ANIMAL CYTOLOGY
AND EVOLUTION

ANIMAL CYTOLOGY
AND
EVOLUTION

M.J.D. WHITE

Professor of Genetics, University of Melbourne

THIRD EDITION

CAMBRIDGE
AT THE UNIVERSITY PRESS
1973

Published by the Syndics of the Cambridge University Press
Bentley House, 200 Euston Road, London NW1 2DB
American Branch: 32 East 57th Street, New York, N.Y. 10022

Library of Congress Catalogue Card Number: 79-190418

ISBN: 0 521 07071 6

First published 1945
Second edition 1954
Third edition 1973

Printed in Great Britain by
William Clowes & Sons Ltd., London, Colchester and Beccles

CONTENTS

CONTENTS

PREFACE TO THE THIRD EDITION

The seventeen years that have elapsed since the publication of the second edition of this book have seen an explosive expansion of cytogenetics. Thus in order to cover the whole field of the earlier volume it was necessary to write an entirely new book, and a much larger one. Nevertheless, the general scope remains the same. No attempt has been made to deal completely or adequately with the field of molecular mechanisms in cytogenetics. In this area the rate of new discoveries is now so great that any comprehensive review would be seriously out of date a year after publication. Thus the earlier chapters of the present book contain only sufficient information on the various aspects of chromosome structure and behaviour at mitosis and meiosis to illuminate the later discussions on chromosomal evolution in its various aspects. The citation of literature ends about the middle of the year 1971, so that some very recent contributions, particularly Francis Crick's latest model of chromosome structure, are not dealt with.

The flood of publications in the field of cytogenetics is now so great that most investigators are only able to read those that seem directly relevant to their own work. This has led to a significant divorce between those cytogeneticists concerned primarily with molecular mechanisms (and who are for the most part philosophically *reductionists*) and those who study the evolutionary consequences of cytogenetic mechanisms at the three levels of the population, the race and the species (the latter being, in general, *compositionists*). One of the aims of this book is to prevent that divorce, since it is clear that discoveries of a fundamental nature will come about in the evolutionary field from the application of the newer knowledge of molecular cytogenetics to evolutionary problems and that an evolutionary perspective is very useful to the investigator of molecular mechanisms. Another consequence of the flood of publications is that many of the younger workers, rightly anxious to be fully informed on all the latest discoveries, seem to have lost historical perspective. I do not, therefore, apologize for the citation of some ancient authors in this book. Our science has a history, and one that we have no reason to be ashamed of. In fact, we can learn much, even from the errors of the past. It is for this reason that I have mentioned and discussed some of these past errors, when it might have been more tactful to keep silent about them. For each aspect of animal cytogenetics I have tried to cite what have seemed to me the most significant publications, regardless of their date. To all who find that their work has not been mentioned, I can only say that any further expansion of the book seemed to me undesirable.

The problem of ignorance of related fields is, of course, not confined to cytogenetics. Many workers in the area of evolutionary theory, speciation and population studies are very ill-informed concerning chromosome biology and some whole schools of evolutionary studies seem to exist in total isolation from cytogenetics. If the present book helps to re-establish chromosomal mechanisms in the centre of the evolutionary stage, the labour of writing it will not have been in vain.

I am grateful to a large number of people who have aided in the writing of this book in one way or another. Most of it has been written at the University of Melbourne, but some sections were written in 1968 and 1970, when I was a guest at the Museum of Comparative Zoology, Harvard University and the Department of Animal Genetics, University College London. Professors Ernst Mayr and Hans Grüneberg most kindly provided favourable atmospheres for writing in their respective institutions. To a number of secretaries, artists and photographers in Melbourne who helped in the production of the manuscript I am especially indebted – particularly to Geraldine Spilman and Florence Deeley, who did most of the typing. My colleague Dr J. A. Thomson read a number of the chapters in manuscript and made helpful suggestions for their improvement.

Melbourne   M. J. D. White
November 1971

I

CHROMOSOMES AS SUPERMOLECULAR SYSTEMS

Corporibus caecis igitur natura gerit res.

LUCRETIUS: *De Rerum Natura*

The chemical basis of organic heredity in all forms of living matter consists of nucleic acid molecules, which are polymers of a relatively small number of different kinds of nucleotides (each composed of a purine or pyrimidine base plus a phosphorylated pentose sugar). These carry, in coded form, the specification of the amino acid sequence in all the enzymes and, more generally, all the proteins which go to make up the individual or the species (on the principle one gene or cistron to one polypeptide chain). This is the central fact of biology, from which both the immanent and historical aspects of life (Simpson 1960) take their origin. The actual operation of the coding mechanism, with its two levels ('transcription' and 'translation') is outside the scope of this book. In all 'complete' organisms, which reproduce themselves without involvement in the genetic system of another form of life, the primary nucleic acid is desoxyribonucleic acid (DNA) in its two-stranded (Watson–Crick) form. In some viruses, however, which are 'incomplete' organisms in that they need to draw on the coding mechanisms of the cells which they parasitize, ribonucleic acid (RNA) is the only one present, DNA being absent. And in a few viruses (e.g. the bacteriophage ϕX174) a single-stranded form of DNA exists (Sinsheimer 1959).

Theoretically, we might imagine an organism in which the genes or 'cistrons' which determine the assemblage of polypeptide molecules, via RNA as an intermediate, were unlinked physically or genetically. In that case the apparatus of coding would be a population of independent cistrons, like a bowl containing a large number of slips of paper on each of which a word is written. But, in fact, it seems that no such organism exists, since even in the simplest viruses the 'words' seem to be joined together in a 'sentence', 'linkage group' or 'chromosome' (Benzer 1955; Luria 1959; Streisinger and Bruce 1960).

PROKARYOTIC AND EUKARYOTIC GENETIC SYSTEMS

Two levels of genetic organization seem to exist in living organisms. The first is represented by the viruses, bacteria and blue-green algae, in which there is no nuclear membrane and the 'chromosome' is simply a nucleic acid molecule without intimately associated basic protein. Such organisms may be designated *prokaryotic*, since they lack true nuclei. All higher organisms, including the algae (other than Cyanophyceae) and fungi are *eukaryotic* (Lwoff 1943) and have chromosomes of a higher order of complexity, in which much protein is associated with the DNA. Whitehouse (1965) speaks of *chromonemal* and *chromosomal* categories of organisms in the same sense as prokaryotic and eukaryotic. If the basic proteins of the chromosome are responsible for regulation of genic activity (switching on and off of biosynthetic function) it is probable that the chromosomal or eukaryotic level of organization had to be attained before any form of cellular differentiation could be attained. Microbial geneticists continue to use the term *chromosome* for the DNA molecules of prokaryotes; but it is, perhaps, rather doubtful whether we should continue to use the same term for two kinds of structures which differ so much in their levels of complexity. It might be better to speak simply of the genomes or genophores (W. K. Baker 1965) of microorganisms.

The DNA 'chromosomes' of those prokaryotes that have been studied seem to be almost invariably circular molecules. Circularity of the bacterial chromosome was first inferred from genetic data on *Escherichia coli* by Jacob and Wollman (1957). It has subsequently been confirmed for this organism by direct autoradiography (Cairns 1963*a*, *b*, *c*) and for numerous bacteriophages such as T2, T4 and ϕX174 by genetic and other means (MacHattie and Thomas 1964; Streisinger, Edgar and Denhardt 1964; Fiers and Sinsheimer 1962). Electron micrographs of the DNA molecules of a number of viruses also show a circular structure (e.g. Hayashi, Hayashi and Spiegelman 1964). However, it is probable that in most or all these cases the 'chromosome' goes through a linear phase in its life cycle. Thus although the linkage map of T4 is circular, the DNA molecule as it exists in the mature phage is linear (Rubenstein, Thomas and Hershey 1961) and in the case of the lambda phage it is also linear although its ends tend to join up to form a ring under certain conditions (Hershey, Burgi and Ingraham 1963; Edgar 1965). The genetic maps of the bacteria *Escherichia coli* and *Salmonella typhimurium* are both circular (Demerec 1964, 1965; Demerec and Ohta 1964; A. L. Taylor and Thoman 1964; K. E. Sanderson and Demerec 1965). Moreover the sequence of homologous genetic loci seems to be the same in these two species. *Streptomyces coelicolor* also has a circular map (Stahl 1967), but it is uncertain as yet whether all bacterial chromosomes are circular.

Increasing size of the chromosome is shown in the series viruses–bacteria–higher organisms. The chromosome of ϕX174 contains about 5500 nucleotides, which might correspond to about 5 cistrons, while that of T2 and T4 have about 200 000 nucleotide pairs (equivalent to 100 or more cistrons). By comparison, bacteria have about 3 000 000 to 6 000 000 nucleotide pairs (*Escherichia coli* has 4 500 000 pairs) and mammalian sperm nuclei about 3 000 000 000 nucleotide pairs.

We shall not be further concerned in this book with the genetic systems of prokaryotes, but in order to stress the basic differences between the cytogenetic mechanisms of pro- and eukaryotes we reproduce below (Table 1.1), in somewhat modified form, a synopsis due to Westergaard (1964).

The chromosomes of higher organisms are complex structures in which DNA, RNA and several types of proteins are intimately associated. They undergo a complicated series of changes in appearance (and presumably in structure and composition) throughout the mitotic cycle. Moreover, in certain organisms and in certain tissues they develop into giant structures which exhibit a very special morphology (Chs. 4–5). All these considerations make it very difficult to give a satisfactory account of chromosome structure in general. The protein constituents of the chromosomes are (i) *histones* (or *protamines*, in the case of some sperm nuclei), simple basic proteins with molecular weights ranging from 14 000 to 20 000 and (ii) more complex

TABLE 1.1. *Cytogenetic mechanisms in bacteria and higher organisms* (based on Westergaard 1964, modified)

Bacteria	Higher organisms
A single, circular DNA chromosome. No centromere or telomeres.	Several nucleoprotein chromosomes. Telomeres probably always present; an individualized centromere usually present in each chromosome.
No mitotic apparatus, division by DNA replication.	Mitotic apparatus present, mitosis follows on DNA replication after an interval of time.
No nuclear membrane.	Nuclear membrane present.
Partial transfer of genetic information (merozygosis) by sexual reproduction, transformation and transduction.	Complete transfer of genetic information (holozygosis) by sexual reproduction. Transformation and transduction not found.
Reduction (haploidization) through crossing-over in the merozygote whereby one of the two products of recombination, being lethal, is eliminated.	Reduction through meiosis, all four products of which are initially (at least) viable.

TABLE 1.2. *Percentage dry-weight composition of HeLa human chromosomes* (from Huberman and Attardi 1966)

	Interphase	Metaphase
DNA	31	16
RNA	5	10
Histone	34	31
Non-histone protein	31	42

non-basic proteins of higher molecular weight (Stedman and Stedman 1947; Alfert 1956; Mirsky and Ris 1949). Some idea of the percentage composition of chromosomes can be obtained from Table 1.2. There are usually five different histone fractions, F_1 (lysine-rich), F_2b (slightly lysine-rich), F_2a1, F_2a2 and F_3 (all arginine-rich) (Pardon and Richards 1972). A regular supercoil structure is very liable to form in nucleohistone molecules and one or more of these types of histones may be responsible for forcing the DNA double helices into this regular tertiary configuration.

In some species of animals a lysine-rich chromosomal histone is replaced by an arginine-rich one during the conversion of the spermatid into the mature sperm. Such a change occurs in the snail *Helix aspersa* (Bloch and Hew 1960), *Drosophila melanogaster* (Das, Kaufmann and Gay 1964) and the grasshopper *Chortophaga viridifasciata* (Bloch and Brack 1964).

In life the chromosomes lie in a clear fluid, the *nuclear sap*, or *nucleoplasm* surrounded by the nuclear membrane or envelope. The nuclear membrane seems to be fundamentally similar to the endoplasmic reticulum of the cytoplasm, with which it is in connection at various points. At the end of prophase in mitosis, it appears to break down, being re-formed (from the endoplasmic reticulum) at the end of the telophase stage.

When fully extended, the chromosomes of eukaryotes always seem to exhibit a *chromomeric* structure, in which local aggregations of DNA-containing material (the *chromomeres*) alternate with inter-chromomeric regions containing much less DNA. This basic interpretation has been generally accepted since the time of Belling (1928). Chromomeric structure is seen especially well in the giant polytene and lampbrush chromosomes (Chs. 4 and 5), but is also evident at various stages of mitosis and meiosis; it is not evident at metaphase and anaphase (even in electron micrographs – see p. 52) when the chromosomes are maximally condensed and contracted. It is now fairly clear that chromomeres are regions where the DNA is coiled into a helix (or perhaps several orders of helical coiling) which is superimposed on the helical structure of the actual duplex molecule itself.

At prophase and metaphase each chromosome consists of two strands or *chromatids* which are roughly parallel to one another, but may be wound round one another in a loose *relational spiral*. In most organisms the two chromatids appear to be firmly held together at one region, the *centromere*. When the spindle is formed at premetaphase it is only the centromere which becomes directly connected with it, mechanically and functionally, the rest of the chromosome being either free-lying in the cytoplasm at metaphase, or else passively included in the spindle. In certain organisms, however, distinct centromeres are not detectable, and in such cases (see pp. 14–17) the chromosomes seem to be associated mechanically and functionally with the spindle throughout their entire length.

It is characteristic of most chromosomes that their cycle of condensation is not uniform but varies along their length, so that certain regions may be precocious in condensation during prophase or may remain condensed throughout interphase, when the rest of the chromosome is diffuse. This phenomenon of differential condensation has been called *heteropycnosis* and regions (or whole chromosomes) which exhibit it are said to be *heterochromatic* or composed of *heterochromatin*.

Chromosomes in prophase, metaphase and anaphase frequently show so-called *constrictions*, which are regions where the diameter of the chromosome is less than elsewhere and where the DNA content is low per unit of length. The centromere usually appears as a constriction, but there are other so-called *secondary constrictions* which have no centromeric properties. One or more of these secondary constrictions may bear, during interphase and prophase, bladder-like RNA-containing structures, the *nucleoli*.

Having defined the various visible structures in the chromosome we shall proceed to consider each of them in detail. This task is especially difficult because we are forced to deal simultaneously with universal properties of eukaryote chromosomes on the one hand and properties which are distinctive of particular types of chromosomes on the other. A number of issues cannot be fully clarified until we have considered the structure of two kinds of giant chromosomes – the *polytene* chromosomes found in certain tissues of dipterous flies (Ch. 4) and the *lampbrush* chromosomes of the oocyte nuclei in Amphibia and some other groups.

CHROMONEMA AND CHROMOMERES

The term *chromonema* is applied to the fundamental thread which maintains a continuous existence throughout the entire mitotic cycle. In interphase the chromonema is maximally extended, while in metaphase it is condensed and coiled into a tight, compact helix. The chromomeres are best seen during the prophase stages, especially the prophase of the first meiotic division, and in the two types of giant chromosomes referred to earlier. There may be several hundred chrommomeres per chromosome, each 0.5 to 2.0 μm in diameter and hence several thousands in the chromosome complement as a whole. But as the cycle of condensation progresses towards metaphase the chromomeres tend to fuse or condense together into blocks of material, and by the time metaphase has been reached they have disappeared.

The compact, smooth-outlined appearance of metaphase chromosomes is thus a deceptive one. Numerous investigations have shown that each chromatid is a highly compressed helix at this stage. Thus one element in the long condensation process that occurs during prophase is a gradual spiralization of the chromatids. Similarly, in the de-condensation of the telophase chromosomes, as they return to the interphase stage, a process of despiralization takes place.

Spiral structure has been studied in natural mitotic cycles or in nuclei that have been exposed to various chemicals, such as ammonia vapour or cyanide, which seem to have the property of slightly relaxing the gyres of the metaphase helix, so as to render them visible.

The spirals of one mitotic cycle, although relaxed in the succeeding interphase, are not always completely lost, and may show up as loose 'relic coils' in the prophase of the next division. At the first meiotic division the metaphase chromosomes have been shown by various workers (Kuwada and Nakamura 1934; Darlington 1935a; Oura 1936; Huskins 1941) to have a 'coiled coil' structure, with a major helix superimposed on a minor one, which will form the major helix of the second meiotic division. Darlington (1958) has pointed out that when chromosomes are under-condensed at metaphase they possess far more gyres of smaller diameter than usual. The process of condensation during prophase presumably consists of a diminu-tion in the number of coils, with an increase in their diameter, which must depend on the development of an internal minor coil, as at meiosis.

A helix may, of course, be either right-handed or left-handed. Darlington (1935a), after studies on the chromosomes of the plant *Fritillaria* (in which the individual chromosomes were not distinguished) concluded that for any particular chromosome arm the direction of coiling was fixed, being predetermined by its molecular structure. He did not admit the possibility that the direction of coiling might be reversed at interstitial points other than the centromere (Nebel, 1932, has shown that frequent reversals of the direction of coiling occurred at the centromere). Huskins (see Huskins and Smith 1935; Huskins and Wilson 1938; Huskins 1941) observed numerous reversals at the centromere and other interstitial points in *Trillium*. He concluded that the direction of coiling was not constant even for individual regions. White (1940*a*) showed that the direction of coiling was random in the case of the *X*-chromosomes of some long-horned grasshoppers (Tettigoniidae), where approximately equal numbers of right-handed and left-handed *X*-chromosomes were found in spermatogonia from the same testis-cyst. More recently, Darlington and Vosa (1963) have studied the direction of coiling in a telocentric chromosome of *Tradescantia*, which was found to be entirely random (equal numbers of right-handed and left-handed chromosomes). They also found that in metacentric chromosomes of *Tradescantia* the direction of coiling in the two arms was about 3.5 times as often in the opposite as in the same direction (the combinations LL:LR:RR being approximately in the ratio 1:7:1).

Usually, the direction of coiling is the same in the corresponding chromatids of a somatic chromosome, at any one locus. However, this is not always so, and Manton and Smiles (1943) claimed that in the fern *Osmunda* corresponding regions in the two chromatids of a single somatic chromosome may be coiled in opposite directions. Ohnuki (1968) has shown that the direction of coiling is random in human chromosomes, that sister chromatids may be coiled in opposite directions and that interstitial reversals of direction of coiling occur.

It may be necessary to point out here certain optical principles affecting the apparent direction of coiling, as seen under the microscope. In the first place, most light microscopes produce an inverted image, in which right-handed spirals appear as left-handed ones and vice versa. Secondly, and independently of this fact, an optical section below the mid-plane of a spiral which is lying horizontally will give the impression of a right-handed spiral where a left-handed one is actually present, and vice versa, because the eye interprets out-of-focus parts of the image as further away, when they may be closer. Great care must consequently be used in the interpretation of photographs of chromosomal spirals, although direct visual inspection of microscopic images should lead to correct interpretations, provided that the above principles are borne in mind.

CENTROMERES

Most chromosomes possess a special region which plays an essential role in connection with the organization of the developing spindle at premetaphase of mitosis and with the separation of the daughter chromosomes at anaphase. This region, chromosomal organelle or locus has been variously called the *achromite, spindle attachment, primary constriction, spindle fibre locus, fibre attachment, kinetochore* or *centromere*. Some of these names indicate functions which have been ascribed to it. We shall adopt the term 'centromere', since it is short and adjectives (e.g. *acentric*, lacking a centromere; *poly-centric*, having many centromeres) can conveniently be formed from it. The concept of the centromere as a self-perpetuating organelle which cannot arise *de novo* in chromosomal arrangements and chromosomal evolution was first clearly formulated by Navashin (1932).

Although the functional role of the centromeres is most evident at metaphase and anaphase, they are permanent and autonomous segments of the chromosomes, which can be seen at other stages of the mitotic cycle in suitable material. The morphology of the centromere varies a good deal from one organism to another but it usually appears as a non-staining gap or constriction at mitosis and meiosis. In certain cases a dark-staining chromo-mere has been observed in the middle of the unstained region and it has been suggested that this is the true centromere or kinetochore, the un-stained material on either side of it being some kind of accessory material. In many cases, however, no such granule or chromomere can be seen and it seems more logical to assume that in those instances where such a chromo-mere does exist, it separates two centromeric segments which function in a coordinated manner at mitosis.

Numerous observations (Mather and Stone 1933 and all later workers) have shown that chromosome fragments which, as a result of fragmentation (whether induced by radiation or by other means) have lost their centromeres – so-called acentric fragments – do not become attached to the spindle. They float about freely in the cytoplasm at mitosis, so that there is no mechanism for ensuring the regular passage of daughter-fragments to opposite poles at anaphase. On the other hand, it has been repeatedly shown (Mather and Stone 1933; McClintock 1951) that in dicentric

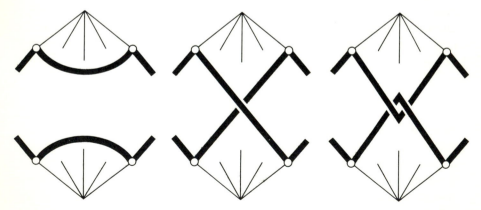

Fig. 1.1. The three ways in which a dicentric chromosome may behave at a mitotic anaphase.

chromosomes the two centromeres in each chromatid may pass to opposite poles at anaphase, the region between them being stretched on the spindle. In such cases the separation of the daughter nuclei may be prevented or else the stretched chromatids between the centromeres may be broken. It is thus easy to understand why chromosomes normally possess only a single centromere rather than none or several. However, a few cases of transmissible dicentric chromosomes have been recorded in the literature. Sears and Camara (1952) described an instance in wheat and Angell, Giannelli and Polani (1970) have reported several cases of supposedly dicentric Y-chromosomes in man.

It is rather generally agreed that the centromeres 'divide' later than the rest of the chromosome at mitosis. Whether the DNA or the proteins of the centromere region actually replicate tardily in the mitotic cycle is still unproven. Certainly, however, most mitotic chromosomes present the appearance of pairs of chromatids which are only held together very loosely except at the centromere, where they are closely united. Three categories of observations reinforce this general conclusion: (1) at the first meiotic division the centromeres do not normally divide at all; (2) in somatic chromosomes from cells that have been treated with such drugs as colchicine the chromatids diverge quite widely, but remain attached at the centromere (Fig. 12.19); (3) under various experimental conditions so-called *diplochromosomes* may be produced, which have undergone an extra replication between mitoses (White 1935*a*, *b*; Barber 1940); such diplochromosomes show incomplete or delayed separation of the four parallel chromatids in the centromere region.

It would be in accordance with modern views of chromosome structure to regard the centromere as built up of hundreds or thousands of repeated cistrons having very special properties. Clearly it can be fragmented into two segments, each of which will then exhibit centromeric properties. Fragmentation of the centromere region in univalents at meiosis or, as it has been called, *mis-division*, was observed by Upcott (1937), Koller (1938) and Darlington (1939*c*, 1940*a*) in plant material. It has been studied especially in wheat by Sears (1952*b*, 1966) and Steinitz-Sears (1963). We shall discuss the whole question of structural alterations of the centromere in detail in Chapter 7.

Chromosomes which have the centromere approximately half-way along their length, so that the two arms are approximately equal, are called *metacentric*. Those in which it is very close to one end, so that one arm is many times the length of the other are said to be *acrocentric*. Chromosomes with strictly terminal centromeres are referred to as *telocentric*. There is no doubt that telocentric chromosomes have been obtained experimentally in a variety of different types of material, following breakage through the centromere region or immediately adjacent to it, and that under certain circumstances they may be stable and capable of passing through a number of successive mitoses. A controversy has arisen, however, as to whether strictly telocentric chromosomes form part of the normal karyotype of any species of higher organism.

The writer (White 1954b, 1957d; White, Blackith, Blackith and Cheney 1967) has adopted the view that all naturally-occurring 'rod-shaped' chromosomes are acrocentric, i.e. that there is always a minute 'second arm', even if this cannot always be seen by conventional techniques. This is in accordance with the early, but careful, studies of Prokofieva-Belgovskaja (1935a, b, c, 1938) on the X-chromosome of *Drosophila melanogaster* and on the chromosomes of a number of fish species. It is supported by the fact that minute second arms can be seen on the proximal ends of the rod-shaped chromosomes of a great many species of animals and some higher plants.

Nevertheless, there have been repeated claims that in particular species the chromosomes, or some of them, were truly telocentric. Such a claim was made by Cleveland (1949) for the chromosomes of the hypermastigote flagellate protozoan *Holomastigotoides* and by Melander (1950a) for the crustacean *Ulophysema öresundense*. It has been put forward for all chromosomes of the mouse karyotype (Tjio and Levan 1954) and all except two chromosomes of cattle (Melander and Knudsen 1953; Melander 1959). And more recently John and Hewitt (1966b, 1968) and John and Lewis (1968) have claimed that, although the rod-shaped chromosomes of some species of grasshoppers are acrocentric, those of other species (and sometimes closely related ones) are truly telocentric. Obviously this difference of opinion as to the occurrence or non-occurrence of telocentric chromosomes in the normal karyotypes of wild species owes its importance mainly to the fact that different mechanisms of chromosomal evolution are implied on the two hypotheses.

There are some species of grasshoppers such as *Locusta migratoria* (White 1935b; John and Hewitt 1966b) in which the short arms of the acrocentrics can be seen at metaphase in good preparations of spermatogonial or somatic metaphases. In other species, however, the second arms can only be detected occasionally, and in many metaphases the rod-shaped chromosomes appear to terminate proximally in a pointed or slightly rounded end. If we were to trust solely to direct visual observation, we might conclude that the chromosomes were acrocentric in some cells of an individual and telocentric in others, which is clearly illogical.

The difficulty here does not seem to be that the short arms are actually below the limit of resolution of the light microscope; it is more a question of the degree of condensation of the metaphase chromosome, which may lead to the short arms being actually absorbed into the proximal end of the main limb, the gap between them (i.e. the actual centromere, on our view) being invisible.

John and Lewis (1968) have claimed that the rod-shaped chromosomes of the grasshoppers *Myrmeleotettix maculatus* and *Chorthippus brunneus* are truly telocentric because the chromomeric region at their proximal end (one to three chromomeres) seems to correspond exactly to one half of the corresponding region in the middle of the metacentric chromosomes of the same species. Lima-de-Faria (1949*a*, *b*, 1952, 1956), who made a detailed study of the centromeres of rye, described a compound chromomeric structure which appeared as a reversed repeat (whether it really is a repeat in any exact molecular sense is uncertain). The centromeres of the *Myrmeleotettix* and *Chorthippus* metacentrics do, indeed, appear to have essentially the same structure as those of rye. But there is no reason to believe that this structure is universal; in fact most animal centromeres appear to be simple unstainable 'constrictions', with no chromomeres in the middle.

The question is, then, whether the essential properties of attachment to the spindle reside in the unstainable region or regions or in the associated chromomeres. If it be true that in many species of animals special chromomeres are not present in the middle of the unstained region, then it seems logical to ascribe the essential properties of the centromere to the unstained segment. The only alternative to this conclusion would seem to be the assumption that centromeric properties reside in chromomeres flanking the constriction on either side.

It is claimed by John and Hewitt (1968) that:

a telocentric chromosome is expected to contain centric chromomeres and there can be little doubt that these have frequently been taken to represent a second arm. This is an error we have ourselves committed . . .

However, it seems to be at least equally probable that the error is the other way round, i.e. that short arms have been mistaken for centromeric chromomeres.

One piece of evidence relied upon by John and Hewitt to support their case seems to point entirely in the opposite direction. They cite a chromosomal polymorphism in the grasshopper *Oedaleonotus enigma*, where a chromosome regarded by them as telocentric has given rise to one which is subacrocentric. Numerous similar cases are known in the North American trimerotropine grasshoppers. Discounting the possibility of terminal inversions, John and Hewitt state:

If, as generally believed these shifts (of centromere position) are two-break pericentric inversions then one of the breaks must have occurred within the centromere.

It will be obvious that a rearrangement of the type they postulate would give rise to a dicentric chromosome which would almost certainly show some degree of instability at mitosis, sometimes giving bridge configurations at anaphase.

There can be very little doubt that, by the criteria of John and Hewitt (1966b) the X and fourth chromosome of *Drosophila melanogaster* would be regarded as telocentric rather than acrocentric. Yet, for at least thirty years, these chromosomes have been considered to be acrocentric by *Drosophila* workers. In the words of Muller (1938):

> ... the minute observations of mitotic chromosomes carried out by Prokofyeva and by Heitz have strongly suggested a non-terminal, that is, a sub-terminal, position for it [i.e. the centromere]. And agreeing with this is Prokofyeva's observation of small 'heads', as she calls them, on the right distal ends of both the X and the fourth chromosomes in the salivary glands ... This conception of the existence of another terminal chromocentral region, beyond the centromere of the apparent rods ... also affords a ready explanation, consistent with the other known facts regarding chromosome rearrangements, for the frequency with which, in translocations, sections of other chromosomes can be attached beyond the centromere of the X and the fourth chromosome, to give rise to chromosomes that are obviously J- or V-shaped.

Kaufmann (1933, 1934), Prokofieva (1935a) and Cooper (1959) all recognized the short, right limb of the *D. melanogaster* X-chromosome.

If we are correct in believing that no 'natural' chromosomes have strictly terminal centromeres, there must be some reason for this. Presumably no breaks through the centromere give rise to absolutely stable chromosome ends, i.e. a single end cannot exhibit both centromeric and telomeric (see p. 17) properties, perhaps because these functions depend on different types of nucleotide sequences.

The unsatisfactory nature of John and Hewitt's distinction between acrocentric and so-called telocentric rod-chromosomes is further illustrated by the fact that they regard the chromosomes of *Locusta migratoria* as acrocentric, since they have quite evident 'short arms', but those of *Humbe tenuicornis* as telocentric. The genera *Locusta*, *Gastrimargus* and *Humbe* form a closely related taxonomic group, so it is clearly illogical to suppose that a fundamental difference in chromosome structure exists.

John and Hewitt (1966b) have also concluded from an examination of the photographs in White, Carson and Cheney (1964) that '*Moraba*' *viatica* has telocentric chromosomes. Yet the very closely related species 'P45c' has 'short arms' of quite significant and measurable length (White, Blackith, Blackith and Cheney 1967) which can be shown to be late-labelling by tritiated thymidine autoradiography (Webb 1973). In *M. viatica* itself the short arms could not be seen as separate entities in material from the

Keith area of South Australia, but were clearly visible (and measurable) in individuals from the Fleurieu Peninsula (White, Blackith, Blackith and Cheney 1967). Whether the difference between the two geographic populations of this species is due to the second arms being actually longer in one than the other is uncertain; it could simply be due to a difference in the degree of condensation and contraction of the proximal ends of the chromosomes. However, the example of P45c, referred to above, suggests that the first explanation is more likely to be correct. A similar instance in the morabine grasshoppers concerns the parthenogenetic species *M. virgo*, which by John and Hewitt's criteria has acrocentric chromosomes and its closest bisexual relative, P151, which they would consider to have telocentrics (White and Webb 1968).

It is certainly possible that the second arms of acrocentric chromosomes have some role in relation to the spindle, i.e. that they have some properties which we may call centromeric. But it seems highly illogical to suppose that two fundamentally different kinds of rod-shaped chromosomes co-exist in closely related species of the same group. The kinds of chromosomes which John, Hewitt and Lewis regard as 'acrocentric' and 'telocentric' represent extremes in a continuum rather than clearly separable types. Any other standpoint makes the interpretation of chromosomal evolution hopelessly confused, and although this is not, strictly speaking, an argument for the interpretation put forward here (because chromosomal evolution may, after all, be more complex than has been suspected until now) it is, on Occam's principle, a reason for rejecting the more complex view until new lines of evidence, such as electron microscopy, provide decisive evidence one way or the other. Perhaps the term acrocentric should be restricted to chromosomes whose short arm is less than 1/20th of the total length (i.e. with an arm ratio >19); chromosomes with an arm ratio between 19 and 9 might be called *subacrocentric*, those with an arm ratio between 9 and 1.25 being *heterobrachial metacentrics* and ones with an arm ratio between 1.25 and 1 being *isobrachial metacentrics*. Levan, Fredga and Sandberg (1965) have proposed a terminology according to which chromosomes with an arm ratio greater than 7 are called *t* chromosomes, those in which it is between 7 and 3 *st* chromosomes, those in which it is between 3 and 1.7 *sm* chromosomes and those in which it is less than 1.7 *m* chromosomes (these abbreviations apparently stand for *acrocentric, subtelocentric, submetacentric* and *metacentric*). Chromosomes with an arm ratio of exactly 1.0 are referred to as *M*-chromosomes and those with an arm ratio equal to infinity are called *I*-chromosomes in the terminology of Levan *et al.*

Some animal species have a karyotype consisting entirely of acrocentric or metacentric chromosomes, while in other species some chromosomes are of one type and some of the other. In the latter case the distinction may be an entirely clear-cut one, or there may be many subacrocentrics and heterobrachial elements in the karyotype. Whole groups of the animal kingdom are known in which strictly acrocentric chromosomes are not known at all (a situation which exists in many groups of higher plants).

HOLOCENTRIC CHROMOSOMES

In certain groups of animals and plants the chromosomes do not seem to show any localized or individualized centromeres. Either the whole chromosome or (in the case of some nematodes of the family Ascaridae) a long region of each chromosome, is attracted to the spindle at metaphase and behaves as if endowed with centromeric properties. Experimental studies on a number of such organisms have shown that if their chromosomes are fragmented by X-rays each portion becomes attached to the spindle in subsequent mitoses, there being no distinction between centric and acentric fragments (Hughes-Schrader and Ris 1941). Chromosomes of this kind have been studied especially in plants of the genus *Luzula* by Nordenskiöld (1951, 1961) and La Cour (1953).

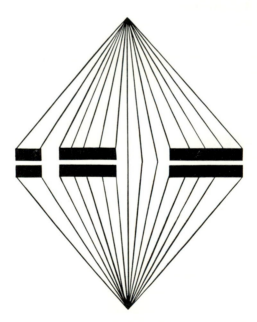

Fig. 1.2. Characteristic mode of orientation of holocentric chromosomes on the spindle at mitosis.

There have been two interpretations of chromosomes of this type. The Schraders (Hughes-Schrader and Ris 1941; Hughes-Schrader 1948a; Schrader 1947, 1953; Hughes-Schrader and Schrader 1961) have spoken of a *diffuse kinetochore* or *diffuse kinetochore activity*, implying that attachment to the spindle in such cases is evently distributed along the entire length of the chromosome. Others have imagined that numerous, but discrete and localized, centromeres are present, and have spoken of *polycentric chromosomes*. Bayreuther (1955) and Scholl (1955) use the term *holokinetic chromosomes* (from *holos*, whole). It must be admitted that we have at present no means of deciding between the two interpretations. It may be, as Schrader (1947) suggested, that there exists more than one type of chromosome lacking localized centromeres, and that the insect cases exhibit diffuse centromeric activity while in the nematode case we are dealing with a truly 'polycentric' condition. For the present we shall use the term holocentric (rather than polycentric) to cover all such cases. This is in agreement with adoption of the term centromere, rather than kinetochore; if it should be shown later that some chromosomes are actually polycentric rather than holocentric then the two terms can be used to distinguish the two types.

Holocentric chromosomes are generally present throughout the insect orders Heteroptera and Homoptera. They seem to be likewise found in the biting and sucking lice (Mallophaga and Anoplura) according to Scholl (1955) and Bayreuther (1955); these orders may be fairly closely related to the Homoptera. Piza (1958) has unconvincingly argued that the chromosomes of the Heteroptera have two terminal centromeres. Whether the chromosomes of the Lepidoptera are holocentric is still somewhat uncertain. Federley (1943, 1945) thought that they were, and was followed in this view by Suomalainen (1953, 1954b); Schrader (1947) also supported the holocentric interpretation in the case of the Lepidoptera. Throughout this order of insects the chromosomes are usually quite short and approximately isodiametric at the metaphase of both mitotic and meiotic divisions. While the morphological evidence for holocentric chromosomes in the Lepidoptera is moderately convincing, it is not right to argue, as Suomalainen has done, that the great cytotaxonomic differences in chromosome number in certain lepidopteran genera point to the occurrence of 'simple' fusions and fragmentations in evolution and hence to holocentric chromosomes (see p. 417). The chromosomes of earwigs (Dermaptera) are almost certainly holocentric (Ortiz 1969; White 1971a), in spite of some statements in the literature to the contrary (Callan 1941b, Henderson 1970). In the dragonflies (Odonata) Schrader (1947) suggested that the chromosomes were holocentric, but the evidence of Oksala (1943–52) makes it fairly clear that they are of the ordinary monocentric type, although Kiauta (1967–9) still regards them as holocentric.

In a few groups such as the scorpions some species seem to have holocentric chromosomes while in others a monocentric condition occurs (see p. 419). This definitely raises the question of the evolutionary relationship between the two types of chromosomes.

In centipedes (Chilopoda) the chromosomes seem, in general, to have localized centromeres; thus rather clearly metacentric chromosomes have been figured by Ogawa (1954, 1961*d, e*) in the case of *Otocryptops sexspinosus, O. capillipedatus* and *Lithobius pachypedatus*. But the same author (1952, 1955*a, c*, 1961*d*) has claimed that in *Thereuonema hilgendorfi, Thereuopoda clunifera* and *Esastigmatobius longitarsis* a pair of giant chromosomes in each case are holocentric. This pair of elements are regarded as X- and Y-chromosomes in the two former species but are not necessarily sex chromosomes in the last-named form. Ogawa states that in *Esastigmatobius* there are two huge holocentric chromosomes which are surrounded by eighteen smaller metacentrics at mitosis. He claims to have proved the holocentric nature of the giant chromosomes in both *Thereuonema* and *Esastigmatobius* by a study of the mitotic behaviour of radiation-induced fragments (Ogawa 1955*a, c*). But it is a little difficult to tell just how critical his evidence was; and one should perhaps be cautious in accepting the somewhat paradoxical idea of holocentric and monocentric chromosomes in the same karyotype.

We shall later deal (pp. 490 and 493) with the mechanism of meiosis in holocentric chromosomes and with the principles of chromosomal rearrangement and karyotype evolution in groups whose chromosomes are of the holocentric type.

Although in monocentric chromosomes the centromere is strictly localized and other regions of the chromosome do not exhibit any centromeric activity as a rule, at least one exception is known. That is in maize, where Rhoades and Vilkomerson (1942) showed that plants with an abnormal type of chromosome 10 have chromosomes which, in addition to their normal centromeres, show centromeric activity in certain other regions as well (so-called neo-centromeric segments).

Bauer (1967) has argued from the low frequency of zygotic lethality following irradiation in Lepidoptera (by comparison with the much higher percentage lethality in the case of Diptera, such as *Phryne* and *Drosophila*, with monocentric chromosomes) that lepidopteran chromosomes are most probably holocentric. The argument rests on the fact that in organisms with monocentric chromosomes half of the mutual translocations produced will give inviable dicentric and acentric chromosomes; whereas in the case of holocentric chromosomes all the translocation products should be cell-viable.

In eukaryotic organisms the chromosome ends do not exhibit any tendency to fuse together spontaneously. They differ in this from artificial chromosome ends caused by breakage of normal chromosomes (whether caused by radiation, stretching on the spindle or other means). Such artificial ends always seem to have some capacity for re-fusion, i.e. they behave as if they were 'sticky'. The natural ends, on the other hand, are 'non-sticky'. This difference in behaviour caused Muller (1932b, 1938) to apply the term *telomere* to these natural chromosome ends. Various authors have claimed from time to time to have identified the telomeres as visible structures in chromosomes of various types; it seems probable, however, that in most cases they were studying structures such as terminal blocks of heterochromatin, much larger than the true telomeres, which are most likely molecular configurations not visible under the light microscope. However, Berendes and Meyer (1968) have demonstrated structures which they interpret as telomeres in electron micrographs of the giant salivary gland chromosomes of *Drosophila* (see Ch. 4).

Muller's original formulation of the telomere concept implied that telomeres were self-perpetuating structures which could never, by rearrangement, come to occupy an interstitial position in a chromosome; and that no interstitial locus could take on the functions and properties of a telomere. Whereas ordinary genetic loci were 'bipolar', telomeres were 'unipolar'. The first proposition seems to have survived virtually unchallenged, i.e. there are no clear instances of a telomere becoming interstitial, e.g. as a result of a truly terminal inversion of a chromosome segment (a type of rearrangement which is probably impossible). The second proposition seems, however, to be subject to various exceptions, i.e. there are a number of conditions under which freshly broken ends can undergo 'healing', thus taking on at any rate some of the properties ascribed to telomeres. One striking exception is in the nematode worms of the family Ascaridae in which the terminal rejoins of the germ-line chromosomes are cast off in the nuclei which will give rise to the somatic tissues of the adult – and especially in *Parascaris equorum*, where, in addition to the loss of the terminal portions the middle segment of each germ-line chromosome undergoes spontaneous fragmentation into a large number of pieces in the somatic nuclei, each of which persists as a separate chromosome through a large number of mitotic divisions. Another case of 'healing' of freshly broken ends occurs in the maize sporophyte, but not in the gametophyte or in the endosperm (McClintock 1939, 1941a). Hsu (1963) has claimed that a particular region of chromosome 1 of the Chinese Hamster is preferentially broken after administration of the drug BUDR and that centric fragments resulting from such breakage behave as if they had a 'built-in telomere'.

There does not seem to be any general evidence that the natural ends of holocentric chromosomes are in any way different from those of monocentric ones, and the meiotic behaviour of holocentric chromosomes in certain Hemiptera (see p. 490) strongly indicates that the natural ends show specialized telomeric properties. Several investigations, however, suggest that the freshly broken ends of holocentric chromosomes are stable or 'self-healing' (e.g. Lin 1954, in the case of *Ascaris* and Nordenskiöld 1962, in that of the plant *Luzula*). There is some evidence that in *Drosophila* self-healing may sometimes take place in heterochromatic regions, although it probably never occurs in euchromatin (Muller and Herskowitz 1954).

RING CHROMOSOMES IN HIGHER ORGANISMS

In no higher organisms are any of the chromosomes naturally ring-shaped. It has proved possible, however, to produce by radiation and other agencies chromosomes which are circular and lack telomeres and such chromosomes must arise spontaneously with a high frequency as a result of chromosomal rearrangements in which both ends of a chromosome are lost.

Ring chromosomes were first studied in maize by McClintock (1932, 1938, 1941b) and in the case of the so-called closed X-chromosomes of *Drosophila melanogaster*, by L. V. Morgan (1933) and Schultz and Catcheside (1937). In maize the ring chromosomes are unstable since they are liable to give rise to double-sized rings with two centromeres at mitosis. When these become stretched on the spindle at anaphase, with one centromere passing to each pole, they will become broken at two random points along their length. Thus each daughter cell receives a metacentric chromo-

a   *b*

Fig. 1.3. Somatic metaphases of *Drosophila melanogaster* females carrying ring *X*'s. *a*, from a fly with one ring *X* and one acrocentric *X*; *b*, from a fly with two ring *X*'s. From L. V. Morgan (1933).

some with two freshly broken ends which unite to form a new ring, larger or smaller than the original one. Maize ring chromosomes are thus compelled to go through a type of 'breakage–fusion–bridge' cycle, which leads to much mosaicism in the tissues in which it occurs. The formation of double-sized rings at mitosis implies that some process analogous to sister-strand crossing-over has occurred during or subsequent to chromosomal replication.

The closed X-chromosomes of *D. melanogaster* have arisen in experiments in various ways, some by a rearrangement in an attached X-chromosome, the two breakage points being in opposite arms, one very close to the centromere, the other very close to the tip. Other ring X's have arisen by crossing-over in an attached X with a long inversion in one arm. Some closed X's do form double-sized rings in the cleavage mitoses of the embryo and hence give rise to breakage–fusion–bridge cycles and mosaicism; but they do not do so very frequently and it is hence possible to maintain laboratory stocks of flies carrying ring X's. Brown, Walen and Brosseau (1962) found that such flies showed mosaicism both as a result of somatic crossing-over and due to elimination of the ring; and Walen (1964) showed that the frequency of somatic crossing-over was positively correlated with the length of the heterochromatic segments in the ring. Levitan (1952) found a wild individual of *Drosophila robusta* carrying a ring autosome; it is unlikely that the ring element would have maintained itself in the population for long.

There have now been numerous recorded instances of human individuals with congenital malformations, heterozygous for ring chromosomes (Smith-White, Peacock, Turner and Den Dulk 1963; Cooke and Gordon 1965; Gripenberg 1967). In some of these instances the X-chromosome was probably the one that formed the ring; in other cases (e.g. those of Aula *et al.* 1967) one of the medium-sized autosomes was involved while in Cooke and Gordon's case the ring was probably formed from one member of the no. 1 pair (largest chromosome). Most, at any rate, of these individuals showed chromosomal mosaicism, presumably due at least in part to breakage–fusion–bridge cycles of the ring. Cooke and Gordon have presented some evidence that in the case they studied stable metacentric chromosomes could arise from the rings, presumably as a result of 'healing' of two freshly-broken ends.

NUCLEOLI

In almost all higher organisms, one or more bladder-like *nucleoli* are prominent structures in interphase and prophase nuclei. These are connected with secondary constrictions located on particular chromosomes; the constrictions appear to function as *nucleolar organizers*. In many species of diploid plants and animals there is a single pair of nucleolar chromosomes, but in others several or even many chromosome pairs may engage in nucleolus formation. Thus in the human species the prophase nuclei of the primordial germ cells have a large nucleolus with which most or all of chromosome pairs 13, 14, 15, 21 and 22 are associated; apparently the short arms of the acrocentric chromosomes all bear nucleolar organizers (Ohno 1965). In bird karyotypes there are numerous very small micro-chromosomes which participate in forming a common nucleolus (Comings and Mattocia 1970). In the midges of the family Sciaridae there are numerous small nucleoli in the salivary gland nuclei (Poulson and Metz 1938; Jacob and Sirlin 1963), while in a midge belonging to a different family, *Chironomus tentans*, there are two major nucleoli, one arising from a nucleolar organizer on the second chromosome, the other from one on the third chromosome (Pelling and Beermann 1966). In crickets the *X*-chromosome is responsible for nucleolus formation, but may contain several nucleolar organizers (Sotelo and Wettstein 1966). In *Drosophila melanogaster* the nucleolar organizer lies in the immediate neighbourhood of the *bobbed* locus on the *X*- and *Y*-chromosomes, between the scute-4 and the scute-8 breakage points (Cooper 1959).

True nucleoli must be regarded as extensions of the chromosomes from which they arise; they frequently exhibit a strong tendency to fuse together, so that, for example, many nuclei containing a pair of nucleolar chromosomes exhibit a single 'fusion nucleolus'. Nucleoli usually shrink during prophase and have disappeared by the time the metaphase stage is reached, being re-formed again after telophase. About 80–90% of the dry weight of nucleoli is protein, but they also contain RNA fibrils about 50 Å in diameter, RNA granules and threads of nucleohistone arising from the nucleolar organizer.

The first critical evidence that nucleoli were localized products of biosynthetic activity at specific chromosomal loci came from the work of Heitz (1931) and McClintock (1934, 1940) on plant materials. Later, it was shown by Caspersson and Schultz (1939), Brachet (1940) and Vincent (1965) that nucleoli were very rich in RNA. More recent work, involving molecular hybridization experiments, has shown that the genes which code for ribosomal RNA lie in or very close to the nucleolar organizer in *Drosophila melanogaster* (Ritossa and Spiegelman 1965; Ritossa, Atwood, Lindsley and Spiegelman 1966) and in the South African clawed toad *Xenopus laevis* (Birnstiel, Wallace, Sirlin and Fischberg 1966; Wallace and Birnstiel 1966); this is presumably true of all higher organisms.

Ribosomes consist of two subunits, a larger one containing a molecule of 28s RNA and one of 5s RNA together with proteins, and a smaller subunit which contains a molecule of 18s RNA and associated protein. The 28s and

18s RNA is formed in the nucleolus by cleaving of 45s RNA molecules; the cleavage process is complex and involves methylation of both molecules (Maden 1968). The genes for 28s and 18s rRNA are apparently arranged alternately in the nucleolar DNA of *Xenopus* oocytes, but are probably separated by nucleotide sequences different from either (Brown and Weber 1968b). The haploid chromosome complement of *Xenopus* contains about 450 of the 28s and 18s cistrons. Apparently the guanine + cytosine content of the 45s, 28s and 18s RNA molecules is unusually high. The 5s RNA component of the ribosomes originates independently outside the nucleolus (in *Drosophila melanogaster* from a locus on chromosome 2, according to Wimber and Steffensen 1970) and is presumably coded for by cistrons which are not included in the nucleolar organizer.

In ordinary somatic cells the number of nucleolar organizers will be constant per diploid genome. In the oocytes of several species of Amphibia, however, it has been shown that the nucleolar organizer undergoes a differential replication (by companion with the rest of the genetic loci) during pachytene, so that there are eventually 1000 or more copies of it in the nucleus. In the toad, *Bufo* the total amount of DNA increases from the 4C quantity to about 10C in the course of the pachytene stage (Gall 1968), while in *Xenopus* the increase is apparently from 4C to about 14C (Macgregor 1968). The thousands of short DNA sequences apparently separate from the original nucleolar organizer; in *Xenopus*, they form a compact mass at one stage, but this breaks up later to form hundreds of granules (each of which is probably a circular DNA molecule). The hundreds of nucleoli which appear at a slightly later stage apparently have one such granule each associated with them. The extra DNA is only synthesized during pachytene, but is stable and persists throughout the later stages of oogenesis in association with the nucleolar structures. The number of ribosomal cistrons, which is approximately 1600 in somatic nuclei, is increased to 4000000 in the oocyte (Perkowska, Macgregor and Birnstiel 1968).

In the primitive tailed frog *Ascaphus*, Macgregor and Kezer (1970) have shown that the last three oogonial mitoses do not involve a cytoplasmic division, so that the oocyte at the beginning of meiosis contains eight 'lampbrush' nuclei, seven of which degenerate late in prophase. Whereas late prophase oocyte nuclei of *Xenopus*, *Bufo* and *Necturus* contain about 27–30 picograms of extrachromosomal nucleolar DNA, the single post-pachytene oocyte nuclei of *Ascaphus* have less than 5 picograms of this material. Thus *Ascaphus* seems to undergo the same degree of amplification of its nucleolar genes as other amphibia, but it does so by having multi-nucleate oocytes, within each of whose nuclei the differential replication of the nucleolar organizer is only on a small scale.

A mutant in which the nucleolar organizer is deleted or inactivated is known in *Xenopus* (Elsdale, Fischberg and Smith 1958; Fischberg and Wallace 1960). Normal individuals have two nucleoli in their diploid nuclei, one derived from each nucleolar organizer (in some somatic nuclei only one double-sized nucleolus is present and it has been assumed, but not proved this is due to fusion of two nucleoli, one derived from each nucleolar organizer – but see Barr 1966). In the cells of the heterozygotes for the anucleolar mutation there is never more than a single nucleolus. The homozygous anucleolar condition is lethal at about the time of hatching; embryo homozygotes show no organized nucleoli in the earlier stages (Wallace 1960), but do develop them later (Barr 1966). A similar situation is found in the midge *Chironomus tentans*, in which embryos lacking nucleoli fail to develop (Pelling and Beermann 1966). *Xenopus* apparently has an efficient control mechanism which ensures that the nucleoli of heterozygotes are as large as the fused double nucleoli of normal individuals (Barr and Esper 1963; Esper and Barr 1964) and synthesize normal amounts of ribosomal RNA, although they have about half the normal amount of ribosomal DNA. The heterozygotes lack the nucleolar constriction in one member of a pair of chromosomes (Kahn 1962), so that the most plausible, although not the only, explanation of the anucleolar mutation in *Xenopus* is that it is an actual cytological deletion. It has been shown by Brown and Gurdon (1964) that in the lethal homozygotes for the anucleolar mutant there is no synthesis of 28s and 18s ribosomal RNA or of high molecular weight ribosomal RNA precursor molecules. Deletion of the nucleolar organizer does not affect the cistrons for 4s and 5s RNA, which must be located elsewhere in the karyotype (Brown and Weber 1968*a*; Wimber and Steffensen 1970).

The special mechanism for magnification of the synthesis of ribosomal RNA in the amphibian oocyte is undoubtedly only one example of a general phenomenon, and we must expect similar mechanisms to operate during oogenesis in many and perhaps all species of animals. But the exact morphological appearances vary from case to case. In insect oocyte nuclei three different DNA-containing bodies have been shown in recent years to be composed of hundreds or thousands of separated replicates of the chromosomal nucleolar organizing region which codes for ribosomal RNA.

The first is the structure which appears in the oogonial nuclei of various species of craneflies of the genus *Tipula* and persists until the diplotene stage in the oocytes, when it disintegrates and disappears (Bauer 1931, 1933, 1952; Bayreuther 1952, 1956; Lima-de-Faria 1962; Lima-de-Faria and Moses 1966). This body is only present in the female sex and first makes its appearance in close association with the sex chromosomes in the oogonia four divisions before meiosis. At each of the oogonial mitoses the DNA body passes to one pole; it increases in size during the premeiotic growth phase. All oocytes possess the DNA body, but only a few of the nurse cells have one. Apparently it contains thousands of copies of the cistrons of the nucleolar organizing region. When fully developed it contains about 60% of the DNA of the nucleus, and has the RNA-rich nucleoli embedded in its substance. At diplotene the DNA body abruptly disappears, releasing its contents into the nucleoplasm or the cytoplasm.

A somewhat similar body which makes its appearance in the oogenesis of the water beetle *Dytiscus* was studied as early as 1901 by Giardina, who named it the '*anello cromatico*' (chromatic ring). It has been recently reinvestigated by Urbani and Russo-Caia (1964), Urbani (1969), Bier, Kunz and Ribbert (1967) and Gall, Macgregor and Kidston (1969). The latter authors also studied a similar body in another beetle, *Colymbetes*. These bodies contain DNA which can be proved by molecular hybridization techniques to code for ribosomal RNA but there is also a considerable quantity of other DNA, whose function is unknown; presumably the ribosomal RNA amplification mechanism is not the only genetic magnification system in the egg development of these insects.

The DNA body of the cricket *Acheta domesticus* (Bier, Kunz and Ribbert 1967; Heinonen and Halkka 1967; Lima-de-Faria, Nilsson, Cave, Puga and Jaworska, 1968; Lima-de-Faria, Birnstiel and Jaworska 1969; Jaworska and Lima-de-Faria 1969; Kunz 1969) resembles those of *Tipula* and *Dytiscus* in many respects. It originates in the oogonia at interphase and is present in all oocytes, eventually reaching a diameter of 6 μm. At the pachytene and diplotene stages of meiosis it has an outer shell of RNA and an inner core of DNA (the chemical nature of these regions has been clearly demonstrated by staining with azure B and Feulgen stain and by incorporation of tritiated thymidine and tritiated uridine). Many DNA-containing fibrils radiate into the RNA zone, giving the DNA body an appearance similar to that of a 'puff' in a dipteran salivary gland chromosome (see p. 106). The DNA body of *Acheta* is apparently formed by one of the autosomes and contains the nucleoli; its DNA has a guanine + cytosine content of 56% and clearly represents hundreds of replicates of the nucleolar organizing cistrons. No comparable body exists in the spermatogenesis of *Acheta*, and the amount of 56% G + C DNA relative to 'ordinary' DNA in the ovary at pre-meiotic interphase is 18 times the amount found in the testis. At diplotene the DNA body disintegrates and liberates its DNA, histone and RNA into the nucleus. As in the dytiscid beetles, there are amplified DNA sequences in the *Acheta* DNA body which do not code for rDNA.

A highly peculiar feature of the *Acheta* DNA body is that it contains groups of packages of multiple synaptinemal complexes, similar to those formed between the homologous chromosomes at meiosis (see p. 156) (Jaworska and Lima-de-Faria 1969). It is uncertain whether these structures are formed between the replicated cistrons of the maternal and paternal homologues or between the extra copies of the ribosomal genes of the individual chromosomes.

Three different intranuclear constituents have been studied by Halkka and Halkka (1968) and Halkka, Halkka and Nyholm (1969) in dragonfly oocytes. The main or primary nucleolus is responsible for ribosome biosynthesis, as in other organisms. The much smaller secondary nucleolus is associated with a different chromosome pair and seems to synthesize sRNA and arginine-rich proteins. In addition to these structures dragonfly oocyte nuclei contain peculiar lamellate bodies of unknown function.

Gall (1968) has clearly pointed out that these peculiar gene-amplification mechanisms in both amphibia and insects must have arisen in evolution to cope with a unique biosynthetic problem faced by large yolky oocytes – they must supply RNA to a volume of cytoplasm which would at other stages contain thousands of nuclei. Since polyteny or polyploidy of the oocyte nucleus as a whole is incompatible with a normal meiotic mechanism, two different biosynthetic systems have been evolved – the nurse-cell mechanism, whereby highly polyploid or polytene nurse-cells (in reality non-functional oocytes) transfer RNA to the oocyte through intercellular bridges of cytoplasm (Bier, Kunz and Ribbert 1969) and the gene-amplification mechanisms whereby particular genetic loci are replicated thousands of times within the oocyte nucleus. As far as insects are concerned, the nurse-cell mechanism seems to exist only in the 'higher' orders with so-called meroistic ovaries (particularly those of the polytrophic type). In the more primitive insects with panoistic ovaries there are no nurse cells. Thus, of the three examples of insects with giant DNA bodies in their oocyte nuclei which we have discussed, both *Tipula* and *Dytiscus* also have the nurse-cell mechanism, but *Acheta* does not.

The essential nature of nucleolar organizer amplification in oocyte nuclei was well understood thirty years ago by Painter and Taylor (1942) who wrote: 'If these nucleolar organizers are genetically the same as those which form nucleoli in ordinary somatic cells, then we may say that the germinal vesicle of the toad is highly polyploid in nucleolar organizers, but otherwise lampbrush chromosomes are normal meiotic structures.' Ribosomal DNA amplification mechanisms have now been demonstrated in the oocytes of such diverse organisms as the teleost *Roccus* (Vincent, Halvorson, Chen and Shin 1968), the mollusc *Spisula* and the echiurid worm *Urechis* (D. D. Brown and Dawid 1968), *Locusta migratoria* (Kunz 1967) and several species of fleas (Bayreuther 1957).

HETEROCHROMATIN, HETEROPYCNOSIS, LATE-REPLICATION AND GENETIC INERTNESS

It has been recognized from the earliest period of chromosome cytology that certain chromosomes or chromosome regions exhibit a condensed and (in fixed material) heavily-staining structure at stages of the mitotic and meiotic cycles when the rest of the karyotype is in a diffuse, weakly-staining condition. This behaviour was referred to as *heteropycnosis* by the earlier cytologists (Gutherz 1907). When seen at prophase it has sometimes been called *precocious condensation*, a term that seems inapplicable when the chromosomes have never been uncondensed, or when the phenomenon is seen in interphase nuclei which may not be destined to undergo further division. Heitz (1928, 1929) introduced the concept of *heterochromatin* as a chromosomal material which exhibits the property of heteropycnosis at particular stages of the mitotic cycle; he contrasted this material with *euchromatin*, which does not show heteropycnosis. Most chromosomes seem to contain both euchromatic and heterochromatic segments, the latter being frequently adjacent to the centromeres and/or telomeres; but some chromosomes, such as the *Y*-chromosomes in many species, seem to be composed entirely of heterochromatin (whether any chromosomes consist entirely of euchromatin is more doubtful). In many interkinetic nuclei large or small condensed masses which stain with basic dyes can be seen (review in Tschermak-Woess 1963). These structures, which have been called *chromocentres*, are composed of heterochromatin; in some instances they represent whole chromosomes, but more frequently they are composed of short segments of chromosomes, frequently proximal or distal ones. To a considerable extent the different appearances of interkinetic nuclei, in ordinary histological preparations ('coarsely granular', 'finely granular' etc.) are due to the number, size and distribution of the chromocentral masses.

An important distinction has been drawn (S. W. Brown 1966) between so-called *constitutive heterochromatin* which is heteropycnotic in all types of cells, and especially in every interphase nucleus, and *facultative heterochromatin*, which only manifests heteropycnosis in special cell types or at special stages. Extreme types are easy to assign to these categories, but there are examples known of chromosomes or chromosome regions which seem to occupy an intermediate position between these theoretically distinct types.

Typically, heterochromatin is more condensed than euchromatin; some authors have spoken of 'over-condensed' segments. But in certain cases a peculiar reversal of this behaviour occurs. Thus in all species of true

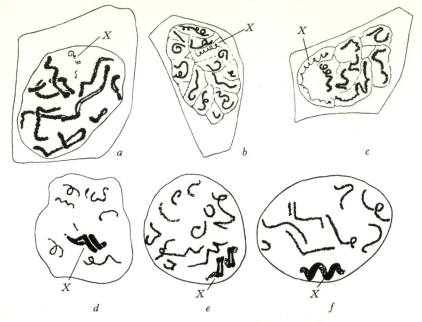

Fig. 1.4. Early spermatogonial prophase nuclei in various species of grasshoppers (*a–c*, Acrididae; *d–f*, Tettigoniidae; in the former the *X*-chromosome is negatively heteropycnotic, in the latter it is positively heteropycnotic). *a, Schistocerca gregaria; b, Melanoplus femur-rubrum; c, Chortophaga viridifasciata; d, e, Platycleis grisea; f, Metrioptera brachyptera*. Relic spirals showing in most of the nuclei. From White (1940*a*).

grasshoppers (suborder Caelifera of the order Orthoptera) the *X*-chromosome is very heavily condensed and densely staining throughout the prophase of the first meiotic division in the male (i.e. at a time when the autosomes are, for the most part, thin and diffuse). And in the early spermatid nuclei

Fig. 1.5. Negative heteropycnosis of the *X*-chromosome in a spermatogonial prophase and metaphase of *Locusta migratoria*. Drawn from irradiated material, hence various structural rearrangements present. From White (1935*b*).

Fig. 1.6. Spermatogonial metaphases of (*a*) the grasshopper *Monistria vinosa* (one pair of autosomes (*h*) as well as the *X*-chromosome negatively heteropycnotic), (*b*) the gryllacridid *Hadrogryllacris* sp. (the *X*-chromosome is more slender than the autosomes, but has a smooth outline).

the *X* is once more a compact, densely-staining mass. At certain other stages in spermatogenesis, however, the same chromosome exhibits precisely the reverse type of behaviour. Thus in the mitoses of the primordial spermatogonia the *X* is thin, diffuse and weak-staining at metaphase, when the other chromosomes are condensed, dark-staining and have smooth outlines. It may also show, in certain species, the same diffuse condition at the first metaphase of meiosis in the male. There is thus a cyclical reversal of behaviour; *negative heteropycnosis* in the spermatogonial metaphases is followed by *positive heteropycnosis* in the prophase of meiosis and a further reversal may occur before the spermatid is formed. It is interesting to note that in those same insects the *X*-chromosomes do not show either type of heteropycnosis in the oogonia or the female meiotic divisions, and they do not seem to do so in the somatic mitoses of either sex. Another interesting point is that many species of grasshoppers possess conspicuous blocks of heterochromatin in certain of their autosomes and some of them also possess heterochromatic supernumerary chromosomes (Ch. 10). Yet, with only very rare exceptions, these autosomes and supernumerary chromosomes never exhibit negative heteropycnosis. Thus we can observe two different kinds of heterochromatin in the same nucleus, one exhibiting a reversal of heteropycnosis, the other not. It is perhaps worth pointing out that the reversible type of heteropycnosis is characteristic of certain groups of organisms and may never be found in other, related, ones. Thus the primordial spermatocytes of crickets (Grylloidea) show the same kind of negative heteropycnosis as those of the true grasshoppers (Acridoidea and Eumastacoidea). But in the long-horned grasshoppers (Tettigonioidea) the negative type of heteropycnosis is quite unknown and the *X* only shows

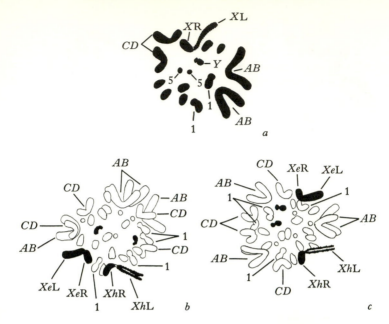

Fig. 1.7. A diploid and two tetraploid spermatogonial metaphases of the Australian morabine grasshopper species 'P52a'. In the case of the tetraploid metaphases the autosomes are shown in outline, the sex chromosomes in black. This is a species in which an evolutionary fusion between the X and an autosome has produced a neo-XY sex-chromosome mechanism. The Y is quite small. The X-chromosome consists of a 'left' limb (XL), which represents the original X, and a 'right' limb, which was originally autosomal. In diploid spermatogonia the XL limb shows a rather variable degree of negative heteropycnosis. In tetraploid spermatogonia only *one* XL limb (XhL) is negatively heteropycnotic; the other one (XeL) is condensed to the same extent as the autosomes and XR. From White (1970b).

positive heteropycnosis (in spermatogonial mitoses, at the meiotic prophase in spermatogenesis and again in the spermatid nuclei).

In tetraploid grasshopper spermatocytes both the X's show the positive type of heteropycnosis (White 1933). It is hence not the single nature of the X in the diploid spermatocytes which determines the positive heteropycnosis. But in tetraploid spermatogonia (at any rate in one species of eumastacid grasshopper) only one of the two X's shows the negative heteropycnosis, the other being condensed to the same degree as the autosomes (White 1970b). Since the two X's must be genetically identical in this case (unless they have undergone some kind of secondary change, e.g. unequal somatic crossing-over), some very special mechanism must exist which determines that one shall undergo condensation while the other remains negatively heteropycnotic. There is a clear analogy to the differential behaviour of the two X-chromosomes in somatic cells of female mammals (see p. 594).

The idea that negative heteropycnosis is the reverse state of positive heteropycnosis has been opposed by Lima-de-Faria and Jaworska (1968). These authors refer to the data of Woodard and Swift (1964) and Woodard Gorovsky and Swift (1966), who showed that normal and cold-treated nuclei and individual chromosomes of the plant *Trillium* showed the same DNA content, in spite of the fact that the cold-treated chromosomes had long regions negatively heteropycnotic (i.e. exhibiting 'nucleic acid starvation' in Darlington and La Cour's (1940) unfortunate and misleading terminology). Woodard *et al.* conclude reasonably enough, that the negatively heteropycnotic segments have 'localized uncoiling' and show a modified RNA and protein metabolism. By way of contrast to these observations Lima-de-Faria and Jaworska cite the work of Lima-de-Faria (1959*b*) in which he claimed to have shown that positively heteropycnotic chromosomes contain 'two to three times more DNA per unit area' than non-heteropycnotic ones (the comparison used here was the grasshopper X-chromosome at prophase of male meiosis versus autosomes at the same stage). It should be obvious that in one case the comparison is between the DNA content of *whole* chromosomes (negatively heteropycnotic in certain regions or not), while in the other case it is on a *per unit area* basis. The two sets of observations, each separately valid, are simply not comparable. We believe, therefore, that it is still justifiable to regard negative and positive heteropycnosis as reverse states, the first being due to a relative degree of uncoiling, the latter to some kind of supercoiling. The claim of Lima-de-Faria and Jaworska would only

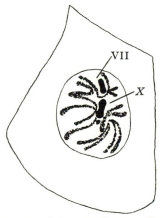

Fig. 1.8. Second spermatocyte of the grasshopper *Stethophyma grossum* in interkinesis (i.e. between the first and second meiotic divisions), showing the positive heteropycnosis of the X and the 'megameric' VIIth autosome.

be valid if it could be shown that the grasshopper X-chromosome showed a higher DNA content when positively heteropycnotic (e.g. in the prophase of male meiosis) than when non-heteropycnotic (e.g. in female somatic cells). Apparently measurements of this kind have never been made. Lima-de-Faria's (1959b) conclusions were based not only on microphotometric measurements but also on grain-counts following tritiated thymidine autoradiography; these counts, however, merely demonstrate the now well-known late-labelling phenomenon and have no relevance as far as DNA content is concerned.

One general feature of heterochromatin is a frequent tendency for different heterochromatic regions to enter into non-specific associations, of a temporary or permanent nature, in various types of nuclei, and at various stages of mitosis or meiosis (Schrader 1941b; Slack 1938b). This is shown in many interphase nuclei by fusion of chromocentral masses. In the polytene nuclei of *Drosophila* (see Ch. 3) the heterochromatic regions adjacent to the centromeres of all the chromosomes become permanently fused to form the so-called chromocentre, and many intercalary heterochromatic segments may show so-called ectopic pairing (Slizynski 1945; Kaufmann and Iddles 1963). And in many species the heterochromatic regions of different chromosomes may adhere temporarily, during the prophase of meiosis. The older workers often referred to such associations as *non-homologous* pairing. It seems likely that it is due, in the main, to true homology, i.e. the repetitiveness of the nucleotide sequences in the heterochromatin. But there are some unexplained features of these associations. For example, in many grasshopper species there are temporary adhesions or associations between blocks of autosomal heterochromatin and the X-chromosome, during the prophase of the male meiosis; it seems unlikely that the nucleotide sequences which are repeated in these two types of heterochromatin are the same, although there may be sufficient similarity to lead to a force of attraction between them.

An association between heterochromatin and genetic inertness was earlier suggested by the situation in *Drosophila melanogaster*. In this species the whole of the Y-chromosome, the proximal half of the X and the regions around the centromeres of the metacentric second and third chromosomes are both heterochromatic and genetically inert and so is a substantial part of the little fourth chromosome. A detailed description of the heterochromatin in the sex chromosomes has been given by Cooper (1959), who has published a detailed map of the heterochromatic region of the X, which includes the minute right arm and four main blocks, separated by constrictions (in the case of the two middle blocks, the constriction between them is the site of the nucleolar organizer).

It is probable that we should distinguish between two kinds of genetic inertness corresponding to the two main types of heterochromatin which we have called *constitutive* and *facultative*. Constitutive heterochromatin, which is probably made up largely or entirely of repeated sequences of satellite DNA, is probably inert in the strict sense of not coding for any protein or only for some very simple protein without much specificity. Facultative heterochromatin, on the other hand may be regarded as genetic material which is biosynthetically active in certain tissues or at certain stages of mitosis and meiosis, but which is liable to be inactivated or switched off in other tissues or stages.

It is not easy to say whether the *X*-chromosomes of grasshoppers should be regarded as composed of constitutive or facultative heterochromatin. If we considered only the male sex we would probably regard them as made of constitutive heterochromatin. But in female somatic, oogonial and meiotic nuclei they seem to be generally, and perhaps universally, non-heteropycnotic; so that it would seem more proper to regard them as made of a special category of facultative heterochromatin, whose behaviour is sex-controlled.

In the somatic nuclei of most female mammals a small heteropycnotic mass can be seen which is absent from the corresponding nuclei of male individuals. This is the so-called sex chromatin or Barr body, first reported by Barr and Bertram (1949). We shall discuss this structure in detail later (Ch. 16) but it is necessary to state here that it represents *one* of the two *X*-chromosomes, which becomes heteropycnotic in the course of development, and that in some cells this is the paternal *X* and in others the maternal *X*.

Obviously, the formation of the Barr body by one of the *X*'s is a classical example of what is meant by facultative heterochromatin. There is now good evidence that heteropycnosis of one of the *X*'s in this case is associated with genetic inactivation or switching-off of certain of the loci in that chromosome (Lyon 1961, 1962, 1963, 1966*a*, *b*, 1968; Russell 1961, 1963, 1964*a*, *b* – see Ch. 16 for fuller discussion).

Another example of an association of heteropycnosis and genetic inactivation is provided by certain male scale insects (S. W. Brown and Nur 1964). Here the entire haploid set of chromosomes derived from the father is heteropycnotic in the nuclei of the male offspring, from an early stage of embryology onwards. We shall discuss later (p. 507) the genetic evidence that this set is also inactivated (because doses of radiation which would induce dominant lethals in an active set, and which do give rise to lethals in the female offspring, do not have a killing effect on the males). Berlowitz (1965*a*) has also shown, by tritiated uridine autoradiography, that the heteropycnotic set of chromosomes in these male scale insects have almost completely discontinued RNA synthesis. There is apparently some difference between the histones of the heteropycnotic and non-heteropycnotic chromosome sets (Berlowitz 1965b).

Another rather special case of facultative heterochromatin occurs in the gall midges (see p. 532 for details). Here the germ-line nuclei (spermatogonia and oogonia) contain a large number of chromosomes and the somatic nuclei a much smaller number. The chromosomes represented in the soma are referred to as S-chromosomes, the extra ones found only in the germ-line being called E-chromosomes. It is a universal rule in this group that the S-chromosomes are positively heteropycnotic in the germ-line nuclei of both sexes, while the E-chromosomes are diffuse in the interphase and prophase nuclei. In this instance, if we regard strong positive heteropycnosis as an indication of genetic inactivation (as we must do, by analogy with other cases), the chromosomes which are the only ones represented in the somatic tissues would be 'switched off' as far as biosynthesis is concerned, in the germ-line, from an early embryonic stage onward.

In recent years it has been shown that the replication of the chromosomal DNA during the so-called S phase of interphase (Howard and Pelc 1951) is not exactly synchronous over the whole karyotype. Certain chromosomes and chromosome segments are late-replicating by comparison with the rest of the karyotype. In general, late-replicating segments correspond to heterochromatic ones, but the relationship is not an entirely simple one.

The exact chronology of replication has been studied by means of tritiated thymidine autoradiography. Thus when we speak of a chromosomal region as late-replicating, we really mean that it is late-labelling. It is possible to determine the 'labelling patterns' of particular cell types by examination of labelled metaphases at varying periods of time after exposure to the radioisotope. This is most conveniently carried out in cultured cells *in vitro*, but in the case of small invertebrates one may also use in-vivo techniques.

The earliest investigations of this type were carried out on spermatocytes of the grasshopper *Melanoplus differentialis* by Lima-de-Faria (1959a, b). The observations, carried out at the zygotene and pachytene stages of meiosis, indicated that the positively heteropycnotic X-chromosome is late-labelling, i.e. that it both starts and finishes DNA replication later than the autosomes. Thus if tritiated thymidine is supplied to the nucleus for a short time early in the S phase it will only be incorporated into the autosomes, if it is available somewhat later in the cycle it will be incorporated into both euchromatin and heterochromatin, while if it is only available at the end of the S phase it will be incorporated mainly or exclusively into the X. Lima-de-Faria also studied autoradiographs of rye chromosomes which also indicated late-labelling of heterochromatin, in this case the heterochromatic segments around the centromeres.

Asynchronous replication of the chromosomal DNA was next studied in cultured somatic cells of the Chinese Hamster, *Cricetulus griseus*, by J. H. Taylor (1960*a*, *b*). The mean time for a complete nuclear cycle was about 14 hours; the S phase lasted about 6 hours and stopped 2–3 hours before metaphase. Taylor found that several pairs of autosomes had late-labelling segments. In male cells the long arm of the X-chromosome and the whole of the Y underwent DNA replication during the latter half of the S phase. The short arm of the X, on the other hand, was replicated during the first half of the S phase. In female cells one whole X was late-replicating, while the other behaved as in the male. Taylor's observations were more satisfactory than the earlier ones of Lima-de-Faria because they were carried out on metaphases, so that localization of labelling pattern to definite chromosomal segments was possible.

The late-replication of one of the two X's in somatic cells of female mammals has now been shown to be present in a large number of species and presumably occurs in all of them, (except for such unusual cases as *Microtus oregoni* and *Ellobius lutescens* – see p. 600). It parallels the presence of a condensed body, the so-called sex-chromatin or Barr body in the interphase nuclei, which is apparently derived from one of the X-chromosomes, which may be the paternal X in some patches of tissue and the maternal one in other patches (see p. 594). The sequence of replication in the human karyotype has been studied by Lima-de-Faria, Reitalu and Bergman (1961), Gilbert *et al.* (1962), Schmid (1963) and German (1964*a*, *b*); the replication patterns of man and the chimpanzee have been compared by Low and Benirschke (1969).

The replication pattern in somatic metaphases of *Drosophila melanogaster* has been studied by Barigozzi *et al.* (1966) and Halfer *et al.* (1969), using tritiated thymidine autoradiography on cells in culture. Heterochromatin was consistently late-replicating by comparison with euchromatin. In stocks with translocations involving the Y, the replication pattern of translocated heterochromatin is modified in some cases but not in others.

The replication patterns of four species of snakes were studied in leucocyte cultures by Bianchi *et al.* (1969). Their most noteworthy finding was that in all species the sex chromosomes finished their replication before any of the other chromosomes.

In birds there are always a large number of small microchromosomes in the karyotype. In the quail these are heterochromatic, nucleolus-organizing and late-labelling (Comings and Mattoccia 1970). On the other hand, the microchromosomes of the chicken are neither heterochromatic nor late-labelling (Schmid 1962; Bianchi and Molina 1967), although they may be nucleolus-organizing (Ohno, Christian and Stenius 1962).

Since positive heteropycnosis is associated with late-labelling, one might expect that negative heteropycnosis would be accompanied by 'early-labelling'. The evidence on this point is not clear, however. Nicklas and Jaqua (1965) applied autoradiographic techniques to the spermatogenesis of the grasshopper *Melanoplus differentialis* and concluded that 'in all cell generations in which the X-chromosome is negatively heteropycnotic, its replication is synchronous with autosomal replication . . .'. However, they do not seem to have actually identified the beginning and end of the X and autosomal replication periods to determine whether they are precisely synchronous.

In addition to heteropycnotic and late-labelling segments, the chromosomes of higher organisms are now known to possess special segments that fluoresce with dyes such as atebrin and quinacrine mustard (Caspersson *et al.* 1968; Caspersson, Zech, Johansson and Modest 1970; Pearson, Bobrow and Vosa 1970; Vosa 1970). Quinacrine mustard binds to DNA in two ways, by its reaction with guanine and by intercalation in the double helix. Bands or regions that fluoresce strongly with quinacrine mustard should hence be guanine-rich. In the human karyotype the distal half of the long arm of the Y, the short arms of chromosomes 13, 14 and 15 and a short region near the centromere of chromosome 3 fluoresce especially strongly. By making photoelectric recordings of the fluorescence patterns along the length of the chromosomes, Caspersson *et al.* (1970) have been able to show that each of the human chromosomes has its own characteristic 'fluorescence profile'. These workers speak of this technique as differentiating between heterochromatic and euchromatic segments. This is certainly an oversimplification. It is probable that only certain types of heterochromatin are guanine-rich. The whole of the human Y-chromosome is heterochromatic and late-labelling, but only the distal part of the long arm fluoresces strongly with quinacrine mustard (Pearson, Bobrow and Vosa 1970; George 1970); and although one X-chromosome is consistently heteropycnotic and late-labelling in female somatic cells it does not fluoresce at all strongly. Strong fluorescence may be characteristic of constitutive heterochromatin but is not likely to be shown by facultative heterochromatin; but we do not really know yet whether all constitutive heterochromatin fluoresces. In at any rate some Australian aborigines the Y is significantly shorter than in Europeans, the difference being apparently due to the much smaller size of the fluorescent segment (Angell, unpublished).

SATELLITE DNA'S

It has been known for the past ten years that higher organisms possess so-called satellite DNA's which can be separated from the main DNA by density gradient centrifugation, because they differ from it in buoyant density. The two strands of satellite DNA, when separated by heating, reassociate much faster than do strands of the main DNA. These properties suggest that satellite DNA's consist of relatively short sequences of bases, repeated many thousands or even millions of times in the genome. For this reason, they have sometimes been called repetitive or repetitious DNA's. Britten and Kohne (1968) give a list of fifty-one species of animals and plants in which repetitious DNA's have been shown to exist; the implication is that they occur universally in eukaryotic organisms. Coudray, Quetier and Guille (1970) have also tabulated extensive data on satellite DNA's and Arrighi, Mandel, Bergendahl and Hsu (1970) have provided data for a large number of species of mammals. There are apparently no differences between the various tissues of the body in regard to the repetitious DNA (McCarthy and Hoyer 1964; Britten and Kohne 1968).

The satellite DNA of the mouse, which has been extensively studied, is rich in A and T nucleotides and correspondingly lacking in G and C ones (G + C content about 30%). It is consequently lighter than the main DNA fraction and constitutes about 10% of the total DNA (Kit 1961; Hennig and Walker 1970). The 5-methylcytosine content of this satellite DNA is also higher than that of the main fraction (Salomon, Kaye and Herzberg 1969). In man, the satellite DNA is likewise lighter than the main fraction, but is only about 2% of the total quantity (Corneo, Ginelli and Polli 1968). Some species have fast-reassociating DNA fractions, which are presumably repetitive, but which lack the difference in base composition from the main DNA fraction which is necessary to give a distinct satellite on a caesium chloride density gradient (McLaren and Walker 1969). And in other species it is possible to distinguish between two fractions of repetitive DNA, one with a fast rate of reassociation and one with an intermediate rate (Hennig and Walker 1970).

Almost all the satellite DNA's of rodents seem to be lighter than the main DNA fraction (Hennig and Walker 1970). Certain heavy satellites reported by Arrighi, Mandel, Bergendahl and Hsu (1970) may have been due to microbial contaminants. Some decapod Crustacea have both a heavy and a light satellite fraction (Skinner, Beattie, Kerr and Graham 1970) and the American Lobster has a heavy satellite DNA but no detectable light satellite. It is highly probable that these heavy satellites are nucleolar in origin. Guinea-pig DNA contains at least three different satellites. Whereas earlier calculations based on the rate of reassociation had suggested a repeat length of 300–600 base pairs for the mouse and guinea-pig satellites, E. M. Southern (1970) has shown by primary structural analysis that the repeating sequence of the guinea-pig α-satellite DNA is probably only six base pairs in length and has the structure:

$$\text{C C C T A A}$$
$$\text{G G G A T T}$$

The guinea-pig genome would contain 10^7 replicates of this sequence. The light satellite of mouse DNA seems to be composed of a somewhat longer sequence (8–13 base pairs) which is known to contain:

$$\text{T T T T T C}$$
$$\text{A A A A A G}$$

The discrepancy between the repeat length predicted from reassociation kinetics and that found by sequence analysis is most probably due to the fact that some of these short sequences contain anomalous bases, due to 'mutation'.

In order to explain the origin of millions of replicates of these short sequences in evolution, Britten and Kohne (1967, 1968) suggested that they arise by a process of saltatory replication and are then spread through the genome of the species. The mechanism of evolutionary replication is possibly the same as that revealed by Keyl's studies (1964, 1965a, b) of *Chironomus thummi* and *C. piger* (see p. 364). E. M. Southern (1970) calculates that the 10^7 copies of the guinea-pig and satellite sequence must contain more than 10^6 'mutations'. But it is not necessary that all these should have been independent events subsequent to the saltatory replication; it seems much more probable that many of the mutations were copied and propagated in the course of the saltatory replication.

The satellite DNA's of closely related species are in some instances remarkably different in density and in amount. Thus the densities of the satellite DNA's of *Apodemus sylvaticus*, *A. flavicollis* and *A. agrarius* are all significantly different (Hennig and Walker 1970).

A general relationship undoubtedly exists between satellite DNA's and the so-called constitutive heterochromatin. But this is certainly not a simple relationship. Part of the difficulty arises from uncertainty as to just which chromosome sequents should be regarded as constitutive heterochromatin and which as facultative heterochromatin (see p. 25).

In the mouse and guinea-pig the satellite DNA's are located in the condensed fraction of isolated interphase chromatin (Yasmineh and Yunis 1970; Yunis and Yasmineh 1970) and in the mouse it has been shown by in-situ reassociation experiments that the satellite DNA is located near the centro-meres (K. W. Jones 1970; Pardue and Gall 1970). The latter observations are particularly demonstrative of an association with heterochromatin, since it is known that mouse chromosomes have proximal heterochromatic blocks. If there are also segments of constitutive heterochromatin adjacent to the telomeres of mouse chromosomes (Ohno, Kaplan and Kinosita 1957), these either do not contain satellite DNA sequences or only contain them in small numbers. Moreover Pardue and Gall report that one chromosome in their preparations consistently failed to bind satellite DNA, and this they believe to be the Y-chromosome, which is certainly composed of constitutive DNA. And the vole *Microtus agrestis* in which the giant sex chromosomes form about 20% of the total chromosome length has no obvious satellite DNA (by density gradient centrifugation) nor any 'fast' reassociating DNA fraction (it does have about 7% of its DNA with an 'intermediate' speed of re-association). In *Drosophila melanogaster* there is no difference in the base-composition of DNA from XO, XX, XY, XXY and XYY individuals (Perreault, Kaufmann and Gay 1968).

In the grasshopper *Myrmeleotettix maculatus* there is no distinct satellite DNA corresponding to the X or to autosomal heterochromatin, but heterochromatic supernumerary chromosomes, which are present in some populations but not in others, contain a light satellite (Gibson and Hewitt 1970). In this species the density of the main DNA appears to be different in populations from Wales and East Anglia and there is likewise a difference (in the same direction) between the densities of the satellite DNA's in these areas.

It is extremely unlikely that the satellite DNA's composed of as few as six to thirteen nucleotide pairs per unit of repetition code for any protein (if they did so it would have to be an immensely long series of repeating di- or tri-peptides). Their chemical structure is hence compatible with two of the classical properties of heterochromatin – promiscuous 'non homologous' pairing and genetic 'inertness'. But clearly long sequences of repeats that have been built up in the course of evolution cannot be functionless and various suggestions have been made according to which they are responsible for the 'integrity' of the chromosome, for 'packaging' or for 'housekeeping' (whatever these metaphors mean). Clearly there has been a very complex evolution of satellite DNA's in rodents (Hennig and Walker 1970), in Crustacea (Skinner, Beattie, Kerr and Graham 1970) and in grasshoppers (Gibson and Hewitt 1970). This is a field in which further remarkable discoveries may be expected in the near future – after which the implications of satellite DNA's and heterochromatin for evolutionary processes may become clearer.

AMOUNT OF DNA IN THE KARYOTYPE

It is now possible to measure the amount of DNA in single nuclei, by various microspectrophotometric techniques, with considerable accuracy. Fluorometric methods have also been used on purified preparations of uniform populations of cells such as erythrocytes or sperms, the cell concentration having been determined by a Coulter Counter or haemocytometer. Many microspectrophotometric studies have been aimed at comparisons of the relative amount of DNA in the nuclei of different tissues in the same material, or at different stages in the mitotic cycle, i.e. they have been designed to detect polyploid nuclei or to throw light on the relation of DNA replication to the nuclear cycle. Other investigations have been carried out to determine the relative amount of DNA in the karyotypes of different species of organisms. In many of these studies the results are only given in arbitrary units. But in other cases, by using as a control a type of nucleus whose DNA value is already known from chemical determinations, the absolute amount of DNA per nucleus [measured in picograms (pg), i.e. $gm \times 10^{-12}$] has been determined. Up to the present time there have been very few determinations of the amount of DNA in individual chromosomes, but they are perfectly possible, and Bailly (1967) has shown that the relation between amount of DNA and chromosome length is linear in the salamander *Pleurodeles*. Almost all the microspectrophotometric determinations of DNA values have been carried out on nuclei stained with Feulgen dye. The quantity of DNA per haploid genome (e.g. in the sperm nucleus) is referred to as the 1C amount. Thus ordinary somatic nuclei contain the 2C amount, while prophase nuclei will have the 4C quantity.

One of the most important applications of these studies has been the determination of the absolute amounts of DNA in the karyotypes of different species. These may be expressed in picograms or converted to the number of nucleotide pairs in the haploid karyotype (1 picogram of DNA being equivalent to about 9.5×10^8 nucleotide pairs). In Table 1.3 we present a summary of the available determinations of the DNA values of animal species. There is some uncertainty about the accuracy of a number of these determinations, especially the early ones, but the data as a whole are probably reliable enough to base some conclusions on. We have not included the very low value of 0.17 pg of DNA in Anlage nuclei of *Drosophila melanogaster* given by Kurnick and Herskowitz (1962) in this table, because of serious doubts as to its reliability. Approximately ten-fold differences in DNA values occur within such groups as molluscs and fishes. The lowest values occur in some marine groups such as coelenterates and tunicates. It has sometimes been stated or implied that there is a progressive increase in DNA values throughout the vertebrate evolutionary series. In actual fact there is nothing of the kind; the most primitive chordates such as tunicates and *Amphioxus* have very low DNA values, but some, at any rate, of the cyclostomes and elasmobranchs show values which are almost as high as those of the mammals and higher than those of the reptiles and birds.

Ohno, Wolf and Atkin (1968) have speculated that the first vertebrate had a DNA content of about 1.4 pg per diploid genome and that 'ancient crossopterygian fishes' increased this amount, not only by regional duplication but also by 'tetraploidization'. They support this conclusion (and the assumption of other tetraploidizations in vertebrate evolution) by evidence derived from the loci coding for the light and heavy chains of gamma-globulin and those which code for the subunits of the enzyme lactate dehydrogenase. Their ingenious argument seems to collapse if one takes into account that duplication of genetic loci in the karyotype can come about by translocation as well as by unequal exchanges between homologous chromosomes or sister chromatids of the same chromosome. Ohno *et al.* present a photograph of meiosis in the Rainbow trout, *Salmo irideus*, with three quadrivalents, as evidence that this species is a tetraploid. The chromosome number is 2n = 68 and the DNA value is rather high for a teleost. The case is superficially convincing, but two alternative interpreta-

TABLE 1.3. *DNA content per nucleus (diploid) in various species of animals*

Phylum or class	Species	DNA content (pg)
Porifera		
Orange sponge	*Dysidea crawshagi*[1]	0.11
Coelenterata		
Jelly-fish	*Cassiopeia*[1]	0.66
Nematoda		
Horse roundworm	*Parascaris equorum*[2]	5.00
Echinodermata		
Sea-urchin	*Arbacia*[3]	1.34
	Paracentrotus[3]	1.40
	Echinometra[3]	1.96
	Lytechinus[1]	1.80
	Strongylocentrotus purpuratus[4]	1.78
Sea-cucumber	*Stichopus diabole*[1]	1.80
Mollusca		
	Fissurella barbadensis[1]	1.00
	Tectarius muricatus[1]	1.34
	Ilyanassa obsoleta[5]	6.50
Squid[1]		9.00
Annelida		
Nereid worm[1]		3.00

TABLE 1.3 — *continued*

Phylum or class	Species	DNA content(pg)
Crustacea		
Cliff crab	*Plagusia depressa*[1]	2.98
Goose barnacle[1]		2.92
Insecta		
	Locusta migratoria[6]	13.42
	Schistocerca gregaria[6]	19.00
Four species of morabine grasshoppers[7]		7.49 to 9.28
	Dytiscus marginalis[8]	
	Drosophila hydei[18]	0.4
	Chironomus tentans[19]	0.5
Tunicata		
	Ciona intestinalis[9]	0.41
	Ascidia atra[1]	0.32
	Amphioxus lanceolatus[9]	1.22
Cyclostomata		
	Lampetra planeri[9]	2.80
Lamprey	*Petromyzon*[1]	5.00
Hagfish	*Eptatretus stoutii*[9]	5.70
Elasmobranchii		
Shark	*Carcharias obscurus*[1]	5.46
Teleostei		
> 200 species[4]		0.80 to 8.8
	Tetraodon fluviatilis[4]	0.80
	Corydoras aeneus[4]	8.8
Shad	*Alosa*[1]	1.82 to 2.01
Brown trout[1]		5.79
	Barbus (2 spp.)[10]	1.49, 161
Carp and Goldfish[1,9]		3.04 to 3.83
Crossopterygii	*Latimeria chalumnae*[21]	5.6
Dipnoi		
Lungfish	*Protopterus*[1]	100
Amphibia	*Lepidosiren paradoxa*[22]	204
	Amphiuma means[23]	192
	Necturus[1]	48.4
'Frog'[1]		15.0 to 15.7
	Bufo (12 spp.)[20]	8.9 to 14.6
	Ascaphus truei[12]	7.1 to 8.2
	Xenopus[13]	6.3
Chelonia		
'Green turtle'[1]		5.12 to 5.27
	Gopherus agassizi[14]	6.10
	Amyda ferox[14]	5.64
Crocodilia		
	Caiman sclerops[14]	5.88
Sauria		
	Anolis carolinensis[14]	4.37 to 5.23
	Gerrhonotus multicarinatus[14]	4.61
	Xenodon merremi[14]	4.52 to 5.75
	Boa constrictor[14]	4.25 to 4.35
	Bothrops jararaca[14]	4.50 to 4.90
	Drimarchon corais[14]	3.91 to 4.02

TABLE 1.3—*continued*

Phylum or class	Species	DNA content (pg)
Aves		
Chicken	*Gallus domesticus*[1]	2.88
Pigeon	*Columbia livia*[14]	3.45
Canary	*Serinus canarius*[14]	3.71
Parakeet	*Melopsittacus undulatus*[14]	2.83
Mammalia		
Echidna	*Tachyglossus aculeatus*[24]	6.4
Platypus	*Ornithorhynchus anatinus*[24]	6.8
Kangaroo	*Macropus rufus*[15]	6.26
Possum	*Trichosurus vulpecula*[15]	6.00
Bandicoot	*Perameles nasuta*[15]	9.25
Dog	*Canis familiaris*[14]	5.86
Mouse	*Mus musculus*[6, 14]	6.52 to 7.93
Vole	*Microtus oregoni*[14]	5.06 to 5.75
Golden hamster	*Mesocricetus auratus*[14]	7.79
Aardvark	*Orycteropus afer*[16]	9.89
Horse	*Equus caballus*[14]	6.36
Man	*Homo sapiens*[25]	7.30

References

[1] Mirsky and Ris 1951*a*; Ris and Mirsky 1949.
[2] Nigon and Bovet 1955.
[3] Vendrely and Vendrely 1948, 1949.
[4] Hinegardner 1968.
[5] Collier and McCann Collier 1962.
[6] Fox 1970*a*.
[7] White and Webb 1968.
[8] Gall, Macgregor and Kidston 1969.
[9] Atkin and Ohno 1967.
[10] Ohno, Muramoto, Christian and Atkin 1967.
[11] Gall 1969.
[12] Macgregor and Kezer 1970.
[13] Brown and Dawid 1968.
[14] Atkin *et al.* 1965.
[15] Rendel and Kellerman 1955.
[16] Benirschke, Wurster, Low and Atkin 1970.
[17] Allfrey, Mirsky and Stern 1955.
[18] Mulder, van Duijn and Gloor 1968.
[19] Daneholt and Edström 1967.
[20] Bachmann 1970.
[21] Vialli 1957.
[22] Ohno and Atkin 1966.
[23] Callan 1972.
[24] Bick and Jackson 1967*c*.
[25] Bachmann 1972

Notes on the Table: The results of Atkin *et al.*, Atkin and Ohno and Ohno, Muramoto, Christian and Atkin have been converted to picograms on the assumption that the human species has a DNA value of 7.30 pg and the chicken one of 2.88 pg.

tions should be borne in mind. It is possible that the individual was hetero-zygous for translocations or that the species has duplicated regions at the ends of the chromosomes. The tettigoniid grasshopper *Metrioptera brachyptera* (White 1940*c*; Southern 1967), the cockroach *Periplaneta americana* (John and Lewis 1958, 1959) and the newt *Triturus helveticus* (Mancino and Scali 1961) all show rings or chains of four chromosomes at meiosis, but are certainly not tetraploids. We shall return to this whole question of the evolutionary significance of DNA values later (p. 402) in connection with the controversial question of the number of DNA strands per chromatid.

Darlington (1932*b* and later papers) considered that the size of the chromo-somes in an organism was under 'genotypic control'. This concept was put forward before there was any understanding of the molecular composition or structure of chromosomes. The idea was that gene mutations, without any structural rearrangements, could produce overall changes in the size of the chromosomes in an organism.

In the light of modern knowledge of chromosome structure, it seems probable that this concept has only limited validity, and may even be totally invalid. A gene mutation cannot change the amount of DNA in the karyo-type (except by the imperceptible amount represented by one or two nucleotide pairs); thus if there is any 'genotypic' effect on chromosome size, it would have to operate on the amount of protein and RNA in the chromosomes, relative to the amount of DNA. While this may be the case, to some extent, it seems unlikely that it plays more than a very minor role in determining differences in chromosome size between related species. In general, chromosomes retain their normal dimensions in the karyotypes of interspecific hybrids (Pegington and Rees 1967; White and Cheney 1966), a fact which argues against the whole hypothesis of genotypic control of chromosome size.

REPLICATION OF THE CHROMOSOMAL DNA AND THE PROBLEM OF 'STRANDEDNESS'

The time and mode of replication of chromosomal DNA has presented cytologists with a series of challenging problems. Two main quantitative techniques have been employed. The total amount of DNA in the nucleus at various stages may be measured, using a suitable photometric system, after Feulgen staining. The time of replication may best be studied by autoradiographic techniques, based on the fact that there is no exchange between fully polymerized DNA and the nucleotide pool in the surrounding medium. Thus nucleotides are incorporated into DNA only at the time of replication and by supplying the living cell with thymidine labelled with a radioisotope (tritium in the case of all recent experimental work) we may distinguish in the autoradiograph between old and new DNA strands (i.e. between those synthesized before or after the administration of the isotopically labelled nucleotide).

The mitotic cycle involves a complicated series of changes in the physical appearance of the chromosomes. In interphase they are at their most diffuse, highly hydrated and difficult to see by conventional microscopical techniques. At metaphase and anaphase the chromosomes reach their maximum degree of condensation, with firm, smooth outlines. Intermediate degrees of condensation occur between interphase and metaphase (i.e. in prophase) and between metaphase and interphase (i.e. in telophase). Rather little is known of the exact nature of the processes of condensation and de-condensation, but in all cases there seem to be differences between the physical states of the chromosomes at prophase and telophase, i.e. the stages of de-condensation are not simply those of condensation in the reverse order.

A prophase or metaphase chromosome clearly consists of two genetically equivalent *chromatids*. These may lie strictly parallel, with or without a visible split between them. In many cases the chromatids are loosely wound round one another in a *relational coil*. Many modern techniques of chromosome study involve the use of drugs such as colchicine which tend to separate the chromatids of prophase and metaphase chromosomes, so that they lie much farther apart than in a normal cell.

At anaphase the two chromatids pass to opposite poles of the spindle; once they have separated from one another they are referred to as daughter chromosomes. The nature of the spindle mechanism and the forces involved in anaphase separation will be discussed in the next chapter.

Some of the problems presented by the DNA replication mechanism may perhaps be formulated as follows: (1) At what stage in the mitotic cycle does DNA replication occur? (2) How is the replication of the other constituents of the chromosome related to that of the DNA? (3) How is the newly synthesized DNA distributed to the chromatids of the next mitosis? (4) How many DNA strands are present in each visible chromatid?

It is now known that DNA synthesis (i.e. the replication of the chromosomes) always takes place during interphase. The period when it occurs is known as the S phase, while the interval between the end of telophase and the beginning of S is the G_1 phase, and that between the end of S and the beginning of prophase is the G_2 phase. Naturally the duration of these stages, in minutes or hours, varies from one cell type to another. The most rapid cell cycles are in early embryos. In cleavage divisions of the toad

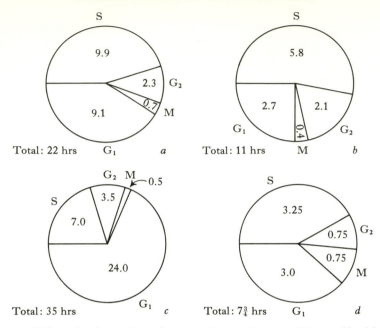

Fig. 1.9. Cell cycles in various tissues and organisms. *a*, Mouse fibroblast; *b*, Chinese Hamster fibroblast; *c*, HeLa cells; *d*, *Paramecium caudatum* micronucleus. Data from Dewey and Humphrey 1962, Hsu, Dewey and Humphrey 1962, Toliver and Simon 1967 and Rao and Prescott 1967.

Xenopus the entire cycle lasts 15 minutes, the duration of the G_1, S, G_2 and mitotic stages being 0, 10.2, 2.4 and 2.4 minutes respectively (Graham and Morgan 1966). In mouse fibroblasts in culture the corresponding periods are 546, 594, 138 and 42 minutes (Dewey and Humphrey 1962). The duration of the G_1 phase appears to be especially variable. The S phase may commonly take about 7–10 hours but lasts as long as 30 hours in the mouse ear epidermis (Bullough and Laurence 1966).

Watson and Crick (1953) suggested that the replication of the DNA molecule is 'semi-conservative', i.e. that each molecule consists of one nucleotide chain directly derived from the parent molecule and a complementary (and anti-parallel) strand which has been synthesized during the replication process. Thus each molecule consists of one 'old' and one 'new' chain. Direct confirmation of this hypothesis has come from experiments, on a wide variety of organisms, in which the 'new' strands are labelled with an isotope such as ^{15}N or ^3H. In the classical work of Meselson and Stahl (1958) bacteria had their DNA labelled with the heavy isotope ^{15}N. After one generation in a medium containing 'light' ^{14}N all their DNA was of

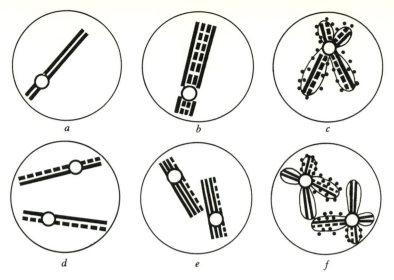

Fig. 1.10. Diagram of the experiment of Taylor, Woods and Hughes. *a*, early interphase (G_1) – on the unineme model the two strands are the constituent nucleotide chains of a Watson–Crick duplex; *b*, late interphase (G_2) after replication in the presence of tritiated thymidine; *c*, first C metaphase after labelling, in the presence of colchicine (both chromatids equally labelled); *d*, second G_1 following labelling (tetraploid nucleus); *e*, second G_2 following labelling. (All tritiated thymidine used up in the first mitotic cycle); *f*, second C metaphase after administration of tritiated thymidine (unequal labelling).

intermediate density (i.e. it consisted of one light and one heavy strand). But after a further generation in ^{14}N medium half the molecules are still 'hybrid' while the other half consist of two 'light' ^{14}N-containing strands. This experiment furnished a decisive proof of the 'semi-conservative' hypothesis at the molecular level. But even earlier (1957) Taylor, Woods and Hughes had shown by tritiated thymidine autoradiography that somatic chromosomes of the broad bean, *Vicia faba* replicated semi-conservatively. Their evidence showed that at the first division after labelling both chromatids of each chromosome had the label, whereas at the next division (in the absence of any further tritiated thymidine) only one chromatid of each chromosome was, in general, labelled.

Taylor, Woods and Hughes used colchicine in their experiments to keep all the daughter chromosomes resulting from the first post-labelling mitosis (X_1) together in the same nucleus until the second post-labelling division (X_2). In some cells it was observed at the tetraploid X_2 division that a few of the chromatids were labelled in one region only; but in such cases the corresponding region in the sister chromatid was unlabelled. This situation was assumed to be the result of sister-strand crossing-over, but it was uncertain whether this was normal or induced by the radiation from the tritium atoms which are necessary for demonstrating the phenomenon. In octoploid X_3 metaphases one half of the chromosomes were entirely unlabelled, while the other half had one chromatid labelled and the other unlabelled.

Essentially similar results indicating a semi-conservative model of replication were obtained in other plants such as *Bellevalia* and *Crepis* (Taylor 1958*a*, *b*) as well as in the Chinese Hamster (Taylor 1960*b*; Prescott and Bender 1963) and the Rat Kangaroo, *Potorous* (Walen 1963, 1965). Discordant results were, however, obtained by La Cour and Pelc (1958, 1959) who repeated the experiment of Taylor, Woods and Hughes both with and without colchicine. In the experiment without colchicine it was estimated that 16% of the metaphases seen 48 hours after labelling were in the X_1 division and 84% in the X_2 mitosis; in spite of this 'most metaphase chromosomes' were labelled over both chromatids (implying that some X_2 chromosomes were labelled throughout both chromatids, which would be contrary to the results of Taylor, Woods and Hughes). In a second experiment in which colchicine was used, La Cour and Pelc obtained metaphases with twelve chromosomes (i.e. X_1 mitoses) 20 and 24 hours after labelling; only a minority of these should labelling along both chromatids in all chromosomes. At the X_2 division (with twenty-four chromosomes) many metaphases contained a few chromosomes labelled along both chromatids in addition to those with only one chromatid labelled.

La Cour and Pelc interpreted their results as indicating – contrary to the conclusions of Taylor *et al.* – that the chromatids are not single-stranded and that colchicine influences the segregation of 'new' and 'old' strands, perhaps by an effect on the centromere.

Woods and Schairer (1959) performed two experiments similar in principle to those of La Cour and Pelc; their results were in accordance with the original findings of Taylor *et al.*, but seem to have been based on only fifty chromosomes (i.e. the equivalent of only four metaphases). Prescott and Bender (1963) working on cultured cells of the Chinese Hamster, also obtained results in accordance with those of Taylor *et al.*, but Deaven and Stubblefield (1969) obtained an entirely different result, which suggested that the chromatids were multi-stranded.

Peacock (1963) carried out four experiments on *Vicia faba*. The first was similar to that of Taylor *et al.*, in the second and third no colchicine was used, the fourth was a repetition of one of La Cour and Pelc's experiments. No effects of colchicine on sister chromatid exchanges or segregation of strands were observed. As in the original experiments of Taylor *et al.* all chromatids were labelled at the X_1 division. In the X_2 division most (923 of 1164) chromosomes showed segregation of the label, and of these about half showed 'isolabelling', i.e. a label over the whole of both chromatids. These results are not easily compatible with the 'unineme' model of chromatid structure favoured by Taylor (1957, 1958a, b, 1963a, b, 1966), who believes that two strands are indeed present in each chromatid but that these are the two polynucleotide chains of a single DNA duplex. Sister-strand crossing-over is supposed to occur, but subject to the restriction that only strands of like polarity can join together. From this model Taylor (1958) calculated that there should be twice as many 'single exchanges of label' at the X_2 division as 'twin exchanges' (i.e. exchanges having an identical location in two homologous chromosomes). As a result of an ingenious experiment involving dicentric Chinese Hamster chromosomes, Brewen and Peacock (1969a) have obtained additional evidence for the principle of polarity in rejoining and, incidentally, for the concept of a single DNA duplex per chromatid.

Certain difficulties remain however. There is now plenty of evidence for iso-labelling at the X_2, but some uncertainty regarding segments showing iso-non-labelling.

Three hypotheses concerning the 'strandedness' of 'ordinary' somatic chromosomes may be considered. According to the *unineme* model each chromatid at prophase or metaphase, consists of a single spirally coiled chromonema consisting of a Watson–Crick DNA duplex together with associated RNA and protein. The bineme model has the chromatids made up of two DNA strands or *subchromatids*. Finally, the *polyneme* model regards chromatids as multi-stranded and hence not fundamentally different from the polytene chromosomes present in some (but not all) of the somatic tissues of the Diptera (see Ch. 4). The latter are, however, in a permanent interphase stage and will never again divide by mitosis.

It is, of course, not logically necessary to suppose that all the cells of an individual show the same degree of strandedness or that corresponding tissues of different species do so. It is quite conceivable that differences exist, for example, between the chromosomes of meristematic and endosperm tissue in plants and that they may also exist between those of such histologically diverse cell types as grasshopper spermatogonia, *Drosophila* neurocytes and mammalian leucocytes.

Five main techniques have been employed to investigate this question of 'strandedness': (1) direct visual observation, (2) electron microscopy, (3) studies of chromosomal rearrangements, especially ones resulting from irradiation, (4) the autoradiographic studies with tritiated thymidine which we have discussed earlier, (5) determinations of DNA content of nuclei.

Evidence that has been regarded as favouring the bineme or polyneme models has come from several sources in addition to the autoradiographic results of Peacock (1963) cited above. Certain workers have claimed that chromatids could be seen to be divided longitudinally into sub-chromatids by ordinary light microscopy, following various pre-treatment techniques. Wolff (1969) has published some particularly clear photographs by J. Wahrman, showing sub-chromatid structure in a grasshopper and Vosa (1968) has published an equally clear photograph of sub-chromatid structure in the plant *Crocus*, following acetic-alcohol fixation, hydrolysis in 1N HCl and squashing in aceto-orcein. Unfortunately, he gives no indication as to the regularity with which such structure can be demonstrated, and the single cell illustrated could be simply an example of diplochromosomes, i.e. of endoreduplication, in the sense of Levan and Hauschka (1953) and Walen (1965). However, Trosko and Wolff (1965) were able to obtain clear visual evidence of sub-chromatid structure in metaphase chromosomes of the bean *Vicia faba*, following air-drying and trypsin digestion. Similar evidence was obtained by Trosko and Brewen (1966) on Chinese hamster chromosomes after treatment with hypotonic solutions, chelation with ethylene diamine tetra-acetic acid (EDTA) and air-drying. These are relatively drastic treatments and the chromosomes no longer have a life-like appearance after their application. But Giménez-Martín *et al.* (1963) have demonstrated sub-chromatid structure in *Scilla* after 8-oxyquinoline treatment followed by conventional fixation and Bajer (1965) has even observed it in living chromosomes of *Haemanthus* endosperm, especially at anaphase. Maguire (1968), using a Nomarski interference contrast optical system obtained good evidence for bineme chromatids at the diplotene stage of meiosis in maize.

So-called side arm bridges at mitotic metaphase or anaphase following X-irradiation have been studied by a number of workers. Some of the 'half-chromatid lesions' observed by Nebel (1936) in *Tradescantia* may have been of this type, but his account is unclear. Swanson (1947) almost certainly observed sub-chromatid rearrangements in *Tradescantia*, but did not illustrate details of the configurations seen. Very clear examples of 'two-side arm bridges' in endosperm divisions of *Scilla* were figured by La Cour and Rutishauser (1954) and interpreted in terms of breakage and reunion of subchromatids. In the commonest and simplest type only one chromatid is involved and the two side arms are the same length. Peacock (1961) obtained extensive data on sub-chromatid rearrangements, following irradiation, in somatic mitoses of *Vicia*. In addition to the two-side arm bridges, obtained by all workers in this field, he occasionally observed (1) one-side arm bridges and (2) Y-shaped chromatids at anaphase with a deficient chromatid at the opposite pole. Östergren and Wakonig (1954) had earlier attempted to interpret these rearrangements in terms of stickiness of the 'matrix' rather than actual sub-chromatid breakage, but in view of modern work on chromosome ultrastructure (see next section) which makes it fairly clear that no matrix exists as a separate structural entity, this interpretation must now be rejected. More recent studies on side arm bridges at mitotic metaphase and anaphase, including some ultrastructure studies of critical configurations, have been published by Heddle and Bodycote (1968), Brinkley and Humphrey (1969) and Humphrey and Brinkley (1969).

Side arm bridges have also been studied at meiosis (usually in first anaphase) by a number of workers. They were seen in the plant *Gasteria* by Darlington and Keffalinou (1957); in grasshoppers by John, Lewis and Henderson (1960) and Lewis and John (1966); and in the plants *Podophyllum* and *Agave* by Newman (1967) and Brandham (1969). All these observations were made on normal, untreated material. Very clear examples of side arm bridges following irradiation were observed at first anaphase of *Lilium* by Crouse (1954, 1961a) and similar results were obtained in *Trillium* by Wilson, Sparrow and Pond (1959). There seems little difference in principle between the side arm bridges and similar configurations obtained at mitosis and meiosis. Almost all the investigators have interpreted such bridges as the result of sub-chromatid breakage and reunion, according to the scheme shown in Fig. 1.11. But critical evidence is obviously lacking and other interpretations are possible. Side arm bridges at mitosis and meiosis seem to provide good evidence that chromatids contain two parallel strands but there seems to be no way of ruling out the possibility that these are the constituent strands of a single DNA duplex, on Taylor's unineme model of chromatid structure.

Kihlman and Hartley (1967) have expressed doubts as to whether sub-chromatid rearrangements are really such at all, because in an experiment performed by them, apparent sub-chromatid rearrangements were followed in the next division by chromosome rearrangements rather than chromatid rearrangements. However, it is not really clear what criteria they adopted to distinguish apparent sub-chromatid rearrangements from chromatid rearrangements in the first cell generation. There is now so much evidence for sub-chromatid rearrangements, in such a variety of materials, both at mitosis and meiosis, that it is impossible not to accept it as a genuine phenomenon, whatever the precise interpretation.

After carefully considering all the evidence on this question of 'strandedness', John and Lewis (1969) concluded that on the whole it favours the bineme model in which each chromatid consists of two sub-chromatids, each of which contains a single DNA duplex. The existence of a higher

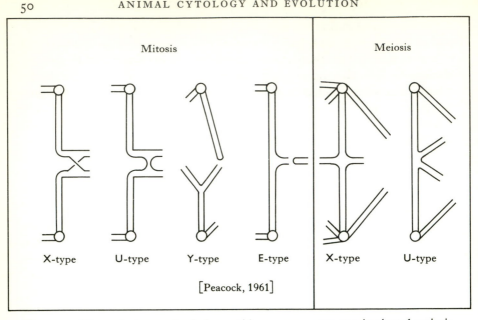

Fig. 1.11. The main types of subchromatid rearrangements at mitosis and meiosis, interpreted on the bineme model of chromosome structure.

degree of strandedness seems to be definitely excluded as far as the lamp-brush chromosomes are concerned by measurements of the thickness of the interchromomeric fibres (about 50 Å) and of the axial fibres of the lateral loops (about 30 Å). And a lower degree of strandedness (i.e. a single DNA duplex per chromatid) seems to be ruled out as far as the somatic chromosomes of *Vicia faba* are concerned by light microscopy, results of X-irradiation, tritiated thymidine autoradiography (Peacock 1965) and trypsin digestion of air-dried chromosomes (Trosko and Wolff 1965). The X-irradiation experiments of Evans and Savage (1963) and Wolff and Luippold (1964) on *Vicia* as well as those of Monesi, Crippa and Zito-Bignami (1967) on Chinese Hamster cells show that the chromosomes right at the beginning of the S phase behave as if they were bineme. It might be suggested that the two strands could be the two antiparallel halves of a DNA duplex which had separated. But Heddle and Trosko (1966) were quite unable to detect single DNA strands, which fluoresce red in the ultraviolet following acridine orange staining (whereas DNA duplexes fluoresce yellow-green).

Wolff (1969) has recently reviewed all the evidence bearing on this vexed question of strandedness and ended by rejecting the unineme hypothesis.

We must consequently conclude that at the present time it is impossible to decide with certainty between the unineme and the bineme models of chromatid structure, but that, with the exception of the evidence from electron microscopy and that from the lampbrush chromosomes (see Ch. 5), the majority of the evidence seems to favour the bineme hypothesis.* What do seem to be ruled out are all hypotheses of polyneme chromatids in ordinary somatic or germ line chromosomes. It follows that theories of 'evolutionary polyteny' according to which large-scale differences in the 'DNA values' of related species were ascribed to different degrees of strandedness of their chromosomes must be decisively rejected.

If no germ-line chromosomes have a structure more complex than the bineme one, the reason, we would suggest, is that many-stranded chromosomes, although they can exist in the polytene nuclei of some Diptera, Collembola and Protozoa, are incompatible with the normal mechanism of meiosis. However, there does not seem to be any evidence for the existence of polyneme chromatids in the germ-line of animal species which have totally abolished meiosis in favour of a mechanism of apomictic parthenogenesis (see Ch. 19).

When fixation has been inadequate, various artefacts may be produced in metaphase chromosomes. Following acetic fixation chromosomes may have a swollen, bloated and vacuolated appearance. If penetration of the fixative into the moribund tissue is delayed, one may get 'clumping' of the chromosomes, with fusion of some of them into aggregates. Frequently associated with clumping is the appearance of stainable connecting strands between the chromosomes. It should be unnecessary to point out that all these types of appearances are essentially inconstant and variable from cell to cell, and that they are not seen in life or in material that has been prepared by the best techniques. Nevertheless such artifacts have frequently trapped inexperienced or careless workers into conclusions which are obviously false. The latest of these is the extraordinary suggestion of Du Praw (1970) that all the chromosomes in a given karyotype may be merely segments of a continuous circular DNA molecule. Some of the evidence put forward in support of this idea, if it were genuine, i.e. non-artefactual (which it is not) would actually seem to be incompatible with it, e.g. lateral as well as terminal connections between the chromosomes in the spermatogonial metaphases of coreid Heteroptera (Du Praw's Fig. 11.3B, taken from Wilson 1925, Fig. 397) and strands apparently joining the centromeres (which are not terminal) of the oogonial chromosomes of the opossum *Didelphis* (Du Praw's Fig. 11.5C).

* Since the above was written, Laird (1971) has shown that most of the nucleotide sequences in mouse, *Drosophila* and *Ciona* (tunicate) DNA are present only once per sperm. This seems to be strong evidence for the unineme interpretation.

ULTRASTRUCTURE OF CHROMOSOMES

The preparation of chromosomes for electron microscopy has presented many technical difficulties. The size and mode of organization of chromosomes, both in the interphase nucleus and at metaphase has rendered them rather unsuitable for thin sectioning methods. Most of the modern work in this field has consequently been carried out on isolated chromosomes. Mirsky and Ris (1947a, b, 1951b) isolated threads resembling metaphase chromosome from many types of tissues, in a saline citrate medium, using a Waring blender. There has been considerable controversy, however, as to whether these structures really were chromosomes. More recently, Gall (1963c, 1966) has developed a technique which involves the spreading of animal nuclei on a water surface, picking them up on a carbon-coated grid and fixing and drying them in various ways. This technique seems more fruitful for studying chromosome ultrastructure than the older methods involving sectioning. None of the modern work on the ultrastructure of chromosomes supports the idea of a 'matrix' or pellicle as an independent constituent, and it now appears certain that the RNA and protein constituents must be intimately complexed with the DNA chromonema rather than present as a distinct layer.

Wolfe and John (1965) studied the ultrastructure of male meiotic chromosomes of the bug *Oncopeltus* using a technique which employs nuclei isolated on a Langmuir trough. They found these chromosomes to be composed of microfibrillar units 250 Å in diameter, but could reach no conclusion as to whether each chromatid was multi-stranded or not (the alternative is that each fibril is folded or coiled, perhaps several times, so as to simulate a multi-stranded condition).

Similar results have been obtained on mitotic chromosomes by Gall (1966). All modern workers on chromosome ultrastructure have found that metaphase chromosomes are built up of a great mass of tangled fibrils usually about 200–300 Å in diameter. According to Ris (1966, 1967), who has studied the erythrocyte nuclei of salamanders, the fibrils are 200–250 Å in diameter if spread on a water surface, but only 80–100 Å in diameter if a solution of sodium citrate or EDTA is used instead. It is noteworthy that chromosomes prepared by these techniques show, in general, no trace of spiral structure, sub-chromatids or chromomeres, nor any differentiation into heterochromatic or euchromatic regions; on the other hand the position of the centromere is in most cases clearly marked, as a region where the two chromatids are joined together (Du Praw 1970). As John and Lewis (1969) point out: 'there is no doubt that the spreading technique generates considerable surface forces. These lead to considerable lateral displacement.

This, in turn, produces an over-complicated, even confused, picture of chromosome organization.'

Eventually, no doubt, there will be a reconciliation between pictures of chromosome structure presented by ultrastructure studies and those revealed by the light microscope on living and fixed material. But that reconciliation is certainly not possible at the present time and may not be easily arrived at in the near future.

There has been considerable argument as to whether the DNA in a chromatid is in the form of a single molecule or consists of a number of molecules placed end-to-end and connected by 'linkers' of some other substance. One thing seems certain, however, namely that when the chromosome undergoes replication, this is initiated at a number of points along its length. Certainly the simplest assumption would be that each of these initiation points is the end of a DNA molecule. Using electron microscopy, Solari (1965) studied DNA fibres up to 93 μm long from sea-urchin sperm. Much longer DNA fibres have been detected, using the autoradiographic technique of Cairns (1963c). Huberman and Riggs (1966), using cultured Chinese Hamster cells, obtained many fibres more than 800 μm in length and some as long as 1800 μm. The relationship between these and actual DNA molecules is not clear, however, and the latter may be longer or shorter than the fibres visualized by autoradiography. Since these hamster cells contain about 7 pg of DNA and an average of twenty-three chromosomes, one can calculate that each chromosome contains about 9 cm of DNA. If this replicated from a single initiation point at the same rate as bacterial DNA (100 μm per minute), the total time for replication would be about 15 hours, whereas it is known to take only about 6 hours, or 1$\frac{1}{2}$ hours in the case of the X-chromosome (Hsu 1964). These arguments, and especially a consideration of the very rapid mitotic cycles of the early cleavage divisions in many embryos, virtually necessitate the assumption of multiple points of initiation of replication.

In later work, Hubermann and Riggs (1968) have shown that the DNA of mammalian chromosomes is made up of a large number of tandemly arranged replication units. Mid-way along each of these is a point where replication is initiated, and from which it spreads in each direction at a Y-shaped growing point. Replication is completed when the growing points of all the replication units meet at the boundaries or termini between them. Most replication sections are shorter than 60 μm, many being so short that they are difficult to resolve by tritiated thymidine autoradiography. The rate of DNA replication from each point of initiation was 2.5 μm per minute or less, and was apparently constant throughout the S phase. Using a different technique, Taylor (1968) arrived at essentially the same result, namely that DNA synthesis proceeds at 1 to 2 μm per minute, and that the maximum length of a unit of replication is from 90 to 180 μm.

Callan (1972) has shown that in newts of the genus *Triturus*, which have about ten times as much DNA per nucleus as *Xenopus*, the replicating units are much longer (possibly ten times longer) but replication proceeds from the initiation sites at about three times the speed. He has also shown that in the male germ-line of *Triturus* the S phases become longer and longer in the later spermatogonial mitoses and the final premeiotic S phase occupies eight to nine days as compared with about one day for a typical somatic metaphase. The long duration of this S phase is apparently due to the much smaller number of initiation sites and hence to the much greater length of the replicating units; the rate of replication per unit of length is about the same as in somatic metaphases.

REGULAR, SPONTANEOUS CHROMOSOME BREAKS AND FUSIONS

There are a number of instances in which spontaneous chromosomal breaks or fusions occur more or less regularly in the course of development. The exact mechanism in all these cases is more or less obscure.

The best-known examples of regular fragmentation of chromosomes occur in the cleavage nuclei of nematodes belonging to the family Ascaridae. In general, the chromosomes of the Ascaridae seem to cast off their terminal segments in certain of the cleavage divisions in the embryo, but only in those nuclei which are destined to pass into the somatic tissues. The details were described by Bonnevie (1902) in the case of *Ascaris lumbricoides* and in a number of other species by Walton (1924). A more complicated case is that of the horse nematode, *Parascaris equorum* (= *Ascaris megalocephala*), studied by so many of the early chromosome cytologists (Boveri 1887, 1888, 1892, 1909; O. Hertwig 1890; Brauer 1893). There are two 'varieties' (more probably full species, according to modern concepts) of this animal. One (var. *univalens*) has only one pair of chromosomes in the spermatogonia and oogonia, while the other (var. *bivalens*) has two pairs. It has been known, however, since the time of the early investigators that in both forms a peculiar fragmentation of the chromosomes takes place during certain of the cleavage divisions. The zygote nucleus contains either two or four long chromosomes, each of which consists of a central region that is either holocentric or polycentric (interpretations vary) and two end segments which are somewhat club-shaped and apparently heterochromatic. The first cleavage division is a simple mitosis, but the second division is different in the two nuclei. In one it is a normal mitosis, but in the other it involves the casting out of the chromosome ends (which are apparently left without

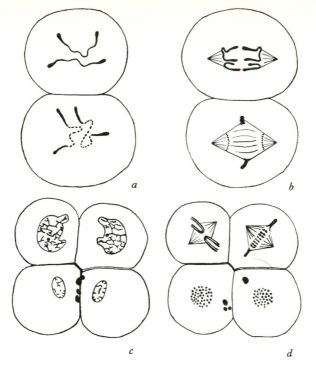

Fig. 1.12. *Parascaris equorum* var. *univalens*. Diagram of the second and third cleavage divisions. *a*, metaphase and *b* anaphase of second cleavage. In *b* the chromosomes are undergoing diminution in the lower cell but not in the upper one, which will give rise to the germ-line. *c*, telophase of second cleavage, with eliminated chromatin in the two lower cells. *d*, metaphase of third cleavage division; the top left-hand cell will go on to give rise to the germ-line at the next cleavage division. Elimination is about to take place in the top right-hand cell. From Boveri, redrawn.

centromeres and undergo degeneration) and the fragmentation of the central segments of each germ-line chromosome into about 30 small portions each of which behaves henceforth as a separate chromosome. At the next cleavage division in the embryo the two nuclei derived from the 'unreduced' cell behave similarly – one divides normally, while the other undergoes reduction. This behaviour is repeated at the fourth and fifth cleavage divisions but then stops: so that at the 32-celled stage the embryo consists of two 'unreduced' cells, which will give rise to the germ-line and thirty reduced ones, which are destined to form all the somatic cells. Boveri's (1909, 1910) famous studies on centrifuged and double fertilized *Ascaris* eggs demonstrated that whether or not a nucleus underwent diminution depended on the type of cytoplasm in which it lay.

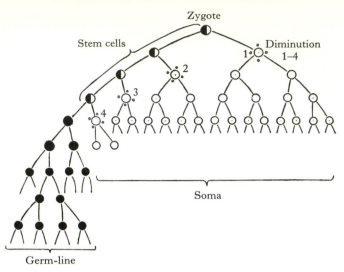

Fig. 1.13. *Parascaris equorum.* Diagram to show the divisions at which diminution takes place. Undiminuted cells half black; diminuted ones white. Germ-line cells black. Diminution actually occurs in the cells marked 1, 2, 3, 4. From Wilson (1925).

From the general standpoint one of the most significant aspects of this case is the fact that freshly-broken chromosome ends seem to undergo a completely satisfactory 'healing' process, as if they had been converted into telomeres.

Another possible instance of elimination of terminal heterochromatic segments occurs in the copepods of the genus *Cyclops*, where it takes place at the fifth or sixth cleavage division in the embryo (S. Beermann 1959). However, Stich (1962) has put forward an alternative interpretation according to which the eliminated material would be DNA of chromosomal origin rather than an integral part of the chromosome itself; in which case this may represent a type of gene amplification mechanism.

Mutations which produce 'fragile sites' (i.e. ones liable to frequent spontaneous breakage) in human chromosomes have been reported by a number of authors, e.g. Lejeune *et al.* 1968; Gooch and Fischer 1969 and Magenis, Hecht and Lovrien 1970. A 'fragile site' of a somewhat different kind has been described in the spermatogenesis of a grasshopper by White (1966); in this case one of the chromosomes (but not its homologue) underwent spontaneous breakage at a locus adjacent to the centromere at a particular stage of meiosis, just prior to first metaphase.

ELIMINATION OF RIBONUCLEOPROTEIN MASSES FROM CHROMOSOMES

In a few instances distinct ribonucleoprotein masses are left behind in the middle of the spindle as the daughter chromosomes pass to the poles at anaphase. A classical case occurs in the mite *Pediculopsis graminum* (Cooper 1939). Here the elimination bodies, which are Feulgen-negative, are seen during both meiotic divisions in the egg, as well as during the early cleavage mitoses. The details are very clear, on account of the low number of chromosomes (2n = 6). In the closely related species *Pediculoides ventricosus* no ribonucleoprotein bodies are visible (Pätau 1936).

In many species of Lepidoptera (Seiler 1914, 1923; Kawaguchi 1928; Fogg 1930; Ris and Kleinfeld 1952) and Trichoptera (Klingstedt 1931) Feulgen-negative bodies of a similar nature are seen in the equatorial region of the spindle between the separating chromosomes at meiosis in the egg. Owing to the large number of chromosomes (usually about thirty) the details are not so clear, and all the elimination bodies may appear clumped together in a great pancake-shaped mass between the two groups of anaphase chromosomes.

It is possible that the helical components of the *Amoeba* interphase nucleus, whose ultrastructure has been studied by Wolstenholme (1966a) are similar to the elimination bodies of *Pediculopsis* and the Lepidoptera; they do not seem to incorporate tritiated thymidine in autoradiographic experiments. It has been shown that they may be either right-handed or left-handed helices, and that the direction of a single helix may undergo reversal; there is thus some reason to believe that they have been derived from the chromosomes and retain the coiled structure of the latter. The true chromosomal material of the *Amoeba* interphase nucleus appears to consist of fibres about 150 Å in diameter.

Whether the bodies seen between the separating chromosomes at anaphase in *Pediculopsis* and the Lepidoptera bear any relationship to the 'synaptinemal complexes' seen in electron micrographs of meiotic prophase bivalents (and which have been seen by Gassner (1969) by light microscopy in the mantid *Bolbe*) must remain uncertain for the present; probably they do not.

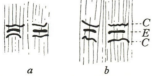

a   *b*

Fig. 1.14. Cleavage divisions in the mite, *Pediculopsis graminum*, in (*a*) early and (*b*) mid anaphase. *C*, daughter chromosomes, *E*, Feulgen-negative elimination bodies. From K. W. Cooper, redrawn.

2

THE MITOTIC MECHANISM IN EVOLUTION

Kräfte deren stilles Treiben in der elementarischen Natur,
wie in den zarten Zellen organischer Gewebe, jetzt noch
unseren Sinnen entgeht . . .

ALEXANDER VON HUMBOLDT: *Kosmos*, vol. 2

A full discussion of modern experimental work on the mitotic mechanism
would be outside the scope of this book, which is concerned primarily with
evolutionary changes in chromosomes and in chromosome behaviour.
Nevertheless, some consideration of the mitotic mechanism is in order,
more particularly since the various processes involved in it do affect and
limit in various ways the kinds of chromosomal changes that can be tolerated
in the evolution of karyotypes.

The shape of the interphase nucleus clearly depends in part on the chromosomes inside it and in part on the cytoplasm outside it. A particularly clear
demonstration of the role of the chromosomes occurs in the polymorphonuclear leucocytes of primates, including the human species; those of females
(with two X-chromosomes) frequently show a knobbed extension, which
has been called a *drumstick*, while those of XY males do not. Studies on
individuals with abnormal numbers of X-chromosomes leave no room for
doubt that one of the X-chromosomes is directly responsible for the formation of the drumstick, and presumably lies within it.

Most interphase nuclei are approximately spherical, slightly ovoid or
somewhat flattened structures. The smallest are about 2 μm in diameter, the
largest being about 500 μm across (i.e. the largest have a volume which is
several million times that of the smallest ones). Nuclear shape is frequently
related to the overall shape of the cell, e.g. flattened cells have flattened
nuclei and spindle shaped ones have long ovoid nuclei. More irregular shapes
such as lobed or branched nuclei are relatively uncommon, but many
striking instances are known. Lobulated nuclei, ranging from ones that are

faintly bilobed or trilobed to some that are elaborately dendritic, are found in some vertebrate leucocytes and in certain specialized secretory cells such as those of the spinning glands in lepidopteran and caddis-fly larvae (Vorhies 1908). These spinning gland cells are undoubtedly highly polyploid. Very similar are the branched polyploid nuclei in the salivary gland cells of some Heteroptera such as *Gerris* (Geitler 1937, 1938, 1940). The highly polyploid macronuclei of some ciliates such as *Condylostoma* and *Ephelota* are likewise moniliform or dendritic. On the other hand, many highly polyploid nuclei in both insect secretory cells and in Protozoa are spherical or nearly so.

In interphase cells the nucleus is enclosed in a double-layered *nuclear membrane*, the outer layer of which is often seen to be continuous with the endoplasmic reticulum. At the end of the prophase stage of mitosis the nuclear membrane disintegrates, to be re-formed at telophase, probably from the endoplasmic reticulum. The nuclear membrane shows many octagonal 'pores' (Gall 1967), which are places where the outer and inner layers are continuous; they are surrounded by a thickened annulus. In amphibian oocyte nuclei they are very numerous, crowded together on the surface of the membrane and about 650–800 Å in diameter. There can be little doubt that nucleo-cytoplasmic exchanges of macromolecules take place through the pores, but the latter are probably not the simple holes they appear to be at first sight; in some instances they have been described as blocked by 'plugs'.

Very rapid mitoses may take less than an hour (less than ten minutes in some embryonic cleavage divisions). More usually the whole process takes 2–3 hours. There is no constant ratio between the duration of the various stages. Usually, but by no means invariably, prophase is the longest stage and anaphase the shortest; but there is a great deal of variation in this respect. Metaphase, the stage when most observations on chromosomes are made, may last several hours or be completed in 20 seconds (Milovidov 1949). In a wide variety of organisms, all stages of mitosis are shorter at higher temperatures (Mazia 1961, Table VII).

The first stage of mitosis, prophase, is a period of increasing condensation of the chromatids, between which a visible split is always apparent. In general prophase is a continuous stage and only in the special case of the meiotic prophase is it liable to be interrupted by periods of stasis. It ends with a relatively brief period which has been variously called *premetaphase* or *prometaphase* (the former seems etymologically correct and will be used here). This is a period when two major changes occur in the cell – the nuclear membrane breaks down and the spindle is formed. Only in certain Protozoa do we apparently have a type of mitosis in which the nuclear membrane persists throughout the whole process.

The spindle is a relatively solid and apparently gelatinous body, whose precise shape is characteristic of particular tissues and organisms. In animals it is almost always an *amphiastral* structure, with a centriole at each pole, surrounded by a region known as the centrosome from which astral rays usually (but not always) radiate. *Anastral* spindles occur in higher plants and in some cases in animals. The fact that such anastral spindles can exist argues against the centrioles or asters playing an essential role in the spindle mechanism. In accordance with this view, it was shown by Dietz (1959, 1966) that in the male meiotic divisions of the craneflies *Pales crocata* and *P. ferruginea* the centrioles, centrosomes and asters, although normally present, are not essential for the functioning of the spindle or for cell division. In the normal spermatogenesis of these insects huge asters come to occupy diametrically opposite positions close to the nuclear membrane so that when the latter breaks down and the spindle forms the centrioles and asters become associated with its poles. By experimentally flattening these spermatocytes, Dietz was able to prevent the asters from migrating to the nuclear membrane; this led to cells in which the asters were entirely independent of the spindle. In this way he was able to obtain sister second spermatocytes, each with a normal haploid set of chromosomes, but one with four centrioles and asters and one with none. The latter type of second spermatocytes formed normal bipolar spindles, while in the former tripolar spindles were organized. Dietz thus regards the association of the centrioles with the spindle poles as a mechanism which ensures the regular distribution of the centrioles to the daughter cells.

In most cases the centrioles are minute granules. But in some groups they are large elongated structures. In the flagellate *Holomastigotoides* the spindles are very small compared with the huge chromosomes (Cleveland 1949); at each end is an elongated centriole which is attached at one end to one of the flagellar bands. The centromeres of the chromosomes (which Cleveland believed to be strictly telocentric) are directly attached to the centrioles in this case, so that there is nothing resembling a metaphase plate stage; the whole division cycle takes place within an intact nuclear membrane. Essentially similar mitotic cycles occur in some related genera of flagellates such as *Trichomonas* and *Joenina*.

Giant centrioles of a somewhat different type occur in the male meiotic divisions of neuropterous insects (Asana and Makino 1937; Hirai 1955c, Friedländler and Wahrman 1966). In the spermatogonia they are minute, but by the time the pachytene stage of meiosis is reached, they have grown into V-shaped bodies which migrate from the cell membrane to the surface of the nuclear membrane. There is a special mechanism whereby the parent centriole produces a daughter centriole from a minute procentriole which

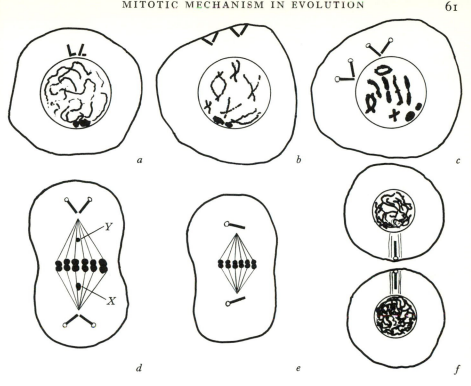

Fig. 2.1. The centrioles at male meiosis in a neuropteran such as *Myrmeleon*. Based on Friedländler and Wahrman (1966). *a*, pachytene (centrioles and growing daughter centrioles near the nuclear membrane); *b*, diplotene (parent and daughter centrioles at the cell membrane, where they acquire centriolar vesicles); *c*, diakinesis (centrioles have elongated and return to vicinity of nucleus); *d*, first metaphase (centrioles at poles); *e*, second metaphase; *f*, second telophase (centrioles in spindle remnant (*Stemmkörper*)).

grows until it is equal in length to the parent. After that both grow in length simultaneously, to produce the V-shaped bodies which move to the poles of the spindles at premetaphase. A single V is passed into each second spermatocyte and its arms then separate and pass to opposite poles of the second meiotic spindle. Each spermatid thus receives a half-V. The mechanism of centriole replication from a procentriole which grows in length while maintaining a constant angle to the parent centriole is possibly found in all organisms possessing centrioles. It was figured as early as 1905 by Schreiner and Schreiner in spermatocytes of the hagfish *Myxine* and has been illustrated in the flagellate *Barbulanympha* by Cleveland (1957).

The centriole undoubtedly plays a special role in the sperm, so that the growth of giant centrioles in spermatogenesis is probably to be understood in the light of this function rather than as related to the spindle mechanisms of the meiotic divisions (Friedländler and Wahrman 1970). These workers (1966) have interpreted the evidence of many earlier investigators as indicating that in insects a single centriole always passes into the sperm while in Coelenterata, Nematoda, Annelida, Mollusca, Crustacea and Vertebrata two invariably do so.

The centriolar mechanism has been studied in a number of Lepidoptera and molluscs in which so-called *apyrene* or *oligopyrene* sperms are formed alongside the normal ones in the testis. These atypical sperms contain no chromosomes or fewer than the normal haploid karyotype and do not function in genetic fertilization. Pollister (1939) and Pollister and Pollister (1943) studied the development of the giant multiflagellate oligopyrene sperms of the snail *Viviparus*. The diploid number was eighteen and in the cells which are going to form normal sperms nine bivalents are present in the first meiotic division; one centriole is present at each pole of the spindle. In the atypical spermatogenesis the chromatids fall apart, so that there are thirty-six of them in the nucleus; only four behave more or less normally, the remaining thirty-two (corresponding to sixteen chromosomes or eight pairs of homologues) seem incapable of association with the spindle. The two centrioles each develop into a mulberry-like cluster of nine bodies. The Pollisters claimed that the eight extra centrioles at each pole were transformed centromeres which had migrated from the abnormally-behaving chromosomes to the poles of the spindle. A reinvestigation of this material by Gall (1961), using electron microscopy, has not confirmed this remarkable hypothesis but neither has it decisively disproved it.

In suitable materials, entire mitotic spindles, with asters at their poles, can be isolated from the rest of the cell by appropriate experimental techniques (Mazia and Dan 1952; Mazia 1961). Most of this work has been carried out on sea-urchin eggs; the isolation is obtained by chemical means, which can be applied on a mass scale. The earlier techniques depended on treatment of the cells with 30–40% ethanol at −10 °C, after which the cytoplasm could be dispersed by detergents, digitonin or ATP. Later studies involved 'protection' of S—S bonds in the mitotic apparatus by an excess of the S—S compound dithiodiglycol. The main advantage of this technique is that it has enabled some studies to be carried out on the chemical composition of the mitotic apparatus (spindles plus asters). Not surprisingly, about 90% of the mitotic apparatus turns out to be protein. The average molecular weight of the proteins of the mitotic apparatus is about 315000, and they have an isoelectric point around pH 5.5. These data prove that spindles and asters do not contain any large quantities of histones or pro-

Typical　　　　　　　　　　　Atypical

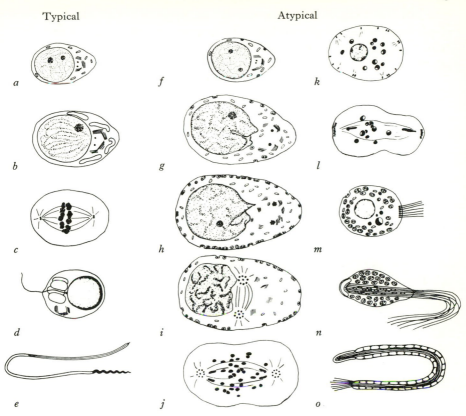

Fig. 2.2. Centriolar behaviour in the typical and atypical spermatogenesis of the snail *Viviparus* (semidiagrammatic, some cytoplasmic bodies omitted in *j*, *k* and *l*). *a–e*, the typical spermatogenesis (*a*, *b*, early and late primary spermatocytes; *c*, first metaphase, showing nine bivalents; *d*, spermatid; *e*, mature sperm). *f–o*, the atypical spermatogenesis (*f–i*, prophase of meiosis – note formation of multiple centriole clusters; *j*, first anaphase with 4 normal, 32 lagging chromatids; *k*, interkinesis – the chromatids which lagged on the spindle are enclosed in micronucleoli; *l*, second anaphase – one chromatid passes to each pole and the centrioles are unequally distributed to a large and a small spermatid; *m–o*, spermatids and mature atypical sperm with multiple flagella. From Gall (1961).

tamines. A certain amount of RNA is also present in the spindle, but its significance is not understood.

Spindles were shown by classical microscopical studies to be fibrous bodies, and various categories of spindle fibres were recognized (see Schrader 1953 for a review of the earlier literature). It was later shown that, when viewed by polarization microscopy, living spindles are strongly bifringent (Inouye 1964); this is a satisfactory proof that their fibrous structure is not an artifact. The two main categories of spindle fibres are the *chromosomal fibres* which run from the centromeres to the poles and the *continuous fibres* which run from pole to pole (Nicklas 1970 calls these *interpolar fibres*).

Electron microscopy (Harris 1962; Roth and Daniels 1962; Bajer and Jensen 1969; Bajer and Mole-Bajer 1969) has revealed the presence of numerous approximately parallel microtubules in the spindles of a wide variety of animal and plant species; these rather obviously correspond to the spindle fibres seen by classical light microscopy. There is the same distinction between *chromosomal* and *continuous* microtubules. In some cases, however, most of the latter appear to run only from the pole to the equator of the spindle where their ends overlap with those of the microtubules connected to the opposite pole. Apparent cross connections have been seen between adjacent parallel microtubules in some instances. Microtubules are about 200 Å in diameter and have a complex structure. There is a good deal of material present in the spindle which is not microtubular, and which fills the space between the microtubules; very little is known about it, but it must be mainly protein, with some RNA, lipids and polysaccharides (Forer 1969). The mitotic apparatus (spindle plus asters) of the sea-urchin egg contains about 12% of the total cell protein, i.e. about fifty times as much protein as occurs in the resting nucleus (Zimmerman 1960). There may be one main protein which is the principal structural component of the spindle and several other minor protein constituents.

The premetaphase stage, called *metakinesis* by Mazia (1961) is usually a brief one, which includes the breakdown of the nuclear membrane, the final condensation of the chromosomes to the metaphase condition (with a smooth surface), the establishment of the spindle and the attachment of the chromosomes to it, on the equator, by their centromeres. The formation of the spindle involves, in most animal material, the migration of the centrioles to its poles. It is usually said that the centromeres orientate exactly on the equator, equally connected to both poles by chromosomal microtubules, and this orientation is termed *amphitelic*, a condition that is contrasted with the *syntelic* orientation at the first meiotic division, when the two centromeres of each bivalent orientate on either side of the equator and equidistant from it. Amphitelic orientation may be a property of centromeres that are already functionally divided, syntelic behaviour being manifested by functionally undivided centromeres. But until we know more about the ultrastructure of the centromere and its relationships to the rest of the chromosome and to the microtubules the concepts of 'functionally single' and 'functionally divided' are little more than a restatement of the observed facts.

The orientation or congression of the chromosomes onto the equator of the spindle involves irregular movements of the chromosomes (at a velocity of 0.25–2.0 μm min^{-1}) which eventually cease when the centromeres reach the equator. The very irregular pathways following by the centromeres during the formation of the spindle have been illustrated by Bajer and Mole-Bajer (1956). It seems probable that only the centromeres are active in these movements, the rest of the chromosome being passively pulled in one direction or another by the centromere. Irradiation of the centromere by u.v. or proton microbeams stops its congression at premetaphase; but if other parts of the chromosome are hit by the beam, movement is unaffected, even though visible damage is caused (Zirkle 1957). In holocentric chromosomes, which have been rather inadequately studied by modern techniques as far as the mitotic mechanism is concerned, the whole chromosome is presumably active in congression. Fragments of chromosomes lacking centromeres (e.g. ones that have been produced by radiation) always seem to be inert as far as the spindle is concerned and usually lie out in the cytoplasm without making any contact with it. Presumably there is a specific molecular binding between certain elements in the centromere (most likely special nucleotide sequences in the centromeric DNA, which may be repeated many times in a long centromere) and subunits of the spindle proteins that may be actually microtubules or closely associated with them.

The movements of the chromosomes at premetaphase have been most thoroughly studied in the case of the meiotic bivalents of crane-flies (Dietz 1956, 1969; Bauer et al. 1961) and grasshoppers (Nicklas 1961a, 1967). Various hypotheses regarding the forces responsible for these movements have been put forward, e.g. the force-equilibrium hypothesis of Östergren (1950) and the alternative model of McIntosh, Hepler and Van Wie (1969) which involves pushing forces between the poles and the centromeres. But there is no rigorous proof of any of the various models that have been put forward. In some groups of animals the meiotic bivalents seem to be subjected to extreme tension during premetaphase and may become extremely elongated during this 'premetaphase stretch' stage; nothing quite comparable to this is known in ordinary mitoses, and in most groups the amount of stretching is only moderate even at meiosis.

Metaphase seems to be essentially a period of stasis, during which the orientation, relative positions and dimensions of the chromosomes do not undergo any change. Only in a few exceptional cases is it possible to distinguish between nuclei in early and late metaphase by direct visual examination. Presumably, invisible changes are occurring which eventually lead to the termination of the static condition and the sudden onset of anaphase.

Although the overwhelming majority of mitotic spindles are symmetrical bipolar structures, other types can occur, either in the peculiar meiotic divisions which are normal for certain groups of organisms, or in abnormal or pathological conditions. Thus in the midges of the families Sciaridae and Cecidomyidae (see Ch. 14) the first meiotic division in spermatogenesis is characterized by a unipolar spindle, or at least one with one end acuminate and the other diffuse. In certain scale insects of the tribe Llaveini (pp. 503–5) the primary spermatocytes at first metaphase show parallel spindles, one for each chromosome, flared out at their ends without any poles in the ordinary sense at all (Hughes-Schrader 1931). Rather similar spindles are present in the egg of the Strepsipteran *Acroschismus* (Hughes-Schrader 1924).

Even among bipolar spindles of the usual type, many variations of shape are known. Some spindles are elongated and sharply acuminate at the poles; others are broadly barrel-shaped. Most of the usual staining techniques do not define the outlines of the spindle clearly, so that often its precise shape is more a matter of conjecture than of accurate observation. In consequence, very little is known about abnormalities of spindle formation under various physiological conditions or, for example, in species hybrids.

Unipolar spindles may occur in sea-urchin eggs treated with various chemicals or following mechanical injury. Such divisions were studied extensively by embryologists such as R. Hertwig and Th. Boveri about the beginning of the century; the whole subject is discussed in detail by Wilson (1925). Multipolar spindles (tripolar and quadripolar) may also be obtained in double-fertilized sea-urchin eggs (Baltzer 1908). Such multipolar spindles are also seen quite commonly in cancer cells, especially where the chromosome number is larger than the normal one. It is usual for such spindles to have several 'equatorial' plates which intersect one another; the chromosomes are orientated amphitelically on these equators. Multipolar spindles also occur in the polyploid megakaryocytes of mammals (Bessis, Breton-Gorius and Thiéry 1958).

The spatial arrangement of the chromosomes on the metaphase plate has been the subject of an extensive literature. Of course, when all the members of the karyotype are indistinguishable in size, regularities of arrangement may remain undetected. Where, on the other hand, there is a considerable range in size, the larger elements usually lie on the periphery of the metaphase plate so that only the centromere regions are actually within the spindle, the long arms projecting out into the cytoplasm; the smaller chromosomes lie in the central area of the metaphase plate and are entirely included in the spindle. This arrangement is particularly well shown when the chromosomes are relatively thin at metaphase; it is very well seen in many reptiles and grasshoppers (Fig. 2.1). On the other hand, when the chromosomes are

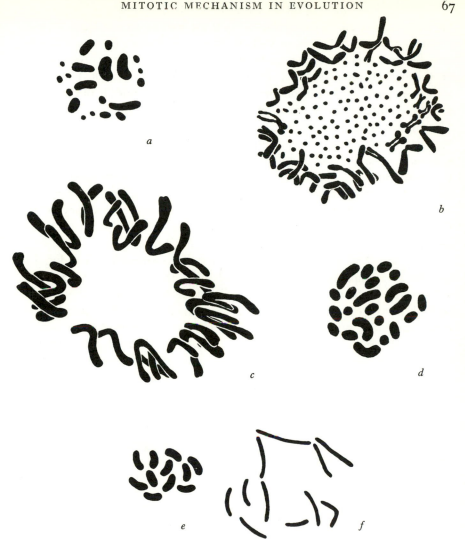

Fig. 2.3. Mitotic metaphases in various species of animals, to show orientation of chromosomes. *a*, spermatogonial metaphase of the tree-cricket *Oecanthus longicauda*, with a very irregular orientation of large and small chromosomes (Makino 1932); *b*, octoploid spermatogonial metaphase of the lizard *Tupinambis teguixin*, showing a very regular orientation of microchromosomes and macrochromosomes (Matthey 1933); *c*, spermatogonial metaphase of the urodele *Diemyctylus pyrrhogaster* (Sato 1932, from Matthey 1949) showing a 'hollow spindle'; *d*, oogonial metaphase of the pentatomid bug *Dinidor rufocinctus* to show irregular (non-radial) orientation of holocentric chromosomes (Schrader 1947); *e, f*, respectively egg follicle cell mitosis and embryonic mitosis of the scale insect *Leucaspis loewi*, showing differences in degree of chromosome condensation (S.W. Brown 1965). All figures redrawn.

very condensed, so that they appear as thick rods or almost isodiametric, the tendency is usually not so well marked, even though the range in size may be considerable (Fig. 2.3). In some species, where the chromosomes are particularly long and thin (as they are in many of the urodeles) they may all occupy peripheral positions around the spindle, the middle of the metaphase plate being empty. Some statistical studies (e.g. Barton, David and Merrington 1965; Irwin 1965) have shown that even after all the deformation and distortion inherent in the making of a squash preparation (as opposed to one made by sectioning) the arrangement of the chromosomes is non-random. Thus Barton *et al.* demonstrated that in the case of the human karyotype, most of the smaller chromosomes tend to occupy a central position, but that the sex chromosomes have a tendency to be peripheral. These positions clearly reflect the natural location of the chromosomes on the metaphase plate, before crushing.

Where the chromosomes are holocentric they are entirely enclosed in the spindle at metaphase and their arrangement is usually fairly random, without any particular tendency to a radial or any other kind of orientation. In such cases large chromosomes may sometimes occupy the centre of the metaphase plate, surrounded by smaller ones (see Fig. 2.3).

In no species are the relative positions of the different chromosomes constant; that is to say they always vary to some extent from cell to cell. But Dobzhansky (1936*a*) showed that in *Drosophila*, with a low number of chromosomes of very different sizes, the relative positions tend to remain the same from one division to the next, or at any rate that sister cells show the same arrangement more frequently than can be accounted for by chance.

Usually there is no particular tendency for homologous chromosomes to lie close together or far apart, in somatic metaphases; or if such a tendency does exist it is not strongly marked and is hence difficult to demonstrate convincingly. But the flies of the order Diptera constitute a most striking exception since in members of this group it is usual for homologous chromosomes to lie next to one another and closely parallel throughout their length in interphase; as a rule somatic pairing becomes less close by metaphase, so that the homologues are somewhat separated in the spindle. It has been assumed that somatic pairing is due to forces of mutual attraction between the chromosomes – forces which are identical with, or similar to, the ones responsible for meiotic synapsis.

Various cytologists have claimed to have observed somatic pairing in other groups of animals, but its existence has never been conclusively demonstrated outside the Diptera. Some early workers also thought they had found evidence of a phenomenon which would be, in a sense, the antithesis of somatic pairing – namely a tendency of homologous chromosomes to become orientated in opposite halves of the metaphase plate (Shivago 1924; Werner 1927). Such a tendency, which has been called gonomery, in those cases where it exists, is probably confined to the first few cleavage divisions in the embryo, in which the chromosome sets derived from the male and female pronuclei may not have become thoroughly mixed. In a first cleavage metaphase of a turbellarian with $2n = 34$, Costello (1970) demonstrated a remarkable instance of gonomery, which was interpreted as indicating that the seventeen chromosomes in the female pronucleus were in the same linear sequence as those in the sperm, due to some specific end-to-end attraction or attachment.

It has been claimed by Lark, Consigli and Minocha (1966) and Lark (1967, 1969) that there is a non-random segregation of daughter chromosomes at anaphase, i.e. that there is a tendency for chromatids synthesized during a particular division to segregate as a group to the same pole. The studies of these workers were carried out on mammalian material and on both diploid and polyploid higher plants. Their conclusions have been challenged by Heddle, Wolff, Whissel and Cleaver (1967), and S. Wolff and Heddle (1968) who found a random distribution of radioactive chromosomes at the second and third mitosis following labelling. The controversy is still unresolved, but Lark's experimental approach is the more direct one, less liable to complications such as sister-strand exchanges. Also, Lark's later data on polyploid wheat seem particularly convincing, since he obtained non-random segregation except when a high dose of the 5D chromosome (known to control meiotic synapsis and segregation) was present.

The chromosomes on the metaphase plate keep at a distance from one another and never touch or approach very close in a normal cell. They behave, that is to say, as if they were repelled from one another because of a surface charge. Under abnormal circumstances, however, they may fuse or clump together. But this is definitely a pathological manifestation and probably always irreversible. Clumping invariably seems to occur a short while after cell death if it is not prevented by coagulation as a result of cytological fixation. But it may also happen after irradiation with X-rays or at abnormal temperatures. Some degree of clumping may also be seen in the meiotic divisions of some species hybrids. Clumping has frequently been ascribed to 'stickiness' of the chromosomes, a cytological concept for which there is not much physical evidence (but which may nevertheless be genuine).

In much modern work on chromosomes it is a matter of routine to treat the cells with an agent such as colchicine or colcemid (desacetyl, dimethyl colchicine) before fixation. Such substances have several effects. They inhibit spindle formation so that nuclei with the chromosomes in a stage equivalent to metaphase accumulate in the tissue and are unable to proceed into anaphase. This is a convenience for the investigator, particularly if the tissue is one with a low 'mitotic index' (i.e. one in which cells in metaphase are normally scarce). These drugs also cause the two chromatids of each chromosome to separate more widely than is normal in untreated cells. Acrocentric chromosomes come to look like V's and metacentrics like crosses, the centromere being at the apex of the V or at the centre of the cross. This effect is also useful to the investigator because it enables the position of the centromere to be defined more precisely, and makes it easier to measure the lengths of the chromosome arms. The use of colchicine and colcemid has become such a universal routine in certain forms of cytogenetics, particularly in the study of vertebrate chromosomes (and above all human chromosomes) that some young workers in this field may never have seen normal chromosomes that have not been treated in this manner! The effects of these drugs seem to be broadly similar in all animals and plants.

The onset of anaphase is always abrupt. Whereas the chromosomes in a nucleus may reach the equator at different times in the premetaphase stage, they always seem to leave it synchronously at the beginning of anaphase. The daughter centromeres suddenly separate from one another and the chromatids are torn asunder in their immediate neighbourhood. The daughter centromeres separate further and further until eventually even the ends of the chromatids are pulled apart (at this stage we may begin to call the chromatids daughter chromosomes). And, at about this time the spindle itself begins to elongate, as a result of extension in the equatorial region.

It is probable that it is the necessity for synchronous initiation of anaphase movement by all the members of the karyotype which has led to the fact that all the chromosomes of a species generally seem to have the same centromere morphology. A karyotype in which some chromosomes had very large centromeres and others small ones might well be inefficient. Synchronous anaphase separation is not always seen, however, at the first meiotic division, where we have complications due to different numbers and location of chiasmata in the various bivalents (see p. 163) which are not present at mitosis. Thus at the anaphase of the first meiotic division we may have certain chromosomes which pass precociously to the poles or ones which lag on the spindle, remaining in the equatorial region after the others have completed their poleward movement.

In different organisms and tissues either the poleward movement of the chromosomes or the elongation of the equatorial region of the spindle may be predominant. But usually they seem to be combined, with the poleward movement of the chromosomes being initiated before any perceptible spindle elongation has begun. It is highly probable that the fact that many of the so-called continuous microtubules actually terminate in the equatorial region, overlapping with the ends of microtubules originating at the opposite pole (see p. 64), leads to the whole equatorial region being less rigid and hence a region of weakness when elongation occurs. Eventually the region of the spindle between the two groups of separating daughter chromosomes becomes converted into the so-called spindle remnant, which may persist long after the nuclei have entered the telophase stage and even after the cell has undergone cytoplasmic division into two daughter cells.

A variety of mechanisms to account for both types of anaphase movement have been proposed, but will not be reviewed in detail here. Most of them involve some kind of sliding of one set of fibres or microtubules on another. An early model of this type was put forward by Bělař (1929); more recent hypotheses which take account of our modern knowledge of spindle structure have been put forward by McIntosh, Hepler and Van Wie (1969) and Nicklas (1970). The model of McIntosh *et al.* assumes that the microtubules are polarized and have polarized binding sites for diagonal cross-bridges. Where bridges are between microtubules of the same polarity no force will be exerted, but bridges between adjacent microtubules of opposite polarities may exert forces which will cause sliding between the microtubules. Stasis at metaphase is assumed to be due to linkers of some kind between sister chromatids whose sudden disappearance initiates the anaphase movement. This particular aspect of the model seems unsatisfactory, since a well-marked and quite static metaphase stage occurs in the second meiotic division, when the chromatids generally diverge quite widely from one another; thus if linkers prevent the onset of anaphase for a while, it is probable that they are strictly confined to the centromere itself.

An alternative sliding microtubule model of chromosome movement has been put forward by Nicklas (1970). All such proposed models are analogous to the contraction mechanism in striated muscle; there seems no immediate prospect of proving one rather than another.

The forces needed to move the chromosomes are certainly very small; Nicklas (1965), working on grasshopper divisions and E. W. Taylor (1965) using newt chromosomes, have calculated from the known velocities and viscosities that the energy sufficient to move one chromosome could be provided by terminal dephosphorylation of about 20 ATP molecules.

Anaphase is followed by telophase, a stage in which nuclear membranes are re-formed around the two daughter nuclei and in which the chromosomes return to the interphase condition, undergoing a progressive decondensation. At the beginning of telophase the chromosomes are orientated in a kind of bouquet. This is, however quite different to the well-known bouquet orientation seen in the early stages of the meiotic prophase, since it involves a bringing together of the centromeres on one side of the nucleus

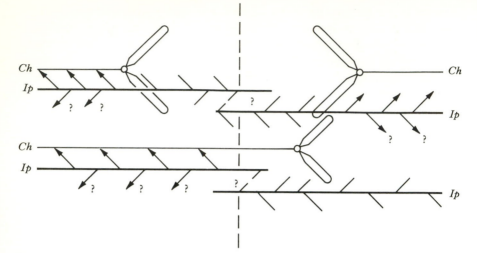

Fig. 2.4. Nicklas' sliding microtubule model of chromosome movement. Mid-anaphase is shown, with two normal daughter chromosomes and one daughter chromosome displaced to the right of the equatorial plane (broken vertical line). Polarized interpolar fibres (Ip) with attached bridges (diagonal lines or arrows) produce force only when adjacent to appropriately oriented, polarized chromosomal fibres (Ch). The question marks denote two unspecified features of the model; bridge interactions between overlapping interpolar fibres of opposite polarity (centre) and possible activity of 'free' bridges in regions of force production between chromosomal and interpolar fibres (right and left). From Nicklas (1970), *Advances in Cell Biology*, vol. 2. David M. Prescott, Lester Goldstein and Edwin McConkey, editors. Copyright 1971. Reprinted by permission of Appleton-Century-Crofts, Educational Division, Meredith Corporation.

rather than a gathering together of chromosome ends (i.e. telomeres). Telophase chromosomes, of course, consist of only one chromatid, and the general appearance of the various stages of their de-condensation always seems to be somewhat different from the stages of condensation that occur during prophase; this may be associated with the fact that telophase nuclei are almost always considerably smaller than prophase ones in the same tissue.

A number of abnormal circumstances may be sufficient to inhibit anaphase separation or prevent it being completed. In some cases in spermatocytes an acentric and a dicentric chromatid (resulting from crossing-over in a paracentric inversion – see p. 216) may do so. And in certain species of mantids a single univalent on the spindle (resulting from asynapsis or desynapsis) may do so (Callan and Jacobs 1957). The mechanisms underlying these inhibitory effects are not understood.

Mitotic non-disjunction (i.e. the passage of both daughter chromosomes to the same pole at anaphase) is a very rare accident of cell division; so rare that it has hardly ever been actually observed *in vivo*, although its effects may be observed in genetic experiments and in various types of somatic mosaicism. Directed non-disjunction, namely the more or less regular passage of both daughter chromosomes to a particular pole (where the two poles are distinguishable because the cytoplasmic division is unequal) occurs particularly in the case of certain supernumerary chromosomes (see p. 326) and in that of the X-chromosome of midges of the genus *Sciara* at the second meiotic division in spermatogenesis. The exact mechanism that operates in these instances is not understood.

Mechanisms of cytoplasmic division (cytokinesis) lie outside of the scope of this book. However, the general significance of the distinction between equal and unequal cytokinesis may be pointed out. In most cases cell division leads to the production of two equal and equivalent daughter cells. In special cases, however, the two products of division are cytoplasmically unequal and have different fates. In most such instances the spindle is bipolar and symmetrical, but located excentrically in the cell. The plane of cytokinesis is then at right angles to the mid-point of the spindle (Mazia 1961). An example that has been studied experimentally is the grasshopper neuroblast (Carlson 1940, 1952) which produces a small ganglion cell and a large daughter neuroblast at each division cycle. Here Kawamura (1960) was able to obtain a variety of abnormal modes of cytokinesis by displacing the spindle in the cell by means of micro-needles.

The factors which induce cells to enter on mitosis are poorly understood. Some types of cells have permanently lost the power to divide. These include most of the highly polyploid or polytene types of cells in insects and other groups (see Chs. 3 and 4); but they also include such diploid cells as the nucleated avian erythrocytes (and presumably those of lower vertebrates as well), which have also lost the power to synthesize protein and RNA (Cameron and Prescott 1963).

In other tissues individual cells divide from time to time, so that we can speak of the mitotic rate or mitotic index of the tissue (i.e. the proportion of cells in division at any one time); there exists a whole literature on the effect of various substances, particularly hormones, on mitotic indices. In other cases, particularly in the early cleavage divisions of embryos, or in the cysts of the testis in insects, we may have strictly synchronous division of all the nuclei and cells, suggesting, in the words of Mazia (1961) that 'rate-determining events may be governed by underlying changes in the intracellular environment that can be shared.'

3

ENDOMITOSIS AND ENDOPOLYPLOIDY

In every living cell, the bearers of all possible reactions, the chromosomes, strove always, unless compelled to some other substituted function, after the one thing – reproduction.

R. BRIFFAULT: *Europa*

The statement made in some elementary accounts of histology, that all the somatic cells of an organism contain the same number of chromosomes, is true only to a limited extent. In all multi-cellular animals and plants certain tissues regularly consist wholly or in part of polyploid cells, whose nuclei contain multiples of the basic number of chromosomes. This phenomenon has been studied especially in certain insect and plant tissues (Geitler 1937, 1938, 1939a, b, 1941, 1953; see general review in Tschermak-Woess 1963). The insect embryo consists of a large number of cells (diploid in most instances, but haploid in male Hymenoptera and some other groups) which have become differentiated into the various kinds of tissues by the time of hatching from the egg. During larval life the volume of the insect increases many times, but mitosis occurs only rarely in some tissues and not at all in others. Thus larval growth involves, in general, a great enlargement of the individual cells and only a slight increase in their number. Examination of the nuclei of these enlarged cells shows that they have become polyploid and that by the time the adult stage is reached their chromosome number is 4n, 8n, 16n . . . etc., the degree of polyploidy being characteristic of each type of tissue (although most such tissues are mosaics of cells of different 'ploidies', i.e. *mixoploid* tissues in the terminology of some authors).

Determination of the chromosome number in these insect somatic cells which will never divide again is rendered possible in some instances by the fact that the chromosomes in the resting nuclei are visible as separate, irregularly-shaped bodies which fix and stain sufficiently well to be countable if the degree of polyploidy is not too high. A good example of this state of

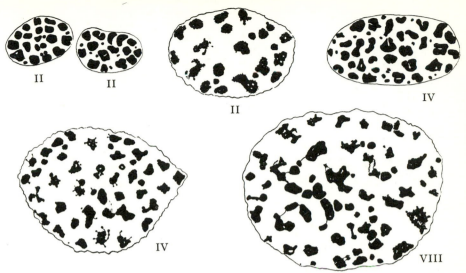

Fig. 3.1. Endopolyploidy in somatic nuclei of the testis sheath in the locust *Schisto-cerca gregaria*. The smallest nuclei (II) are diploid and have 23 chromosomes, the medium sized ones (IV) have 46, the largest ones (VIII) are octoploid, with 92 chromosomes. *X*-chromosomes not distinguishable by any special behaviour.

affairs is the investing epithelium of the testis in orthopteroid insects, a tissue in which the nuclei are flattened pancake-shaped bodies containing irregular masses of chromatin corresponding to the diploid, tetraploid and octoploid numbers for the species (23, 46 and 92 in most of the common species of grasshoppers). Where the chromosome numbers of such endo-polyploid cells are much higher (in the thousands or tens of thousands), as in the nuclei of certain insect secretory cells, it may not be possible to determine their degree of polyploidy directly, although an estimate may be attempted on the basis of the nuclear volume or quantity of DNA present. However, determinations of DNA values by photometric methods become very difficult and inaccurate in the case of extremely large nuclei.

Geitler (1937a, 1938b, 1939a, b, 1941) made a special study of somatic polyploidy in various bugs of the order Heteroptera, particularly the pond-skaters of the genus *Gerris*. Here the diploid number is twenty-one in the males, the odd chromosome being an *X*, which is conspicuously heteropyc-notic in the somatic nuclei of some species such as *Gerris lateralis*, but not in those of others such as *G. lacustris*, the difference possibly depending on the extent of the heterochromatic segments in the *X*.

Owing to the heteropycnosis of the *X*'s in the somatic resting nuclei of *G. lateralis* it is possible to distinguish them from the autosomes and count

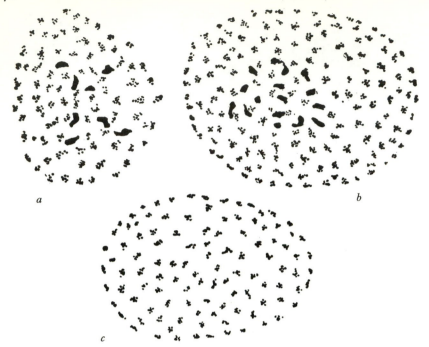

Fig. 3.2. Endopolyploidy in the pond-skaters *Gerris lateralis* and *G. lacustris*. *a*, octoploid nucleus from a testis-septum cell of *G. lateralis*; *b*, 16-ploid nucleus from a similar cell; *c*, 16-ploid nucleus from testis septum of *G. lacustris*. The *X*-chromosomes are distinguishable by their heteropycnosis in *a* and *b*, but not in *c*. From Geitler (1937).

them. Obviously, the number of haploid sets of chromosomes in the nucleus is twice the number of *X*'s. In this way Geitler was able to show that the giant nuclei in the salivary gland are highly polyploid; some of them are 512-ploid, having undergone eight cycles of endoreduplication, while the largest were estimated from their volume to be 1024- and even 2048-ploid (it is not possible to actually count the *X*'s in these largest nuclei). These highly polyploid nuclei contain many nucleoli, but the latter show a strong tendency to fuse together, so that the number of nucleolar structures is not a reliable index of the number of sets of chromosomes present. In some other Heteroptera such as *Lygaeus saxatilis* the *X*-chromosomes are not heteropycnotic in the somatic nuclei, but the *Y*-chromosomes do show conspicuous heteropycnosis, so that it is possible to determine the degree of polyploidy of male somatic cells by counting the number of *Y*'s. However, in certain polyploid tissues all the *Y*'s in a nucleus fuse together in a single mass. The much-branched nuclei in the spinning gland cells of the Lepidoptera and

Fig. 3.3. Endopolyploid nucleus (128- or 256-ploid) from the salivary gland of a male *Lygaeus saxatilis* (Heteroptera). The large black mass represents many fused heterochromatic *Y*-chromosomes. From Geitler (1939*a*).

Trichoptera (Vorhies 1908) probably have a similar structure to those of the salivary glands of *Gerris*.

The terms endomitosis, endopolyploidy and endoreduplication have been used in connection with this phenomenon. The prefix 'endo' indicates that the events take place within the nucleus, without any breakdown of the nuclear membrane. Associated with each endoreduplication is a cycle of chromosomal condensation and de-condensation. The stages of this cycle have been termed endoprophase, endometaphase, endoanaphase and endotelophase. Endoprophase involves increasing condensation; the chromosomes are visibly split into chromatids. Endometaphase is simply the stage of maximum condensation; there is no sign of a spindle or any orientation of the chromosomes onto a metaphase plate. Endoanaphase is a stage when the daughter chromosomes appear to be falling apart; it may resemble a diplochromosome nucleus. In endotelophase the daughter chromosomes have separated but without passing to any poles; they may also be undergoing decondensation. The interphase is a stage when the chromosomes become diffuse, although the extent to which they do is highly variable.

Some of these endopolyploid nuclei occur in flat pancake-shaped cells and are themselves very much flattened. It is possible that a true mitosis with a spindle mechanism would be incompatible with the dimensions of such nuclei and cells. However, it is unlikely that cellular dimensions have caused insects to develop the mechanism of endopolyploidy; it is probable that the relationship is the other way around, namely that the existence of endopolyploidy has permitted the evolution of certain types of cellular mechanisms and histological architecture which would have been incompatible with a mitotic mechanism.

The situation where the number of chromosomes can be directly counted in endopolyploid nuclei is relatively uncommon. Thus in many instances the degree of ploidy attained can only be estimated. A more satisfactory method of investigation consists in determining the total amount of DNA in the nuclei.

The DNA values of the nuclei in a number of endopolyploid tissues of *Dermestes* beetles were studied by Fox (1969b). The same author had earlier shown (Fox 1969a) that six species of this genus, with very similar karyotypes, showed a considerable range of DNA values in the spermatids, the highest value being 2.7 times the lowest one; DNA values were roughly proportional to chromosome volume.

In a number of endopolyploid somatic tissues the DNA values form a doubling series. Thus in the abdominal fat body there are 2C and 8C cells, with a few 4C and 16C ones (C is the amount of DNA in the haploid spermatid nuclei). But in other tissues a doubling series is not found (Fox 1970b). All somatic tissues investigated showed some 2C nuclei, but a few nuclei showed less than the 2C amount in *D. maculatus* and *D. lardarius*, suggesting that some chromosome loss had occurred. The Malpighian tubule nuclei seem to fall into 4C, 8C and 16C classes in *D. ater*, but approximate to a continuous series in some other species or show peaks at intermediate values (e.g. a peak between 8C and 16C in *D. maculatus*). Autoradiographic studies have shown that the intermediate values are not due to the nuclei being in S phase at the time (the frequency of nuclei in S phase is very low in these tissues – from 1 in 1000 to 1 in 10000 nuclei). It seems probable that, as in the larval brain ganglia of *Drosophila hydei* (Berendes and Keyl 1967), the non-doubling series in certain somatic tissues is due to differential replication of euchromatin and heterochromatin. This hypothesis has been confirmed by making detailed density maps of Feulgen-stained nuclei in order to estimate the proportion of the total DNA which is heterochromatic and determine the replication-status of the heterochromatin and euchromatin in different nuclei of the same tissue (Fox 1972). It is fairly obvious that wherever we have differential replication of heterochromatic and euchromatic DNA the nucleus must have, at least in part, a polytene structure. But many of these somatic nuclei are not easily amenable to analysis by ordinary microscopical techniques.

In locusts (*Schistocerca gregaria* and *Locusta migratoria*) Fox (1970a) found that the DNA values formed a discontinuous series in all tissues but except for the first two classes in *Schistocerca* testis wall and ovariole and male fat body, this is not a doubling series. He suggests that at each cycle of DNA replication a fraction of the total genome is omitted, and that this fraction increases with each successive S phase. Such a mechanism could theoretically play an important and significant role in tissue differentiation in insects.

Although it seems clear that endoreduplicated chromosomes have lost the power to divide by mitosis, exceptions undoubtedly exist. Thus in mosquito larvae with $2n = 6$, the cells of the iliac epithelium apparently undergo a series of endomitoses whereby the number of chromosomes per nucleus is increased to 12, 24, 48, 96 and sometimes 192. These cycles of replication take place during larval life in nuclei which show a typical interphase appearance. It is only when the pupal metamorphosis begins that these cells once more acquire the ability to divide by mitosis (Berger 1936, 1937, 1938; S. M. Grell 1946a, b). As they pass into prophase, it can be seen that in each nucleus there are three large bundles of chromonemata which result from somatic pairing of the homologous chromosomes and repeated replication and splitting of each of the paired elements.

As these cells reach metaphase, the bundles of chromonemata become dispersed, but apparently the mitoses are reductional, i.e. the chromosomes play the role normally assumed by daughter chromatids. As a result of a series of such divisions (the later ones only differing from the first in that bundles of chromonemata are not formed at prophase) the number of cells in the iliac epithelium is greatly increased, their size is correspondingly diminished and their chromosome number is reduced from $32n$ or $64n$ to $2n$. Although there does not seem to be any doubt that this process of somatic reduction actually occurs in mosquito development, some of the details are not entirely clear. These mitoses have not been satisfactorily illustrated and the investigation was carried out before accurate measurement of DNA values by microdensitometry became possible.

A somewhat similar, but different, phenomenon occurs in the development of the larval ganglionic cells in *Drosophila*. It has been known for many years that the mitoses of the ganglion cells of the last instar larvae of *Drosophila* show particularly large chromosomes, compared with those seen in spermatogonial, oogonial and embryonic mitoses, for example. They have consequently been routinely employed for studies of *Drosophila* karyotypes, e.g. by Clayton and Ward (1954) and Clayton and Wasserman (1957). Swift (1962) found differences in DNA values in cerebral ganglion interphase nuclei of *Drosophila virilis* larvae, some of which were found to have the 16C amount. More recently, Gay, Das, Forward and Kaufmann (1970) found that in *D. melanogaster* the ganglionic metaphases of first instar larvae show the 4C quantity of DNA while those of third instar larvae have the 8C amount (and perhaps occasionally the 16C quantity): The large metaphase chromosomes in the ganglion cells of the third instar larvae are both longer and thicker than those of the first instar larvae; they do not show any obvious multi-stranded or diplochromosome structure.

These phenomena in somatic cells of mosquitoes and *Drosophila* seem to be rather special ones, perhaps associated with the existence of the somatic-pairing phenomenon characteristic of the Diptera. It would certainly be wrong, at the present stage, to use them as an argument for any general occurrence of multi-stranded, polyneme metaphase chromosomes in other groups of organisms.

Nur (1968*a*) has studied endomitosis in the Malpighian tubule nuclei of the female mealybug, *Planococcus citri*. The cells of the Malpighian tubules are regularly binucleate in this insect. Nuclei with the shortest chromosomes and with the chromatids closely approximated were considered to be in 'endometaphase'. Ones with short chromosomes but with the chromatids more widely separated were regarded as 'endoanaphase' stages. During endoanaphase and endotelophase there is a tendency for the daughter chromosomes to remain attached at their ends. The same worker (Nur 1966) had earlier shown that in males of *Planococcus* (which have the paternal chromosome set in a permanently heteropycnotic state) the paternal heterochromatic chromosomes in the nuclei of the testis sheath and in most of the oenocytes do not replicate their DNA while the euchromatic set undergo several rounds of endoreduplication.

The endomitotic cycle of the developing hypopharyngeal gland cells of worker honey bees, which will secrete the so-called royal jelly at a later stage, was studied by Painter and Biesele (1966*a*), with special reference to the nucleolar cycle which has been rather neglected in other investigations on endomitosis. In these bee gland cells there are a large number of nucleoli in the interphase nuclei. Nuclei presumed to be diploid have up to eight nucleoli and the number is doubled with each endomitotic cycle. The units of which the nucleoli are composed contain ribosomes embedded in a mass of protein. During early endoprophase when the structure of the chromosomes is unclear the nucleoli undergo fragmentation releasing ribosomes and quantities of newly synthesized protein into the nuclear fluid. The ribosomes then enter the nuclear pores where they are organized into polysomes and passed into the cytoplasm. By the time the endometaphase stage has been reached the nucleoli have disappeared and the chromosomes are visible in bundles. In the next interphase period a new set of nucleoli make their appearance and the whole cycle is repeated. Painter and Biesele regard these repeated endomitotic cycles as a mechanism for increasing polysome synthesis preparatory to the secretory phase of the hypopharyngeal cells. In a later paper (Painter and Biesele 1966*b*) they describe a similar cycle in the nurse cells of *Drosophila* oocytes.

We have already referred (p. 9) to the occasional occurrence of nuclei containing 'diplochromosomes' in various types of material following

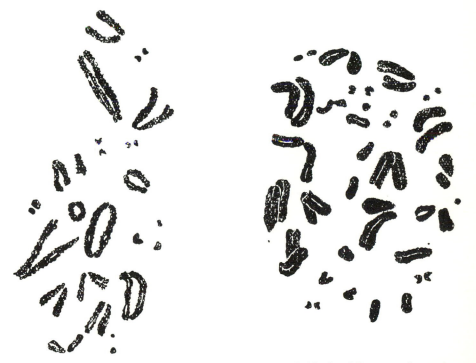

Fig. 3.4. Spermatogonial nuclei from irradiated individuals of *Locusta migratoria* showing diplochromosomes (these were the first instances of diplochromosome formation recognized). From White (1935*b*).

irradiation or various other types of treatment (White 1935). There does not seem to be any fundamental difference between experimental diplo-chromosome formation and natural endomitosis, although diplochromo-somes tend to be held together by functionally undivided centromeres, whereas in naturally occurring endoreduplication the chromosomes seem usually to separate completely. If we get three rather than two successive replication cycles without mitosis so-called 'quadruplochromosomes' may result.

Several cytologists have used tritiated thymidine autoradiography to study the pattern of endoreduplication in the formation of diplochromo-somes. It is clear from the work of Walen (1965), Schwarzacher and Schnedl (1966) and Herreros and Giannelli (1967) that when diplochromosomes are obtained following two cycles of DNA replication after labelling with tritiated thymidine, only the outer chromatids are labelled. In the case of quadruplochromosomes, one replication cycle later, it is the inner chroma-tids of the outer 'chromosomes' that are labelled. These results are con-

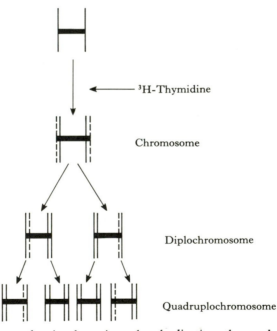

Fig. 3.5. Diagram showing how, in endoreduplication, the newly-formed DNA strands are always on the outside. The 'hot' strand formed at the first replication (in the presence of tritiated thymidine) is indicated by a dashed line. The cross-bar represents the centromere region, where the strands are held together. Based on the experiments of Walen (1965), Schwarzacher and Schnedl (1966) and Herreros and Giannelli (1967).

sistent with the view that when endoreduplication exists, the newly synthesized DNA strand is always on the outside. The reason for this is totally obscure.

The highest degree of endopolyploidy hitherto recorded is probably that of the nucleus of the giant neuron in the abdominal ganglion of the mollusc *Aplysia*. According to Coggeshall, Yaksta and Swartz (1970) this nucleus may reach a volume of 7 million μm^3 in adult individuals and contain 75 000 times the haploid amount of DNA. Such nuclei may be as much as 220 μm long and 70 μm wide. They may have undergone sixteen or seventeen cycles of endoreduplication, and there is some evidence that these are synchronous for all the chromosomes in the nucleus, since the amount of DNA appears to go up in a number of steps in the course of development, rather than continuously.

ENDOPOLYPLOIDY IN PROTOZOA

Two theoretically important kinds of endopolyploid nuclei are known in Protozoa, the macronuclei of the ciliates and the so-called *primary nuclei* of certain Radiolaria. Modern work on ciliate macronuclei has been reviewed in detail by Raikov (1969). This author has made a study of certain lower ciliates (mainly marine) which have micronuclei and macronuclei (several or many of each) but in which the macronucleus is diploid. These forms, of which *Loxodes* may be considered as an example, seem to be one stage further advanced than *Stephanopogon*, which has no differentiation of micronuclei and macronuclei. *L. striatus* has two micronuclei and alongside those, two spherical macronuclei. *L. magnus* has about thirty nuclei of each kind. The macronuclei of these lower ciliates arise from micronuclei but once they have grown in size they never themselves divide again, either mitotically or amitotically. They do, however, show a most peculiar cycle of nucleolar activity, which is presumably their main *raison d'être*. These peculiar forms have been referred to as 'karyological relics' by K. G. Grell (1962); they belong to several different and not very closely related families.

In the higher ciliates the macronucleus is always very large, highly polyploid and divides amitotically when the individual undergoes fission. It may be spherical, ovoid or elongate; in some cases it acquires a filamentous, sickle-shaped or branched form, and in a number of species it appears like a string of beads. Some Ciliata have multiple or fragmented macronuclei. In all cases the DNA content of polyploid ciliate macronuclei seems to be very high. In most species the internal structure of the macronucleus is difficult to make out; it appears to be densely packed with chromosomes and nucleolar structures. The nuclear fluid is usually very much reduced in amount.

There may be many small nucleolar bodies, a few large ones or (in a few species) a single very large one; they may be distinguished from the 'chromatin bodies' because they are Feulgen-negative. It is uncertain in most cases whether the latter are to be looked up as chromosomes, or as portions or segments of chromosomes.

Two types of macronuclei were distinguished by Fauré-Fremiet (1957). In *homomerous* ones there are no separate zones and the entire nuclear cavity is more or less uniformly filled with chromosomal and nucleolar bodies. In *heteromerous* macronuclei, which occur in relatively few groups of ciliates, there are two segments or zones, the *orthomere* and the *paramere*, which differ in their contents.

In all the higher ciliates with endopolyploid macronuclei the macronucleus develops in the first instance from a diploid micronucleus by a series of endomitotic replications. In *Stylonychia mytilus* (Ammermann 1963), *S. muscorum* (Alonso and Pérez-Silva 1965) and *Nyctotherus* (Golikowa 1965) the chromosomes within it go through a typical polytene stage in which they have a transversely banded structure just like the polytene chromosomes of the Diptera and some Collembola (see Ch. 5). In *Ephelota gemmipara* the chromosomes in the macronucleus do not pass through a banded polytene stage, but they do form bundles, as in the ileal cells of mosquitoes, the chromosomes being especially closely associated at one end, where the centromere is perhaps located (K. G. Grell 1949, 1950, 1952).

In *Nyctotherus*, according to Golikowa, all the chromosomes appear to be attached end-to-end during the polytene stage, which lasts about two weeks; there is apparently just a single giant polytene element in the nucleus at this time, with two free ends. The situation is different in *Stylonychia*. In *S. mytilus* there are about 250 chromosomes in the nucleus during the prepolytene stage which join together to form one or two long chains at the time when they acquire a polytene structure. In *S. muscorum*, on the other hand, there seem to be fifty to sixty free polytene chromosomes in the macronucleus. The polytene stage is only temporary in these macronuclei; Ammermann has described a process of transverse fragmentation of the polytene chromosomes in *S. mytilus*, with each Feulgen-positive band becoming transformed into a chromatin granule, while in *Nyctotherus* the giant polytene element seems to split longitudinally into many non-polytene chromosomes.

The degree of ploidy has been determined for a number of ciliate macronuclei by comparing the amount of DNA they contain with the amount present in the micronucleus. In *Paramecium bursaria* the macronucleus only contains the 16n amount of DNA; in *P. caudatum* different strains have

52n to 160n amounts, while *P. aurelia* has an 860n macronucleus (Cheissin and Ovchinnikova 1964; Cheissin, Ovchinnikova and Kudriavtsev 1964; Woodard, Gelber and Swift 1964). The highest ploidy recorded in a ciliate macronucleus is an estimated 13 150n in *Spirostomum ambiguum* (Ovchinnikova, Selivanova and Cheissin 1965). In some cases the degree of ploidy is a power of two (e.g. *P. bursaria* and a strain of *P. caudatum* with a 128n macronucleus); but most frequently it is not, indicating some form of asynchrony, either of whole chromosomes or chromosome regions, in replication.

Most ciliate macronuclei divide amitotically into equal halves when the individual undergoes fission; usually the amitotic division of the macronucleus is accompanied or preceded by a mitotic division of the micronucleus. But in some ciliates, especially Suctoria, small daughter macronuclei are budded off from the main macronucleus. The DNA replication of several types of macronuclei has been studied by modern methods. The S phase in the macronucleus usually lasts much longer than in the micronucleus. In some cases either the G_1 or the G_2 period may be shortened, so that it is practically absent; in *Stentor* both G_1 and G_2 are virtually absent.

In some ciliate macronuclei DNA replication occurs simultaneously throughout the nucleus. But in many forms, especially ones with heteromerous macronuclei, so-called reorganization bands in which replication occurs pass across the nucleus during the S phase. In *Euplotes* the reorganization bands (sometimes referred to as replication bands) make their appearance simultaneously at opposite ends of the sickle-shaped macronucleus and migrate towards one another until they meet in the middle region and disappear (Turner 1930). This process takes $8-12\frac{1}{2}$ hours. It was shown by Gall (1959), using tritiated thymidine autoradiography, that DNA replication takes place only in the bands. It is, in fact, restricted to the leading edge of the posterior zone of the reorganization zone (Prescott and Kimball 1961; Prescott 1962). Protein synthesis accompanies DNA replication in the reorganization band (Ringertz and Hoskins 1965; Prescott 1966) and the synthesis of histones at any rate seems to be closely associated with DNA replication. But RNA synthesis is continuous in the *Euplotes* macronucleus *except* in the replication bands (Kluss 1962; Prescott 1966), suggesting that replicating DNA is actually incapable of RNA synthesis. Complex changes in the chromatin granules and nucleoli as the reorganization band passes over them; the chromatin granules break down into fibrils in the band but after this has passed over them they become once more converted into granular form, but the granules are now approximately eight times their previous volume (Ringertz, Ericsson and Nilsson 1967).

The division of the macronucleus is generally described as amitotic; but in some heteromerous macronuclei the paramere appears to play a spindle-like role. Usually the macronucleus degenerates during the sexual conjugation process and a new one is formed from a derivative of the zygote nucleus. In some, but not all, species the degeneration of the macronucleus involves fragmentation. In *Paramecium aurelia* it was shown by Sonneborn (1940) that fragments of the macronucleus can be experimentally induced to regenerate an entire macronucleus; presumably there is a regulatory process which ensures that regenerated macronuclei contain equal numbers of replicates of all the chromosomes in the basic karyotype.

Protozoan nuclei which are capable of multiple division into several or many genetically equivalent products were called *polyenergid* by Hartmann (1909). In ciliates such multiple division or 'budding' of the macronucleus occurs in *Ephelota*, *Tachyblaston* and *Anoplophrya* (Grell 1952).

The phenotype of ciliates is determined by the macronucleus alone, as was shown by Sonneborn's (1946, 1947, 1954) experiments in which individuals of *Paramecium* were obtained with a macronucleus and a micronucleus of different origin and hence of different genetic constitution. The inactivity of the micronucleus is apparently not due simply to the fact that the macronucleus contains far more copies of all the genes. In fact, the micronucleus usually synthesizes no RNA; in *Tetrahymena* it contains about 70% more histones per unit of DNA than the macronucleus (Alfert and Goldstein 1955).

Sonneborn (1947) suggested that the macronucleus of the ciliates consisted of a number of diploid 'subnuclei', each containing a complete genome and capable of independent reproduction. At macronuclear division these subnuclei would be distributed at random to the daughter nuclei. Electron microscopy does not support this concept in its original form. A different genetic model to explain polyenergid nuclei was put forward by K. G. Grell (1953, 1956), on the basis of his work on the very large 'primary nucleus' of the radiolarian *Aulacantha scolymantha*. This contains over a thousand chromosomes, and arises through a series of endomitotic cycles, each involving an endoprophase, endometaphase, endoanaphase and endotelophase. During swarmer-formation the nucleus breaks up into a large number of secondary nuclei. At division both chromatids of each chromosome pass into the same daughter nucleus.

Grell believes that in this case all the chromosomes in a genome or karyotype are connected end-to-end so as to form *composite chromosomes*. These chains of chromosomes would consequently be the units of heredity and aneuploidy would be impossible, as in Sonneborn's model. There is some morphological evidence for the existence of composite chromosomes, not only in *Aulacantha*, but also in the macronuclei of a number of ciliates.

4

GIANT CHROMOSOMES — THE POLYTENE TYPE

...un cordon cylindrique diversement enroulé sur lui-même a la manière d'un intestin.

E. G. BALBIANI

In the year 1881 Balbiani published the first description of the giant chromosomes in the salivary gland nuclei of midges of the genus *Chironomus*. These structures were also studied by Korschelt (1884) and Carnoy (1884), who published a remarkably accurate figure of their morphology, but who made the mistake of supposing that the same type of nucleus occurred in other insect orders. These early workers did not recognize the giant cross-banded structures as chromosomes and tended to believe, erroneously, that they formed a continuous 'spireme' (i.e. an endless ring tangled up within the nucleus). All of them described the banded appearance of the chromosomes, but early in the twentieth century a number of cytologists misinterpreted this as gyres of a spiral. The modern work on the giant chromosomes of the Diptera dates from the work of Heitz and Bauer (1933), Painter (1933) and King and Beams (1934), who clearly interpreted them as chromosomes, realized the significance of the bands and pointed out their importance for detailed cytogenetic investigations. According to Beermann (1962) only Rambousek (1912) among the earlier workers clearly realized the chromosomal nature of the polytene elements, but his work was published in Czech and remained unnoticed.

After 1933, it was soon found that chromosomes of this type, although widespread in the two-winged flies (order Diptera) were not confined to the larval salivary gland – although they usually attain their largest size in that organ – but occur in many other somatic tissues. The term salivary gland chromosomes is hence inappropriate as a general designation for these structures and the term *polytene* (i.e. many-stranded) chromosomes, introduced by Koller (1935) seems preferable. Beermann (1962) prefers *giant chromosomes* (German: *Riesenchromosomen*), but we shall use this term in a more general sense, to include also the 'lampbrush' chromosomes of the amphibian and other oocytes (Ch. 5), which are also very large, but not many-stranded.

Naturally, the nuclei in which the polytene chromosomes lie are generally very large. In various Chironomidae those of the larval salivary gland may reach 100 to 150 μm in diameter, while those of *Drosophila* and most *Sciara* species are considerably smaller (about 50 μm in diameter).

The work of Heitz and Bauer was carried out on the salivary gland nuclei of *Bibio* larvae. They showed that the worm-like elements really were separate chromosomes and not a continuous 'spireme' as some earlier workers had assumed. They also showed that these bodies were present in the haploid number, each of them being formed by intimate side-by-side pairing of two homologous chromosomes, presumably as a result of the somatic pairing phenomenon, which is so characteristic of dipterous chromosomes. In some instances the homologues appear to be actually fused, while in other cases they are in close contact but still distinguishable as separate entities; some elements may show intimate fusion in certain regions and quite loose pairing in other sections.

Painter (1933, 1934*a, b, c,* 1935) was the first to apply the salivary gland technique to *Drosophila*, and its use in cytogenetic analysis dates from that time. He was able to confirm the idea that the giant strands of the salivary gland nuclei actually were synapsed chromosomes by using stocks of *Drosophila* (provided by W. S. Stone) known, on genetic grounds, to be heterozygous for structural rearrangements such as inversions and trans-locations; it was found that even in such structural heterozygotes the homologous regions were still paired together, band for band, giving rise to complicated configurations.

The relationship between the salivary gland cells (which will never undergo mitosis) and the normal mitotic elements was taken up by C. B. Bridges (1935*a*) and Koltzov (1934), who independently put forward the view that the salivaries were similar in structure to prophase chromosomes, but that they had uncoiled and replicated many times, without any separation of the resulting strands (i.e. that they had a cable-like structure).

We can thus say that the polytene chromosomes of the Diptera represent a very special case of the much more widespread phenomenon of endopoly-ploidy. Some species of Diptera such as the cecidomyids of the genus *Miastor* (see p. 524) do not seem to show polytene chromosomes in any of their tissues, even those with the largest nuclei exhibiting an endopolyploid condition, the individual strands having separated after replication. Polytene nuclei do not seem to have been reported for any tissues of the primitive

Fig. 4.1. The four polytene elements from a salivary gland nucleus of *Chironomus thummi*, showing the banding. Small arrows indicate the approximate positions of the centromeres. The shortest chromosome has its longer arm largely heterochromatic and the other three elements have short heterochromatic regions at the tips. From Bauer (1935).

family Tipulidae, in which the larval salivary gland nuclei are certainly endopolyploid.

The dark-staining bands in the polytene chromosomes clearly correspond to the chromomeres seen at various stages of mitosis and meiosis, and especially to those seen in the 'lampbrush' chromosomes of the amphibian oocyte (see Ch. 5). Each band is in reality a disk, i.e. it goes right through the thickness of the chromosome and is formed by more or less intimate fusion of chromomeres corresponding in number to the number of strands or chromonemata. In the thinner bands the chromomeres are smaller and tend to be separate; in such cases their number has been assumed to be a power of two, depending on the number of rounds of replication the original one has undergone. Larger chromomeres tend to be fused together to form vesiculate bands. In *Chironomus* there may be more than 16000 strands in the polytene chromosomes, resulting from thirteen to fourteen successive replications.

There can be no doubt that the chromomeres contain much-folded or spiralized DNA, while in the interband regions the DNA is in an extended state. This would explain why it has proved impossible to demonstrate DNA at all in the interbands by techniques such as Feulgen staining, although this has now been done by fluorescence microscopy (Wolstenholme 1965, 1966b). The interband regions are disrupted by DNAase, but not by RNAse or proteases (Lezzi 1965).

Most of the larval organs of the Diptera undergo lysis at the time of pupation, the adult organs being developed from the imaginal buds of the larva. This is the reason why larval tissues have been most extensively used for studies of polytene chromosomes. Nevertheless, some tissues of the pupa and imago do possess polytene chromosomes and pupal footpad cells and trichogen cells of the thoracic bristles have been studied by Whitten (1965, 1969), Ribbert (1967) and Thomson (1969). These particular types of cells, however, undergo death (autocytolysis) about the time of emergence of the imago (although there seems to be a difference between the trichogen cells of *Sarcophaga* which undergo cytolysis and those of *Calliphora*, which do not). We shall consider the causes of this 'programmed cell death' later (p. 110).

The diameter of the polytene chromosomes is by no means uniform. Almost all polytene chromosome limbs (except very short ones) show one or more 'waists' or narrow constrictions. Other regions are characteristically thick and swollen. Such constantly swollen regions, which may include a dozen or more bands, are entirely different from the 'puffs' and Balbiani rings discussed later, which are products of the activity of single loci. Variation in diameter is especially striking in some of the polytene chromosomes that have recently been discovered in Collembola (see Fig. 4.16). In salivary gland nuclei of certain gall midges (Cecidomyidae) there is one short chromosome that is very much thicker than any of the other chromosomes and appears to have a large number of characteristically granular bands (Fig. 4.3); the significance of this chromosome is not understood (White 1948a); Ashburner (1970) has suggested that it has undergone an extra cycle of replication by comparison with the other chromosomes. A similar state of affairs has been seen in some other fungus-feeding Cecidomyidae and something of the same kind has been illustrated by Melland

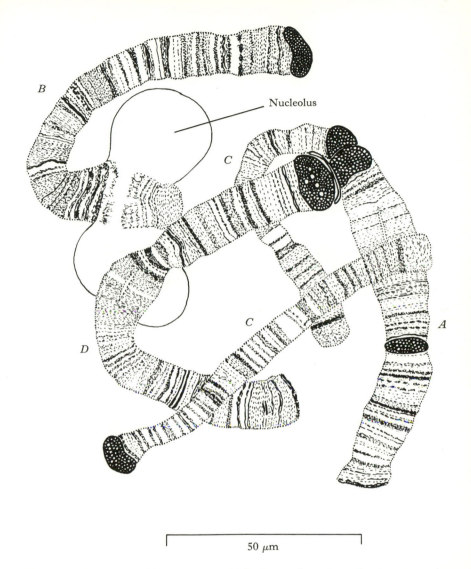

Fig. 4.2. Salivary gland chromosomes of '*Cecidomyia*' *serotinae*, showing the nucleo-lar chromosome (*B*) and the tendency of the condensed heterochromatic regions at the ends of the chromosomes to fuse together. Chromosome *A* possesses, in addition, an interstitial heterochromatic region. In this particular cell the two *C*-chromosomes were unsynapsed. The nucleolus arises from a region composed of loose hetero-chromatin. From White (1948*b*).

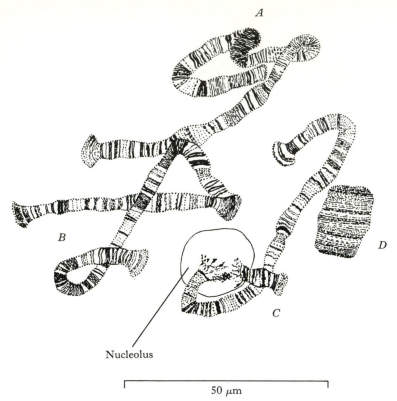

Fig. 4.3. Salivary-gland chromosomes of *Camptomyia* sp. (Diptera, Cecidomyidae), showing the nucleolar chromosome (*C*) and the thick chromosome (*D*), which has probably undergone an additional replication cycle. Note the association of loose heterochromatin with the nucleolus. From White (1948*b*).

(1942) in the case of the chironomid *Anatopynia varius*. An even more peculiar condition has been reported in the cells of the salivary gland 'reservoir' in the cecidomyid *Mikiola fagi* by Kraczkiewicz and Matuszewski (1958); in this case there are four polytene elements, two of which are 'normal', with prominent heterochromatic regions around the median centromeres. The other two chromosomes are ribbon-like, with the bands running *longitudinally*. It would appear that we are dealing here with very short chromosomes which have become polytene, but in which the multiple strands lie essentially in one plane rather than forming a compact bundle. The mode of replication in these ribbon-like chromosomes may well be peculiar and different from the usual one.

In many Diptera the ends of the polytene chromosomes, or some of them, are flared out in trumpet-like expansions. They must terminate in special telomeric regions, but it is only in *Drosophila hydei* that these have been subjected to a detailed study by electron microscopy (Berendes and Meyer 1968).

A convincing proof of the polytene interpretation was provided by the autoradiographic experiments of Beermann and Pelling (1965) who labelled the cleavage division chromosomes of *Chironomus* embryos with tritiated thymidine. In fully-grown larvae, after the chromosomes of the salivary glands have become polytene and undergone twelve to fourteen cycles of replication single, double or quadruple labelled strands could still be seen in the giant chromosomes, running continuously from end to end. A different type of experiment leading to the same conclusions was carried out by Slizynski (1950), who irradiated embryos of *Drosophila* (in which salivary glands were already present but in which they were very small) and obtained structural rearrangements involving only a part (presumably one-quarter, one-eighth, etc.) of the entire thickness of the chromosome.

Polytene chromosomes have been extremely important in cytogenetics for two main reasons. On the one hand, studies of their detailed structure and especially of the DNA replication cycle and the puffing phenomenon (see p. 106) have led to new insights on fundamental problems such as the nature and mode of action of genes. On the other hand, comparison of banding sequences of different individuals, populations and species have been of great significance in the analysis of evolutionary cytogenetic processes. We shall be considering some of the results of the latter type of analysis later (in Chs. 8 and 11). In the present chapter we shall be principally concerned with the role of the polytene chromosomes in genetic processes at the molecular and cellular level.

In the salivary gland nuclei of triploid individuals of *Drosophila* it is usual for all three chromosomes of each kind to 'pair' together, although some regions or chromosomes may remain unpaired, as happens occasionally, even in diploid nuclei. Where small regions are reduplicated in otherwise diploid nuclei, so that they are present three or four times in the karyotype, they may sometimes fuse together in threes or fours, one of the double-polytene chromosomes being attached to another, laterally. Where a region is reduplicated within a single chromosome, as in the sequence *ABCDEFG HIDEFJK*. . . . the chromosome may loop around, so that the two *DEF* regions are paired with one another. It was long ago pointed out by C. B. Bridges (1935), and has been confirmed by other workers since, that the normal polytene chromosomes of *Drosophila melanogaster* contain a number of regions in which the banding pattern is similar or identical. These regions, known as repeats, are presumably duplications of small segments which have become established in the course of evolution. Some of them must be of considerable antiquity, since the same ones are present in both *D. melano-gaster* and *D. simulans* and must ante-date the evolutionary separation of these species. A similar situation occurs in *Sciara*, where Metz and Lawrence (1938) and Metz (1938*b*, 1941) have shown that the *X*-chromosomes of *S. ocellaris* and *S. reynoldsi* contain the same triple repeat. That repeats do really represent duplications is clear from the fact that such regions are occasionally paired with one another in salivary nuclei. It is now possible to carry out computerized 'matching' of sections in photometrically-scanned polytene chromosomes (Castleman and Welch 1968), removing the sub-jective element from the detection of repeats.

If a duplicated segment is represented by the letters *PQ*, then sequences such as *PQPQ* and *PQQP* are referred to as *tandem repeats* (the former being a *direct* tandem repeat, the latter a *reversed* one). Direct tandem re-peats are unstable, since by unequal crossing-over they occasionally give rise to the unrepeated sequence *PQ* and the triplicated one *PQPQPQ*. The reversed type, however, should be stable, since if unequal crossing-over occurs at all it will only give rise to inviable dicentric and acentric chromatids (Lewis 1945). It is consequently not unexpected that all the tandem repeats which are present in the normal karyotype of *Drosophila melanogaster* seem to be of the reversed type.

Where individuals are heterozygous for inverted segments in their chro-mosomes we must expect to find 'reversed loop' pairing in their polytene nuclei, unless the inversion is too short to form a loop. The observation of inversion loops is an important technique in the study of the population genetics of *Drosophila* and other Diptera (see Ch. 8).

It is probable that in life the distal ends of many of the chromosomes are attached together in the polytene nuclei of *Drosophila* species, but that these are rather weak, unstable connections, readily broken during the making of a squash preparation. Such terminal adhesions were observed by some of the earlier workers such as Bauer (1936*b*) and Prokofieva-Belgovskaya (1937*b*). They were subjected to a detailed study by Hinton and Atwood (1941), Hinton and Sparrow (1941), Hinton (1945) and Kaufmann and Gay (1969). In all probability they involve fusions between the actual telomeres, identified as compact structures in the electron micrographs of Berendes and Meyer (1968). This would imply that all the telomeres are 'homol-ogous', or in biochemical terms, that the nucleotide sequence constituting the telomere is a unique one. The non-randomness of the associations (those between certain ends are much commoner than others) could be due to the overall pattern of orientation of the chromosome arms in the nucleus, or to the synaptic properties of the genetic loci immediately adjacent to the telomeres. However, Berendes and Meyer report that in *D. hydei* the telo-

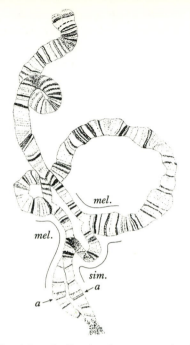

Fig. 4.4. 'Reversed loop' pairing, indicating heterozygosity for a long inversion in the left limb of the third chromosome in an F_1 hybrid from the cross *D. melanogaster* × *D. simulans*. From Pätau (1935).

meres appear to consist of fibrils (presumably DNA) 100–125 Å in diameter covered with a thick protein coat which is removed by protease digestion. It is conceivable that the DNA fibrils of the telomeres are bent back on one another in a U-shaped manner. These special regions at the ends of the chromosomes differ somewhat in size, that of chromosome 2 being the largest and that of chromosome 4 the smallest. The frequency with which the telomeres of the various chromosomes participate in associations with one another seems to be directly proportional to their size in *D. hydei*.

NUMBER OF BANDS

In C. B. Bridges (1938) map of the *X*-chromosome in *Drosophila melanogaster* 1024 bands were identified. This was an increase of 299 on his earlier (1935*b*) figure of 725. The difference was mainly due to the much larger number of 'doublets' which were recognized in the later work, the great majority of which had been regarded as single bands in 1935. The later maps of Bridges and Bridges (1939), P. N. Bridges (1941*a*, *b*, 1942) and

TABLE 4.1. *Numbers of bands in the salivary gland chromosomes of* Drosophila melanogaster (data from C. B. Bridges 1938; C. B. and P. N. Bridges 1939; P. N. Bridges 1941*a*, *b*, 1942; Slizynski 1944*, Cooper 1959)

Chromosome arm	Total number of bands	Number of chromo-central bands	Number of 'doublets'
X	1024	17	313
II L	803	25	196
II R	1137	15	282
III L	884	29	196
III R	1178	0	263
IV L	25	?	9
IV R	110	?	40
Total	5161	86+	.1299

* Hochman (1971) could not confirm Slizynski's report of over 100 bands in IV R, and states that only about 50 bands can be seen. The small, thin IV L is probably essentially heterochromatic.

Slizynski (1944), for the other chromosome elements seem to have been modelled on the 1938 map of the X-chromosome and bring the total number of bands in the entire karyotype of this species to 5161 (Table 4.1). As analysed by Cooper (1959), somewhat more than 86 would be embedded in the chromocentre and hence rather indistinct (see p. 115). It has been assumed by most cytogeneticists that this total of 5161 bands was a minimum estimate, liable to be increased by future work, perhaps by as much as 10 or 20%. However, Berendes (1970) has carried out an electron microscopical study of the 1A1 – 4D7 region of the X-chromosome of *Drosophila melanogaster*. Somewhat surprisingly, the number of bands detected was actually *less* than the number shown on Bridges' 1938 map, which was based, of course, on observations with the light microscope (116 bands as compared with 174). The difference seems to be accounted for mainly by the fact that Bridges regarded fifty-three of the bands as doublets, whereas Berendes regards all but four of these as single (but Berendes has fourteen 'granular regions', some of which might be doublets). There can be little doubt that a number of bands which appeared to be doublets by Bridges' technique are actually single. If this conclusion can be applied to the whole karyotype of *Drosophila melanogaster* the total number of bands would be only 3441 (5161 × 116/174). Less detailed polytene chromosome maps showing about 2000 bands in the case of each species have been published for *D. virilis* (by Patterson, Stone and Griffen 1940 and Hsu, in Patterson and Stone 1952), *D. repleta* (Wharton 1942, reprinted in Patterson and Stone), *D. pseudoobscura* (Kastritsis and Crumpacker 1967), *D. paulistorum* (Kastritsis

1966*a*), *Sarcophaga bullata* (Whitten 1969), *Calliphora erythrocephala* (Ribbert 1967), *Lucilia cuprina* (Childress 1969), *Culex pipiens* (Dennhöfer 1968), *Anopheles gambiae* (Coluzzi and Sabatini 1967) *Chironomus tentans* and *C. pallidivittatus* (Beermann 1952*a*, 1955), *Lundstroemia parthenogenetica* (Porter 1971) and *Sciara coprophila* (Gabrusewycz-Garcia and Kleinfeld 1966).

The system of numbering the bands originally introduced by Bridges (1935*b*) has been adopted, with minor modifications, by subsequent workers. The chromosome set of *D. melanogaster* was first of all divided into 102 segments of approximately equal length (1–20 in the X, 21–60 in the IInd chromosome, 61–100 in the IIIrd and 101–102 in the IVth. Each numbered section was then divided into subsections designated by capital letters. Thus segment 1A is the distal tip of the *X*-chromosome and 42C lies near the middle of chromosome II. Within each subsection the individual bands are indicated by numbers, 1A1 being the most distal band in the *X* and 102F17 the one at the tip of the right limb of the IVth chromosome (Slizynski 1944). Similar systems of numbering the bands have been adopted in the case of *D. pseudoobscura* by Tan (1935, 1937), Dobzhansky and Tan (1936*b*) and Dobzhansky and Sturtevant (1938), in *D. willistoni* by Dobzhansky (1950*b*), in *D. virilis* by Patterson, Stone and Griffen (1940) and Hsu (1952), and in *Chironomus tentans* by Beermann (1952*a*).

THE RELATIONSHIP BETWEEN GENES AND BANDS

In the earliest period of work on the polytene chromosomes of *Drosophila*, Muller and Prokofyeva (1935) put forward the concept of a one-to-one correspondence between genes and bands. It is certain that ideas on the nature of genes have changed significantly since that time. Nevertheless, in the second edition of the present work (1954) the one-gene–one-band hypothesis was accepted. Since that time much more evidence has accumulated and ideas on the fundamental nature of the genetic units have clarified. It is now certain that there can be a number of different mutational sites or loci within a single functional unit or cistron, corresponding to the 'gene' of classical genetics, and that recombination can occur within cistrons as well as between them. The available evidence seems to be generally compatible with the one-band–one-cistron hypothesis. Nevertheless, a number of so-called complex loci (see below) complicate the picture somewhat.

By the use of deficiencies and duplications for recombination studies Judd (1962) and Kaufman, Shen and Judd (1969) identified 112 different genetic loci (recessive lethals or semi-lethals) between the genes *zeste* and *white* on the *X*-chromosome of *Drosophila melanogaster*, a region between bands 3A2 and 3C3 of Bridges. These 112 loci could be assigned to thirteen functional units or cistrons. On Bridges' map there are twelve bands in this region, but Berendes' (1970) ultrastructure studies suggest that the real number is thirteen. Rayle and Green (1968) found five separable loci between the bands 3B2 and 3C3 (four bands on Bridges' map, five on Berendes' ultrastructure map). Hochman (1971) has detected a total of forty loci in the right arm of chromosome IV, compared with an estimated fifty bands in the polytene chromosome. Thus in certain regions at any rate, there seems to be a satisfactory one-to-one relationship between genes (cistrons) and bands. In some other cases, however, it does appear that there is a deficiency of bands by comparison with the known number of genes. However, it is possible that in these instances resolution of granular structures by the electron microscope may eventually increase the number of recognizable bands.

A number of so-called complex loci or pseudoallelic series are now known in *Drosophila*. These are clusters of functionally related cistrons lying very close together in the chromosome, each cistron being responsible for coding for a single polypeptide. Complex loci have frequently been interpreted in evolutionary terms as the result of duplication, triplication or quadruplication of an originally single cistron, followed by divergent mutational change. However *vermilion*, which appears to be a complex locus in *Drosophila melanogaster*, seems to be represented by two unlinked loci in *D. virilis* (Green 1955); proponents of the evolutionary duplication hypothesis would regard the *D. melanogaster* condition as the more primitive one. The usual test for defining cistrons and complex loci is the so-called *cis–trans* one. If the heterozygotes for two recessive mutations in coupling (the *cis* arrangement, $ab/++$) and for the same mutations in repulsion (the *trans* arrangement, $a+/+b$) both exhibit the wild type, then we are dealing with different cistrons; conversely, if they give rise to different phenotypes, the *trans* being mutant and the *cis* wild type, we have a single cistron. Exceptions to this simple principle are known, however, and they are of two kinds. On the one hand we have cases of 'intraallelic complementation' where alleles in the same cistron give identical phenotypes regardless of whether they are in the cis or the trans configuration. The other type of exception is where we have 'position pseudoallelism' in which two cistrons forming part of a complex locus show different cis and trans phenotypes. We shall return to such cases later, when considering position effects (p. 101).

The whole series of *white* mutants, which lie in the X-chromosome of *D. melanogaster* at 1.3 map units from the distal end and seem to occupy the band 3C1 and the adjacent doublet 3C2.3, could be considered as falling within the limits of a single cistron on the basis of the cis–trans test. However,

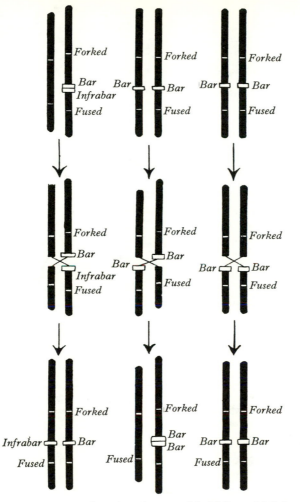

Fig. 4.5. Equal (right-hand column) and unequal (middle and left-hand columns) crossing-over at the *Bar* region in the *X*-chromosome of *Drosophila melanogaster*. From Dobzhansky (1936*d*).

they show some functional diversity (Lewis 1967), and it is probable that three or four pseudoallelic loci are involved (Green 1963, 1965; Judd 1964; Gersh 1967).

The rough eye mutants of the *Lozenge* series (Oliver 1940; Green and Green 1949, 1956; Green 1961) seem to fall into four different loci which lie in the 8D section of the polytene *X*-chromosome, but it is uncertain how many bands correspond to this complex locus. The *Notch* series of mutants

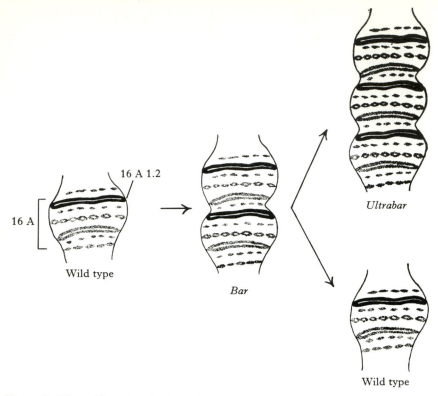

Fig. 4.6. The 16A region in the salivary *X*-chromosomes of wild type, *Bar* and *Ultrabar* flies. Bar is located in the 16A 1.2 doublet. Based on the work of C. B. Bridges (1936) and Sutton (1943).

(Welshons 1958, 1965; Welshons and von Halle 1962) appear to occupy at least eleven 'sites', but these are regarded by Welshons as lying in a single cistron which corresponds to a single band 3C7.

The complicated case of the *bithorax* mutants has been extensively studied by Lewis (1951, 1955, 1963, 1964, 1967); these mutants, which fall into five groups, seem to be located in two doublet bands in segment 89E of chromosome arm 3R.

In summary, and without discussing all these various cases in full detail, there seems no cogent evidence at the present time for rejecting the one-cistron–one-band model, which should, however, continue to be regarded as a working hypothesis rather than as an established fact. One argument that has been urged against it is that of Daneholt and Edström (1967) who calculated that the average polytene band of *Chironomus tentans* contained about 10^{-4} picograms of DNA, which they regard as an improbably high number

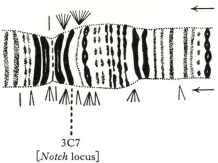

← 15 breaks with recipient section in euchromatin

← 25 breaks with recipient section in heterochromatin

3C7
[*Notch* locus]

Fig. 4.7. A portion of the salivary *X*-chromosome in *Drosophila melanogaster*, showing the 3C7 band (*Notch* locus) and the positions of 40 breaks involved in rearrangements giving a *Notch* phenotype. After Demerec (1942), redrawn.

for a single cistron (since it would code for 30 000 amino acids) unless bands 'either represent operational units of great complexity or . . . a large part of the DNA is non-functional in the informational sense'. Since there are some fairly powerful arguments in favour of the latter possibility, it does not seem that this quantitative consideration is a serious objection to the one-cistron–one-band hypothesis.

POSITION EFFECTS

In the very early days of genetics it was supposed that the genes were entirely discrete and independent entities, which could be separated or rearranged without affecting their genetic properties (i.e. without changing the phenotype). From these early investigations it appeared that crossing-over only took place between (rather than within) genes and that there were no instances where the functioning of a gene was altered by a change in the allele present at a neighbouring locus.

The first case which seemed to constitute a clear exception to these principles was that of *Bar* eye in *Drosophila melanogaster* (Zeleny 1921; Sturtevant 1925), which gave rise to the concept of *position effect*. *Bar* is a sex-linked character which affects the development or the eyes, *Bar* flies having narrower eyes with fewer ommatidia than normal. Heterozygous females are intermediate in phenotype between homozygotes and the wild type. The *Bar* condition is somewhat unstable, since by rare 'unequal crossing-over' it gives rise to a more extreme condition known as *Ultrabar* and to the original wild type. *Ultrabar* is likewise unstable, reverting occasionally to *Bar* and wild type. That these transformations of *Bar* and its derivatives were actually due to a special type of crossing-over was suggested by the fact that they occurred only in females. The hypothesis was proved to be correct by a genetical study of *Bar* chromosomes carrying the flanking markers *forked* and *fused*; it was found that each apparent mutation of *Bar* was accompanied by a cross-over between *forked* and *fused*.

Bar was found by Bridges (1936) and Muller, Prokofieva and Kossikov (1936) to be a direct tandem duplication of the 16A region of the *X*-chromosome, including five bands; later work (Sutton 1943) has suggested that it is the 16A 1.2 band which is involved in the causation of *Bar* phenotypes. *Ultrabar* is hence caused by a triplication of this region.

Since female flies which are heterozygous for *Ultrabar* and those which are homozygous for *Bar* both contain the same number of 16A regions (i.e. four per nucleus), one might expect that they would be phenotypically indistinguishable. But this is by no means so – the number of eye facets is very significantly lower in the heterozygous *Ultrabars*. Thus 'extra' *Bar* regions seem to be more potent when present in the same chromosome than when located in different ones. It was to this phenomenon that the term 'position effect' was first applied by Sturtevant. The early literature on position effects was reviewed by Dobzhansky (1936c, d); more recent work has been discussed by E. B. Lewis (1950, 1951, 1963, 1964, 1967) and Carlson (1959).

Two main types of position effects have been recognized in *Drosophila*. On the one hand there are cases where the phenotypic effect is 'stable', i.e. manifested in all those cells which are developmentally competent to manifest it. This category has been termed *S*-type position effects by Lewis (1950). They are believed to arise when both chromosome breaks involved in the rearrangement are located in euchromatin.

The other category are the so-called V-type position effects, which are somatically unstable so that they give rise to variegation or mosaicism. In general these seem to arise when one break is in euchromatin and one in heterochromatin, so that a genetic locus which was formerly situated in euchromatin is brought into close proximity to heterochromatic material.

Apart from the *Bar* case not many examples of *S*-type position effects are known in *Drosophila*, although the fact that many translocations are lethal in the homozygous condition might be taken as an indication that many of them exist but are inviable. However, Lea and Catcheside (1945) showed that the quantitative data do not require such an interpretation and that the lethality of many induced translocations may be due simply to the fact that a lethal point mutation has been induced near one of the breakage points.

Nevertheless, a number of *S*-type position effects have been reported in *Drosophila* (Demerec 1941a, b). Lewis (1950) suggested two models to account for such phenomena, where they involve closely linked loci which might be due to an evolutionary duplication (as in the *Star-asteroid* case, where the genotype *S*, *ast*/++ and *S*+/+ *ast* exhibit different phenotypes). The first model would depend on the two loci competing for a common substrate; the second would be based on the individual loci controlling separate steps in a series of reactions $S \rightarrow A \rightarrow B$. These models were put forward before the modern era of molecular biology had elucidated the main principles of gene action, and neither looks a particularly probable interpretation today. On the other hand it is now well-known that the puffing of individual bands may affect whole chromosome regions, involving perhaps twelve or fifteen bands, and it may well be that this is the basis of some position effects. If so we might have an explanation for the fact that, with very few exceptions, position effects have not been reported in organisms lacking polytene chromosomes.

If *S*-type position effects are to a large extent confined to organisms with polytene chromosomes, the rarity of evolutionary translocations in *Drosophila*, by comparison with other organisms (p. 433) might be explained in part by the greater prevalence of such effects in *Drosophila*, rather than by the population genetics considerations advanced by Wright (1941). However, position effects have not prevented the establishment of many thousands of inversions in the phylogeny of *Drosophila* (Stone 1949, 1955, 1962).

V-type position effects have been studied almost exclusively in *Drosophila*; certain cases known in the mouse (Russell and Bangham 1959, Cattanach 1961, Lyon 1966) are all concerned with *X*-linked loci and may be associated with the well known phenomenon of inactivation of one of the *X*-chromosomes in female somatic cells (see p. 594). The literature has been comprehensively reviewed by Baker (1968). A V-type effect occurs when a rearrangement in which a gene normally located in euchromatin is brought into close proximity to a heterochromatic segment; a complete proof being available if the effect is reversed when the gene is returned to its normal position by a crossover between it and the adjacent breakpoint (Dubinin and Sidorov 1935; Panshin 1935; Baker 1954; Judd 1955).

In almost all cases of position-effect variegation, the variegation is suppressed (in the direction of a uniform wild type phenotype) by the addition of a *Y*-chromosome (Gowen and Gay 1933 and many later workers). An exception is the light (*lt*) gene in *D. melanogaster*, where addition of *Y*-chromosomes increases the mutant tissue in the variegated eyes, whereas deletion of *Y*'s suppresses the mutant regions (Schultz 1936, Baker and Rein 1962).

Position effect variegation may extend over chromosome segments of considerable length, and in some instances genes fifty bands away from the heterochromatic break may be affected (Demerec 1941). It has been claimed in several instances that rearrangements in which one break is in the heterochromatin and one in euchromatin may lead to facultative hetero-chromatinization of the euchromatic regions thus brought into the proximity of heterochromatin (Caspersson and Schultz 1938, Prokofieva-Belgovskaya 1941, 1947). But, as Baker (1968) points out, there is a complete lack of critical evidence concerning such heterochromatinization in patches of tissue exhibiting mutant and wild type phenotypes. However, a causal connection between heterochromatinization and variegation is very prob-able. Thus in *Drosophila melanogaster* the degree of heterochromatinization of the *yellow-achaete* region in the *scute*[8] chromosome can be expressed as follows, according to Prokofieva-Belgovskaya (1947):

$$\frac{sc^8}{\widehat{XY}} > \frac{sc^8}{Y} > \frac{sc^8}{y \; ac \; v} > \frac{sc^8}{sc^8}$$

and this series also corresponds to the extent of mosaicism (Noujdin 1944). Addition of extra Y's and abnormal temperatures (Hartmann-Goldstein 1967) also have parallel effects on heterochromatinization and variegation.

DNA REPLICATION IN POLYTENE CHROMOSOMES

The polytene chromosomes develop from the ordinary mitotic chromosomes of the embryo by a series of endomitotic replications without separation of the resulting strands; at the same time a great elongation of the chromosome takes place, undoubtedly involving despiralization of much of the DNA (that in the interband regions).

The details of DNA replication in polytene chromosomes have been studied by tritiated thymidine autoradiography, using either injection or an in vitro technique. The earlier literature on the subject is slightly confused, because it was not realized initially that there is an important difference between the polytene chromosomes of the Sciaridae and those of *Chironomus* and *Drosophila*, namely that the former develop special puffs in which extra DNA is synthesized, whereas the puffs of *Drosophila* and *Chironomus* polytenes are RNA-puffs, in which there is no extra DNA. It is still uncertain whether the extra DNA in the puffs of the Sciaridae consists of extra chromonemata or is present in some other form.

It was shown by Keyl and Pelling (1963) that following a short incubation of *Chironomus* salivary glands in medium containing tritiated thymidine one may observe either a general labelling of almost all the bands, or a restriction of the label to a few specific bands. Apparently, in each cycle of replication, all the bands begin to replicate at about the same time, but they take different times to complete the process, so that a number of them (including those with the highest DNA content) are late-labelling.

The situation in *Drosophila* seems somewhat different. According to Plaut and Nash (1964), Howard and Plaut (1968) and Plaut (1968) the sequential pattern is quite complex. Quite different results were obtained with *D. melanogaster* and with members of the *willistoni* group. In the former species, following a 10- to 20-minute pulse of tritiated thymidine, about 95% of the nuclei which show any label at all show interrupted labelling, with some chromosomal segments unlabelled. In members of the *willistoni* group, under similar conditions, less than 10% of the labelled nuclei show the interrupted pattern, more than 90% showing all the chromosomes labelled from end to end.

Some idea of the replication sequence in the *X*-chromosome may be obtained from Table 4.2, which is taken from Plaut's data. It will be seen that there is no simple distinction between late-labelling and non-late-labelling segments. A somewhat more detailed study by Arcos-Terán and Beermann (1968) has revealed about twenty-four late-labelling 'spots' distributed in a rather random manner along the length of the *X*-chromosome. Five of these are extremely late-labelling. There seems to be a general pattern of coincidence between late-labelling spots and the regions

TABLE 4.2. *Replication sequence in the X-chromosome of* D. melanogaster
(from Plaut 1968)

Regions of the chromosome showing label												% Frequency of occurrence
—	—	—	—	—	—	—	—	9	—	—	—	8.0
1	—	—	—	—	—	—	—	9	—	—	—	1.7
1	—	—	—	—	—	—	—	9	—	—	12	10.5
1	—	3	—	—	—	7	—	9	—	—	12	0.7
1	—	3	—	—	—	7	—	9	10	11	12	0.9
1	—	3	—	—	—	7	8	9	10	11	12	3.7
1	2	3	—	—	—	7	8	9	10	11	12	0.7
1	2	3	4	5	6	7	8	9	10	11	12	9.5
1	—	—	—	—	—	7	—	9	10	11	12	0.2
1	—	—	—	—	—	7	—	9	—	—	12	0.9
1	—	—	—	—	—	—	—	—	—	—	12	1.1
1	—	—	—	—	—	—	—	—	—	—	—	0.9
—	—	—	—	—	—	—	—	—	—	—	—	61.0
												99.8

Each row represents an observed labelling pattern; the numbers indicate labelled sections, the dashes unlabelled segments. The vertical dimension of the array represents time, but the direction is undetermined. The sequence of rows has been determined on the assumption that DNA synthesis in each section is uninterrupted until completed; but this is not the only possible sequence of rows which complies with this restriction.

shown by Kaufmann *et al.* (1948) to exhibit non-homologous (ectopic) pairing and to be especially liable to break in salivary gland preparations.

The replication of the heterochromatic regions in the polytene chromosomes seems to be independent of that of the euchromatic bands. On the basis of cytophotometric measurements Berendes and Keyl (1967) have suggested that in *Chironomus* polytenes there are fewer cycles of replication in heterochromatin than in euchromatin, while Rudkin (1965*a*, *b*) has claimed that in *Drosophila melanogaster* most of the heterochromatin simply does not replicate at all during the formation of the salivary gland chromosomes. This is supported by the observations of Dickson, Boyd and Laird (1971), according to which *Drosophila hydei* DNA from polytene nuclei contains only about 5% fast-renaturing sequences (believed to be mainly located in centromeric heterochromatin) compared with the 20% found in embryonic and pupal DNA's.

Synthesis of proteins in the polytene chromosomes has been studied by Cave (1968), using tritiated tryptophane to label the non-histone proteins (since histones do not contain tryptophane) and tritiated lysine (histones being lysine-rich). The results suggested that histone synthesis takes place simultaneously with DNA replication but that there is no increase in the synthesis of the non-histone proteins during DNA replication.

PUFFING

At particular stages in development, certain of the genetic loci in the polytene chromosomes undergo a spectacular change in appearance – they become converted into large swellings or puffs. It has been generally believed that the process of puffing is due to or permits the biosynthetic activity of the particular loci concerned, hence the great importance of the phenomenon for the understanding of gene action. The whole subject of puffing has recently been reviewed by Ashburner (1970) and cannot be considered in detail here. In numerous instances it has been possible to trace these puffs back to their origin in single bands or pairs of adjacent bands ('doublets'). Some of the very large specialized puffs are referred to as Balbiani rings (from the name of the cytologist who first observed them, ninety years ago); but there does not seem to be any fundamental difference between these and the smaller puffs.

Puffing involves an unravelling or unfolding of the DNA in the chromomeres of the particular band concerned. In some instances the process has been observed to start at the edge of a band and spread towards the centre (Pelling 1964, 1966). A particular band of *Chironomus thummi* which contains eight times as much DNA as the corresponding band in the closely-related *C. piger* (Keyl 1965*a*) may produce a puff, either from a terminal 1/8th subunit or from the whole band. Very large puffs such as the Balbiani rings naturally lead to dispersion and swelling of bands adjacent to those from which they originate, so that a whole segment of up to ten bands becomes involved (Bauer and Beermann 1952).

Not all the bands undergo puffing, even at the totality of stages and in the totality of tissues that are available for investigation. In *Chironomus tentans* about 2000 main bands have been mapped (Beermann 1952*a*); but of these only about 15% undergo visible puffing in the salivary gland (Pelling 1964, 1966). It has been suggested by some workers that puffing is not necessary for gene activity. This point of view has been especially developed by Goodman, Goidl and Richart (1967) who have claimed that they were able to inhibit all puff-formation by cortisone in the salivary gland chromosomes of *Sciara coprophila* without preventing normal development. However, this result has been challenged by Crouse (see p. 114) and it is in any case doubtful whether we should expect the synthesis of the salivary proteins to be essential for development.

Puffing is associated with an intense local synthesis of RNA, which can be demonstrated by tritiated uridine autoradiography or by special staining techniques (Pelling 1959, 1964, 1966). It was early suspected that this was mRNA, a view which was strengthened when it was found that Balbiani ring RNA had a specific base composition, which differed from that of ribosomal RNA in its high adenine content (Edström and Beermann 1962; Beermann 1967).

Pelling showed that the puffs of *Chironomus tentans* stained reddish with toluidine blue while unpuffed regions stained blue. This metachromasia seems to be due to the presence or absence of ribonucleoprotein, since in-vivo experiments showed that tritiated uridine was only incorporated into the red-staining regions. However there were a good many bands that showed some incorporation of the label in spite of the fact that they were not visibly puffed. It is possible that there are many minor puffs or slightly puffed sites which ordinarily escape detection.

The RNA synthesized in the puffs is believed to leave its location in the chromosomes, as ribonucleoprotein, in the so-called puff granules, electron-dense structures 300–500 Å in diameter (Beermann and Bahr 1954; Sorsa 1969). These are probably transferred to the cytoplasm in small pouches or blebs of the nuclear membrane (Gay 1956, 1959). According to Berendes (1968) all the newly synthesized RNA in puffs which have been prevented from further RNA synthesis by actinomycin D is released to the nuclear sap within 30–35 minutes.

Berger (1940) and Slizynski (1950) first clearly demonstrated that in different tissues of the same individual (salivary gland, mid-gut, fat body, etc.), the banding pattern was the same, and this was confirmed by Beermann (1952a, 1956a) who showed, however, that bands which were puffed in one tissue might be quite unpuffed in other tissues; for example, the three huge Balbiani rings in the fourth chromosome of *Chironomus tentans* are only seen in the salivary gland nuclei and not at all in other tissues. In later work (Beermann 1961) it was shown that both *C. tentans* and the closely related *C. pallidivittatus* possess at the base of the salivary gland a group of four special cells which, in the second species, but not in the first, produce a special kind of secretion granules ('SZ' material). In F_1 hybrids between the two species it was found that an intermediate amount of SZ material is synthesized. The presence of the SZ material is correlated with the appearance of a specific puff, BR4(SZ) which is not seen at all in *C. tentans*, and only in the special cells of *C. pallidivittatus*. In the special cells of the hybrids it can be seen that there is heterozygosity for the presence of this particular puff which is present in the *pallidivittatus* homologue, but not in the *tentans* one (the two homologues can also be distinguished on other grounds, since they differ by inversions).

It was concluded from this work that the largest puffs represent genetic loci coding for some of the non-enzyme proteins, such as those of the salivary secretion, which are synthesized in very large amounts in the course of development. More recently, the puffing patterns of *Chironomus tentans* and *C. pallidivittatus* have been studied by Grossbach (1968, 1969) in relation to the synthesis of specific silk-like proteins which are produced in large amounts in the cells of the salivary glands. In *C. tentans* this secretion consists of five protein subunits; in *C. pallidivittatus* an additional subunit (no. 6) is present. The synthesis of subunit no. 6 seems to be controlled by a locus in the distal region of chromosome 4 which forms a tissue-specific Balbiani ring. In *C. pallidivittatus*, but not in *C. tentans*, an additional protein subunit (no. 7) is formed in these cells; it is controlled by another locus in chromosome 4 which forms a Balbiani ring which is not present in the ordinary gland cells.

Although differences in puffing patterns between the various tissues of the body undoubtedly exist, Clever (1966) has concluded that, when allowance is made for developmental changes and some physiological variation between cells of the same tissue, very few puffs seem to be rigorously tissue-specific. Thus some workers have been led to criticize the earlier models of puffing as a model for genetic regulation in higher organisms, particularly since polytene chromosomes are restricted to very few groups of organisms. However, Thomson (1969) has pointed out that studies of polytene chromosomes have been carried out on cell types in which the giant chromosomes are most favourable for analysis (i.e. most compact). These are mostly 'terminal' elements such as trichogen cells and those of the salivary gland, specialized for the synthesis in very large amounts of a small range of proteins. Study of a tissue such as the larval fat body, in *Calliphora*, leads to a very different picture. This organ is, of course, one of the main biosynthetic tissues of the larva, producing large quantities and many types of proteins. On the fourth day of development long sequences of bands are visible in the fat body polytene chromosomes, with only short segments in a dispersed state; but from the fourth to sixth day, when biosynthesis is much more active, most of the chromosomes are in a diffuse state and only short compact banded regions can be seen. Thomson thus concludes that puffing or the assumption of a diffuse structure is normally associated with, but does not actually cause, genic activity; it may, however be essential for complexing and protection of the transcribed RNA. A model of puffing involving a dynamic equilibrium between RNA synthesis, RNA processing and ribonucleoprotein liberation from the puff has been put forward by Yamamoto (1970) on the basis of ultrastructure studies of puffs after recovery from inhibition by heat shocks and actinomycin D treatment.

The normal 'puffing pattern' in the autosomes and X-chromosome in the salivary gland nuclei of *Drosophila melanogaster* has been described by Ashburner (1967, 1969*a*). A total of eighty-three autosomal loci and twenty-one in the X-chromosome undergo puffing at specific stages of larval and prepupal development. The puffing patterns of the X-chromosome loci

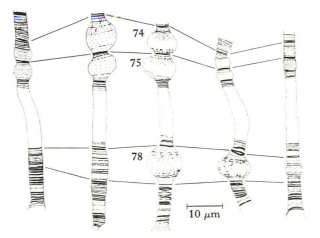

Fig. 4.8. Formation and regression of puffs in sections 74, 75 and 78 of chromosome arm 3L of *Drosophila melanogaster* at the end of the third larval instar up to the formation of the puparium (right). From Becker (1959).

were similar in the two sexes, but a group of puffs were active for a longer time in the male than in the female. There are two main peaks of puffing activity – at the time of puparium formation and in the prepupa, four hours before the pupal molt. The puffing patterns of the sibling species *D. melanogaster* and *D. simulans* are similar (Ashburner 1969*a*, *b*). In the *X*-chromosome *D. simulans* shows two puffs which are not present in *D. melanogaster*. There are also a number of differences in the precise timing of puffing and in the size of the puffs. *D. melanogaster* has one autosomal puff (46A) which is not present in *D. simulans*; hybrids between these species show a heterozygous puff. Differences in time of puffing and size of puffs also exist between different strains of *D. melanogaster* (Ashburner 1969*c*) and certain puffs are active in some strains but are not seen in others. Thus puff 64C is present in strain vg6 but not in larvae of the Oregon strain. In F_1 heterozygotes between these two strains this puff is expressed 'homozygously' only when the region is synapsed; if it is unsynapsed only one homologue shows the puff. But in heterozygotes between vg6 and an Oregon stock carrying an inversion including the locus of the puff 64C forms a 'heterozygous' puff even when the homologues are synapsed.

The puffing pattern of the pupal footpad cells of *Sarcophaga* has been described in detail by Whitten (1969). The sequence of puffing is very orderly and regular, and most puffs exist only for a limited time. Thus a particular puff 'B9c-10a' swells up quite suddenly on day 6 of pupal development, when layers two, three and four of the cuticle are being formed and regresses equally rapidly a few hours later. On the other hand puff D10a lasts for about five days, reaching a maximum at day 8 and then regresses on the ninth day. A few puffs seem to be present in both the trichogen and footpad cells (possibly ones concerned with common processes involved in cuticle formation), but most of them are specific to one or other cell type. In some instances there is an apparent movement of a puffed region along the chromosome, suggesting sequential activation of successive loci; similar observations were made by Mechelke (1961) in the case of the salivary gland chromosomes of *Acricotopus*, but this phenomenon does not seem to be very general in polytene chromosomes. A series of large puffs which appear suddenly right at the end of the life of the footpad cells are regarded by Whitten as possibly concerned with the molecular mechanism of 'programmed death' of these cells.

Detailed studies of the puffing sequences in footpad cells of *Sarcophaga* have also been carried out by Bultmann and Clever (1966), who have demonstrated differences in the puffing pattern in the cells from the three pairs of legs, suggesting an influence of specific intracellular factors rather than any general effect of hormone titre.

Structural rearrangements can affect the puffing pattern, i.e. puffing can be manifested as a position-effect. Thus Kaufmann and Gay (1969) have shown that in *Drosophila melanogaster* a translocation between chromosome 4 and the tip of arm 2L may lead to pronounced puffing in the segment of 2L (21A-B) adjacent to the point of fusion; this is particularly evident in translocation heterozygotes where the distal region of 2L fails to synapse and the homologue with the rearrangement shows the puff while the untranslocated one does not.

Polymorphism for puffing patterns has been described in natural populations of a number of species. Hsu and Liu (1948) described polymorphism for a puff in a Chinese species of *Chironomus*; in heterozygotes a large puff in one homologue is represented by four conspicuous bands and one or two thin ones in the other chromosome. In *Sciara* species, Metz (1935, 1937) described a number of instances of heterozygosity for the size or appearance of single bands. Pavan and Perondini (1967) have reinvestigated *Sciara ocellaris*, one of the species studied by Metz. They found a number of cases where homologous bands undergo their puffing cycles asynchronously, but with an overlap in timing. Thus at the beginning and end of the puffing period the puffed region appears asymmetrical, while in the middle of this period there is no visible asymmetry. They suggest that in these instances heterosis could be due, either to the genetic activity being maintained over a longer period than in homozygotes or to a form of genetic complementation.

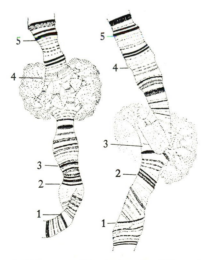

Fig. 4.9. Formation of different Balbiani rings in different cell types of *Trichocladius vitreipennis*: the same chromosome segment in cells of the ordinary salivary gland (left) and in the special granular cells (right). From Beermann (1952*b*).

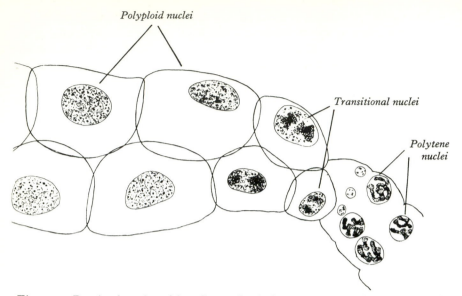

Fig. 4.10. Proximal portion of the salivary gland of *Dasyneura affinis* (Cecidomyidae), showing the transition between the polytene nuclei of the 'basal reservoir' and the endopolyploid ones in the gland proper. From White (1948*b*).

It has been shown by a number of workers (Clever and Karlson 1960; Clever 1961; Berendes 1967) that numerous puffs can be experimentally induced at an earlier developmental stage than that at which they would normally occur, by administration of the insect steroid moulting hormone ecdysone. The degree of puffing and its duration depend on the concentration of the hormone. The effect is a remarkably rapid one in some instances; thus one puff (I18C1) begins to form 15 minutes after the hormone has been administered and another follows 30 minutes later. A number of other environmental conditions such as heat treatment (Ritossa 1962; Berendes, van Breugel and Holt 1965; Yamamoto 1970) are also capable of inducing specific puffs.

Kroeger (1963*a*, *b*, 1964) studied especially the influence of sodium and potassium ions on puffing *in vitro* and claimed that the effect of ecdysone was not a direct one on the genetic loci which were induced to puff but an indirect one, via the sodium–potassium equilibrium. He proposed that potassium ions led to an anticipation of puffing patterns normally characteristic of later stages and that sodium ions caused a return to an early condition ('rejuvenation' of the puffing pattern). Later work has not supported this rather simplistic model. In the studies of Berendes *et al.* (1965) on *Drosophila hydei* it was shown that twenty changes in the puffing pattern normally occur just before puparium formation, when ecdysone action would be at a maximum. Only seven of these could be induced at earlier stages by KCl, and the others could not be stimulated by any known agency. Potassium chloride treatment not only failed to induce many of the normal puffs – it led to the appearance of three puffs never before seen in the salivary chromosomes of *D. hydei*.

The effect of ecdysone is not simply to cause bands to puff. Rather, it is to induce a whole pattern of puffing which may involve the regression or disappearance of certain puffs. Thus, in prepupae of *Acricotopus lucidus*, Panitz (1964) and later work referred to by Ashburner (1970) has demonstrated regression of two large Balbiani rings in the cells of the anterior lobe of the salivary gland following a rise in ecdysone titre. The experimental conditions under which this occurred included transplantation of the glands into prepupae of *Calliphora*, incubation in the presence of *Calliphora* ring glands and injection of ecdysone into *Acricotopus* larvae.

In other experiments, Kroeger (1960) transplanted isolated salivary gland nuclei of *Drosophila busckii* into a medium obtained by crushing eggs of *D. melanogaster*. If pre-blastoderm eggs were used (rather than later embryos) a large puff (II R/22) developed in the *D. busckii* second chromosome, which is never normally observed, and which does not make its appearance when *D. busckii* eggs or blastoderm eggs of *D. melanogaster* are used for making the medium. Presumably, the *D. melanogaster* egg cytoplasm, at a specific stage, contains a substance which evokes the activity of a particular locus in the *D. busckii* chromosome.

DNA PUFFS

In midges of the family Sciaridae the polytene chromosomes develop puffs in which an intense DNA synthesis occurs at the time of puffing (Breuer and Pavan 1955; Rudkin and Corlette 1957; Swift 1962). Thus although these structures resemble the well-known RNA puffs and Balbiani rings of the Drosophilidae and Chironomidae, they are biochemically quite different. RNA puffs also occur in the Sciaridae, in which there is no disproportionate DNA synthesis (gene amplification); and of course DNA puffs also synthesize RNA in large quantities (Pavan and Da Cunha 1969) as well as histone and non-histone protein (Ficq and Pavan 1961; Swift 1962). It has been proposed that these sciarid DNA puffs are the site of specific gene amplification mechanisms that are important in differentiation and development of these insects. However, Goodman, Goidl and Richart (1967) claimed that by feeding cortisone to the larvae they could inhibit DNA puff formation without preventing development and metamorphosis.

These authors criticized the whole concept of puffing as a mechanism of genic activity. Their conclusions have, however, been challenged by Crouse (1968), who found that the moulting hormone ecdysone, when injected into the larvae, would produce a rapid induction of the DNA puffs at the usual sites. Injection of ecdysone plus cortisone was usually followed by puffing but larvae injected with cortisone alone did not develop puffs. However, pupation was delayed in larvae that had been fed cortisone, and Crouse believes that the larvae which achieved metamorphosis in the experiments of Goodman *et al.* would have shown puffs if they had been dissected shortly before pupation. The DNA puffs of *Rhynchosciara* seem to be very highly tissue-specific, no individual puff having been observed in more than a single tissue (Breuer and Pavan 1955).

The nature of the DNA puffs in *Sciara* has been investigated spectro-photometrically by Crouse and Keyl (1968). They found that the DNA increase in the puffs is not merely disproportionate (compared with bands which do not puff); it takes place step-wise, as if by a series of extra rounds of replication. In this respect it is different from the production of RNA in an RNA puff, which is a continuous process.

In each of three species of *Sciara* investigated by Gabrusewycz-Garcia (1971) there are about twenty DNA puffs per nucleus, of which about nine are large. A pair of puffs adjacent to one another appear to be homologous in two sibling species. The DNA puffs of *Sciara* do not make their appearance until late in the last larval instar (Breuer and Pavan 1955, Gabrusewycz-Garcia 1964).

In *Sciara coprophila* Gabrusewycz-Garcia and Kleinfeld (1966) have observed the formation of 'micronucleoli' in the polytene nuclei; apparently the RNA synthesized in the DNA puffs and in some heterochromatic chromosome regions is shed from the chromosomes in these small globules. In this case the micronucleoli apparently do not receive any of the 'extra' DNA, which remains in the chromosome when the puffs undergo regression. In *Hybosciara fragilis*, however, some DNA is extruded into the micro-nucleoli (Da Cunha, Pavan, Morgante and Garrido 1969; Da Cunha 1972).

DNA amplification mechanisms have been found in a number of other species of sciarid midges, by Da Cunha (1972). Numerous micronucleoli having a DNA core surrounded by ribonucleoprotein are formed and set free from at least forty distinct short segments in the polytene chromosomes of the salivary glands in *Hybosciara fragilis* and *Trichosia pubescens*. In *Plecia* sp. there is a slightly different mechanism; in the salivary gland nuclei and especially in those of the Malpighian tubules, the polytene chromosomes show lateral loops which break off and are released into the nuclear sap.

A special phenomenon has been described in diapausing larvae of

Chironomus melanotus by Keyl and Hägele (1966). Masses of DNA and protein were extruded from heterochromatic regions around the centromeres during the fourth instar. These structures are never seen in normal, non-diapausing larvae, and it is possible (although not certain) that they represent a special type of DNA amplification mechanism. Some DNA globules are apparently also shed from certain loci in the polytene chromosomes of the foot-pad cells of *Sarcophaga* (Whitten 1965; Roberts 1968).

HETEROCHROMATIN IN POLYTENE CHROMOSOMES

The heterochromatic regions of the polytene chromosomes are relatively very much shorter than they are in ordinary mitotic chromosomes. Thus in *D. melanogaster* the proximal one-third of the X-chromosome is hetero-pycnotic at mitosis; but the same region forms less than one twentieth of the total length of the salivary-X chromosome. This heterochromatic proximal segment of the X contains only about six indistinct bands in the salivary gland nuclei. These bands are made up of 'heterochromomeres' which stain faintly and which seem to be associated in a rather irregular manner (Muller and Prokofieva 1935). The Y-chromosome has a similar structure in the salivary nuclei, being so short that in the early work on the polytene chromosomes of *Drosophila* it was altogether overlooked. Later, Prokofieva-Belgovskaya (1935a, b, c; 1937a) showed that it was a small body containing about eight indistinct bands. The heterochromatic part of

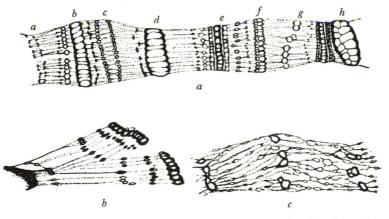

Fig. 4.11. Detailed structure of salivary chromosomes in *Simulium virgatum*. *a*, surface view of a euchromatic region, showing the appearance of the bands, some of which are made up of small, distinctly staining chromomeres, while others are composed of large, crowded 'vesicles'; *b*, a portion of the same, crushed in mounting; *c*, a piece of the heterochromatic 'spread out' region. From Painter (1939).

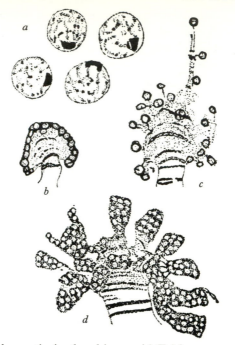

Fig. 4.12. Heterochromatin in the chironomid *Trichotanypus pectinatus*. *a*, small somatic nuclei showing chromocentral masses; *b–d*, ends of salivary chromosomes III, IV and V, showing the terminal heterochromatin. From Bauer (1936*a*).

chromosome II which lies between the centromere and the secondary constriction forms about one-tenth of the whole chromosome at mitosis, but is represented only by a single band in the salivary chromosome (Hinton 1942).

In *Drosophila* species all the heterochromatic regions round the centromeres are fused in the polytene nuclei to form a common mass known as the chromocentre. The chromomeres of these regions seem to attract one another in a rather non-specific manner, so that they are all irregularly paired in a rather disorganized way. The *Y*-chromosome is completely included in the chromocentre. Thus if we squash a salivary nucleus of *D. melanogaster* the chromosomes consist of six strands radiating out from a common mass in the centre – something like a six-limbed ophiuroid. One of these limbs is very short and represents the little IVth chromosome, while another represents the *X*. The four remaining strands are the 'right' and 'left' limbs of the two metacentric chromosome pairs II and III. A definite chromocentre is present in all species of *Drosophila*, although its size is quite variable. A similar structure is also present in the polytene nuclei of the mosquito

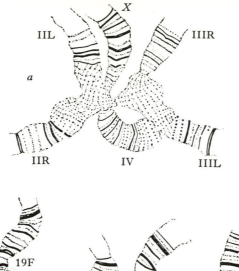

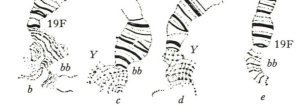

Fig. 4.13. Structure of the chromocentre in *Drosophila melanogaster*. *a*, chromocentre in a female salivary nucleus, with the IVth chromosome attached to it at both ends; *b–e*, parts of the chromocentre from male nuclei, showing the pairing of the *Y* with the proximal region of the *X* and the locus of *bobbed* gene (*bb*). From Prokofieva-Belgovskaya (1935*a*).

Anopheles atroparvus, but chromocentres are generally absent in Sciaridae, Bibionidae, Simuliidae and Chironomidae, in which all the polytene chromosomes are free or only very loosely attached by their ends.

A number of workers have described the presence of more than one type of heterochromatin in the polytene chromosomes of *Drosophila* and other Diptera. Heitz (1934) distinguished an α- and a β-heterochromatin in the salivary gland nuclei of *D. virilis*. The α-heterochromatin forms a compact body in the chromocentre, while the β-heterochromatin is more diffuse and forms part of the chromocentre as well as the proximal segments of the chromosome arms arising from it. In many species we can draw a distinction between 'compact' heterochromatin, in which the chromomeres still form regular bands, and 'loose' heterochromatin, in which the regular arrangement of the chromomeres is more or less lost. Bauer (1936) showed that in *D. pseudoobscura* it was possible to break up the chromocentre by squashing,

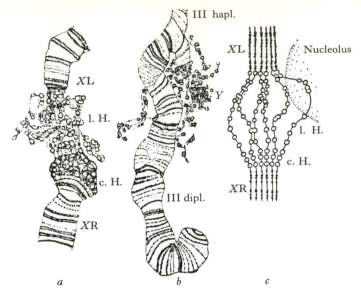

Fig. 4.14. Structure of the *X*- and *Y*-chromosomes in salivary gland nuclei of *Drosophila pseudoobscura*. *a*, proximal region of the *X*, showing the 'loose heterochromatin' (l.H.) and the 'compact heterochromatin' (c.H.). *b*, a translocation between the IIIrd chromosome and the *Y*, showing the loose heterochromatin of the latter. *c*, diagram of the proximal part of the *X*, showing its relation to the nucleolus. From Bauer (1936*b*).

so that the two arms of the metacentric *X*-chromosome were left attached to one another by heterochromatin on either side of the centromere. Part of this segment is composed of compact heterochromatin and seems to be the proximal part of *X*R. The remainder is loose heterochromatin and represents the proximal region of *X*L. Similar observations have been made by Pavan (1946) on *D. nebulosa* and by Hsiang (1949) in *D. tumiditarsus*, in both of which species there is a 'compact' type of heterochromatin whose length in the salivary chromosomes seems to bear about the same relation to its mitotic length as in the case of the euchromatic segments; and a loose heterochromatin which is very massive in the mitotic nuclei but reduced to a few irregularly arranged heterochromomeres in the polytene elements. In all these species the *Y* seems to be composed mainly or entirely of loose heterochromatin.

In the cecidomyids a very clear distinction between the loose and compact types of heterochromatin exists in some species such as *Cecidomyia serotinae* (White 1948*a*). In these insects the nucleolus always seems to arise from a region of loose heterochromatin; blocks of compact heterochromatin of very striking appearance are present on the ends of some of the chromosomes and may be present in intercalary situations (probably around the centromeres) in some of them (Fig. 4.2). The view of Rudkin (1965a, b) and Berendes and Keyl (1967) that the heterochromatin does not replicate during the growth of the polytene chromosomes probably applies only to the loose type and not to the compact heterochromatin.

The satellite DNA's of *Drosophila hydei, D. neohydei* and *D. pseudo-neohydei* and the localization of these repetitive nucleotide sequences in the polytene chromosomes of these species have been studied by Hennig, Hennig and Stein (1970). Approximately 15% of the total DNA is repetitive. The banding pattern of the three forms is essentially identical. In spite of this, the satellite DNA's identified by density gradient centrifugation were very different. In-situ molecular hybridization experiments demonstrated that the distribution of these repetitive DNA fractions over the karyotype is highly diverse. In some cases a particular satellite DNA is widely distributed over all the chromosomes; other satellite DNA's are apparently restricted to the centromeric regions of individual chromosome elements. In general, satellite DNA's seem to be located in regions composed of α-heterochromatin; none of them seems to be included in the heterochromatin of the Y-chromosome. Similar studies have been carried out on *Rhynchosciara hollaenderi* by Eckhardt and Gall (1971); satellite complementary RNA underwent molecular hybridization with the centromeric heterochromatin of all four chromosome elements and with certain dark-staining bands in the telomeric regions of two of them. Complementary RNA prepared from main-band DNA annealed to many loci distributed over the whole karyotype.

As briefly mentioned in the last chapter, the ganglionic metaphases of *Drosophila* larvae seem to have polynemic chromosomes (Swift 1962, Rudkin 1965, Gay, Das, Forward and Kaufmann 1970). These elements differ from polytene chromosomes in two respects: they have not lost the power to divide by mitosis and the replication of the heterochromatic segments seems to proceed synchronously with that of the euchromatic ones instead of being out of step, as it always seems to be in true polytene chromosomes. Banded structure is not evident in the ganglionic chromosomes.

EFFECTS OF INFECTIVE ORGANISMS ON POLYTENE CHROMOSOMES

The effects of infection of dipteran larvae by microsporidian parasites (particularly members of the genus *Thelophania*) on the structure and functioning of the polytene chromosomes has been studied by Diaz and Pavan (1965), Pavan and Basile (1966), Pavan (1967), Pavan and Da Cunha (1968) and Diaz, Pavan and Basile (1969). Infected salivary gland cells of

Rhynchosciara undergo hypertrophy of the nuclei and polytene chromosomes. Usually the enlarged nuclei fall into classes having 2, 4, 8, 16 or 32 times the DNA content of normal, non-infected cells (Roberts, Kimball and Pavan 1967), indicating that up to five extra cycles of replication have occurred. These hypertrophied polytene chromosomes continue to synthesize RNA actively and may show a condition that has been described as 'generalized puffing'. However, chromosomes from different infected tissues show some specific differences in their puffing patterns (Pavan and Basile 1966). The DNA puffs which normally develop at the end of larval life in *Rhynchosciara* do not make their appearance in cells infected by microsporidia.

The effects of microsporidian infections on polytene cells of *Sciara ocellaris* have also been studied by Pavan and Perondini (1966) and Pavan, Perondini and Picard (1969); they are generally similar to those that occur in *Rhynchosciara*. The normal puffing pattern of the salivary glands is altered and the autocytolysis which the gland cells normally undergo at the end of larval life may be indefinitely delayed, so that gland cells may persist even into the imaginal stage of infected individuals. Heavy infections may almost completely destroy the cytoplasmic constituents of the cells, while the polytene chromosomes become very large and are very active in RNA- and DNA-synthesis.

Some effects of infection by a gregarine in another sciarid, *Trichosia* sp. have also been described by Da Cunha, Morgante, Pavan and Garrido (1968). In this case the polytene chromosomes of the intestinal caeca grow to a giant size and may show abnormal constrictions and breaks. They then undergo dissociation into their constituent chromonemata, which fall apart; this process begins first of all in the *X*-chromosomes, the autosomes dissociating at a somewhat later stage.

Hypertrophy of polytene chromosomes as a consequence of microsporidian infections is also known in other families of Diptera. It was recorded for a simuliid by Debaisieux and Gastaldi (1919) and for *Chironomus anthracinus* by Keyl (1960c).

These protozoan parasites are, of course, cytoplasmic. Infection of *Rhynchosciara* by an intranuclear polyhedral DNA virus leads to somewhat different effects to those observed in the case of microsporidian infection (Diaz, Pavan and Basile 1969). Hypertrophy of the chromosomes is caused equally by virus infections, but in this case puff formation is apparently suppressed.

Various mutations are known in *Drosophila* which affect rate of larval development and may also lead to alterations in the polytene chromosomes. In *D. melanogaster* larvae homozygous for the *tu-h* gene ('tumorous head') have their metamorphosis blocked and the duration of larval life greatly increased; their polytene chromosome have at least one extra round of DNA replication and are much wider than normal (Rodman 1967). Another gene, *ltl* (lethal-tumorous) seems to produce similar effects (Kobel and van Breughel 1968). But some other mutants which have an extended larval period, e.g. *lgl* (lethal giant) show normal polytene elements (Ashburner 1970). 'Supergiant' polytene chromosomes which have undergone extra rounds of DNA replication can be produced artificially by culturing salivary glands of larval *Drosophila* in the abdomens of adult flies for several weeks (Hadorn, Ruch and Staub 1964). In all these conditions it is probable that the cells are exposed to abnormally high ecdysone levels (Ashburner 1970).

NUCLEOLI IN POLYTENE NUCLEI

Most fully mature polytene nuclei contain one or more large bladder-like nucleoli which do not seem to be different from those of ordinary mitotic nuclei in any essential respect. They arise from localized nucleolar organizers in one or more of the chromosome elements. In some cases (e.g. *Drosophila virilis* studied by Heitz 1934) the nucleolus seems to be physically separated from the chromosome, to which it is attached by a thread. More frequently (Fig. 4.1) the nucleolar organizer appears as a gap in the polytene chromosome surrounded by the nucleolus; in this case the bands in the immediate vicinity of the nucleolar organizer are somewhat disrupted and the chromonemata on either side of the organizer appear to extend like a complex branching system of roots into the actual nucleolus. Species of *Drosophila* have a single nucleolus, the nucleolar organizer being located in the proximal 'pairing segment' of the sex chromosomes. In some species of *Chironomus*, however, more than one nucleolus is present.

Unlike the puffs, the nucleolus is present in all tissues and in all stages of development, and its location in the karyotype is the same in both polytene and mitotic nuclei, i.e. the same nucleolar organizer or organizers seem to be active in all cell-types. The haploid genome of *D. melanogaster* contains 130–190 copies (*in tandem*) of the cistrons coding for the 18S and 28S rRNA species; these are located in the nucleolar organizers on the X- and Y-chromosomes (Ritossa, Atwood and Spiegelman 1966); on the other hand, the genes coding for 5S rRNA are located in section 56EF of chromosome arm 3R (Wimber and Steffensen 1970).

The various 'nucleolar' ribonucleoprotein bodies in polytene nuclei from various tissues of *Calliphora* have been studied by Thomson and Gunson (1970). Three distinct classes of inclusion bodies were distinguished (type 1 – 'amorphous'; type 2 – fibrous core surrounded by a cortex; type 3 –

coarsely granular) which are probably stages in a complex developmental sequence differing from tissue to tissue. Thomson and Gunson suggest that these various ribonucleoprotein bodies contain different classes of mRNA which are being processed and stabilized in relation to the synthesis of the ribosomes. A certain amount of DNA seems to be regularly present in the nucleoli of polytene nuclei. This has been studied in various species of *Drosophila* by Barr and Plaut (1966), using fluorescent staining after digestion with RNAse or DNAse; it usually appears as a number of small flecks, together with a larger irregularly-shaped mass.

A particularly elegant cytobiochemical technique for studying nucleolar function in polytene nuclei has been developed by Pardue, Gerbi, Eckhardt and Gall (1970). These workers prepared radioactive ribosomal RNA and used a technique of 'molecular hybridization' to anneal it *in vitro* to the DNA cistrons coding for ribosomal RNA in the polytene nuclei. In *Drosophila hydei* the ribosomal cistrons were found to be located within the nucleolus. *Rhynchosciara hollaenderi* has ribosomal genes on one end of the X-chromosome and on one end of an autosome; in addition the ribosomal RNA hybridizes with DNA in most or all of the scattered micronucleoli. In *Sciara coprophila* the ribosomal cistrons are located in the nucleolus organizer region at the proximal end of the X-chromosome and also in the micronucleoli. Apparently the extra DNA synthesized in the DNA puffs of these Sciaridae does not code for ribosomal RNA.

COMBINATIONS OF POLYTENY AND ENDOPOLYPLOIDY

Several interesting conditions intermediate between polyteny and endopolyploidy, or combining features of both systems, have been described in the salivary gland nuclei of various Cecidomyidae by White (1946*a*, 1948*a*), Matuszewski (1964, 1965) and Henderson (1967*a*, *b*). In the salivary glands of *Lestodiplosis* sp. (White 1946*a*) there is one 'super-giant' cell in each gland whose polytene nucleus is apparently 32-ploid; the other cells of the gland are polytene but not polyploid. In the super-giant cell pairing of homologues is apparently restricted to twos (unlike the situation in triploid *Drosophila* larvae, in which the homologues fuse together in threes, as shown by Painter, 1934*b*). The embryology of the super-giant cells of *Lestodiplosis* sp. has not been studied, so that it is not known whether endopolyploidy precedes polyteny in development or vice versa. There are four chromosomal elements in the somatic nuclei of *Lestodiplosis* sp, which may be referred to as *A*, *B*, *C* and *D*. In the super-giant nucleus the *A* and *B* elements seem to consist, in general, of synapsed homologues, while the *C*- and *D*-chromosomes are unsynapsed.

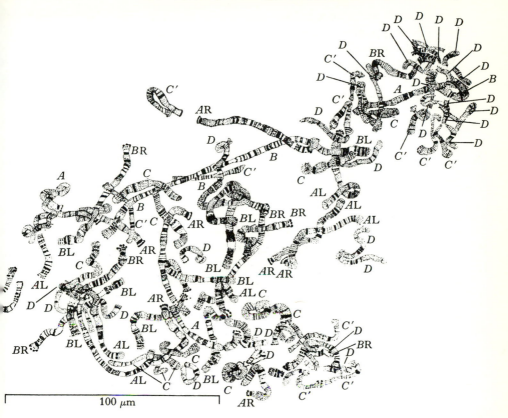

Fig. 4.15. Complete set of chromosomes from one of the polyploid-polytene nuclei in a 'super-giant' salivary gland cell of *Lestodiplosis* sp. (Cecidomyidae). There are four different chromosomal elements, designated *A*, *B*, *C*, *D*. Most of the *A*, *B* and *C* elements are synapsed, while the *D*'s are unsynapsed. *AR* and *AL*, the 'right' and 'left' ends of the *A* chromosome, *BR* and *BL* the corresponding ends of the *B* element. From White (1946*a*).

Henderson (1967*b*) has described a similar case in a cecidomyid which may have been *Lestodiplosis pisi*. In this instance two super-giant cells were present at the base of each gland, and the degree of ploidy was probably half that in White's case. Henderson therefore suggests that in the development of *Lestodiplosis* sp. a fusion of two super-giant cells similar to those of 'L. *pisi*' occurs. A study of the development of the super-giant cells of 'L. *pisi*' suggested that they develop from ordinary polytene nuclei by repeated splitting of the chromosomes.

Matuszewski (1964) has shown that giant polyploid polytene nuclei develop in the cecidomyid *Dasyneura urticae* following infection by a hymenopteran parasite. These nuclei were estimated at approximately 64- or 128-ploid. The origin of these giant cells is not known in detail, but it is clear that they are derived from the tissues of the host rather than those of the parasite.

A large number of cecidomyid species show one or two super-giant cells in each salivary gland (White 1948*a*), but the nuclei of these cells are generally *either* polytene *or* endopolyploid, and it is only in the genus *Lestodiplosis* that the two conditions have been found to occur normally in combination. In general, in Diptera, it is usual for the largest nuclei of the body to show polytene chromosomes, while the smaller ones either do not do so, or show lower degrees of polyteny (i.e. thinner chromosome elements). But there are some significant exceptions to this rule: thus in the cecidomyid *Dasyneura affinis* (White 1948*b*) the relatively small nuclei in the proximal division of the salivary gland contain polytene chromosomes, while in the very much larger nuclei in the distal part of the gland the chromosomes have split repeatedly with separation of the resulting strands, so that these nuclei are endopolyploid but not polytene in the ordinary sense (Fig. 4.10). However, Matuszewski (1965), who has made a detailed study of the situation in *D. urticae*, whose larval salivary gland has the same histology as that of *D. affinis*, concludes that the endopolyploidy of the nuclei in the gland proper results from 'progressive splitting of the polytene chromosomes into numerous fibrils of a low degree of polyteny (oligotene fibrils)'. Apparently, these oligotene fibrils do not undergo any significant contraction after they are liberated from the polytene chromosomes. White pointed out that in 'transitional nuclei' located between the basal and the main part of the gland the disintegration (splitting) of two of the chromosomal elements has proceeded further than in the case of the other two and Matuszewski has shown that splitting begins in different chromosomes in the same nucleus at different stages of development. These observations have been extended to *Dasyneura crataegi* by Henderson (1967*a*), who suggests that possibly the disintegration of the polytene chromosomes is associated with a burst of RNA synthesis and is an alternative to the puffing phenomenon (the one chromosome in *D. crataegi* which shows large puffs is the last to under disintegration). In *Mikiola fagi* and *Rhabdophaga saliciperda* Kraczkiewicz and Matuszewski (1958) have claimed that the chromosomes of the gland proper undergo transverse fragmentation as well as longitudinal splitting. And in *Perrisia ulmariae* the endopolyploid nuclei of the main part of the gland, which have gone through the 'splitting phase' of development later become secondarily polytene (Ashburner and Henderson, cited by Ashburner 1970).

Special types of behaviour are characteristic of the chromosomes in the ovarian nurse-cells of the higher Diptera. It has been shown by Bauer (1938), Bier (1957, 1958, 1959, 1960a, b) and Ribbert and Bier (1969) that in *Calliphora*, *Lucilia* and *Musca* these chromosomes become polytene and then fall apart into their constituent strands. Late in development they once more become polyploid and assume a banded appearance, with puffs and Balbiani rings. Studies with tritiated uridine have shown that all loci of these 'secondarily polytene' chromosomes, and not merely those which are visibly puffed, are actively synthesizing RNA. This is probably what we would expect, in view of the presumed functions of the nurse-cells – to supply the developing ovum with a very wide selection of different proteins which will be needed in the development of an embryo.

In mosquitoes, or at any rate in the genus *Anopheles*, the chromosomes of the ovarian nurse-cells remain polytene, without any dissociation into their constituent strands (Coluzzi 1968). In fact such chromosomes have been used to prepare maps of the banding pattern of certain mosquito species (Coluzzi, Cancrini and Di Deco 1970); as we should perhaps expect, the puffing patterns of the salivary cells and the ovarian nurse-cells seem to be very different.

POLYTENE CHROMOSOMES IN ANIMALS OTHER THAN DIPTERA

Very striking polytene chromosomes have recently been described in the salivary gland nuclei of a number of species of Collembola belonging to the family Neanuridae by Cassagnau (1966, 1968a). Actually, the first report of polytene chromosomes in Collembola appears to be that of Prabhoo (1961) in a species of *Womersleya*. The appearance of these collembolan polytene nuclei is essentially as in the Diptera. In *Neanura monticola* certain polytene chromosomes may reach a length of 400 μm (in an insect only 4–5 mm long!). In *Bilobella massoudi* the polytene nuclei contain fourteen chromosomes ranging in length from 45 to 170 μm; these consist of seven pairs of homologues which are generally unsynapsed, although occasionally lateral synapsis or end-to-end adhesion of homologues has been observed. The banding patterns and details of the individual bands seem to be as in the Diptera. No chromocentral formations occur, but heterochromatic regions are seen and there are a number of giant puffs and structures similar to Balbiani rings. No differences between the polytene nuclei of the two sexes were observed. This seems to exclude an XO condition in either sex; presumably one sex is XY with minimal differences between the two sex chromosomes. The existence of polytene chromosomes in the Collembola, a group not known to show 'somatic pairing' suggests that the association of

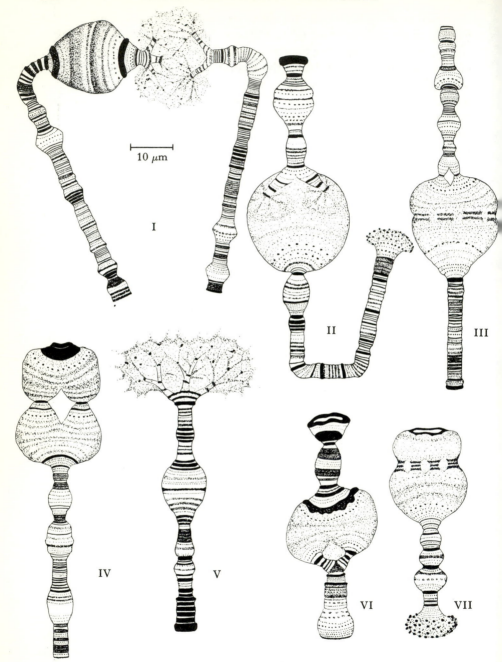

Fig. 4.16. Polytene chromosomes of the collembolan *Bilobella massoudi*. From Cassagnau (1968).

polyteny and somatic pairing in Diptera is at any rate largely fortuitous. In *Bilobella matsakisi* polymorphism for pericentric inversions has been found in two chromosome pairs by Cassagnau (1970) in Grecian populations. Since the polytene chromosomes of *Bilobella* are generally unsynapsed, no inversion loops of the dipteran type are seen. Polyteny is not generally present in the salivary gland nuclei of Collembola, being apparently confined to the members of the tribe Neanurini (Cassagnau 1968*b*). There have been no reports of polyteny in collembolan tissues other than the salivary gland.

There have been a few reports of genuine polytene chromosomes in animals other than arthropods. One is in the macronuclear *Anlage* of *Stylonychia muscorum*, a hypotrichous ciliate (Alonso and Pérez-Silva 1965). In these very large nuclei, there are about fifty to sixty giant banded chromosomes 15–55 μm in length and 0.8 to 2.5 μm in width. There is no chromocentre and the large number of chromosomes leads to the nucleus having a rather different appearance to a dipteran salivary nucleus; but there seems no doubt that these chromosomes are comparable in all respects to the polytene elements of the Diptera. It is not clear, however, whether the *Stylonychia* elements consist of two synapsed chromosomes or not. Polytene chromosomes have also been studied in the development of the macronucleus in *Euplotes* by Narasimha Rao and Ammermann (1970).

5

GIANT CHROMOSOMES — THE LAMPBRUSH TYPE

> Hope may also be derived from the fact that some of those
> things which are already discovered are of such a kind as,
> previous to their discovery, would not have easily occurred to
> anyone: they would simply have been rejected as impossible.
>
> FRANCIS BACON: *Novum Organum*

A second type of giant chromosome occurs in the mid-prophase of meiosis in the oogenesis of certain animals. Oocyte nuclei are almost always large, and perhaps especially so in groups with yolky eggs. It is apparently when such nuclei occur in species with high DNA values that we are apt to get giant chromosomes of the 'lampbrush' type – so called from their imagined resemblance to the brushes used to clean old fashioned oil-lamp chimneys. It is fairly clear, however, that these chromosomes do not really constitute an entirely distinct type, but are merely extreme examples of the more usual type of mid-prophase chromosome in oogenesis, and perhaps in spermatogenesis too. Thus many kinds of intermediate types between typical lampbrush and 'normal' oocyte chromosomes occur in various groups (Callan 1957).

The first lampbrush chromosomes to be studied were those of sharks (Rückert 1892), but almost all the recent work has been on those of urodeles, especially members of the genus *Triturus*. The pioneer work of Duryee (1941, 1950), Dodson (1948) and Guyénot and Danon (1953) has been followed by the detailed studies of Tomlin and Callan (1951), Gall (1952, 1954, 1955, 1956, 1958, 1963*a*, *b*), Callan (1952, 1955, 1957, 1963), Callan and Lloyd (1956, 1960*a*, *b*), Callan and Macgregor (1958) and Gall and Callan (1962), using improved techniques not available to the earlier workers. The finest details can be seen in unfixed, i.e. 'living', chromosomes in saline by phase-contrast microscopy in preparations with the nuclear membrane ruptured, using an inverted optical system. Such preparations may also be fixed in formaldehyde or osmic vapour.

There seems to be some kind of association between the amount of yolk or at any rate the total cytoplasmic volume attained by the oocyte and the degree of development of the lampbrush elements; the latter are well seen, for example, in the oocyte nuclei of reptiles and birds, even though the birds have quite low DNA values. Among invertebrates, lampbrush chromosomes have been seen in the arrow-worm *Sagitta*, the squid *Sepia*, the crustacean *Anilocra* and in an echinoderm (Callan 1957).

The lampbrush chromosomes are in the diplotene stage of meiosis, i.e. they are bivalents. In the amphibian oocyte they remain in this stage for at least six months before maturation is completed. In *Triturus viridescens*, with 2n = 22, the oocyte nuclei reach a diameter of nearly 500 μm at one stage, and the individual bivalents range from about 400 μm–1 mm in length (Gall 1954; Miller 1965). Thus if all the eleven lampbrush elements of a single nucleus were placed end-to-end without stretching they would

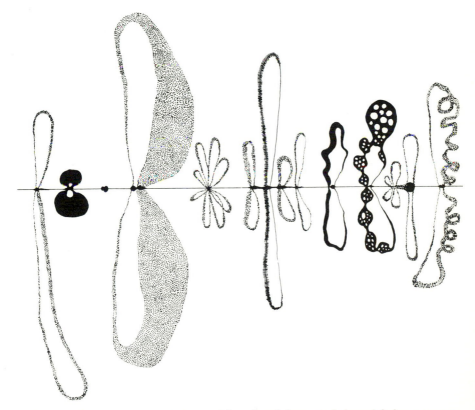

Fig. 5.1. Some characteristic types of lampbrush loops and the axial chromomeres from which they arise. From Callan (1955).

extend for more than half a centimetre (5.8 mm in *T. cristatus* according to the most accurate measurements).

Each bivalent consists of two homologues associated at one or more points, most of which at any rate correspond to chiasmata. A single homologue consists of a row of Feulgen-positive chromomeres, about 1 μm in diameter and 1–5 μm in length, connected by an extremely thin continuous chromonema (30–50 Å in diameter according to the electron micrographs of Miller 1965). In addition to DNA, the chromomeres contain basic protein (Gall 1954). Most of the chromomeres bear pairs of lateral loops; either a single pair or several. It is the hundreds of loops arising from each bivalent which give it the characteristic fuzzy or hairy 'lampbrush' appearance under low magnification or in fixed preparations. Members of each pair of loops are alike in length and morphology; and the loops on the corresponding chromomeres of the two homologues are, in general, alike (although important and significant exceptions occur, see p. 142); but loops arising from different, non-homologous chromomeres differ greatly in length and appearance. A few chromomeres bear quite minute loops and the majority are 5–10 μm in length, but certain giant loops may extend out laterally for over 100 μm. Some loops are straight, while others are characteristically twisted or spirally contorted. In the anuran *Xenopus* the longest loops are only about 10 μm long and the majority are no more than 1 μm (Callan 1963). In *Triturus cristatus*, Mancino (1963) counted fifty chromomeres in a short section of chromosome 12 and calculates that the total number in the haploid complement would be about 3300; but, as he admits, this is probably a minimum which would be increased if better optical techniques were used. Callan (1963) states that there are of the order of 5000 chromomeres per haploid complement.

Since we are dealing with diplotene bivalents, we would expect that the chromonema would be visibly two-stranded and that the chromomeres, which are localized enlargements or swellings of the chromonema, should also show indications of doubleness. In spite of this, both chromonema and chromomeres ordinarily appear unsplit, as if they had either not undergone

Fig. 5.2. Postulated structure of a chromomere and its attached pair of loops. From Gall (1956). Brookhaven National Laboratory.

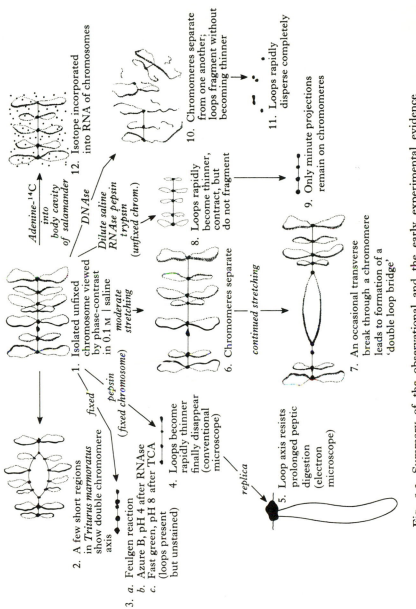
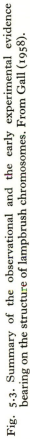

Fig. 5.3. Summary of the observational and the early experimental evidence bearing on the structure of lampbrush chromosomes. From Gall (1958).

replication at all, or as if the two strands were physically fused throughout their length. They even appear single in the best electron micrographs that have been obtained. It is nevertheless fairly clear that these structures really are double. Doubleness may, in fact, be evident in the neighbourhood of chiasmata and there are two regions of the shortest bivalent of *Triturus marmoratus* and *T. cristatus* which show doubleness of both chromonema and chromomeres very clearly; the split or dissociated chromomeres of these regions only give rise to *one* loop each (Fig. 5.3). Ullerich (1966, 1967, 1970) has carried out careful observations on the ultrastructure of anuran lampbrush chromosomes in relation to the DNA values of the species. He concludes that the large amounts of DNA in Amphibia and the range of variation in DNA values from species to species are not due to polynemy (many-strandedness), since in all species the number of DNA strands in the chromosomes is the same.

As meiosis proceeds towards first metaphase, the loops undergo regression and are apparently eventually resorbed into the chromosomes, which shorten and thicken, being finally converted into normal compact meiotic bivalents with smooth outlines; earlier suppositions that the loops were cast off in their entirety were apparently unfounded. As the loops are withdrawn the chromomeres become larger and adjacent ones tend to fuse together.

Each lateral loop consists of a very thin axis surrounded by a variable amount of 'matrix' material. The loops appear to be Feulgen-negative in stained preparations. The characteristic differences between the individual loops depend to a large extent on the amount and kind of matrix. In many loops the matrix material seems to consist of numerous fine fibres projecting out radially from the axis. The chromonema, chromomeres and loop axes are all resistant to proteolytic enzymes and to RNAse, but are broken or disintegrated by DNAse. There is thus good reason to believe that these structures are all composed of DNA, or at any rate that their continuity depends on DNA. Gall (1963a) has carried out kinetic studies of enzymic digestion of lampbrush chromosomes. His results suggest that when an interchromomeric connection is broken enzymatically four polynucleotide strands are fragmented. Enzymatic breakage of loops suggests that their axis consists of only two polynucleotide strands (i.e. a *single* Watson–Crick duplex). The matrix material of the loops is apparently ribonucleoprotein. The chiasmata are always between regions of the interchromomeric chromonema and never involve the loops, although 'non-homologous' fusions between the matrix material of some of the loops do sometimes occur. Callan (1955) has shown that when lampbrush chromosomes are mechanically stretched, they break transversely through the middle of a chromomere,

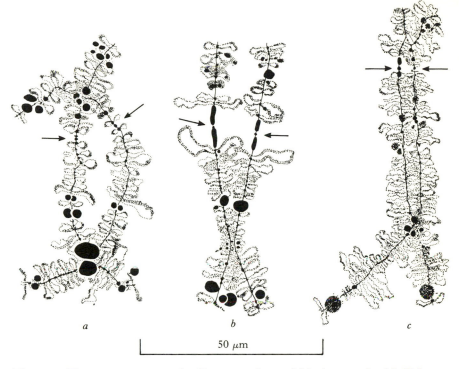

Fig. 5.4. The centromeres and adjacent regions of bivalent 10 in (*a*) *Triturus cristatus carnifex*, (*b*) *T.c. karelinii*, (*c*) the hybrid between them. Centromeres indicated by arrows. Some chiasmata are visible in these bivalents. From Callan and Lloyd (1960*a*).

the half-chromomeres so formed being still connected by the lateral loops, which are stretched between them. This suggests that the chromomeres have a sort of symmetrical duplex structure, one of the bases of each loop arising from one half of the duplex and one from the other (Fig. 5.2).

The essential picture of the lampbrush chromosome as analysed by modern workers is, hence, that each homologue consists of two immensely long DNA fibres (one per chromatid) which are closely paired in the inter-chromomeric sections, and presumably also in the chromomeres, where they must be compacted, probably by extensive folding or spiralization. Between the chromomeres the DNA chromonemata are in an extended and 'naked' state. From the chromomeres they run out into the lateral loops, forming their axes; but in the loops they are covered by a considerable amount of RNA and protein material, presumably synthesized under the control of genic DNA. It is to be presumed that each lateral loop synthesizes

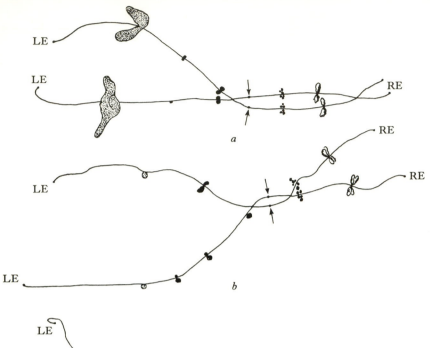

a

b

c

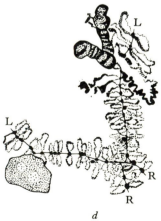

d

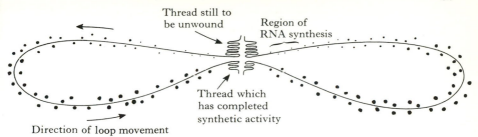

Thread still to
be unwound

Region of
RNA synthesis

Thread which
has completed
synthetic activity

Direction of loop movement

Fig. 5.8. Interpretation of tritiated uridine experiment in terms of a conveyer belt model of lampbrush loops. Modified from Gall and Callan (1962).

(Fig. 5.8). Gall and Callan have given reasons for rejecting alternatives (1) and (2). Experiments with tritiated phenylalanine suggest that when it has finished synthesizing RNA the active region begins to synthesize protein. More recently Gall (see Miller 1965) has broken loops by X-irradiation and then used labelling with tritiated uridine to demonstrate RNA synthesis: somewhat unexpectedly this was found to occur on both sides of the induced breaks. However, the essential correctness of interpretation (3) has been supported by the experiments of Snow and Callan (1969), in which they used actinomycin D to inhibit RNA synthesis in the lateral loops. The accessory ribonucleoprotein matrix of some of the loops is periodically cast off in the form of masses of material which float in the nuclear sap, undergoing various changes in appearance. These sloughed-off bodies, of which there may be as many as a thousand in a single nucleus, are generally spherical refractile bodies, the majority of which are peripherally situated, just inside the nuclear membrane. They obviously represent products of intense biosynthetic activity under direct genic control. Earlier (e.g. Gall, 1955) they were referred to as nucleoli, a term that seems misleading and undesirable. Gall believed at one time that in *Triturus viridescens* and *Ambystoma tigrinum* all the extrachromosomal masses of ribonucleoprotein were formed at the locus of the single nucleolar organizer, but it appears from Callan's work on *T. cristatus* that they arise from many of the loops rather than from any one locus. The nature of the proteins formed by the loops has not been studied in any detail.

Fig. 5.7. Polymorphism for loops in *T.c. carnifex. a*, bivalent 10, homozygous for presence of the giant loops; *b*, the same, homozygous for their absence; *c*, the same, heterozygous for giant loops; *d*, portion of bivalent 11 with giant loops fused together in one homologue but not in the other (typical of this particular individual). LE and RE, left and right ends, respectively. In *c* the two right ends are fused together. From Callan and Lloyd (1960*b*).

It does not seem to be definitely known whether the sequence of 'thick' and 'thin' loop-ends along the length of the chromosome is random or whether some ordered sequence exists (e.g. the thick ends might all be proximal or distal to the thin ends of the same loop). As far as individual loop pairs are concerned, polarity is constant, e.g. the giant loops on bivalents 10 and 11 of *T. cristatus carnifex* have their thin ends on the side towards the centromere, while the giant loops on the left arm of bivalent 12 in *T. c. carnifex* and *T. c. cristatus* have their thick ends on the side directed towards the centromere (Callan and Lloyd 1960*b*).

Earlier interpretations of lampbrush chromosomes envisaged some kind of polytene or multi-stranded structure (Ris 1945, 1956, 1957). It is now clear from the work of Callan and Gall that all such interpretations must be decisively rejected. There are apparently just four Watson–Crick duplex DNA strands in each lampbrush bivalent, corresponding to the four chromatids or strands of the classical genetic picture. The electron micrographs of Miller (1965) show that the interchromeric axis (2 fused chromatids) is 30–50 Å in diameter – a dimension which definitely excludes any kind of polyteny and makes it certain that each chromatid consists essentially of a single DNA molecule of enormous length with associated RNA and protein in the loops.

Gall and Callan (1962) have presented some calculations as to the total length of the DNA thread in a loop and its associated chromomere. The total traverse of a giant loop takes 10 days, and the lampbrush stage lasts at least 180 days. This implies eighteen traverses, i.e. the total length is eighteen times the length of the loop. Since this is about 50 μm the total 'site length' is about 1 mm. The sum of all the loops on a haploid chromosome set of *T. viridescens* or *T. cristatus* is estimated at 50 cm. If each site contains a thread eighteen times the length of the loop, the total length of the DNA in the haploid chromosome set would be about 9 metres. Photometric measurements have shown the total amount of DNA in the haploid set to be about 30×10^{-12} gm. Calculating this as a Watson–Crick double helix gives a length of 9.9 metres. The agreement between the two calculations is remarkable and confirms the view that there is only a single Watson–Crick double helix in each chromatid. It seems unlikely that any folding or spiralization of the DNA interchromomeric axis exists when the lampbrush chromosomes are at their maximum size, but the DNA in the chromomeres must certainly be coiled.

The various experiments in which lampbrush chromosomes were treated with enzymes (Macgregor and Callan 1962) have not demonstrated any 'linkers' in the DNA chromonema. There is, in fact, no evidence from any of these enzymatic studies for any kind of bonding in the continuous axis of the chromosome other than the phosphodiester linkages of the DNA strand. On the other hand it would be going too far to say that the enzymatic experiments that have been performed necessarily exclude the possibility of linkers of some kind, perhaps in the chromomeres.

If we are right in accepting the Callan–Gall interpretation of the DNA of the interchromeric axis, chromomeres and loops as essentially continuous, it nevertheless remains that there must be some general chemical difference between the DNA's composing these structures to account for their different behaviour. At least this must be so in the case of the DNA of the interchromomeric axis (the DNA of the loops may be merely a part of the chromomeric DNA, temporarily extended). Since the interchromomeric DNA is apparently never extended into loops it may be inactive as far as RNA synthesis is concerned. Nevertheless, we might imagine that even if the whole of the chromonema between the chromomeres were 'non-genic' (i.e. carrying no coded information or perhaps carrying only 'nonsense') the exact length of these segments, i.e. their *spacing* on the chromosome, as distinct from their *sequence*, might not be irrelevant to genic functioning.

The nature of the free nucleolar structures in the nuclear sap was not elucidated in the early work on *Triturus* spp. It has now become much clearer as a result of work by Kezer, Wimber and Peacock (see Peacock 1965a) on the oocyte nuclei of plethodontid salamanders. In these nuclei there are 200 to 300 free nucleolar structures each of which is a ring of DNA (the chemical nature has been confirmed by Feulgen staining and enzyme tests). The rings show a considerable size range, from a few micra up to 400 μm in circumference. At certain stages the rings have a 'beaded necklace' appearance, apparently due to local accumulation of RNA and protein materials on the circular DNA molecule.

These nucleolar rings seem to originate from a peculiar region of one of the bivalents. In this section there are a number of single unpaired loops which are presumed to arise from only one of two sister chromatids by Peacock (Fig. 5.9). An alternative interpretation would be that the apparently single loops have arisen by an intimate pairing of two identical loops, as in the case of the apparently single (but in reality double) interchromomeric chromonema. It is believed by W. J. Peacock (1965a) that the nucleolar rings are liberated from the unpaired loops, following on some kind of replication process; but it is unclear whether the chromosomal loops are really circular DNA molecules or whether the newly liberated rings are linear but immediately become circular by fusion of their ends. O. L. Miller (quoted by W. J. Peacock, 1965a) has found that the free nucleolar structures of *Triturus* are also rings, although frequently appearing compact so that their true nature is not revealed. There is a possibility that the ring structures

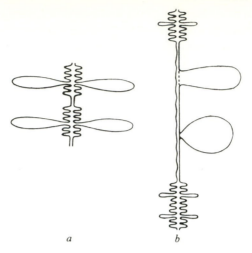

Fig. 5.9. *a*, diagrammatic representation of a normal portion of a lampbrush chromosome showing paired loops; *b*, a comparable diagram for the nucleolar-ring loop region in *Plethodon cinereus*, with single loops. From Peacock (1965).

may undergo further replication in the nuclear sap, thereby increasing their number. Assuming that they are molecular copies of the DNA in the chromosomal loops, they may perhaps represent a special device for greatly increasing the production of messenger RNA of certain types.

Callan and Lloyd (1960*b*) have carried out a detailed comparison between the lampbrush chromosomes of several 'subspecies' of *Triturus cristatus* (probably these would more appropriately be considered as distinct species). The centromeres of all the chromosomes in a subspecies karyotype are alike, but they are characteristically different in the lampbrush chromosomes of the four subspecies. *T.c. carnifex* centromeres are spherical chromomeres without lateral loops and slightly larger than most of the ordinary chromomeres. The centromeres of *T.c. cristatus* are similar but smaller, while those of *T.c. danubialis* are the smallest of all. *T.c. karelinii* differs from the other forms in having its centromeric chromomeres flanked by thick blocks or bars of Feulgen-positive material (probably to be interpreted as heterochromatin) without lateral loops (Fig. 5.4). It is characteristic of the oocyte nuclei of *karelinii* that homologous centromere regions are frequently fused together and that even triple non-homologous fusions of the flanking blocks occur; fusions of centromere regions have not been observed in the other subspecies. Oocyte nuclei of hybrids between *karelinii* and *carnifex* or *danubialis* show the centromere-flanking bars on one chromosome of each bivalent but not on the other (Fig. 5.4*c*). In *T. italicus* the centromeres are single rather large chromomeres without lateral loops (Mancino and Barsacchi 1969), but in *T. alpestris apuanus* they are difficult to detect (Mancino and Barsacchi 1965), but consist in some cases of a sequence of several fused chromomeres without loops. The lampbrush elements of *Triturus italicus* show rather numerous bars without loops, but these do not bear any relationship to the centromeres (Mancino and Barsacchi 1969). At the ends of the chromosome arms, in *Triturus cristatus*, the telomeres are morphologically visible as small spheres of distinctive structure (Callan, 1963) which do not bear loops. They are definitely larger in *T.c. cristatus* than they are in *T.c. karelinii*. In the former, but not in the latter, fusions of telomeres (on homologous or non-homologous chromosome ends) are common, but do not persist until metaphase, so that they cannot be due to terminalized chiasmata.

Apart from differences in centromere morphology, there are numerous conspicuous loops which are present in one subspecies but cannot be recognized in the others. Callan and Lloyd have accordingly constructed 'working maps' of the lampbrush chromosomes of *cristatus*, *carnifex* and *karelinii* in which the more conspicuous loops are indicated. Similar working maps have been published for *Triturus vulgaris meridionalis* by Mancino (1966) and Barsacchi, Bussotti and Mancino (1970), for *T. alpestris apuanus* by Mancino and Barsacchi (1965), for *T. italicus* by the same authors (1969) and for *T. helveticus helveticus* by Mancino (1965) and Mancino and Barsacchi (1966).

Maps have also been published for two species of *Pleurodeles*, *P. waltlii* (Portugal) and *P. poireti* (Algeria) by Lacroix (1968a); there are very great differences between the loop patterns of these presumably very closely related forms. Maps for *Salamandra salamandra* have been published by Mancino, Barsacchi and Nardi (1969); in this species five of the eleven bivalents show regions bearing numerous long loops and eight bivalents bear one or more 'spheres' (the latter structures seem to exist also in members of the genera *Triturus*, *Pleurodeles* and *Euproctus*, but are less numerous than in *Salamandra*). All these maps only show the more conspicuous landmarks; no cytologist has yet succeeded in constructing really detailed maps of lampbrush chromosomes comparable to those that now exist for the polytene chromosomes of many species of Diptera. However, those that have been worked out have proved adequate for the detection of individuals with natural or induced rearrangements such as translocations, inversions and deletions (Gallien, Labrousse and Lacroix 1966; Mancino, Nardi and Barsacchi 1970).

Normally in amphibian oocyte nuclei the loop pairs on homologous chromosomes are similar in size, location and morphology. But exceptions exist, i.e. some bivalents may exhibit heterozygosity for presence or absence of loops. In all four subspecies of *T. cristatus* the longest bivalent has a long segment in which chiasmata are never formed and in which the loops of the two homologues do not correspond. This is presumably the *XY* sex-chromosome bivalent and the region for which all females seem to be heterozygous is a 'differential segment'. Similar differential segments have been found in chromosome 4 of *Pleurodeles waltlii* by Lacroix (1968b) and in *P. poireti* by Lacroix (1970). In the latter species it appears as if the *Y*-chromosome may bear an extra region with loops of special type, which is absent in the *X*. Individual females of *T.c. carnifex* may show different loop patterns in these differential segments. It thus appears (on the hypothesis outlined above) that this subspecies, at any rate, is polymorphic for several kinds of *X*'s and also for several kinds of *Y*'s, differing in their more conspicuous loops.

Populations of *Triturus cristatus* are polymorphic for presence or absence of certain giant loops on chromosomes 10 and 12. A sample of twenty-two individuals of *T.c. carnifex* included representatives of seven of the expected nine genotypes, showing that, in the case of both chromosome elements, homozygotes for deficiency of the loops, as well as for their presence, are viable. Heterozygotes show a loop on one homologue of the bivalent but not on the other (Fig. 5.7). Unfortunately, Callan and Lloyd do not seem to have determined whether a chromosome without a giant loop still bears an axial chromomere which fails to produce a loop. Callan and Lloyd (1956) have interpreted these polymorphisms for presence or absence of a loop as 'allelic differences' but if the chromomere is also lacking in loopless chromosomes, it might be better to think of them as due to minute deletions.

As we have already suggested, it is extremely likely that all chromosomes have a 'lampbrush' structure during the meiotic prophase, both in oogenesis and spermatogenesis. Thus the fuzzy, irregular outline of the bivalents in the diplotene and diakinesis stages (see p. 187) is almost certainly due to the presence of thousands of loops, arising from the chromatid axes and more or less covered with a 'matrix' of ribonucleoprotein. In many meiotic bivalents, however, the two chromatids of which each chromosome is composed are clearly separated and not intimately fused as they are in the amphibian oocyte.

It must be admitted, however, that the dimensions of the chromosomes in the meiosis of Orthoptera, Heteroptera and plants such as *Trillium* and *Tradescantia* (all organisms with exceptionally large chromosomes) are not such as to permit analysis of the postulated loops under the light microscope. Neither have techniques of electron microscopy been successfully applied as yet to such material. There is, however, one other type of chromo-

TABLE 5.1

Segment	Name of paired loop structure
Short arm (Centromere)	Schlinge (nooses)
1	Keulen (clubs)
2 or 3	Tubuli-Bänder (tubular ribbons)
4 5 6 7	{ No loop structures in this region
8 or 9	Pseudonucleolus
9 or 10	Faden (threads)

some in which an undoubted 'lampbrush' structure, albeit of a somewhat different type to that in the amphibian oocyte has been demonstrated. Meyer (1963), Hess and Meyer (1963a, b), Hess (1965) and Meyer and Hess (1965) have carried out a series of investigations, using phase-contrast and electron microscopy, on the nuclei of *Drosophila* spermatocytes during the prophase of the first meiotic division. In *D. hydei, D. repleta* and a number of other species these nuclei contain very large and conspicuous paired structures which have been shown to arise as loops from a lampbrush-type Y-chromosome. In *D. hydei* there are five pairs of giant loops which differ greatly in appearance (Figs. 5.10 and 5.11). Hess has studied stocks with translocations and other rearrangements which have enabled him to localize these structures on a cytological map of the Y-chromosome as in Table 5.1 (the long arm is considered to be subdivided into ten equal segments).

Males with Y-chromosomes which lack any of the loop-forming segments are invariably sterile, sperm-formation being inhibited at different stages, depending on which pair of loops is wanting.

Although the basic interpretation of these giant structures as loops arising from the Y-chromosome is not in doubt, the precise relation between the axes of the loops and those of the chromatid between the successive loops has not been determined and various possibilities exist, which are shown diagrammatically in Fig. 5.11.

The detailed morphology of the paired structures is somewhat different in *D. hydei* and *D. neohydei* and if a Y-chromosome of the former is transferred, by interspecific crossing, into the genotype of the latter it continues to form paired structures of the original type. Y-linked mutations are also known which affect the appearance of the paired structures and spermatocytes containing two Y's of different genetic constitution show both kinds of structures. Injection of the flies with actinomycin (an inhibitor of RNA

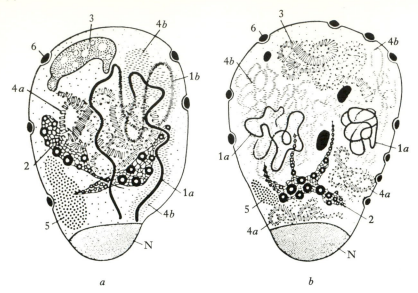

Fig. 5.10. Spermatocyte nuclei of (a), *Drosophila hydei* and (b), *D. neohydei*, showing the structures visible under phase-contrast in the prophase of the first meiotic division. 1*a*, the compact and 1*b*, the diffuse section of the 'threads'; 2, the 'clubs'; 3, the 'pseudonucleolus' in *D. hydei* and the distal coil in *D. neohydei*; 4*a*, the proximal coils, 4*b*, the covering of the threads; 5, granular material; 6, invaginations of the nuclear membrane with 'secretion'; N, nucleolus. From Meyer (1963).

synthesis) causes a regression and disintegration of the material composing the loop structures, but regeneration occurs later. This experiment may be interpreted as indicating that the bulk of the loop material is a ribonucleo-protein matrix surrounding a very fine (and therefore invisible) DNA axis, as in the loops of the amphibian lampbrush nuclei.

The *Y*-chromosomes of *D. hydei* and its relatives are totally heterochromatic by all the usual criteria. It is interesting that the *X*-chromosome of *hydei* has one heterochromatic arm but that this does not form any looped structures in the primary spermatocyte; it may be partially homologous to the long segment in the middle of the *Y* which does not form any giant looped structures.

At the end of the meiotic prophase in *D. hydei*, as the *Y*-chromosome condenses to assume its metaphase shape, the matrix material of the different loops clumps together and is released into the developing spindle. It can still be seen after completion of the first meiotic division. There is no significant development of looped structures corresponding to those of the

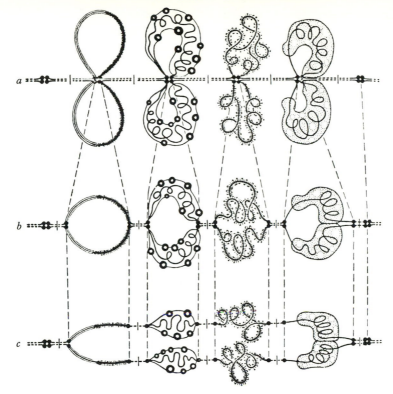

Fig. 5.11. Three alternative interpretations of the paired structures produced by the Y-chromosome of *Drosophila hydei*. From Meyer (1963).

meiotic prophase in the nuclei of the secondary spermatocytes and spermatids of *D. hydei*.

Spermatocytes of a number of other species of *Drosophila* have been examined by Meyer and Hess, but most of them only show rather ill-defined structures that could be associated with the Y-chromosome. In *D. melanogaster* structures referred to as 'tubuli' and 'reticular elements' are present in the spermatocyte nuclei of XY individuals, but not of XO ones (Beermann, Hess and Meyer 1963). However, tubuli do not seem to arise from a single locus on the Y-chromosome because they are present in nuclei containing only the short limb or only the long limb of the Y.

A general reconciliation between the superficially very different structures of the lampbrush and polytene types of giant chromosomes seems possible. The chromomeres of the lampbrush elements clearly correspond to the bands of the polytene chromosomes – or, more strictly, to the elementary chromomeres which fuse to form the bands in the latter case. The inter-chromomeric chromonemata of the lampbrush elements are equivalent to the interband regions of the polytene chromosomes – strictly speaking, to the DNA strands which traverse those spaces (in both cases it would seem that spiralization of the interchromomeric DNA strands is minimal or non-existent).

As first realized by W. Beermann (1952a), the loops of the lampbrush bivalents are obviously represented in the polytene chromosomes by the various 'puffs'. And here certain major differences are apparent. A 'puff' must consist of hundreds or thousands of identical DNA loops, similar to those in the amphibian oocyte, each with its matrix and each arising from a chromomere at the puff-producing locus. And, whereas in the lampbrush chromosome all (or almost all) chromomeres have formed loops, in the polytene element only a few do so at any one stage of development, or in any particular tissue.

The latter point of difference is just what we might have expected if (as seems overwhelmingly probable) loop formation, or puffing, is associated with intense biosynthetic activity (formation of messenger RNA) by genetic loci. In an oocyte which is shortly going to undergo cleavage to form an embryo we might expect that all, or almost all, of the genetic loci would be 'biosynthetically' active, leading to the presence in the cytoplasm of the mature egg of all the thousands of protein species needed for morphogenesis and differentiation. In the secretory cells of the Diptera, on the other hand, we have highly differentiated elements, specialized for the production of what is probably a very limited range of proteins. It is thus not at all un-expected that the number of genetic loci showing visible evidence of bio-synthetic activity should be so much lower in the case of the polytene chromosomes. Like the puffs, but less obviously, the lampbrush loops each have their own cycle of development and regression with a very precise timing that differs slightly from loop to loop.

The detailed analysis of the lampbrush chromosomes of the amphibian oocyte does not provide any immediate explanation of the very high DNA values met with in urodeles. It definitely excludes all forms of polynemy or polyteny as the explanation. There are no conspicuous heterochromatic segments in *Triturus* spp., although certain short sections do seem to be undercondensed following cold treatment (Callan 1942). Therefore the possible explanation of high DNA values in urodeles as due to accumulation of large blocks of heterochromatin seems untenable. It seems virtually certain that the individual chromomeres of the urodele lampbrush elements contain far more DNA than the comparable structures in other classes of vertebrates.

A cistron in phage T4 may consist of several hundred nucleotide pairs (Benzer 1961). It is perhaps a DNA fibre 0.3 μm in length (Gall 1963). On this basis a lampbrush loop 50 μm in length might contain 100 to 200 cistrons. It seems difficult to avoid the conclusion that the chromosome-loop units of the lampbrush chromosomes must contain numerous duplicated and re-duplicated copies of the same basic cistron. The method of repeated evolutionary duplication which has given rise to this situation would presumably be the one which Keyl has demonstrated in the polytene chromosomes of *Chironomus thummi* (see p. 364).

6

THE MECHANISM OF MEIOSIS

From the fact of sexual reproduction, which brings together
equal amounts of paternal and maternal germ-plasm at each
fertilization, I inferred not only the composition of the germ-
plasm out of a number of units, the ancestral plasms, but also
the necessity of a reduction of the germ-plasm each time to
one half of its bulk, as well as a reduction of the number of the
ancestral plasms contained in it.

A. WEISMANN: *The Germ-Plasm: a Theory of Heredity*

As with other cytological phenomena, the mechanism of meiosis may be
investigated at various levels. There are many aspects that can be studied
and analysed at the level revealed by ordinary microscopy. Other and
perhaps more basic elements of the whole process lie in the fields of ultrastructure and molecular biology and are as yet imperfectly understood.

Although a final interpretation of meiosis in biochemical terms is not
possible as yet, and although a number of aspects are still controversial, the
cytogenetic analysis during the 1920s and 1930s of this, the essential
machinery of the genetic system in higher organisms, represents a major
achievement of biological science, and was historically necessary for the
development of the modern 'synthetic' theory of evolution. Authoritative
reviews of the meiotic mechanism from the classical standpoint have been
published by Rhoades (1961) and John and Lewis (1965a); the molecular
mechanisms of genetic recombination have been reviewed by Emerson
(1969) and Whitehouse (1970).

In the great majority of higher organisms, whether animals or plants, the
main outline of the meiotic divisions is essentially uniform. On the other
hand, there are certain groups in which these processes have been more or
less profoundly modified. In this chapter we shall deal only with the 'normal'
type of meiosis, in organisms with monocentric chromosomes, deferring

consideration of the various 'anomalous' types (which are nevertheless 'normal' for the groups in which they occur) until Chapter 12. In viruses and bacteria, as pointed out in Chapter 1, no phenomenon strictly analogous to meiosis occurs.

An understanding of the mechanism of meiosis is basic for the interpretation of the role of recombination in population genetics and evolution. Cytological analysis may enable us to distinguish between species with much recombination and those with very little ('recombophil' and 'recombophobe' species) even in the absence of actual genetic data. Historically, it was not until the basis of crossing-over had been worked out at the cytogenetic level (even if not at the chemical one!) by Janssens (1909, 1924) and especially Darlington (1930, 1931, 1932a, b, 1936, 1940b) that the principles governing the inheritance of inversions, translocations and other structural changes of the chromosome set could be understood. Many types of chromosomal rearrangements are unable to establish themselves in natural populations because they are unable to pass through the meiotic divisions without giving rise to gametes carrying broken chromosomes, duplications or deficiencies and hence lead to partial sterility. Moreover, since even 'normal' meiosis varies to some extent in the details of the various stages and processes, from one group to another, rearrangements which would lead to a high degree of sterility in one group may hardly lower the fertility of the organism at all in another. Thus when interpreting the chromosomal evolution of a population, species or group, it is first of all necessary to know the details of the particular type of meiosis which is present.

Except in the flagellates *Oxymonas*, *Saccinobaculus*, *Leptospironympha*, *Notila* and *Urinympha* (Cleveland 1950a, b, 1951) in which a type of one-division meiosis occurs, meiosis invariably consists of *two* nuclear divisions which follow on one another, usually with only a short resting stage, or no resting stage at all, in between. Typically, each division is followed by cytokinesis, but in certain cases this may be omitted or delayed. In multicellular animals the meiotic divisions are the last ones to take place during the process of formation of the gametes (spermatogenesis or oogenesis, as the case may be); in the former process they lead to the production of four sperms from each primary spermatocyte, while in the latter one ovum and three non-functional polar bodies or polar nuclei are produced from each primary oocyte. In a few groups of insects, reduction from the diploid to the haploid condition occurs at an embryonic division (see p. 506); it is really a question of semantics whether such processes should be termed *meiosis*.

In general, the nuclear phenomena of meiosis seem to be similar in the two sexes of a species (see, however, p. 169), but the cytoplasmic processes,

the details of which are outside the scope of this book, are of course profoundly different (in some cases there is no cytokinesis accompanying the
two female meiotic divisions, the polar nuclei remaining within the egg
cytoplasm and eventually degenerating). For technical reasons (abundance
of meiotic cells, absence of a large mass of yolky cytoplasm) spermatogenesis is much easier to study than oogenesis, so that most of our knowledge
of meiosis is based on the process as it occurs in spermatogenesis (or micro-
sporogenesis, in the case of the higher plants). In some groups where the
mechanism of meiosis is highly 'anomalous' in the male it may be relatively
or entirely normal in the female (see Ch. 13); this is the case in *Drosophila*,
where it has long been known that genetic crossing-over occurs in the female,
but not in the male.

Each of the meiotic divisions consists essentially of the same stages
(prophase, premetaphase, metaphase, anaphase and telophase) as an ordinary
somatic division; but these stages are modified in various respects. The
prophase of the first meiotic division is usually very protracted (it may
often last for weeks or months) and it involves a complicated sequence of
phenomena that do not occur in an ordinary mitosis. For this reason a
special terminology has been developed in order to describe the various
substages into which it may be divided. Several variants of this terminology
are in use, the two most common being as follows:

(This book)		(Swanson, 1957)
Leptotene	=	Leptonema
Zygotene	=	Zygonema
Pachytene	=	Pachynema
Diplotene	=	Diplonema
Diakinesis		

The leptotene stage follows on the last premeiotic spermatogonial or
oogonial division; it corresponds to the very earliest part of an ordinary
mitotic prophase. Leptotene chromosomes are usually very long slender
threads and in some organisms can be seen to bear very numerous minute
chromomeres distributed along their length. Most observers have reported
that the leptotene threads are visibly single, and Darlington (1932b, 1937b)
laid considerable stress on the singleness of the leptotene chromosomes as a
prerequisite for the pairing (synapsis) which occurs in the zygotene stage.
However, Huskins and Smith (1935), Nebel and Ruttle (1936) and some
other cytologists claimed to have observed split leptotene chromosomes,

especially in plant material. Since it is well known that the undoubtedly double chromosome axes of newt lampbrush chromosomes appear as single strands throughout most of their length (Callan and Lloyd 1960b) it is obvious that observations of 'single' leptotene strands are not evidence for true singleness. The leptotene stage is probably one in which the chromosomes reach their maximum elongation; Walters (1970) has determined that in *Lilium* they are more condensed and spiralized in the preleptotene period and subsequently elongate as they enter on leptotene.

Autoradiographic studies on a variety of animal and plant species are in substantial agreement that DNA replication occurs during or before leptotene (Swift 1950, Taylor and McMaster 1954 and Hotta, Ho and Stern 1966 on the plants *Lilium, Trillium* and *Tradescantia*; Wimber and Prensky 1963 and Callan and Taylor 1968 on newts of the genus *Triturus*; Henderson 1966 and Peacock 1970 on locusts and grasshoppers and Monesi 1962 on the mouse). The replication phase is of considerable duration (about 14 hours in the mouse, according to Monesi, and nine to ten days in *Triturus*, according to Callan and Taylor). It is not quite synchronous for all chromosomes and chromosome regions; thus in the newt there are certain heterochromatic segments which start and finish replication about one day later than the rest of the karyotype.

Premeiotic replication results in a change from the 2C to the 4C amount of DNA in the nucleus (Swift 1950 and all subsequent workers). No further replication takes place between the two meiotic divisions, so that the DNA is reduced to the 2C amount in the first meiotic division and to 1C in the second division. The male and female pronuclei undergo replication to give the 2C amount in each, so that after fertilization, i.e. at the metaphase of the first cleavage division the 4C amount is present, as at any diploid somatic metaphase.

Wimber and Prensky (1963) claimed that, in addition to the premeiotic DNA replication, there was a further synthesis of a small quantity of DNA (about 2% of the total) at zygotene–pachytene. Callan and Taylor have criticized this conclusion, claiming that such a small synthetic activity could not be reliably demonstrated by ordinary autoradiographic procedures. But, using more sophisticated techniques, Hotta, Ho and Stern (1966) have demonstrated that a very small amount of synthesis (about 0.3% of the total) does actually take place at pachytene in *Lilium* and *Trillium*; the DNA synthesized at this time has a different base-ratio to that synthesized at the premeiotic stage. This observation is of critical importance because it is now believed that crossing-over (chiasma-formation) occurs at pachytene and because the special DNA synthesized at this time may play a critical role in genetic recombination.

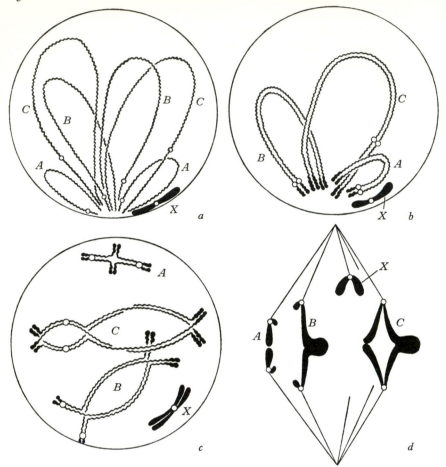

Fig. 6.1. General diagram of certain stages of meiosis in an imaginary organism with three pairs of autosomes and an *XO* sex-chromosome mechanism. *a*, leptotene; *b*, pachytene; *c*, diplotene; *d*, first metaphase, in side view. The *X*-chromosome is shown as heterochromatic; in most species of animals it would probably be folded or doubled on itself, instead of stretched out straight (as shown here).

Henderson (1966), working with the locust *Schistocerca gregaria*, kept at 30 °C, showed by tritiated thymidine autoradiography that the pre-meiotic DNA replication occupied about two days and occurred thirteen to fourteen days before first metaphase, presumably during the leptotene of preleptotene stage. Using heat shocks (40 °C) to inhibit chiasma formation it was found that the time interval between the heat-sensitive period and first metaphase was nine to ten days. The results were interpreted as demonstrating unequivocally that chiasma formation must take place during zygotene or early pachytene at least several days after premeiotic DNA synthesis is completed and that true breakage and reunion of chromatids (rather than a copy-choice mechanism occurring during DNA replication) must take place. Peacock (1970) has carried out similar experiments on the grasshopper *Goniaea australasiae*, which indicated that chiasma formation occurred in early pachytene, ten days before first metaphase (at 37 °C); the premeiotic DNA replication took place fourteen days before first metaphase.

Biosynthesis of the histone components of the chromosomes undoubtedly takes place at the same time as the replication of DNA (Bloch and Goodman 1955; Ansley 1957; Gall 1959).

SYNAPSIS

During the zygotene stage, homologous chromosomes become closely approximated, side-by-side, as a result of the process of synapsis. In a diploid organism this results in each pair of homologous chromosomes forming a *bivalent*, which is necessarily twice as thick as the original leptotene chromosomes. In polyploids, one chromosome may undergo synapsis with two others in different parts of its length so as to give rise to a *multivalent* association (*trivalent* if composed of three chromosomes, *quadrivalent* if it consists of four chromosomes, etc.). But, in general, synapsis is restricted to two chromosomes at any one locus (Darlington 1930); apparent exceptions to this principle will be considered later (p. 650).

The exact nature of the mechanism responsible for synapsis is still, to a large extent, mysterious. Riley and Law (1965) speak of this as 'probably the only biological process, of universal occurrence in higher organisms, whose causal processes are completely unknown' (echoing the statement of Digby (1910) that 'synapsis . . . faces one as an impenetrable wall.').

Two main viewpoints have been put forward. One stresses the fact that highly specific forces of attraction must be exerted over distances of several micra (Fabergé 1942). The second argues that these forces need not really extend over such considerable distances, because a polarized orientation of the chromosomes in leptotene brings all the ends together in a very small volume, on the inner surface of the nuclear membrane, opposite to the centrioles.

The truth probably lies somewhere between these viewpoints. Certainly a polarized or so-called bouquet orientation of the chromosomes is very generally present at leptotene–zygotene. And it is equally clear that, in general, synapsis seems to begin at the closely approximated chromosome ends and then spread along the length of the chromosomes zipper-like, from both extremities. This polarized orientation seems to be very generally present in animal cells, and occurs regardless of whether the chromosomes are metacentric or acrocentric; it has been studied in such diverse organisms as planarians (Gelei 1921, 1922) and salamanders (Kezer 1970). In plants the leptotene chromosomes are not arranged in a definite bouquet, but may form a tangled mass, the *synizetic knot* of some early cytologists, to one side of the nucleus.

It is, however, equally true that when chromosomal rearrangements such as inversions, duplications and translocations are present, each region somehow manages to 'find' its partner, in general; and this 'finding' argues either for some kind of long range forces or for chromosomal movements on a scale which would bring *all* regions into very close approximation with one another at least once during the period when the chromosomes are capable of synapsis. One piece of evidence suggesting that chromosomal synapsis depends on some prior spatial relationship is the work of Driscoll and Darvey (1970) who found that in wheat the drug colchicine will reduce the chiasma frequency of the ordinary chromosomes to about half the normal level (presumably by somehow disrupting the conditions necessary for synapsis) but has no effect on the chiasma frequency of an isochromosome, where two homologous chromosome limbs are held in close approximation at one end through being attached to the same centromere. Another line of evidence indicating special spatial relationships of the chromosomes prior to synapsis is the rarity of interlocking of bivalents; if orientation of the leptotene chromosomes was anything like random and if (as seems certain) synapsis is usually initiated at both ends of the chromosome and proceeds towards the middle, it would frequently happen that one chromosome would be caught between two synapsing homologues, so leading to interlocking of the bivalents at a later stage. It is true that such interlocking does occasionally occur; but it must be regarded as an extremely rare accident of meiosis

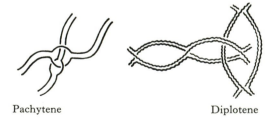

Pachytene   Diplotene

Fig. 6.2. Diagram showing how interlocked bivalents occasionally arise through accidental interlocking during synapsis.

whose frequency of occurrence is two or three orders of magnitude lower than would be the case on the hypothesis of random leptotene orientation.

Walters (1970) has concluded from her very detailed studies of the pre-leptotene period in *Lilium* that there is no spatial association of homologues at this stage. However, Maguire (1967), making use of heterochromatic knobs to identify the chromosomes of maize, claims that some degree of pairing between homologues exists before leptotene.

Suggestions regarding the nature of the forces responsible for synapsis range from the hypothesis of Fabergé (1942), based on Guyot-Bjerknes hydrodynamic effects, to Delbrück's (1941) suggestion, based on the reduction of peptide bonds and the formation of resonance bonds between pairing peptides. The more recent model of Comings and Okada (1970b) will be considered in the next section.

True synapsis seems to be restricted to homologous chromosome regions. But where a pair of chromosomes are homologous throughout part of their length but non-homologous in another region (as a result of a chromosomal rearrangement) 'true' synapsis may be continued (as a kind of 'pseudo-synapsis') in the non-homologous segment.

This non-homologous pairing, first detected in maize by McClintock (1933), is now known to occur in grasshoppers, midges (Chironomidae) and a number of other organisms. Whereas true synapsis may be followed by genetic crossing-over and chiasma formation, there is no evidence that chiasmata can ever be formed between segments that are synapsed non-homologously.

As soon as synapsis has been completed the nucleus may be said to have entered the pachytene stage. The bivalents are now at least twice as thick as the leptotene threads were and they continue to shorten and thicken during pachytene. In most organisms some shortening of the chromosomes seems to take place between leptotene and pachytene, but there seems to be a good deal of variation from group to group in this regard. Thus Rhoades (1950) believes that in maize the leptotene chromosomes are several times the length of the pachytene bivalents, while Manton (1945b) states that no contraction takes place in the fern *Osmunda* between leptotene and pachytene. Pachytene bivalents have a distinctly 'hairy' appearance; it is probable that this is due to hundreds or thousands of paired loops similar to those present in the giant lampbrush chromosomes, only shorter.

In most species of animals the bouquet orientation of the chromosomes is retained throughout zygotene and is still present in the early pachytene nuclei, although it may be lost in the later pachytene stages, as the bivalents shorten and become more condensed.

THE SYNAPTINEMAL COMPLEX

Synapsis is accompanied by the formation of linear axial bodies, the so-called *synaptinemal complexes* or *cores*, between the pairing homologues (Moses 1956, 1958). These bodies consist of parallel strands, usually three in number (Moses 1969). They were originally detected by electron microscopy in crayfish spermatocytes, but have subsequently been detected in the meiotic bivalents of plants (orchids and lilies), insects (*Drosophila*, crickets, cockroaches) and vertebrates (fishes, birds and mammals). Synaptinemal complexes do not occur at mitosis and they are not present in polytene chromosomes or in the case of somatic pairing in Diptera. In grasshoppers (Acrididae) the univalent X of XO species does not have a complex, but in crickets occasional synaptinemal complexes have been found within the univalent X-chromosome or between the latter and the nucleolus to which it gives rise (Sotelo and Wettstein 1966; Wettstein and Sotelo 1967). Menzel and Price (1966) observed occasional synaptinemal complexes in haploid tomato pollen mother cells; presumably these were formed in segments that were synapsed, either homologously or non-homologously. On the other hand Moses and Coleman (1964) found no complexes in the spermatogenesis of the normally haploid male scale insect *Steatococcus*.

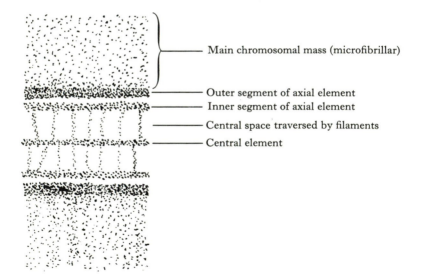

Main chromosomal mass (microfibrillar)

Outer segment of axial element
Inner segment of axial element
Central space traversed by filaments
Central element

Fig. 6.3. Diagram of a frontal section (electron micrograph) of a synaptinemal complex, following fixation with glutaraldehyde and osmium tetroxide. Based on Moses (1968).

The synaptinemal complexes usually seem to separate from the bivalents at diplotene and either disperse or become associated with one another in bundles or stacks that have been called polycomplexes. These structures have been studied in the mosquito *Aedes* by Roth (1966), in an opilionid by Wettstein and Sotelo (1965), and in hemipteran insects by Maillet and Folliott (1965) and Barker and Riess (1966). On the other hand, Moens (1968b, 1969) described the synaptinemal complexes of grasshoppers (*Chorthippus longicornis*) as being released into the cytoplasm at first anaphase and then aggregating into polycomplexes. If the lateral elements of the synaptinemal complexes are really part of, and integrated with the synapsed chromosomes it is difficult to see how the complexes could persist as such after the beginning of diplotene, when the synapsed chromosomes separate from one another except where they are held together by chiasmata.

Structures similar to polycomplexes have been observed in certain spermatids and Moens (1968) states that the polycomplexes in grasshopper spermatogenesis persist through the second division into the cytoplasm of the spermatid.

The synaptinemal complex is not ordinarily visible by conventional light microscopy, presumably because it is dispersed by the usual cytological fixatives. But Gassner (1969) has demonstrated what appear to be synaptinemal complexes (or at any rate remnants of synaptinemal complexes) by phase-contrast microscopy, between the separating homologues of the mantid *Bolbe nigra* at first anaphase. The material in this case had been treated as for electron microscopy, with glutaraldehyde fixation and embedding in epoxy resin.

Meyer (1964) has shown that synaptinemal complexes are not present in the meiosis of Diptera which do not form chiasmata. They are not seen in the spermatogenesis of *Drosophila*, *Tipula caesia* and *Phryne fenestralis* but are visible in the chiasmatic meiosis of the male *Tipula oleracea*. They occur in the normal oogenesis of *Drosophila* but not in that of females homozygous for Gowen's gene (C3G) which suppresses synapsis and crossing-over. In heterozygous females both synapsis and the occurrence of synaptinemal complexes is much reduced.

This correlation between achiasmatic meiosis and absence of the synaptinemal complex does not seem to hold outside the Diptera, however. Gassner (1967) has shown that synaptinemal complexes are present in the spermatogenesis of the scorpion fly *Panorpa*, in which chiasmata are absent (Ullerich 1961). And in two species of mantids with achiasmatic spermatogenesis, *Bolbe nigra* and *Holaptilon pusillulum* synaptinemal complexes are clearly seen (Gassner 1969; J. Wahrman, pers. comm.). Also, according to Meyer (quoted by Moses 1968) they are present in both oogenesis and spermatogenesis of the Lepidoptera *Pieris*, *Ephestia* and *Galleria*, although chiasmata are believed to be lacking in the female meiosis. Menzel and Price (1966) have shown that in F_1 hybrids between the plants *Lycopersicon esculentum* (tomato) and *Solanum lycopersicoides*, synapsis occurs between the chromosomes of the parent species and synaptinemal complexes are present, but no chiasmata are formed. Thus the formation of synaptinemal complexes appears to be a necessary but not a sufficient precondition for genetic crossing-over. Heat treatments which interfere with chiasma formation in grasshoppers (Peacock 1970) do not prevent synapsis or the formation of the synaptinemal complex.

A theory of the mechanism of synapsis based on the role of the synaptinemal complex has been put forward by Comings and Okada (1970a, b). These workers have used whole-mount electron microscopy to study the complexes in mouse and quail spermatogenesis. They were shown to be resistant to digestion by deoxyribonuclease (which removes the chromatin fibres) and ribonuclease, but were destroyed by treatment with trypsin or urea. These results conclusively establish the protein nature of the synaptinemal complex, including both the central and the lateral elements, which latter were shown to contain two fibres, presumably one for each chromatid. The space between the central element and the lateral elements is traversed by protein filaments said to number 1000 or 2000 per *Drosophila* bivalent (Meyer 1964) and assumed by Comings and Okada to be loops.

Comings and Okada postulate that in the formation of the synaptinemal complex the lateral elements synthesize loops of protein, in a regular temporal sequence, from one end of the chromosome and that these come together to form the central element. Thus the chromosomes are envisaged as being zipped together because the corresponding elements of the zipper are formed simultaneously – one loop 5 synapses with the other loop 5 because the two no. 4 loops are already synapsed and the no. 6 loops are not yet formed. Apart from this specificity of time of synthesis, it is not assumed that there is anything specific about the individual loops. Comings and Okada have published a diagram showing how their model could account for reversed loop pairing in the case of an individual heterozygous for an inversion.

The main objection to this model, however, arises when we consider the situation in individuals homozygous for chromosomal rearrangements, and especially ones homozygous for numerous structural changes. In many instances the karyotype of species *B* is homozygous for a large number of inversions, translocations, deletions and duplications that have occurred since its evolutionary separation from species *A* or from a common ancestor. Consequently, if the temporal order of formation of a linear sequence of successive synaptinemal loops is 1, 2, 3 . . . in species *A* or in the common ancestor, it will be essentially chaotic in species *B*. Thus this particular feature of the Comings–Okada model of synapsis requires the subsidiary assumptions of a special evolution of the time sequence of synaptinemal loop formation – an evolutionary process for which there is absolutely no evidence.

Although, for the reasons stated above, we find it necessary to reject the Comings–Okada model of the mechanism of synapsis there can hardly be any doubt that the synaptinemal complex does play an essential and important role in synapsis. In triploid pollen mother cells of *Lilium tigrinum* no more than two chromosomes become synapsed at any one level; the trivalent has a single synaptinemal complex which is continuous through a change of partners (Moens 1968*a*). And in triploid *Drosophila* females only the two synapsed chromosomes have a synaptinemal complex, the third, unpaired, one not showing one (Meyer 1964). It is thus quite conceivable that the restriction of meiotic synapsis to pairs of chromosomes (Darlington 1930) depends on an inability of a chromosome to form synaptinemal complexes with more than one other chromosome, at any particular locus.

CHIASMA FORMATION

It will be realized that the pachytene bivalents must consist of four parallel chromatids; that is, however, not how they appear under the microscope. The two chromatids of each chromosome are so closely associated (being presumably spiralized jointly) that no split is visible and the bivalent appears to consist of only two parallel strands. Darlington (1935*b*, 1939*b*, 1940*b*) claimed that the two chromosomes are spirally wound round one another at this stage, and attached considerable importance to this. Although 're-lational' coiling of this kind is undoubtedly present in many organisms at pachytene, it is certainly not universal.

At the end of pachytene two striking visible changes occur, either simultaneously or in very rapid succession: (1) the appearance of a visible split between the chromatids of each homologue, (2) a separation begins to take place between the two homologues. These changes do not take place quite simultaneously in all the bivalents of a nucleus, or in all regions of the same chromosome, but they quickly spread over the entire nucleus to produce the diplotene stage. If (1) precedes (2) we have a very short distinct four-stranded pachytene period, the 'tetramite' stage of McClung (1941), when there are two splits in the pachytene bivalent in planes perpendicular to one another. As the transition to diplotene proceeds one of the two splits becomes much wider, while the other remains unchanged. It has been generally assumed that the 'opening-out' of the bivalents is due to a sudden replacement of the synaptic attraction by a force of repulsion as the homologues become visibly split into chromatids. It may, however, be due to the shedding or dissolution of the synaptinemal complex.

This 'opening-out' of the bivalents would cause them to fall asunder into separate chromosomes, were it not for the fact that the four strands remain held together at certain points where two of them cross over reciprocally from one chromosome to the other, forming an X. These chromatid exchanges are known as *chiasmata* (singular : *chiasma*) and in normal meiosis there is always at least one in each bivalent, and there may be several.

Whereas diplotene is usually a relatively short stage in spermatogenesis, in oogenesis it may be extremely prolonged. Thus in urodele oocytes it may last up to six months (see p. 129), while in the human ovary the oocyte nuclei remain in diplotene from the time of birth onwards (Ohno, Makino, Kaplan and Kinosita 1961).

Obviously, there are two topological possibilities with regard to chiasmata. The *classical* theory of some early cytogeneticists assumed that they were due to the 'opening-out' occurring in the plane of the primary split (i.e. the one between the chromosomes) on one side of the chiasma and in that of the secondary split (the one between the chromatids) on the other side. The *partial chiasmatype* interpretation of Darlington (1930), based on earlier hypotheses of Janssens (1909, 1924), assumed that each chiasma is a place where two out of the four chromatids, one derived from the 'paternal' chromosome and one from the 'maternal' one, have undergone genetic crossing-over, i.e. where they have broken at exactly corresponding loci and re-fused reciprocally. Thus according to the chiasmatype theory the 'opening-out' is in the primary or *reductional* plane on *both* sides of the chiasma and each visible chiasma corresponds to a genetic cross-over which has occurred at some earlier stage (on the classical theory it was assumed that crossing-over took place at some stage between diplotene and first anaphase, without there being necessarily a one-to-one relationship between chiasmata and genetic cross-overs).

Several proofs of the partial chiasmatype interpretation have been put forward. A convincing genetic demonstration is possible whenever two chromosomes with appropriate genetic markers have also cytological markers at each end (e.g. length differences). Such evidence was provided by Stern (1931) in the case of *Drosophila melanogaster* : an X-chromosome lacking the distal end and marked with the mutations *carnation* and *Bar* underwent crossing-over with an X carrying the wild type alleles and having a piece of Y-chromosome translocated to its proximal end. The proof was obtained when it was found that there was a correspondence between phenotype and karyotype in the offspring. An exactly similar demonstration of the correctness of the chiasmatype theory was obtained by Brink and Cooper (1935) in the case of a maize chromosome.

Topological proofs of the partial chiasmatype theory which depend on cytological observation rather than genetical experiment were put forward by Darlington (1930), Mather (1933a, 1935a, b, 1938) and others. Mather (1935a) described cases in lilies essentially similar to the *Drosophila* and maize examples cited above – a very short chromosome is synapsed with an interstitial segment of a much longer one; at diplotene–diakinesis it can be seen that chromatids of identical length are paired on *both* sides of the chiasma.

Darlington's (1930) proof is based on observations of trivalents of the type shown in Fig. 6.4 in triploid or tetraploid plants; it depends on acceptance of the proposition that synapsis of three chromosomes in the same segment is impossible (a proposition which seems to be generally true, although probably not in the case of the sex chromosomes of *Rhodomantis* – see p. 650). If Darlington's argument is accepted, trivalents of the kind illustrated in Fig. 6.4 would, on the chiasmatype theory, be consistent with two-by-two pairing at zygotene–pachytene, whereas the classical interpretation would involve the assumption that one chiasma includes chromatids derived from three different chromosomes. Trivalents showing the critical configuration have been observed repeatedly in triploid hyacinths and lilies.

An alternative proof (Mather 1933a) depends on the observation of a 'double interlocking' between two bivalents, a rare accident of meiosis

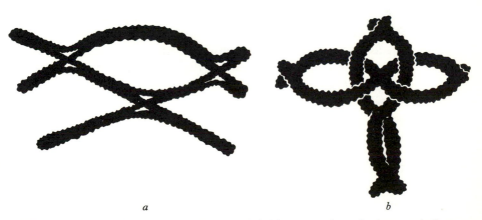

a   *b*

Fig. 6.4. Two topological proofs of the partial chiasmatype hypothesis. *a*, a trivalent of a type that would be impossible on the classical interpretation of chiasmata; *b*, two 'doubly interlocked' bivalents. Further explanation in the text. Based on Darlington (1930) and Mather (1938).

which has actually been seen in only three cells, two in *Lilium* and one in *Eremurus*. Here again the classical interpretation leads to a *reductio ad absurdum*, namely 'identical or sister chromatids on opposite sides of the intruded arm of the second bivalent in one loop of the first bivalent' (Mather 1938). This proof would be absolutely convincing if only the critical configuration had been observed more often.

A very conclusive proof of the partial chiasmatype interpretation has been provided by the work of Kayano (1960*a*, *b*) on translocation heterozygotes in the liliaceous plant *Disporum sessile* (see also Lewis and John 1963, pp. 122–7).

On the chiasmatype theory it is possible to calculate 'map distances' from chiasma frequencies and vice versa, a bivalent with a chiasma frequency of 1.0 (i.e. two single cross-over chromatids and two non-cross-over ones) having a map length of fifty units (50% crossing-over). Unfortunately the only higher organisms for which really adequate genetic maps are available are species of *Drosophila*, in which cytological determinations of chiasma frequencies are hardly feasible, on account of observational difficulties. In maize, Darlington (1934*b*) found that the combined chiasma frequency of the ten bivalents was about 27.0. This would indicate a total genetic length of 27 × 50 or 1350 cross-over units. The latest linkage maps (Neuffer, Jones and Zuber 1968; Vetturini 1968) add up to a total of about 1100 units. Since the end segments of most of the chromosomes are presumably still unmarked by genes which have been used in linkage studies, the data are in good agreement with the chiasmatype theory, and we may anticipate that future genetic work will eventually increase the known map lengths until their combined total approaches closely to the predicted length of 1350 units (unless Darlington's data are inaccurate or not characteristic of all maize varieties). Of course, if in any organism the genetic length of the chromosomes could be shown to be greater than the length calculated from the chiasma frequency, there would be a powerful argument against the chiasmatype theory; but no such case is known.

Various attempts have been made to test the hypothesis of a 1:1 correspondence between chiasmata and genetic cross-overs by comparing the number of visible chiasmata at diplotene and the number of 'switch points' between labelled and unlabelled chromatids at second metaphase following tritiated thymidine autoradiography (Taylor 1965, Moens 1966). The results have been inconclusive because of difficulties in determining the frequency of sister-strand exchanges and particularly because of the possibility that the presence of tritium may increase the frequency of sister-strand exchanges (Olivieri and Brewen 1966; Geard and Peacock 1969; Brewen and Peacock 1969*b*).

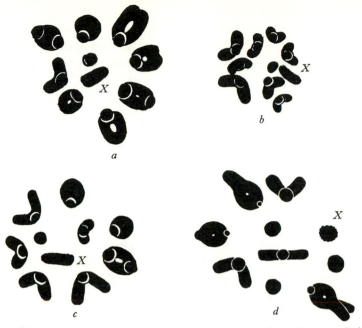

Fig. 6.5. First metaphases (polar views) in four species of grasshoppers belonging to the family Pyrgomorphidae (X-chromosome and nine autosomal bivalents in each case). (a) *Aularches miliaris* (7 ring bivalents with two chiasmata each); (b) *Pyrgomorpha bispinosa* (no ring-bivalents, i.e. each bivalent with only one chiasma); (c) *Chrotogonus* sp., showing 3 ring bivalents; (d) *Psednura pedestris* (two ring bivalents and one with 3 chiasmata). a–c from Rao (1937).

The number of chiasmata per bivalent varies from one to about thirteen. The latter number was observed occasionally by Maeda (1930) in the case of the longest chromosome in the bean *Vicia faba*; such very high numbers of chiasmata are very rare, and do not seem to have been unequivocally demonstrated in higher animals. A general relationship exists between chromosome length and the number of chiasmata formed. Thus very short chromosomes invariably show a single chiasma and never 0 or 2, while medium-sized ones may show 1 to 4, the number varying in homologous bivalents from cell to cell. Direct proportionality between the length of the chromosome and the number of chiasmata formed in it is not invariably found, however, and instances are even known where one or more of the shorter chromosomes of the karyotype exhibit, on the average, more chiasmata than some of the longer chromosomes. The mean number of chiasmata is known as the *chiasma frequency* (of the bivalent, or of the whole karyotype).

The appearance of the bivalents at diplotene naturally depends on the number of chiasmata they contain. Thus homologous bivalents may look quite unlike in different cells because they have different numbers of chiasmata. A bivalent with a single chiasma has the form of a cross, each of the four limbs being composed of two chromatids paired side by side, while one with several chiasmata appears as a series of loops (Fig. 6.1) with an incomplete half-loop at each end.

In early diplotene the successive loops in a bivalent with several chiasmata seem to be in the same plane, or nearly so. But as the chromosomes condense in diplotene and diakinesis the successive loops come to lie in planes that are perpendicular to one another, as in a piece of stretched chain. This process of 'rotation' is a universal phenomenon in meiotic bivalents with a single chiasma, which may be said to have two 'half-loops', one on either side of the chiasma.

As a result of this phenomenon the three-dimensional bivalents have a very regular orientation when they later become attached to the metaphase spindle. Since the two centromeres are attached in the axial or vertical plane, at equal distances 'above' and 'below' the equator, the loop or half-loop in which the centromere is situated is likewise axial. The next loop or half-loop on either side is necessarily horizontal, i.e. in the equatorial plane, and so on, if there are still more chiasmata. Thus acrocentric bivalents with one proximal and one distal chiasma appear as signet-ring structures in the equatorial plane and metacentric bivalents with one chiasma near each end are seen as vertical rings.

Four different relationships may exist between two successive chiasmata in the same bivalent. If we call the chromatids A, A', B and B', then if A and B cross-over at one chiasma, the chromatids which cross-over in the second chiasma may be A and B, A and B', A' and B or A' and B'. These different relationships and the terminology used to designate them are shown in Fig. 6.6. Unfortunately it is only rarely possible to distinguish reciprocal, complementary and diagonal pairs of chiasmata under the microscope, owing to the difficulty of following the course of the chromatids between the successive chiasmata. From a genetical standpoint a pair of reciprocal chiasmata give rise to two double cross-over chromatids and two non-cross-overs ('two-strand double exchange'), a pair of complementary chiasmata produce four single cross-over chromatids ('four-strand double exchange'), while diagonal chiasmata give rise to one double cross-over, two single cross-over and one non-cross-over chromatid ('three-strand double exchange').

The possibility that the four types of relationship between two successive chiasmata might not occur at random has been called *chromatid interference* (Mather 1933*b*). The studies of Emerson and Beadle (1933) on crossing-over in attached X-chromosomes of *Drosophila melanogaster* (isochromosome

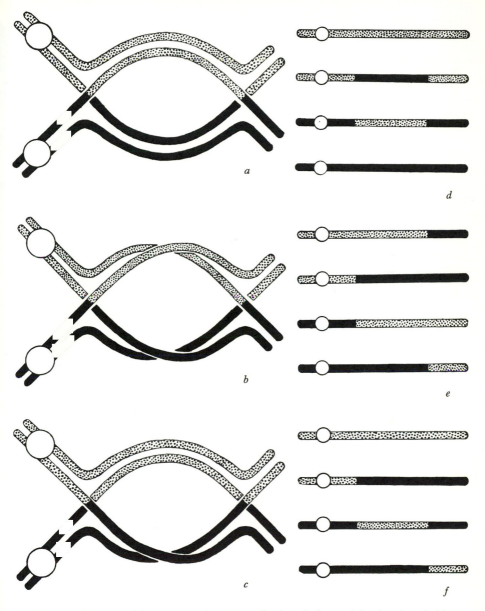

Fig. 6.6. Reciprocal (*a*, *d*), complementary (*b*, *e*) and diagonal (*c*, *f*) pairs of chiasmata. The reciprocal and complementary types of relationship are sometimes described as *compensating*, diagonal pairs of chiasmata being *non-compensating*.

X's which form chiasmata between their two limbs) proved that there was no significant chromatid interference in that organism (Emerson, 1969, after a detailed consideration of all the evidence concludes that 'the available attached X data are insufficient to distinguish between the complete absence of chromatid interference, on the one hand, and strong negative chromatid interference over short map distances, on the other'.). In the plant *Lilium formosanum*, Brown and Zohary (1955) obtained similar results in an experiment making use of inversion heterozygosity. On the basis of this somewhat meagre evidence it has been generally concluded that chromatid interference does not exist in higher animals and plants, whatever the situation in fungi, where the evidence is somewhat contradictory.

Chiasmata, from their very nature, are interstitial structures, i.e. they cannot arise terminally. Thus observations of very early diplotene bivalents almost always reveal the chromosome ends as free structures and the chiasmata as interstitial. Only in the lampbrush chromosomes of *Triturus* spp. (Callan and Lloyd 1960b) and a few other cases do we find that in the case of some bivalents the ends of the two homologues are attached at this stage.

As diplotene and diakinesis proceed, however, there is an undoubted tendency, in many organisms, for chiasmata which were originally close to the chromosome ends to 'slip along' so that they may even reach the end. This is the process of *terminalization* (Darlington 1932b). What moves, of course, is the visible chiasma, not the genetic cross-over. Thus, although we may say that a chiasma is the 'outward and visible sign' that a cross-over has occurred, the two do not necessarily correspond exactly in position, in the diakinesis and metaphase bivalents. Henderson (1969) considers that the terminalization phenomenon is probably less widespread than the earlier workers believed and that 'many cases of presumed terminalization are likely to be examples of distal localization'. But the existence of completely terminalized chiasmata, in many organisms, proves that it is a common and widespread phenomenon.

When a chiasma reaches the end of the chromosome arm, the two homologues do not fall apart but are, on the contrary, firmly held together in an end-to-end association which is called a *terminal chiasma* (Darlington 1932b). In many species of organisms all the chiasmata become completely terminalized by first metaphase. Bivalents with a terminal chiasma in each arm will appear as rings, symmetrical if they are equal-armed metacentrics, more or less D-shaped at first metaphase if they are unequal-armed. Some metacentric and many acrocentric bivalents show a terminal chiasma in one arm only at first metaphase, the other arm lacking a chiasma.

The nature of the association between the chromosome ends which are held together in end-to-end contact has been somewhat controversial. It is presumably due in some way to the special properties of the telomeres. Darlington (1932b) believed that there is a special affinity or attraction between the chromosome ends to which a chiasma terminalizes. This leaves many unanswered questions – for example, why does this attraction not manifest itself in the absence of a chiasma ? Callan (1949b) suggested that the telomeres may not replicate at the same time as the rest of the chromosome and may be still in an unreplicated condition up to the beginning of first anaphase. Taken literally, this suggestion seems biochemically awkward. But there seems no reason to object to some kind of sister-strand union at the chromosome end, between the telomeres of the two chromatids (White 1959b). The fine strands which are usually seen between the ends of chromosomes united by a terminal chiasma, may be taken as evidence of some kind of sister-strand union, no doubt due to a special type of chemical bonding.

Darlington has claimed that several chiasmata may terminalize to the same chromosome end, so that there is an apparent decrease in the chiasma frequency from diplotene to metaphase (counting each terminal association as a single chiasma). But this decrease is not very well-documented; in some of the published data which purport to support it, there is a possibility that the counts of diplotene 'chiasmata' include some overlaps, and are hence in excess of the true number.

On Darlington's hypothesis of the nature of terminal chiasmata, a pair of reciprocal or complementary chiasmata (two-strand or four-strand doubles) can terminalize to the same chromosome end without 'cancelling out' – i.e. when they reach the end a terminal chiasma will still be produced. But if the hypothesis of Callan and White is true, this would not happen. It seems significant that almost all species with extreme chiasma terminalization

a   b

Fig. 6.7. Terminal chiasmata. Maternal chromatids white, paternal ones black. On the interpretation put forward here, all terminal chiasmata are of type a; on Darlington's interpretation a configuration like b can arise through terminalization of a pair of compensating chiasmata to the same chromosome end. From White (1959b).

Fig. 6.8. First metaphase (side view) in a male individual of the grasshopper *Trimerotropis gracilis* which was heterozygous for the *Standard* and *Oreana* sequences of chromosome 7. A rare accident of meiosis has apparently occurred in chromosome 4 – the 'cancelling out' of two reciprocal chiasmata in terminalization, thus leading to precocious separation on the spindle. From White (1959*b*).

are ones in which there is never more than one chiasma per chromosome arm, while conversely, in species with many chiasmata per arm, terminalization is absent or not pronounced. In many species with two or more chiasmata per arm the distal one alone terminalizes, the others remaining interstitial. Clearly, it 'would not do' for the only two chiasmata in a bivalent to cancel out, so that the bivalent fell apart into its constituent chromosomes at or about the premetaphase stage, as this would be liable to lead to non-disjunction. The mechanism of terminalization and of the terminal chiasma must hence have been perfected and moulded into a rigidly canalized system by natural selection.

What happens to 'terminal chiasmata' at first anaphase? One might imagine that it was the mechanical tension imposed by stretching on the spindle which breaks the association. But there are various reasons for believing that this is not so, or at any rate that it is not the whole of the explanation, and that some specific chemical action (a 'telomerase'?) is involved. In the first place, 'terminal chiasmata' are not broken through by the violent stretching that goes on at premetaphase on some organisms such as certain phasmatids, mantids, earthworms, gastropods, etc. (see p. 188). And in the second place, chromosomes which have had a chiasma in one arm only regularly seem to be X-shaped at telophase and in interkinesis, i.e. the chromatid ends which are not stretched on the spindle separate just as readily, and at the same time, as those that are violently pulled apart.

In certain Hemiptera, with holocentric chromosomes, whose bivalent separation at first anaphase is 'reductional' at the end where the chiasma is (see p. 490), the two chromatids of each chromosome are held together end-to-end between the two meiotic divisions by 'half a terminal chiasma'

(Fig. 13.9). In these species the special bonding which we believe exists between sister chromatid ends is clearly *not* broken by chemical action at first anaphase, as it probably is in other cases.

The significance of chiasmata is a two-fold one. On the one hand they contribute to genetic recombination, without which truly progressive evolution is difficult. In the great majority of higher organisms, however, they also perform an essential mechanical function, by holding together the homologues of which the bivalents are composed, from the end of pachytene up to first metaphase. In such organisms a bivalent which forms no chiasma is resolved into two univalents before metaphase, an accident which is liable to lead to non-disjunction and hence to aneuploid gametes. Thus in species with chiasmate meiosis there is a very rigid canalization which ensures that *each* bivalent forms *at least one* chiasma – i.e. that bivalents with no chiasmata do not occur, even though the mean number of chiasmata per bivalent may be 1.0 or only slightly higher. Such canalization must be the result of natural selection which has perfected the meiotic system of each species, from the earliest times.

Credit for realizing the fundamental importance of this mechanical role of the chiasmata must go to Darlington (1932, etc.); however, the situation is not quite as simple as originally supposed by him. In the first place, it is now clear that in the male *Drosophila* and in fact in most or all of the 'higher' Diptera (and also in some 'lower' ones such as the Mycetophilidae, Phryneidae, Bibionidae, Scatopsidae and Blepharoceridae) the male meiosis is achiasmate, the homologues being held together during the critical period from pachytene to the end of first metaphase by some kind of synaptic attraction apparently similar to that which operates in the somatic cells of the Diptera (see p. 68). And in *Drosophila*, although the female bivalents do show chiasmata, these are not so necessary for regular disjunction as in most other organisms, since inversions which suppress them do not lead to anything like the theoretically-anticipated rate of non-disjunction (Cooper 1945).

Achiasmate meiosis, confined to one sex, seems to be of rather widespread sporadic occurrence in the animal kingdom (see p. 476). It is certainly found in a number of forms which do not show any chromosomal pairing in their somatic nuclei; in most cases it can be described as resulting simply from a prolongation of the pairing relationships characteristic of pachytene to a later stage, i.e. until first anaphase.

These facts demonstrate that certain organisms have, as it were, invented alternative mechanical devices which hold the chromosomes together in meiotic bivalents, even when chiasmata are absent. And in some African grasshoppers belonging to the Eumastacidae, subfamily Thericleinae, although chiasmata are present in the male meiosis, they do not become manifest until mid-metaphase or even anaphase (White 1965*a*), so that their mechanical function is probably almost vestigial (two species of Thericleinae appear to have a genuinely achiasmate type of meiosis).

The primary reason for the retention of chiasmata in the meiotic mechanisms of most organisms is therefore the genetic one, i.e. their role in genetic recombination. But if chiasmata are retained because they lead to recombination (a second-order effect), their frequency is canalized to produce regularity of disjunction (a first-order effect).

Some degree of localization of chiasmata is probably universal, i.e. chiasmata are probably never formed entirely at random along the length of the chromosome. But the degree of localization varies enormously; in some species the chiasmata are almost all formed in the proximal regions of the chromosomes, or very close to the distal ends, while in yet others they are formed near both ends of the chromosome but seldom or never in the middle regions. More frequently, perhaps, localization is less pronounced, so that it may be difficult to demonstrate by direct observation, although a statistical analysis might show that it was present to a slight degree.

Localization of chiasmata obviously affects the overall level of genetic recombination. It does so because the closer a chiasma is to a chromosome end, the less recombination it causes; thus species which have pronounced terminal localization of chiasmata will show little recombination. Certain species of grasshoppers with acrocentric chromosomes seem to have gone to extraordinary lengths to render the chiasmata that are formed in spermatogenesis genetically ineffective, either through extreme proximal or extreme distal localization.

Ordinarily, acrocentric chromosomes do not form chiasmata in their tiny 'second arms', but in certain acrocentric bivalents these short arms have a chiasma frequency out of all proportion to their minute size. In extreme cases a large proportion of the chiasmata are confined to the short arms of acrocentric chromosomes, and there are no chiasmata in the main arm, or only a distally localized one. McClung (1928) first recognized bivalents of this type in three species of the grasshopper genus *Stethophyma* (= *Mecostethus*) and called them *ditactic* chromosomes. Such bivalents appeared strikingly different from the usual type at first metaphase, since their long arms were orientated in the axial rather than in the equatorial plane of the spindle. The members of the *pallidipennis* group of the grasshopper genus *Trimerotropis* also show a ditactic bivalent which always forms a proximal chiasma, either in the short arm or the long arm, but never in both. Other striking examples of bivalents with short arm chiasmata occur in the Texas grasshopper *Chloroplus cactocaetes* (Fig. 6.16) and the Australian species

Austroicetes vulgaris (White 1958); in these cases a second, distal chiasma in the long arm is frequently present as well.

John and Hewitt (1966*b*) regard most or all bivalents with what we call short arm chiasmata as having the chiasma located within the centromere region itself. But in all other bivalents regular orientation on the spindle and disjunction at anaphase depends on the two centromeres being situated at equal distances above and below the equatorial plane of the spindle. It is difficult to believe that two centromeres intimately held together by a chiasma (which might presumably be located anywhere within the centromere region) would function as efficiently as they undoubtedly do in 'ditactic' bivalents in ensuring normal orientation and segregation; in other words a chiasma within the centromere would seem incompatible with the regular syntelic orientation (Dietz 1969) which undoubtedly occurs in ditactic bivalents (White 1949*a*, 1958). Shaw (1970) has claimed that the ditactic bivalents of *Stethophyma* lack a chiasma altogether; he also believes the chromosomes to be strictly telocentric in this genus. Both conclusions seem to arise from an inability to appreciate the fact that forms of structure and organization which are not revealed by ordinary light microscopy after the usual cytological techniques, may exist in chromosomes.

In cases such as these the mechanical function of the chiasmata is, of course, unimpaired. But their genetical function is clearly minimal. It has been argued by a number of cytogeneticists that such rigid systems of chiasma localization must be the result of natural selection for close linkage. In fact, there exists a whole theoretical literature about this supposed process of selection for close linkage, where epistatic interactions exist between polymorphisms on the same chromosome (Fisher 1930; Darlington 1939*a*; Darlington and Mather 1949; Sheppard 1953, 1955; Kimura 1956*a*; Bodmer and Parsons 1962).

In a careful laboratory experiment involving the *black–purple–vestigial* segment close to the centromere of chromosome 2 of *Drosophila melanogaster*, Parsons (1958) obtained a significant increase in recombination in the black–purple interval after nine generations of selection. Because of the breeding system used, there was a concomitant decrease in recombination in the purple–vestigial segment, so that there was little or no change in recombination over the whole black–vestigial interval, i.e. selection had probably modified the distribution of crossovers. Acton (1961) was unsuccessful in reducing the amount of crossing-over between *cinnabar* and *vestigial* even after twenty-three generations of selection.

Thus it is possible that natural selection for close linkage may have been operative in some instances. But it is very doubtful whether the cytologically **known** examples of rigid chiasma localization furnish evidence for such a process, since in most of these species localization seems to be confined to one sex, the other exhibiting little or no localization, or even an opposite type of localization. Thus Watson and Callan (1963) have shown that in the newt *Triturus helveticus* there are about twenty-two distally localized chiasmata in the male, while in the female there are approximately twenty-five chiasmata which are not restricted to any particular regions of the chromosomes. In *T. cristatus* the relationship is reversed. Thus in *T. helveticus* there is far more genetic recombination in the female than in the male, while in *T. cristatus* most of the recombination occurs in spermato-genesis. In two species of the urodele genus *Euproctus*, Mancino (1965) and Mancino and Barsacchi (1966) have shown that males have about twenty-four distally localized chiasmata, while females show 28 to 29.5 proximal and interstitial chiasmata. The Australian grasshopper *Keyacris scurra* does not have extreme distal localization in the male, but there is a strong tendency for the chiasmata in the large metacentric '*AB*' bivalent and in the '*CD*' bivalent, when heterozygous for an inversion, to be distally located. In the female, on the other hand, the chiasmata in these bivalents are situated more proximally. The European grasshopper *Stethophyma* (*Mecostethus*) *grossa*, with all its chromosomes acrocentric, is a classic example of extreme proximal localization of chiasmata in the male; but no such localization is evident in the oocyte, many of the chiasmata being interstitial or distal. The same lack of correlation between the patterns of chiasma distribution in the two sexes was found by Fogwill (1958) in the plants *Fritillaria meleagris* and *Lilium* spp. As she concluded: 'There are two systems of recombination, one on the male and the other on the female side within each species.'

This is not the kind of situation that would be produced by selection for tight linkage *per se*. On the other hand, we are not suggesting that 'unisexual localization', where it occurs, has not been evolved as a result of natural selection. It is such a rigidly canalized mechanism that it clearly has been moulded by intense selective pressures. But what type of pressures would be capable of producing quite different patterns of chiasma distribution in the two sexes? Clearly, selection pressures related to the efficiency of the meiotic disjunction mechanism, which is necessarily somewhat different in the oocyte and the spermatocyte, due to differences in the dimensions of the spindle and in the properties of the cytoplasm surrounding it. It is quite possible that if, in a species such as the grasshopper *Stethophyma grossa*, the oocyte bivalents showed the same proximal chiasma localization as occurs in the spermatocytes much non-disjunction would result. Thus we conclude that most cases of strict chiasma localization result from the operation of natural selection in respect of the mechanical functions of the chiasmata rather than their genetic functions.

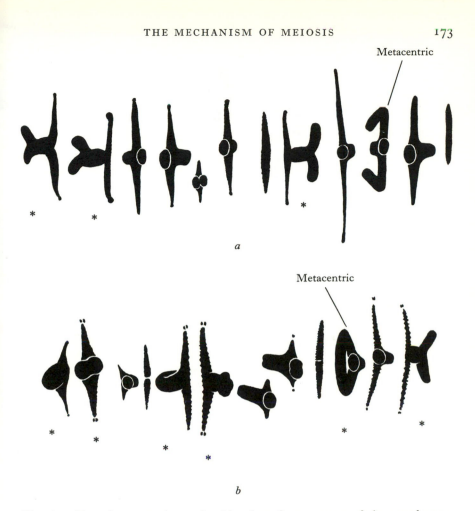

Fig. 6.9. Two first metaphases, in side view, from oocytes of the grasshopper *Austroicetes pusilla*. One bivalent is metacentric, the remaining eleven acrocentric. Bivalents with two chiasmata indicated by an asterisk; the remainder have one chiasma each. Compare the appearance of the proximal ends of the chromosomes in *a* and *b*, and the clear doubleness of the short arms in *b* (this is evidence that they are not centromeres). This is a species in which the distribution of the chiasmata is approximately the same in both sexes.

An interesting comparison between the chiasma frequencies of spermatocytes and oocytes in the planarian *Dendrocoelum lacteum* was carried out by Borragán and Callan (1952). Since this animal is a hermaphrodite, they were able to compare the chiasma frequencies in oogenesis and spermatogenesis of the same individual. It was found that the chiasma frequency of the oocytes was consistently higher, but there was considerable inter-individual

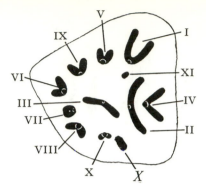

Fig. 6.10. First metaphase (polar view) in spermatogenesis of the grasshopper *Stethophyma grossa*, with extreme proximal localization of chiasmata. All the chromosomes are acrocentric. Each bivalent except no. VII shows a single chiasma quite close to the centromere. VII has two chiasmata in this cell and consequently forms a small ring bivalent.

variation and a strong correlation between the values found in the spermatocytes and oocytes of the same individual.

Where chiasmata are distally localized, i.e. where they are all remote from the centromeres, the bivalents will be elongated in an axial direction on the first metaphase spindle, and will not project far into the cytoplasm. On the other hand, when they are proximally located the bivalents will extend out from the equatorial plane laterally, into the cytoplasm. In general, distal localization of chiasmata should tend to facilitate anaphase separation. But if the most proximal chiasma is abnormally far from the centromeres we may, under certain circumstances, get failure of centric coordination, i.e. the two centromeres of the bivalent may arrange themselves amphitelically on the equator, instead of equidistant from it (White 1954, fig. 64). All these considerations help to understand the different patterns of chiasma localization in males and females of the same species.

Discussion of the actual mechanism of chiasma localization is unsatisfactory, in the absence of a completely acceptable explanation of the molecular basis of genetic crossing-over. Darlington (1935*b*, 1937*a, b*, 1958) claimed that localization was simply due to incomplete or intermittent pairing; White (1954) has questioned this interpretation as far as the grasshopper *Stethophyma grossa*, with extreme proximal localization of chiasmata, is concerned. Henderson (1969) has described observations on two other species *Chrysochraon dispar* and *Parapleurus alliaceus*, which in his opinion support Darlington's interpretation. But his illustrations of incomplete synapsis are all of early diplotene, where such incompleteness is to be expected

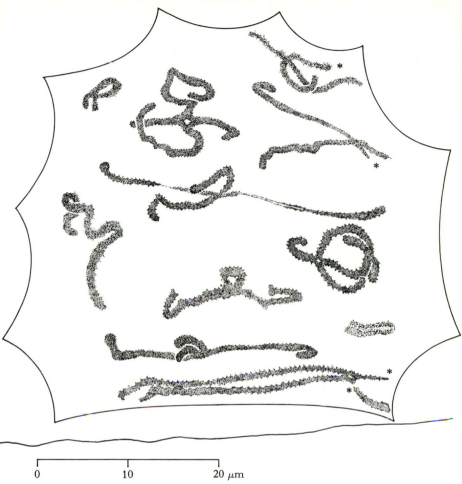

0   10   20 μm

Fig. 6.11. A diplotene nucleus from a sectioned oocyte of the grasshopper *Stetho-phyma grossa*; the nucleus is somewhat collapsed (as can be seen from its irregular outline) and the chromosomes are broken or cut at some points (indicated by asterisks); but it can be clearly seen that many of the chiasmata are interstitial.

anyhow, owing to slight asynchrony in the 'opening out' of the different bivalents. He makes the surprising statement that the pachytene stage does not occur in the spermatogenesis of these species, and regards zygotene nuclei as 'irresolvable'. The whole matter deserves a critical re-investigation. Chiasma localization is such a widespread and, indeed, universal phenomenon that it would be surprising if it were always due to incomplete synapsis. On

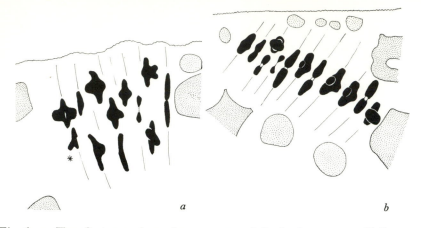

Fig. 6.12. Two first metaphases from oocytes of *Stethophyma grossa*. Yolk masses stippled. There are many interstitial chiasmata, so that the bivalents appear quite different to the corresponding bodies in spermatogenesis (compare Fig. 6.10).

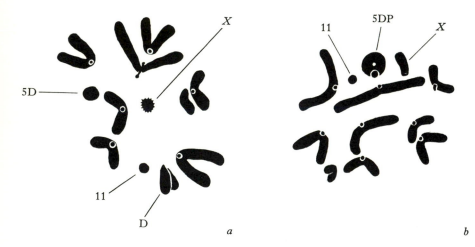

Fig. 6.13. Two first metaphases (polar view) of the grasshopper *Bryodema tuberculatum* (male), showing strong proximal localization of chiasmata. In *a* chromosome 5 has formed a distal chiasma, while in *b* it has formed both distal and proximal chiasmata. The bivalent labelled D in *a* is a 'ditactic', i.e. it has only a short-arm chiasma. The smallest chromosome (11) normally forms a distal chiasma in this species; the other bivalents (with the exception of no. 5) normally form a single chiasma very close to the centromere. From White (1954).

the other hand, in organisms like the grasshoppers of the genera *Stetho-phyma* and *Bryodema* (White 1954) with an extreme development of chiasma localization in the male (we do not know the situation in female *Bryodema*) it is certainly conceivable that there could have been an evolutionary loss of synapsis in segments where it was of no genetic significance.

It was long ago suggested by White (1945) that where chiasma localization is extreme the chiasmata are usually formed in the immediate neighbourhood of transition points between heterochromatin and euchromatin; and this was used as evidence for what must now be regarded as a rather primitive hypothesis of crossing-over, according to which the breaks were supposed to be caused by these 'frontiers'. More recently, Rees and Evans (1966) have shown that in the plant *Scilla campanulata* the chromosomal regions having the highest chiasma frequencies are late-replicating, and have suggested a causal connection.

It is clear that chiasma-formation normally involves two non-sister chromatids out of the four strands. May not crossing-over also occur between

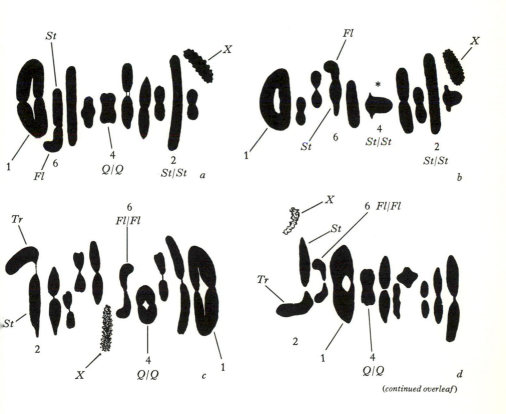

(*continued overleaf*)

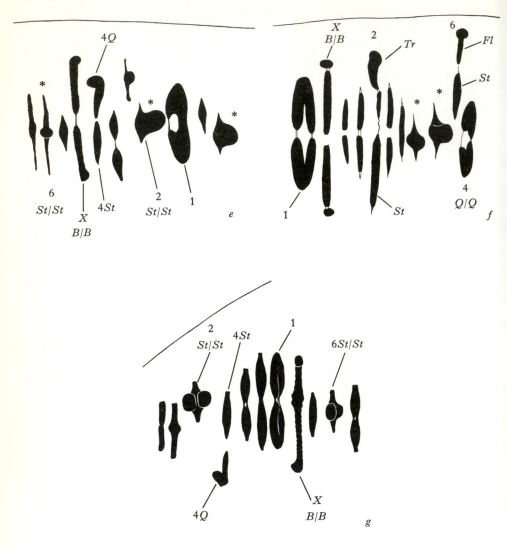

Fig. 6.14. First metaphases (side views) in spermatogenesis and oogenesis of the grasshopper *Austroicetes interioris*, an example of distal chiasma localization in both sexes. *a–d*, in spermatogenesis, *e–g* in oogenesis. $2n\male = 21$, $2n\female = 22$. The largest autosome (1) is invariably metacentric. The species is polymorphic for pericentric inversions; chromosome 2 may be *Standard* (*St*) or *Trangie* (*Tr*), chromosome 4 may be *Standard* or *Quorn* (*Q*), chromosome 6 may be *Standard* or *Flinders* (*Fl*). The *X*-chromosome may be *Standard* or *Buronga* (*B*). In each case the standard sequence is acrocentric. The genetic constitution of the seven individuals illustrated is indicated. Bivalents marked with an asterisk have a proximal as well as a distal chiasma. *a–d*, from White and Key (1957).

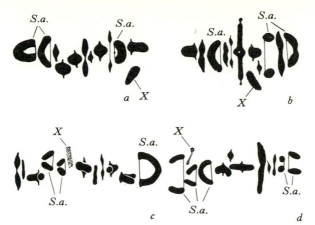

Fig. 6.15. Four first metaphases, in side view, from spermatocytes of the grass-hopper *Austroicetes vulgaris*. Note bivalents with short arm chiasmata (*S.a.*); some of the bivalents also show a distal chiasma in the main arm, others do not. Compare with other species of the same genus (Figs. 6.9 and 6.14). From White (1958).

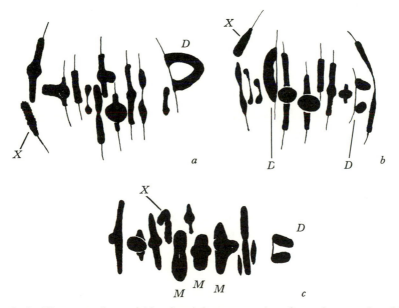

Fig. 6.16. First metaphases (side views) in two species of grasshoppers in which chiasmata are regularly formed in the short arms of some of the bivalents. Such 'ditactic' bivalents are designated *D* in the figure; some of them have a second chiasma in the long arm, while others do not. *a, b, Chloroplus cactocaetes; c, Tri-merotropis pallidipennis*, a species in which there are three large metacentric bivalents (*M*).

sister-strands? This would not lead in the ordinary way to genetic recombination or to cytologically visible chiasmata, but might be detectable in ring chromosomes.

In maize, Schwartz (1953, 1955) studied first and second meiotic divisions in heterozygotes for a normal and a ring-shaped chromosome 6. In the absence of sister-strand crossing-over and chromatid interference (believed not to occur in maize) there should have been equal numbers of double bridge chromatid configurations at first anaphase and single bridge configurations at second anaphase, whereas the numbers actually observed were 81 out of 620 (13%) and 166 out of 475 (35%). Single bridges at first anaphase should have been three times as frequent as double bridges at first anaphase but were somewhat more than four times as frequent (368/620 compared with 81/620). Finally, 10% double bridges were observed at second anaphase compared with an expectation of 0% from single and double cross-overs and only 6.25% from triple cross-overs.

Schwartz explained these results by assuming the occurrence of sister-strand crossing-over. In the case of *Drosophila* ring *X* chromosomes, however, there is no evidence for meiotic sister-strand crossing-over. But since occasional sister-strand crossing undoubtedly takes place in ring chromosomes at mitosis, its occurrence at meiosis cannot be conclusively ruled out (see p. 19).

(see p. 19)

CHIASMA INTERFERENCE

The early *Drosophila* geneticists discovered that the occurrence of a cross-over in a particular chromosome segment sharply decreases the probability of another cross-over occurring in its immediate neighbourhood. This is the now well-known phenomenon of *chiasma interference*, which has been found in all organisms in which it has been looked for. Haldane (1931) pointed out that if there were no chiasma interference each pair of chromosomes should form bivalents with 0, 1, 2 ... chiasmata with frequencies governed by the Poisson distribution:

$$\frac{e^{-m} m^x}{x!}$$

In actual fact the observed distributions are always much narrower than those calculated from the Poisson formula, and the extent of the deviation between the observed and the calculated Poisson distributions can be taken as an approximate measure of the 'strength' of chiasma interference. Many chromosomes invariably form a single chiasma, never 0 or 2.

In *Drosophila* (Mather 1936, 1938; W. L. Stevens 1936) and maize (Rhoades 1941) genetical evidence suggests that there is no interference across the centromere region in metacentric chromosomes. Cytological evidence points to the same conclusion in such plants as *Fritillaria chitralensis* (Bennett 1938) and *Uvularia perfoliata* (H. N. Barber 1941). But there are numerous species in which cytological evidence which is highly significant statistically exists which seems to show that interference *does* extend over the centromere region. This is so in the Diptera *Culex pipiens* (Pätau 1941; Callan and Montalenti 1947) and *Dicranomyia trinotata* (Pätau 1941). Interference across the centromere region also seems to exist in the crustacean *Asellus aquaticus* (Vitagliano 1947) and in the earwig *Forficula auricularia* (Callan 1949a). However, Owen (1949) has pointed out a logical fallacy in this argument, namely it rests on the assumption that all chiasmata have an equal initial probability of formation. He has suggested that if the first chiasma is essential on account of its mechanical function in holding the chromosomes together in a bivalent, the probability of this chiasma being formed may have reached a high value (close to 1.0) as a result of selection. If so, the chiasma in the other arm may have a much lower probability of formation, so giving a spurious appearance of interference. Owen's hypothesis of a 'primary' chiasma with a high probability of occurrence and 'secondary' chiasmata with a much lower probability of formation is reasonable in a formal statistical sense, but seems plausible biologically only if the primary and secondary chiasmata are localized in different chromosome arms, a restriction which he did not make. Thus Owen's objection probably does not apply in the case of the 'ditactic' bivalents (i.e. acrocentric bivalents with a chiasma in the short arm) in certain grasshoppers such as *Trimerotropis pallidipennis*, *Chloroplus cactocaetes* and *Austroicetes vulgaris* (White 1949, 1958); statistical data on the actual frequencies of chiasmata in the long and short arms of individual chromosomes have not been compiled in these species, but one certainly has the impression that these cases are not to be explained on the basis suggested by Owen.

THE MOLECULAR MECHANISM OF CROSSING-OVER

There have been a number of attempts to interpret the actual mechanism of crossing-over, but it is only in recent years that these have been framed in definitely molecular terms. Even the latest models, however (Holliday 1962, 1964; Whitehouse 1963a, b, 1965, 1966; Whitehouse and Hastings 1965) only consider crossing-over in the DNA of the chromosome, the protein components being disregarded. If most models seem somewhat unsatisfactory as complete explanations, it must be remembered that we are dealing with a phenomenon which is not yet open to direct experimental approaches. Genetic experiments (e.g. Catcheside 1968) do, indeed, represent an indirect approach; but it appears that some new types of mechanical, biochemical or biophysical experiments are needed before final answers to a number of questions can be given.

We may distinguish between two types of theories of crossing-over. According to the first type, crossing-over occurs during or at the time of DNA replication and as a consequence of it. On the second type of hypothesis crossing-over occurs at some time after DNA replication. Westergaard (1964) refers to hypotheses of the first type as 'replication crossing-over' hypotheses. The prototype of such hypothetical mechanisms is the one proposed by Belling (1931, 1933), which may be described as a *copy-choice* or *switch synthesis* hypothesis. Belling attached particular importance to the 'chromomeres' which he had observed in liliaceous plants and which he believed corresponded to 'genes'. His work was, of course, carried out before any modern knowledge of the DNA molecule existed. Belling supposed that chromosomal replication took place at pachytene and that this occurred in two stages: first a replication of the chromomeres (genes) and then the formation of new intergenic connections. At each locus and in

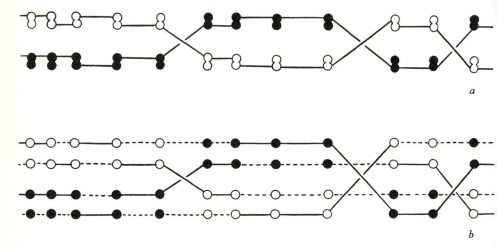

Fig. 6.17. Belling's copy-choice hypothesis of crossing-over. *a*, two-strand stage, in which the chromosomes (but not the chromonema) have replicated; *b*, four-strand stage. Maternal chromosomes white, paternal ones black. 'Old' interchromomere connections represented by solid lines, new connections by dashed ones. The diagram is a two-dimensional representation of what would be a three-dimensional structure.

each homologue the two daughter chromomeres were supposed to be exactly equivalent, and it was supposed to be a matter of chance which of them was connected to the next chromomere on each side. This would be equivalent to a large number of sister-strand cross-overs in addition to cross-overs of the usual type. The latter were supposed by Belling to be due to chance overlaps of the original strands, the new intergenic connections taking the shortest route.

This hypothesis did not seem to be incompatible with any of the data of classical *Drosophila* and maize genetics, including the data on attached -*X* and ring *X*-chromosomes. But neither, as Westergaard (1964) points out, did any of the data of classical genetics *require* acceptance of a replication crossing-over hypothesis. There was never, of course, any evidence for a different time of replication in the chromomeres and intergenic connections. However, all variants of the copy-choice model of crossing-over must now be rejected, since the discovery that both isolated DNA and intact chromosomes replicate semiconservatively rather than conservatively, and that there are hence no 'old' and 'new' chromatids after replication, each chromatid being 'half old' and 'half new' (Taylor, Woods and Hughes 1957; Meselson and Stahl 1958). Also, it is now clear that DNA replication, and probably that of histones as well, precedes synapsis instead of following it.

Models of crossing-over which are not based on the copy-choice principle but involve breakage and rejoining must involve enzymatic action. A minimum of two enzymes would be required, one for breakage and one for re-union. In earlier considerations of crossing-over, it was always difficult to understand why the two strands were broken at exactly corresponding loci. A possible explanation would be that breakage is due to enzyme molecules with two active sites, the distance between which is equal to the distance between the paternal and maternal chromatids which are broken.

Three recent molecular models of crossing-over will be considered here. These are due to Uhl (1965a, b), Whitehouse (1963a, b, 1965, 1966) and Whitehouse and Hastings (1965) and Holliday (1964, 1968). Full discussion of these models will not be attempted, but the salient differences between them will be pointed out.

Uhl's model of crossing-over assumes the existence of linkers (presumably protein) in the chromonema, each linker being bonded on both sides to both strands of the DNA duplex. Meiotic synapsis is assumed to occur after DNA replication but before incorporation of new linkers into the chromatids. Slightly later it is assumed that some new linkers come to reciprocally connect DNA helices from different chromatids, thus leading to cross-overs. The weakest part of this model is the fundamental assumption of the existence of linkers. As pointed out on p. 53, there is no real evidence for the existence of such linkers in chromosomes and if any are present they are almost certainly not protein and may be only a special type of nucleotide sequence in the DNA.

Whitehouse (1963a, b) and Whitehouse and Hastings (1965) have proposed a 'polaron hybrid DNA' model of crossing-over. This assumes breakage of single nucleotide chains of two homologous DNA molecules at the same locus, followed by a re-joining process which involves (1) synthesis of two new strands over a short segment, (2) formation of 'hybrid' (i.e. new–old) strands over this segment, (3) destruction of the old, unpaired nucleotide segments over the same region. In this model the two primary breaks are in chains having opposite polarities. The fact that the two breaks occur at precisely corresponding loci may plausibly be accounted for by supposing that they are due to an endonuclease molecule with two active sites which spans the distance between the two chains. On this model, if a mutant site is included in the hybrid sequence (i.e. if one chromatid contained AT and other CG before recombination) the hybrid strands will each contain two mis-matched identical bases (e.g. TT and CC).

Holliday (1964, 1968) has proposed a rather similar model, in which the primary breaks are in strands having the same polarity. In this case the mis-matched base pairs will be different (e.g. AC and TG). A cross-over can occur without any turnover of DNA, but if the hybrid DNA region includes a heterozygous site (i.e. if there are mis-matched bases) it is assumed that a repair process occurs which involves enzymically controlled excision, single strand degradation, repair synthesis and rejoining of polynucleotide chains.

It does not seem possible to decide at present between these different models of genetic crossing-over or other similar models which can be imagined. Basically, however, the idea of crossing-over as due to the action of endonuclease molecules with two active sites, followed by DNA synthesis and degradation over short regions, is probably sound. The models of Whitehouse and Hastings and Holliday were put forward to account, not only for inter-allelic recombination (crossing-over) but also for intra-allelic recombination not involving crossing-over of flanking markers (the 'gene conversion' phenomenon). The extent to which gene-conversion occurs in organisms above the level of the fungi is still somewhat controversial. The reader is referred to Whitehouse (1970) for an up-to-date discussion of the problems involved.

Whatever the precise mechanism of crossing-over, it is probable that it depends on some kinds of key molecules (enzymes or DNA) which are in short supply in the nucleus. Such a situation would explain the well known Schultz-Redfield effect (Schultz and Redfield 1951; Steinberg 1936, 1937;

A *Uhl*

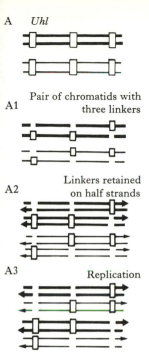

A1 Pair of chromatids with three linkers

A2 Linkers retained on half strands

A3 Replication

A4 Change of partners

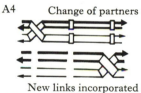

A5 New links incorporated at missing sites
(Those at upper left and lower right give rise to crossovers.)

B *Whitehouse and Hastings*

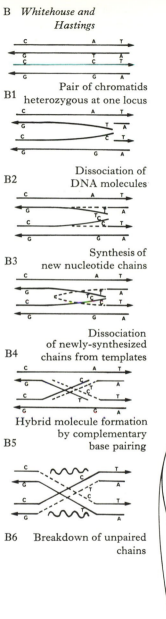

B1 Pair of chromatids heterozygous at one locus

B2 Dissociation of DNA molecules

B3 Synthesis of new nucleotide chains

B4 Dissociation of newly-synthesized chains from templates

B5 Hybrid molecule formation by complementary base pairing

B6 Breakdown of unpaired chains

C *Holliday*

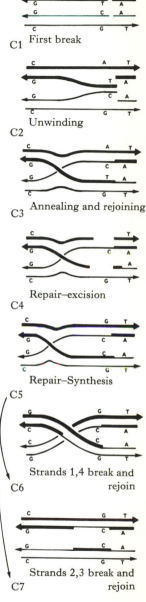

C1 First break

C2 Unwinding

C3 Annealing and rejoining

C4 Repair–excision

C5 Repair–Synthesis

C6 Strands 1,4 break and rejoin

C7 Strands 2,3 break and rejoin

Fig. 6.18. Models of crossing-over of Uhl, Whitehouse and Hastings, and Holliday. Explanation in text.

Steinberg and Fraser 1944; White and Morley 1955) according to which inversions that suppress crossing-over in one bivalent or in a part of it automatically increase the chiasma frequency of other bivalents in the same nucleus, or other sections of the same bivalent (of course some of the suppression of crossing-over by inversions in *Drosophila* is only apparent, due to the elimination of cross-over chromatids in the polar body nuclei, but some of it is entirely genuine). A limited number of endonuclease molecules or of desoxyribonucleotides in the pachytene nucleus would also explain the data of Mather and Lamm (1935) who showed that in rye and *Vicia* the variation of chiasma frequency among cells was less than would be expected on the basis of the chiasma frequencies among bivalents (i.e. that there is a negative correlation between the chiasma frequencies of the different bivalents, a reduction in chiasmata in some bivalents being compensated for by an increased number of chiasmata in others).

To reconcile the facts concerning chiasma localization with the above speculations it seems necessary to suppose that chromosomes possess some kind of specific mechanism for attracting and capturing DNA-breaking endonuclease molecules. Such a mechanism might explain the absence of bivalents with no chiasmata, in organisms with chiasmate meiosis. It seems possible that chiasma interference could also be interpreted in terms of enzyme capturing mechanisms provided that such mechanisms could be either distributed over a considerable length of the chromosome (in cases where there is no strict chiasma localization) or more sharply localized (where extreme chiasma localization occurs).

Grell (1962a, b, 1964a, b, 1965, 1967, 1969) has put forward a novel interpretation of the sequence of meiotic events in the female *Drosophila melanogaster*. She starts with the observed fact that non-random segregation of non-homologues may occur in the case of chromosomes which failed to undergo crossing-over, and postulates two kinds of pairing, one occurring before crossing-over, the other (distributive pairing) following the time of crossing-over. The evidence suggests that 'distributive pairing' is extremely dependent on the size of the chromosomes, with elements of similar size participating preferentially. It is also claimed (Grell 1965) that distributive pairing is the mechanism causing regular segregation of the fourth chromosome, whose cross-over frequency is very low.

Novitski (1964) has suggested a simpler alternative to Grell's distributive pairing hypothesis. He supposes that all the chromosomes are polarized in a bouquet or chromocentral type of orientation prior to synapsis, thus facilitating the synaptic process. Those chromosomes which were unable to successfully take part in bivalent formation (because of absence of a homologue, interference with synapsis by a rearrangement such as an inversion or failure to form a chiasma) would be left behind to take part in heterologous associations after the bivalents had separated from the chromo-central association. Novitski postulates that the production of some inviable zygotes as a result of non-homologous pairing 'is a price that the organism must pay for the use of a chromocentral association as a pre-requisite to normal pairing of homologues'. But Grell insists, in spite of the complete absence of direct cytological evidence, that 'distributive pairing' occurs under normal conditions and is then restricted to homologues and provides a mechanism for ensuring their regular segregation when chiasma formation has failed to occur. Whatever its precise nature, distributive pairing in *Drosophila* is probably associated with the somatic pairing so characteristic of that order of insects, and is unlikely to be important in other groups of organisms.

Long ago Agar (1911) described meiosis in the male of the South American lungfish, interpreting it in terms of a first and second pairing, with an intermediate stage in which de-synapsis occurred. It is fairly clear from an examination of the original figures that the nuclei illustrated by Agar as showing the desynaptic stage were abnormal or pathological and not part of the normal sequence.

DIPLOTENE TO FIRST METAPHASE

There is no fundamental difference between the diplotene and diakinesis stages, and it might be better to replace the latter term by 'late diplotene'. The only universal difference is the greater degree of condensation and contraction of the bivalents in diakinesis. Right up to the end of diakinesis the bivalents retain their 'hairy' appearance, presumably due to large numbers of loops, as in the lampbrush bivalents of the amphibian oocyte. Rotation of the bivalents (see p. 164) and terminalization of chiasmata (if it occurs in the organism concerned) is completed during the diakinesis stage.

The nuclear membrane breaks down at the end of diakinesis and, as at an ordinary somatic mitosis, the spindle develops rapidly during a brief premetaphase stage (in the work of Dietz (1969) on *Tipula* spermatocytes this lasts for about $1\frac{1}{2}$ hours).

Autoradiographic studies with tritiated uridine on grasshoppers by Henderson (1964) and on the cranefly *Pales* by Petzelt (1970) have shown that RNA synthesis is carried out by all uncondensed chromosomes throughout the meiotic prophase, but ceases during the metaphase and anaphase stages of both meiotic divisions, when the chromosomes are condensed. The positively heteropycnotic X-chromosomes of grasshoppers are completely inactive, as far as RNA synthesis is concerned, during the meiotic prophase.

ORIENTATION OF CHROMOSOMES ON THE FIRST METAPHASE SPINDLE

In many species of animals rather violent movements occur at the meiotic premetaphase and the bivalents are violently stretched on the developing spindle and may become extremely drawn out and attenuated. This premetaphase stretch is particularly well shown in certain phasmatids and mantids (Schrader 1944; Hughes-Schrader 1947a); it is also seen in the oogenesis of some gastropods (Staiger 1954), earthworms (Omodeo 1952) and marsupials (McIntosh and Sharman 1953). The premetaphase stretch phenomenon is of very sporadic occurrence throughout the animal kingdom and many groups do not show it to any extent; it does not seem to occur in plants.

Whether or not a distinct premetaphase stretch occurs, the two centromeres of each bivalent orient themselves on the spindle more or less equidistant from one another on opposite sides of the equatorial plane. This is known as *syntelic* orientation and the centromeres are said to be co-orientated. Co-orientation may also occur where there are three centromeres, in a trivalent, or four, in a quadrivalent. The centromeres at the first meiotic division are either actually single and undivided or behave as such, in which case they may be described as functionally undivided.

Time-lapse cinematography has been employed by Nicklas (1961a, 1963, 1967), Nicklas and Staehly (1967), Nicklas and Koch (1969), Henderson and Koch (1970) and Henderson, Nicklas and Koch (1970) to study the mechanisms of orientation of grasshopper chromosomes at meiosis. Bauer, Dietz and Röbbelen (1961) and Dietz (1956, 1969) have carried out important studies on the meiotic orientation of chromosomes in craneflies of the genus *Tipula*, Bajer, Hansen-Melander, Melander and Mole-Bajer (1961) have studied meiosis in living snail spermatocytes and Seto, Kezer and Pomerat (1969) have carried out similar investigations of salamander spermatocytes.

Nicklas showed that bivalents or the univalent X-chromosome that have been completely detached from the spindle by micromanipulation and moved to any position in the cell will travel back to the spindle, centromeres foremost, and reattach in a normal position. It is possible to experimentally induce malorientation, i.e. orientation of both half-bivalents towards the same pole; but such bivalents always undergo reorientation before the beginning of first anaphase. Stability of bivalents on the spindle results from the tension imposed by spindle forces towards opposite poles; this tension is lacking when both centromeres are oriented towards the same pole, and reorientation ensues. Nicklas and Koch have shown that one can induce stability of malorientated bivalents by imposing tension on them by micromanipulation. Similar results were obtained by Henderson and Koch, who made use of bivalents that had been experimentally interlocked by micromanipulation.

The behaviour of the chromosomes in cranefly meiosis (Dietz 1956, 1969) differs in detail but not in principle from their orientation in grasshopper spermatocytes. The movements of the centromeres on the spindle are rather unpredictable. The univalent X- and Y-chromosomes in this case show a tendency to orientate amphitelically on or close to the equator and may reverse their orientation several times during premetaphase and metaphase. Eventually, at first anaphase, they travel to opposite poles, but their poleward movement does not begin until some time after that of the centromeres of the autosomal bivalents.

Obviously, bivalents in which the intercentromeric distance is too long for the spindle to be able to impose tension on the system will be unstable and may be unable to achieve a bipolar orientation. In some cases this factor may have prevented extra-long chromosomes, arising by structural rearrangement, from establishing themselves in a natural population.

Univalent X-chromosomes in the spermatocytes of XO grasshoppers attach themselves to the spindle syntelically and then make several pole-to-pole movements, with the centromere leading toward the pole, usually without any pause or reversal of direction (Nicklas 1961a). These trips from pole to pole occur during both premetaphase and metaphase in *Melanoplus differentialis*, but are always completed before anaphase begins. Such reorientations of univalents signify that we cannot state that a given univalent will pass at anaphase to a particular pole because it is closer to it than to the other pole at first metaphase.

Both genetic and cytological evidence indicates that there is in general, no tendency for the 'paternal' or 'maternal' centromeres of the different bivalents to pass to the same or opposite poles of the spindle at the first meiotic division. The different pairs of autosomes thus segregate independently of one another and of the sex chromosomes at anaphase (Carothers 1913, 1917). In *Drosophila melanogaster*, with four bivalents, all the 'paternal' centromeres will pass to the same pole (and all the 'maternal' ones to the opposite pole) in one out of every sixteen first meiotic divisions, while in the butterfly *Lysandra atlantica*, with 223 bivalents (DeLesse 1970), this event will happen once in 2^{223} divisions.

Certain exceptions to this independent segregation do undoubtedly occur, however. In the mouse, Michie (1953), Wallace (1953, 1958) and Parsons

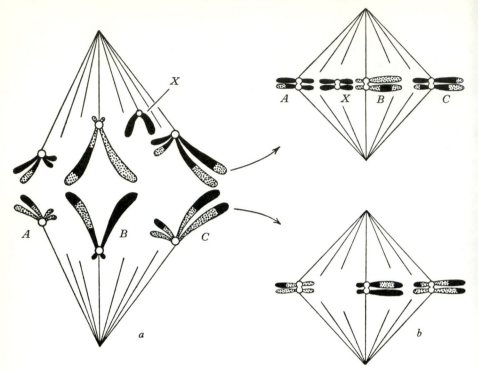

Fig. 6.19. Genetic consequences of meiosis (continuation of Fig. 6.1). Maternal portions of chromosomes black, paternal parts stippled. *a*, first anaphase, *b* second metaphases.

(1959) have presented statistical evidence showing that both in crosses between the 'species' *Mus musculus* and *M. bactrianus* and in laboratory crosses involving only *M. musculus* certain marker genes of similar ancestry but located on different chromosomes tend to segregate together at meiosis. This phenomenon, first termed quasi-linkage, was interpreted as due to a tendency for centromeres of similar ancestry to pass to the same pole at meiosis – a phenomenon which has been called *affinity*. There can be little doubt as to the genuineness of non-random combinations of markers in the offspring of those crosses, but the proposed explanation seems to be open to various criticisms. John and Lewis (1965a) have pointed out that the linkage maps of the mouse are still inadequate to establish close linkage between the markers that have been studied and the centromeres. Wallace (1958) localized the agency responsible for quasi-linkage of chromosome 5 in an interstitial position between the loci of *fi* (fidget) and *Sd* (Danforth's short tail). If so, in view of the strictly acrocentric nature of all the chromosomes in the mouse karyotype, it is most unlikely to be the centromere. Wallace claimed that she had excluded both erratic and systematic viability inter-action as possible causes of the observed ratios. But 'affinity', if defined as a property of centromeres or centromeric regions also seems a very improbable explanation.

A very special case of non-random segregation in the mole-cricket *Gryllo-talpa* (or *Neocurtilla*) *hexadactyla*, in which one member of an unequal bivalent always passes to the same pole as the X, at first anaphase, will be discussed later (p. 385).

The separation of the chromosomes at first anaphase takes place as a result of a poleward movement of the centromeres and an elongation of the equatorial region of the spindle. It has been shown by Dietz that in craneflies the centrosomes are not essential to poleward movement of the centromeres, since even in their absence anaphase separation takes place normally.

During first anaphase the chiasmata are pulled along the chromosomes toward the distal ends; it is as if the process of chiasma terminalization were resumed. The final separation of the bivalents into two chromosomes takes place quite suddenly, as if the terminal associations were instantaneously broken.

MEIOSIS IN POLYPLOIDS

A good deal has been learned about meiotic mechanisms by studying the behaviour of the chromosomes in polyploids (Darlington 1932*b*, 1937*b*, 1939*a*; Mather 1938; Levan 1940; Riley 1958, 1960). In animals there are so few polyploid species that there is not much opportunity for studies of this kind. On the other hand, all animal species seem to show occasional poly-ploid spermatocytes in otherwise normal testes. Such polyploid spermato-cytes are of two different types. On the one hand there are those that have arisen through failure of spermatogonial divisions. Such cells are usually tetraploid; they may occur singly or in groups of two, four, eight or some higher power of two, surrounded by diploid cells. On the other hand, under abnormal conditions which are not understood, several primary spermato-cytes may fuse together, at any stage from leptotene onwards, to give rise to giant polyploid cells which may be tetraploid, hexaploid, octoploid, deca-ploid, etc. John and Lewis (1965*a*, Fig. 117) have illustrated a binucleate grasshopper spermatocyte with two diplotene nuclei which would obviously have fused to give a tetraploid first metaphase.

Since the first type of polyploid spermatocyte is already polyploid before the onset of zygotene there is a possibility that the chromosomes may synapse in threes or fours as well as in twos. In other words, such a cell may show

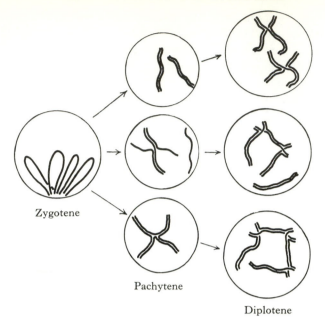

Fig. 6.20. Diagram showing how four homologous chromosomes in a tetraploid may behave in several different ways at meiosis, forming bivalents, trivalents or quadrivalents.

trivalents and quadrivalents as well as bivalents. The second type of cell ordinarily arises after synapsis has been completed, so that no multivalents can be formed in it. An example of the latter type, a decaploid primary spermatocyte in a grasshopper, containing fifty-five bivalents and five separate X-chromosomes, has been illustrated by White (1951*b*, Fig. 2).

Some tetraploid first metaphases from a cyst of 256 such cells in an individual of the long-horned grasshopper *Tettigonia viridissima* are illustrated in Figs. 6.21 and 6.22. In such cells each autosome will be represented four times, while sex chromosomes (X's in this case, X's and Y's in some other organisms) will be represented twice (White 1933; Moffett 1936; Klingstedt 1937*a*; Makino 1939*b*; Hughes-Schrader 1943*a*; Ray Chaudhuri and Bose 1948 and Henderson 1965). Each set of four autosomes can behave in three different ways at zygotene; synapsis may result in (1) a quadrivalent (association of four chromosomes), (2) a trivalent (association of three) and a univalent, or (3) two separate bivalents. Such theoretical possibilities as the formation of a bivalent and two univalents or complete failure of synapsis (four univalents) hardly ever seem to occur in practice. At pachytene chiasmata can then be formed in any of the paired regions; but if they fail to

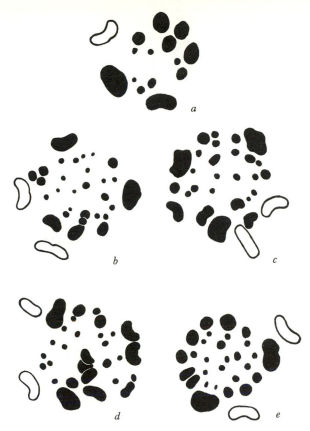

Fig. 6.21. A diploid and four tetraploid first metaphases (in polar view) from a tetraploid cyst in *Tettigonia viridissima*. Not counting *X*-chromosomes (shown in outline) which are always unpaired at this stage, *b* and *d* each show 27 bivalents and 2 univalents, *c* shows one quadrivalent and 26 bivalents, while *e* shows 28 bivalents.

arise between any two chromosomes, the latter will then fall apart. Unfortunately, almost all observations on polyploid spermatocytes have been limited to post-pachytene stages. In such cells the shorter chromosomes are usually present as bivalents. Whether this is due to their having synapsed as bivalents at zygote, or to quadrivalents having fallen apart into bivalents at the beginning of diplotene, can hardly be decided in those cases that have been studied. Obviously, no multivalents can be maintained after pachytene in the case of chromosomes whose physical length is less than the length over which total interference operates, i.e. chromosomes which are incapable of forming more than a single chiasma. Longer chromosomes with higher chiasma frequencies can theoretically form trivalents and quadrivalents, and

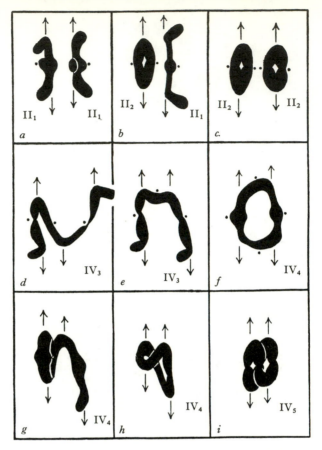

Fig. 6.22. Behaviour of the largest chromosome in nine tetraploid spermatocytes of *Tettigonia viridissima* (first metaphases in side view). In *a, b, c* the four chromosomes have formed two separate bivalents (II, is a bivalent with one chiasma, II_2 one with two chiasmata). In *d–i* the four chromosomes have formed quadrivalents (IV_3, IV_4 and IV_5 being ones with 3, 4 and 5 chiasmata respectively). The arrows indicate the positions of the centromeres and the dots (in *a–f*) those of the chiasmata. *d* and *e* are 'chain quadrivalents', *f* is a 'ring quadrivalent'.

often do so in practice. But it is unlikely that length and chiasma frequency are the only factors determining whether multivalents will be formed. The nature of the bouquet orientation at leptotene and the frequency with which synapsis is initiated at interstitial points in ordinary diploid cells are likely to be critical factors. Henderson (1969, Fig. 1) has illustrated a pachytene quadrivalent in the locust *Schistocerca gregaria* which shows so many changes of partner that synapsis must have been initiated at six different points. The

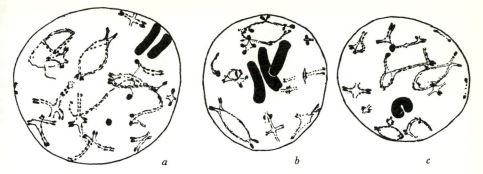

Fig. 6.23. Tetraploid spermatocytes of the locust *Schistocerca gregaria* in early diplotene. *a*, a cell with the full tetraploid number of chromosomes (four sets of autosomes and two *X*'s); *b* and *c* are a pair of cells of which the first contains three *X*'s, the second only one (the result of non-disjunction at the last premeiotic division). Not all the autosomes are drawn – most of them have conspicuous heterochromatic segments at both their proximal and distal ends.

speed at which synapsis spreads along the length of the chromosome may also play some part in determining the frequency of multivalent formation in a polyploid cell (if synapsis were instantaneous throughout the nucleus no multivalents could be formed, while if it were very slow and could be initiated at many points along the length of the chromosomes, every chromosome would be part of a multivalent at pachytene, although some might fall asunder at diplotene, owing to failure of chiasma formation between them).

In the coreid bug *Archimerus* all ordinary autosomes are associated as bivalents in post-pachytene tetraploid spermatocytes (Wilson 1932). The situation seems to be essentially the same in those Orthoptera which have few chiasmata, even in the longer chromosomes (e.g. *Podisma*, studied by Makino 1939*b*). On the other hand, in grasshoppers such as *Schistocerca gregaria* (White 1933) and *Chrysochraon* (Klingstedt 1937*a*) the longer chromosomes form two to four chiasmata in the diploid spermatocytes, so that there is a fair chance that these chromosomes will be represented by multivalents in post-pachytene tetraploid cells, which is what is actually found. *Tettigonia viridissima* occupies an intermediate position; the only chromosome which regularly forms a quadrivalent in a proportion of the tetraploid nuclei is a big metacentric element which has a chiasma frequency of about 1.0 per arm in ordinary diploid spermatocytes (Figs. 6.21 and 6.22).

The behaviour of the sex chromosomes in these tetraploid nuclei is instructive. *Archimerus* and the grasshoppers mentioned above have *XO* males, there being no Y. The X is strongly positively heteropycnotic during the prophases of meiosis, and this heteropycnosis is just as strongly expressed in tetraploid spermatocytes and even in hypo- and hyper-tetraploid nuclei, with one and three X's respectively (Fig. 6.23). Presumably, in all these cases the X's do not have synaptinemal complexes. But at pachytene the X's may be very closely associated, or even apparently fused together; whether this indicates that they have undergone true zygotene synapsis is uncertain, but probably it does not. However, in all cases they fail to form chiasmata and consequently fall apart after the end of pachytene and are present as univalents in later stages. This is, of course, in sharp contrast to what happens in normal diploid oocytes, where the X's are not heteropycnotic and form a true bivalent in which they are held together by chiasmata (Fig. 6.14). Presumably the intense positive heteropycnosis which occurs in the spermatocyte nuclei in some way prevents chiasma formation, perhaps because it interferes with true synapsis or with the formation of a synaptinemal complex or both.

Trivalents and quadrivalents orientate themselves on the spindle at premetaphase in accordance with the general principle that orientations under tension are stable, whereas those that are not are liable to undergo reorientation. This may lead to a statistical excess of final orientations in which all adjacent centromeres are orientated towards opposite poles (alternate type of orientation) as compared with ones in which some adjacent centromeres in the multivalent configuration are orientated towards the same pole.

The polyploid spermatocytes which occur as aberrations in otherwise diploid testes may go on to produce diploid or polyploid sperms, but it is extremely doubtful whether these are ever functional, so that the occurrence of such cells is probably of no significance in the genetic system. Abnormal triploid individuals arising occasionally in normally diploid species (e.g. *Homo sapiens*) are likely to have two maternal chromosome sets and one paternal one; alternatively, if they have two paternal sets, these may have been derived from different sperms.

THE SECOND MEIOTIC DIVISION

The transition from the end of the first meiotic division to the beginning of the second is invariably brief and relatively direct. In most instances a distinct interphase or interkinesis occurs, during which the chromosomes undergo de-condensation. But in some organisms the telophase of the first division seems to be followed immediately by the prophase of the second one. No DNA replication takes place between the two meiotic divisions.

It is usual for the chromatids of each chromosome to diverge widely in interkinesis, being only attached at the centromere; this is not only due to the chiasmata having been pulled along the length of the chromosome during anaphase, since the chromatids diverge widely even in chromosome limbs which have not contained a chiasma, and in the segments between the centromere and the most proximal chiasma.

The second metaphase resembles a normal somatic metaphase except that the chromosomes are present in the haploid number and their constituent chromatids diverge much more widely. The DNA is, of course, present in the 2C amount, i.e. there is half as much as in a somatic or gonial metaphase. The centromeres become orientated amphitelically, on the equatorial plane, but the orientation of the divergent chromatids is often irregular. At the second anaphase the centromeres divide (strictly synchronously, as a rule) and the half-chromosomes pass to the poles.

Such, in outline, is the process of meiosis. Before the general acceptance of the chiasmatype theory there was much discussion as to whether the first or the second division was the 'reductional' one (the other being conceived of as an 'equational' division, similar to an ordinary mitosis). On the chiasmatype theory the disagreement is meaningless if one thinks in terms of whole chromosomes. The concept of a reductional division is one that separates 'paternal' strands from 'maternal' ones. (Whereas an equational division is one that separates a paternal and a maternal strand from a maternal and a paternal one, at the first division – and which separates strands of identical origin at the second division.) Consequently, the first division (according to the chiasmatype theory) is always reductional for segments between the centromere and the most proximal chiasma in each arm. Between the first chiasma and the second one in each arm the first division is always equational, but distal to the second chiasma the situation depends on whether the first two chiasmata are compensating or non-compensating. Regions which separate reductionally at the first division obviously separate equationally at the second one and vice versa.

7

CHROMOSOMAL REARRANGEMENTS –
OCCURRENCE AND MECHANISMS

Nascono, è vero, qualche volta Mostri, qualche volta la
materia guidata da certa necessità si svia, ma questo sviamento
non è senza legge, ed hanno i suoi termini prefissi anche gli
errori.

ANTONIO VALLISNERI: *Considerazioni ed Experienze
Intorno alla Generazione dei Vermi*

Except for evolutionary polyploidy, which is very rare in the animal kingdom
(see pp. 454–63), the visible differences between the karyotypes of different
species and, where these exist, between individuals in a chromosomally
polymorphic species, depend on chromosomal rearrangements that have
occurred in the recent or in the more remote past. Recent work has led to the
conclusion that 'mutations' are of three main types: (1) *point mutations,*
invisible in the polytene chromosomes and involving rearrangements or
substitutions at the level of the individual nucleotide pairs in the DNA
molecule, (2) *gene duplication* of the type studied by Keyl (1963*b*, 1964,
1965*a, b*) in *Chironomus* species, (3) large scale chromosomal rearrangements,
involving microscopically visible segments of the chromosome containing
several or many genetic loci (i.e. in general, thousands or tens of thousands
of nucleotide pairs). It has long been recognized that the first and third of
these categories are different (see especially Muller, 1956); but it is only
recently that the second category has been identified as a kind of mutation
distinct from the other two. The discovery that the satellite DNA's of
closely related species may be very different (see p. 36) forces us to the
view that structural rearrangements involving duplications and deletions of
this material must have occurred on a scale that has been quite unappreciated
in the past. Such rearrangements may, of course, arise through some process
of unequal crossing-over.

The great geneticists of the recent past (e.g. R. A. Fisher, J. B. S. Haldane, H. J. Muller) hardly considered chromosomal rearrangements as an important or significant factor in evolution. For Fisher (1930) 'the evolutionary possibilities of these kinds of change are evidently extremely limited compared to those of the type of change to which the term gene-mutation is applied, . . . only in very special cases could they contribute appreciably to the genetic diversity of an interbreeding population'. Sewall Wright (1941) calculated that selection against translocations must be so severe that they would hardly ever manage to establish themselves. H. J. Muller seems to have regarded inversion polymorphism as a kind of temporary imperfection of the evolutionary process, destined to be soon superseded by improvements based on point mutations. And even Haldane, who probably had a deeper knowledge of all aspects of the evolutionary process than any of his contemporaries, never seems to have quite appreciated the evolutionary role of chromosomal rearrangements, in all its complexity. With this neglect by geneticists, it is hardly surprising that chromosomal rearrangements have been left out of account by many non-genetical evolutionists.

In the next two chapters (8 and 9) we shall consider the evidence which now forces us to take account of chromosomal rearrangements to a much greater extent than did the founders of the synthetic or neo-Darwinian theory of evolution. And in Chapter 11 we shall deal with the special part played by chromosomal rearrangements in speciation phenomena.

In the present chapter we shall be concerned with the third category of genetic changes, namely the large-scale structural rearrangements, whose role in evolutionary processes is much better understood than that of the gene duplications investigated by Keyl. In accordance with the general scope of this book, the molecular mechanisms underlying chromosomal rearrangement will not be discussed in detail, but the various factors which prevent, permit or encourage the survival of rearrangements at the cellular and populational levels will be considered.

Chromosomal rearrangements may occur within single chromosomes or may affect two or more members of the karyotype simultaneously. They seem to depend on two kinds of events: breakage of chromosomes and reunion in a way which is different from the original one. Almost all the evidence favours the view that breakage without reunion (i.e. simple fragmentation) does not give rise to stable alterations of the karyotype and that reunion without previous breakage (i.e. simple fusion) does not occur at all (apparent exceptions to these generalizations and claims to the contrary will be discussed later).

We speak of the chromosomal rearrangements that have occurred in evolution, or which we observe in single (i.e. unique) individuals in natural populations, or in the tissues of chromosomal mosaics, as *spontaneous*. This is simply a confession of ignorance as to their causation. Almost certainly they arise from many causes, acting singly or in combination. There is reason to believe that only a very small fraction of them are caused by ionizing radiation, whether from the earth's crust, cosmic rays or other sources. Many may arise from abnormal biochemical conditions in the cell or as a result of virus infections (Aula, Nichols and Levan 1968). In *Drosophila robusta*, Levitan (1963) and Levitan and Schiller (1963) have studied a maternal factor (presumably cytoplasmic) which breaks paternal chromosomes in the offspring. It is known that when chromosomes are mechanically stretched on the spindle they may be broken, especially at anaphase, and it may be that some breaks are due to this cause, while others may be due to temperature shocks. It is fairly certain that freshly broken chromosome ends are capable of rejoining in a way different from the original one, so as to produce a chromosomal rearrangement. But if the exchange hypothesis of Revell (see below) is correct, this is not the usual way in which radiation-induced rearrangements occur. There is still complete uncertainty as to the actual mechanisms involved in spontaneous chromosomal rearrangement and it is by no means clear how far we can legitimately use the results of experiments in which the chromosomes have been artificially broken in order to analyse the changes in gene sequence which have occurred in the course of evolution and which have produced the karyotypes found in wild individuals, populations and species. The study of 'spontaneous' rearrangements in tissue culture now occupies the attention of many investigators; its much higher frequency in such situations may be taken as an indication that even under the best conditions the life of the cell in tissue culture is biochemically abnormal by comparison with its existence in the intact organism. The actual molecular mechanism of chromosome breakage is not understood, but the fact that it can be produced by mechanical stretching suggests that an enzymatic mechanism is not always involved.

The general principles which govern the rearrangement of chromosome parts were originally worked out on *Drosophila*, maize and the plant *Tradescantia*. Later, extensive data on many other species of animals and plants, including *Vicia*, wheat, grasshoppers and the human species, suggest that these principles apply, with only minor modifications, throughout the animal and plant kingdoms. This is what we should expect, if, as is generally believed, the molecular architecture of the chromosomes is fundamentally the same in all eukaryotic organisms. Some differences undoubtedly occur, however, and, in particular, we must not expect the principles of chromosomal rearrangement to be quite the same in groups with holocentric chromosomes.

When a 'population' of cells (either in culture or in the intact organism) including all stages of the mitotic cycle is irradiated, the first ones to come into metaphase will show so-called sub-chromatid rearrangements (see p. 48). These are ones which were in prophase or premetaphase at the time of the irradiation. The next group of cells to enter on metaphase will show chromatid rearrangements, resulting from breaks and reunions of single chromatids during the S and G_2 phases. The main types of chromatid rearrangements are shown in Fig. 7.1. The last group of cells to come into

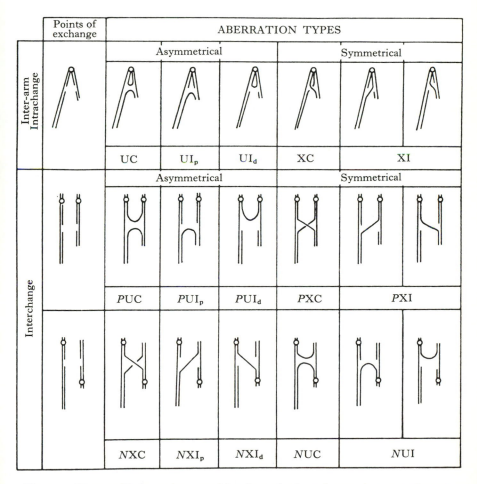

Fig. 7.1. Chromatid aberrations resulting from single exchanges between chromosomes and chromosome arms. *P* and *N*, polarized and non-polarized chromosomes. U and X are types of exchange which are complete (C) or incomplete proximally (I_p) or distally (I_d). From Evans (1962).

mitosis will be those that were in G_1 or in the previous division at the time of irradiation. These will show so-called chromosome rearrangements, in which the breaks and reunions affect both chromatids at the same level (because they occurred before replication, and have hence been copied in the replication process). There is a brief period at the beginning of the S phase when replication is incomplete and one may get the simultaneous occurrence of both chromosome and chromatid aberrations, either in different cells of the same 'population' or, rarely, even within the same cell (Sax and Mather 1939). The rearrangements that we observe in individuals (regardless of whether the whole body carries the rearrangement or is a mosaic for it) are of course all chromosome rearrangements; but there seems no reason to doubt that they originated, at the cellular level, in chromatid rearrangements.

Irradiation of cells which have already entered on mitosis leads, in many types of material, to a generalized clumping of the chromosomes which frequently assume abnormal and fantastic shapes. This 'primary' effect of irradiation (Alberti and Politzer 1923) is apparently irreversible. It has been ascribed to a general 'stickiness' of the chromosomes, but this term is more a description of the condition than an interpretation, and it must be admitted that the biochemical basis of the phenomenon is not understood, although it has been plausibly ascribed to depolymerization and cross-linking of the chromosomal DNA. Various agencies other than irradiation can also lead to so-called 'stickiness' which may even be manifested in species hybrids (White and Cheney 1971). Whatever the precise nature of stickiness, it is essentially a pathological condition, which probably has nothing to do with the processes of chromosomal rearrangement which are of evolutionary significance (Fox 1966a).

There are at the present time two interpretations of the mechanism of structural change in chromosomes. These hypotheses have been put forward especially in connection with radiation-induced rearrangements and it is not certain that 'spontaneous' rearrangements occur in precisely the same manner. The classical model of chromosomal rearrangement assumed that chromosome breaks were produced first, and that the freshly broken ends then rejoined in ways different from the original one (if they joined up in the same manner as before we should merely have a restitution). This model was put forward by Stadler (1931), Sax (1938, 1941) and Muller. The alternative interpretation is the exchange hypothesis of Revell (1955, 1959, 1963, 1966), which has been supported by Evans (1966) and Brewen and Brock (1968). We shall discuss the evidence for these opposing hypotheses later (p. 210); for the time being it will be convenient to discuss the various kinds of rearrangements in terms of the classical hypothesis.

On the classical hypothesis, the first general principle governing the occurrence of chromosomal rearrangements is that before any stable rearrangement can take place, at least two breaks must be present in the same nucleus, either in the same chromosome or in different ones. The formulation of this principle was mainly due to Muller (1932*b*, 1938, 1940*a*, *b*; see also Muller and Herskowitz 1954). If only a single break is present, this can lead to a terminal deletion, i.e. loss of a terminal acentric segment of a chromosome. But, as a general rule, at any rate, the centric fragment will undergo *sister-strand reunion* between the two chromatids at the site of breakage. The consequences of such an event are disastrous, because at the next mitosis a continuous chromatid will inevitably be stretched between the two separating daughter centromeres. Either the chromatid bridge will prevent the completion of cell division or it will break during telophase or when cytokinesis occurs. Undoubtedly the precise details differ from one type of cell to another. But if cytokinesis does occur and the chromatid breaks, then further sister-strand reunions are liable to occur again in the daughter cells, followed by more breakage.

Although chromosomes with freshly broken ends do not seem to be stable in *Drosophila* (and, as a general rule in other animals, although some holocentric chromosomes and those of the Ascaridae may constitute exceptions) because of sister-strand unions, it was shown by McClintock (1941*a*, *b*) that in maize, freshly broken ends, although unstable in the gametophyte and the endosperm, undergo a kind of 'healing' in the sporophyte and will not subsequently fuse with other broken ends or undergo sister-strand union. Just what molecular changes are involved in this process of healing is quite unknown, and it is uncertain whether true telomeres arise in the course of healing.

A special category of simple breaks are ones through the centromere. Grasshoppers heterozygous for a break through the centromere of a metacentric have been recorded by White (1956) and Southern (1969). Such individuals possess two telocentrics which undergo regular synapsis with the limbs of the unbroken metacentric at meiosis. Obviously the telocentrics must have been stable throughout the spermatogonial mitoses. But there is no evidence that they are stable in the somatic divisions or in oogenesis. The individuals in which they have been found were quite possibly mosaics, i.e. there is no proof that the telocentrics were present in the somatic nuclei as well as in the testis. Thus the existence of such individuals should not be claimed as evidence for the view that evolutionary replacement of metacentric chromosomes by two rod-shaped elements is always or even usually due to simple breakage through the centromere.

Although almost all types of chromosomal aberrations (e.g. ring chromosomes and isochromosomes of various types) have now been recorded from human patients, there seem to be absolutely no records of any telocentrics having been found; since they must surely arise from time to time, this strongly indicates that they are not transmissible, either mitotically or meiotically. Neither have telocentrics arising by fragmentation of meta-centrics through the centromere been recorded in any species of *Drosophila*.

Thirty years ago, Rhoades (1940) after a most careful study of the in-heritance of a telocentric chromosome in maize, wrote:

The production of secondary or isochromosomes at meiosis from the telocentric chromosome and its loss or modification in somatic tissue show that a terminal centromere is unstable. Such a telocentric chromosome would tend to be eliminated by natural selection. This instability may apply to all telocentric chromosomes and account for the fact that telocentric chromosomes are rarely, if ever, found in the normal chromosome complement of any organism.

In spite of all the detailed studies that have been carried out since that time (in part reviewed by Marks 1957) there seems no reason to question or modify these conclusions today.

The centromere is undoubtedly a functionally compound body, in the sense that fragments of centromere still possess centromeric properties. Presumably it consists of a particular nucleotide sequence (or sequences) repeated many times. Lima-de-Faria (1955) studied a rye chromosome which had a centromere region about one-third of the normal length (pre-sumably due to a deletion). Its ability to pass to the poles of the spindle at mitotic and meiotic divisions was stated to be hardly impaired at all (however *any* impairment would be likely to prevent such a chromosome establishing itself in a natural population).

It is thus not disputed by anyone that breakage through the centromere can lead to two chromosomes each endowed with some degree of centromeric activity. Darlington (1939c) adopted the term *misdivision* for a type of acci-dent which sometimes occurs at the first anaphase of meiosis and leads to the replacement of a metacentric by two telocentrics. In some cases telocentrics convert themselves into isochromosomes, presumably by sister-strand union at the freshly broken end. In maize, Rhoades (1938, 1940) found a telocentric to be mitotically unstable, undergoing loss or reduction in size in the somatic tissues; at the pollen mitosis they sometimes became iso-chromosomes. In wheat telocentrics, Sears (1952a, b) has likewise reported the same two types of instability.

It is obvious that mitotically unstable chromosomes cannot be expected to survive in evolution, even if the instability is only occasionally manifested. Thus the opinion of John, Lewis, Hewitt and Southern that simple fission through the centromere can convert a metacentric into two acrocentrics, and that this process has occurred often enough in evolution to be significant, could only be true if (1) some telocentrics are much more stable than those of maize and wheat, or (2) some unstable telocentrics are capable of undergoing rapid changes to a stable condition. For the time being we shall continue to adhere to the opinion expressed earlier (White 1957a, 1969) that evolutionary replacements of metacentric chromosomes by two acrocentrics occur by dissociation rather than by simple centric fission.

Equal-armed metacentrics whose two arms are homologous and nontandem are referred to as *isochromosomes* (Darlington 1940a). They have the structure ABC . CBA, where the point indicates the centromere. A classical example is provided by the attached *X*-chromosomes in laboratory stocks of *Drosophila melanogaster*. Isochromosomes can arise in three different ways: (1) through centric fusion between two homologous acrocentrics, (2) through replacement, by two successive translocations, of the arms of a heterochromatic chromosome such as the *Y* of *D. melanogaster*, by homologous acrocentrics, (3) by sister-strand reunion, following breakage through or immediately adjacent to the centromere ('mis-division' in the terminology of Darlington and his school).

Isochromosomes can, and usually do, undergo internal auto-synapsis at meiosis, with formation of chiasmata between their arms. There have been many genetic studies of crossing-over between the limbs of attached *X*-chromosomes in *Drosophila*. When a single chiasma is formed in an isochromosome, a ring univalent will be seen at meiosis; such structures are regularly formed in the case of certain supernumerary chromosomes (see p. 319). It is easy to see that autosynapsis will prevent isochromosomes from becoming normal members of a diploid karyotype which depends on regularity of synapsis between pairs of chromosomes to perpetuate it from generation to generation. Hypotheses which (like those of Henderson 1965, in regard to the *Y*-chromosome of some mantids) postulate isochromosomes as members of the normal chromosome set are hence to be rejected.

It is generally accepted that dicentric chromosomes cannot function as members of the karyotype of a wild species of organism. However, the existence of groups with holocentric chromosomes should perhaps make us rather cautious and undogmatic on this point. Some dicentric chromosomes are known to be transmissible in experimental organisms, even if their behaviour is not entirely normal.

Thus Sears and Camara (1952) have reported on a transmissible dicentric chromosome in wheat; it had an apparently normal submedian centromere and a weak subterminal one. At the first meiotic division the two centromeres normally orientated themselves syntelically. Although transmissible, this dicentric is liable to be broken, both at mitosis and meiosis.

So-called 'dicentric Y-chromosomes' have been described several times in sexually abnormal human patients (Ferrier, Ferrier and Bill 1968; Angell, Giannelli and Polani 1970). These are transmissible and seem to consist entirely of Y-chromosome material; they are probably isochromosomes, but it is not certain whether they have some kind of double or compound centromere or two separate centromeres.

A distinction exists between dicentric chromatids resulting from sister-strand reunion and true dicentric chromosomes. The former necessarily form a chromatid bridge at the next division; this becomes stretched on the spindle and then breaks abruptly (Bajer 1963). There may be some types of cells in which a dicentric chromatid does not break but prevents the completion of anaphase separation; however they are probably not common.

Dicentric chromosomes do not necessarily break at mitosis. They may behave in three ways at anaphase (Darlington and Wylie 1953; Giles 1954; Kaufmann 1954; Swanson 1957; Wolf 1961; Bajer 1963). These may be called *parallel, criss-cross* and *interlocked* orientation (Fig. 1.1). The first gives rise to free separation without the formation of chromatid bridges. Criss-cross orientation gives rise to two bridges and, as a rule, two breaks; occasionally, however, it can lead to three or four breaks, if one or both chromatids are broken close to a centromere and one or two of the long stretched limbs is caught in the middle of the spindle when cytokinesis occurs (Bajer 1963). Theoretically, the two freshly broken ends in each cell resulting from a criss-cross anaphase could fuse; but in practice they usually seem to remain unfused and undergo sister-strand reunion in the next interphase, following replication.

In a number of experiments on various types of material it has been found that heterochromatic chromosomes are less likely to participate in reciprocal translocations than euchromatic ones. Thus Westerman (1968) obtained a total of 163 different reciprocal translocations in a radiation experiment on male locusts (*Schistocerca gregaria*), but not one of these involved the heterochromatic X-chromosome. A similar result was obtained by Helwig (1938) in an experiment on the grasshopper *Trimerotropis verruculatus*. It is, of course, known that such X-chromosomes have on a number of occasions undergone evolutionary fusions (a type of translocation) to autosomes (see p. 608). Westerman (1968) has suggested that fusions of this kind may have established themselves more readily in the female germ-line, in which the X is not heteropycnotic. However, there does not seem to be any evidence as to the frequency of various types of induced rearrangements in female grasshoppers. The insects irradiated in Westerman's experiments must have had their germ cells in the spermatogonial stage, when the X-chromosomes would have been negatively heteropycnotic. A very different result was obtained by Manna and Mazumder (1967b), who obtained numerous X-autosome translocations in experiments in which grasshopper spermatocytes in meiotic prophase were irradiated. It is at least probable that the difference is due to the fact that in the spermatocytes the X-chromosomes are positively heteropycnotic.

It has been generally assumed that since spontaneous breaks at a particular locus occur with a frequency of about 1 in 10^6 a particular two-break rearrangement could only be expected to recur with a frequency of less than 1 in 10^{12} individuals (bearing in mind that the two breaks would have to be in the same nucleus within a limited period of time). Thus in practice recurrence of particular rearrangements would not be observed. Nevertheless, Grüneberg (1937) observed a single, apparently perfect, spontaneous reinversion of the roughest[3] inversion in *Drosophila melanogaster* (which produces a distinctive phenotype of roughened eyes). And Novitski (1961) obtained six such reinversions in 32 000 male offspring of irradiated females.

THE EXCHANGE HYPOTHESIS OF CHROMOSOME REARRANGEMENT

The 'exchange hypothesis' of the mechanism of chromosomal rearrangement was put forward by Revell (1955, 1959, 1963, 1966); it postulates that all chromatid rearrangements produced by irradiation are the result of exchanges where two strands cross one another. Such exchanges may be complete (in which case both re-joins occur) or incomplete (only one re-join occurs). Chromatid breaks and isochromatid breaks are thus believed to result from so-called intrachanges, exchanges which are supposed to occur at small loops in the chromosome which may be related to gyres of the chromonemal helix at interphase.

The exchange hypothesis has received strong support from a number of later workers. But it is entirely uncertain whether, even if it is true for radiation-induced rearrangements, it applies also to spontaneous ones. And the work of McClintock (1938, 1941a, b) on the chromosome- and chromatid-types of breakage–fusion–bridge cycles in maize proves conclusively that true chromatid breaks can take place by stretching on the spindle and subsequent growth of a cell wall which cuts through the 'bridge'; and that freshly broken ends produced in this way, do join up in the manner postulated in the classical 'breakage first' hypothesis. Furthermore, McClintock (1942) showed that chromosomes do not have to be in contact at the time of breakage in order for fusions to take place between the broken ends, since fusions can occur in the zygote or early embryonic nucleus between one broken end derived from the male gamete and another contributed by the female gamete.

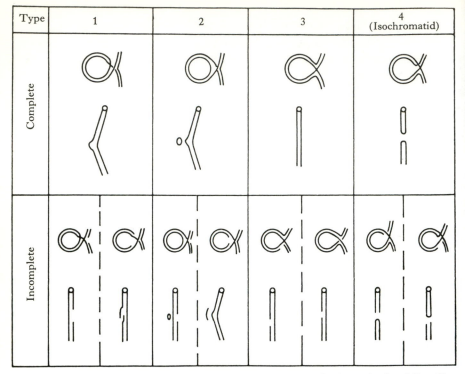

Fig. 7.2. Types of chromatid intrachanges according to the hypothesis of Revell. From Evans (1962).

A fundamental postulate of the classical hypothesis is that large numbers of chromatid breaks are produced by irradiation, most of which then rejoin ('restitute'). Of those that fail to restitute, some rejoin in novel ways (leading to rearrangements), while others appear as true breaks at the following metaphase. Such breaks were classified as chromosome breaks (induced prior to replication, i.e. in G_1), chromatid breaks (produced after replication) and isochromatid breaks (breaks in both chromatids at precisely the same level occurring after replication).

There has always been considerable difficulty in scoring the number of chromatid breaks seen at metaphase, in experimental material, and different schools of workers have reported quite discrepant results, even on the same organism. The difficulty seems to result from a confusion between true chromatid breaks and non-staining gaps or lesions produced by irradiation which are similar in appearance to the naturally-occurring secondary constrictions. The nature of these achromatic lesions has given rise to much controversy. They are not complete breaks in the sense of interrupting the continuity of the chromosome, although they are certainly mechanically weak places, where the chromosome may break if stretched (H. J. Evans 1961). Some authors have interpreted these achromatic gaps as sub-chromatid breaks. However, Evans (1962) has pointed out that if this explanation was correct they should yield a larger number of chromatid breaks at the second post-radiation mitosis, which is not in fact observed. It is possible that these gaps are a heterogeneous group of aberrations, arising from several causes. It has been suggested that they are due to localized despiralization of the DNA, but if so the mechanism of despiralization is unknown, and it is not clear whether gaps involve mutational changes at the site of the observable change. Achromatic gaps may be produced by radiation at the end of the interphase stage or in prophase. In some species of animals similar 'elastic constrictions', which become enormously stretched on the spindle at anaphase, have been occasionally observed in material collected from natural populations (White 1957c).

The evidence from irradiation experiments on the Broad Bean, *Vicia faba* (Neary and Evans 1958; Evans, Neary and Williamson 1959; Revell 1959) has suggested that the frequency of true chromatid breaks may be only one tenth or one twentieth of that earlier given by Thoday (1951). Such a low frequency of breaks, relative to rearrangements, is a necessary requirement for the exchange hypothesis of Revell, which would be incompatible with the much higher estimates given by some earlier workers.

In the earlier work it was claimed that, whereas the frequency of isochromatid breakage rose approximately in proportion to the 1.5 power of the gamma- or X-ray dose, the relationship in the case of chromatid breaks was a linear one, indicating that they were due to single 'hits', whereas in the case of the isochromatid breaks there would be a combination of a one-hit and a two-hit component. The more recent estimates of chromatid break frequency (excluding 'gaps') do suggest the same '1.5 dose power' relationship as in the case of isochromatid breaks and chromatid interchanges. This seems to be so both in plant material (Revell 1963) and in locust embryo cells (Fox 1967b). On the other hand the frequency-dose relationship in the case of the non-staining gaps is a linear one.

Revell has proposed that there are four types of 'intrachanges' which occur with equal frequencies, because the two chromatids have equal probabilities of undergoing an exchange. Corresponding to each of these four types there are two possible types of 'incomplete' intrachanges (Fig. 7.2, bottom row), so that there are eight possible types of incomplete intrachanges which are also believed to arise with equal frequency (but the two sub-types of type 3 would be indistinguishable). Complete isochromatid intrachanges will have both proximal and distal sister-strand union (SU), while incomplete ones will be of two equally frequent types with proximal and distal non-union respectively (NU_p and NU_d). Failure of sister-strand union both proximally and distally (NU_{pd}) is very rare in *Vicia faba*, but significantly more frequent than complete sister-unions (SU) in the locust embryos irradiated by Fox (1967).

Type 1 intrachanges lead to a small duplication (of the direct tandem type) in one chromatid and a deletion in the other. When 'complete' they can be recognized by a small bulge in one chromatid only; when 'incomplete' there will either be only one chromatid for a short distance, or three chromatids.

Type 2 intrachanges produce a minute deletion from one chromatid. If the rearrangement is 'complete' the deleted fragment will be a ring, if it is 'incomplete' the fragment may either be a ring or a rod.

Type 3 intrachanges give a small inversion in one chromatid. If 'incomplete' the inversion will be adjacent to a chromatid break (either proximal or distal to it).

The type 4 intrachanges are responsible for the isochromatid breaks, earlier supposed to be due to ionizing radiation causing two independent breaks at corresponding loci in the two chromatids. But on the exchange hypothesis the sister-unions are *not* between identical loci.

Apparently, in *Vicia faba* these chromatid rearrangements can be induced only during interphase, and cease when the prophase stage begins (Revell 1959). There seems a strong possibility that the single-locus duplications described by Keyl (1963*b*, 1964, 1965*a*, *b*) are due to type 1 intrachanges. The configurations actually observed at metaphase following irradiation of G_2 interphase nuclei seem, in general, to be of the types expected on the exchange hypothesis, i.e. minute 'buckles' (due to a small duplication in one chromatid or a deletion in the other), minute fragments and gaps marking the position of small deletions.

In addition to the 'intrachanges', chromatid aberrations also include translocations or exchanges. Revell (1955) originally claimed that a disproportionate number of these appear to be between homologous chromosomes and at apparently corresponding loci (i.e. they resemble meiotic chiasmata). This suggested that they must arise at a time when there is a polarized orientation of the chromosomes involving some kind of somatic pairing. Fox (1966*b*), in his studies of chromatid interchanges following irradiation of locust embryos, found a marked excess of so-called U-type exchanges over X-type ones (159 versus 34). There were also almost twice as many 'polarized' orientations as 'non-polarized' ones (128 versus 65). These findings seem more easily interpretable on the basis of the exchange model of rearrangement than on the classical one. Fox suggests that the chromosomes of the interphase nucleus have a polarized orientation and lie more or less parallel to one another; he also believes that U-type exchanges are fundamentally different in nature from the X-type.

Fox (1967a) did, however, find certain types of rearrangements which did not fit into Revell's original scheme. These were intrachanges involving insertion of short lengths of paired sister chromatids into a chromatid break; both complete and incomplete types were observed. A modification of Revell's original hypothesis was proposed to account for these types of rearrangements, which were not encountered by the earlier workers on *Vicia faba*. Revell himself has discussed (1966) the origin of so-called triradial configurations as due to a chromatid interchange and a type 4 intrachange occurring close together. He gives the following dose exponents for the various types of aberrations observed after irradiation of *Vicia faba* cells:

Gaps (achromatic lesions – not true breaks)	1.07 ± 0.08
Real chromatid discontinuities	1.71 ± 0.09
Isochromatid discontinuities	1.48 ± 0.04
Chromatid interchanges	2.00 ± 0.03
Triradials	2.92 ± 0.21

The dose exponent suggests that the origin of triradials is more complicated than is apparent at metaphase.

The exchange hypothesis of chromatid rearrangements has been extended to chromosome rearrangements by Brewen and Brock (1968) in their work on irradiated cultured cells of the swamp wallaby, *Wallabia bicolor*. They found that terminal chromosomal deletions showed two-hit kinetics, whereas on the classical breakage-first hypothesis their frequency should have been a linear functional of dose (with the possible addition of a small two-hit component due to incomplete exchange).

It seems difficult to see how some complex rearrangements involving three or four breaks which have arisen spontaneously (e.g. those reported by White 1963 and White and Cheney 1971 in individual morabine grasshoppers from natural populations) could have arisen by the exchange method. Since most spontaneous rearrangements are probably due to chemical mutagens or to virus infection, it will be interesting to see whether careful quantitative studies of chromatid rearrangements induced by such means yield the same picture as those induced by irradiation. Such comparative studies would be most revealing if they could be carried out on the same biological system, since the differences between the aberrations induced by radiation in *Vicia faba* root-tip cells, grasshopper embryos and mammalian cells in culture are fairly significant.

There does not seem to be any particular difficulty in explaining the origin of large inversions on the exchange hypothesis; they would be similar to exchanges between different segments of the same chromosome. But it is difficult to understand how the reversed repeats observed in *Drosophila* following irradiation or treatment with chemical mutagens could have arisen by the exchange mechanism (see discussion in Revell 1963).

A peculiar type of chromosomal rearrangement has been observed in somatic cells of the plant *Crepis*, following irradiation, by Dubinin and Nemtseva (1970). This phenomenon, which has been called 'vesting' seems to result from two breaks in the same chromosome, in the G_1 phase prior to replication. These lead to the production of three fragments, of which the middle one proceeds to form a ring which encircles the chromosome resulting from the fusion of the other two fragments. Following replication, two sister chromatids are encircled by two sister rings; but in addition to this 'expected' configuration, some more complicated and unexpected ones were seen. These included figure-of-eight rings 'vesting' a chromatid or a chromosome twice and 'vesting' of sister chromatids, individually, by simple rings. These patterns were interpreted as due to a combination of mitotic crossing-over and the slipping out of the internal chromatids from the rings through breaches when crossing-over occurs. Ten different configurations can be predicted on this hypothesis, of which six were actually observed. Dubinin and Nemtseva suggest that chromosomal rearrangements are really of two types, with and without contact–exchange processes. Thus far, there have been no studies of 'vesting' in organisms other than *Crepis*. The relationship between Dubinin and Nemtseva's views, on the one hand, and those of Revell and Evans, on the other, is not entirely clear.

Although the telomere concept was originally put forward by Muller in the context of the classical hypothesis of chromosome rearrangement, the exchange hypothesis does not seem to require any fundamental modification of it. But we might say that telomeres cannot participate in chromosomal rearrangements for two separate reasons: (1) both intrachanges and exchanges are, by their very nature, interstitial phenomena, (2) telomeres are 'non-sticky' ends, in the same sense as on the classical hypothesis.

The basic types of two break chromosomal rearrangements are (1) translocations (strictly speaking, mutual translocations or interchanges), (2) inversions, (3) tandem duplications, (4) deletions. On the view adopted throughout this book, the various kinds of rearrangements decreasing or increasing the chromosome number (fusions and dissociations) are simply special types of translocations in which the breakage points are very close to the centromeres or telomeres.

Mutual translocations (interchanges), resulting from a break in each of two non-homologous chromosomes, should theoretically be of two equally frequent types, called symmetrical and asymmetrical by Lea (1946). In a symmetrical translocation the two new types of chromosomes will both be monocentric, and hence able to pass through mitosis unchanged and unharmed. In an asymmetrical translocation, on the other hand, one of the resulting chromosomes is a dicentric and the other an acentric. Thus, in general, asymmetrical translocations will be cell-lethal and unable to survive (because the acentric will be lost and the dicentric will form chromatid bridges at mitotic anaphases). A possibility exists, however, that in some instances evolutionary fusions of acrocentric chromosomes (to produce to a metacentric) may have been due to asymmetrical translocations which produced a metacentric with two centromeres so close together that they functioned as one, and a minute acentric fragment containing only dispensible heterochromatin. In *Vicia faba* Heddle (1965) found that the frequencies of symmetrical and asymmetrical interchanges induced by radiation at the G_1 stage were equal, as expected on theoretical grounds.

More complex types of translocations are possible, of course, when three or more breaks have occurred in a cell. Some of these are *insertional* translocations, in which a piece is deleted in one chromosome and inserted into the gap created by a break in a second chromosome.

Inversions are simply chromosome segments that have been turned around through 180°, so that their position in the sequence has been reversed. Single inversions are obviously two-break rearrangements. If both breaks were situated on the same side of the centromere they will be *paracentric* inversions; if they were on opposite sides of the centromere the inversion is said to be *pericentric*. Paracentric inversions may be identified in certain cases at meiosis and they can always be detected in polytene chromosomes. But unless they change the relative sequence of conspicuous heteropycnotic and euchromatic segments they will not be detectable in mitotic divisions. They are thus overlooked in most studies of induced rearrangements in somatic cells, whether in culture or in tissues such as plant root-tips. Pericentric inversions, on the other hand, change the arm-ratio of the chromosome, unless the two breaks are precisely equidistant from the centromere or from the chromosome ends, so that they are, for the most part, detectable in somatic divisions as well as at meiosis.

It seems highly unlikely that terminal inversions can occur. Such a rearrangement would involve a telomere becoming interstitial through fusion with a freshly-broken end. And, in addition, the other freshly-broken end would have to undergo a form a healing, i.e. it would have to take on the functions and properties of a telomere. There is considerable doubt as to whether the first process is possible at all, and the second may be possible only under very special circumstances, as seems to happen in the maize sporophyte.

Supposedly terminal inversions were described in the polytene chromosomes of several species of *Drosophila* by a number of the early workers. Such rearrangements were recorded in unirradiated individuals of *D. ananassae* by Kaufmann (1936), Kikkawa (1938) and Dobzhansky and Dreyfus (1943), in *D. nebulosa* by Pavan (1946) and in *D. robusta* by Carson and Stalker (1947). They were also claimed in various species of *Sciara* (Carson 1944; McCarthy 1945). More recently three terminal inversions and three terminal translocations were reported in *Chironomus riparius* by Blaylock and Koehler (1969). Unfortunately, none of these cases has been subjected to a really critical analysis to determine whether a minute region (perhaps including one or two bands) may not exist beyond the limits of the inversion (if such a region were sufficiently minute, the observational difficulty of distinguishing between a truly terminal and a subterminal inversion might be insuperable). In view of the vast amount of work that has been done on inversions in polytene nuclei, and assuming that breakage points are distributed approximately at random along chromosomes, it was virtually inevitable that there should be some published 'discoveries' of terminal rearrangements based in reality on subterminal ones.

The situation with regard to terminal deletions is somewhat different. There is no doubt that this category of rearrangements can *occur*; the doubt is as to whether they can survive. It is true that Sutton (1940) recorded five terminal deletions of the *X*-chromosome in *Drosophila melanogaster*, but no really critical evidence that the telomere (in the sense of Berendes and Meyer 1968) had actually been removed was presented.

It is generally accepted that, in all chromosomal rearrangements, directional polarity of DNA strands is conserved (Taylor 1959; Brewen and Peacock 1969*a*). This has certain important consequences at the molecular level. If we call the two anti-parallel DNA strands 1 and 2, then in the case of an inversion the inverted segment of strand 1 will be inserted, after rotation through 180°, into strand 2, and vice versa, the inverted segment of strand 2 will be inserted into strand 1. As a result, neither strand will have the same total-base composition, after inversion, as it did before. Presumably, each cistron contains within itself, or closely adjacent, the instruction as to which strand is to be responsible for transcription as well as an instruction specifying the polarity of the cistron itself (i.e. whether it is to be read from 'left' to 'right' or from 'right' to 'left'). And both kinds of instructions must, in general, be unaffected by structural rearrangements.

On the classical breakage-first hypothesis of chromosomal rearrangements, the relative frequency of spontaneous or induced translocations and inversions in a particular species will depend on the number and sizes of the chromosomes which make up the karyotype. Considering the simple case where all the chromosomes are the same length, in a species with n = 2, inversions and translocations should be equally frequent, while in one with n = 3 translocations should be twice as frequent as inversions. In species with high chromosome numbers, of course, inversions should be very rare by comparison with translocations. In an irradiation experiment with the locust *Schistocerca gregaria*, with 2n = 23 acrocentrics, Westerman (1968) records 163 translocations, but no inversions; it is uncertain, however, what fraction of the total number of paracentric inversions produced would have been detected in this case.

On the exchange hypothesis of chromosomal rearrangement, assuming some polarization of the chromosomes in the interphase nucleus, it does not seem possible to make any prediction as to the expected ratios of inversions and translocations in different karyotypes. But one might expect that in all karyotypes, even those with low numbers of chromosomes, inversions should be infrequent by comparison with translocations.

In the case of equal-armed metacentric chromosomes (arm-ratio 1.0) we may expect the frequency of spontaneous (or induced) paracentric and pericentric inversions to be equal. On the other hand, as the arm-ratio gets smaller, paracentric inversions will become commoner and pericentric ones rarer. In the case of truly acrocentric chromosomes, with very minute short arms, breaks through the latter will necessarily be very rare, and pericentric inversions correspondingly infrequent. But on the exchange hypothesis all inversions will result from formation of chromosome loops at interphase and the sizes and types of loops (i.e. their relation to the centromeres and telomeres) would not be predictable on any general principles but would have to be determined empirically.

HETEROZYGOUS REARRANGEMENTS AT MEIOSIS

At meiosis, in an individual heterozygous for an inversion, complete synapsis will lead to a 'reversed loop' at pachytene, similar to those seen in salivary gland nuclei (Fig. 4.4). If the inversion is paracentric, the centromeres will be outside the loop, while if it is pericentric they will be inside. If the inversion is sufficiently short we might expect that the chromosomes might be too rigid to form a 'reversed loop' and in such cases synapsis between the mutually inverted regions might not occur.

In a bivalent with an inversion loop, chiasmata may be formed either in the loop or outside it. The genetical consequences will depend on the positions of these chiasmata, their relation to one another (reciprocal, complementary, or diagonal pairs of chiasmata) and on whether the inversion is pericentric or paracentric. In a bivalent heterozygous for a pericentric inversion, chiasmata outside the loop, in either or both of the free arms, will not lead to any structurally changed chromatids. A single chiasma within the loop will, however, lead to the production of two chromatids which lack one end region and have the other end in duplicate (i.e. each of these strands will carry both a deficiency and a duplication). Two reciprocal chiasmata within the loop will not lead to the production of such deficiency-duplication chromatids, but diagonal or complementary pairs of chiasmata will do so.

In a bivalent which is heterozygous for a paracentric inversion (i.e. in which the centromeres are outside the reversed loop) the consequences of crossing-over will be quite different. A single chiasma within the loop will again give rise to two chromatids, each with a terminal deficiency and a terminal duplication, but one of these will be dicentric and the other acentric. At first anaphase the former will be stretched on the spindle between the two centromeres, forming a 'bridge', and will either prevent the separation of the two daughter nuclei or will break under the strain. The acentric chromatid will usually remain passively in the equatorial part of the spindle, without passing to either pole.

If two chiasmata are formed within the limits of a paracentric inversion the result will depend on their relation to one another. If they are reciprocal, no acentric or dicentric chromatids will be produced. On the other hand, a pair of complementary chiasmata within the loop will give rise to two dicentric chromatids and two acentrics. The dicentrics will later form a 'double bridge' at the first anaphase. A pair of diagonal chiasmata will produce a dicentric and an acentric chromatid in the same way as a single chiasma within the loop does.

More complicated situations arise when chiasmata occur between the loop and the centromere as well as within the loop (Fig. 7.9). It will be seen that in some cases loop chromatids (i.e. ones with both ends attached to the two halves of the same centromere) are produced. These will pass to the pole as such at first anaphase but will give rise to dicentric bridges at the second meiotic division.

Two inversions in the same chromosome may be either *independent*, *included*, *overlapping* or *tandem*; the first three combinations arise (on the classical hypothesis) from four breaks (simultaneously or two at one time and two at another). Tandem inversions result from three breaks in the same chromosome and seem difficult to interpret on the exchange hypothesis. They may be either direct or reversed with respect to one another, i.e. the sequence ABCDEF may give rise to ACBEDG (direct tandem inversions) or to AEDCBF (reversed tandem inversions).

Included and *overlapping* pairs of inversions, when heterozygous, give rise to characteristic 'double loop' configurations in polytene nuclei (Fig. 8.14). Theoretically, they might produce the same configurations at meiosis, but the extent to which they actually do so is not really known. Overlapping pairs of inversions are especially interesting, since it is often possible to trace the phylogenetic history of the different chromosomal sequences by an analysis of those that occur in natural populations. The method, due in principle to Sturtevant and Dobzhansky (1936b), is as follows:

When three sequences are known in a particular chromosome, (1) ABCDEFGHI, (2) AB*FEDC*GHI, (3) ABFE*HGCD*I (the inverted segments being italicized for the sake of clarity), then (1)/(2) and (2)/(3) heterozygotes show single loops in polytene nuclei while (1)/(3) heterozygotes show a double loop, which permits accurate mapping of the limits of the inversions. From an evolutionary standpoint, either (1) gave rise to (2) and that to (3), or (3) gave rise to (2) and that to (1), or else (2) give rise, on different occasions, to both (1) and (3). But (1) cannot have arisen directly from (3), or vice versa, unless all four breaks took place in the same cell and at the same time and the pieces rejoined in such a manner as to simulate a pair of overlapping inversions (an event with a probability of occurrence so low that it can be safely dismissed).

Although all symmetrical mutual translocations are alike in principle, their genetic consequences are very different according to the positions of the breaks in relation to the centromeres and chromosome ends. The lengths of the chromosome regions on either side of the breaks and the chiasma frequencies of those regions will determine the mechanics of the translocation at the first meiotic division. However, the shape and properties of the meiotic spindle itself certainly vary from one species to another, so that structurally similar translocations probably behave differently in species with dissimilar spindle mechanisms.

We shall consider first the behaviour of mutual translocations in the heterozygous condition. If complete synapsis occurs it will lead to a cruciform configuration at pachytene. The four arms of the cross will theoretically be all different in length; but these differences, or some of them, may not be detectable by the eye. In the case of a symmetrical translocation the centromeres will, of course, be located in opposite arms of the cross (in an asymmetrical one they would be in adjacent arms). The centromeres need not be equidistant from the centre of the cross. Some failure of synapsis may be expected at the intersection of the arms.

Chiasmata may be formed, theoretically, in any or all of the four arms. And in the arms containing the centromeres chiasmata may be formed in the

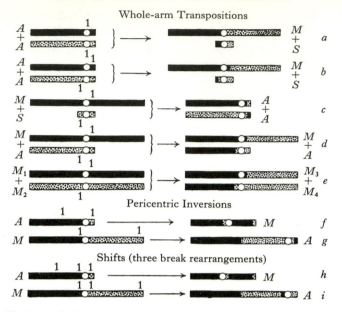

Fig. 7.3. Various whole-arm transpositions and rearrangements which alter the position of the centromere in a chromosome. Positions of breaks indicated by short vertical lines. *A*, an acrocentric; *M*, a metacentric; *S*, a small chromosome fragment. *a* and *b* are different types of centric fusion, *c* is a dissociation.

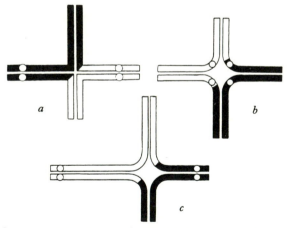

Fig. 7.4. Pachytene pairing in translocation heterozygotes. Synapsis is probably always interrupted in the middle of the cross, as in *b* and *c* (not as in *a*, which represents a condition that probably never occurs). The breakage points may be close to the centromeres (as in *b*) or far away from them (as in *c*). Chiasmata may be formed later in any or all of the four arms of the cross.

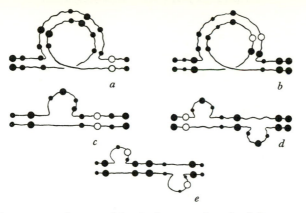

Fig. 7.5. Pachytene or polytene pairing in the case of a pair of chromosomes hetero-zygous for (*a*) a paracentric inversion, (*b*) a pericentric inversion, (*c*) a deletion, (*d*) a paracentric shift, (*e*) a pericentric shift. Centromeres indicated by hollow circles.

'external segment' distal to the centromere or in the 'internal' one between the centromere and the centre of the cross. Once chiasmata have been formed, terminalization may lead to their slipping along to the ends of the arms. In general, it seems probable that most chromosome regions retain their original chiasma frequencies, following a translocation, with only minor changes.

If the chromosomes taking part in a mutual translocation are not capable of forming more than a single chiasma, there is no possibility of the associ-ation of four chromosomes being maintained after the end of pachytene – it will as a rule fall apart into two separate bivalents. No doubt this is how many translocations behave in species with small chromosomes. Since there will be nothing to compel the two bivalents to orientate themselves on the spindle of the first meiotic division in a 'disjunctional' manner they will probably do so in a 'non-disjunctional' way in about 50% of cells. Thus individuals heterozygous for translocations which regularly or frequently form two bivalents (rather than a chain or ring of four chromosomes) will produce a large proportion of aneuploid gametes and will be semi-sterile, so that natural selection against such translocations will be very severe and they can hardly be expected to establish themselves in natural populations or contribute to the evolutionary process, unless the aneuploidy is in respect of minute or heterochromatic segments so that zygotes trisomic for one segment and monosomic for another are viable.

Where the chromosomes taking part in a translocation are capable of forming two or more chiasmata, we may have chiasmata in three of the arms

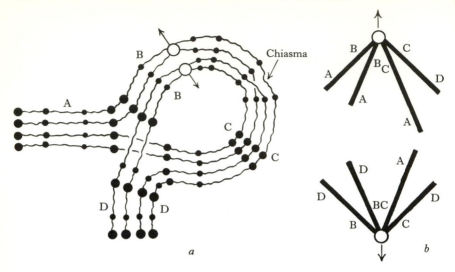

Fig. 7.6. Diagrams showing how a single chiasma within a pericentric inversion leads to the production of gametes with deficiencies and duplications. *a*, end of pachytene; *b* first anaphase.

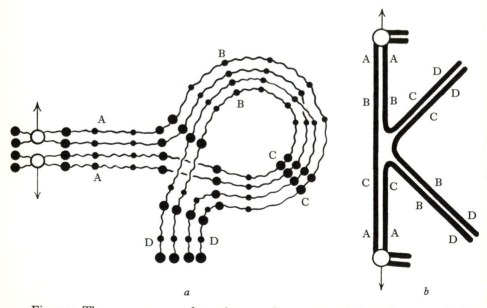

Fig. 7.7. The consequences of crossing-over in a paracentric inversion. *a*, end of pachytene; *b*, first anaphase. A single cross-over gives rise to a dicentric 'bridge' and an acentric chromatid.

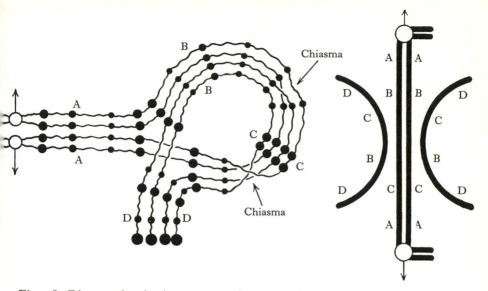

Fig. 7.8. Diagram showing how two complementary chiasmata within a paracentric inversion loop give rise at anaphase to a double 'bridge' and two acentric chromatids.

of the cross, or in all four of them. In the former case we get a chain of four chromosomes, in the latter case a ring. Rings and chains of four may become orientated on the meiotic spindle in either a zig-zag, 'disjunctional' manner or in either of two 'non-disjunctional' ways (adjacent-1 and adjacent-2). One might think that these three kinds of orientations would occur with equal frequency, but this is apparently seldom or never so, there being a general tendency for disjunctional orientations to occur more frequently than non-disjunctional ones. In Bedichek-Pipkin's (1940a) 2–3 translocation in *Drosophila melanogaster* disjunctional segregation occurred in about 59% of cases in both sexes.

Translocations between very long chromosomes are likely to lead to very unwieldy rings and chains that have great difficulty in accommodating themselves to the dimensions of the spindle, especially if some of the intercentromeric distances in the configuration are longer than any that occur normally in the species. Such very awkward configurations as the one shown in Fig. 9.6b are never likely to be successful in establishing themselves in natural populations. They will rather frequently orientate in such a manner that three chromosomes will pass to one pole at anaphase and one to the other.

The mutual translocations that have the best chance of evolutionary success are probably those which regularly form a single chiasma in each of

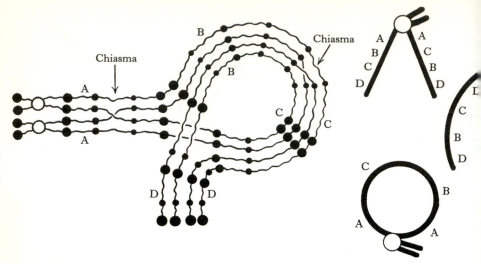

Fig. 7.9. Diagram showing how one chiasma within a paracentric inversion loop and another between the loop and the centromere may give rise to a loop chromatid and an acentric fragment at first anaphase.

the four arms (or in three of them at any rate) and in which the chiasmata are either formed near the ends of the arms, or terminalize to the ends later. Chiasmata formed internally to the centromeres are likely to interfere with regularity of disjunction and hence to be inimical to evolutionary success. On the other hand a ring or chain of four elements of similar size held together by means of chiasmata that are close to the ends of the chromosomes may show very nearly 100% disjunctional orientation on the spindle. It is on translocations of this type that the ring-forming genetic systems of

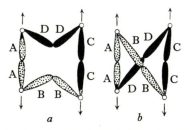

Fig. 7.10. Diagrams showing how a ring of four chromosomes in a translocation heterozygote may arrange itself in a 'square' or a 'zig-zag' orientation at first metaphase of meiosis. Actually, there are two alternative 'square' orientations (in the other one the two A segments go to one pole and the two C segments to the other pole). Square configurations give rise to duplication-deficiency gametes; zig-zag orientations to ones having a complete haploid set of genes.

certain evening primroses (*Oenothera*) and some other plants (*Rhoeo discolor*, *Hypericum punctatum*, etc.) are based.

Since, as we have already pointed out, chiasmata formed 'internal' to the centromeres in translocation configurations will be inimical to evolutionary success, it follows almost automatically that translocations involving acrocentric or near-acrocentric elements are hardly ever likely to be successful (except for 'centric fusions' – Fig. 7.12).

Translocations in which the breakage points are adjacent to centromeres or telomeres and in which the regions interchanged are very small or very unequal constitute special categories whose properties seem to be quite different from those previously discussed. In the first place we may consider the type of translocation in which two chromosomes interchange their telomeres with perhaps only one or two adjacent genetic loci. A heterozygote for such a translocation will almost certainly form two bivalents at meiosis rather than a cruciform configuration. About half the gametes will carry a minute duplication and a minute deficiency. In some instances this may lead to a lethal or sublethal phenotype in the trisomic–monosomic zygote. But if the regions interchanged are sufficiently small it may rather frequently be perfectly compatible with viability. Such translocations of minute terminal regions may even be expected, on occasion, to lead to situations in which even the tetrasomic–nullisomic zygotes are viable. There is a good deal of evidence that in a number of species all or many of the telomeres in the karyotype are homologous (perhaps with minute regions adjacent to them) and such homology implies a common origin by translocation of very short terminal segments (see p. 564 for further discussion of this point).

Translocations that involve very unequal regions also have special properties. We may consider first of all the type of translocation in which two acrocentric chromosomes break, one of them just distal to the centromere (in the long arm) the other proximal to it (in the short arm). A symmetrical translocation will give a large monocentric metacentric and a very minute one. If the latter is not essential for viability it will soon be lost (its chiasma frequency will be close to zero in most instances so that it can hardly be inherited in any regular manner). This type of translocation will be referred to as a 'centric fusion' since its effect is to fuse two acrocentric elements to form a double-sized metacentric. In heterozygotes for centric fusions a chain of three chromosomes will be formed at meiosis, the two acrocentrics being paired with the corresponding limbs of the metacentric. Provided that chiasmata are not formed close to the centromeres (i.e. in the proximal regions) such chains may undergo almost invariable disjunctional orientation on the spindle, so that the two acrocentrics pass to the same pole at anaphase and the metacentric to the opposite pole. Proximal localization of

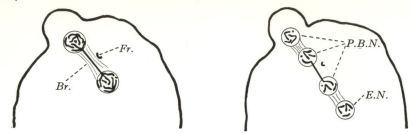

Fig. 7.11. Orientation of the products of the meiotic divisions in a *Drosophila* egg, showing the fate of the dicentric and acentric chromatids resulting from a single cross-over in a paracentric inversion loop. *Br.* bridge; *Fr.* acentric fragment; *P.B.N.* polar body nucleus; *E.N.* egg nucleus.

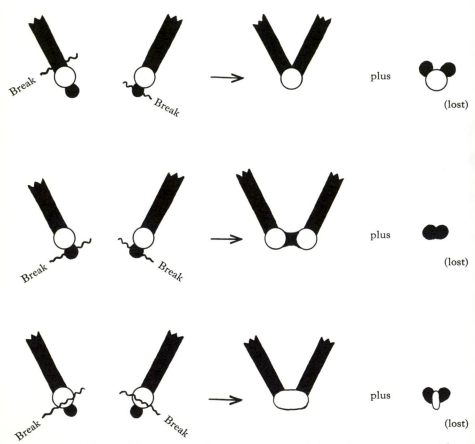

Fig. 7.12. Three different ways in which centric fusions between acrocentric chromosomes may occur. From White (1957*d*).

chiasmata is thus highly inimical to the success of centric fusions and distal localization favourable to it. The symmetry of the chain is also important; centric fusions between acrocentrics of quite different lengths will give lopsided configurations which are apt to undergo non-disjunction.

Although we have stated above that centric fusions result from 'a break proximal to one centromere, and one distal to the other centromere' a possibility exists that some may arise from two breaks actually through the centromeres or perhaps from two breaks both of which are distal to the centromeres, i.e. in the short limbs. The latter type of centric fusion would give a dicentric metacentric with the two centromeres so close to one another that they might function efficiently as one and an acentric chromosome which would necessarily be lost from the karyotype. It is rather uncertain, however, whether a chromosome with two centromeres, however close together, would be sufficiently stable for evolutionary success.

Centric fusions have been of major importance in the karyotype evolution of many groups of animals. They may be present in the heterozygous condition in some animal populations, but in many evolutionary lineages they seem to have rather rapidly reached fixation in the homozygous condition, the original acrocentric elements disappearing from the population (see p. 372).

In addition to centric fusions, there can be no doubt that 'tandem fusions' between acrocentrics (to give a double-length acrocentric) or between an acrocentric and a metacentric, can become established in evolution. Such fusions result from a proximal break in one chromosome and a distal one in the other. But tandem fusions, when heterozygous, will produce 50% aneuploid gametes if a chiasma occurs regularly between the centromere and the point of fusion (Fig. 7.13), so that selection against them will be fairly severe. Well authenticated instances occur in the morabine grasshopper species 'P45b' (see p. 378), where an acrocentric X and a small acrocentric autosome have fused and in a number of species of chironomid midges, in which tandem fusions between a large metacentric and a small acrocentric have occurred (see p. 364). A few 'double-length' acrocentrics that have been reported in various *Drosophila* species (Patterson and Stone 1952) may have arisen by tandem fusion, but they may also have been produced by a pericentric inversion following a centric fusion.

A type of translocation which is, in a sense, the opposite of a centric fusion may be called a 'dissociation'. This occurs when a metacentric and a minute 'donor' chromosome both break in the immediate neighbourhood of their respective centromeres. The 'donor' may be a very small member of the regular karyotype. Alternatively it may be an acrocentric from which almost the entire genetically active region has been deleted, leaving almost nothing

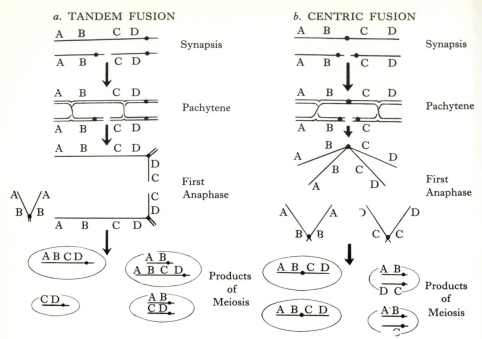

Fig. 7.13. Meiotic segregation in the case of an individual heterozygous for (*a*) a tandem fusion and (*b*) a centric fusion. In the latter case all gametes will be euploid if a trivalent is regularly formed, chiasma-formation is regular and the orientation of the centromeres on the spindle is disjunctional. In the case of a tandem fusion 50% of the gametes are expected to carry a duplication or a deficiency. From White, Blackith, Blackith and Cheney (1967).

but the centromere and two telomeres. And, in some instances, small supernumerary chromosomes (see p. 329) may have functioned as donors.

The effect of a dissociation is that a metacentric is replaced by two acrocentrics. If the 'donor' is not a member of the regular karyotype a dissociation will lead to an increase in the amount of genetic material present and an increase in chromosome number. This seems to be what has happened in the karyotype evolution of many groups. Dissociations which leave the number of chromosomes unchanged (i.e. in which a minute chromosome which is part of the original karyotype loses its separate identity) may occur but are not definitely known as evolutionary events.

It would seem that, in addition to dissociations of metacentrics another type of dissociation should be possible whereby an acrocentric chromosome breaks in a position somewhere mid-way along its length, the freshly broken ends being 'capped' by minute telomere-bearing and centromere-bearing segments derived from the breakage of a small donor fragment. The effect of such a dissociation would be the replacement of an acrocentric by two acrocentrics each half the length of the original one. This type of dissociation is in a sense the opposite to a tandem fusion. The reason why more examples of this kind of dissociation are not known is, we believe a mechanical one; namely that even if a trivalent is regularly formed in such cases and segregates normally at meiosis, one quarter of the gametes will be deficient in an extensive chromosomal region, while another quarter will carry it in duplicate (Fig. 7.13).

Matthey (1939b, 1945a, 1948, 1963a) has criticized the idea that centric fusions are usually or always due to reciprocal translocations. His arguments appear to be three in number: (1) centric fusions have been 'extraordinarily frequent' in evolution, whereas translocations with the breakage points very

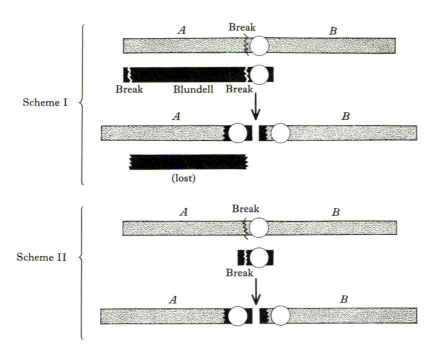

Dissociation of the *AB*

Fig. 7.14. Two alternative mechanisms for the dissociation of a metacentric into two acrocentrics. Based on the case of the grasshopper *Keyacris scurra* (White 1957a) where the dissociation of the large '*AB*' metacentric seems to have been effected through a translocation in which a '*CD*' chromosome carrying the Blundell inversion acted as a donor of a centromere and two telomeres. On Scheme I there was a 3-break rearrangement; on Scheme II the Blundell *CD* chromosome underwent a large deletion first, and this was followed by a 2-break translocation.

close to the centromeres of acrocentric chromosomes must be very rare, (2) in certain special cases such as the cleavage divisions of the ephemerid *Chloeon dipterum* temporary and reversible terminal adhesions between acrocentric chromosomes occur, (3) in two female hybrids between the rodents *Acomys minous* and *A. cahirinus* two acrocentrics from the former species were converted into a metacentric in the cytoplasm of *A. cahirinus*; no such fusion was found in two male hybrids.

It will be appreciated that the first objection is the only general one, the second and third being based on very special cases, however they are interpreted. The first argument collapses when we consider the actual data on the frequency of spontaneous translocations. In the grasshopper *Keyacris scurra* at least one individual per thousand is heterozygous for a newly arisen translocation; the rate in some other grasshopper species may possibly be a good deal higher. The frequency of viable translocations in the human species must be of the same order of magnitude. This implies that in some animal species, at any rate, thousands or tens of thousands of translocations must be occurring in every generation. It is thus not unreasonable to suppose that from 10^7 to 10^{11} translocations probably occur in the entire phylogenetic history of most species of higher animals (these figures refer, of course, only to translocations which occur in the germ-line and are actually transmitted to one or more zygotes at fertilization). It is clear that even if only one in 10^3 of these translocations is of a type which gives rise to a centric fusion, Matthey's main objection to the breakage-plus-reunion hypothesis of fusions is invalid.

The *Chloeon dipterum* case seems to us irrelevant to the mechanism of centric fusion, because the adhesion which occurs between the X and Y is clearly not a centric fusion, since it undergoes spontaneous dissociation later in development. Most probably it is simply a temporary synapsis between heterochromatic segments, similar to those which have been reported in the phorid *Aphiochaeta xanthina* (= *Megaselia scalaris*) by Barigozzi and Semenza (1952).

The *Acomys* case is an extraordinarily confused one. The work was carried out on mitoses in the spleen, and there were unexplained variations in chromosome number in the spleen cells of single individuals of *A. minous*. To what extent these were due to non-disjunction, chromosomal rearrangement or simply to inadequate technique is quite uncertain. The matter is complicated by the fact that *A. minous* either shows chromosomal polymorphism for chromosome number or has two wild 'races' differing in chromosome number. A reinvestigation using more individuals, different somatic and germ-line tissues and a variety of techniques is obviously desirable. Another peculiar case that deserves reinvestigation is that of the

Japanese cyprinid fish *Acheilognathus rhombea*, in which Nogusa (1955*b*) has described three karyotypes: 2n = 44 (including four metacentrics), 2n = 46 (including two metacentrics) and 2n = 48 (no metacentrics). According to Nogusa's account two or all three karyotypes may be found in the same individual. It is not clear why karyotypes with odd numbers of metacentrics were not encountered. A somewhat similar case was reported by Beçak, Beçak and Ohno (1966) in the Green Sunfish (*Lepomis cyanellus*); three genotypes were found (2n = 48, no metacentrics; 2n = 47, one metacentric; 2n = 46, a pair of metacentrics). But the heterozygotes, with one metacentric, also showed some cells with two metacentrics and some with none. Beçak, Beçak and Ohno interpret this as a case of somatic segregation. An even more complex case has been described by Ohno, Stenius, Faisst and Zenzes (1965) in the Rainbow Trout (*Salmo irideus*), in which the chromosome number was found to range from 2n = 58 to 2n = 65; as many as seven different karyotypes occurred in the somatic cells of one individual.

An objection to the translocation hypothesis of centric fusion that has been urged by B. E. Wolf (1960) is based on the finding by Wahrman and O'Brien (1956) that the DNA content of nuclei in the mantid *Ameles heldreichi* is no lower in individuals with fusions than in ones without them, thus indicating that loss of the small product of translocation cannot have occurred. But since the small product is assumed to be in most cases less than a hundredth of the total chromosome length of the entire karyotype, and since the DNA values determined by Wahrman and O'Brien may easily have been in error by as much as 5%, this objection is obviously invalid. In such cases as *Chilocorus stigma*, where J-shaped chromosomes have undergone centric fusion with the visible loss of much longer segments (see p. 300), DNA determinations have not been carried out.

Belief in simple adhesion between chromosome ends as a cause of centric fusion has sometimes been linked with the idea that 'rod-shaped' chromosomes are strictly telocentric, rather than acrocentric, i.e. that their centromeres are actually terminal. Now, it is universally admitted that the 'short arms' of most acrocentrics are usually very difficult to demonstrate, and that there is no universally successful technique for revealing them.

In the heterozygous condition a dissociation will have the same properties as a centric fusion, i.e. at meiosis two acrocentrics will synapse with the corresponding limbs of the metacentric. Only in the case of a centric fusion the metacentric is the 'new' chromosome, while in the case of a dissociation the metacentric is the original element. Just as in the case of centric fusions, proximally located chiasmata will tend to prevent the success of dissociations.

Centric fusions and dissociations which have occurred in evolution are referred to by many authors, following Matthey, as 'Robertsonian changes' after the American cytologist W. R. B. Robertson, who, in 1916, pointed out that a metacentric chromosome in one species, race or individual may correspond to two acrocentrics in another. It does not seem that Robertson ever dealt with any cases of dissociation in the course of his work, so that the term seems somewhat inappropriate.

The view that evolutionary replacements of a metacentric chromosome by two rod-shaped elements (acrocentric or telocentric) are usually or always due to 'dissociation' has been challenged by John and Hewitt (1968), John and Lewis (1968) and Southern (1969). These authors contend that simple fission through the centromere of a metacentric chromosome may sometimes give rise to two stable telocentrics. Their views are consequently bound up intimately with their opinion that many animal species do, indeed, possess truly telocentric chromosomes, rather than acrocentric ones.

In all discussions of the evolutionary role of chromosomal rearrangements it is necessary to bear in mind that the number of spontaneous rearrangements is vastly greater than the number which become established in phylogeny, either as 'floating' polymorphisms, or in the 'fixed' monomorphic condition. Those that survive the various hurdles, at the level of the cell, the individual and the population are probably well below one millionth of those that occur, in the germ-line of all the individuals comprising a species. Thus no direct correspondence between the types of chromosomal rearrangements that occur and those that survive need be expected. Certain kinds which are of very frequent occurrence may have been of negligible importance in the evolution of particular groups; while others whose origin is quite rare and exceptional may have played a major role in evolution.

8

CHROMOSOMAL POLYMORPHISM IN NATURAL POPULATIONS. I. INVERSIONS

The grosser forms of mutation . . .

R. A. FISHER: *The Genetical Theory of Natural Selection*

Many species of animals appear to be cytologically monomorphic, while others show various kinds of chromosomal polymorphism. In the first type of species we may examine very large numbers of individuals and find that they are cytologically indistinguishable or only show the chromosomal differences associated with maleness and femaleness (X- and Y-chromosomes). Such uniformity may be spurious in some instances, i.e. it may simply result from our inability to demonstrate minute cytological differences of various kinds which remain concealed from the observer. This is especially likely to be the case in those groups of animals where the chromosomes are very small. But to some extent this difficulty exists even in organisms whose chromosomes are large and favourable for study. Even where polytene salivary gland chromosomes are present, certain types of chromosomal rearrangements, particularly ones situated in heterochromatic segments, may escape detection.

In spite of these strictures, it seems likely that the natural populations of some and indeed many species are cytologically monomorphic; at least in some species of *Drosophila* such as *D. simulans, D. virilis, D. novamexicana* and *D. repleta* where the polytene chromosomes are large and favourable for study, complete chromosomal uniformity (except for sex-differences) seems to prevail (Wasserman and Wilson 1957; Stone, Guest and Wilson 1960). However, even in such species it is probable that a study of a very large number of individuals would reveal a few that were carrying chromosomal rearrangements of some kind. And we might increase the probability of discovering these individuals if we concentrated our search on abnormal-looking members of the population. In the grasshopper *Keyacris scurra* at least one individual in a thousand seems to be heterozygous for some kind of unique, and presumably newly arisen chromosomal rearrangement, usually a translocation (White 1961*b*, 1963). In the human species, cytological surveys of inmates of mental institutions and hospital patients with congenital malformations have now revealed a number of individuals carrying translocations or rearrangements of various kinds, in addition to the now well known trisomic types such as mongols and the *XO*, *XXY*, *XYY*, *XXX*, and *XXXY* sex types (see p. 580).

These karyotypes are clearly pathological and give rise to obviously non-adaptive syndromes. Considerable uncertainty still exists, however, as to the adaptive significance of the so-called minor variants of the human karyotype such as an increased length of one of the *D* group chromosomes, or an unusually large satellite on one of the *D* or *G* chromosomes. There are a great many different kinds of these minor chromosomal polymorphisms, which may affect many of the heterochromatic segments; their total frequency was estimated at 39% in a sample of Caucasian infants and 52% in a Negro sample (Lubs and Ruddle 1971). These figures include variations in the length of the *Y*-chromosomes, earlier reported by Cohen, Shaw and Maccluer (1966) and Unnérus, Fellman and de la Chapelle (1967). No evidence exists for the hypothesis of E. B. Ford (1964) according to which the human species would be polymorphic for numerous short inversions. Polymorphism for small pericentric inversions may, however, be much commoner in certain groups than has previously been suspected; thus Yosida and Amano (1965) and Yosida, Nakamura and Fukaya (1965) have recorded this kind of polymorphism in both *Rattus rattus* and *R. norvegicus*.

THE PRINCIPLES OF GENETIC EQUILIBRIUM IN NATURAL POPULATIONS

By a cytologically or chromosomally polymorphic population or species we mean, then, one that normally and regularly contains several different kinds of individuals whose karyotypes are visibly different (excluding, obviously, chromosomal differences between the sexes). Such a population may be in equilibrium or not. If it is in perfect equilibrium, then the frequencies of the alternative types of chromosomes will not change significantly from one generation to another. The chromosomal inversions present in some natural grasshopper populations seem to show this kind of perfect equilibrium (White 1958; Lewontin and White 1960; White, Lewontin and Andrew 1963), although it is probable that if we were able to examine larger samples or to study them over a great number of generations we should find that the equilibrium was less perfect.

A different kind of equilibrium, which we may designate imperfect, seems to occur in many natural populations of *Drosophila* species and is likely to be found in other organisms that have several generations in a year. The frequencies of alternative types of chromosomes may undergo fluctuations throughout the year, but these seasonal cycles are, to a considerable extent, repeated from year to year. Thus the composition of the January and July populations in any one year may be quite different, but the composition of the January populations over a series of years will be very similar. Such seasonal cycles are good evidence for the action of selective forces on the cytogenetic composition of the population.

Imperfect equilibria are not necessarily limited to organisms having several or many generations a year; but they will be more difficult to observe in species whose life cycle is considerably longer. In this kind of imperfect equilibrium, as a result of natural selection, the relative frequencies of two or more alternative types of chromosomes are changing but will eventually reach a new stable equilibrium after many generations. The frequency of the Pikes Peak inversion sequence has been increasing in Californian populations of *Drosophila pseudoobscura* (Dobzhansky 1956, 1958a, b, 1963b) but there seems little likelihood that it will replace all the other sequences.

Populations in genetic disequilibrium will manifest transient polymorphism, i.e. they are destined to become monomorphic after a certain number of generations of natural selection. Since most environments change with time it may be difficult to predict whether a particular trend that can be observed in a natural population is evidence of a complete disequilibrium or will slow down and become arrested before the population becomes monomorphic.

The various conditions leading to genetic equilibrium are now fairly well understood from a mathematical standpoint (Kimura 1956b, Haldane 1962). But theoretical understanding has greatly outstripped factual knowledge in this field, and it is still difficult to tell how far various mathematical models correspond to situations that actually occur in natural populations.

It is customary to assume that natural populations are large, so that inbreeding and the effects of genetic 'drift' can be neglected; and that they are panmictic, i.e. that mating is at random, as far as genotype is concerned. The first condition is true for many species, but rather obviously not for very rare ones, which undoubtedly form a considerable fraction of the total number. Even in the case of species which form relatively large populations, restricted vagility may lead to a slight degree of inbreeding (sib-matings, cousin-matings, etc. more frequent than if the whole population were randomized). Lewontin and Dunn (1960) have been led to the view that the house mouse may have such small genetic isolates in nature that a very significant amount of inbreeding occurs. The same may be true for the cytologically polymorphic shrew *Sorex araneus* (see p. 301). Studies of such species may reveal very significant departures from panmixia when areas of a hectare or so are considered. But they are very difficult, since for statistical purposes we require large samples, and these may simply not exist in species with very small local populations.

It is commonly assumed that the selective values of particular genotypes are the same in both sexes, and that they are independent of the frequencies of those genotypes in the population. These are convenient assumptions for mathematical exercises but perhaps unlikely to apply in nature.

The simplest conditions for genetic equilibrium in a large panmictic population are (in the case of an autosomal polymorphism) where the selective values are neither sex-influenced nor frequency dependent and where the heterozygote has a higher selective value than either homozygous genotype ('positive heterosis'). Symbolically, $AA < Aa > aa$. Under such circumstances we may designate the selective values of the three genotypes as $1 - S_A$ (AA): 1 (Aa):$1 - S_a$ (aa) where S_a and S_A are the selection coefficients. It is easy to prove that the equilibrium frequencies of A and a will be:

$$\frac{S_a}{S_a + S_A} \quad \text{and} \quad \frac{S_A}{S_a + S_A} \quad \text{respectively}$$

Thus the point of equilibrium is determined by the relative selective values of the two homozygotes and not by the degree of superiority shown by the heterozygote.

Where there are three or more alternative types of chromosomes, analogous equilibria will be established, provided that each heterozygote has a higher selective value than the corresponding homozygous genotypes (Kimura, 1956b).

In the above instances there is a unique point of equilibrium. Where the selection coefficients are different in the two sexes, however, there may be two alternative points of equilibrium (Owen 1953). No actual case seems to be known in which a natural polymorphism exemplifies this principle (i.e. in which some demes show one equilibrium frequency and others show the alternative frequency).

Where the selective values of the three genotypes AA, Aa and aa are frequency-dependent, so that each declines as the genotype in question increases in frequency, an equilibrium may be established even though at equilibrium the heterozygote does not show any superiority and may even be inferior to both homozygotes (Teissier 1954). The same will be true in a population whose individuals have the choice of two different ecological niches, in one of which AA is favoured while in the other aa is superior (Levene 1953; Li 1955). This is the principle of *annidation* (Ludwig 1950), namely the adaptation of specific genotypes in a polymorphic population to alternative habitats within the same environment. Actually, there may be no difference in practice between the frequency-dependent model and the annidation one, since the latter assumes that when each 'preferred' niche has been occupied by the genotype adapted to it, there will be a 'spill-over' into other niches to which the particular genotype is less well adapted.

The use of the term 'heterosis' to cover cases where the equilibrium depends on frequency-dependent selective values or annidation rather than on a simple selective superiority of the heterozygote regardless of time and place is perhaps permissible so long as we appreciate the essential differences involved. Thus, in a polymorphic population at equilibrium where 'simple' heterosis exists, and where natural selection acts at some time between fertilization and maturity, an examination of adult individuals should reveal an excess of heterozygotes over the number to be expected on the binomial square rule, and a corresponding deficiency of the homozygous genotypes. But no such excess of heterozygotes need occur, and a deficiency may even be observed, if the equilibrium is maintained by frequency-dependent selective values or annidation. Neither will there be an excess of heterozygotes in the adult population if the heterotic superiority depends on an increase of fecundity rather than of viability.

The situation with regard to sex-linked genetic polymorphisms is more complex. We may consider first of all the situation where the Y is non-existent or irrelevant to the situation. For convenience we suppose that the male is the heterogametic (XY or XO) sex. For a gene or inversion located on the X-chromosome we have to consider the selective values of the five genotypes:

$$(X_A X_A, X_A X_a, X_a X_a, X_A Y \quad \text{or} \quad X_A O, X_a Y \quad \text{or} \quad X_a O)$$

Although some authors have assumed that the selective values of $X_A X_A$ and $X_A Y$ would be the same (and similarly for $X_a X_a$ and $X_a Y$) this is clearly unjustified (Bennett 1958). This author points out that a selective advantage of the heterozygote over both homozygotes in the homogametic sex is neither a necessary nor a sufficient condition for genetic equilibrium. Nor does an excess of heterozygotes in the homozygous sex over 50% necessarily indicate a selective superiority of heterozygotes, as Da Cunha (1953) has supposed. For example, where the selective values are:

Genotypes	$X_A X_A$,	$X_A X_a$,	$X_a X_a$,	$X_A Y$,	$X_a Y$
Selective values	0.33	1	2	6	1

an equilibrium will be established with 64% of the adult females $X_A X_a$ (Bennett 1958).

An even more complex situation is where a given gene or inversion can be either on the X or on the Y. In that case we have to consider the selective values of all seven genotypes:

$$X_A X_A, X_A X_a, X_a X_a, X_A Y_a, X_A Y_a, X_a Y_A, X_a Y_a$$

This situation seems to actually occur in the natural populations of certain midges (Chironomidae) where A and a are cytological inversions and the X and Y cytologically indistinguishable, so that we cannot tell $X_A Y_a$ and $X_a Y_A$ apart (Acton 1957b; Martin 1962). As in the simpler case considered by Bennett one may have considerably more than 50% of the females heterozygous without any heterosis in that sex.

In practice, almost all the chromosomal polymorphisms which are encountered in natural populations seem to be in equilibrium, although the frequencies of the alternative chromosome types may undergo seasonal fluctuations or show changes over a period of years. This is not really surprising. A large number of chromosomal rearrangements which are deleterious in the heterozygous condition undoubtedly arise, but they will never reach high frequencies in the population and hence are unlikely to be observed unless very large numbers of individuals are examined. The situation in such cases may be symbolized as $S^O S^O > S^O S^N$ where S^O is the original sequence and S^N the new one; in a large panmictic population the selective value of the $S^N S^N$ genotype will be irrelevant, since there will be no such homozygotes so long as the frequency of S^N is very low. S^N will be eliminated from the population under these circumstances; such instances are the very rare or unique rearrangements mentioned earlier as existing in all species.

The opposite type of 'transitory polymorphism' would be where a newly arisen rearrangement steadily increases in frequency until it reaches fixation (0% to 100% in a few generations). The only circumstances under which this is at all likely, from a theoretical standpoint, would be where the hierarchy of the selective values is:

$$S^O S^O < S^O S^N < S^N S^N$$

Instances of this kind of situation may occur occasionally, even in the case of newly arisen rearrangements (e.g. where the S^N sequence increases viability when present in either single or double dose, but has a deleterious effect on fecundity when heterozygous). It is also a situation that may eventually develop as a result of a long evolutionary process in the course of which the genetic properties of S^O and S^N undergo modification. The selective value of $S^N S^N$ in a population with balanced polymorphism may gradually increase until it exceeds that of $S^O S^N$. Thus equilibrium will gradually change until it overbalances and the polymorphism is replaced by a new monomorphism. Change of the environment may, of course, play a role in changing the selective values, even without evolution of the genetic properties of the cytological sequences.

The opposite process may, of course, be expected to take place occasionally, The selective value of the $S^O S^O$ genotype may gradually increase until it exceeds that of $S^O S^N$; thus, theoretically, a new sequence may arise, may increase in frequency in the population and then later decrease once more, to disappear eventually.

There certainly exists another way in which new sequences may spread and reach fixation in natural populations. That is through drift in small and perhaps rather strongly isolated local populations (demes). This will be especially likely to occur on the periphery of the species distribution. We may imagine a situation in which we have $S^O S^O > S^O S^N < S^N S^N$ with $S^O S^O >$ or $< S^N S^N$ (homozygote superiority or 'negative heterosis'). In this case the S^N sequence may establish itself partly by drift and partly by selection. Since certain types of rearrangements reduce the fertility of heterozygotes by causing 'meiotic accidents', this kind of situation may not be too uncommon, especially in rather sedentary organisms with limited powers of locomotion.

All the above considerations about genetic equilibria are based on the assumption that the newly arisen rearrangement does not enjoy any special segregational advantage ('meiotic drive') in the heterozygote. It is at least possible (see p. 379) that a significant fraction of the chromosomal rearrangements that have established themselves in evolution may have been ones which tended to be included in the egg nucleus rather than in a polar body nucleus.

We have thus far considered only single cytological polymorphisms which may be compared to an allelic series: a chromosome 'A' exists in a series of alternative forms A_1, A_2, A_3, ... so that we can have diploid genotypes $A_1 A_1$, $A_1 A_2$, $A_2 A_2$, $A_1 A_3$ ... The rules governing the attainment of genetic equilibrium in such cases are fairly well understood.

Where a second polymorphism B_1, B_2 ... coexists with the first in the same population it may be located in the same chromosomal element or in a different one. The genetic relationship between the A and B polymorphisms may be a single additive one, or may involve an epistatic interaction. The consequences of linkage and epistasis in the attainment of genetic equilibrium in such situations has been explored theoretically by Lewontin and Kojima (1960). In actual populations epistatic interactions between different polymorphisms have been studied in *Drosophila* by Levitan (1955, 1958) and Levitan and Salzano (1959); and in the grasshopper *Keyacris scurra* by White (1957*b*), Lewontin and White (1960), White and Andrew (1962) and White, Lewontin and Andrew (1963). We shall consider the evidence in such cases later (pp. 260–3).

We must draw a distinction between cytologically *polymorphic* species, in which many or all demes contain several different karyotypes; and cytologically *polytypic* ones, which consist of two or more chromosomal races differing in karyotype. In the latter case, if the races have a zone of contact or overlap the species will be chromosomally polymorphic in that zone. We shall consider such situations later, as potentially leading to speciation. Some species are known which may be described as both polymorphic and polytypic. The difficulty here lies in the definition of a 'cytological race'. Should we, for instance, regard two populations which contain the same chromosome sequences, but in very different proportions, as constituting cytological races? We give, in Table 8.1, a tentative classification of the main types of cytological races, with a suggested terminology. In a cytologically polymorphic species with an extensive geographic distribution virtually every local colony or deme will be a microcytorace, some being more strongly demarcated than others.

Chromosomal polymorphism has been studied especially in Diptera, where the polytene chromosomes can be used in its investigation (Dobzhansky 1937*c*–70; Stone 1949, 1955, 1962; W. Beermann 1952*a*, 1955; Carson

TABLE 8.1. *Kinds of cytological races*

Populations *A* and *B* differ only in the relative proportions of certain gene sequences.	'Microcytoraces'
Population *B* contains one or more sequences not present in *A* (a special case of this situation is where one population is cytologically monomorphic, the other polymorphic).	'Mesocytoraces'
Population *A* is monomorphic for one sequence, population *B* is monomorphic for an alternative one. If a zone of overlap occurs the population in that zone will be polymorphic for the rearrangements in question.	'Macrocytoraces'

and Blight 1952; Patterson and Stone 1952; Kitzmiller 1953; Mainx, Kunze and Koske 1953; Brncic 1954, 1961*a*; Acton 1955–62; Carson 1955*a*, *b*, 1958*a*; Da Cunha 1955, 1960; Mainx, Fiala and von Kogerer 1956; Rothfels 1956; Rothfels and Dunbar 1953; Rothfels and Fairlie 1957; Mainx 1958; Stone, Guest and Wilson 1960; Kitzmiller and French 1961; Martin 1962; Krivshenko 1963; Pasternak 1964; Stalker 1964*a*, *b*; Petrukhina 1966; Grinchuk 1967). The families of Diptera in which paracentric inversion polymorphism has been recorded include the Limoniidae (Wolf 1941; Mainx 1951*a*), Ptychopteridae (Mainx 1951*a*), Culicidae, Chironomidae, Simuliidae, Sciaridae (Carson 1944), Cecidomyidae (White 1946*a*; Abramowicz 1964), Liriomyzidae (Mainx 1951*b*) and Drosophilidae. Inversion polymorphism has now been found also in the Collembola which possess polytene chromosomes (see p. 125).

Some other groups in which sufficient information is beginning to accumulate from studies of meiosis to enable us to estimate the role of chromosomal polymorphism of one kind or another in the natural populations include the grasshoppers and some other orthopteroid insects (White 1941*a*, 1951*a*, *b*, *c*, 1956, 1957*a*, *b*, *c*, 1958; Wahrman 1954*a*; White and Andrew 1960, 1962; Lewontin and White 1960; White, Lewontin and Andrew 1963), and certain genera of beetles (Manna and Smith 1959; Smith 1959, 1960*b*, 1962*a*, *b*, *c*, *d*, 1963). In many of these cases extensive numerical data have been published. Less complete investigations or isolated observations only have been carried out on various forms of cytological polymorphism in a neuropteran (Klingstedt 1934), several species of roaches (Lewis and John 1957*a*; John and Lewis 1958, 1959; John and Quraishi 1964), a number of species of scorpions (Piza 1947*a*, *b*, *c*, 1948*a*, *b*, *c*, 1950*a*, *b*, 1952; Sharma, Parshad and Joneja 1959), a marine isopod (Staiger and Bocquet 1956), the prosobranch mollusc *Thais lapillus* (Staiger 1954, 1955), the shrew *Sorex araneus* (Ford, Hamerton and Sharman 1956; Ford and Hamerton 1958; Meylan 1960; Matthey and Meylan 1961), certain African species of mice (Matthey 1963*b*, *c*, 1966*b*, *c*), the North American deer mouse *Peromyscus maniculatus* (Ohno, Weiler, Poole, Christian and Stenius 1966), *Rattus rattus* (Yosida, Nakamura and Fukaya 1965), *Rattus norvegicus* (Yosida and Amano 1965) and *Cercopithecus* spp. (Chiarelli 1968).

Variations in chromosome number have been recorded in populations of several species of Lepidoptera (Lorkovic 1941; Maeki 1958*a*, *b*), but the cytological conditions in this group are not very favourable for a precise analysis. In the Hymenoptera, with haploid males, a case of variation in chromosome number has been reported in *Tenthredo acerrima* by Waterhouse and Sanderson (1958). In recent years an increasing number of chromosomal polymorphisms have been found in mammals and it seems likely that even more would be found if a special search were to be made, since in the case of many species only one or two individuals have been investigated cytologically. There seems no doubt that eventually instances of cytological polymorphism will be found in all the major groups of animals; but it is difficult from the existing evidence to estimate what proportions of the species in the various groups will eventually prove to be cytologically polymorphic. At present it looks as if a very high percentage of species of Diptera, a lower percentage of mammalian species and an intermediate proportion of grasshopper species show chromosomal polymorphism as a regular feature of their natural populations. In many groups, of course, the chromosomes are too small to permit the ready detection of rearrangements or otherwise unsuitable for cytological study of a really thorough type.

PERICENTRIC AND PARACENTRIC INVERSIONS

The main types of cytological polymorphism that have been studied are listed in Table 8.2. We may begin by considering inversions. In the dipterous genera *Drosophila* (Sturtevant and Beadle 1936), *Sciara* (Carson 1946), *Chironomus* (Acton 1956; Beermann 1956b) and *Dicranomyia* (Wolf 1941) mutually inverted chromosome segments apparently may twist round,

TABLE 8.2. *The main types of chromosomal polymorphism in natural populations, and ways of detecting them*

	Polytene chromosomes	Ordinary somatic chromosomes	Meiosis
Inversions			
Paracentric	+	−	(+)[1]
Pericentric	+	+[2]	+
Translocations			
Mutual	+	(+)[3]	+
Fusions	(+)[4]	+	+
Dissociations	(+)[4]	+	+
Duplications	(+)[5]	+	+
Deletions	(+)[5]	+	+
Supernumerary chromosomes	(+)[5]	+	+

[1] Provided that an inversion loop is formed and a chiasma is formed within it.
[2] Except for 'symmetrical' types.
[3] Detectable only if the regions interchanges are unequal in length or carry distinctive markers, e.g. a nucleolar organizer.
[4] Detectable only with difficulty in forms with a chromocentre.
[5] Duplications and deficiencies of heterochromatic regions are not detectable (or only with great difficulty) in polytene chromosomes.

producing reversed loops at zygotene, so that chiasmata can be formed between them. Presumably this is a general feature of dipteran genetic systems; but it is certain in some species of *Chironomus* that loops are not formed with the same frequency at meiosis as in polytene nuclei (Martin 1967). A single chiasma within an inversion loop will lead to the formation of an acentric and a dicentric chromatid if the inversion is paracentric or to two monocentric chromatids each carrying a deficiency and a duplication if it is pericentric. All such chromatids must be expected to have a lethal effect in the zygote. In *Drosophila* and *Sciara* there is no chiasma formation in the male so that 'lethal' chromatids will only be formed in the female. However, in both genera there seems to be a meiotic system in oogenesis which ensures that acentric and dicentric chromatids pass into the polar body nuclei rather than into the egg nucleus (Sturtevant and Beadle 1936; Carson 1946) so that fecundity is not reduced. Deficiency-duplication monocentrics such as are produced in heterozygotes for pericentric inversions will, however, be just as likely to get into the egg nucleus as into any of the other three products of meiosis.

If two chiasmata are formed within a paracentric inversion loop the consequences will depend on which chromatids are involved. 'Two strand double' crossing-over will produce four monocentric chromatids, three strand double crossing-over will yield a dicentric and an acentric in addition to two monocentrics. Four strand double crossing-over gives two dicentric chromatids and two acentrics. In the latter case the egg nucleus will not receive any chromatid from the bivalent in question and an inviable zygote (or a patroclinous XO male if the inversion is in the X) will result. A three strand double exchange with one chiasma between the inversion loop and one between the loop and the centromere will give rise to an acentric and a ring chromatid at first anaphase. The latter will form a dicentric bridge at the second anaphase; in consequence half the eggs will be expected to be inviable.

Double crossing-over within a pericentric loop will give rise to deficiency-duplication chromatids except in the case of a two strand double exchange. In view of these properties of pericentric inversions it is not surprising that almost all the inversions that occur naturally in fly populations are of the paracentric type. On theoretical grounds we should expect natural selection to eliminate all newly arisen pericentric inversions rather rapidly, and no doubt this is what happens in dipteran populations in general. A few pericentric inversions have, however, been recorded in *Drosophila*. Two are known in *D. robusta*, in the United States, one of which (3 L-R) is common in certain areas (Carson and Stalker 1947); apparently the explanation is that chiasmata are only infrequently formed within the inversion loop. Pericentric inversions have also been recorded in *D. ananassae* in Brazil

(Freire-Maia 1952), where the explanation is probably the same. In *D. algonquin* Miller (1939) found a pericentric inversion which coexisted with a paracentric one that largely overlapped it, so minimizing the rate at which duplication-deficiency eggs would be produced. Alexander (1952) tested two artificially produced pericentric inversions in the second chromosome of *D. melanogaster* (Plum[2] and Glazed) and found that heterozygotes laid about 7–10% inviable eggs, by comparison with controls.

Some evidence that pericentric inversions have occurred in the phylogeny of a group may be obtained simply by comparing the karyotypes of the species which have survived. Stone (1955) has listed thirty-two pericentric inversions which he believes have occurred in the history of the genus *Drosophila*, of which the best known is the one that established itself in the second chromosome of the *montana* section of the *virilis* group, converting that element from an acrocentric to a metacentric. This figure compares with over 35 000 paracentric inversions which are believed to have become established in the phylogeny of the genus.

For a long while the occurrence of paracentric inversions in natural populations of Chironomidae, Culicidae, Simuliidae, Limoniidae and Ptychopteridae seemed paradoxical, since in these families of Diptera chiasmata are formed in spermatogenesis. One explanation seems to be that where a dicentric chromatid is formed by crossing-over the 'bridge' between the two nuclei at first anaphase prevents their separation, so that a giant functionless cell results which does not develop into a sperm (Beermann 1956*b*). The number of sperms formed by inversion heterozygotes is slightly reduced (which is unimportant) but no aneuploid sperms which would be lethal to the egg they fertilized are formed. Thus these 'chiasmate' Diptera avoid paying the penalty for paracentric inversion heterozygosity, without having given up chiasma formation altogether in the male, as the Sciaridae, Drosophilidae and many other families of 'higher' Diptera have done. Another factor which must increase the number of viable sperms produced by chironomids heterozygous for inversions is that there are frequently fewer inversion loops present at pachytene than in the polytene nuclei of the same individual, i.e. not all the potential loops are actually formed at zygotene, presumably due to non-homologous synapsis (Martin 1967). A third factor which may operate to reduce the number of dicentric chromatids in some chironomid species is that there is distal localization of chiasmata, most of the inversions being located proximally.

INVERSIONS IN GRASSHOPPER POPULATIONS

In a number of species of grasshoppers and crickets belonging to several different groups, pericentric inversions are regularly present in the natural populations. Since polytene chromosomes are not found in these insects, such rearrangements are usually detected in the meiotic chromosomes, although they can also be studied in somatic or gonial mitoses. Most of them convert acrocentric chromosomes into metacentrics, or vice versa, but in a few instances a pericentric inversion has converted a metacentric into a different type of metacentric. Pericentric inversions that were perfectly symmetrical (i.e. with the breakage points situated an equal distance on either side of the centromere) would be very difficult to detect in grasshoppers.

The genetic properties of inversions, whether paracentric or pericentric, depend on the fact that they act as *cross-over suppressors*. In the case of the pericentric inversions of grasshoppers this suppression is absolute, i.e., chiasmata are simply not formed in the mutually inverted segment, in heterozygotes of either sex. In all probability, reversed loops are not formed at zygotene (Coleman 1948; White, unpublished observations), the regions which are inverted with respect to one another undergoing instead a kind of non-homologous pairing similar to that which sometimes occurs in maize chromosomes (McClintock 1933). There are only three known exceptions to the principle that chiasmata are not found between mutually inverted chromosomes in grasshoppers. One (Figs. 8.1 and 8.2) is a newly-arisen pericentric inversion in *Keyacris scurra* (White 1961*b*); another is a rather long pericentric in *Trimerotropis sparsa* (White and Morley 1955). In both cases the 'exceptional' chiasmata were quite rare. Nur (1968*b*) found a single individual of *Camnula pellucida* which was heterozygous for a paracentric inversion extending over about 10% of one of the longest chromosomes. Non homologous 'straight' pairing occurred in 88% of the pachytene nuclei, inversion loops in about 4% and asynapsis in approximately 8%. About 8% of the cells in first and second anaphase–telophase showed acentric fragments and dicentric chromatids (evidence of crossing-over within an inversion loop).

Suppression of crossing-over may be effectively complete in the case of single short inversions in *Drosophila* species where only two sequences coexist in the population. Loops will be formed at zygotene and chiasmata will be formed within them, but since these are eliminated in the polar body nuclei they are genetically ineffective. In the case of longer inversions suppression of crossing-over will be incomplete, since two strand double crossing-over within the loop will permit some transfer of genetic material between the mutually inverted segments. Incomplete genetical isolation of mutually inverted segments may also exist where several cytologically overlapping inversion sequences coexist in the same population.

Where genetical isolation between mutually inverted sequences is absolute or near-absolute, different alleles will accumulate on them, so that

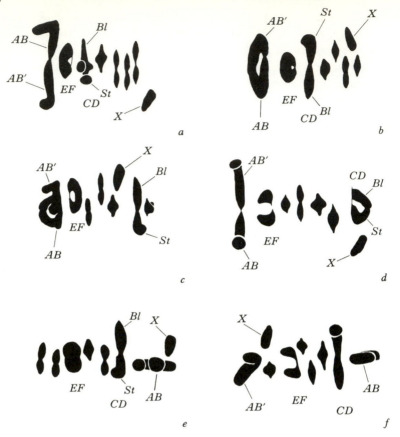

Fig. 8.1. First metaphases (side views) in an individual of the grasshopper *Keyacris scurra* (15-chromosome race) that was heterozygous for a spontaneous pericentric inversion in the *AB* chromosome (this chromosome is normally structurally homozygous). *a, d*, with a single chiasma in the *AB* bivalent; *b*, with one in both limbs; *c*, with one within the inverted segment and one beyond it; *e*, a cell in which the two centromeres of the *AB* bivalent have orientated amphitelically and *f*, a cell in which separate *AB* and *AB'* univalents have both orientated amphitelically. The *CD* chromosome was heterozygous Blundell/Standard in this individual and in *d* there is an anomalous terminal chiasma between the proximal end of the Blundell chromosome and the short limb of the Standard homologue. From White (1961*b*).

the genetic properties of the inversions are likely to become more and more different and distinctive in the course of time. Genetic divergence is limited, however, by the fact that homozygotes for both sequences have to be functioning, viable individuals, i.e. the two sequences must control substantially the same biosynthetic functions. Haldane (1957) has concluded,

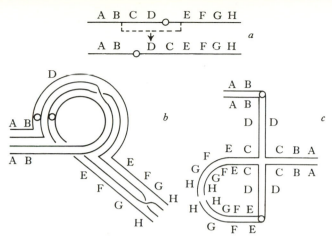

Fig. 8.2. The approximate extent of the pericentric inversion in Fig. 8.1 and the mode of origin of the bivalent shown in Fig. 8.1c. From White (1961b).

from a mathematical analysis, that in the case of autosomal inversions the only alleles that stand any chance of being added to an inversion in the course of evolution are ones which produce a cumulatively heterotic effect. If this were the whole story, we should expect that inversion polymorphisms would go on becoming more and more heterotic until they eventually became balanced lethal systems, the homozygotes being completely inviable.

The answer to this apparent paradox (since balanced lethal systems are apparently very uncommon in natural populations – see p. 283), is, in all probability, that in a species with inversion polymorphism the genetic evolution of the rest of the karyotype is such as to counteract this tendency to increased heterosis which would result from the evolution of the mutually inverted regions.

The main groups of grasshoppers in which pericentric inversions are relatively common are the trimerotropine grasshoppers of North America (one section of the genus *Trimerotropis*, the entire genus *Circotettix* and the single species of *Aerochoreutes*) and the morabine grasshoppers of Australia. In addition they have been recorded in an Indian species of *Orthacris* (Rao 1934), in the Australian *Cryptobothrus chrysophorus* (White, unpublished) and in another Australian species, *Austroicetes interioris* (White and Key 1957; Nankivell 1967). The earlier hypothesis that these could be centromere shifts (i.e. 3-break rearrangements) rather than pericentric inversions can be discarded now that it is known that many inversions regularly pair non-homologously at zygotene rather than forming reversed loops.

The great majority of grasshopper species, of course, lack these rearrangements in their natural populations. It is quite uncertain whether in these a mechanism of non-homologous pairing of inverted segments is lacking so that reversed loop pairing would occur if they did possess inversions, with dicentric and acentric chromatids as a consequence. Rearrangements which appear to be of the same nature as the pericentric inversions of the trimerotropine and morabine grasshoppers have been recorded in the mantids *Tenodera aridifolia* and *Dystacta* sp. (White 1941*a*, 1965*b*), in the neuropteran *Hemerobius stigma* (Klingstedt 1934) and in the mammals *Peromyscus maniculatus* (Ohno, Weiler, Poole, Christian and Stenius 1966) and *Mastomys natalensis* (Matthey 1966*a*).

The cytological polymorphism of the Trimerotropi has been studied by Carothers (1917, 1921, 1926, 1931), Wenrich (1917), King (1923), Helwig (1929), Coleman (1948) and White (1949, 1950*b*, 1951*a, b, c, d, e, f*, 1958). The group is especially characteristic of the arid regions of western North America, and contains about fifty-four species. *Trimerotropis* (about forty-six species) is probably a composite assemblage which should be split up into several genera. Six species occur east of the Mississippi, the remainder being western in distribution (one is endemic in Chile, all the others being North American, although *T. pallidipennis* occurs in South America as well). *Circotettix* includes seven species, all western in distribution, while *Aerochoreutes* has a single western species.

On the basis of karyotypes and genetic mechanisms, we may subdivide the genus *Trimerotropis* into two sections. In the more primitive of these (A) all the chromosomes (apart from some supernumerary elements that occur in certain species) are acrocentric; while in section B certain of the chromosomes have become metacentric as a result of pericentric inversions. Apart from supernumerary chromosomes and chromosome regions, which are known in a number of species, the members of section A do not show chromosomal polymorphism. In section B there are likewise some species in which the pericentric inversions have all reached fixation, the number of metacentric and acrocentric elements being constant for the species. Thus the widespread *T. pallidipennis* regularly shows a metacentric *X*-chromosome, three pairs of metacentric autosomes and eight pairs of acrocentrics, and so does the closely related *T. diversellus*, confined to a small area of Wyoming and Montana. *T. saxatilis* (southeastern United States) and *T. schaefferi* (Gulf Coast of Texas) are likewise cytologically monomorphic, but have four pairs of metacentric autosomes, in addition to the metacentric *X*, and only seven pairs of acrocentrics.

In the members of this group polymorphic for pericentric inversions the number of metacentric chromosomes is not fixed but varies from one

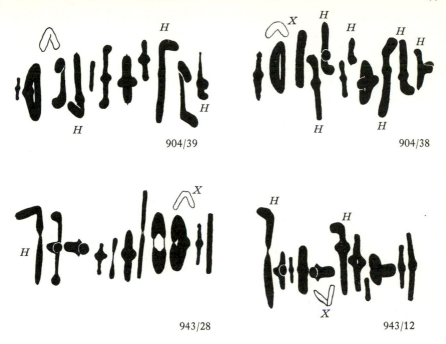

904/39   904/38

943/28   943/12

Fig. 8.3. First metaphases (side view) from four different individuals of the grass-hopper *Trimerotropis suffusa*. H, structurally heterozygous ('heteromorphic') bivalents. Individuals 904/38 and 904/39 are from a population at Truckee, California, with a mean of about 3.0 heteromorphic bivalents per individual. Individuals 963/12 and 943/28 are from a population in Archuleta county, Colorado, with a mean of about 0.3 heteromorphic bivalents per individual.

individual to another. Thus a particular chromosome pair is sometimes represented by two acrocentrics, sometimes by an acrocentric and a meta-centric and sometimes by two metacentrics. In a few instances a chromosome may have two different metacentric sequences in addition to the original acrocentric one (instances of this may be seen in Figs. 8.4, 9.4). The extent of all this polymorphism is highly variable, both from species to species and also from population to population within the individual species. Eight species in section B of *Trimerotropis*, five species of *Circotettix*, the single species of *Aerochoreutes* and a form known as '*Circotettix*' *maculatus* (almost certainly not a true *Circotettix*) all show pericentric inversion polymorphism. The morphological diversity of the members of this assemblage suggests that the group as a whole is of considerable antiquity. It has been suggested that *Aerochoreutes* and *Circotettix* are related to the Palaearctic group Bryodemae, and an Old World origin has been claimed for these two genera. But *Bryodema tuberculatum* from Bavaria (Fig. 6.13) and *B. semenovi* from Kazakhstan show an entirely different picture, with no meta-centric chromosomes in the karyotype, no cytological polymorphism and a type of chiasma localization unlike anything known in the Trimerotropi (White 1954 and unpublished studies). It therefore seems probable that the evolution of the structurally heterozygous trimerotropine grasshoppers took place in Tertiary times in western North America, a few members of the group having subsequently extended their range beyond that area (*Trimerotropis verruculata*, which has reached the mountains of New England and *T. ochraceipennis* (Fig. 8.5), whose ancestors somehow reached Chile). Pre-sumably the cytologically monomorphic members of section B (*pallidipennis, diversellus, saxatilis, schaefferi* and perhaps *huroniana*) passed through a stage in their evolution when they were cytologically polymorphic; they must have 'experimented with' inversion polymorphism and later abandoned it, becoming once more cytologically monomorphic. Naturally, we cannot tell whether this 'experiment' was of long or short duration, or whether it occurred in many ancestral species or in a single one.

In some of the cytologically polymorphic species only one chromosome pair has an acrocentric and a metacentric sequence. This seems to be the case in *Circotettix coconino* (northern Arizona) and *C. crotalum* (confined to two mountain ranges in southern Nevada). At the other end of the scale, the population of *Trimerotropis sparsa* at Whitewater, Colorado is polymorphic for inversions in at least seven chromosome pairs out of eleven (White 1951c) and in the Californian *T. thalassica* inversion polymorphism seems to occur in as many as ten chromosome pairs (King 1923; White, unpublished observations). In these demes there should be, respectively, 3^7 ($= 2187$) and 3^{10} ($= 59\,049$) cytologically different types of individuals; some of these may, however, be so rare that they may never yet have occurred in the history of the species.

There seems to be no direct or inverse correlation between the extent to which chromosomal polymorphism is developed in these species and the area occupied by that form at the present time. For example, *T. thalassica* occupies an area of California which is about equal in extent to the area of Arizona occupied by *Circotettix coconino*. Most of the species which have some 'polymorphic' chromosomal pairs also have some which are invariably acrocentric and others which are invariably metacentric. Thus such species as *C. crotalum* and *C. coconino*, which have four pairs invariably metacentric probably showed a greater degree of cytological polymorphism at some time in the past, or are descended from more polymorphic forms, and are perhaps on their way to becoming cytologically monomorphic.

Throughout this whole group (Section B of *Trimerotropis* + *Circotettix* +

Fig. 8.4. *Trimerotropis sparsa. a,* spermatogonial metaphase with a metacentric X and six metacentric autosomes (Lovelock, Nevada); *b,* one with an acrocentric X and only five metacentric autosomes (Cherry Creek, Nevada); *c,* first metaphase (side view) in an individual from Duchesne, Utah which was heterozygous for a centric fusion and showed a trivalent (T) as well as two heteromorphic bivalents (H); *d,* first metaphase from an individual from Olathe, Colorado with three heteromorphic bivalents, one of which is composed of two metacentrics which differ in the position of the centromere. From White (1951*f*).

Aerochoreutes + '*Circotettix*' *maculatus*) the X-chromosome is always metacentric, except in a few populations of *Trimerotropis gracilis, T. sparsa* and *T. thalassica*, in which some males have acrocentric X's (presumably a reversion to the primitive condition). Some females in these populations will necessarily have one X acrocentric and the other metacentric.

In the species of Trimerotropi characterized by extensive cytological polymorphism the chromosomal constitution varies greatly from one population to another, sometimes over distances of only ten to twenty miles (White 1951*c*). These species, in fact, show a strong tendency to form micro-geographic races, distinguishable by the amount of structural heterozygosity, relative numbers of acrocentric and metacentric chromosomes,

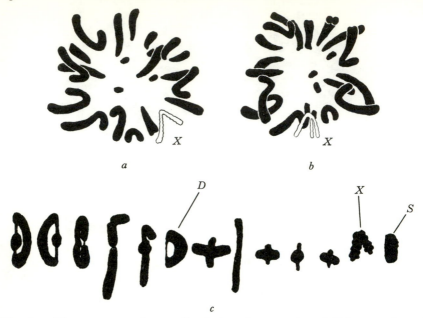

Fig. 8.5. Chromosomes of some South American species of *Trimerotropis. a, b,* spermatogonial metaphases of an individual of *T. ochraceipennis* from Chile (metacentric *X* and seven metacentric autosomes). *c,* a first metaphase of an individual of *T. pallidipennis* from 11 400 ft in the Argentinian Andes with 2 heteromorphic bivalents and a supernumerary chromosome (*S*). *T. pallidipennis* (if, indeed, it is the same species) is not known to be chromosomally polymorphic in North America; but note that both South and North American populations (compare Fig. 6.16) show short arm chiasmata in one bivalent (*D*). *c* from Mesa (unpublished).

presence or absence of supernumerary chromosomes, etc. Thus in *T. sparsa* (a species of the 'shadscale' vegetational zone) the mean number of metacentric autosomes per individual is 19.05 at Whitewater, Colorado, while at Delta (nineteen miles away) it is 11.40 and at Mack (thirty-four miles in the opposite direction) the corresponding figure is 11.81 (White 1951*c*). This is an extreme example, but less striking differences in the cytological composition of local populations seem to occur throughout this whole group of grasshoppers. No monomorphic populations are definitely known, however, in any of the species which show chromosomal polymorphism, although the degree of polymorphism may vary enormously. Thus in *T. suffusa*, populations from pine forests near Lake Tahoe, California, have a mean of about 4.0 structurally heterozygous bivalents per individual, while in southern Colorado the corresponding figure is approximately 0.28 (Fig. 8.3). Thus within the same species we have some populations in which almost every individual is a structural heterozygote (most of them being heterozygous for several chromosome pairs) and other populations in which the overwhelming majority of the individuals are structurally homozygous.

The population of *Trimerotropis gracilis* at Limerick Canyon, Humboldt Range, Nevada has been studied over the period 1949–57 (White 1958). There are pericentric inversions in chromosomes 5, 6, 7, 8 and 9. In each case the acrocentric type of chromosome is referred to as Standard, the metacentric being given an arbitrary name (derived from the geography of the area). In the case of chromosome 5, there is a significant excess of heterozygotes over the number expected on the binomial square rule, so that we may conclude that there is a heterotic effect on viability (Table 8.3). In the case of chromosome 7, there seems to be a more complex situation (Table 8.4), with the most common heterozygous genotype apparently showing a low viability. As calculated, these selective values do not indicate genetic equilibrium. Nevertheless, there is good evidence that the composition of the population (as far as chromosomes 5 and 7 are concerned) remained unchanged between 1949 and 1957. The reason for this discrepancy is not known; possibly the selective values of the various genotypes are to some extent frequency-dependent.

Trimerotropis sparsa, gracilis, cyaneipennis and *suffusa* all show supernumerary chromosomes in some of their populations, and so does *Circotettix undulatus* (see p. 322). At least four of the cytologically polymorphic trimerotropine species include races which possess a centric fusion (i.e. in which $2n\male = 21$) as well as populations lacking the fusion ($2n\male = 23$). Thus in *Trimerotropis gracilis sordida* from Vancouver Island there is no fusion (Coleman 1948); but in material of *T. g. gracilis* from Colorado, New Mexico,

TABLE 8.3. *Trimerotropis gracilis* (Limerick Canyon population 1949–57)

Chromosome 5. Observed numbers of individuals and numbers expected on the Hardy–Weinberg formula*

	Observed	Expected
Standard/Standard	1237	1255
Standard/Humboldt	890	854
Humboldt/Humboldt	127	145
	2254	2254

* Based on White 1958

TABLE 8.4. *Trimerotropis gracilis* (Limerick Canyon population 1949–57)

Chromosome 7. Observed and expected numbers of individuals and calculated selected values of the genotypes*

Genotype	Observed number	Expected number (Hardy–Weinberg)	Calculated selective value
Standard/Standard	29	33.44	0.70
Standard/Oreana	324	317.44	0.82
Standard/Rochester	151	148.68	0.82
Oreana/Oreana	790	753.40	0.85
Oreana/Rochester	626	705.76	0.72
Rochester/Rochester	204	165.28	1.00
	2124	2124.00	

* Based on White 1958

Utah, Nevada and California such a fusion seems to be invariably present (White, unpublished). A similar situation exists in *T. cyaneipennis*, where a fusion is present in all the material collected between western New Mexico and Oregon, but is absent from populations in central New Mexico. In *T. sparsa* a fusion is present in certain populations in northwestern Colorado and the adjacent part of Utah, but is absent from most of the range of the species (White 1951*f*). In this case two populations were encountered in which the fusion was present in some individuals but not in others, one individual being a fusion heterozygote in which a characteristic trivalent was present at meiosis (Fig. 8.4*c*). A similar situation exists in populations of *Circotettix undulatus* in the Truckee Valley, Nevada, in which fusion heterozygotes have also been found (Evans 1954). Helwig (1933) illustrated a similar fusion heterozygote in a population of *Trimerotropis suffusa* at Crescent Lake, Washington.

The pericentric inversions studied by Nankivell (1967) in natural populations of the Australian grasshopper *Austroicetes interioris* differ in two respects from those of the genus *Trimerotropis*. In the first place, the mutually inverted sequences in *A. interioris* contain blocks of heterochromatin which differ in size and position; in other words duplications or deficiencies of heterochromatic segments have been superimposed on the original inversion polymorphism (Fig. 8.6). Secondly, in certain classes of multiple inversion heterozygotes there are 'homoeologous' associations of chromosome ends, leading to the formation of chains of four chromosomes at meiosis. These associations are presumed to be chiasmatic. Since alternate disjunction of the chains of four is more frequent than would be the case if

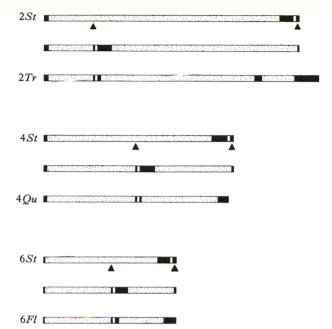

Fig. 8.6. The main heterochromatic regions of the structurally polymorphic chromosomes 2, 4 and 6 in the grasshopper *Austroicetes interioris*. Heterochromatin black, euchromatin stippled, centromeres unshaded, breakage points indicated by triangles. In each case the acrocentric *Standard* sequence is shown above (presumed ancestral condition), while below the sequence that would result from a single pericentric inversion is indicated. The lowest sequence of each chromosome (*Trangie* in the case of no. 2, *Quorn* in that of no. 4 and *Flinders* in that of no. 6) represents the arrangement of euchromatic and heterochromatic segments which is actually found. From Nankivell (1967).

orientation were at random, there is some degree of segregation distortion as far as the inversions are concerned. The inversion polymorphisms in chromosomes 2, 4 and 6 all show an excess of heterozygotes over the Hardy–Weinberg expectation, i.e. the data are suggestive of heterosis. This extremely complex system of chromosomal polymorphism seems to exist throughout the geographic range of the species.

In the Australian morabine grasshoppers about 10% of approximately 180 species that have been examined have their populations regularly polymorphic for pericentric inversions. These seventeen or eighteen species are not closely related, so that a general tendency to develop this kind of genetic polymorphism seems to be present throughout the group. Most of these species have inversions in only one, two or three chromosome pairs.

The number of cytologically distinct genotypes is hence three, nine or twenty-seven.

The genetic properties of these grasshopper pericentric inversions have been investigated in some detail in *Keyacris* (formerly *Moraba*) *scurra*

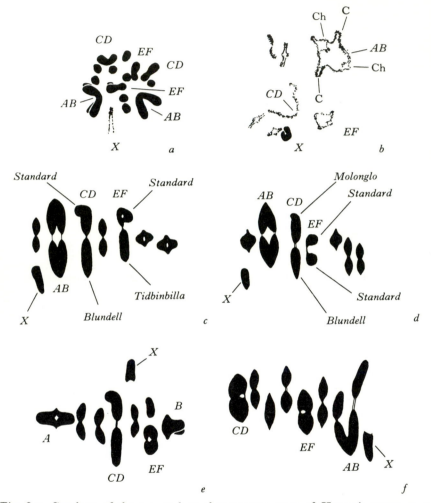

Fig. 8.7. Cytology of the 15- and 17-chromosome races of *Keyacris scurra*. *a*, a spermatogonial metaphase from an individual of the 15-chromosome race with the constitution *St/St*, *St/St*; *b*, diplotene, showing the positions of the centromeres (c) and the chiasmata (Ch) in the *AB* bivalent; *c*, first metaphase (side view) of a *St/Bl*, *St/Tid* individual of the 15 chromosome race; *d*, a similar figure from a *Mol/Bl*, *St/St* individual; *e*, a first metaphase from a *St/St*, *St/St* individual of the 15-chromosome race which was heterozygous for a 'broken' *AB* chromosome. From White (1956).

where over 19000 individuals from natural populations have been studied (White 1956, 1957a, b, c; White and Andrew 1960, 1962; Lewontin and White 1960; White, Lewontin and Andrew 1963). In this species the inversions are present in two chromosome pairs, the '*CD*' pair and the '*EF*' pair. In the case of the *CD* element four naturally-occurring sequences are known, which have been named *Standard, Blundell, Molonglo* and *Snowy*, but the last has only been found at a single locality, and the distribution of *Molonglo* is also quite restricted, so that most of the work has been carried out on *Standard* and *Blundell*. In the *EF* chromosome only two sequences are known, *Standard* and *Tidbinbilla*.

The Standard sequences of both elements are metacentric. There is some reason to believe that this was the ancestral condition in both of them. The *CD* element is generally metacentric throughout the morabine grasshoppers and is certainly so in the species most closely related to *K. scurra*; while the *EF* element arose by a fusion between '*E*' and '*F*' elements that are separate in these related species.

The *Blundell, Snowy* and *Tidbinbilla* sequences are all acrocentric, while *Molonglo* is a J-shaped element with one limb much shorter than the other. *Snowy* and *Blundell* are indistinguishable from one another in a mitotic division and even at meiosis a *Snowy/Snowy* individual cannot be distinguished from a *Blundell/Blundell* one. In *Snowy/Blundell* heterozygotes, however, it is very clear from meiotic divisions that the centromere is at opposite ends of the chromosome in the two sequences (i.e. if one is $a.bcdefg$ $hijk$, the other is $ajihgfedcb.k$, where the point represents the centromere). The phylogeny was probably

$$Standard \rightarrow Blundell \rightarrow Snowy$$
$$\rightarrow Molonglo$$

less probably

$$Standard \rightarrow Blundell \rightarrow Snowy$$
$$\rightarrow Molonglo$$

In demes that have been studied over periods of several years (there is only one generation a year) no significant changes in the frequencies of the various cytological sequences have occurred. The genetic equilibrium accordingly appears to be a very stable one, little or not at all affected by seasonal variations in amount of rainfall or other meteorological factors. On the other hand, demes only five miles apart may differ very considerably in cytogenetic composition. This may indicate a relationship between the

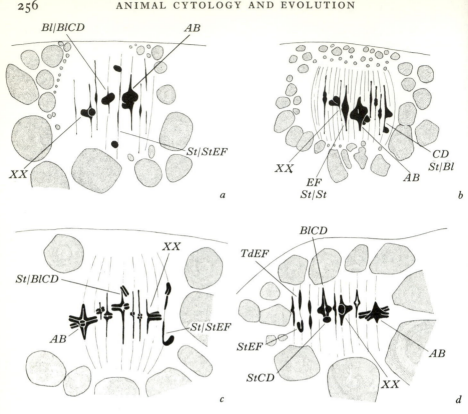

Fig. 8.8. First metaphases (side views) in oocytes of the 15-chromosome race of the grasshopper *Keyacris scurra* (compare with male first metaphases in Figs. 8.7 and 8.9). The constitution of the *CD* and *EF* bivalents is indicated. Note the interstitial chiasmata, particularly in the *AB* bivalent and in structurally heterozygous *CD* bivalents – in the male these bivalents usually show only terminal chiasmata at first metaphase. Yolk masses stippled.

genetic properties of the inversion sequences and the prevalence of various food plants or shelter plants; alternatively it may result from a relationship to the soil type (since the eggs are laid in the soil and embryonic mortality is probably high).

The main types of demes that have been encountered in *Keyacris scurra* are shown in Table 8.5. Type 6, which is the common one throughout a large geographic area, has been thoroughly investigated. In most such demes the frequency of the *Blundell* sequence is well over 50% and reaches over 92% at a locality known as Tarago Swamp. However, at Michelago, N.S.W., the *Standard CD* sequence greatly exceeds the *Blundell* one in frequency.

TABLE 8.5. *Types of demes in Keyacris scurra*

CD chromosome	EF chromosome	Localities
(1) *St* sequence only	*St* sequence only	Bright, Wodonga, Benalla
(2) *St* sequence only	Mainly *St*, (*Td*)	Yackandandah, Beechworth
(3) *St*, *Bl*	*St* only	Murringo, South Gundagai
(4) *St*, *Bl*, (*Mol*)	*St* only	Taemas Bridge
(6) *St*, *Bl*	*St*, *Td*	Wombat, Wallendbeen, Murrumbateman, Hall, Tarago Swamp, Royalla
(7) *Bl*, *Sn*	*St*, *Td*	Berridale

A simple relationship seems to exist between these inversion sequences and the overall size of the insects (White and Andrew 1960, 1962; White, Lewontin and Andrew 1963). The *Standard CD* and *EF* chromosomes are size-increasing elements, the *Blundell* and *Tidbinbilla* sequences being size-decreasing. Thus in a population of type (6) the *St CD/St CD*, *St EF/St EF* individuals will be, on the average, larger than any other genotype and the *Bl/Bl*, *Td/Td* insects will be the smallest. Overall size shows considerable geographic variation, however, so that the *Bl/Bl*, *Td/Td* individuals at one locality may be larger than the *St/St*, *St/St* ones at another.

The biometric effects of these *K. scurra* inversions clearly indicate that there has been a differential accumulation of size-increasing and size-decreasing alleles on the mutually inverted cytological sequences. Since the biometric effects are essentially the same over an area of hundreds of square miles, part of which is inhabited by the 15-chromosome race and part by the 17-chromosome race (see p. 259), they seem to have shown amazing evolutionary stability, probably over thousands of years. However, we only know about the biometric effects of the inversions in demes of type (6) and they may be different in the other types.

The biometric effects of the *Keyacris scurra* inversions seem to be simply additive, and if any dominance or epistasis exists, it is not very evident. The situation with regard to the effects on viability are somewhat more complex, however. In populations of type (6), containing the chromosomes *Standard CD*, *Blundell CD*, *Standard EF* and *Tidbinbilla EF*, there will be nine genotypes, which we may designate by the shorthand notations *SS-SS*, *SB-SS*, *BB-SS*, *SS-ST*, *SB-ST*, *BB-ST*, *SS-TT*, *SB-TT*, *BB-TT*. However a statistical study of contingency tables reveals slight but consistent deviations from independence between the *SS*, *SB* and *BB* genotypes for the *CD* chromosome and the *SS*, *ST* and *TT* genotypes for the *EF* element. In most populations of type (6), the commonest genotype is *BB-SS* and the

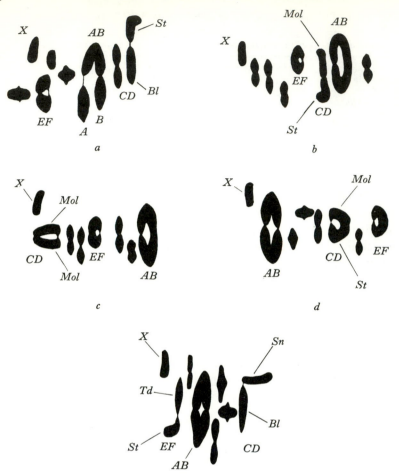

Fig. 8.9. Male first metaphases (side views) of *Keyacris scurra* showing hetero-zygosity for various chromosomal rearrangements. *a*, in an individual heterozygous for the dissociation of the *AB* chromosome (the offspring of a cross between the 15-chromosome and the 17-chromosome races) – this individual was heterozygous for the *Standard* and *Blundell* sequences of the *CD* chromosome; *b, d* in individuals of the 15-chromosome race heterozygous for the *Standard* and *Molonglo* sequences of the *CD* chromosome (in *b* the heterozygous bivalent shows only a distal chiasma, in *d* it has both proximal and distal chiasmata); *c*, in an individual homozygous for the *Molonglo* sequence (both proximal and distal chiasmata present). All these individuals homozygous for the *Standard* sequence of the *EF* chromosome; *e*, an individual heterozygous for the *Blundell* and *Snowy* sequences of the *CD* chromo-some (both acrocentric, but with the centromere at opposite ends, so that the chromo-somes are indistinguishable and labelling is arbitrary); this individual was hetero-zygous *Standard/Tidbinbilla*. Partly from White (1957*a*).

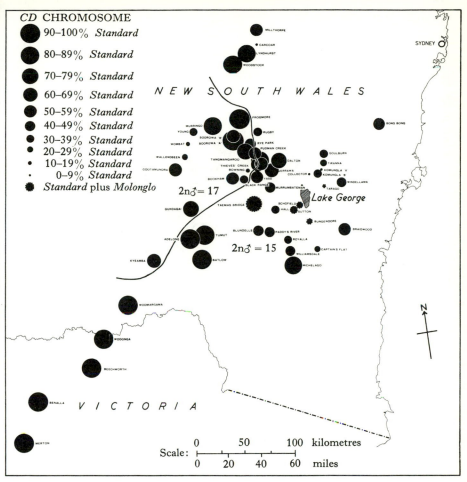

Fig. 8.10. Regularities and irregularities in the geographic distribution of the *Standard* sequence of the *CD* chromosome in the Australian grasshopper *Keyacris scurra*. The alternative sequence is the *Blundell* inversion, but in a few localities near Lake George a third sequence, *Molonglo*, is present in low frequency. The boundary between the $2n\male = 15$ and $2n\male = 17$ populations is also shown. From White (1957*b*).

rarest *SS-TT*. As a matter of fact, only about a dozen individuals of the latter genotype have been recorded, and they presumably have a very low viability in most populations. If we calculate relative viabilities from the ratio of observed individuals of a particular genotype to the number expected on the Hardy-Weinberg formula (binomial square rule) we find that in populations of this type, although *BB-SS* individuals may form the bulk

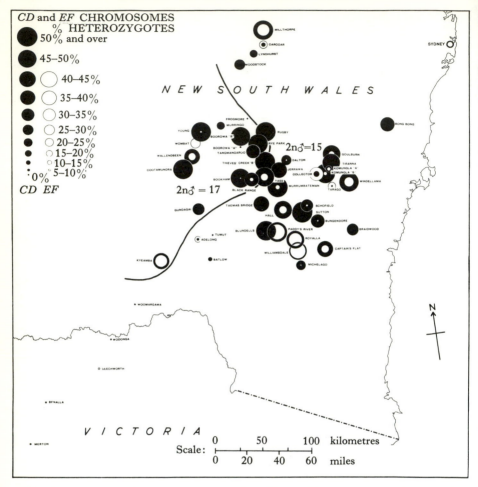

Fig. 8.11. Extent of heterozygosity for the *CD* and *EF* chromosomes in natural colonies of *Keyacris scurra*. All-black symbols represent colonies in which the *Tidbinbilla* sequence is not known to occur; black symbols with a white centre those where the percentage of heterozygosity for the *CD* pair exceeds that for the *EF* pair; and white symbols with a black centre the colonies where that relationship is reversed. At Wombat the percentage heterozygosity for the two chromosome pairs is almost exactly equal. From White (1957*b*).

of the population, their relative viability is less than that of the *SB-SS* heterozygotes. All categories of *Tidbinbilla* homozygotes probably have quite low viabilities, but because of sampling error these are difficult to calculate with any degree of accuracy. In general, each chromosome pair exhibits heterosis as far as viability is concerned, but there is a powerful

CD chromosome

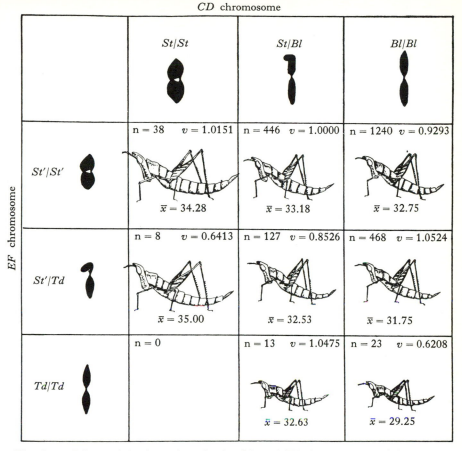

Fig. 8.12. Effects of the inversions in the *CD* and *EF* chromosomes of the grass-hopper *Keyacris scurra* on size and relative viability in a sample of 2315 male individuals from Wombat, New South Wales. The numbers of individuals of each of the 9 theoretically possible genotypes is shown in the top left hand corner of each square. *v*, relative viabilities, calculated from the ratios of observed numbers to numbers expected on the binomial square rule (the viability of *SB*, *SS* being arbitrarily set at 1.0000). The *Standard* sequences of both chromosomes are size-increasing and the *Blundell* and *Tidbinbilla* sequences are size-decreasing (as shown, in somewhat exaggerated manner by the silhouettes of grasshoppers – the mean live weights ($\bar{x}$) in mg, are shown under each individual). From White and Andrew (1962).

epistatic interaction between the two systems which results in double heterozygotes (*SB-ST*) having a lower viability than the single heterozygotes cited above. Obviously, as far as natural selection is concerned, the two polymorphisms are interlocked. By taking the calculated selective values of the nine genotypes in individual demes it was possible to compute 'topographies' for each of them, i.e. contour diagrams in which the vertical dimension represents the mean relative viability ($\overline{W}$) of hypothetical populations having any arbitrary frequencies of the *Blundell* and *Tidbinbilla* sequences, from 0 to 100%. These topographies (Fig. 8.13) seem to be all essentially similar in form; they have an oblique ridge running across the surface, with a low point or saddle somewhere near its middle. The existing composition of each population corresponds to a point on the saddle. Each deme is thus not (as we might have expected) on an 'adaptive peak'; nor, for that matter, is it in a valley. If we accept as valid the estimated relative viabilities, the situation is paradoxical and incomprehensible, since the composition of the populations should change under natural selection until

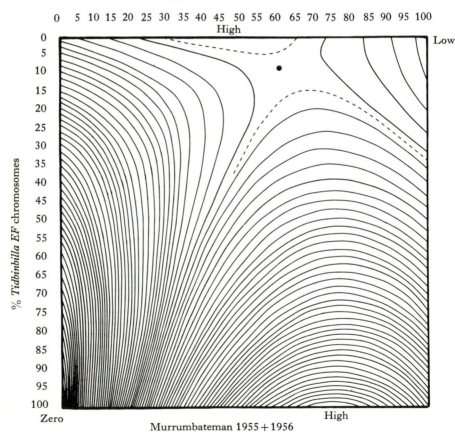

% *Blundell CD* chromosomes

Murrumbateman 1955 + 1956

a

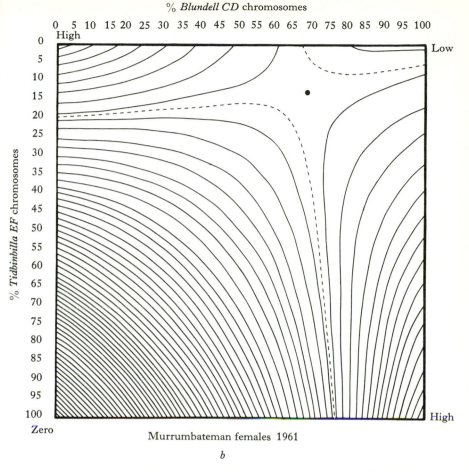

Fig. 8.13. Computed 'topographies' for mean relative viability in the Murrumbateman, New South Wales colony of *Keyacris scurra*. *a*, for the male population in 1955 and 1956, *b*, for the female population in 1961 (showing that the effects of the inversions on viability are essentially the same in the two sexes). Further explanation in text. *a* from White, Lewontin and Andrew (1963).

they climb up from the saddle to one of the peaks. The most probable explanation is that the relative viabilities are frequency-dependent, i.e. not constant.

Paracentric inversions *appear* to be lacking in grasshopper species, with one or two exceptions, such as the one studied by Nur (see p. 243). Such rearrangements could theoretically be detected (*a*) by an altered sequence of chromomeres or blocks of heterochromatin in prophase chromosomes, (*b*) by observing reversed loops at pachytene, (*c*) by observing dicentric and acentric chromatids in diplotene, diakinesis or first metaphase. But since it now appears that pericentric inversions do not in general form reversed loops in grasshopper meiosis it is quite possible that paracentric ones may behave in the same manner, in which case they would be virtually undetectable, since method (*a*) is hardly practicable at present. Their apparent absence from the genetic systems of grasshopper species may hence be spurious and they may, in reality, be as widespread as the pericentric type, or even more so.

There is an undescribed species of morabine grasshopper from central Australia in which two kinds of individuals exist. In one kind the large equal-armed metacentric *AB* bivalent shows a chiasma in each limb; in the other kind this bivalent shows a chiasma in one arm only. In each case 'exceptional' cells are only about 1% of the total. It is possible that the individuals with only one chiasma in the *AB* bivalent are heterozygous for one or more paracentric inversions which effectively suppress crossing-over in one limb. Many thousands of first metaphases have been examined in this species without any bridge or fragment chromatids being found.

INVERSIONS IN DROSOPHILA AND OTHER DIPTERA

The great majority of the studies which have been carried out so far on chromosomal polymorphism in natural populations have been on members of the insect order Diptera (especially the genera *Drosophila*, *Sciara*, *Chironomus*, *Simulium*, *Liriomyza*, *Anopheles*), making use of the salivary gland technique. In some species of *Drosophila* paracentric inversions are extremely common, there being a large number of alternative gene sequences for one or more chromosome members of the karyotype. The 'champion' species from this point of view are undoubtedly *D. willistoni*, a tropical American species, and *D. subobscura*, a European one, in each of which over forty inversions are now known (Da Cunha, Burla and Dobzhansky 1950; Dobzhansky, Burla and Da Cunha 1950; Townsend 1952; Da Cunha and Dobzhansky 1954; Da Cunha 1955; Dobzhansky 1957*b*; Da Cunha, Dobzhansky, Pavlovsky and Spassky 1959; Cordeiro, Townsend, Petersen and Jaeger 1959 for *willistoni* and Mainx, Koske and Smital 1953; Stumm-Zollinger 1953; Kunze-Mühl and Sperlich 1955; Richter and Hündler 1955; E. Goldschmidt 1956*a*, *b*, 1958; Mainx 1958; Stumm-Zollinger and Goldschmidt 1959; Sperlich 1961, 1964; Prevosti 1964; Götz 1965*a*, *b*, 1967; Pentzos-Daponte and Sperlich 1966; Sperlich and Feuerbach 1966 for *subobscura*).

Other species of *Drosophila* which show the same type of chromosomal polymorphism, although not to the same extent, include *D. pseudoobscura* and *D. persimilis* (Dobzhansky and Queal 1938; Dobzhansky and Sturtevant 1938; Dobzhansky 1939*b*, 1941*a*, 1943, 1948*a*, 1956, 1958*b*, 1963*b*; Dobzhansky and Epling 1944; Epling, Mitchell and Mattoni 1953, 1955, 1957; Brncic 1954; Epling and Lower 1957; Spiess 1957, 1961; Dobzhansky, Hunter, Pavlovsky, Spassky and Wallace 1963; Mohn and Spiess 1963) *D. algonquin* (Miller 1939), *D. athabasca* (Novitski 1946), *D. azteca* (Dobzhansky and Socolov 1939; Dobzhansky 1941*b*), the *obscura* group in general (Buzzati-Traverso and Scossiroli 1955), *D. ananassae* (Kaufmann 1936, 1937; Kikkawa 1938; Dobzhansky and Dreyfus 1943), *D. prosaltans* (Cavalcanti 1948), *D. nebulosa* (Pavan 1946), *D. robusta* (Carson and Stalker 1947; Carson 1953, 1958*a*), *D. guarú* (J. C. King 1947), *D. melanica* (Ward 1952), *D. pavani* (Brncic 1957, 1961*a*, *b*; Brncic and Solar 1961), *D. tropicalis* (Dobzhansky and Pavlovsky 1955), *D. rubida* (W. B. Mather 1961, 1963) *D. guaramunu* (Brncic 1953; Da Cunha, Brncic and Salzano 1953; Levitan and Salzano 1959), *D. paramelanica* (Stalker 1960*a*, *b*), *D. immigrans* (Freire Maia *et al.* 1953, Brncic 1955; Toyofuku 1957, 1958), the *D. paulistorum* complex (Dobzhansky and Pavlovsky 1962; Kastritsis 1967) and *D. euronotus* (Stalker 1963, 1964*b*). The whole subject of cytogenetic polymorphism in *Drosophila* has been comprehensively reviewed by Sperlich (1967). Populations of *D. melanogaster*, a species which has become distributed all over the world by human activity, and is associated with garbage dumps and similar habitats, also show inversion heterozygosity, in Russia (Dubinin, Sokolov and Tiniakov 1936, 1937), North America (Warters 1944; Ives 1947), Egypt (Mourad and Mallah 1960) – and no doubt elsewhere.

On the other hand, certain species of *Drosophila* such as *D. simulans*, *D. virilis* (Patterson, Stone and Griffen 1940, 1942; Warters 1944; Patterson and Stone 1952; Stone 1955, 1962) and *D. novamexicana* (Hsu 1952; Stone 1962) are not known to possess inversions in their natural populations, although the related species *D. americana* and members of the *montana* complex do so (Hsu 1952; Stone, Guest and Wilson 1960; Stone 1962). The members of the large *repleta* species group seem, in general, to show few or no inversions. Only about 144 have survived in the 46 species studied by Wasserman (1963) and of these only 52 are polymorphic, the remainder having undergone fixation. There is hence an average of 1.1 polymorphic inversions per species in this group, many species being cytologically monomorphic.

Where a large number of different inversions are present in a species they may be apparently distributed at random over the karyotype, or there may be a concentration of them on a single chromosome. Thus in the *repleta* group referred to above, there is a marked concentration of inversions in the second chromosome, and this applies both to inversions that have undergone

TABLE 8.6. *Inversions in the repleta group of
Drosophila* (from Wasserman 1963)

Chromosome	Total inversions	Polymorphic
X (=1)	9	1
2	103	39
3	18	7
4	4	3
5	10	2
	144	52

fixation and those that have not (Table 8.6). In *D. pseudoobscura* and *D. persimilis* the great majority of the polymorphic inversions are on the third chromosome. In the *melanica* species group (*D. melanica, D. paramelanica* and *D. euronotus*) there is a strong tendency for the natural inversions to be in the second chromosome (Ward 1952; Stalker 1960a, b, 1961, 1963, 1964b).

There have been two kinds of explanations put forward to account for such non-random distributions of inversions. We may refer to these as the *mechanical* and the *coadaptation* hypotheses. The mechanical hypothesis was first put forward by Novitski (1946) who suggested that, because of the presumed formation of a reversed loop in cells of inversion heterozygotes (probably with an asynaptic region at the ends of the inverted section) one inversion would tend to lead to the formation of others, with closely similar breakage points. One can imagine asynapsis favouring breakage or, alternatively, preventing restitution (and thereby increasing the likelihood of a rearrangement). Bernstein and E. Goldschmidt (1961) carried out an irradiation experiment which provided some support for the hypothesis that inversion heterozygosity tends to prevent restitution in its neighbourhood. Rothfels and Fairlie (1957) likewise favoured Novitski's hypothesis, on the basis of their work on the midge *Tendipes decorus*, in which the inversions are concentrated in the second chromosomes (fifty-four out of a total of seventy identified breakage points), and, within that chromosome, in certain regions rather than others.

The coadaptation interpretation is that the concentration of inversions in a particular chromosome or chromosome region is a result of natural selection; inversions occur with an approximately equal frequency in other chromosomes or regions, but are eliminated by selection. As Wasserman (1963) puts it:

Survival of a newly arisen inversion is dependent upon its fortuitous occurrence in a chromosomal region which is preadapted for a recombinant suppressor, that is, the region contains co-operating loci, among which recombination would be disadvantageous ... given the conditions for preadaptation, any one of a large number of distinguishable rearrangements could equally serve the purpose.

At present it does not seem possible to decide definitely between these two interpretations. The mechanical hypothesis would seem plausible if the frequency of spontaneous rearrangement were very low, so that non-randomness in their occurrence would play a significant role in determining the final pattern. On the other hand, if spontaneous rearrangements are far more common, so that only one in 10^3, 10^4 or 10^5 becomes established in a *Drosophila* or *Tendipes* species, then it seems more likely that the final pattern would be determined by natural selection. Work on grasshoppers (see p. 228) suggests that the frequency of spontaneous rearrangements is quite high (of the order of 1 per 10^3 gametes). The evidence thus seems to favour the coadaptation hypothesis, according to which the 'clustering' of inversions in certain species would be a consequence of the 'genetic architecture' rather than due to a chain reaction depending on an initial rearrangement. Carson (1969) has drawn attention to the fact that six closely related species of the *Drosophila grimshawi* group in Hawaii show inversion polymorphism in chromosome 4. In four of these very similar but not identical structural polymorphisms exist. He interprets this situation as indicating that certain sections of the middle and proximal regions of chromosome 4 tend to produce high fitness when heterozygous; thus inversions covering these regions would be particularly likely to give rise to balanced polymorphisms.

INVERSION PHYLOGENIES

We have already discussed (p. 217) the construction of inversion phylogenies, based on the occurrence of overlapping inversions. In such phylogenies, each sequence differs from the next by a single inversion, but it is not possible to state unequivocally which sequence was the earliest. Wasserman (1963) has discussed possible sources of error in conclusions based on this principle (due to cytological parallelism and convergence); he concludes that they are not likely to be at all frequent.

As an example of this method we may take the gene sequences in the long arm of chromosome 'A' of *D. azteca*, studied by Dobzhansky and Socolov (1939). Here seven sequences are now known, which have been named *alpha, beta, gamma, delta, epsilon, zeta* and *eta*. These may be arranged in a linear phylogenetic sequence, except for *delta*, which forms a side branch (Fig. 8.15). It is interesting to note that sequence *zeta* was actually predicted by this method and only found later (Dobzhansky 1941b), thus providing a striking justification of the method.

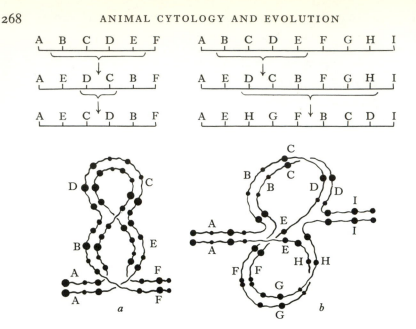

Fig. 8.14. Diagrams of pairing of chromosomes heterozygous for (a) included inversions, (b) overlapping inversions. Based on the work of Dobzhansky.

In a case like this we have no means of knowing which of the seven sequences was the original one, so that the diagram is only 'phylogenetic' in a rather limited sense. The original sequence may have been one of the terminal ones or it may equally well have been an interstitial member of the series. Sequences *alpha* and *beta* have only been found in Guatemala and Mexico, *gamma* and *delta* are present in central Mexico, while *eta* is only known from Arizona and California. Thus if it could be shown that the species originated in Central America and spread northward, *alpha* or *beta*

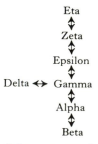

Fig. 8.15. Phylogenetic chart of the sequences in the *A*-chromosome of *Drosophila azteca*. Each pair of sequences separated by a double-headed arrow differ by a single paracentric inversion.

would probably be the most archaic sequence; while if the species originated in the United States and spread southward, *eta* was probably the ancestral sequence. But it is also possible that the species spread both northward and southward from a centre of origin in Mexico, in which case *gamma* or even *delta* may be the original sequence.

In the sibling species *D. pseudoobscura* and *D. persimilis* (which belong to the same species group as *azteca*) the third chromosome (an acrocentric autosome) is far more variable than any of the other members of the karyotype (Sturtevant and Dobzhansky 1936*b*). In *D. pseudoobscura*, twenty-one different sequences are now known for the third chromosome, while in *D. persimilis* eleven have been discovered (Crumpacker and Kastritsis 1967; Kastritsis and Crumpacker 1967). Only one sequence ('*Standard*') is common to both species. The geographic distribution of these sequences has been studied in great detail by Dobzhansky (see Dobzhansky 1939*a*, 1948*a*, 1951; Dobzhansky and Epling 1944; Spiess 1950; Epling and Lower 1957). Some have been found only at a single locality, while others are widespread. Some local populations of *D. pseudoobscura* contain as many as eight third chromosome sequences, while others have only three or four. In an area which includes northern Arizona and New Mexico and southwestern Colorado one sequence (Arrowhead) is very much commoner than any of the others. Apart from this exception, all populations show a considerable degree of polymorphism for third chromosome sequences. The percentage of heterozygotes for third chromosome inversions in the natural populations ranges from 6.8% at Grand Canyon, Arizona, to 76.2% at Mather, California, and one locality in Colorado (Anderson, Dobzhansky and Kastritsis 1967). The typical population of *D. pseudoobscura* or *persimilis* thus consists of a mixture of homozygotes and heterozygotes for third chromosome inversions; in some populations the homozygotes outnumber the heterozygotes while in others the reverse is the case. There is no marked decrease of inversion polymorphism in geographically peripheral populations of these species, as there is, for example, in *D. willistoni* and *D. robusta*; the isolated population in Guatemala shows four sequences and an even more isolated and marginal one in Colombia has two (Dobzhansky, Hunter, Pavlovsky, Spassky and Wallace 1963).

The sequences in the third chromosome of these two species may be placed on a branching phylogenetic chart (Fig. 8.16), in which each sequence differs from those connected to it by an arrow by a single inversion. In order to complete the phylogenetic 'tree', one sequence that has not actually been found in nature (*Hypothetical*) has had to be added (i.e. *Standard* differs from *Santa Cruz* by two overlapping inversions). Presumably the hypothetical sequence existed at one time but has now become extinct. The possibility

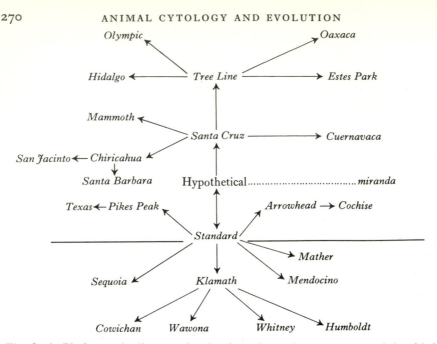

Fig. 8.16. Phylogenetic diagram showing how the various sequences of the third chromosome of *Drosophila pseudoobscura* and *D. persimilis* are related. Sequences connected by a single arrow differ from one another in respect of a single paracentric inversion. Sequences above the horizontal line occur only in *pseudoobscura*, those below it only in *persimilis* (*Standard* is found in both). From Dobzhansky (1951), modified.

that it may yet be found appears remote in view of the extensive collecting that has been carried out. Actually, it is hardly surprising that the phylogenetic chart contains a gap – it is rather remarkable that it has only *one* gap. Obviously, every inversion sequence which becomes established in the third chromosome of these species (i.e. neglecting those which are eliminated almost at once from the population) stands an excellent chance of persisting for a significant period of geologic time.

There has been considerable discussion as to the antiquity of the various inversion sequences in *D. pseudoobscura* and *persimilis*. Since it occurs in both species, it seems likely that *Standard* is the most archaic sequence, although *Hypothetical* may have been still older. Obviously, there is an a-priori likelihood that the sequences on the central axis of the phylogenetic chart are more ancient than the peripheral ones and that such sequences as *Texas, Santa Barbara, San Jacinto* and *Cochise*, which are removed two steps from the central axis, are as a class, more recent than those which are only one step removed from *Standard* and *Santa Cruz*. This deduction is not only in accordance with their position on the phylogenetic chart, but is also supported by their very restricted distribution. We may suggest that the sequences two steps removed from the main axis have not been in existence for long enough to have become spread over a wide area (we have in mind here, not merely the rate of migration of a newly established sequence, but also the fact that it can only spread by displacing already established sequences). The *Thomas* and *Vandeventer* sequences, both rare and geographically restricted, are probably recently derived from *Standard*.

But although we can get an approximate idea of the *relative* antiquity of the various sequences, it is unlikely that we shall ever be able to do more than guess at their *absolute* antiquity. Epling (1944) attached considerable importance to the fact that *Tree Line* and *Santa Cruz* (both presumed to be relatively ancient from their position on the main axis) occur in three disjunct areas: California, Central Mexico and Guatemala. He argued that these sequences must have arisen at a time when the species could cross the Colorado–Sonoran desert and the isthmus of Tehuantepec, and that climatic conditions which would permit dispersal across these barriers have not existed since mid-Tertiary times. However, the barriers to the dispersal of *D. pseudoobscura* in these areas are probably not as wide as the published evidence suggests; Stebbins (1945), who supported Epling's viewpoint on other grounds, admitted that the discovery of Santa Cruz in the unstudied areas of Baja California, Sonora and Sinaloa (and, we may add, in the highlands of Chiapas) would considerably reduce the force of the argument for a Miocene or earlier origin of these sequences.

ADAPTIVE SIGNIFICANCE OF INVERSION POLYMORPHISM

We have already considered (pp. 232–7) the conditions which may lead to genetic equilibria in natural populations. It remains to consider just what biological role the inversion polymorphism characteristic of so many species of Diptera actually plays in the life of the species.

It was generally assumed by the early workers that chromosomal inversions were 'adaptively neutral'. This idea seemed to be supported by the fact that individuals of *Drosophila* heterozygous for inversions did not look externally different from structural homozygotes and showed no *obvious* indications of heterosis such as larger size; it also appeared to be in accordance with the finding (Dobzhansky and Epling 1944) that inversion heterozygotes and homozygotes were present in natural populations of *D. pseudoobscura* in frequencies which were in good statistical agreement with the values to be expected (on the binomial square rule) if random mating and equal viability existed.

The concept of selective neutrality had to be abandoned, however, when it was found that the frequencies of the different sequences underwent temporal fluctuations in the wild populations (Dobzhansky 1940, 1943; Dubinin and Tiniakov 1945, 1946a, b; Strickberger and Wills 1966). A particular sequence which was common in one month might become much rarer a few months later and then recover its original frequency. If these temporal changes had been entirely random, they might have been ascribed to 'drift' – although most *Drosophila* populations are far too large for 'drift' to be a significant factor in determining their composition. If the changes had been entirely in one direction (i.e. if a particular sequence went on getting rarer and rarer until it was eliminated, while another increased in frequency and reached fixation), we might have been dealing with some type of transient polymorphism. In point of fact, however, the changes were not unidirectional and were far from random, but exhibited seasonal regularity, certain sequences being always common at particular seasons of the year and rare at other seasons. Thus the changes in the composition of any particular population tend to be repeated from year to year; of course the repetition is never exact, but neither are the meteorological and ecological conditions ever exactly repeated from year to year. Such seasonal fluctuations do not occur in all *Drosophila* species with inversion polymorphism, and they sometimes occur in one area and not in another. Thus Carson (1958a) found rather insignificant seasonal changes in the frequency of the inversions in *D. robusta* populations in Missouri; Levitan (1951, 1957), on the other hand, found quite marked seasonal fluctuations in the case of some populations of the same species in Virginia. Most grasshopper populations have only one generation a year, so that seasonal fluctuations in inversion frequencies of the *Drosophila* type are not to be expected, although it may well be that carriers of different cytological sequences differ in longevity (the difficulty here is to obtain large enough samples for examination at the end of the season, when a large mortality has occurred).

The fact that considerable seasonal fluctuations in inversion frequencies were common in *Drosophila* populations (even if not universal) was sufficient to disprove the 'selective neutrality' hypothesis. It was, however, reinforced by evidence from population cage experiments in the laboratory. These studies on artificial populations demonstrated considerable changes in the frequencies of inversion-sequences, over periods of a few generations, and showed that such changes were reproducible, being due to selection and not to drift (of course, if the population is small enough a significant amount of drift will modify the outcome of the experiment). In a particular experiment of Dobzhansky a population which initially contained about 11% *Standard* third chromosomes and 89% with the *Chiricahua* sequence had at the end of a year reached a situation in which there were about 70% *Standard* chromosomes and only 30% *Chiricahua* ones. These frequencies seemed to represent equilibrium values for the population in question, since only minor fluctuations occurred thereafter. Thus, under the admittedly artificial environment of the population cage, the cytological polymorphism is maintained in equilibrium from generation to generation, just as it is in the natural environment (although the equilibrium frequencies of the two sequences are not necessarily the same in the cage as they are in any natural environment which the species is known to inhabit).

This type of experimental result shows clearly 'natural' selection is operating in these artificial populations to change the frequencies of the different sequences, which are hence not adaptively neutral; but it does not explain just how selection operates. By taking samples of eggs from the population cages and raising the larvae which hatched under optimum conditions, so that mortality was minimal, it was determined that the zygotes were being produced in frequencies which were in statistical agreement with the expectation calculated from the binomial square rule. However, when the adult flies were removed from the cage (i.e. after considerable mortality had occurred) a significant excess of the structural heterozygotes and a corresponding deficiency of the homozygotes were encountered. The inversion-polymorphisms 'transplanted' from nature to population cages thus show a type of *heterosis*.

There is, however, a complication in such calculations, which has been pointed out by Wallace (1958) and Novitski and Dempster (1958). In essence it is this: in order to calculate the expectations, we need to use certain values of q and $(1 - q)$, the frequencies of the alternative sequences. But the frequencies which should be used for this purpose are those obtaining in the population of gametes which gave rise to the generation we are studying. If the population is in genetic equilibrium, so that the values of q and $(1 - q)$ are the same in successive generations, we can legitimately use values derived from the sample itself in order to calculate the expected frequencies of genotypes. But in *Drosophila* populations, whether in cages or in nature, equilibrium is uncommon, and hence the calculation of expected genotypic frequencies is usually legitimate only when we have evidence of the composition of the previous generation (assumed to produce gametes in the same frequencies). In the grasshopper populations we have previously considered, this particular difficulty either does not arise (because the population is in perfect genetic equilibrium) or can be simply overcome (by studying several successive and discrete generations and using the q and $(1 - q)$ values of each generation to calculate the expected composition of the next one).

Before this situation was generally recognized, Dobzhansky and Levene (1948) claimed to have evidence that even in wild populations of *Drosophila pseudoobscura* structural heterozygotes were more frequent, and homozygotes less frequent than would have been expected on the basis of adaptive neutrality. This is indeed clearly so in the case of certain inversions in *D. willistoni*, where the heterozygotes exceed 50% (since *no* values of q and $(1 - q)$ will give an expectation of over 50% for the heterozygotes). But it does not necessarily follow from Dobzhansky and Levene's data, using the type of calculation they employed, even though the balance of the evidence strongly favours heterosis as the overall reason for the maintenance of inversion polymorphism in natural populations of *Drosophila* and other animals. The same objection applies likewise to the arguments of Epling, Mitchell and Mattoni (1953, 1955, 1957) who failed to find consistent excesses of heterozygotes either in nature or in population cage experiments. These authors put forward an entirely different interpretation of inversion heterozygosity, namely that it plays a role in natural populations in virtue of its effects on the level of genetic recombination in the other chromosomes (the 'Schultz–Redfield effect' – see p. 184). Space does not permit a discussion of their argument, but it should be obvious, on general grounds, that genetic equilibria based on such a second order effect would be expected to be very unstable.

In the meanwhile, Dobzhansky (1948*b*) had shown that the 'heterosis effect' in population cage experiments is obtained consistently only when the two chromosome sequences come from the same region. Thus, when the *Standard* chromosomes were from Mather, California and the *Chiricahua* ones from Mt. San Jacinto, about 550 km to the south, instead of obtaining an apparent excess (i.e. higher viability) of heterozygotes, it was found that the heterozygotes had a calculated viability intermediate between that of the two homozygous genotypes. In this case no equilibrium existed and the *Chiricahua* chromosomes would be expected to become extinct after a few generations. In some experiments involving chromosomes from California and Mexico, the heterozygotes were even inferior in adaptive value to both homozygous genotypes (Dobzhansky 1950*a*).

It follows from the above experiments that the genetic properties shown by the various inversion sequences are not due to the inversions themselves, but to the genes contained in them, which must be expected to become coadapted, as a result of natural selection, in any particular population. Thus the genetic properties of the *Standard* sequence at Mather will be coadapted to those of the *Chiricahua* sequence at the same locality, but not to those of the *Chiricahua* chromosome at Mt. San Jacinto. This genetic evolution of the different sequences is made possible by the fact that crossing-over between mutually inverted regions is effectively suppressed, so that the different sequences in any population may be expected gradually to acquire different complexes of alleles, as a result of mutation. As a matter of fact, it was shown by Dobzhansky and Epling (1948) that the suppression of crossing-over is by no means confined to the inverted region itself, but extends a considerable way along the chromosome, both proximally and distally (probably because the inversion prevents effective synapsis for some distance on either side). Thus the genes responsible for the specific biological properties of the inversion do not necessarily all lie within it; some of them may be a short distance outside.

Since this early pioneer work of Dobzhansky and his associates on the biological role of the inversions of *Drosophila pseudoobscura* additional possibilities have become apparent. Thus it now appears quite probable that, in addition to heterosis, we may be dealing with a certain degree of *annidation* (the term of Ludwig 1950), i.e. an adaptation of different cytological genotypes to different niches within the general environment occupied by the species. Frequency-dependent selection has been demonstrated in the case of certain inversions in *D. ananassae* populations by Tobari and Kojima (1967).

The adaptive chromosomal polymorphism of *D. pseudoobscura* and other species of *Drosophila* is probably a mechanism whereby the individual populations are 'buffered' against seasonal changes in the environment and against alterations in the microenvironments within a particular habitat. The regular seasonal changes in the frequencies of the various sequences within particular populations may be accounted for by the operation of natural selection, operating differently on the successive generations throughout the year. In some populations and species of *Drosophila*, however, such seasonal fluctuations are slight, absent or irregular (Carson and Stalker 1949; Spiess 1950). Where a species of *Drosophila* occupies a considerable altitudinal range (e.g. *D. pseudoobscura* and *D. persimilis* in the Sierra Nevada of California and *D. robusta* in the Great Smoky Mountains of Tennessee) regular gradations or altitudinal clines in the frequencies of the inversions are found (Dobzhansky 1947; Stalker and Carson 1948; Carson and Stalker 1949; Spiess 1950; Carson 1967b).

Apart from short-term fluctuations in the frequencies of inversion sequences in natural populations, some major long-term trends have been detected. Thus over the period 1940 to 1958 the Pikes Peak sequence increased its frequency in numerous Californian populations from about 0.02% to between 5.0 and 12% (Dobzhansky 1958a, b, 1963b; Dobzhansky, Anderson and Pavlovsky 1966). This is a 400-fold increase in about 100 generations of the fly. The increase in the frequency of the Pikes Peak sequence has been mainly at the expense of the *Chiricahua* one. It has occurred equally in habitats disturbed by human activity and in those removed from such interference; the localities affected extend from southern California to north of San Francisco. The most probable interpretation would be that the Pikes Peak sequence was transported to California by automobile traffic some time during the 1930's and that it then acquired new genetic properties (possibly as a result of crossing-over) which greatly increased its selective value, and enabled it to partially displace the *Chiricahua* sequence.

It was pointed out by Wallace (1953) that there exist series of sequences (called by him *triads*) such that *A* has given rise to *B* by a simple inversion and *B* to *C* by an overlapping inversion. So long as only two sequences are present in a population the mutually inverted segments will be genetically isolated. But if all three were to coexist, serial transfer of genes would occur, by crossing-over, thus disrupting coadapted gene complexes or preventing them from developing. Wallace consequently argued that such triads should not coexist at the same locality in nature, or if they do so, one sequence should be very rare. Data on the distributions of sequences in nature seem to support Wallace's hypothesis in the case of *Drosophila pseudoobscura*. But the data for *D. robusta* do not fit the hypothesis (Levitan, Carson and Stalker 1954), so that it seems to be of rather limited applicability.

INVERSION-POLYMORPHISM IN CENTRAL VERSUS PERIPHERAL POPULATIONS

It was pointed out by Da Cunha, Burla and Dobzhansky (1950) and Da Cunha and Dobzhansky (1954) that the degree of inversion polymorphism in *Drosophila willistoni* is greatest in the centre of its range (in the equatorial regions of Brazil) and declines rather regularly towards the periphery. Additional data have been provided by Townsend (1952) and Dobzhansky (1957b). In some localities in central Brazil the average female individual may be heterozygous for more than nine inversions, and the males for over six. In the extreme south of Brazil and in the Antilles and Florida the mean number of heterozygous inversions per individual falls to one, two or three (the most nearly monomorphic population being the one on the island of St Kitts, with about 0.2 heterozygous inversions per fly).

Similar data have been obtained by Carson (1955a, b, 1958a, b, 1959) and Carson and Heed (1964) in the case of *D. robusta* in the United States. Marginal populations in Nebraska, Vermont and Florida are mainly homozygous for gene sequence, whereas centrally located populations in Missouri and the Appalachian region are highly heterozygous for inversions. Some other species of *Drosophila* which reach their southern or northern limits in

Florida also seem to show a reduced level of inversion polymorphism there. *D. subobscura* seems to show some falling-off of its chromosomal polymorphism at the periphery of its distribution. It has the highest degree of inversion polymorphism in the Mediterranean region, with reduced polymorphism in Scotland, Norway, Persia and (perhaps) Tangier (Götz 1967).

Dobzhansky has argued that the amount of chromosomal polymorphism in all these cases is related to the diversity of ecological niches occupied by the population. This is probably true to a considerable extent; the species may be imagined to occupy the greatest number of different ecological niches in the centre of its range, and may be supposed to be precariously exploiting only a single niche at the very edge of its area of distribution. But there are obviously many complicating factors, and not all species show a decline in chromosomal polymorphism at the edge of their range, nor should we expect them to. The distribution limits of the grasshopper *Trimerotropis gracilis* in the western United States seem to be imposed by the natural distribution of its food-plant, namely sagebrush (*Artemisia tridentata sens. lat.*); its degree of inversion polymorphism does not seem to decline at the periphery, where it may be encountering as much environmental diversity (e.g. in types of soil in which the eggs develop) as in the centre of the range. Even where there is a marked fall-off in chromosomal polymorphism at the periphery of the range, it is by no means probable that Dobzhansky's explanation applies in every instance. For example, in some cases a species may only have extended its range into certain peripheral areas very recently; in which case not all the genetic polymorphism of the species may have reached the newly-occupied territory. Some species seem to show fairly regular clines in degree of inversion polymorphism, without any definite central–peripheral pattern. Thus the grasshopper *Trimerotropis suffusa* seems to show a gradual decrease in inversion polymorphism as one goes eastward from California to Colorado.

NON-RANDOM ASSOCIATIONS OF INVERSIONS

One might expect that independent inversions in the same chromosome – and *a fortiori* ones in different chromosomes – would be combined at random in the individuals comprising a natural population. However, we have already seen that, in the case of the inversions of the *CD* and *EF* chromosomes of *Keyacris scurra* slight but consistent and significant departures from randomness occur. Much more extreme non-random associations of inversions located in the *same* chromosome have been found in certain species of *Drosophila* and *Chironomus* (Levitan 1954, 1955, 1958; Martin 1962, 1965).

TABLE 8.7. *Non-random association of inversions in the second chromosome of D. robusta at two localities in Virginia* (Levitan 1958)

Linkage chromosome	In Females		In Males	
	Observed	Deviation from random association	Observed	Deviation from random association
S–S	522	− 5.7	651	−23.8
S–1	44	+ 5.7	84	+23.9
1–S	423	− 4.9	537	−19.5
1–1	36	+ 4.9	69	+19.5
2–S	300	− 4.7	411	+ 6.5
2–1	27	+ 4.7	30	− 6.5
3–S	736	+15.3	959	+36.9
3–1	37	−15.3	45	−36.9
Total	2125		2686	

$\chi^2_{(3)}$ (females) = 7.60; $\chi^2_{(3)}$ (males) = 38.00**.

We may take Levitan's data on the second chromosome inversion sequences of *Drosophila robusta* as a first example (Table 8.7). Apparently the departure from randomness is caused by the excess of certain types of double heterozygotes (especially among males) such as S–1/3–S and 1–1/3–S and corresponding deficiencies of other types such as S–S/3–1 and 1–S/3–1, having the same inversions in a different configuration. The non-randomness seems to be caused by natural selection acting in opposition to crossing-over between the inversions, which would otherwise lead to random equilibrium.

A second type of non-random association is seen in the case of the X-chromosome inversions of *D. robusta*. Here the departure from randomness is in the opposite direction at Blacksburg, Virginia and Gatlinburg, Tennessee (Table 8.8). This example seems to demonstrate that the non-randomness must depend on natural selection and that the genetic properties of the inversions are not the same in Virginia and Tennessee. The alternative possibility – namely that the non-randomness is some kind of historical accident which depends on the phylogeny of the X-chromosome is excluded by the fact that the departure from randomness is in the opposite direction in the two populations.

A far more extreme departure from random equilibrium has been demonstrated by Levitan and Salzano (1959) in the case of the E and H inversions of the fourth chromosome of the Brazilian *Drosophila guaramunu*. Only 1.7% of the larvae were heterozygous for one inversion and homozygous

TABLE 8.8. *Non-random associations of inversions in the X-chromosome of D. robusta with deviations from expectation* (Levitan 1961, simplified from original table)

Chromosome	Blacksburg	Gatlinburg
S–S	444 (+69.4)	54 (−48.5)
S–2	956 (−69.4)	190 (+48.5)
1–S	435 (−69.4)	224 (+48.5)
1–2	1450 (+69.4)	194 (−48.5)
Total	3285	662

$$\chi^2_{(1)} = 30.6. \qquad \chi^2_{(1)} = 62.7.$$

for the other (*EEHh, eeHh, EeHE, Eehh*); 73.8% were double homozygotes (*EEHH, eeHH, EEhh, eehh*) and 24.5% were double heterozygotes (*EeHh*). Interpretation of the results is complicated by the fact that the two homozygous arrangements *HH* and *hh* were not distinguished in the original data, but in any case the interpretation seems to be in terms of natural selection.

There are now a number of other records of non-random associations of inversions in species of *Drosophila*. Particularly noteworthy are the studies of Stalker (1960*a*, 1964*b*) on *D. paramelanica* and *D. euronotus* and those of W. B. Mather (1961, 1963) on *D. rubida*. The phenomenon is certainly not confined to *Drosophila*, since associations of this type have been recorded in *Chironomus tentans* by Acton (1955) and in *Simulium tuberosum* by Landau (1962). They are also evident in the data of Rothfels and Fairlie (1957) on *Tendipes* (= *Chironomus*) *decorus*.

In the Australian *Chironomus intertinctus* Martin (1962, 1965) has made a careful study of two non-random associations. In the third chromosome, which constitutes an *XY* pair, the Corio inversion is closely linked to the sex determining region. Different selective values in the two sexes result in the Standard sequence being much commoner in the *Y*-chromosomes while the Corio sequence is commoner in the *X*'s. In the second chromosome of the same species the Barwon and Lonsdale inversions (situated in opposite arms of the chromosome) tend to occur together, and so do the corresponding Standard sequences. This results in $St^{Lo}/St^{Lo}\ St^{Ba}/St^{Ba}$, $St^{Lo}/Lo\ St^{Ba}/Ba$ and $Lo/Lo\ Ba/Ba$ genotypes being present in excess, while the $Lo/Lo\ St^{Ba}/St^{Ba}$ and $St^{Lo}/St^{Lo}\ Ba/Ba$ ones are deficient (in the case of the other four genotypes there is no consistent excess or deficiency in the data). The association of the two Standard sequences breaks down when the Hume inversion is superimposed on Barwon; in this case an association of St^{Lo} with Hu seems to be favoured by natural selection instead.

The animal species with the most complex system of inversions hitherto described is undoubtedly the blackfly *Simulium vittatum*, with 134 different paracentric inversions known (Pasternak 1964). It is probable that non-random associations of inversions occur in this species also, as they do in *S. tuberosum*. This enormous number of rearrangements is distributed over the chromosome complement, with no obvious tendency to concentration in one chromosome. In spite of the great number of inversion sequences, the mean number of heterozygous inversions is no higher than in *Drosophila subobscura* or *D. willistoni* (1.90 to 6.01 per individual in different populations investigated by Pasternak).

GENETIC LOADS DUE TO NATURAL POLYMORPHISM

The general view of cytological polymorphism that has developed in recent years is that it is an adaptive phenomenon, i.e. an advantage to the species under the particular conditions of the environment. Yet cytogenetic polymorphisms imply the existence of some less well adapted genotypes which die prematurely or have their fecundity reduced in some manner.

The concept of a genetic load or burden in a population is that of a certain proportion of genetically handicapped individuals being produced, generation after generation, because of some inherent genetic property of the population. By 'genetically handicapped', we mean individuals which fall below the optimum that the species is physiologically capable of, as far as biological or 'Darwinian' fitness is concerned. Three main kinds of load may be defined and some other rarer or less generally important types may be imagined (Muller 1950a; Morton, Crow and Muller 1956; Dobzhansky 1957a; Haldane 1957, 1958; Crow and Morton 1960; Kimura 1960; Crow 1963; Brues 1969). The whole subject has recently been reviewed by Wallace (1970).

In the first place we have mutations arising in natural populations which lower the fitness of the individual even when heterozygous (and which may be even more deleterious when homozygous). Chondrodystrophic dwarfism, Retinoblastoma and Multiple Neurofibromatosis (von Recklinghausen's disease) in the human species will serve as examples, and, in all probability, so will most cytological deletions in diploid organisms. These mutations are entirely deleterious, with no compensatory advantages. They exist in the population at an equilibrium frequency which depends on the rate at which they arise by mutation and the rate at which they are eliminated by natural selection. A limiting case is where a mutation is obviously recessive but deleterious when homozygous. We may refer to this as the *mutational load*.

In the second place, we have a *segregational load*, due to the segregation of inferior homozygotes in populations containing genes leading to heterosis in the heterozygotes. The S-haemoglobin gene in many African populations exposed to malaria is an example. AA individuals are 'normal' but liable to suffer from malaria, AS individuals are resistant to malaria and have a higher fitness, while SS individuals have a severe and usually fatal anemia (Allison, 1955). Segregation of inferior genotypes in other types of balanced polymorphism (see p. 234) is similar in principle. In the case of cytological polymorphisms the segregational load is defined as the loss due to homozygotes of poor viability or fecundity. An additional load may be due to the production of some gametes with deficiencies or duplications, but this may be referred to as the *recombination load* and is strictly speaking, a distinct type of genetic burden.

Finally, there is the *substitution load* due to evolutionary changes in gene frequency occurring in the course of adaptation to a changing environment; this is the 'cost of evolution' as Haldane has named it. This would be minimal in a genotypically bradytelic species, maximal in a genotypically tachytelic one (definition of species or groups as bradytelic or tachytelic by evolutionists rests, of course, on observations of phenotypes – we have no means of estimating the extent of the genetic changes that may have taken place in a species whose phenotype has remained virtually unchanged over a long period of geological time).

In most wild species, living in natural environments to which they are presumably well adapted, the substitution load is probably small compared to the other two. A controversy has, however, arisen as to the relative importance of the mutational and segregational genetic loads in natural populations including the human species. Muller (1950a) stressed especially the mutational load; Dobzhansky on the other hand, while not denying the mutational load, has tended to stress especially the balanced genetic load which results from heterotic polymorphisms. Direct evidence has been sought in both man and *Drosophila*, but with somewhat inconclusive or contradictory results (Morton, Crow and Muller 1956; Dobzhansky 1958a; Dobzhansky, Krimbas and Krimbas 1960). Muller (1959) went so far as to regard heterotic polymorphisms as 'temporary makeshifts that arose in the stress of comparatively rapid evolutionary flux and that are due to be rectified ultimately, when a longer term natural selection repairs its short-term imperfections and miscarriages'.

The assumption implied in this statement is that eventually a biologically adaptive (i.e. heterotic) heterozygote a_1a_2 will be replaced by an even superior homozygote a_3a_3. That genetic polymorphism is a makeshift, i.e. a compromise, is easily admitted. But all life is based on compromises and many of them seem to have endured for long periods of geological time. It seems inherently improbable that the human species was ever monomorphic for the blood group genes and for genes determining haptoglobins, transferrins and ability to taste or excrete a great variety of substances (Williams 1956).

As far as inversions are concerned, Muller admitted their heterotic nature but argues that this is irrelevant since it has been built up by selection and that, anyhow, polymorphism is precluded in most groups by the existence of crossing-over in the males. The matter cannot, however, be disposed of in this way. Cytological polymorphisms of one kind or another has been found in every group of animals that has been seriously examined by cytogeneticists – in grasshoppers, crickets, mantids, roaches, beetles, neuropterans, butterflies, isopods, scorpions, molluscs, shrews, rodents, as well as many groups of Diptera, some of which *do* have chiasmata in the males. Thus even though many and perhaps most species of animals may lack such mechanisms, they are sufficiently widespread to constitute one of the great general phenomena of natural populations of animals. We cannot therefore escape the conclusion that many species have quite a significant genetic load from this cause alone. One might imagine that this might be related to the 'innate rate of increase' of the species (Andrewartha and Birch 1954) and that species of low prolificity might lack such mechanisms. However, the grasshopper *Keyacris scurra*, a conspicuous example of cytological polymorphism, lays a maximum of about forty eggs (with an average of perhaps twenty-five) and the prolificity of the cytologically polymorphic shrew *Sorex araneus* (Ford, Hamerton and Sharman 1957) is probably even lower.

As far as spontaneously arising cytological aberrations are concerned the mutational load seems to be small. In *Keyacris scurra* about one individual in a thousand carries a unique chromosomal aberration, usually a translocation. Since these rearrangements do not seem to establish themselves in natural populations they are presumably deleterious in the heterozygote and will be eliminated from the population, after a few generations, by natural selection. The picture in human populations does not seem to be essentially dissimilar. In both species a larger number of chromosomal rearrangements are presumably lethal in the early embryonic stage. One difference between *Keyacris scurra* and *Homo sapiens* seems to be that trisomic individuals have never been detected in the former, whereas certain types of trisomics (namely those for some of the shorter chromosomes) exist in human populations, although their fitness is much reduced (see p. 465). Presumably trisomics in *Keyacris scurra* are inviable at a very early embryonic stage.

The above conclusions seem to apply generally in animal populations. No species is known in which any large fraction of the population are handicapped by recently arisen chromosomal rearrangements. There is no evidence that more than a small fraction of the lethals present in *Drosophila* populations are cytological deficiencies or duplications. On the other hand, the heterotic load due to segregation of adaptively inferior structural homozygotes may obviously be considerable in populations of Diptera and Orthoptera polymorphic for chromosomal rearrangements. It would be especially heavy if the hypothesis of 'pure heterosis' (see below) were largely valid.

The most extreme type of segregational load is, of course, where we have a balanced lethal mechanism, both homozygous genotypes being inviable. Rather understandably, such balanced lethal systems do not seem to be common in natural populations. Dobzhansky and Pavlovsky (1955) have described a situation in a central American population of *Drosophila tropicalis* where the viability of the inversion homozygotes is very low. Other examples in animal species are more hypothetical.

'PURE HETEROSIS' VERSUS 'ANNIDATION'

There are two opposed views as to the role of heterotic polymorphisms in natural populations. The first may be referred to as the 'pure heterosis' viewpoint. According to this hypothesis, heterozygotes *Aa* are inherently fitter than either homozygote *AA* or *aa*. Natural selection will establish an equilibrium, the relative frequencies of *A* and a depending on the relative fitness of *AA* and *aa*. The superior fitness of the *Aa* genotype is believed to exist in a wide variety of ecological niches, or even in all niches. As Carson (1959) puts it:

The heterozygotes are selected not because they confer specific adaptive features on the organism in response to selection for a specific niche, but rather because the heterozygote exhibits a general vigor, a heterotic buffering, which does not relate to any specific component of the environment. Thus the heterozygous organism is . . . selected because of its good performance in all of the niches a population has mastered.

Carson calls this type of selection *heteroselection* and contrasts it with *homoselection*, i.e. selection for a homozygous genotype adapted to a specific ecological niche. The terms are useful ones, even if we reject some other parts of Carson's argument.

Opposed to the 'pure heterosis' hypothesis is the view that in a population composed of *AA*, *Aa* and *aa* individuals, each genotype may be superior in some ecological niche. The overall superiority of *Aa* individuals is then due to the fact that they are superior in the predominant niche or niches or over a wider range of niches than *AA* or *aa*. This is the 'annidation model' (see p. 235). As rightly pointed out by Cain and Sheppard (1954): 'Observations on relative coefficients of selection of different polymorphs within a population give no evidence on the ability of any one polymorph to survive as a pure stock.'

The general trend of modern research into the genetics of natural populations seems to lay more stress on the annidation and frequency dependent principles than on 'pure heterosis'. As we have seen, there are examples of natural populations in proven genetic equilibrium which show deficiencies of heterozygotes, by comparison with the expectation on the binomial square rule (see Table 8.4). It is quite probable, however, that *Drosophila* populations whose natural genetic equilibria depend on annidation and frequency-dependency may show pure heterosis in the simpler environments of laboratory population cages, where the ecological niches to which the various homozygous genotypes are adapted do not occur.

It would seem, however, that systems of frequency-dependent selective values based on the principle of annidation can only operate if individuals are free to move from one niche to another or if consumable requisites (e.g. prey) or pathogens can similarly move. Thus if we consider a polymorphic population of a grasshopper species with considerable mortality in the egg stage which is passed underground, the soil may consist of a variety of ecological niches (depending on its water-holding capacity, aeration, organic content, pH etc.). But it is difficult to understand how a genetic equilibrium based on frequency dependent selective values could exist under such circumstances unless pathogens or egg predators were involved (which, of course, is not improbable).

9

CHROMOSOMAL POLYMORPHISM IN NATURAL POPULATIONS. II. TRANSLOCATIONS AND OTHER TYPES OF REARRANGEMENTS

These are all mutations in the wide and primitive meaning of the term.

R. A. FISHER: *The Genetical Theory of Natural Selection*

Very few natural polymorphisms involving mutual translocations (other than fusions and dissociations) are known in animals. The reasons are fairly obvious (Wright 1941), since heterozygotes for translocations will show rings or chains of four chromosomes at meiosis and if these orientate on the spindle in a non-disjunctional 'square' instead of a disjunctional 'zig-zag' manner, aneuploid gametes carrying a duplication and a deficiency will be produced. Thus selection against the translocations that are constantly arising in natural populations is expected to be extremely severe unless the interchanged regions are so minute that the genetic unbalance involved in aneuploidy is not deleterious or unless the orientation is, for mechanical reasons, regularly zig-zag. The latter is only likely to be the case when the chromosomes have interchanged virtually entire chromosome arms and have distally localized chiasmata. Other ways in which aneuploid gametes may be produced by translocation heterozygotes include (1) formation of two bivalents (instead of a chain or ring of four) in some cells; the bivalents may then segregate non-disjunctionally, (2) occasional formation of a chiasma between a centromere and the point of translocation, in an association of four chromosomes. In spite of all these theoretical, a-priori expectations some instances of translocation heterozygosity do exist in natural animal populations. Thus Lewis and John (1957*a*) and John and Lewis (1957, 1958, 1959) have reported translocation heterozygosity in populations of the roach *Periplaneta americana* from coal mines and laboratory cultures, and in laboratory colonies of another species of roach, *Blaberus discoidalis*. These authors believe that the populations in which these translocations were encountered had been inbred and that selection under such enforced inbreeding is likely to favour structural heterozygosity. Although the proposition is questionable and the extent of the alleged inbreeding in the case of *P. americana* and its duration in *B. discoidalis* were probably insufficient to affect the situation, there can be no doubt as to the reality of the chains and rings of metacentric chromosomes observed at first metaphase. Some males of *P. americana* even showed three different rings or chains of four. Orientation of these configurations on the spindle is almost invariably of the zig-zag (disjunctional) type, there being hardly any 'square' orientations which lead to non-disjunction. Some individuals also carried rings of six chromosomes, being presumably heterozygous for two serial translocations (of the type *AB, BC, CD, DE, EF, FA*). Orientation of these configurations is likewise usually disjunctional. Some small samples of *P. americana* from Pakistan had relatively more structural homozygotes than the British populations; but they also showed more rings of 6, and even (in one individual) a ring of 8 (John and Quraishi 1964).

A disturbing and peculiar feature of this case is the frequency with which, in individuals heterozygous for a translocation, the ring is replaced by two bivalents. On the assumption that these segregate at random with respect to one another, half the sperms resulting from first metaphases with bivalents should carry a duplication and a deficiency. In some of the individuals heterozygous for several translocations, over 50% of the sperm should be aneuploid, on this basis. Such a situation would be expected to lead to extraordinarily strong selection against translocation heterozygosity. The paradox could, in theory, be resolved if we supposed that a number of the chromosome arms were genetically quasi-inert and more or less equivalent, so that duplication-deficiency zygotes were viable. But it would be necessary also to postulate that chiasmata between genetically active arms were invariable, those between quasi-inert arms facultative. These are very special assumptions, which may not correspond to reality. But the case of *Periplaneta* is certainly inexplicable without special assumptions of some kind.

It has been claimed (Sharma, Parshad and Sehgal 1956, 1959) that the chromosomes of *Periplaneta* are acrocentric rather than metacentric (as maintained by Lewis and John) and that no chiasmata are formed at meiosis in the male. These conclusions have been satisfactorily refuted by Lewis and John (1957*b*) and John and Lewis (1960*a*). Chiasmata are admittedly not visible in this species as cross-shaped configurations of chromatids. But the indirect evidence for their existence is cumulative; particularly convincing is the clear distinction between rod and ring bivalents and the corresponding difference between chains and rings of four chromosomes in the translocation heterozygotes.

The translocations reported by Piza (1939, 1940, 1943a, c, 1944, 1946b, 1947a, b and c, 1948a, b and c, 1949, 1950a, b) in various Brazilian scorpions (species of *Tityus* and *Isometrus*) are especially interesting, since certain individuals are heterozygous for several translocations, leading to the formation of multiple rings of chromosomes at meiosis, similar to those of the plant *Oenothera*.

Although in some non-buthid scorpions such as *Opisthacanthus* (Wilson, 1931) the chromosomes seem to be monocentric, those of *Tityus* and *Isometrus* do not show any constrictions which might indicate the presence

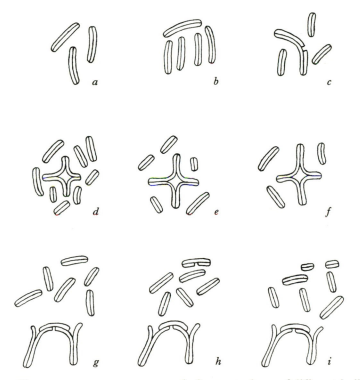

Fig. 9.1. Chromosome sets, as seen at male first metaphase, of different individuals of the scorpion genus *Tityus*. (a) *T. bahiensis*, 'normal' set (2n = 6, three bivalents); (b) *T. bahiensis* (2n = 10, five bivalents); (c) *T. bahiensis* (2n = 9, three bivalents and a trivalent); (d) *T. mattogrossensis* (2n = 20, eight bivalents and an association of four); (e) *T. trivittatus* (2n = 14, five bivalents and an association of four); (f) *T. bahiensis* (2n = 10, three bivalents and an association of four); (g) (h) (i) three individuals of *T. bahiensis* from the same locality, with 2n = 17, 18 and 19 (each with an association of five chromosomes). From Piza, redrawn and modified.

of an individual centromere, and at both mitotic and meiotic anaphases the daughter chromosomes remain straight and parallel to the equatorial plane, as if they were attached to the spindle along their whole length. In some cells the ends of the chromosomes may be slightly bent up towards the poles at early anaphase, and Piza has concluded from this that there may be one centromere at or near each end. There is no critical evidence in favour of this 'dicentric' interpretation and the appearance of these chromosomes is, in fact, quite unlike that of dicentrics in other organisms. Thus one must agree with Brieger and Kerr (1949) who consider these chromosomes to be holocentric (or possessing a 'diffuse centromere activity'), particularly in view of the observations of Rhoades and Kerr (1949) on chromosome fragments produced by irradiation.

The second peculiar feature about the cytology of these Brazilian scorpions is that there are apparently no chiasmata formed during spermatogenesis. It is true that Brieger and Graner (1943) claimed to have seen some appearances suggestive of chiasma formation, but Piza's evidence that the homologous chromosomes simply remain paired throughout their length from pachytene up to the beginning of first anaphase seems sound. Whether absence of chiasmata indicates a real absence of genetic crossing-over in this case is not known, nor is there any evidence as to whether chiasmata are formed during oogenesis.

Tityus bahiensis, the species that has been most extensively studied, has a variable chromosome number. At Piracicaba most individuals have 2n = 6, at Batatais 2n = 10 and at Ouro Preto 2n = 17 to 19. Clearly, fusions or dissociations (or both) have taken place repeatedly during the history of the species and, in fact, chromosome number heterozygotes occur in some populations (Fig. 9.1). The mechanisms of such fusions and dissociations would probably not be essentially different from those obtaining in organisms with monocentric chromosomes, except that the breaks would not have to take place adjacent to the centromeres. Thus, in a sense, fusions and dissociations should take place more readily in a species with holocentric chromosomes than in one with monocentric ones. However, in most groups of animals with holocentric chromosomes it would be very difficult for trivalents to be maintained up to metaphase (because of low chiasma frequency) and the orientation of any trivalents that were formed would be irregular and liable to lead to frequent non-disjunction. *Tityus* appears to have circumvented this barrier to fusions and dissociations by its peculiar mechanisms of achiasmate meiosis and parallel disjunction at first anaphase (what happens in the female is, of course, still an open question). Thus, whereas in other groups with holocentric chromosomes fusions and dissociations have probably had to establish themselves in spite of heterozygote inferiority ('negative heterosis'), as in the morabine grasshoppers (see p. 374), in *Tityus* they may enjoy positive heterosis in some instances, since they clearly exist in a state of balanced polymorphism in certain populations.

Apart from chromosome number heterozygotes, many of the individuals of *Tityus* in natural populations are translocation heterozygotes. At meiosis a single cross-shaped configuration may be present (Fig. 9.1*d*) or there may be two separate crosses. In some individuals even more complex configurations occur, indicating the presence of multiple translocations. Thus in the individual shown in Fig. 9.2*a* there is apparently a ring of seven chromosomes and five bivalents, while in those shown in Fig. 9.2*c* all the chromosomes are joined together in a complex multivalent at meiosis, certain chromosomes having regions homologous to parts of three other members of the complement. Such complex meiotic configurations seem to be the rule rather than the exception, in *Isometrus maculatus* (Fig. 9.2*d–e*). It is hardly possible, on the basis of Piza's published illustrations, to understand some of the more complicated configurations he has observed, but it

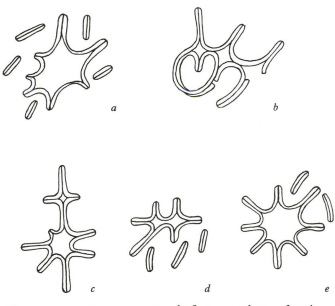

Fig. 9.2. Chromosome sets, as seen at male first metaphase, of various species of scorpions. (*a*) an individual of *Tityus bahiensis* (2n = 17, five bivalents and a ring of seven chromosomes); (*b*) *T. bahiensis* (2n = 7, all chromosomes associated in a complex configuration); (*c*) *T. bahiensis* (2n = 9, all chromosomes associated in a complex configuration); (*d*) and (*e*) *Isometrus maculatus* (2n = 12 in both individuals, but there is a complex configuration of six in (*d*) and a ring of eight in (*e*)). From Piza, redrawn and modified.

seems clear that rings and other configurations made up of odd numbers of chromosomes (e.g. seven and nine) can occur. These must give rise to gametes with deficiencies and duplications, and the individuals with such meiotic configurations must themselves have been derived from gametes with deficiencies and duplications. Possibly there are extensive heterochromatic regions in these chromosomes which can be dispensed with or be present in duplicate without serious effects on viability and fecundity. But various other possibilities seem to exist. One is that the male is the heterogametic sex (this is not known for certain in any species of scorpion) and that the structural heterozygosity of *Tityus* and *Isometrus* species is somehow involved in sex determination, in which case the degree of structural heterozygosity might be much less. Another is that, since *Tityus serrulatus* is known to be a thelytokous species without males (Matthiesen 1962), the other species may likewise reproduce parthenogenetically, the males (which only form about 25% of the population in *T. bahiensis* and *T. trivittatus*) being functionless. Perhaps the most likely explanation is pseudogamous reproduction (i.e. a form of parthenogenesis in which sperms are necessary for the initiation of development in the egg but do not contribute any chromosomes to the offspring).

The kind of translocation heterozygosity seen in *Tityus* and *Isometrus* spp. in Brazil seems to be a rather fundamental feature of the genetic system of the buthid scorpions, since it has been reported in two Indian species by Sharma, Parshad and Joneja (1959). In *Buthus doriae odonturus* (2n = 22) one male (out of five) carried a ring of ten chromosomes. In some cells,

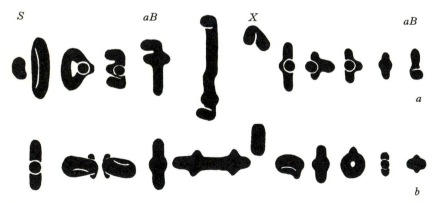

Fig. 9.3. Chromosome sets at male first metaphase in (*a*) an individual of *Trimerotropis suffusa*, (*b*) one of *T. citrina*. In *a* there is a supernumerary chromosome (*S*), two heteromorphic bivalents (*aB*) and a chain of four chromosomes resulting from a translocation. In *b* there is a similar chain of four chromosomes but no heteromorphic bivalents or supernumeraries. From Helwig (1933).

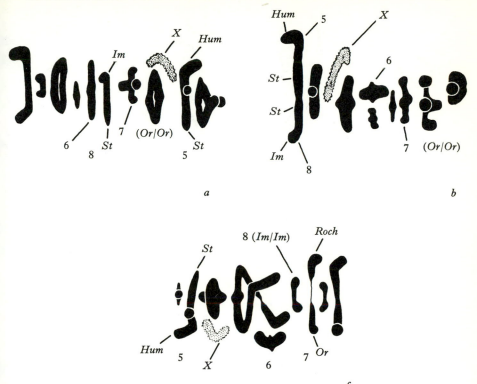

Fig. 9.4. Three male first metaphases (side views) of the grasshopper *Trimerotropis gracilis* (Humboldt Range, Nevada). *a, b* from the same individual which was heterozygous for chromosome pairs 5 (*Standard/Humboldt*), and 8 (*Standard/Imlay*) and homozygous for chromosome 7 (*Oreana/Oreana*). In *b* the *Standard* 5 and *Standard* 8 chromosomes are associated end-to-end, presumably by a terminal chiasma. Cell *c* is from another individual, which was heterozygous for a reciprocal translocation and shows a chain of four chromosomes (some other cells showed a ring of four); this individual had the constitution *Standard/Humboldt, Oreana/Rochester, Imlay/ Imlay*. From White (1961*a*).

however, a ring of six and a ring of four were present. In a single individual of *Buthus tamulus sindicus* associations of three or four chromosomes were also observed.

Translocations occurring in single individuals of a population cannot be regarded as forming part of an adaptive polymorphic system. We illustrate in Figs. 9.4, 9.5, 9.6 and 9.7 several examples in the grasshoppers *Trimerotropis gracilis, Cryptobothrus chrysophorus, Keyacris scurra* and an unnamed morabine species. Helwig (1933) has figured similar translocations in *Trimerotropis suffusa* and *T. citrina* (Fig. 9.3). In these cases few or no

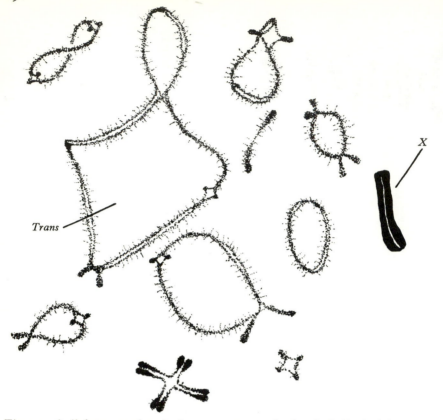

Fig. 9.5. A diplotene nucleus in the spermatogenesis of an individual of the grass-hopper *Cryptobothrus chrysophorus* heterozygous for a translocation. A configuration (*Trans*) of four chromosomes held together by five chiasmata is present.

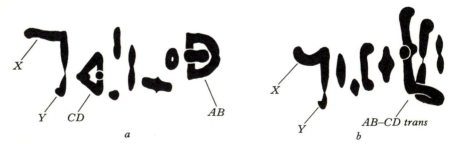

Fig. 9.6. Two male first metaphases (side views) from individuals of the Australian morabine grasshopper 'P52a'. *a*, normal; *b*, from an individual which was hetero-zygous for a 3-break rearrangement involving the *AB* and *CD* chromosomes (a translocation plus a pericentric inversion in the *CD* chromosome). From White and Cheney (1972).

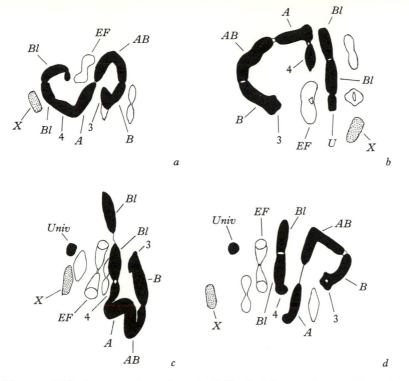

Fig. 9.7. Male first metaphases in an individual of the grasshopper *Keyacris scurra* caught in the wild which was carrying a translocation involving four different chromosomes. *a, b* configurations without univalents – in *a* there is a chain of 8 chromosomes, in *b* there is a chain of 5 and one of 3. *c*, a configuration with a chain of 7 and a univalent; *d* a cell with a chain of 4, one of 3 and a univalent. From White (1963).

chiasmata were formed in the segments between the centromeres and the point of interchange. However, John and Hewitt (1963) have described a spontaneous translocation in the grasshopper *Chorthippus brunneus* in which there was extensive chiasma formation in this interstitial segment. White (1963) has reported an individual of *Keyacris scurra* heterozygous for a complex translocation involving breaks in four different chromosomes (Figs. 9.7, 9.8) and Coleman (1947) described one of *Chorthippus longicornis* heterozygous for two translocations and a pericentric inversion. Such complex rearrangements would almost certainly lead to complete sterility, so it is highly unlikely that they were inherited from a previous generation; the most plausible explanation in both cases would be that the breaks arose in one of the gametes which produced the individual in question. Such complex

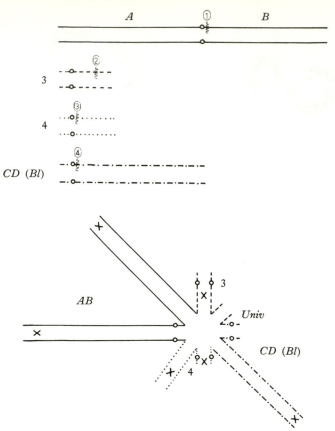

Fig. 9.8. Presumed breakage points and mode of synapsis in the individual illustrated in Fig. 9.7. Crosses indicate regions where chiasmata are usually formed. From White (1963).

spontaneous rearrangements (four breaks in one instance and six in the other) do not seem to have been observed in animals other than grasshoppers.

POLYMORPHISM FOR FUSIONS OR DISSOCIATIONS

Cytological polymorphism in respect of chromosome number may be due to several quite distinct causes. In many instances it may be due to super-numerary chromosomes, present in varying numbers in addition to the 'basic' set (see Ch. 10). In other cases it may be due to chromosomal fusions or dissociations which have established themselves in natural populations but have not reached fixation. In such instances a metacentric chromosome may be genetically equivalent to two acrocentrics so that there are three types of individuals in the population: MM, M/A^1A^2 and A^1A^2/A^1A^2. The heterozygotes will show a characteristic trivalent at meiosis, in which the two acrocentrics are paired with the corresponding limbs of the metacentric; provided that the orientation of this trivalent on the meiotic spindle is regular (i.e. provided that the three centromeres are arranged in an isosceles or equilateral triangle) disjunction will be fairly regular.

The question whether heterozygosity for fusions or dissociations can be regarded, like inversion heterozygosity, as cross-over suppressing, is not easily answered. Polymorphisms involving fusions or dissociations seem only able to persist in natural populations (1) where the metacentric element is approximately equal-armed, (2) where each arm of the metacentric has a chiasma frequency of 1.0 or only very slightly higher, and (3) where the chiasmata show a considerable degree of distal localization. These conditions are needed if the trivalents in the heterozygotes are to be symmetrical, so that disjunction will be regular.

The evidence thus suggests that a considerable degree of suppression of crossing-over in proximal regions precedes the establishment of this type of structural polymorphism. However, the suppression may be absolute in the case of the structural heterozygotes and only relative in the homozygotes. If so, opportunity will exist for divergent genetic evolution of the proximal regions of the metacentric and acrocentric elements and the whole system will function genetically in very much the same manner as an inversion polymorphism.

The pre-condition that in such systems the arms of the metacentric must be approximately equal in length, seems to apply fairly strictly in all those cases discussed below where fusions or dissociations exist in a state of flux in natural populations. It is less strictly observed, apparently, in some instances where we have related species or races differing cytotaxonomically in respect of fusions or dissociations, but where there is no evidence that an intra-population polymorphism has been either long-lasting or geographically widespread. Thus, in the phylogeny of the Australian morabine grasshoppers the CD chromosome, which is an unequal-armed element, has undergone dissociation on a number of occasions (although significantly less frequently than the AB element, which is equal-armed) and fusions between acrocentrics of very different length have established themselves in several species.

It is difficult to estimate how many of the numerous instances of variation in chromosome number that have been recorded in groups such as the Lepidoptera with dot-like chromosomes and an absence of visible centromeres should be regarded as falling into the 'fusion and dissociation'

TABLE 9.1. *Variations of chromosome number in Lepidoptera probably due to fusions or dissociations*

Species	n	Author
Argynnis ino	12–13	Federley 1938
Gonepteryx rhamni	31–32	Federley 1938
Leptidia sinapis	28–41	Lorkovic 1950; Maeki 1958b
Pieris rapae	25–26	Federley 1938; Lorkovic 1941; Maeki and Remington 1960b
P. bryoniae	25–28*	Lorkovic 1968
Polyommatus icarus	22–23	Federley 1938; Lorkovic 1941
Agrodiaetus hamadensis	21–22	De Lesse 1959c
A. sennanensis	28–30	De Lesse 1959c
A. dama	41–42	De Lesse 1959c
Lysandra hispana	84–85	De Lesse 1952, 1956
L. dolus	124–125	De Lesse 1954
L. argester	131–151	De Lesse 1954
L. nivescens	190–191	De Lesse 1954
Melanitis phedina	28–30	Maeki 1953

* Plus o to 5 small supernumerary chromosomes; there seem to be two different types of chromosome number variation in this species. A similar situation seems to occur in *P. napi pseudorapae* from the Caucasus, although *P. napi napi* invariably shows n = 25.

category. Probably most of them are of this type, but in Table 9.1 we have only listed those instances that seem to be clearly due to fusions or dissociations.

Whether any particular chromosome number polymorphism has been due to fusion or dissociation can, as a rule, only be decided by comparison with related populations or species. However, as we have already pointed out, dissociation leads to duplication of minute regions adjacent to the centromeres and fusion to deletion of similar regions. In the grasshopper *Keyacris scurra*, it was possible to prove the existence of the duplicated (i.e. tetrasomic) regions in the race homozygous for the higher chromosome number and hence to prove that a dissociation rather than a fusion had occurred (White 1957a).

Obviously, there is a difference in principle between species which exhibit a chromosome number polymorphism over a wide geographic area or even throughout the entire range of the species and instances where there are two chromosome number races, each cytologically monomorphic throughout most of its range, but with a narrow zone of overlap in which chromosome number heterozygotes occur. The distinction will be clear when an extensive investigation involving a study of many local populations has been carried out, but not if sampling has been less extensive. Where only a few individuals of a species or even a single one have been examined and found to be fusion or dissociation heterozygotes there is a high probability, but no certainty, that the polymorphism is a geographically extensive one. This is the case in the grasshoppers *Cornops aquaticum* and *Aularches miliaris*, a tettigoniid of the genus *Jamaicana* and the lizards of the genus *Gerrhonotus*.

Complicated systems of chromosomal polymorphism involving fusions or dissociations have been described by Smith (1962a, d), and Manna and Smith (1959) in the Canadian bark weevil *Pissodes approximatus* and in the Californian *P. terminalis*. In the former, a metacentric A may be represented by two acrocentrics A^l and A^r. The highest frequencies of A occur in eastern Ontario populations, the lowest (i.e. the highest frequencies of A^l and A^r) in Manitoba populations referred to $P.$ *canadensis*, a form which is probably at most racially distinct from $P.$ *approximatus*. In individual colonies the frequencies of the three kinds of individuals (A/A, A/A^lA^r and A^lA^r/A^lA^r) are in good agreement with the expectation on the binomial square rule and the data are consistent with a stable genetic equilibrium. Manna and Smith tentatively suggest, after comparison with related species, that dissociation has occurred in this case, i.e. that the metacentric AA represents the primitive condition. But the situation could also be due to a fusion.

Two other cytological polymorphisms occur in $P.$ *approximatus*. A rather obscure inequality in length of two homologues may be ascribed to a supernumerary chromosome region, probably heterochromatic. A much more complex and interesting polymorphism occurs in a chromosome element referred to as B. This exists in four different 'forms'. Two of these are J-shaped metacentrics which are cytologically indistinguishable but apparently differ by a pericentric inversion – we may call them B^1 and B^2. Both seem to have undergone dissociation into 'left' and 'right' limbs at some time in the history of the species, so that we have acrocentric chromosomes B^{1l}, B^{1r}, B^{2l}, B^{2r} all floating in the population. There will altogether be nine euploid genotypes

$$\left[\frac{B^1}{B^1}, \frac{B^2}{B^2}, \frac{B^1}{B^{1l}, B^{1r}}, \frac{B^1}{B^{2l}, B^{2r}}, \frac{B^2}{B^{2l}, B^{2r}}, \frac{B^2}{B^{1l}, B^{1r}}, \right.$$

$$\left. \frac{B^{1l}, B^{1r}}{B^{1l}, B^{1r}}, \frac{B^{1l}, B^{1r}}{B^{2l}, B^{2r}}, \frac{B^{2l}, B^{2r}}{B^{2l}, B^{2r}} \right]$$

in the population, but only six of these will be cytologically distinguishable. In addition, simple recombination should yield six additional genotypes

$$\left[\frac{B^1}{B^{1l}, B^{2r}}, \frac{B^2}{B^{1l}, B^{2r}}, \frac{B^1}{B^{2l}, B^{1r}}, \frac{B^2}{B^{2l}, B^{1r}}, \frac{B^{1l}, B^{2r}}{B^{1l}, B^{2r}} \text{ and } \frac{B^{1r}, B^{2l}}{B^{1r}, B^{2l}}\right]$$

which are all heteroploid (respectively, monosomic, monosomic, trisomic, trisomic, nullisomic and tetrasomic for the c region). These heteroploid genotypes, which should be cytologically distinguishable, were not encountered by Manna and Smith in 934 individuals examined; it is not clear whether their absence is due to gametic mortality, selective fertilization or zygotic inviability.

In the Californian *P. terminalis* there is a polymorphism for the 'C' chromosome which is similar to that for the A-chromosome in *P. approximatus*, but all of 218 males were C/C^lC^r heterozygotes while forty-one females were C/C homozygotes. There is an easily recognizable XY bivalent in the males of these species and this is orientated at random with regard to the C trivalent at first metaphase and therefore presumably segregates at random at first anaphase. Selective fertilization accounts for the absence of the expected C/C males and C/C^lC^r females (Smith and Takenouchi 1962, Smith 1962*d*). This unique incompatibility system persists even in crosses between males of *P. terminalis* and females of *P. yosemite*, a species which is C/C in both sexes. Smith and Takenouchi believe, following Drouin *et al.* (1963) that *P. terminalis* arose in evolution by hybridization between *P. yosemite* and a species such as *P. strobi* that was homozygous for the C^lC^r condition.

P. terminalis has a polymorphism in the B-chromosome which seems to be of the same type as that of *P. approximatus* although not all the expected genotypes have been encountered; it also shows dissociation-polymorphism for a third chromosomal element A, all three genotypes being present in this case.

It is interesting that there are two quite different polymorphisms with 'missing genotypes' in these weevils, the B polymorphism and the C polymorphism of *P. terminalis*. This suggests that the same mechanism of gametic mortality or selective fertilization may be involved. Zygotic inviability seems unlikely, because of the very considerable genetic load it would imply. Mortality of gametes with unbalanced combinations of acrocentric chromosomes and selective fertilization are probably involved in the case of chromosome C.

The cytological polymorphism of the coccinellid beetle *Chilocorus stigma* (Smith 1956*a*, 1957*a*, 1958, 1959, 1960*b*, 1962*b*) is similar in some respects to that of the *Pissodes* weevils. All populations of this species have an X_1X_2Y sex-chromosome mechanism which has resulted from a Y-autosome fusion that is absent in related XY species. Florida populations lack fusions between autosomes and have $2n\male = 25$, but ones from Maine and Nova Scotia have an autosomal fusion referred to as the *Kentville* fusion which exists in a state of flux in these regions. This fusion involves chromosomes '1' and '2'. Ontario populations also contain the Kentville fusion (in somewhat higher frequency) and a second fusion known as the *Vineland* one, between chromosomes '4' and '5'. In Manitoba, the Kentville and Vineland fusions are present in still higher frequencies and a third *Morden* fusion makes its appearance; thus $2n\male$ varies from nineteen to twenty-five. Finally, a population in Saskatchewan was homozygous for the Kentville fusion and possessed the other two in a state of flux (Fig. 11.20). The cytogenetic status of populations throughout the greater part of the eastern and middle-western states of the United States of America is still unknown.

Smith believes that increasing chromosomal polymorphism permitted the species to extend its distribution northward and westward after the last glaciation. The idea of the most polymorphic populations of a species occupying the advancing frontier of its range is a very different one from the views of Dobzhansky (based on *Drosophila willistoni*) and Carson (based on *D. robusta*) according to which peripheral or invading populations should be more homozygous than those of areas where the species has been longer established. This does not mean that Smith's interpretation is necessarily incorrect, but it does suggest that more populations of *C. stigma* should be studied before the history of the species can be regarded as fully understood. All the other species of *Chilocorus* lack the Y-autosome fusion that is present in all populations of *C. stigma*, i.e. they are XY species; three species (*C. cacti, orbus* and *fraternus*) have $2n = 22$, i.e. they are homozygous for two fusions. The first is from the southern United States and the Caribbean, while the other two are Californian. *C. tricyclus* ($2n = 20$) from British Columbia is homozygous for three fusions and *C. hexacyclus* ($2n = 14$) from Alberta for six fusions.

Studies of interspecific hybrids have shown that one of the two fusions present in *C. orbus* involves the same two chromosome elements (1 and 2) as the Kentville fusion in *C. stigma*. The second *orbus* fusion involves elements 3 and 4, i.e. it is a fusion not present in *stigma*. *Tricyclus* and *hexacyclus* possess both the *orbus* fusions plus additional ones of their own. Up to a point, these species support the idea that in this genus chromosomal fusions have assisted in expansion of the range northward and westward, but not that chromosomal polymorphism as such has done so (since these species are structurally homozygous).

The fusions that have occurred in *Chilocorus stigma* and other species of the genus are peculiar in that the elements that have participated in them, instead of being acrocentric, as is usually the case where centric fusions have occurred, are J-shaped elements with the shorter limb heterochromatic. These heterochromatic limbs are lost in the process of fusion, but in some instances may be represented in the karyotype by heterochromatic supernumerary chromosomes, of which as many as five may be present in the same individual.

It is characteristic of the male meiosis of *Chilocorus* that the euchromatic chromosome arms always form a single chiasma, while the heterochromatic ones never form chiasmata at all. Thus thirteen chiasmata are invariably found at meiosis, regardless of the number of fusions present and disjunction is entirely regular. Meiosis in the female has now been studied by Smith, who has shown that the heterochromatic arms do not form chiasmata in the oocyte either.

A very elaborate system of chromosomal polymorphism based on fusions or dissociations has been described in the littoral mollusc *Thais* (= *Purpura*) *lapillus* by Staiger (1954). On the coast of Brittany certain populations show $2n = 36$, all the chromosomes being rod-shaped (not strictly acrocentric, for the most part). These demes exist in localities sheltered from wave action where the food supply is limited. In localities exposed to strong wave action, with an abundant supply of food (mostly mussels) a race with $2n = 26$ is found, having eight pairs of rod-shaped chromosomes and five pairs of metacentrics. In stations which are ecologically intermediate, cytologically polymorphic populations occur in which many of the individuals have chromosome numbers between twenty-six and thirty-six and show from one to five trivalents at meiosis. Only the twenty-six chromosome race occurs on the Atlantic coast of North America (Mayr 1963).

Assuming that the twenty-six chromosome race arose by five centric fusions, then certain relatively short chromosome arms, probably heterochromatic, were lost in the process, just as in *Chilocorus stigma*. But this may well be a case where the thirty-six chromosome race arose by five dissociations, rather than vice versa.

Staiger assumes that, in the case of each of the five polymorphisms, the hierarchy of selective values is:

$$\frac{M}{M} > \frac{M}{A^1, A^2} > \frac{A^1, A^2}{A^1, A^2} \qquad \text{in exposed localities}$$

$$\frac{M}{M} < \frac{M}{A^1, A^2} > \frac{A^1, A^2}{A^1, A^2} \qquad \text{in intermediate localities}$$

$$\frac{M}{M} < \frac{M}{A^1, A^2} < \frac{A^1, A^2}{A^1, A^2} \qquad \text{in sheltered localities}$$

Heterosis is thus confined to ecologically intermediate situations. Accepting this interpretation, it is most interesting that the adaptive properties of five different, non-homologous, fused chromosomes are all similar – they all help to adapt the individual to 'exposed' environments.

About 1% of the individuals of *Thais lapillus* were heterozygous for reciprocal translocations of several kinds and showed chain or ring multivalents at meiosis (Staiger, 1955).

Since centric fusions always involve the loss of small proximal regions adjacent to the centromeres, there does not seem to be any difference *in principle* between 'ordinary' centric fusions of acrocentrics and the situation in *Chilocorus* and *Thais*, where the regions lost are much more extensive than is usual. But it is certainly interesting that in two outstanding instances of evolutionary success of centric fusions as heterotic polymorphisms, losses of large heterochromatic segments should be involved; whereas in the numerous cases where fusions have established themselves and rapidly reached fixation in a race or a species (i.e. where there is no evidence of selective superiority of the heterozygotes) losses of heterochromatic material have been on a much smaller scale.

Another case where a fusion-polymorphism has probably involved loss of a substantial heterochromatic region is that of the marine isopod *Jaera albifrons*, described by Staiger and Bocquet (1956). In one locality about 45% of the individuals showed an autosomal trivalent at meiosis; but this consisted of a metacentric paired with one acrocentric and a J-shaped chromosome. In all probability the shorter limb of the latter was deleted during the formation of the metacentric. *Jaera syei* has a series of populations with $2n\delta = 28$, 26, 24, 22, 20 and 18 from the Baltic to the Basque coast; these differ by chromosome fusions. A hybrid zone exists between the 24- and 22-chromosome populations but colonies of this species are not, in general, polymorphic for chromosome number (Lécher 1967).

Chromosome number polymorphism due to a total of six fusions or dissociations exists in the European shrew *Sorex araneus* (Bovey 1949; Sharman 1956; Ford, Hamerton and Sharman 1957; Meylan 1960, 1964, 1965a; Matthey and Meylan 1961; Ford 1970). Both this and the sibling species *Sorex gemellus* (Ott 1968) possess a sex trivalent of the XY_1Y_2 type in the male, the X being a big metacentric, Y_1 a small acrocentric and Y_2 a medium-sized subacrocentric. *S. gemellus* is cytologically monomorphic and invariably has $2n\delta = 23(2n\female = 22)$; it is a West European form which occurs in the Alps, Jura Mountains, France, Belgium, Holland and the island of Jersey. *S. araneus* is chromosomally polymorphic and has $2n\male = 21$ to 33; it occurs in England, Holland, Scandinavia, Germany and also in the Jura Mountains and Alps above about 900 metres. Apparently it exists and is polymorphic as far east as Siberia (Kozlovsky 1970). There is a broad zone of overlap between the two species which runs from Holland to the Alps. The karyotypic differences between the two siblings seem to be extensive and even a 23-chromosome male of *S. araneus* differs from *gemellus* by several inversions, translocations or other structural changes.

If we consider the males of *araneus* with thirty-three chromosomes as representing the 'primitive' karyotype, *araneus* would be polymorphic for six fusions; an individual with $2n = 21$ would be homozygous for all of these. However, no population is known to be polymorphic for all these fusions and most only seem to carry one or two of them. It is probable that many local populations are cytologically monomorphic, but this is naturally difficult to prove conclusively.

The relationship between the two sibling forms is not entirely clear. Both possess the same complex sex chromosome mechanism, which must hence have evolved before they diverged from one another.

This example differs from most of the invertebrate cases we have considered in the very small prolificity and the restricted population size and low degree of vagility of the species. Colonies of *Sorex araneus* are probably quite highly inbred. It is thus not at all surprising if the species has become cytologically monomorphic in many areas – the wonder is that it has remained so highly polymorphic in others. A species with such a low prolificity (by comparison with most insects, arachnids and crustaceans) is surely ill-equipped to bear the heavy segregational load which such complex polymorphisms imply, especially in the presence of a significant degree of inbreeding. Further studies on this most interesting species will be eagerly awaited.

Other species of *Sorex* have much higher chromosome numbers ($2n = 38$ in *S. minutissimus*; $2n = 42$ in *S. minutus*, *S. caecutiens*, *S. shinto saevus*, *S. unguiculatus* and *S. isodon*; $2n = 58$ in *S. alpinus*; $2n = 66$ in *S. fumeus* and *S. cinereus*); all these are $XX:XY$ species (Meylan 1965*b*; Orlov and Alenin 1968; Halkka and Skarén 1964; Skarén and Halkka 1966; Fredga 1968*b*; Takagi and Fujimaki 1966; Shimba and Itoh 1969; Halkka, Skarén and Halkka 1970). It is thus probable that there have been a whole series of chromosomal fusions and pericentric inversions in the lineage leading to the siblings included under the name *S. araneus*; and that the polymorphism of *araneus* is for fusions, not dissociations. Polymorphism for a single fusion has been encountered in Canadian populations of the shrew *Blarina brevicauda* by Meylan (1967).

Clear cases of fusion- or dissociation-polymorphism have also been reported by Matthey (1963a, b, c; 1964b, d; 1965b, c; 1967) in four taxa of African mice, *Mus* (subgenus *Leggada*) *minutoides minutoides* (2n = 18–20), *M. (L.) minutoides musculoides* (2n = 30–34), a subspecies of *M. triton* (2n = 20–22) and *M. (L.) tenellus*. The first of these is polymorphic for one fusion (and for a pericentric inversion as well), the second is polymorphic for two fusions and the latter for a single one. No polymorphic systems have been encountered in a number of other races and species belonging to this taxonomically complex group of forms. Some of these African 'Leggadas' appear to have the *X*- and *Y*-chromosomes (primitively acrocentric in *Mus musculus*, *M. indutus* and *M. setulosus*) fused with the members of a pair of autosomes. Extreme variability of chromosome number (from 2n = 48 to 2n = 59) occurs also in the field mouse *Apodemus giliacus* in Hokkaido, but has not been analysed in detail (Hayata *et al.* 1970).

Polymorphism for several centric fusions has been reported by Gustavsson and Sundt (1968, 1969) in two subspecies of the Asiatic deer *Cervus nippon*. However, since their material came from European deer parks there is a possibility that it was of racially mixed origin; the chromosomal situation in truly wild populations of the species has not been investigated.

Chromosome number polymorphism is known in three species of the mantid genus *Ameles* from the eastern Mediterranean region (Wahrman 1954a, b; Wahrman and O'Brien 1956). In populations of *A. heldreichi* from Israel and Turkey one chromosome element may be represented by a metacentric (M) or by two acrocentrics $(A^1$ and $A^2)$; the former condition being much the commoner. Wahrman considers that the metacentric condition was the primitive one in this genus, so that we would be dealing with dissociation-polymorphism. *A. cypria* from Cyprus has a similar polymorphism and an undescribed species of the same genus from Israel seems to be polymorphic for as many as six dissociations, although it is not clear whether all this polymorphism is found at any one locality. Several other cases of chromosome number polymorphism are known in mantids. One is the European *Iris oratoria* (Wahrman 1954a); others occur in the South African genera *Hoplocorypha* and *Antistia* (White 1965b).

The observations of Mesa (1956) on a single individual of the South American grasshopper *Cornops aquaticum*, which was heterozygous for three different fusions suggests that this is another species that is highly polymorphic for chromosome number.

Although polymorphism for fusions or dissociations will undoubtedly be found in many more groups of animals in the future, it is clearly a rare phenomenon, dependent on very special pre-conditions. In the morabine grasshoppers about sixty different fusions and dissociations are known to

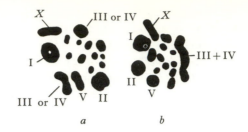

Fig. 9.9. Male first metaphases (polar views) in the tettigoniid grasshopper *Metrioptera brachyptera*. *a*, a cell with fifteen bivalents; *b*, one with a chain of four chromosomes. There are two possible interpretations of *b*, heterozygosity for a reciprocal translocation between chromosomes III and IV or reduplication of a minute terminal segment, as is probably the case in Fig. 9.4*b*.

have established themselves in the phylogeny of the group, but no populations polymorphic for such arrangements are known, other than those which must exist in the zones of contact or very narrow geographic overlap between races differing in chromosome number. It is regrettable that only in the case of *Thais* do we have any evidence as to the biological significance of the polymorphism in terms of adaptation to specific ecological niches. Multiple polymorphisms of the *Chilocorus* and *Sorex* type should furnish excellent material for the study of epistatic interactions between polymorphisms, but this aspect has not been tackled as yet.

A most unusual case of fusion heterozygosity has been reported in the rodent *Sigmodon minimus* by Hsu (1969); six individuals (out of a total of seven) from a population in New Mexico were heterozygous for a fusion between two homologous acrocentrics, i.e. for an isochromosome. Theoretically this should produce complete sterility in matings with normals and 50% sterility in matings between heterozygotes; but since the condition was found in the majority of the individuals collected in 1967 and 1968 it may be present as a balanced polymorphism in the population.

DUPLICATIONS AND DEFICIENCIES IN NATURAL POPULATIONS – 'UNEQUAL BIVALENTS'

Two broad categories of deficiencies and duplications have been encountered in natural populations. On the one hand we have deletions or duplications of single bands or very short regions, visible in dipteran polytene chromosomes. Such changes were studied especially by Metz (1939, 1941, 1947) in *Sciara*. On the other hand we have deficiencies or duplications of much longer regions, visible in ordinary mitotic or meiotic chromosomes.

As many as seventeen minute deficiencies or duplications are known in *Sciara ocellaris*. In some instances a band is present in one homologue but

Fig. 9.10. *Drosophila ananassae*. Distal ends of some of the salivary gland chromosomes, showing the minute terminal deficiencies or duplications which are present in certain stocks. *a* and *e*, homozygous deficient types; *b, c, f, g* heterozygous types; *d* and *h*, individuals homozygous for the presence of the terminal band. From Kikkawa (1937).

not in the other (so that the band appears to go half way across the double polytene element), while in other cases a thick band in one homologue is opposite a thin one in the other. No phenotypic differences have been associated with any of these minute rearrangements, and their role in the dynamics of natural populations of *Sciara* species is not known. These minute rearrangements are known in nine different species of *Sciara*, but paracentric inversions are also quite common in wild populations of some members of this genus.

In the genus *Drosophila*, this kind of minute rearrangement is uncommon in natural populations but some species such as *D. ananassae* are polymorphic for minute duplications or deficiencies (of one or two bands in the polytene elements) at or very close to the ends of the chromosomes (Kikkawa 1938; Dobzhansky and Dreyfus 1943).

The second category of duplication-deficiency that has been encountered in natural populations is where a particular chromosome may either carry or lack a segment of sufficient length to be easily visible under the microscope. In these cases heterozygotes will show so-called 'unequal bivalents' composed of two homologues, one of which is markedly longer than the other. Instances of this type have been studied in many species of grasshoppers and in a few other insects and molluscs. A particularly clear case exists in the rat flea *Nosopsyllus fasciatus*, where four different kinds of unequal bivalents have been figured by Bayreuther (1969).

We must presume that in all these cases we are dealing with balanced polymorphisms, which depend on heterozygote advantage, or on frequency-dependent selective values, with the homozygotes (long/long and short/short) adapted to particular ecological niches. There is, however, the possibility that in certain instances a deleterious chromosome (either the long or the short homologue) might be kept in the population by meiotic drive.

The 'extra' region that is present in the longer homologue is probably in all instances a heterochromatic segment. In many cases the published evidence does not show whether all three expected genotypes exist in the population; we may presume, however, that in most cases they do, and that neither of the homozygous ones is actually lethal. 'Extra' heterochromatic regions which are not essential to life are clearly akin to supernumerary chromosomes (Ch. 10). There is, however, an essential difference. Super-numerary chromosomes frequently and perhaps always display unorthodox

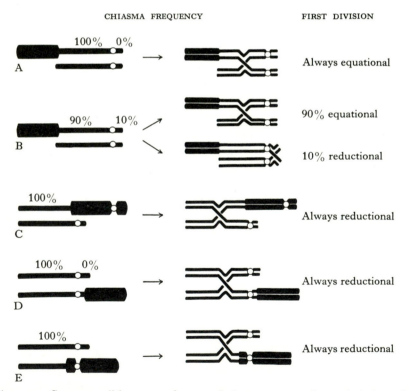

Fig. 9.11. Some possible types of unequal chromosome pairs and their various modes of segregation at the first meiotic division, depending on the relative positions of centromeres, chiasmata and inequalities. All these are terminal inequalities.

behaviour at meiosis in one or other sex and may show 'accumulation' mechanisms at various stages of the germ-line; thus their inheritance deviates in various ways from the Mendelian rules. Supernumerary regions which form part of regular members of the karyotype are likely to be inherited in the conventional manner although, as pointed out above, the possibility of their showing meiotic drive in some instances cannot be ruled out at present.

Where we have a polymorphism for a supernumerary chromosome segment we may regard the longer chromosome as carrying a duplication or the shorter one as bearing a deletion. Which interpretation is true in an evolutionary sense can usually not be determined; the most we can do is to demonstrate whether the extra region is interstitial or terminal. A chromosome with the sequence of regions *ABCDE* may give rise to one with the sequence *ADE* by deletion. But a pair of *ABCDE* chromosomes may also give rise to *ABCBCDE* and *ADE* elements by translocation; in which case we shall have all three chromosomes present in the same population, and a fourth type *ABCBCBCDE* may be expected to arise by unequal crossing-over. Where the extra region in the longer homologue is terminal, the situation is likely to have arisen by translocation between non-homologous chromosomes, one of which may in some instances have been a supernumerary.

There are two types of cases where we can be reasonably sure that the supernumerary region is interstitial and one where we can be certain that it is terminal (distal). It will be simpler to consider the situation where both members of the unequal pair are acrocentric. If, in such a case, we have two kinds of unequal bivalents in the same heterozygous individual each with a single chiasma in the main arm, but in one instance with the chiasma proximal to the inequality and in the other distal to it, we may be certain that the inequality is interstitial. An example of an unequal bivalent has been found in the Australian grasshopper *Cryptobothrus chrysophorus*, which falls in this category. It can be seen that bivalents with a proximal chiasma will be 'equational' at the first division, segregation of the inequality being postponed until the second anaphase. On the other hand, bivalents with a distal chiasma will show a 'reductional' first anaphase. A case just like the *Cryptobothrus* one was studied by Nur (1961) in the grasshopper *Calliptamus palaestinensis*. In *Cryptobothrus* there are several other types of unequal bivalents in the natural populations, one of which corresponds to category D or E of Fig. 9.11. John and Hewitt (1966a) have described an interstitial supernumerary chromosome segment in the British grasshopper *Chorthippus parallelus* which behaves just like that of *C. palaestinensis* at meiosis.

The unequal pair in *Calliptamus* and the commonest type found in

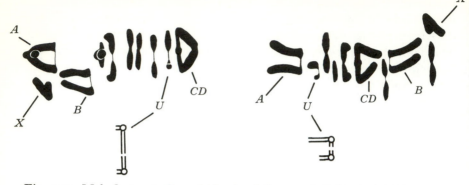

Fig. 9.12. Male first metaphases (side views) from an individual of the Australian morabine grasshopper 'P146', heterozygous for an unequal bivalent. In this species the usual *AB* chromosome has undergone dissociation into *A* and *B* acrocentrics which seem to invariably form a chiasma in the short arm (a chiasma may be present in the long arm as well). The unequal bivalent (*U*) consists of two J-shaped chromosomes, the inequality being in the longer limb. Two kinds of 'reductional' *U*-bivalents occur. In the first (frequency about 25%) the chromosomes are terminally associated by their longer (unequal) limbs; in the second (frequency about 75%) they are similarly associated by their shorter limbs. The inequality is clearly interstitial and must have arisen by deletion.

Cryptobothrus both form univalents in a certain proportion of cases. Such cells may be expected to give rise to aneuploid gametes in about 50% of cases. If the genetic equilibrium depends on heterosis in these instances, it must be strong enough to outweigh the less of fecundity due to this cause. The unequal chromosome pair found in the Tasmanian grasshopper *Exarna includens* by Sharman (1952) seems to fall into the same category as the *Calliptamus* one. Once again, some cells with univalents were observed.

There is a second type of unequal pair in which we can be sure that the inequality is interstitial. This is illustrated in Fig. 9.12. A case has been found in an undescribed species of morabine grasshopper from northern Australia. In this case there are two different types of reductional bivalents at first metaphase, one with a distal chiasma in the long arm, the other with a chiasma in the short arm.

Unequal bivalents which are invariably pre-reductional, due to a single distal chiasma (i.e. in which the large and small elements invariably separate at the first anaphase) may be interpreted as ones in which the inequality is proximal and probably includes the centromere and the short arm (i.e. the two centromeres are not truly homologous). Another type is where the chiasma is invariably proximal and the bivalent always post-reductional; in such a case we could conclude that the inequality was terminal and distal.

The heterochromatic supernumerary segments present in some populations of the grasshopper *Chorthippus parallelus* in Britain and France are of this type. They occur in the two smallest chromosome pairs and, when present in either the heterozygous or the homozygous state, raise the chiasma frequency of the other bivalents significantly (John and Hewitt 1966a). In the case of chromosome $S8$ the three genotypes are present in the Hardy–Weinberg proportions, while in the case of chromosome $M7$ there is a significant deficiency of heterozygotes (Hewitt and John 1968, 1970b; John and Hewitt 1969; Westerman 1969).

There can be little doubt as to the evolutionary interconvertibility of the three main types of autosomal heterochromatin met with in grasshoppers (the homozygous heterochromatic segments, the blocks of heterochromatin in the larger members of unequal pairs of chromosomes and the heterochromatin of supernumeraries). The origin of an 'unequal bivalent' polymorphism is hence probably in some cases a translocation between two autosomes, which may or may not be homologous (one or both, in either case, having a terminal block of heterochromatin). However, some 'unequal bivalent' systems may have arisen by translocations involving a supernumerary chromosome. It is rather unlikely that the heterochromatin of the X-chromosome has become implicated in any of these transformations (at least in the grasshoppers); it always shows a characteristic negative heteropycnosis (Figs. 1.4, 1.5) in spermatogonial divisions, which is never shown by supernumerary chromosomes or chromosome regions. Any meiotic synapsis between X- or supernumerary chromosomes and autosomes will be liable to lead to non-disjunction of the autosomal pair and will hence be strongly opposed by natural selection. It is hence to be expected that, in cases where both supernumerary chromosomes and supernumerary chromosome regions coexist in a population, synapsis between them will be very rare or non-existent, as in the case of *Calliptamus palaestinensis*, studied by Nur (1961).

Almost all species of grasshoppers possess in their karyotypes one pair of medium-sized or small autosomes with a very conspicuous heterochromatic segment, or with several such segments, and it is probably from such a pair of homologues that the unequal pairs are derived, by duplication, deletion or translocation. The heteropycnotic autosome is so obvious at meiosis in many grasshopper species that it has received various special names (the *megameric* chromosome of Corey 1938, 1939; the *precocious* chromosome of Darlington 1936; and the *dyade compagnon de l'X* of Janssens 1924, since it is often loosely associated with the X-chromosome during the meiotic prophase, although this association is almost certainly non-chiasmatic and does not persist into metaphase.

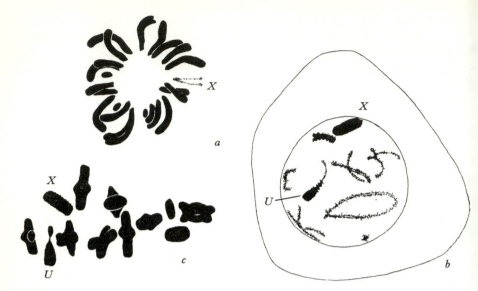

Fig. 9.13. Cytology of an individual of the grasshopper *Trimerotropis bilobata* which has an unequal pair of chromosomes. *a*, spermatogonial metaphase (the two shortest pairs of chromosomes consist of three small acrocentrics and one medium-sized one); *b*, diplotene, showing the heterochromatic segment in the larger member of the unequal bivalent (U); *c*, first metaphase, showing the unequal bivalent in side view.

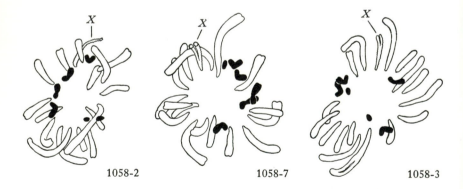

Fig. 9.14. Spermatogonial metaphases of three different individuals of the very rare grasshopper *Pedioscirtetes nevadensis*. The three small autosomal pairs are shown in black, for clarity. In 1058-2 two of these pairs are unequal, in 1058-7 one of them is unequal, while 1058-3 is structurally homozygous. From White and Nickerson (1951).

Fig. 9.15. Three male first metaphases (side views) of the mantid *Paratenodera sinensis*. In *b* and *c* there is an unequal bivalent (*U*) which may form chiasmata on one side of the centromere or on both sides (in *c*). The three sex chromosomes X_1, X_2 and Y form a characteristic trivalent. From White (1941*a*).

The chromosomal polymorphism of the cricket *Pteronemobius taprobanensis* (Manna and Bhattacharjee 1964) is apparently due to 'extra' heterochromatic arms which may be present or absent from chromosomes 4 and 6 (when present, these chromosomes are metacentric; when absent they are acrocentric). In the case of chromosome 4, the heterozygotes appear to be significantly above 50% of the population, indicating a rather high degree of heterosis.

Polymorphisms for duplications or deficiencies in the Y-chromosome are known in a number of species. In *Drosophila pseudoobscura* there are a number of different types of Y's known, some of which are longer than others (Dobzhansky 1935*a*, 1937*a*). Several types of Y are also found in *D. athabasca* (Miller and Roy 1964) and in *D. birchii* (Baimai 1969). In man, also, the size of the Y-chromosome is variable (Cohen, Shaw and MacCluer 1966), although it is not clear whether this should be interpreted as a true balanced polymorphism.

Intrapopulational variation in size of the X is less common. In an African species of mouse, *Mus (Leggada) triton*, Matthey (1966*b*) has studied a sample of twelve individuals from Bukavu (Congo); three of the females were heterozygous for a large deletion in the X-chromosome, converting a metacentric element into a subacrocentric one. It is not known in this case whether males carrying deleted X's or females homozygous for the deletion are viable.

10

SUPERNUMERARY CHROMOSOMES

Supernumeraries thus arising may therefore lead a kind of
wandering life in the species . . .

E. B. WILSON: *The Cell in Development and Heredity*

In natural populations of certain species of animals and plants, supernumerary chromosomes (i.e. ones additional to the normal karyotype and not homologous, or only partly homologous, to members of the regular set) are present in some individuals but not in others. In some cases the majority of the population may carry supernumeraries, while in other instances the frequency of individuals carrying them is very low. In all cases, there should theoretically be a series of genotypes with 0, 1, 2, 3 . . . supernumeraries, but where the extra bodies are sufficiently rare, individuals with more than one of them may not be found, in practice, in natural populations. Frequently, a species which shows supernumeraries in certain geographic areas may lack them in others.

Most supernumerary chromosomes are heterochromatic; some have been described as wholly or partially euchromatic, but one may well question whether they would be so by all criteria (e.g. by tritiated thymidine autoradiography, which might show them to be late-labelling). Many supernumeraries are smaller than the members of the regular chromosome set, but some are quite large (Fig. 10.1). In general, their presence or absence does not seem to affect the appearance of the individual in any noticeable way although a variety of subtle biometric effects on viability and fecundity undoubtedly exist and have been studied especially in plants such as rye (Müntzing 1963; Moss 1966) and *Lilium callosum* (Kimura and Kayano 1961; Kimura 1962) (see reviews by Müntzing 1958 and Battaglia 1964). These effects are, however, such that it is not possible to determine from the external phenotype whether an individual carries supernumeraries or not.

Moss (1966) has shown that in Rye, supernumerary chromosomes increase the variance of the population for a number of biometrical characters.

We have defined the term 'supernumerary chromosome' somewhat rigorously, since it was used very loosely by some early authors to cover a variety of cases where the chromosome number is not constant. Some instances of so-called supernumerary chromosomes recorded in the literature may have been due to trisomy for a member of the regular karyotype, to polymorphism for fusions or dissociations or to occasional asynapsis at the first metaphase; none of these phenomena having anything in common with true supernumeraries.

True supernumerary chromosomes have been found in numerous species of higher plants and have been studied especially in maize (Longley 1927, 1956; Darlington and Upcott 1941; Randolph 1941; Roman 1947a, b; Roman and Ullstrup 1951; Blackwood 1956) *Lilium callosum* (Kayano 1956a, b, 1957, 1962), *Trillium grandiflorum* (Rutishauser 1960), *Plantago serraria* (Fröst 1959), various grasses (Östergren 1945, 1947; Bosemark 1954, 1956, 1957; Müntzing 1957, 1959), rye (Müntzing 1945, 1950, 1958, 1959, 1963; Lima de Faria 1948) and *Clarkia* (H. Lewis 1951). In bryophytes they have been investigated by Vaarama (1953) and others. Most plant cytologists refer to them as *B*-chromosomes. Melander (1950b) has used the term 'accessory chromosomes', a term which indicates their essential nature.

In animals, supernumerary chromosomes of the classical type have been described in flatworms (Melander 1950b), snails (Evans 1960), Isopoda (Rocchi 1967), grasshoppers, scale insects, Heteroptera, Lepidoptera, beetles and some Diptera. They seem to be very rare in vertebrates, but have been found in the urodele genus *Oedipina* (Kezer, unpublished) and in the marsupial Glider *Schoenobates volans* (Hayman and Martin 1965c). Morescalchi (1968c) studied two female individuals of the primitive New Zealand frog *Leiopelma hochstetteri*, one of which had 2n = 23, the other having 2n = 34. It appears as if there are twenty-two large chromosomes and a varying number of small ones which may be supernumeraries. There has been a second report of supernumerary chromosomes in a species of frog by Nur and Nevo (1969), in the North American *Acris crepitans*. General reviews of the occurrence and behaviour of supernumerary chromosomes have been published by Battaglia (1964) and Müntzing (1966).

Most supernumeraries obviously exist in a state of genetic equilibrium in the natural populations in which they occur. There have been two main interpretations of these equilibria. The first type of model assumes that single supernumeraries (or low numbers of them) confer an adaptive advantage on the individual, but that higher numbers of them are deleterious. We may call this the 'heterotic' interpretation. since an individual with a single supernumerary may be regarded as a heterozygote. According to the second type of explanation the supernumeraries are deleterious, regardless of their number, and owe their continuous existence in the population to an 'accumulation mechanism' which operates at meiosis in one or other sex or at some of the mitotic divisions in the germ-line. On this second interpretation we might speak of the 'parasitic nature' (Östergren 1945) of supernumeraries. A thorough investigation of each case of supernumeraries will obviously involve a detailed study of their meiosis in both sexes and their behaviour at other stages in the germ-line and in the somatic tissues as well as an analysis of their effects on viability and fecundity. It is hence not surprising that such complete data hardly exist for any concrete instances of organisms carrying supernumeraries and that our knowledge of most of the recorded cases is extremely fragmentary and unsatisfactory.

One can state in a general way that supernumerary chromosomes must have arisen in evolution from members of the regular karyotype. However, since they do not generally synapse with the other chromosomes at meiosis (some types may undergo a temporary adhesion to sex chromosomes during the prophase of meiosis, but this seems to be almost always resolved before first metaphase) they must have undergone numerous changes since their origin. In a general way, one can imagine very small fragment chromosomes (consisting of a centromere and minute adjacent segments) arising through various kinds of deletions and translocations. Such fragments may then become converted into true supernumeraries as a result of successive duplications and perhaps also by translocations from other chromosomes. But the modes of origin of supernumerary chromosomes are necessarily conjectural and the process is hardly open to direct investigation. They are perhaps especially liable to arise from members of the regular karyotype which contain extensive heterochromatic segments adjacent to the centromeres and telomeres.

An important distinction may be drawn between supernumeraries which are mitotically stable and those which show more or less frequent non-disjunction; the usual result of which is that a cell with one supernumerary gives rise to daughter cells with 0 and 2 supernumeraries. As far as animals are concerned, we know rather little about the mitotic stability or instability of supernumeraries in the somatic tissues, since almost all the studies that have been made were confined to spermatogenesis.

Accumulation mechanisms may be of several kinds, and may operate at mitosis or at meiosis. Melander (1950b) discovered mitotically unstable supernumerary chromosomes in one population of the triclad *Polycelis tenuis*. The normal type of supernumerary is metacentric, with a satellite on the end of one arm. However, both satellited and non-satellited acrocentrics were also seen, having presumably arisen by some kind of breakage through or adjacent to the centromere. The supernumeraries of *P. tenuis* tend to be lost from the somatic cells of fully-grown animals. Their number in the meiotic nuclei is considerably greater, but variable. In the male first metaphases of one individual, nine cells had no supernumerary, sixteen had one, thirty had two, fifteen had three and seven had four. This is presumably due to non-disjunction in the spermatogonia. Apparently many supernumeraries degenerate and are lost during the anaphase and telophase of the second meiotic division of spermatogenesis. Melander claimed that in the female meiosis the supernumerary chromosomes undergo an endomitotic reduplication in the premeiotic resting stage in oogenesis. If true, this would constitute a powerful accumulation mechanism, which would be in equilibrium with the loss of supernumeraries that occurs at the second meiotic division and as a result of natural selection against the relatively inviable individuals carrying supernumeraries.

This interpretation of the supernumeraries of *P. tenuis* as entirely deleterious elements which are only maintained in the population as a result of a supplementary endomitotic replication appears extremely dubious. It seems to be based on the postulate that the genotype of the species is unable to evolve in a way that would get rid of the 'parasitic' supernumeraries, or render them innocuous.

A different type of accumulation mechanism has been reported for the supernumerary chromosomes of the snail *Helix pomatia* by Evans (1960). In two populations of this species a large proportion of the individuals carried one or more small supernumeraries. The number of supernumeraries varied considerably from one first spermatocyte to another in a way which suggested that non-disjunction had occurred in the spermatogonia. But there was a deficiency of first metaphases carrying no supernumeraries, as compared with those carrying two or more, which suggested that cells with supernumeraries divided faster than ones lacking them. Where two supernumeraries were present they formed a bivalent in most cells and the meiotic behaviour seems to be normal.

Another species of snail which probably carries supernumerary chromosomes in its natural populations is *Triodopsis fraudulenta*, investigated by Husted and Burch (1946). The normal chromosome number in this species is $2n = 58$, but individuals with 59, 60, 61 and 62 were also found. Husted and Burch regarded the extra chromosomes as homologous to members of the regular karyotype (i.e. they considered the individuals with more than 58 chromosomes as polysomics), but it seems more probably that they were dealing with true supernumeraries. It is noteworthy that unequal bivalents (due to extra heterochromatin in the larger member of a pair of chromosomes) also occur in some individuals of this species.

In at least seven species of grasshoppers, *Camnula pellucida* (Carroll

1920; Nur 1969a), *Locusta migratoria* (Itoh 1934), *Neopodismopsis abdominalis* (Rothfels 1950), *Patanga japonica* (Sannomyia 1962), *Atractomorpha australis* (White, unpublished), *A. bedeli* (Sannomyia and Kayano 1968) and *Calliptamus palaestinensis* (Nur 1963b) different follicles or cysts in the same testis commonly contain different numbers of supernumeraries, suggesting that mitotic non-disjunction occurs in the germ-line. Nur has carried out a careful mathematical analysis of the *Calliptamus* case; in the population studied by him 12.3% of the males carried supernumeraries. About 11% were presumed to have started life with a single supernumerary. Such individuals usually showed some follicles without a supernumerary and some with two and sometimes more. If simple non-disjunction were responsible the number of cysts or follicles with 0 and 2 supernumeraries would be expected to be equal, whereas in *Calliptamus* those with 2 outnumbered those with 0 in the ratio 15 : 1. Nur consequently concludes in favour of the hypothesis of differential non-disjunction at embryonic divisions in which future germ-line cells are segregated from future somatic cells. This mitotic non-disjunction presumably functions as an 'accumulation mechanism'. Although the evidence for the other species mentioned above is less complete, it seems probable that the mechanism is the same in all of them. The meiotic behaviour of these species is essentially normal in the male, but has not been studied in the female. Univalent supernumeraries lag on the first anaphase spindle in the male and divide tardily in *Locusta* (Rees and Jamieson 1954) and *Atractomorpha australis*. In *Camnula pellucida*, Nur (1969a) found that 105 out of 780 males at a certain locality in California carried supernumeraries. The number of supernumeraries per cell in such individuals varied from 0–4 in different follicles. A study of the number of supernumerary chromosomes in the cells of the gastric caeca suggested that all but three of the 105 individuals started life with a single supernumerary. Males which originally possessed a single supernumerary show a mean of about 1.37 in the spermatocytes, suggesting an accumulation of 37% in the development of the testis. There is a loss of supernumeraries amounting to about 1%, in the form of an occasional elimination in microspermatids.

A supernumerary chromosome with an accumulation mechanism will increase in frequency in the population unless the accumulation is counteracted by elimination of individuals with supernumeraries through natural selection. However, it must be pointed out that in the cases considered above we have no idea how the supernumeraries behave in the female meiosis (i.e. we do not know whether they have an equal chance of getting into the four products of meiosis). Neither do we know whether there may be some gametic selection at some stage. In the absence of such information it seems rather futile to speculate as to the adaptive significance of this type of polymorphism.

A large number of species of grasshoppers belonging to the tribe Trimerotropi in western North America are known to carry supernumerary chromosomes in some or all of their populations. Acrocentric supernumeraries have been found in *Trimerotropis diversellus*, *T. cyaneipennis*, *T. gracilis*, and *Circotettix thalassinus*. Equal-armed metacentric supernumeraries which are probably isochromosomes occur in *T. latifasciata* and *C. rabula*, while both types coexist in *T. suffusa*, *T. sparsa* and *C. undulatus*. Most of these species are ones which also have elaborately developed systems of inversion polymorphism in their natural populations (see p. 246).

In all these species the supernumerary chromosomes are largely heterochromatic and show strong positive heteropycnosis during the prophase of meiosis in the male. However, they frequently show some euchromatic segments, or at any rate, ones which do not undergo heteropycnosis at the same time as the rest of the chromosome. Mitotic non-disjunction does not seem to take place in the male germ-line, so that all the spermatocytes of a single individual have the same number of supernumeraries.

In individuals with a single acrocentric supernumerary, the latter attaches itself to the first meiotic spindle near one of the poles rather than in the plane of the equator. It passes undivided to one pole at first anaphase, but splits in the second meiotic division, so that half the sperms contain the extra element. In individuals carrying two acrocentric supernumeraries, these invariably form a bivalent in *T. sparsa* and *C. thalassinus*. These supernumerary bivalents show a single chiasma which is probably situated in one of the euchromatic segments; in the case of the acrocentric supernumeraries of *Trimerotropis sparsa* the position of the chiasma seems to be somewhat variable (White 1951*f*), while in those of *Circotettix thalassinus* the chiasma seems to be rather strictly localized very near to the distal end. A peculiarity of these bivalents composed of two acrocentric supernumeraries is that, in both the species mentioned above, they rather frequently orientate themselves on the spindle of the first meiotic division with both centromeres close to the same pole; one would expect that this behaviour would frequently lead to the production of second spermatocytes with two supernumeraries and with none. If so, the number of individuals in the populations with different numbers of supernumeraries would depart from that expected on the basis of Mendelian inheritance, but without any increase in the mean number of supernumeraries per individual. However, the behaviour of the supernumeraries in the female is not known in any trimerotropine grasshopper.

Metacentric supernumeraries in the Trimerotropi show a very character-

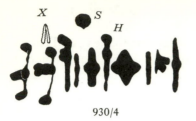

930/4

930/3

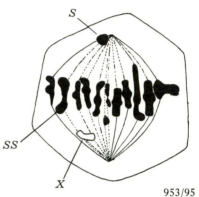

H H   953/114

953/95

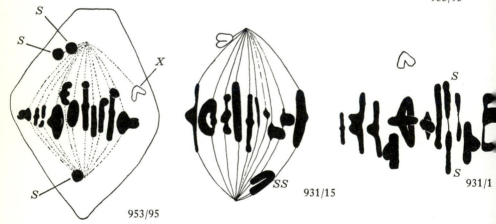

953/95

931/15

931/1

Fig. 10.1. First metaphases in individuals of *Trimerotropis sparsa* with supernumerary chromosomes (side views). No. 930/4 has one metacentric supernumerary, 930/3 has two metacentric supernumeraries; 953/95 has three metacentric supernumeraries (which may either form a bivalent and a univalent or three univalents); 953/114 has one acrocentric supernumerary and 931/15 has two acrocentric supernumeraries that regularly pair to form a bivalent (which is sometimes orientated in the form of a U with both its centromeres directed toward the same pole). *H*, a 'heteromorphic' bivalent; *S*, a single supernumerary; *SS*, a bivalent composed of two supernumerary elements. From White (1951*b*).

istic type of behaviour at meiosis in the male. The two arms are precisely equal in length and in individuals with a single such supernumerary a chiasma is regularly formed at meiosis between the distal regions of the two limbs. It therefore seems clear that these elements must be interpreted as isochromosomes. At diplotene and diakinesis they appear as rings which are largely heteropycnotic but have a subterminal euchromatic segment in each arm; it is apparently in this segment that the chiasma is formed. At first metaphase they are thick rings with a minute hole in the middle and at first anaphase they ordinarily pass undivided to one pole, dividing at second anaphase.

Where two or more isochromosome supernumeraries are present in the same individual they sometimes pair to give bivalents or chains of three, held together by chiasmata in the same subterminal euchromatic regions; however, in the majority of the first spermatocytes they are present as univalent rings with 'internal' chiasmata. Thus individuals with two iso-chromosome supernumeraries produce about 20% sperms with no super-numeraries and 20% with two. In individuals of *Trimerotropis sparsa* with both acrocentric and metacentric supernumeraries, no pairing between the two types has been seen, thus suggesting that no very close degree of homo-logy or phyletic relationship exists between them. None of these trimero-tropine supernumeraries show any tendency to pair with the X at meiosis, and they are non-heteropycnotic in the spermatogonial divisions, in contrast to the X, which is negatively heteropycnotic at this stage.

In *Trimerotropis inconspicua* a type of supernumerary has been found which may be interpreted as an isochromosome from which a large part of one limb has been deleted or, alternatively, as one in which a pericentric inversion has occurred. This chromosome is J-shaped, but at meiosis it forms a ring with a subterminal chiasma between the two arms, like the 'perfect' isochromosome supernumeraries with arms of equal length (Fig. 10.3).

It is characteristic of these trimerotropine grasshoppers that the fre-quency of the supernumerary chromosomes in the populations shows considerable geographic variation (Table 10.1). And, although individuals with two and three supernumeraries may be reasonably frequent in some populations, the frequency of individuals in the population carrying super-numeraries never seems to exceed 30%. No interpretation of the role of supernumerary chromosomes in any trimerotropine population is possible until the meiotic behaviour in the female has been studied. One must strongly suspect that meiosis in the egg includes an 'accumulation mechanism' and the natural selection operates to eliminate individuals with more than a certain number of supernumeraries; the two opposing forces leading to the establishment of a stable equilibrium.

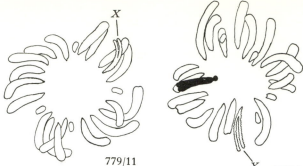

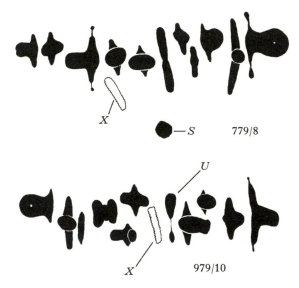

Fig. 10.2. Spermatogonial metaphases and first meiotic metaphases in *Trimerotropis inconspicua*. 979/11, spermatogonial metaphase from an individual without supernumeraries or unequal bivalents; 979/4, spermatogonial metaphase from an individual with a J-shaped supernumerary (shown in black); 979/8, first metaphase in an individual similar to 979/4; 979/10, first metaphase in an individual with no supernumerary but with an unequal bivalent (U).

It seems probable that supernumerary chromosomes have been present throughout the evolution of the trimerotropine grasshoppers and that they have been handed down to such present-day species as *T. gracilis*, *T. sparsa*, *T. suffusa*, etc. from some common ancestor of those forms (no doubt undergoing structural changes in the course of phylogeny, thus accounting for their present-day differences in shape and size).

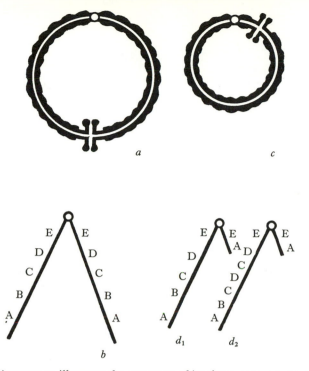

Fig. 10.3. Diagrams to illustrate the structure of isochromosome supernumeraries. *a*, diplotene configuration of a 'perfect' isochromosome supernumerary which has formed a chiasma with itself; *b*, genetic interpretation of the same (such supernumeraries have been found in three species of *Trimerotropis* and two species of *Circotettix*); *c*, the type of isochromosome supernumerary which has been found in *Trimerotropis inconspicua* (diplotene); d_1 and d_2, two alternative genetic interpretations of the same (d_1 assumes a deletion in one arm, d_2 has a pericentric inversion).

The supernumerary chromosomes of the British grasshopper *Myrmeleotettix maculatus* have been studied by Barker (1960, 1966) and also by John and Hewitt (1965a, b) and Hewitt and John (1967), who have disproved Barker's conclusions in some respects. There are two types of supernumeraries in this species – metacentrics (B^M) and submetacentrics (B^{SM}). Most populations in the southern half of the British Isles carry one or both types in a high proportion of the individuals (15 to 70%). In northern England, in Scotland and at high elevations in the south of the British Isles, supernumeraries seem to be generally absent. Apart from individuals lacking supernumeraries, the following genotypes have been found in wild populations: $1B^M$, $1B^{SM}$, $2B^M$, $2B^{SM}$, $1B^M + 1B^{SM}$, $3B^M$, $2B^M + 1B^{SM}$. Both types of supernumeraries are mitotically stable; they apparently do not

TABLE 10.1 *Supernumerary chromosomes in trimerotropine grasshoppers*[1]

Species	Locality	Percentage of individuals having			No. of individuals studied
		One super-numerary	Two super-numeraries	Three super-numeraries	
T. latifasciata (?)	Roswell, New Mexico	11.54	–	–	26
T. sparsa	Craig, Colorado	6.85	5.48	–	73
	Mesa, Garfield, Delta and Montrose Counties, Colorado	0.47	–	–	212
	Pojoaque, New Mexico	12.23	1.06	–	188
	Great Basin (nine localities)	9.42	–	0.53	199
T. gracilis	Humboldt Range, Nevada	0.36	–	0.52	559
T. suffusa	Truckee, California	9.52	–	–	189
	Archuleta County, Colorado	9.30	2.33	–	43
	Crescent Lake, Washington	15.56	8.89	–	45[2]
T. diversellus	Yellowstone National Park	10.70	–	–	56
C. undulatus	Ashton, Idaho	17.91	7.14	3.57	28
C. rabula[3]	Carbondale, Colorado	–	–	–	25
	Cowles, New Mexico	13.04	4.35	–	23
	Cloudcroft, New Mexico	4.79	–	–	313
	Collyer, Kansas	16.22	10.81	–	37[2]

[1] Mostly unpublished data of M. J. D. White. No distinction is made in this table between acrocentric and metacentric (isochromosome) supernumeraries.
[2] Data of Helwig (1933)
[3] The Kansas material was referred to erroneously as C. undulatus by Helwig.

possess any mechanism of accumulation or elimination on the male side, but the situation in the female has not been studied. At meiosis in the male the supernumeraries of *Myrmeleotettix* form chiasmata between the arms of the same chromosome and (in cells with several supernumeraries) between chromosomes. The B^M and B^{SM} chromosomes must be largely homologous, since chiasmata are formed between them.

Both types of supernumeraries present in *Myrmeleotettix* occur in populations living in warm, dry environments; they are rare or absent in habitats which are colder or wetter. Individuals with supernumerary chromosomes show a higher chiasma frequency than individuals from the same population lacking them; and the frequency of supernumeraries in the populations is positively correlated with the characteristic chiasma frequencies of these populations.

The populations of *M. maculatus* on the mainland of Europe that have been studied lack supernumeraries and Hewitt and John (1970a) have concluded from this that these chromosomes arose since 8000 years B.P. (and perhaps since 3000 B.P.), i.e. since the formation of the Straits of Dover after the end of the Pleistocene glaciation. It may be questioned whether enough Continental European populations have been examined for this to be quite certain. Gibson and Hewitt (1970) have shown that the DNA of the supernumerary chromosomes has a different base composition to that of the remaining DNA of the karyotype. Their data are somewhat complex and different to interpret, however (Table 10.2), since there are differences between the base ratios of the main peaks of the four populations. Somewhat surprisingly, the heterochromatin of the X does not form a subsidiary peak of its own.

Probably we can regard the supernumeraries as having been built up by repeated duplication of a small segment of autosomal heterochromatin characterized by a relatively low percentage of G + C. The rather remote possibility that they might have been derived by introgressive hybridization from another species cannot be totally excluded however.

TABLE 10.2 *Base composition of DNA's of Myrmeleotettix maculatus* (% G + C)

	Main peak, due to autosomes and X-chromosomes	Subsidiary peak, due to supernumerary chromosomes
Wales population A	37	–
B	40	33
East Anglia population C	48	–
D	46	38

An interesting type of supernumerary has been described in the Australian grasshopper *Phaulacridium vittatum* by Jackson and Cheung (1967). In the particular population studied 11.12% of males carried a single acrocentric supernumerary and 0.14% carried two supernumeraries. The supernumerary does not seem to be entirely homologous to the *X*, since it does not show negative heteropycnosis in spermatogonial divisions (at a time when there is marked, but variable, negative heteropycnosis of the *X*). Nevertheless, in the first meiotic division of the males, the *X* and the supernumerary are closely associated and may be united by a chiasma, although this is not certain. They pass to opposite poles at first anaphase in 70% of cases; there is thus an increased transmission to sons and a decreased transmission to daughters. The frequency of the supernumerary in the female population and its behaviour in female meiosis have not yet been studied.

Other species in which heterochromatic supernumeraries are associated with the *X*-chromosome during the male meiotic prophase include the grasshoppers *Neopodismopsis abdominalis* (Rothfels 1950), *Phaulotettix eurycercus* (Fig. 10.4) and the grouse locust *Tettigidea parvipennis* (Robertson 1917). There have been no detailed studies of segregation in these cases, so that we do not know whether a mechanism similar to that of *Phaulacridium*

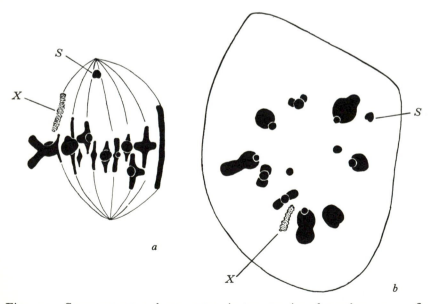

Fig. 10.4. Supernumerary chromosomes in two species. of grasshoppers. *a*, first metaphase (side view) in an individual of *Phaulotettix eurycercus*; *b*, *Netrosoma fusiforme*. The exact form of these supernumeraries has not been determined, but they are probably both small metacentrics.

operates. In general it seems likely that the association in these instances is not maintained into the metaphase stage, and is hence definitely non-chiasmate. Whether any parts of the X-chromosome have been translocated to the supernumerary in any of these species is uncertain, but probably unlikely. It is at any rate clear that supernumerary chromosomes do not lead to any evident degree of intersexuality in any species of grasshopper.

In spite of the widespread occurrence of supernumerary chromosomes in groups such as the Acrididae, there is no evidence that a supernumerary has ever become stabilized, as a member of the regular karyotype, in any species of animal. They could hardly be expected to do so in cases where they are mitotically unstable, but in such forms as the trimerotropine grasshoppers, one might perhaps have expected they might have evolved into normal members of the karyotype in some species. However, we do not know (1) whether

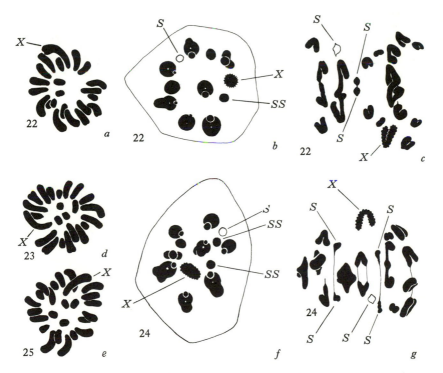

Fig. 10.5. Supernumerary chromosomes in the pyrgomorphid grasshopper *Atractomorpha australis*. Chromosome number variable in different follicles. *a–c*, from an individual with 22 chromosomes ($18A + X + 3S$); *d–g* from an individual with a variable chromosome number ($18A + X + 4$–$6S$). *a, d, e*, spermatogonial metaphases; *b, f*, first metaphases in polar view; *c, g*, first anaphases in side view. S, a supernumerary chromosome; SS, a bivalent composed of two supernumeraries.

the behaviour of the supernumeraries in the somatic mitosis is entirely normal and (2) whether their inheritance in the female meiosis deviates in any way from Mendelian expectation. If, as seems possible, the centromeres of *all* supernumeraries exhibit peculiar forms of behaviour (accumulation mechanisms, a tendency to non-disjunction etc.), then such chromosomes could hardly be expected to evolve into members of the regular karyotype. However it would still be possible for translocations to occur between supernumeraries and ordinary chromosomes, and some of these may have been involved in the origin of the supernumerary chromosome regions discussed in Chapter 9. It is possible, but rather unlikely, that some supernumeraries have acted as donor chromosomes in the dissociation of metacentrics to give acrocentrics (see p. 226).

The supernumerary chromosomes in the rat flea *Nosopsyllus fasciatus* studied by Bayreuther (1969) regularly form bivalents when two or four are present. When there are three of them, they form a bivalent and a univalent, being apparently incapable of forming a trivalent. However, occasionally, trivalents seem to be formed by the synapsis of a supernumerary with an autosomal bivalent. Non-disjunction of the supernumeraries occurs at the second meiotic division in the male; there is an accumulation mechanism in oogenesis.

A particularly interesting accumulation mechanism for supernumerary chromosomes has been studied in the mealy bug *Pseudococcus obscurus* by Nur (1962*a*, *b*, *c*). In this insect each primary spermatocyte gives rise to two functional sperms and two non-functional products of meiosis. The supernumeraries segregate preferentially into the functional nuclei; actually a male transmits about 88% of his supernumeraries to his sperms. Since females transmit close to 50% of their supernumeraries to their eggs, the calculated increase in the number of supernumeraries per generation will be about 33%. Nur found, however, that the number of supernumeraries per individual stayed relatively constant from year to year; so that natural selection must operate against individuals with large numbers of supernumeraries. The mean number of supernumeraries is considerably less than 1 per individual in some populations and in these it is probable that even a single one is deleterious, but in those where large numbers are present the deleterious effect may only be manifested in individuals with several supernumeraries. In one population the calculated relative viabilities declined from 1.0 in individuals with no supernumeraries to 0.1 in ones with five supernumeraries. It has been established in this case that the supernumeraries have one significant biometric effect – they increase the length of the hind tibiae in the female, especially when several of them are present. We shall describe the cycle of heteropycnosis in these supernumeraries later (p. 510), when considering the mechanism of meiosis in these coccids.

In later studies, Nur (1966a, b, 1969b) found that the supernumeraries have little or no harmful effect in females. By studying the progeny of 114 females which had mated in the wild it was possible to calculate the constitution of the males with which they had copulated. The 114 females had a mean of 1.73 supernumeraries per individual and 15 of them had none. The males with which these females had mated had a mean of 1.41 supernumeraries per individual. There was no significant difference between the frequency of supernumeraries in the mothers and their offspring. It was concluded that the supernumeraries have little or no effect on the fitness of females, but that the fitness of males with 0, 1, 3, 3 and 4 supernumeraries was in the ratio 1.0, 0.52, 0.63, 0.49 and 0.23. In a later experiment 213 females without supernumeraries were inseminated by wild males. The relative fitness of males with 0, 1, 2, 3 and 4 supernumeraries was calculated to be 1.0, 0.64, 0.56. 0.38, 0.20 in this experiment. Males with supernumeraries had significantly fewer sperm cysts in their testes than ones lacking supernumeraries. Nur considers that the supernumeraries of *P. obscurus* are only maintained in the natural populations by their accumulation mechanism, in the absence of which they would be eliminated by natural selection. It is difficult to imagine, however, that they would have been preserved in the species through all the vicissitudes of inter-deme selection, unless they possessed some adaptive significance.

The supernumeraries of *Pseudococcus* seem to be mitotically stable in the germ-line of both sexes. Those of another scale insect, *Nautococcus* are unstable in both the embryonic somatic cells and in the spermatogonia (Hughes-Schrader 1942); they do not pair at meiosis, either with one another, or with any of the members of the regular karyotype.

It is characteristic of supernumeraries that they tend to vary greatly in number from one locality to another. This almost certainly implies that their frequency varies from generation to generation, within the same deme or local population. There have been hardly any investigations on this aspect, however. A population of *Pseudococcus obscurus* at Berkeley, California showed an increase in the number of supernumeraries over a six-year period (Nur 1966a).

One of the earliest reports of supernumerary chromosomes in natural populations was that of Stevens (1908a) on two chrysomelid beetles now regarded taxonomically as *Diabrotica u. undecimpunctata* and *D. u. howardi*. The number of supernumeraries in both races ranged from 0 to 4, with a mean of about 0.72 per male individual. Smith (1956c) investigated the situation in 1954 in two populations of the former race and found it to be 1.29 and 1.50. Since the collection stations were not the same, one cannot be certain whether the difference was a geographic or a temporal one, but there is certainly some prima-facie evidence for supposing that there has been an

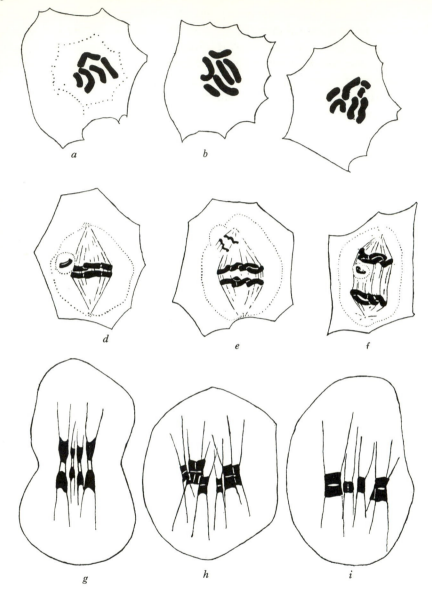

Fig. 10.6. Mitotic and meiotic divisions in the scale insect *Nautococcus schraderae*. *a*, normal somatic set of male; *b*, same with one large supernumerary; *c*, same with two small supernumeraries; *d, e* and *f*, mitotic anaphases showing irregular behaviour of the supernumerary elements, leading to variation of their number in the tissues of a single individual; *g*, first metaphase with a single supernumerary; *h*, same with two supernumeraries that have not paired together; *i*, same with one large supernumerary. From Hughes-Schrader (1942).

approximate doubling in the frequency of these supernumerary chromosomes in the course of about fifty years. One might speculate that this may have been the result of selection following the introduction of agricultural practices involving insecticides, weedicides and other chemicals. The formal genetics of these *Diabrotica* supernumeraries has not been thoroughly studied. There seems to be no accumulation mechanism in the male, but the situation in the female meiosis is unknown. In spermatogenesis of individuals with several supernumeraries prophase adhesions may take place between them, but these do not seem to involve chiasma formation (at least not usually). This contrasts with the situation in the beetle *Tribolium madens*, in which Smith has shown that supernumerary chromosomes can form true bivalents in the male. In the scarabeid beetles *Epicometis hirta* and *Oryctes nasicornis* very small supernumerary chromosomes occur which seem to undergo some kind of prophase association in the male meiosis which is resolved before first metaphase (Virkki 1954).

It is perhaps natural that in groups characterized by high numbers of very small chromosomes, supernumerary chromosomes, when present, should likewise be small and numerous. A number of instances have been described in Lepidoptera, especially by de Lesse (1960). The clearest is perhaps that of the closely related species *Coenonympha arcania* and *C. gardetta*, which have a 'basic' chromosome number of n = 29, but which may show up to fifty-two bodies at first metaphase, owing to the presence of small supernumeraries. Some populations lack such elements, or only show small numbers of them. Their meiotic history has not been followed in detail; probably they are always univalents. Intra-individual variation in number of supernumerary bodies at first metaphase is only slight, and may be due to occasional mitotic non-disjunction. The situation in *Reverdinus floccifera* and *R. boeticus* seems to be similar, but *R. stauderi* probably show only the 'basic' number of n = 30. These cases of genuine supernumerary chromosomes in Lepidoptera need to be distinguished carefully from instances of species that are polymorphic for fusions or dissociations (see p. 417) which likewise show variations in the number of chromosomal bodies at first metaphase. The distinction is not easy, because of the great difficulty of distinguishing univalents, bivalents and multivalents in the Lepidoptera with their dot-like chromosomes, in which centromeres cannot be detected.

There are a few other species of animals in which large numbers of minute supernumerary chromosomes are sometimes present. Boyes and van Brink (1967) have reported that whereas material of the syrphid fly *Xylota nemorum* from eastern Canada show five pairs of large chromosomes, material from western North America had, in addition, about twenty microchromosomes. No further details have been published, and it is not clear whether this is a case of true supernumerary chromosomes.

The first report of supernumerary chromosomes in a mammal appears to be that of Hayman and Martin (1965c) for the marsupial *Schoinobates volans*. Seven individuals from the same locality all showed, in addition to twenty-one (♂) or twenty-two (♀) large chromosomes in the diploid karyotype, a variable number (two to six) of much smaller chromosomes. One of the latter is presumably the *Y* in the case of the male. Meiosis was not studied, so that the interpretation of this case remains somewhat uncertain. Shell-hammer (1969) has found from o to 7 minute supernumerary chromosomes in the North American harvest mouse *Reithrodontomys megalotis*.

It will be noted from Table 10.3 that there are a number of groups in which supernumerary chromosomes have never been recorded, in spite of extensive cytological studies on a large number of species. There seem to be no genuine records of their occurrence in any species of *Drosophila*, the individuals of *D. calloptera* (Metz 1916) and *D. putrida* (Wharton 1943) formerly cited as having supernumeraries being now interpreted differently. Neither is there any record of a supernumerary in any species of mantid. In the later case it is possible that the male meiotic mechanism is simply not

TABLE 10.3 *Species of animals in which supernumerary chromosomes have been recorded*

PLATYHELMINTHES
Turbellaria
    Polycelis tenuis   Melander (1950b)
    Rhynchodemus terrestris   Melander (1950b)

MOLLUSCA
    Helix pomatia   Evans (1960)
    Triodopsis fraudulenta   Husted and Burch (1946)

CRUSTACEA
Isopoda
    Asellus coxalis   Rocchi (1967)
Amphipoda
    Gammarus pulex   Orian and Callan (1957)
    Marinogammarus marinus   Orian and Callan (1957)
    M. pirloti   Orian and Callan (1957)

INSECTA
Orthoptera
    Gryllidae
    Eneoptera surinamensis   Claus (1956); Mesa & Bran (1964)
    Cyrtoxiphus ritsemae   Ohmachi (1935)
    Euscyrtus sp.   S. G. Smith (1953a)
    Gryllotalpidae
    Gryllotalpa africana   Makino, Niiyama and Asana (1938); Asana,
                                  Makino and Niiyama (1940)
    G. gryllotalpa (Romanian race) Steopoe (1939)

TABLE 10.3—*continued*

Rhapidophoridae	
Ceuthophilus sp.	Stevens (1912)
Tettigoniidae	
Theudoria melanocnemis	White, Mesa and Mesa (1968)
Tetrigidae	
Tetrix	Henderson (1961)
Tettigidea lateralis	Robertson (1917)
Eumastacidae	
Moraba carissima	White (unpublished)
Moraba viatica	White, Carson and Cheney (1964)
Moraba 'P24'	White, Blackith, Blackith and Cheney (1967)
Acrididae	
Acrida nasuta	Minouchi (1934)
Stauroderus scalaris	Corey (1933)
Myrmeleotettix maculatus	Barker (1960, 1966); John and Hewitt (1965a, b); Hewitt and John (1967)
Cryptobothrus chrysophorus	Sharman (1952); White (unpublished)
Neopodismopsis abdominalis	Rothfels (1950)
Aioloplus sp.	Ray-Chaudhuri and Manna (1951)
Acrotylus humbertianus	Manna (1954)
Camnula pellucida	Carroll (1920)
Locusta migratoria	Itoh (1934); Rees and Jamieson (1954); Nur (1969)
Phlaeoba infumata	Hsu, Liu and Hsiang (unpublished)
Trimerotropis latifasciata (?)	White (1949)
T. inconspicua	White (unpublished)
T. sp. (Beatty, Nevada)	White (unpublished)
T. diversellus	White (unpublished)
T. cyaneipennis	King (1923); White (unpublished)
T. sparsa	White (1951b, f)
T. gracilis	White (unpublished)
T. suffusa (= *fallax*)	Carothers (1917); White (unpublished)
Circotettix rabula	Evans (unpublished); Helwig (1933)
C. undulatus	White (unpublished)
C. thalassinus	White (unpublished)
Aerochoreutes carlinianus	White (unpublished)
Hesperotettix viridis viridis	McClung (1917)
Phaulotettix eurycercus	White (unpublished)
Bibracte bimaculata	Powers (1942)
Melanoplus sanguinipes	Wasserman (unpublished)
M. borealis stupefactus	Wasserman (unpublished)
M. differentialis	Singleton (unpublished)
Podisma motodomariensis	Helwig (1942)
P. sapporoensis	Helwig (1942)
Miramella mikado	Helwig (1942)
Tristria pulvinata	Manna and Mazumder (1967a)
Scotussa delicatula	Mesa (1964b)
Scotussa sp.	Mesa (1960)

TABLE 10.3—*continued*

Calliptamus palaestinensis	Nur (1963*b*)
Netrosoma fusiforme	White (unpublished)
Atractomorpha australis	White (unpublished)
A. bedeli	Sannomyia and Kayano (1968)
Patanga japonica	Sannomyia (1962)
Pyrgomorpha kraussi	Lewis and John (1959)
Phaulacridium vittatum	Jackson and Cheung (1967)
Coleoptera	
Lampyridae	
Pyractomena angulata	Smith and Maxwell (unpublished)
Cerambycidae	
Elaphidion sp., prob. *parallelus*	Smith (1953*b*)
Chrysomelidae	
Diabrotica duodecim-punctata[1]	Stevens (1908*a*);
	Smith (1956*c*)
D. soror[1]	Stevens (1908);
	Smith (1956)
Coccinellidae	
Chilocorus stigma	Smith (1956, 1957*a*, 1959)
Mulsatina sp. 4	Smith (1960)
Exochomus uropygialis	Smith (1965)
E. lituratus	Smith (1965)
Tenebrionidae	
Tribolium madens	Smith (1956)
Bruchidae	
Spermophagus (Zabrotes) subfasciatus	Minouchi (1935)
Scarabeidae	
Epicometis hirta	Virkki (1954)
Oryctes nasicornis	Virkki (1954)
Homoptera	
Aleurodidae	
Aleurodes proletella	Thomsen (1927)
Coccoidea	
Nautococcus schraderae	Hughes-Schrader (1942)
Pseudococcus obscurus	Nur (1962*a, b*; 1966*a, b*)
Antonina pertiosa	Nur (1962)
Heteroptera	
Coreidae	
Acanthocephala (= *Metapodius*)	
terminalis[2]	Wilson (1907, 1910)
A. femorata[2]	Wilson (1907, 1910)
A. granulosa[2]	Wilson (1907, 1910)
Lygaeidae	
Dieuches uniguttatus	Manna (1951*b*)
Pentatomidae	
Banasa calva	Wilson (1907)
Cimicidae	
Cimex lectularius[3]	Slack (1939*a, b*);
	Darlington (1939*b*);
	Ueshima (1967)
Neuroptera	
Ascalaphidae	
Ascalaphus libelluloides	Naville and de Beaumont (1933)

TABLE 10.3—*continued*

Lepidoptera	
Coenonympha arcania	De Lesse (1960)
C. gardetta	De Lesse (1960)
Reverdinus floccifera	De Lesse (1960)
R. boeticus	De Lesse (1960)
Diptera	
Tipulidae	
Tipula paludosa	Bauer (1931)
Phryneidae	
Phryne cincta	Wolf (1950)
Syrphidae	
Xylota nemorum	Boyes and van Brink (1967)
Tachinidae	
Ceromasia auricaudata	Boyes and Wilkes (1953)
Aphaniptera	
Nosopsyllus faciatus	Bayreuther (1969)
VERTEBRATA	
Urodela	
Oedipina sp. (Costa Rica)	Kezer (unpublished)
Reptilia	
Anolis cristatellus (Puerto Rico)	Gorman (1969)
Anura	
Acris crepitans	Nur and Nevo (1969)
Mammalia	
Schoenobates volans	Hayman and Martin (1965c)
Reithrodontomys megalotis	Shellhammer (1969)

[1] Both now regarded as subspecies of *D. undecimpunctata*
[2] Supernumeraries possibly homologous to the *Y*-chromosome
[3] Supernumerary chromosomes probably derived from the *X*-chromosome

capable of dealing with univalents of any kind, which inhibit anaphase separation at the first division (Callan and Jacobs 1957); but it is not certain whether this is the true explanation. In some other groups the absence of supernumeraries may simply be due to the absence of blocks of autosomal heterochromatin capable of generating them – for example, all the heterochromatic segments may contain too many active genetic loci. A comparison of the cytology of the Coleoptera (Smith 1953*b*, 1960*a*) with the situation in the grasshoppers shows that there are far fewer species with supernumerary chromosomes in the former group. No group seems to be known, however, in which polymorphism for 'extra' heterochromatic segments exists without the simultaneous presence of supernumerary chromosomes, either in the same or closely related species. This argues strongly for the adaptive role of these two cytological phenomena being similar, in spite of the differences in the mechanism of inheritance.

II

CHROMOSOMAL REARRANGEMENTS AND SPECIATION

The close genetic ties which bind species together into single bodies bring into relief the problem of their fission – a problem which involves complexities akin to those that arise in a discussion of the fission of the heavenly bodies, for the attempt to trace the course of events through intermediate states of stability seems to require in both cases a more detailed knowledge than does the study of stable states.

R. A. FISHER : *The Genetical Theory of Natural Selection*

The detailed cytogenetic analysis of speciation in animals is a field which has only recently been developed in a serious manner. The words quoted above recognize its existence as an independent branch of genetics. Yet, as Lewontin (1965) points out, the section of R. A. Fisher's book devoted to speciation (from which they are taken) is only three and a half pages in length.

The recognition during the 1930's that the speciation process is, in general, a gradual one, in which the essential feature is the acquisition of genetic isolating mechanisms, paved the way for the 'biological' species concept that developed during the next two decades. Even now, however, the genetic analysis of speciation phenomena is a somewhat neglected field. Most students of speciation are not primarily geneticists and some population geneticists seem rather unaware of the special problems in the genetic study of speciation.

Mayr (1940, 1942, 1947, 1954, 1963) performed the important task of formulating a consistent interpretation of the process of speciation, based primarily on studies of taxonomy, geographic variation and distribution. Rejecting the earlier concept of 'sympatric' speciation (i.e. the origin of new species from 'ecological' or 'biological' races, coexisting in the same geographic area), Mayr regards all speciation as 'allopatric', that is, preceded by complete geographic isolation. In his own words: 'The essential factor . . . is the interruption of gene flow by extrinsic barriers.' The basic features of the allopatric model are contained in the well-known dumbbell diagram (Mayr 1942, Fig. 16), often presented in elementary accounts of evolution (e.g. Stebbins 1966) as if it embodied some kind of final truth on the speciation problem.

Critics of Mayr's standpoint have based themselves on several grounds. Some (e.g. Thorpe 1930, 1940, 1945) have argued in favour of sympatric speciation from the existence of so-called 'biological races' or 'host races' of phytophagous insects, adapted to live on different food plants. Many supposed cases of 'host races' have, however, been shown to be actually quite distinct species, although with minimal morphological differences (i.e. sibling species). But Bush (1966a), in his careful study of the dipteran genus *Rhagoletis*, concludes that, after all: 'The possibility of two species arising sympatrically from a single population of flies, therefore, is a definite possibility and needs further investigation . . .' Other zoologists and entomologists (e.g. J. W. Evans 1962) have claimed that, in the groups they are familiar with, too many closely-related species coexist geographically (i.e. are sympatric at the present time) for the allopatric hypothesis to be at all credible. This argument has also been used in relation to the species-flocks of certain freshwater animals in ancient lakes such as Baikal and Nyasa (although most modern authors such as Brooks 1950 consider that speciation in ancient lakes has been allopatric). Mayr's interpretation of all species-flocks and similar situations is that where related or sibling species are now sympatric they were formerly allopatric and that the descendant populations have only come to overlap geographically at a later date, after the acquisition of genetic isolating mechanisms. An entirely different type of argument in favour of sympatric speciation has been put forward by Thoday (1958), Thoday and Boam (1959) and Thoday and Gibson (1962) who consider that the results of their experiments on disruptive selection in *Drosophila* make the origin of genetic isolating mechanisms in a continuous population (i.e. without 'interruption of gene flow by extrinsic barriers') plausible.*

The distinction between the sympatric and allopatric models of the speciation process was a vitally important *Fragestellung* in the development

---

* The best discussion of the possibility of sympatric speciation from a theoretical and mathematical standpoint is probably that of Maynard Smith (1966) who concludes: 'The crucial step in sympatric speciation is the establishment of a stable polymorphism in a heterogeneous environment. Whether this paper is regarded as an argument for or against sympatric speciation will depend on how likely such a polymorphism is thought to be, and this in turn depends on whether a single gene difference can produce selective coefficients large enough to satisfy the necessary conditions . . .'.

of evolutionary ideas. But both models seem to be based on implicit acceptance of the idea that gene mutations, and especially ones with rather slight
quantitative effects, are the only real basis of speciation, and that chromosomal rearrangements such as inversions and translocations are either unimportant in the speciation process or play a part which is similar to that of
the point mutations. The alternative, of course, is that certain types of
chromosomal rearrangements may, in some groups of animals at any rate,
have played a particular role in speciation, as primary causes of new genetic
isolating mechanisms. Both for this and for other reasons, it seems important
to examine the fundamental question: are the sympatric and allopatric
models of speciation the only conceivable ones? Even if we admit that Mayr
(1947, 1963) has effectively demolished the earlier arguments in favour of
sympatric speciation, it does not seem to follow that the allopatric model
(in the sense of the dumbbell diagram) is necessarily the only alternative,
equally valid for all groups of animals. At any rate, the time seems ripe for a
detailed examination of the actual genetic processes involved in the origin
of isolating mechanisms, which is the really essential core of the speciation
process. The whole subject is almost certainly much more complex than has
been appreciated in the past and its investigation is far from easy. In such a
situation dogmatism should be avoided at all costs (White 1970*d*).

A great variety of types of population structure exist in different groups
of animals. On the one hand, we have wide-ranging species of unrestricted
vagility, able to occupy a wide variety of ecological habitats. On the other
hand, there are the ecologically specialized species with interrupted distributions and restricted vagility. *A priori*, it does not seem very likely that the
same model of the speciation process should apply equally to oceanic birds,
pelagic fishes, small rodents, land molluscs, *Drosophila*, monophagous gall-
forming insects, hermaphroditic and parasitic flatworms, etc. Of course, the
great variety of population structure is not denied by Mayr and those who
adopt his standpoint. But it is implied that these differences are only of
minor significance in determining the mode of speciation.

Fig. 11.1. Chromosome phylogeny of 82 species of Hawaiian *Drosophila*. Under
the name of each species is an indication of the island, or islands, on which it occurs.
5*R* 1*D*, a karyotype composed of five acrocentric rods and one dot chromosome;
4*R* 1*V* 1*D*, one composed of 4 acrocentric rods, 1 metacentric V and one dot. The
inversion formulae for the individual species should be read additively by following
the line from the *Standard* chromosome sequences for the five large chromosomes
(box, upper centre). For example, *D. liophallus* has the formula *Xh* 2 3*i* 4*b* 5*d*. Single
letters represent fixed inversions, diagonal strokes in the inversion formula indicate
chromosomal polymorphism. From Carson (1971).

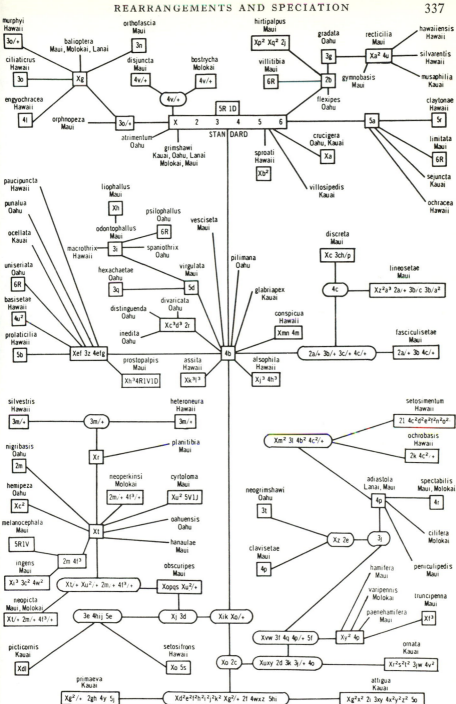

We shall discuss this general question at the end of the present chapter, when we have considered the kinds of cytogenetic differences between closely related races and species in a number of representative groups; and we shall return to the same topic in Chapter 21. It is one of the major problems of biology, and no final answers are likely to emerge from the discussion for several decades yet.

In most groups of animals that have been studied by cytogeneticists in detail, it has been found that even the most closely related species differ cytologically, i.e. their karyotypes can be distinguished by a difference in chromosome number, shape, size or other features. In such groups as *Drosophila*, *Chironomus*, grasshoppers, beetles, mammals and many others, cytotaxonomic differences seem to be almost invariably present. However, in the genus *Drosophila*, there are some instances of species whose karyotypes are indistinguishable, even after detailed studies of banding pattern in the polytene chromosomes. This is so for the three *repleta*-group species, *mulleri*, *aldrichi* and *wheeleri*, from Texas and northern Mexico (Wasserman 1960). It is possible that wherever there has been very rapid differentiation of species in the course of an adaptive radiation into empty ecological niches, 'genic' rather than 'chromosomal' evolution may have been predominant. The Hawaiian Drosophilidae seem pertinent here. An estimated 700 species of *Drosophilidae* exist in the Hawaiian islands, morphologically quite diverse, but possibly descended from a single or at most two or three, introduced species (Hardy 1965; Carson, Hardy, Spieth and Stone 1970). Karyotype differences seem to be minimal in this assemblage, and in many instances quite different species appear to have identical banding patterns in their polytene chromosomes, with no fixed inversion differences such as commonly exist between continental species of the genus. Carson, Clayton and Stalker (1967) and Carson and Stalker(1968) have described a number of Hawaiian species complexes which they designate as *homosequential* (having the same gene sequence).

In groups which lack polytene chromosomes, it is at present impossible to prove the existence of homosequential species pairs or species complexes. The most that we can say is that two or more species appear to have identical karyotypes. But such apparent similarity may conceal inversions or other differences which do not affect the gross morphology of the karyotype. Although lampbrush chromosomes are more difficult to analyse in detail than polytene elements, it should be possible to use them in the same kind of way for the analysis of cytotaxonomic differences, even though such investigations are very laborious. In one instance in which this has been done, namely the 'geographic races' of *Triturus cristatus* (possibly better regarded as distinct species) many minute cytological differences **were** detected in the appearance of the loops (Callan and Lloyd 1960b), but the exact nature of these differences is somewhat uncertain, and it is not clear whether some or most of them are due to the 'Keyl phenomenon' (see p. 364) or to minute structural rearrangements of other kinds. Similar analyses have been carried out by Mancino (1965, 1966) and Mancino and Barsacchi (1965, 1966) on several other European species of the genus *Triturus*.

It is fairly clear that many of the cytotaxonomic differences between closely related or 'sibling' species must have arisen in chromosomal differences between racial populations which had not reached the level of full species. However, it is quite possible that in some instances cytotaxonomic differences may have arisen after speciation has been completed, through chromosomal rearrangements arising in a 'young' species, or in the ancestral one, spreading through its entire population and eventually reaching fixation. However, in many well-investigated groups of insects and mammals the evidence points rather clearly to cytological races as the main source of incipient species; or, to put it somewhat differently, the facts suggest that most of the geographic races capable of evolving into distinct species are at the same time chromosomal races. This does not mean, of course, that all chromosomal races will eventually succeed in becoming full species; some will simply become extinct after a relatively brief career. And, certainly, the existence of some homosequential species groups in the genus *Drosophila* warns us against the facile assumption that *all* full species arise from chromosomally distinguishable races.

The cytological races which are especially likely to become incipient species are surely the 'macrocytoraces' of our classification (Table 8.1), although in some instances 'microcytoraces' or 'mesocytoraces' may be on the way to becoming incipient species. But these kinds of chromosomal races will only exist in species whose populations are cytologically polymorphic, at least in part of their geographic range; and it is still uncertain to what extent chromosomally polymorphic species occur in most groups of animals.

The extent to which chromosomal races of all kinds occur may be expected to vary with the vagility of the organisms we are considering. Very wide-ranging species, individuals of which may traverse in a single life-cycle the whole territory occupied by the species, will almost always be incapable of giving rise to distinct races of any kind. Less vagile organisms such as small mammals may frequently do so. It is, however, among very sedentary forms such as wingless terrestrial arthropods, that 'raciation' is most highly developed, and it is probably among these that the most instructive examples of cytological races can be found.

We shall select for special consideration in this chapter certain groups which have been subjected to particularly detailed cytogenetic analysis – the dipterous families, Drosophilidae, Simuliidae (Black flies), Chironomidae (midges) and Culicidae (mosquitoes); the grasshoppers, mole crickets and certain families of beetles; and, among the vertebrates, the rodents of the family Muridae. In the dipteran families the evidence is based largely on the giant polytene chromosomes which do not exist in the other groups.

CHROMOSOMAL REARRANGEMENTS AND SPECIATION IN DROSOPHILA

The evolutionary literature on *Drosophila* is now so large that only a few aspects can be considered here. This literature was last reviewed in a comprehensive manner by Patterson and Stone (1952). Since that time, important contributions have been made by Wasserman (1954, 1960, 1962*a*, *b* *c*, *d*, 1963, 1968), Stone (1955, 1962), Carson (1959), Stalker (1960, 1966), Stone, Guest and Wilson (1960), Dobzhansky and Pavlovsky (1962), Throckmorton (1962, 1965), Kastritsis (1966*a*, *b*, 1967), Carson, Clayton and Stalker (1967), Carson, Hardy, Spieth and Stone (1970), Brncic (1970) and many others.

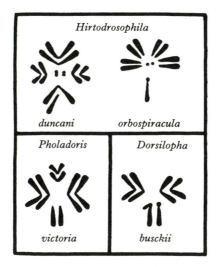

Fig. 11.2. Male karyotypes of some members of the subgenera *Hirtodrosophila*, *Pholadoris* and *Dorsilopha* of *Drosophila*. Sex chromosomes at the bottom of each figure (*orbospiracula* is an *XO* species). After Wharton (1943) and Patterson and Stone (1952), but redrawn.

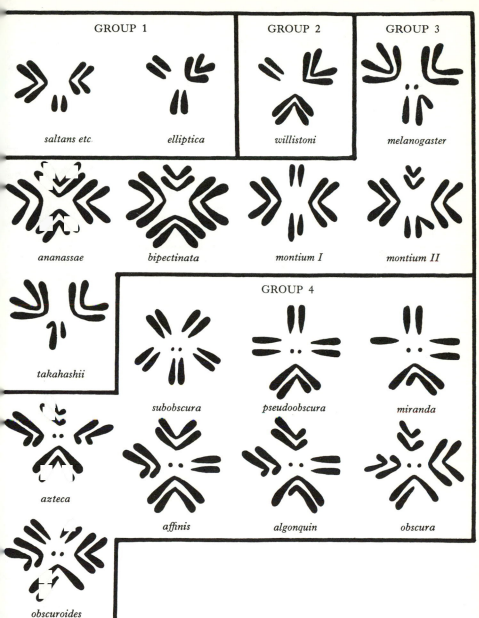

Fig. 11.3. Male karyotypes of some members of the subgenus *Sophophora* of *Drosophila*. Several different karyotypes have been reported for *montium* and *takahashii*. Based on various authorities.

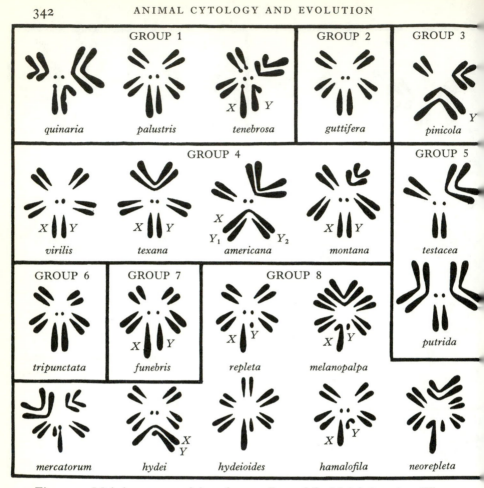

Fig. 11.4. Male karyotypes of the subgenus *Drosophila*. Mainly based on Wharton (1943) and Patterson and Stone (1952) but redrawn.

We shall discuss in Chapter 12 the total number of chromosomal rearrangements that have occurred in the genus *Drosophila*; concentrating our attention for the present on a detailed account of the cytotaxonomy of some representative species groups. It is, in general, characteristic of *Drosophila* that a great many of the cytotaxonomic differences between closely related species have arisen through fixation of paracentric inversions in phylogeny. Since paracentric inversions are also frequently found in an 'unfixed' (i.e. floating and polymorphic) condition in natural populations of *Drosophila* species, there is an evident parallelism between interspecific and intraspecific chromosomal variation. However, the fusions and whole-arm transpositions which distinguish some *Drosophila* species are not present as floating intraspecific polymorphisms.

(1) The *virilis* group

This is essentially a Holarctic group of species, although *D. virilis* itself has become largely cosmopolitan, probably through human agency. Hsu (1952), Patterson and Stone (1952) and Stone (1962) have worked out the phylogeny of the group, based on the fact that some members have inversion sequences in common which are lacking in other forms.

D. virilis has a karyotype of five pairs of acrocentric chromosomes and a pair of dot chromosomes. This is regarded by Patterson and Stone as the primitive karyotype for the genus *Drosophila* as a whole, and *virilis* seems to be close to the common ancestor of the species group which bears its name. It is supposed that *virilis* gave rise to two hypothetical forms 'Primitive II' and 'Primitive III'. These may have been distinct species or geographic races or strains of some of the known species.

Primitive II differed from *virilis* by two inversions in the X and one in chromosome 2. It gave rise to the three North American forms – *novamexicana*, *americana* and *texana*. The first of these retains the basic karyotype of five acrocentrics and a dot chromosome; it differs from Primitive II by six inversions and is a cytologically monomorphic species like *virilis*. *Americana* and *texana* are probably best regarded as geographic races or subspecies; they both have a fusion of chromosomes 2 and 3 and are polymorphic for several inversions. *Americana* has a fusion between the X and chromosome 4 so that the male is, in effect, XY_1Y_2, the unfused fourth chromosome being confined to the male line, so that it may be designated Y. *Americana* and *texana* hybridize in a broad zone where their distributions overlap (Carson and Blight 1952). *Americana* shares some inversions with *texana* and others with *novamexicana*; Patterson and Stone (1952) originally interpreted this situation as indicating that it arose by hybridization between these forms. It seems more probable, however, that the ancestral population was polymorphic for several inversions which are still 'floating' in *americana*, but absent in *texana* and fixed in *novamexicana*.

All descendants of 'Primitive III' are homozygous for a pericentric inversion in chromosome 2, which has converted this element into a metacentric. This lineage includes both Palaearctic and Nearctic forms. The Old World species *ezoana* and *littoralis* both carry a considerable number of additional fixed inversions, especially in the X-chromosome. The Y in these species has been converted into a metacentric. *Littoralis* (but not *ezoana*) has a fusion between chromosomes 3 and 4.

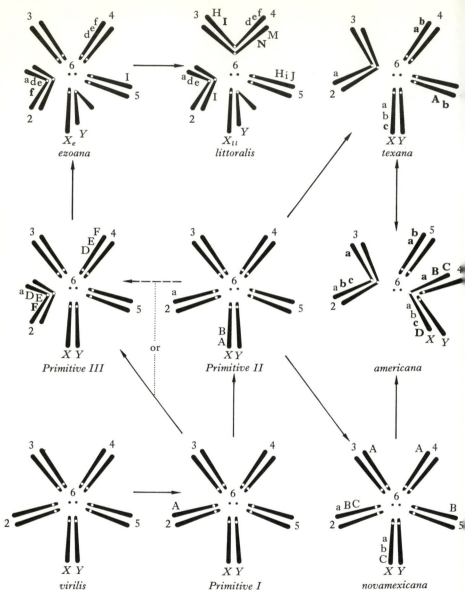

Fig. 11.5. Chromosome phylogeny of part of the *virilis* group of *Drosophila*. Inversions are designated by capital letters on the first occasion when they appear in the phylogeny, by lower case letters thereafter. Inversions designated by bold letters are ones for which the species is polymorphic (i.e. ones which have not yet reached fixation). X_{li} and X_e, the *littoralis* and *ezoana* X's, which differ from those of the other species by complex structural changes, not fully analysed. From Stone (1962).

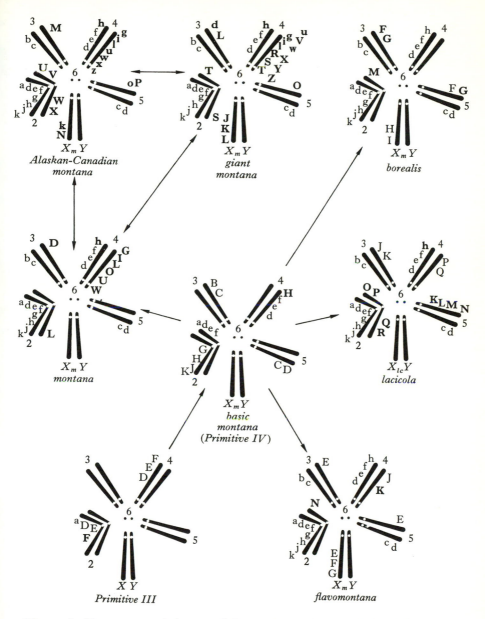

Fig. 11.6. Chromosome phylogeny of the *montana* subgroup of the *virilis* group. Lettering as in Fig. 11.2. Double-headed arrows designate uncertainty as to the direction of the phylogenetic relationship. X_m, X_{lc}, the characteristic X-chromosomes of *montana* and *lacicola*, which differ by multiple structural changes. From Stone (1962).

'Primitive III' also gave rise to the North American *montana* complex, including *montana*, *borealis*, *lacicola* and *flavomontana*. These are all species carrying many additional inversions, both 'fixed' and 'floating'. Inversion polymorphism reaches a maximum in populations of *montana* from the Pacific northwest which have at least twenty-one floating inversions. There has been a strong tendency for inversions to undergo fixation in the X-chromosomes of the descendants of Primitive III; *ezoana*, *littoralis*, *montana* and *lacicola* all possess distinctive X-chromosomes which have undergone rather radical repatterning (*borealis* and *flavomontana* possess X's of the *montana* type).

Montana consists of three geographic races: one from the Rocky Mountain area, one from the Pacific northwest ('Giant *montana*' of Moorhead 1954) and one from Canada and Alaska; these share some inversion polymorphisms, but each is also polymorphic for some inversions which are lacking in the other two.

The cytological evolution of the nine species of the *virilis* group has hence involved well over a hundred paracentric inversions (the number in the X-chromosome is not known, because the X's of *ezoana*, *littoralis*, *montana* and *lacicola* have not been fully analysed). In addition, there have been three centric fusions, a pericentric inversion and some undefined structural changes in the Y. Some of the inversions are of particular phylogenetic interest. Thus in chromosome 2, inversion F must have been floating in Primitive III (i.e. both F and F^+ were present). *Ezoana* retains this polymorphism but *littoralis* is homozygous for F^+, while *montana*, *borealis*, *lacicola* and *flavomontana* are all monomorphic for F. Thus the two alternative sequences present in Primitive III and *ezoana* have each undergone fixation in different lines of descent. In chromosome 4, *lacicola* and all geographic races of *montana* are polymorphic for inversion H, but this sequence has undergone fixation in *borealis* and *flavomontana*, the alternative sequence (H^+) being lost.

(2) The *repleta* group

The *repleta* group is much larger than the *virilis* group, since it includes over sixty species. However, the number of inversional changes that have occurred in the phylogeny of the *repleta* group is relatively smaller (Wasserman 1954–63). These two species groups consequently represent extremely different patterns of cytogenetic evolution within the genus *Drosophila*. Altogether, forty-six species of the *repleta* group have been studied cytologically (out of a total of sixty-eight known). Three of the forty-six (namely *mulleri*, *aldrichi* and *wheeleri*) are homosequential, but all the others are chromosomally unique. Wasserman (1960, 1963) states that a total of 144 inversions (92 fixed in phylogeny and 52 floating in a polymorphic condition in present-day populations) occur in the forty-six investigated species; but these figures really represent minimum estimates, because there are about six more fixed inversions in the *fasciola* subgroup, and some additional floating inversions would undoubtedly be found if more extensive material of some of the species, covering their entire geographic range, were to be studied. These rearrangements have not been randomly distributed over the several chromosome elements. The fact that only four inversions (3% of the total) have been incorporated in the fourth chromosome and only

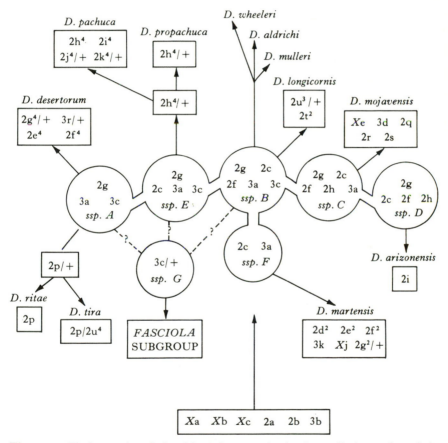

Fig. 11.7. Phylogenetic relationships of 12 species in the *mulleri* complex of the *repleta* group of *Drosophila* and the ancestor of the *fasciola* subgroup. All these are interpreted as having arisen from a large cytologically polytypic species with seven chromosomal races or subspecies. *D. wheeleri*, *D. aldrichi* and *D. mulleri* are homosequential species. From Wasserman (1963).

ten (7%) in the fifth chromosome shows that these elements have been extremely stable throughout the phylogeny of forty-six species. On the other hand, no less than 103 inversions (70% of the total number) have established themselves in the second chromosome (estimated to be 23% of the total length of euchromatin in the karyotype). It is extremely unlikely that the second chromosome has a higher rate of breakage, and much more probable that, for some unknown reason, second chromosome inversions have a much higher chance of survival in evolution.

The primitive karyotype of the *repleta* group undoubtedly consisted of an acrocentrix X and Y, four pairs of acrocentric autosomes and a pair of dot-autosomes. In *D. nigrohydei*, *D. hydeoides* and *D. anceps* the dot chromosomes have been replaced by rod-acrocentrics, as a result of additions of hetero-chromatin, and in *D. hydei* the X-chromosome has become metacentric as a result of the addition of a heterochromatic limb equal in length to the euchromatic one. In *D. meridiana rioensis*, *D. promeridiana* and *D. meridion-alis*, chromosomes 2 and 3 have fused to give a metacentric. Apart from these rearrangements, the metaphase karyotypes are relatively uniform in the *repleta* group.

Wasserman has constructed detailed phylogenies of the various sub-groups of the *repleta* group, based on the inversions which have undergone fixation. There are three instances where pairs of species share the same inversion polymorphism. Thus *mercatorum* and *paranaensis* are both poly-morphic for the $3g$ inversion and the corresponding standard sequence; *melanopalpa* and *canapalpa* are both polymorphic for the $5p$ inversion; and *pictilis* and *pictura* are both polymorphic for the $2n^3$ inversion.

In the phylogeny postulated for the *repleta* group, there have been several occasions when a highly polytypic species, assumed to be carrying different inversions in different parts of its range, gave rise to a number of descendant species. Wasserman (1963) states:

It is believed that the ancestral species consisted of a number of small, semi-isolated populations among which some gene exchange was possible. The species was poly-typic for combinations of inversions: each inversion had its own distribution, over-lapping the ranges of one or more of the others . . . Fragmentation of this ancestor into seven distinct forms, several of which later split into two or more species, resulted in the twelve species (of the *mulleri* complex) which share the six inversions.

(3) The *obscura* group

This group includes a number of Holarctic and North American species of *Drosophila*. *D. pseudoobscura* and *D. persimilis* are a pair of sibling species from western North America which have been studied from almost every conceivable aspect by Dobzhansky and numerous collaborators. *D. subobscura* is a species whose range extends from Scotland to North Africa, Israel and Persia; its very extensive chromosomal polymorphism has been studied by numerous workers (see Ch. 8). Other members of the *obscura* group have been less thoroughly investigated. *D. subobscura* has the most primitive karyotype, with an acrocentric *X*, four pairs of acrocentric autosomes and a dot chromosome. In all other members of the group, the original *X* has become fused to chromosome arm *D*, in the terminology of Muller (1940a) – i.e. the element corresponding to arm 3L of *Drosophila melanogaster* – to form a metacentric. In consequence, this limb has become

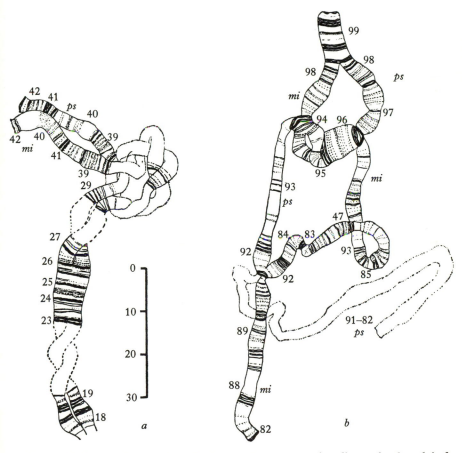

Fig. 11.8. Incomplete pairing of polytene chromosomes in salivary gland nuclei of hybrids between *Drosophila miranda*, *D. pseudoobscura* and *D. persimilis*. *a*, *X*R in a *miranda* × *persimilis* hybrid; *b*, chromosome 4 in a *miranda* × *pseudoobscura* hybrid. Scale of μm alongside. From Dobzhansky and Tan (1936b).

haploid in the male karyotype, the previously autosomal genetic loci in it being now inherited in a sex-linked manner. It is probable that in the evolution of the *obscura* group karyotype from the original *subobscura* condition there was an intermediate stage in which a population or species was XY_1Y_2 in the male (like *D. americana*), the 'Y_2' being simply the unfused *D* element. It is fairly certain, however, that this population or species is now extinct and that the Y_2 chromosome has been lost from the karyotype in all members of the group which possess the *A–D* fusion (i.e. all except *subobscura*). Most probably, this loss of the Y_2 took place through its fusion with the original $Y (= Y_1)$ and subsequent heterochromatinization.

Some aspects of the extensive inversion polymorphism of *D. pseudo-obscura* and *D. persimilis* have already been discussed (p. 269). These behave in nature as entirely distinct species and hybrids do not seem to occur in the wild, although they can easily be obtained in the laboratory. Apart from the numerous inversions in the third chromosome, for which both species are polymorphic (*Standard* being the only sequence found in both of them), *pseudoobscura* and *persimilis* differ in respect of an inversion in the second chromosome and one in the 'left' limb of the X (Dobzhansky and Epling 1944). No rearrangements other than paracentric inversions seem to have occurred since the evolutionary bifurcations which gave rise to these species, except perhaps in the Y-chromosome, where some duplications or deficiencies seem to have established themselves, leading to a situation in which there are several types of Y (differing in size and in centromere position) in both species.

Drosophila miranda is obviously closely related to the *pseudoobscura – persimilis* stock, (but there has been a further fusion involving the Y-chromosome, so that it possesses a multiple sex-chromosome mechanism of the X_1X_2Y type (see p. 657)). Hybrids between *miranda* and either *pseudoobscura* or *persimilis* show rather incomplete pairing of their salivary gland chromosomes. This situation was interpreted by Dobzhansky and Tan (1936*a*, *b*) as indicating that numerous shifts and translocations, involving some 50 to 100 chromosome breaks, had taken place since the separation of *miranda* from the *pseudoobscura – persimilis* stock. In view of later work on the salivary chromosomes of many other hybrid combinations in *Drosophila* and other genera of Diptera, it is now clear that lack of synapsis does not always indicate structural non-homology or translocational changes. By analogy with the $5B$ factor in wheat, which inhibits meiotic pairing between 'homoeologous' chromosomes (Riley 1966), we might conclude that there are specific genes in many *Drosophila* species which cause localized asynapsis in hybrid polytene nuclei. Thus, until the situation in the *miranda* hybrids has been reinvestigated, the conclusions of Dobzhansky and Tan must remain in doubt; it seems unlikely that the amount of chromosomal repatterning has been anything like as extensive as they suggested.

Members of the North American *affinis* subgroup of the *obscura* group have the same A–D fusion, but differ from *pseudoobscura* and *persimilis* in that two pairs of autosomes have become J-shaped (submetacentric), presumably as a result of the establishment of pericentric inversions in phylogeny. *D. azteca* and *D. athabasca* have gone one stage further, in that three pairs of autosomes are submetacentric; but since the number of chromosome arms in the polytene nuclei is the same as in the *affinis* subgroup, it is probable that a substantial block of heterochromatin has been added to the karyotype, and that there has been no additional pericentric inversion. All these species have a pair of small dot chromosomes.

In contrast to the cytologically primitive *D. subobscura*, the European representatives of the *obscura* group (*D. ambigua*, *D. bifasciata* and *D. tristis*) have three pairs of metacentric autosomes, but show eight long arms and a short one (corresponding to the dot chromosome pair) in their polytene nuclei. They have presumably acquired three pericentric inversions in their autosomes. Buzzati-Traverso (1950) was able to obtain a few hybrid larvae from matings between females of *ambigua* and males of *pseudoobscura* and *persimilis*. Their polytene nuclei showed thirteen long chromosome arms (eight from one parent species, five from the other), there being no synapsis at all.

(4) The *paulistorum* complex

Drosophila paulistorum is a 'superspecies' consisting of seven forms, some of which have reached a degree of reproductive isolation which enables them to coexist sympatrically in certain areas, without merging (Dobzhansky and Spassky 1959; Ehrman 1960, 1962, 1965; Dobzhansky and Pavlovsky 1962; Carmody *et al.* 1962; Dobzhansky 1963*a*; Dobzhansky, Ehrman, Pavlovsky and Spassky 1964). The range of the superspecies as a whole is from Guatemala southward through Central America to Bolivia and southern Brazil. The seven races or incipient species are the Centro-American, Amazonian, Andean–South Brazilian, Orinocan, Calypso, Transitional and Guianan. The latter has later been regarded as a full species and named *D. pavlovskiana* (Kastritsis and Dobzhansky 1967). Laboratory crosses between these taxa frequently give no progeny; sometimes they give fertile daughters and sterile sons; and occasionally fertile hybrids of both sexes were obtained. The latter outcome was obtained especially in crosses involving the Transitional race (Dobzhansky *et al.* 1964).

In general, each form has its own chromosomal inversions, for which it is polymorphic, and which are not found in other races. This does not apply to the Transitional race, which has approximately the same sequences as the widespread Andean–South Brazilian race. The complex as a whole shows more than sixty-three different inversions of which no less than thirty-two are in the third chromosome (Kastritsis 1967). Some of these are confined to a single race and some have only been found in one strain of a particular race. But it is characteristic of the *paulistorum* complex that many inversion polymorphisms are present in two, three or four races. Thus the Transitional, Andean, Centro-American and Orinoco races are all polymorphic for inversion XL-II, while the Amazonian race and *D. pavlovskiana* are monomorphic for the corresponding Standard sequence. In fact, *D. pavlovskiana* is not known to share any polymorphisms with the other forms, but the analysis is difficult because of poor pairing of its polytene chromosomes with those of the other forms in laboratory hybrids (Kastritsis 1966a, 1967).

Dobzhansky and Pavlovsky (1962), having found that laboratory-produced interracial hybrids were heterozygous for more inversions than the parental strains, concluded that most of the races had arisen from cytologically monomorphic (or relatively monomorphic) peripherally-isolated populations of pre-existing races. They accordingly regarded the situation in the *paulistorum* superspecies as resulting from classical allopatric speciation with 'peripheral homoselection' (Carson 1955a, b, 1959), as in *D. robusta* and *D. willistoni*. The situation has been changed, however, by Kastritsis' (1966a, 1967) discovery that the incipient species share rather a large number of inversion polymorphisms. The most probable interpretation is that populations of the ancestral *paulistorum* occupying peripheral areas became more or less completely isolated genetically and later became sympatric as a result of range-extension. The Guianan race (= *D. pavlovskiana*) may be a result of peripheral homoselection; but there is no evidence as to whether it acquired its genetic isolating mechanism before or after it was cytologically monomorphic.

It is in any case unlikely that the mechanisms of genetic isolation in the *paulistorum* complex are directly related in any way to the chromosomal inversions. Where male hybrids between the various taxa are sterile, this seems to be an inheritance from the mother via the egg cytoplasm. Ehrman and Williamson (1965) and Williamson and Ehrman (1967) have demonstrated the existence of a factor causing male sterility which can be transmitted by injection as well as through the egg cytoplasm. This factor is now known to be a symbiotic microorganism (Kernaghan and Ehrman 1970). Dobzhansky and Pavlovsky (1967) have made the interesting suggestion that speciation in the *paulistorum* group may have resulted from the establishment of new symbiotic relationships between the fly and a microorganism which may not be compatible with the genes of other populations of the species, so that the hybrids are sterile.

(5) The *melanica* group

There are seven species of this North American group. The cytogenetics of one of these has not been studied, but the other six have been investigated by Ward (1949, 1952) and Stalker (1960*a*, *b*, 1961, 1964*a*, *b*, 1965, 1966).

D. micromelanica has an acrocentric X, but in five other species, the X is metacentric as a result of a fusion with an autosome. The relation of *D. micromelanica* to the rest of the *melanica* group is hence analogous to that of *subobscura* to the remaining members of the *obscura* group, i.e. it is a cytologically primitive species which still retains the originally acrocentric X. It is remarkable for possessing four visibly different karyotypes. These differ in presence or absence of a fusion between two acrocentric autosomes, in the shape of chromosome 4 (which may be either an acrocentric or a metacentric) and in the size of the small sixth chromosome, which may appear either as a dot or a short rod, presumably indicating loss or gain of a heterochromatic segment. The geographic distribution of these karyotypes is not well known and it is not clear whether two or more of them may occur at a single locality.

The six species of the *melanica* group that have been investigated show about sixty-one fixed inversions; they have at least eighty floating inversions. The mean number of inversions of both kinds per species is 23.5, compared with 3.1 for the *repleta* group, 9.6 for the *cardini* group and 13.0 for the *virilis* group. There is a great concentration of both fixed and floating inversions in chromosome 2, a total of fifty-six out of the eighty floating inversions and about twenty-five of the sixty-one fixed inversions being in this chromosome. There is no case known of two species of the *melanica* group being polymorphic for the same inversion, which is remarkable in view of the overall high level of inversion polymorphism (the *repleta* group, with a much lower level of inversion polymorphism is known to show three such cases). Altogether, the speciation mechanisms of the *melanica* and *repleta* groups seem to be significantly different. Whether the species of the *melanica* group have actually gone through an initially cytologically monomorphic stage cannot be regarded as certain – incipient species of this group may have lost their 'old' inversion polymorphisms *after* having acquired new polymorphisms of their own. Stalker (1966) has discussed the various possible explanations for the concentration of inversions in the second chromosome of the *melanica* group, but no definite conclusion can be reached at present.

(6) The *tripunctata* group

This is one of the largest species-groups in the genus *Drosophila*, with at least fifty-four species, mainly in the neotropical region. The chromosomes of ten species have been studied by Kastritsis (1966*b*). The group is probably fairly closely related to the *melanica* group, and is characterized by an acrocentric *X*, which corresponds to *X*R of *melanica*, the third chromosome of the *tripunctata* group (an acrocentric autosome) being homologous to *X*L of *melanica*.

Only three of the ten species examined had the *X* polymorphic for inversions. The second chromosome is highly polymorphic for inversions in most of the species. There is a suggestion that in the phylogeny of the group, inversions have tended to undergo fixation in certain chromosomes (e.g. the fifth) which show very few floating inversions – while in certain other chromosomes which show much more inversion polymorphism there has been less fixation of rearrangements. If this tendency is a real one, it indicates a difference from the *melanica* group, in which both the fixed and the floating inversions are heavily concentrated in the second chromosome.

(7) The Hawaiian species of *Drosophila*

Space will not permit a full consideration here of the extremely significant studies on the karyotypes and evolutionary patterns of the Drosophilidae of the Hawaiian archipelago carried out by Carson and Stalker (1968, 1969), Carson, Hardy, Spieth and Stone (1970), Carson (1970) and several other collaborators on the taxonomy, behaviour and biochemistry of these flies. The Hawaiian islands have a very rich fauna of Drosophilidae consisting of 650–700 species, all but 17 of which are endemic. There are two main cosmopolitan genera, *Drosophila* (324 species) and *Scaptomyza* (121 species), together with a number of smaller endemic genera which seem to have evolved from these ancestral genera. Ten 'homosequential' species groups of *Drosophila* have been recognized in the Hawaiian fauna (i.e. groups of species having an identical banding pattern in the polytene chromosomes. One of these include five species, two include four species each, four include three species each and three consist of pairs of species. Five pairs of species are known to share a common inversion polymorphism. The majority of the Hawaiian species of *Drosophila* retain the primitive karyotype of five acrocentric pairs and a pair of dot chromosomes. But four

Hawaiian species seem to possess chromosome fusions and many of them have undergone changes in the extent and distribution of blocks of heterochromatin.

From the general standpoint the most interesting conclusion that can be drawn from these investigations is the large number of species that have arisen since the origin of this group of islands. The oldest island (Kauai) is about five million years old, while the youngest (Hawaii) is only about one million years old. It is uncertain whether the whole endemic drosophilid fauna of the archipelago can be traced back to a single ancestral immigrant species (in which case the genus *Scaptomyza* would have diverged from *Drosophila* in the Hawaiian chain of islands) or to two immigrant forms, one ancestral to the native species of *Drosophila* and related genera, the other to the *Scaptomyza* complex. In any case, the evolution of the Hawaiian drosophilid fauna in a mere five million years or so must be characterized as 'explosive'.

CHROMOSOMAL REARRANGEMENTS IN BLACK FLIES (SIMULIIDAE)

Detailed studies of the polytene chromosomes of numerous species of Simuliidae (mainly Canadian and European) have been carried out by Rothfels and Dunbar (1953), Rothfels (1956), Dunbar (1958, 1959, 1965, 1966) Basrur (1959, 1962), Rothfels and Basrur (1960), Landau (1962), Pasternak (1964), Carlsson (1966) and Ottonen (1966). Unlike *Drosophila*, this is a family in which there are chiasmata in the male meiosis.

Simulium vittatum was studied by Rothfels and Dunbar (1953) and, in great detail, by Pasternak (1964). It is one of the two members of the subgenus *Neosimulium*, the other being *S. argus*. Both have three pairs of metacentric chromosomes (the usual karyotype in Simuliidae). The long arm of chromosome 1 and the short arms of chromosomes 2 and 3 appear to show exactly the same banding pattern in the two species. The short arm of chromosome 1 also has the same banding pattern, but there are structural differences in the centromeric region. The long arm of chromosome 3 has at least two fixed inversion differences and there are more profound differences in the long arm of chromosome 2 (presumably the sex-determining limb). *S. vittatum* always shows very loose pairing of the chromosomes in the salivary gland chromosomes, regardless of the degree of polyteny (Rothfels and Dunbar 1953). The species shows an extraordinarily complex system of paracentric inversion polymorphism, a total of 127 simple inversions and 7 complex overlapping inversions (two-step rearrangements) being known (Pasternak 1964). As in a long series of *Simulium* and *Eusimulium* species, chromosome 3 bears the nucleolus and Balbiani ring.

'*Simulium tuberosum*' seems to consist of four or five sympatric forms in southern Ontario, whose taxonomic rank is somewhat doubtful (Landau 1962). They differ in respect of floating inversions and in respect of the sex-determining chromosome arm (the short arm of chromosome 2). The species or superspecies is widely distributed in North America and northern Europe.

Seven types of 2S arms were identified by Landau. These are referred to as Standard, A, AB, CDE, CKL, FG, FGH; each letter representing a simple inversion difference from Standard. Standard and CDE are almost or essentially confined to males, and may be interpreted as *Y*-linked sequences. A, AB and FG are presumably *X*-linked These 2S arms were found to occur as follows:

Sibling (1) Males A/CDE, females A/A
   (2) Males mainly AB/St, females AB/AB
   (3) Males mainly FG/St, females FG/FG
   (4) Both sexes FGH/FGH

A rare and doubtful fifth sibling may have CKL/A males and CKL/CKL females.

There were no fixed inversion differences between the sibling forms of *S. tuberosum*, other than those in the 2S chromosome limb. However, a total of eighty-three floating inversions were found by Landau. Seven inversion polymorphisms are shared by two or three of the sibling forms.

The evolutionary implications of this most interesting case are somewhat unclear. Probably a much more extensive investigation of this super-species, throughout a much larger part of its geographic range, would be necessary for a full understanding of the method of evolutionary differentiation of the various forms. It does appear, however, as if speciation was proceeding as a result of a number of alternative sex-chromosome mechanisms.

The African *Simulium* '*damnosum*' appears to be a superspecies which includes at least four sibling forms (Dunbar 1966). All these have three pairs of metacentric chromosomes. They have the same basic banding pattern except for a few 'fixed' interspecific inversions. There are not many 'floating' inversions, but those that exist are different in each of the four species. In one species the centromeres of the three polytene chromosomes are tightly fused, but in the other three they are free. All four have a nucleolar organizing region in the shorter arm of chromosome 1; one species has a second nucleolar organizer in the long arm of chromosome 3.

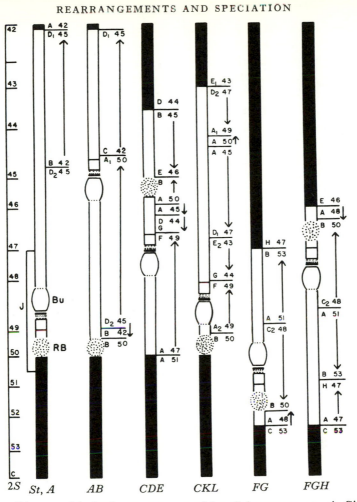

Fig. 11.9. Diagram of the basic arrangements of the 2*S* chromosome arm in *Simulium tuberosum*. Bu, Bulge; RB, Balbiani ring. Black areas indicate extent of terminal homology with St, except in the first column. Arrows denote orientation of inverted segments. J is a floating inversion. From Landau (1962). [Reproduced by permission of the National Research Council of Canada from the *Canadian Journal of Zoology*, **40**, 921–39 (1962).]

No translocations of pericentric inversions have occurred in this super-species – the only chromosomal changes that seem to have established themselves are paracentric inversions and alterations in the minute structure of the centromere and nucleolar-organizing system.

The superspecies *Eusimulium aureum* includes seven presumptive sibling species which have been recognized by Dunbar (1958, 1959). These forms differ from other simuliids in having n = 2 instead of 3. Apparently the two shorter chromosomes (2 and 3) present in other members of the family have been combined to form one extremely long chromosome. It has been suggested (Dunbar 1958) that this occurred by pericentric inversions of chromosomes 2 and 3 (giving acrocentrics), followed by a centric fusion. The species of the *aureum* complex characteristically show very tight pairing of the polytene chromosomes.

Of the seven forms recognized by Dunbar, two (*A* and *B*) occur sympatrically near Toronto and Montreal, two others (*C* and *D*) are sympatric in Manitoba, and two more (*E* and *F*) occur near Leningrad, U.S.S.R.; a seventh (*G*) is found in northern California. Most of these forms differ from the others by one or more fixed inversions for which they are homozygous; in addition, each of them carries a different set of floating inversions, for which it is polymorphic. *D* differs from *C* only by its floating polymorphisms, a situation which also exists between forms *E* and *F*.

The comparative cytology of some species of *Prosimulium* has been studied by Basrur (1959, 1962), Rothfels and Basrur (1960) and Carlsson (1966). The genus consists of two sections which differ in their sex-determining mechanism; in the European species and the North American *P. fontanum* chromosome element 1 (the longest) is the *XY* pair in the male (*XX* in the female), while in the remaining North American species (*mixtum, fuscum*) the role of sex determination has been taken over by chromosome 3. There are over fifty species known in this genus.

Basrur (1959) has studied the cytology of seven forms of *Prosimulium* in which the centromere region of chromosome 1 is considerably expanded, with heavy bands. *P. fuscum* and *P. mixtum* lack this 'transformed' centromere region. Among the forms which possess it, *P. fontanum* and the European *P. hirtipes*, *P. inflatum* and *P. vigintiquaterni* are clearly distinct species which differ in respect of ordinary morphological characters and in many fixed inversions. They also differ in their specific inversion polymorphisms. Three European forms related to *P. inflatum* are of much more uncertain taxonomic status. *P. fontanum*, *P. inflatum* and the forms related to the latter all have a chromocentre in the polytene nuclei, but this is not present in *P. hirtipes* and *P. vigintiquaterni*.

In a later paper Basrur (1962) has dealt with a group of species of *Prosimulium* from Alaska and British Columbia which form a natural unit characterized by the possession of the fixed inversion 2L2. He has published a hypothetical phylogeny of these species according to which they are descended from an ancestral form that was polymorphic for the 'transformed' and 'standard' centromeric regions of chromosome 1 (and in which some heterozygotes for the type of centromere region would have occurred). The

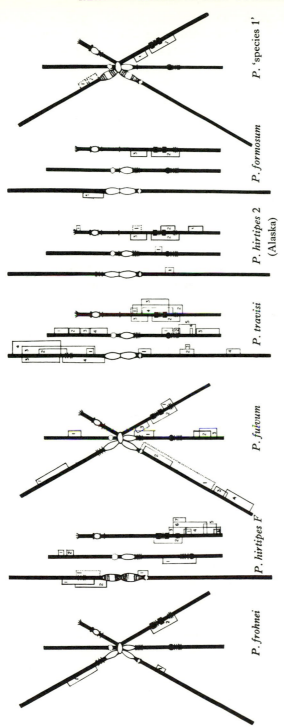

Fig. 11.10. Idiograms of the seven species of *Prosimulium* which carry inversion III L-2. Interspecific inversions shown by brackets to the left of chromosome arms, floating inversions by brackets to the right, X inversions in broken line, Y inversions in dotted lines. *P. frohmei*, *P. fulvum* and *P.* 'species 1' show chromocentral associations. In all species chromosome I has a swollen region on either side of the centromere, with the nucleolar organizer adjacent to it in the long arm. The short arm of chromosome III ends in a 'frazzle'. Various other landmarks (the Balbiani ring, the 'para-Balbiani', the 'Blister', the 'Shield' and the 'Triad') shown conventionally. From Basrur (1962). [Reproduced by permission of the National Research Council of Canada from the *Canadian Journal of Zoology*, **40**, 1019–33 (1962).]

same concept occurs in the work of Ottonen (1966) on another group of six North American species of *Prosimulium*, one of which (*P. dicum*) has the centromere 'transformed'. No individual heterozygous for centromere type has actually been seen, and in both instances it would seem more probable, on general grounds, that the ancestral species were polymorphic for inversions rather than for the type of centromere region in chromosome 1, i.e. that the proposed phylogenies should be amended. It is not clear on how many occasions a 'transformed' centromere region has been acquired in chromosome 1 in the genus *Prosimulium* (Ottonen implies that it arose independently in *dicum*). Two species of *Prosimulium* (*magnum* and *multidentatum*) are homozygous for a reciprocal translocation between chromosomes 1 and 2; the breakage points leading to this translocation were adjacent to the centromere, so that the longest chromosome in these species consists of limbs 1L and 2L, while the second longest is made up of 1S and 2S.

Ottonen makes the extremely important point (already foreshadowed by Rothfels, 1956) that all known instances of incipient speciation in the Simuliidae seem to have been accompanied by divergent changes in the sex-chromosome mechanism. The essential sex-determining mechanism in Simuliidae seems to be a very small chromosome segment close to the centromere. Males are XY, females XX. In many species a differential segment with failure of pairing in the polytene nuclei has been developed and superimposed on the primitive mechanism. In still more highly evolved sex-chromosome mechanisms in the Simuliidae distal inversions have been added to either the X or the Y or both. These presumably contain genetic complexes coadapted with the primary sex locus.

In *P. magnum* chromosome 3 is the sex-determining element. The primitive sex chromosomes are designated X_0 and Y_0. These do not differ visibly. Some populations contain only these sex chromosomes. Others contain a variant type of X designated X_c which differs from X_0 in the appearance of bands 82*e* and 83*e*. These populations contain Y_c, which differs from Y_0 by changes in the 82–83 region. The differences between X_c and Y_c in male larvae are conspicuous. Other types of Y-chromosomes in *P. magnum* are referred to by Ottonen as Y_1, Y_3 and Y_5; these differ from Y_c by various inversions. Another type of X was designated *Epd*; it differs from X_0 by having an extra band between 81*a* and 81*b* and by having band 83*e* weak-staining (these two characteristics can also occur separately in X-chromosomes designated *Ep* and *Ed*). Chromosomes X_0 and Y_0 appear to be coadapted and X_c and Y_c are likewise coadapted. In some areas both X_0Y_0 and X_cY_c populations seem to coexist without hybridization; but in one locality X_0X_c and X_0Y_c heterozygotes were encountered.

In the closely related *P. multidentatum* the function of sex determination has been taken over by chromosome limb 2L (which is part of the longest chromosome). Presumably, *P. multidentatum* has passed through a stage when it possessed sex-determining mechanisms on two different chromosome pairs, as in the case in certain populations of *Chironomus tentans* in northern Europe.

The polytene chromosomes of three species of *Twinnia*, a North American genus closely related to *Prosimulium*, have been studied by Rothfels and Freeman (1966). *T. tibblesi* has the standard association of chromosome arms, but the other two species both have a whole-arm interchange so that they possess the metacentric chromosomes 2L3L and 2S3S instead of the usual 2L2S and 3L3S. The genera *Twinnia* and *Gymnopais* share an inversion '1L–1' and two inversions in chromosome arm 3L. The position of the nucleolus is variable in *Twinnia*, four different locations being known in the three species; in no case was the nucleolar shift accompanied by any visible rearrangement and it is an open question whether there has been a shift of a single locus (the alternative being that one locus has lost its nucleolar-organizing function and that another has gained this property).

CHROMOSOMAL REARRANGEMENTS AND SPECIATION IN CHIRONOMID MIDGES

The mechanisms of chromosomal evolution in the Chironomidae seem to be broadly similar to those of the Simuliidae. Detailed cytogenetic investigations on European chironomids have been carried out by Bauer (1945), Acton (1957*a*, *b*, *c*), Keyl (1957*b*, 1960*a*, *b*, 1961*a*, *b*, 1962, 1963*a*, 1964, 1965*a*, *b*). In North America, Rothfels and Fairlie (1957) studied *Chironomus attenuatus* (= *decorus*), Basrur (1957) compared Canadian and German representatives of *Glyptotendipes barbipes*, Acton (1958, 1959, 1962, 1965) compared Canadian and European material of *C. tentans* and Wülker, Sublette and Martin (1968) have published a general review of the cytology of the neoarctic forms. The work of Blaylock (1963, 1965, 1966) on a species from Tennessee is faulty in that the material he worked with was not *C. (Camptochironomus) tentans*, as stated, but a member of the *thummi* group of the subgenus *Chironomus*. A number of Australian species complexes in the Chironomidae have been investigated by Martin (1962, 1963, 1964, 1965).

Keyl (1962) constructed a phylogenetic tree of twenty European species of the genus *Chironomus* on the basis of the inversions and translocations which they carry: this scheme has since been extended to include a number of species from other continents. It is generally possible to homologize chromosome arms throughout the genus by observing the banding pattern

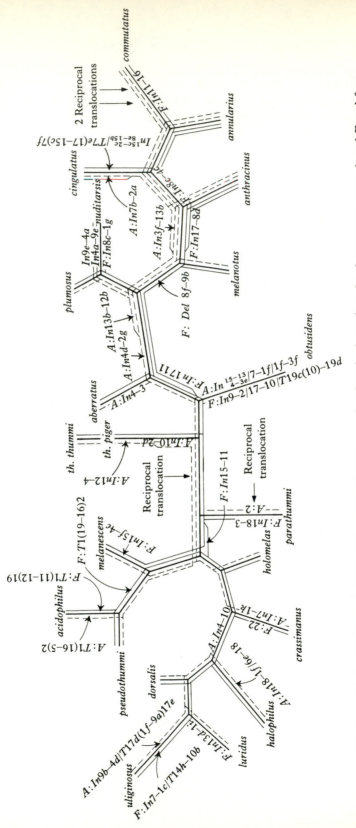

Fig. 11.11. Phylogeny of 20 species of the genus *Chironomus*, showing the inversions in chromosome arms A and F and four reciprocal translocations. From Keyl (1963).

and specific markers in the polytene chromosomes; but this is frequently difficult and in the case of the North American *C. attenuatus* and the European *C. salinarius*, the banding pattern has been so much altered that the homologies are uncertain.

The primitive chironomid karyotype includes seven chromosome arms which have been called *A, B, C, D, E, F, G*. Two widespread species groups show three metacentric chromosomes and a short acrocentric, which bears the nucleolus; these are the *thummi* group (configuration $AB \frown CD \frown EF \frown G$, where the yoke sign indicates an association of chromosome arms to form a metacentric) and the *pseudothummi* group (configuration $AE \frown CD \frown BF \frown G$). Obviously, either of these karyotypes (1 and 2 in Table 11.1) could have given rise to the other by a mutual whole-arm translocation. Karyotype (3), in *parathummi*, can likewise be derived from (2) by a further whole-arm translocation and the karyotype of the subgenus *Campto-chironomus* (5) is also derivable from the *parathummi* one by yet another translocation or from either of karyotypes (1) and (2) by two translocations.

Chironomus commutatus has only three pairs of chromosomes. Its karyotype is *AD, BC, GEF*; apparently it has been derived from the *thummi* group by a mutual translocation between the *AB* and *CD* chromosomes and a tandem fusion between the acrocentric *G* and the *E* limb of the *EG* metacentric. A group of North American species including *staegeri* and

TABLE 11.1 *Associations of chromosome arms in Chironomus**

(1) $\widehat{AB}$ $\widehat{CD}$ $\widehat{EF}$ G	(2) $\widehat{AE}$ $\widehat{CD}$ $\widehat{BF}$ G	(3) $\widehat{AC}$ $\widehat{DE}$ $\widehat{BF}$ G
thummi (E)	*pseudothummi* (E)	*parathummi* (E)
piger (E)	*acidophilus* (E)	
aberratus (E)	*melanescens* (E)	
obtusidens (E)	*holomelas* (E)	
plumosus (E)	*crassimanus* (E)	
melanotus (E)	*dorsalis* (E)	
cingulatus (E)	*halophilus* (E)	
anthracinus (E)	*luridus* (E)	
annularius (E)	*uliginosus* (E)	
nuditarsis (E)	Seven Australian and one New Zealand species	
(4) $\widehat{AD}$ $\widehat{BC}$ $\widehat{GEF}$	(5) $\widehat{AB}$ $\widehat{CF}$ $\widehat{DE}$ G	(6) $\widehat{AB}$ $\widehat{CD}$ $\widehat{GEF}$
commutatus (E)	*Ch. (Camptochironomus) tentans* and *pallidivittatus*	*staegeri* (NA) *crassicaudatus* (NA)
(7) $\widehat{BF}$ $\widehat{CD}$ $\widehat{AEG}$	(8) $\widehat{AF}$ $\widehat{BE}$ $\widehat{CD}$ G	
duplex (Aus)	Two species from California and Ontario	

* (E) – European; (NA) – North American; (Aus) – Australian.

crassicaudatus have the karyotype *AB, CD, GEF*, but these forms are probably derived from the *thummi* group. An Australian species (*C. occidentalis*) has a *BF, CD, AEG* karyotype and appears to be derived from the *pseudothummi* group. It is thus rather uncertain how many different tandem fusions between the *G* and *E* elements have taken place in the phylogeny of the genus, although obviously several have occurred.

A further tandem fusion was found in some species of the genus *Sergentia* by Bauer (1945). *S. longiventris* has the usual chironomid karyotype of three metacentric elements and one subacrocentric but *S. coracina* has only three chromosome pairs, the subacrocentric being fused with one of the metacentrics. Two further fusions of considerable interest have been reported by Bauer in European members of the genus *Cryptochironomus*; in species '1' n = 3, while in species '2', n = 2. In the common ancestor of these species, the subacrocentric fourth chromosome (*G*) must have become fused with one of the longer chromosomes; the karyotype of species '2' must have been derived by at least one additional pericentric inversion and a further fusion. There is a North American species of the subgenus *Chironomus* which likewise has n = 2 (Wulker, Sublette and Martin 1968), but its karyotype has not been analyzed in detail.

A large number of paracentric inversions have occurred in the phylogeny of the chironomid midges. Some of these have reached fixation and now serve to distinguish species or higher categories, while others are still present in a floating polymorphic condition in the populations of many species. Keyl (1962) recorded thirty-nine inversional changes in the phylogeny of the twenty European species considered by him at that time, but it is certain that this number is greatly exceeded by some American and Australian groups of species. Thus the subgenus *Kiefferulus* includes three quite distinct Australian species, whose polytene chromosomes show little resemblance in banding pattern (Martin 1963, 1964). All show considerable inversion polymorphism; *intertinctus* has sixteen known inversions, *paratinctus* and *martini* eight each.

A special kind of cytotaxonomic difference has been discovered in *Chironomus* by Keyl (1964, 1965a, b, 1966a, b) who has studied the DNA content of corresponding bands in the salivary gland chromosomes of *Chironomus thummi* and *C. piger* (referred to by him as two subspecies of *C. thummi*, but best regarded as sibling species). The observations were made on individual bands in Feulgen-stained preparations of F_1 hybrid individuals, with the aid of an integrating scanning microspectrophotometer; unsynapsed chromosome segments were used in order to ensure accuracy in the determinations.

It is presumed (although there does not seem to be absolutely critical evidence for this) that *piger* is the ancestral form and *thummi* the derived one. The bands in *thummi* are either as thick as the corresponding ones in *piger* (in which case the DNA content is the same) or thicker (in which case the DNA content is twice, four times, eight times or sixteen times that of the

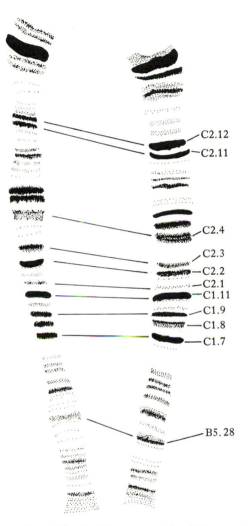

Fig. 11.12. Comparison of the banding pattern in chromosome arm IIR of *Chironomus piger* (left) and *C. thummi* (right) in the hybrid between them. Bands C2.11 and C2.4 show a two-fold increase in DNA content in *thummi*, band C2.12 shows a four-fold increase, bands C2.2, C2.1 and C1.9 show the same amount. In the case of bands C2.3, C1.11, C1.8, C1.7 and B5.28 the increase is variable. From Keyl (1965).

piger band). No other relationships were ever observed, i.e. no band of *thummi* contains less than the corresponding band of *piger*, and where the *thummi* band contains more DNA, the multiplication factor is always a power of two. There does not seem to be any special order or sequence of duplicated and non-duplicated bands along the length of the chromosome. The whole phenomenon clearly has nothing to do with differential polyteny and in fact the degree of polyteny must be assumed to be the same in the two species. Both polytene and spermatocyte nuclei of *thummi* contain approximately 27% more DNA than the corresponding nuclei of *piger*, as we should expect, since many of their genetic loci have undergone duplication, sometimes repeatedly.

A number of single locus duplications seem to exist in a floating condition in populations of *thummi*, presumably as adaptive polymorphisms. That this kind of evolutionary duplication of individual bands or loci may be quite widespread in Diptera, at any rate, is suggested by the earlier data of Metz on *Sciara* species.

Keyl (1965*a*) has considered the possible mechanisms which could underly the single locus duplications. If such events were due to unequal crossing-over, the duplications would be of the tandem rather than the reversed type. Subsequent unequal cross-overs would, moreover, be expected to produce bands with three times the original DNA content – which were never found. Keyl accordingly has put forward a molecular hypothesis according to which the DNA in the bands is organized in the form of rings or ring-like structures, the successive rings being connected by 'linkers' or interband connections. A supplementary replication of one ring is supposed to be followed by end-to-end attachment of the two identical rings. This will also lead to a tandem duplication, but with the essential difference that the two identical halves of the doubled locus will be contin-uous and not separated by a 'linker'. If we accept the idea that ordinary meiotic crossing-over can only occur in linkers, or inter-band connections, it is natural that bands containing three times, five times . . . the original amount of DNA could not be produced by unequal crossing-over. This hypothetical mechanism may be brought into relationship with Callan's (1967) hypothesis of 'master' and 'slave' genes, if we suppose that the master locus is in the linker.

That evolutionary losses of bands have occurred in some Chironomidae, is strongly suggested by the work of Martin (1963) on *Chironomus martini*. A number of individuals of this species were found to be heterozygous for interstitial 'deletions or duplications' of up to ten bands in one case; the banding pattern in this instance seems to rule out a tandem or a reversed duplication in the longer homologue, so that a deletion in the shorter one is the most probable explanation, although we do not know whether deletion homozygotes are viable.

CHROMOSOMAL REARRANGEMENTS AND SPECIATION IN MOSQUITOES (CULICIDAE)

Like the Simuliidae and Chironomidae, the mosquitoes are a family of 'lower' Diptera which show chiasmata at meiosis in the males. Except for the genus *Corethra* with 2n = 8, all species that have been investigated have three pairs of chromosomes. In *Anopheles* there are two pairs of large metacentric autosomes and a pair of smaller sex-chromosomes, consisting of two X's in the female and an unequal-sized XY pair in the males. The sizes of the sex chromosomes differ somewhat in different species of *Anopheles* (Frizzi 1953; Rishikesh 1959*a*, *b*; Rai and Craig 1961; Rai 1963; Breland 1963. Sex chromosomes have not been detected in other genera of Culicidae.

The earliest work on mosquito cytogenetics aimed at an understanding of speciation mechanisms was that of Frizzi (1947*a*, *b*, *c*, 1949, 1950, 1951, 1952) on the sibling species of the *Anopheles maculipennis* complex (long regarded as races of a single species but distinguishable on minor morphological characters of the eggs, larvae and adults). There are six main taxa: *typicus*, *subalpinus*, *messeae*, *labranchiae*, *atroparvus* and *elutus*. *Messeae* is a species which is polymorphic for inversions, but the others seem to be structurally homozygous species. If the polytene chromosome map of *atroparvus* is taken as a standard, *labranchiae* seems to have the same sequences in all its chromosomes, since no rearrangements could be detected in the polytene nuclei of laboratory hybrids between them; on the other hand the salivary chromosomes of the hybrids do show long asynapsed regions which are not seen in the parent species, so that the possibility of some minute rearrangements having occurred cannot be excluded. Hybrids between *typicus* and *atroparvus* likewise show long asynapsed regions in their polytene nuclei, particularly near the chromosome ends; but in this case there is a long inversion difference in chromosome limb 3L. *Subalpinus* does not seem to differ cytogenetically from *typicus*. *Elutus* differs from the standard *atroparvus* karyotype by a long inversion in 3L and a small one in the *X*. *Messeae* is a complex species, since there are two kinds of *X*-chromosomes, one with the standard sequence and one with a complex rearrangement involving at least five breaks. In the vicinity of Pavia, Italy, these differ in ecological and seasonal distribution (Frizzi 1951). Very few heterozygotes were found, by comparison with the number expected on the Hardy–Weinberg formula. It seems likely that this was due to non-random mating between incipient races rather than to differential viability.

In North America there are several species of mosquitoes that are considered to belong to the *maculipennis* complex. A polytene chromosome map of *A. freeborni* has been published by Kitzmiller and Baker (1963); this species was also studied by Frizzi and de Carli (1954), who noted a general similarity of the banding pattern to that seen in the European *atroparvus*. *A. freeborni* is polymorphic for two autosomal inversions and sometimes shows a tandem duplication of nine bands in chromosome arm 2L. *Anopheles earlei* (Kitzmiller and Baker 1965), *A. punctipennis* (Baker and Kitzmiller 1963, 1964*a*), *A. aztecus* (Baker and Kitzmiller 1964*b*), *A. quadrimaculatus* (Klassen, French, Laven and Kitzmiller 1965), *A. neomaculipalpus* (Kitzmiller, Baker and Chowdaiah 1966), and *A. vestipennis* (Chowdaiah, Baker and Kitzmiller 1966) all seem to show considerable resemblances in banding pattern. These are species from North and Central America. *Neomaculipalpus* has a polytene karyotype that includes several reverse repeats. Chromosome arm 2R differs from the corresponding limb of *vestipennis* by one inversion and an extensive duplication which has increased its length. The polytene chromosomes of *A. algeriensis*, from the Mediterranean region, have been mapped by Kitzmiller (1966); there appear to be similarities in banding pattern between this species and *atroparvus*, on the one hand and the North American *freeborni*, on the other. It is now clear that anopheline mosquitoes represent extremely interesting material for studies of chromosomal evolution, but it will probably be necessary to study the polytene chromosomes of interspecific hybrids in order to carry the analysis to a satisfactory conclusion. To some extent this has been done for the North American species by Kitzmiller, Frizzi and Baker (1967) who conclude that between ten and twenty major inversions may have occurred in the phylogeny of *freeborni*, *occidentalis*, *aztecus*, *quadrimaculatus*, *earlei* and *punctipennis*. The last-named species is also polymorphic for at least nine different floating inversions. Schreiber and Memoria (1957) have also recorded inversion polymorphism in several Brazilian species of *Anopheles*.

Recent work has shown that the African mosquito formerly known as *Anopheles gambiae*, is actually a group of at least five sibling species (Davidson 1962, 1964; Coluzzi 1964, 1966; Davidson, Paterson, Coluzzi, Mason and Micks 1967). These include the saltwater species *A. melas* (West African) and *A. merus* (East African) and three cryptic freshwater species '*A*', '*B*' and '*C*'. The polytene chromosomes of species *A* and *B* have been studied by Coluzzi and Sabatini (1967), who also examined the salivary gland chromosomes of laboratory hybrids between them. In both species the karyotype consists of a subacrocentric X and two large pairs of metacentric autosomes; the Y of the male is an acrocentric.

The banding pattern seems to be very similar in the autosomes of species A and B and there is good synapsis of the autosomes in both meiotic divisions and polytene nuclei of the hybrids. But the banding pattern of the X-chromosome is profoundly different in the two species and the two X's are largely asynaptic in the salivary gland nuclei of female hybrids. Species B is polymorphic for six autosomal inversions, five of which also occur in species A, a situation which is reminiscent of that which occurs in the 'races' of Drosophila paulistorum. Species C is cytogenetically intermediate between species A and B (Coluzzi and Sabatini 1968). The West African A. melas has an X-chromosome which has a long heterochromatic limb, so that it is very much longer than the X's of the freshwater species and merus in the mitotic karyotypes; but the euchromatic segment of the melas X is almost indistinguishable from that of species C (Coluzzi and Sabatini 1969). The autosomes of melas differ from those of the freshwater species by a number of fixed paracentric inversions. The East African A. merus has an X-chromosome which has an extra segment, characteristically swollen, by comparison with the freshwater species. Neither of the saltwater species seem to be polymorphic for intraspecific inversions. Melas and merus seem to have arisen independently; the former is most closely related to species C, the latter to species A.

The role of paracentric inversions in chironomid, simuliid and culicid populations and phylogeny is probably very similar to the part which they play in Drosophila. We have explained (p. 242) the reasons why heterozygosity for such rearrangements does not lead to the production of aneuploid sperms in Chironomus, and it is probable that the same mechanisms operate in the Simuliidae and Culicidae. However, inversions in simuliids probably show a higher frequency of reversed-loop pairing at pachytene than they do in chironomids, as demonstrated by the frequent occurrence of bridge and fragment chromatids at first anaphase of simuliid male meiosis. In all these families selection against pericentric inversions seems to be very severe, so that very few cytotaxonomic differences which can be ascribed to the fixation of this type of rearrangement have been recorded. We have, however, pointed out a few such instances in the Chironomidae, and in Prosimulium decemarticulatum chromosome 3 has an extreme arm ratio, suggesting that a pericentric inversion has become fixed (Rothfels 1956). And in P. mixtum and P. fuscum the X- and Y-chromosomes differ by a pericentric inversion (a different one in each of these species) according to Rothfels (see also Basrur 1959).

The frequency of mutual whole-arm translocations between metacentrics has undoubtedly been much higher in the phylogeny of the Chironomidae than in any of the other dipteran groups we have considered. And in the

Chironomidae we also find that a highly unexpected type of rearrangement – tandem fusion of an acrocentric to a metacentric – has successfully established itself on several occasions. Such rearrangements must be expected to give rise to extremely 'awkward' configurations at meiosis in heterozygotes i.e. ones liable to give rise to aneuploid gametes. It is hence very unlikely that they have established themselves by virtue of heterozygote superiority or were ever present in a population in a balanced polymorphic state.

CHROMOSOMAL REARRANGEMENTS AND SPECIATION IN GRASSHOPPERS

The four families of Diptera discussed above consist of winged insects. Even if they are not strong fliers, individuals of *Drosophila*, *Simulium*, *Chironomus* and *Anopheles* are liable to be carried considerable distances by air currents. In contrast to these genera of flies, many grasshoppers are wingless, or at any rate flightless. The influence of vagility on speciation is very clearly seen in the North American grasshopper genus *Melanoplus*. Those sections of this genus which include fully-winged forms show relatively few, geographically widespread species. The sections which include brachypterous, flightless insects seem to have undergone speciation on a much more extensive scale, the areas occupied by the individual species being much smaller. A comparison of cytogenetic mechanisms of speciation in the two sections of the genus might be interesting.

In the grasshoppers belonging to the large family Acrididae (to which *Melanoplus* belongs) an apparent karyotypic uniformity is very evident, the great majority of the species having $2n\male = 23$ acrocentric chromosomes, the X being a medium-sized chromosome and the autosomes showing a considerable size range. Three reservations to this concept of karyotypic uniformity must, however, be made. First of all, there are quite a number of species in which the chromosome number has been reduced, by fusions between autosomes or between the X and one of the autosomes; these species will be discussed in Chapter 12. Secondly, there seem to be fairly large differences between related species in the extent and distribution of heterochromatic blocks; there is frequently one autosomal element which is largely heterochromatic, and its size varies considerably. Thirdly, as pointed out by John and Hewitt (1966b) the DNA values of species of Acrididae show quite a large range and there has undoubtedly been a considerable amount of 'chromosomal repatterning' even in species retaining the ancestral 23-chromosome karyotype; the 'karyotypic uniformity' is thus somewhat illusory. Nevertheless, the fact that many hundreds of species (undoubtedly many thousands, if all of them were investigated) possess twelve pairs of acrocentrics and that these species belong to several subfamilies and include morphologically diverse forms, some winged and others wingless, undoubtedly implies fairly severe restrictions on the kind of rearrangements that can establish themselves in the phylogeny of this group of insects.

The grasshoppers belonging to the group Trimerotropi considered in Chapter 9 are fully-winged insects. Section A of the genus *Trimerotropis* does not seem to differ from hundreds of other genera of Acrididae in its cytogenetic picture; but section B of the same genus and the genera *Circotettix* and *Aerochoreutes* are highly polymorphic for inversions, some of which are floating while others have undergone fixation in phylogeny. These insects seem to be generally similar to species of *Drosophila* in their population structure and in the relation of chromosomal rearrangements to speciation, except that the inversions present in most of the species are pericentric rather than paracentric. Thus the Trimerotropi do not seem to teach us any important lessons which cannot also be derived from studies of Drosophilidae, Simuliidae and Chironomidae.

The situation with regard to the wingless grasshoppers belonging to the endemic Australian subfamily Morabinae (one of about twenty subfamilies of the tropicopolitan family Eumastacidae) is considerably different. These are insects with quite feeble powers of locomotion, which are usually restricted to a limited range of foodplants (Blackith and Blackith 1966); individuals are unlikely to travel more than a few metres from their birthplace in the course of an entire life-cycle. As with other animals having the same general type of population structure, the actual distributions of the species are not continuous, but consist of a series of populations occupying favourable habitats; the effect of human interference with the environment over the past 150 years has been to render the distributions still more discontinuous. Many species have now been reduced to a few completely isolated colonies, surviving precariously for the time being.

Approximately 250 species of morabine grasshoppers are now known, although only 35 have been described in the taxonomic literature. The total number (including yet undiscovered forms) may be about 300 species. The generic classification of the group as a whole has not yet been established and the species are provisionally assigned to 'species groups', many of which will eventually be recognized as genera. Undescribed species are referred to for the time being by symbols consisting of letters and numbers. The karyotypes of about 190 species have been studied and detailed investigations have been carried out on several of the species groups.

There are some species groups (e.g. the tropical *misilliformis* group and the desert genus *Warramunga*) in which the ranges of the species overlap to a considerable extent. In others, such as the *viatica* group, most of the taxa (species or races – it is not always easy to decide) have 'contiguous' or 'parapatric' distributions that meet in very narrow zones of overlap. Where such situations have been carefully studied, the zone of coexistence has been found to be only a few hundred metres wide. In the cases that have been investigated hybridization either definitely occurs, or is suspected to occur, with diminished fecundity or viability of the hybrids.

It is tempting to conclude that the '*viatica*' and '*misilliformis*' types of distribution represent two different stages of evolutionary differentiation, the separation of the various taxa being relatively recent in the *viatica* group and more ancient in the *misilliformis* group. Morphological studies seem to confirm this view, the male genitalia being more different in the case of species which are sympatric over wide areas than they are in forms showing the 'parapatric' type of distribution.

About sixty evolutionary changes of chromosome number (fusions and dissociations) are known to have occurred in the morabine grasshoppers (Table 11.2). The exact number that can be postulated depends to some extent on phylogenetic interpretations; in Table 11.2 it has been assumed that where similar fusions or dissociations (e.g. a dissociation of a particular metacentric chromosome) are present in different species groups they are to be interpreted as different evolutionary events, but that when they occur in several members of the same species group, only one evolutionary event should be postulated.

The whole of this interpretation of chromosomal evolution in the morabine grasshoppers depends on the concept of a primitive karyotype with $2n\male = 17$ (an X-chromosome, a pair of large equal-armed 'AB' autosomes, a pair of smaller unequal-armed 'CD' autosomes and six pairs of small autosomes). It is supposed that the common ancestor of the whole group possessed this karyotype. However, even if it did not, it must have had a

TABLE 11.2 *Fusions and Dissociations in morabine grasshoppers*

Fusions		Dissociations	
Between X and an autosome	11	Of the AB chromosome	16
Between the Y and an autosome	6	Of the CD chromosome	6
Tandem fusion of X or Y and an autosome	1		
Between two long autosomes	3		
Between a long and a short autosome	8		
Between two short autosomes	10		
	39		22

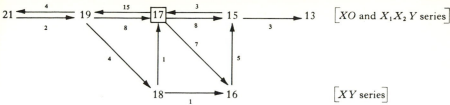

Fig. 11.13. Evolution of chromosome numbers (2n♂) in the Australian morabine grasshoppers, starting from the primitive karyotype of 2n♂ = 17. The numbers alongside the arrows indicate the number of times a particular type of fusion or dissociation has taken place in phylogeny. Horizontal arrows represent fusions or dissociations of autosomes, diagonal arrows X-autosome fusions and vertical arrows Y-autosome fusions. From White (1969).

relatively similar one (possibly with 2n♂ = 19 or 15), and the count of fusions and dissociations would only have to be slightly modified. One argument for a primitive karyotype of 2n♂ = 17, with AB and CD metacentrics, is that such a karyotype has also been found in a member of the related subfamily Biroellinae (White 1968*b*).

How many chromosomal rearrangements other than fusions, dissociations and translocations have taken place in the morabine grasshoppers is very uncertain. A considerable number of pericentric inversions have undoubtedly occurred, and some species such as *Keyacris scurra* (see p. 254) are polymorphic for this type of rearrangement. The majority of the species probably have five of the small chromosome pairs acrocentric, one pair being metacentric. But the size of these small autosomes varies a good deal throughout the group and there are species in which all the small autosomes are metacentric and ones in which they are all acrocentric. To what extent these changes in the small chromosomes have been caused by pericentric inversions and to what extent by duplications and deletions of genetic material is very uncertain. In the species with XO males (about 170 out of the 190 examined) the X may be either acrocentric or metacentric, the difference between these two types being undoubtedly due to pericentric inversions which have repeatedly converted metacentric X's into acrocentric ones, and vice versa.

About fifteen species of morabine grasshoppers are known which consist of two chromosomal races which differ in respect of a fusion or dissociation. In those instances that have been thoroughly investigated, there is always a zone of overlap and hybridization a few hundred metres wide. This is so in the case of the 15- and 17-chromosome races of *Keyacris scurra* (White and Chinnick 1957), the 19- and 17-chromosome races of '*Moraba*' *viatica* (White, Carson and Cheney 1964) and the XO and XY races of a form known as 'P24' (White, unpublished). A similar situation exists where the 17-chromosome race of *viatica* comes in contact with the XY race of P24 on Kangaroo Island (White, Key, André and Cheney 1969). In all these cases, hybridization occurs freely in the zone of overlap, although the fecundity of the hybrids is somewhat reduced in the first two instances; ethological isolating mechanisms are either absent altogether or poorly developed. Numerous other cases of 'parapatric' distributions of chromosomal races or closely related species are known in the morabine grasshoppers, but these are the only ones in which detailed field work has confirmed the existence of natural hybrids. There is one other case in the morabine grasshoppers where a very narrow zone of overlap exists, without any adult hybrids being present. This is where the ranges of the 19-chromosome race of *viatica* and the XY race of P24 overlap on Kangaroo Island (White, Key, André and Cheney 1969). These forms will produce hybrids in the laboratory, but these grow very slowly and do not reach the adult stage; one immature hybrid was collected in nature.

The narrowness of the hybrid zones in all these cases indicates clearly that there is strong selection against the chromosome number heterozygotes ('negative heterosis'); an alternative way of expressing the situation would be to say that there is strong selection against fused chromosomes in 'unfused territory' and, vice versa, against unfused chromosomes in 'fused territory'. However, even this state of affairs would not lead to such extremely narrow zones of hybridization, were it not for the very low degree of vagility of these insects. A similar situation seems to exist in some coccinellid beetles (Zaslavsky 1963), but these insects are liable to be carried by air currents and the zones of hybridization are 20–30 km wide. In the case of the $2n\male = 22$ and $2n\male = 24$ races of the isopod crustacean *Jaera syei* on the coast of Brittany the zone of overlap seems to be about 80 km wide; but chromosome number heterozygotes are always fewer than would be expected on the Hardy–Weinberg formula (Lécher 1967).

The causes of the reduced selective value of the interracial hybrids are rather diverse in the morabine grasshoppers. In several instances (but these are mainly ones where the parental forms seem to have reached the level of full species) the whole development may be inhibited or the growth of the gonads may be blocked at an immature stage. In the case of the interracial hybrids, however, and in some between forms here considered to be specifically distinct, it is usual for development to be outwardly normal, although the viability of the chromosome number heterozygotes has not been accurately compared with that of non-hybrid individuals. In such hybrids, however, we do find relatively minor abnormalities of meiosis (asynapsis, malorientation of bivalents or multivalents) which lead in various ways to the production of aneuploid gametes and hence to a reduction in fecundity.

The first case to be investigated was that of the 15- and 17-chromosome races of *Keyacris scurra* (White 1957a; White and Chinnick 1957). A single hybrid individual was found in nature, but numerous others were obtained by laboratory crosses. In the western, 17-chromosome race the big '*AB*' metacentric has undergone dissociation into two acrocentrics, '*A*' and '*B*'. In interracial hybrids a trivalent is usually formed at meiosis, the two acrocentric elements being associated with the corresponding limbs of the metacentric. In male hybrids, at any rate, the orientation of the trivalent on the spindle is extremely regular. However, in individuals which happen to be heterozygous for the 'Blundell' pericentric inversion in the *CD*-chromosome, a chain of five chromosomes is sometimes formed instead of a trivalent (see p. 258). Malorientation of these chains is frequent and undoubtedly leads to the production of some aneuploid sperms. It is probably the main cause of natural selection against chromosome number heterozygotes, since in populations close to the zone of overlap, a large proportion of the population are heterozygous for the Blundell inversion.

In the *viatica* group of morabine grasshoppers there are five coastal 'forms' (*viatica* itself, 'P24', 'P25', 'P45b' and 'P45c'). The exact status of these taxa as races or species is somewhat uncertain. Each of them is, in turn, broken up into several chromosomal races. There are, in addition, two or three inland 'forms' whose relationship to the coastal ones is not understood. One has the impression that an ancestral species which we may call *proto-viatica* has undergone fragmentation, probably within the last one or two million years, into about seven incipient species, which show little morphological differentiation. The distribution of the coastal forms in South Australia is shown in Figure 11.14.

The 17-chromosome race of *viatica* has a fusion between chromosomes '*B*' and '6' which the 19-chromosome race lacks. This *B* + 6 fusion is also present in P24, and must hence have arisen prior to their evolutionary separation. Natural hybrids between the two races of *viatica* have been found in their zones of overlap; in this case there are no chains of five chromosomes and in most spermatocytes there is a trivalent, formed by pairing of the '*B*' and '6' chromosomes of the 19-chromosome race with the unequal-armed *B* + 6 metacentric derived from the 17-chromosome form (Fig. 11.15). The trivalent orientates correctly in most of the cells in spite of the fact that it is lopsided. But in about 10% of the spermatocytes it is maloriented and in a further 10% of first metaphase the little sixth chromosome is a univalent (whether as a result of asynapsis or de-synapsis following pachytene is uncertain). Thus the chromosome number heterozygotes probably produce 10–15% aneuploid sperms, and this is presumably the reason for their low selective value (i.e. for the narrowness of the zone of hybridization). The meiotic behaviour of the female hybrids has not yet been studied.

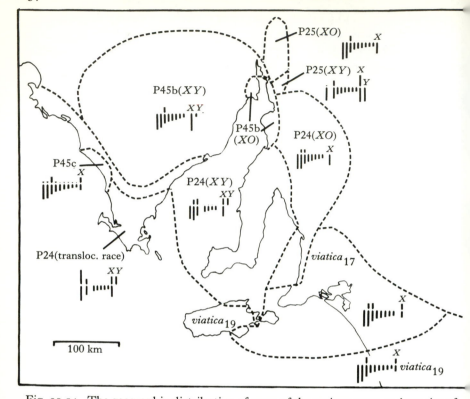

Fig. 11.14. The geographic distribution of some of the various races and species of the *viatica* group in South Australia. The Eyre, Yorke and Fleurieu Peninsulas are shown, with Kangaroo Island across the entrance to St Vincent's Gulf, between the latter two. The ranges of some of the taxa are shown as if they extended beyond the present coastline, because the evolutionary divergence of these forms is believed to have taken place mainly during the Pleistocene when the coastline was some distance south of Kangaroo Island and the whole of Spencer's and St Vincent's Gulfs were dry land. Small ideograms indicate the karyotypes of the various taxa (haploid, except that in the case of forms with XY males, both sex chromosomes are shown). No ideogram is shown for the XO race of 'P45b'; this form has a karyotype which is indistinguishable from that of the XO race of P25. The zones of natural overlap between contiguous taxa are all, as far as known, narrower than the dashed lines. From White (1968a) but brought up-to-date as far as Kangaroo Island is concerned.

The most primitive race of P24 has XO males, but the X is acrocentric, unlike that of *viatica*, which is an equal-armed metacentric. Presumably a pericentric inversion in the X reached fixation in the phylogeny of P24, P25 and P45b, all of which have an acrocentric sex chromosome as the basic, primitive condition. In the XY race of P24 the X has undergone a centric

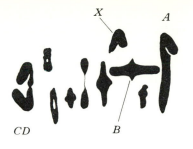

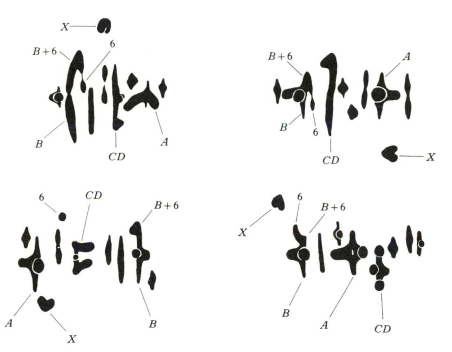

19-chromosome race   17-chromosome race

Interracial hybrids

Fig. 11.15. Male first metaphases (side views) in the 19- and 17-chromosome races of the grasshopper *Moraba viatica* and in natural hybrids collected in the zone of overlap near Keith, South Australia. The individual of the 19-chromosome race was heterozygous for a pericentric inversion in the *A* bivalent and the one belonging to the 17-chromosome race was carrying a supernumerary chromosome (sup). The inter-racial hybrids are heterozygous for the *B* + 6 fusion. A trivalent is usually formed, and generally orientates in a disjunctional manner on the spindle, but sometimes the small number 6 chromosome is a univalent. In two of the hybrid metaphases the *CD* bivalent is heterozygous for a pericentric inversion. From White, Carson and Cheney (1964).

fusion with the largest of the small autosomes (no. 1). A narrow hybrid zone between the XO and XY races occurs north of Adelaide, both types of male being present in the population. Some female hybrids have been found, with trivalents (1–fused X–unfused X) at meiosis in their eggs.

A third cytological race of P24, which exists in the southern part of the Eyre Peninsula, is derived from the XY race, but is homozygous for a translocation between the 'A' chromosome and the '$B + 6$' one. Hybrids between the translocation race and the ordinary XY race (which also occurs on the Eyre Peninsula) may exist, but have not yet been found; one would expect that their fecundity would be rather severely reduced.

Species P25 lacks the $B + 6$ fusion. It has an XO race (the X being acrocentric) and an XY one – but in this case the centric fusion is between the X and the B-chromosome, since in laboratory-bred male hybrids between females of the XY race of P25 and males of P24 (any race) we find one limb of the neo-X of P25 synapsed with the $B + 6$ chromosome of P24.

The zone of overlap between the XO and XY races of P25 has not been found yet, so that natural hybrids have not been studied. Neither has any zone of overlap between P24 and P25 been discovered. Laboratory-bred male hybrids between these forms show about 70–80% asynapsis between the little sixth chromosome of P25 and the $B + 6$ element of P24. Their fecundity must be much more seriously reduced than that of the natural hybrids between the two races of *viatica*.

'P45b' is likewise a species with XO and XY races, the karyotype of the former being like that of the XO race of P25. But in the XY race this has undergone a tandem fusion (i.e. not a centric one) with autosome 6; the original X occupying the distal position in the fused chromosome. At meiosis in the XY race the neo-Y (i.e. the unfused sixth chromosome) synapses with the proximal segment of the long acrocentric neo-X. Since a single chiasma is always formed between them, this is a post-reductional sex-chromosome mechanism, the X- and Y-chromosomes separating at the second anaphase instead of at the first one, as they do in the XY races of P24 and P25. Laboratory-bred male hybrids show a great deal of asynapsis between the little sixth chromosome of the XO race and the basal segment of the X derived from the XY form, thus proving that there are genetic differences between the sixth chromosome as it exists today in the XO form and the 'derived' neo-Y of the XY race. In some male hybrids between these two races there is also much degeneration of the spermatocytes and few or no sperms are produced. Female hybrids have not been studied, but because of the nature of tandem fusions (Fig. 7.13) they must be expected to produce at least 50% aneuploid eggs. Altogether, the 'races' of P45b seem to show strong genetic isolation and to be, in fact, biological species rather than races.

'P45c' has a karyotype which is basically like that of $viatica_{19}$, except that the short limbs of most of the acrocentric autosomes have become much longer; these limbs are late-labelling following tritiated thymidine auto-radiography, which probably indicates that they are heterochromatic and that a number of duplications have established themselves in the short arms since the evolutionary divergence of P45c from its nearest relatives.

The elaborate chromosomal evolution of the *viatica* group should tell us something about the mechanisms of speciation that have operated. White, Blackith, Blackith and Cheney (1967) and White (1968) have given reasons for believing that the original ancestor, *proto-viatica*, with $2n\delta = 19$ and a metacentric X, occupied approximately the same area as the descendant forms do today. This was considerably more than 10000 years ago (the approximate date at which Kangaroo Island became separated from the mainland). Two chromosomal rearrangements must have occurred at an early stage in the central part of the range of *protoviatica* – a pericentric inversion in the X (converting it into an acrocentric) and the $B + 6$ fusion. It is assumed that these spread out from their respective points of origin. In combination, they led to the effective evolutionary separation of the two races of *viatica* and of P24 from *viatica* to the south and from P25 to the north (Fig. 11.16). The three X-autosome fusions and the translocation in the XY race of P24 would have occurred later and spread in essentially the same manner.

This concept of a species with a wide and essentially continuous range undergoing fragmentation into a number of descendent races and species as a result of chromosomal rearrangements which spread out from their point of origin until their expansion is arrested and the narrow zone of hybridization, which has been undergoing a secular movement across the terrain, becomes stabilized, has been called *stasipatric speciation* (White, Blackith, Blackith and Cheney 1967; White 1968a). It seems to be essentially different from both the generally rejected sympatric model of speciation and the classical allopatric model (as represented by the dumbbell diagram). But the reason for the spread of the rearrangements concerned in the process (which are certainly not heterotic, since they diminish the fecundity of the heterozygotes) is not clear. It is possible that the hierarchy of selective values is $OO > ON \ll NN$ in the area where the new chromosome arrangement N is spreading, and that fixation of N occurs in one small colony after another, by a combination of genetic drift and selection, in spite of the inferiority of the heterozygotes. But it is also possible that the rearrangements which have been evolutionary successes (perhaps 1 in 10^4 or more of the total number that occurred in the phylogeny of the *viatica* group), were ones that happened to enjoy a segregational advantage of 'meiotic drive' in the heterozygous female, passing into the egg nucleus more frequently than

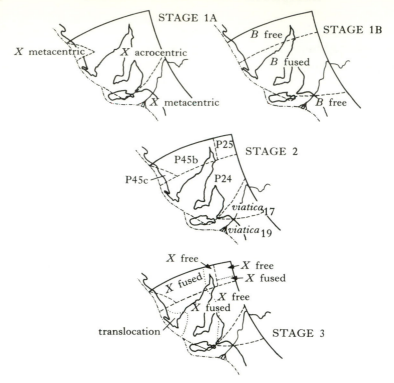

Fig. 11.16. 'Stasipatric' model of speciation in the *viatica* group of morabine grasshoppers in South Australia. It is assumed that in an original *proto-viatica* population two events occurred: a pericentric inversion which converted the originally metacentric *X*-chromosome into an acrocentric and a fusion between the *B*- and 6-chromosomes. The combination of stages '1A' and '1B' (it is not known which occurred first) gives stage 2; the further addition of three different *X*-autosome fusions and a translocation in the southern part of the Eyre Peninsula produces stage 3, which is virtually the situation at the present day. From White (1968a).

into the polar body at first anaphase. An attempt to interpret raciation and speciation in the *viatica* group on the allopatric model led to various logical difficulties and apparent conflicts between the cytological evidence and bioclimatic data. Bazykin (1969) has considered, from a theoretical standpoint, a model of speciation which is quite similar to the 'stasipatric' one of White; he regards this as essentially different from the sympatric and allopatric models, but in some respects intermediate.

No evidence is available, of course, as to the adaptive properties of the various chromosomal rearrangements that have established themselves in the *viatica* group. And even if their effects on viability, in heterozygotes and homozygotes, could be determined in various environments, there is no guarantee that these effects were the same at the time when the rearrangements first occurred. The narrow zone of overlap between the two races of *viatica* is clearly a 'zone of tension' (Key 1968) but it does not seem to correspond to any ecological or geographical discontinuity. The present situation may be one of stabilized tension and may not be leading to increasing genetic isolation. Alternatively, the zone of overlap may be moving across the terrain, perhaps at the rate of a few metres a year. The degree of isolation between P24 and P25 seems to be much greater than that between the chromosome number races of *viatica*. In both cases the reduction in fecundity results from asynapsis and malorientation of chromosomes involved in a fusion; none of the other chromosomes show any abnormalities of synapsis or segregation. It is fairly obvious that the *viatica* case and the P24–P25 one represent two stages in an evolutionary process. If this view is correct, it is probably a process which is not confined to morabine grasshoppers, but has occurred in other groups having the same type of population structure. We might imagine that, in the case of *viatica*, genetic changes in the chromosomes involved in the fusion might eventually build up (by mutation) to such an extent that the adaptive value of the natural hybrids in the zone of overlap would be even more drastically reduced. This may be so in the zone of overlap between $P24(XY)$ and $viatica_{19}$ on Kangaroo Island, although these forms differ in respect of three chromosomal rearrangements ($B + 6$ fusion, pericentric inversion in X, $X + 1$ fusion). One might imagine that once the selective value of the hybrids had fallen to a sufficiently low level, selection for pre-mating isolating mechanisms in the zone of overlap might begin to occur.

Alternatively, one might imagine that in some cases chromosomal races will be unable to evolve into full species until climatic, ecological or geographical changes have produced complete geographic isolation between them, with *no* zone of overlap. Certainly, however, the morabine grasshoppers provide no evidence for this rather extreme hypothesis.

The stasipatric model involves the geographic spread of a chromosome with a new rearrangement through territory already occupied by populations of the species. The allopatric model, as applied to cases where a rearrangement gives rise to a genetic isolating mechanism, would seem to involve the origin of a rearrangement in a geographically peripheral deme and its spread into territory previously unoccupied by the species. Clearly, we are dealing with secular processes, and since the former distributions of species can only be known in the most general way, a definite decision between the two models will be almost impossible to arrive at, in any actual case, on the basis of present evidence. Various intermediate models of speciation that combine features of the stasipatric and the allopatric models seem to be possible.

No other group of grasshoppers has been so thoroughly investigated, from the cytogenetic standpoint, as the *viatica* group. In most groups of grasshoppers belonging to both the Acrididae and the Eumastacidae,

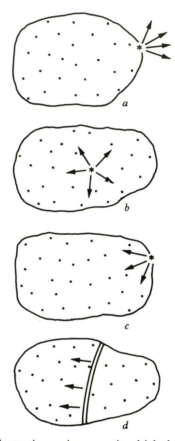

Fig. 11.17. Diagram illustrating various ways in which chromosomal rearrangements which do not give to heterosis might extend their geographic range. *a*, the chromosomal rearrangement manages to establish itself in a geographically isolated peripheral deme and spreads into territory previously unoccupied by the species (one form of the allopatric model); *b*, the rearrangement establishes itself in a non-peripheral local colony and spreads through the range of the species on an advancing front (the stasipatric model); *c*, the rearrangement establishes itself in a peripheral colony and thus spreads through the existing species population (modified form of the stasipatric model); *d*, the result of *b* or *c* – a narrow hybrid zone showing a slow secular movement across the territory occupied by the species until it is arrested in some way. From White (1968*a*).

fusions, dissociations and pericentric inversions do not seem to have occurred to the same extent and hence cannot have played the same major role in speciation. But small changes in the amount and distribution of heterochromatin have been extremely common (see Ch. 9) and it is possible that in many groups these may have played the same kind of role in speciation as the more conspicuous types of rearrangements seem to have done in the morabines.

CHROMOSOMAL RACES OR SIBLING SPECIES IN MOLE CRICKETS

The mole crickets of Europe and the Mediterranean region have been placed in a single species *Gryllotalpa gryllotalpa* L. by taxonomists. At least seven cytologically distinguishable forms are now known to be included under this name; in the absence of any really critical modern taxonomic work, it is quite uncertain how many of these should be regarded as cytological races (analogous to those found in several species of morabine grasshoppers) and how many as distinct species. They seem to be largely allopatric and it is clear that their evolutionary relationships will not be understood until the situation in their zones of contact or overlap has been examined; experimental hybridization may also be necessary to clear up some points. Zones of contact or overlap should occur in Spain, Italy, Macedonia and Palestine and perhaps in Yugoslavia and Hungary. There has been a good deal of speculation (some of it unwarranted) as to the cytotaxonomic relationships and status of the forms of *G. gryllotalpa*; detailed studies of the karyotypes by modern methods might clear up some doubtful points, but natural or artificial hybrids seem to be essential for a complete analysis.

The seven forms, whose chromosome numbers ($2n\male$) range from twelve to twenty-three are shown in Table 11.3, with their geographic distribution as far as known. There has been some debate as to whether the forms with low or high chromosome numbers should be regarded as the more primitive, cytologically (i.e. whether the cytological evolution in this group has been preponderantly an increase or a decrease in chromosome number). Various authors (e.g. de Winiwarter 1927, 1937; Barigozzi 1933a) concluded, at a time when the principles of chromosomal evolution were not properly understood, that the lowest chromosome number should be regarded as the most primitive, partly because the form with $2n = 12$ seems to be the most widespread, geographically (surely, a most unconvincing argument).

Four other species of mole crickets, namely *G. africana* (Asana, Makino, and Niiyama 1940; Makino, Niiyama and Asana 1940; Kushnir 1952), *G.*

TABLE 11.3. *Members of the 'Gryllotalpa gryllotalpa' species group*

2n ♂	Distribution	Authors
23 (*XO*)	Israel (a single locality on the shores of the Dead Sea)	Kushnir 1948, 1952
19 (*XO*)	Israel (widespread), Rhodes, Greece (including Pelleponesus)	Kushnir 1948, 1952; Krimbas 1960
18 (*XY*)	Northern Italy	Barigozzi 1942; Tosi 1959
17 (*XO*)	Southern Spain, probably also central Italy	Ortiz 1951, 1958; Senna 1911
15 (*XO*)	Southern Italy	Payne 1916; de Winiwarter 1927, 1937; Barigozzi 1933*a, b*
14 (*XY*)	Turkey, Rumania, Thrace, Macedonia, Mytilene, Samos, Crete	Steopoe 1939; Kushnir 1952; Krimbas 1960
12 (*XY*)	France, Belgium, Germany, northern Spain, northern Italy	Payne 1916; de Winiwarter 1927, 1937; Ortiz 1951, 1958

himalayana (Bhattacharjee and Manna 1967), *G. hexadactyla* (= *G. borealis*) (Payne 1912*b*, 1916; White 1951*b*) and *Scapteriscus tetradactylus* (Dreyfus 1942) all have 2n♂ = 23, although some individuals of the first have supernumerary chromosomes in addition to the regular set. Some of these are species with a vast geographic range (e.g. *G. africana* occurs throughout Africa, a large part of Asia and some regions of Australia and the distribution of *G. hexadactyla* extends throughout North and South America). It thus seems reasonable to assume that chromosomal evolution in the *G. gryllotalpa* group of forms has been predominantly (although not necessarily

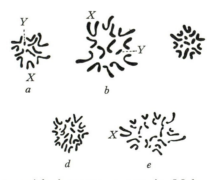

Fig. 11.18. Spermatogonial chromosome sets in Mole crickets of the genus *Gryllotalpa*. (*a*) *G. gryllotalpa*, north European race with 2n♂ = 12 (after de Winiwarter); (*b*) Romanian race with 2n♂ = 14 (after Steopoe); (*c*) southern Italian race with 2n♂ = 15 (after Payne); (*d*) *G.* (*Neocurtilla*) *hexadactyla* with 2n♂ = 23 (after Payne); (*e*) *G. africana* with 2n♂ = 23 (after Asana, Makino and Niiyama).

exclusively) in a downward direction, by chromosomal fusions. North-western Europe is clearly a peripheral area for mole crickets, a group whose main evolution has undoubtedly been in warmer areas (in spite of the fact that all the known fossils of this group are from western and central Europe).

Centric fusions alone do not explain the cytological evolution of this group, since the autosomes are probably always metacentric or J-shaped. Presumably there must have been a number of pericentric inversions, followed by fusions. No heterozygosity for pericentric inversions has, however, been recorded in any mole cricket. Alternatively, it is possible that some of the chromosomal limbs in forms with the higher numbers are (or were) heterochromatic, as in the ladybird genus *Chilocorus* (see p. 386), so that evolutionary loss of such limbs could occur without necessarily diminishing viability. Kushnir (1952) interpreted the 19-chromosome form in Palestine as equivalent to the 15-chromosome race from southern Italy with the smallest chromosome pair reduplicated (i.e. in the hexasomic condition). His views rest on rather superficial comparison of the karyotypes of the two forms and on the fact that four of the small elements in the nine-teen chromosome form may be associated as a quadrivalent at meiosis and may show various forms of lagging at first anaphase (as an alternative to the polysomy hypothesis one might suggest that they are sometimes hetero-zygous for a translocation, possibly involving heterochromatic regions). Kushnir assumed that further reduplication of the small chromosomes led to the 23-chromosome Dead Sea 'race'. This latter form is distinct in external morphology so that there does seem to be a case for regarding it as a distinct species.

The primitive sex-chromosome mechanism in the mole crickets was undoubtedly an XO one with the X a large metacentric element. In *G. hexadactyla* there is a peculiar mechanism with an unequal-sized pair of chromosomes, the larger member of which invariably passes to the same pole as the X at first anaphase, in spite of there being no visible physical connection between them; this mechanism has been found in material from Indiana (Payne 1912b, 1916), North Carolina and Texas (White 1951b). Clearly this unique mechanism, whatever its precise nature and significance, is irrelevant to the evolution of the XO and XY systems in the *G. gryllotalpa* group of forms. It may be that in the *gryllotalpa* complex the X underwent a pericentric inversion at some stage, followed by a centric fusion, creating a neo-X: neo-Y system. But other possibilities exist, since supernumerary chromosomes occur in the 14-chromosome race in Rumania and Greece; a translocation between one tip of a metacentric X and a supernumerary chromosome (with a homologous supernumerary becoming the Y) is not unthinkable in this case. But chromosomal evolution in this group has undoubtedly been complex and it is not impossible that some of the XO forms such as the one from southern Italy may have been secondarily derived from XY forms, perhaps by fusion of the Y with the X as Barigozzi suggested. Y-chromosomes in *G. gryllotalpa* seem to be small acrocentric or subacrocentric elements.

CHROMOSOMAL REARRANGEMENTS IN BEETLES

Cytogenetic studies on speciation mechanisms in Coleoptera have been carried out especially by S. G. Smith (1956a, b, 1957a, b, 1958, 1959, 1962, 1965a, b, 1966), who has worked on various genera of the Coccinellidae (especially *Chilocorus* and *Exochomus*) and Curculionidae (genera *Pissodes* and *Hylobius*). A number of other families of beetles have also been investigated in recent years by Virkki (1961, 1962a, b, 1963, 1964, 1965a, b, 1967a, b, c, 1968a, b, 1970; Virkki and Purcell 1965).

In the ladybird genus *Chilocorus* certain cytologically primitive species such as *similis* have most of their autosomes acrocentric. Actually, *similis* has eight pairs of acrocentric autosomes and one pair of metacentrics, which has presumably arisen through centric fusion (Smith 1965b); the acrocentric bivalents show a single chiasma, while the metacentric one always has a chiasma in each arm. The sex chromosomes consist of a metacentric X, one limb of which is associated at meiosis with an acrocentric Y. This is probably a neo-XY system, which has arisen by a fusion between an original X and one member of a pair of autosomes; but no member of the genus *Chilocorus* is known which lacks this fusion.

The majority of the species of *Chilocorus* differ from *similis* in that each of the acrocentric autosomes has become metacentric through the 'addition' of a heterochromatic limb which may be almost as long as the euchromatic one (the possibility that *C. similis* is not cytologically primitive and has lost these heterochromatic arms cannot, however, be excluded). Smith (1966) calls these metacentric chromosomes with one arm heterochromatic and the other euchromatic *diphasic*; at meiosis in the male, they always form a single chiasma between the euchromatic arms, there being never a chiasma between the heterochromatic limbs. *C. kuwanae* has one pair of *monophasic* metacentrics, like *similis*, with both arms euchromatic, and eight pairs of diphasic autosomes.

A series of species of *Chilocorus* (*fraternus, cacti, bipustulatus, orbus*) have two pairs of monophasic autosomes and eight pairs of diphasic ones; they consequently have $2n = 20 + XY$ instead of $18 + XY$ as in *similis* and *kuwanae*. A further development of this group is seen in *tricyclus* with three pairs of monophasic metacentrics and six pairs of diphasics ($2n = 18 + XY$);

it has presumably arisen from the $2n = 20 + XY$ stock by a further centric fusion in the course of which two heterochromatic arms are lost. The final stage is represented by *C. hexacyclus*, in which all the autosomes are mono-phasic metacentrics ($2n = 12 + XY$).

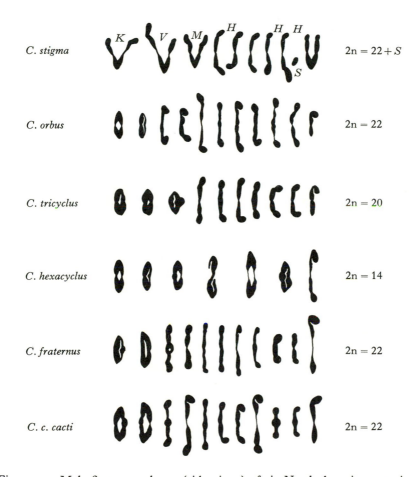

Fig. 11.19. Male first metaphases (side views) of six North American species of *Chilocorus*. In all cases the sex chromosomes (X_1X_2Y in *C. stigma*, XY in the other species) are on the extreme right. The individual of *C. stigma* is heterozygous for the *Kentville*, *Vineland* and *Manitoba* fusions and consequently shows three tri-valents; it is also heterozygous for deletions in the heterochromatic arms of three chromosome pairs (H) and carries a small supernumerary chromosome (S). Ring-bivalents are formed by metacentric chromosomes which have both limbs euchro-matic; bivalents with a chiasma in one limb only have the other limb heterochromatic. From Smith (1962).

Fig. 11.20. The geographic gradient in the frequency of fusion chromosomes in *Chilocorus stigma* (expressed as a percentage of the maximum possible six fusion chromosomes). The percentages are Florida, New York and Connecticut (0.00%), Maine and New Hampshire (2.19%), Nova Scotia and New Brunswick (2.30%), southern Ontario (4.40%), central Ontario (27.87%), western Ontario (44.79%), Manitoba (47.15%), Saskatchewan (58.53%) and Alberta (80.00%). From Smith (1962).

We have already described in Chapter 9, the situation in the chromosomally polymorphic species *C. stigma*, in which there are three 'floating' fusions between diphasics to give monophasics. This species is also unique in the genus in having a Y-autosome fusion, which has converted the sex-chromosome mechanism into an $X_1X_2Y(\male)$ one. In *C. stigma* the trivalents which are formed at meiosis in the fusion-heterozygotes regularly orientate on the spindle with the centromeres in a triangle. Thus no aneuploid gametes are formed and the structural heterozygotes do not have their fecundity reduced.

Hexacyclus (homozygous for six autosomal fusions) and *tricyclus* (homozygous for three fusions) are very closely related allopatric species, the former occurring in Saskatchewan and southern Alberta, the latter being restricted to British Columbia and the state of Washington, i.e. an area entirely west of the Rocky Mountains. A hybrid population occurs in the Crowsnest

River Valley on the east side of Crowsnest Pass in the Rocky Mountains (Smith 1966). This situation is apparently the result of a yearly invasion of a *hexacyclus* population by individuals of *tricyclus* carried through the Pass by westerly winds. The hybrid population contains individuals showing every chromosome number between 2n = 14 and 2n = 20. In the male hybrids, at any rate, synapsis is complete and all spermatocytes show thirteen chiasmata, like the parent species. However, the trivalents in these hybrids, unlike those of the naturally polymorphic *C. stigma*, often show malorientation on the spindle at the first meiotic division (linear rather than triangular configuration of the centromeres). This kind of malorientation occurred with a frequency of 25–50% in the different categories of heterozygotes.

C. hexacyclus predominates in the Crowsnest Valley population. Out of 389 individuals analysed cytologically, 226 (156♂♂, 70♀♀) had the *hexacyclus* chromosome number (2n = 14). But not all of them were pure *hexacyclus*, as 24 of the 156 males carried Y-chromosomes derived from *tricyclus* (the shape of the Y differs in the two parent species). In eight males having the *tricyclus* chromosome number (2n = 20), all in fact had a *tricyclus* Y. A total of 75 individuals had 2n = 17, i.e. might have been F_1 hybrids (some were undoubtedly the result of back-crossing and repeated inter-crossing).

It is fairly clear that the situation in the case of these two species of ladybird beetles is very similar to that found in the morabine grasshoppers – strong selection against fused chromosomes in 'unfused' territory and vice versa. The hybrid zone is considerably wider, because individuals of *Chilocorus* are carried by air currents for considerable distances, which is not known to occur in morabine grasshoppers. Apparently the whole of the reduction in fecundity which occurs in the *Chilocorus* hybrids is attributable to malorientation of the trivalents, whereas in the morabines some asynapsis usually seems to occur in chromosome number heterozygotes. The loss of heterochromatic arms in the formation of 'monophasic' metacentrics in *Chilocorus* is an unusual feature, but only as far as the length of the regions deleted is concerned, since *all* centric fusions involve loss of some proximal regions of the acrocentrics which fuse.

The cytotaxonomic situation in some species of another coccinellid genus, *Exochomus*, in India, is very similar to that encountered in *Chilocorus* (Smith 1965a), although in this case hybrid populations have not been found (they may well occur, however).

Exochomus 'uropygialis' consists of three sibling species, two of which occur sympatrically at Rawalpindi while the third has been found at Dalhousie, Punjab. '*Uropygialis* 1' has six pairs of acrocentric autosomes and one pair of monophasic metacentrics (2n = 14 + XY). '*Uropygialis* 2' has six pairs of diphasic autosomes and a similar pair of monophasics. '*Uropygialis* 3' has eight pairs of acrocentric autosomes. The *Y*'s of the first and the third species are diphasic metacentrics, while that of *Uropygialis* 2 is acrocentric. The latter species carries supernumerary chromosomes in some individuals.

Exochomus '*lituratus*' consists of two sibling species. '*Lituratus* 1' has eight pairs of acrocentric autosomes and a diphasic *Y*. '*Lituratus* 2' shows seven pairs of diphasic autosomes, one pair of acrocentrics and an acrocentric *Y*. The *X* is a diphasic metacentric in both species.

Another coccinellid genus, *Mulsantina*, includes three species in Canada, with 2n = 20, 18 and 12 respectively (Smith 1962c). Unlike the situation in *Chilocorus*, two species of *Mulsantina* are sympatric in many areas, without any hybridization; ethological isolation is presumably complete.

It is apparent that one aspect of chromosomal evolution in these ladybird beetles is increase or decrease in the size of the heterochromatic chromosome limbs (we assume that even the acrocentric chromosomes have very short heterochromatic second limbs).

The karyotypes of the weevils of the genus *Hylobius* are quite diverse (Smith 1956b, 1962). *H. piceus* and *H. pinicola* both have 2n = 40, all the chromosomes being presumably acrocentric. *H. warreni* has 2n = 32, including two very large pairs of metacentric autosomes. It seems likely that either centric fusion plus tandem fusion or (more probably) centric fusion plus pericentric inversion have produced this karyotype.

Five other species of *Hylobius* have 2n = 22. Two of them have all their autosomes acrocentric, two others have nine pairs metacentric and one pair acrocentric, and a fifth has five pairs of acrocentric autosomes and five pairs of metacentrics. Unlike *H. warreni*, which has acrocentric *X*- and *Y*-chromosomes, the 22-chromosome species have a metacentric *X* and a very minute *Y*.

There is no report of entirely heterochromatic arms in *Hylobius*, and it must be presumed that the metacentric chromosomes in this genus are all of the 'monophasic' type. The difference between acrocentric and metacentric autosomes in the 22-chromosome species has almost certainly come about as a result of pericentric inversions, but there is insufficient evidence as to which type of chromosome gave rise to which in the evolution of these species of weevils. We have already dealt in Chapter 9 with the processes of chromosomal evolution in the bark-weevils of the genus *Pissodes*, where the picture is complicated by peculiar systems of chromosomal polymorphisms.

Other studies on the cytotaxonomy of numerous genera and families of beetles are not sufficiently detailed to provide critical evidence on mechanisms of chromosomal evolution and speciation. A few families, of which the Alticidae are an outstanding example, show an extreme range of karyotypes – in this case from $2n = 6 + XY$ in *Homoschema nigriventre* to $48 + XY$ in *Disonycha alternata* (Virkki 1964, 1970; Virkki and Purcell 1965; Manna, in Smith 1960a). There are a great variety of different sex-chromosome mechanisms in the Alticidae (see p. 625) which provide additional evidence of the karyotypic instability in this family. Some other families or genera of beetles show a very uniform karyotypic picture, e.g. twenty-two species of *Carabus* have $2n = 26 + XY$ (Weber 1966, correcting earlier erroneous records).

CHROMOSOMAL REARRANGEMENTS AND SPECIATION IN MAMMALS

In general even very closely related species of mammals seem to show considerable differences in karyotype, and frequently in chromosome number. For the class as a whole, the extremes of chromosome number are $2n♀ = 6$, $2n♂ = 7$ in the deer *Muntiacus muntjak* (Wurster and Benirschke 1970) and $2n = 84$ (presumably in both sexes) in the African Black Rhinoceros (Hungerford, Sharat Chandra and Snyder 1967)[*] (two other species of Rhinoceroses show $2n = 82$, according to Wurster and Benirschke 1968a). In general, pericentric inversions and centric fusions have been invoked by most workers to explain evolutionary changes in mammalian karyotypes; but Todd (1970), whose views will be criticized later (see p. 400) has ascribed extraordinary importance to a process termed by him 'karyotypic fissioning' in the phylogeny of the Carnivora. Obviously, the generally high chromosome numbers in most groups of mammals and the difficulty of examining the karyotypes of large numbers of individuals, especially in the case of rare species, have hampered critical analyses of the role of chromosomal rearrangements in mammalian speciation.

The most primitive true mammals, the didelphid marsupials, show a considerable variety of karyotype, with chromosome numbers ranging from $2n = 14$ to 22 (Reig and Bianchi 1969). Considerable karyotypic variation also exists in the Australian kangaroos and wallabies (Hayman and Martin 1969).

In the insectivores, we have described the situation in the shrews of the genus *Sorex*, where one species shows a considerable level of polymorphism for centric fusions (see p. 302). Other insectivores also show large scale karyotypic differences between related species. Thus six species of tree shrews of the genus *Tupaia* have $2n = 52$ to 68 (Arrighi, Sorenson and

---

[*] Since the above was written, Gardner (1971) has shown that the Peruvian rodent *Anotomys leander* has $2n = 92$.

Shirley 1969); only two of them seem to have indistinguishable karyotypes. Borgaonkar and Gould (1968) have postulated a reciprocal translocation as a basis for speciation in the Madagascan tenrec genus *Microgale*, but their evidence is not certain.

Mammalian chromosomes are usually metacentric, but many karyotypes include acrocentric or subacrocentric chromosomes as well. In a few species all the chromosomes are acrocentric. This seems to be the case as far as the autosomes are concerned in cattle (*Bos taurus*) although in this species the X and Y are metacentric (Melander 1959; Makino 1944). In the house mouse (*Mus musculus*) all the chromosomes including in this case the X and Y are acrocentric or subacrocentric (Levan, Hsu and Stich 1962; Crippa 1964). It is not always easy to say how these aberrant karyotypes with acrocentric chromosomes have arisen; dissociation of metacentrics, pericentric inversions and deletions in the short arms of J-shaped chromosomes represent the three obvious possibilities.

The most extensive investigations on mammalian karyotypes, and the most significant from the standpoint of speciation mechanisms, have been carried out on rodents, especially the cricetine, microtine and murine mice. Matthey (1953-70) has carried out very detailed studies on the karyotypes of the numerous races and species of the rats and mice belonging to the African genera *Acomys*, *Praomys*, *Mastomys* and *Leggada* (the latter is regarded by many as a subgenus of *Mus*, but will be cited here as if it was of generic rank).

There seem to be numerous cryptic or sibling species of these mice, distinguishable on their karyotypes, but not recognizable on ordinary taxonomic criteria. The situation in this respect is similar to what has been found in many groups of insects and other invertebrates; until very recently vertebrate zoologists did not recognize the existence of such morphologically indistinguishable but biologically distinct species, although often willing to name 'subspecies' on very minor biometrical characters.

In the genus *Leggada*, there are a number of species with a 'primitive' karyotype of $2n\male\female = 36$ acrocentrics (e.g. *L. setulosus* from central Africa, *L. tenellus* from Ghana, *L. indutus* from South Africa and *L. bufo* from the Congo). *L. triton* from the Congo ('subspecies 1') shows $2n = 32$; the X has become metacentric as a result of a fusion with an autosome (the homologous autosome has probably fused with the Y, but the latter is quite short, so that deletions have presumably occurred). 'Subspecies 2' of *L. triton*, from Tanzania and Malawi, has a variable chromosome number ($2n = 20$ to 22). The meagre data show that this form is homozygous for five fusions and polymorphic for a sixth one.

The most complex 'superspecies' is *L. minutoides* in which at least six chromosomally distinct 'subspecies' have been recognized by Matthey. A primitive 'subspecies 1' from the Congo shows $2n\male\female = 36$ acrocentrics. Its X- and Y-chromosomes are unfused. In a race of *minutoides* from Rhodesia, the Y has become fused with an autosome; this form is consequently X_1X_2Y in the male (see p. 640) and has $2n\male = 35$. In the remaining subspecies, there has been a further X-autosome fusion (i.e. 'X_1' and 'X_2' have fused). In *L. minutoides musculoides* from central Africa the Y is still metacentric ($2n = 30$ to 34) but in *L. minutoides minutoides* from South Africa, it has once more become acrocentric, presumably by loss of one arm. This latter form has $2n = 18$ to 20, being homozygous for seven fusions between acrocentric autosomes and polymorphic for an eighth one. *L. indutus* with $2n = 36$ acrocentrics (regarded by some taxonomists as a subspecies of *minutoides*) coexists with a form of *L. minutoides* having $2n = 18$ metacentrics near Kirkwood, South Africa, apparently without hybridization.

The main lines of chromosomal evolution in these pygmy mice is clear enough, but the overall evolutionary situation is obscure. What isolating mechanisms exist and what happens in zones of overlap between closely related forms is still unknown.

The karyotype of the house mouse *Mus musculus* is well known to consist of $2n = 40$ acrocentrics. A population which has been named *M. poschiavinus* Fatio 1869, known only from a single locality, Poschiavo in eastern Switzerland, has $2n = 26$ (14 metacentrics, 12 acrocentrics). It is consequently homozygous for seven chromosomal fusions (Gropp, Tettenborn and von Lehman 1970; Gropp, Tettenborn and Léonard 1970). Laboratory hybrids between *M. musculus* and *poschiavinus* have seven trivalents at meiosis, each composed of a metacentric and two acrocentrics. These trivalents show much non-disjunction, so that many aneuploid gametes are formed and the F_1 hybrids have their fecundity markedly reduced. Apparently, *M. musculus* and *M. poschiavinus* coexist in nature without hybridization. The case of *M. poschiavinus* is an extremely interesting one, since it seems probable that this species must have evolved *within* the distribution range of *M. musculus*, rather than allopatrically. Pathak (1970) has reported that the Indian *Mus platythrix* has $2n = 22$, but in this case the number of major chromosome arms (in the female) is only twenty-eight.

Karyotype evolution has undoubtedly been complex in the genus *Rattus*. The species *R. rattus* includes forms having $2n = 42$ and $2n = 38$, the latter being recorded from South America, Italy, Egypt, New Guinea, Australia and New Zealand (Capanna and Civitelli 1969; Bianchi, Paulete-Vandrell and de Vial Rioja 1969; Paulete-Vandrell 1970; Yosida and Tsuchiya 1969; Capanna, Civitelli and Nezer 1970); it is rather clear that fixation of two fusions has occurred in the ancestry of the 38-chromosome race (or races). But Badr and Badr (1970) report that near Cairo, Egypt no less than three forms of *R. rattus* exist – the 42- and 38-chromosome biotypes recorded from other continents and, in addition, a form with approximately 54 chromosomes. This is certainly a very puzzling case, which cannot be fully interpreted on that basis of the published evidence. It will be recalled that, at least in Asia, the longest chromosome in the karyotype of this species is polymorphic for a pericentric inversion, and the number 9 and number 13 chromosomes show a similar polymorphism in some populations (Yosida, Moriguchi, Kang and Shimakura 1967, Yosida and Tsuchiya 1969).

Yong (1968, 1969, 1970) has studied the karyotypes of the numerous wild Malayan species of the genus *Rattus*, which show a range of chromosome numbers from 2n = 36 to 52 and of major chromosome arms (the *nombre fondamental* of Matthey 1945a) from 54 to 83. In Australia, it would appear that an ancestor with a karyotype like that of *Rattus lutreolus*, with 2n = 42 (itself very similar to that of *R. norvegicus* and the Malayan *R. tiomanicus jalorensis*) has given rise to the karyotypes of *R. fuscipes*, *R. greyi* and *R. assimilis* (all with 2n = 38) and *R. conatus* (2n = 32) by centric fusions (Kennedy 1969; J. H. D. Martin 1969; Dartnall 1970). But *R. villosissimus* has 2n = 50, including two pairs of very large metacentrics, so that there have probably been dissociations of some of the smaller metacentrics in its ancestry.

Chromosomal reorganization on a fairly large scale seems to have occurred in the evolutionary differentiation of the rodent genus *Pitymys* (Matthey 1964c). Three European and one North American species have metacentric *X*-chromosomes, but *P. tatricus*, from the Tatra mountains, has an acrocentric *X*, possibly as a result of a pericentric inversion. Two species have 2n = 62 (thirty pairs of acrocentric autosomes), one has 2n = 54 (twenty-five pairs of acrocentric autosomes and one pair of metacentrics) and one has 2n = 32 (eight pairs of acrocentrics and seven pairs of metacentrics).

In the North American cricetine genus *Peromyscus* Hsu and Arrighi (1968) found that nineteen species all had the same chromosome number (2n = 48), but there was much interspecific variation in karyotypes, apparently as a result of fixation of pericentric inversions in phylogeny. Certain 'species' such as *P. truei*, *P. maniculatus*, *P. leucopus* and *P. difficilis* seem to be complexes which include a number of allopatric populations (possibly sibling species) differing in karyotype; some, at least, of these may also show intrapopulation polymorphism for chromosomal rearrangements, particularly pericentric inversions (see also Ohno, Weiler, Poole, Christian and Stenius 1966; Singh and McMillan 1966; Sparkes and Arakaki 1966; Arakaki, Veomett and Sparkes 1970). The overall range in number of major chromosome arms (*nombre fondamental*) is from fifty-six to ninety-six in the genus *Peromyscus*.

In another genus of North American cricetine rodents, *Neotoma*, 2n ranges from 38 to 56 (Baker and Mascarello 1969); six species showed 2n = 52, but their karyotypes were not identical; as in *Peromyscus*, several species are chromosomally polymorphic. In the marmots, Rausch and Rausch (1965) have shown that two 'subspecies' of *Marmota caligata* are, on the evidence of their chromosome numbers, entirely distinct species.

Some other large cricetine and microtine rodent genera show even more variation in chromosome number. Thus the voles of the genus *Microtus* show a range from 2n = 17 in *M. oregoni* (see p. 600) and 2n = 24 in *M. montanus* to 2n = 62 in *M. socialis* (Matthey 1958*b*). Fedyk (1970) considers that in the case of the subarctic species of the *gregalis* group of *Microtus*, the karyotype evolution of the palaearctic species involved centric fusions whereas that of the nearctic species occurred through pericentric inversions. *Tatera* has species with 2n = 42, 44 and 72. In jirds (*Meriones*), also, the chromosome numbers range from 42 to 72; *M. shawi* and *M. lybicus* differ by at least three autosomal rearrangements and a remodelling of the *X*-chromosome that involved either duplication or deletion (Lay and Nadler 1969).

Gerbillus pyramidum includes a number of allopatric forms that differ greatly in chromosome number (Wahrman and Zahavi 1955, 1958; Zahavi and Wahrman 1957). The exact status of these as races or sibling species is uncertain. A form from the Negev has 2n = 66 (6 pairs of metacentrics, 27 pairs of acrocentrics) while one from the coastal plain of Palestine has 2n = 52 (11 or 12 pairs of metacentrics, 14 or 15 pairs of acrocentrics) and one from Algeria has 2n = 40 (19 pairs of metacentrics, 1 pair of acrocentrics). In some, but apparently not all, Algerian populations, the *XY* pair is associated with an autosomal bivalent, in some cells, to form a chain of four chromosomes, presumably as a result of a mutual translocation (Matthey 1952*b*, 1953*a*). Except for a report of some individuals from the Negev with 2n = 62 (Wahrman and Zahavi 1955) no natural populations polymorphic for chromosome number seem to have been encountered. A laboratory hybrid between a female from the Negev and an Algerian male had 2n♂ = 53 and showed three bivalents, 9–11 trivalents composed of our Algerian metacentric paired with two Negev acrocentrics and 3–4 chain configurations of 5–7 chromosomes at meiosis. This suggests that some of the fusions present in the Negev are different from ones in the Algerian race (e.g. if the Algerian form has fusions *AB*, *CD*, *EF*, *GH*, *IJ*..., where each letter represents a chromosome arm, the Negev population may have such fusions as *BC*, *DE*, *FG*). The evidence clearly indicates that fusions rather than dissociations have occurred (since one acrocentric can fuse with any one of a number of other acrocentrics, while a metacentric can only undergo dissociation into the same two acrocentrics). Thus the Negev race, with the highest number of acrocentrics, would be the most primitive of the three, although a hypothetical form with an even higher number of acrocentrics is indicated as a common ancestor.

Similar conditions exist in the spiny mice of the genus *Acomys*. Matthey (1965*a*, 1968*a*) has studied ten species with 2n = 36, 38, 38, 38, 38–40, 50, 60, 60, 64, 66. The corresponding numbers of major chromosome arms are 68, 68, 70, 68, 68, 66, 70, 76, 66, 66. Here it is obvious that there have been numerous fusions or dissociations, the species with the higher chromosome numbers having mainly or exclusively acrocentric chromosomes, while those with lower chromosome number have many metacentrics. Both chromosomal studies and hybridization experiments have shown that there are more biological species in this genus than had been recognized by morphological taxonomists using traditional procedures. In one case in this genus Wahrman has found a zone of natural hybridization 20–30 km wide on the west side of the Gulf of Aqaba, in which heterozygotes for a single centric fusion or dissociation occurred. As in the genus *Gerbillus*, different centric fusions seem to have occurred in several allopatric taxa of *Acomys*. Thus laboratory hybridization between two forms, one from Palestine, the other from Cyprus, but both with 2n = 38, gave rise to F_1 males with a chain of twenty-eight chromosomes, a chain of four, two autosomal bivalents and an XY bivalent at meiosis (Wahrman, unpub.). One would think intuitively that such sequences of fusions as those responsible for a chain of twenty-eight at meiosis could not have come into existence by chance and that some special explanation is required; but a statistical consideration shows that this is probably not so, provided we postulate a common ancestor with all its chromosomes acrocentric and a strong tendency for fusions to establish themselves in both lineages. In two South African species of *Acomys* the *X*-chromosomes are of giant size, which sets them apart from the rest of the genus, in which *X*-chromosomes are of 'normal' length.

All these rodents we have been discussing are probably forms of low vagility which form local colonies and do not move far in the course of an entire life-cycle. Certain burrowing, subterranean rodents probably have even less mobility. The cytotaxonomy of the fossorial genus *Spalax* in Palestine has been studied by Wahrman, Goitein and Nevo (1969*a*, *b*). There are four taxa of '*S. ehrenbergi*' with chromosome numbers of 2n = 52, 54, 58 and 60, whose geographic ranges seem to be essentially contiguous (parapatric). Natural hybrids between the last two forms have been found at two localities and two hybrids between the 52- and 54-chromosome

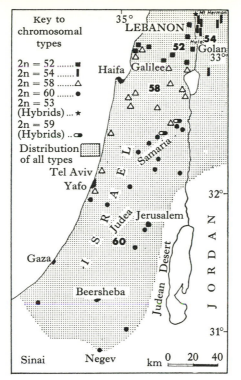

Fig. 11.21. Distribution of the chromosomal races of the mole-rat *Spalax ehrenbergi* in Israel. From Wahrman, Goitein and Nevo (1969), *Science*, **164**, 83. Copyright 1969 by the American Association for the Advancement of Science.

forms were also collected in nature. In the European *Spalax leucodon*, Soldatović, Živkovic, Savić and Milosevié (1967) found four different karyotypes, ranging from $2n = 26$ to $2n = 48$. Two biotypes of this super-species from Romania showed $2n = 50$ and $2n = 56$, while *S. microphthalmus* showed $2n = 62$ (Raicu, Bratosin and Hamar 1968).

The burrowing rodents of different continents seem to show a similar cytogenetic picture. Thus the karyotypes of eleven species of 'tuco-tucos' belonging to the South American genus *Ctenomys* seem to be very diverse (Reig and Kiblisky 1969); the chromosome numbers range from $2n = 22$ to $2n = 68$ and the number of major chromosome arms from 44 to 122. Obviously there have been numerous chromosomal fusions (possibly also dissociations) and pericentric inversions in the phylogeny of the genus. Chromosome polymorphism seem to occur in three of the species. Reig and Kiblisky have calculated that a minimum of thirty chromosomal rearrangements must have occurred during the evolutionary differentiation of the closely related species *C. talarum* and *C. porteousi*, which both have 2n = 48, but with very different karyotypes.

A similar picture seems to exist in the African genus *Praomys*, studied by Matthey (1965*d*), where three species have 2n = 28, 34 and 42, the number of major chromosome arms being, respectively, 30, 34 or 36 and 44 or 46. Clearly, the relationship of these karyotypes is not interpretable in terms of single fusions or dissociations; most probably pericentric inversions have also occurred. But it is also possible that losses or gains of heterochromatic chromosome arms have been involved in these considerable evolutionary changes.

In the cases considered earlier of natural hybridization in *Spalax* and *Acomys* the parental forms seem to be less than full species. But an instance of natural hybridization between mammalian taxa which have undoubtedly differentiated to full specific level and whose karyotypes differ somewhat more than those of the forms of *Spalax ehrenbergi* has been described by Patton and Soule (1967) in the case of the pocket mice *Perognathus penicillatus pricei* (2n = 46) and *P. pernix rostratus* (2n = 52) in one area of sympatry in Mexico. This is almost certainly a zone of 'secondary' hybridization, in contrast to the 'primary' hybrid zones in *Acomys* and *Spalax*. Three different karyotypes are known in six subspecies of *Perognathus penicillatus*. One of these, in California and Arizona, is regarded as primitive by Patton (1969*b*); southward a different karyotype has evolved by fixation of a single pericentric inversion, while to the southeast in Mexico three, different pericentric inversions have established themselves.

Six different chromosomal races of the Pocket mouse species *Perognathus goldmani* have been found in western Mexico by Patton (1969*a*). These differ in respect of four different chromosomal fusions and a couple of pericentric inversions, one of which is in the *X*. These taxa are parapatric, like those of *Spalax ehrenbergi* and the forms of the *viatica* group of morabine grasshoppers. But a peculiar feature of the *P. goldmani* case is that where two races are contiguous they are separated by rivers or streams, and the riparian vegetation fringing these is inhabited by a different species of *Perognathus* – *P. artus*. Patton rejects White's (1968*a*) stasipatric model as not applicable to *P. goldmani* and adopts an allopatric interpretation based on range expansion of populations homozygous for new chromosomal rearrangements into previously unoccupied territory. Only three natural hybrids between chromosomal races α and δ of *P. goldmani* were found by Patton and none between any of the other races, the low frequency of hybridization in this case being presumably due in part to the physiographic and ecological barriers (rivers and riparian vegetation) between the races. Chromosome numbers range from 2n = 50 to 56 in *P. goldmani. Perognathus*

S. hispidus

S. arizonae

XY

XY

Fig. 11.22. Karyotypes of the Cotton Rats, *Sigmodon hispidus* (2n = 52) and *S. arizonae* (2n = 22). Note the differences in the sex chromosomes. From Zimmerman and Lee (1968), redrawn.

baileyi, with 2n = 46 to 56, seems to be highly polymorphic for chromosome number in most populations, but the details have not been worked out yet.

The most extraordinary case of extreme chromosomal differentiation in rodents is perhaps that of two species of Cotton Rats *Sigmodon hispidus* and *S. arizonae*, which have 2n = 52 and 2n = 22 respectively (Zimmerman and Lee, 1968; Zimmerman 1970). Kiblisky (1969*b*) has shown that the 52-chromosome form extends as far south as Venezuela and an isolated population exists at Yuma, Arizona. *Sigmodon fulviventer*, with 2n = 28 to 30, is polymorphic for chromosome number (Lee and Zimmerman 1969).

Although the evidence is certainly insufficient for any final conclusion, it undoubtedly suggests that the speciation mechanisms in some of these small mammals may have been more similar to those of the morabine grasshoppers than to those of *Drosophila* species, with much greater population size and potential vagility. But rodents are not the only order of mammals in which sibling species exist which differ greatly in karyotype. Thus striking karyotypic differences exist between the living species of horses and zebras (see p. 448) i.e. mammals whose vagility is much greater than that of small rodents and insectivores.

Chu and Swomley (1961) have reported two instances where 'subspecies' of Madagascan lemurs have quite different chromosome numbers (2n = 58 and 54 in two 'subspecies' of *Hapalemur griseus* and 2n = 60 and 48 in 'subspecies' of *Lemur fulvus*. It is probable that in both cases we are dealing with pairs of true biological species which are affectively isolated reproductively. The variation in chromosome number within the subspecies *Lemur fulvus collaris* is due to fusions or dissociations (Rumpler and Albignac 1969). On the other hand, *Lemur catta* seems to show no geographic variation in karyotype (Rumpler 1970).

In bats, on the other hand, the careful chromosomal studies of Baker (1967), Baker and Patton (1967) and Kiblisky (1969*a*) as well as the extensive studies of Capanna, Civitelli and Conti (summarized in Capanna and Civitelli 1964, 1970) have revealed hardly any instances of races or sibling species differing in karyotype; although there is the case of *Macrotus water-housii* (Nelson-Rees, Kniazeff, Baker and Patton 1968), where two geographic populations have 2n = 46 and 2n = 40, as a result of three fusions present in the latter but not in the former. In this order species that are sufficiently closely related to be placed in the same genus by systematists generally seem to have similar or indistinguishable karyotypes, but there are rather radical differences between karyotypes at the generic level. It is certainly tempting to speculate that these apparent differences from the cytogenetic picture in other mammalian orders are due to the flying habits of bats (speaking of another group of flying vertebrates, Piccini and Stella (1970) state that 'our data confirm the great stability of the karyotypes within the various orders of birds'). Taylor, Hungerford, Snyder and Ulmer (1968) comment on the karyotypic uniformity of the Camelidae.

The Carnivora seem to be a 'normal' mammalian order as far as cyto-taxonomic differentiation is concerned. The cats (genus *Felis*) of the northern hemisphere all seem to show 2n = 38, but there are a number of different karyotypes in the species that have been investigated, depending on the number of acrocentric and metacentric chromosomes present (Hsu, Rearden and Luquette 1963; Wurster 1969). Several South American species have 2n = 36, due to a fusion. The situation in this genus does not seem to be greatly different from that found in the rodent genus *Peromyscus*. All true dogs (genus *Canis*) seem to have the very high chromosome number 2n = 78 and an apparently uniform karyotype; the closely related genera *Dusicyon*, *Atelocynus*, *Chrysocyon* and *Lycaon* have the very similar chromosome numbers 74, 74 or 76, 76 and 78 respectively (Gustavsson and Sundt 1966; Wurster and Benirschke 1968a, 1969).

Todd (1970) has argued that the high chromosome numbers of some Canidae and Ursidae are derivative rather than primitive and that ancestral carnivores had chromosome numbers in the vicinity of 2n = 38. Thus far his conclusions are reasonable. But to explain them he invokes an entirely hypothetical evolutionary process: ' . . . simultaneous massive increase in diploid number through centric mis-division, here called karyotypic fissioning. This event occurs in a single individual and subsequently spreads through a population depending on chance, meiotic compatibility, dynamics of the population and natural selection. The possibility that fusions also occur sporadically is not precluded, although such events are not seen as abundant.' He then goes on to relate 'presumed episodes of karyotypic fissioning with known periods of explosive speciation and adaptive radiation.' Todd's extreme and unbalanced hypothesis is at variance with almost every careful cytotaxonomic investigation that has ever been carried out and, we believe, with every single one discussed in this chapter. It is still extremely doubtful, in spite of the case described by Southern (1969), whether simple breakage through the centromere of a metacentric can produce two fully functional and stable telocentric chromosomes, capable of persisting indefinitely. To suppose that all the chromosomes of a karyotype would undergo this process simultaneously is equivalent to a belief in miracles, which has no place in science.

GENERAL CONCLUSIONS

The instances we have discussed in this chapter have been mainly ones where an ancestral species has split up into a number of relatively 'young' sibling species whose geographic ranges are essentially contiguous or 'parapatric'. In such cases, it may be difficult to determine the exact degree of evolutionary divergence, and designation of a particular form as a race, semi-species or full species may be somewhat arbitrary, in the absence of complete information concerning the strength of the various genetic isolating mechanisms involved. Nevertheless, as we have seen, karyotypic divergence can be quite extensive in many cases where separation of the taxa by classical morphological methods is barely or not at all possible.

The exact nature of the cytotaxonomic differences between sibling forms varies greatly in the cases we have discussed in this chapter, and it seems unlikely that the mechanisms of speciation have been the same in all of them. In most of the *Drosophila* species groups, and perhaps especially in the *paulistorum* complex, there seems little reason to believe that inversion polymorphisms have played a primary role in generating the isolating mechanisms. In the *viatica* group of morabine grasshoppers, in *Chilocorus* and in *Leggada*, chromosomal fusions and other types of rearrangements may well have played a major part, although in some of these instances the original chromosomal isolating mechanism has almost certainly been reinforced or overlain by additional isolating mechanisms not due to chromosomal rearrangement. The available information is insufficient for us to be able to hazard guesses as to the extent to which speciation in *Chilocorus* or *Leggada* has followed the allopatric or the stasipatric model. The vagility of Diptera such as the chironomid and simuliid midges is such that classical allopatric speciation seems more probable. Possibly burrowing forms like mole crickets and fossorial mammals, (e.g. *Spalax*, *Ctenomys*) are likely to show 'stasipatric' speciation mechanisms. There seems to be a strong suggestion that in a few groups (e.g. the Simuliidae and the *Anopheles gambiae* complex) speciation follows on situations where there are two alternative sex-determining mechanisms, or two different types of X-chromosomes coexisting in an ancestral population.

Whatever the precise genetic and cytogenetic mechanisms involved in speciation in various groups of animals, it is obvious that we are dealing in all cases with a gradual process. If races (or some of them) are incipient species, then raciation is the early stage of speciation. Some evolutionists have focussed their attention especially on the genetic or cytogenetic changes which, by creating a new isolating mechanism, are responsible for the initial divergence between two taxa. Others have felt that it is the final rather than the early stages of speciation which are the most significant. Cytogenetic evidence seems to suggest that in most cases where chromosomal rearrangements have played a direct causative role in speciation, this has been exerted from an early stage in evolutionary divergence onward. The final stage of speciation is more likely to involve the accumulation of polygenic changes, quantitatively increasing the strength of genetic isolation until it becomes complete. An evolutionist who thinks of speciation solely or mainly in terms of the last stage of the whole process (and this is the view of many working taxonomists) is more likely to doubt or deny the significance of chromosomal rearrangements in speciation.

It is unfortunate that (with the exception of the Hawaiian Drosophilidae) no chromosomal studies appear to have been carried out on groups which have undergone explosive speciation on a large scale in very limited geographic areas (e.g. the gammarids of Lake Baikal and the forty-three species of the beetle genus *Microcryptorhynchus* on the 4×5 mile Pacific island of Rapa, reported by Zimmerman 1938). It is probably in such cases, if anywhere, that one might eventually obtain critical (as opposed to purely circumstantial) evidence of sympatric speciation.

The view which we have tacitly adopted in this chapter is that all visible differences between the karyotypes of related species are due to chromosomal rearrangements, including not merely such major changes as inversions and translocations, but also duplications and deletions of short regions, especially heterochromatic ones. According to this viewpoint, differences in the DNA values of related species, apart from the special case polyploidy, would be entirely the result of duplication or loss of single loci or longer segments of chromosomes.

This conclusion necessarily follows from the 'mononeme' hypothesis of chromosome structure, discussed in Chapter 1. If, on the other hand, it could be established that ordinary mitotic and meiotic chromatids were multi-stranded, other possibilities would exist. Accepting the multi-stranded model, the Schraders, in a series of papers (Schrader and Hughes-Schrader 1956; Hughes-Schrader and Schrader 1956; Hughes-Schrader

1957, 1958) sought to introduce the principle of 'differential polyteny' as a major factor in evolutionary cytogenetics. The concept is simply that of changes in the number of strands per chromatid occurring from time to time in the evolution of particular groups. The exact adaptive significance of such quantum steps was left vague, but certainly one cannot deny *a priori* that they might be adaptive. It must be pointed out, however, that even if chromatids are multi-stranded, it does not necessarily follow that evolutionary changes in the degree of strandedness have occurred – the number of strands could be the same throughout major taxonomic groups, or even in all higher organisms.

The Schraders measured the relative amount of DNA in primary spermatocytes of twenty species of bugs belonging to the tribe Pentatomini. Six species of *Thyanta* (four with n = 7, one with n = 8 and one with n = 14) showed a range of DNA values (in arbitrary units) from 0.89 to 1.10; thirteen other species belonging to twelve other genera showed values of 1.43, 1.69, 1.80, 1.82, 2.19, 2.24 (two species), 2.27, 2.32, 2.39, 2.52, 2.58 and 3.13; four additional species in the genus *Acrosternum* showed values from 3.6 to 4.0. These differences rather naturally correspond to visible differences in the size of the mitotic chromosomes, the species with the highest DNA values having the longest and thickest chromosomes in a spermatogonial metaphase. Sixteen of these species have n = 7, one has n = 6.

It will be obvious that these data, although certainly compatible with the hypothesis of differential polyteny, do not provide any really critical evidence in favour of it. The relatively few changes in chromosome number in this group can surely be attributed to fusions and dissociations, as in other groups. And the differences in DNA values may plausibly be attributed to duplications and deletions of chromosome segments (perhaps in the main heterochromatic ones) and possibly also single locus duplications of the type demonstrated by Keyl in *Chironomus thummi*. The occurrence of duplications and deletions of heterochromatic segments in the phylogeny of *Drosophila* is well documented and universally admitted; it is also obvious in the chromosomal evolution of *Chilocorus* (see p. 387). And in grasshoppers it is clear that there has been an elaborate evolution of the amount of heterochromatin, involving numerous duplications and deletions; many grasshopper species in addition are known to show polymorphism for duplications or deficiencies of large heterochromatic segments (see pp. 307–10).

In spite of this situation, the Schraders put forward the hypothesis that the germ-line chromosomes of pentatomid bugs are multi-stranded and that the number of strands in the case of *Thyanta* species is half that in the species with DNA values ranging from 1.69 to 2.58; they suggested that the species with a DNA value of 3.13 and some related forms may have four times as many strands in their chromosomes as the species of *Thyanta*. They also thought that the single species of *Thyanta* with n = 14, which has 0.93 units of DNA and shows an $X_1 X_2 Y$ mechanism in the male, might have undergone a permanent separation of its chromatids (i.e. polyploidy combined with a halving of the number of strands per chromatid). Possibly their ideas on the subject were influenced by the finding of a genuine split between 'sub-chromatids' in the coccid *Llaveiella* (see p. 504) – a split which is not visible in related forms.

Apparently the Schraders were willing to concede that differences of DNA value within the range 1.69 to 2.58 (which obviously cannot be due to doubling of the number of strands) might be due to duplications or deletions of chromosome segments, but were unwilling to extend this interpretation to forms having slightly higher or lower values. If a species with a DNA value of, say, 1.35 were to be discovered, their case that species with 1.69 to 2.58 units of DNA possess twice as many strands in their chromosomes as *Thyanta* species with 0.89 to 1.10 units would obviously be weakened.

Hughes-Schrader (1958) also determined the DNA values of four species of the mantid genus *Liturgousa*. Extensive chromosomal rearrangements must have occurred in the evolutionary differentiation of these species, since their chromosome numbers are widely different and cannot be explained by simple fusions or dissociations. The data are as follows: *L. maya* ($2n\male = 17$) DNA value = 1.00; *L.* sp. *n.* ($2n\male = 21$) DNA value = 1.50; *L. actuosa* ($2n\male = 23$) DNA value = 1.43; *L. cursor* ($2n\male = 33$) DNA value = 0.94. Obviously, 'differential polyteny' cannot be put forward as an explanation of this case, since the highest DNA value is less than twice the lowest one. A correlation exists, of course, between the subjectively estimated size of the chromosomes at metaphase and the accurately determined DNA value. Thus, among the mantids, the chromosomes of *Pseudomiopteryx infuscata* are quite unusually small for the group. It is therefore no surprise to find that their DNA value is 0.74, compared with one of 1.59 for *Stagmomantis heterogamia* (Hughes-Schrader 1958). In another mantid genus, *Ameles*, Wahrman and O'Brien (1956) determined the DNA values of individuals of the same species differing in chromosome number as a result of fusions or dissociations. No significant differences in DNA value were found, indicating that the minute gains and losses of material around the centromeres which must have occurred in such cases are too small to be demonstrated by this method.

Christensen (1966) has determined the DNA values of forty-five different biotypes (in many cases known polyploid species or races) of enchytraeid worms. In diploid forms the overall range in DNA values is from 0.40 to

1.12 arbitrary units. In the genera *Lumbricillus* and *Enchytraeus*, with a range of diploid numbers (2n = 26 to 34 in the first case and 32 to 42 in the second) there is also a range of DNA values, while some genera and species groups which are conservative in chromosome number also show little variation in DNA values. Two cases, one in the genus *Enchytraeus* and one in *Buchholzia*, are tentatively interpreted by Christensen in terms of differential polyteny, but the doubling of DNA values in each case is only approximate, so that the evidence is not really critical.

We must draw the general conclusion from this discussion that the evidence for evolutionary changes in degree of strandedness of germ-line chromosomes is extremely weak and will not withstand critical examination. On the other hand, evolutionary changes in the degree of polyteny or endo-polyploidy attained by specific categories of somatic cells in the course of ontogeny have clearly occurred in insects (and probably in many other groups), so that evolutionary changes in polynemy are certainly not cyto-logically impossible, although probably inapplicable to the germ-line, for the reasons given in Chapter 1.

12

EVOLUTION OF CHROMOSOME NUMBERS
AND CHROMOSOME FORM

> Chromosome numbers are a readily quantified aspect of a fundamentally important adaptive mechanism of almost all organisms. It is remarkable how little we understand of their general significance.
>
> GEORGE C. WILLIAMS: *Adaptation and Natural Selection*

In the early period of cytogenetics it appeared to some investigators unlikely that chromosome numbers and form as such were of any particular adaptive or evolutionary significance. Thus Morgan, Bridges and Sturtevant (1925) wrote that 'To a geneticist many of these comparisons (i.e. between the karyotypes of different species) will seem of little significance, because to him it is not the shapes and sizes of chromosomes which are important, but the genes contained in them' and Wilson (1925) considered that 'Both cytological and genetic evidence prove that the chromosomes are compound bodies containing many different components. So long as the sum total of these remains the same, or nearly so, it seems immaterial whether they be grouped to form few or many aggregates'.

Such views were expressed in the heyday of atomistic genetics when genes were regarded as entirely independent entities imagined as 'beads on a string'. They are out of date for a number of reasons. We are today very much interested in genetic systems in all their manifestations, and not merely in the properties of single genetic loci, and since the karyotype of a species is the physical basis of its genetic system we are necessarily interested in the evolution of karyotypes. Species with high chromosome number will show considerably more genetic recombination than ones with low numbers of chromosomes, and this may be connected with their systems of population genetics and reflected in their evolutionary patterns. Since evolutionary fusions of chromosomes involve the loss of some genetic material while dissociations involve duplication of centromeric and telomeric regions it is obvious that such rearrangements may have more or less significant genetic consequences and that the sum total of hereditary material does not remain precisely the same when an evolutionary change in chromosome number occurs.

The karyotypes of animal and plant species must be adapted to the cellular dimensions of the organisms in which they occur. The mechanisms of mitosis and meiosis are highly precise ones in which the length of the metaphase spindle, the diameter of the equatorial plate and the degree of spindle elongation at anaphase are all functionally related to the number, shape and size of the chromosomes. And in histologically complex organisms these functional interrelations must exist for a great variety of different types of cells, large and small. Thus, for example, chromosomes whose limbs were too long would not be pulled completely apart at anaphase and might be trapped or caught by the cleavage-furrow when cytokinesis occurred; while if the chromosomes were too numerous to be accommodated on the metaphase plate various abnormalities of the mitotic mechanism might result.

Documented lists of animal chromosome numbers have been published by several authors. That of Makino (1951, 1956) covers the whole animal kingdom, but is seriously incomplete, even for work published before 1950. Later, more satisfactory lists for particular groups are those of Smith (1953b, 1960a), for Coleoptera, Matthey (1958b) for eutherian mammals, Hayman and Martin (1969) for marsupials, Bender and Chu (1963) for primates, Borgaonkar (1965, 1966a, b, 1967) for insectivores and primates and Niiyama (1959) for several groups of Crustacea. For the insect order Heteroptera, the list of Takenouchi and Muramoto (1969) should be consulted. Nogusa (1960) has reviewed the chromosome cytology of fishes, Hughes-Schrader (1950) and White (1965b) that of mantids, Maeki and Remington (1960–1) and de Lesse (1960) that of butterflies. Inaba (1953, 1959a, b), Nishikawa (1962) and Burch (1960a) have tabulated much information on the chromosome numbers of various groups of molluscs.

Makino's list included data on 3317 species of animals, i.e. less than 0.4% of the total number of described species, believed to be now over a million. Possibly an up-to-date and complete list might include about 6000 species. This would include at least 846 species of beetles (Smith 1960a; Virkki 1964) and at least 600 species of butterflies (Maeki and Remington 1960a, b, 1961a, b; de Lesse and Condamin 1962, 1965; de Lesse 1966b, 1967). No doubt some of the determinations of chromosome numbers, particularly the earlier ones, are erroneous, but the majority are probably accurate enough to base general conclusions on. The lists, however, only cite chromosome numbers; other cytological details such as centromere positions, the number of major chromosome arms in the karyotype (the *nombre fonda-mental* of Matthey 1945*a*, 1949*d*), the extent and location of heterochromatic segments of various kinds, nucleolar organizers and other special structures, being recorded only for a much smaller number of species. In most groups of animals, however, such detailed information is not likely to be available for a long while, so that we are forced to base our tentative conclusions about the chromosomal evolution on what meager data exist.

The haploid numbers of most species of animals lie between six and twenty, numbers above and below these limits being rare except in a few groups. The only species known to have a single pair of chromosomes is the nematode *Parascaris equorum* var. *univalens*, but this instance is complicated by the fact that the number of chromosomes in the somatic cells is very much higher, because the chromosomes undergo fragmentation during certain of the cleavage divisions (see pp. 54–5). Species with n = 2 include *P. equorum* var. *bivalens*, a number of rhabdocoel Turbellaria, a few chiro-nomid midges (Bauer 1945) and some simuliids (Dunbar 1958, 1959) and the coccids of the tribe Iceryini; while species with a haploid number of three occur in quite a number of groups (Turbellaria, scorpions, some mites, a few Heteroptera, some species of *Drosophila*, mosquitoes, simuliids and a few Chironomidae). The highest number known in any metazoan is 217–223 in the lycaenid butterfly *Lysandra atlantica* (de Lesse 1970); other very high numbers being those of the crayfish *Astacus trowbridgei* with n = 188 (Niiyama 1962), the hermit crab *Eupagurus ochotensis* with n = 127 (Niiyama, 1951) and the geometrid moth *Phigalia pedaria* with n = 112 (Regnart 1933). It is not quite clear why the range of chromosome numbers should be so restricted in animals, particularly since some plant species have much larger numbers (up to n = *ca.* 510 in the fern *Ophioglossum petiolatum*, which is certainly a high polyploid, according to Manton and Sledge, 1954). Possibly in most groups of animals the mechanism of mitosis is only adapted to accomplish the separation of a moderate number of bodies, the spindle being unable to deal efficiently with too few or too many chromo-somes (in which case the more simple histological differentiation of plants might be less restrictive and permit higher chromosome numbers).

The extent of variation in chromosome number which exists in various groups of animals can be appreciated from an inspection of the histograms in Figs. 12.1–12.4. It will be seen that in some groups nearly all the species have the same chromosome number, while in others there is a great range of variation. Some authors have assumed that the modal chromosome number (sometimes called the 'type number') is necessarily ancestral for the group in question, all the other numbers having been derived from it, by fusions and dissociations. While this may be true for some or even for many groups, it is unlikely to be so for all. In most cases it is an unproven, and perhaps unprovable, hypothesis. In many cases the modal number of a group is simply the chromosome number characteristic of a subgroup that is numerically dominant at the present time. For example, the histogram for the praying mantids ranges from $n = 8$ to $n = 20$, with a very clear mode at 14. But this is simply the number characteristic of the large sub-family Mantinae – it almost certainly has no ancestral significance for the order Mantodea as a whole.

Generally speaking, it becomes more and more difficult to distinguish clearly defined type numbers the higher one goes in the systematic hierarchy;

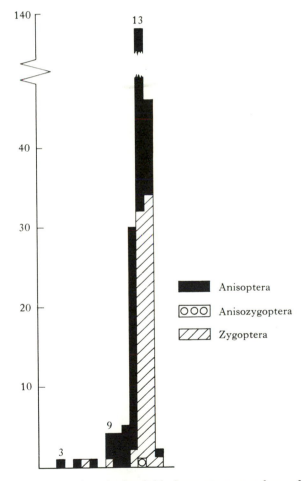

Fig. 12.1. Histogram showing the haploid chromosome numbers of 233 species of dragonflies (Odonata). Data from Kiauta (1967).

on the other hand, in the lower categories (families, subfamilies, tribes, genera) the species which have been studied by cytologists usually do not represent a random sample, particular sections of the group having been investigated more extensively than others.

In many groups below the rank of class or subclass, however, the concept of a type number is probably a useful one. To speak of a type number for the Insecta, the Vertebrata or even the Mammalia, would clearly be absurd. But to regard thirteen as typical for the dragonflies (Odonata) seems legitimate, since it is the commonest number in the suborder Anisoptera and the second commonest in the Zygoptera and most of the other numbers do not deviate far from it (Fig. 12.1); it also seems to occur in the primitive suborder Anisozygoptera. However, Kiauta (1967) has argued rather unconvincingly for $n = 9$ being the most primitive number in the Odonata, from the phylogenetic standpoint. Some other groups which show fairly well defined type numbers are the Coleoptera ($n = 10$ occurs in many different families) and the long-horned grasshoppers (Tettigoniidae), in which $n = 16$ is the commonest number in several large subfamilies, although higher or lower numbers are characteristic of other subfamilies.

Many groups appear to have more than one type number. Thus the histogram for the Heteroptera, whose complete range is from $n = 2$ (in *Lethocerus* sp., according to Chickering 1932; Chickering and Bacorn 1933) to $n = 25$ in the mirid *Dicyphus stachydis* (Leston 1957), shows peaks at 7, 12 and 17. Each of these peaks represents the modal number of one of the dominant families of which the order is composed. Thus the peak at seven is due to the Pentatomidae and Lygaeidae, that at twelve to the Corixidae, while that at seventeen is mainly due to the Miridae (Manna 1951b, 1958, 1962; Leston 1957, 1958). But the shape of a histogram such as this clearly depends a great deal on the extent to which the various sub-groups have been studied by cytologists.

We may consider the decapod Crustacea, the Lepidoptera and the birds (Fig. 12.3) as outstanding examples of groups in which chromosome numbers are unusually high. Yet the pattern of chromosomal evolution seems to have been rather different in these three groups, as can be seen even from an inspection of the three histograms and consideration of the means, variances and skewness of the frequency distributions. A more detailed study of the data, considering individual genera or families of the three groups, confirms the preliminary impressions and conclusions.

In the decapod Crustacea (Fasten 1918; Niiyama 1959, 1962, 1966) there is a great range of chromosome number (from $n = 41$ to $n = 188$ in 37 species) and a very high variance. If more species were studied a distinct modal number or numbers might become evident, but this is not certain.

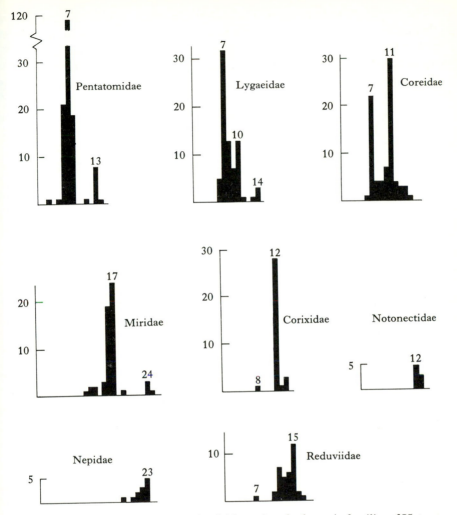

Fig. 12.2. Histogram showing the haploid numbers in the main families of Heteroptera. These are calculated for the female sex, i.e. a species with $2n\male = 12 + X_1X_2Y$ is considered to have n = 8. Data mostly from Manna (1951, 1956, 1962) and Leston (1957).

The crabs (tribe Brachyura) have lower numbers than the crayfish (Astacura) and the hermit crabs (Paguridea). Clearly the decapods have efficient evolutionary mechanisms increasing chromosome numbers, and probably also mechanisms leading to decreases (although this is less obvious from the data, since there are no species which have clearly undergone an evolutionary decrease in chromosome number).

The Lepidoptera have a greater range in chromosome number than any other group of animals (from n = 7 to n = *ca.* 220). But they show a strongly marked mode at 29–31 so that the variance is much lower than in the decapod Crustacea. A subsidiary mode at 23–24 is due to the family Lycaenidae ('Blue' butterflies), the only family of Lepidoptera with a clearly marked mode of its own, different from that of the group as a whole. The distribution

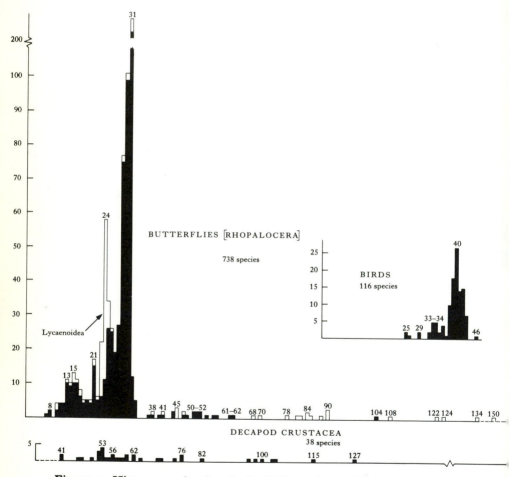

Fig. 12.3. Histograms showing the haploid numbers of three groups of animals with high chromosome numbers. In the case of the butterflies, the Lycaenoidea (families Lycaenidae and Riodinidae) are shown in white, the remaining families in black. Where two races have different chromosome numbers they have been shown separately, but species whose populations are clearly polymorphic for chromosome number have been excluded. Data from Maeki and Remington (1960*a*, *b*, 1961*a*, *b*) and later publications by Maeki and by de Lesse.

of chromosome numbers in the Lepidoptera is clearly skewed and asymmetrical about the mode, there being far more species with numbers below twenty-nine than there are ones with numbers above thirty-one. In other words fusions seem to have been considerably more frequent (or more successful in an evolutionary sense) than dissociations, in this group of insects. Individual genera of Lepidoptera include some in which all or almost all the species have the same chromosome number. Thus about fifteen species of *Papilio* have $n = 30$, only a few having other numbers such as $n = 27$ or 31 (Maeki and Remington 1960a), almost all members of the tribe Nymphalini have $n = 31$ and all members of the Limenitini have $n = 30$ (Maeki and Remington 1961b).

On the other hand we have many genera and tribes of Lepidoptera in which extreme variations in chromosome number occur. Thus the three species of the remarkable Giant Skippers (Megathymidae) that have been investigated have haploid numbers of 21, 27 and 50 (Maeki and Remington 1960b). In the genus *Erebia*, the primitive chromosome number was probably about 28 or 29, as in other genera of the family Satyridae (Maeki and Remington 1961a). But some species have very much lower numbers such as $n = 7$ in *E. aethiopellus* and $n = 8$ in *E. calcarius*; and some have considerably higher numbers such as $n = 40$ in *E. ottomana* and $n = 51$ in *E. iranica* and *E. dromulus* (Federley 1938; Lorković 1941; de Lesse 1955, 1959b). In this genus, apparently, both fusions and dissociations can establish themselves in evolution. In the lycaenid genus *Lysandra*, however, (Lorković 1941; de Lesse 1953, 1954, 1959a, 1969, 1970) dissociations seem to have been spectacularly successful, but there is no evidence that fusions have occurred at all. This genus includes a few cytologically primitive species with $n = 24$ such as *L. syriaca* (the usual number in the family Lycaenidae) and a series of species with the haploid numbers 45, 82, 84, 88, 90, 124, 131–150, 190–191 (in *L. nivescens* from Spain) and 217–223 in *L. atlantica* from Morocco (the animal species with the highest chromosome number known). The very closely related lycaenid genus *Agrodiaetus*, however, includes species such as *A. posthumus* ($n = 10$) and *A. araratensis* ($N = 13$) in which fusions have established themselves and others like *A. phyllis* ($n = 79$–82) in which many dissociations have occurred (de Lesse 1957, 1959c).

It is not only in these Old World Lycaenidae that spectacular evolutionary changes in chromosome number have taken place. Such New World members of the Lycaenidae as *Hemiargus hanno* ($n = 14$), *Calephelis virginiensis* ($n = 45$) and *Lycaena heteronea* ($n = 68$) show wide deviations from the modal number, and so do such Japanese species as *Taraka hamada* ($n = 15$) and *Ussuriana stygiana* ($n = 47$) (de Lesse 1967; Maeki and Remington 1961a). In the family Pieridae, the Euchloini have a modal number of 31 and the Pierini one of 25–26. However, a few divergent species occur in this family also. Thus *Pieris brassicae* has n = 15 and an African *Leptosia* n = 12 (Maeki and Remington 1960*b*). In the Nymphalidae, most species of *Vanessa*, including Japanese individuals of *V. indica*, have n = 31 (Maeki 1961), but Indian material of this species shows n = 15 according to

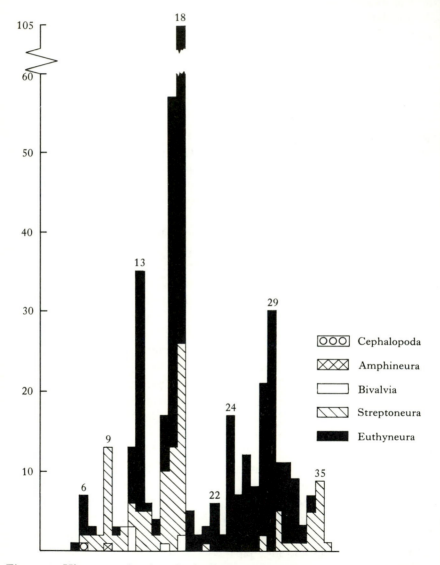

Fig. 12.4. Histogram showing the haploid numbers of 430 species of Mollusca. Based on Butot (1967).

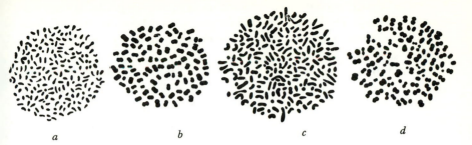

Fig. 12.5. Chromosomes of some decapod Crustacea. *a, Cambaroides japonicus*, spermatogonial metaphase (196 chromosomes); *b*, the same, first meiotic metaphase (98 bivalents); *c, Paralithodes camtschatica*, spermatogonial metaphase (208 chromosomes); *d*, the same, first meiotic metaphase (104 bivalents). From Niiyama (1934, 1936).

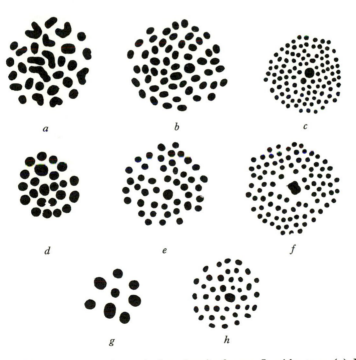

Fig. 12.6. Male first metaphases (polar views) of some Lepidoptera. (*a*) *Leptidea sinapis* (29 elements, some certainly multivalents); (*b*) *L. morsei* (n = 54); (*c*) *L. duponcheli* (n = 104); (*d*) *Lycaena icarus* (n = 23); (*e*) *Lysandra bellargus* (n = 45); (*f*) *L. coridon* (n = 90); (*g*) *Erebia tyndarus* (n = 18); (*h*) *E. ottomana* (n = 40). From Lorković (1941, 1949).

Y. Gupta (1964); probably the two forms should be regarded as distinct species. Two species of *Actinote* (subfamily Acraeinae) from Bolivia show the widely different chromosome numbers n = 14 and n = ca. 150 (de Lesse 1967).

In a number of lepidopteran species with high chromosome numbers numerical variations have been recorded. Most of the work has been done on first metaphases, so that what has been counted are the visible bodies (no doubt mostly bivalents but in some instances possibly univalents or multivalents) rather than single chromosomes. Unfortunately, there seems to be no means of distinguishing between univalents, bivalents and multivalents in lepidopteran spermatogenesis – they all look like small spheres or isodiametric bodies in which no structure is observable. Some of the numerical variation in these species is no doubt due to the inherent difficulties of the work, i.e. it is due to less than perfect technique. Thus, when de Lesse reports that *Lysandra argester* from various localities in Spain and Savoie show a variation from n = 147 to n = 151 we may legitimately suppose that the apparent variation is due to imperfections in technique, although this conclusion does not necessarily follow and it may be that some real variation occurs. When, however, he adds that four individuals of the same species from high elevations in the Sierra Nevada of southern Spain show n = 131–134 it is clear that he has discovered a geographic race with a lower chromosome number. Different populations of *Agrodiaetus dolus* from Italy, southern France and Spain show n = 108, 122 and 124 (de Lesse 1966a).

Spectacular evolutionary changes in chromosome number have occurred

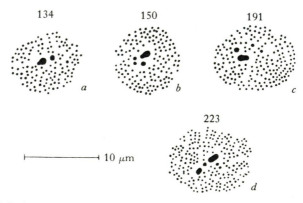

Fig. 12.7. Male first metaphases (polar views) of four closely related species of the butterfly genus *Lysandra*. (a) *L. golgus* (n = 134) from the Sierra Nevada, Spain; (b) *L. argester* (n = 150) from Teruel, Spain; (c) *L. nivescens* (n = 191) from Teruel, Spain; (d) *L. atlantica* (n = 223) from the Middle Atlas, Morocco. From de Lesse (1970).

in moths as well as butterflies. Thus Suomalainen (1963, 1965) has investigated forty species of the large geometrid genus *Cidaria*. Thirteen of them had n = 31, and 32 had numbers between n = 28 and n = 32; but species with 19, 17 and 13 chromosomes also occurred. The chromosomes are obviously larger in the species with lower chromosome numbers and photometric measurements show that the total amount of DNA is approximately the same in spite of the great differences in chromosome number. This shows clearly that we are not concerned with polyploidy. More extensive and accurate determinations of the DNA values of species that have undergone large scale increases or decreases of chromosome number in the course of phylogeny are badly needed; it is to be expected that species whose chromosome numbers have increased (e.g. *L. nivescens* and *L. atlantica*) will show higher DNA values and ones with diminished chromosome numbers (e.g. *Agrodiaetus posthumus*) lower than average DNA values, if we are correct in assuming that these evolutionary changes in chromosome number have been due to dissociations and fusions.

On a somewhat different footing are those instances where genuine variations in chromosome number have been recorded within demes or within individuals. Intrademic variation without intra-individual variation is likely to be entirely genuine, and in some instances both phenomena may really coexist. Just what the explanation is in species like *Leptidea sinapis*, where Lorković (1941) has recorded n = 26–41, is uncertain. Supernumerary chromosomes might be involved, but it seems more probable that the species is polymorphic for a number of chromosomal fusions or dissociations, since *L. morsei* shows n = 54, *L. amurensis* n = 61 and *L. duponcheli* n = 104 (Lorković 1941; Maeki 1958a, b). It seems useless to speculate about such cases until they have been critically reinvestigated, with determinations of the total amount of DNA present.

In considering the processes which have been involved in the evolution of unusually high or low chromosome numbers in the Lepidoptera, we may first of all rule out polyploidy, since there is no real suggestion of polyploid series in such genera as *Erebia, Lysandra, Agrodiaetus*. Furthermore, as pointed out earlier (White 1946c) many of the species with very high numbers have one bivalent which is much bigger than the others, or several large bivalents which differ visibly in size – i.e. karyotypes that exclude any possibility of straightforward polyploidy. And, although DNA values have not been determined, several authors (e.g. Y. Gupta 1964) have noted that where two closely related species or geographic races of Lepidoptera differ greatly in chromosome number, the chromosomes of the one with the lower number are much larger, so that the overall volume of the metaphase chromosomes is similar – which is what we should expect if fusions or dissociations have occurred. It is, of course, possible to maintain that polyploidy has occurred but has then been complicated by further karyotypic changes that have obscured the evidence for it. But there is no justification for such unnecessarily complicated hypotheses. On the contrary, it now seems clear that lepidopteran species with extremely high or low chromosome numbers represent the end products of evolutionary processes, each step of which was an increase or decrease of the chromosome number by one element. The extreme case is *Lysandra atlantica*, which has presumably accumulated 193 to 199 dissociations since the lineage to which it belongs diverged from the main stock of the 24-chromosome lycaenid butterflies. There is still some doubt as to whether localized centromeres are present in lepidopteran chromosomes and it may be that a holocentric condition is usual. But in any case each dissociation would have required a donor chromosome (to provide telomeres, even if centromeres were not needed), and the final result would be a species carrying much duplicated genetic material, perhaps mainly heterochromatic.

The caddis-flies (Trichoptera), an order of insects phylogenetically close to the Lepidoptera, show a similar degree of variation in chromosome number (Pchakądze 1930); the histogram extends from n = 6 to n = 30; but *Agrypnetes crassicornis* has about fifty chromosomes in the haploid set, according to Klingstedt (1931).

The birds, with a clear mode at 39–42, are another group with very high chromosome numbers. In contrast to the decapods and the Lepidoptera there is an extreme size range in the karyotype, the 'macrochromosomes' being very much larger than the extremely small 'microchromosomes' which occupy the center of the metaphase plate. Technical difficulties prevented the earlier investigators from arriving at accurate chromosome counts, but in recent years these obstacles seem to have been overcome to a very considerable extent (Yamashima 1950, 1951; Udagawa 1952, 1953, 1955, 1957; Makino, Udagawa and Yamashima 1956; van Brink 1959). The claims of Newcomer (1957) that the smaller elements in the metaphase plates of the chicken are not chromosomes at all or that the chicken is heterozygous for various kinds of structural rearrangements should not be taken seriously (van Brink 1959 and all later workers).

The really remarkable thing about frequency-distribution of bird chromosome numbers (Fig. 12.3) is the smallness of its variance. The range is from n = 25 to n = 43, but the great majority of the species have chromosome numbers between 33 and 43, only a few auks, parrots and gulls having lower numbers. Thus, by comparison with the decapods and Lepidoptera the birds have been extremely conservative as far as chromosome numbers are concerned, both increases and decreases having been very infrequent in evolution. This is definitely *not* an automatic consequence of the considerable range of size of the chromosomes; lizards show a similar size range but are much less conservative as far as chromosome numbers are concerned. That the amount of DNA per karyotype is probably rather uniform in birds is suggested by the chromosome measurements of Ohno *et al.* (1964).

The scorpions are another group which seem to show a striking range in chromosome number. However, the family Buthidae, which show moderate or low numbers, probably separated off from the phylogenetic line which includes the remaining families (which have high chromosome numbers) as early as the Silurian (Kraepelin 1905). In the Buthidae, *Tityus bahiensis* usually has n = 3 (Piza 1939, 1940), *T. serrulatus* has n = 6 (Piza 1947*a*) and various species of *Buthus* have n = 11–12 (Sokolov 1913; Sato 1936; Sharma, Parshad and Joneja 1959; Guénin 1961). These species of Buthidae appear at first sight to have holocentric chromosomes, since in the metaphases and especially at the first metaphase of meiosis in the male the chromosomes are stiff rods lying parallel to the equatorial plane. A considerable polemic developed on this question between Piza, who claimed that the chromosomes of *Tityus* had a centromere at each end (Piza 1939 and later papers), and his critics (Brieger and Kerr 1949; Rhoades and Kerr 1949) who supported the holocentric interpretation. Guénin (1961) has, however, produced some evidence that in *Buthus occitanus*, at any rate, the chromosomes are in reality monocentric after all, the centromeres being subterminal and only detectable at mitotic anaphases.

In the non-buthid scorpions the centromeres are much more easily detected and there can be no doubt that the chromosomes are truly monocentric. In *Pandinus imperator* n = 60, the two largest chromosomes being metacentric and the remainder acrocentric or subacrocentric (Guénin 1957). Three species of *Palamnaeus* have n = 32, 43 and 56 (Sharma, Parshad and Handa 1962; Venkatanarasimhaiah and Rajasekarasetty 1964); once again the chromosomes are clearly monocentric. The family Vejovidae possibly have the highest numbers since Wilson (1931) estimated the diploid chromosome number of two species at approximately 100. The Bothriuridae, with n = 18 in *Bothriurus* sp. (Piza 1957*a*) may have the lowest numbers in the non-buthid group of families.

By way of contrast to groups such as the decapods, Lepidoptera, birds and scorpions, with their extreme evolutionary instability of chromosome number, we may consider the short-horned grasshoppers (suborder Caelifera of the order Orthoptera) a group in which a very large number of species have been studied cytologically, so that the published information is very extensive and detailed. We give, in Table 12.1, the modern classification of this group and their recorded chromosome numbers. The degree of 'karyotypic conservatism' varies considerably from one family to another (White 1972). Obviously we cannot say anything about such families as the Trigonopterygidae, Pneumoridae, Xyronotidae and Charilaidae, in each of which only a single species has been investigated cytologically.

The tropical and subtropical family Eumastacidae includes a number of rather diverse groups which have been regarded as subfamilies, but are perhaps in some cases rather remotely related. They may represent survivals of a Mesozoic group which radiated out from Gondwanaland. Only fifteen of the twenty-one subfamilies have been studied cytologically; as we should perhaps expect from their varied morphology, their karyotypes are fairly diverse and it does not seem possible to establish a 'typical' or 'primitive' karyotype for the Eumastacidae as a whole. A group of Old World subfamilies (Chorotypinae, Eruciinae, Pseudoschmidtiinae and Mastacideinae) seem to have a primitive karyotype consisting of $2n\male = 21$ acrocentrics, but the

TABLE 12.1. *Classification of the Orthoptera (suborder Caelifera), with typical chromosome numbers and deviant numbers (in parentheses)*

	$2n\male$	No. of major chromosome arms in typical $\male$ karyotype	Sex chromosomes
Superfamily (1) *Eumastacoidea*			
Family *Eumastacidae*			
Subfamily *Chorotypinae*[1,2]	21 (19)	21	XO, X_1X_2Y
Eruciinae[1,2]	21	21	XO
Chininae[3]	14*	19	XY
Thericleinae[4,5]	19 (17, 16)	19	XO, XY
Euschmidtiinae	?	?	?
Pseudoschmidtiinae[1,4,5]	21 (19)	21	XO, X_1X_2Y
Miraculinae[5]	25		XO
Mastacideinae[5]	21	21	XO
Eumastacinae[5]	21*	23	XO
Episactinae	?	?	?
Parepisactinae[5]	19*	32	XO
Paramastacinae[5]	19*	31	XO
Morseinae[1]	23*	23	XO
Teicophryinae[1]	17*	?	XO
Espagnolinae	?	?	?
Pseudomastacinae	?	?	?
Temnomastacinae	?	?	?
Gomphomastacinae[3]	19	20	XO
Biroellinae[2]	17*	34	XO
Morabinae[6]	17 (21, 19, 18, 16, 15, 13)	22	XO, XY, X_1X_2Y
Family *Proscopiidae*[7]	19 (17)	19	XO

TABLE 12.1—*continued*

	2n♂	No. of major chromosome arms in typical ♂ karyotype	Sex chromosomes
Superfamily (2) *Trigonopterygoidea*			
Family *Trigonopterygidae*[8]	23*	?	*XO*
Superfamily (3) *Pneumoroidea*			
Family *Pneumoridae*[8]	23*	?	*XO*
Tanaoceridae	?	?	?
Xyronotidae[8]	23*	?	*XO*
Superfamily (4) *Acridoidea* (true grasshoppers, 'short-horned grasshoppers')			
Family *Charilaidae*[9]	23*	23	*XO*
Pamphagidae[10]	19	19	*XO*
Ommexechidae[11]	23 (25)	25	*XO*
Pyrgomorphidae[12]	19 (18, 17, 15, 11)	19	*XO, XY*
Lentulidae[9]	23 (21, 20, 19)	23	XO, XY, X_1X_2Y
Lathiceridae	?	?	?
Pauliniidae[13]	23	23	*XO*
Acrididae[14]	23 (22, 21, 20, 19, 18, 17, 16, 15, 14, 13, 12, 8)	23	XO, XY, X_1X_2Y
Superfamily (5) *Tetrigoidea* ('Grouse locusts')			
Family *Tetrigidae*[15]	13	13	*XO*
Superfamily (6) *Tridactyloidea*			
Family *Tridactylidae*[16]	13	24	*XO*
Cylindrachetidae	?	?	?

[1] Helwig, in Rehn 1948
[2] White 1971*b*
[3] White 1968*b*
[4] White 1965*a*
[5] White 1970*c*
[6] White 1956, 1966*a* White and Cheney 1966, 1971 White, Blackith, Blackith and Cheney 1967 White and Webb 1968
[7] De Castro 1946 Piza 1943*b*, 1945 Dasgupta 1968
[8] Helwig 1958

[9] White 1967
[10] Granata 1910
[11] Mesa 1961, 1963*a*, 1964*a*
[12] McClung 1932 Rao 1937 Momma 1943 White unpublished
[13] Mesa 1961
[14] Numerous authors
[15] Robertson 1916, 1931 Rayburn 1917 Harman 1920 Misra 1937*b* Henderson 1961
[16] Ohmachi 1935

* Only a single species has been studied cytologically.

very distinct African Thericleinae have either 2n♂ = 19 or 17 acrocentrics. The Australian Morabinae and Biroellinae seem to have a primitive karyotype of 2n = 17, but with two large pairs of metacentric autosomes; the complex evolution of the karyotype in the Morabinae has been discussed in Chapter 11. The neotropical Proscopiidae, a family of grasshopper-like insects related to the Eumastacidae, show 2n♂ = 17 and 2n♂ = 19, with

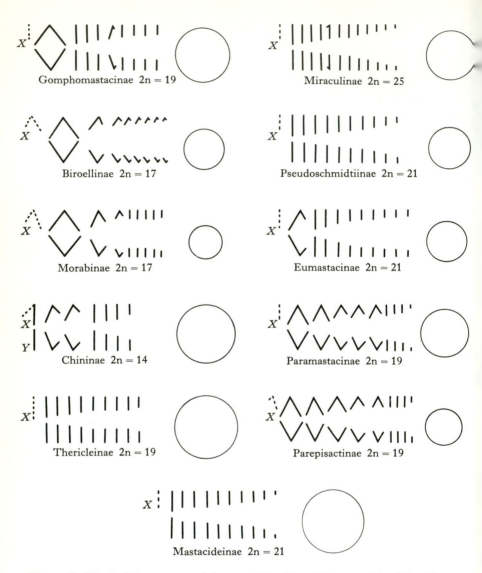

Fig. 12.8. 'Typical' karyotypes of eleven subfamilies of Eumastacidae (Biroellinae, Chininae, Eumastacinae, Paramastacinae, Parepisactinae each based on a single species; Gomphomastacinae, Miraculinae, Mastacideinae, each based on only two species). The circles alongside the karyotypes represent the mean size of the pachytene nuclei. From White (1970c).

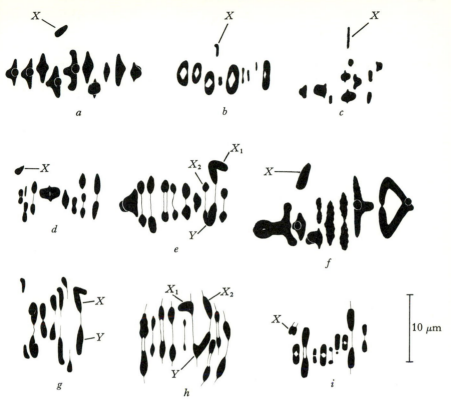

Fig. 12.9. Male first metaphases (side views) of tropical Eumastacid grasshoppers. (a) *Eumastax salazari* (Eumastacinae) from Colombia; (b) *Parepisactus carinatus* (Parepisactinae) from Colombia; (c) *Mastacides* sp. (Mastacideinae) from India; (d) *Apteropeoedes elegans* (Pseudoschmidtiinae) from Madagascar; (e) *Xenomastax wintreberti* (Pseudoschmidtiinae) from Madagascar; (f) *Manowia alca* (Thericleinae) from Malawi; (g) *Mnesicles* sp. (Chininae) from Manus Is.; (h) *Erianthus guttatus* (Chorotypinae) from Malaysia; (i) *Erucius dimidiatipes luteipes* (Eruciinae) from Malaysia. Partly from White 1968.

most or all of the chromosomes acrocentric (Piza 1943*b*, 1945; De Castro 1946; Dasgupta 1968; Albizu de Santiago 1968).

By comparison with the Eumastacidae, the larger families of the super-family Acridoidea seem to be karyotypically conservative. Particularly in the very large family Acrididae, the general impression that one gets after examining the karyotypes of many hundreds of species is of great cytogenetic uniformity. But, as John and Hewitt (1966*b*) have rightly emphasized, this uniformity is more apparent than real. All the usual processes of chromo-somal evolution have been at work in the Acridoidea, even if they have not led to such striking variations in karyotype and in chromosome number as we find in other groups.

The Acridoidea are a group that are, in general, characterized by acrocentric chromosomes. Metacentric chromosomes do occur, however, in many genera and species; some owe their origin to fusions between originally acrocentric chromosomes while others have arisen by pericentric inversions. In the former case we find a reduction in chromosome number, while in the latter case the original number is retained. The families Pneumoridae, Xyronotidae, Trigonopterygidae, Charilaidae, Lentulidae, Pauliniidae and Acrididae all seem to show a 'typical' karyotype of 23 acrocentrics. Information concerning the cytology of the first three of these families consists of a mere mention of the chromosome number by Helwig (1958) with no other details. These are primitive families of grasshoppers, represented at the present time by only a few species, restricted to small areas of the tropics and southern hemisphere. It seems probable that $2n\male = 23$ acrocentrics was the primitive chromosome number in the evolutionary stock which gave rise (presumably in Mesozoic times) to the three superfamilies Trigonopterygoidea, Pneumoroidea and Acridoidea. It occurs in only a slightly modified form in the South American family Ommexechidae which have, in general, $2n\male = 23$, with one chromosome pair J-shaped, presumably as a result of a pericentric inversion that occurred early in the phylogeny of this family (Mesa 1963a); a few genera lack the J-shaped element (Mesa 1964a) and in the Chilean genus *Conometopus* (with three species) $2n\male = 25$, presumably as a result of a dissociation of the J-shaped chromosome (Mesa, unpublished).

Two families of the Acridoidea, the Pamphagidae and the Pyrgomorphidae, which do not seem to be particularly closely related, are unusual in having $2n\male = 19$ acrocentrics in the vast majority of their species. Morphologically aberrant Pyrgomorphidae such as the West African *Chapmanacris* and the Australian Psednurini betray their relationships by possessing the same chromosome number as typical members of the family. And several American genera of Acrididae, formerly classified in error as Pamphagidae (an exclusively Old World family) on superficial characters, can be excluded, since they have the typical Acridid number of $2n\male = 23$.

Deviations from $2n\male = 19$ occur in some Pyrgomorphidae, exclusively, as far as known, due to fusions. Thus three South African species show $2n\male = 18$ (the result of an X-autosome fusion), 15 (two autosomal fusions) and 11 (four autosomal fusions).

The 19-chromosome karyotypes of the Pamphagidae and Pyrgomorphidae are far more likely to be derivative than primitive in the Acridoidea. Presumably the reduction in chromosome number took place independently in two phyletic lineages, in each case by two fusions followed by pericentric inversions (since the chromosomes of the 19-chromosome karyotypes are all acrocentric).

The great majority of deviations from the typical chromosome numbers of the various Acridoid families are due simply to chromosomal fusions, i.e. they are changes which leave the number of major chromosome arms unaltered. In addition, it is clear that very numerous deletions and duplications of chromosomal material have taken place in the Acrididae, since the DNA values are quite variable (John and Hewitt 1966*b*; Fox 1970*a*; Kiknadze and Vysotskaya 1970), the extent of heterochromatic segments differs from species to species, and the relative lengths of the chromosomes which make up the karyotype are far from constant (e.g. some species have two pairs of small chromosomes, others have three, etc.). Kiknadze and Vysotskaya found a more than twofold difference in DNA values in the Acrididae, from 301 arbitrary units in *Eirenephilus longipennis* to 682 in *Bryodema tuberculatum* and 763 in *Angaracris barabensis*; a certain correlation with the amount of heterochromatin seen at pachytene is evident in the nine species they studied from this point of view, but it is an imperfect one.

Approximately 90% of the species belonging to the huge family Acrididae show the typical 23-chromosome karyotype. Starting with this condition, a single fusion between acrocentric autosomes will reduce the karyotype to $2n\male = 21$ (19 acrocentrics and 2 metacentrics). Two such fusions will produce $2n\male = 19$ (15 acrocentrics and 4 metacentrics). Members of the European genera *Chorthippus*, *Stauroderus*, *Omocestus* and *Stenobothrus* and the North American genera *Chrysochraon*, *Chloealtis* and *Napaia* show three autosomal fusions and have a karyotype of eleven acrocentrics and six metacentrics. Rothfels (1950) inclined to the view that the three fusions are truly homologous (i.e. monophyletic) throughout this assemblage, but this is not certain.

A fusion between the X and an autosome occurring in a 23-chromosome species will convert it into one with $2n = 22$ in both sexes, while a similar transformation in a 21-chromosome species will give $2n = 20$. The ultimate reduction that can be obtained in the Acrididae by fusion of acrocentrics, without other changes, is $2n\male = 12$, which is what is found in the Mexican *Philocleon anomalus* (Helwig 1941). There exists, however, a South American acridid grasshopper, *Dichroplus silveiraguidoi*, with only four pairs of chromosomes, i.e. with $2n\male = 8$ (Saez 1956, 1957). Presumably this species must have acquired a total of one X-autosome fusion, seven fusions between autosomes and six pericentric inversions – the minimum number of rearrangements needed to convert the usual $2n\male = 23$ karyotype into that of *D. silveiraguidoi* (Fig. 12.10), unless we postulate tandem fusions (see p.

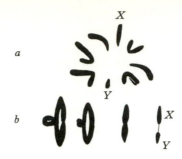

Fig. 12.10. *a*, Spermatogonial metaphase and *b*, first metaphase (side view) of the South American grasshopper *Dichroplus silveiraguidoi*, with n = 4 (From Saez 1956, redrawn).

225). Some species of the genus *Dichroplus* retain the primitive 23-chromosome karyotype while *D. bergi* has an *X*-autosome fusion (2n♂ = 22) and some forms assigned to *D. pratensis* have 2n♂ = 18; one of these is said to have two pairs of metacentric autosomes, which is what we would expect if it has acquired an *X*-autosome fusion and two autosomal fusions. Another form of '*D. pratensis*' apparently has no metacentric autosomes, so that it has presumably become homozygous for two pericentric inversions in addition. These forms seem to represent intermediate stages, so to speak, on the route that *D. silveiraguidoi* has travelled. No populations polymorphic for fusions or inversions seem to have been recorded in the genus *Dichroplus*. Similarly extreme reductions in chromosome number to n = 4 have occurred in the alticid beetle *Homoschema nigriventre* (Virkki and Purcell 1965) and the carabid *Graphipterus serrator* (Wahrman 1966).

There are a number of other species of Acrididae in which the chromosome number has been reduced below 2n♂ = 23, without there being any metacentrics to account for the reduction. Thus in the tribe Podismini the genera *Miramella*, *Zubowskya* and *Niitakacris* all have 2n♂ = 21 acrocentrics (Makino 1939*b*; Helwig 1942; Momma 1943; S. G. Smith 1944). The same is true of *Indopodisma*, a genus in which the short arms of all the chromosomes are unusually large by comparison with other Acrididae. It seems likely that a fusion and a pericentric inversion occurred in the phylogeny of this group of genera, there having been at one stage a pair of metacentric chromosomes (probably J-shaped with very unequal arms) which later become acrocentric by pericentric inversion. The same has probably happened in the neotropical grasshopper *Coscineuta virens*, the Mexican *Perixerus squamipennis* (where an *X*-autosome fusion has also occurred) and *Machaerocera mexicana*, where three autosomal fusions and one *X*-autosome one have also occurred (Helwig 1942). In the grasshopper *Dactylotum bicolor* $2n\male = 17$ acrocentrics, i.e. no less than three pairs of autosomes have disappeared as independent entities (Helwig 1942; Powers 1942).

It is fairly clear that there is a special barrier which operates against the establishment of metacentric chromosomes in the Acridoidea. Although this barrier has broken down in a few genera this does not alter the fact that the vast majority of the species in these groups have only acrocentric chromosomes.

Now there are, of course, two ways in which metacentric chromosomes can arise in such a karyotype – by centric fusion and by pericentric inversion. The fact that *both* types of transformation have been very rare in the 'acridoid' grasshoppers argues strongly in favour of the view that it is against metacentric chromosomes, as such, that the barrier operates (rather than against one or other type of rearrangement). It is also noteworthy that in certain of the trimerotropine grasshoppers and the genus *Dichroplus*, in which the barrier seems to have been ineffective, *both* fusions *and* pericentric inversions have succeeded in establishing themselves. There is no indication that a special barrier against either metacentric or acrocentric chromosomes operates in the Morabinae, although there probably is a barrier against metacentrics in some other subfamilies of eumastacid grasshoppers such as the Thericleinae (White 1965a). Very short pericentric inversions may have established themselves quite frequently in many evolutionary lineages of grasshoppers, leading to changes in the length of the short arms of the acrocentric chromosomes. The restriction against pericentric inversions thus seems to operate increasingly strongly against longer inversions. Where fusions involving short chromosomes have presumably occurred (e.g. in *Dactylotum*) they must have been followed in many cases by pericentric inversions, thereby restoring the acrocentric condition. Only in a few species such as the Ommexechidae, and the Acrididae *Oedaleus nigrofasciatus* and *Austroicetes pusilla* do we have J-shaped chromosomes which have arisen by pericentric inversion fixed in the karyotype, although such rearrangements are quite common in one section of the North American genus *Trimerotropis* and in the related genera *Circotettix* and *Aerochoreutes* (see Ch. 9). In *A. pusilla* there is one pair of medium-sized metacentrics in a 23-chromosome karyotype (Fig. 6.9); polymorphism for pericentric inversions occurs in the related *A. interioris* (White and Key 1957; Nankivell 1967), but *A. pusilla* itself is not a cytologically polymorphic species. In *O. nigrofasciatus* there are four pairs of J-chromosomes which have presumably arisen by pericentric inversion (Nolte 1939); no polymorphism for pericentric inversions is known in this genus.

In a group such as the Acridoidea it seems clear that increases in chromosome number (e.g. from $2n\male = 23$ to $2n\male = 25$) can only come about through pericentric inversions followed by dissociation of the metacentric or submetacentric chromosome so produced into two acrocentrics. Since both types of rearrangement are rare it is hardly surprising that there should appear to be a barrier against increases in chromosome number. One exception, the Chilean ommexechid genus *Conometopus* with three species showing $2n\male = 25$, has already been mentioned (p. 424). No such exceptions are known in the families Pamphagidae, Pyrgomorphidae and Acrididae, in which many hundreds of species have been studied cytologically, i.e. there are no known species of the former two with $2n\male = 21$ or of the latter with $2n\male = 25$. It might have been expected that one or more species of the North American trimerotropine grasshoppers, with numerous metacentric chromosomes which have arisen by pericentric inversion, would have undergone an increase of the chromosome number by dissociation; but no such species is known.

The Tetrigidae (grouse locusts) are only remotely related to the other groups of Caelifera. They are an example of extreme karyotypic conservatism, since all the species that have been studied have $2n\male = 13$ acrocentrics (Robertson 1916, 1931; Rayburn 1917; Harman 1920; Misra 1937*b*; Henderson 1961). There is no good reason to suppose, with Helwig (1958) that this low chromosome number has any ancestral significance for the Caelifera as a whole. Possibly related to the Tetrigidae are the little Tridactylidae, one member of which also shows $2n\male = 13$ (Ohmachi 1935); but although the chromosome number is the same, the karyotype of the single species that has been studied is entirely different.

The long-horned grasshoppers (Tettigonioidea) seem to be somewhat intermediate between the Eumastacoidea and the Acridoidea in degree of karyotypic conservatism. The chromosome numbers ($2n\male$) range from 20 to 35 (*Saga pedo* has 68, but it is a parthenogenetic tetraploid – see p. 715). Some subfamilies of the Tettigoniidae have a rather uniform karyotype, but the uniformity is probably never as great as in the Acrididae. Thus most of the Old World Decticinae have $2n\male = 31$ acrocentrics, but the New World decticine genera *Atlanticus* (White 1941*b*), *Steiroxys* (Davis 1908) and *Anabrus* (McClung 1914), with one or more pairs of metacentrics, diverge from this standard karyotype. Bianchi (1966) has shown that there are two quite different karyotypes ($2n\male = 29$ and 23) among Mediterranean species of the Ephippigerinae, which are not simply derivable from one another by fusions or dissociations. The first is found in the genera *Ephippiger*, *Ephippigerida* and *Callicrania*; both karyotypes have been found in different species of *Steropleurus* (Bullini and Bianchi-Bullini 1969). The Australian

Fig. 12.11. Spermatogonial metaphases of various Tettigoniidae. (*a*) *Platycleis grisea*; (*b*) *Metrioptera brachyptera*; (*c*) *Atlanticus pachymerus* (all subfamily Decticinae); (*d*) *Tettigonia viridissima* (Tettigoniinae); (*e*) *Insara tolteca*; (*f*) *Insara gracillima*; (*g*) *Microcentrum* sp.; (*h*) *Leptophyes punctatissima* (all Phaneropterinae); (*i*) *Neoconocephalus* sp. (Copiphorinae). From White 1941*b*.

genus *Yorkiella* (subfamily Listroscelinae) includes two species with $2n\text{\male} = 31$ and another (*Y. picta*) which has $2n\text{\male} = 20$ and an *XY* mechanism in the male (White, Mesa and Mesa 1967; Ferreira 1969). *Y. picta* has presumably acquired one *X*-autosome fusion, five fusions between autosomes and a number of pericentric inversions since its evolutionary divergence from the species with the presumed ancestral karyotype for the genus. Two other neo-*XY* Tettigoniidae are known. One of them (*Isopsera* sp.) from India shows no fusions between autosomes (Dave 1965), while the other (a race of *Polichne parvicauda* from New South Wales) shows two autosomal fusions as well as the one involving the *X* (Ferreira 1969). *Letana atomifera* from India has an X_1X_2Y sex-chromosome mechanism, but apparently shows no fusions between autosomes (Dave 1965).

The subfamily Phaneropterinae shows a considerable variability in chromosome numbers ($2n\text{\male}$) which range from thirty-three in *Amblycorypha* spp. (Pearson 1929) and *Microcentrum* (McClung 1902) down to nineteen in *Torbia viridissima* (Ferreira 1969); since the latter species has all its chromosomes acrocentric, pericentric inversions as well as fusions have presumably occurred. Some members of this subfamily have the *X* metacentric, while others have an acrocentric *X*, the situation in this respect paralleling that in the Morabinae and Trimerotropi, where the *X* has repeatedly changed its shape by pericentric inversion. Generally speaking, the species of Tettigoniidae with relatively low chromosome numbers (e.g. *Neoconocephalus* sp. with $2n\text{\male} = 25$) have several pairs of metacentric autosomes which have clearly arisen by centric fusion, but *Torbia* and some other forms are exceptions.

The rather heterogeneous Gryllacridoidea, a superfamily allied to the Tettigoniidae and to the crickets, show a very considerable range of chromosome numbers.

The Cave and Camel crickets of the family Rhaphidophoridae mostly show high chromosome numbers. In the European fauna Baccetti (1958) found four species of *Dolichopoda* had $2n\text{\male} = 31$ and one had $2n\text{\male} = 35$; these were all species with *XO* males. Saltet (1959, 1960, 1967) has studied six additional *XY* species of this genus with $2n\text{\male} = 28, 30, 30, 32, 32, 34$. In the subfamily Macropathinae, Mesa (1965) showed that four species of the South American genus *Heteromallus* had $2n\text{\male} = 45$. The same basic chromosome number occurs in the Australian and New Zealand representatives of this subfamily (Mesa, 1970b). Reductions of chromosome number, as a result of fusions, have occurred in a number of species, however. $2n\text{\male} = 45$ occurs in about four-fifths of the New Zealand species, but there are also forms with 43, 41, 39, 37, 35, 33, 32 (a neo-*XY* species), 31 and 27. In the Australian fauna, also, $2n\text{\male} = 45$ seems to be the primitive chromosome number, but in the genus *Cavernotettix* there are two XY species with $2n = 44$ and 34 and three $X_1 X_2 Y$ ones with 43, 43 and 39 (Mesa, Ferreira and de Mesa 1968, 1969; Mesa 1970a). The Mesas' work emphasizes the karyotypic conservatism of this undoubtedly ancient group with a typically 'antarctic' distribution. Two species of the large North American genus *Ceuthophilus* have $2n\male = 37$ (Thompson 1911; Stevens 1912). Species of *Tachycines* and *Diestrammena* show the very high chromosome number of $2n\male = 57$ (Makino 1931; Mohr and Eker 1934; Schellenberg 1913).

The families Gryllacrididae and Schizodactylidae seem to show quite low numbers such as $2n\male = 11$, 14, and 17 (Ohmachi 1935; Heberer 1937; McClung and Asana 1933); but an Australian species of *Hadrogryllacris* has $2n = 31$. Finally, the Stenopelmatidae are possibly a heterogeneous family since one species has $2n\male = 15$ and another $2n\male = 47$. (Piza 1947d; Stevens 1909). Altogether, the Gryllacridoidea show a far greater range of chromosome numbers than any other group of Orthoptera Saltatoria.

The crickets (superfamily Grylloidea) have not been studied by cytologists to quite the same extent as the other main groups of orthoperoid insects, but the karyotypes of a number of species have been figured by Baumgartner (1929), Ohmachi (1935), Piza (1946a), Dutt (1949), Smith (1953a), Ohmachi and Ueshima (1955, 1957a, b), Claus (1956), Ueshima (1957), T. Sharma (1963) and others. We have already referred (p. 383) to the cytology of the mole crickets (Gryllotalpidae) in connection with the problem of sibling species of *Gryllotalpa* in Europe. The number of chromosomes and major chromosome arms seem to vary more than in the Tettigoniidae but certainly less than in the Gryllacridoidea. There is a general tendency for the X to be a metacentric element, although acrocentric X's are known in *Cyrtoxiphus ritsemae* and *Homoeogryllus japonicus*. In a number of cricket species, centric fusions or dissociations which have not reached fixation are responsible for variations in chromosome number within the species. Thus in all three sibling species of *Loxoblemmus* from Japan formerly included under the name *L. arietulus* the chromosome number is variable (Ohmachi and Ueshima 1955); pericentric inversions as well as fusions or dissociations have been involved, since the number of major chromosome arms varies from 20 to 22 in two of the species. $2n\male$ ranges from 13 to 17 in the three species, but the whole range probably does not occur in any single one of them. The situation in *Brachytrupes portentosus* seems to be similar, with $2n\male$ ranging from 13 to 17 (Ohmachi 1935; Momma 1941). In *Anaxipha pallidula*, also, males with 17, 18 and 19 chromosomes exist, as a result of fusions or dissociations (Ueshima 1957). Considerable cytotaxonomic differences exist between the species of such large genera of crickets as *Pteronemobius* and *Nemobius*. The subfamily

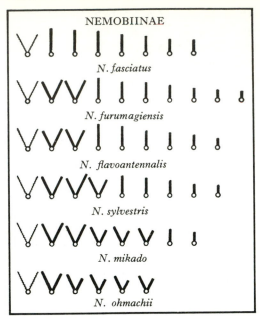

Fig. 12.12. Karyotypes in some members of the cricket genus *Nemobius*, based on the work of Baumgartner (1929) and Ohmachi (1935). *X*-chromosomes on left.

Gryllinae shows a range of chromosome numbers from 2n = 11 to 2n = 29. Recent work by Indian cytologists on this group has been reviewed by Manna (1969).

It is perhaps instructive to compare the processes of chromosomal evolution in the orthopteroid groups with the situation in the Diptera, an outstanding example of a large taxonomic group characterized by low chromosome numbers. The complete range in this order of insects is from 2n = 4 to 20, with peaks at 6 and 8 (due to several families of 'lower' Diptera, i.e. Nematocera, such as the midges and mosquitoes) and 12 (a huge peak, due to various acalypterate and calypterate families, including the Drosophilidae). It is probably legitimate to conclude that there is some kind of special barrier which operates against increases of chromosome number in the Diptera; large numbers are present in the germ-line in certain Nematocerous groups (Cecidomyidae and Orthocladiinae – see p. 525), but most of these are not represented in the somatic tissues.

The chromosome numbers in the genus *Drosophila* range from n = 3 to n = 7. For a long time it appeared as if no species of *Drosophila* had more than six pairs of chromosomes, but *D. trispina* was eventually found to possess an additional pair of small chromosomes which are mainly or entirely heterochromatic (Patterson and Stone 1952), and since that time several other species with a similar karyotype have been found; these species may be compared with the Chilean grasshoppers of the genus *Conometopus* which have likewise increased their chromosome number above the normal upper limit for the group (see p. 424).

Patterson and Stone (1952) and Stone (1955, 1962) have interpreted karyotype evolution in *Drosophila* in terms of changes which have occurred in a 'primitive configuration' consisting of five pairs of acrocentric chromosomes (including the sex chromosome pair) and a pair of small dot-like elements which ordinarily lie in the middle of the metaphase plate. This primitive karyotype may be compared with the '17 chromosome' one found in the morabine grasshoppers (see p. 372) and the '23 chromosome' one found in most families of acridoid grasshoppers (p. 424). It still occurs in *Drosophila orbospiracula* and *D. longala* (in which, however, the *Y* has disappeared so that males are *XO*), *D. subobscura*, numerous members of the *quinaria* group, *D. guttifera*, *D. virilis* and *D. novamexicana*, and (in various slightly modified forms) in several species belonging to the *pinicola*, *funebris*, *repleta*, *melanica*, *macroptera* and *guarani* groups.

According to the analysis of Patterson and Stone, based on the very extensive work of Wharton (1942), Ward (1949, 1952), Wasserman (1954, 1960, 1962*a*, *b*, *c*, *d*), Clayton and Wasserman (1957) and others, the modifications of the primitive karyotype which are detectable by an examination of the metaphase chromosomes (i.e. excluding the very numerous paracentric inversions) have been (1) centric fusions, (2) pericentric inversions, (3) changes in the overall amount of distribution of heterochromatin and (4) translocations. Stone (1962) considered that there was evidence of fifty-eight fusions, thirty-two pericentric inversions, thirty-eight 'additions of heterochromatin' and only three translocations which have reached fixation in the karyotypes of approximately 300 species of *Drosophila* that had been examined cytologically up to that time. The number of paracentric inversions that have undergone fixation in the karyotypes of these species is not known, but Stone estimated that there are about fourteen 'floating' in the population of the average *Drosophila* species, and that 1.5 times this number have undergone fixation; on this basis our sample of about 300 species would include 21 × 300 or 6300 'fixed' paracentric inversions.

The proposition that there have been no dissociations of metacentric chromosomes into acrocentrics in the genus *Drosophila* seems questionable. At least the species with n = 7 have almost certainly evolved from ones with n = 6 by dissociations of small metacentric elements; and some dissociations of larger metacentrics may be suspected as well. However, Stone was probably correct in thinking that fusions have greatly outnumbered dissociations in the phylogeny of the group, i.e. that the evolution of chromosome numbers has been preponderantly in a downward direction.

The karyotypes of a number of other families of higher Diptera have been studied by Boyes and collaborators (Boyes 1953, 1954, 1958, 1963, 1964; Boyes and Wilkes 1953; Boyes and Naylor 1962; Boyes, Corey and Patterson 1964; Boyes and van Brink 1964). In the Sarcophagidae almost all species have six chromosome pairs; the autosomes are invariably metacentric, but there is great variation in the size and shape of the X and Y. The Muscidae mostly show six pairs of chromosomes including a very small heterochromatic X and Y; but three species studied by Boyes, Corey and Patterson (1964) had only five pairs, the sex chromosomes being apparently missing (they have possibly been fused with one of the larger chromosome pairs).

EVOLUTION OF CHROMOSOME NUMBERS AND CHROMOSOME FORM IN VERTEBRATES

So much is known concerning the evolutionary history of the Vertebrates that their comparative karyology is necessarily of particular interest. It is to be hoped that in a few years we shall have far more information concerning DNA values, genetic mechanisms of sex determination and other cytogenetic questions, particularly in the lower vertebrates, to supplement the presently available data on chromosome numbers and shapes. The most recent review of this field is that of Morescalchi (1970).

A single species of the Cephalochordata, *Branchiostoma belcheri*, has n = 16 acrocentrics and an XY sex-chromosome pair in the male, according to Nogusa (1957, 1960); the chromosomes are very minute. In the Cyclostomata, two species of Pacific hagfishes investigated by the same author had n = 23 and 24 acrocentrics; they both showed a most peculiar association of several bivalents at both meiotic divisions. Lampreys have much higher numbers, e.g. *Petromyzon marinus* has 2n = 168 (Potter and Rothwell 1970). According to Howell and Duckett (1971) two species of Australian lampreys have 2n = 76, while five Northern Hemisphere species have 2n = 142, 146, 164, 164 and 168. The chromosomes of these Northern Hemisphere species of *Lampetra*, *Petromyzon* and *Ichthyomyzon* are all acrocentric, while many of those in the karyotypes of the Australian species of *Mordacia* are metacentric and about twice the size of the chromosomes of the Northern Hemisphere forms.

Elasmobranchs seem, in general, to have high chromosome numbers. The skate *Raja meerdervoortii*, with n = 52, has one of the highest chromosome numbers known up to the present time in vertebrates. *Squalus sucklei* and

Scyllium catula also show a high number (n = 31), while *Mustelus manazo* has n = 36 and the sting ray *Dasyatis akajei* has n = 42 (Nogusa 1960).

Most teleost fishes show relatively large numbers of small chromosomes. Detailed studies of the karyotypes have hardly begun, and will obviously be far from easy. Certain families appear, superficially at least, to have rather uniform karyotypes. For example, the gobies mostly seem to have n = 22 or 23 acrocentric chromosomes with one or more elements very much shorter than any of the others (Nogusa 1960). Relative uniformity also seems to prevail in the Cyprinodontidae, which show n = 19 to n = 24 acrocentrics, according to Wickbom (1943). In both these families centric fusions and pericentric inversions seem not to have established themselves at all frequently, although we must suppose that they have done so occasionally, in order to explain the range of chromosome numbers found. Karyotypic variation is somewhat more pronounced in the Cyprinidae, where a few metacentric chromosomes occur in a number of the species. There is a possibility (but no proof) that polyploidy may have occurred in the goldfish, *Carassius auratus* and the carp *Cyprinus carpio*, both of which have 2n = 104 (approximately), whereas two species of *Barbus* have 2n = 50 and 52, with about half as much DNA (Ohno, Muramoto, Christian and Atkin 1967). In the Salmonidae, species of *Oncorhynchus* have n = 50 or 54 (the latter number, in *O. nerka* being the highest known for any teleost). *Salmo* shows a range of chromosome numbers from n = 30 to 42, with 5 to 8 pairs of metacentrics (Prokofieva 1934; Svärdson 1945). The Atlantic Salmon (*S. salar*) with n = 30 has almost exactly the same DNA value as the Brown Trout (*S. trutta*) with n = 40 (Rees 1964).

The Dipnoi differ fundamentally from the teleosts in the large size of their chromosomes; they resemble the urodeles in this respect and have high DNA values. *Lepidosiren paradoxa* has n = 19 metacentrics (Agar 1911), *Protopterus annectans* has n = 17 and *Epiceratodus forsteri* about the same number (Wickbom 1945). The chromosome cytology of the only living coelacanth fish *Latimeria chalumnae* does not appear to have been studied as yet; however this species has quite a small DNA value according to Vialli (1967). It must be admitted that the cytogenetic evidence at present available would seem to point to a dipnoan origin of the Amphibia rather than the crossopterygian one indicated by the palaeontological evidence. Possibly the dramatic increase in the DNA content of the karyotype, from the teleost level of the dipnoan–amphibian one, took place independently in two phyletic lineages, as suggested by Morescalchi (1970).

The urodeles seem to show rather uniform karyotypes within the individual families, although there are great differences between them. The Salamandridae almost all have n = 12, but species of *Ambystoma* have

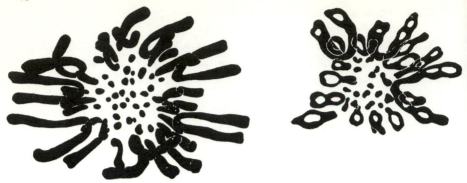

Fig. 12.13. Spermatogonial metaphase and first meiotic metaphase in the urodele *Cryptobranchus alleghaniensis* n = 31; the chiasma frequency of the longest metacentric chromosomes seems to be about 8 or 9. From Makino (1935).

n = 14 and *Triturus viridescens* has n = 11. The Plethodontidae have n = 13 and n = 14 in seventeen genera studied by Kezer (unpublished). *Necturus maculosus* and *Proteus anguinus* both have n = 19 (Seto, Pomerat and Kezer 1964), while the Hynobiidae and Cryptobranchidae mostly have n = 28, although some species have n = 13 and n = 32 (Makino 1934, 1935, 1939*a*; Iriki 1932). Most urodele karyotypes show a graded series of chromosomes but in *Cryptobranchus alleghaniensis* there is a clear distinction between the larger chromosomes and the numerous small microchromosomes which occupy the centre of the metaphase plate (Makino 1935). Metacentric chromosomes are the rule in this order. All species of urodeles seem to have very high DNA values, ranging from about 58 picograms (2C) in *Triturus cristatus* to 192 picograms in *Amphiuma* (Callan 1972).

Three members of the burrowing limbless order Apoda (Gymnophiona) have been studied cytologically by Seshachar (1937, 1939, 1944). They have 2n = 30, 36 and 42, but the number of major chromosome arms is 50 or 52 in all of them, the karyotypes including metacentrics, acrocentrics and short 'dot chromosomes'. One member of this order has been shown by Beçak *et al.* (1970) to have a very high DNA value (in the urodele range); however, Goin, Goin and Bachmann (1967) report quite a low value for another species of this order.

In the frogs and toads (order Anura) there is a very general constancy of chromosome numbers within the larger genera and even within families. Thus most species of *Bufo* have 2n = 22 (but some have 2n = 20), most species of *Hyla* 2n = 24 (a few have 2n = 22 or 2n = 30) and all frogs of the genus *Rana* 2n = 26 (Stohler 1928; Makino 1932*b*; Galgano 1933*a*, *b*; Saez, Rojas and de Robertis 1936; Bushnell, Bushnell and Parker 1939;

Wickbom 1945, 1949, 1950*b*; Matthey 1947*c*; Morescalchi 1962, 1963*a*, 1967*a*, *b*, 1968*a*, *b*; Saez and Zorilla 1963; Ullerich 1966; Bogart 1968). Most Leptodactylidae have 2n = 22, but a few have 2n = 26. Since the chromosomes appear to be generally metacentric in frogs and toads, karyotypic conservatism is not unexpected. The primitive frog *Ascaphus truei* from the northwestern U.S. has 2n = 44 (twelve large metacentrics and thirty-four acrocentric microchromosomes) according to Morescalchi (1967*a*) and the South American *Pipa pipa* has 2n = 22, the size range being also very great (Wickbom 1950*a*); while *Discoglossus pictus* has 2n = 28 (Morescalchi 1964) and *Xenopus laevis* has 2n = 36 (Weiler and Ohno 1962; Morescalchi 1963*b*). *Xenopus* does not show the extreme range in chromosome size that is shown by *Ascaphus* and *Pipa*. The karyotypes of

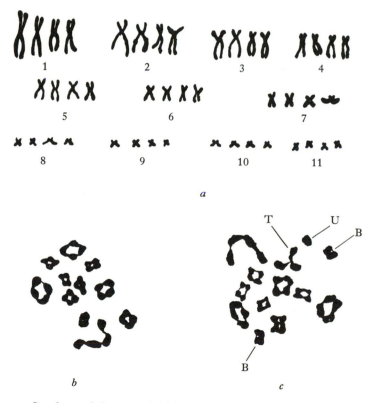

Fig. 12.14. Cytology of the tetraploid frog *Odontophrynus americanus. a*, somatic karyotype of female, with chromosomes arranged in eleven groups of supposed homologues; *b*, and *c* male 'first metaphases' (perhaps in reality diakinesis nuclei). *b* shows 11 quadrivalents, *c* has 9 quadrivalents, 1 trivalent, 2 bivalents and a univalent. From Beçak, Beçak and Rabello (1966), redrawn.

the genera *Bufo, Rana, Hyla, Discoglossus, Pelobates* all show a considerable size-range, but all the chromosomes are metacentric or J-shaped, unlike the situation in the primitive genera *Ascaphus* and *Pipa*. A useful summary of the karyotypes of anurans is provided by Rabello (1970).

In the South American Ceratophrydidae some very much higher chromosome numbers are encountered (Saez and Brum 1959, 1960). In the genus *Odontophrynus* six species have 2n = 22 and seven others have 24, 30, 42, 44, 50 and 60; while in the single 'species' *Ceratophrys ornata* Saez and Zorilla (1963) found four different chromosome numbers: 26, 92, 96 and 108. Beçak, Beçak and Rabello (1966) and Bogart (1967) have published incontrovertible evidence that *Odontophrynus americanus*, with 2n = 44, is a tetraploid, since it forms eleven quadrivalents at male meiosis (occasionally a few trivalents, bivalents and univalents are seen). The other members of this family with high chromosome numbers are most likely also polyploids, since Saez and Brum-Zorilla (1966) appear to have seen multivalents in *Ceratophrys ornata*. *C. dorsata*, with 2n = 104, is an octoploid, since some octovalents were seen at meiosis. The DNA values of *O. cultripes* (2n = 22), *O. americanus* and *C. dorsata* are in the ratio 1:2:4 (Beçak, Beçak, Lavalle and Schreiber 1967). In one small area of Brazil, however, Beçak, Beçak and Vizotto (1970) have found a diploid (2n = 22) population of *O. americanus*. Bogart (1967) has argued, without good evidence, that the polyploidy of these species arose through 'chromatid autonomy' (Schrader and Hughes-Schrader 1958).

In a different family of Anura, the Hylidae, A. Wasserman (1970) has recently shown that the North American species *Hyla versicolor* is almost certainly a tetraploid, but the meiosis is not described in the preliminary account and the DNA value has not been determined. Another hylid, *Phyllomedusa burmeisteri*, is also a tetraploid and forms both quadrivalents and bivalents at meiosis (Beçak, Denaro and Beçak 1970).

The genetic systems of these polyploid Amphibia are somewhat difficult to understand; it is not clear whether they form a significant proportion of aneuploid gametes, and it is not known how sex determination operates. It has been suggested by Schultz (1969) that polyploidy may have been able to establish itself in these bisexual species via an intermediate stage in which they reproduced parthenogenetically; but natural parthenogenesis is not known to occur in the Anura.

The Anura show a considerable range of DNA values, from '43% of that of placental mammals' (in *Leptodactylus ocellatus*) to 147% of the same value (in *Hyla pulchella prasina*); even within the genus *Hyla* there is a considerable range of variation, since *Hyla nana* has a value 54% of the mammalian figure. The octoploid *Ceratophrys dorsata* has 181% of the mammalian figure, which is more, but not much more, than the diploid *Hyla pulchella prasina* (Beçak, Beçak, Schreiber, Lavalle and Amorim 1970).

Reptiles in general show lower DNA values than Amphibia. Relatively low chromosome numbers (n = 15 to 21) occur in the Crocodilia (Cohen and Gans 1970), much higher ones in the order Chelonia (Huang and Clark 1967; Sasaki and Itoh 1967); ranging up to n = 32 in *Amyda japonica* (Oguma 1937a); much lower numbers (n = 13–14) have been found in species of the Amazonian genus *Podocnemis* (Ayres, Sampaio, Barros, Dias and Cunha 1969). *Sphenodon*, the sole living representative of the order Rhynchocephalia, shows n = 18 (Keenan 1932).

The karyotypes of numerous species of lizards were studied by Matthey (1931, 1932a, b, c, 1933, 1939b, 1945a, 1949d), using traditional sectioning techniques. His survey was extended by Asana and Mahabale (1940, 1941), Margot (1946), Makino and Momma (1949), Cavazos (1951), Nogusa (1953), Werner (1956) and Bhatnager (1958). More recently, numerous species have been investigated using modern methods by Gorman (1965, 1968), Gorman and Atkins (1966, 1967), Gorman, Atkins and Holzinger (1967), Huang, Clark and Gans (1967), Gorman, Thomas and Atkins (1968), Dallai and Vegni Talluri (1969), Pennock, Tinkle and Shaw (1969), Gorman and Gress (1970), Cole, Lowe and Wright (1969), Cole (1970) and Ivanov and Federova (1970). There is usually a great range of size of chromosomes

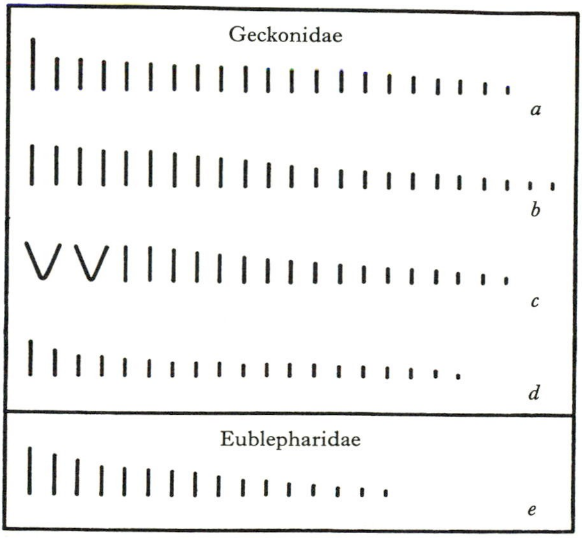

Fig. 12.15. Ideograms of the haploid karyotypes of some lizards. (*a*) *Tarentola mauretanica*; (*b*) *Hemidactylus bowringi*; (*c*) *Gekko japonicus*; (*d*) *Hemidactylus milliousi*; (*e*) *Eublepharis variegatus*. Based on the work of Matthey (1931, 1933).

in this group, and in many species there is a break in the middle of the series, so that there are two rather sharply defined types, the 'macrochromosomes' (which are arranged around the periphery of the spindle at mitosis and meiosis) and the 'microchromosomes' (which occupy the central region of the spindle). This condition is also found in the birds (see pp. 418–9), but there the two types are not so sharply defined and tend to intergrade. Morescalchi (1970) has put forward the view that high chromosome numbers and possession of a distinct category of microchromosomes were characteristics of primitive Amphibia and reptiles and that these features have been lost in the modern families of urodeles, the Anura and the marsupials and Eutheria, but not in the birds and monotremes.

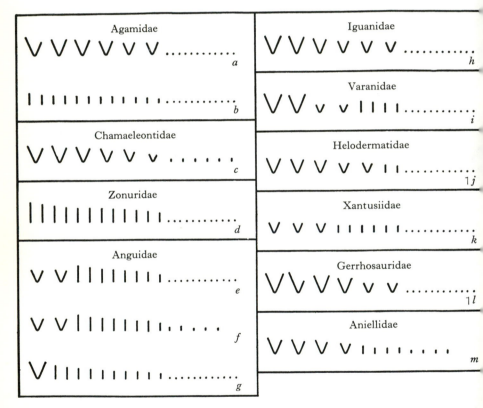

Fig. 12.16. Ideograms of the haploid karyotypes of some more lizards. (a) *Agama stellio* and *Uromastix hardwicki*; (b) *Japarula swinhonis*; (c) *Chamaeleo vulgaris*; (d) *Zonurus catephractus*; (e) *Pseudopus apus*; (f) *Ophisaurus ventralis*; (g) *Gerrhonotus scincicauda*; (h) *Anolis carolinensis*; (i) *Varanus gouldi*; (j) *Heloderma suspectum*; (k) *Xantusia henshawi*; (l) *Gerrhosaurus flavigularis*; (m) *Aniella pulchra*. Based on the work of Matthey (1931, 1933).

On the basis of their karyotypes, Matthey divided the Lacertilia into three main groups: (1) the *'complexe geckonoïde'*, which includes the geckos and the Eublepharidae, (2) the *'complexe iguanoïde'*, which consists of the families Agamidae, Iguanidae, Zonuridae, Anguidac, Hclodcrmatidae, Varanidae, Xantusiidae, Aniellidae and Chamaeleontidae, and (3) the *complexe scincolacertoïde'*, with the families Scincidae and Lacertidae. In the first of these groups there is usually no sharp distinction between the larger and the smaller chromosomes; the number of major chromosome arms in the haploid karyotype ranges from 19 to 32 and there are relatively few metacentrics (*Gekko japonicus* has two pairs (Nakamura 1932); six other species have none).

In the *'complexe iguanoïde'* there is usually a sharp distinction between macro- and microchromosomes. Typically, the macrochromosomes include twelve major arms in the haploid set, while the microchromosome part of the karyotype consists of an equal number of major limbs, but centric fusions may have occurred among either the macrochromosomes or the microchromosomes, or both. This basic karyotype runs through a number of families and is almost world-wide. Thus *Heloderma suspectum* from Arizona has five pairs of large metacentrics, two pairs of large acrocentrics

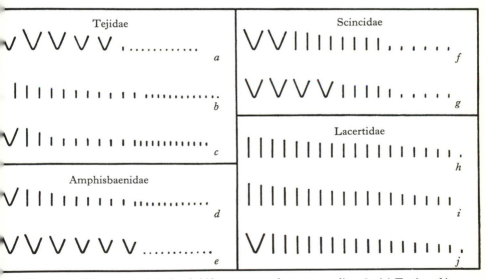

Fig. 12.17. Ideograms of the haploid karyotypes of some more lizards. (*a*) *Tupinambis teguixin*; (*b*) *Ameiva surinamensis*; (*c*) *Cnemidophorus sexlineatus*; (*d*) *Rhineura floridana*; (*e*) *Trogonophis wiegmanni*; (*f*) *Scincus officinalis*; (*g*) *Chalcides tridactylus*; (*h*) *Lacerta* (most species), *Psammodromus hispanicus* and *Tachydromus* spp.; (*i*) *Lacerta vivipara*; (*j*) *Lacerta ocellata*. Based on the work of Matthey (1931, 1933).

and twelve pairs of small acrocentric microchromosomes. Typical iguanids such as the North American *Anolis carolinensis* and *Phrynosoma cornutum* have all their macrochromosomes (but none of their microchromosomes) metacentric. Gorman (1965) and Gorman and Atkins (1966, 1967) have shown that a taxonomic subdivision of the large genus *Anolis* into two sections is supported by differences in karyotype.

Various Japanese workers, e.g. Oguma (1934) and Makino and Asana (1948) claimed that various species of lizards had one chromosome less in the male sex and that there was consequently an $XO(\male):XX(\female)$ sex-determining mechanism. Matthey (1949*d*) and Matthey and van Brink (1956*a*) denied the validity of these claims and strongly insisted that no morphologically differentiated sex chromosomes existed in any reptiles. Later work has not confirmed either of these standpoints. It is true that no credible accounts of visibly different X- and Y-chromosomes exist for turtles (Chelonia), Crocodilia, some groups of lizards and snakes. But Kobel (1962), Beçak, Beçak and Nazareth (1962) and a number of later workers have demonstrated morphologically different X- and Y-chromosomes in many female snakes (i.e. female heterogamety) and Gorman and Atkins (1966) and other recent investigators have similarly demonstrated male heterogamety in several groups of lizards (whereas Ivanov and Federova (1970) have claimed to have demonstrated female heterogamety in *Lacerta strigata*).

In some recent publications, Matthey (1957, 1961; Matthey and van Brink 1956*b*, 1960) has taken up the cytotaxonomy of the chamaeleons of Africa and Madagascar. The haploid numbers range from 18 (six pairs of 'macros' and twelve pairs of 'micros') down to 10, the histogram being a bimodal one with peaks at 12 and 18. Matthey divides the karyotypes of twenty-four species into two types. The 'continental' type, which is widespread in the African species, is characterized by a clear size difference between the six pairs of macrochromosomes and the microchromosomes, there being no elements of intermediate size; and by the fact that the microchromosomes represent only about 15% of the total chromosome length, all the microchromosomes being acrocentrics of about the same size. The 'insular' type of karyotype, which is predominant in Madagascar, has a more or less even gradation of chromosome length, there being chromosomes which are intermediate in size between 'macros' and 'micros'. The chromosomes shorter than the six longest elements are heterogeneous in size and shape and constitute about 32% of the total chromosome length. Species with the typical 'continental' karyotype include *Brookesia stumpfi* and *Chamaeleo fischeri*, *C. cristatus*, *C. oweni*, *C. johnstoni* and *C. parsonii*. The first and the last of these, however, are Madagascan species. The 'insular' karyotype is a good deal more variable. In some cases we could formally derive 'insular' karyotypes from 'continental' ones by centric fusions of originally acrocentric microchromosomes to give metacentrics. This might account for the n = 10 to 12 karyotypes with all or most of the smaller elements metacentric. But it will not account for the 'insular' karyotype of C. *cephalolepis* (Comoro Islands) with n = 14, all except the two smallest chromosomes being acrocentric. The basic phenomenon in the origin of the 'insular' karyotypes seems to have been the increase in size of the microchromosomes by duplications of chromosomal material (or, alternatively, the origin of the 'continental' karyotypes must have involved repeated deletions from the microchromosomes). The distribution of heterochromatic segments has not been studied in this group, but one may suspect that the difference between the two kinds of karyotypes resides in the extent of heterochromatic regions in the smaller chromosomes.

In the '*complexe scinco-lacertoïde*' there are typically eighteen pairs of acrocentric 'macros', together with a single pair of 'micros'. This is the condition in most of the species of the European genus *Lacerta*, but *L. vivipara* has lost the microchromosome pair (Oguma 1934; Matthey 1934a; Margot 1946), presumably as a result of its fusion with one of the larger autosomes. In the Scincidae, *Scincus officinalis* has 14 acrocentrics and 2 metacentrics in the haploid set, while *Chalcides tridactylus* has 10 acrocentrics and 4 metacentrics.

The karyotypes of snakes also show a sharp distinction between macrochromosomes and microchromosomes (Nakamura 1927, 1928, 1935; Bhatnager 1957, 1959, 1960a, b; Beçak and Beçak 1969; Itoh, Sasaki and Makino 1970; Raychaudhuri, Singh and Sharma 1970). The usual haploid set consists of 8 pairs of 'macros' and 10 pairs of 'micros'; the former range from several pairs of large metacentrics to 3 or 4 pairs of medium-sized acrocentric or subacrocentric elements. In the primitive family Boidae no differentiated sex chromosomes have been found, but in members of the families Colubridae, Viperidae, Elapidae and Crotalidae morphologically distinguishable X and Y (or Z and W, according to the terminology adopted) chromosomes have been described in the female sex (Kobel 1962, Beçak, Beçak and Nazareth 1962, 1963; Beçak, Beçak, Nazareth and Ohno 1964). Thus, whereas lizards (with the possible exception of *Lacerta strigata*, referred to earlier) have male heterogamety, snakes have female heterogamety. Apart from the sex chromosomes, the general uniformity of the karyotype throughout the Ophidia contrasts with the diversity found in most families of lizards. However, some Colubridae have aberrant karyotypes such as n = 25 in *Clelia occipitolutea* (presumably due to 6 dissociations) and n = 15 in *Xenodon merremi* (Beçak, Beçak, Nazareth and Ohno 1964).

We have dealt with the evolution of karyotypes in birds in an earlier section (p. 418). In general, bird karyotypes seem to be of a 'reptilian' type, but with considerably higher numbers, which make detailed analysis very difficult. All birds appear to have female heterogamety, but any close evolutionary relationship to snakes seems very unlikely, in spite of the arguments of Beçak, Beçak, Nazareth and Ohno (1964) and Ohno (1967). Evolutionary switches from male to female heterogamety and vice versa seem to have occurred several or even many times in the phylogeny of fishes, amphibia and reptiles, so that too much significance should not be read into the common possession of female heterogamety by these two groups. Birds have low DNA values so that considerable losses of genetic material must have occurred in their phylogeny.

The karyotypes of the monotremes (van Brink 1959; Bick and Jackson 1967a, b) have been stated to be of the same general reptilian type, but appear to be somewhat peculiar. In the Echidna (*Tachyglossus aculeatus*) there is a graded series of 31 autosomal pairs, the larger chromosomes being J-shaped and the smaller ones metacentric. In the platypus (*Ornithorhynchus anatinus*) the karyotype is similar, but there are only 26 pairs of autosomes. The monotremes seem to have male heterogamety, since they have an uneven chromosome number in the male soma (63 in *Tachyglossus*, 53 in *Ornithorhynchus*). The X is a large chromosome in both; but it is not known whether the males are XO, X_1X_2Y or XY with a mechanism similar to that present in the bandicoots (see p. 603) whereby the Y is eliminated from the somatic tissues.

In the true mammals (marsupials and Eutheria) we find a considerable range of chromosome numbers (Fig. 12.18) and a great variety of karyotypes. But although both large and small chromosomes coexist in many species, a clear distinction between macrochromosomes and microchromosomes such as exists in many lizards and other reptiles is never found. Male heterogamety is universal and XO species do not occur, apparently because sex determination depends on a 'dominant Y' principle (see p. 579).

Fig. 12.18. Histogram showing the female haploid chromosome numbers of 924 mammals (based on the list of Matthey (1958) together with the eight supplements by the same author published in the *Mammalian Chromosomes Newsletter* down to January 1970; data for marsupials from Hayman and Martin (1969); determinations made after Matthey's eighth supplement not included, except for *Muntiacus muntjak* with n = 3). Where a species has several subspecies or races with the same chromosome number, it has been counted only once; but where several races have different chromosome numbers, they have all been included. Doubtful cases and species whose populations are regularly polymorphic for chromosome number have been excluded.

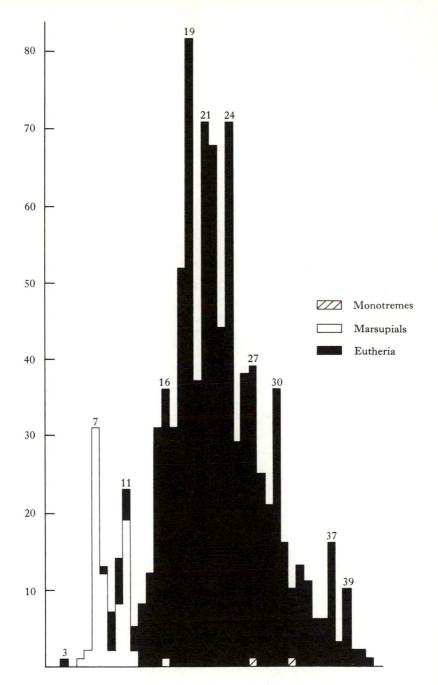

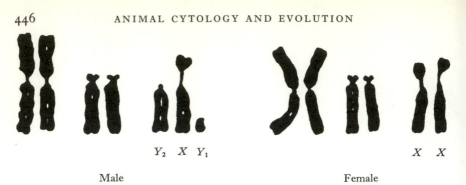

$Y_2 \ X \ Y_1$ $X \ X$

Male   Female

Fig. 12.19. Karyotypes of the male and female of the Indian Muntjac deer, the mammal with the lowest chromosome number known. From Wurster and Benirschke (1970), redrawn.

There is a general difference between the karyotypes of marsupials and Eutheria, the former having, in general, lower chromosome numbers and larger chromosomes. It is rather doubtful, however, whether the DNA values of marsupials are, on the whole, higher than those of placental mammals, as has been stated, for example, by Morescalchi (1970). Certain families of marsupials (e.g. Dasyuridae and Peramelidae) show the very low chromosome number n = 7 in most or all of their species. Others, such as the Didelphidae, Phalangeridae and Macropodidae have a range of chromosome numbers from n = 7 to n = 11 (Reig and Bianchi 1969; Hayman and Martin 1969; Gunson, Sharman and Thomson 1968), with two species of Macropodidae having n = 5 and n = 6 (in the females – the males are XY_1Y_2 and have 2n = 11 and 13 respectively).

The eutherian mammals have, in general, much higher chromosome numbers than the marsupials. However, the mammal with lowest chromosome number is the Muntjac deer, with n = 3 in the female (the male is XY_1Y_2 and has 2n = 7). Karyotype studies on mammalian species have been greatly developed in recent years, using modern methods based on cell culture techniques. However, the relatively large chromosome numbers in most groups are a severe limitation on detailed studies of karyotype evolution.

The extreme variability of mammalian chromosome numbers seems to occur in all the main orders. It is shown equally by macropodid marsupials, shrews, rodents, horses, artiodactyls and primates. Thus, from the beginning of their phylogeny, mammals seem to have been particularly liable to acquire fusions, dissociations and pericentric inversions in their chromosomes. One certainly gets the impression that even the most closely related species of mammals usually differ in karyotype to some extent, and frequently differ very greatly. Thus there are very few pairs or groups of species that *could* be homosequential in the sense of Carson, Clayton and Stalker (1967). It is true that in the cats (genus *Felis*) relative karyotypic uniformity prevails, perhaps in part because the population structure of these wide-ranging hunting carnivores is not conducive to the establishment of chromosomal rearrangements. Thus eight species of *Felis* and three belonging to the closely-related genera *Panthera* and *Lynx* all show $2n = 38$. However, *F. wiedii* and *F. pardalis* are aberrant in having $2n = 36$ (Hsu and Rearden 1965; Hsu, Rearden and Luquette 1963). And even within the group of species with $2n = 38$ the karyotypes are not quite uniform and some pericentric inversions or other rearrangements affecting centromere position have occurred (Maloef and Schneider 1965). The bears and weasels likewise seem to show fairly uniform karyotypes ($2n = 74$ in the former group and usually 38, 40 or 42 in the latter). On the other hand the foxes, which presumably have the same general type of population structure as these other groups of the Carnivora, have a wide range of chromosome numbers ($2n = 38, 40, 64$ and 78 in four species). We have already discussed (Ch. 11) a number of instances of karyotypic diversity and karyotype evolution in various genera of rodents where restricted vagility and the existence of populations with many small local colonies may be especially favourable to the establishment of chromosomal rearrangements. However, one rodent genus in which chromosome numbers seem to be invariable is *Eutamias* (family Sciuridae), in which all the species appear to have $2n = 38$ (Nadler and Block 1962).

Matthey (1957, 1968*b*; see also Matthey and van Brink 1960) has suggested, in his work on the karyotypes of chamaeleons and mice, that cytotaxonomic data might throw light on some general problems of faunistics and zoogeography. If the native rodents of islands such as Madagascar or Australia proved to be cytotaxonomically rather uniform one might incline to think they were descended from a single invading stock; whereas if they were chromosomally diverse, one might conclude in favour of 'multiple colonization'. But the cytotaxonomic diversity of many rodent genera is such that any such conclusions would be unjustified. Thus even if considerable cytotaxonomic diversity were to be found in the rodent faunas of Madagascar or Australia, this would not be convincing evidence of multiple colonization.

Since so much is known of the phylogeny of the horse family, it is interesting to note that the few surviving members of this group (all included in the genus *Equus*) show great karyotypic diversity (Table 12.2). We may at least surmise that the evolution of the Equidae in the past was accompanied by numerous changes in chromosome number and other structural rearrangements of the karyotype. On the other hand, the karyotypes of the

TABLE 12.2. *Chromosome numbers of living members of the Horse family (mainly from Benirschke 1964 and Benirschke et al. 1964, 1965)**

	2n
Equus caballus caballus (Horse)	64
E. przewalskii (Przewalski's Horse)	66
E. onager (Asiatic Onager)	54
E. asinus (Ass)	62
E. grevyi (Grevy's Zebra)	46
E. zebra hartmannae (Mountain Zebra)	34
E. burchelli boehmi (Grant's Zebra)	44

* See also Makino, Sofuni and Sasaki 1963.

Indian and African Elephant seem to be virtually indistinguishable (Hungerford, Chandra, Snyder and Ulmer 1966), so it is probable that the phylogeny of the Proboscidea involved far fewer chromosomal rearrangements than that of the Equidae.

The comparison between the human karyotype and those of the anthropoid apes is naturally of particular interest, in view of the light which it might throw on problems of human origin and evolution. Unfortunately, in a group like the primates, with relatively high chromosome numbers, the techniques available, even today, are inadequate for anything like a complete interpretation of all the cytotaxonomic differences, in terms of the chromosomal rearrangements that have occurred in phylogeny.

The human chromosome number, formerly believed to be 2n = 48, is now known with certainty to be 2n = 46 (Tjio and Levan 1956; Ford and Hamerton 1956; Denver Report 1960; London Conference 1963; Turpin and Lejeune 1965), except for occasional trisomics, intersexes and mosaics. There are no differences in chromosome number between the major racial subdivisions of mankind and the only karyotypic differences that have been reliably reported are minor ones affecting the size of the Y-chromosome and perhaps especially the size of its fluorescent segment. However, it should occasion no particular surprise if, in the future, some other minor cytological differences are discovered. Such differences may perhaps be found in the case of populations such as Bushmen, Australian Aborigines, Ainus, etc. which are now, or have been in the past, strongly isolated. Even if more differences (i.e. ones additional to the size-differences in the Y) are eventually found, there will be no reason to doubt that *Homo sapiens* is, biologically, a single species.

We show, in Figure 12.20, a diagrammatic 'ideogram' of a standard

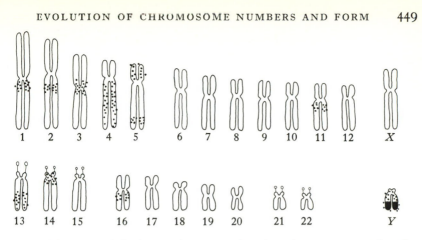

Fig. 12.20. Ideogram of the human karyotype, based on the one presented at the 1966 Chicago Conference on 'Standardization in Human Cytogenetics', with the main late-replicating DNA segments shown (stippled) and the approximate location of the fluorescent segment in the Y-chromosome (black).

human karyotype. There are certain chromosomes such as nos. 1 and 18, that are quite distinctive; but even with the best technical methods it is very difficult to distinguish between elements 13, 14 and 15 and between chromosomes 21 and 22. There are five subacrocentric elements, 13, 14, 15, 21 and 22, which all bear 'satellites' on their short arms. The Y is similar in dimensions to elements 21 and 22, but does not seem to bear a satellite.

The great apes apparently all have 2n = 48 (Yeager *et al.* 1940, Young *et al.* 1960, Chu and Bender 1961 for the Chimpanzee; Chiarelli 1961 for the Orang Utan; Hamerton *et al.* 1961 for the Gorilla). There is a close general resemblance between the karyotypes of these three species and between them and the human karyotype; but many differences of detail are present. There seem to be more definitely metacentric chromosomes in the Gorilla and Chimpanzee than are present in *Homo*. The Gibbons (Hylobatinae) show 2n = 44 and 2n = 50. The lowest chromosome number known is in the Catarrhini, 2n = 42, found in most Cercopithecidae; the highest being 2n = 72, found in two species of *Cercopithecus* (Bender and Chu 1963). The range of chromosome numbers in the Platyrrhini (2n = 34 to 62) and in the Lemurs (2n = 38 to 66) is similar (Bender and Mettler 1958; Chu and Giles 1957). *Tarsius bancanus* has 2n = 80, one of the highest chromosome numbers known in the Mammalia (Klinger 1963). The evidence thus indicates that all Primates have fairly high chromosome numbers. Man is likely to be descended from a 48-chromosome stock in which a centric fusion became established at some stage. The Primates seem to be rather more karyotypically conservative than some other mammalian orders such as the Rodentia. We have already referred, however (p. 399), to the existence of 'races' (probably in reality sibling species) of Lemurs with quite different chromosome numbers.

KARYOTYPIC ORTHOSELECTION

Almost any survey of the chromosome cytology of a group of animals demonstrates evolutionary trends which in earlier writings we have referred to as the *principle of homologous change*, but for which the term *karyotypic orthoselection* is perhaps more appropriate. Even a casual glance at the chromosome sets of a few related species will show that in most of them the lengths and shapes of the chromosomes are not at random. In many species we have *symmetrical karyotypes* (Stebbins 1950) in which all the chromosomes are metacentric and about the same size. In other groups we encounter *asymmetrical karyotypes*, in some of which there may be two size classes of chromosomes, each with definite characteristics. These conditions would not be found with such regularity if the structural changes which become established in phylogeny were of all possible types. It seems, rather, that in many groups chromosome after chromosome has undergone the same type of structural change, so that they have all retained a similar morphology. This is particularly clear in such striking instances as *Lysandra atlantica*, which has acquired about 195 dissociations, the Algerian race of *Gerbillus pyramidum* which has accumulated 19 fusions, the species of trimerotropine grasshoppers which have incorporated one pericentric inversion after another in their genetic systems and such grasshoppers as *Indopodisma* in which the short arms of *all* the chromosomes are much longer than in related forms. It is, however, a much more widespread and even universal principle than these few outstanding instances by themselves suggest. In one lineage, fusions have repeatedly established themselves, in another dissociations, in others paracentric or pericentric inversions or duplications or deletions of heterochromatin. The result is an order and symmetry in the karyotypes of the great majority of animal species which would be entirely lacking if it were not for the operation of this all-pervading principle of karyotypic orthoselection.

It is certain that we do not need to invoke any mystical inherent tendencies to explain these facts: we are dealing with the effects of orthoselection in Simpson's (1944) sense, not orthogenesis in the sense of the earlier palaeontologists. There are, however, two distinct ways in which selection may be imagined to canalize the types of structural changes occurring in chromosomes. On the one hand, there is the possibility that changes of the same structural type, occurring in a species will all have the same kind of physiological effect. Thus the Blundell and Tidbinbilla inversions in *Keyacris scurra* (each of which converted a metacentric chromosome into an acrocentric) are both size-decreasing today, and may have been so when they first arose. And, if we regard the 2n = 36 'race' of *Thais lapillus* as having given rise to the 2n = 26 race by five fusions, each of these events helped to adapt the individual to life in a particular type of environment characterized by increased wave action and a relative abundance of food.

Orthoselection imposed by the environment, because structurally similar chromosome rearrangements tend to cumulatively adapt the organism to new ecological opportunities, may be contrasted with orthoselection caused by the need for chromosomal shapes and sizes to remain harmonious with the dimensions of the spindles and the cells in all the tissues of the body. Control in the first case is external to the body, in the second internal. We may imagine that invasion of a more arid environment, for example, in some way favours a particular type of structural change. Or we may imagine that the development of some new type of cell, smaller or flatter or more elongated than any that existed in the body before, has the same effect. There are many types of unitary control in the cell about which we know next to nothing; why is it, for example, that all the chromosomes of a karyotype are usually about the same thickness at metaphase?

There is another and more subtle way in which karyotypic orthoselection operates. In general, but with certain minor exceptions, the distribution of chiasmata tends to be similar in the different members of the chromosome complement. Some species show conspicuous proximal localization, others distal localization, and such patterns may be common to groups of related species. Thus three species of the grasshopper genus *Stethophyma* (=*Mecostethus*) have pronounced proximal localization of chiasmata in the male, all beetles of the genus *Chilocorus* distal localization and most spiders seem to show extreme proximal localization. But sometimes we find striking differences between related forms, e.g. the grasshopper *Moraba curvicercus* has its chiasmata distributed in a fairly random manner over its chromosome set in spermatogenesis, while in some closely related members of the '*curvicercus* group' extreme distal localization is universal, in the male.

These regularities may be regarded as resulting from a form of karyotypic orthoselection. But they also help to determine further orthoselection, as far as structural rearrangements are concerned; because the distribution of chiasmata is similar in different chromosome elements, the types of rearrangements that manage to establish themselves in phylogeny will tend to be similar.

Groups which originally possessed only acrocentric chromosomes and which still retain this type for the most part (e.g. the acridoid grasshoppers) will be ones in which far more fusions than dissociations will have established themselves; the histogram of their chromosome numbers will be strongly skewed to the left (i.e. towards the lower numbers) and they will exhibit a 'maximum chromosome number'. Groups in which the original chromosomes were all metacentric will, on the other hand, show dissociations in some species; the histogram will be skewed to the right and such groups will tend to have a minimum chromosome number. But what are we to say of groups like the Mammalia in which both acrocentric and metacentric chromosomes exist, and in which the histogram is symmetrical and approximates to a normal curve? Two possibilities seem to exist. The first is that dissociations have actually been as frequent as fusions in the phylogeny of such groups; a rather unlikely state of affairs, since the process of dissociation requires a pre-existing chromosome fragment to serve as a 'donor' of a centromere and telomeres (see p. 225). The other possibility seems to be that dissociations have been less frequent than fusions in the phylogeny of such groups, but that more species with fusions than ones with dissociations have proved to be 'dead ends' in evolution. Although centric fusion is an 'easy' type of structural rearrangement where acrocentric chromosomes are present, it must not be forgotten that it involves the permanent loss of small proximal regions of two chromosome pairs. There may be some, or even many, groups in which there are a sufficient number of biologically important genetic loci close to the centromeres to make centric fusion distinctly hazardous, either in the short-term or the long-term sense.

The problem of karyotypic orthoselection is intimately bound up with the question whether we can really recognize a 'primitive karyotype' in certain groups, from which all the aberrant karyotypes encountered in the group have been derived. 'Primitive karyotypes' can *apparently* be found in such diverse groups as the genus *Drosophila* (Patterson and Stone 1952; Stone 1962), the morabine grasshoppers (see p. 372), and the 'iguanoid' families of lizards (see p. 441). But to what extent are these karyotypes really primitive – may they not, perhaps, represent 'equilibrium conditions' to which karyotypically aberrant lineages revert repeatedly in the course of phylogeny? Probably the answer to this question depends on the antiquity of the 'group' in question and the number of species included in it. We may obviously consider a particular karyotype as the primitive one for a 'species group' of drosophilids or morabine grasshoppers. But can we legitimately speak of a primitive karyotype for the genus *Drosophila* as a whole (over 1000 species), the morabine grasshoppers (about 250 species) or the iguanoid families of lizards? Because the phylogeny of these larger groups can never be known in detail, the extent to which an apparently primitive karyotype (e.g. the '17-chromosome karyotype' of the morabines) has been repeatedly attained in them is impossible to determine. But the two concepts of a 'primitive karyotype' and an 'equilibrium karyotype' are not necessarily mutually exclusive. Thus we can imagine that in a particular group we may have an original ancestral karyotype which is periodically reacquired in phylogeny. In such groups it might eventually be possible to distinguish those species which have retained the primitive karyotype (never having lost it) and those which have reacquired it after having for a time 'experimented' with various aberrant karyotypes.

The histogram of chromosome number for the mammalia shows modes at $n = 19$, 21 and 24; the mean chromosome number is 21.8. The mean for the Eutheria is 23.0. In previous editions of this book the histogram for the mammalia showed a mode of 24, and many authors, from Painter (1925) to Matthey (1955, 1958b) have suggested that $n = 24$ was the primitive or ancestral chromosome number for eutherian mammals. The change in the modes between 1958 and 1970 has been caused partly by the determination of many chromosome numbers in the order Primates, in which $n = 21$ is undoubtedly the commonest chromosome number. The Rodentia, by far the most numerous order of mammals, still show a mode at $n = 24$. It seems quite possible that if as many Artiodactyla as Primates were studied cytologically, the histogram for the eutherian mammals as a whole might have an entirely different shape, with a mode at or near $n = 30$. The weakness of the case for regarding any particular chromosome number as ancestral for the Eutheria in general should be obvious.

In spite of these strictures, it is fairly clear that almost all eutherian orders probably evolved from ancestors having chromosome numbers between $n = 19$ and $n = 30$. Eutherian mammals with chromosome numbers above and below those limits certainly represent for the most part relatively recent derivatives from stocks with chromosome numbers within that range. There is one important fact about the histogram for the mammalia which was emphasized by Matthey (1958b) and which is still true today, in spite of the change in mode – namely that the graph for the Eutheria is a fairly good approximation to a normal curve. We interpret this to mean that increases in chromosome number (by dissociation) have been about as frequent in mammalian phylogeny as decreases (by fusion). On the other hand there certainly occur groups in which the histogram is strongly skewed to the left, i.e. ones in which there seem to have been far more fusions than dissociations. The grasshoppers of the family Acrididae seem to furnish an extreme example of this type. Another group whose histogram is very much skewed to the left is the Lepidoptera (Fig. 12.3), only in this case there are a few species in which really spectacular increases in chromosome number seem to have occurred.

The question remains, however, as to what we actually mean if we say that there have been about as many dissociations as fusions in the phylogeny of the eutherian mammals. Could it not be that there have been more fusions, but that species with low chromosome numbers have been less successful in evolution and have left fewer descendent species?

The answer one gives to this question depends in part on his view as to whether chromosomal rearrangements have a causal relation to speciation or whether, alternatively, they are merely concomitants of the speciation process. On the former view one might think of a mammalian species with many acrocentric chromosomes as one in which large numbers of different fusions could occur; it would hence be one with the potential for producing large numbers of new species, usually as geographically peripheral isolates. On the other hand a species with all its chromosomes metacentric would be one in which no further fusions could occur; it might hence be unable to give rise to cytogenetically isolated, geographically peripheral incipient species.

However, the answer to this question will certainly not be the same for all groups of animals. Thus in many groups of grasshoppers where the chromosomes are acrocentric, fusions seem to be incapable to establish themselves (see p. 424). And in the Hawaiian Drosophilidae speciation seems to have occurred on a very considerable scale without any assistance from chromosomal rearrangements of any kind (see p. 338). Statements on the relationship between chromosomal rearrangements and speciation consequently seem best restricted to particular groups in which the evidence is particularly complete and especially to ones in which studies have been carried out on critical cases of populations that are on the threshold of specific status.

POLYPLOIDY IN ANIMALS

In the higher plants, as is now well known, polyploidy has played a major role in speciation and evolution, particularly since it has made possible the establishment of many fertile genotypes of hybrid origin (allopolyploids). It is quite possible that more than half the species of higher plants are polyploids or have been derived from polyploid ancestors, and in the grasses this figure is probably over 70%. However the extent to which polyploidy has occurred in the major groups of plants varies a great deal. It seems hardly to have played a significant role at all in the fungi and the gymnosperms, but has been very important in the ferns, psilotales and horsetails (Manton 1950; Manton and Sledge 1954; Walker 1962). It has not occurred in certain large genera of woody plants such as *Quercus* and *Eucalyptus*, but has been frequent in some others such as *Salix* and *Casuarina*.

In animals it is admitted that evolutionary polyploidy has occurred in a number of species whose reproduction is parthenogenetic (see Ch. 18). These polyploid parthenogenetic species include a number of Diptera, Lepidoptera, Coleoptera, Turbellaria, Oligochaeta, Crustacea and isolated representatives of other groups, so that there is clearly no *general* barrier to the development of evolutionary polyploidy in animals with parthenogenetic methods of reproduction (barriers may exist in specific instances, since not *all* parthenogenetic forms are polyploids).

On the other hand, it is clear that a powerful barrier, or barriers, to evolutionary polyploidy does operate in the case of sexually reproducing animals, and that polyploidy has consequently played a far smaller role in animal evolution than in that of the higher plants. This much is agreed by all authorities, but there has been considerable argument as to how absolute this barrier (or barriers) has been.

From the reproductive standpoint we may distinguish three types of sexual reproduction in animals: (1) self-fertilizing hermaphroditism, (2) cross-fertilizing hermaphroditism, (3) bisexuality. The first two categories undoubtedly intergrade, i.e. there are probably a number of hermaphroditic forms which are capable of both self- and cross-fertilization. However, we have very little precise information as to the extent to which self-fertilization does actually occur in such groups of hermaphroditic animals as the flatworms, pulmonate Mollusca, etc. One has the impression that it is probably rare, exceptional and in some species only takes place under abnormal circumstances. If this is so, our category (1) may be numerically unimportant.

The first suggestion to account for the rarity of polyploidy in animals as compared with plants was that of Muller (1925), who pointed out that in bisexual organisms even-numbered polyploidy will inevitably upset the sex-chromosome mechanism, since tetraploidy, hexaploidy, etc. will automatically abolish the heterogamety upon which sex determination depends. Thus a tetraploid of an originally $XY:XX$ form would be $XXYY:XXXX$ and nearly all the gametes of the originally heterogametic sex would be XY, due to the two X's and the two Y's forming separate bivalents at meiosis. No one, of course, expects that odd-numbered polyploidy (triploidy etc.) could establish itself in sexually-reproducing species which necessarily depend on a regular meiotic mechanism, although it is known that triploidy *has* been successful in many parthenogenetic forms.

Muller's argument was first put forward at a time when it was believed that the 'genic balance' principle of sex determination which operates in

Drosophila was general or universal. Now that it is known that in some groups such as urodeles and mammals the 'switch mechanism' controlling sex differentiation is simply the presence or absence of a Y-chromosome (see Ch. 15), Muller's argument loses a good deal of its force. It has been criticized, on the basis of the work of Westergaard (1940, 1953, 1958), who was able to produce a stable tetraploid strain of the normally bisexual plant *Melandrium album*. He obtained XXXX (♀) and $XXYY$ (♂) plants experimentally, and from them an $XXXY$ (♂): XXXX (♀) stock was built up, the male-determining power of the Y being strong enough to override the three X's (see also Warmke 1946). This may be what has happened in the evolution of polyploid species of dioecious willows and some other plants. Several authors have wondered whether it may not also have occurred in bisexual animals.

The second type of barrier that can be imagined as operating against evolutionary polyploidy in animals is simply the prevalence of cross-fertilization which is obviously obligatory in bisexual species and may also be so in the majority of hermaphroditic ones. A single tetraploid individual that has arisen accidentally will find only diploid mates and consequently will leave only sterile triploid offspring – and even if several tetraploid individuals are produced in a large population the chances of two of them mating will be infinitesimal. In a sense this type of barrier to the development of polyploidy is one which arises from the general argument against all types of sympatric speciation. Several tetraploid individuals are only likely to coexist at the same time in a *very* large population of individuals (of the order of many millions or billions) – and it is precisely in such a population that geographic or ecological isolation is least possible. Parenthetically, it may be remarked that the only type of polyploid which can really be expected to arise accidentally in natural populations with any appreciable frequency are triploids, which have arisen as a result of the union of a haploid and a diploid gamete, the latter having originated by failure of nuclear division (Böök 1940, found such a triploid individual in a natural population of newts).

A third type of barrier to evolutionary polyploidy in animals which has been suggested by some authors such as Stebbins (1950) is what we may call the *histological* one. Animals are histologically more complex than plants and it might be imagined that a doubling of the chromosome number, with various effects on cellular dimensions, might lead to difficulties in certain types of cells, perhaps especially in specialized ones such as those of the nervous system, sensory epithelia such as the retina, etc. The main argument – and a very powerful one – against this particular type of 'barrier' being a real or effective one is based on the high frequency of evolutionary polyploidy in parthenogenetic animals (see pp. 709–10).

We shall indicate at this stage our conviction that the main barrier that has operated against polyploidy in sexually reproducing (i.e. non-parthenogenetic) animals is the second one, i.e. the prevalence of obligatory or near-obligatory cross-fertilization; with 'Muller's principle' playing an additional role in certain groups where the 'genic balance' system of sex determination exists. These, however, are conclusions which arise from a survey and analysis of the actual state of affairs regarding the occurrence of evolutionary polyploidy in various groups of animals; so that we shall now proceed to discuss the evidence.

Taking the hermaphroditic groups first, in the cestodes and the digenetic trematodes there is no evidence for polyploidy. In the former group Jones (1945) studied fifteen species whose haploid numbers ranged from 5 to 8; while in the latter the data of Britt (1947) on forty-nine species include no plausible instances of evolutionary polyploidy.

A considerable variety of reproductive mechanisms exist in the Mollusca, but at least in the pulmonates and opisthobranchs hermaphroditism is universal. Self-fertilization is probably rare but has been observed to take place in *Limnaea auricularia, Bulinus contortus, Arion ater, Omalogyra* (references in Burch 1960a and Morton 1964) and may occur in many more, either as a regular or an occasional phenomenon. Thus there exists a theoretical possibility that evolutionary polyploidy might have occurred in these groups.

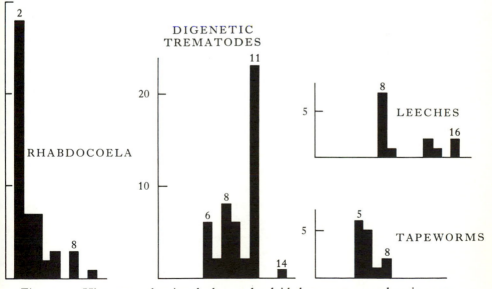

Fig. 12.21. Histograms showing the known haploid chromosome numbers in some groups of hermaphroditic animals (for euthyneurous Mollusca, see Fig. 12.4).

The Pulmonata include the aquatic Basommatophora and the terrestrial Stylommatophora, which are best considered separately. Following the earlier work of Inaba (1953) and Le Calvez and Certain (1950), Burch (1960a, b, 1961, 1962) has reviewed the evidence in the case of the Basommatophora, of which about fifty species have now been studied cytologically. The great majority of the species have n = 18, and this was almost certainly the ancestral number for the group as a whole. *Gyraulus circumstriatus*, with n = 36, (*G. deflectus* has n = 18, *G. triomantium* is stated to have n = 16) is a probable, but not certain tetraploid. In the African genus *Bulinus*, *B. tropicus* has n = 18; *B. natalensis* has 19–21 bodies at first metaphase, there being apparently some supernumerary chromosomes. The *truncatus* group, on the other hand, which are important as intermediate hosts of *Schistosoma haematobium*, are tetraploids with 36 pairs of chromosomes and *B. sericinus* is an octoploid, with 72 pairs (Burch 1963). The ancylids *Ferrissia parallela* (n = 30) and *Ancylastrum costulatum* (n = ca. 59) *may* be polyploids but too little is known about the cytology of other members of the family Ancylidae for us to be certain of this. The common, widespread species of *Lymnaea*, *Physa* and *Planorbis* are definitely *not* polyploids and it seems that evolutionary polyploidy has been quite rare in this group.

The most recent summary of the situation in the land snails (Stylommatophora) is that of Inaba (1959a), but there are more recent contributions by Beeson (1960) and Burch and Heard (1962). The haploid numbers of about 80 species range from n = 17 to n = 34; there is not a single form that could plausibly be interpreted as a polyploid. Burch (1964) has later shown that two species of the succineid snail genus *Catinella* have n = 5 and n = 6; these are presumably forms that have undergone a reduction in chromosome number and they do not suggest that all other snails have a polyploid ancestry.

We may, perhaps, compare the above data for the hermaphroditic gastropods with the situation in the generally bisexual prosobranch families. Nishikawa (1962) has determined the chromosome numbers of numerous species and reviewed the findings of earlier workers. In general, these marine molluscs seem to be rather conservative in their chromosomal evolution. Thus nine species of Limpets (Patellacea) have n = 9, eleven out of twelve species of Trochidae and Turbinidae that have been studied have n = 18 and seven species are likewise known to show n = 18. The Littorinidae, with n = 15, 17, 17, 18 in four species, seem to show rather more variation in chromosome number. The Neogastropoda, which mostly have higher numbers (n = 13 to 36) show considerably more variability; within the genus *Thais* (*sens. lat.*) there are forms with n = 13, 18 and 30. Altogether, we cannot say that the bisexual groups of gastropods show more or less variation in chromosome number than the hermaphroditic ones.

In the hermaphroditic Turbellaria there are a number of possible, and some probable, polyploids. Thus in the rhabdocoel genus *Mesostoma* there are species with n = 2, 4, 5 and 8 (Valkanov 1938; Ruebush 1938; Husted and Ruebush 1940); it is possible that those with 4 and 8 chromosomes may be tetraploid and octoploid, respectively, but there is no proof of this. The evidence is somewhat better in the case of *Macrostomum hustedi* (Jones 1944), where n = 6 (ten other species of the genus have n = 3), since in this species multivalents occur at meiosis. But since one of the multivalents was an association of six chromosomes, one may wonder whether translocations as well may not have occurred. In the genus *Phaenocora* there are species with n = 2, one with n = 3 and one with n = 6 (Ruebush 1938; Cognetti de Martiis 1922; Valkanov 1938); again there is a possibility, but no proof, that the species with n = 6 is a polyploid. In the large genus *Dalyellia* there is no evidence for polyploidy, since the eleven species which have been studied all have n = 2 (Ruebush 1937, 1938).

The Tricladida have, in general, higher chromosome numbers than the rhabdocoels. *Dendrocoelum infernale*, with n = 16, appears to be a genuine tetraploid by comparison with D. *lacteum*, with n = 8 (Aeppli 1951, 1952). We shall deal later (p. 735) with the complicated situation in certain species of *Dugesia* and *Polycelis* in which polyploidy is combined with parthenogenesis. Self-fertilization seems to be rare in triclads, but occurs regularly in the marine species *Procerodes lobata* and the freshwater *Cura foremanii*; and sporadically in *Polycelis nigra* (references in Benazzi 1962).

Two groups of Annelida, namely the leeches (Hirudinea) and the earthworms (Oligochaeta) are basically hermaphroditic, although a number of species have developed parthenogenetic modes of reproduction (spermatogenesis may be abnormal or entirely suppressed in such forms). In the leeches there are a few possible polyploids, but the evidence (Wendrowsky 1928) is inconclusive and there have been no recent studies. The earthworms do, however, show a number of undoubted polyploid species and races, at any rate in the families Lumbricidae (Muldal 1948, 1949, 1952; Omodeo 1951a, b, c, d, e, f, 1952, 1953a, b, 1955) and Enchytraeidae (Christensen 1959, 1960, 1961). Most of these are parthenogenetic (see p. 732) but at least eight out of forty-two species of sexually reproducing Lumbricidae and twenty-seven out of seventy-three Enchytraeidae seem to be polyploids (see Table 19.3). Some of these forms, especially in the Lumbricidae, are members of species groups or complexes that also include parthenogenetic forms. It rather looks as if polyploidy in this group was closely linked with the parthenogenetic methods of reproduction developed in many of the species or biotypes, even though there are some polyploid forms which reproduce exclusively by sexual means at the present time. The large genus

Lumbricus appears to consist entirely of diploid, sexual species (Muldal 1952).

Experimental and spontaneous triploid and tetraploid individuals (as opposed to regularly polyploid populations or races) have been studied in numerous species of animals. Such anomalous triploids are usually viable and may show some degree of fertility in both Urodeles (Fankhauser and Humphrey 1950; Kawamura 1951a) and anurans (Kawamura 1940, 1951b). A triploid individual of *Triturus vulgaris* was even found in the wild by Böök (1940). Numerous triploid, tetraploid and pentaploid urodele larvae, together with a few aneuploid ($2n - 2$ and $2n - 3$) individuals were obtained by Griffiths (1941), Fankhauser (1945), Böök (1945) and Fischberg (1948), following exposure of the eggs to extreme temperatures. The triploid and tetraploid ones were sometimes able to complete metamorphosis, but the aneuploids were very abnormal and all died before or during metamorphosis. In all these experiments a small percentage of haploid larvae were also obtained; they usually seem to be viable only up to the time of metamorphosis (it will be recalled that in *Drosophila* haploidy seems to be lethal in the early embryo). Haploid frogs have, however, been reared to the adult stage by Miyada (1960). Not very much is known as to the precise mode of origin of the abnormal chromosome numbers in the 'heteroploid' newt larvae. The haploid individuals are probably ones which have developed from eggs in which the male or the female pronucleus (or both) started to cleave without previous fusion. Triploid larvae may arise from fusion of a haploid sperm with a diploid (unreduced) egg nucleus, but the mode of origin of the triploids and pentaploids is not clear. In the human species triploids seem to be absolutely inviable (Edwards *et al.* 1967); they have a frequency of about 4.5% in spontaneous abortions (Carr 1965). Edwards (1970) suggests that about 1% of human conceptions are triploids; they include *XXX*, *XXY* and (probably) *XYY* types. In the chicken, an adult triploid was found by Ohno, Kittrell, Christian, Stenius and Witt (1963).

From time to time suggestions have been made that polyploidy has occurred in the phylogeny of some group consisting entirely of bisexual species. Thus Slack (1938a) claimed that it had occurred in certain Heteroptera, Lorković (1941, 1949) argued in favour of polyploidy in several genera of butterflies, and Svärdson (1945) suggested that some salmonid fishes were polyploids. More recently, Sachs (1952) and Darlington (1953) claimed that the Golden Hamster, *Mesocricetus auratus* ($n = 22$) arose as an allopolyploid between two species such as *Cricetus cricetus* and *Cricetulus griseus* (both with $n = 11$).

Apart from the case of the South American Anura of the family Ceratophrydidae and two species of hylid frogs (see p. 438) in which multivalents occur at meiosis, and the possible cases of the Goldfish and the Carp

(p. 435), the evidence for such views has never been of a conclusive kind and has been criticized by White (1946*c*, 1959*a*) and Matthey (1953*b*). Svärdson's view that polyploidy has occurred in the salmonid fishes has been widely quoted, and the additional data of Kupka (1948) on the cytology of some Swiss coregonids was considered by some to support it, but Matthey (1949*d*) and Bogart (1967) are rightly sceptical concerning it, since even where the evidence of chromosome numbers appears to support the hypothesis of evolutionary polyploidy, the number of major chromosome arms (*nombre fondamental*) is in disagreement with it. Actually, Svärdson's and Kupka's arguments in favour of polyploidy in the coregonids are not even in agreement with one another, Svärdson claiming that the species with 2n = 80 (*Coregonus laveretus* and *C. albula*) are octoploids with a 'basic' haploid number of 10, while Kupka regards the forms with 2n = 70 or 72 (*C. schinzii duplex*, *C. wartmanni caeruleus* and *C. exiguus albellus*) as tetraploid by comparison with *C. asperi maraenoides* (2n = *c*. 36), i.e. as having a 'basic' haploid number of 18. In a later paper Kupka (1950) illustrated a cleavage mitosis of *C. wartmanni* in which 36 chromosomes are more darkly stained than the other 36, which appear as less condensed. On the basis of chromosome length and form he interprets these two 'sets', not as the maternal and paternal karyotypes, but as half the maternal and half the paternal one in each case. This he regards as evidence for the allopolyploid nature of *C. wartmanni*. His interpretation seems insecurely based, since no critical karyotype has been worked out, the arm lengths of the individual chromosomes being unknown. Simon and Dollar (1963) report that five species of the salmonid genus *Oncorhynchus* have chromosome numbers ranging from 2n = 52 to 2n = 74; the *nombre fondamental* is 102–4 in four species and 112 in the fifth. They also report that *Salmo gairdneri* and *S. clarki lewisi* have 2n = 60 and 64 respectively (*nombre fondamental* 104 and 106). Boothroyd (1959) reported that Canadian *Salmo salar* has 2n = 56, compared with 2n = 60 in European populations, but the *nombre fondamental* is the same. Finally, Rees (1964) has found that the DNA value of diploid nuclei in European *Salmo salar* and *S. trutta* (2n = 80) is almost precisely the same, a finding which is not easily reconcileable with Svärdson's view that *salar* is a hexaploid and *trutta* an octoploid. Altogether, the evidence that has accumulated since the early work of Svärdson and Kupka seems to weaken considerably the case for polyploidy and suggests that the evolution of chromosome numbers in the salmonid fishes is far more likely to have been by fusions and dissociations. However, Suomalainen (1958*a*) reports that the Pike-perch *Lucioperca sandra* has 2n = 48 in Finland, and 2n = 24 in Sweden, which looks more like a genuine case of polyploidy, although no details of the karyotypes have been published.

In the *Mesocricetus* case the polyploidy hypothesis entirely fails to explain why *auratus* has a single pair of sex chromosomes rather than two pairs (if *auratus* really were an $XXXX:XXXY$ species, there should be cytological evidence of this). Actually, there are a whole series of palaearctic cricetine rodents with chromosome numbers intermediate between n = 11 and n = 22 (Matthey 1952a, 1953a, 1957, 1959; Makino 1949). The alleged allopolyploid origin of *auratus* is not anatomically plausible; neither is it supported by Moses and Yerganian's (1952) data on the DNA values of *griseus* and *auratus*, the nuclear dimensions of the species concerned (Weiss 1958) and the genetic evidence (Robinson 1958). In fact, it is clear that it is the cricetine rodents with *low* chromosome numbers (i.e. 2n = 20 or 22) which are the aberrant members of their taxonomic group and which hence stand in need of explanation (such an explanation must surely be in terms of chromosomal fusions and other types of rearrangements, especially pericentric inversions, but possibly including some others such as deletion of heterochromatic arms). Mayr (1963) concludes categorically: 'No case of [evolutionary] polyploidy is known among mammals' – and this is still the case.

Certain authors such as Bauer (1947) and E. Goldschmidt (1953a) have put forward the view that evolutionary polyploidy has occurred in groups of organisms with multiple sex-chromosome mechanisms (e.g. X_1X_2Y in one sex, $X_1X_1X_2X_2$ in the other). We shall consider later (Ch. 17) the ways in which such multiple sex-chromosome mechanisms have arisen; there is not a single one that cannot be much more plausibly interpreted on the basis of chromosomal rearrangements without any polyploidy: in fact it is in general absolutely essential to the proper functioning of such mechanisms that the X_1 and X_2 chromosomes should be *non* homologous (if they had arisen by polyploidy, they would be homologous, at least in origin). Bauer argued in favour of polyploidy in the Dermaptera since at the time three species with low chromosome numbers (2n = 12 and 14) were known to be XY in the males, while several with $2n\male = 25$ had X_1X_2Y males and *Prolabia arachidis* with $2n\male = 38$ had $X_1X_2X_3Y$ males.

Henderson (1970) has even suggested that allopolyploidy has occurred several times in the Dermaptera, and that it has involved hybridization between diploid and tetraploid species as well as between diploid species having different chromosome numbers. There is certainly no solid evidence for these fanciful speculations and the supposed correlation between high chromosome numbers and multiple sex-chromosome mechanisms disappears when we take into account the additional data of Srikantappa and Rajasekarasetty (1969–70), Srikantappa and Aswathanarayana (1970), Ortiz (1969) and G. C. Webb (unpublished).

It may well be that there is a general tendency in some groups for multiple

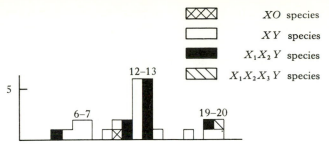

Fig. 12.22. Histogram of the known female haploid chromosome numbers of Earwigs (order Dermaptera, including Hemimerina).

sex-chromosome mechanisms to occur in the species with larger numbers of autosomes, but any such tendency must surely be attributed to the operation of the principle of karyotypic orthoselection, favouring dissociations in both types of chromosomes in certain evolutionary lineages. In Spiders X_1X_2 species seem to have higher numbers of autosomes than XO ones, and $X_1X_2X_3$ ones still higher numbers.

The whole topic of the occurrence of evolutionary polyploidy in bisexual animals seems to be one which restraint, objectivity and critical judgment are especially necessary. This applies equally to the protagonists on both sides of the argument. Because the overwhelming majority of the claims for polyploidy that have been made in the past were obviously rash, ill-considered guesses based on uncritical evidence, we must not conclude that genuine cases of evolutionary polyploidy in bisexual animals do not exist. Those of the ceratophrydid amphibia, *Hyla versicolor* and *Phyllomedusa burmeisteri* (see p. 438) are undoubtedly genuine and there are several highly suggestive cases in teleost fishes. But it is clear that any such instances are so extremely rare that they are quite insignificant in the general picture of animal evolution.

One way in which chromosome numbers might theoretically undergo evolutionary increase is through reduplication of existing chromosomes, i.e. through polysomy. Trisomic individuals must be constantly arising in all species, with a certain frequency, and if two trisomics mate some tetrasomic offspring are likely to be produced. A stabilized tetrasomic population will have one more pair of chromosomes than formerly; it might be expected to show an occasional trivalent or quadrivalent at the first metaphase of meiosis if the chiasma frequency is high enough for this (but not if the chromosomes are incapable of forming more than a single chiasma).

Such a hypothetical mode of increase of chromosome number may seem plausible in view of the evidence from 'repeats' in *Drosophila* polytene chromosomes that evolutionary duplications of genetic material have occurred in that genus (although not duplications of whole chromosomes). However, there is really no evidence that simple polysomy of whole chromosomes has been a significant factor in the chromosomal evolution of animals. In the first place, species or races showing an occasional multivalent such as one would expect to result from a tetrasomic condition are not known. Secondly it is a matter of general experience that individuals trisomic or tetrasomic for an autosome have a very low viability in most species of animals. In *Drosophila melanogaster* this is true even for triplo-4 individuals,

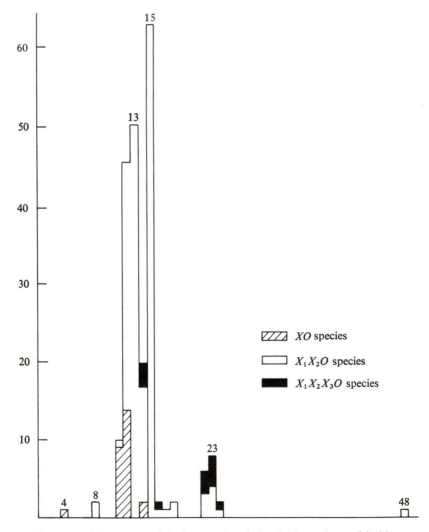

Fig. 12.23. Histogram of the known female haploid numbers of Spiders.

in which the little fourth chromosome (about 2% of the total chromosome length) is present in triplicate. In the grasshopper *Keyacris scurra*, in which many thousands of individuals have been examined cytologically, trisomic individuals have not been found, which suggests (since they must surely arise from time to time) that they are inviable. In the related *viatica* group of species trisomic embryos have been encountered in the offspring of backcrosses involving two different species, but such trisomic individuals do not seem to survive to the adult stage. Various workers (e.g. Callan, 1941*a*; Lewis and John 1959; Sharman, Parshad and Bedi 1963; Hewitt and John 1965; Sharma, Gupta and Randhawa 1967) have found individual grasshoppers trisomic for various autosomes, but in none of these cases was it established that the trisomy was present in the somatic tissues as well as in the testis; some at least were karyotypic mosaics.

The situation in vertebrates is similar. In the urodele *Pleurodeles waltlii*, Lacroix (1967) obtained a fertile female which was trisomic for a chromosome constituting about 10% of the total chromosome length. Fredga (1968*a*) has reported a water vole (*Arvicola terrestris*), caught in the wild, that was trisomic for the smallest autosome. Several trisomic individuals have now been recorded for the laboratory mouse (Cattanach 1964; Griffen and Bunker 1964, 1967); they were phenotypically normal males but sterile or semi-sterile.

About four or five different syndromes in the human species have been associated with trisomy for specific autosomes (see review of the earlier literature by Fraccaro 1962). Some difficulties in their interpretation arise from inherent technical obstacles (e.g. the difficulty of identifying the separate elements of the karyotype, and the possibility that some of the individuals in question may have been cytological mosaics and that others may have been carrying duplicated fragments, as a result of translocations or deletions, instead of being simple trisomics).

The first trisomic syndrome to be recognized was the well-known condition of Down's Syndrome (Mongolism), which occurs with a frequency of about 1 in 700 births. This is usually due to the presence of chromosome no. 21 in triplicate (Lejeune, Gautier and Turpin 1959; C. E. Ford *et al.* 1959; Böök *et al.* 1959; Harnden, Miller and Penrose 1960). Most such individuals have 47 chromosomes instead of the usual 46. Mongolism involves poor viability, complete sterility in the male and very low fertility in the female. It has been shown that in male mongols a small trivalent is usually present at meiosis (Sasaki 1965). Some mongols have only 46 chromosomes but carry translocations involving chromosome 21 which lead to trisomy for a part of this element (Fraccaro *et al.* 1960; Penrose *et al.* 1960; Polani *et al.* 1960). It is now well known that the incidence of trisomy mongolism in the offspring is strongly correlated with maternal age at the time of conception, so that non-disjunction (probably at meiosis) is more frequent in the oocytes of older women. Naturally, there is no such correlation with maternal age in the case of translocation mongolism. McClure *et al.* (1969) have described a case of autosomal trisomy in a chimpanzee with a phenotype resembling that of Down's syndrome. Heterozygosity for a deletion of part of chromosome 21 leads to a syndrome which has been called anti-mongolism (Reisman *et al.* 1966).

Trisomy for chromosome no. 18 has been reported by a number of authors (e.g. Edwards *et al.* 1960; Pätau *et al.* 1960, 1961) in a number of patients showing a characteristic syndrome of mental retardation, hypertonicity, flexion of fingers, low-set, malformed ears, small mandible, etc. Trisomy for chromosome 13 has been reported by several workers (Pätau *et al.* 1960; Therman *et al.* 1961). The patients in which this condition occurred showed mental retardation, deafness, myoclonic seizures, eye defects and other abnormalities; the condition leads to early death. Trisomy for chromosomes 16 and 22 has also been reported by various workers (e.g. F. J. W. Lewis *et al.* 1963 and Uchida *et al.* 1968).

Obviously, all these types of trisomy in man are sublethal; a considerable literature on their pathogenesis exists (Rohde 1965). We may suspect that in the case of most of the 22 human autosomes the trisomic condition is absolutely inviable at the embryonic stage, so that these genotypes are only seen as spontaneous abortions. And this is probably especially true of the autosomes longer than number 13. Hardly any autosomal monosomics are viable in the human species;* as far as the *X*-chromosome is concerned, *XO* individuals seem to be reasonably healthy, although sterile 'sub-females'. The fact that *XXY* and even *XXXY*, *XXXXY* and *XXXXX* individuals are fairly viable presumably depends on the fact that all but one of the *X*'s are genetically inactivated, to a considerable extent, in early development (see p. 596).

Polysomy might appear particularly probable, as an evolutionary mechanism, in groups such as birds, where some of the chromosomes are extremely minute and only contain a very small amount of the total DNA of the nucleus. Nevertheless, as we have seen, the range of chromosome numbers in birds is rather small and certainly does not suggest that they possess a mechanism for increasing the chromosome number which is absent in other groups of animals.

---

* A number of cases of apparent monosomy for chromosome 21 or 22 have been described. All these patients were severely affected physically or mentally. Some of them may have been mosaics, i.e. the monosomy may not have been present in all cells.

13

THE EVOLUTION OF THE MEIOTIC MECHANISM

> We ought now to be used, at least in biology, to hypotheses
> which are not like physical laws, but more like Michael
> Ventris' reading of the Linear B script – the test of them is
> how far they give consistent translations of empirical fact
> when we take proper precautions against circular fallacies . . .
>
> ALEX COMFORT: *Darwin and Freud* (*Lancet* 1960)

The general course of meiosis is surprisingly constant throughout multicellular animals and plants. In the vast majority of diploid, sexually reproducing organisms the same sequence of pairing, chiasma formation and
segregation occurs with complete regularity, the variations which are
observed being concerned with such details as the timing of the individual
stages, the dimensions and degree of elongation of the spindles, the number
of chiasmata per bivalent, their localization in particular regions and so on.
The close similarity between the meiotic processes of organisms so widely
different as lilies, grasshoppers, urodeles and mammals impressed itself
forcefully on some of the early investigators (e.g. Janssens 1900) and has
also been emphasized by more recent investigators (see Darlington and
La Cour 1942, p. 18). It is still rather uncertain how far this similarity, in the
case of the higher plants and animals, reflects a common ancestry and how
far it is due to convergence.

The existence of a fundamental similarity between the meiotic mechanisms
of most multicellular organisms should not blind us, however, to the complex
evolution which meiosis has undergone and the numerous detailed modifications which occur. Some of these modifications are of a more profound
type, and may involve complete suppression of chiasma formation, for
example. We shall discuss such types of meiosis in the present chapter; in
the following one we deal with the even more radical modifications that have
evolved in a few groups, usually in association with particular types of

cleavage mitoses involving elimination of certain chromosomes from the soma or germ-line or both.

Anomalous meiotic mechanisms are usually confined to one sex, usually but not invariably the male. However, in Dragonflies the meiotic mechanism in both sexes shows a rather fundamental difference from the usual process (see p. 494) and in the Gall Midges (Ch. 14, pp. 523–38) radical evolutionary changes have occurred in the meiotic machinery of both spermatogenesis and oogenesis. In some of the cases considered in Chapter 14 only one or two sperms are formed from each primary spermatocyte, instead of the usual four, the other meiotic products degenerating in the same way as the polar bodies do in oogenesis.

Anomalous types of meiosis usually affect all the chromosomes of the karyotype in the same way, or at any rate all the autosomes. There are, however, a few instances in which only one chromosome pair behaves anomalously. Thus in the Heteroptera of the family Coreidae there is always a pair of so-called m-chromosomes which fail to pair at zygotene and remain as univalents throughout the prophase of the first meiotic division. In *Archimerus calcarator* and *Pachylis gigas* the m-chromosomes are exceedingly minute, while in *Protenor belfragei*, *Leptoglossus phyllopus* and *Anasa tristis* they are not much smaller than the ordinary autosomes (Wilson 1905a, b, 1909a, b, 1911). In all these forms their behaviour is essentially the same; they are heteropycnotic during the prophase of the first meiotic division and show no tendency to pair until premetaphase, when they arrange themselves in the centre of the developing spindle, one above the other (Wilson 1911). Thus at first anaphase they regularly pass to opposite poles, although they have never formed a true bivalent. It is not known whether they have in the same manner in oogenesis, or whether they undergo synapsis and chiasma formation in the female sex. Their mode of orientation within the developing first division spindle in the male is clearly a very special mechanism which ensures a regular disjunction in spite of the absence of synapsis. In individuals of *Acanthocephala* (*Metapodius*) which are trisomic for the m-chromosome the three m's arrange themselves in a row in the centre of the spindle (Wilson 1910), and a similar orientation occurs in tetraploid spermatocytes of *Archimerus*, in which a chain of four m's is formed (Wilson 1932).

MUTATIONAL CHANGES IN THE MEIOTIC MECHANISM

We must imagine the various evolutionary changes which the meiotic mechanism has undergone as having had a mutational origin. It is, of course, not necessary to assume that any of them arose suddenly in its completed condition; many of them may have arisen in a somewhat irregular and inefficient form at first and been gradually perfected under the influence of natural selection. Stated in this way, we seem to be postulating an intermediate evolutionary stage in which the meiotic processes were mechanically unsatisfactory. It is, of course, highly unlikely that any mutation which rendered meiosis inefficient, in a sexually reproducing organism, would stand any chance of establishment. But there are several instances of two types of meiosis occurring in the same gonad, in different lobes, cysts or other divisions (see pp. 512–14). It is thus quite possible that some of the more drastic evolutionary modifications of the meiotic mechanism may have been 'tried out' first in special gonadal regions rather than in the gonad as a whole.

A number of mutations are known in various species of *Drosophila* which affect meiosis or some other part of the chromosome cycle. Most of them seem to be limited in their affects to the female, and do not alter spermatogenesis in any way. One of these mutations, '*Gowen's gene*' (a third-chromosome recessive) prevents crossing-over in *D. melanogaster* and also leads to a considerable amount of non-disjunction (Gowen 1928, 1933). These effects are presumably due to asynapsis of the chromosomes in oogenesis, but there has apparently been no cytological investigation of this case.

In *D. simulans* another third-chromosome mutant, *claret*, in addition to affecting eye colour, leads to extensive disturbances of meiosis in the eggs (distorted spindles at the first meiotic division, failure of the second division with reorganization of the daughter chromosomes into a large number of micronuclei, each containing very few of them) and also to widespread irregularities of cleavage (Sturtevant 1929; Wald 1936). Fertilization and cleavage fail in most eggs of homozygous *claret* females, but a few offspring do develop, some of which are XO/XX gynandromorphs. An apparently homologous mutation in *D. melanogaster* affects the colour of the eyes in a similar manner, but has no cytological effects, although the allele ca^{nd} (claret non-disjunctional) causes nondisjunction of the X and chromosome 4 (Lewis and Gencarella 1952; Davis 1963).

The case of the *claret* gene in *D. simulans* suggests that some mutations affecting chromosome behaviour in spermatogenesis or oogenesis may have established themselves in evolution because of pleiotropic effects on the phenotype, e.g. on eye colour. In *D. subobscura*, Fahmy (1952) studied the effects of a *cross-over suppressor* mutation which prevents chiasma formation in oogenesis and leads to some asynapsis (or desynapsis?) The second meiotic divisions and cleavage are more nearly normal than in the case of the *simulans claret*.

These three mutations are all strictly sex-limited in their effects, the heterozygous and homozygous males being normally fertile, and they are completely recessive in the female. It would hardly be possible to maintain laboratory stocks of such mutants if they were partially dominant or affected both sexes.

Sandler, Lindsley, Nicoletti and Trippa (1968) recovered a total of fourteen meiotic mutants from two natural populations of D. *melanogaster* near Rome. In all cases mutants affecting the first meiotic division were limited to one sex or the other in their effects, emphasizing the fact that the genetic control of the first meiotic division is to a large extent independent in the two sexes. On the other hand one mutant caused a high rate of non-disjunction at the second meiotic division in both sexes (Nicoletti 1968).

MEIOTIC DRIVE MECHANISMS

Several types of cytogenetic systems are known, both in plants and animals, in which a heterozygote Aa produces more than 50% of one kind of gamete (say A). Such systems are consequently exceptions to the ordinary Mendelian ratios. If the A-chromosome itself determines the mechanism producing unequal segregation it will automatically increase in frequency in the population. Sandlér and Novitski (1957) applied the term *meiotic drive* to such mechanisms. The basis for deviations from the expected 1:1 segregation ratio is fairly obvious in the case of the female since both meiotic divisions in oogenesis involve an unequal cytoplasmic division and mechanisms causing unequal chromosomal segregation at either of the meiotic divisions are consequently to be expected. The situation in the case of spermatogenesis is different since (except in such organisms as the midges of the families Sciaridae and Cecidomyidae, dealt with in the next chapter) the two meiotic divisions are believed to give rise to four functional sperms. Zimmering, Sandler and Nicoletti (1970) have considered six possible mechanisms leading to unequal transmission of the members of a pair of homologous chromosomes in a heterozygote. These are: (1) a supplementary replication of the 'driven' chromosome, with a concomitant loss or degeneration of its homologue, (2) impairment of sperm function by abnormal chromosome behaviour (e.g. chromosome breakage), with sperms carrying the favoured chromosome remaining unimpaired, (3) and (4) sperm competition or dysfunction, i.e. differential sperm behaviour depending on the genetic constitution, (5) a system in which only one of the products of the first meiotic division is destined to give rise to functional sperms, with preferential segregation of the favoured homologue to the functional pole at first anaphase, (6) differential 'acceptance' by the ovum of the two kinds of sperms. It will be seen that only two of these conceivable systems, namely

(1) and (5) are meiotic in the strict sense, so that to call the others mechanisms of meiotic drive seems rather inappropriate (although (2) could be a meiotic mechanism in some instances). The consequences of all of these possible mechanisms would, however, be similar in terms of population genetics.

It has been shown by Prout (in Dunn 1953) that if a heterozygote produces two kinds of gametes with frequencies k and $(1 - k)$ respectively, and if the heterozygote has a reproductive fitness f_{het}, then the chromosome causing meiotic drive will become established in the population if $2f_{het}k > 1$. Fixation of the chromosome will be expected when f_{hom} (the fitness of the homozygote) $> 2f_{het}(1 - k)$.

The so-called sex-ratio condition in the *obscura* group of *Drosophila* is a classical example of meiotic drive. 'Sex ratio' males carry an X-chromosome which causes them to produce more daughters than sons, the deviation from equality not being due to zygotic mortality. The phenomenon is only superficially similar to another sex-ratio condition, known especially in the *willistoni* group of *Drosophila*, in which infection by a cytoplasmically transmitted microorganism leads to death of Y-carrying zygotes.

As far as the *obscura* group is concerned, *sex-ratio* conditions were found in wild populations of *D. affinis* by Gershenson (1928). They are also known in *D. pseudoobscura, D. persimilis, D. azteca* and *D. athabasca* (Sturtevant and Dobzhansky 1936a; Dobzhansky 1937c; Darlington and Dobzhansky 1942; Wallace 1948). In all of these cases *sex-ratio* males produce over 90% daughters in their offspring, irrespective of the genotype of the mother. *Sex-ratio* is thus sex-limited in expression and sex-linked in inheritance, behaving as if it were a gene located in the right arm of the X-chromosome (actually it is more probable that it depends on several genetic loci in that arm). At least in *pseudoobscura, persimilis, affinis* and *azteca*, X-chromosomes carrying *sex ratio* are cytologically distinguishable by the presence of inversions, from X's lacking this gene or complex of genes. The normal or standard XR of *pseudoobscura* is cytologically indistinguishable from the *sex-ratio* XR of *persimilis*; and in *pseudoobscura* the standard and sex-ratio chromosomes differ by three independent inversions, while in *persimilis* the corresponding chromosomes differ only by a single inversion, which is not identical with any of the three in *pseudoobscura* (Fig. 13.1).

The spermatogenesis of sex-ratio males of *pseudoobscura* was studied by Sturtevant and Dobzhansky (1936a), whose conclusions have been widely cited; but a reinvestigation by Novitski, Peacock and Engel (1965) has shown that the mechanism is quite different from that postulated by the earlier workers. The X and Y regularly pass to opposite poles at the first anaphase. The Y then degenerates, the time of degeneration being somewhat

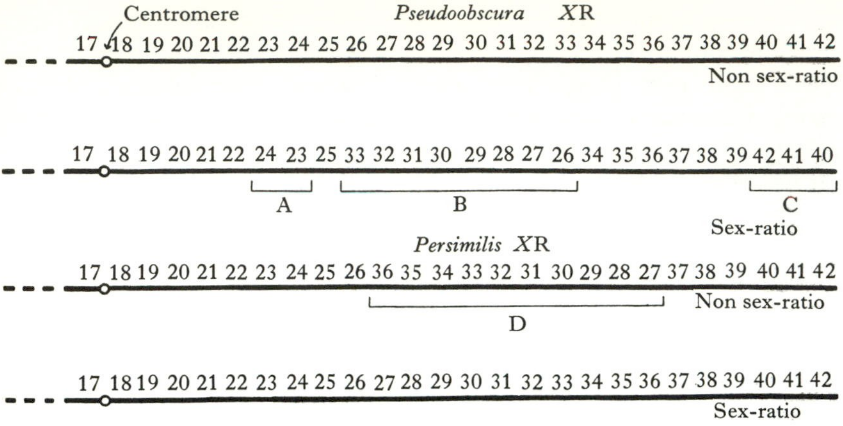

Centromere Pseudoobscura XR

17 18 19 20 21 22 23 24 25 26 27 28 29 30 31 32 33 34 35 36 37 38 39 40 41 42

Non sex-ratio

17 18 19 20 21 22 24 23 25 33 32 31 30 29 28 27 26 34 35 36 37 38 39 42 41 40

A B C
Sex-ratio

Persimilis XR

17 18 19 20 21 22 23 24 25 26 36 35 34 33 32 31 30 29 28 27 37 38 39 40 41 42

Non sex-ratio
D

17 18 19 20 21 22 23 24 25 26 27 28 29 30 31 32 33 34 35 36 37 38 39 40 41 42

Sex-ratio

Fig. 13.1. Sequence of regions in the 'non sex-ratio' and 'sex-ratio' XR chromosome limb in *Drosophila pseudoobscura* and *D. persimilis*. In the former species the two chromosomes differ in respect of three inversions, A, B and C; while in *persimilis* they differ by a single inversion, D. The numbers 17–42 correspond to the sections of the salivary chromosome map.

variable. Spermatids in which Y-chromosomes have degenerated do not form functional sperms, so that in sex-ratio males each primary spermatocyte ordinarily produces only two sperms capable of fertilizing eggs. Although the number of spermatids per cyst is normal, the number of mature sperms per bundle is only about half this, so that approximately half the sperms fail to complete their development (Policansky and Ellison 1970). The X does not undergo an extra replication cycle, as supposed by Sturtevant and Dobzhansky. Apparently, in some cases sperms are formed containing both an X and a Y, a Y alone or no sex chromosome at all. The last two classes give rise to the rare sons, which form less than 10% of the progeny. Some of these sons are XY, others XO, the latter being, of course, sterile.

Since the progenies of sex-ratio males are of approximately normal size, it follows, as first pointed out by Gershenson (1928) in the case of *D. obscura*, that the frequency of *sr* X-chromosomes should increase at each generation (at the expense of the Y and *standard* X-chromosomes, both of which would tend to disappear from the populations), unless natural selection counteracted such an increase. Since a population consisting entirely of females (presumably incapable of parthenogenesis) would inevitably become extinct, it seemed clear that the frequency of *sr* chromosomes must be kept in check by some means.

Sex-ratio X-chromosomes are widespread in natural populations of several species of *Drosophila*. In *D. pseudoobscura sr* seems to be absent from the most northern part of the distribution area of the species (British Columbia, Washington, Oregon) but reaches a frequency of about 30% in certain areas of southern Arizona and northern Mexico. All populations from the southwestern United States and Mexico seem to contain some *sr* X-chromosomes (usually about 5–15%). The situation in *persimilis* seems to be basically similar, the most northerly populations lacking *sr* chromosomes.

It was found by Wallace (1948) that in population cages an equilibrium is established between *sr-X* and *standard-X*-chromosomes at 16.5 °C (*sr* attaining a frequency of 6–10%). At 25°C, however (under the conditions of the experiment), *sr* was eventually eliminated from the population. Wallace's analysis makes it clear that, in addition to the fact that *sr* chromosomes are reproducing nearly twice as fast as *standard X*'s in the male, they also tend to be kept in the population by the high adaptive value of the heterozygous *sr/standard* combination in the female. On the other hand, the frequency of *sr* is kept down by the fact that homozygous *sr/sr* females have a very low viability. Thus we have in the sex-ratio phenomenon a most interesting combination of a heterotic mechanism with one which disturbs the usual sex ratio of the species in a direction (excess of females) which raises the innate rate of increase of the population. But a high price is paid for these advantages, in the form of the near-lethality of the *sr/sr* females. This genetic load may be insupportable in the northern populations, liable to pass through narrow 'bottle necks' in the winter.

Since the three inversions of the *pseudoobscura* chromosome are independent, one might assume that they would become separated by crossing-over; but there is evidence that they only do so extremely rarely (Wallace 1948). It is probable that the combination of the three inversions effectively suppresses crossing-over between them as well as within them, and that the complex of genes responsible for the sex-ratio effect (and possibly for the heterotic effect as well) may be located between the inversions as well as (or instead of) within them. Since the *standard* sequence in *pseudoobscura* is identical with the *sex ratio* sequence in *persimilis*, it seems unlikely that sex ratio is a position effect. Novitski (1947) discovered an autosomal recessive gene in *D. affinis* which reverses the action of *sr*, so that *sr* males which are homozygous for it have only male offspring. Stalker (1961) has studied a similar *sex-ratio* condition in *D. paramelanica*. *Sex-ratio* X-chromosomes in this species differ from standard X's by two inversions in the 'right' arm and two in the 'left' arm. There are two kinds of sex-ratio X's, referred to by Stalker as 'Northern' and 'Southern'. Northern-type sex-ratio X's have the effect suppressed by southern-type Y-chromosomes; southern-type sex-ratio X's are not suppressible by any type of Y. As in *pseudoobscura*, the highest frequencies of sex-ratio X's are attained in southern populations of the species, northern populations having far fewer.

A very different type of sex-ratio distortion, due to a factor called 'D', has been described in the mosquito *Aedes aegypti* by Craig, Hickey and VandeHey (1960) and Hickey and Craig (1966*a*, *b*). This insect has XY males, there being a long homologous pairing sequent in the sex chromosomes. Males that are Xd/YD produce a preponderance of male offspring. There is no effect in XD/YD or XD/Yd males or in females. It seems probable that this system does depend on unequal segregation at meiosis; there is some evidence that Xd/YD males produce less than the normal number of sperms.

Another example of 'meiotic drive' is the so-called *segregation-distorter* (S.D.) locus found in certain stocks of *Drosophila melanogaster* from Wisconsin and Baja California (Sandler, Hiraizumi and Sandler 1958; Sandler and Hiraizumi 1959, 1960*a*, *b*; Hiraizumi, Sandler and Crow 1960; Mange 1961). SD is located in the second chromosome, near the proximal heterochromatin. Males heterozygous for SD transmit the SD-bearing chromosome to the great majority of their offspring, usually 90% or more. Segregation is normal in females heterozygous for SD and in homozygotes of both sexes (even in homozygotes where one SD locus was of Wisconsin origin and the other from a Baja California population, 3200 km away). Since the effect is confined to the male it is obviously not connected with crossing-over; but the fact that it is not shown when the SD locus is covered by a heterozygous inversion may suggest that close synapsis is a *sine qua non* for its expression. Crow, Thomas and Sandler (1962) originally believed that the SD locus caused actual breakage of the homologous chromosome. However Peacock and Erickson (1964, 1965) showed that this was not so. These workers put forward the astonishing hypothesis that even in the normal spermatogenesis of *Drosophila* only one of the cells formed at the first meiotic division gives rise to functional sperms and that the other produces two sperms which are incapable of fertilizing an egg. They suggested that the basis of the segregation–distortion phenomenon was a directed segregation of the SD^+ chromosome to the 'non functional' pole at the first meiotic division. The same interpretation was used by Peacock (1965*b*) to explain the segregation of the X- and Y-chromosomes in a stock where the X was carrying a large deficiency in the heterochromatin.

There were three lines of evidence which seemed to point to this conclusion. In the first place, there can be no doubt that in certain strains of *D. melanogaster* the primary spermatocytes do contain in their cytoplasm masses of symbiotic microorganisms, probably rickettsial in nature (Erickson and Acton 1969) and that these are unequally distributed to the two daughter cells at first anaphase; they were consequently assumed to be 'markers' of an inherently polarized cytoplasmic division. In addition, Peacock and Erickson claimed that SD and normal males produced approximately equal numbers of total progeny and that in both cases the ratio of progeny to sperm transferred in copulation was about 0.5.

These statistical ratios have now, however, been challenged by a number of other workers. Nicoletti, Trippa and De Marco (1967) found that the fertility of heterozygous SD males was about half the normal level. Hartl, Hiraizumi and Crow (1967) and Hiraizumi and Watanabe (1969) have also shown that the number of progeny of SD/SD^+ males is diminished by comparison with normal males and, further, that in different crosses the number of progeny is negatively correlated with the extent of segregation distortion. Also, Zimmering, Barnabo, Femino and Fowler (1970) have found very variable progeny: sperm ratios in matings of SD males with females of different genotypes, due either to different relative efficiencies with which the females utilize sperm or specific interactions between the genotype of the sperm and that of the female or her eggs. Finally, Hartl (1968) has shown that males which carry two SD chromosomes form motile sperm which is transferred in copulation, but such males have their fecundity strongly reduced (in extreme cases to zero), as if each SD chromosome caused the sperm carrying the other to be non-functional. Taken together, all this evidence strongly suggests that Peacock and Erickson's original hypothesis was incorrect and that the true explanation of the SD case is simply that the sperms carrying the SD^+ chromosome are in most cases non-functional. The actual mechanism which is the immediate cause of the non functioning of the + spermatozoa has been investigated by Tokuyasu, Peacock and Hardy (1971) who have shown that in the testes of SD/SD^+ males 0 to 32 of the 64 spermatids in each cyst fail to undergo the normal 'individualizing' process which strips away the syncytial bridges, excess cytoplasm and unneeded organelles from the developing gametes. Only sperms that undergo 'individualization' are released from the testis.

Chromosomes bearing SD may also carry lethal factors; several of those found in wild populations were associated with inversions. Linkage with a lethal will tend to counteract the tendency of SD to increase in frequency in the population, and will lead to an equilibrium. We agree with Mayr (1963) that although these examples of meiotic drive are extremely interesting as cytogenetic mechanisms and as 'natural experiments' in population genetics, there is (in spite of the suggestions of Sandler and Novitski 1957) no real evidence that they are in any way relevant to progressive phyletic evolution. It is possible, however, that meiotic drive may in some instances have helped chromosomal rearrangements, which would not otherwise have been able to do so, to establish themselves in evolution. If this is so it may, in some instances, have played a role in speciation. It is also conceivable that meiotic drive mechanisms may have been contributory causes to the extinction of some species; and this possibility will have to be examined if work now in progress shows that they can be used for the control or elimination of pest species.

Although it has now been shown that the mode of action of SD does not depend on chromosome breakage there is one case known in which a meiotic drive mechanism does involve breakage of a chromosome caused by a specific locus in the homologous chromosome. This is in the 'RD' strain of *Drosophila melanogaster* (originally derived from an irradiation experiment), in which males with the RD factor on their X-chromosome produce an excess of X-bearing over Y-bearing sperm. Erickson (1965) has shown that in this line the Y undergoes fragmentation following the second meiotic division, but apparently not at a fixed locus. The Y fragments apparently have a cell-lethal effect during spermiogenesis. White (1966a) has reported an instance of spontaneous chromosome breakage, but in this case at a specific locus, occurring at the first meiotic division in a grasshopper; the exact mechanism is not understood, however, and it is not known whether this is a single-locus phenomenon (i.e. a chromosome with a 'suicidal' mutation) or a two-locus one ('murder' of one chromosome by another as in the case of RD).

ACHIASMATIC MEIOSIS

Forms of meiosis without chiasmata have been evolved in a number of groups of animals. In all cases except one, the abolition of chiasmata is confined to one sex, usually the male. It is to be presumed that absence of chiasmata is accompanied by absence of genetic crossing-over, but it is only in the males of *Drosophila* sp. and *Phryne fenestralis* (Diptera, Nematocera) that this has been confirmed genetically.

Achiasmatic meiosis has been reported in several Protozoa such as *Actinophrys sol* (Heliozoa) and *Aggregata eberthi* (Coccidia) (Belar 1923b, 1926). It is definitely present in some Foraminifera (Le Calvez 1950; K. G. Grell 1958) and also in the flagellate *Trichonympha* (Cleveland 1949) and in the gregarine *Stylocephalus longicollis* (Grell 1940). Some other Protozoa show quite clear chiasmata at meiosis, however. Until recently no instances of achiasmatic meiosis were known in plants, but recently Noda (1968) has reported a species of *Fritillaria* with achiasmatic bivalents in the pollen mother cells (numerous other species of this genus show chiasmata, which are proximally localized in some forms).

In certain hermaphroditic enchytraeid worms (Oligochaeta) chiasmata seem to be absent in both oogenesis and spermatogenesis (Christensen 1961), but this is the only case in which the cytological evidence suggests that intra-chromosomal recombination has been entirely abolished in a sexually reproducing species – a state of affairs which may be regarded as 'half-way to parthenogenesis'.

Other invertebrates in which chiasmata seem to be absent in spermatogenesis are the mollusc *Sphaerium corneum* (Keyl 1956), a few species of mites (Keyl 1957*a*; see also the figures of Sokolov 1945, 1954) and certain species of scorpions, especially those belonging to the genera *Tityus* and *Isometrus* (family Buthidae) (Piza 1939, 1943*a*, *b*, 1947*a*, *b*, *c*). In a number of species of copepods (a group in which the female is the heterogametic sex – see p. 605) chiasmata are absent in the female meiosis (W. Beermann 1954; S. Beermann 1959; Ar-Rushdi 1963).

In insects it is well known that spermatogenesis without chiasmata occurs in many (probably all) families of 'higher' Diptera (suborder Brachycera) and also in certain families of the more primitive suborder Nematocera

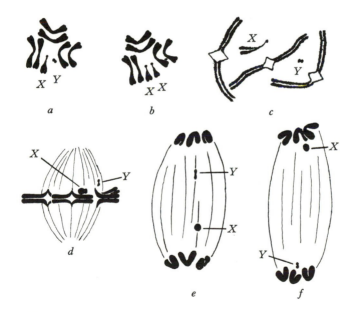

Fig. 13.2. The chromosome cycle of *Fungivora guttata* (Diptera, family Mycetophilidae). *a*, spermatogonial metaphase, showing the acrocentric *X* and the very small *Y*; *b*, female somatic metaphase; *c*, diplotene; *d*, first metaphase in side view; *e* and *f*, early and late first anaphase stages. The autosomal bivalents are achiasmatic and the *X* and *Y* are never associated to form a bivalent. Based on Le Calvez (1947), redrawn and modified.

such as the Phryneidae, Bibionidae, Thaumaleidae, Scatopsidae, Blepharoceridae and Mycetophilidae (Fig. 13.2). One might well add the Sciaridae and Cecidomyidae here, but in these families the male meiosis has been even more profoundly modified, so that it is best considered separately (Ch. 14). In several other families of Nematocera such as the Culicidae (Fig. 13.3), Dixidae, Chironomidae, Simuliidae and Psychodidae chiasmata seem to be universally present in the males.

Apart from these Diptera, the main examples of achiasmatic meiosis in insects are the Scorpion flies of the genus *Panorpa* (Ullerich 1961) and certain mantids (White 1938, 1965*b*, and later unpublished work; Hughes-Schrader 1943*a*, 1950, 1953*b*; Gupta 1966*b*). In the Orthoptera (Saltatoria) the only species known to have achiasmatic spermatogenesis are two species of African eumastacid grasshoppers belonging to the sub-family Thericleinae (White 1965*a*).

Since it is quite clear that achiasmatic forms of meiosis have arisen polyphyletically in many groups it is likely that the details will be somewhat different in the various cases; the mode of evolutionary origin may also have been different. Christensen (1961) has described several types of achiasmatic meiosis in the worms *Buchholzia*, *Fridericia*, *Enchytronia*, *Marionina* and *Henlea*; they seem to differ mainly in the presence or absence of premetaphase stretch. Most of these enchytraeid species with achiasmatic meiosis are diploids; but a number of tetraploid biotypes exist. These latter show almost exclusively bivalents, but occasional quadrivalents are seen. The majority of species of enchytraeids have perfectly typical chiasmate bivalents in oogenesis and probably in spermatogenesis as well.

Grasshoppers are classical material for the study of 'normal' meiosis but in the eumastacid grasshoppers of the African subfamily Thericleinae (twenty-four species have been studied) the diplotene and diakinesis stages are entirely absent in spermatogenesis, the transition from pachytene to first metaphase being quite direct, without any 'opening-out' of the bivalents

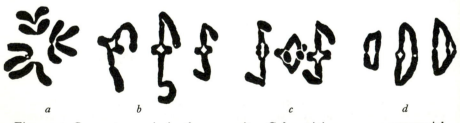

    a   *b*   *c*   *d*

Fig. 13.3. Spermatogenesis in the mosquito, *Culex pipiens*. *a*, spermatogonial metaphase; *b*, *c* and *d*, first metaphases in side view with 3, 4 and 6 chiasmata respectively. It is not certain which of the three pairs of chromosomes is the *XY* pair. From Moffett (1936).

(White 1965a). In these cases the *early* first metaphase bivalents are cylindrical worm-like structures, slightly forked proximally, where the centromeres are pulling the chromosomes asunder; apart from this proximal fork, no trace of a reductional split can be seen (Fig. 13.4). Mid-way through the

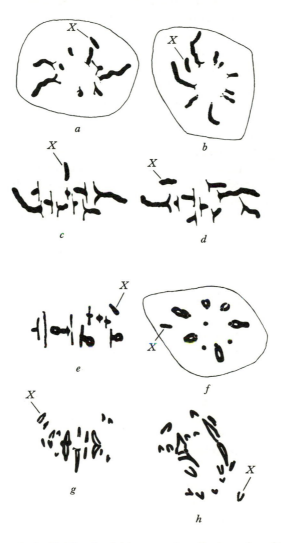

Fig. 13.4. First meiotic division (male) in an undescribed species of *Thericles* with 'cryptochiasmatic' meiosis. *a, b*, very early first metaphases in polar view (no chiasmata visible); *c, d*, early first metaphase in side view; *e, f*, later first metaphases in side and polar view (chiasmata now visible); *g, h*, first anaphases in side view. From White (1965a).

first metaphase a dramatic change occurs in the bivalents; the delayed opening-out occurs and it is now obvious that chiasmata are present after all. In some other species of this group the appearance of the chiasmata is even more delayed, and definite cruciform configurations only become evident very briefly at first anaphase. But in such species the first anaphase chromosomes have the chromatids widely divergent, as if they had been separated in the final terminalization of proximal or interstitial chiasmata. This 'cryptochiasmate' type of meiosis does not occur in the other subfamilies of the Eumastacidae which have been investigated cytologically – it is apparently restricted to the Thericleinae. The rapid condensation of the pachytene chromosomes in the Thericleinae may be the antithesis to the 'diffuse' diplotene seen, for example, in the Grouse Locusts (Henderson 1961). It is possible that in the latter insects much synthesis of proteins needed for sperm formation takes place between pachytene and first metaphase, while in the Thericleinae these substances have for the most part already been synthesized before the end of pachytene (White 1965a).

In these Thericleinae it is fairly clear that the apparent absence of chiasmata in the early first metaphase bivalents is deceptive. Crossing-over has occurred, but because of the delayed 'opening-out', chiasmata do not become visible until much later. In two species of Thericleinae, however (*Thericles whitei* and an undescribed species), chiasmata seem to be genuinely absent at all stages of the male meiosis. White (1965a) described achiasmatic meiosis in three species of Thericleinae, but his species '*V*' was probably conspecific with *T. whitei*. In these species, in sharp contrast to those previously referred to, a very evident 'reductional split' can be seen throughout the whole of premetaphase and first metaphase (Fig. 13.5). Moreover, the chromatids of the early first anaphase chromosomes are closely paired, so that no split can be seen between them. There is thus critical evidence that no chiasmata have been forced along to the distal ends of the chromosomes during anaphase separation. Oogenesis has been studied in *T. whitei*; quite typical chiasmata were seen in the female metaphase bivalents of this species. The genetic system is thus presumably *Drosophila*-like (White 1965a).

These unique grasshoppers seem to present a most instructive evolutionary series. It is fairly obvious that in those Thericleinae which only reveal their chiasmata very briefly at the end of metaphase or in anaphase, the 'mechanical' function of the chiasmata is vestigial. Such forms are hence preadapted for the final loss of the chiasmata which has occurred in the two species which show a clear reductional split at all stages. This loss could have occurred in gradual stages in these thericleines; it is not necessary to assume that it did so by a single cytological mutation, although it may have done.

The meiotic mechanism in these achiasmatic Thericleinae seems to be fundamentally different from that in the oogenesis of the copepod *Ectocyclops* (W. Beermann 1954). In the latter case the chromosomes seem to keep parallel as they separate at first anaphase, almost as if they had diffuse centromeres (in fact, the *Ectocyclops* chromosomes are acrocentric, as can be seen from the mitotic divisions). At the second metaphase and anaphase of *Ectocyclops*, however, the centromeres precede the rest of the chromosome in the poleward movement.

The situation in the Praying Mantids seems to be basically similar to that in the Thericleinae. The majority of the 90-odd species of these insects that

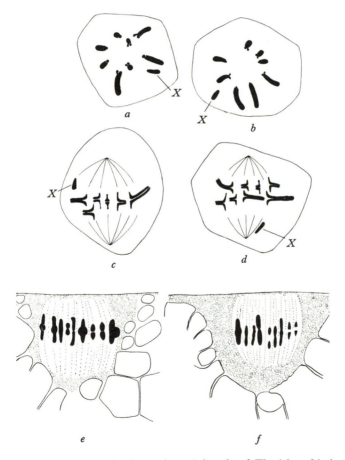

Fig. 13.5. First metaphases in the male and female of *Thericles whitei. a, b*, male first metaphases in polar view; *c, d*, the same in side view (in *c* only six of the eight bivalents shown); *e, f*, first metaphases in the egg, in side view. From White (1965*a*).

have been investigated (see Tables 17.2 and 17.3) show chiasmata in the males at first metaphase, but there are varying degrees of suppression of the diplotene and diakinesis stages; and in a number of forms the bivalents do not reveal their chiasmata at premetaphase but only after the first metaphase is well established.

In at least fourteen genera of mantids genuinely achiasmatic mechanisms of spermatogenesis occur. These are *Callimantis* (White 1938; Hughes-Schrader 1943*a*), *Humbertiella* (Hughes-Schrader 1948*b*), *Promiopteryx* (Hughes-Schrader 1953*b*), *Pseudomiopteryx* (Hughes-Schrader 1950), *Amorphoscelis* and *Haldwania* (M. L. Gupta 1966*b*), *Holaptilon* (J. Wahrman, unpub.), *Harpagomantis* (White 1965*b*), and the Australian genera *Bolbe*, *Kongobatha* and *Nanomantis* (subfamily Iridopteryginae) and *Glabromantis* and *Cliomantis* (subfamily Perlamantinae) (White 1965*b* and unpub. data). The general appearance of the bivalents at first metaphase in *Harpagomantis* is shown in Fig. 13.6.

Some difficulty in interpreting the meiotic mechanisms of mantids was encountered because of the existence of a number of species, e.g. *Miomantis* sp. (White 1941*a*) and *Acontiothespis* sp. (Hughes-Schrader 1950) which have chiasmatic bivalents at first metaphase but in which no chiasmata can be seen at premetaphase, and in which the premetaphase bivalents exhibit a 'reductional split'. In such species the transition from premetaphase to metaphase seems to be quite sudden and involves a violent 'opening out' of two 'splits' at right angles to one another.

There seem to be two possible interpretations of the meiotic mechanism of these species. Either the 'reductional' split seen in the premetaphase is in reality reductional only on one side of an invisible cross-over and is

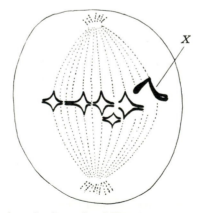

Fig. 13.6. First metaphase in the male of *Harpagomantis tricolor* (side view). Only five of the twelve bivalents shown. From White (1965*b*).

equational on the other side; or the actual process of crossing-over does not take place in these bivalents until the moment when premetaphase 'explodes' into metaphase. The first interpretation leaves one with the uncomfortable feeling that if cross-overs can exist and be concealed throughout a premetaphase with an apparent reductional split, there is no reason why they should not exist in a concealed state throughout all stages of the first meiotic division in *Callimantis*, and the other ten genera previously mentioned. It was this consideration which led White (1941*a*) to conclude that the genetic mechanism of all male mantids was probably fundamentally similar, with only varying degrees of concealment of chiasmata. In the light of all the work that has been done on achiasmatic meiotic mechanisms in the past two decades it now seems more likely that the eleven genera mentioned above really do lack genetic crossing-over in the males. The most rigorous test would seem to be the appearance of the chromosomes at first anaphase; the studies on the thericleines seem to demonstrate that if crossing-over has occurred it *must* reveal itself at anaphase, even if it has been concealed throughout metaphase. Meiosis of individuals heterozygous for various kinds of structural rearrangements needs to be studied for a complete understanding of the genetic implications of these achiasmatic bivalents.

The second possibility considered above is unorthodox in assuming that crossing-over can occur in chromosomes which have practically or entirely reached the metaphase degree of condensation. Most cytogeneticists will probably reject such an idea on the ground that the various mechanisms of crossing-over that have been proposed (see Ch. 6) all require the chromosomes to be uncondensed at the time of crossing-over. But we know so little for certain of the actual molecular process that the hypothesis is perhaps not too fanciful to be considered, although it seems very improbable.

The main difference between the 'cryptochiasmate' thericleines and mantids (e.g. *Miomantis*, *Acontiothespis*) with chiasmata whose appearance is delayed until 'full metaphase' or anaphase is that in the thericleines a 'reductional split' is *not* visible at premetaphase or is only visible in some regions, while in the mantids the 'reductional split' is clear and continuous.

Whatever the interpretation of such forms as *Miomantis*, there is obviously a general tendency in mantids for the 'mechanical' role of the chiasmata to be rendered obsolete by the 'telescoping' or virtual suppression of the stages between pachytene and first metaphase. This makes them, so to speak, preadapted for the evolutionary loss of chiasmata that has occurred in the fourteen genera cited above. This loss has probably occurred independently at least five or six times in the phylogeny of the mantids (Fig. 13.7) i.e. in the ancestors of the Caribbean *Callimantis*, the neotropical *Miopteryginae*, the African *Harpagomantis* and the Australian Perlamantinae and

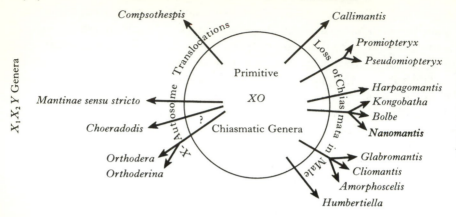

Fig. 13.7. Main lines of chromosomal evolution in mantids (see pp. 482–4). The origin of X_1X_2Y genera by X-autosome translocations is discussed in Ch. 17.

Iridopteryginae and possibly also in an ancestor of *Humbertiella*. The oogenesis of these mantids with achiasmatic spermatogenesis has not been studied as yet, so that we do not know whether they have chiasmata in the females, as we should expect, both on general principles and by analogy with *Thericles whitei*.

Many species of mantids show a rather pronounced and apparently violent premetaphase stretch stage, in which the bivalents are drawn out into very long slender shapes on the developing spindle. This is only in the chiasmatic genera, however – the achiasmatic forms conspicuously lack a premetaphase stretch, which, if it occurred, would lead to complete separation of the chromosomes before the spindle was ready to undergo its anaphase elongation.

The thericleines and mantids with achiasmatic males do not seem to have been conspicuous evolutionary successes. The total number of known species in the fourteen genera of achiasmatic mantids and closely related genera which may be suspected of being achiasmatic is less than 100. This compares with about 1800 species now known for the order as a whole. Of course it is probable that additional achiasmatic genera exist and will eventually be discovered; but it is already clear that all the really large genera and tribes of mantids have chiasmatic bivalents. The achiasmatic species of mantids and thericleine grasshoppers are all small insects. This may be a fortuitous relationship, but looks genuine and significant. Possibly the true correlation is between achiasmatic meiosis and a low 'innate capacity for increase' (Andrewartha and Birch 1954); since the size of the eggs in orthopteroid insects varies far less than that of the adults, it inevitably follows that small species will lay fewer eggs per life cycle. In such species we might possibly expect that natural selection would favour low rates of genetic recombination.

The achiasmatic spermatogenesis of the higher Diptera (suborder Brachycera) and especially that of *Drosophila* spp. has been studied and discussed by many authors. It is, of course, well known that crossing-over does not normally occur in *Drosophila* males, although occasional recombinations do occur in the spermatogonia (Friesen 1936; Whittinghill 1937, 1947, 1955).

The early studies on *Drosophila* spermatogenesis (Stevens 1908*b*; Metz 1914, 1916*a*, *b*; Huettner 1930) showed that the meiotic process was not grossly anomalous but did not explain the difference between the genetic behaviour of the two sexes. However, the work of Guyénot and Naville (1929) did suggest that there were differences between the prophase stages of meiosis in oogenesis and spermatogenesis. It is important to remember in any interpretation of *Drosophila* meiosis that homologous chromosomes show 'somatic pairing', not only in somatic nuclei but also in the gonial divisions in both sexes.

Realizing that the absence of genetic crossing-over in the male *Drosophila*, in spite of the presence of bivalents at meiosis, consituted an apparent exception to his 'chiasmatype' theory of meiosis, Darlington (1931) made the ingenious suggestion that the chiasmata might be restricted to inert segments and always present as reciprocal pairs so that they would be genetically undetectable. He later (1934*a*) investigated the situation in *D. pseudoobscura* and found that, as far as the autosomes were concerned, there were no traces of chiasmata at any stage, the four chromatids of each bivalent being paired throughout their length up to the beginning of anaphase separation, with no exchanges of partner. On the other hand, the association of the X and Y was confined to a 'pairing segment' around the centromeres and seemed to be more intimate in this region than in the case of the autosomes. Thus, so far as the sex bivalent was concerned, Darlington continued to cling to his 1931 hypothesis of reciprocal chiasmata in an inert region, which he had himself shown to be untenable in the case of the autosomes.

Kaufmann (1934) showed that *apparent* chiasmata were present in the synapsed chromosomes in the neuroblasts of *Drosophila* larvae. Similar 'pseudo-chiasmata' are sometimes visible in *Drosophila* spermatogenesis (Cooper 1949, 1950). These seem to be 'exchanges of partners' among the very loosely associated chromatids of the bivalents; they may be chiasmata according to semantic definition, but they do not really *look* like chiasmata. The whole matter has been reinvestigated by Slizynski (1964*a*) who concludes that they are not true chiasmata. It seems unfortunate that Cooper's observations should have been regarded by some as evidence against that general principle which is the cornerstone of cytogenetic theory – namely that each chiasma is the visible sign that a genetic cross-over has occurred.

Doubt was cast on Darlington's interpretation of the XY bivalent of *Drosophila* by Cooper's (1941a) investigation of the parasitic hippoboscid fly, *Melophagus ovinus*, in which the autosomes are only paired in short proximal regions, in which the association is very intimate, but clearly does not involve the presence of chiasmata. Cooper, therefore, implied that the meiotic association in the autosomal bivalents of *Melophagus* was similar in nature to that which occurs between the X and Y in *Drosophila* spp., and that neither involved chiasma formation. In *Melophagus* no synapsis between the X and Y takes place at any stage, but these chromosomes regularly pass to opposite poles at first anaphase. Additional evidence against Darlington's hypothesis of reciprocal chiasmata between the X and Y was provided by Cooper's (1944a, b) work on another hippoboscid fly, *Olfersia bisulcata*, in which he showed that the X and Y do pair in one short 'conjunctive segment', but that this pairing was initiated, not at zygotene, but much later, at diakinesis. The bivalent so formed is strikingly similar to the XY bivalent of *Drosophila pseudoobscura*, but since it is generally agreed that crossing-over is a process which takes place at the end of pachytene, it is impossible that it should possess any chiasmata.

We may accordingly conclude that there are no chiasmata in the male *Drosophila*, either in the autosomes or in the sex chromosomes, the evidence from cytology and genetics being mutually confirmatory. Since (with one notable exception) the bivalents of all the 'higher' Diptera (suborder Brachycera) that have been studied (Stevens 1908b; Metz 1961a, 1922; Metz and Nonidez 1921, 1923, 1924; Keunecke 1924; Neville 1932; Emmert 1935; Ribbands 1941; Cooper 1941b; Perje 1948; Ullerich 1963) resemble those of *Drosophila* in structure, it seems likely that achiasmatic meiosis is found throughout the whole suborder, with the single exception of the family Phoridae, in one member of which (*Megaselia scalaris = Aphiochaeta xanthina*) Barigozzi and Semenza (1952) and Barigozzi and Petrella (1953) have found chiasmatic bivalents. Tokunaga and Honji (1956), Mainx (1959, 1962) and Burisch (1963) have all reported genetic crossing-over in the male of this species, although with a much lower frequency than in the female.

In the Diptera, the 'somatic pairing force' which is manifested to some extent in all species and in all types of cells seems to persist throughout the first meiotic division and holds the homologous chromosomes in the male together in bivalents until they are torn apart at anaphase. This pairing force seems to lead to especially close synapsis in localized and probably heterochromatic regions (tentatively designated conjunctive segments by Cooper 1944*b*). The fact that the pairing segments of the *Drosophila X* and *Y* are much more closely associated than the autosomal chromosomes is probably due to the extensive heterochromatic segments in this region of the chromosome. And in *Melophagus* it seems that *only* the regions around the centromeres are paired during the spermatogonial divisions and again at diakinesis and first metaphase.

In the spermatogenesis of *Olfersia* (Cooper 1944*b*) the longest bivalent at diplotene consists of two and a half loops (i.e. it *appears* to show three chiasmata, one of which is terminal). At diakinesis the loops are mutually at right angles, just as in a bivalent with several chiasmata, but by first metaphase they all come to lie in the same plane. It thus appears from Cooper's careful analysis that the three points of association between the homologues are not chiasmata but 'conjunctive segments', like the proximal pairing segments of the *Melophagus* chromosomes.

In certain families of the 'lower' Diptera (suborder Nematocera) perfectly typical chiasmata occur in spermatogenesis. This is the case in most of the crane flies (Tipuloidea) that have been studied (but not in *Tipula caesia*

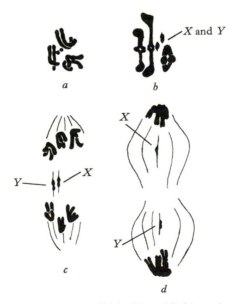

Fig. 13.8. Spermatogenesis in the tipulid fly *Dictenidia bimaculata. a,* spermatogonial metaphase, showing three pairs of large metacentric autosomes and the small *X* and *Y; b,* first metaphase, in side view, showing the chiasmata in the three autosomal bivalents; *c,* first anaphase; *d,* first telophase, showing the passage of the *X* and *Y* to opposite poles. From Wolf (1941), redrawn.

and *T. pruinosa*, which have achiasmatic bivalents in the male). Chiasmata also occur in male Culicidae (Moffatt 1936; Pätau 1941; Callan and Montalenti 1947), Dixidae (Frizzi and Contini 1962), Chironomidae (Philip 1942; Acton 1956; W. Beermann 1956*b*; J. Martin 1967), Psychodidae (Montalenti 1946; Sarà 1950) and Simuliidae (Montalenti 1947). On the other hand, certain families of the Nematocera such as the Phryneidae (Bauer 1946; Wolf 1946, 1950), Bibionidae, Scatopsidae and Thaumaleidae (Wolf 1941), Blepharoceridae (Wolf 1946) and Mycetophilidae (Frolova 1929; Le Calvez 1947; Fahmy 1949*b*) seem to have a spermatogenesis of the *Drosophila* type, without chiasmata. In some of the Tipuloidea such as *Dictenidia bimaculata* (Fig. 13.8) the X and Y do not undergo synapsis to form a bivalent but regularly pass to opposite poles at first anaphase. In most of these nematocerous families with chiasmata in the males X and Y chromosomes are not distinguishable cytologically; where, as in some Culicidae, a distinction can be seen, they form a bivalent in spermatogenesis. The work of Gilchrist and Haldane (1947) on a sex-linked gene in a species of mosquito seems to demonstrate that genetic crossing-over between the X and Y does actually occur.

The evolutionary origin of achiasmatic meiosis in the Diptera has probably been somewhat different to that of achiasmatic meiosis in the other groups referred to above, since throughout the whole order Diptera we find a strong development of the somatic-pairing phenomenon which seems to be completely lacking in grasshoppers, mantids, etc., including even those species in which achiasmatic meiosis has been evolved. It is quite possible that in the evolution of the Diptera achiasmate meiosis was not preceded by a 'cryptochiasmate' stage (involving the suppression of diplotene and diakinesis) as in mantids and thericleine grasshoppers.

It is quite possible, and indeed probable, that achiasmatic male meiosis has been acquired independently on a number of occasions in the Diptera. One instance where it has certainly done so is in the crane flies *Tipula caesia* and *T. pruinosa* (Bauer, in Bayreuther 1956) which completely lack chiasmata in the male, differing in this respect from 15 other species of the genus (Bauer 1931, Dietz 1956, John 1957. Henderson and Parsons 1963).

The case of *Megaselia scalaris*, referred to above, almost seems to suggest that chiasmatic meiosis can be gained (or rather regained) as well as lost in evolution, since the Phoridae are most likely descended from flies with an achiasmatic meiosis in the male. However, the possibility that they have evolved from nematocerous ancestors quite independently of the remaining Brachycera should not be lost sight of. Further cytological studies on other members of this family may enable the whole matter to be cleared up.

According to Ullerich (1961) the bivalents of male scorpion flies of the genus *Panorpa* (order Mecoptera) are achiasmatic. Representatives of two other families of Mecoptera, the Bittacidae (Matthey 1950b; Atchley and Jackson 1970) and the Boreidae (Cooper 1951) seem to show typically chiasmate bivalents in spermatogenesis.

The achiasmatic bivalents seen in the spermatogenesis of certain mites and scorpions differ from those considered above in that the chromosomes are holocentric. The case of certain water-mites (Keyl 1957a) is especially clear. In *Eylais setosa*, with $2n = 4$, there are only two bivalents at meiosis; the 'reductional split' is quite wide and traces of an equational split can be seen as well in the prophase stages of meiosis. Keyl has proved the holocentric nature of the chromosomes in this species by a study of radiation-induced fragments (which all attach to the spindle): but at first anaphase in the male one end of each of the two bivalents separates and passes to the pole first, leaving the other and dragging behind. This seems to indicate that, although the chromosomes are holocentric, there is some kind of gradient of centric activity from one end of the chromosome to the other. Sokolov (1945) figured achiasmatic bivalents in the spermatogenesis of some tyroglyphid mites but the status of the centromeres is uncertain in that group.

The adaptive significance of achiasmatic meiosis is not entirely clear. Obviously it has been evolved repeatedly in at least ten major groups of animals. In those cases where achiasmatic meiosis is confined to one sex, we have far too little information as to the chiasma frequency in the other sex. If it were generally true that in the 'chiasmatic sex' there is only one chiasma per bivalent, we might assume that selection for a low level of genetic recombination *per se* had led to total loss of chiasmata in one sex. If, on the other hand, absence of chiasmata in one sex is generally accompanied by a fairly high chiasma frequency in the other, this explanation obviously would not hold. Only evidence from species that had acquired achiasmatic mechanisms fairly recently in their phylogeny would be significant from this point of view (*Tipula caesia* would be relevant – *Drosophila* spp. would not). The only evidence on this point seems to be the grasshopper *Thericles whitei*, with no chiasmata in the male and a very low chiasma frequency in the female; it obviously supports the hypothesis of selection for low levels of recombination.

A possible alternative explanation for the evolution of achiasmatic meiosis might be that it would facilitate paracentric inversion heterozygosity. Many species of chironomid midges seem to manage to combine inversion heterozygosity and chiasmatic spermatogenesis through the three mechanisms described earlier (p. 242). But one might imagine that it would be more satisfactory to evolve an achiasmatic spermatogenesis if the genetic system relied heavily on inversion-heterozygosity. However, none of the mantids, roaches, enchytraeids, etc., with achiasmatic meiosis show any 'reversed loops' in the achiasmatic sex; neither does *Thericles whitei* show any bridges and fragments in the female meiosis.

MEIOSIS IN GROUPS WITH HOLOCENTRIC CHROMOSOMES

Obviously, we must expect some unusual and aberrant types of meiosis in organisms with holocentric chromosomes. The interpretation of such systems has been attempted by Schrader (1940*a*), Ris (1942), Hughes-Schrader (1948*b*), Helenius (1952), Rao (1957, 1958), Halkka (1959) and Parshad (1958) in the case of various Heteroptera and Homoptera. The status of centromeres in the chromosomes of Lepidoptera, and Centipedes (Chilopoda) is still somewhat doubtful (see pp. 15–16): important papers on the meiotic mechanisms of these groups have been published by Suoma-lainen (1953, 1965) for Lepidoptera and Ogawa (1950–54) for Chilopoda. The chromosomes of the Mallophaga and Anoplura seem to be undoubtedly holocentric, but meiosis in these groups is entirely aberrant in the male (see p. 541) and has not been studied in the female. Battaglia and Boyes (1955) have reviewed and discussed meiotic mechanisms in organisms with holocentric chromosomes.

In the Heteroptera and Homoptera with holocentric chromosomes syn-apsis and chiasma formation appear to take place quite normally, at least in the autosomes (with the exception of certain small '*m*-chromosomes' in the Coreidae, see p. 468). Terminalization of chiasmata is usually complete by first metaphase. Some species invariably show only one chiasma per bivalent, while in others bivalents with one chiasma and two chiasmata occur alongside one another. Bivalents with three or more chiasmata do not seem to have ever been seen in these two insect orders (Halkka 1964), a fact which may be significant for an understanding of their meiotic mechanisms. Where two chiasmata are present they invariably terminalize in opposite directions, to give a ring-bivalent at metaphase. The occurrence of chiasma terminali-zation in these holocentric chromosomes proves that any material correspon-ding to centromeres which may be present must be fully divided by the diplotene–diakinesis stage. Where a single cross-over has occurred, the metaphase bivalent will consist of two pairs of chromatids connected end-to-end by a terminal chiasma; on each side of the chiasma one chromatid will be a cross-over strand while the other will be a non-cross-over strand. In the case of ring bivalents with two terminal chiasmata, there will theoretic-ally be three genetic types, according to whether the pair of chiasmata are *reciprocal, complementary* or *diagonal* (see p. 165).

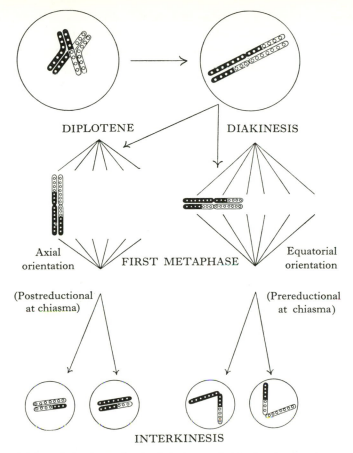

Fig. 13.9. Meiotic behaviour of a single holocentric bivalent in the Heteroptera and Homoptera. Paternal chromosome in black, maternal one in white. It is assumed that the chromatids have multiple centromeres (indicated by small circles). A single chiasma is shown, which undergoes complete terminalization between diplotene and diakinesis. After diakinesis orientation may be either axial or equatorial (see text).

The meiotic mechanisms of the Heteroptera and Homoptera seem to depend on obligate chiasma terminalization. The restriction of the chiasma frequency to a maximum of two per bivalent probably depends on this fact, since it is generally true that in monocentric chromosomes with strong chiasma terminalization there is a maximum of one chiasma per chromosome arm (see p. 167).

Clearly, the terms *reductional* and *equational*, as applied to the two meiotic divisions, are meaningless in this case, as far as the chromosome as a whole is concerned. All that we can say is that the first division will be reductional (separating two paternal strands from two maternal ones) on one side of each point of crossing-over, and equational (separating a paternal and a maternal strand from a maternal and a paternal one) on the other side; and that regions which separate reductionally in the first division will do so equationally in the second, and vice versa. But from a cytological standpoint, there is a distinction between a first meiotic division in which the two chromatids on the same side of the terminalized chiasma pass to the same pole and one in which they pass to opposite poles. In the first case the bivalent is orientated *axially* with regard to the spindle and there is a post-reduction (i.e. second division separation of maternal and paternal strands) at the chiasma. In the second case the bivalent is orientated *equatorially* with regard to the spindle and there is pre-reduction at the chiasma. Darlington (1937*b*), Darlington and Mather (1949), Helenius (1952) and Lewis and John (1963) speak of *co-orientation* of bivalents, where we use the term axial orientation, and *auto-orientation* where we speak of equatorial orientation (Darlington's terms were originally defined quite rigorously in relation to the behaviour of discrete centromeres, so that it is not desirable to extend them to holocentric chromosomes, where centromeres cannot be seen and *sensu stricto*, may not exist). In axially orientated holocentric bivalents that have completely terminalized before metaphase, the first anaphase separation only tears asunder the terminal connections between telomeres; in equatorially orientated bivalents the synaptic association is torn apart at first anaphase (Fig. 13.9).

The above distinctions between two possible kinds of bivalent orientation in Hemiptera and Homoptera seem theoretically clear, and even their application to ring-bivalents with two chiasmata does not present insuperable difficulties. But in practice, and especially with the short, condensed metaphase chromosomes which are usual in these insects, great difficulties are encountered in determining with certainty which type of bivalent we are dealing with. Ris (1942) and Hughes-Schrader (1940, 1942, 1948*b*) concluded that in the aphids and the more primitive coccids (tribe Llaveini and genus *Puto*) equatorial orientation occurs at the first division. In the case of some of the Llaveine coccids (see p. 504) a 'tertiary split' (i.e. one preparatory to the first cleavage division in the embryo) is already visible during the prophase of the first meiotic division, so that there are really eight chromatids per bivalent instead of four.

Where holocentric bivalents are orientated equatorially at first metaphase, the question arises as to what holds the two chromatids of each chromosome together during interkinesis. In certain coccids they probably dissociate completely and reassociate during the second division. But the more usual thing is for the two chromatids to be held together end-to-end by 'half a terminal chiasma' (Fig. 13.9).

The Schraders have referred to meiosis with the bivalents equatorially orientated as the aphid–coccid type. There is evidence, however, that it is considerably more widespread in the Heteroptera and Homoptera. Rhoades (1961) suggested that all holocentric chromosomes always undergo equatorial orientation, but this is now known not to be so. Within the Homoptera, most members of the suborder Sternorhyncha seem to show equatorial orientation in spermatogenesis and oogenesis, while in the suborder Auchenorhyncha axial orientation is the rule in spermatogenesis and apparently also in oogenesis (Helenius 1952; Halkka 1959). Whitten (1965) has provided a definite proof of axial orientation in the male of the leafhopper *Alodeltocephalus longuinquus* (Auchenorhyncha), where individuals heterozygons for fusions or dissociations occur in natural populations; prereduction of the associations of three chromosomes occurred regularly; similar observations have been made on *A. draba* (Whitten and Taylor 1969). On the other hand, the Psyliidae (Jumping Plant Lice), included in the Sternorhyncha, also show axial orientation of their bivalents at the first division in the male (Suomalainen and Halkka 1963); in the female of one species the bivalents showed equatorial orientation. In the Heteroptera axial orientation seems to be the rule in spermatogenesis; but in the oogenesis of some lygaeids equatorial orientation apparently occurs (Helenius 1952; Halkka 1956). Ueshima (1963) has provided a neat proof of axial orientation in the spermatogenesis of Bed Bug hybrids with eight unequal bivalents consisting of a large and a small chromosome; the first anaphase is always reductional for the inequality, which is clearly at the opposite end of the chromosome to the single terminalized chiasma.

According to Suomalainen's (1953) interpretation, the chromosomes of all Lepidoptera are holocentric and the bivalents usually show two terminalized chiasmata in the male, but only one in the female. In later publications, however (Suomalainen 1963, 1965) he claims that lepidopteran bivalents in oogenesis are achiasmatic. His conclusions are based on his own studies of *Cidaria* spp. and on the observations of earlier workers such as Dederer (1907), Seiler (1914), Buder (1915), Cretschmar (1928), Federley (1931, 1945) and Bauer (1939, 1953). He believes that the bivalents are orientated axially at first metaphase in male Lepidoptera, equatorially in the female. The interpretation of metaphase bivalents in lepidopteran oogenesis is complicated by the presence of the 'elimination chromatin' between the two halves of the bivalent. Suomalainen claims that in order to see the achiasmatic structure of the bivalents clearly it is necessary to use Feulgenstained material, in which the elimination chromatin is unstained. We believe that he has made out a strong case for achiasmatic oogenesis in the genus *Cidaria*; but with the examples of the mantids, thericleine grasshoppers and Tipulidae before us, we should be unwise to assume without further evidence that *all* Lepidoptera have achiasmatic oogenesis. De Lesse (1964) has published what appear to be clear figures of achiasmatic bivalents in the oogenesis of the butterfly *Erebia*, without calling attention to their significance. Suomalainen (1969b) has also figured achiasmatic bivalents and sex-chromosome trivalents in the oogenesis of several species of moths. Elsewhere (1966b) he has presented evidence that the oogenesis of the caddis-flies (Trichoptera), an order closely related to the Lepidoptera, is also achiasmatic.

MEIOSIS IN DRAGONFLIES

There seem to be a number of unique features about the meiosis of the Dragonflies (Odonata), according to the analysis of Oksala (1943, 1945, 1948, 1952) but at least in the genus *Aeschna* the chromosomes seem to be metacentric, with a single, localized centromere. It is probable, however, that the two daughter centromeres in each chromosome undergo auto-orientation at the first meiotic division, just as in mitosis. The orientation of the bivalents is thus equatorial at first metaphase, both in spermatogenesis and oogenesis. Chiasma formation seems to be normal in both sexes; in the males of *Aeschna* spp. the bivalents almost invariably form a single chiasma which terminalizes before metaphase, but in the females of many Odonata the bivalents regularly seem to show two chiasmata which terminalize to opposite poles so that a ring bivalent is produced. However, because of the equatorial orientation of these bivalents, they may appear as two parallel rods when seen in polar view at metaphase (Fig. 13.10), the circular space in the centre of the ring being apparently compressed into a narrow slit. The single chiasmata in the male bivalents are probably an indication of strong interference across the centromere (Oksala 1952).

The proof of the auto-orientation of the centromeres in Dragonfly meiosis comes from the behaviour of the sex chromosomes. Most species have *XO* males, the *X* dividing in the first meiotic division and passing undivided to one pole at the second. In several species of *Aeschna*, however, the *X* has become fused to an autosome, so that we have a neo-*X*:neo-*Y* system (see p. 608). These fusions are probably 'tandem' rather than 'centric'. Although the single chiasma in the sex bivalent is at the opposite end to the differential *X*-segment, the latter divides and passes to both poles at first anaphase, which could not happen in an orthodox meiosis unless there were a chiasma between the centromere and the differential segment,

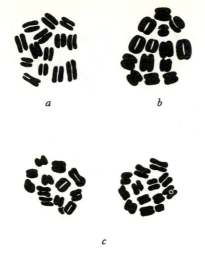

a   *b*

c

Fig. 13.10. Oogenesis in dragonflies. *a*, first metaphase in *Ischnura elegans* (polar view); *b*, the same in *Enallagma cyathigerum*; *c*, sister anaphase groups of the first meiotic division in *Leucorrhinia dubia*. From Oksala (1945).

which is obviously not the case here (Fig. 13.11). The neo-X and neo-Y separate from one another at second anaphase (Fig. 13.11*j*, *k*).

Seshachar and Bagga (1962) have claimed that in the Indian Dragonfly *Hemianax ephippiger* there is a neo-X:neo-Y system in which the X and Y differential segments pass to opposite poles at first anaphase. If this is so, the mechanism would differ from that in the XY species of *Aeschna*; but the evidence is not really clear.

Interkinesis in Dragonflies, at any rate in male *Aeschna*, is quite unique: the chromosomes ('double chromatids' in Oksala's terminology) show three 'constrictions', the central one being 'half a terminalized chiasma', the other two being the centromeres (Fig. 13.11*i*). At the second meiotic division the two centromeres of each chromosome undergo co-orientation, equidistant above and below the equator. Thus, as far as the behaviour of the centromeres is concerned, the normal sequence of the first and second divisions is reversed in the Odonata – they divide in the first division but not in the second. The suggestion of Schrader (1947) that the chromosomes of Dragonflies might be holocentric does not seem to be in accordance with the facts; the centric constrictions are clearly visible at many stages of mitosis and meiosis. However, Kiauta (1967, 1968, 1969*a*, *b*) continues to regard them as holocentric.

From the standpoint of single-locus genetics it does not seem to matter much whether the bivalents orientate axially or equatorially at meiosis.

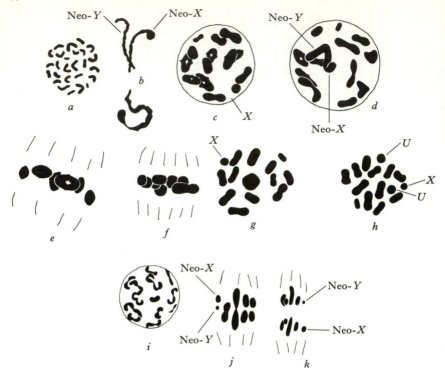

Fig. 13.11. Spermatogenesis in dragonflies. *a*, spermatogonial metaphase of *Aeschna juncea* (2n = 26); *b*, the sex-chromosome bivalent of *A. grandis* in diplotene; *c*, diakinesis in *A. crenata* (13 bivalents and an *X*-chromosome); *d*, diakinesis in *A. grandis* (12 autosomal bivalents and a neo-*XY* bivalent); *e*, first metaphase (side view) in *A. crenata*; *f*, the same in *A. grandis*; *g*, first metaphase (polar view) in *A. crenata*; *h*, the same, with two univalents; *i*, interkinesis in *A. juncea*; *j, k*, second meiotic division in *A. coerulea*, showing the separation of the sex chromosomes. From Oksala (1943).

But differences between the meiotic mechanism in various groups with holocentric chromosomes and in organisms with monocentric chromosomes may be vitally significant from the standpoint of chromosomal evolution. We have almost no information about the occurrence of inversions, translocations and other types of chromosomal rearrangements in groups with holocentric chromosomes, except for the case of the buthid scorpions (p. 287), where meiosis seems to be achiasmatic in the males. We do not know nearly enough about what happens in the meiosis of Dragonflies, Heteroptera, Homoptera and Lepidoptera heterozygous for inversions, translocations, fusions or dissociations. A dragonfly heterozygous for a fusion or dissociation should show at second metaphase chains of three chromatids joined together end-to-end by 'half terminal chiasmata' and we may well doubt whether the three centromeres would regularly orientate themselves in a disjunctional triangle on the second meiotic spindle. This may, in fact, be the main reason why the Dragonflies do not seem to have undergone many evolutionary changes in chromosome number (Fig. 12.1). Since the distance between the axially orientated centromeres at the second division is twice the distance between each centromere and the terminal 'half chiasma', one may wonder whether acrocentric chromosomes (except for very short ones) might be unable to undergo regular axial orientation at the second division in Odonata. If so, the general occurrence of metacentric elements in the group is explained.

The Odonata are an archaic and isolated order of insects, a survival from a group of Palaeozoic orders in which wings first developed. We may never know how and when they underwent a fundamental change in the meiotic mechanism of their centromeres, nor what the adaptive significance of the change was at the time. Unlike many other evolutionary changes in the meiotic mechanism, this one did not lead to any change in the level of genetic recombination.

INDICES OF RECOMBINATION

It would obviously be very desirable to have some overall measure of the total amount of genetic recombination in a species, race or individual. Darlington (1937a) proposed that one should use a *recombination index*, obtained by simply adding together the haploid number and the mean number of chiasmata. This index obviously represents the mean number of blocks of genes segregating at meiosis. Where the index differs in the two sexes, the mean between them would be the biologically meaningful quantity.

Although Darlington's recombination index is very simple to determine, it has a number of weaknesses. Where the chromosomes are all the same size and the chiasmata are distributed strictly at random, the index will truly represent the amount of genetic recombination. But, obviously, in an organism with a great size range in its karyotype, the smallest chromosomes will contribute little to the total amount of interchromosomal recombination by comparison with larger elements containing far more genetic loci. And, similarly, chiasmata situated very close to the chromosome ends will be very ineffective in a genetic sense, while those at the mid-point will be most effective. Thus, to be really meaningful, Darlington's recombination index needs to be weighted by factors to take care of variation in chromosome length and non-randomness in the distribution of the chiasmata along the chromosomes.

In species with n > 100 there is probably always a single chiasma per bivalent. Thus the recombination index of the butterfly *Lysandra atlantica* with n = 223 will be 446, and that of the hermit crab *Eupagurus ochotensis* with n = 127 (Niiyama 1951) will be 254. At the other end of the scale, *Drosophila melanogaster*, with no chiasmata in the male and about six in the female will have a recombination index of approximately 7. *Homo sapiens*, with n = 23, probably has a recombination index of about 75 to 80 (see Böök and Kjessler 1964; Evans, Breckon and Ford 1964). The lowest recombination indices in the animal kingdom are probably those of some mites and iceryine coccids with n = 2 or 3 and haploid males. These will have recombination indices of 2.5 to 4 (when computing the recombination index of a species with haploid males, n should be taken to be

$$\left(\frac{(\text{female n}) + 1}{2} \right).$$

Certain Chironomidae and Simuliidae with n = 2 probably have very low recombination indices, in spite of being diploid and having chiasmata in both sexes. Some rhabocoel Turbellaria likewise have very low recombination indices (Ruebush 1938).

Halkka (1964) has calculated the recombination indices of sixty species of leafhoppers and related Homoptera. These were based on the chiasma-frequencies in the males; apparently there is hardly ever more than one chiasma per bivalent in the female in this group. Chromosome lengths and distribution of chiasmata at the time of formation were not taken into account in this investigation, so that the results are of limited value. However, the indices that were obtained ranged from 10 in *Idiodonus cruentatus* (with n = 5) to 31 in *Criomorphus bicarinatus* (n = 15). Members of this group that have undergone evolutionary decreases in chromosome number (no doubt by fusions) have about the same chiasma frequencies per bivalent as related species with much higher chromosome numbers. Thus when two chromosomes fuse to form a double-sized one the chiasma frequency of the latter is much less than the sum of those of the original chromosomes. The same seems to be true of the grasshopper *Dichroplus silveiraguidoi*, which has undergone an evolutionary decrease in chromosome number from n = 12 to n = 4 (Saez 1956, 1957). But some exceptions seem to exist: the large metacentric chromosome in an undescribed genus of acridid grasshopper from New South Wales regularly shows four chiasmata in the female – which suggests that no reduction in chiasma frequency took place when the fusion occurred which gave rise to it.

It was suggested by Mather (1953) that the very low levels of recombination prevalent in the genus *Drosophila* (n = 3 to 6, no crossing-over in the male, only about 6 to 8 chiasmata in the female) may be adaptive to the type of life cycle found in these insects. The argument is that the genetic stability produced by a genetic system with a low recombination index is suitable for species with many generations a year (multivoltine) – preventing them from evolving cold-adapted races every winter (which would be ill-adapted when the spring comes) and heat-adapted strains every summer (which would be handicapped as soon as winter arrived). The argument is a plausible one and derives additional support from the fact that mites (Acarina), which mostly have very low chromosome numbers (Sokolov 1954) and, presumably, low recombination indices (many of them have haploid males and some, at any rate, that have diploid males show achiasmatic bivalents in spermatogenesis) have many generations a year. But in the case of the Diptera, it is not only the genus *Drosophila* which shows low levels of recombination – the whole order does so, including the primitive Craneflies (Tipulidae), with much longer generation-times than the more 'advanced' forms like *Drosophila*. Mather's interpretation of the adaptive significance of the dipteran 'recombophobe' genetic systems consequently remains somewhat hypothetical, although plausible. It is possible that low levels of recombination first evolved in the Diptera for what we may call reasons of cell-mechanics (see p. 451) and that they then facilitated the evolution of multivoltine life cycles. Collembola, which have several or many generations per annum, have low chromosome numbers (n = 4 to 7) according to Núñez (1962).

14

THE EVOLUTION OF ABERRANT GENETIC SYSTEMS

Das Unbeschreibliche
Hier ist's getan

GOETHE : *Faust*

In the present chapter we shall describe the chromosome cycles of some animal groups in which the usual meiotic mechanism has been profoundly modified in one or both sexes, i.e. in which the modifications are of a more radical kind than those described in the last chapter. In most cases these highly aberrant meiotic mechanisms are combined with unusual mitotic phenomena during the cleavage divisions in the embryo.

These unique and bizarre chromosome cycles occur in the Scale Insects (Coccoidea), in the Biting and Sucking Lice (Mallophaga and Anoplura) and in three groups of 'lower' Diptera, the families Sciaridae and Cecidomyidae and the subfamily Orthocladiinae of the Cecidomyidae.

We shall consider later (Chs. 17, 18) those aberrant chromosome cycles in which parthenogenetic reproduction of one kind or another is involved, and in Chapters 15 and 16 various forms of aberrant behaviour by sex chromosomes in organisms whose autosomes undergo normal or only slightly modified meiotic divisions.

Some most interesting aberrant chromosome cycles and anomalous types of meiosis have been encountered in the Scale insects or coccids (super-family Coccoidea of the order Homoptera). The earlier work has been reviewed by Hughes-Schrader (1948*a*); more recent work is mostly by S. W. Brown and his students (Brown and Bennett 1957; Brown 1957, 1958, 1959, 1960*a, b*, 1961, 1963, 1964, 1965, 1967; Brown and Nelson-Rees 1961; Brown and McKenzie 1962; Chandra 1962, 1963*a,b*; Nur 1962*b*, 1963*a*, 1965, 1967, 1970; Nur and Chandra 1963; Brown and Nur 1964). An attempt has been made in Tables 14.1 and 14.2 to summarize the findings. Most species of coccids are bisexual, with extreme sexual dimorphism, but there are a number of parthenogenetic species, in which males are unknown.

TABLE 14.1. *Summary of chromosomal cycles in Scale Insects**

Superfamily Coccoidea	
Section Margaroidae	
Fam. Ortheziidae	Males diploid, meiosis normal; no sex chromosomes visible.
Fam. Margarodidae	
Tribe Llaveini	Males diploid, meiosis modified; males *XO*.
Tribe Iceryini	Males haploid (impaternate). Females in some species converted into haplodiploid sex mosaics.
Tribe Monophlebini	*Aspidoproctus* has a llaveine type of spermatogenesis.
Tribe Coelostomidiini	*Nautococcus* has a llaveine type of spermatogenesis.
Section Lecanoidae	
Fam. Eriococcidae	Males diploid. Most have systems intermediate between the lecanoid and Comstockiella ones.
Fam. Pseudococcidae	*Puto* has diploid males with normal meiosis. Sex chromosomes *XO*:*XX*. Other genera have lecanoid system.
Fam. Dactylopiidae	Lecanoid system.
Fam. Coccidae (= Lecaniidae)	Lecanoid system.
Fam. Asterolecaniidae	Lecanoid system in *Cerococcus*; Comstockiella system in *Mycetococcus*
Fam. Lacciferidae	Lecanoid system
Fam. Conchaspidae	Lecanoid system
Section Diaspidoidae	
Fam. Diaspididae	Most species have the diaspidid system; some have the Comstockiella system; yet others have both the lecanoid and Comstockiella system. A modified Comstockiella system in *Parlatoria*.
Fam. Phoenicococcidae	Some have the Comstockiella system; others a combination of the Comstockiella and lecanoid systems; one species has only the lecanoid system.

* Mainly from S. W. Brown 1959, 1965 and Brown and McKenzie 1962.

TABLE 14.2. *Genetic systems in the Comstockiella–lecanoid Scale Insects* (from S. W. Brown 1967)

System			Examples
L	–	Lecanoid.	*Planococcus citri*
C	–	Typical Comstockiella, one D pair predetermined.	*Comstockiella sabalis*
C var D		Comstockiella with single D pair, but D role played alternately by more than one pair of homologues.	*Ancepaspis tridentata*
CL^M	–	Mixed systems; Comstockiella in some cysts, lecanoid in others.	*Nicholiella bumeliae*
C^C	–	Complete Comstockiella, no D pair.	*Madarococcus totarae*
CL^I	–	Comstockiella–lecanoid intermediates.	*Parlatoria oleae*

All coccid chromosomes seem to be holocentric; the various types of meiosis encountered in the group are probably derivatives of the 'aphid type' in which there is 'equatorial' orientation of bivalents at first metaphase and prereduction at the chiasma (Fig. 13.9). Only three or four species, out of the large number that have been investigated, can be said to have an orthodox chromosome cycle and meiotic mechanism in the male. The first of these is *Puto* sp. (Hughes-Schrader 1944), somewhat surprisingly a member of the specialized section Lecanoïdae, the other members of which have a highly peculiar type of male meiosis (see below). This form has 2n = 14 in the female and 13 in the male, the latter being consequently *XO*. During spermatogenesis there are six autosomal bivalents and a univalent *X*. Both meiotic divisions are bipolar and four sperms are formed from each primary spermatocyte. *Puto albicans*, with $2n\male = 19$, $2n\male = 20$ has a basically similar type of spermatogenesis (Brown and Cleveland 1968). The Australian *Callipappus rubiginosus* has some embryos with 2n = 13, others with 2n = 14, so that an *XO:XX* sex-chromosome mechanism is probably present in this case also (Hughes-Schrader, unpub.).

The other coccid with an apparently orthodox chromosome cycle is *Orthezia* sp., a member of the relatively primitive family Ortheziidae (Brown 1958). Here 2n = 16 in both sexes and eight bivalents are formed in spermatogenesis. Since the coccids were undoubtedly in origin a group with *XO* males, the only thing that is mysterious about *Orthezia* is its sex-determining mechanism, which remains unknown.

Two basically different types of chromosome cycle which we may call the llaveine and the iceryine, occur in the primitive family Margarodidae. In the genera which exhibit the llaveine type of cycle both sexes are diploid and in all the forms that have been studied there are two pairs of autosomes and an *XO:XX* sex-chromosome mechanism (i.e. the somatic sets of the male and female consist of five and six chromosomes respectively). Meiosis in the females is quite regular, three bivalents being formed in each egg.

The spermatogenesis of the Llaveini is characterized by the fact that during the prophase of the first meiotic division the chromosomes are apparently enclosed in separate vesicles rather than in a single nuclear membrane. In *Protortonia primitiva* (Schrader 1931) there are four vesicles; one of these contains the *X*, the other three the autosomes. It seems that the members of one autosomal pair are regularly included in one vesicle, while those of the other pair are in separate vesicles. Both pairs of autosomes are asynaptic, and their chromatids apparently fall asunder inside the vesicles. At the premetaphase of the first division the vesicles become spindle-shaped and closely pressed together. Within each vesicle the chromatids now approach one another again, forming a chain of four in the large vesicle and associations of two in the three smaller ones. The vesicular membranes then disappear and anaphase separation takes place, being presumably equational for all the chromosomes. The second meiotic division in *Protortonia* is even more peculiar, since all the chromosomes align themselves in a chain of five elements which are stretched linearly between the two 'poles', two elements going to one pole and three to the other.

In *Llaveia* (Hughes-Schrader 1931 – see Fig. 14.1) about 5% of the primary spermatocytes show four vesicles as in *Protortonia* but the remainder have only three, two of them containing pairs of autosomes, while the third contains the *X*. The peculiar meiotic spindles of *Llaveia* are 'compound', each autosomal pair and the *X* having a separate mass of gelated spindle substance flaring out on either side of it. In *Llaveiella* (Hughes-Schrader 1940) three vesicles are usually formed, but in some cells either or

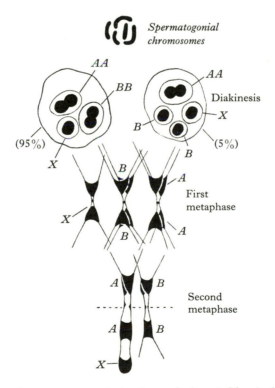

Fig. 14.1. Course of spermatogenesis in the scale insect *Llaveia bouvari*. The appearance of the first metaphase bivalents is the same, regardless of whether the *B*-chromosomes have been in the same vesicle at diakinesis or in different ones. After Hughes-Schrader (1931).

both pairs of autosomes may be enclosed in separate vesicles. *Llaveiella* is also unique in showing a clear split between subchromatids from pre-metaphase onwards, so that the first metaphase bivalents are visibly octo-partite instead of quadripartite, as they are in the other genera (Fig. 14.1). Also the X may dissociate completely into its constituent chromatids, which may go through the prophase stages in separate vesicles. The spermato-genesis of the African *Aspidoproctus* (tribe Monophlebini) is essentially similar to that of the Neotropical Llaveini (Hughes-Schrader 1955) and so is that of *Nautococcus*, of the tribe Coelostomidiini (Hughes-Schrader 1942). These forms show a graded seriation as far as asynapsis is concerned; in *Llaveia*, *Llaveiella* and *Nautococcus* 3–7% of the primary spermatocytes show asynapsis of one (occasionally both) autosomal pairs. This percentage rises to 30% in *Aspidoproctus* and 100% in *Protortonia*. Where bivalents are formed, in these llaveine Scale Insects, they are clearly chiasmatic.

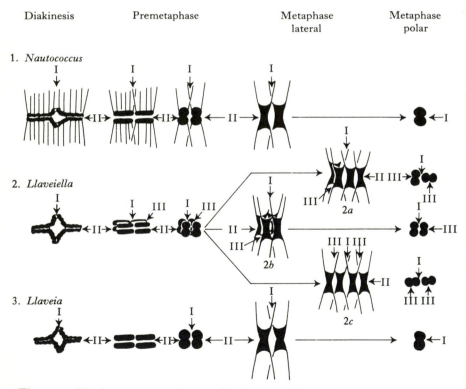

Fig. 14.2. Bivalent structure in male llaveine coccids. *a*, *b*, the 'reductional' and 'equational' splits; *c*, the subchromatid split (only in *Llaveiella*). From Hughes-Schrader (1948*a*).

1. *Nautococcus*

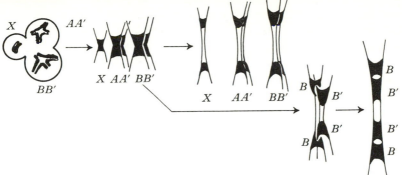

2. *Nautococcus* asynaptic AA'

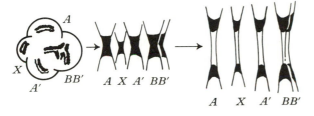

3. *Protortonia*

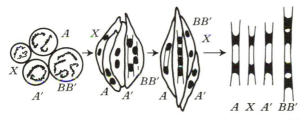

Fig. 14.3. Structure of spindles in male llaveine coccids. *Nautococcus* – top row shows separate spindle units, lower row (right) shows formation of a linear aggregate spindle in the case of bivalent BB'. Asynaptic homologues form separate spindles in both *Nautococcus* and *Protortonia*. From Hughes-Schrader (1948a).

In the tribe Iceryini the males are, as far as is known, always haploid (Hughes-Schrader 1925, 1927, 1930a, b, 1948a; Schrader and Hughes-Schrader 1926, 1931; Hughes-Schrader and Ris 1941). The details of the chromosome cycle seem to be identical in *Icerya littoralis*, *I. montserratensis* and in the genera *Echinicerya*, *Crypticerya* and *Steatococcus*. In all these the females have two pairs of chromosomes (2n = 4) while the males have only two chromosomes in their spermatogonia and in their embryonic somatic cells (in adult individuals, of course, some tissues are endopolyploid, as in all insects). Two quite normal bivalents are formed in oogenesis, so that the egg nucleus is haploid. Spermatogenesis involves only a single 'equational' maturation division, so that two haploid sperms are produced from each spermatocyte. The Australian *Auloicerya acaciae* also has $2n\male = 2$, $2n\female = 4$ (Hughes-Schrader, unpublished).

Icerya purchasi, a well known pest of citrus trees, has the 'females' transformed into functional self-fertilizing hermaphrodites, which produce both eggs and sperms. Externally, these hermaphrodites look like the females of the other Iceryini, but they possess an ovotestis in which the centre is testicular, the cortex being ovarian. True males also occur in *I. purchasi*, but are very rare in most localities; they are haploid like those of related species.

The cytology of the hermaphrodites is quite unique; they are chromosomal mosaics, the testicular part of the gonad being haploid, the ovarian part and the somatic tissues diploid. The germ cells of the future hermaphrodite are apparently all diploid up to the time of hatching from the egg, but shortly thereafter haploid nuclei make their appearance in the gonad. Hughes-Schrader and Monahan (1966) have shown that in both *Icerya purchasi* and *I. zeteki* the haploidization process involves the degeneration and elimination of one haploid chromosome set in early gonia of a sexually undifferentiated female. This process occurs during prophase, and none of the details are in any way reminiscent of a true meiosis. It is not known whether the set that is eliminated is 'paternal' or 'maternal' or may include one chromosome derived from the sperm and one derived from the egg (it is really inappropriate to speak of 'paternal' and 'maternal' chromosome sets in the case of self fertilization). The African *I. bimaculata* shows hermaphroditism like *I. purchasi*, but some pure females probably exist as well (Hughes-Schrader 1963). The haploid cells of the hermaphrodite iceryas form the core of the gonad, from which the testis later develops. The spermatogenesis of the hermaphrodites follows the same course as in the true males, there being only one maturation division.

The hermaphrodites of *I. purchasi* seem to be normally self-fertilizing, but they are capable of copulating with the occasional males which develop from eggs that have somehow escaped fertilization. The hermaphrodites are incapable of fertilizing one another, and their behaviour patterns are entirely female.

The haplodiploidy of the Iceryini clearly arose in evolution from a chromosome cycle basically like that of the Llaveini. The ancestors of the Iceryini were presumably *XO* in the males; at one stage in the history of the group one set of autosomes was lost in the male sex, with the establishment of haploid parthenogenesis. Thus in the cytological evolution of the Iceryini there were three main steps: (1) the partial disappearance of bivalent formation in the males, which has already happened in the Llaveini, (2) the establishment of male haploidy, (3) the transformation of the female into a self-fertilizing hermaphroditic mosaic (a step which has only taken place in a few species of *Icerya*). The last step was presumably due to a mutation or mutations which produced the special type of reduction division that occurs in the early part of the germ-line. The males of the coccids in general are very fragile and short-lived in the adult stage, so that any reproductive device which rendered them unnecessary, while retaining occasional genetic recombination, might possess considerable selective advantage. *Kuwania oligostigma* has $2n♀ = 16$; males are almost certainly diploid, but their precise karyotype is unknown (Hughes-Schrader 1963).

Except for the cytologically primitive genus *Puto* the 'lecanoid' families of coccids possess a highly peculiar type of chromosome cycle in which one haploid set of chromosomes is heteropycnotic in the somatic cells of males, from the blastula stage onwards; in females both sets are euchromatic. The basic facts were discovered by Schrader (1921, 1923b) and interpreted by Schrader and Hughes-Schrader (1931). These authors suggested that it was the paternal set which became heteropycnotic in male embryos and that this involved a functional inactivation, so that the males would be physiologically haploid, although actually carrying two sets of chromosomes in their nuclei. Both of these hypotheses have been confirmed experimentally by Brown and Nelson-Rees (1961) who have shown that (1) irradiation of fathers is followed by the appearance of breaks in the heteropycnotic chromosomes of their sons, (2) irradiation of mothers leads to the appearance of breaks in the euchromatic set of their sons, (3) after irradiation of fathers with high doses dominant lethality is present among the daughters, but not in the sons. Irradiation of mothers leads to dominant lethality in offspring of both sexes. If still higher doses are employed for paternal irradiation the number of both sons and daughters is reduced, due to chromosome bridges and general abnormalities of development.

The heteropycnosis of the paternal chromosome set in male lecanoids is thus a classic and unique example of functional inactivation of entire chromosomes, presumably by histones. It may be compared with the inactivation of one X in the somatic tissues of female mammals (see p. 594), although in that instance it may be either the maternal or the paternal chromosome which is inactivated by heteropycnosis. Baer (1965) has shown that, as we should expect, the DNA of the heteropycnotic chromosome set of *Pseudococcus* is late-replicating in the male by comparison with the non-heterochromatic set.

It is logical to suppose that the heteropycnosis of the paternal chromosome set in the lecanoid scales is determined by its having passed through the sperm, but that this heteropycnosis is only expressed in genetically male zygotes. Nur and Chandra (1963) have shown that in hybrid male embryos heteropycnosis of the paternal set can occur in the cytoplasm of another species; the mechanism is thus not species-specific. The mechanism of sex determination in these insects is not understood. There are no detectable sex chromosomes and the sex ratio is highly variable from one female to another and under different environmental conditions; it is very characteristic that male and female embryos tend to be produced at different periods of the oviposition cycle. All these facts suggest that some mechanism of histological differentiation or cytoplasmic segregation in the ovary produces male and female eggs.

Certain lecanoid scales such as *Pulvinaria hydrangeae* and *P. mesembryanthemi* are parthenogenetic (Nur 1963a; Pesson 1941). In the former, meiosis occurs in the developing oocyte; diploidy is restored by fusion of the first two cleavage nuclei so that all individuals must be completely homozygous. In spite of this, two kinds of embryos are produced – 95% with both chromosome sets euchromatic and 5% with one haploid set euchromatic and the other heterochromatic. The latter are presumably male but adult males are not known; perhaps the embryos with a heterochromatic set die, or perhaps the heteropycnosis is only temporary and disappears later in development. In *P. mesembryanthemi* males are apparently produced parthenogenetically, but never mate.

The case of *P. hydrangeae* proves that heteropycnosis can occur in a haploid set which has never passed through a sperm, and which is genetically identical with a set that remains euchromatic.

Heteropycnosis is not necessarily associated with sex in coccids. For example, in many species a pentaploid mycetome tissue, containing intracellular symbionts and analogous to a plant endosperm is formed by fusion of a diploid cleavage nucleus with the first (2n) and second (n) polar body nuclei (Schrader 1923a); three of the haploid sets in certain species undergo heteropycnosis in both male and female embryos (Brown and Nur 1964) and may fuse to form a persistent chromocentre, leaving only two haploid sets diffuse.

We may take the chromosome cycle of *Phenacoccus acericola* (Hughes-Schrader 1935) as typical of the lecanoids. In this species 2n = 12 in both sexes, all the chromosomes being alike in size and none of them being recognizable as sex chromosomes. Oogenesis is normal, with formation in bivalents. In spermatogenesis the heteropycnosis of the paternal chromosomes occurs in the spermatogonia and during the prophase of the first

Lecano–diaspidid systems of chromosome behaviour

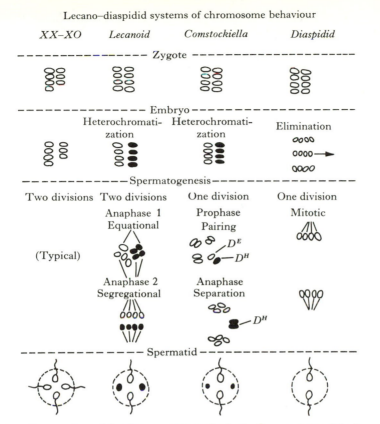

Fig. 14.4. Comparison of the chromosome cycles in the four systems of the lecanoid–diaspidoid evolutionary sequence. The first is only represented by *Puto*. From S. W. Brown (1963).

meiotic division. The heteropycnotic chromosomes form a group to one side of the nucleus, the non-heteropycnotic (diffuse) ones being distributed throughout the rest of the nuclear cavity.

No synapsis takes place during spermatogenesis, so that there is no possibility of genetic crossing-over in the male. By the time first metaphase has been reached the two sets of chromosomes have reached the same degree of condensation and can no longer be distinguished. All twelve univalents are divided into their chromatids, which separate at anaphase in a normal mitotic manner and pass to the poles. During interkinesis one haploid set of six chromosomes (presumably the same one as before) becomes heteropycnotic again. The spindle of the second meiotic division is a unipolar structure which is formed in connection with the heteropycnotic group of chromosomes; it separates them from the non-heteropycnotic group which remain diffuse and weak staining.

The second meiotic division is thus a 'reductional' one, for all the chromosomes. The cytoplasm is not completely divided at either division, so that a syncytial mass of cytoplasm is left with four haploid nuclei, two containing heteropycnotic chromosomes, the other two formed from the diffuse ones. Only the latter nuclei give rise to sperms, the heteropycnotic nuclei degenerating in the mass of cytoplasm which is sloughed off from the developing spermatids.

The chromosome cycle of three species of Mealy bugs (genus *Planococcus* or *Pseudococcus*) follows almost exactly the same course as that of *Phenacoccus*, except that $2n\male = 10$ (Schrader 1923*b*). The same kind of cycle occurs in the genera *Lecanium* (Thomsen 1927), *Dactylopius* (Hughes-Schrader 1948*a*), *Cerococcus*, *Tachardiella* and *Conchaspis* (Brown 1959).

Although the heteropycnotic chromosome set in the mealy bugs is genetically inactivated to a large extent it certainly retains some more or less general functions (Brown and Nur 1964). Thus following on very heavy irradiation of the fathers (60 000 to 90 000 rep) a few sons still survive. But only individuals with an approximately normal *amount* of heteropycnotic material reach the adult stage. Also, although with lower doses (30 000) all the sons reach adulthood, many of them are sterile. Finally, interspecific hybridization studies have shown that the heteropycnotic chromosome set of one species cannot substitute for another. In some cases the eggs hatch but development of the hybrids does not proceed beyond the first instar; in such cases the male embryos have one haploid set heteropycnotic, thus proving that heteropycnosis can occur in a foreign cytoplasm. It is obvious that we are only at the beginning of the analysis of the genetic functions of these 'inactivated' chromosome sets.

In *Pseudococcus obscurus* there may be one or several supernumerary chromosomes in addition to the regular karyotype (Nur 1962*a*). They are heteropycnotic in some tissues of females and behave like members of the heteropycnotic set in males (regardless of whether they have been derived from the father or the mother) up to the beginning of the male meiosis. They then undergo a reversal of heteropycnosis, becoming even less condensed than the non-heteropycnotic set. They divide equationally, like all the other chromosomes, in the first meiotic division and segregate preferentially with the non-heteropycnotic set in the second division. In somatic tissues of the male *Planococcus citri* the nuclei are endopolyploid, but only the non-heteropycnotic chromosomes have undergone endoreduplication (Nur 1967).

Nur (1970) has studied the spermatogenesis of irradiated males of

Planococcus in which translocations had been induced between members of the euchromatic and heterochromatic sets in the first instar. The boundary between the two regions in *E–H* translocation chromosomes was quite sharp and in some cases they formed bridges in the second meiotic division.

In the majority of the Armoured Scales (Diaspidoidea) the males are true haploids, with half as many chromosomes in their nuclei as the females. But they arise from fertilized eggs, the male embryos eliminating the paternal chromosomes during certain of the late cleavage divisions, i.e. at precisely the stage when heteropycnosis first makes its appearance in the lecanoid system. The details were first worked out in *Pseudaulacaspis pentagona* (Brown and Bennett 1957, Bennett and Brown, 1958) and have subsequently been confirmed for a large number of other species of diaspidoids (Brown 1958; Brown and De Lotto 1959). Experiments in which fathers had been irradiated so that some of the chromosomes in the sperms were broken showed that it is the paternal chromosomes which are eliminated from the male embryos.

This type of chromosome cycle necessarily raises the question as to what is the switch mechanism that determines whether a given embryo will or will not eliminate the paternal chromosome set. Although this question is not fully resolved, there are some indications from *P. pentagona* that the switch may be a cytoplasmic one; in this species the eggs containing female embryos are laid first and are coral coloured, while the eggs containing the pinkish-white male embryos are deposited later.

A type of chromosome cycle which is clearly related to the lecanoid one is found in a number of genera of diaspidoid scale insects such as *Comstockiella, Aonidia, Ancepaspis, Parlatoria, Odonaspis* and *Nicholiella* (Brown 1963; Brown and Nur 1964). In this 'Comstockiella type' of genetic system the males are diploid and one haploid set (presumably the paternal one) is heteropycnotic from the blastula stage onwards as in the lecanoid system. We may take *Parlatoria oleae* ($2n = 8$) as an example (Nur 1965; Kitchen 1970). Here the heterochromatic chromosomes degenerate and are lost prior to meiosis, except for one heteropycnotic element 'D_H' and its homologous non-heteropycnotic partner 'D_E'. There is a single meiotic division, at which the euchromatic univalents divide and the daughter and the daughter chromosomes pass to opposite poles. The 'D_H' element is eliminated, while the 'D_E' chromosome divides equationally, contributing a daughter chromosome to each of the separating anaphase groups. In most species showing the *Comstockiella* system the D chromosome pair is a fixed one, but in some species such as *Ancepaspis tridentata* and *Phoenicococcus marlatti*, in which the chromosomes are strikingly different in length, one can see that the 'D' chromosome is a long chromo-

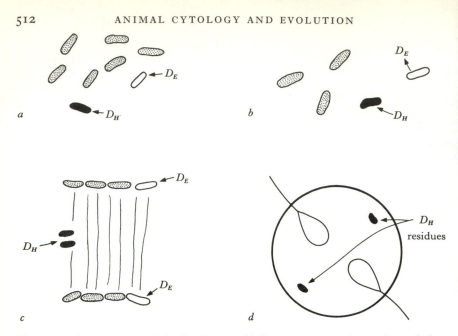

Fig. 14.5. Spermatogenesis in the Comstockiella system. *a*, early prophase; *b*, late prophase; *c*, anaphase of the single meiotic division; *d*, post-meiotic stage. From S. W. Brown (1964), but modified according to the data and interpretation of Kitchin (1970).

some in one cyst, a short one in a second and so on – i.e. any one of the members of the karyotype can play a '*D* role' (Brown 1963, 1965). This strange behaviour is reminiscent of the peculiar sex-determining mechanism of the phorid fly *Megaselia scalaris*, in which a 'wandering' sex factor may be located on any of the chromosome elements (see p. 619). This 'variable *D*' system has probably arisen in several different evolutionary lineages. The diaspidoid system almost certainly arose (possibly more than once) from either the lecanoid or the Comstockiella system (Brown and McKenzie 1962; Brown 1963).

In at least six species of armoured scale insects belonging to no less than four different tribes we have the extraordinary situation that the Comstockiella and lecanoid mechanisms coexist in the same individual male, but in different cysts of the testis (Brown 1963, 1964). The significance of this most peculiar situation is not understood – it must result in the production by each male of two kinds of sperms, one derived from a Comstockiella type of meiosis, the other from a lecanoid type. *Platycoccus tylocephalus* (family Phoenicococcidae) is the only true diaspidoid known to shown only the lecanoid type of spermatogenesis (Brown 1965); it may be a species

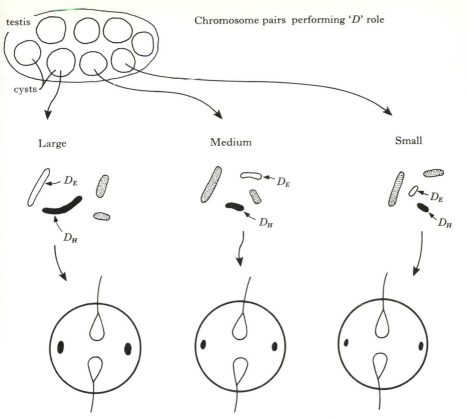

Fig. 14.6. The 'variable D' Comstockiella system in the scale insect *Ancepaspis tridentata*. From S. W. Brown (1964), but modified according to Kitchin (1970).

which is descended from ancestors having a 'mixed' type of testis but which has lost the Comstockiella-type cysts in the course of its evolution.

In the family Eriococcidae, Brown (1967) has found a variety of genetic systems intermediate between the Comstockiella and lecanoid ones. In these species the number of chromosomes which remain heteropycnotic during spermatogenesis is variable and frequently unstable from cyst to cyst within the same individual. Some of these species (e.g. *Apiomorpha pharetrata*) have no second meiotic division as in the Comstockiella system, while others (e.g. *Scutare lanuginosa*) clearly have two meiotic divisions, as in the lecanoid system. The Australian *Ascelis schraderi* and the New Zealand *Madarococcus totarae* show what Brown calls a 'complete Comstockiella system'. Although one haploid set is heteropycnotic in the somatic

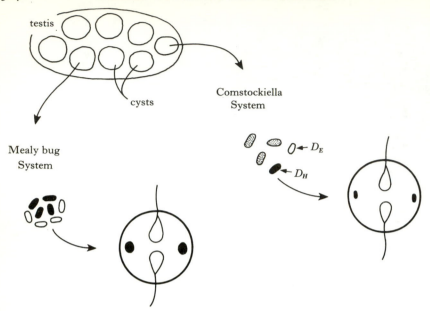

Fig. 14.7. The coexistence of the lecanoid ('mealy bug') system and the Comstockiella system in different cysts of the same testis. This situation occurs in some scale insects such as *Nicholiella bumeliae*. From S. W. Brown (1964), but modified according to Kitchin (1970).

cells of the males, it becomes euchromatic in spermatogenesis; synapsis occurs and bivalents are formed, but there is no second meiotic division.

In *Gossyparia spuria* Schrader (1929) long ago described a unique type of spermatogenesis obviously related to the lecanoid and Comstockiella systems but different from either. His account has been corrected in various respects by Nur (1967). This species has $2n = 28$ chromosomes and these form fourteen bivalents in the egg. Fourteen chromosomes are heteropycnotic in the somatic cells of the male. Before the beginning of the first meiotic division in spermatogenesis, some of the heteropycnotic chromosomes become euchromatic. The first prophase nuclei (i.e. stages corresponding to diplotene-diakinesis) show fourteen euchromatic bodies and from one to over nine heterochromatic chromosomes. Apparently some of the euchromatic bodies are univalents while others are bivalents, consisting of a maternal chromosome and a paternal one which has undergone 'deheterochromatinization' prior to meiosis. All the chromosomes divide in first anaphase, i.e. the univalents derived from the heteropycnotic and non heteropycnotic chromosomes divide equationally while the homologues of which each bivalent is composed separate from one another. After an interkinesis in which there is differential uncoiling, there is a second meiotic division of the lecanoid type in which heteropycnotic and non heteropycnotic chromosomes separate from one another. In the first meiotic division of *G. spuria* the bivalents are probably unreplicated and achiasmatic. White's suggestion in the second edition of this work, that *G. spuria* might be tetraploid, was based on Schrader's original account, and is certainly wrong.

From the genetic standpoint male coccids with the lecanoid and Comstockiella systems will be phenotypically haploid and will also breed as haploids. In species with the diaspidoid system the males will be true haploids and will also breed as such.

Brown (1963, 1964) has argued that the several steps in the evolutionary series: *Puto* mechanism → lecanoid → Comstockiella → diaspidoid were facilitated by what he calls *automatic frequency response* ('AFR'). The essential nature of this hypothetical principle may be understood by considering a species with half its testis cysts showing the lecanoid mechanism and half the Comstockiella one. Now imagine a mutation l → L which leads to all the cysts being of the lecanoid type. There will be four kinds of males in the population, ll, lL, lL, LL, where the allele in heavy type is in the heterochromatic set. If L is active in the euchromatic but not in the heterochromatic set we shall have the following situation:

ll males have equal number of lecanoid and Comstockiella cysts. They transmit 100% l

lL males have equal numbers of lecanoid and Comstockiella cysts. The former transmit only l, the latter equal numbers of l and L. Thus these males transmit 67% l, 33% L

lL males have all cysts lecanoid. They transmit 100% L

LL males have all cysts lecanoid. They transmit 100% L

Thus the L mutation will automatically increase, without any natural selection in the ordinary sense being involved.

Similarly, in the origin of male haploidy, one can imagine a mutation h → H causing the unfertilized eggs of hH and HH females to develop successfully as haploids. An 'automatic frequency response' will occur (in a population containing both haploid and diploid males) because the offspring of the haploid males are all females which, since they will carry in greater frequency the H gene responsible for the production of haploid males, produce more of them than in the previous generation.

It seems premature to speculate on the relative importance of AFR and natural selection of the orthodox type in the evolution of the genetic systems of the coccids. AFR is not exactly a type of meiotic drive (see p. 470) but is in some ways analogous.

As far as sex determination is concerned, the coccids appear to show an orthodox male heterogamety in the primitive forms such as *Puto* and the llaveines; in the more 'advanced' scale insects sex determination appears to be non-genetic and probably due to the females producing two kinds of eggs which are cytoplasmically different. However, the possibility exists that the differentiation of two kinds of eggs is preceded by some kind of nuclear event, e.g. a switching on or off of a special genetic locus controlling the differentiation of the egg cytoplasm. Analogous nuclear mechanisms are probably responsible for (1) heteropycnosis versus non heteropycnosis of one haploid set, (2) assumption by one chromosome pair or another of the '*D* role' in the 'variable *D*' Comstockiella mechanism, (3) the choice between the lecanoid and Comstockiella types of spermatogenesis in the testis cysts of a species such as *Nicholiella bumeliae*. As far as the second and third of these phenomena are concerned the determining events seem to occur at the time when the primordia of the testis cysts are formed from the embryonic germ-line.

Three groups of the 'lower' Diptera (suborder Nematocera) are characterized by genetic systems which are in many ways more complicated and bizarre than even those of the coccids. These groups are the families Sciaridae (fungus gnats) and Cecidomyidae (gall midges) and the subfamily Orthocladiinae of the Chironomidae. The Sciaridae and Cecidomyidae are fairly closely related and have both probably originated from primitive Mycetophilidae, a family in which the chromosome cycles are apparently orthodox except that the bivalents are achiasmatic in spermatogenesis.

The entirely unique chromosome cycle of the Sciaridae is fortunately known rather completely, both from the cytological and the genetic standpoint (Metz, Moses and Hoppe 1926; Metz 1926–41, 1959; Metz and Schmuck 1931*a*, *b*; Du Bois 1932*a*, *b*, 1933; Schmuck 1934; Smith-Stocking 1936; Berry 1939, 1941; Crouse 1939, 1943, 1947, 1960*a*, *b*, 1961*b*, 1965; McCarthy 1945; Fahmy 1949*a*; Metz and Armstrong 1961). The genus *Sciara* includes several hundreds of species of fungus-feeders. Since the cytological picture is of the same general type in all the species which have been investigated (about twenty in all) we may take *S. coprophila* (the best-known form) as the type. In this species there are seven chromosomes in the somatic cells of the male and eight in those of the female, one pair of chromosomes being metacentric, the other three acrocentric. The male soma is thus *XO*, the female soma *XX* (the *X* being one of the acrocentric elements, which are all about the same length). The spermatogonia and

Fig. 14.8. Chromosomes of *Sciara coprophila*. *a*, male somatic set; *b*, female somatic set; *c*, oogonial or spermatogonial set. *X*-chromosomes in each case at the bottom of the figure. l, 'limited' chromosomes. From Du Bois (1933).

oogonia, however, both contain ten chromosomes (occasionally 9 or 11). This germ-line karyotype is made up of the four pairs already mentioned, together with a large pair of metacentrics which are not represented in the soma. Both oogonia and spermatogonia are thus *XX*, in spite of the fact that the male somatic cells are *XO*.

The large chromosomes which are confined to the germ-line were referred to as the 'limited chromosomes' (L's for short) by Metz and his school. Usually a pair, they may be present singly or in triplicate without affecting the outward phenotype of the fly. However, individuals of *S. coprophila* have never been found without any limited chromosomes, although some other species of the genus such as *reynoldsi* and *ocellaris* do lack them altogether. In *coprophila* the L's may vary in size, so that those present in natural populations may carry duplications or deficiencies.

All these facts suggest that the limited chromosomes are largely inert, genetically, although they undoubtedly do have some genetic functions in those species of *Sciara* (probably the majority) in which they are found. The L's are heteropycnotic at some stages in the mitotic cycle, but since they are never present in somatic tissues it is not known what they would look like in salivary gland nuclei. Their invariable presence in the species in which they do occur prevents us from regarding them as supernumerary chromosomes, although they do have many of the properties of super-numeraries.

The meiosis of the female *Sciara* follows a normal course; bivalents are formed and there is genetic evidence that crossing-over takes place (Schmuck and Metz 1932). It is not definitely known whether the L's undergo synapsis and crossing-over, but if two of them are present they segregate from one another regularly at meiosis. Thus the female pronucleus comes to contain a full haploid set of chromosomes (three acrocentrics, one small metacentric and a large L). In individuals which have one or three L's in the oogonia the eggs may sometimes carry zero or two L's.

The spermatogenesis of *Sciara* is of a very anomalous type (Metz, Moses and Hoppe 1926; Metz 1933). Whereas in the somatic cells the homologues show the usual dipteran 'somatic pairing', in the spermatogonia there is no obvious approximation of homologues. No synapsis occurs during the meiotic prophase, so that all the chromosomes remain univalent, the L's being clearly distinguishable from the rest by their heteropycnosis. When the spindle of the first meiotic division forms it is a 'unipolar' cone-shaped body, the chromosomes being all attached in an irregular manner to the base of the cone, there being no regular 'equatorial plate'. At first anaphase the L's, together with one member of each of the other pairs, pass to the pole of the cone, the remaining four chromosomes moving in the opposite direction into a small mass of cytoplasm which is then expelled from the cell and eventually degenerates (Fig. 14.9). There is genetic evidence that the chromosomes which are got rid of in this way are

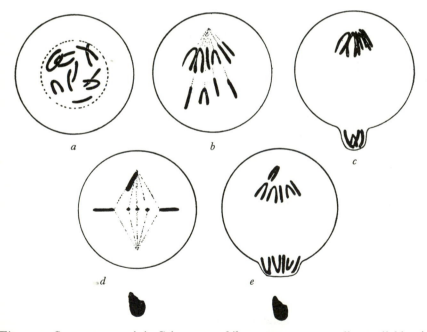

Fig. 14.9. Spermatogenesis in *Sciara coprophila*. *a*, stage corresponding to diakinesis (all chromosomes unpaired); *b*, anaphase of first meiotic division, *c*, telophase (four paternal chromosomes being eliminated); *d*, metaphase of second meiotic division: *e*, second anaphase, showing the split *X* at the upper pole of the spindle. In *d* and *e* the degenerating 'bud' is seen at the bottom of the figure. From Du Bois (1933).

the set derived from the father, the only paternal chromosome retained in the sperm being the L.

The second meiotic division is a bipolar one in which all six elements (two L's, three autosomes and the X) become attached to the spindle. The X is not attached in the equatorial region but near one of the poles; it splits into two, but both daughter chromosomes remain in the same half of the spindle and pass into the same spermatid. The daughter chromosomes which pass to the other pole are cast off in a small bud of cytoplasm which degenerates, like the one formed at the first division. Thus only one sperm is produced from each primary spermatocyte; it contains the L chromosomes (usually two in number), two X's and a complete haploid set of autosomes. All the 'ordinary' chromosomes (autosomes and X's) in the sperm are of maternal origin, so that the male *Sciara* only transmits genes derived from the mother. Although the male soma is XO the males are actually homogametic, all their sperms having the same chromosome set.

Crouse (1960a, b) has studied nine reciprocal translocations between the X and autosomes of *S. coprophila*. The aberrant behaviour of the X at the second meiotic division was found to be controlled by a heterochromatic segment in the short arm of the acrocentric X. When this segment became translocated (without its centromere) to the distal end of a second chromosome the rearranged chromosome behaved like an X at the second meiotic divisions, in spite of having an autosomal centromere.

The fertilized egg of *S. coprophila* contains three pairs of autosomes, three X's and a variable number of L's (usually three). One of the X's has been derived from the egg pronucleus, the other two came from the sperm nucleus. During the cleavage divisions a number of the chromosomes are eliminated both from the somatic cells and from those of the germ-line, so restoring the karyotypes which are characteristic of the adult tissues. These elimination divisions were described by Du Bois (1932a, b, 1933) and Berry (1941).

The first four cleavage divisions are normal mitoses. Separation of the future germ-line and soma is accomplished before the sixth cleavage. At the fifth or the sixth cleavage division the L's are eliminated from the future somatic cells; they go through an apparently normal prophase and become attached to the spindle, but at anaphase it is obvious that their chromatids are unable to separate in the normal manner except proximally. As the ordinary chromosomes pass to the poles, the L's are left in the middle part of the spindle, stretched between their daughter centromeres. Here they remain and eventually degenerate in the egg cytoplasm.

A further series of eliminations occurs during the later cleavage divisions. At the seventh or the eighth mitosis the somatic nuclei, which still contain

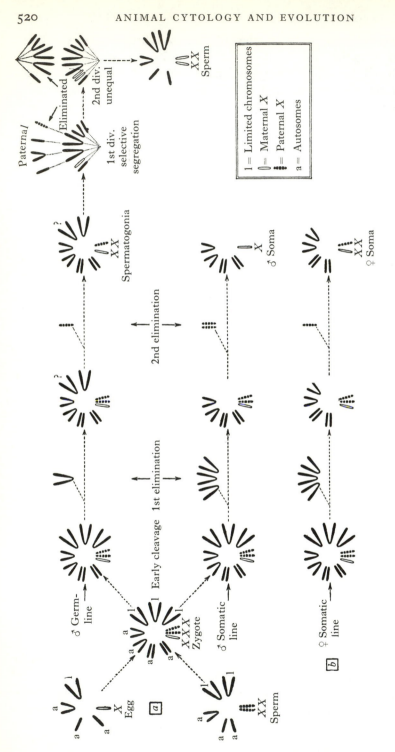

Fig. 14.10. The chromosome cycle of *Sciara coprophila* during fertilization, cleavage and spermatogenesis. The 'maternal' X-chromosomes, which go into the sperm become, of course, the 'paternal' XX at fertilization. The number of limited chromosomes actually varies from one to three. Diagram *a* will serve equally well to represent the conditions in the female, except as regards gametogenesis, which is normal, and as regards the second chromosome elimination from the soma. From Metz (1938*a*).

Fig. 14.11. Elimination of X-chromosomes at the seventh cleavage division in *Sciara coprophila. a*, four divisions in a female embryo (only one X eliminated); *b*, a division in a male embryo (two X's eliminated). From Du Bois (1933).

three X chromosomes, extrude one or two of these. In embryos which are destined to become females only one X is eliminated, while in those which will become males two are got rid of (Fig. 14.11). The genetic evidence shows that the X's which are eliminated at this stage are paternal, i.e. in females one of the two sister X's brought in by the sperm is thrown out, while in males both sister X's are cast away. In the germ-line a single paternal X is got rid of at a slightly later stage, irrespective of the sex of the embryo. According to Berry (1939, 1941) the elimination of the X from the germ-line nuclei takes place in a totally different manner from the earlier elimination processes, the X-chromosome simply migrating through the nuclear membrane and degenerating in the cytoplasm.

Presumably, one or more L's must be eliminated from the primordial germ cells as well, since in spite of the fact that each sperm transmits the full number present in the primary spermatocyte the number of L's does not increase with each generation; but it is not clear exactly when and how this elimination of L's from the germ-line nuclei takes place (Metz, 1938*a*).

The whole of the extraordinary series of processes in the chromosome cycles of *Sciara* spp. obviously bear some relationship to the mode of sex determination. Certain species of *Sciara* invariably produce unisexual progenies (all female or all male), other produce bisexual progenies, while in yet others different strains may give either bisexual or unisexual families. In all cases the sex of the offspring seems to depend on the genetic constitution of the mother (in the species which have unisexual progenies there are two kinds of females, respectively male-producing and female-producing). The genotype of the father seems to have no influence on the sex of the progeny.

In species such as *S. coprophila* and *S. impatiens*, which nearly always produce unisexual progenies, there are no phenotypic or visible cytological differences between the two types of females, which apparently differ in an invisible genetic factor. Female-producing females are assumed to be heterozygous (XX'), male-producing ones being homozygous (XX). Normally, male- and female-producers are present in the population in approximately equal numbers, so that a sex ratio of about 1:1 is maintained. Actually, a good many of the 'unisexual' families are not completely so, since occasionally individuals of the wrong sex are produced. These exceptional flies have the sexual constitution corresponding to their phenotype, i.e. their soma is XO if they are males, XX if they are females. Thus they seem to result from a failure of the genetic constitution of the mother to control the elimination of the X's. It is possible that the two alleles (X and X') postulated by Metz are not the only ones and that actually a series of alleles of varying potency exists. The actual sex of the individual is apparently determined by the elimination of one or two X's during cleavage; but it is the genetic constitution of the mother (operating, no doubt, through the egg cytoplasm) which determines whether one or two X's are lost.

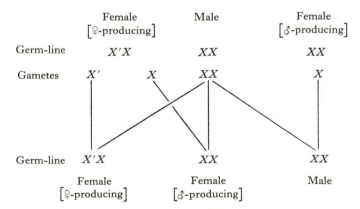

Fig. 14.12. Sex determination in a species of *Sciara*, such as *S. coprophila*, which produces unisexual progenies.

Among the species of *Sciara* which normally produce bisexual families are *S. pauciseta* and *S. prolifica*. The difference between these and a species like *coprophila* would appear to be one of degree rather than kind; in *coprophila* the 'exceptional' individuals are few in number, while in *pauciseta* and *prolifica* they make up 50% of the offspring. Even in *coprophila* 'bisexual' strains are occasionally encountered (Reynolds 1938).

An interesting situation exists in *S. ocellaris*, in which there are 'unisexual' and 'bisexual' strains which differ in that one of the autosomes is acrocentric in the former, metacentric in the latter (Crouse, 1939). In the polytene nuclei of hybrids between the two races, the acrocentric and metacentric elements are synapsed throughout their entire length, with no inversion loop. Either the rearrangement responsible for the difference in chromosome shape was a centric shift (3-break arrangement) or a pericentric and a paracentric inversion with almost coincident breakage points must have occurred. Whether the difference in the sex-determining mechanism in the two races is really a position effect resulting from the rearrangement is still an open question. In the closely related species *reynoldsi* (which always produces bisexual broods) the chromosome in question is invariably metacentric.

Since the chromosomal constitution of the germ-line in *Sciara* is always the same in the two sexes, what is it that causes the gonad of a female to become an ovary and that of a male to become a testis? The obvious answer would seem to be that the chromosomal constitution of the surrounding soma must determine the development of the embryonic gonad in one direction or the other. Lawrence and Crouse (see Crouse 1943) studied some gynandromorphic hybrids between *S. ocellaris* and *S. reynoldsi* in which some somatic tissues were XX and other XO, and found numerous cases in which an ovary and a testis were present in the same individual, which is what one would expect if the sex of the gonad were determined by the chromosomal constitution of the surrounding somatic tissues. Crouse (1960b, 1965) has studied patroclinous males that were XO in both soma and germ-line. In these the single paternal X is eliminated at first meiotic anaphase, and there is no X at the second divison; such males are sterile.

Plastosciara pectiventris seems to have a chromosome cycle essentially like that of the species of *Sciara* investigated by Metz and his school; there are one to four limited chromosomes in this species (Fahmy 1949a). *Rhynchosciara angelae* has been used extensively for studies on its giant polytene chromosomes (Pavan 1965) but its meiosis and cleavage divisions have not been studied as yet.

The gall midges (family Cecidomyidae), which are fairly closely related to the Sciaridae, possess even more aberrant chromosome cycles. In certain genera of the primitive subfamilies Heteropezinae and Lestremiinae, such as *Miastor, Heteropeza* (= *Oligarces*) and *Mycophila*, reproduction is mainly by larval parthenogenesis (pedogenesis), although sexual genera-tions occur under certain conditions. In *Tecomyia populi* and *Henria psalliotae* it is the pupae which reproduce pedogenetically (Nikolei 1961, Wyatt 1961). The overwhelming majority of species of Cecidomyidae, however, reproduce exclusively by sexual means. Not all Cecidomyidae are gall-making insects; the genera which we have mentioned above are fungus feeders and members of some other genera such as *Lestodiplosis* are predators in the larval stage.

As far as the pedogenetic forms are concerned, the cytology of *Miastor* was studied by Kahle (1908) and Hegner (1914), whose classic studies were amplified and corrected by Huettner (1934), Kraczkiewicz (1935*a, b*, 1937, 1938), White (1946*b*) and Nicklas (1959). Certain differences in chromosome number exist between European material of this genus (referred to as *M. metraloas*) and the North American species studied by Nicklas. *Heteropeza pygmaea* (= *Oligarces paradoxus*) has been studied by Reitberger (1939), Hauschteck (1959, 1962), Camenzind (1966) and Panelius (1968, 1971), while *Mycophila speyeri* has been investigated by Nicklas (1960).

In most of these forms the number of chromosomes in the germ-line (oogonia and spermatogonia) is the same in the two sexes; it is in all cases very much higher than in those somatic cells which are not endopolyploid. There are, in fact, a large number of 'E-chromosomes' which are present in the zygote nucleus, but are destined to be eliminated from the future somatic nuclei during certain of the cleavage divisions.

In addition, there are a number of so-called *S*-chromosomes in the germ-line cells, which are also represented in the somatic nuclei (Table 14.3). *Heteropeza pygmaea* differs from the other cecidomyids that have been investigated in having a higher number of chromosomes in the germ-line of the female than in that of the male (Hauschteck 1959; Camenzind 1966; Panelius 1968, 1971).

In most of the species of Heteropezinae that have been studied the female somatic nuclei contain a diploid set of *S*-chromosomes while the male somatic nuclei only carry a haploid set. However, in *Heteropeza* some females have a haploid set of chromosomes in the soma. In pedogenetic larvae of *M. 'metraloas'* (which are all of the female sex) there are twelve chromosomes (six pairs of homologues) in the somatic nuclei. The oogonial nuclei, as the other hand, contain forty-eight chromosomes. White (1946*b*) supposed that these germ-line nuclei were octoploid, but later (1950*a*) concluded, on the basis of studies on members of the subfamily Cecido-myinae, that no homology existed between the *E*- and *S*-chromosomes. In the light of later work by Camenzind and Panelius it seems probable

TABLE 14.3. *Chromosome numbers in Cecidomyidae*

Species	Chrom. in Germ-line	E-chromosomes	S-chromosomes ♀ soma	S-chromosomes ♂ soma	Egg nucleus	Sperm nucleus	Author
Miastor 'metraloas' (U.K.)	48	36	12	6	42(?)	6	White 1946b
Miastor sp. (U.S.A.)	38–40	30–32	8	4(?)	–	–	Nicklas 1959
Heteropeza pygmaea			see text				Camenzind 1966 / Hauschteck 1959, 1962 / Reitberger 1939
Mycophila speyeri	29	23	6	3	14–15	3	Nicklas 1960
Taxomyia taxi	40	32	8	6	–	4	White 1947a
Phytophaga celtiphyllia	36(?)	–	12	–	–	–	White 1950a
Rhopalomyia sabinae	ca.25	ca.17	8	6	21(?)	5–8	White 1950a
Oligotrophus pattersoni	34	26	8	6	30(?)	10	White 1950a
Lasioptera rubi	40(?)	32(?)	8	6	–	4	Kraczkiewicz 1950
L. asterspinosae	–	–	–	–	–	4	White 1950a
Trishormomyia helianthi	24	16	8	6	20(?)	4	White 1950a
Monarthropalpus buxi	ca.50	–	8	6	–	28	White 1950a
Wachtliella persicariae	40	32	8	6	–	–	Geyer-Duszyńska 1959
Mikiola fagi	24	16	–	–	20	–	Matuszewski 1960, 1962
Walshomyia texana	–	–	8	–	–	–	White 1950a
Lestodiplosis spp.	–	–	8	–	–	–	White 1946
Rhabdophaga batatas	40	32	8	–	–	4	Geyer-Duszyńska 1961
R. salicidiperda	–	–	–	–	–	–	Kraczkiewicz 1966
Mayetiola destructor	ca.40	ca.32	8	8*	–	–	Bantock 1961

* Based on Bantock's statement. No information was given as to the stage or tissue investigated.

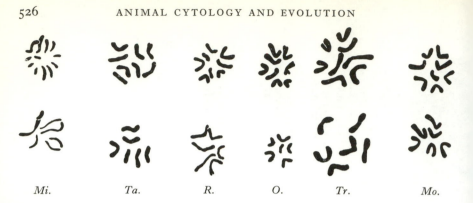

Mi.   *Ta.*   *R.*   *O.*   *Tr.*   *Mo.*

Fig. 14.13. Somatic metaphases in females (upper row) and males (lower row) of several species of Cecidomyidae. From left to right the species are: *Miastor* sp., *Taxomyia taxi, Rhopalomyia sabinae, Oligotrophus pattersoni, Trishomomyia helianthi, Monarthropalpus buxi*. From White (1946b, 1950a).

that both viewpoints are over-simplified and that the real situation is very complex. In the species of *Miastor* studied by Nicklas there are eight chromosomes in the female somatic nuclei and thirty to thirty-two *E*-chromosomes, making thirty-eight to forty altogether, in the germ-line nuclei. The somatic set of the American species could have been derived from that of *M.* '*metraloas*' by two centric fusions.

The oogonia of the pedogenetic larvae develop into oocytes whose nuclei undergo a single non-reductional maturation division. The polar body nucleus passes to the periphery of the egg and may divide again, but no polar body as such is given off. Soon after the maturation division is completed the eggs begin to undergo cleavage and develop into embryos and eventually into daughter larvae within the body of the mother larva which is parasitized (and finally killed) by her own offspring.

During the third and fourth cleavage divisions of the pedogenetic embryo a peculiar 'elimination process' occurs in those nuclei which are destined to pass into the somatic cells of the future individual. This process consists in the *E*-chromosomes remaining in the equatorial region of the mitotic spindle, while the *S*-chromosomes divide normally, pass to the poles and form the daughter nuclei. Apparently the *E*-chromosomes at the elimination divisions have undergone normal replication and consist of two parallel chromatids at metaphase, but fail to move poleward as a result of non-development or under-development of the usual mid-anaphase tension (Nicklas 1959). Probably this peculiarity of spindle physiology is to be ascribed to non-functioning or partial functioning of the centromeres of the *E*-chromosomes (Geyer-Duszyńska 1959). It seems to be a general characteristic of embryonic development in all species of Cecidomyidae. The eliminated chromosomes clump together in pycnotic masses which remain in the yolky cytoplasm for a considerable part of the embryonic development, eventually breaking up and degenerating, their final destruction being presumably brought about by the action of cytoplasmic enzymes. One nucleus alone escapes undergoing the elimination process and retains the full number of chromosomes; this is the primordial germ cell nucleus which enters a special morphologically posterior region of the egg cytoplasm, the so-called polar plasm, prior to the third cleavage division and divides by a normal mitosis at the fourth cleavage division.

Under certain adverse environmental conditions the pedogenetic larvae of *Miastor*, *Heteropeza* and *Mycophila* give rise (again by pedogenesis) to sexual larvae which are of two types, respectively male and female. Unlike the pedogenetic larvae, the sexual ones eventually pupate and develop into winged midges which copulate, the females subsequently depositing small numbers (six to eight in *M. metraloas*) of eggs which are much larger than the 'internal' eggs of the pedogenetic larvae. The fertilized eggs develop into pedogenetic larvae, thus completing the cycle.

In *M. metraloas* the female sexual larvae, although differing from the pedogenetic ones in external morphology, show the same cytological picture, i.e. they have twelve chromosomes in their somatic nuclei and forty-eight ($12 + 36$ E-chromosomes) in their oogonia. The male sexual larvae, however, possess only six chromosomes in their somatic nuclei, which are accordingly haploid by comparison with female larvae (endopolyploidy occurs in certain of the somatic cells in both sexes, but these chromosome numbers are present in embryonic somatic cells and in the nervous tissue of older individuals). The number of chromosomes in the spermatogonia is forty-eight and there seem, in fact, to be no chromosomal differences between spermatogonia and oogonia.

Just how the male somatic nuclei in *M. metraloas* come to contain only six chromosomes is not known, since the elimination divisions in male embryos have not been studied in detail.

The spermatogenesis of the male *Miastor* (which occurs in the prepupa and pupa) is highly peculiar. No synapsis occurs between any of the chromosomes and the prophase stages are not greatly different from those of a spermatogonial mitosis. The spindle which develops at premetaphase is, however, peculiar in that its two poles are quite different, one being acuminate, the other flared out (originally this spindle was described as 'unipolar', but the interpretation given above seems preferable). Geyer-Duszyńska (1961) believes that no true spindle is formed at the first meiotic division in the cecidomyids, the structure regarded by other workers

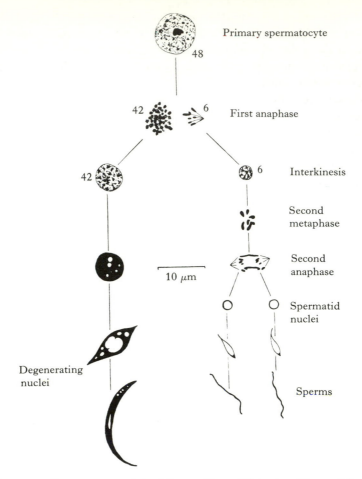

Fig. 14.14. Spermatogenesis in *Miastor*. From White (1946*b*), modified.

as a spindle being simply the nuclear sap contained in a persistent nuclear membrane. This difference of opinion may turn on the rather narrow question as to how far the spindle substance has undergone gelation or is still in a liquid state. Geyer-Duszyńska's observations on the absence of birefringence certainly suggest that the latter is the case. However, we may continue to speak of the acuminate pole, which corresponds to the 'centre of the nucleoplasmic island' in Geyer-Duszyńska's terminology. Panelius (1971) could find no difference between the structure of the spindles at the first and second meiotic divisions in *Heteropeza*. Six chromosomes (three in *Mycophila speyeri* and presumably four in the species studied by Nicklas), which can be seen by their characteristic sizes and shapes to correspond to a haploid set of S-chromosomes, now orientate themselves with their centromeres directed toward the acuminate pole, while the remainder (i.e. one haploid set of S-chromosomes and all the E-chromosomes) remain in an unorientated group on the other side of the nuclear area. This stage may be regarded as the first metaphase, although there is, strictly speaking, no metaphase 'plate'.

First anaphase and telophase consist simply in the separation of the two groups of chromosomes (six to the acuminate pole and forty-two to the diffuse one in M. 'metraloas' and corresponding numbers in the other species). Two kinds of secondary spermatocytes are accordingly formed, of very different sizes (since the cytoplasmic division at first telophase is an unequal one, the two cells receiving amounts of cytoplasm roughly proportional to their chromosome number). The larger category of cells never divide again and merely undergo a series of degenerative changes, but the smaller cells go through a second division (which is a simple mitosis) and form two spermatids that proceed to transform into mature spermatozoa. Thus in M. 'metraloas' each primary spermatocyte with forty-eight chromosomes forms two sperms with six chromosomes and a 'residual' cell with forty-two.

Both the life-cycle and the chromosome cycle of *Heteropeza pygmaea* are even more complex than those of *Miastor* spp. (Reitberger 1939; Hauschteck 1959, 1962; Ulrich 1962; Camenzind 1962, 1966; Panelius 1968, 1971). There are essentially three modes of reproduction in this species; pedogenesis, parthenogenesis and sexual reproduction.

In pedogenetic reproduction a gynogenic larva (in German, *Weibchenmutter*) gives rise to daughter larvae, by a non-sexual process. The eggs of such females undergo only a single equational maturation division; different races or populations show fifty-five, sixty-six or seventy-seven chromosomes in the pedogenetic germ-line. Such larvae have ten chromosomes in their somatic nuclei. Panelius (1968) has shown that the primordial germ cell only gives rise to oocytes (a maximum of sixteen in each ovary); the nurse-cells (trophocytes) are derived from the soma.

Two other kinds of *Heteropeza* larvae reproduce pedogenetically. These are the androgenic larvae (*Männchenmutter*) which give birth to male imago-larvae and the amphogenic larvae which can produce both ordinary daughter larvae and male larvae. Female imago-larvae are also formed by pedogenesis. The male and female imago-larvae eventually pupate and produce winged midges. The eggs of female midges may develop after fertilization (sexual reproduction) or without being fertilized (parthenogenesis); but in both cases they give rise to ordinary gynogenic larvae.

In the development of the eggs of androgenic larvae (i.e. eggs which are going to give rise to males) there is a true meiosis. Panelius (1971), who worked on 55-chromosome material, states that 27–28 bivalents are formed, but that multivalents are sometimes present; the chromosome number of the egg nucleus seems to be reduced (after two meiotic divisions) to 28, 33 or 38 (?39), according to the race. The three polar body nuclei do not degenerate, but become supplementary cleavage nuclei. There are two (or possibly more) small somatic nuclei in the egg cytoplasm, and these fuse with the egg nucleus before the first cleavage division. Such somatic nuclei in the oocyte cytoplasm almost certainly do not exist in all cecido-myids, but were seen by White (1950a) in *Monarthropalpus buxi* (they were called 'enigmatic bodies' at that stage). Similar somatic nuclei seem to occur also in *Miastor* (Kraczkiewicz 1936). Camenzind considers that this fusion of the 'small nuclei' with the egg nucleus, after reduction, restores the chromosome number and he argues from this and from the meiotic behaviour that in *Heteropeza* few, if any, of the chromosomes are really confined to the germ-line and do not exist in the soma. This is similar to the earlier view according to which the germ-line was supposed to be essentially polyploid (White 1946b), the *E*-chromosomes consisting of a number of 'sets' of autosomes. But since it is fairly certain that 'small nuclei' do not contribute to the germ-line of the developing embryo in most of the gall-forming cecidomyids, it may be that the *Heteropeza* condition is a primitive one, in an evolutionary sense, and that the *E*-chromosomes of the primitive cecidomyids do not consist of a number of 'sets' of *S*-chromosomes, while in the more advanced forms the homology between *S*-chromosomes and *E*-chromosomes has been partially or totally lost.

The usual cecidomyid elimination processes occur in *Heteropeza* in the second to the fourth cleavage divisions, reducing the chromosome number in the future somatic nuclei to six. A second elimination process occurs in the sixth to the eighth cleavage division, at which a single chromo-some is lost, so that the chromosome number is finally five in the male soma. Similar events occur in gynogenic eggs and in the pedogenetic cycle, except that the chromosome number in the somatic nuclei is reduced by the first elimination process to eleven and by the second elimination process to ten.

The eggs of the imagines may develop with or without fertilization, but most of the larvae which develop parthenogenetically die. In both cases the imaginal eggs undergo meiosis so that the chromosome number is halved. It is restored in the case of the fertilized eggs by fusion of several sperm nuclei and several 'small nuclei' with the egg nucleus. In the case of the eggs which develop parthenogenetically, there are of course, no sperm nuclei, and only the 'small nuclei' are added to the egg nucleus. Camenzind has shown that both the female larvae that have arisen by the sexual process and those that have arisen by parthenogenesis (of which few survive) have only five chromosomes in the soma. It was earlier supposed that sex determination was dependent on whether five or ten chromosomes were present in the soma. But since female larvae that have arisen by pedo-genesis have ten chromosomes in their somatic cells while ones that have arisen from eggs laid by imagines have only five, this is clearly not so, and sex determination must depend on some unknown mechanism.

Spermatogenesis in *Heteropeza* follows the usual cecidomyid pattern. In the spermatogonial divisions the five or six S-chromosomes form a central group, easily distinguishable from the rest. Panelius (1971) has discovered the extraordinary fact that there are actually four different kinds of males, which produce different kinds of sperm. In two of these the sperm nuclei contain five chromosomes, while in the other two they contain six. The sperm karyotypes may be represented as follows:

Type	Chromosomes
(5a)	I, II, III, IV, Va
(5b)	I, II, III, IV, Vb
(6a)	I, II, III, IV, Va, VI
(6b)	I, II, III, IV, Vb, VI

Each of these chromosomes represented by roman numerals can be recognized on the basis of length and centromere position. Chromosomes Va and Vb are, in all probability, partially homologous, because they seem to show some degree of somatic pairing in female somatic metaphases.

Panelius has proposed an extremely complex genetic model to explain the chromosome cycle of *Heteropeza*. Using the symbol S to designate the set of chromosomes I, II, III and IV, the karyotype of the pedogenetic germ-line would be:

$$55 = 5 \, (\text{VI} + \text{Va} + \text{Vb} + 2 \, S).$$

After the first elimination in the future somatic cells there would be left:

$$11 = \text{VI} + \text{Va} + \text{Vb} + 2 \, S.$$

And after the second elimination:

$$10 = \text{Va} + \text{Vb} + 2 \, S \text{ (the female somatic karyotype)}$$

The four post meiotic nuclei in male-production would be of two types, each with six subtypes:

$$27 = (2 \, \text{VI} + 5 \, \text{V}^* + 5 \, S) \quad \text{and} \quad 28 = (3 \, \text{VI} + 5 \, \text{V}^* + 5 \, S)$$

where $5 V^*$ indicates any of the six combinations fron $5 Va$, $0 Vb$ to $0 Va$, $5 Vb$.

After 'regulation' by two small somatic nuclei the male germ-line nuclei would be $47 = (2 VI + 9 V\S + 9 S)$ and $48 = (3 VI + 9 V\S + 9 S)$ where $9 V\S$ designates any of the six combinations from

$$7 Va, 2 Vb \quad \text{to} \quad 2 Va, 7 Vb$$

Male somatic nuclei would be of two types: $5 = (Va + S)$ and $5 = (Vb + S)$ and spermatids of the four types previously mentioned.

In the great majority of the members of the subfamily Cecidomyinae there is no pedogenesis and reproduction is exclusively sexual. The number of E-chromosomes is variable (Table 14.3). In most of the species there are six chromosomes in the somatic cells of males and eight in those of females. In these cases the salivary gland nuclei show four elements, regardless of the sex of the individual, so that we may designate the somatic sets as:

$$X_1 X_1 X_2 X_2 \ A_1 A_1 A_2 A_2 \ (\female) : X_1 X_2 \ A_1 A_1 A_2 A_2 \ (\male)$$

where X_1 and X_2 represent non-homologous sex chromosomes and A_1 and A_2 the autosomes. In *Wachtliella persicariae* it has been shown that the E-chromosomes are eliminated at the fourth cleavage division and that in male embryos two more chromosomes (X_1 and X_2) are eliminated at the seventh division (Geyer-Duszyńska 1959). *Mayetiola destructor* is said to have eight chromosomes in the somatic cells of both sexes (Bantock 1961).

Spermatogenesis in the Cecidomyinae follows, in general, the same course as in *Miastor*. In such forms as *Taxomyia taxi* (White 1947) and *Lasioptera rubi* (Kraczkiewicz 1950) the S-chromosomes show strong positive heteropycnosis in the resting spermatogonia and oogonia, and may form a compact group on one side of the nucleus, while the E-chromosomes are diffuse and distributed throughout the rest of the nuclear cavity (Fig. 14.15). Spermatogenesis in these forms is of the same type as in *Miastor*, a haploid set of four S-chromosomes ($X_1 X_2 A_1 A_2$) passing to the 'pole' at the first meiotic division. Thus in *Taxomyia*, *Lasioptera* and probably the great majority of the gall-making cecidomyids the sperm only transmits four chromosomes to the zygote.

In a species such as *Miastor* '*metraloas*' with forty-eight chromosomes in the germ-line, but only six in the sperm nucleus, we should expect the nucleus of the 'sexual' egg to contain forty-two chromosomes and we should expect the maturation divisions of the egg to accomplish a reduction from 48 to 42. Whether this is what actually happens in that species is uncertain. But this seems to be essentially what happens in the gall-making forms

Trishormomyia helianthi and *Oligotrophus pattersoni* (White 1950a) and in *Mikiola fagi* (Matuszewski 1960), although the chromosome numbers are different (in *T. helianthi* and *M. fagi* there are 8 *S*- + 16 *E*-chromosomes in the cells of the germ-line, while in *O. pattersoni* there are 8 *S*- + 26 *E*-chromosomes).

In the oogenesis of *Trishormomyia*, *Oligotrophus* and *Mikiola* it can be seen from pachytene onwards that only the *S*-chromosomes have paired to form four normal-looking chiasmate bivalents; the *E*-chromosomes remain in a univalent condition. In *Mikiola* (and probably in the other species as well) the *E*-univalents divide once only in the course of the female meiotic divisions and the egg nucleus receives 4 *S*-chromosomes and a complete set of *E*-chromosomes. Thus the zygote receives a haploid complement of *S*-chromosomes from each parent, and these will have undergone

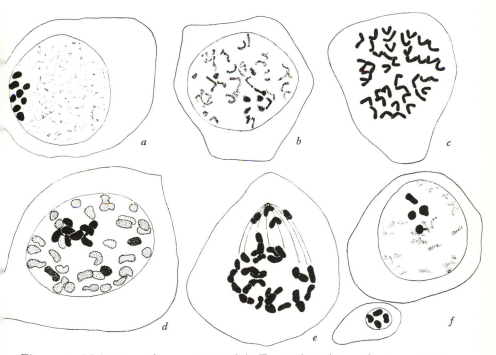

Fig. 14.15. Main stages of spermatogenesis in *Taxomyia taxi*. *a*, resting spermatogonium, showing positive heteropycnosis of the eight *S*-chromosomes; *b*, prophase and *c*, metaphase of a spermatogonial mitosis (40 chromosomes); *d*, prophase of first meiotic division, showing the clumped *S*-chromosomes in black and the 32 *E*-chromosomes, stippled; *e*, anaphase of first meiotic division with 4 *S*-chromosomes on the spindle, the other 4 being left with the 32 *E*-chromosomes at the 'base' of the spindle; *f*, interkinesis, showing one of the small cells with 4 chromosomes and one of the large residual ones with 36. From White (1947a).

crossing-over in the mother but not in the father; the *E*-chromosomes are inherited solely from the mother and never undergo crossing-over.

In *Wachtliella persicariae*, diakinesis in the egg has a very strange appearance (Kunz, Trepte and Bier 1970). The 4 *S*-bivalents, showing typical chiasmata, occupy the centre of the nucleus and are surrounded by concentric lamellae. Outside the lamellae, but within the nuclear membrane, are the 32 *E*-chromosomes which are diffuse and grouped into four loose bundles. At first metaphase is approached, the lamellae disappear, the *S*- and *E*-chromosomes eventually attain the same degree of condensation and

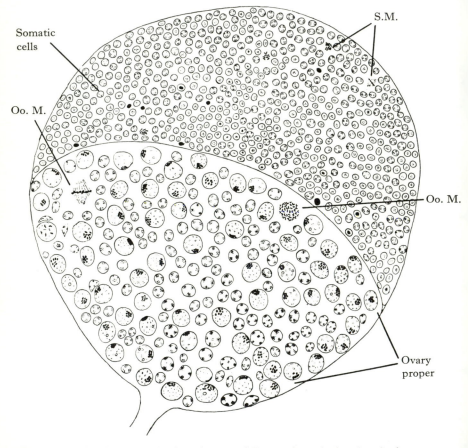

Fig. 14.16. Section through a larval ovary of *Taxomyia taxi*; showing the large mass of somatic cells not belonging to the germ-line. Some of these are dividing (S.M., somatic metaphases showing eight chromosomes). In the ovary proper some oogonial metaphases (Oo.M.) show 8 heteropycnotic *S*-chromosomes and an undeterminable number (presumably 32) of diffuse *E*-chromosomes. From White (1950*a*).

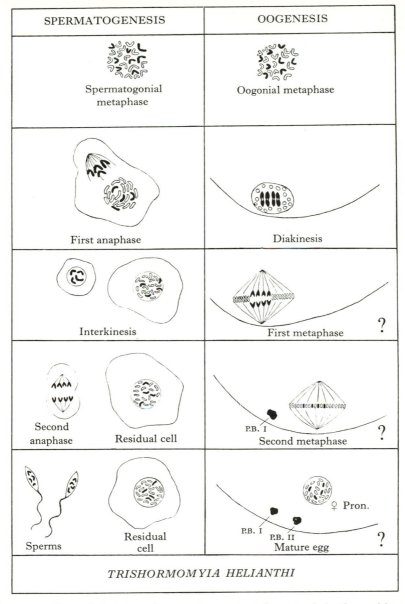

SPERMATOGENESIS	OOGENESIS
Spermatogonial metaphase	Oogonial metaphase
First anaphase	Diakinesis
Interkinesis	First metaphase ?
Second anaphase	Residual cell	Second metaphase ?
Sperms	Residual cell	♀ Pron. Mature egg ?

TRISHORMOMYIA HELIANTHI

Fig. 14.17. General diagram of spermatogenesis and oogenesis in the cecidomyid *Trishormomyia helianthi*. *S*-chromosomes in solid black, *E*-chromosomes in outline. The three lower figures of the right hand side are hypothetical as these stages have not actually been observed. From White (1950*a*).

it becomes evident that there are 16 *E*-bivalents as well as the 4 S-bivalents. Whether the *E*-bivalents are chiasmate is not entirely clear.

The above scheme seems to be the most widespread type of genetic system in the Cecidomyidae, but modifications exist in a number of genera. Thus in *Mycophila speyeri*, which has twenty-nine chromosomes in the germ-line and three in the sperm-nucleus the oogenesis is 'normal', i.e. the twenty-nine chromosomes form fourteen chiasmate bivalents and one univalent (Nicklas 1960). The female pronuclei contain fourteen or fifteen chromosomes and it is not clear how the original germ-line number of twenty-nine is restored, although we may suspect that the female pro-nucleus fuses with a sperm nucleus and two 'small' somatic nuclei each containing six chromosomes.

In some Cecidomyinae such as *Oligotrophus pattersoni* and *Monarthro-palpus buxi* the sperms carry a certain number of *E*-chromosomes as well as the haploid set of S-chromosomes although the spindle mechanism at the first meiotic division is essentially of the *Miastor–Taxomyia* type and a residual cell and two sperms are formed from each primary spermatocyte. In *Oligotrophus pattersoni* some *E*-chromosomes are presumably eliminated from the germ-line nuclei at some stage.

In some cecidomyids such as *Thomasiniana* spp. and *Rhabdophaga* spp. (Barnes 1935, 1944), unisexual broods are produced, while others, such as *Lasioptera asterspinosae* have bisexual progenies (White 1950a). In the Hessian Fly (*Mayetiola destructor*) most progenies are unisexual but some show both sexes, there being usually a great preponderance of one or other sex (R. H. Painter 1930).

The genetic significance of the *E*-chromosomes in the cecidomyids is just as mysterious as that of the *L*-chromosomes in the Sciaridae, which they resemble in many respects. Nicklas (1959), on the basis of chromosome measurements, concluded that at least half of the germ-line chromosomes in his species of *Miastor* could not be homologous to the S-chromosomes; but this is questioned by Panelius (1971). On the whole it seems at present likely that in the evolution of the cecidomyids the *E*-chromosomes arose as a number of replicate sets of S-chromosomes, and that the condition of a polyploid germ-line is retained (at least to a considerable extent) in *Hetero-peza*. But in that organism the basis for its retention seems to be the 'regula-tion' process whereby the small somatic nuclei fuse with the egg nucleus, thereby recruiting S-chromosomes to the germ-line at each sexual genera-tion. It is certainly doubtful whether this mechanism exists in most of the 'higher' cecidomyids.

It was shown by Geyer-Duszyńska (1961), working on *Wachtliella persicariae*, that one can cause *all* the embryonic nuclei to undergo the

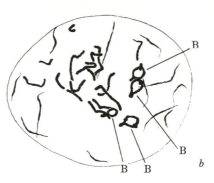

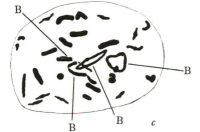

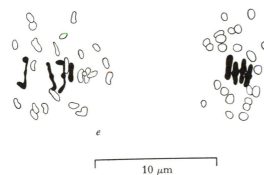

10 μm

Fig. 14.18. Stages of the first meiotic division in the egg of the cecidomyid *Oligo-trophus pattersoni*. *a*, early diplotene; *b–c*, mid diplotene, *d*, diakinesis; *e–f*, pre-metaphase. In *a* the four bivalents formed by the *S*-chromosomes can be seen clearly (not all the univalents are drawn in this nucleus). In *b* there are 4 ring bi-valents (B) and 26 univalents. In *c* there are 3 ring bivalents, one rod bivalent and 26 univalents. In *d–f* the bivalents are shown in solid black, the univalents in outline. *d* appears to show only 23 univalents, *e* shows 24 and *f*, 25 (numbers of univalents below 26 being probably due to 'clumping' in the making of the preparation). From White (1950a).

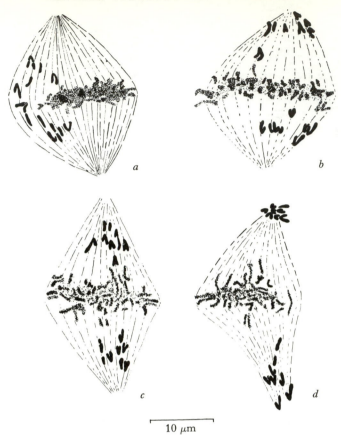

Fig. 14.19. Elimination mitoses from an embryo of the cecidomyid *Monarthropalpus buxi* undergoing the fifth cleavage division. The *E*-chromosomes are clumped in an irregular mass lying across the equatorial region of the spindle, while the *S*-chromosomes (shown in solid black) have divided and are passing to the poles. From White (1950*a*).

elimination of the *E*-chromosomes by preventing the posterior nucleus entering the polar plasm (this can be achieved, either by centrifugation or ligation). Such embryos develop normally, in spite of the fact that their germ cells contain eight instead of forty chromosomes; however oogenesis and spermatogenesis do not occur normally and the germ-cells degenerate.

Differences in chromosome number between germ-line and soma also occur in members of the subfamily Orthocladiinae of the family Chironomidae (Bauer and Beerman 1952*b*, *c*). In eleven species belonging to six genera which they investigated the number of chromosomes in the somatic

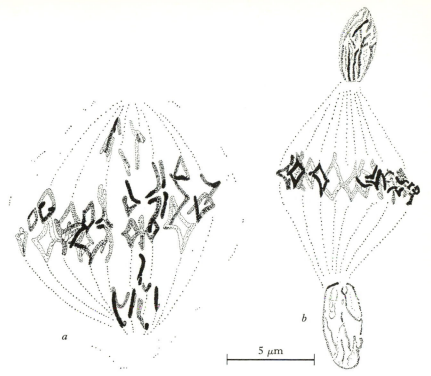

Fig. 14.20. Elimination mitoses in the embryo of *Miastor* sp. *a*, mid-anaphase; *b*, telophase. From Nicklas (1959).

cells is four or six (there being two or three double elements in the salivary gland nuclei), but the number of chromosomes in the oogonial and spermatogonial nuclei is very much higher. The 'extra' chromosomes are eliminated from the future soma during the fifth to the seventh cleavage divisions, as in *Sciara* and the cecidomyids. The *E*-chromosomes of the Orthocladiinae seem to be variable in number within the species; thus *Metriocnema cavicola* has twenty-four to fifty-two of them, while *M. hygropetricus* has two to eight.

Elimination of half the *E*-chromosomes takes place during the oogonial and spermatogonial divisions. But this is compensated for at a later 'differential mitosis' in both oogenesis and spermatogenesis. At these differential mitoses the *E*-chromosomes split but both daughter *E*'s pass to the same nucleus, which becomes the functional oocyte or spermatocyte nucleus as the case may be, the other cell (which has received only a diploid set of *S*-chromosomes) becoming a nurse-cell in the female and an aberrant

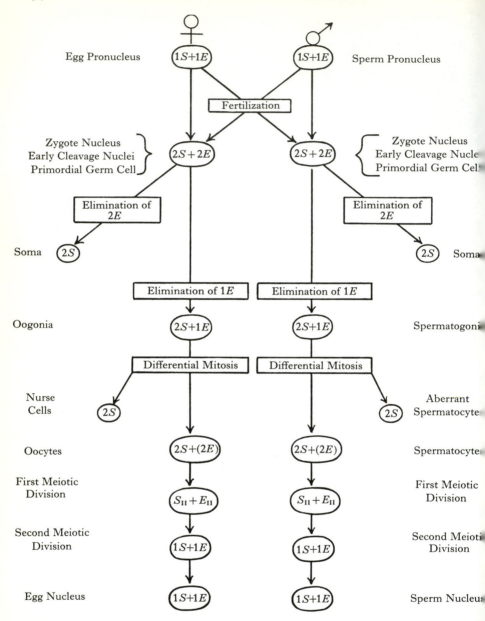

Fig. 14.21. Chromosome cycle of the Orthocladiinae, based on the work of Bauer and Beermann. $1S$ = a haploid set of S-chromosomes, $1E$ = a haploid set of E-chromosomes; S_{II} and E_{II} designate sets of S- and E-bivalents. Further explanation in text.

non-functional spermatocyte in the male. In the functional spermatocytes the *E*-chromosomes mostly form bivalents (usually without chiasmata) although some univalents may also be present.

The Chironomidae are not closely related to the sciarid–cecidomyid stock of Nematocera, and although there may be a general analogy between the anomalous chromosome cycles of the Orthocladiinae and those of the Sciaridae and Cecidomyidae, so that an entirely independent evolutionary origin seems probable.

In a few insects which have diploid, biparental males, the reduction to the haploid number of chromosomes takes place in the primordial spermatogonia, so that the secondary spermatogonia and spermatocytes are haploid, and no further reduction takes place before the formation of the sperms. This is definitely the case in the testicular part of the gonad in the hermaphrodite coccid *Icerya purchasi* and some related species (see p. 506). A somewhat analogous condition occurs in the Biting and Sucking Lice (Mallophaga and Anoplura). The early work was carried out on the human louse, *Pediculus humanus* (Doncaster and Cannon, 1919; Cannon 1922; Hindle and Pontecorvo 1942), but conditions in the Hog louse *Haematopinus suis* (Bayreuther 1955) and the Biting Lice *Goniodes* (Perrot 1934) and *Gyropus* (Scholl 1955) seem to be similar. Throughout this group the chromosomes seem to be generally holocentric. In *P. humanus* the somatic (2n) number is 12 in both sexes. Oogenesis has not been studied in detail, but is probably normal. In the male, however, the primordial spermatogonia undergo a reduction division, so that the secondary spermatogonia only contain six chromosomes. Bivalents are formed at this reduction (see especially Bayreuther's observations on *Haematopinus*), but there do not seem to be any chiasmata, the four chromatids all lying parallel, as in *Drosophila* and *Callimantis* (but not separated at a localized centromere).

Apparently both haploid products of the reduction division remain in the same cyst, and then undergo five synchronous divisions which are ordinary mitoses and yield 4, 8, 16, 32 and 64 secondary spermatogonia. Then follows a growth phase which converts the small spermatogonia into much larger spermatocytes. This is followed by a single maturation division which is an unequal one as far as the cytoplasm is concerned, a small nucleated 'bud' containing six chromosomes being cut off from the main cell. Only the latter form sperms, the buds degenerating. There is no second maturation division, each cyst giving rise to a bundle of 64 sperms and 64 small pycnotic masses.

The sex-determining mechanisms in this group are somewhat mysterious. No sex chromosomes have been seen in any of the species. In *Pediculus* pair matings may give rise to almost unisexual progenies (Hindle 1919),

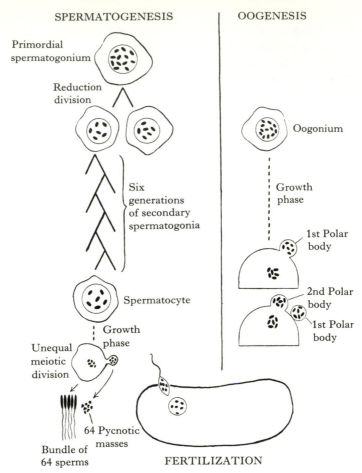

Fig. 14.22. Chromosome cycle in the louse. Explanation in text. Based on the work of Cannon (1922) and Hindle and Pontecorvo (1942).

although females mated to several males usually produce approximately equal numbers of sons and daughters (Buxton 1940). Thus there are probably genetic differences between the males, some being male-producers, others female-producers. This situation contrasts with the one that obtains in *Sciara* spp. and some other groups, where the sex of the progeny depends on the genetic constitution of the mother, there being only one type of male, from the standpoint of sex determination.

In *Gyropus* there are only two chromosomes in the haploid set, and one fewer spermatogonial division, so that the sperms are in bundles of 32, but otherwise the details of the chromosome cycle are very similar.

It is interesting to find such a close similarity between the cytogenetic system of the Anoplura and Mallophaga; groups that differ considerably in external morphology but are probably more closely related than they appear to be.

Apart from groups such as the coccids, sciarids, cecidomyids and others, where anomalous meiotic mechanisms have become part of the normal genetic mechanism of many or all of the species, there are a number of groups of animals in which a 'degenerate' type of meiosis coexists with a normal type in the same testis, and leads to the production of non-functional sperms alongside the normal ones.

A relatively 'mild' form of sperm polymorphism seems to be very wide-spread in the pentatomid bugs; in many species with seven lobes in the testis the spermatocytes in all except lobes 4, 5 and 6 are of normal size, while those in lobes 4 and 6 are much larger and those in lobe 5 are much smaller. The situation has been investigated especially in *Arvelius albopunctatus* by Schrader and Leuchtenberger (1950). The nuclear volumes of the three categories of spermatocytes are approximately 200, 400 and 1600 μm^3, but their DNA content is the same. The nucleolus is much larger in the largest class of cells and contains more protein and RNA, while the cytoplasmic volume is also much greater and the amount of cytoplasmic protein and RNA correspondingly increased. The nuclear phenomena of meiosis are the same in all three classes of cells, but three different kinds of sperms are produced. Apparently the acrosomes of the very large sperms contain much larger quantities of a polysaccharide which may be associated with the enzyme hyaluronase. It is not known whether these 'large' sperms function genetically; they certainly contain a normal haploid chromosome complement. Polyspermy (penetration of the egg by several sperms) is normal in many insects, so it is possible that the giant sperms contribute a significant quantity of protein and RNA to the egg cytoplasm, their role being similar to that of nurse-cells.

In a few pentatomid bugs, almost all tropical members of the tribes Pentatomini, Halyini and Discocephalini, the fifth lobe, which in many other species produces the smallest class of sperms, contains spermatocytes whose meiotic divisions are so anomalous that only aneuploid sperms are produced which obviously cannot function in a genetic sense. This is what Schrader (1945*a*, *b*, 1946*a*, *b*, 1960*a*, *b*) called a 'harlequin lobe'.

In the genera *Loxa* and *Mayrinia* (Schrader 1945*a*,*b*) the primary spermatocytes in the harlequin lobe fuse together to form giant polyploid cells which undergo a highly irregular meiosis and give rise to sperms containing all sorts of aberrant chromosome numbers (up to more than 100 in *Loxa*). *Brachystethus rubromaculatus* has quite a different type of 'degenerate meiosis' in its harlequin lobe (Schrader 1946a). In this species the spermatocytes do not fuse and the prophase stages of the first meiotic division are quite normal. Just before the first metaphase, however, all the autosomal bivalents become clumped together. They do not divide in either meiotic division, but remain as a single aggregate. The X- and Y-chromosomes, on the other hand, behave normally, splitting in the first division and passing undivided to opposite poles at the second. Thus, in the main, four types of sperms are produced by the fourth lobe: (1) with an X but no autosomes, (2) with a Y but no autosomes, (3) with an X and a diploid set of autosomes, (4) with a Y and a diploid set of autosomes. It seems certain that none of these types of sperms ever function in a genetic sense, but it is not known whether they may penetrate the eggs.

A somewhat similar type of degenerate meiosis occurs in some penta-tomids of the tribe Discocephalini such as *Mecistorhinus* spp. and *Neodine* (Schrader 1946b), where it is confined to the fifth lobe of the testis. In this type no synapsis takes place; all the autosomes become clumped at the first

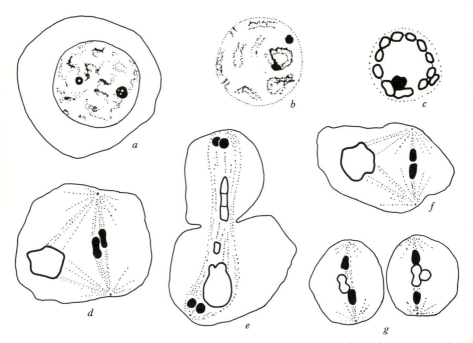

Fig. 14.23. Spermatogenesis in the 'harlequin lobe' of the testis in the pentatonid *Mecistorhinus sepulcralis. a*, post-pachytene stage; *b*, diakinesis; *c*, first metaphase in polar view, X and Y associated with large autosomes; *d*, first metaphase in side view; *e*, late first anaphase; *f*, second metaphase in one of the large cells; *g*, second metaphase in one of the small cells. From Schrader (1946).

metaphase as in *Brachystethus*, but one large autosome separates from the aggregate at the first anaphase and passes to the opposite pole from the rest of the mass. The sex chromosomes behave normally, as in *Brachystethus*. At the second meiotic division the single autosome almost always passes to the same pole as the X, so that the following types of sperms are produced: (1) with an X and a large autosome, (2) with the Y alone, (3) with the X and a variable number of autosomes. Nicklas (1961*b*) has shown that in *Mecistorhinus* and *Neodyne* the premeiotic DNA replication is normal in the harlequin lobe.

A variety of other types of abnormal spermatogenesis occur in the harlequin lobes of the neotropical pentatomids *Schraderia*, *Alitoris*, *Macropygium*, *Moncus*, *Melanodermus*, *Architas*, *Ablaptus* and *Pseudoevoplitas* (Schrader 1960*a*). There is a general tendency for chromosomes, such as the X and Y (and certain autosomes as well) which are mainly or entirely heteropycnotic in the prophase of meiosis to behave relatively normally in the meiotic divisions, the abnormal behaviour being shown by the euchromatic autosomes. Schrader (1960*b*) points out that all the species with 'harlequin lobes' are tropical forms, none of the temperate North American genera possessing such a development. On the other hand, many neotropical genera undoubtedly lack harlequin lobes; information for other continents is unfortunately lacking.

Somewhat similar types of degenerate spermatogenesis occur in many of the prosobranch Mollusca (Meves 1902; Kuschakewitsch 1913; Ankel 1930; Morita 1932), in which *apyrene* sperms (i.e. ones lacking a nucleus altogether) and *oligopyrene* ones (ones containing very few chromosomes) are produced. The details were worked out in four species of viviparid snails by Pollister (1939) and Pollister and Pollister (1943). No synapsis occurs in the 'atypical' spermatocytes, and the sister chromatids fall asunder at a stage corresponding to diakinesis. In *Viviparus malleatus* the nucleus contains thirty-six separate chromatids at the premetaphase of the first meiotic division ($n = 9$ in this species). Four of these chromatids behave normally during the meiotic anaphases, so that each spermatid receives one of them. The remaining thirty-two chromatids behave as if they were acentric; at the first anaphase they separate into two roughly equal groups which pass into the two daughter cells, but at the second division they all remain at one end of the spindle, so that two kinds of spermatids are formed, both with a single 'normal' chromosome but only the larger kind having the rest of the chromosomes, which are now obviously degenerating. At the same time as the 'abnormal' chromatids begin to behave as if they were acentric the normal number of two centrioles at each pole of the spindle is augmented by sixteen centriole-like bodies, making a total of eighteen apparent centrioles at each pole. The suspicion that the extra centrioles are actually derived from the centromeres of the chromatids which behave abnormally was confirmed by a study of three other species; all these have different chromosome numbers, but in each case the number of supplementary centrioles corresponds exactly to the number of chromatids which behave as if acentric.

Apyrene and oligopyrene sperms are also formed in some Lepidoptera (Meves 1902; Gatenby 1917; Bowen 1922) but the chromosomal details have not been worked out as completely as in the pentatomids and prosobranch molluscs.

15

THE CYTOLOGY OF INTERSPECIFIC HYBRIDS

On the theory of natural selection the case is especially important, inasmuch as the sterility of hybrids could not be of any
advantage to them, and therefore could not have been acquired
by the continued preservation of successive profitable degrees
of sterility.

C. DARWIN: *Origin of Species* (1st edition)

Comparison of the cytogenetic systems of related species is greatly facilitated if interspecific hybrids can be obtained. Such hybrids occur occasionally in nature, the isolating mechanisms between the parent species being
incomplete. In some such cases the F_1 hybrids are completely sterile, so that
no 'leakage' or 'introgression' (Anderson and Hubricht 1938; Anderson
1949) of genetic material from one species into the other is taking place,
while in other cases a slight amount of gene transfer does appear to occur, in
one or both directions. Mayr (1963, Ch. 6) has described in detail the
various ways in which isolating mechanisms may break down and has cited
numerous instances which we shall not discuss, since cytological information
on them is lacking. He concludes that as far as animals are concerned, the
evolutionary role of interspecific hybridization has been small, since even
where fertile hybrids are produced, there is severe selection against the
genetically unbalanced genotypes resulting from introgression. Mayr's
views on the relative unimportance of hybridization in animal evolution are
shared by the present author, but it is only fair to point out that they are not
universally accepted.

Many species which never seem to produce hybrids in nature can be
induced to do so in captivity, when they are not supplied with mates of
their own species, or when they are kept under crowded or otherwise
abnormal conditions. However, certain pairs of species which are regarded

by taxonomists as very closely related have never been successfully hybridized, either because they will not mate or because the cross is completely sterile. In such cases the isolating mechanism is complete, so that its genetic basis can only be studied by indirect methods, such as crossing both forms to a third one. In other instances it has proved possible to obtain hybrids, but only by using techniques of artificial insemination.

Whenever the isolating mechanism or complex of mechanisms which normally keep two species separate is incomplete or can be broken down by experimental means, some kind of an F_1 generation can be obtained. This may consist mainly of a few sickly embryos that never develop beyond a certain stage or of vigorous individuals that become fully fertile adults. In some cases the F_1 progeny of an interspecific cross are all of one sex, the offspring of the opposite sex being inviable. In other instances the sexual development of the hybrids is disturbed so that they become intersexes or undergo sex reversal.

Experimental hybridization may be directed toward revealing the nature and extent of some of the isolating mechanisms between the species concerned and, when combined with cytological analysis, to show the kinds of structural rearrangements which have taken place in their chromosomes since they diverged in evolution (see Ch. 11). It does not follow, however, that either the strength and extent of the genetic isolating mechanisms or the degree of karyotypic divergence is in any particular instance an accurate index of the degree of phyletic relationship between the species being studied. Some very closely related species may be separated by strong isolating mechanisms and show considerable karyotypic differences – and vice versa. 'Phyletic relationship' can only be assessed on the basis of *all* the evidence, biochemical and biometrical as well as genetic and chromosomal.

A great many species hybrids (i.e. F_1 individuals resulting from interspecific crosses) are less fertile than normal individuals of the parent species, the loss of fertility ranging from near normality down to complete sterility. Obviously, hybrid sterility does not, in itself, prevent interspecific matings; it is, however, an effective post-mating isolating mechanism.

The whole subject of the causation of hybrid sterility is a very complex one, involving endocrinological and morphogenetic aspects as well as genetic and cytogenetic ones. Each of the parents in a species cross probably contains within its chromosomes a very large number of genes controlling the development of the gonads, the course of oogenesis and spermatogenesis, the relations of the future germ-cells to various 'nutritive' tissues and cellular elements, the development of the very specialized sperm cell from the spherical or polygonal spermatid, the motility of the sperm and its ability to survive in the female genital tract and a number of other physiological processes essential to fertile reproduction. A careful comparative study of meiosis in any group of organisms would certainly show differences between the species in regard to the timing of the stages, the rate and extent of chromosome condensation. If the genes controlling all these processes are sufficiently alike in the two parent forms they may build up a compatible system in the hybrid, so that the latter has an approximately normal gametogenesis and is wholly or partially fertile. On the other hand, if there is a lack of harmony between the genetic systems of the parent species we may have a general disturbance of gametogenesis, a failure of synapsis or of chiasma formation at meiosis or the suppression or modification of some essential stage or process in meiosis or the later stages of gametogenesis. The male germ-cells, in particular, seem to be rather precariously balanced in respect of some essential physiological processes, in many species, so that they are very liable to become pycnotic and die prematurely. This happens in the testes of many non-hybrid animals, but it is particularly liable to occur in those of hybrid constitution.

In some species hybrids the gonads are quite small and imperfectly formed and contain only spermatogonia or oogonia and no later stages. In such instances there is no question of meiosis occurring. The whole situation may be regarded as a result of some specific endocrine disturbance, but very little is known of the actual factors involved. In the well-known hybrids between *Drosophila melanogaster* and *D. simulans* the ovaries and testes are very small and gametogenesis is arrested before meiosis, so that sterility is complete (Kerkis 1933). Similarly, in the testes of male hybrids between the chicken (*Gallus gallus*) and the pheasant (*Phasianus colchicus*) the spermatogonial mitoses seem to be normal but the spermatocytes degenerate during the prophase of meiosis (Yamashina 1943). The same is probably true of the mule (Makino 1955), although the old work of Wodsedalek (1916) suggests that meiotic divisions are sometimes present. In this case the horse *Equus caballus* is known to have 2n = 64, while the donkey *E. asinus* has 2n = 62; there are considerable differences between the karyotypes of the two species, suggesting that a fairly large number of chromosomal rearrangements have occurred since the evolutionary divergence of these species (Trujillo *et al.* 1962; Makino, Sofuni and Susaki 1963; Mukherjee and Sinha 1964). Even more profound differences exist between *E. asinus* and the Grevy zebra, *E. grevyi*, which apparently has 2n = 34; in spite of this sterile hybrids between these species can be obtained (Benirschke *et al.* 1964). In spite of alleged cases of fertile mule mares which have been reported in the literature, it seems likely that all mules and hinnies are completely sterile (Benirschke, Low, Sullivan and Carter 1964). On the other hand, hybrids between *E. caballus* and *E. przewalskii* (2n = 66) are said to be fully fertile (Benirschke, Maloef, Low and Heck 1964); the X-chromosomes of these species are apparently indistinguishable, whereas those of the horse and donkey are markedly different. Some examples of fertile species hybrids in bears, Camelidae and deer are described by Benirschke (1967).

It is rather unfortunate that a great many of the most significant studies of hybrid sterility have been carried out on forms whose meiotic divisions are, for one reason or another, hard to study. Thus all the excellent work of Federley, Suomalainen, de Lesse and others on lepidopteran species hybrids suffers from the fact that the chromosomes of the Lepidoptera are, in general, too small and numerous (and also too similar in size) to permit a really detailed analysis. The structure of the meiotic bivalents in Lepidoptera is not easily discerned and trivalents and quadrivalents cannot be clearly distinguished from bivalents. The very extensive studies of many workers on hybridization in the genus *Drosophila* (see especially Patterson and Stone 1952) have been carried out on the polytene chromosomes and, with certain exceptions to be noted later, there have been no detailed studies of the meiotic divisions of the hybrids, because of the technical difficulties involved. Investigations of various fish hybrids resemble those on Lepidoptera and suffer from the same defects, inevitable in view of the nature of the material.

More satisfactory investigations on the cytologic basis of hybrid sterility have been carried out on forms such as Orthoptera and Amphibia, where the chromosomes are large enough for detailed studies. Among those species hybrids in which meiosis occurs at all (i.e. excluding those in which gametogenesis is arrested before meiosis begins or at an early prophase stage) we may distinguish various types of meiotic anomalies, each of which may occur in various grades, from mild to severe. Broadly speaking, we may classify the kinds of meiosis in animal hybrids as follows: (1) complete synapsis; spindle morphology and anaphase separation normal, (2) incomplete synapsis; meiotic divisions otherwise normal, i.e. except for the presence of some univalents. (3) complete synapsis but physiology of meiosis abnormal, i.e. chromosomes 'sticky', spindles incompletely formed or abnormal in shape, anaphase separation difficult, irregular or impossible, (4) incomplete synapsis and spindle mechanism abnormal, (5) no synapsis, meiotic spindles normal, (6) no synapsis, spindles abnormal. Usually, it is only hybrids belonging to types (1) and (2) which show some degree of fertility, either *inter se* or in backcrosses to one or other of the parent species, hybrids in categories (3) to (6) being almost invariably sterile.

When we find univalents at the first meiotic division of a species hybrid it is, of course, quite uncertain whether they are due to zygotene asynapsis or to 'desynapsis' at a later stage, perhaps as a result of failure of chiasma formation. Some of the earlier cytogeneticists assumed that asynapsis in hybrids must be due to structural chromosomal differences between the parent species. It is fairly obvious that if two species have sustained very great structural changes in their chromosomes since they diverged in the course of phylogeny, then the hybrid between them would be liable to show some degree of asynapsis. But it is known that in *Drosophila* a large number of heterozygous inversions may be present in a non-hybrid individual without leading to a significant amount of non-disjunction such as would indicate that asynapsis was occurring. In this connection one must point out, however, that we have no right to base our ideas as to the extent of synapsis at meiosis on observations of pairing in the polytene nuclei. There is ultimately only one way of studying synapsis at meiosis and that is by a microscopical examination of the meiotic nuclei themselves, at the zygotene–pachytene stage; but in *Drosophila* this is extremely difficult and the evidence obtainable is of doubtful value.

Dobzhansky (1937c, 1951) has drawn a distinction between *chromosomal sterility*, caused by structural differences between the karyotypes of the parent species, and *genic sterility*, where the impediment to pairing is physiological and not merely mechanical. Although Dobzhansky's distinction may be theoretically valid it is difficult to apply in practice. Many instances of hybrid sterility may depend on a combination of structural or mechanical and physiological impediments to synapsis. It must be said, however, that there is rather little evidence for chromosomal sterility in animal species hybrids, although it may play a more important role in plants (Dobzhansky 1951); the more severely abnormal types of gametogenesis and meiosis in hybrids seem to be always of the genic type. If there is little or no synapsis in a diploid hybrid of constitution AB but synapsis is restored in the tetraploid AABB it has been sometimes assumed that the asynapsis of the diploid hybrid is caused by structural differences such as inversions or translocations. But this is not necessarily the case; it is possible to imagine that the chromosomes of species A and B have the same gene sequence but differ in some general respect such as their cycle of condensation, so that their pairing is inhibited in the hybrid. It is, of course, important to realize that failure of synapsis is only one of the causes of hybrid sterility and that even where synapsis is completely normal, gametogenesis may become abnormal at a later stage, so that few or no viable eggs or sperms are produced.

Many cases of species hybrids whose meiosis is believed to be essentially normal are known in groups whose cytology is not suitable for a detailed study of synapsis or chiasma formation. An early example is provided by the hybrids between the pentatomid bugs *Euschistus variolarius* and *E. servus* (Foot and Strobell 1914). Here bivalent formation in the F_1 males was apparently normal and the hybrids were fertile, so that a second generation could be reared.

In the hybrids between the bed bug species *Cimex lectularius* and *C. columbarius* pairing of the autosomes is likewise complete (Darlington 1939*b*) and in the *Cimex pilosellus* complex Ueshima (1963) has shown that the hybrids between two species with n = 11 and n = 12 regularly show nine bivalents, a trivalent and the *X*- and *Y*-univalents at the first meiotic division. Complete synapsis at meiosis may hence be quite usual in heteropteran species crosses. The hybrid crickets studied by Cousin (1934, 1941) and those between the grasshoppers *Trimerotropis maritima* and *T. citrina* obtained by Carothers (1941*a*, *b*) probably also belong in this category, although no detailed cytological study was made, since in both cases an F_2 was obtained. A similar situation is known to exist in some hybrid birds. Thus the Japanese copper pheasant *Syrmaticus soemmeringii* and the golden pheasant *Chrysolophus pictus* both seem to have n = 41 and in the male hybrid between them pairing seems to be complete at meiosis so that forty-one bivalents are regularly formed (Yamashina 1943). Meiosis is normal, although a few spermatocytes do degenerate, and the hybrids are quite fertile.

In all the above instances except the *Cimex* species studied by Ueshima the chromosome number is the same in the two parent species and the gene sequences of the chromosomes are probably very similar. Almost

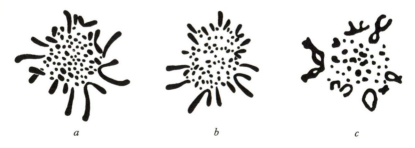

a   *b*   *c*

Fig. 15.1. Chromosomes of two species of pheasants, *Syrmaticus soemmeringi* and *Chrysolophus pictus* and the hybrid between them. *a*, spermatogonial metaphase of *Syrmaticus*; *b*, spermatogonial metaphase of *Chrysolophus*; *c*, first metaphase in spermatogenesis of the hybrid between them, showing apparently complete synapsis and evidence of a high chiasma frequency in the larger bivalents. From Yamashina (1943), redrawn.

complete pairing can occur, however, even when the chromosome number of the parent species is different. Thus in the hybrids between the moths *Bombyx mori* ($n = 28$) and *B. mandarina* ($n = 27$) studied by Kawaguchi (1928) twenty-six bivalents and a trivalent composed of two *mori* chromosomes and one *mandarina* one were observed at meiosis. Associations of more than two chromosomes are also seen at meiosis in the hybrids between the moths *Orgyia antiqua* ($n = 14$) and *O. thyellina* ($n = 11$) studied by Cretschmar (1928).

Trivalents in species hybrids may be expected wherever one parent species has a fusion or dissociation not present in the other. But the trivalent may be realized in only a minority of the cells, the remainder having a bivalent and a univalent. This is the state of affairs in hybrids between the morabine grasshopper species 'P24' and 'P25' (White, Blackith, Blackith and Cheney 1967). P24 has a fusion between a long acrocentric and a short one. In P25 the fusion is lacking and, in addition, the chromosome corresponding to the short limb of the fusion chromosome of P24 is a little metacentric. Perhaps because of the latter fact, the small chromosome of P25 is a univalent in about 80% of the spermatocytes in the hybrids; in the remaining 20% a chain trivalent is formed.

Multivalents will also be expected in species hybrids where one parent species carries a translocation which the other lacks. They will also occur in hybrids between two species B and C where both are derived from an ancestral form A, if one set of fusions occurred in the evolutionary pathway leading to B and a different series of fusions in that leading to C (e.g. the ancestral acrocentrics a, b, c, d, e, f may be combined as metacentrics ab, cd, ef in species B and as fa, bc, de in species C). The best example of the latter situation is that of the hybrids between gerbils from Algeria and the Negev studied by Wahrman and Zahavi (1958).

The situation where wide differences in chromosome numbers does not prevent complete pairing at first metaphase clearly depends on regular zygotene synapsis, but in organisms with chiasmatic meiosis it is also dependent on the ability to form two chiasmata in the chromosomes of the species with the lower chromosome number. This seems to be rather frequently the case in lepidopteran species hybrids. Thus in male hybrids between *Dicranura vinula* ($n = 21$) and *D. delavoiei* ($n = 31$) twenty-one bodies are frequently found at first metaphase, ten of these being made up of two *delavoiei* chromosomes paired with a single large *vinula* chromosome (Federley 1939). In the F_1 of certain crosses between *vinula* stocks from Finland and central Europe (both parent species having $n = 21$) only twenty bodies were present at first metaphase. Presumably one of these bodies was an association of four chromosomes, but it is not possible to

distinguish multivalents from bivalents at first metaphase in the spermatogenesis of most Lepidoptera.

In many lepidopteran species hybrids, on the other hand, pairing is very
incomplete, regardless of whether the chromosome numbers of the parent
species are the same or not. Thus most of the chromosomes are univalents
at the first meiotic division in hybrids between *Dicranura erminea* (n = 28)
and *D. vinula* (Federley 1943) and in the hybrids between *Cerura* (*Dicranura*)
bifida (n = 49) and *C.* (*D.*) *furcula* (n = 29) fifty-two to seventy bodies
are seen at first metaphase, so that here likewise pairing is incomplete
although some chromosomes obviously do manage to associate. Hybrids
between *C. bifida* and *C. bicuspis* (n = 30) show thirty to forty-three bodies
at first metaphase so that pairing may be complete in some cells but not in
others. Incomplete pairing occurs also in the hybrids between the moths
Saturnia pavonia (n = 29) and *S. pyri* (n = 30) studied by Pariser (1927).
In the F_1 hybrids there are about eight to fourteen bivalents. These hybrids
produced some sperms with slightly less than the diploid set of chromosomes; when backcrossed to either of the parental species they gave rise
to intersexual subtriploid individuals.

A large number of interspecific and intergeneric crosses have been carried
out in the sphingid moths (Federley 1915, 1916, 1928, 1929; Bytinski-Salz
and Günther 1930, Bytinski-Salz 1934). In the genera *Celerio* and *Pergesa*
the extent to which the chromosomes of the F_1 hybrids pair at meiosis
seems to be directly correlated with the degree of relationship between the
parent species, as judged by ordinary taxonomic criteria. Thus all hybrids
between *Celerio euphorbiae*, *C. galii*, *C. hippophaes* and *C. vespertilio* show
complete pairing at first metaphase, and so do hybrids between *Pergesa
elpenor* and *P. porcellus*. In hybrids between species of *Pergesa* and species of
Celerio, however, there is very little pairing (only four or five bivalents
being formed, out of a potential twenty-nine). *C. lineata* seems to be more
distantly related to the other *Celerio's* than they are to one another, since in
the hybrids between it and *euphorbiae*, *galii*, *hippophaes* and *vespertilio*
pairing is incomplete, although more bivalents are formed than in the
intergeneric hybrids. In none of these lepidopteran hybrids does one find
the gross abnormalities of meiosis which occur in many *Drosophila* hybrids.
Apparently the genetic control of the physiology of meiosis is less delicately
poised than in *Drosophila*, so that it is not so easily upset by the presence of
the chromosomes of two different species within the same nucleus.

All the above data refer, of course, to the male hybrids. In female hybrids
of *Celerio* and *Pergesa* the histology of the ovaries is normal in the offspring
of interspecific crosses, but in the female intergeneric hybrids the ovarioles
do not contain any fully developed eggs. In some of these sphingid species
crosses the female hybrids never emerge from the pupa.

By way of contrast to the sphingid hybrids we may consider those between the moths *Pygaera curtula* (n = 29) and *P. pigra* (n = 23) obtained by Federley (1913, 1931). Whereas female hybrids show almost complete pairing at meiosis, in the spermatogenesis of the male hybrids there are few or no bivalents at first metaphase. The univalent chromosomes were said to divide in both divisions by Federley. At any rate, spermatids are produced with the diploid chromosome set. There has been some discussion as to whether the failure of synapsis in these *Pygaera* hybrids is 'genic' or 'chromosomal'. Dobzhansky (1937*d*) suggested that a true chromosomal sterility was involved because in triploid *Pygaera* hybrids with two chromosome sets from one species and one from the other the two homologous sets formed bivalents, leaving the third set unpaired. This is a different situation to that found in triploid *Drosophila* hybrids with two *melanogaster* sets and one *simulans* set, where pairing of the *melanogaster* sets does not occur (Schultz and Dobzhansky 1933). The fact that pairing is almost complete in diploid female *Pygaera* hybrids strongly suggests, however, that failure of synapsis in the male hybrids is physiological, i.e. genic rather than chromosomal in origin. The pairing which takes place in the triploid *Pygaera* hybrids cannot really be regarded as an argument in favour of chromosomal sterility, since the genic equilibrium is obviously different in a diploid and in a triploid hybrid.

The general picture which emerges from all this work on lepidopteran hybrids is thus one of relative normality of spermatogenesis in the F_1 of crosses between fairly closely related forms, the only abnormality being in most instances the presence of a certain number of univalents. In some species of sphingid moths such as *Smerinthus ocellatus*, however, hybrids between forms currently regarded as conspecific show rather incomplete pairing (Federley 1914, 1915), so that the degree of taxonomic relationship which is associated with failure of synapsis (or failure of chiasma formation) obviously varies from group to group.

There is usually a good deal of variability in the cytological picture presented in species hybrids; at least this is so where abnormalities are moderate in degree, involving some asynapsis. This variability is strikingly shown in the hybrids between the grasshoppers *Circotettix verruculatus* and *Trimerotropis suffusa*, studied by Helwig (1955). Actually these species are closely related and should obviously be included in the same genus. Some F_1 hybrids had over 90% of the spermatocytes with complete synapsis while in other hybrid individuals less than 10% of the spermatocytes have complete synapsis. In general, asynapsis was commonest in some of the medium-sized chromosomes and rarest in the longest ones. But a particular chromosome pair such as no. 3 showed 0.5% asynapsis in one individual and 67.8% in another. The factors leading to this extreme and

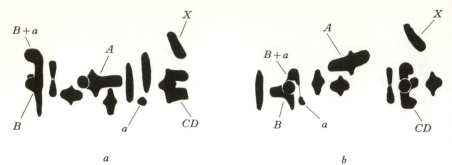

Fig. 15.2. First metaphase in an F_1 hybrid between the morabine grasshopper species 'P24' and 'P25'. In *a* the small chromosome (*a*) derived from P25 is an unsynapsed univalent; in *b* there is a trivalent, which is segregating disjunctionally.

chaotic variability of behaviour are quite obscure. Variability seems to be much less in cases where the abnormalities are much more extreme. Thus in certain hybrids between an Australian and a New Zealand species of the grasshopper genus *Phaulacridium* there is a fairly uniform picture involving degeneration of all the spermatogonia before meiosis has begun (White unpublished).

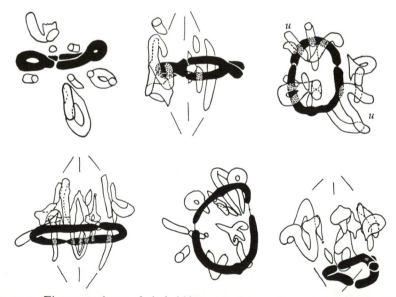

Fig. 15.3. First metaphases of a hybrid between the grasshopper species *Chorthippus bicolor* and *C. biguttulus*, showing the abnormal behaviour of bivalents with a lower chiasma frequency than usual, which frequently form huge loops surrounding the spindle. *u*, univalents in one of the metaphases. From Klingstedt (1939).

As an example of the third type of species hybrid (those in which pairing is complete, but meiosis is physiologically abnormal) we may take the hybrids between the European grasshoppers *Chorthippus biguttulus* and *C. brunneus* (= *bicolor*) studied by Klingstedt (1939). These species are difficult to distinguish, but there can be no doubt as to their specific rank; where they overlap geographically they occasionally hybridize, and one of the hybrid individuals investigated by Klingstedt was caught in the wild.

In the spermatogenesis of the male hybrids all the chromosomes usually pair and the chiasma frequency is not much below normal. At the first metaphase the bivalents are thinner than usual and their orientation is irregular. Sometimes one large ring bivalent surrounds the whole spindle, so that the axis of the latter passes through the loop. It seems likely that the spindle itself is abnormal, but this is not certain. Anaphase separation seems to take place with great difficulty and many chromosomes break or fail to separate. It is not quite clear how far this is due to abnormalities of the spindle or to disturbances in the condensation and separation of the chromatids.

Both the parent species have three pairs of large metacentric chromosomes in addition to five pairs of acrocentric autosomes. Since the former always form three bivalents in the hybrids, usually with chiasmata in both arms, the arrangement of the chromosome limbs must be the same in both species (and not, for example *AB*, *CD*, *EF* in one and *AB*, *CE*, *DF* in the other). Some chromosome 'bridges' were observed in the hybrids at first anaphase, but it is uncertain whether they were due to crossing-over in inversions, or to the difficulty which the chromosomes experience in separating from one another. An 'unequal bivalent' composed of a large and a small acrocentric was observed in the natural hybrid. Although Klingstedt's experiment was not carried beyond the first generation, it seems likely that the hybrids would have been entirely sterile, in spite of the complete pairing of the chromosomes.

Crosses between species and races of salamanders provide a graded series, in which the magnitude of the meiotic anomalies is roughly proportional to the apparent degree of evolutionary divergence of the parental forms. A considerable amount of work has been carried out on the species *Triturus cristatus* and *T. marmoratus* (White, 1946d; Lantz 1947; Spurway 1953; Lantz and Callan 1954). The first of these is a polytypic species which has a wide distribution in northern, central and eastern Europe and is represented by distinct subspecies (*T. c. carnifex*, *T. c. karelinii* and *T. c. danubialis*) in Italy, the Black Sea–Caucasus area and the Danube valley. *T. marmoratus* is confined to the Iberian peninsula (from which *T. cristatus* is absent) and western France. The two species have a wide zone of overlap in central France, within which some natural hybrids (originally described as a distinct species 'T. blasii') occur (Vallée 1959). In the region of overlap recognizable hybrids form about 5% of the population, the parent species being about equally abundant.

Laboratory F_1 hybrids between the various subspecies of T. cristatus and T. marmoratus show a considerable degree of fertility in the females, but male hybrids are entirely sterile. Their chiasma frequencies are much lower than those of the pure species and there are many univalents at first metaphase (Fig. 15.4). To what extent these are a consequence of asynapsis (failure of zygotene pairing) and to what extent they result from desynapsis (falling apart at a later stage) is not known. Owing to the presence of univalents and the frequent spindle abnormalities at first anaphase, the second meiotic divisions very seldom possess a correct assortment of chromosomes. In any case, the spermatids are in most cases highly abnormal and undergo degeneration without forming sperms. The sterility of the male hybrids thus depends on two factors: (1) failure of pairing of the chromosomes (or, at any rate, failure to remain paired) and (2) degeneration of the spermatids. Genetical and cytological studies on backcross and 'double backcross' hybrids (i.e. ones with five-eighths of their chromosome set from one species and three-eighths from the other) suggest that the sterility of the male hybrids is due to a number of genes, as in the case of the *Drosophila pseudo-obscura* × *persimilis* cross (see p. 568).

Minor disturbances of meiotic pairing occur even in hybrids between the various races of *Triturus cristatus*, which may perhaps be regarded in some cases as approaching the species level. As we have already seen (p. 141) there are numerous differences between the lampbrush chromosomes of the races in respect of the detailed appearance of the individual loops and even in the nature of the centromere regions. Members of this group of forms show very little chiasma localization; the chiasma frequency is highest in subspecies *karelinii*, lowest in *carnifex*. The male F_1 hybrids show about half as many chiasmata as their parents (15–21 instead of 30–40) and some univalents are present (Spurway and Callan 1950; Callan and Spurway 1951). These anomalies of meiosis are, however, not as severe as in hybrids involving *marmoratus*.

In the genus *Triturus* it is not possible to obtain hybrids between members of different species groups by natural mating; however, by using artificial insemination, Benazzi and Lepori (1947) were able to obtain hybrids of both sexes from the cross *Triturus c. carnifex* × *T. vulgaris meridionalis*. The spermatogenesis of the male hybrids showed the same types of anomalies as those seen in the *cristatus* × *marmoratus* hybrids, but in an even more exaggerated degree; a few bivalents were, however, still present. Benazzi and

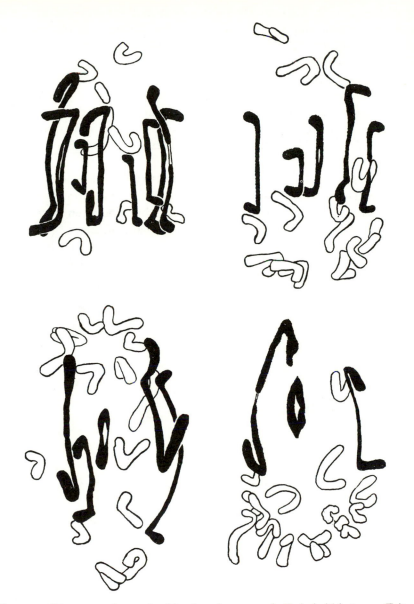

Fig. 15.4. First metaphases, in side view, from a male F₁ hybrid between *Triturus cristatus carnifex* and *T. marmoratus*. Bivalents in black, univalents in outline. From White (1946c).

Lepori (see also Benazzi 1948*b*) believe that in these hybrids zygotene pairing is normal, and that the chromosomes undergo desynapsis after pachytene, presumably owing to the fact that in most bivalents no chiasmata are formed.

The high degree of genetic incompatibility between the European species of *Triturus* in hybridization experiments seems to be paralleled by the situation in the Japanese salamander genus *Hynobius* (Kawamura 1952, 1953). On the other hand, F_1 interspecific hybrids in the Californian genus *Taricha* show little or no reduction of fertility in either sex (Twitty 1964).

In lizards, 'natural' hybridization occurs between two species of *Anolis* on the island of Trinidad, where they have been accidentally introduced by human agency (Gorman and Atkins 1968). F_1 hybrids between *A. trinitatis* (native to the island of St Vincent) with $2n = 36$ and *A. aeneus* (from the Grenada bank) with $2n = 34$ show incomplete pairing at the male first metaphase and a considerable accumulation of first metaphases, presumably due to some kind of block or inhibition of first anaphase.

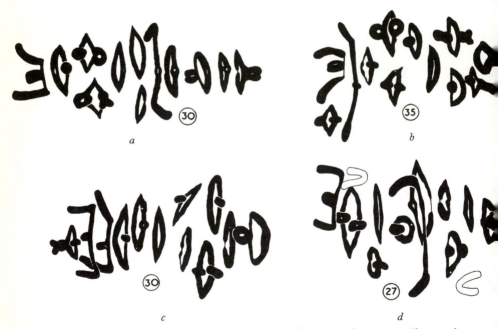

Fig. 15.5. Four first metaphases in the male of *Triturus cristatus carnifex*, to show numbers and distribution of chiasmata (numbers of chiasmata shown in small circles). Compare with the figures of the same stage in the interspecific and inter-racial hybrids. *T. cristatus cristatus* and *T. c. karelinii* have even higher numbers of chiasmata. From White (1946*c*).

Another case of natural hybridization, analogous to the '*Triturus blasii*' one, that has been investigated cytologically concerns the butterfly genus *Lysandra*. *L. bellargus* (n = 45) and *L. coridon* (n = 88 in most areas, but n = 87 in peninsular Italy) are very clearly distinct species. They are sympatric over a large part of Europe, but it is only in a few areas that natural hybrids have been detected in small numbers. In the Abruzzi these received the name *italoglauca*, while in the Pyrenees they are called *polonus*. De Lesse (1960) has shown that these forms show a variable number of bivalents and univalents at male first metaphase (51–70 bodies altogether) and other abnormalities of meiosis (lagging univalents at first anaphase, chromosomal bridges between telophase nuclei, etc.). Thus the hybrid status of these forms has been confirmed cytologically. It must be presumed that they are quite sterile, at least in the male sex, and that consequently no introgression is occurring (if 'Haldane's rule' were obeyed the female hybrids would be at least equally sterile, and in any case the great difference in chromosome number between the present species would suggest that female hybrids are unlikely to be fertile). It seems likely that some natural hybridization between *L. coridon* and *L. hispana* (n = 84) occurs at various localities in southern Europe (de Lesse 1969); laboratory hybrids between these species have a normal spermatogenesis and a fairly high degree of fertility (de Lesse 1956).

F_1 hybrids between geographic races or geographically remote populations of the same species may in some instances provide evidence of a special kind with regard to cytogenetic processes in evolution. Such hybrids may, of course, show an absolutely normal meiosis in both sexes. Frequently, however, they show various degree of asynapsis, abnormal synapsis and

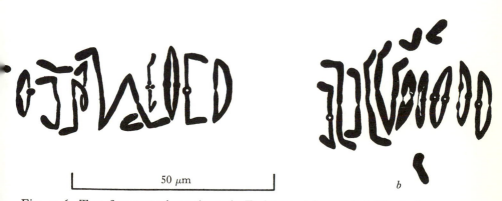

50 μm

b

Fig. 15.6. Two first metaphases in male F_1 interracial newt hybrids. *a*, from a *T. c. karelinii* × *T. c. cristatus* hybrid; *b*, shows 16 chiasmata, a chain of 3 chromosomes and 3 univalents. From Callan and Spurway (1951).

multivalent formation. The interpretation of the cytological picture presented has given rise to some controversy.

As examples of hybrids of this type we may take those between individuals of the grasshopper *Keyacris scurra* from populations about 270 km apart reared by White (1957*a*), those between the geographic races of the European newt *Triturus cristatus* (Callan and Spurway 1951) and those between geographic races of the African grasshopper *Eyprepocnemis plorans* studied by John and Lewis (1965*b*). Multivalents in which the association is by terminal chiasmata have been observed in a non-hybrid individual of *Triturus helveticus* (Mancino and Scali 1961) and in material of *Salamandra maculosa* (see Hartmann 1953 p. 633). And in *Triturus helveticus* apparent chiasmata occur between interstitial regions of different lampbrush bivalents in the oocyte nuclei (Mancino and Barsacchi 1965); their significance is not understood, and it is not known whether they persist until first metaphase. In *K. scurra* there are no major chromosomal differences between the parental populations, but in the F_1 hybrid chains of three, four and five chromosomes were observed at first metaphase. These involved the large '*AB*' metacentric, the '*CD*' pair which is heterozygous for a pericentric inversion, the '*EF*' pair and at least one of the smaller chromosome elements. The end-to-end associations were interpreted as terminal chiasmata. But the situation is certainly not to be explained by heterozygosity for mutual translocations. In a diploid translocation heterozygote each chromosome end can be paired with only one another, whereas in the *Keyacris scurra* interracial hybrids there was evidence that certain chromosome ends could be associated with any one of four different chromosome ends.

In the *Triturus* hybrids chains of three and four metacentric chromosomes were originally interpreted by Callan and Spurway as evidence for translocation heterozygosity, but this interpretation is not supported by detailed study of the giant lampbrush chromosomes of the females. Chains of chromosomes have also been seen in hybrid material of the Californian salamander genus *Taricha* (Callan – see White 1961*a*).

Fig. 15.7. First metaphases in F_1 male hybrids between individuals of the grasshopper *Keyacris scurra* from populations at two localities about 270 km apart. These show various types of anomalous associations (presumably due to terminalized chiasmata) between the *AB* chromosome, the *CD* and the 4th. These individuals were heterozygous for the *Standard* and *Blundell* sequences of the *CD* chromosomes. In *a* two univalent 4th chromosomes can be seen, and in *b*, *e* and *f* there are single univalent 4th chromosomes. From White (1957*a*).

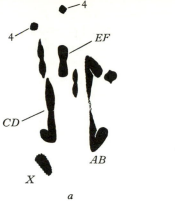

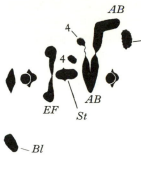

a

b

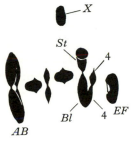

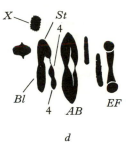

c

d

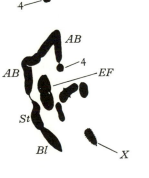

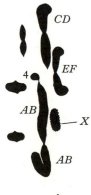

e

f

The situation in the *Eyprepocnemis* hybrids is a little different. The chromosomes of these grasshoppers are all acrocentric. Numerous unequal bivalents were seen in the hybrids, these being apparently always pre-reductional, i.e. the inequality is proximal to the chiasma. Numerous chains of up to ten chromosomes were observed; in some cases non-terminal chiasmata between heterologous chromosomes were present. One chromosome (S_9) was observed to be associated with five others in different nuclei (whereas on the hypothesis of diploidy, with reciprocal translocations it should have showed associations with only two others, since its proximal and distal ends could not be distinguished).

'Anomalous' associations of chromosome ends of the type seen in these grasshopper and urodele hybrids are also found in non-hybrid individuals of a number of grasshopper species that are heterozygous for pericentric inversions. In such cases the terminal associations are always between two bivalents that are structurally heterozygous. Isolated instances have been seen in *Keyacris scurra* and in *Trimerotropis gracilis* (White 1961a, Fig. 1), while in *Austroicetes interioris* they are quite frequent (Nankivell 1967).

The conclusion to which these observations lead seems inescapable – namely that in both grasshoppers and newts a considerable number of the chromosome ends are sufficiently alike for synapsis (and presumably also chiasma formation) to occur between them whenever the normal synaptic pattern is somewhat disrupted. In the interracial hybrids the chiasma frequency is reduced, which probably indicates a partial inhibition of synapsis. In the heterozygotes for pericentric inversions true synapsis between the mutually inverted regions is inhibited, since 'reversed loops' are not seen at pachytene.

Homology between many of the chromosome ends of a species implies that numerous mutual translocations between minute terminal regions have occurred in the past, and that this has frequently been followed by extinction of one of the translocated chromosomes and a non-translocated heterologous chromosome. Such a process, if it occurs repeatedly, will lead to karyotypes in which certain chromosome ends are represented four, six, eight or more times. No conversion of interstitial material into telomeres is involved, and no new principle has to be invoked. On general grounds it does not seem at all improbable that loss of minute terminal segments (and their replacement by other terminal segments) will sometimes be tolerated. In other words, duplication-deficiency karyotypes are usually inviable when the chromosome regions concerned are of considerable length, but if they are sufficiently short it is fairly certain that the homozygous duplication-deficiency karyotype will be viable and may even in some instances be adaptively superior. Thus it is actually difficult to avoid the conclusion that mutual translocations of minute terminal segments must be frequently successful in evolution. The phenomenon is not ordinarily detected in completely homozygous material – some special kind of heterozygosity is needed to partially inhibit the complete synapsis between homologous chromosomes, thereby leaving them free to synapse at the ends with heterologous elements having minute homologous segments at the ends.

John and Lewis (1965b) disagree with the above interpretation. They believe that the anomalous terminal associations in hybrids or multiple pericentric inversion heterozygotes are due to synapsis between non-homologous regions and that the associations are non-chiasmate. Their evidence for this view is that some terminal associations between the X and an autosome were observed in the *Eyprepocnemis* hybrids, and that they believe the X to be incapable of forming chiasmata in the male. It seems quite clear, however, that the terminal associations between heterologous chromosomes which we are considering are extremely firm ones which persist through the rather violent movements involved in spindle formation at premetaphase. A mere 'attraction' would be unlikely to do this and we feel reasonably certain that terminal chiasmata are present.

There remain the interstitial chiasmata between heterologous chromo-somes observed in the *Eyprepocnemis* hybrids. These are interpreted by John and Lewis as evidence of heterozygosity for mutual translocations between acrocentric chromosomes which are supposed to have arisen during the divergence of *E. plorans meridionalis* and *E. plorans ornatipes*. This seems an unlikely hypothesis; it implies the repeated evolutionary success of chromosomal rearrangements which would lead to semi-sterility in the heterozygote. It seems more likely that in the evolution of *Eyprepocnemis plorans* proximal heterochromatic segments have undergone duplications (possibly of the tandem rather than the reversed type), to a varying degree in different races and populations; and that such duplications are respon-sible for the unequal bivalents observed in the hybrids, as well as the sub-terminal (rather than strictly terminal) chiasmata between otherwise non-homologous chromosomes. The situation where two geographic races or sibling species differ in the extent of terminal heterochromatic segments in a number of chromosome elements seems a relatively common one; a good example is provided by the two siblings of the *Cimex pilosellus* complex studied by Ueshima (1963). Such differences will, of course, lead to unequal bivalents in the hybrids (assuming that synapsis and chiasma formation both occur).

Our interpretation of the *Keyacris scurra*, *Austroicetes interioris* and *Triturus cristatus* cases is therefore that we are dealing with a situation of the following type in the hybrids ('t' is supposed to represent the telomere and letters followed by an apostrophe allelic differences between the races):

$$\text{Chromosome I} \begin{cases} tabcdefgh'i \dots \text{race A} \\ tabcd'efghi \dots \text{race B} \end{cases}$$

$$\text{Chromosome 2} \begin{cases} tabpqrst'uv \dots \text{race A} \\ tabpqrstu'v \dots \text{race B} \end{cases}$$

On the other hand we imagine the *Eyprepocnemis* situation to be like this (the *ab* region being assumed to be heterochromatic and relatively inert genetically):

$$\text{Chromosome I} \begin{cases} tSA.abababcdef'ghi \dots \text{Race A} \\ tSA.abcd'efghi \qquad \dots \text{Race B} \end{cases}$$

$$\text{Chromosome 2} \begin{cases} tSA.ababpqrst'uv \quad \dots \text{Race A} \\ tSA.abpqrstu'v \qquad \dots \text{Race B} \end{cases}$$

(SA represents the short arm and the full stop the centromere)

The tandem duplications of heterochromatin we postulate would presumably have arisen by translocation – perhaps mainly between homologous chromosomes. The difference of opinion between John and Lewis, on the one hand, and the present author, on the other, is thus partly, but not entirely, a semantic one. Evidence as to which opinion is correct could be obtained by a study of the DNA values of the two *Eyprepocnemis* races or species: on the interpretation put forward here they should be significantly different, while on John and Lewis' hypothesis they would be identical or very similar.

We have dealt with this question at some length because of its general importance. Certain authors have been willing to postulate a high frequency of establishment of mutual translocations in animal evolution – and moreover ones involving acrocentric chromosomes, i.e. of a type particularly liable to impair fecundity in the heterozygotes. They have further suggested at times that such rearrangements may play an important role in race and species formation. Theoretical considerations render this highly unlikely and the evidence from studies of polytene chromosomes in *Drosophila* is entirely opposed to it. Even in *Chironomus*, where translocations may have been rather more frequent (Keyl 1962) those that have established themselves seem to have involved virtually entire arms of metacentric elements, in which case the reduction in fecundity may not have been considerable.

Most of the interspecific hybrids that have been obtained by laboratory crossing in the genus *Drosophila* seem to show fairly severe abnormalities of meiosis. Those between *Drosophila pseudoobscura* and *D. persimilis*, which were studied in considerable detail by Dobzhansky and Boche (1933) and Dobzhansky (1934) fall into our categories (3) and (4). It had earlier been shown by Lancefield (1929) that the hybrid males from the cross *pseudoobscura* ♀ × *persimilis* ♂ have testes of normal size, while in those from the reciprocal mating they are very much reduced in size. In either case, however, the male hybrids are entirely sterile. Corresponding to the difference in testis size, the cytology of the reciprocal hybrids is quite different. The matter is, however, complicated by the fact that within each species there are 'strong' and 'weak' strains which give different-sized testes when crossed ('strong' strains giving small testes, 'weak' ones producing testes of normal or only slightly subnormal size).

In *pseudoobscura* ♀ × *persimilis* ♂ hybrids the early stages of spermatogenesis are normal. In some instances all the chromosomes pair to form bivalents during the meiotic prophase, while in other cases only some of them, or none at all, do so. In general, hybrids between 'strong' races show few or no bivalents, while those between 'weak' races have a higher frequency of pairing. At the anaphase of the first meiotic division in these hybrids the whole spindle elongates to an extraordinary extent, so that it becomes

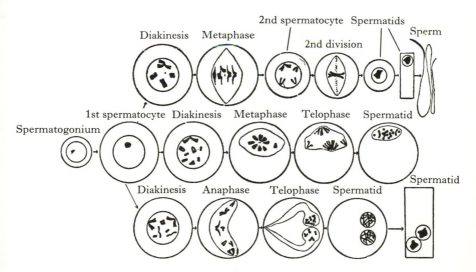

Fig. 15.8. Spermatogenesis in *Drosophila pseudoobscura* (top row), in *pseudoobscura* ♂ × *persimilis* ♀ hybrids (middle row) and in *pseudoobscura* ♀ × *persimilis* ♂ hybrids (bottom row). From Dobzhansky (1934).

bent round within the cell into a horseshoe shape. No division of the cytoplasm occurs, so that both telophase nuclei become included in a single cell. The second meiotic division never occurs and giant worm-like spermatids are formed, which never develop into functional sperms.

The spermatogenesis of the *persimilis* ♀ × *pseudoobscura* ♂ hybrids follows an entirely different course. There are very few spermatogonia and spermatocytes in the testis and the number of spermatogonial divisions is often reduced. The greater part of the mature testis is filled with cellular debris produced by the degeneration of the spermatids. The chromosomes are usually unpaired at meiosis and the anaphase of the first meiotic division is abnormal, although there is no marked elongation of the spindle, as in the reciprocal hybrids. No cytokinesis occurs and there is no second meiotic division as a rule, but in 'strong' *persimilis* ♀ × *pseudoobscura* ♂ hybrids a number of supernumerary mitoses may occur after meiosis, leading to the production of multinucleate spermatids.

In neither type of hybrid are viable sperms produced, thus explaining the complete sterility of the F_1 males. The genic incompatibility which gives rise to hybrid sterility in this cross is probably fairly complex, so that different degrees of incompatibility occur, according to the 'strength' of the strains used.

Female hybrids between *D. pseudoobscura* and *D. persimilis* are not completely sterile when backcrossed to the parent species. Thus Dobzhansky (1936*b*, 1937*c*) was able to obtain individuals bearing various combinations of *pseudoobscura* and *persimilis* chromosomes. These varied in degree of fertility, and it was shown that the sterility of the male hybrids depended on numerous genes located in all the chromosomes except the *Y* and the minute fifth (dot) chromosome. However, the cytoplasm also plays some part in determining the size of the testes in the hybrid males, since two individuals carrying the same set of chromosomes may differ in testis size according to the history of the cytoplasm (i.e. whether the mother of the individual in question was a hybrid or a pure *pseudoobscura* or *persimilis* fly).

Hybrids between *D. pseudoobscura* and *D. miranda* were obtained by Dobzhansky and Tan (1936*a*, *b*) and Dobzhansky (1937*a*) and were studied cytologically by Kaufmann (1940). Whereas the male hybrids have very small testes in which no sperms are formed, the F_1 female hybrids produce numerous eggs in which some kind of meiosis takes place (it is stated that there are no striking irregularities of meiosis, but it is uncertain how far synapsis and chiasma formation are normal). Such eggs may be fertilized by sperms from either parent species, and cleavage may proceed as far as the stage when there are 16 nuclei. Kaufmann pointed out that a chance distribution of chromosomes at meiosis should give some egg pronuclei containing only *pseudoobscura* chromosomes or only *miranda* ones and that a few of the zygotes will consequently have all their chromosomes derived from one species. Since all these backcross zygotes died at an early embryonic stage he believed that there was a lethal interaction of the chromosomes with the hybrid cytoplasm.

The hybrids between *D. melanogaster* and *D. simulans* are entirely sterile (Kerkis 1933), so that it has not been possible to analyse the causes of their sterility in the same way. Nevertheless, Muller and Pontecorvo (1940*a*, *b*, 1941) found an ingenious way of overcoming this difficulty. They crossed triploid *melanogaster* females, which produced eggs with some extra chromosomes with *simulans* males that had some of their chromosomes incapacitated by a heavy dose of X-rays. In this manner they were able to obtain offspring with various combinations of *melanogaster* and *simulans* chromosomes and to prove that the sterility effect, like that of the hybrids between *pseudoobscura* and *persimilis*, depends on numerous genes scattered over all the chromosomes. Males which possess a *simulans* Y, but in which all the other chromosomes are derived from *melanogaster*, are sterile just like XO males. Individuals having all their chromosomes from *melanogaster* except the two fourth (dot) chromosomes were fairly viable, fertile if female but sterile if male (the male sterility in this case probably depends on a single gene in the fourth chromosome). Hybrids between *D. affinis* and *D. athabasca* were obtained by Miller (1941); their spermatogenesis was grossly abnormal, giant multinucleate cells being formed in the testis, so that complete sterility resulted.

Haldane (1922) long ago enunciated the rule that when, in the offspring of an interspecific cross, one sex is rare, absent or sterile, that sex is usually the heterogametic one (i.e. the female in birds and Lepidoptera, the male in most other groups). 'Haldane's rule' has so many exceptions that it can no longer be regarded as generally valid, even though it is probably true that significantly over half the known cases support it. Inviability or sterility of the heterogametic sex but not the homogametic one in the progeny of a species cross may be due to a deleterious interaction between the Y of one species and the X, autosomes or cytoplasm of the other. This explanation, however, will obviously not apply in the case of XO forms. In some cases the X of one species may carry recessive alleles that react, when hemizygous, with the autosomes or cytoplasm of the other species in such a way as to lead to lethality, poor viability or reduced fertility.

In the genus *Drosophila*, Haldane's rule seems to apply in some cases, but there are many clear-cut exceptions. Thus in crosses between *D. mulleri* and certain strains of *D. aldrichi* the progeny are almost all males, the female offspring being for the most part killed by a gene in the *aldrichi* X-chromosome which acts as a semi-lethal in the hybrid, although it has no such effect in the pure species (Crow 1942; Patterson 1942). A rather similar situation occurs in crosses between *D. texana* and *D. montana*; whereas the mating *texana* ♀ × *montana* ♂ gives sons and daughters in approximately equal numbers, the reciprocal mating gives all male offspring, the daughters being killed as a result of a lethal interaction between the *texana* X-chromosome and the *montana* egg cytoplasm (Patterson and Griffen 1944).

Many lepidopteran species crosses exemplify 'Haldane's rule'. Here, of course, the female is the heterogametic sex (in some cases XY, in others XO). The cross between the European species *Pergesa elpenor* ♀ × *P. porcellus* ♂ gives caterpillars of both sexes, but the females die in the pupal stage. The same seems to be true of the cross *Celerio euphorbiae* ♀ × *C. galii* ♂ although the reciprocal mating gives adult hybrids of both sexes (Federley 1929). In the *elpenor* × *porcellus* cross the X of the former and the Y of the latter seem to combine to give a lethal effect in the pupal stage. Similar results were obtained by Federley (1931) in some of his *Pygaera* crosses. Thus in the cross between *P. curtula* ♂ × *P. anachoreta* ♀ the combination $X^c Y^a$ is semi-lethal, and similarly in the cross between *anachoreta* and *pigra* the combination $X^p Y^a$ leads to the death of most of the female larvae.

In a few instances the absence of one sex in the offspring of an inter-specific cross may be due, not to differential mortality but to a complete sex reversal. Thus Federley (1933) found that the cross between the European moths *Smerinthus populi* ♀ × *S. ocellata* ♂ gave only male off-spring. He was able to eliminate the possibility of a selective mortality, since all the larvae which died were dissected or sectioned and found to be males. It would appear that the *ocellata* X-chromosome is sufficiently 'strong' to overcome the female-determining genes of the *populi* Y or autosomes. Thus both $X^p X^o$ and $X^o Y^p$ individuals are males. This situation seems to be distinctly unusual, although it may occur elsewhere in groups such as the cyprinodont fishes, Amphibia and birds, in which complete sex reversal can be effected by an alteration in the endocrine balance of the organism.

Partial sex reversal takes place in some bird hybrids. Thus in the cross between the domestic duck (*Anas platyrhyncha*) and the Muscovy (*Cairina moschata*) the progeny of the mating *Cairina* ♀ × *Anas* ♂ are sterile males and females, while those of the reciprocal mating are sterile males and intersexes which outwardly resemble males but are actually masculinized females with rudimentary ovaries. The sex ratio is not disturbed in this case, the males and intersexes being produced in approximately equal numbers. In some other bird species crosses, however, there may be a genuine deficiency of female offspring (due to differential mortality), as in the case of the pigeon–ring dove hybrids studied by Painter and Cole (1943).

The spermatogenesis of the *Anas* × *Cairina* hybrids has been studied by Yamashina (1942). In spite of the fact that the chromosome numbers of the parent species seem to be the same and the sizes and shapes of the chromosomes very similar, most of the bodies seen at diakinesis and first metaphase are univalents. It is not clear whether this is due to failure of synapsis at zygotene or of chiasma formation at the end of pachytene. Degeneration of the spermatocytes sets in during the first metaphase, the chromosomes clumping together to form irregular masses. There is consequently no first anaphase or telophase and all the spermatocytes degenerate without going through a second meiotic division. There is a general similarity between the spermatogenesis of these hybrids and those between the species of *Triturus* described above (see p. 558), but the degeneration of the spermatocytes takes place at an earlier stage in the duck hybrids.

Although the biological definition of species as breeding units is the only type of species definition that is acceptable today, it is undeniable that occasional hybridization does occur in nature between members of different species populations. There has been considerable discussion as to the general importance and significance of this phenomenon in animal speciation and evolution. Obviously we cannot dismiss it as of no importance in the evolutionary process simply because it is rare; after all, spontaneous mutations and structural rearrangements are rare events. But if the offspring of such interspecific matings are entirely sterile, then no leakage ('introgression') of genetic material from one species into the other can take place. This seems to be the situation in the case of *Drosophila aldrichi* and *D. mulleri* (Patterson and Crow 1940). Occasional hybrids between these two species occur in nature, but owing to their complete sterility there seems to be no possibility of introgression occurring. Even if the F_1 hybrids of a species are partially fertile in backcrosses it may be that because of strong natural selection against the F_2 and backcross individuals introgression may be kept down to a very low level or even prevented altogether. Clearly, *Triturus cristatus* and *T. marmoratus* remain two distinct entities in central France in spite of the fact that about 5% of the combined population are phenotypically recognizable hybrids, the females having a fair degree of fertility.

Differences of opinion on the role of hybridization in animal evolution arise, therefore, from different estimates of two factors. In the first place, we do not know to what extent the undoubted cases of hybridization

between distinct species in freshwater fishes, toads, jays and other vertebrates owe their origin to human interference with natural habitats, creating
more or less artificial conditions and environments. It may well be that in
former times, when the human population was much smaller, or before a
human population existed at all, such instances were very much rarer.
Evidence on this question is bound to be indirect and unsatisfactory, but
when we find that two toad species remain quite distinct in central Texas
except on a golf-course (A. Wasserman 1957) it is surely suggestive.

The other issue in which there is much genuine divergence of opinion
is as to how much natural selection actually operates against the products
of hybridization. W. F. Blair (1951) reviewed the voluminous literature
on natural hybridization in vertebrates, including that of Hubbs (see
Hubbs, Hubbs and Johnson 1943) on fishes and A. P. Blair (1941) on toads.
He concluded at that time that: 'the amount of gene transfer between sympatric species of vertebrates is unknown, but it seems probable that it is
negligible'. Mayr (1963) likewise concludes that: 'the evolutionary importance of hybridization seems small in the better-known groups of animals
and reticulate evolution above the species level plays virtually no role in
the higher animals'. The evidence from *Drosophila* and grasshoppers
seems in complete agreement with that from mammals and birds in this
respect. Patterson and Stone (1952) concluded that *Drosophila americana*
might have arisen by hybridization of *D. texana* and *D. novamexicana*,
because it combined some inversion sequences present in the former but
not in the latter with some known in the latter but not in the former. However, it seems far more probable that a widespread '*proto-americana*' with
different inversion polymorphisms in its eastern and western populations
gives rise to *novamexicana* in the west and *texana* in the east (see Mayr 1957).
It is possible, however, that in fishes interspecific hybridization may have
played a significant role (Hubbs and Miller 1943; Svärdson 1957; Dottrens
1959). Cytological evidence for interspecific hybridization having played a
significant evolutionary role seems to be non-existent in groups with giant
polytene or lampbrush chromosomes.

16

THE EVOLUTION OF SEX DETERMINATION
I. SIMPLE SYSTEMS

> Strange streaks here and there you will find: hair-trigger
> insanities, barely showing, like flaws in ice, but running in a
> steady heavy river, the endless tributary of sex.
>
> L. DURRELL: *The Black Book*

The great majority of the species of higher animals have separate sexes and possess genetic mechanisms which determine them. On the other hand, in several large groups such as the ctenophores, flatworms, Ectoprocta, oligochaetes, leeches, euthyneurous Mollusca and tunicates, hermaphroditism is the rule and bisexuality the exception. Within these groups only a few species such as the marine triclads *Sabussowia dioica* and *Ceryra teissieri*, the trematodes of the family Schistosomatidae, the tapeworm *Dioicocestus* and the tunicates *Distaplia*, *Sycozoa* and *Oikopleura* have reverted to bisexuality (Benazzi 1947). Presumably each such reversion has involved the de-novo origin of a genetic sex-determining mechanism; but except in the Schistosomatidae, where some species seem to lack morphologically distinguishable sex chromosomes, while others possess them (see p. 582) no genetic or cytologic studies have been carried out on any of these forms.

The Mollusca include many bisexual as well as many hermaphroditic species. On the other hand, coelenterates, nematodes, polychaetes and echinoderms are almost all bisexual, although some hermaphroditic species exist in all these groups, having presumably lost the sex-determining mechanisms characteristic of the classes and phyla to which they belong. The arthropods are almost all bisexual except for certain Crustacea (especially barnacles), some phorid Diptera that live in termite nests and a few scale insects of the genus *Icerya* (see p. 506). The Crustacea, with their hormonal control of sexual differentiation (Charniaux-Cotton 1960) seem to have been far more liable to develop hermaphroditism in the course of their evolution than the insects, in which circulating sex hormones seem to be generally absent (but see p. 575). Thus among such typically bisexual groups of Crustacea as the Tanaidacea and the oniscoid Isopoda we have the sporadic appearance of hermaphroditic species such as *Apseudes hermaphroditus* (Lang 1953) and *Rhyscotoides legrandi* (Johnson 1957). Among the vertebrates the only hermaphroditic forms are a few species of fishes (D'Ancona 1945; Benazzi 1947; Clark 1959; Mead 1960).

Although one might perhaps look for vestiges of sex chromosomes in a few hermaphroditic species which are closely allied to bisexual forms, and which may hence be presumed to have become hermaphroditic fairly recently, one must not expect animals such as flukes, land snails and earthworms (which have doubtless been hermaphroditic for vast periods of geologic history) to possess any special sex chromosomes. In these forms the testicular and ovarian tissues are produced by ordinary histological differentiation; oogonia and spermatogonia possess the same set of chromosomes and differ in the same kind of way as, say, the liver cells and kidney cells of the same individual. Some early cytologists who though they had found sex chromosomes in such groups were certainly mistaken, either in their observations or in their interpretation. In many hermaphroditic molluscs and in a few species of fishes the sexual differentiation is consecutive rather than simultaneous, so that the individual passes through male and female phases in its life cycle.

In the present chapter we shall be concerned with the cytogenetically simple sex-determining mechanisms, i.e. those which depend on a single pair of chromosomes, reserving the more complex ones, in which more than a single pair of chromosomes are involved, until Chapter 17. Sex determination by male haploidy is dealt with in Chapter 18.

Where the genetic mechanism of sex determination consists of a single pair of chromosomes, these are homologous throughout their length in one sex (X-chromosomes) and non-homologous or only partly so in the other (X- and Y-chromosomes). Thus the X is present in both sexes, while the Y is represented in one only. In a number of animal species and groups the Y is absent altogether (having presumably been 'lost' in the course of evolution*); in such cases the diploid number in one sex is an uneven one. We can accordingly distinguish between an $XY:XX$ type of mechanism and an $XO:XX$ one, the 'O' indicating merely the absence of a Y.

The XX sex is said to be *homogametic*, since it produces only one kind of gamete, while the XY or XO sex is said to be *heterogametic*, because it produces two kinds of gametes in equal numbers. In most groups of bisexual animals it is the male sex which is heterogametic, so that there are

---

* The 'loss' may have occurred through fusion with the X or with an autosome in some cases.

two kinds of sperms but only one kind of egg. However, in some major groups such as the Lepidoptera and birds and in a few Diptera, Crustacea, fishes, amphibians and reptiles the situation is reversed, the females being heterogametic and the males homogametic. Some authors speak of ZZ and WZ sex chromosomes in the case of female heterogamety, which seems an unnecessary terminological complication except in the special case of the fish *Xiphophorus* (*Playtpoecilus*) *maculatus* where some races have male heterogamety and others female heterogamety. We do not use the ZZ and WZ terminology in this book, and it must be remembered that when we speak of an X-chromosome in birds and Lepidoptera we mean one that is represented twice in the male karyotype and once in the female.

There are a number of groups of marine invertebrates in which sex seems to be rather labile, so that genetic factors either play no role or only a subsidiary one. The classical case is that of the echiuroid worms of the genus *Bonellia* (*B. viridis* and *B. fuliginosa*) in which the minute male is parasitic on the female, either on her proboscis or in her uterus. Here isolated larvae develop into females which secrete masculinizing substances (ecto-hormones) into the sea water, inducing larvae that are influenced by them to develop into parasitic males. Even here, however, it is not certain that genetic sex-determining factors are entirely absent. A very similar case is that of the bopyrid Isopod crustacean *Ione thoracica* in which the small males are likewise ectoparasitic on the females (Reverberi and Pitotti 1942).

We may regard all kinds of genetic sex-determining systems as mechanisms which switch the pattern of development onto one or other of two alternative tracks. The basic distinctions drawn by Goldschmidt (1931*a*) between *hermaphroditism* (the normal and functional coexistence in one individual of maleness and femaleness), *gynandromorphism* (mosaicism for sex chromosomes) and *intersexuality* (phenotypes intermediate between maleness and femaleness) remain valid, although Goldsmidt's particular interpretation of intersexuality in terms of rates of sexual differentiation (see p. 578) no longer seems acceptable. Clear instances of gynandromorphism have been recorded in almost all orders of insects, in spiders (Bonnet 1934) and ticks (Campana-Rouget 1959; Loomis and Stone 1970). They may be regarded as evidence against the existence of circulating sex hormones in those species in which they have been recorded. However even in insects there is now at least one case (the beetle *Lampyris noctiluca*) where an androgenous hormone produced by the testis under the influence of a neurosecretary hormone is capable of masculinizing female larvae (Naisse 1969).

Although sex-chromosome mosaicism undoubtedly occurs as an occasional aberration in such groups as Crustacea and vertebrates, with circulating sex hormones, its phenotypic expression is necessarily modified in those groups, in which clearly defined gynandromorphs, with a sharp separation of male and female parts, do not seem to occur.

It is frequently assumed that the effect of an $XY:XX$ or $XO:XX$ sex-chromosome mechanism is to produce equal numbers of the sexes at fertilization, and it has been claimed that this is an adaptive situation. It may be that in many groups the numbers of the sexes at the time of fertilization are in fact approximately equal. In mammals it was earlier claimed that this was far from being so; estimates of the sex ratio at conception ranged from 120 to 150 males per 100 females for the human species, the pig and the Golden Hamster (Parkes 1963). But later work (Stevenson and Bobrow 1967; McKusick 1970) suggests that morphological sexing of early human foetuses was unreliable in some of these studies and that the true primary sex ratio in the human species is very close to 1.0.

Two basically different genetic types of XY switch mechanisms are known. We may refer to these as the 'genic balance' mechanism and the 'dominant Y' system. The first was demonstrated as early as 1925 in *Drosophila* (Bridges 1925a, b, 1930, 1932); in this case the sexual phenotype depends on a balance between female-determining genes in the X-chromosome and male-determining genes in the autosomes. The Y-chromosome plays no part in the determination of the externally visible sex phenotype, so that XO individuals are males, although sterile because of abnormal spermatogenesis, and XX individuals with one or more Y's in addition are phenotypically normal and fertile females. Bridges demonstrated that experimentally obtained triploid individuals of *Drosophila* with three sets of autosomes but only two X's are intersexual, the extra set of autosomes having shifted the equilibrium in the male direction. Strictly triploid ($3A:3X$, whose A represents a haploid set of autosomes) and tetraploid ($4A:4X$) flies are females with poor viability. Haploid ($1A:1X$) *Drosophila* have never been obtained in experiments and are presumably non-viable. Flies of the constitution $2A:3X$ and $3A:1X$ flies have been repeatedly obtained in experiments: they are sterile 'super-females' and 'super-males' ('meta-females' and 'meta-males' of some authors, who emphasize that their sexual characteristics are not exaggerated).

A number of autosomal mutations are known in various species of *Drosophila* that transform chromosomally female individuals into intersexes (Sturtevant 1920, 1921; Lebedeff 1938; Dobzhansky and Spassky 1941, Newby 1942; Stone 1942; Morgan, Redfield and Morgan 1943; Spurway and Haldane 1954). The recessive third chromosome mutant *transformer* (Sturtevant 1945) actually changes females of *D. melanogaster* into apparently normal but sterile males. On the other hand, the only clear instance of a mutation in *Drosophila* that transforms males into intersexes is the third chromosome gene *doublesex* which also transforms females into intersexes of extremely similar phenotype (Hildreth 1965).

Such mutations might seem to suggest that perhaps the normal sex determination in *Drosophila* depends on a single genetic locus or at any rate on the action of relatively few genes. As far as the X is concerned, however, this is certainly not the case. Various authors (Dobzhansky and Schultz 1931, 1934; Patterson, Stone and Bedichek 1937; Bedichek-Pipkin 1940b) have studied $3A:2X$ intersexes in which short regions of the X were present only once or in triplicate (due to deficiencies or duplications). They found that all regions of the X were female-determining, so that any extra fragment shifted the phenotype in a female direction, the extent of the shift being roughly proportional to its length. This suggests that the number of female determining loci is large, and that they are distributed fairly uniformly along its length, no single one of them being much more powerful than the rest.

As far as the autosomes of *Drosophila melanogaster* are concerned, the second and third chromosomes are apparently male determining, and Gowen and Fung (1957) have presented some evidence which has been interpreted as indicating the existence of a major male-determining locus in the third chromosome.

The strength of the sex genes is not necessarily the same in different species of *Drosophila*. Thus Sturtevant (1946) showed that *D. neorepleta* carries an autosomal gene which, when present in single dose in a female hybrid from a cross with *D. repleta* confers a tendency toward maleness on her eggs. This tendency is only partly overcome by two *repleta* X's and male-like intersexes are produced. Sturtevant concluded that both the male-determining activities of the autosomal genes and the female-determining potency of the X-chromosomes are stronger in *neorepleta* than in *repleta*.

In the Gypsy moth *Porthetria* (formerly *Lymantria*) *dispar*, R. Gold-schmidt (1931a, b, 1932, 1934) carried out a large number of interracial crosses with a view to elucidating the genetic mechanism of sex determination. *Porthetria* is, of course, heterogametic in the female sex, so that the X (or Z) is a male-determining chromosome. The location of the female-determining factors has been much debated; theoretically they could lie either in the Y or in the autosomes, or even be cytoplasmic; the evidence, although not entirely conclusive, favours the view that they are located in the Y.

The importance of *Porthetria* lies in the fact that the potency of the sex genes in the X differs in the numerous geographic races into which the species may be subdivided. Since sexual dimorphism is very pronounced in *Porthetria*, the material was well suited for an analysis of sex determination. Goldschmidt designated an individual as weak if its X bore sex genes of low potency, 'strong' races being ones in which the X is very actively male-determining. It is not clear in this case whether the X bears a single sex locus with a number of alleles of varying potency or an additive system of many sex-determining loci as in *Drosophila*.

Porthetria dispar occurs throughout the Palaearctic region from Europe to Japan. Most of the European races and those from the island of Hokkaido are 'weak', while those from the southern island of Japan are 'medium' or 'strong'. In purebred races the male- and female-determining factors are in equilibrium so that no intersexes are produced, but in interracial crosses the balance may be upset. Thus when females of a 'weak' race are crossed with males of a 'strong' one, the F_1 consists of males and intersexes, the latter being XY individuals whose X is too strong for the female-determining genes they have inherited from their mother. The reciprocal cross gives rise to XX intersexes in the F_2. All degrees of intersexuality may be obtained by using parents of the approxiate 'strength'.

Goldschmidt was able to distinguish as many as eight races from different parts of the Japanese archipelago, which could be arranged in a series, according to the 'strength' of the X. The difference between the end members of the series was so great that when females of the weakest race were crossed with males of the strongest one, the offspring were all completely male in appearance, although half of them were XY and half XX. The difference between the race *hokkaidoensis* and other races is so great that it might almost be considered as a distinct species.

The female-determining factors in *Porthetria* crosses seem to be inherited in a strictly maternal fashion, so that they cannot be located to any considerable extent in the autosomes. Goldschmidt concluded that they were cytoplasmic, but it seems more probable on general principles that they are carried on the Y-chromosome. Goldschmidt himself regarded the *Porthetria* results as a proof of his 'time law of intersexuality', according to which an intersex begins to develop in the direction of one sex and then switches over at a certain point in development to differentiation in the direction of the other sex. The time of the switch was supposed to regulate the degree of intersexuality. Goldschmidt's views were criticized by Newby (1942) and especially by Seiler (1949b, 1958, 1965); they are not accepted by Stern (1965) in a balanced summary of the whole controversy. Apparently, in insects, intersexes are essentially mosaics of small patches of tissue, karyotypically uniform, but which have differentiated in a male and female direction, respectively (Stern 1965).

Even if the 'time law of intersexuality' is abandoned, however, important difficulties still remain. No arithmetical balance of male- and female-determining factors acting simultaneously (rather than sequentially) will explain the radically different phenotypes associated in *Porthetria* with male intersexuality (i.e. *XX* individuals, partially sex-reversed) and female intersexuality (i.e. intersexes with an *XY* genotype). Regardless of the degree of intersexuality, the former type is, in general, characterized by a coarse sexual mosaicism, the latter by a much finer one. It seems conceivable that some kind of inactivation of sex chromosomes, analogous to that which occurs in female mammals (see p. 594) may be the basis for the mosaicism of the sexual phenotype in the intersexes of *Porthetria*.

The significance of the *Porthetria* work from the evolutionary standpoint is by no means clear. Why, after all, should the strength of sex-determining genes be in an active state of evolution in a particular species? It is clear that *P. dispar* is a polytypic species which has differentiated into a number of geographic races, and perhaps we should regard its sex genes merely as part of a much larger system of balanced relationships which must evolve in harmony with the whole if they are to continue to produce coadapted phenotypes.

The 'dominant *Y*' system of sex determination occurs in mammals (with male heterogamety) and the Axolotl (with female heterogamety). In the former there is evidence from both man and the mouse that the *Y* is a powerfully male-determining chromosome. Thus *XO* individuals of *Homo sapiens*, instead of being sterile males (as they are in *Drosophila*) are sterile females with 'ovarian dysgenesis' or 'Turner's syndrome'. *XYY* human individuals are virtually normal males but may be unusually tall, mentally subnormal and aggressive (Jacobs *et al.* 1965), while *XXY*, *XXXY*, *XXYY* and *XXXXY* individuals are all intersexes of a particular type referred to as Klinefelter's syndrome, instead of being normal females as they would be in *Drosophila*. It will be apparent thet the phenotypes of both the *XO* and Klinefelter individuals provide evidence for the male-determining properties of the *Y* in the human species. The *X* on the other hand, must play some role in stimulating the development of the ovaries, since these organs fail to develop normally in an *XO* individual.

Although, in general all human individuals with one or more *Y*'s are males or male-type intersexes, it has been shown that individuals with an isochromosome for the long arm of the *Y* are always females (Jacobs and Ross 1966). This suggests that the male-determining locus is situated in the short arm of the *Y*. A syndrome known as *testicular feminization* (Morris 1953), due apparently to a single gene, converts an *XY* individual into a female-type intersex with undescended testicles (Court Brown *et al.* 1964; McKusick 1964). *XX* males or male-type intersexes have occasionally

TABLE 16.1. *Sex chromosomes in Homo sapiens**

Sex chromosomes	Phenotype	No. of late-replicating X's	Maximum no. of peripheral sex chromatin bodies
XO	Turner's syndrome	o	o
XY	Male	o	o
XYY	Male	o	o
XX	Female	1	1
XXX	Female	2	2
XXXX	Female	3	3
XXY	Klinefelter's syndrome	1	1
XXXY	Klinefelter's syndrome	2	2
XXYY	Klinefelter's syndrome	1	1
XXXYY	Klinefelter's syndrome	2†	2†
XXXXX	Female ('Penta X syndrome')	4	4
XXXXY	'Tetra X-Y syndrome'	3	3
$\widehat{XX}$	Female	1 (Iso-X)	1 large
$XO/\widehat{XX}$ mosaic	Female	1 (Iso-X)	1 large
$XO/XX/XXX$ mosaic‡	Female	o–2	o–2

* Based on Atkins *et al.* 1963 and Sasaki and Makino 1967.
† Presumably – not studied by the original author.
‡ Even more complex human mosaics are on record, e.g. *XXY/XX/XY/XO* and *XXXY/XXXXY/XXXXXY*.

been recorded in the human species, but Mittwoch (1967) considers it probable that these individuals commenced their development with a *Y* which was subsequently lost.

The frequency of *XO* individuals in the human species is about 0.03 % of all female births, but as many as 4% of spontaneously aborted conceptuses seem to be *XO*, i.e. the condition is usually lethal before birth. The single *X* may be either patroclinous or matroclinous, but studies involving the X_g blood group system have shown that the latter condition is far more common (Ferguson-Smith 1966). The frequency of *XYY* and *XXY* individuals is about 0.1 to 0.2 of all male births, according to the latest data (Sutherland 1970). Genetic studies on *XXY* individuals have not provided any very clear evidence as to their mode of origin, although it is known that in some cases both *X*'s are of maternal origin. *XXX* females have a frequence of about 0.1% of all female births (Sasaki and Makino 1967). The total frequency of sex-chromosome aneuploids in the human species is thus close to 0.3% of all births.

An *XXY* intersex in cattle, analogous to the human Klinefelter intersexes, has been reported by Scott and Gregory (1965). *XXY* mice are apparently sterile males with small testes but without obvious signs of intersexuality (Catlanach 1961*b*; Russell and Chu 1961; Slizynski 1964*b*). In marsupials,

Sharman (1970) has briefly described the intersexual phenotypes of an *XXY* and an *XO* wallaby (*Macropus eugenii*).

Another organism in which sex seems to depend on the 'dominant *Y*' principle is the silkworm *Bombyx mori* (Tazima 1954, 1964; Tanaka 1953). All types of diploid, triploid and tetraploid individuals with one or more *Y* (= *W*) chromosomes are females; even *XXXY* individuals with four sets of autosomes not showing any sign of intersexuality. It is therefore obvious that any male-determining factors in the *X* and autosomes must be extremely weak; the whole genetic mechanism of sex determination seems to be very different from that of *Porthetria* (see p. 577). In the moths of the genus *Solenobia* both parthenogenetic and sexually reproducing females are usually *XO* (*XY* females also occur in some populations), so that the genetic mechanism of sex determination seems to be quite variable in the Lepidoptera, although female heterogamety always occurs. In the tipulid fly *Pales ferruginea XXY* individuals are fertile males (Ullerich, Bauer and Dietz, 1964), so that here also sex determination seems to depend on a dominant-*Y*. Dominant *Y* mechanisms seem to exist also in the calliphorid flies *Phormia regina* and *Lucilia cuprina* in which *XXY* individuals are males and *XO* ones normal and fertile females (Ullerich 1963).

It is fairly obvious that sex determination must depend on genic balance rather than on a dominant *Y* throughout those groups where many of the species are *XO* in the male sex. Thus the great majority of the species in the orthopteroid insects (i.e. roaches, mantids, phasmatids and Orthoptera (Saltatoria), including the crickets, tettigoniids and grasshoppers) and Odonata (dragonflies) have *XO* males and all other types of sex-determining mechanisms in these groups are obviously derivative and recent in an evolutionary sense. We must accordingly conclude that sex determination throughout at least one or two main branches of the phylogeny of winged insects was originally, and remains in most or all of the species, of the *Drosophila* type, depending on a balance between female-determining *X*-chromosomes and male-determining autosomes. It may well still be of this type, even in species that have acquired neo-*Y*'s or complex X_1X_2Y systems (see pp. 608–15). There are a sufficient number of species with *XO* males in both Heteroptera and Homoptera to make us suspect that these orders of insects also have genic balance sex-switch mechanisms, and the same is probably true of the Coleoptera (see p. 622).

Other groups in which *Y*-chromosomes are generally or universally absent include the spiders (see p. 669) and the nematodes (p. 675).

Particular interest naturally attaches to instances of newly-arisen sex-chromosome mechanisms, such as we may find in dioecious members of otherwise hermaphroditic groups. One case of this kind that has been studied by a number of workers is that of the trematodes of the family Schistosomatidae. Various early workers reported sex chromosomes and male heterogamety in *Schistosoma haematobium* and *S. japonicum*. More recent work on the latter species (Ikeda and Makino 1936) and on *S. mansoni* (Niyamasena 1940; Short 1957) failed to demonstrate sex chromosomes in either sex, the somatic number being 16 in both males and females. In *Schistosomatium douthitti*, on the other hand, Short reports female heterogamety with the female XY and the male XX (the X is a large metacentric, the Y a medium-sized acrocentric). One feature of interest is that short found two triploid snail infections, one male, the other female (the first was XXY, the second XYY). This suggests that sex in schistosomes is determined in the *Drosophila* manner (only with the roles of the X and Y reversed since we are dealing with female heterogamety) rather than in the Axolotl manner, with the Y strongly female-determining.

DOSAGE COMPENSATION

It has been known in a general way, at least since the early work of Stern (1929) and Muller (1932a) that there exists, in *Drosophila*, a mechanism tending to equalize the effects of sex-linked genes in the male (with one X-chromosome) and the female (with two X's). This phenomenon of *dosage compensation* seems to operate in respect of most, but not all, mutant alleles in the X. It does not exist, for example, in the case of the *eosin* and *facet* phenotypes; and *a priori* it cannot occur in the case of the actual sex-determining loci (or there would be only one sex rather than two). Its adaptive and evolutionary significance is clearly in relation to the normal wild type alleles in the X. By a study of flies hyperploid for sections of the X it has been shown that for many genetic loci, not merely is one 'dose' in the male functionally equivalent to two in the female – two in the male are approximately equivalent to three in the female.

It is believed that dosage compensation is due to numerous systems of genic modifiers on the X (the 'dosage compensators') which either act as enhancers of biosynthetic activity in single-X genotypes or as repressors in two-X individuals. Thus, as shown long ago (Mohr 1923) in the case of the *White–Notch* region of the X, a short deletion of the main genetic locus may cause a more extreme change in the phenotype than loss of the entire X, as represented by the normal male genotype (including the dosage compensators as well as the main gene).

From an evolutionary standpoint it seems reasonable that dosage compensation systems should have been built up, by natural selection, in groups where the phenotypes of the two sexes are basically similar, evolving *pari passu* with the sex-determining mechanism itself. However, in groups where sexual dimorphism is extreme (e.g. scale insects, angler fishes, etc.) it may be that there are no dosage compensation mechanisms. Muller (1950b) has claimed that in *Drosophila pseudoobscura* the limb of the X which is of fairly recent autosomal origin (see p. 349) has less perfect dosage compensation than the other limb (the original X). Studies on the expression of the *facet* mutant in XX individuals transformed into males by the autosomal recessive gene *tra* (Sturtevant 1945, see also Stern 1960a) suggest that dosage compensation is not dependent on sex as much.

The molecular mechanism of dosage compensation is of considerable interest. It has long been known that in the polytene chromosomes of *Drosophila* the single X of the male has a diameter which is close to that of the two synapsed X's in the female (Offerman 1936). This is especially clear in the hybrids between *D. tropicalis* and *D. insularis* studied by Dobzhansky (1957c) in which all the chromosomes are unsynapsed in the polytene nuclei, and the single X of the male is about twice as thick as any of the others (Fig. 16.1). However, this is not due to a different level of polyteny in the male X, since the amount of DNA is not increased, although the amount of protein is about 11% greater than in one of the two X's in the female (Rudkin 1964). Dosage compensation apparently operates at the level of RNA synthesis in the chromosome, since studies of Mukherjee and Beermann (1965) involving quantitative authoradiography following labelling of RNA with tritiated uridine have shown that the male X operates in RNA synthesis at a level close to that of the two X's in the female. Comparison of grain counts in similar lengths of the X and autosomes suggests that dosage

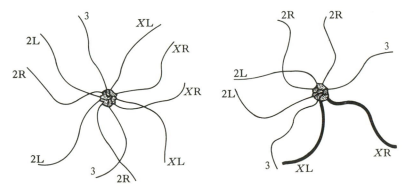

Fig. 16.1. Salivary gland chromosomes of female (left) and male (right) hybrids from the cross *Drosophila insularis* × *D. tropicalis*. No synapsis has occurred, except in the chromocentre (dotted). Note the greater thickness of the arms of the X-chromosome in the male, as compared with the female. From Dobzhansky (1957).

compensation involves enhancement of activity in the male rather than repression in the female.

A basically different mechanism of dosage compensation which involves genetic inactivation of large sections of one of the X's in the female apparently operates in the Mammalia (see p. 594). What mechanisms of this kind exist in other groups is not known. In many groups the need for dosage compensation may not arise, either because the X and Y are homologous for most of their length, differing only in a small sex-determining section, or because the sex chromosomes, although large, are heterochromatic and hence genetically sub-inert throughout their entire length. Cock (1964) has presented evidence that a number of sex-linked loci in birds do not show dosage compensation. We shall discuss later (see p. 686) the question whether dosage compensation is likely to occur in groups with haploid males.

SEX DETERMINATION IN VERTEBRATES

There have been very few credible accounts of morphologically distinguishable sex chromosomes in fishes. In the goby *Mogrunda obscura*, Nogusa (1955a) has described an XY bivalent in the male; some doubt however must be expressed as to whether this was not merely an autosomal bivalent heterozygous for a rearrangement such as a pericentric inversion. However, a very clear XY pair (the presumed X being the largest element in the karyotype and the presumed Y, the smallest) has been found in males of the deep-sea fish *Bathylagus wesethi* by Chen and Ebeling (1966); a similar sex chromosome pair apparently exists also in two other species of this genus and in some species of *Fundulus* (Chen and Ruddle 1970). An undescribed Mexican cyprinodont fish appears to have an X_1X_2Y multiple sex chromosome mechanism, with a very large metacentric Y (Uyeno and Miller 1971).

In the cyprindont *Lebistes reticulatus* sex chromosomes cannot be detected cytologically; nevertheless there is abundant genetic evidence that this is normally an XY ($\male$):XX ($\female$) species. Both the X and Y carry sex-linked genes and crossing-over occurs between them (Winge 1923, 1932, 1934; Winge and Ditlevsen 1947). Presumably, these chromosomes are homologous for most of their length, the differential region, on which sex determination depends, being probably very short. Winge was able, by selection, to obtain a stock of *Lebistes*, all the individuals of which were XX, but in which an autosomal pair of genes had come to control sex determination; this stock showed female heterogamety, the newly arisen mechanism giving femaleness when heterozygous. Winge interpreted these results as proving the existence of multiple male- and female-determining genes which are sorted out by selection. XX males when mated with normal XX females gave all-female offspring and YY males mated with XX females produced all-male progenies.

In the medaka, *Oryzias latipes*, genetic evidence has shown that the male is XY, the female XX (Aida 1921). As in *Lebistes*, there are body colour genes on the sex chromosomes which show crossing-over from the X to the Y. Yamamoto (1963, 1964a, b) has been able to obtain sex reversal from genetic males to phenotypic females by the use of oestrogens. Mating two XY individuals gives rise to some YY individuals, which are males. The R allele (giving orange-red body colour) is usually homozygous lethal when carried by the Y-chromosome. Yamamoto apparently obtained an XXY individual which was a fertile male, thus suggesting that sex determination depends on the 'dominant Y' principle. Nevertheless it was possible, by the use of oestrogens to induce complete sex reversal even in YY individuals.

The situation in another cyprinodont, *Xiphophorus* (*Platypoecilus*) *maculatus* with regard to sex determination is also interesting. 'Domesticated' stocks of this species (Gordon 1937; Kosswig 1939; Breider 1942) show female heterogamety, but wild material from the rivers of the east coast of Mexico show male heterogamety (Gordon 1946, 1947, 1951). The latest investigation is that of Kallman (1965). Sex chromosomes have not been identified cytologically in this species, but can be inferred from certain dominant pigment pattern alleles which exhibit sex-linked inheritance. In British Honduras females appear to be WY and males YY. In two Mexican rivers (the Papaloapan and Coazacoalcos), however, females are XX and males XY. In about 60% of the known range of the species WY, WX and XX females coexist with XY and YY males, there being presumably a balanced polymorphism. It is suggested that *X. maculatus* arose from an $XX:XY$ form by the acquisition of a strongly female-determining W-chromosome.

The general picture that emerges from this work is that the sex-determining mechanism in these cyprinodont fish is still in a very 'primitive' state, the sex chromosomes being very little differentiated from autosomes, and hence not distinguishable morphologically. Such 'primitive' mechanisms of sex determination seem to be unstable in several different ways. Thus in *Lebistes* selection of 'minor' sex-determining genes in the autosomes may lead to a new sex-determining mechanism (but one which may break down unless the selection is maintained). And in *Xiphophorus maculatus* instability of a different kind (due to mutation of the 'main' sex genes) seems to have occurred in evolution. Finally, in *X. helleri* intersexuality and sex reversal seem to be not uncommon, so that it has even been suggested that sex determination is controlled environmentally.

In the urodeles, also, both male and female heterogamety occur. In the Axolotl (*Ambystoma mexicanum*) Humphrey (1942, 1945) mated females which had been experimentally converted into males with normal females. The offspring consisted of approximately 75% females and 25% males, suggesting that females normally have the constitution XY, so that the mating $XY \times XY$ gives $1XX:2XY:1YY$ (i.e. one-quarter males, if YY individuals are viable and female. The existence of these YY individuals in the F_1 (indistinguishable phenotypically from normal XY females) was confirmed by rearing a second generation (approximately one-third of the F_1 females gave all-female progeny after mating with normal males). Later work (Fankhauser and Humphrey 1950) suggested that experimentally produced triploid axolotls are female if XXY or XYY, but male if XXX. Sex determination hence depends on the dominant-Y principle, with the Y female-determining.

Female heterogamety was demonstrated by the same genetic techniques in the salamander *Pleurodeles waltlii* by Gallien (1954). In neither of these two species have sex chromosomes been demonstrated cytologically, so that any visible differences which may exist between the mitotic X- and Y-chromosomes must be very minute. In *Triturus cristatus*, however, Callan and Lloyd (1960b) have demonstrated that there is an extensive region in the longest bivalent which is invariably heterozygous for loop-pattern in the lampbrush oocyte nuclei. This may be considered as a cytological demonstration of an XY pair in the female *Triturus*, even though the two chromosomes have not been shown to be distinguishable in somatic metaphase karyotypes. It also proves that in cases such as this, where the X and Y are the same size and shape, so that they cannot be distinguished by conventional techniques, the differential segments are not necessarily minute – in fact they may be quite extensive. No chiasmata are ever formed in this differential region, apparently.

Three genera of plethodontid salamanders from Central America, *Oedipina*, *Thorius* and *Chiropterotriton* definitely show an XY pair in the males (Kezer, unpub., see Mancino and Barsacchi 1966b), there being a large X and a very small Y. No cytologically visible sex chromosomes seem to exist in *Necturus maculosus* and *Proteus anguinus* (Seto, Pomerat and Kezer 1964; Kezer 1962) or in the families Cryptobranchidae and Hynobiidae with high chromosome numbers (Makino 1934, 1935, 1939a; Iriki 1932).

In the Anura it likewise seems probable that both male and female heterogamety coexist. Gallien (1956) established the existence of female heterogamety in *Xenopus laevis* by sex-reversal experiments; a small J-shaped *X*-chromosome and a large metacentric *Y* were demonstrated in female metaphases by Weiler and Ohno (1962), whose results were confirmed by Morescalchi (1963*b*). *Discoglossus pictus* likewise shows an *XY* pair in the female, but in this case both elements seem to be acrocentric, the *Y* being larger (Morescalchi 1964). The evidence for male or female heterogamety in the genera *Rana*, *Bufo* and *Hyla* is very conflicting, although the work of Witschi (1923) on the progeny of a hermaphrodite *Rana temporaria* and the more recent study of Kawamura and Yokota (1959) on the progeny of sex-reversed females of *Rana japonica* probably indicate that the Ranidae have male heterogamety. However, the work of Ponse (1942) on *Bufo vulgaris* suggests that Bufonidae have female heterogamety. No cytologically distinguishable sex chromosomes seem to be present in either of these genera (Morescalchi 1962, 1963*a*; Makino 1932*b*; Galgano 1933*a*, *b*; Saez, Rojas and de Robertis 1935, 1936). On the other hand Yosida (1957) has claimed to have demonstrated cytological male heterogamety in *Hyla arborea japonica*; however, his evidence does not seem convincing.

In the reptiles sex reversal has not been possible, as an experimental technique, so that the evidence for male or female heterogamety rests solely on the cytological data, Most of the earlier workers on reptile chromosome were unable to demonstrate *X*- and *Y*-chromosomes in either sex. And this is still the situation as far as the orders Chelonia and Crocodilia are concerned (Huang and Clark 1967; Sasaki and Itoh 1967; Ayres, Sampaio, Barros, Dias and Cunha 1969; Cohen and Clark 1967; Cohen and Gans 1970). The chromosomes of *Sphenodon*, the only living representative of the Rhynchocephalia, have not been studied by modern methods. However, there is now good evidence for female heterogamety in at least three families of snakes. Kobel (1962) found an *XY* pair in the female of *Vipera berus* and Beçak *et al.* (1962) reported an unequal pair of sex chromosomes in the female of *Bothrops jararaca*, although they were unable to demonstrate any sex chromosomes in *Boa constrictor*. Apparently morphologically differentiated sex chromosomes cannot be detected in the family Boidae, but can generally be found in members of the families Colubridae and Crotalidae (Beçak 1967); the *Y*-chromosome is much larger than the *X* in *Clelia occipitolutea*. Matthey and van Brink (1957) could find no evidence of sex chromosomes in twenty-three embryos of the snake *Natrix rhombifera* (presumably including both sexes); they were also unable to find any differences between the male and female crocodilian *Caiman sclerops*.

In a number of Antillean species of the lizard genus *Anolis* Gorman and Atkins (1966) have demonstrated male heterogamety. There are two sections of this genus, the so-called *alpha* and *beta* sections. In the alpha section most species have 2n = 36 and no distinguishable sex chromosomes (Gorman and Atkins 1966, 1968), but members of the *bimaculatus* group of species have $2n\delta = 29$, $2n\female = 30$ and an X_1X_2Y sex-chromosome mechanism. In the beta section *A. conspersus* apparently has an *XY* pair in the male, with one member acrocentric, the other metacentric; while in *A. biporcatus* there is an X_1X_2Y mechanism like that of the *bimaculatus* group, but probably independently evolved.

In another genus of the family Iguanidae, *Uta*, there is an *XY* micro-chromosome pair in the males, with the *Y* extremely minute (Pennock, Tinkle and Shaw 1969); and in some species of *Sceloporus* X_1X_2Y mechan-isms have been reported (Cole, Lowe and Wright 1967; Cole 1970). At least in the genus *Cnemidophorus* sex determination seems to depend on the 'domi-nant *Y*' principle, since *XXXY* tetraploid individuals are males (see p. 742). Most species of this genus lack easily distinguishable sex chromosomes, but *C. tigris* has a visibly detectable *XY* pair in the male (Cole, Lowe and Wright 1969). In the Russian lizard *Lacerta strigata* Ivanov and Fedorova (1970) have claimed that female heterogamety occurs, the *Y* being a microchromosome.

In the birds there is ample genetic evidence for female heterogamety in the chicken and several other species. The cytology of birds is exceptionally difficult, on account of the very high chromosome numbers and the small size of many of the individual members of the karyotypes; but the careful work of Yamashina (1943, 1944) has shown that in the chicken the *X*-chromosome is a metacentric which is fifth in order of size (2n = 78) many of the microchromosomes being less than 0.5 μm in length).

It was not known until recently whether female birds were *XO* or *XY*. There is now convincing evidence for the existence of a *Y* in the Budgerigar (Rothfels, Aspden and Mollison 1963), the chicken (Frederic 1961; Schmid 1962), the pigeon and the Canary (Ohno, Stenius, Christian, Beçak and Beçak 1964). In the Budgerigar the *X* is the fifth chromosome in order of size (van Brink 1959; Rothfels *et al.* 1963), while the *Y* is probably the tenth in order of size; it shows negative heteropycnosis in some of the somatic metaphases.

In the case of the chicken it was found by Schmid that in all female nuclei one of the larger microchromosomes exhibits 'late labelling' after treatment with tritiated thymidine; such behaviour is never seen in male cells. This is clearly the *Y*-chromosome, first definitely reported by Frederic. It is, of course, uncertain whether all bird species will turn out to have *XY* females; nor is there any definite evidence as to whether sex determination follows the *X*: autosome genic balance scheme or the 'dominant *Y*' scheme. Neither is it known what mechanism of dosage-compensation (if any) exists in birds. 'Sex chromatin' has been reported in resting nuclei of female chickens by several workers, but it could not be confirmed by Miles and Storey (1962) and Schmidt's account makes it clear that autoradiography provides no support for the extension of the 'fixed inactivation hypothesis' (see p. 594) to birds. Recent work by Abdel-Hameed and Shoffner (1971) has shown that triploid chickens of the constitution $3AXXY$ are intersexual (externally mainly female but with well-developed testes). This seems to exclude the possibility of the Y being strongly female-determining and suggests a *Drosophila*-like mechanism with the roles of the X and autosomes reversed.

Birds seems to differ from mammals in the large number of sex-linked loci that have been encountered; the $X (= Z)$ is probably more euchromatic than in most mammals, where it is largely heterochromatic, so that only a few sex-linked loci (e.g. those for haemophilia, colour blindness, the blood group X_g and glucose 6-phosphate dehydrogenase deficiency in man and the Tabby gene in mice) are known.*

It is to be presumed that the X-chromosome has had a continuous evolutionary history throughout the Mammalia, so that we should expect genetic loci that are sex-linked in one species of mammal to show sex-linkage in others as well. In fact, the glucose 6-phosphate dehydrogenase locus has been shown to be X-linked in the mule (Mathai, Ohno and Beutler 1966) and in a cross between two species of hares (Ohno, Poole and Gustavsson 1965). In doves, on the other hand, it is inherited autosomally, thus suggesting that there is no evolutionary continuity between the sex chromosomes of mammals and birds.

In monotremes an XO mechanism possibly exists in the Echidna, *Tachyglossus aculeatus*, according to the work of Bick and Jackson (1967a). The same authors (1967b) report that a similar mechanism occurs in *Ornithorhynchus*. These conclusions were based on the diploid karyotypes and until the male meiosis has been studied there remains a distinct possibility that the males are X_1X_2Y rather than XO.

In all marsupials male heterogamety is evident, and is especially clear on account of the low number of the chromosomes. Apart from a few species which have XY_1Y_2 or X_1X_2Y males (see p. 638) all the other marsupials that have been studied cytologically (Painter 1922; Agar 1923; Drummond 1933, 1938; Matthey 1934b; Koller 1936a, b; McIntosh and Sharman 1954; Sharman 1961; Hayman and Martin 1969) have an XY pair of chromosomes in the male, and these two elements are associated to form a sex bivalent during the prophase stages of meiosis, the association being maintained

---

* Levitan and Montagu (1971) list 81 inherited conditions in the human species for which good evidence of X-linkage now exists. It is uncertain how many genetic loci are involved, but almost certainly more than the number of conditions.

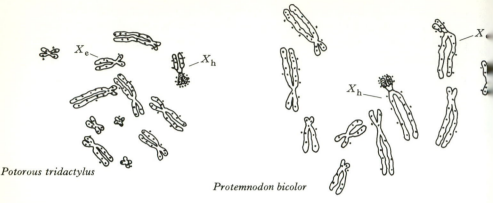

Fig. 16.2. Female karyotypes of the marsupials *Potorous tridactylus* and *Protemnodon bicolor*, following tritiated thymidine autoradiography. In each case one of the two *X*-chromosomes (X_h) is showing late-labelling in its short arm, while the other (X_e) is not. Based on Hayman and Martin (1965).

until it is broken at first anaphase, as the spindle lengthens. The sex chromosomes are usually quite small, the *X* being a metacentric or J-shaped element about the same length as the smallest of the autosomes and the *Y* a minute dot.

The *X*- and *Y*-chromosomes of male mammals undergo synapsis to form an *XY* sex bivalent at meiosis (Fig. 16.3). And, with the sole exception of the mouse genus *Apodemus* (see below) the *X* and *Y* separate to opposite poles at first anaphase (pre-reduction). There has been considerable controversy, however, as to the nature of the *XY* bivalent, and particularly as to whether a chiasma is formed between the two sex chromosomes in the male. According to the interpretation of Koller and Darlington (1934) and Koller (1936–41*b*) the *X*- and *Y*-chromosomes in mammals invariably possess a common homologous pairing segment, and in addition each of these elements has a segment which is not represented in the other ('differential segment'). Thus, at zygotene the pairing segments are supposed to undergo synapsis, while the two differential segments have nothing to pair with; the sex bivalent at pachytene would hence be (at least theoretically!) a triradiate structure, one of the rays being formed by the synapsed pairing segments while the other two would consist of the differential segments of the *X* and *Y*, which would be unequal in length. Koller and Darlington supposed that one or more chiasmata were formed in the pairing segment, leading to the maintenance of the association of the *X* and *Y* until first metaphase. Koller seems to have supposed that in marsupials the centromeres were invariably situated in the differential region, so that the *XY* bivalent would invariably undergo prereduction, regardless of the number of chiasmata formed in the pairing segment. On the other hand, he and Darlington supposed that in eutherian mammals the centromeres lay in the pairing segments, or so near the junction of the pairing and differential segments that one could not be sure which they lay in. In the former case (centromeres in the pairing segment) there would theoretically be a possibility of a chiasma being formed between the centromeres and the differential segments, thus leading to postreductional bivalents, i.e. ones in which the X and Y differential segments would not segregate from one another until second anaphase. Koller and Darlington (1934), Koller (1937, 1938) and Ahmed (1940) claimed that in actuality prereductional and postreductional XY bivalents could be found alongside one another in the same testis (depending on the position where the chiasma was located) in the rat, man, the Golden hamster and the sheep. These claims formed the cytological basis for Haldane's (1936) concept of partial sex-linkage in the human species.

It is now fairly certain that this interpretation, in its entirety, is not supported by objective evidence. Much of the trouble in interpreting the XY bivalents of the Mammalia results from the fact that the sex chromosomes are attached to, and usually embedded in, a large nucleolar structure, the so-called sex vesicle, during the prophase stages of meiosis. Thus triradiate pachytene bivalents, pairing segments and chiasmata cannot, in general, be directly observed. What is fairly certain is that the sex chromosomes emerge from the nucleolar mass at metaphase, firmly attached end-to-end. However, in such cases as the mouse, where the X and Y are acrocentric or nearly so, there has been some dispute as to whether they are attached at first metaphase by their short arms or their long ones.

Reinvestigation of male meiosis in the species in which Koller claimed that some XY bivalents were postreductional has not confirmed this interpretation. Thus Makino (1943) found 100% prereduction in the sheep and Matthey (1952a) could not find any postreduction in the hamster. In all probability the alleged instances of postreductional XY bivalents observed by the earlier workers were in most cases merely ones that had not yet properly orientated themselves on the spindle.

There remains the question whether the association between the X and Y in mammals is chiasmate at all. Certain authors (e.g. Makino 1941; Matthey 1949d, 1953a, and Geyer-Duzyńska 1963) seem to feel that because chiasmata cannot be unequivocally demonstrated in diplotene–diakinesis, as cruciform configurations, they must necessarily be absent. If so, there would be no compelling need to believe in the existence of pairing segments in the X and Y at all; the attachment of the X and Y would be due simply to some kind of terminal adhesion. Ford and Hamerton (1956) considered that the terminal association between the human X and Y at meiosis could be due to a terminal chiasma. But Reitalu (1970) studied human sex bivalents which he interpreted as showing an association between the middle part of the long arm of the Y with the distal region of the short arm of the X. However, using fluorescent staining, Pearson and Bobrow (1970) have definitely shown that it is the short arm of the Y which is associated with the X during meiosis. Chen and Falek (1969) also conclude that the association is between the short arms of both chromosomes.

In the absence of direct visual evidence, we are forced to rely on analogy. In a great variety of mammalian species belonging to all the main orders, the X and Y have been seen at first metaphase, firmly connected by a thread between their ends. Where similar connections are seen in bivalents of grasshoppers, urodeles or even higher plants, it is almost universally conceded that they are due to terminalization of chiasmata that have been formed very close to the ends of the chromosomes. Particular instances that seem relevant in this connection are the threads between the short arms of some acrocentric chromosomes (see p. 170) and those between the X and Y in certain 'neo-XY' systems in orthopteroid insects. That the above connections really are due to terminalized chiasmata is shown most clearly by the fact that when a subterminal chiasma *can* be seen in such bivalents the terminal connection is invariably lacking. General considerations thus strongly suggest that the association between the X and Y in mammals is due to a single terminalized chiasma originally formed in what must be a very short terminal pairing segment in most species. This is the conclusion reached by Ohno, Kaplan and Kinosita (1959a) for the mouse, in which the chiasma is probably formed between the very short second arms of the X and Y. In the goat, also, Datta (1970) has shown that a chiasma is formed between the X and Y in minute terminal segments. The alternative interpretation of an achiasmatic association necessarily postulates the existence of a very strong force of attraction or adhesion between the chromosome ends which can withstand considerable stretching on the spindle – i.e. a force which is unprecedented and unexplained. Unlike the case of the 'parachute' XY bivalents of Coleoptera, there seems to be no possibility that persistent nucleoli hold the X- and Y-chromosomes of mammals together at first metaphase. In fact, in man the X and Y seem to be present as unpaired univalents in about 27% of the first metaphases (Sasaki and Makino 1965).

If we are right in supposing that true pairing segments do exist in the X- and Y-chromosomes of mammals, partial sex linkage would be theoretically possible. But it seems likely that in most species the pairing segments are heterochromatic and so short that the chances of finding genetic loci exhibiting partial sex linkage will be small indeed. Thus the alleged instances of alleles crossing-over between the X and Y in *Homo sapiens* studied by Haldane (1936) and others are almost certainly to be interpreted in other ways, as concluded by Stern (1960b).

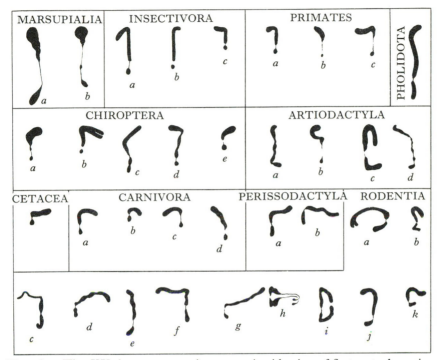

Fig. 16.3. The XY chromosome pair, as seen in side view of first metaphase, in a number of species of mammals. MARSUPIALIA: (*a*) *Trichosurus vulpecula*; (*b*) *Pseudochirus peregrinus*; INSECTIVORA: (*a*) *Talpa europaea*; (*b*) *Neomys fodiens*; (*c*) *Erinaceus europaeus*; PRIMATES: (*a*) *Pithecus entellus*; (*b*) *Macacus rhesus*; (*c*) *Homo sapiens*; PHOLIDOTA: *Manis pentadactyla*; CHIROPTERA: (*a*) *Plecotus auritus*; (*b*) *Barbastella barbastellus*; (*c*) *Pipistrellus pipistrellus*; (*d*) *Pteropus dasymallus*; (*e*) *Rhinolophus ferrum equinum*; ARTIODACTYLA: (*a*) *Bos taurus*; (*b*) *Capra hircus*; (*c*) *Rangifer phylarchus*; (*d*) *Sus scrofa*; CETACEA: *Phocoenoides dallii*; CARNIVORA: (*a*) *Felis leo*; (*b*) *Felis domestica*; (*c*) *Vulpes vulpes*; (*d*) *Mustela itatsi*; PERISSODACTYLA: (*a*) *Equus asinus*; (*b*) *Equus caballus*; RODENTIA: (*a*) *Peromyscus maniculatus blandus*; (*b*) *Chinchilla laniger*; (*c*) *Mesocricetus auratus*; (*d*) *Nesokia nemorivaga*; (*e*) *Mus musculus*; (*f*) *Rattus fulvescens*; (*g*) *Rattus losea*; (*h*) *Apodemus speciosus*; (*i*) *Microtus kikuchii*; (*j*) *Myocaster coypus*; (*k*) *Cavia cobaya*. The small chromosome (assumed to be the Y) is at the bottom in each case. The first division is 'reductional' for the sex chromosome pair in all except *Apodemus speciosus*. The figures of marsupials are from Koller (1936a), and those of insectivores from Bovey (1949b). The remainder are from Makino (various papers) except those of *Plecotus*, *Barbastella* and *Pipistrellus* (Bovey 1949b), *Macacus* and *Homo* (Painter 1924, 1925) and *Mesocricetus* (Matthey 1952). All figures redrawn; magnification varies.

As mentioned previously, postreduction of the XY bivalent does definitely occur in the mice of the genus *Apodemus* with the exception of *A. geisha*, which is said to show invariable prereduction. Matthey (1936) and Koller (1941*a*) earlier claimed that both prereduction and postreduction of the sex bivalent occurred in *A. sylvaticus*, but this is somewhat uncertain (Matthey 1947*a*). However, in *A. mystacinus* from Palestine, pre- and postreduction of the XY definitely coexist side by side (Wahrman and Ritte 1963). In all probability some type of structural rearrangement (possibly tandem fusions between the X and Y and an autosomal pair?) has taken place in the sex chromosome pair, at some stage in the phylogeny of *Apodemus*, leading to the establishment of postreduction as the normal method of segregation of the X and Y. Whereas in other mammals the sequence of the sex bivalent is chiasma–centromere–differential segments, in *Apodemus* spp. (except *A. geisha*, *A. mystacinus* and possibly *A. sylvaticus*) it is presumably centromere–chiasma–differential segments.

It has been shown in recent years that the two X-chromosomes exist in a different state in the somatic cells of female mammals (M. L. Barr and Bertram 1949; Barr 1959; Ohno *et al.* 1959*b*; Barr and Carr 1960, 1962; Taylor 1960*b*; Cattanach 1961, 1963; Lennox 1961; Lyon 1961, 1962, 1963, 1966*a, b*, 1968; Ohno and Makino 1961; Russell 1961, 1963, 1964*a, b*; Atkins *et al.* 1962, 1963; Beutler *et al.* 1962; German 1962; Gilbert *et al.* 1962; Grumbach and Morishima 1962; Grumbach *et al.* 1962, Morishima *et al.* 1962; Vandenberg, McKusick and McKusick 1962; Moorhead and Defendi 1963; Ohno 1963; Reed, Simpson and Chown 1963; McKusick 1964; Mukherjee and Sinha 1964; Mukherjee 1965; Thompson 1965). One of the X's in each cell undergoes a special type of transformation early in development; it is now clear that this transformation affects the maternal X in some cells and the paternal one in others, but that once it has occurred it is transmitted to all the descendants of that particular X-chromosome in later cell generations.

The 'transformation' involves (1) the development of precocious condensation (positive heteropycnosis) during prophase and formation of a peripherally situated heteropycnotic body (the 'sex chromatin' of Barr) in resting nuclei (Fig. 16.4); (2) delayed replication of the DNA during the synthetic phase (S stage) of interphase; (3) inactivation or 'switching off' of the genetic loci, or some of them, in the transformed X.

The sex chromatin phenomenon (i.e. the appearance of a Barr body in somatic interphase nuclei of XX females, but not in those of XY males) is apparently not seen in all tissues, and has not been observed at all in the mouse (Grüneberg 1969).

We shall refer to this transformation as the 'fixed inactivation pheno-

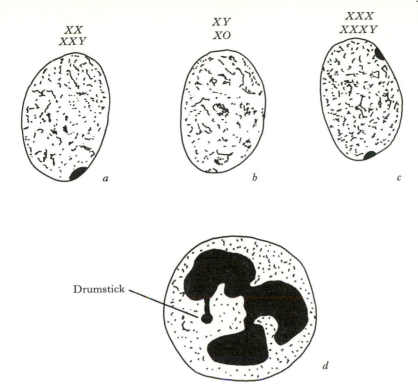

Fig. 16.4. Barr bodies ('sex chromatin') in mammalian somatic nuclei (e.g. human buccal mucosa). *a*, with one Barr body (*XY* or *XO*); *c*, with two Barr bodies (*XXX* or *XXXY*); *d*, the 'drumstick' seen in approximately 5 % of the polymorphonuclear leucocytes of the human female (never seen in the male).

menon', which is a genetic term – the cytologic equivalent might be 'fixed heterochromatinization phenomenon'. By some authors it has been called the 'Lyon hypothesis' after one of its discoverers.

The phenomenon has been studied mostly in human material and in the mouse and Chinese Hamster. But since late replication of one *X* has now been demonstrated by tritiated thymidine autoradiography in female marsupials (Fig. 16.2) by Walen (1963) and by Hayman and Martin (1965*a*) and Martin and Hayman (1966), in the armadillo by Grinberg, Sullivan and Benirschke (1966) and in the mule (Mukherjee and Sinha 1964), it is fairly certainly characteristic of mammals in general. Heteropycnosis of the small *Y* (in the male) and *one* of the much larger *X*'s (in the female) has been observed in somatic cells of the South American opossum *Philander* by Perondini and Perondini (1966).

The fixed inactivation phenomenon may be viewed, on the one hand, as a mechanism of dosage-compensation and, on the other, as a form of genetic mosaicism, since in some cells it is the maternal X which is inactivated while in others it is the paternal one which is 'switched off'. It results in a situation in which both males and females have effectively only one functioning X in their somatic nuclei. The female 'tortoiseshell' cat is probably an example of an XX constitution, heterozygous for a pair of sex-linked alleles (*black* and *yellow*) only one of which is genetically active in each area of skin, thus giving rise to a mosaic of patches of black and yellow fur. Male tortoiseshell cats have occasionally been reported; most of them have proved sterile. They are probably diploid XXY individuals, for the most part, although one was found to be a 3n XXY/2n XX chimaera (Chu, Thulene and Norby 1964). Earlier reports of male tortoiseshell cats with 2n = 38 (Komai and Ishihara 1956) are probably wrong.

Various workers have studied the somatic nuclei of human individuals with more than two X-chromosomes. The evidence (Table 16.1) indicates that in the cells of such individuals, regardless of whether a Y is present, all but one of the X's undergo heterochromatinization (Fig. 16.5). This is almost certainly the reason why such individuals are viable and in certain cases even approximately normal, phenotypically, whereas trisomics for much smaller autosomes are highly abnormal (see p. 465).

A special aspect of the 'sex chromatin' phenomenon in female mammals is the appearance of a small lobe, the so-called 'drumstick', protruding from the nuclei of some of the polymorphonuclear leucocytes of females (Davidson and Smith 1954). 'Drumsticks' are not normally present in males. The evidence indicates that the drumstick contains the heteropycnotic and inactivated X. Individuals with a XXX or $XXXY$ constitution show two drumsticks in some of their cells (Jacobs, Baikie, Court Brown, MacGregor, MacLean and Harnden 1959; Harnden and Jacobs 1961; Maclean 1962). Drumsticks also occur in some other mammals such as the rabbit (De Castro 1963) and the armadillo (Grinberg, Sullivan and Benirschke 1966).

It is fairly clear that the inactivation of one X in female mammals is not complete. Some sex-determining loci are obviously not inactivated – if they were, the phenotypes of XO and XX individuals would be identical. Certain sex-linked loci, both in *Homo sapiens* and the mouse, do not show mosaicism in heterozygous females. One is the human sex-linked blood group X_g (Read, Simpson and Chown 1963). Another is the *scurfy* gene, in the mouse. Studies on X-autosome translocations in the mouse suggest that inactivation proceeds from a localized 'inactivating centre' in the X. Thus autosomal loci which do not normally show mosaicism in heterozygous females will do so when brought into the neighbourhood of the 'inactivating

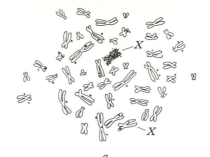

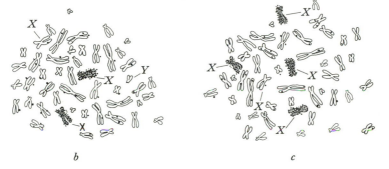

Fig. 16.5. 'Late-labelling' of *X*-chromosomes in human cells, following tritiated thymidine autoradiography. *a*, normal female (*XX*); *b*, an *XXXY* cell; *c*, an *XXXX* cell. Based in part on Grumbach, Morishima and Taylor (1963).

centre' by translocation (Cattanach 1961*a*). Conversely, the *X*-linked *tabby* locus, which normally shows variegation in female heterozygotes does not do so when separated from the inactivation centre as a result of an *X*-autosome translocation (Searle 1962). In the case of an *X*-autosome translocation in the cow (Gustavsson, Fraccaro, Tiepolo and Lindsten 1968) the normal *X* was always late-labelling, the translocated *X* never so. Similarly, in female mice heterozygous for Searle's translocation, the translocated *X* is active, the normal one inactive (it shows positive hetero-pycnosis in 90% of cells). On the other hand, in the case of Cattanach's translocation in the mouse (Ohno and Cattanach 1962; Evans, Ford, Lyon and Gray 1965) which is a non-reciprocal one (i.e. the insertion of an auto-somal segment into the *X*) equal numbers of normal and translocated *X*'s are late-labelling and genetic loci in the inserted segment show mosaicism.

A particularly ingenious demonstration of the fixed inactivation pheno-menon was provided by Mukherjee and Sinha (1964) who carried out autoradiographic studies on cells of a female mule following tritiated thymidine labelling (Fig. 16.6). The *X*-chromosome of the horse is the second largest of the karyotype while that of the donkey is the fourth largest and has a subterminal centromere. The two X's can thus be distinguished in the karyotype of the female mule, and it has found that about half the cells had the horse X late-labelling, while in the other half it was the donkey X which was late-labelling.

The quantitative aspect of this result has, however, been challenged by Hamerton *et al.* (1969) who found that in 80–90% of the cells it was the paternal (donkey) X which was late-labelling. They also studied the glucose 6-phosphate dehydrogenase isoenzymes (which differ in the horse and donkey but are X-linked in both). In one individual mule the paternal G6PD allele was only very weakly expressed, suggesting that the majority of the donkey X's had been inactivated. In the human species R. G. Davidson *et al.* (1963) have demonstrated the existence of two populations of cells in females heterozygous for glucose 6-phosphate dehydrogenase variants.

In human females, although the inactivation of the paternal and maternal X-chromosomes seems to be random when they are both structurally normal, it has been shown in a number of cases that where one of them is structurally abnormal, the abnormal chromosome is invariably the one to be inactivated (Gianelli 1963 ;.Fraccaro and Lindsten 1964; Klinger, Lindsten, Fraccaro, Barrai and Dolinar 1965; Hamerton 1968). In women with isochromosomes for the long arm of the X the sex chromatin is consistently larger than normal. It is probable that in marsupials it is always the paternal X which is inactivated (Sharman 1971).

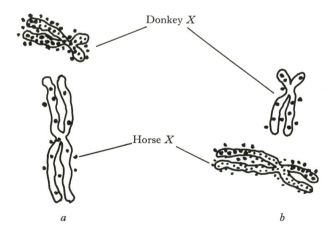

Fig. 16.6. Result of tritiated thymidine autoradiography on the cells of a female mule. In approximately half the cells (*a*) the donkey X is late-labelling, while in the other half (*b*) the horse X is late-labelling. Based on Mukherjee and Sinha (1964).

The X-chromosome of most mammalian species constitutes about 5% of the total haploid chromosome length. In a few mammals, however, a double-sized X occurs. Most of these are rodents, e.g. the Golden Hamster, *Mesocricetus auratus* (Ohno, Beçak and Beçak 1964), *Rattus natalensis* (Huang 1968), *Chinchilla lanigera* (Nes 1963) and the North American Porcupine, *Erithizon dorsatum* (Benirschke 1968). There is a general tendency for the Y-chromosome to be also larger than usual in these species, although not so large as the X. A triple-sized X-chromosome, about 16% of the total haploid chromatin, exists in the Creeping Vole, *Microtus oregoni* (Ohno, Beçak and Beçak 1964) and a quadruple-sized one occurs in *Microtus agrestis* (Matthey 1949c, 1950a). In the order Artiodactyla, *Tragelaphus spekei*, *Antelope cervicapra* and the reindeer, *Rangifer tarandus*, all have double-sized X-chromosomes (Wurster, Benirschke and Noelke 1968).

There are two ways of interpreting these cases. They could be due to some kind of duplication of the original X chromatin itself – either by a number of small duplications or by one large one. Alternatively, they could represent the results of past X-autosome fusions (presumably with associated Y-autosome fusions, since these species do not have multiple sex-chromosome mechanisms). Schmid, Smith and Theiler (1965) have shown that the species with 'original type' X's have relatively extensive late-replicating autosomal segments, that in the hamster some of this late-replicating autosomal material seems to have been transferred to the X and that *Microtus agrestis* has no late-replicating segments in its autosomes. They accordingly conclude in favour of X-autosome translocations as an explanation of 'giant' sex chromosomes in a few species of mammals. The fact that the 'giant' X-chromosome of *Erethizon* has a euchromatic (i.e. non-late replicating) segment and that the Y of the same species also has a region that is not late-replicating seems to point in the same direction. But it is not necessary to conclude that the mode of origin of mammalian giant sex chromosomes has been the same in all cases.

There is a considerable literature on the subject of supposed instances of holandric (Y-linked) inheritance in man. Stern (1957) concluded that the evidence is in all cases equivocal and Penrose and Stern (1958) have reinterpreted one famous pedigree formerly claimed to illustrate holandric inheritance. Since that time Dronamraju (1960, 1964, 1965) has claimed that hypertrichosis of the pinna of the ear is a strictly Y-linked trait. The matter cannot be regarded, however, as finally settled, and there is a possibility that the trait is an autosomal one, sex limited in expression (Stern, Centerwall and Sarkar, 1964). It thus seems best to conclude for the time being that there is no certain case of a Y-linked trait in man. The differential segment of the Y may consequently be regarded as one which only carries sex-determining genes and either has no other genetic activity or has only functions of a subtle, quantitative kind. It is thus not altogether surprising that individual and even racial differences in the length of the human Y-chromosome should have been recorded by several investigators. Thus Cohen, Shaw and MacCluer (1966) have found significant differences between the mean length of the Y in several human racial groups, the Japanese having particularly long Y's. Australian aborigines apparently have shorter Y-chromosomes than Europeans (Angell, unpubl.).

In the mouse, it is now definitely established that the Y-chromosome carries a histoincompatability gene which is responsible for the rejection by females of all skin grafts from males (Hauschka and Holdridge 1962; Celada and Welshons 1963).

An entirely unique mechanism of sex determination exists in the vole *Microtus oregoni*, according to the analysis of Ohno, Jainchill and Stenius (1963) and Ohno, Stenius and Christian (1966). It had earlier been shown by Matthey (1956, 1958*a*) that there were seventeen chromosomes in the male germ-line, including one unpaired sub-acrocentric element and seventeen in the female soma, including an unpaired equal-armed metacentric. Ohno *et al.* have shown that the male soma is XY, with eighteen chromosomes and that the X is lost in the embryology of the male germ-line. As a result of a special process of non-disjunction XXY and OY cells arise, but only the latter form the definitive spermatogonia. The female soma is

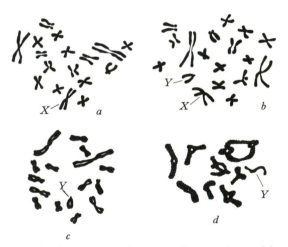

Fig. 16.7. Cytology of *Microtus oregoni*. *a*, somatic metaphase of female (2n = 17, X-chromosome indicated); *b*, somatic metaphase of male (2n = 18, X- and Y-chromosomes indicated); *c*, spermatogonial metaphase (2n = 17, Y-chromosome indicated); *d*, diakinesis in the male (8 bivalents + Y); Based on Ohno, Jainchill and Stenius (1963), redrawn.

TABLE 16.2. *Summary of sex-determining mechanism in Microtus oregoni*

	Male	Female
Zygote and soma of Adult	XY 18	XO 17
Germ-line	OY 17	XX 18
Gametes	O | Y 8 | 9	X | X 9 | 9

XO (seventeen chromosomes), but as a result of a special process of 'selective non-disjunction' the oogonia become XX (Fig. 16.7). The whole scheme may be summarized as in Table 16.2.

It will be seen that the Y-chromosome is still male-determining, as in all other mammals. Whereas in female mammals in general one X is inactivated in the somatic cells, in the female of *Microtus oregoni* only one X is present in the female soma, since female zygotes are derived from O-bearing sperms. It will be noted that although there are two X's at meiosis in the female, these would be molecular copies of one another, so that there would be no genetic recombination in the X, in either sex. The non-disjunction which occurs in the formation of the definitive oogonia must result in XX and OO cells (i.e. ones containing no sex chromosomes) but the latter do not seem to give rise to any germ cells. A peculiar feature is that approximately two-thirds of the very long X is heteropycnotic in female somatic nuclei – it presumably consists of recently acquired heterochromatin.

The Persian Mole-vole *Ellobius lutescens* is another microtine rodent with a highly peculiar sex-chromosome mechanism (Matthey 1953c, 1954a, 1958a, 1962, 1964a). Unlike *Microtus oregoni*, *E. lutescens* has seventeen chromosomes in both the soma and the germ-line of both sexes (Fig. 16.8). This karyotype consists of eight pairs of autosomes and a small equal-armed metacentric (chromosome no. 9) which is obviously a sex chromosome of some kind. There is no visible difference between this sex chromosome in males and females, and its replication pattern is the same in both sexes (Castro Sierra and Wolf 1967). Although meiotic divisions are strangely rare in males it seems fairly certain that the univalent sex chromosome passes undivided to one pole at first anaphase and that there are consequently two kinds of second divisions, with eight and nine chromosomes respectively

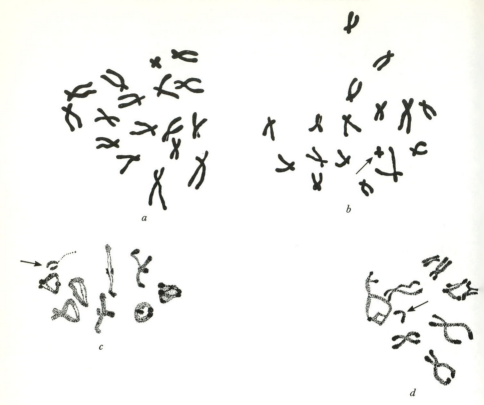

Fig. 16.8. *Ellobius lutescens*. *a*, male somatic metaphase; *b*, female somatic metaphase (both colcemid-treated); *c*, male first metaphase; *d*, female diakinesis. The small arrows indicate the sex chromosomes. From Matthey (1962, 1964).

(and hence sperms with and without the little metacentric sex chromosome). First meiotic divisions in the egg show eight bivalents and the small meta-centric sex chromosome. It is consequently virtually certain that both sexes produce two kinds of gametes in equal numbers, with and without a sex-chromosome. It was suggested (White 1957*e*, 1960) that the sex chromosome of the male *Ellobius lutescens* is an 'attached-XY' element and that of the female an 'attached-XX', the original X and Y before the occurrence of the two fusions having been acrocentrics closely similar in length. The mechan-ism postulated would have to function as a balanced lethal mechanism, the zygotes with no sex chromosome and those with an $\widehat{XX}$ and $\widehat{XY}$ aborting. The establishment of such a mechanism in a mammal which produced more ova than could be reared in any case might not have been opposed by natural selection. However the evidence of Castro-Sierra and Wolf is not favourable to White's earlier hypothesis. The same authors (Castro-Sierra and Wolf 1968) found an inconstant pachytene association between the ninth chromosome and an autosomal bivalent. They favour a hypothesis according to which the original Y, assumed to be very small, was translocated to one member of a pair of autosomes. They suppose that it is this chromosome pair which may have the ninth chromosome temporarily associated with it at pachytene. There would still be a balanced lethal mechanism, although not of the type postulated by White. Whatever the correct interpretation, it seems clear that the superficial similarity between the peculiar sex-chromosome mechanisms of *Microtus oregoni* and *Ellobius lutescens* is deceptive. There is one other species of *Ellobius* in the U.S.S.R., *E. talpinus*, which shows 2n = 52 in some areas and 2n = 54 in others (Vorontsov and Rajabli 1967; Ivanov 1967) so that *E. lutescens* has probably acquired many chromosomal fusions in its phylogeny. *E. talpinus* apparently has acrocentric X- and Y-chromosomes, about the same length. The Japanese *Microtus montebelli* in which Oguma (1937*b*) reported $2n\male = 31$ actually has 2n = 30 in both sexes (Utakoji 1967); the male is consequently XY and there is nothing unusual about the sex-determining mechanism.

There are several species of mammals in which one X is not merely inactivated but is actually eliminated from the somatic nuclei of the female. Hayman and Martin (1965*b*, 1969; Hayman, Martin and Waller 1969) have definitely shown that in three species of bandicoots (marsupials) belonging to the genera *Isoodon*, *Permeles*, *Perorcytes* and *Echymipera* there is an XO condition in some of the somatic tissues of both sexes, although the usual $XY:XX$ condition occurs in the germ-line. They interpret the loss of one X from the female soma as an extreme type of dosage compensation, one stage beyond the genetic inactivation of one X which occurs in other female mammals. But it is not known whether the X that is lost is invariably the paternal or the maternal one, or may be either. The genetic significance of the loss of the Y from the male soma is uncertain; it is not known, for example, whether the interstitial tissue of the testis (which in all mammals produces male sex hormones) is XY or XO. It is certain that only some of the somatic tissues are XO; cultures of skin cells of *Perameles nasuta* obtained by Jackson and Ellem (1968) were apparently XY in constitution and so were corneal cells of *Perameles* spp. and *Peroryctes*; in *Isoodon* spp, however, the corneal cells were XO (Hayman, Martin and Waller 1969). Another species of bandicoot (*Thylacomys lagotis*) has an XX female soma and an XY_1Y_2 male soma (see p. 639). In the New Guinea *Echymipera* there are small supernumerary chromosomes in some individuals, which are present in the germ-line and in the corneal cells but absent from those somatic tissues which are XO in constitution. In the South American field mouse *Akodon azarae* some, but not all, females have partial or total somatic deletion of one of the *X*'s (Bianchi and Contreras 1967).

Another mysterious case, not fully investigated yet, is that of the African mouse *Acomys selousi*, studied by Matthey (1965). The males have a large metacentric *X* and an acrocentric *Y* about as long as one limb of the *X*; apparently this is the condition in both germ-line and soma. The two sex chromosomes form a bivalent in spermatogenesis. Female somatic divisions contain a single *X*; the situation in the female germ-line is not yet known. It is probable that this is another case in which one of the *X*'s is eliminated from the female soma in the course of development. The two-toed sloth, *Choloepus hoffmanni*, also appears to have an *XO* soma in the female (Corin-Frederic 1969); this is a species with an X_1X_2Y constitution in the male as a result of a fusion between the original *Y* and a small autosome, the homologue of which has become X_2. Thus the constitution of the female soma should really be designated $X_1OX_2X_2$. Presumably the female germ-line retains both X_1's, but this has not been confirmed by direct observation.

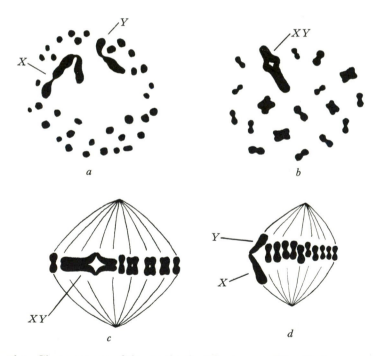

Fig. 16.9. Chromosomes of the centipede, *Thereuonema hilgendorfi*. *a*, spermatogonial metaphase (34 autosomes, *X* and *Y*); *b*, first metaphase in polar view; *c*, the same in side view, showing the *XY* bivalent; *d*, second metaphase in side view, showing the *X* and *Y*. From Ogawa (1950*a*), redrawn.

POSTREDUCTIONAL *XY* MECHANISMS

We have already referred to postreductional XY systems in the mice of the genus *Apodemus*. Where meiosis is chiasmate, such mechanisms will depend on formation of a chiasma in the pairing segment, between the centromere and the differential regions. Such a mechanism has evolved in one species of morabine grasshopper, as a result of a tandem fusion between the X and an autosome, with the X occupying the distal position (see p. 378). Similar mechanisms, but perhaps not due to recent tandem fusions, probably exist in the centipedes *Thereuonema hilgendorfi* and *Thereuopoda clunifera* (Ogawa 1950, 1952). The former species has seventeen pairs of autosomes and two huge sex chromosomes in the male, one of which consists of three distinct regions, the other of two segments (Fig. 16.9). The former is interpreted by Ogawa as the X, the latter as the Y. The positions of the centromeres are uncertain and, in fact, Ogawa has claimed that the chromosomes are holocentric. At meiosis the X and Y form a bivalent, there being apparently terminal pairing segments in which a chiasma is formed. But orientation of the bivalent at first metaphase is apparently equatorial rather than axial (see p. 492), so that each second metaphase shows an X and Y, associated end-to-end by 'half a terminal chiasma'.

SEX CHROMOSOMES AND SEX DETERMINATION IN CRUSTACEA

A considerable variety of sex-determining mechanisms are known in the Crustacea. Bisexual reproduction is the rule in this class, but in barnacles (subclass Cirripedia) and in two families of parasitic Isopoda, the Cymothoidae and Cryptoniscidae, hermaphroditism is found in most species; it also occurs sporadically in some other groups of Crustacea (see Charniaux-Cotton 1960).

Male heterogamety, with a cytologically distinguishable XY or XO mechanism, has been demonstrated in several species of crabs (Niiyama 1937, 1938, 1941). The Ostracoda also have male heterogamety, although here multiple sex-chromosome mechanisms are common (see p. 672). Male heterogamety of the XO type has also been reported in the branchiopod *Chirocephalus nankinensis* (Pai 1949). On the other hand, some at least of the copepods have female heterogamety (W. Beermann 1954; Rüsch 1960; Ar-Rushdi 1963); some species have XY females while in others the female is XO and in yet others sex chromosomes cannot be detected morphologically. Female heterogamety of the XY type occurs in the bisexual biotype of the Brine shrimp *Artemia salina*. In this case Stefani (1963) has shown that the two sex chromosomes are slightly unequal in length. Bowen (1963, 1964,

1965) has studied a partially sex-linked mutation *w* in *Artemia* which causes white eyes; this gene can cross-over from one sex chromosome to the other. We must accordingly suppose that an XY sex bivalent is found at meiosis, with a chiasma in a pairing segment.

Female heterogamety has also been found in some isopods. Some species of *Jaera*, *Janiropsis* and *Janira* do not show any morphologically distinguishable sex chromosomes but probably possess an XY pair consisting of two chromosomes of approximately the same size; the *Jaera marina* species complex, on the other hand, shows a sex trivalent at meiosis in the female (Staiger and Bocquet 1956). This XY_1Y_2 mechanism has almost certainly arisen as a result of a fusion between the original X and one member of a pair of acrocentric autosomes.

In some Crustacea, at any rate, maleness is due to a hormone produced by an androgenic gland which is quite separate from the testis, being attached to the vas deferens (Charniaux-Cotton 1960). An organ of this type seems to be present in all Malacostraca. No similar hormonal mechanisms of sex determination are known in any other arthropods.

In a number of isopod genera the females regularly produce unisexual broods. In *Armadillidium vulgare* the populations contain approximately equal numbers of female-producing (*gynogenic*) and male-producing (*androgenic*) females, while in *A. nasutum* females usually produce both sons and daughters (i.e. they are *amphogenic*) according to Vandel (1941). This species has male heterogamety, but in a strain studied by Lueken (1966) autosomal female-determining factors are present which, in sufficient dosage, cause XY individuals to develop into females. Such females produce progenies with a 1♀:3♂♂ sex ratio, one third of the males being YY and therefore capable of producing sons only. In *Trichoniscus provisorius* gynogenic, androgenic and amphogenic females all exist, but the distinction between them is not clear-cut. Presumably in such cases a number of alleles are present in the population, the various combinations of which determine whether any given female shall be gynogenic, amphogenic or androgenic.

The situation with regard to sex determination in another isopod, *Asellus aquaticus* is complex (Vitagliano-Tadini 1958; 1963, Montalenti 1960, 1961; Montalenti and Vitagliano-Tadini 1960, 1963). The proportion of males in natural sibships ranges from 0 to 95%. Montalenti and Vitagliano-Tadini (1963) assume three sex-determining loci *Aa, Bb* (closely linked) and *Zz* (more loosely linked). Females would have no dominants or only one, while males would have 2, 3 or 4 dominants. Crosses between individuals from mainly gynogenic families (G♂ × G♀) and between individuals from mainly androgenic families (A♂ × A♀) are much less fecund than ones between individuals from androgenic and gynogenic families (A♂ × G♀

and G♂ × A♀). There is obviously a complex adaptive polymorphism and Vitagliano-Tadini and Montalenti believe that one effect of the system is to prevent inbreeding. Similar systems seem to exist in members of the genera *Cylisticus*, *Porcellio* and *Tracheoniscus* (de Lattin 1951, 1952).

THE EVOLUTION OF SEX-CHROMOSOME MECHANISMS IN PRIMITIVE INSECTS

The apterygote groups related to the basic stock of the Insecta all, as far as is known, show male heterogamety. A number of species of Collembola have XO males, according to Núñez (1962) but the Neanuridae studied by Cassagnau (1968a, b, 1970) do not seem to show morphologically differentiated sex chromosomes. The early work of Charlton (1921) suggests that the 'silverfish' *Lepsima domestica* has an X_1X_2O mechanism in the male. The situation in the Protura, Diplura, and in other thysanurans has not been investigated.

All the more primitive orders of exopterygote insects seem likewise to have male heterogamety, usually of the XO type. We must envisage a number of evolutionary lineages such as those represented by the Odonata (dragonflies), the orthopteroid orders and the blattoid–mantid–termite stock as having been characterized by a sex-determining mechanism of the *Drosophila* type, depending on a genic balance between female-tendency genes in the X and male-tendency genes in the autosomes. Within these stocks relatively few evolutionary reversions to XY and X_1X_2Y mechanisms have occurred (see next section). Several other orders of primitive exopterygote insects seem to be characterized by XO ($♂$) systems. These include the Psocoptera (Boring 1913; Wong and Thornton 1966) and the Embioptera (Kichijo 1942; Le Calvez 1949; Hirai 1951; Stefani 1959). Information is lacking for the Zoraptera, and in the case of the termites it is exceedingly scanty (Stevens 1905a; Benkert 1930; Light 1942; Banerjee 1961), although it is known that both sexes in this order are diploid (unlike the situation in the social Hymenoptera, where the males are haploid). The Plecoptera (Stone flies) include both XO and XY species as well as some with multiple sex-chromosome mechanisms (see p. 674). The Ephemeroptera (mayflies) likewise include some species with XY males (Katayama 1939b) and some in which the male is XO (Wolf 1946).

The earwigs (order Dermaptera) differ from the other blattoid–orthopteroid orders in that almost no XO species are known, the males being usually XY, but in some cases X_1X_2Y or XYY (Ortiz 1969; Webb and White 1970); recently, however, an earwig species with XO males has been studied by Webb (unpublished). A species of *Hemimerus*, a genus usually included in the Dermaptera but sometimes regarded as a separate order of insects, has been shown to have X_1X_2Y males and generally earwig-like cytology (White 1971*a*).

Numerous instances are now known in which $XO:XX$ sex-chromosome systems have 'reverted' to $XY:XX$ ones in the course of evolution. The usual way in which such transformations have come about is through centric fusions between the acrocentric X of an $XO:XX$ form and an acrocentric autosome. Such a fusion may be said to create a 'neo-X' chromosome. When the fusion has reached fixation in the population (i.e. when unfused X's no longer occur) the original acrocentric autosome will be confined to the male line (or female line, in groups with female heterogamety) and will constitute a 'neo-Y' (Fig. 16.11).

It is convenient to speak of the limb of the neo-X which is derived from the original X as the 'left' limb (XL); the originally autosomal limb being referred to as the 'right' one (XR). The neo-Y is, then, originally homologous to XR throughout its length, and synapses with it at meiosis, so that a neo-XY bivalent is formed at meiosis in the heterogametic sex and undergoes prereduction. However, as we shall see later, various kinds of secondary genetic and structural changes may take place in both XR and the neo-Y, so that the length of the homologous pairing segments may be gradually reduced.

Evolutionary reversions of this type have been studied most extensively in grasshoppers; but they are also known to have occurred in phasmatids, crickets, beetles and may presumably be found in most groups in which XO karyotypes with acrocentric X's and autosomes occur.

Although almost fifty neo-XY mechanisms in grasshoppers are known to have arisen through centric fusions between X's and autosomes, there is only one instance known where such a mechanism has arisen as a result of a tandem fusion. This is in a rare undescribed species of morabine grasshopper from the Eyre Peninsula, Australia (White, Blackith, Blackith and Cheney 1967). Unlike the other neo-XY mechanisms, which are all prereductional, this one is post reductional like the XY mechanisms found in the rodent genus *Apodemus* (see p. 594); the original X thus occupies the distal position in the fusion chromosome (Fig. 16.12).

Helwig (1941, 1942) stated that neo-XY mechanisms were known in fourteen genera of grasshoppers at that time, but apparently about half of these cases were never published. Since that time many additional instances have been found in North and South American grasshoppers (King 1950;

White 1953; Saez and Díaz 1958; Saez 1960, 1963 and especially Mesa 1956–64) as well as in ones from India (Ray Chaudhuri and Guha 1952), Africa (White 1965*a*) Europe (John and Hewitt 1970) and Australia.

There seem to be two basically different types of neo-XY systems; but these may be just the ends of a continuous series of cytogenetic mechanisms. On the one hand we have neo-XY chromosome pairs in which the Y seems to be still homologous throughout its entire length with the XR limb. The homology is demonstrated by complete synapsis in the prophase of meiosis and by the formation of one or more chiasmata between the neo-Y and XR, which are not distally localized. This is the situation in the Australian tettigoniid *Yorkiella picta* (Fig. 16.10) in which chiasmata may apparently

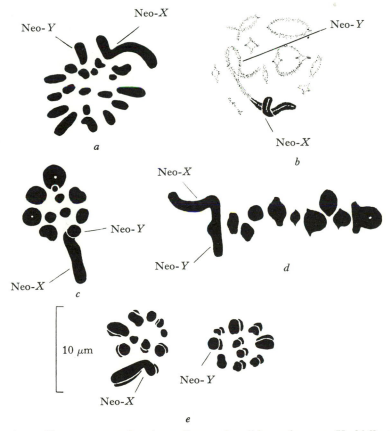

Fig. 16.10. Chromosomes of an Australian tettigoniid grasshopper, *Yorkiella picta*. *a*, spermatogonial metaphase (2n♂ = 20); *b*, diplotene, showing 2 chiasmata between the neo-X and neo-Y; *c*, first metaphase in polar view; *d*, the same, in side view; *e*, second metaphases with neo-X and neo-Y. From White, Mesa and Mesa (1967).

be formed between almost any region of XR and the corresponding segment of the neo-Y (White, Mesa and Mesa 1967). In *Hypochlora alba*, the pairing segment is still long enough for two chiasmata to be formed in it (King 1950) and in the South American *Neuquenina fictor* as many as four chiasmata may occur (Mesa 1961).

In most neo-XY systems, however, the homologous pairing segments are restricted to small distal segments of the Y and XR, the proximal regions being no longer homologous. It is usually found, in such cases, that extensive blocks of heterochromatin are present in the rest of the Y – whereas in neo-XY mechanisms of the first type there has been little or no development of heterochromatic blocks in this chromosome. In both types of mechanism the neo-X and neo-Y are held together by chiasmata at first metaphase, but in mechanisms of the second kind there is only a single chiasma which is formed very close to the distal end and terminalizes completely.

In such grasshopper species as *Mermiria intertexta*, *Hypochlora alba* and *Thisiocetrus pulcher* (King 1950; Ray Chaudhuri and Guha 1952) the greater part of the Y has become heterochromatic, while the originally homologous part of XR has apparently remained euchromatic. In *Paratylotropidia beutenmulleri* (White 1953), on the other hand, some heterochromatic blocks have also developed in the XR limb as well, but their size and distribution shows that they are not homologous to those in the Y. In the South American grasshopper species *Dichroplus bergi* an X-autosome fusion leading to a neo-XY system seems to have been followed by a pericentric inversion in the neo-Y, whose *short* arm is accordingly synapsed with the tip of XR. The pattern of DNA replication in the sex-chromosome

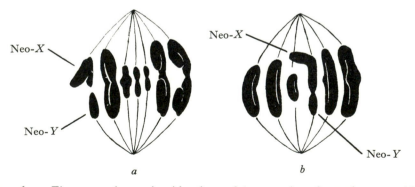

a   *b*

Fig. 16.11. First metaphases, in side view, of two species of grasshopper with neo-XY sex-chromosome mechanisms. *a*, *Machaerocera mexicana*; *b*, *Philocleon anomalus*. The complete complement of autosomal bivalents is shown in both cases. Based on Helwig (1941, 1942), redrawn.

bivalent has been studied by Diaz and Saez (1968); apparently the short pairing segments are early-replicating at meiosis, XL and the long limb of the neo-Y being late-replicating.

Once the neo-Y has undergone evolutionary heterochromatinization one might expect that it would undergo further changes involving, perhaps, deletions or duplications. When chiasmata are no longer formed in its proximal segment, there is no barrier (other than that presented by position effects) to the establishment of pericentric inversions. Presumably this is what has occurred in *Mermiria bivittata*, in which the neo-Y is metacentric but has an overall length which is approximately the same as that of the XR limb (McClung 1917). In *M. maculipennis mcclungi* the Y is likewise metacentric, but the combined length of its two arms is less than that of XR, so that it has presumably undergone one or more deletions as well as a pericentric inversion. Similar changes seem to have occurred in several species of Australian morabine grasshoppers with neo-XY mechanisms. No species of grasshopper is known in which a neo-Y may be presumed to have existed at some time in the past and later been lost; but such a species may one day be discovered. The first type of neo-XY mechanism referred to earlier is clearly more primitive. But whether such mechanisms are necessarily more recent in a chronological sense than ones in which extensive new differential, and usually heterochromatic, segments have developed, is uncertain. It may be that, for some unknown genetic reasons, some neo-XY systems are incapable of developing new differential regions. A good deal may depend on whether the autosome which fuses with the X is one with a fairly high chiasma frequency or one which never forms more than a single chiasma, usually localized distally.

In short-horned grasshoppers the X of the XO species invariably seems to show a special kind of reversible heteropycnosis, being 'under-condensed' in spermatogonial metaphases and 'over-condensed' in the prophase of meiosis (see p. 27). The blocks of heterochromatin which develop in the neo-Y (and occasionally in the XR limb as well) never show this reversible behaviour; they are not under-condensed in the spermatogonial metaphases as XL (= the original X) is.

We do not know in any single case, whether the centromere of a neo-X chromosome has been derived from the original X or from the autosome which fused with it. In a large number of instances, especially in the Australian morabine grasshoppers, an originally acrocentric neo-Y has become metacentric, presumably by pericentric inversion. There is no known instance, however, of a neo-X undergoing pericentric inversion; presumably rearrangements which bring 'reversible heterochromatin' into juxtaposition with euchromatin or non-reversible heterochromatin are eliminated by natural selection, perhaps because they lead to variegation (see p. 103).

The genetic evolution of the neo-XY system in the undescribed species of morabine grasshopper 'P45b' which arose by a tandem fusion has been entirely different (White, Blackith, Blackith and Cheney 1967). In this case the entire length of the neo-X shows negative heteropycnosis, in spite of the fact that its proximal region is clearly autosomal in origin. We may interpret this by saying that in cases of neo-X's that have arisen by centric fusion the centromere prevents in some way the spread of a special type of heterochromatinization (i.e. the type which is manifested as negative heteropycnosis in the spermatogonia) to the previously autosomal limb of the chromosome; whereas in the case of tandem fusions there is nothing to prevent this spread. There is another possible interpretation, however. The tandem fusion which occurred in 'P45b' must have brought 'reversible heterochromatin' into juxtaposition with an autosomal region which did not respond by immediately giving position effects of a gravely deleterious

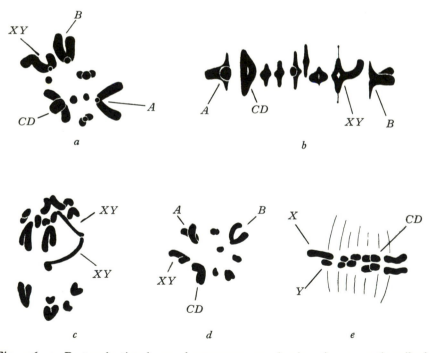

Fig. 16.12. Post-reductional sex-chromosome mechanism in an undescribed species of morabine grasshopper, 'P45b'. a, first metaphase in polar view; b, in side view; c, first anaphase; d, second metaphase in polar view; e, the same in side view. From White, Blackith, Blackith and Cheney (1967).

nature. It may have been this unusual property of an autosomal region which permitted the tandem fusion to establish itself in the first place and later allowed the conversion of the whole proximal segment of the neo-X into reversible heterochromatin.

It has been suggested (White 1957d) that one reason for the successful fixation of X-autosome and Y-autosome fusions in certain species has been that they 'captured' heterotic systems advantageous to the heterogametic sex, but of no consequence or even disadvantageous to the homogametic sex. The argument is as follows: among the innumerable heterotic mechanisms that establish themselves in the evolution of natural species (some 'monogenic', others depending on blocks of genes, inversions, etc.) a few may be expected to affect the sexes differently. If an autosomal heterotic mechanism manifests itself mainly or exclusively in the heterogametic sex $(a_1a_1 < a_1a_2 > a_2a_2)$ and not in the homogametic one $(a_1a_1 \geqslant a_1a_2 \leqslant a_2a_2)$ then a fusion between the X and either the a_1 or the a_2 chromosome will, as it reaches fixation, lead to the elimination of the biologically inferior a_1a_1 and a_2a_2 homozygotes in the XY sex. The XX sex will, of course, become all a_1a_1 or a_2a_2 (as the case may be), but these are *ex hypothesi* not inferior. The same principle may apply in all types of evolutionary alterations of the sex-chromosome mechanism that involve the incorporation of previously autosomal material in X's or Y's, including the various mechanisms by which complex sex-chromosome systems have arisen (see Ch. 17).

It is probable that the XO grasshoppers have very few major genetic loci exhibiting sex-linked inheritance (since the X's are entirely heterochromatic). However, in those species which have neo-XY mechanisms of the second type – those, that is to say, in which synapsis between the X and Y is restricted to minute terminal segments – we must imagine that a considerable number of genetic loci (those in XR) will show X-linked inheritance. Such loci will have passed through an evolutionary stage in which they showed partial sex-linkage (when the neo-XY system was still of the 'first type'). Some Y-linked (holandric) loci probably exist, in species with neo-XY systems of the second kind, but the number and genetic activity of such loci is probably progressively reduced as the Y-chromosomes of such species become more and more completely heterochromatic.

In the grasshopper *Hesperotettix viridis* some individuals have an X-autosome fusion while others lack it; at least in certain areas of west Texas both conditions occur in the same populations (McClung 1917; Helwig 1942). According to King (1950) there are no differences of size of heterochromatic regions between the neo-Y and XR in this case, which is what we should expect to find if the fusion has not yet reached fixation. Conceivably, the existence of two kinds of males (those with the X fused and those with it free) and three kinds of females (homozygous 'fused X', heterozygotes and homozygous 'free X') in the same population may represent an instance of adaptive polymorphism of some kind. Several fusions between autosomes also exist in this species and have likewise not reached fixation. But the whole situation needs reinvestigation before one can decide whether one is dealing with widespread adaptive polymorphisms or a narrow hybrid zone. In *Hesperotettix pratensis* and *H. speciosus* X-autosome fusions have apparently reached fixation; it is uncertain how many different X-autosome fusions have established themselves in the whole phylogeny of this grasshopper genus. Another grasshopper species in which an X-autosome fusion has occurred but has not undergone fixation, so that XO and XY males coexist in the population is *Eyprepocnemis* sp. from India (Manna and Chatterjee 1963).

Fusions between the X with an autosome have also been reported in several species of stick insects (phasmatids). Hughes-Schrader (1947*b*, 1959) has described a clear case of an XY mechanism in the male of *Isagoras schraderi*, although two other species of the genus have XO males. Since all the XO species of phasmatids so far studied (including the two other members of the genus *Isagoras*) have metacentric X's the fusion that has occurred in *I. schraderi* was probably preceded by a pericentric inversion, producing an acrocentric X. A similar XY mechanism is known in the Australian phasmatid *Podacanthus typhon* (Fig. 16.13) and in some of the races of *Didymuria violescens* (Craddock 1970). Two races of the latter insect, however, have XY sex chromosomes which form ring-bivalents at meiosis; they have presumably arisen by two successive fusions of autosomes to the limbs of the original X and a fusion between the homologues of those autosomes.

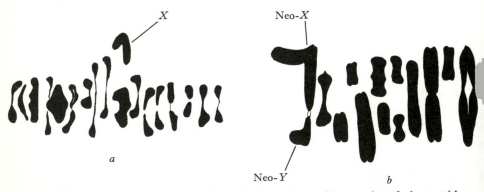

Fig. 16.13. First metaphases (side view) of two Australian species of phasmatids. (*a*) *Podacanthus wilkinsoni* with $2n\male = 35$ (male XO); (*b*) *Podacanthus typhon* with $2n\male = 28$ (neo-XY mechanism).

Once a neo-XY system has arisen by centric fusion it can undergo a further conversion into an X_1X_2Y system by a Y-autosome fusion – that is, if a second pair of acrocentric autosomes is available to become incorporated in the sex-chromosome mechanism. We shall consider the properties of these multiple sex-chromosome mechanisms in the next chapter. It seems worthwhile, however, to point out here an interesting difference between the grasshoppers of the superfamily Acridoidea and the Australian Morabinae. About thirty different neo-XY mechanisms are known to have evolved in the former group; of which six (i.e. one-fifth) have undergone the further transformation into X_1X_2Y systems (in the genera *Paratylotropidia*, *Karruacris*, *Eurotettix*, *Scotussa* and *Dichroplus*).

In the Morabinae there have been eleven independent X-autosome fusions in approximately 200 species that have been studied cytologically, but in no less than six of these instances a further Y-autosome fusion has led to the origin of X_1X_2Y mechanisms. It is unlikely that the greater tendency for Y-autosome fusions to establish themselves in the Morabinae depends on any simple mechanical principle. The autosomes involved in these fusions may be long or short in both groups and in both there are plenty of acrocentric autosomes available to fuse with the Y's of neo-XY systems. We must, in fact, assume that Y-autosome fusions have occurred repeatedly in neo-XY Acridoidea, but that they have only succeeded in establishing themselves in the five genera cited above.

An entirely illogical interpretation of the relationship between XO and XY sex-chromosome mechanisms has been put forward by Manna (1967), according to which the XY condition was primitive in the group, the far more numerous XO species having been derived from XY ancestors by dissociation of the metacentric X. This interpretation entirely ignores the structural differences (and differences in distribution of heterochromatic blocks) which exist between XR and the Y in the XY species. The only way in which a neo-XY system could be expected to give rise to a new XO one would be by gradual decrease in size of the Y and its eventual disappearance. But there is no evidence that this has happened in any of the grasshoppers.

SEX-CHROMOSOME MECHANISMS IN THE PANORPOID COMPLEX OF INSECT ORDERS

The 'Panorpoid complex' includes the orders Mecoptera, Trichoptera, Lepidoptera, Neuroptera (*sens. lat.*), Diptera and Aphaniptera. In this assemblage the Mecoptera seem to represent archaic types, closely related to the ancestral stock of the entire complex. They seem to be mostly XO in the males; at least this is so in several species of *Panorpa* (Naville and de

Beaumont 1934; Ullerich 1961; Atchley and Jackson 1970), one of *Bittacus* (Matthey 1950b), one of *Nannochorista* and one of *Chorista* (G. L. Bush unpublished). On the other hand, *Boreus brumalis* has a multiple sex-chromosome mechanism of the X_1X_2Y ($\male$) type, whose evolutionary origin is not clear (Cooper 1951).

The caddis flies (Trichoptera) and Lepidoptera are undoubtedly descended from Mecopteroid ancestors, but show female heterogamety. At some stage in their phylogeny the old mecopteroid sex-determining mechanism must have been discarded and replaced by an entirely new one. There have been very few really detailed studies of the sex chromosomes in the Trichoptera and Lepidoptera; both XY and XO systems seem to occur in the latter order, but the former condition is certainly the most widespread. An XY bivalent is formed in oogenesis but it is not clear whether it is chiasmatic or not – indeed we do not even know whether chiasmata are formed in the autosomal bivalents in oogenesis (see p. 493).

In a number of species of Lepidoptera with a Y-chromosome the interphase somatic nuclei of females can be distinguished from those of males by the presence of a heteropycnotic body similar to the sex chromatin of mammals (see p. 594); it is clear that this body is the Y-chromosome (Smith 1945; Jucci 1948; Cock 1964).

The sex-chromosome mechanism of the Neuroptera Planipennia is of a highly distinctive type (Fig. 16.14). A large number of species belonging to the families Hemerobiidae, Osmylidae, Chrysopidae, Myrmeleonidae, Ascalaphidae, Coniopterygidae, and Sisyridae have been studied cytologically by Oguma and Asana (1932), Klingstedt (1934, 1937b), Naville and

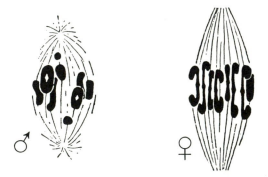

Fig. 16.14. First metaphases in the two sexes of the neuropteran *Macronemurus appendiculatus*. X- and Y-chromosomes lying unpaired on opposite sides of the equatorial plane in the male. In the female (which seems to have a lower chiasma frequency) the two X's form a bivalent which cannot be distinguished from the autosomal ones. From Naville and de Beaumont (1933).

de Beaumont (1933, 1936), Kichijo (1934), Asana and Kichijo (1936), Katayama (1939*a*), Hirai (1955*a*, *b*, 1956), H.O.T. Suomalainen (1952*a*, *b*) and Hughes-Schrader (1969). With the possible exception of *Sisyra fuscata* (Klingstedt 1937*b*), and *Boriomyia nervosa* (Suomalainen 1952*a*), the details of the sex-chromosome mechanism seem to be very uniform. The males of all the species studied are *XY*, and the two sex chromosomes do not form a bivalent at the first meiotic metaphase, but regularly lie in opposite halves of the spindle, so that they segregate from one another at first anaphase. It has been generally assumed that the two sex chromosomes are not associated during the prophase and that their extremely regular orientation at first metaphase is due to something that has been called distance pairing (compare the behaviour of the minute *m*-chromosome of the Coreidae – see p. 468). But there is no complete account of the prophase stages in any neuropteran, and it is possible, as suggested by S. G. Smith (1952) that the *X* and *Y* form a bivalent during the prophase but disjoin precociously as a result of a 'premetaphase stretch'. However, in the Mantispidae studied by Hughes-Schrader (1969) it is clear that the *X* and *Y* do not form a bivalent at any stage. In the females it has been shown by Naville and de Beaumont (1933) that the two *X*'s form a bivalent and are united by chiasmata (Fig. 16.14). The behaviour of the *X* and *Y* in the Raphidoidea seems to be essentially as in the Planipennia (Naville and de Beaumont 1936; Klingstedt 1937*b*), but the situation in the Sialoidea (Itoh 1933*a*, *b*) is not known with certainty. Certain Sisyridae (Klingstedt 1937*b*) and the hemerobiid *Boriomyia nervosa* (H. O. T. Suomalainen 1952*a*) do definitely show *XY* bivalents at meiosis; they may of course be forms in which the original sex chromosomes have undergone fusion to a pair of autosomes.

An unusual variety of sex-chromosome mechanisms occur in the order Diptera. It would appear that on several or even many occasions in the phylogeny of this order an old sex-determining mechanism has been superseded by an entirely new one, frequently embodying a new genetic principle altogether.

The Crane-flies (Tipulidae) are, by common consent, the most primitive or one of the most primitive families of Diptera. In general, Tipulidae have *X*- and *Y*-chromosomes in the male which do not pair at meiosis and are consequently univalents at first metaphase (Bauer 1931; Wolf 1941; Dietz 1956, 1959; Henderson and Parsons 1963). *Dolichopeza albipes*, *Tipula lateralis* and *T. marginata* have large sex chromosomes while in *T. flavolineata* the *X* and *Y* are very small. In all cases the *X* and *Y* regularly pass undivided to opposite poles at first anaphase as if there were some mysterious repulsion between them, even when they lie on opposite sides of the metaphase plate (Henderson and Parsons have corrected an earlier claim by John (1957) that the X and Y divide in the first anaphase of *Tipula maxima*). In spite of the non-paired condition of the X and Y at first metaphase in Tipulidae, they definitely seem to show close somatic pairing in *Dolichopeza*, at any rate.

Tipula caesia and *T. pruinosa* lack distinct sex chromosomes but have the same three pairs of autosomes as other Tipulidae (Dietz 1956; Bayreuther 1956). It seems likely that in these related species the X and Y, having first become quite small, then underwent fusion with the members of a pair of autosomes. If so, one of the three 'autosomal' pairs is in reality a pair of sex chromosomes. These two species are also aberrant in having an achiasmatic meiosis in the male (other Tipulidae show chiasmata in spermatogenesis).

In several species of tipulids there are two kinds of X-chromosomes which differ in size (Ullerich, Bauer and Dietz 1964); since all combinations are viable in *Pales ferruginea*, the difference probably resides in a heterochromatic segment. In this species XXY individuals are fertile males, thus showing that the mechanism of sex determination depends on a 'dominant Y' principle, rather than on the genic balance scheme which is characteristic of *Drosophila*.

The sex-determining mechanisms of the calliphorid flies have been studied by Ullerich (1963). All species examined showed six pairs of chromosomes, including a pair of small, partially or entirely heterochromatic elements. In most of the species this pair consists of equal-sized X's in the female and an unequal XY pair in the male. In *Calliphora erythrocephala*, *Chrysomyia albiceps* and *Chrysomyia rufifacies*, however, the small pair consists of equal-sized elements in both sexes. In *C. erythrocephala* this pair has lost its sex-determining functions which have been taken over by one of the large pairs, i.e. a pair which were formerly autosomal. Male heterogamety has been retained and it is uncertain whether the 'switch' consisted in the translocation of a sex-determining region or in the replacement of the 'old' sex-determining region by a 'new' sex-determining locus. In either case the heterochromatic chromosomes of *C. erythrocephala* must be regarded as ex-sex chromosomes. In *Musca domestica* the sex chromosomes are entirely heterochromatic and, in spite of much genetic work, no lethal or visible mutations have been found in either the X or the Y (Wagoner 1965).

In the two species of *Chrysomyia* unisexual broods occur and females are of two kinds, male-producing and female-producing. Males are of one kind only and have no influence on the sex of the offspring. Apparently daughter-producing females are heterozygous for a dominant F factor, i.e. they are Ff in constitution; son-producing females and all males are ff. The sex of the individual thus depends entirely on the genotype of the mother; the mechanism ensures the production of son- and daughter-producing females in equal numbers in the population.

The phorid fly *Megaselia scalaris* has a most peculiar system of sex determination in which any one of the three chromosomal elements in the haploid karyotype can function as a dominant male-determining chromosome if it bears at one end (always the *same* end, apparently) a factor which Mainx 1959, 1962, 1964a, b) and Burisch (1963) refer to as M, or the 'sex-realizer'. Crossing-over occurs in the male of this fly, although with a much lower frequency than in the female. The M factor may hence be transferred from one chromosome to its homologue by normal crossing-over. It may also be transferred to the tip of a non-homologous chromosome by a process which Mainx calls 'translocation' but which is probably due to 'anomalous chiasmata' of the type we have discussed earlier (p. 564); the frequency of such 'translocations' is about 0.4% when M is on chromosome 3 and 0.05% when it is on chromosomes 1 or 2.

Until recently it was believed that all Diptera showed male heterogamety of one kind or another; female heterogamety has, however, now been demonstrated in certain gall-forming Australian Tephritidae by Bush (1966b) and in the chironomid *Polypedilum nubifer* by Martin (1966).

The situation in the Tephritidae is that all members of the subfamilies Dacinae and Trypetinae that have been examined show either XY or X_1X_2Y males. The subfamilies Tephritinae and Oedaspinae, however, show an unequal pair of sex chromosomes (XY or WZ) in the females – except for *Spathulina* (*Tephritis*) *arnicae* which was stated by Keuneke (1924) to

TABLE 16.3. *Diptera with an XO:XX sex-chromosome mechanism*

Drosophilidae		
Drosophila (*Hirtodrosophila*)	orbospiracula	Clayton & Ward 1954
	longala	Clayton & Wassermann 1957
	pictiventris	
	grisea	
Drosophila (*Drsoophila*)	annulimana	Dobzhansky & Pavan 1943
	mercatorum	Wharton 1944
	(some strains)*	
Asilidae		
Dasyllis grossa		Metz 1922
Tephritidae		
Spathulina (= *Tephritis*) *arnicae*		Keuneke 1924
Ephydridae		
Scatophila unicornis		Heitz 1933

* In this case it is definitely known that the Y is fused with an autosome so that the true situation is $XYY(\male):XXYY(\female)$.

have XO males. Possibly the identification of the latter material, or the cytological work on it, were in error. In any case it appears certain that at a certain point in the phylogeny of the Tephritidae male heterogamety was superseded by female heterogamety in one lineage.

X- and Y-chromosomes without pairing segments are also characteristic of the plant bugs of the order Heteroptera, although a number of species have lost the Y-chromosome (the males being XO) and in certain families (particularly the Reduviidae, Cimicidae and Nepidae) multiple sex-chromosome mechanisms involving several different kinds of X's are of common occurrence (see Ch. 17). Except in a few special cases (where the condition is undoubtedly secondary) the X- and Y-chromosomes of the Heteroptera do not form a true bivalent at meiosis (Fig. 16.15). In some Heteroptera there is no synapsis of the X and Y at all, while in others a temporary pairing occurs, but obviously does not involve chiasma formation, since the two chromosomes separate after diakinesis and behave as separate univalents at first metaphase. With very few exceptions, the sex chromosomes of the Heteroptera divide equationally in the first meiotic division in the male, so that all second metaphases are of one type, with both an X and a Y. Usually the X and Y come together in a temporary association at second metaphase, in the middle of a ring of autosomes, and then separate to opposite poles at anaphase (postreduction). The course of events in most of the XO genera is essentially similar, i.e. the single X-chromosome divides the first meiotic divisions and segregates to one pole at the second. The details were studied in a number of genera (*Anasa, Protenor, Alydus, Pyrrhococoris*) by Wilson (1905*a, b*, 1906, 1907, 1909*a, b, c*, 1912). A few XO species of Coreidae such as *Archimerus calcarator* (Wilson 1905*a, b*, 1909*a*) and *Pachylis gigas* (Schrader 1932) seem, however, to have undergone an evolutionary reversion to the 'usual' prereductional type of behaviour in which the X passes undivided to one pole at the anaphase of the first division rather than the second.

The chromosomes of the Heteroptera are rather ill-adapted for detailed analysis of their structure, and the fact that they seem to be generally or universally holocentric means that evolutionary interpretations of their sex-chromosome mechanisms necessarily rest on rather insecure foundations. The X's and Y's seem to be heterochromatic throughout their whole length except in a few instances, where there is reason to believe that a relatively recent fusion between the original sex chromosomes and a pair of autosomes has occurred. Thus in the pentatomid *Rhytidolomia senilis*, with the very low chromosome number $2n = 6$, the X and Y are associated during the meiotic prophase, and this association is maintained up to and through first metaphase and anaphase, the X and Y dividing equationally in the first division (Wilson 1913; Schrader 1940*a*, *b*). Probably the sex chromosomes of this species are held together by a chiasma formed in a euchromatic pairing segment which has been derived from a pair of autosomes (obviously two separate fusions would be needed to cause the attachment of both the X and Y to the members of a pair of autosomes). At interkinesis and second metaphase the X and Y of *R. senilis* appear to be joined together end-to-end by 'half a terminal chiasma' (see p. 493).

Attachment of the sex chromosome pair to a pair of autosomes seems to have also taken place in one of the giant water bugs (family Belostomatidae) – another form with a very low chromosome number, i.e. one in which many fusions have established themselves. Whereas *Lethocerus americanus* has $2n = 8$ including a small X and Y which behave in the usual heteropteran

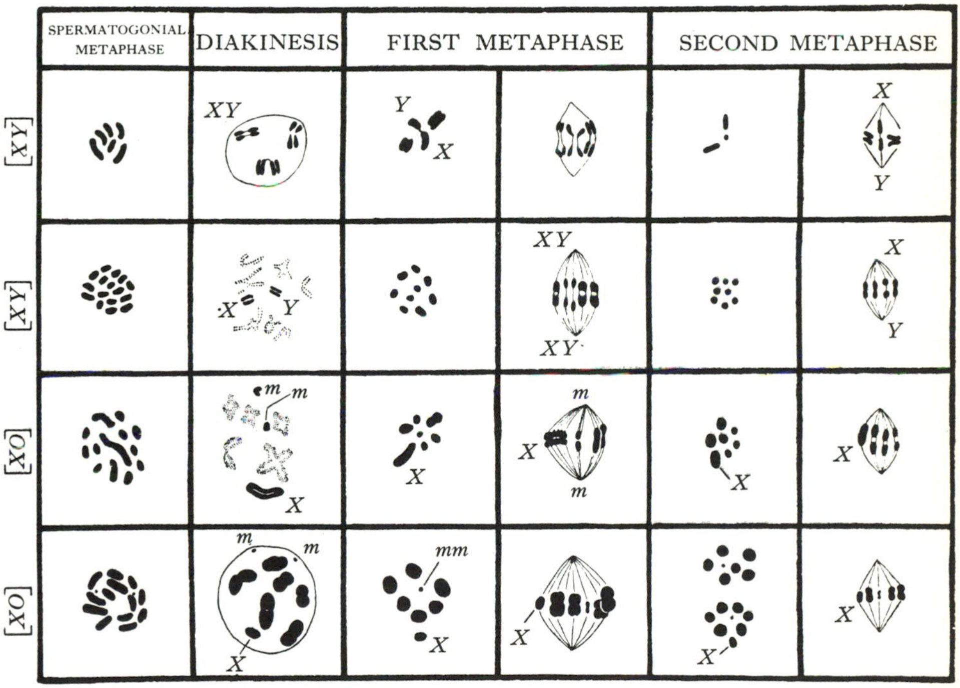

Fig. 16.15. Main stages of meiosis in the males of four representative members of the order Heteroptera. In all except *Archimerus* the segregation of the sex chromosomes takes place at the second division, all the second metaphases being alike. In *Archimerus* there are two types of second metaphases, with and without an X-chromosome. *Protenor* and *Archimerus* belong to the family Coreidae, and have *m*-chromosomes (large in *Protenor*, very small in *Archimerus*). The X and Y form a bivalent in *Rhytidolomia*, but not in *Oncopeltus*. Figures from Wilson (1905*b*, 1912) and Schrader (1940*a*), but redrawn and modified.

manner at meiosis, a related form (sibling species?) from Michigan has
$2n = 4$, there being no trace of the X and Y (Chickering 1927, 1932;
Chickering and Bacorn 1933). At meiosis two bivalents are formed, the
homologues being held together by chiasmata. Presumably what has
happened here is that the small X and Y, or at any rate the parts of them that
are genetically important, have been transferred, by translocations, to one
of the autosomal pairs (an additional fusion between two pairs of autosomes
must also have occurred, to account for the reduction in chromosome
number that has taken place).

SEX CHROMOSOMES IN BEETLES

The careful tabulations of coleopteran cytological data by Smith (1953*b*,
1960*a*), based largely on his own work and that of Virkki, Takenouchi,
Manna, Yosida, Guénin, Suomalainen and earlier workers, enables us to get
a comprehensive picture of the evolution of the sex-chromosome mechanism
in this order of insects, the sex chromosomes having been studied in over
700 species, distributed over forty-three different families (Fig. 16.16).

Six different types of sex-chromosome mechanisms were distinguished
by Smith (1953*b*); one of these ('Xy_c') was based on a misinterpretation, but
an additional category (multiple mechanisms e.g. X_1X_2Y, $X_1X_2X_3Y$, etc.)
should be added. Yet another mechanism has been discovered in a few
species by Virkki (1963, 1964). The aberrant family Micromalthidae
includes a single species with haploid males (Scott 1936, see p. 689) in which
sex chromosomes do not seem to exist.

The first category of sex chromosomes is the XO system, known in
sixteen of the forty-three families, although usually in a minority of the
species. In most instances the X undergoes prereduction, i.e. it divides in the
second meiotic division of the male, but in the fire-flies and some related
families (Lycidae, Lampyridae, Cantharidae) the X is postreductional, i.e. it
divides in the first meiotic division. The Lampyridae are, incidentally, the
only family of beetles in which a considerable number of species have been
investigated and all have had XO males.

We may infer that in the sixteen families of beetles in which XO species
are known the Y is not an essentially male-determining chromosome in the
species in which it occurs. Most probably it functions in a similar manner
to the Y of *Drosophila melanogaster*. Whether this conclusion can safely be
extended to the order Coleoptera as a whole, is, however, rather doubtful.
With the single exception of *Anthonomus bisignifer* (Takenouchi 1963) none
of about 180 species of weevils (family Curculionidae) that have been studied
have XO males. There is no reason to believe that the XO mechanism is a

primitive one in the order Coleoptera as a whole, and it is much more probable that it has arisen repeatedly from the more common XY condition by evolutionary 'loss' of the Y (such 'loss' consisting in the majority of instances of the transfer of most of its substance to the X or to an autosome, in the manner we have suggested for *Drosophila* spp.).

The second category of sex chromosomes in the Coleoptera is an XY system in which one element (assumed by Smith to be always the Y) is very much smaller than the other. At meiosis an XY bivalent is formed, the two elements being united by one arm of each to form a rod-bivalent. Presumably a single chiasma is formed in a very short pairing segment. Smith calls this an Xy_r system, the lower case letter y indicating that the Y-chromosome is indeed very minute. This type of mechanism seems quite uncommon in the order, and a good many of the reported examples should probably be referred to the next type.

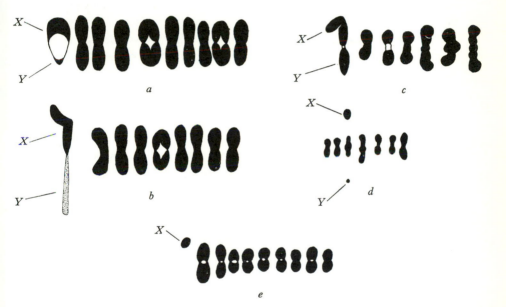

Fig. 16.16. First metaphases, in side view, of various species of beetles. (*a*) *Tribolium castaneum*; (*b*) *Tribolium confusum*; (*c*) *Arthromacra aenea*; (*d*) *Zopherus haldemani*; (*e*) *Pyractomena borealis*. *T. castaneum* shows a 'parachute type' XY bivalent, while in *T. confusum* and *Arthromacra* the X has only one limb associated with the Y (these sex chromosomes have probably arisen through some kind of translocation involving an autosome). *Zopherus* is peculiar in that the X and Y separate precociously into opposite halves of the spindle and are not associated to form a bivalent at first metaphase. *Pyractomena* is XO in the male. From S. G. Smith (1950, 1952c), redrawn.

The third category is an XY system in which the minute Y is associated by *both* its arms with the X. The result is a 'parachute' bivalent (Xy_p in Smith's terminology) in which the Y represents the aviator, the X resembling the open canopy of the parachute. This is by far the commonest type of sex-chromosome mechanism in the Coleoptera; it is recorded from twenty-nine of the forty-three families and may well occur in most of ten other families in which only one or two species have been studied cytologically. Parachute mechanisms seem to be absent, however, from the more primitive suborder Adephaga (Smith, 1960*a*). There have been two attempted interpretations of these parachute bivalents. One is that there are two terminal pairing segments in each sex chromosome, i.e. that the parachute is really a ring bivalent composed of two metacentric chromosomes of quite different sizes held together by terminalized chiasmata. The other interpretation (John and Lewis 1960*b*) assumes that there are no chiasmata, the X and Y being held together by a nucleolar formation which is supposed to occupy the space between them. No really critical evidence exists, since cruciform configurations which would indicate the existence of chiasmata have not been demonstrated in the prophase stages. Nevertheless, as in the case of the mammalian XY bivalents (see p. 592) the hypothesis of a chiasmatic association seems more plausible, in view of the very regular appearance of the parachute bivalents.

The fourth category is an XY system in which the Y is quite large and about the same size as one limb of the X. Such mechanisms are referred to as neo-XY mechanisms by Smith; most of them are probably correctly interpreted as such, and have arisen from XO systems in the same way as in the Orthoptera previously considered. But a word of caution seems in order, since mechanisms of this type could obviously arise from the second category (Xy_r) through an increase in the size of the Y by duplication of some regions, i.e. without the incorporation of any autosomal material into the sex chromosome pair. Presumed neo-XY species have been described in Elateridae (subfamily Pyrophorinae) by Piza (1958) and Virkki (1962*a*).

The fifth category is represented by those species in which the X and Y are both quite large metacentrics, indistinguishable in size and shape ('XY' of Smith). Genuine examples seem to be very rare, although the carabid *Bembidion transversale*, figured by Smith (1960*a*) is a clear case. The weevil *Ceuthorrhynchus lewisi*, studied by Takenouchi (1970) has a peculiar XY mechanism in which both ends of the metacentric X are associated with the same end of the acrocentric Y at first metaphase; Takenouchi states that the assodiation is not chiasmate and that it is not due to a nucleolus. It seems most likely that there are in reality two terminalized chiasmata and that the association is really analogous to those found in the mantids *Rhodomantis* and *Compsothespis* (see pp. 649–52).

In several genera of alticid beetles, *Alagoasa*, *Oedionychus* and *Walterian-ella*, X- and Y-chromosomes of enormous length occur (Virkki 1961, 1963, 1964, 1970). These are asynaptic in the prophase of meiosis, orientate syntelically at first metaphase on a spindle which is separate from the one to which the autosomes are attached (Fig. 16.17) and pass to opposite poles at first anaphase (in the earlier papers they were said to undergo equational division at first anaphase, but this was an error). At second metaphase the single sex chromosome (X or Y, as the case may be) is amphitelic and divides normally. Some species of the related genus *Disonycha* probably have a similar mechanism although their sex chromosomes are quite small. (Virkki

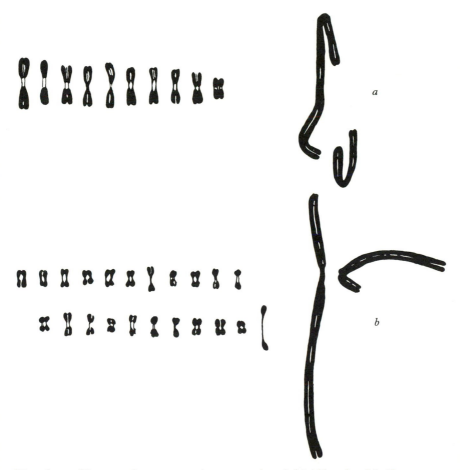

Fig. 16.17. Giant sex chromosomes in two species of alticid beetles. (*a*) *Alagoasa* sp. (2n♂ = 20 + XY); (*b*) *Walterianella venusta* (2n♂ = 44 + XY). In neither case is it known which of the giant chromosomes is the X and which the Y. From Virkki (1964).

interprets these species differently, but the above seems the most probable interpretation).

Lastly, we have several groups of beetles in which multiple sex-chromosome mechanisms have been evolved. The first are the Tiger beetles (family Cicindelidae) in which the North American species are X_1X_2Y in the males (Smith and Edgar 1954), while European ones are $X_1X_2X_3Y$ (Guénin, 1952); earlier work on this family was unreliable and should be disregarded. The second group in which multiple sex-chromosome mechanisms have been evolved is the genus *Blaps* (including *Caenoblaps*), which includes X_1X_2Y, $X_1X_2X_3Y$ and $X_1X_2X_3X_4Y$ species and the extraordinary *Blaps polychresta* in which the male is: $X_1X_2X_3X_4X_5X_6X_7X_8X_9X_{10}X_{11}X_{12}Y_1$-$Y_2Y_3Y_4Y_5Y_6$. Finally we have the coccinellid *Chilocorus stigma*, which shows an X_1X_2Y condition, other species of the genus having a neo-XY system. These cases will be discussed in the next chapter.

The primitive suborder Adephaga includes the Carabidae with approximately equal numbers of XO and neo-XY species and the Cicindelidae with their multiple mechanisms. It seems reasonable to assume that the Adephaga were primitively XO, the other mechanisms being derivative.

The situation in the much larger group of the Polyphaga is less clear. The picture is dominated by the prevalent parachute mechanism which occurs in at least twenty-nine of the thirty-nine families that have been studied cytologically. We might regard it as primitive for the suborder as a while, all the other mechanisms being ultimately derived from parachute systems by one, two, three or more structural changes. Alternatively, parachute mechanisms may have arisen repeatedly in the evolution of the Polyphaga from some other types. Possibly Xy_r mechanisms could have given rise to Xy_p ones through translocation of a minute terminal segment of the non-pairing limb of the X to the non-pairing arm of the Y (or vice versa). But if this were a possible mechanism we also have to account for a change in the centromere position from a subterminal position in the Xy_r system to a submedian one in the Xy_p one.

17

THE EVOLUTION OF SEX DETERMINATION
II. MULTIPLE SYSTEMS

Et des raffinements toujours inépuisés,
Laisse du vieux Platon se froncer l'œil austère.

CH. BAUDELAIRE: *Les Fleurs du Mal*

We have already referred to the existence of sex-chromosome mechanisms in which more than one kind of X or one kind of Y is present. Such systems may be formally classified as X_1X_2Y, $X_1X_2X_3Y$, X_1X_2O, XY_1Y_2, etc., according to the number of non-homologous (or only partly homologous) X's and Y's in the heterogametic sex.

It is, perhaps, necessary to point out that one cannot distinguish between an X_1X_2Y mechanism and an XY_1Y_2 one simply by study of the heterogametic sex of an isolated species – an investigation of the homogametic sex or of related forms is necessary before one can decide between the two possibilities.

Naturally, the only way to understand the origin of multiple sex-chromosome mechanisms is to compare them with the simpler XY or XO conditions met with in related species or genera. Most of them have clearly arisen as a result of translocations involving sex chromosomes and autosomes (Fig. 17.1). Three main types may be distinguished: (1) those due to centric fusions; (2) those due to mutual translocations between metacentric chromosomes; (3) those due to dissociations. Not all the instances that have been described can be unequivocally assigned to one of these three categories, but it is probable that the great majority will eventually be included in one or other of them, although the possibility cannot be excluded that some other types of changes (e.g. tandem fusions) have been involved in a few instances.

The multiple sex-chromosome systems which have been described may also be classified according to the type of meiotic association between the chromosomes in the heterogametic sex, and in particular according to

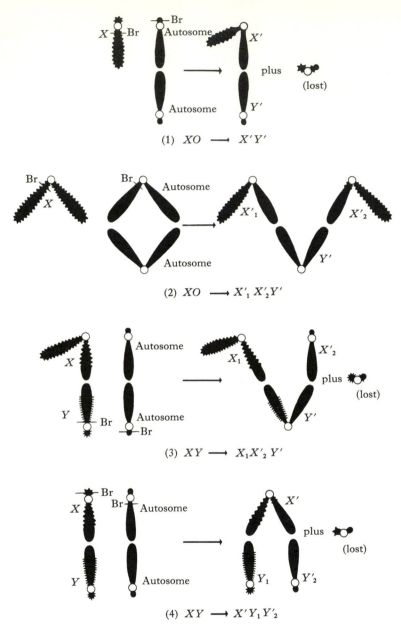

Fig. 17.1. Origin of neo-XY, X_1X_2Y and XY_1Y_2 mechanisms from the simpler XO and XY ones, by centric fusions and (in case (2)) by a translocation. Br, breaks; 'neo' X's and Y's (i.e. ones that were at least in part autosomal before the rearrangement) designated X', Y' etc. From White 1957d.

whether chiasmata are formed between the X's and Y's. Many instances that have been described are, however, somewhat difficult to interpret and in some cases it is uncertain whether chiasmata are formed or not. The effect of the meiotic mechanism must be, however, in all such cases, such as to ensure that all the X's go to one pole and all the Y's to the other, at either the first division (the more usual condition) or the second (in some Heteroptera).

MECHANISMS THAT HAVE ARISEN BY CENTRIC FUSION

One of the most straightforward cases of a multiple sex-chromosome system that has arisen by centric fusion occurs in the North American grasshopper genus *Paratylotropidia* (Figs. 17.2 and 17.3). There are three species of this genus, all of which are rare and localized, brachypterous insects. *P. beutenmulleri* (southern Appalachian Mountains) is a neo-X:neo-Y species with $2n\male = 22$ (White 1953). It has clearly acquired a fusion between the originally acrocentric X and an acrocentric autosome. Its neo-Y is an acrocentric, approximately equal in length to the XR limb, with which it must originally have been totally homologous. There are cytological indications, however, that it contains several blocks of heterochromatin which are not present in XR, so that the homology is now only partial and probably confined to small terminal pairing segments in which chiasmata can be formed.

P. brunneri (middle western states, rather widespread relict populations) and *P. morsei* (confined to the Ouachita mountains of Oklahoma and Arkansas) both have X_1X_2Y males and have $2n\male = 19$, there being a fusion between two autosomes as well as one between an autosome and the neo-Y (King and Beams, 1938; White 1953). In all three species, the number of major chromosome arms (i.e. excluding the minute second arms of acrocentric elements) is the same as in the usual acridid karyotype (23 in the male, 24 in the female). In *brunneri* and *morsei* both sexes have four metacentric chromosomes; two of these represent a pair of autosomes; the other two are the two X_1's in the female, the X_1 and the Y in the male.

At meiosis in males of *brunnei* and *morsei*, a sex trivalent is formed which consists of two metacentric chromosomes (X_1 and Y) and an acrocentric (X_2). The 'free' left limb of X_1, which does not pair with anything at meiosis, clearly represents the original X of a remote XO ancestor. The 'right' limb of X_1 is terminally associated at first metaphase with the 'left' limb of the Y and the acrocentric X_2 is similarly associated with the 'right' limb of the Y. The trivalent becomes orientated with the three centromeres in a triangle, so that X_1 and X_2 pass to the same pole at anaphase, the Y

passing to the other pole. X_1R must originally have been homologous to YL and X_2 must have been homologous to YR. The whole mechanism has clearly arisen by two centric fusions, one (present in *P. beutenmulleri*) between the X and an autosome which we may call 'A', and the other between the homologue of A (i.e. the neo-Y) and a second autosome, 'B'. Thus X_1 is composed of X and A, Y of A and B, while X_2 is the unaltered B-chromosome.

This is not the whole story, however, since at the present time homology between XR and YL and between YR and X_2 seems to be confined to quite small distal pairing segments, in each of which a chiasma is formed at meiosis and then terminalized. The much larger proximal regions of these chromosomes have evolved into new differential segments. Thus the proximal regions of YL and YR show strong positive heteropycnosis in the meiotic prophases, while those of X_1R and X_2 are mainly euchromatic. We are here confronted with the same problem already met with in the neo-X:neo-Y grasshoppers, namely the 'heterochromatinization' of the proximal regions of the Y since the origin of the new sex-chromosome mechanism. Other changes, involving increases and decreases in genetic material (i.e. duplications and deletions of short sections) have also occurred.

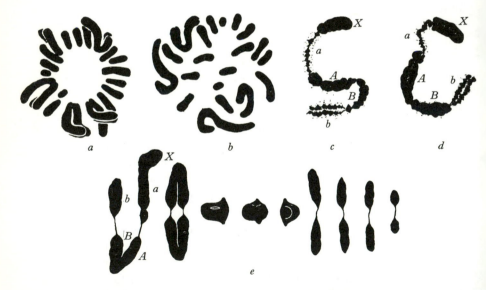

Fig. 17.2. Chromosomes of the grasshopper *Paratylotropidia brunneri*. *a*, spermatogonial metaphase (19 chromosomes); *b*, female somatic metaphase (20 chromosomes); *c*, sex trivalent at diplotene; *d*, the same at diakinesis; *e*, first metaphase in the male, seen from the side. From King and Beams 1938.

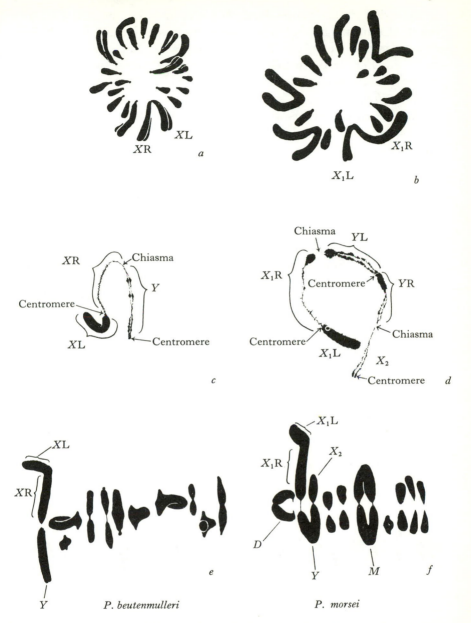

Fig. 17.3. *a*, spermatogonial metaphase of *Paratylotropidia beutenmulleri*; *b*, the same of *P. morsei*; *c*, the sex bivalent of *P. beutenmulleri*, at diplotene; *d*, the sex trivalent of *P. morsei* at diplotene; *e*, first metaphase of *P. beutenmulleri*; *f*, first metaphase of *P. morsei* (both in side view). *M*, the large metacentric bivalent, *D*, the bivalent which forms a chiasma in the short limb. From White 1953.

Somewhat similar to the *Paratylotropidia* case is that of the wingless South African grasshopper *Karruacris browni* (family Lentulidae), which has two geographic races, one XY, with $2n\male = 20$; and one X_1X_2Y, with $2n\male = 19$ (Fig. 17.4). In this case, some 'mixed' populations presumably occur where the two races come into contact but this is not definitely known (White 1967).

The sequence of events which has occurred in the evolution of the sex-chromosome mechanism in *Paratylotropidia* and *Karruacris* is known to have occurred on no less than six separate occasions in the phylogeny of the Australian eumastacid grasshoppers of the subfamily Morabinae (which number about 250 species altogether). On eleven occasions in this group an XO form gave rise to an XY one, as a result of an X-autosome fusion, and in six of these cases the process went one stage further, a second fusion (i.e. a Y-autosome one) giving rise to an X_1X_2Y system (Figs. 17.5 and 17.6). Two species groups with both XY and X_1X_2Y species have been dealt with by White and Cheney (1966, 1972).

In 5 of the 6 instances, one or more representatives of the XY stage (races or species) are still in existence and have been studied, although it cannot be assumed that the XY species are *directly* ancestral to the X_1X_2Y forms. In one instance, the X_1X_2Y form is a geographic race of an otherwise XY species, in three instances we have single X_1X_2Y species which are closely related to one or more XY species and in the remaining case we have a group of (probably) eight X_1X_2Y species with two XY relatives (the *curvicercus* group). Only in this last case can it be said that the X_1X_2Y forms have been conspicuously successful in an evolutionary sense. The other five X_1X_2Y forms of morabine grasshoppers are all apparently confined to quite small geographic areas; this may, however, be simply due to their being of relatively recent origin.

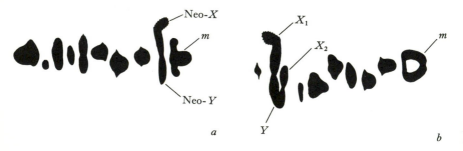

Fig. 17.4. First metaphases in the two races of the South African grasshopper *Karruacris browni*. *a*, the XY race; *b*, the X_1X_2Y race. *m*, a bivalent which is metacentric in both races.

The autosomes involved in fusions with the X and the neo-Y, in the morabine grasshoppers, were not the same ones, in the various cases. Sometimes large elements were involved, sometimes small ones. In the X_1X_2Y species trivalents are formed in the males, chiasmata being apparently restricted to short distal pairing segments. A general tendency to distal localization of chiasmata has probably favoured the occurrence of the fusions, in the species in which they are found. In some of these species structural changes of various kinds (pericentric inversions, heterochromatinization, deletions and duplications) have taken place in the sex chromosomes of the XY and X_1X_2Y species, subsequent to the occurrence of the fusions. Similar X_1X_2Y mechanisms are known in two other subfamilies of eumastacid grasshoppers – *Xenomastax wintreberti* (Pseudoschmidtiinae) from Madagascar (White 1970c) and *Erianthus guttatus* (Chorotypinae) from Malaya (White 1971b).

Fig. 17.5. Chromosomes of the grasshopper *Moraba curvicercus* (northern Australia). *a*, spermatogonial metaphase; *b* and *c*, first metaphases in side view. Note chiasmata in short limbs of some of the bivalents.

Four species of South American Acrididae (*Scotussa daguerrei*, *Eurotettix lilloanus*, *Leiotettix politus* and *Dichroplus dubius*) were shown by Mesa (1962*a*, 1963*b*) and Mesa and de Mesa (1967) to have X_1X_2Y males, although the mode of origin in these cases must surely have been the same as the other grasshopper species cited above. All four species are apparently ones of very restricted distribution. In *S. daguerrei* the pairing segments of X_1R and YL and those of X_2 and YR seem to be quite extensive, while in *E. lilloanus* they are quite minute, as in *Paratylotropidia*. *Leiotettix politus* seems to be normally a species with XY males, but one individual collected at Rio Tacuarí, Uruguay, had an additional Y-autosome fusion and an X_1X_2Y condition.

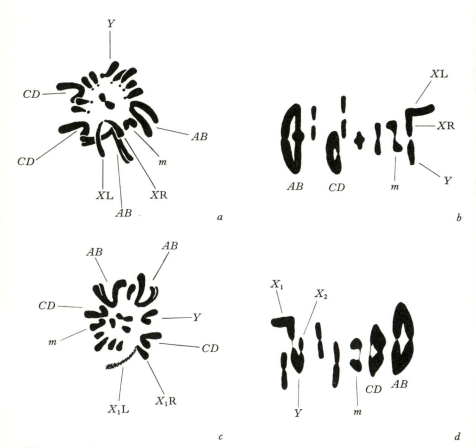

Fig. 17.6. Chromosomes of two races of an undescribed species of morabine grasshopper from the Lake Eyre region of Australia (species 'P32'). *a*, spermatogonial metaphase and *b*, first metaphase of the 'XY race'; *c*, spermatogonial metaphase and *d*, first metaphase of the 'X_1X_2Y race'. *m*, a small metacentric chromosome.

An X_1X_2Y mechanism which appears to be similar to those of *Paratylotropidia*, *Karruacris* and the morabine eumastacids was found by Dave (1965) in the Indian tettigoniid *Letana atomifera*, but in the absence of any details on the prophase stages of meiosis, it is impossible to be certain that it has arisen in the same manner, namely by two successive centric fusions, although the prevalence of acrocentric rather than metacentric autosomes in the Tettigoniidae renders this likely.

Multiple sex chromosomes of the X_1X_2Y type which form chiasmatic trivalents at meiosis have also been described in several species of Australian gryllacridid Cave crickets by Mesa, Ferreira and de Mesa (1968, 1969). Cytologically primitive species of this group have XO males, but in the genus *Cavernotettix* two species have become XY as a result of X-autosome fusions (one of these seems to have acquired five autosomal fusions as well so that the primitive chromosome number of 2n = 45 has been reduced to 2n = 34). A further three species have become X_1X_2Y in the male sex as a result of an additional Y-autosome fusion.

In the above cases, we have called the separate sex chromosomes X_1, X_2 ... in the evolutionary sequence, i.e. 'X_1' is always the original X, X_2 being a phylogenetically younger sex chromosome (i.e. one that was added to the sex mechanism later). In a number of cases to be discussed below, however, the phylogenetic history is not known, and in such cases it seems best to designate X_1, X_2 ... in order of size.

Relatively few multiple sex-chromosome mechanisms are known in beetles. The only fully-understood case is that of the coccinellid *Chilocorus stigma* (S. G. Smith 1956a, b, 1959, 1960b, 1962c) in which a fusion between the Y of an originally XY system and an autosome has produced an X_1X_2Y mechanism. At meiosis the X_1- and X_2-chromosomes are associated with the Y by means of chiasmata.

Somewhat less well understood instances of multiple sex-chromosome systems occur in the Tiger beetles (genus *Cicindela*). Four European species have an $X_1X_2X_3Y$ mechanism (Guénin 1952) while four North American species have X_1X_2Y males (Smith and Edgar 1954). The number of autosomes is the same in both cases, namely nine pairs of metacentrics. Thus, whatever the origin of the X_1X_2Y system, the X_3 was almost certainly *not* acquired by a fusion involving an autosome. In both groups of species a multivalent is formed at meiosis, but whether the association between the X's and the Y is by chiasmata is uncertain. Dasgupta's (1967) account of an Indian species which may have ten pairs of autosomes is somewhat unclear.

Even more complex sex-chromosome mechanisms exist in the tenebrionid genus *Blaps* (Table 17.1 and Fig. 17.7). The early work of Nonidez

(1915, 1920) has been followed up by that of Guénin (1948, 1949, 1953, 1956) and more recently, Lewis and John (1957c) have studied the sex multivalent in one species. *Blaps polychresta*, with twelve different kinds of X's ($X_1 \ldots X_{12}$) and six Y's ($Y_1 \ldots Y_6$) has by far the most complex sex-chromosome mechanism known. It is not entirely clear how these

TABLE 17.1 *Multiple sex chromosomes in beetles*

Families and Species	Sex chromosomes of Male	Comments
Cicindelidae		
Cicindela	$X_1 X_2 Y$	
Four North American species[1]		Nine pairs of autosomes in all species
Cicindela	$X_1 X_2 X_3 Y$	
Four European species[2]		
Coccinellidae		
Chilocorus stigma[3]	$X_1 X_2 Y$	
Tenebrionidae		
Blaps lusitanica[4]	$X_1 X_2 Y$	$2n\male = 19$
B. lethifera[4]	$X_1 X_2 Y$	$2n\male = 37$
B. waltli[5]	$X_1 X_2 X_3 Y$	$2n\male = 34$
B. mortisaga[4] and *B. mucronata*[4,6]	$X_1 X_2 X_3 Y$	$2n\male = 36$
B. gigas[4]	$X_1 X_2 X_3 X_4 Y$	$2n\male = 35$
B. polychresta[7]	$X_1 \ldots X_{12} \ Y_1 \ldots Y_6$	$2n\male = 36$
Caenoblaps nitida[8]	$X_1 X_2 Y$	$2n\male = 35$
Chrysomelidae		
Rhaphidopalpa femoralis[9]	$X_1 X_2 Y$	
Alticidae		
Phenrica austriaca[10]	$X Y_1 Y_2 Y_3 Y_4$	
Omphoita clerica[10,11]	$X Y_1 Y_2$	Interpretation
Asphaera daniela[10,12]	$X Y_1 Y_2 Y_3 Y_4 Y_5 Y_6$	not certain
A. pauperata[10]	$X Y_1 Y_2$	
A. reflexicollis[10]	$X Y_1 Y_2 Y_3 Y_4$	
Altica sp.[13]	$X Y_1 Y_2$	
Cyrsylus volkameriae[14]	$X_1 X_2 Y_1 Y_2$	
Curculionidae		
Diocalandra sp.[15]	$X Y_1 Y_2$	

[1] Smith and Edgar 1954
[2] Guénin 1952
[3] Smith 1956b, 1957a, b, 1959, 1960b, 1962b
[4] Guénin 1949
[5] Nonidez 1915
[6] Lewis and John 1957c
[7] Guénin 1953
[8] Guénin 1956
[9] Goto and Yosida 1953; Yosida 1953
[10] Virkki 1970
[11] Virkki 1967a
[12] Virkki 1968a
[13] Kasturi Bai and Sugandhi 1968
[14] Virkki 1968b
[15] Takenouchi 1969

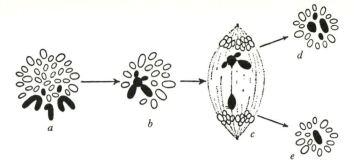

Fig. 17.7. Behaviour of the sex chromosomes in a species of *Blaps* (not *B. lusitanica*, as stated by the original author – perhaps *B. gigas*). *a*, spermatogonial metaphase; *b* and *c*, first meiotic division in polar and side view; *d* and *e*, second meiotic divisions ($X_1X_2X_3X_4$ and Y classes). Sex chromosomes black, autosomes in outline. From Wilson (1925), based on the work of Nonidez.

multiple systems have been built up but since, with the exception of *B lusitanica*, there is a progressive decrease in number of autosomes with increasing numbers of sex chromosomes (the X_1X_2Y species having 17 pairs of autosomes, the three $X_1X_2X_3Y$ species 15, 16 and 16 pairs respectively, the $X_1X_2X_3X_4Y$ species 15 pairs and the $X_1 \ldots X_{12}\ Y_1 \ldots Y_6$ species 9 pairs) it seems most likely that autosomes have become included in the sex-chromosome mechanism, but whether by centric fusion or by reciprocal translocation is not clear. Suggestions by Gates (1953) and Lewis and John (1957) that polyploidy has occurred in the evolution of the genus *Blaps* lack any sound foundation.

Various attempts have been made, without much success, to interpret the sex-chromosome multivalents which are formed in these species. Guénin (1948) believes that in *B. mortisaga* the three X's are associated with the Y by means of chiasmata. Lewis and John (1957c) on the other hand, believe that in *B. mucronata* one of the X's is associated with the Y by a chiasma, while the other two are connected to it by fusion of nucleoli. The situation in *B. polychresta* is not clear, since there does not appear to be a nucleolus, but Guénin (1953) does not believe that chiasmata are present between the eighteen sex chromosomes which form a great mass in the middle of a ring of autosomal bivalents at first metaphase. We cannot avoid a suspicion that multivalents essentially similar to those of *Rhodomantis* (see p. 650), i.e. with 'multiple chiasmata', are present in some of these species of *Blaps*, but the facts are extremely unclear.

The XY_1Y_2 mechanism in the European isopod 'superspecies' *Jaera marina*, with female heterogamety is probably one that has arisen by an

X-autosome fusion, since other fusions are known in this group (Staiger and Bocquet 1956).

About twenty-eight species of mammals are now known to possess multiple sex-chromosome mechanisms (out of about 1000 species that have been studied cytologically). In fifteen certain cases the male is XY_1Y_2. In the Rat-kangaroo or Potoroo, *Potorous tridactylus*, Sharman, McIntosh and Barber (1950) and Sharman and Barber (1952) showed that $2n\male = 13$, $2n\female = 12$. The Y_1-chromosome is very small and apparently forms a single chiasma with the shorter 'left' arm of the X. The Y_2 element which pairs with the longer 'right' limb of the X, with which it forms several chiasmata, has clearly been derived from an autosome. Sharman *et al.* interpreted this case as having resulted from a tandem fusion (rather than a centric one), i.e. they suppose that an acrocentric autosome broke at a point adjacent to the centromere and that a metacentric X broke near the distal end of its 'free' limb (i.e. the limb which did not pair with the Y). Since the limb of the X which pairs with Y_2 has an extensive proximal heterochromatic segment, the hypothesis of such a fusion is reasonable. But there is another possibility, namely that a centric fusion occurred and was followed by heterochromatinization of the proximal segment of XR. The demonstration by Hayman and Martin (1965*a*) that the short arm of one X is late-replicating in the female, while the long arm is not, seems to support the idea of a centric rather than a tandem fusion. Shaw and Krooth (1964), who have studied the somatic karyotype of the Potoroo, call the chromosome designated Y_1 by Australian authors Y_2; since the phylogenetic history is known with virtual certainty in this case, their terminology seems undesirable.

Another case in marsupials is very similar. This is the wallaby *Protemnodon bicolor*, studied by Agar (1923) and reinvestigated by Sharman (1961). Again there is a trivalent formed at meiosis, a metacentric X having one limb associated with a very small Y_1 (the original Y) while the other is associated by 2–3 chiasmata with an acrocentric Y_2. Clearly Y_2 is of autosomal origin and must be presumed to be genetically 'active'. Genetic crossing-over must occur between the X and Y_2 and partial sex linkage must be a reality in *Potorous* and *Protemnodon bicolor*, regardless of whether it is in the human species (see p. 591). Sharman has put forward reasons for believing that the multiple sex-chromosome mechanism of *P. bicolor* also arose by a tandem fusion. Since proximal chiasmata would probably be incompatible with an 'efficient' trivalent their non-formation between XR and Y_2 is expected; distal localization of chiasmata *may* have led to heterochromatinization of the proximal section of XR if Sharman is wrong in assuming a tandem fusion. The length of XR and Y_2 in mitotic metaphases looks to be approximately equal, which would argue in favour of a centric fusion; obviously a firm decision between the two hypotheses is not possible on the basis of the published evidence, but as in the case of *Potorous*, the evidence from autoradiography supports the idea of a centric fusion (Hayman and Martin 1965*a*, 1969). A third species of marsupial in which an XY_1Y_2 mechanism has been briefly reported is the Bandicoot, *Thylacomys* (= *Macrotis*) *lagotis* (Martin and Hayman 1967).

In the sibling species of European shrews *Sorex araneus* and *S. gemellus*, an XY_1Y_2 mechanism likewise exists (Bovey 1948, 1949; Sharman 1956; Ford, Hamerton and Sharman 1957; Meylan 1960, 1964). The X is a metacentric with approximately equal arms, one of which pairs with a small acrocentric Y_1 (presumably the original Y) while the other pairs with a much longer acrocentric Y_2, whose length is approximately equal to that of the arm of the X with which it pairs (two chiasmata are frequently formed between them). This certainly looks like a mechanism that has arisen by centric fusion between an acrocentric autosome and an acrocentric or nearly acrocentric X. It will be recalled that *S. araneus* shows chromosomal polymorphism for autosomal fusions (see p. 301).

An XY_1Y_2 mechanism also exists in the rodent *Gerbillus gerbillus* (Matthey 1954*b*, 1955; Wahrman and Zahavi 1955). The male has 2n = 43, the female 2n = 42. The X is a long chromosome with very unequal limbs, but the shapes of the two small Y elements are somewhat uncertain. A trivalent is formed in spermatogenesis, but its structure is not clear; possibly both Y-chromosomes are associated at first metaphase with the *same* limb of the X.

A multiple sex-chromosome was also reported by Matthey (1952*b*, 1953*a*) in two males of *Gerbillus pyramidum* from Algeria, but Wahrman and Zahavi (1955) could not find such a mechanism in three other Algerian individuals. In Matthey's animals about half the first metaphases had the XY bivalent free, while in the remainder the X had one limb (XL) terminally associated with the Y and the other (XR) connected by a chiasma with an autosomal bivalent, so as to form a chain of four chromosomes. Matthey believes the autosomal pair has taken part in a mutual translocation with the X_1, small terminal segments being interchanged. It is not clear whether the four centromeres in the chain-quadrivalents undergo regular alternate disjunction at first anaphase or not.

There seem to be several possibilities in this case: (1) Matthey's hypothesis according to which the autosomal pair is heterozygous for a translocation is correct; in that case some individuals homozygous for the two alternative forms of this chromosome would be produced, and some of these would be males, others females (it is, of course, uncertain whether they would be viable). The second possibility (2) is that *G. pyramidum* has the autosomal pair in question homozygous for the distal segment of XR (i.e. this region is trisomic in males, tetrasomic in females). No decision between the various possibilities can be reached on the published evidence. Formally, this could perhaps be called an $X_1 X_2 Y_1 Y_2$ mechanism, but if 'X_2' and 'Y_2' are structurally identical (as the second hypothesis proposes) such a nomenclature would be unjustified. In fact, it is a little difficult to decide whether this should really be called a complex sex-chromosome mechanism; it may simply represent a translocation in a small local population. In any case it clearly has nothing to do with the genuine complex sex-chromosome mechanism of $G.$ $gerbillus$.

A total of six species of American bats belonging to the family Phyllostomidae have been shown by Hsu, Baker and Utakoji (1968), Yonenaga, Frota-Pessoa and Lewis (1969) and Kiblisky (1969a) to possess $XY_1 Y_2$ sex-chromosome mechanisms. This is a family in which very numerous chromosomal rearrangements have established themselves, as can be seen from the fact that the chromosome numbers range from $2n = 46$ down to $2n = 16$, probably as a result of a large number of different centric fusions.

It seems likely that $XY_1 Y_2$ mechanisms have arisen three times in this group, in the genera $Choeroniscus$ (one $XY_1 Y_2$ species), $Carollia$ (two $XY_1 Y_2$ species) and $Artibeus$ (three $XY_1 Y_2$ species). However, the possibility that the multiple sex-chromosome mechanisms of $Choeroniscus$ and $Carollia$ have a common evolutionary origin is not entirely ruled out by the evidence. Since the X-chromosomes are metacentric or J-shaped in these genera, one may probably conclude that the multiple sex-chromosome mechanisms arose by tandem fusions of small acrocentric autosomes to an already metacentric X. In the genus $Artibeus$ one species has an XY mechanism which is probably secondary, the single Y having arisen by fusion of Y_1 and Y_2. In a later paper, Baker and Hsu (1970) have reported $XY_1 Y_2$ mechanisms in two additional species of phyllostomid bats, $Enchisthenes$ $harti$ and $Ametrida$ $centurio$.

An $X_1 X_2 Y$ mechanism has been found by Matthey (1965b) in material of the complex species Mus $minutoides$ from Southern Rhodesia. Two different kinds of XY mechanisms are known in this group – a 'primitive' type in which both X and Y are acrocentric and a derived type in which both the original sex chromosomes have undergone centric fusion with the members of an autosomal pair. The Southern Rhodesian race have apparently got the Y-autosome fusion but not the X-autosome one.

Two species of antelopes, studied by Wurster and Benirschke (1968b) and by Wallace and Fairall (1968) likewise seem to have $X_1 X_2 Y$ males. The Indian Muntjac deer ($Muntiacus$ $muntjak$), with the unusually low chromosome number of $2n\male = 7$, $2n\female = 6$, is XY_1Y_2 in the male (Wurster and Benirschke 1970).

Seven species of mongooses of the genus *Herpestes* have X_1X_2Y sex-chromosome mechanisms (Fredga 1965a, b, 1970). The Y-autosome translocation responsible for this system is presumably very ancient. The sizes and shapes of the three sex chromosomes differ slightly in the various species, but the Y is always an acrocentric or subacrocentric. A sex trivalent is formed in spermatogenesis. The Marsh mongoose *Atilax paludinosus*, which also has $2n\male = 35$, $2n\female = 36$, probably has a similar mechanism (Wurster and Benirschke 1968a). We have already described the peculiar multiple sex-chromosome mechanism of the Two-toed sloth (see p. 604). The bat species *Mesophylla macconnelli* has $2n\male = 21$, $2n\female = 22$ and may have a sex-chromosome mechanism of the same general nature (Baker and Hsu, 1970).

Finally a peculiar X_1X_2Y mechanism has been reported in the Hare-wallaby *Lagorchestes conspicillatus* (Martin and Hayman 1966). Although not all the details are clear, it seems likely that there has been an X-autosome fusion, a fusion of the Y to the homologous autosome, and a third fusion between the neo-Y so formed and another autosome. It is perhaps significant that one individual was found to be in addition heterozygous for a fusion between two autosomes; clearly there is a special tendency for fusions to establish themselves in the species, which has only six pairs of autosomes by comparison with ten pairs in *L. hirsutus*.

One may ask why XY_1Y_2 systems should have established themselves in mammals so frequently, since it is the X_1X_2Y type that seem to have been successful in most of the invertebrate groups. It has been suggested (White 1960) that because of the important sex-determining properties of the Y in the Mammalia, small portions of the X may be more dispensible than similar regions of the Y, and that the X is hence more likely to be successfully involved in fusions than the Y.

A number of species of reptiles are now known to possess multiple sex-chromosome mechanisms, and these are all systems which have probably arisen by chromosome fusions. Gorman and Atkins (1966) showed that three species of lizards belonging to the *Anolis bimaculatus* group from the West Indies had an X_1X_2Y mechanism in the males, females being $X_1X_1X_2X_2$. The minute sex chromosomes form a trivalent at meiosis. *A. biporcatus* from Panama seems to have a similar mechanism, but the sex chromosomes are larger. A multiple sex chromosome exists also in the Indian snake *Bungarus caeruleus*, with female heterogamety; females are X_1X_2Y ($=Z_1Z_1W$), males $X_1X_1X_2X_2$ ($=Z_1Z_1Z_2Z_2$) (Singh, Sharma and Ray Chaudhuri 1970).

MECHANISMS THAT HAVE ARISEN BY MUTUAL TRANSLOCATION

Having described a number of instances in which X_1X_2Y or XY_1Y_2 mechanisms have arisen from XY systems by a fusion, we may proceed to consider some cases in which a direct transformation from the XO to the X_1X_2Y condition has taken place by a mutual translocation between a metacentric X and a metacentric autosome, without loss of any genetic material (White 1940b). The classic example is that of certain Mantoidea (Praying mantids). In twenty-four genera of these insects, the great majority of which certainly form a monophyletic group (the Mantinae *sensu stricto*) an X_1X_2Y condition is found (Table 17.3), while in a further forty genera (Table 17.2), an XO condition exists in the male sex. Oguma (1946) called the XO group the Monosomata, the X_1X_2Y group the

TABLE 17.2. *Genera of Mantodea with simple sex-chromosome mechanisms* (XO *males*)

Subfamily and Genus	No. of species investigated	Distribution	2n♂
Perlamantinae			
Amorphoscelis	1	Asia	33
Cliomantis	2	Australia	25
Eremiaphilinae			
Humbertiella[1,2,14,16]	4	Asia	23, 31, 39
Didymocorypha[1,14,16]	2	Asia	15, 17
Mantoida[3]	1	Central America	37
Amelinae			
Ameles[4,5,6,7]	4	Mediterranean region	25, 29*
Apteromantis[8]	1	Africa	29
Ligaria[13]	1	Africa	23
New genus near *Cimantis–*			
Eumantis[16]	1	Asia	25
Schizocephalinae			
Angela[9]	1	Central America	19
Schizocephala[1]	1	Asia	27
Thespinae			
Hoplocorypha[13]	2	Africa	29*, 35
Parathespis[16]	1	Asia	31
Miopteriginae			
Promiopteryx[10]	1	South America (Trinidad)	19
Pseudomiopteriginae			
Pseudomiopteryx[3]	1	Central America	17
Tarachodinae			
Antistia	1	Africa	27, 29

TABLE 17.2.—*continued*

Subfamily and Genus	No. of species investigated	Distribution	2n♂
Oligonicinae			
Oligonyx[3]	1	Central America	17
Thesprotia[10]	2	North America (Florida)	23
Liturgousinae			
Liturgousa[3]	3	Central America	17, 23, 33
Caliridinae			
Leptomantis	1	Asia	39
Photininae			
Iris[4]	1	Europe	25
Brunneria[11]	1	North America	(2n = 28)
Epaphroditinae			
Acanthops[3]	2	Central America	19
Pseudocanthops[3]	1	Central America	19
Hymenopodinae			
Creobroter[1,14,16]	1	Asia	27
Harpagomantis[13]	1	Africa	25
Evantissa[14]	1	Asia	33
Hestiasula[14]	1	Asia	27
Toxoderinae			
Toxomantis[1]	1	Asia	27
Empusinae			
Empusa[4,14,16]	2	Europe, Asia	27
Gongylus[4]	1	Asia	27
Aconthiothespinae			
Acontiothespis[3,4]	3 ?	Central America	15
Tithrone[3]	1	Central America	15
Mantinae (sens. lat.)			
Miomantis[4]	1	Africa	15
Callimantis[12]	1	Antilles	17
Bisanthe[13]	1	Africa	21
Haldwania[14]	1	Asia	15
Vatinae			
Aethalocroa[1]	1	Asia	29
Iridopteryginae			
Bolbe[13]	1	Australia	26
Nanomantis[15]	1	Australia	36

* Chromosome number variable in some species due to polymorphism for centric fusions.

[1] Oguma 1946
[2] Hughes-Schrader 1948*b*
[3] Hughes-Schrader 1950
[4] White 1941*a*
[5] Wahrman 1954*a*
[6] Wahrman 1954*b*
[7] Wahrman and O'Brien 1956
[8] Matthey 1949
[9] Hughes-Schrader 1943*b*
[10] Hughes-Schrader 1953*b*
[11] White 1948*b*
[12] White 1938
[13] White 1965*b*
[14] Gupta, M. L. 1964
[15] White (unpub.)
[16] Gupta, M. L. 1966

TABLE 17.3. *Genera of Mantodea with X_1X_2Y males*

Genus	No. of species investigated	Distribution	2n♂
Compsothespis[3]	2	Africa	23
Mantis[1,2,3]	2	Europe, Australia	27
Statilia[4]	1	Asia	27
Hierodula[4,5]	4	Asia	27
Sphodropoda[3]	2	Australia	27
Sphodromantis[6,7]	4	Africa	27, 23
Polyspilota[3]	1	Africa	27
Tenodera[6,8]	3	Asia, Australia	27
Paratenodera[1,6]	1	Asia	27
Stagmomantis[1,9,10]	2	North America	27
Stagmatoptera[10]	1	Central America	27
Phyllovates[10]	1	Central America	27
Vates[10]	1	Central America	27
Oxyopsis[11]	1	South America	27
Tauromantis[10]	1	Central America	27
Antemna[10]	1	Central America	27
Melliera[10]	1	Central America	27
Archimantis[3]	2	Australia	27
Rhodomantis[12]	2	Australia	27
Orthodera[3,12]	2	Australia	25
Orthoderina[14]	1	Australia	25
Choeradodis[9,10]	1	Central America	31
Cheddikulama[13]	1	Asia	27
Deiphobe[15]	1	Asia	27

[1] King 1931
[2] Callan and Jacobs 1957
[3] White 1965b
[4] Oguma 1946
[5] Asana 1934
[6] White 1941a
[7] Wahrman (*in litt.*)
[8] Oguma 1921
[9] Hughes-Schrader 1943b
[10] Hughes-Schrader 1950
[11] Hughes-Schrader 1953b
[12] White 1962
[13] Gupta, M. L. 1964
[14] White (unpublished)
[15] Gupta, M. L. 1966

Trisomata. However, there are strong reasons for believing that an X_1X_2Y condition may have arisen independently in (1) *Compsothespis*, (2) *Orthodera*, (3) *Choeradodis*, (4) a common ancestor of the remaining eighteen genera (Mantinae *sensu stricto*). Thus use of the term Trisomata in a taxonomic sense seems undesirable.

In all the XO species that have been studied the X is metacentric and in the great majority of them all the autosomes are also metacentric. Thus it seems fairly certain that the conversion of the XO mechanism into the X_1X_2Y one took place directly, without an intermediate XY stage, by a reciprocal translocation between the X and an autosome, both chromosomes being metacentric and the breakage points being situated very close to the

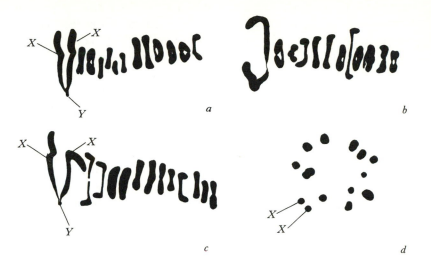

Fig. 17.8. First metaphases in an African mantid, *Polyspilota* sp. *a–c*, side views; *d*, a polar view, with the *X*-chromosomes in optical section. In *b*, the sex trivalent is not properly orientated. X_1 and X_2 indistinguishable in this species, hence both labelled *X*. From White 1965*b*.

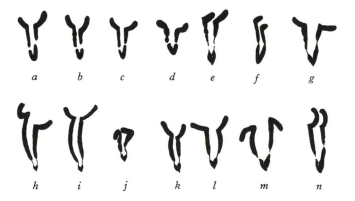

Fig. 17.9. The X_1X_2Y sex trivalent in fourteen species of mantids (first metaphase in side view). (*a*) *Sphodromantis gastrica*; (*b*) *Sphodromantis viridis*; (*c*) *Paratenodera sinensis*; (*d*) *Tenodera aridifolia*; (*e*) *Hierodula* sp.; (*f*) *Mantis religiosa*; (*g*) *Choeradodis rhombicollis*; (*h*) *Melliera brevipes*; (*i*) *Phyllovates tripunctata*; (*j*) *Vates pectinicornis*; (*k*) *Stagmomantis carolina*; (*l*) *Stagmomantis heterogamia*; (*m*) *Tauromantis championi*; (*n*) *Stagmatoptera septentrionalis*. From White 1941*a* and Hughes-Schrader 1950.

centromeres (Fig. 17.1). This probably occurred three or four times in the phylogeny of the mantids (there is a possibility that *Choeradodis* is only an aberrant member of the Mantinae). Various groups such as the New World Vatinae, formerly assigned to separate subfamilies, are $X_1 X_2 Y$ and are clearly only morphologically aberrant members of the Mantinae. Conversely, certain genera such as *Miomantis* (*Calidomantis*) and *Callimantis*, formerly regarded as belonging to the Mantinae, are XO in the males; they should either be excluded from the Mantinae, or regarded as primitive Mantinae lacking the X-autosome translocation.

The Orthoderinae and Choeradodinae are small subfamilies, each including only a few species. The Mantinae *sensu stricto* (i.e. excluding the XO genera) are, however, a very large group, now distributed over Africa, Southern Europe, Asia, Australia and North and South America; there are at least 300 species altogether. We may regard this group as strictly monophyletic in the sense of having been derived from a single species in which the transformation of the sex-chromosome mechanism from the XO to the $X_1 X_2 Y$ one took place. There is not the slightest evidence that the latter system arose by successive fusions in the mantids as it has done in *Paratylotropidia* and the Morabinae, and the generally metacentric nature of mantid chromosomes renders such a mode of origin highly unlikely.

The Y of the $X_1 X_2 Y$ mantids is thus clearly a neo-Y, originally an autosome, but now confined to the male line. The three sex chromosomes in the males form a trivalent at meiosis, being generally associated terminally at meiosis. In certain species, however, minute enlargements or lateral lugs are present at the point of junction of X_1 and X_2 with the arms of the Y, indicating the presence of chiasmata which are not quite terminalized (Callan and Jacobs 1957).

If we call the six chromosome arms in the trivalent X_1L, X_1R, YL, YR, X_2L and X_2R, then the distal ends of X_1R and X_2L pair with the distal ends of YL and YR respectively. X_1L and X_2R remain free and unpaired; they clearly represent the two limbs of the original X-chromosome and betray their origin by manifesting a special heteropycnosis at various stages (negative in spermatogonial divisions, positive in the meiotic prophases). The Y may also be partly heterochromatic, at least in some species, and the three sex chromosomes have a peculiar tendency to lag behind the autosomes in spermatogonial anaphases. No observations have been made on the meiosis of the $X_1 X_1 X_2 X_2$ females, but it must be presumed that two sex bivalents ($X_1 X_1$ and $X_2 X_2$) are formed in oogenesis.

The relative lengths of the arms of X_1 and X_2 vary to some extent throughout the Mantinae (Hughes-Schrader 1950), indicating that other changes have taken place in these chromosomes since the original translocation. It is, in fact, not possible to homologize the X_1 and X_2 elements throughout the Mantinae, although in a fair number of the species examined it seems possible to recognize a shorter element with approximately equal limbs ('X_2') and a slightly longer one with the free limb ('X_1L') shorter than the one which pairs with the Y ('X_1R'). In a species of *Sphodromantis* from Kenya studied by Henderson (1965) X_2L is very short, so that X_2 is practically acrocentric.

The neo-Y is, however, much the most variable of the three sex chromosomes. Relatively large in *Mantis religiosa*, *M. octospilota*, *Choeradodis rhombicollis*, *Orthodera ministralis*, *Archimantis* spp., it is minute in *Melliera brevipes*, *Phyllovates punctata* and an unidentified species of *Polyspilota* from South Africa (Figs. 17.8 and 17.9). Even in the single genus *Hierodula*, four species have Y's of very different sizes. Originally the length of the neo-Y must have equalled the combined length of X_1R and X_2L, but if it has become largely subinert genetically since it has been confined to the male line, it is not surprising that it should have become reduced in size, through deletions, in many of the present-day forms. It still performs an essential mechanical function, however, since by pairing with X_1 and X_2 at meiosis, it ensures that they regularly pass to opposite poles at first anaphase and hence behave as a unit in heredity. It seems to be significant that in the three species mentioned above which have minute Y's, the X_1R and X_2L limbs are unusually long. Two possible explanations of this correlation may be suggested. In the first place, coexistence of long pairing limbs in the X's and the Y is probably impossible, for mechanical reasons (with chiasmata confined to minute terminal segments, the distance between the X-centromeres, on the one hand, and the Y-centromere, on the other, would approach or exceed the maximum permitted by the spindle dimensions). Thus if multiple interstitial duplications have established themselves in the X_1R and X_2L limbs of certain mantid species, deletions of equivalent sections from the Y may have been a *sine qua non*. However, another possibility exists, namely that in these cases we are not concerned with duplications and deletions, but with direct transfers of material to X_1R and X_2L, from the corresponding limbs of the Y, by translocation. In this case the breakage points would have been such that a long segment of the Y was on each occasion exchanged for a short region of X_1 or X_2. It does not seem possible to decide which of these two explanations (the duplication-deficiency hypothesis and the translocation one) is the correct one, in *Melliera*, *Phyllovates* and *Polyspilota* sp.

With only six exceptions the thirty-five X_1X_2Y species of mantids have $2n\male = 27$. The exceptions are *Composothespis anomala* with $2n\male = 23$, two species of the Australian genus *Orthodera* and one of *Orthoderina*

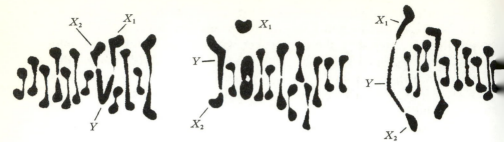

Fig. 17.10. First metaphases of the mantid *Orthodera ministralis*. The one on the left is 'normal', the middle one has an $X_2 Y$ bivalent and a univalent X_1, the one on the right has the sex trivalent malorientated on the spindle. From White 1962*a*.

with $2n\male = 25$ and *Choeradodis rhombicollis* with $2n\male = 31$ (which may not be members of the Mantinae) and one undoubted member of the Mantinae, the Algerian race of *Sphodromantis viridis* ($2n\male = 23$). Palestinian material provisionally assigned to the latter species has $2n\male = 27$ and so does the South African *Sphodromantis gastrica* (White 1941*a* and later observations) so that the change of chromosome number in the Algerian form (by fixation of two fusions, presumably) is probably fairly recent. The almost complete uniformity of chromosome number in the Mantinae *sensu stricto* is a strong argument in favour of their being a strictly monophyletic group, descended from a single population in which the X-autosome translocation established itself; since the chromosome numbers of the XO forms are much more variable (from $2n\male = 15$ to 39).

It is not possible to determine the antiquity of the $X_1 X_2 Y$ mechanism in the Mantinae, but in view of the many genera and species included in the group, their wide distribution in both the Old World and the New, and the considerable morphological divergence between some of them, the translocation which converted an XO species into the ancestor of all the $X_1 X_2 Y$ forms must have taken place a very long while ago – perhaps in early Tertiary times. If the $X_1 X_2 Y$ mechanisms of the Orthoderinae and *Choeradodis* arose independently, they undoubtedly did so in the same manner as the Mantinae, and possibly at a later date. Gupta (1966) has reported that one species of *Deiphobe* has $2n\male = 27$ and an $X_1 X_2 Y$ system, while another has $2n\male = 19$ and an XO mechanism; no further details are given and one must suspect an error of some kind unless an $X_1 X_2 Y$ system evolved quite independently in this genus.

The $X_1 X_2 Y$ mechanisms of many mantid species are less than 100% mechanically efficient. In a significant number of cells in the first metaphase stage either X_1 or X_2 is unpaired, so that instead of a trivalent we have a bivalent and a univalent (White 1941a, 1965b; Callan and Jacobs 1957). Such cells might be expected to give rise in about 50% of cases to aneuploid sperms which would kill the eggs they fertilized. However, Callan and Jacobs have shown that cells containing univalents become blocked in first metaphase and eventually degenerate without forming sperms. Their work was carried out on *Mantis religiosa*, but exactly the same situation occurs in *Orthodera* sp. (Fig. 17.10). It is the *presence* of a univalent which inhibits anaphase movement rather than the absence of a trivalent, since autosomal univalents, which are occasionally observed, also exert the same inhibitory effect. Presumably, the mechanism of first anaphase separation is very delicately poised in mantids, so that inefficiency of the sex-trivalent mechanism is not strongly opposed by natural selection. Callan and Jacobs have made the interesting suggestion that the fact that no supernumerary chromosomes have been recorded in any mantid species is a consequence of their easily-inhibited anaphase mechanism. In nine species of morabine grasshoppers with X_1X_2Y mechanisms failure of pairing has hardly ever been observed in thousands of cells examined, so that chiasma formation in the pairing segments is presumably extremely regular and, in a genetic sense, highly canalized. Possibly in this group, the presence of univalents if they occurred, would not inhibit anaphase separation. Failure of pairing does, however, sometimes occur in the sex-chromosome system of *Paratylotropidia brunneri* and in the South American *Eurotettix lilloanus* (Mesa 1962a) and is quite frequent in the Eumastacid *Erianthus guttatus* (White 1971b).

It is clear that failure of pairing at metaphase in mantids is due to breakage of the trivalent during the premetaphase stretch; but whether this results from breakage of a terminal chiasma during the premetaphase stretch or from failure of chiasma formation at an earlier stage is not known.

In many species of X_1X_2Y mantids a second type of apparent 'inefficiency' of the sex-chromosome trivalent occurs, namely rather frequent malorientation of the trivalent at first metaphase, X_1 being close to one pole and X_2 to the other. It is rather uncertain how many such malorientations lead to the formation of aneuploid cells. Many of them occur at premetaphase but are probably 'corrected' later. Where malorientation persists it probably inhibits anaphase separation, in the same way as univalents do, so that aneuploid sperms are not produced. Malorientation of trivalents is extremely rare in the X_1X_2Y morabines.

In one genus of Mantinae, the Australian *Rhodomantis*, a further and fundamental change has occurred in the X_1X_2Y mechanism (Fig. 17.11). In the two species of this genus that have been studied (there are several more that have not been examined cytologically) the X_1R and X_2L limbs are terminally associated with the *same* limb of the Y at first metaphase, the other limb being free (White 1962, 1965*b*). Presumably the distal segments of X_1R, XL and X_2L are all homologous, so that two chiasmata can be formed between them, holding the three ends together. The males would be trisomic for this segment, the females tetrasomic for it. The condition probably arose from the usual type of X_1X_2Y system by a translocation between the distal ends of X_1R and X_2L, with survival of one of the original chromosomes and one of the translocated ones. There

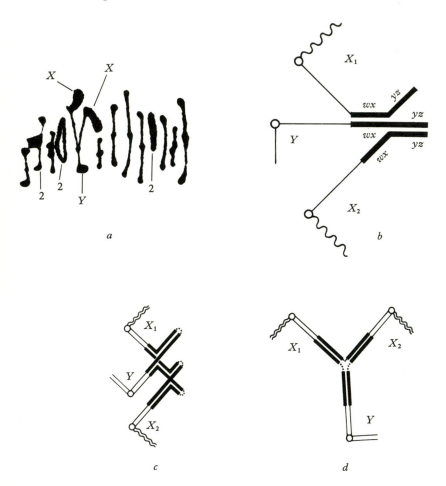

Fig. 17.11. The chromosomes of *Rhodomantis pulchella*. *a*, first metaphase in side view; X_1 and X_2 not distinguishable, hence both labelled X. *b*, presumed mode of zygotene pairing of the sex trivalent, with a triple synaptic segment; *c*, presumed diplotene configuration with two chiasmata; *d*, first metaphase, after terminalization of the chiasmata. From White 1962*a*.

is, however, the alternative possibility that the translocation which gave rise to the *Rhodomantis* mechanism was between the pairing limb of one of the X's (X_1R or X_2L) and the *non-corresponding* limb of the Y. The *Rhodomantis* mechanism is a very 'efficient' one, failure of pairing between one of the X's and the Y being exceedingly rare, by comparison with most of the Trisomata. It is hence possible that the translocation was adaptive if it resulted in the disappearance of a pairing segment that had become so inefficient as to reduce fecundity to some extent.

The X_1X_2Y mechanism of two species of the South African genus *Compsothespis* is likewise quite different from the usual type. The three sex chromosomes are all metacentric and all form a trivalent in the male, but 'X_2' has *both* ends associated (apparently by chiasmata) with 'YR' (Fig. 17.12). Presumably a triple pairing segment is present, as in *Rhodomantis*, but in this instance it consists of the terminal segments of X_2L, X_2R and YR (Fig. 17.13). Such a system could have arisen by a translocation between the tips of XR and X_2R in a 'conventional' X_1X_2Y mechanism that probably arose quite independently of the one in the Mantinae *sensu stricto* (to which *Compsothespis* does not seem closely related).

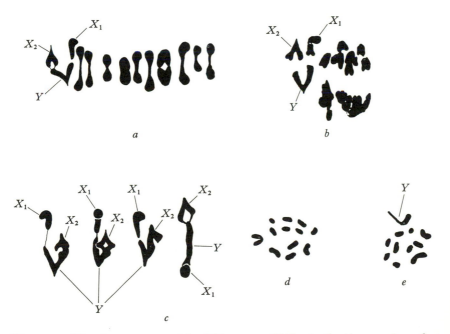

Fig. 17.12. The chromosomes of the African mantid *Compsothespis anomala. a*, first metaphase in side view; *b*, first anaphase; *c*, the sex trivalent at first metaphase from four different cells; *d* and *e*, second metaphases. From White 1965*b*.

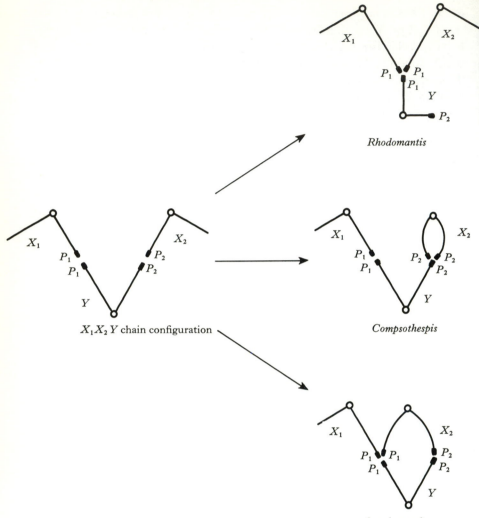

Fig. 17.13. Presumed mode of origin of the *Rhodomantis* and *Compsothespis* sex-chromosome mechanisms (and of a third, still hypothetical, mechanism of the same general type) from the usual X_1X_2Y chain configuration. P_1 and P_2, pairing segments of the original mechanism. In *Rhodomantis* X_1 and X_2, as a result of a translocation both carry the same pairing segment (assumed here to be P_1). In *Compsothespis* both ends of X_2 are assumed to be carrying P_2. It is not implied that *Rhodomantis* and *Compsothespis* had a common X_1X_2Y ancestor – the latest common ancestor of these two genera was undoubtedly an XO species. From White 1965b.

There are various indications that even in mantid species which normally form a chain of three sex chromosomes at meiosis, the tips of X_1R and X_2L are partly homologous. Thus White (1965b) has figured an X_1X_2 bivalent in a diploid spermatocyte of *Polyspilota* sp. and Henderson (1965) has seen such bivalents in tetraploid spermatocytes (in an otherwise diploid individual of *Sphrodromantis* sp.). He likewise observed two X_1YYX_2 chains, proving that a similar homology exists between the two ends of the Y. Such homologies (probably confined to very minute segments) are presumably analogous to those that have been demonstrated in *Triturus* hybrids by Callan and Spurway (1951) and in various grasshoppers (see p. 564). They should not be accepted as evidence for Henderson's far-fetched and improbable speculation that the Y of the Mantinae was originally an isochromosome (see p. 205).

Several species of fleas possess multiple sex-chromosome mechanisms in the males (Table 17.4), but it is uncertain whether these have arisen by fusions or reciprocal translocations. Two species seem to have XY males, one has X_1X_2Y males and the third shows a chain of four chromosomes at meiosis which has been regarded by Bayreuther as an $X_1Y_1X_2Y_2$ quadrivalent (in the female meiosis separate X_1X_1 and X_2X_2 bivalents are formed). The latter mechanism, if it has been correctly interpreted, is likely to have arisen by translocation. This species is also polymorphic for deletions or duplications in some of its autosomes (see p. 305) and shows supernumerary chromosomes in many individuals (see p. 326).

In all probability the X_1X_2Y mechanism of the mecopteran *Boreus brumalis* (Cooper 1951) has arisen by an X-autosome translocation of the mantid type. The evidence, while not conclusive, is twofold: (1) the only other Mecoptera whose spermatogenesis has been studied have XO males (see p. 615); (2) all the chromosomes of *B. brumalis* seem to be metacentric, so that an origin by centric fusion is unlikely.

TABLE 17.4. *Cytology of fleas* (*Aphaniptera*)

Species	2n♂	2n♀	Sex chromosomes
Leptopsylla musculi[1]	22	?	$XY:XX$ (?)
Ctenocephalus canis[2]	14 (in embryo of unknown sex)		$XY:XX$ (?)
Xenopsylla cheopis[3,4]	17	18	$X_1X_2Y:X_1X_1X_2X_2$
Nosopsyllus fasciatus[3,4]	20 to	27*	Probably $X_1Y_1X_2Y_2:X_1X_1X_2X_2$

* Supernumeraries probably present.
[1] Karnkowska 1932 [3] Bayreuther 1954
[2] Kichijo 1941 [4] Bayreuther 1969

In certain Australian ixodid ticks of the genus *Amblyomma* Oliver (1965*a*) and Oliver and Bremner (1968) have described X_1X_2Y mechanisms which may have arisen directly from XO systems by a single translocation; but the evidence is inadequate for a definite conclusion. The primitive condition in this group has almost certainly $XO:XX$. But populations of *A. moreliae* from near Brisbane have XY males (2n♂♀ = 20); while material from near Sydney has X_1X_2Y males (2n♂ = 21, 2n♀ = 22). Oliver and Bremner believe that the XY and X_1X_2Y conditions arose independently from the ancestral XO system (and that the XY condition did not give rise to the X_1X_2Y one).

MULTIPLE SEX-CHROMOSOME MECHANISMS WITHOUT CHIASMATA

In all the above-mentioned cases of multiple sex-chromosome mechanisms a multivalent is formed at meiosis, in which chiasmata are obviously or presumably formed in certain homologous pairing segments and play an essential role in regularizing disjunction at the first meiotic division. In numerous instances of multiple mechanisms, however, chiasmata appear to be entirely absent, disjunction of X's and Y's being nevertheless quite regular. We shall consider first of all a case in crickets.

A multiple sex-chromosome mechanism has been described by Piza (1964*a*), Claus (1956) and Mesa and Bran (1964) in the South American Cricket *Eneoptera surinamensis*. Piza interpreted this as an XY_1Y_2 system, but Claus and Mesa and Bran have shown that he miscounted the chromosomes of the female and that it is actually an X_1X_2Y mechanism. The chromosome number appears to be 2n♂ = 11, 2n♀ = 12, but one pair of chromosomes are minute dots close to the limit of resolution of the microscope, and it is not certain that they are really constant members of the karyotype (Fig. 17.4). The major chromosomes include three pairs of metacentric autosomes, a metacentric X_1, a somewhat larger metacentric Y and an acrocentric X_2. At diplotene–diakinesis the three chromosomes are not connected, but at the first metaphase the two X's regularly lie in one half of the spindle, the Y in the other, so that at anaphase the X's and the Y pass to opposite poles. It is the absence of any definite pairing mechanism and the fact that no sex trivalent is formed which renders this case so strikingly different from those we have previously discussed. We shall deal later (p. 662) with the general problem of the mechanism involved in the meiotic orientation of complex sex-chromosome systems without chiasma formation. Apparently a multiple sex-chromosome mechanism exists also in an Indian cricket *Seychellsia* sp. (Bhattacharjee and Manna 1967).

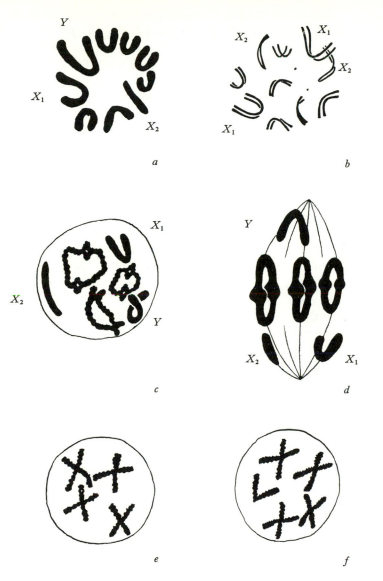

Fig. 17.14. Chromosomes of the Cricket, *Eneoptera surinamensis. a,* spermatogonial metaphase showing three pairs of autosomes and the X_1, X_2 and Y-chromosomes; *b,* somatic metaphase from a female, showing ten chromosomes of normal size and two minute dot chromosomes which are sometimes present; *c,* diakinesis in the male, showing the three autosomal bivalents and the three sex chromosomes; *d,* first metaphase in side view; *e* and *f,* the two categories of interkinetic nuclei. From Piza (1946*a*) and Claus (1956), redrawn; interpretation according to Claus, and Mesa and Bran (1964).

A highly peculiar multiple sex-chromosome mechanism exists in the North American mole cricket *Neocurtilla* (= *Gryllotalpa*) *hexadactyla* (= *G. borealis*), studied by Payne 1912*b*, 1916; White 1951*b* and Camenzind and Nicklas (1968). The species actually extends into South America, but it is not known whether neotropical material is cytologically identical with that from the United States. At least in material from Indiana, Texas and North Carolina there is a metacentric *X*-chromosome which clearly corresponds to the *X* of such *XO* species of the genus as *G. africana* (see p. 384); but there is, in addition, a bivalent consisting of a large and a small acrocentric element which is only present in the male and which regularly orientates itself on the first meiotic spindle in such a way that the larger member invariably passes to the same pole as the *X*. Thus we may formally designate the larger element of this pair as X_2, the smaller one (which is confined to the male line) being a *Y*. But there is absolutely no pairing at first metaphase between the *X* and the unequal bivalent, which may lie quite far apart on the spindle (it is not known whether any pairing relationships exist during the prophase stages of meiosis, which are difficult to study in *Gryllotalpa*).

The origin of the *G. hexadactyla* sex-chromosome system is not entirely clear, but it seems likely that the extra proximal segment in the larger member of the bivalent has been derived, by translocation, from the *X* or from a supernumerary chromosome, that it is heterochromatic and shows some kind of pairing with the *X* during prophase, but is unable to form a chiasma with it because of its heteropycnosis (cf. the behaviour of the two *X*'s in tetraploid grasshopper spermatocytes, p. 196).

In *Eneoptera surinamensis* and *Gryllotalpa hexadactyla* there is no obvious physical connection between chromosomes which regularly segregate to opposite poles at first anaphase, i.e. we have highly efficient complex sex-chromosome systems without formation of a multivalent. In three species of higher Diptera, however (a group in which chiasmata are generally absent in spermatogenesis), a sex-chromosome trivalent is formed, but without any chiasmata. Hughes-Schrader (1969) has described an $X_1X_2X_3Y_1Y_2Y_3$ ($\male$) mechanism in the mantispid *Entanoneura phthisica*. The three *X*'s segregate from the three *Y*'s at the first meiotic division without there being any physical connection between them. Thus the distance segregation mechanism normally exhibited by the *XY* pair in the order Neuroptera (see p. 617) is here extended to a multiple sex-chromosome system. At least, this is Hughes-Schrader's interpretation. There seems to be a possibility, however, that the species is actually *XY* in the male, there being two pairs of small autosomes which are always asynaptic in the male and have a distance-segregation mechanism like the *XY* pair.

That this may be the correct interpretation is suggested by the fact that in some spermatocytes one or two pairs of undoubted autosomes are present as asynaptic or desynaptic univalents which nevertheless undergo regular distance segregation. The supposed sex chromosomes do not form bivalents even in tetraploid spermatocytes, but segregate regularly. *E. phthisica* and the related *E. limbata*, which is undoubtedly XY in the male, both have the same chromosome number ($2n\male\female = 20$).

Two species of *Drosophila* are known to possess multiple sex-chromosome mechanisms which have arisen from the $XY:XX$ mechanisms present in related forms. One of these is X_1X_2Y, the other XY_1Y_2. The first is *D. miranda* (Dobzhansky 1935*b*), which ranges from the states of Washington and Oregon southward to Mt. Whitney in the Sierra Nevada of California. It is closely related to *D. pseudoobscura* and *D. persimilis*, which it closely resembles, and with both of which it is sympatric (Dobzhansky 1941*a*).

The karyotype of the female *miranda* is indistinguishable in metaphase plates from that of *pseudoobscura* and *persimilis*, consisting of a pair of metacentric X's, three pairs of acrocentric autosomes and a pair of dot chromosomes. In the male, however, there is one 'autosome' less, i.e. $2n\male = 9$. The missing element is one member of the third pair of autosomes (element C in Muller's terminology and the chromosome which carries the greater part of the inversion polymorphism of the related species); the male *miranda* thus appears to have become haploid for this chromosome There is both genetic and cytological evidence that the single third chromosome always passes to the same pole as the X in the male. It has become, in effect, a sex chromosome and we may formally designate it as X_2, the original X (which is homologous to that of related species) being referred to as X_1.

Actually, however, the 'haploidy' of X_2 is not complete, since it has been shown that the Y of *miranda* contains some euchromatic segments which are homologous to parts of X_2 and have presumably been derived from the missing third chromosome. Thus in the evolution of the *miranda* sex-chromosome mechanism one member of the third pair of autosomes must have become fused to the Y (probably with loss of one limb of the latter). Subsequently this originally autosomal limb became altered by progressive heterochromatinization and probably structural changes as well (it is not very likely that the euchromatin became broken up and distributed among the heterochromatin by inversions – some of which would have had to be pericentric – as McKnight (1939) seems to have believed). Some genetically active segments still persist in the neo-Y of *D. miranda* – remnants of the third chromosome. Mutations in these segments are inherited exclusively through the male line, from father to son.

During spermatogenesis a trivalent is formed (McKnight and Cooper

1944; Cooper 1946), the X_1- and X_2-chromosomes both pairing with different parts of the Y, so that they regularly segregate to the opposite pole at anaphase (with 97–99% efficiency, according to Cooper). Apparently, the segments in the Y which are homologous to regions of X_2 function as pairing segments. The exact location of these segments (proximal, sub-terminal or distal) is uncertain.

A multiple sex-chromosome mechanism is also present in *D. americana* (subspecies *americana*), where it is presumably of fairly recent origin, since it is not present in the subspecies *texana*. In this case (Spencer 1940*a*, *b*; Stalker 1940; Patterson, Stone and Griffen 1940, 1942; Stone and Patterson 1947) a centric fusion has occurred between one member of the fourth pair of autosomes and the X – thus the other member of the fourth pair has become a neo-Y_2, restricted to the male line. Apparently, very little evolutionary divergence has taken place between the Y_2 element and the arm of the X with which it is homologous; in spite of being confined to the male line and hence unable to undergo crossing-over with its homologue, Y_2 has not lost its genetic activity by 'heterochromatinization'. Just what changes *have* taken place in it, is somewhat uncertain – the question deserves further investigation.

Stone (1955) has listed fourteen instances in the phylogeny of the genus *Drosophila* (in addition to *D. americana*) where fusions between the X and an autosome have occurred. These are all represented by $XY:XX$ species or groups of species at the present day. We have already (p. 350) referred to the situation in the *obscura* group. It is probable that in all these instances a Y-autosome fusion has also occurred, the autosome which became fused to the Y eventually undergoing heterochromatinization so that it is now unrecognizable. Thus in such cases the lineage will have passed through either an X_1X_2Y or an XY_1Y_2 stage (depending on whether the Y-autosome fusion or the X-autosome one occurred first). Thus multiple sex-chromosome mechanisms are apparently rather short-lived in *Drosophila* and those of *D. miranda* and *D. a. americana* are probably very recent.

In the anthomyid fly *Hylemyia fugax* ($2n\male = 13$) Boyes (1952, 1954) has described an X_1X_2Y mechanism which he has compared with that of *D. miranda*. X_1 is a medium-sized acrocentric, X_2 a small acrocentric, while the Y is a metacentric (not an acrocentric as stated in the 1952 paper). It would be natural to interpret this case as having arisen by centric fusion of a small acrocentric autosome with an acrocentric Y similar to those of other *Hylemyia* species (which are XY in the male and have $2n\male = 12$). But since the number of autosomes has not undergone reduction, by comparison with eight other species of the genus that have been studied, the interpretation is somewhat doubtful. Possibly, *H. fugax* was derived from an ancestor with $2n\mathord{\circlearrowleft} = 14$ rather than 12. Apparently no chiasmata are formed between the sex chromosomes at meiosis; the X_2 is associated with the shorter limb of the Y and the X_1 with the longer limb by pairing segments which remain associated up to and through metaphase. Boyes accordingly believes that the sex-chromosome trivalent of *H. fugax* is basically similar to that of *D. miranda* as interpreted by Cooper (1944*a*, *b*, 1946). In the fruit-flies of the family *Tephritidae* Bush (1962, 1966*a*) has reported X_1X_2Y mechanisms in the males of *Anastrepha serpentina* and *Rhagoletis striatella*, which have presumably arisen by Y-autosome fusions.

MECHANISMS THAT HAVE ARISEN BY DISSOCIATION

Multiple sex-chromosome mechanisms that have arisen by dissociation (rather than by centric fusions or mutual translocations) are ones in which an original X or Y has become broken into two portions, both of which have survived, presumably because their broken ends were 'capped' by chromosome ends derived from another chromosome, usually, but perhaps not always, an autosome or autosomal fragment, or a supernumerary. In the case of organisms with monocentric chromosomes, one of the 'capping' fragments must contain a centromere. It is characteristic of multiple sex-chromosome mechanisms derived by dissociation that (unlike those which have arisen by centric fusions or mutual translocation) they do not involve any diminution in the number of autosomes.

We have spoken of these mechanisms as having arisen by dissociation, and that is undoubtedly true of most of the instances in the Heteroptera and Homoptera, discussed below, and perhaps also of some in the earwigs (Dermaptera), nematodes and other groups. But it is impossible to formally exclude the possibility that in some instances multiple sex-chromosome mechanisms have arisen by the addition to the karyotype of a sex chromosome carrying a fairly large deletion, i.e. that the multiple mechanism, at the time of origin, involved hyperploidy for extensive regions of the original X (or Y, as the case may be). But any homology between different X's will be liable to cause complications at meiosis in the homogametic sex. Similarly, any homology between Y_1, Y_2 ... will be liable to lead to the formation of Y_1Y_2 bivalents in the heterogametic sex. Thus the origin of multiple sex-chromosome mechanisms by 'hyperploidy' rather than by dissociation is not very plausible.

In general, the origin of multiple sex-chromosome mechanisms by dissociation is not possible in those groups where the meiotic association of the sex chromosomes in the heterogametic sex is by means of chiasmata. An XY mechanism in which the pairing segment, within which chiasmata are formed, is confined to one arm cannot be expected to successfully convert itself into an X_1X_2Y or XY_1Y_2 system by dissociation of the X or Y.

Classical examples of multiple sex-chromosome mechanisms which appear to have arisen by dissociation occur in numerous species of Heteroptera, belonging to the families Pentatomidae (Wilson 1911; Troedsson 1944; Schrader 1947; Srivastava 1957), Lygaeidae (von Pfaler-Collander 1941; Menon 1955; Parshad 1957; Banerjee 1958; Jande 1959a), Dysodiidae (Schrader 1947), Coreidae (Manna 1951b; Dutt 1957; Parshad 1957; Jande 1959a), Alydidae (Parshad 1957; Jande 1959a), Reduviidae (Payne 1909, 1912a; Schreiber and Pellegrino 1950; Manna 1951a, b; Banerjee 1958; Jande 1959b), Miridae (Leston 1957), Nepidae (Steopoe 1925, 1927, 1931; Manna 1951b, 1958; Dass 1952; Parshad 1956; Das 1958; Banerjee 1958), Pyrrhocoridae (Mendes 1947, 1949; Piza 1947e; Ray-Chaudhuri and Manna

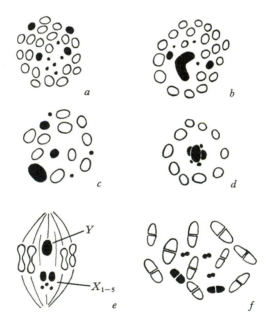

Fig. 17.15. Chromosomes of *Acholla multispinosa* (Heteroptera, Reduviidae). Sex chromosomes in black, autosomes in outline. The largest chromosome is the Y; X_1 and X_2 are medium-sized, X_3, X_4 and X_5 very small. *a*, oogonial metaphase; *b*, spermatogonial metaphase; *c*, first metaphase in the male (polar view); *d*, second metaphase in the male (polar view); *e*, second metaphase in side view, showing the separation of the Y from the five X's; *f*, first metaphase in the female (polar view), showing ten autosomal bivalents and the five sex bivalents X_1X_1, X_2X_2, X_3X_3, X_4X_4 and X_5X_5. From Payne (1910) and Troedsson (1944), redrawn and modified.

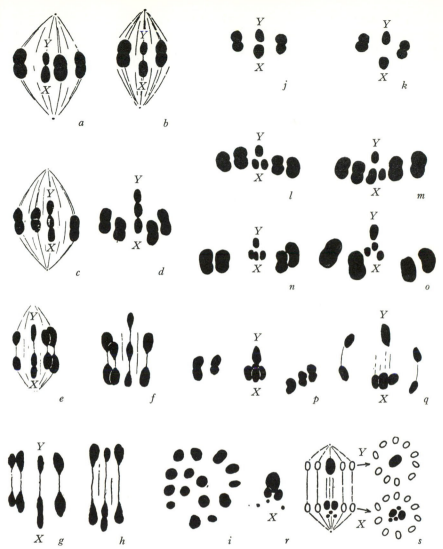

Fig. 17.16. Sex chromosomes of various Heteroptera. *a* and *b*, second metaphases of *Thyanta custator* (Pentatomidae), an *XY* species; *c* and *d*, corresponding stages in *T. calceata* (an X_1X_2Y species); *e–h*, anaphases of the same; *i*, polar view of first metaphase in the same; *j* and *k*, second metaphases, inside view, of *Diplocodus exsanguis* (Reduviidae), an *XY* species; *l*, similar view of *Rocconata annulicornis* (X_1X_2Y); *m*, same stage in *Conorhinus sanguisugus* (X_1X_2Y); *n*, *Sinea diadema* ($X_1X_2X_3Y$); *p* and *q*, metaphase and anaphase of second division in *Gelastocoris oculatus*; *r*, sex chromosomes of *Acholla multispinosa* at second metaphase; *s*, diagram showing the distribution of the sex chromosomes in the second division of *Acholla*. From Wilson (1911).

1955; Battaglia 1956; Sharma 1956; Sharma, Sud and Parshad 1957; Manna 1957; Parshad 1957; Banerjee 1958), Notonectidae (Jande 1959a), Mesoveliidae (Ekblom 1941), Gelastocoridae (Payne 1909, 1912a) and Cimicidae (Slack 1939a, b; Darlington 1939b).

All these mechanisms have evolved from the usual heteropteran sex-chromosome mechanism, characterized in general by failure of synapsis of the holocentric X- and Y-chromosomes, which divide in the first meiotic division and separate from one another at the second division, after a brief 'touch and go' pairing.

The multiple sex-chromosome mechanisms in the Heteroptera range from X_1X_2Y and X_1X_2O (see p. 676) up to $X_1X_2X_3X_4Y$ and $X_1X_2X_3X_4X_5Y$ types (see Table 17.5). There is, in general, no diminution in the number of autosomes as the number of sex chromosomes increases. Thus where we have a genus containing some species with a single XY or XO mechanism and one or more with a complex mechanism, the number of autosomal pairs is either the same or actually higher in the species with the multiple sex chromosomes. Situations of this kind have been described in the genera *Thyanta* (Pentatomidae) by Wilson (1911) and Schrader and Hughes-Schrader 1956, *Sinea* (Reduviidae) by Payne (1909, 1912a), *Triatoma* (Triatomidae or Reduviidae) by Schreiber and Pellegrino (1950) and *Trapezonotus* (Lygaeidae) by von Pfaler-Collander (1941). Although the details are not entirely clear in all instances, the general situation is that the acquisition of a multiple X has not involved the loss of a pair of autosomes and must hence have occurred by some type of fragmentation or dissociation. The chromosomes of Heteroptera seem to be generally holocentric, so that dissociational changes should be somewhat easier than in most groups, the 'donor' chromosome having to provide only two telomeres and not a centromere as well.

This interpretation is strengthened by the meiotic behaviour of these multiple sex-chromosome systems. Since the pairing of the autosomes involves chiasma formation, we should expect that if the multiple mechanisms resulted from the inclusion of one or more pairs of autosomes in the sex mechanism, chiasmata should be formed between some of the elements. In fact, none of the many multiple-X systems which have been found in various Pentatomidae, Cimicidae, Lygaeidae, Nepidae and Reduviidae show any true pairing or chiasma formation between the X's and the Y. Thus at the first meiotic division, in spermatogenesis, all the sex chromosomes divide equationally, while in the second division the X's form a group in one half of the spindle, opposed to the single Y, and the anaphase separation takes place just as in the XY species, all the X's behaving as a unit (Figs 17.16 and 17.17). It will be obvious that this behaviour involves

TABLE 17.5. *Sex chromosomes in Assassin bugs (fam. Reduviidae)*

	No. of pairs of autosomes	Sex chromosomes of ♂	Author
Subfam. Apiomerinae			
Apiomerus crassipes	11	XY	Payne 1912*a*
A. spissipes	11	XY	White (unpublished)
Subfam. Ectrychotinae			
Ectrychotes dispar	14	XO	Manna 1951*b*
Subfam. Harpactorinae			
Acholla multispinosa	10	$X_1X_2X_3X_4X_5Y$	Payne 1909, 1910
Arilus cristatus	11	$X_1X_2X_3Y$	Payne 1909
Harpactor fuscipes	12	$X_1X_2X_3Y$	Manna 1951*b*
H. marginatus	12	$X_1X_2X_3Y$	Banerjee 1958
Sinea confusa	12	$X_1X_2X_3Y$	Payne 1912*a*
S. complexa	12	$X_1X_2X_3Y$	Payne 1912*a*
S. diadema	12	$X_1X_2X_3Y$	Payne 1909
S. spinipes	12	$X_1X_2X_3Y$	Payne 1912*a*
S. rileyi	12	$X_1X_2X_3X_4X_5Y$	Payne 1912*a*
Sycanus sp.	12	$X_1X_2X_3Y$	Manna 1951*b*
Coranus fuscipennis	12	X_1X_2Y	Jande 1959*b*
Staliastes rufus	12	XY	Jande 1959*b*
Sycanus collaris	12	$X_1X_2X_3Y$	Jande 1959*b*
Polididus armatissimus	5	XY	Banerjee 1958
Zelus exsanguis	12	XY	Payne 1909
Fitchia spinulosa	12	X_1X_2Y	Payne 1909
Rocconata annulicornis	12	X_1X_2Y	Payne 1909
Pselliodes cinctus	12	$X_1X_2X_3Y$	Payne 1912*a*
Subfam. Piratinae			
Androclus pictus	10	XY	Jande 1959*b*
Ectomocoris atrox	10	XY	Jande 1959*a*
E. cordiger	10	X_1X_2Y	Jande 1959*a*
E. ochropterus	10	X_1X_2Y	Jande 1959*a*
Sirthenea sp.	13	XY	Jande 1959*b*
Subfam. Microtominae			
Microtomus conspicillaris	14	XY	Piza 1957
Subfam. Acanthaspidinae			
Pasiropsis sp.	12	$X_1X_2X_3X_4Y$	Jande 1959*b*
Reduvius personatus	10	XY	Payne 1912*a*
Subfam. Stenopodinae			
Pnirontis modesta	10	$X_1X_2X_3X_4Y$	Payne 1912*a*
Oncocephalus sp. '1'	10	X_1X_2Y	Jande 1959*a*
O. impudicus	10	X_1X_2Y	Jande 1959*a*
O. sp. '2'	10	$X_1X_2X_3Y$	Jande 1959*a*
Pygolampis foeda (Calcutta)	11	XY	Banerjee 1958
P. foeda (North India)	11	X_1X_2Y	Jande 1959*a*
Subfam. Triatominae			
Triatoma (= *Conorhinus*)			
sanguisugus	10	X_1X_2Y	Payne 1909, 1912*a*
T. rubrofasciatus	11	X_1X_2Y	Manna 1950
T. vitticeps	10	$X_1X_2X_3Y$	Schreiber and Pellegrino 1950

extremely precise mechanisms governing the orientation of the sex chromosomes, if it is not to lead to non-disjunction of some of the sex chromosomes at second anaphase.

We accordingly conclude that, although fusions between sex chromosomes and autosomes may well have occurred in the pentatomid *Rhytidolomia senilis* and the belostomatid *Lethocerus* sp. (Chickering 1932) (which have simple XY mechanisms), they have not taken place in the species of Heteroptera with multiple mechanisms.

Up till now we have been considering multiple sex chromosomes which are constant in the population, the same number of X's or Y's being present in all individuals of the heterogametic sex. In the Bed bug (*Cimex lectularius*) the number of X's varies greatly from one individual to another (Slack 1939*a*, *b*; Darlington 1939*b*). The related species *C. rotundatus*, *C. columbarius* and *C. stadleri* seem to be X_1X_2Y, but in *lectularius* a large number of supplementary X's of various sizes are usually present, so that the total number of X elements in the male individuals varies from two to fifteen (Fig. 17.17). The females have not been studied so extensively, but presumably show a corresponding variation (i.e. from four to thirty). Since no intersexes are apparently produced, it seems likely that only the first two X elements carry major sex-determining loci, the remainder being possibly X's from which the sex-determining regions have been deleted. As in the case of true supernumeraries (see Ch. 10) it is unlikely that the supplementary X's or X fragments of *C. lectularius* are entirely inert, genetically, since in some populations there is an average of nine of them in the males.

At the first meiotic division in these Bed bugs it is usual for all the sex

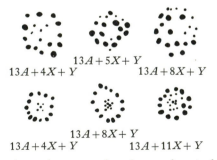

Fig. 17.17. First metaphases (upper row) and second metaphases (lower row) of the bed bug, *Cimex lectularius* showing the variable number of sex chromosomes (there are always 13 pairs of autosomes). In the first metaphases the arrangement of the chromosomes is apparently haphazard, while in the second metaphases the sex chromosomes form a group in the centre of the spindle, surrounded by a ring of autosomes. The Y cannot be identified with certainty.

TABLE 17.6. *Sex chromosomes in Water scorpions (fam. Nepidae)*

	No. of pairs of autosomes	Sex chromosomes of ♂	Author
Subfam. Nepinae			
Nepa cinerea	14	$X_1X_2X_3X_4Y$	Steopoe 1925, 1931
Laccotrephes griseus	19	X_1X_2Y	Manna 1951*b*
L. maculatus (North India)	19	X_1X_2Y	Sharma and Parshad 1955
L. maculatus (South India)	19	$X_1X_2X_3X_4Y$	Dass 1952
L. rubra	19	$X_1X_2X_3X_4Y$	Manna 1951*b*
Subfam. Ranatrinae			
Ranatra linearis	19	$X_1X_2X_3X_4Y$	Steopoe 1927, White (unpublished)
R. elongata (Bangalore)	19	$X_1X_2X_3X_4Y$	Dass 1952
R. elongata (Allahabad)	19	$X_1X_2X_3Y$	Das 1958
R. sordidula	19	$X_1X_2X_3Y$	Parshad 1956

chromosomes to divide equationally, but where the number of X's is high a certain amount of non-disjunction may occur, leading to intra-individual variation in the number of X's in later stages of spermatogenesis. At the second meiotic division, regardless of the number of the X's, they invariably arrange themselves in the centre of the ring of autosomes, opposed to the single Y. Thus, as a rule, all the X's pass to one pole, the Y to the other; but where the number of X's is high a certain amount of non-disjunction seems to occur, as at the first anaphase. Although the size of the Y seems to be somewhat variable (possibly Y's carrying deletions or duplications occur in natural populations) every male seems to carry a single Y, individual lacking one or possessing several being unknown.

The problem of *why* there is this precise coorientation of X's and Y at the second meiotic division is one which exists in the case of every heteropteran with a multiple sex-chromosome mechanism. Clearly, Darlington's (1939*b*) explanation in terms of 'strong' and 'weak' centromeres is inadequate.

The counterpart to the *Cimex lectularius* situation is provided by several species of coreid bugs belonging to the American genus *Acanthocephala* (= *Metapodius*), in which the number of Y's is variable, although there is only one X (Wilson 1907, 1909*a*, 1910). In some individuals of *A. femorata* the Y may be entirely absent, but most males have from one to five Y's, some of which are smaller than others. These multiple Y's generally arrange themselves in a chain at the second meiotic division (instead of in a 'plate' as the multiple X's of *Cimex* and other Heteroptera do); but the final result is the same, namely regular segregation of the Y's from the X.

In the neuropteran *Ascalaphus libelluloides* Naville and de Beaumont (1933) found that the number of apparent sex chromosomes varied considerably in both sexes. Probably they were dealing with a case similar to *Cimex lectularius*, i.e. with supernumerary chromosomes of the same general nature as the sex chromosomes, and ultimately derived from them, but lacking any strong sex-determining properties. Other species of the genus have a simple XY mechanism in the male. An X_1X_2Y sex-chromosome mechanism with distance-pairing has also been recorded in the neuropteran *Plethosmylus decoratus* by Hirai (1956).

In cases where two closely related species or geographic races of Heteroptera have different multiple sex-chromosome mechanisms (e.g. X_1X_2Y and $X_1X_2X_3Y$) it is possible that the one with the higher number has undergone an additional dissociation. But it is also quite probable that in some instances the lower number is the derived one, a fusion having occurred.

In the Lepidoptera, with female heterogamety, what are probably XY_1Y_2 trivalents (apparently achiasmatic) have been described in the oogenesis of four species (Suomalainen 1969*b*). In all these cases the combined length of Y_1 and Y_2 seems to be very close to that of the X.

Unusual multiple sex-chromosome mechanisms have been reported for several species of alticid beetles by Virkki (1967*a* 1968*a*, *b*, 1970). These are neotropical forms belonging to the tribe Oedionychini, in many members of which there are giant X and Y chromosomes in the male which do not synapse at meiosis but regularly pass to opposite poles at first anaphase. In *Omophoita clerica* from northern Brazil there are three sex chromosomes in the male, but it is not known whether the constitution is X_1X_2Y or XY_1Y_2. There is no synapsis between these chromosomes and no trivalent is formed. At first anaphase a J-shaped chromosome regularly goes to one pole while the acrocentric and metacentric members of the trio go to the opposite pole. In *Asphaera daniela* from Venezuela males have one large sex chromosome and six small ones. This is interpreted as an $XY_1Y_2Y_3Y_4Y_5Y_6$ system by Virkki. Again, there is no synapsis, and the six small sex chromosomes all pass as a group to the same pole, the large sex chromosome going to the other. *Haltica caerulea* appears to have XY_1Y_2 males and a similar meiotic mechanism (Kasturi Bai and Sugandhi 1968). *Phenrica austriaca* has $XY_1Y_2Y_3Y_4$ males; at meiosis the X and Y_1 form a bivalent but the other Y's are univalents.

An entirely different type of multiple sex-chromosome mechanism seems to have been derived from the usual 'XY parachute' mechanism in a number of species of beetles through dissociation of the minute Y into two even more minute Y elements. Such $X_{y_1y_2}$ systems seem to be present in two species of *Dermestes* (John and Shaw 1967), in a coccinellid (Takenouchi 1968) and in two species of Curculionidae (Takenouchi 1969a, b).

A number of multiple sex-chromosome mechanisms have been recorded for various species of earwigs (order Dermaptera). Centromeres have not been detected with certainty in this group, and it seems fairly sure that the chromosomes are holocentric (Webb and White 1970; White 1971a). Certain members of this group have XY males (W. P. Morgan, 1928; Asana and Makino, 1934; Misra 1937a; Bauer 1947). In these species both sex chromosomes are strongly heteropycnotic during the prophase stages of meiosis; they form an XY bivalent, but it is not clear whether there are any chiasmata between them. Three species of *Anisolabis* and *Forficula smyrnensis* have X_1X_2Y or XY_1Y_2 males (W. P. Morgan, 1928; Sugiyama 1933; Schrader 1941a; E. Goldschmidt 1953a); the three elements form a trivalent at meiosis, but it is equally impossible to say whether their metaphase association is by means of chiasmata. Disjunction of the two X's from the Y at first anaphase seems to be regular. Finally, *Prolabia arachidis* has $X_1X_2X_3Y$ or $XY_1Y_2Y_3$ males (Bauer 1947).

The situation in the European *Forficula auricularia* is somewhat more complex. Both 24-chromosome (XY) and 25-chromosome males exist in this species (Figs. 17.18 and 17.19) and the two types usually coexist in the same population, although some colonies appear to consist entirely of 24-chromosome individuals. Usually the 25-chromosome males are in a minority, although Morgan reports that in Switzerland the two types were found in equal numbers. The 25-chromosome males (Fig. 17.18) were assumed to be X_1X_2Y by Callan (1941b), Ortiz (1969) and Henderson (1970). If so, there should be several kinds of females, with different

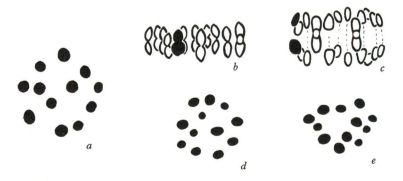

Fig. 17.18. Meiosis in a 24-chromosome male of *Forficula auricularia. a*, first metaphase in polar view; *b*, the same in side view; *c*, first anaphase; *d* and *e*, second metaphases in polar view. From Callan (1941b).

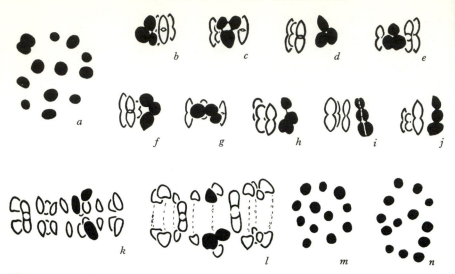

Fig. 17.19. Meiosis in a 25-chromosome male of *Forficula auricularia* (XY_2Y_2). *a*, first metaphase in polar view; *b–j*, same in side view showing various modes of association between the two Y's and the X; *k–l*, first anaphases; *m–n*, second metaphases showing 12 and 13-chromosomes respectively. From Callan (1941*b*) but reinterpreted.

chromosome numbers. Webb and White (1970) have shown that in Australian material of this species all females have twenty-four chromosomes and are XX. The polymorphism is thus a Y-chromosome polymorphism, and there are four main types of males, XY_1, XY_2, XY_2Y_2 and mosaics XY_2/XY_2Y_2. In certain individuals cells with as many as five Y_2-chromosomes may be present, as a result of some kind of accumulation mechanism. The adaptive significance of this peculiar sex-chromosome polymorphism deserves further study. E. Goldschmidt, Ortiz, Bauer and Henderson have all tried to argue that the multiple sex-chromosome mechanisms of the Dermaptera have arisen by polyploidy [since the XY forms have $2n\male = 12$, 14, 14, 24, 24, 24, the X_1X_2Y ones have $2n\male = 21$, 25, 25, 25, 25, and the single $X_1X_2X_3Y$ species has $2n\male = 38$ (but *Nala lividipes*, with $2n\male = 37$, is X_1X_2Y or XY_1Y_2)]. Although there may be a tendency for the species with multiple sex-chromosome mechanisms to exhibit higher numbers of autosomes this is far more easily explicable as an expression of a common tendency, in both types of elements, to undergo dissociation in certain phyletic lineages. As in other similar cases (see p. 461) nothing like a true polyploid series has been demonstrated in the Dermaptera.

X_1X_2O, $X_1X_2X_3O$ ... MECHANISMS

Thus far we have been considering multiple sex-chromosome mechanisms where both X and Y elements are present. In several groups of the animal kingdom, however, mechanisms are known in which the X is represented by several separate elements $(X_1, X_2, X_3 \ldots)$ but there is no Y. The principal interest of these cases lies in the seemingly mysterious way in which the different X's pass regularly to the same pole at the first meiotic division, and never to opposite poles.

The principle involved in mechanisms of this type appears to be that several heterochromatic chromosomes are held together or in close proximity by a general pairing attraction throughout the meiotic prophase without precise synapsis and hence without chiasma formation. Presumably there is enough similarity between their nucleotide sequences (without there being an absolute identity) to produce this generalized attraction. The situation is somewhat analogous to that which occurs in the case of the two X's in a tetraploid grasshopper spermatocyte (see p. 196).

The most noteworthy example of a group characterized by an X_1X_2O mechanism is that of the spiders. Of approximately 240 species which have been studied cytologically by Painter (1914), Hard (1939), Revell (1947), Pätau (1948), Hackman (1948), Bole-Gowda (1950, 1952, 1958), Suzuki (1950, 1951, 1952, 1954), Suzuki and Okada (1950), G. P. Sharma (1950), G. P. Sharma and Singh (1957), G. P. Sharma, Jande, Grewal and Chopra (1958), G. P. Sharma, Jande and Tandon (1959), G. P. Sharma, Tandon and Grewal (1960), Mittal (1960), T. Sharma (1961) and Sokolov (1960, 1962a), all except thirty-eight seem to be X_1X_2O in the male. In some of these species the females have been shown to be $X_1X_1X_2X_2$, and such a constitution may be inferred in the remainder.

In the X_1X_2O species of spiders the X's are invariably acrocentric. Usually X_1 and X_2 are about the same length, but in some forms one is slightly longer than the other, and this inequality is quite marked in some Philodromidae (Hackman 1948), in *Hersilia savignyi* (Bole-Gowda, 1958), and in certain species of *Lycosa* (Suzuki 1952). The X's show strong positive heteropycnosis during the prophase stages of meiosis in the male and in X_1X_2O and $X_1X_2X_3O$ species alike they are closely paired together from zygotene until first metaphase. No chiasmata are formed between them in the male, and it seems unlikely that there is any strict homology between them. As the first meiotic spindle develops, the X's become orientated, still closely parallel to one another, near one of the poles, to which they pass as a group at anaphase. The orderly segregation of the X's to the same pole is probably facilitated by the precocious 'polarization' of the nucleus during the prophase stages of meiosis (Revell 1947). In the $X_1X_1X_2X_2$ females the sex chromosomes are not heteropycnotic in meiosis and form two separate bivalents (Hackman 1948; Pätau 1948).

Of the thirty-eight aberrant species of spiders, twenty-six (belonging to the families Dysderidae, Sparassidae, Thomisidae, Oxyopidae and Salticidae) are simply XO in the male sex, while three species of the family Agelenidae, five of the Sparassidae, two of the Linyphiidae and one each of the Selenopidae and Tetragnathidae are $X_1X_2X_3O$.

Since the X_1X_2O mechanism has been found in the primitive Liphystiomorphae and Mygalomorphae as well as in the overwhelming majority of the more 'advanced' Arachnomorphae (Labidognatha) – including representatives of twenty-two families – there can be no doubt that it was the primitive sex-determining mechanism of the whole group, which has been handed down from Palaezoic times. The amazing evolutionary stability of this unusual cytogenetic system consequently demands an adequate explanation. In fact, the twenty-six XO species seem to represent the result of perhaps as few as six independent evolutionary transformations from the original X_1X_2O condition. Similarly, the twelve $X_1X_2X_3O$ species may have arisen by five separate $X_1X_2O \rightarrow X_1X_2X_3O$ transformations.

It is to be presumed that the origin of XO mechanisms from the X_1X_2O one has been by fusions, and in all probability centric rather than tandem ones. However, only one of the twenty-six XO species is known with certainty to have a metacentric X. That is *Heteropoda sexpunctata*, a species with $2n\male = 21$, all the chromosomes being metacentric except for one pair of acrocentrics (Bole-Gowda 1952). Two other species of *Heteropoda* are $X_1X_2X_3O$ in the males and have $2n\male = 41$, all the chromosomes being acrocentric. It seems likely that *H. sexpunctata* arose from an X_1X_2O species like *Olios lamarcki*, with $2n\male = 42$ by a total of eleven centric fusions, one being between X_1 and X_2. Another XO species of spider which has undergone a spectacular evolutionary decrease in chromosome number is *Ariadna lateralis*, which has the lowest number known in the group, namely $2n\male = 7$ (three pairs of giant metacentric autosomes and an acrocentric X according to Suzuki, 1954).

The twenty-four species of Thomisidae, Salticidae and Oxyopidae with XO mechanisms have acrocentric X's. If they have lost their X_1X_2O mechanisms by four different ancestral centric fusions (as seems likely), then in each instance a pericentric inversion or deletion of almost all of one limb must have converted the metacentric X so formed into an acrocentric element. These species do not have metacentric autosomes.

In order to understand the evolutionary stability of the peculiar sex-chromosome mechanism found in about 85% of all species of spiders, and its occasional breakdown, certain general facts about the cytology of the group must be made clear. The vast majority of the species have all the chromosomes acrocentric and show very strong proximal localization of chiasmata in the male, there being only one in each bivalent, situated very close indeed to the centromere, but in the long arm. In a few species of spiders (about nine out of the 240 that have been studied) all or some of the autosomes are metacentric and in most of these cases the chiasmata in the male are distal or interstitial rather than proximal. Thus we may conclude that proximal localization of chiasmata in spermatogenesis has imposed a major barrier to the establishment of centric fusions in spider phylogeny (for the reasons explained on p. 295). Many species with acrocentric chromosomes have, however, undergone evolutionary decreases in chromosome number. In these, presumably, centric fusions led to the production of metacentric elements which were later converted into acrocentrics by pericentric inversions (the pattern of chiasma distribution must have evolved as follows: proximal → distal → proximal localization).

Meiosis in female spiders has not been adequately studied, but the work of Pätau (1948) on *Aranea reaumuri* suggests that this species shows distal rather than proximal localization in the egg. The pattern of chiasma localization in the male could, of course, not influence the probability of centric fusions between X_1 and X_2 establishing themselves in phylogeny, only the chiasma distribution in the female being significant in this connection. Thus until more is known about chiasma distribution in the female meiosis of a number of species, the fewness of the XO species remains somewhat mysterious. We may be sure that hundreds of thousands or even millions of X_1–X_2 fusions must have occurred in the long history of this group, since Palaeozoic times; but the reason why all but a very few were unsuccessful in establishing themselves is still obscure; it may, of course, have been a reason not dependent on chromosome mechanics.

The mode of origin of the $X_1X_2X_3O$ systems is not entirely clear, but it is rather unlikely that centromere misdivision was involved, as some earlier authors supposed. More probably, in each case the third X-chromosome established itself from a small supplementary fragment derived from X_1 or X_2 by deletion of most of its substance. In those species in which X_1, X_2 and X_3 are subequal in size the third X may have subsequently increased in length, by duplications.

No species of spiders with a Y-chromosome are known, and there is no evidence of any X-autosome fusions having established themselves in the group.

There are a few instances where species of spiders seem to have undergone spectacular increases in number of autosomes, without any change in the sex-chromosome mechanism. One is *Aranea ventricosa*, with twenty-two pairs of acrocentric autosomes, whereas related species have eleven pairs, likewise acrocentric (Suzuki 1951). One would almost suspect polyploidy here, were it not for the fact that it is hardly possible to imagine a form of autosomal polyploidy that did not involve the sex chromosomes. Suzuki assumed that the high chromosome number of *A. ventricosa* was primitive and that the lower numbers of other Argiopidae were derived from it, which hardly seems likely. The form referred to by Suzuki (1954) as *Linyphia montana* may represent a similar case; but Sokolov (1960) gives an entirely different account of the cytology of this species (misidentification of his material by one or other worker seems probable).

There has been no experimental work on sex determination in spiders, so that the relative roles of X_1 and X_2 are not known, and we do not know, for example, in the case of any of the common X_1X_2O species, what the phenotypes of X_1O, X_2O, $X_1X_1X_2$ and $X_1X_2X_2$ genotypes would be. In at least eighteen species of spiders gynandromorphic individuals have been recorded (Bonnet 1934; Exline 1938; Knülle 1954). Unfortunately, in no case was their chromosomal constitution determined; but their existence suggests an absence of circulating sex hormones, as in insects.

MULTIPLE SEX CHROMOSOMES IN OSTRACODS

An interesting series of multiple sex-chromosome mechanisms occur in ostracods (Bauer 1940; Dietz 1954, 1955, 1958). The male is the heterogametic sex in this group in contrast to some other groups of Crustacea such as the copepods and isopods, which have female heterogamety. Of the species of ostracods that have been studied (Table 17.7) some members of the genus *Cyclocypris* show relatively primitive conditions. *Cyclocypris globosa* has *XO* males, the *X* being a metacentric; there are six pairs of acrocentric autosomes in this species, which seems to be the maximum number in these ostracods.

In *C. laevis* an *X*-autosome fusion seems to have taken place, so that there is a neo-*X*, neo-*Y* system and five pairs of autosomes. At the first meiotic division the acrocentric *Y* is associated terminally with one limb of the *X*, presumably by a terminal chiasma.

Cyclocypris ovum includes a number of different kinds of males. There are two series of these, one with five pairs of autosomes and without an *X*-autosome fusion, the other with only four pairs of autosomes and with an *X*-autosome fusion like that of *C. laevis*. But in both types of individuals

TABLE 17.7. *Complex sex-chromosome mechanisms of Ostracods*

Subfamily and Species	Sex chromosomes	2n♂
Cyprinae		
Notodromas monacha[1,2]	X_1X_2O	18
Cypris, four spp.[3]	$X_1X_2X_3Y$	19
Heterocypris incongruens[2,4]	$X_1X_2X_3X_4X_5X_6Y$	15
Scottia browniana[2,3]	$X_1X_2X_3O$	15
Platycypris baueri[3]	$X_1X_2X_3O$	19
Cyclocyprinae		
Cyclocypris globosa[2,3]	XO	13
C. ovum[2,3]	$X + 3$–6 Y's	15–18
C. laevis[2,3]	XY	12
Cypria exsculpta[2,3]	$X_1X_2X_3Y$ and $X_1X_2X_3X_4Y$	18
C. opthalmica[2,3]	$X_1X_2X_3X_4Y$	17
Physocypria kliei[3]	$X_1X_2X_3O$	17

[1] Dietz 1954   [3] Dietz 1958
[2] Dietz 1955   [4] Bauer 1940

there are in addition three to six small supplementary Y's which are univalents at meiosis. Presumably they have arisen by some kind of dissociation, perhaps with reduplication of some regions.

In the genus *Cypris*, instead of multiple Y's we find multiple X's. *C. exsculpta* has three of them which are fused at the first meiotic division into a triradiate group which passes to the opposite pole from the single Y. This species, however, also includes individuals in which a Y-autosome fusion has occurred, and this seems to be the regular condition in *C. ophthalmica*.

Four species of *Cypris* have $X_1X_2X_3Y$ males. From Dietz' figures it seems likely that one of the X's is paired with the Y by means of a chiasma, and that all the X's are associated in a group without any chiasmata between them. The Y is possibly a neo-Y. More primitive conditions seem to occur in *Notodromas monacha*, with X_1X_2O males and *Scottia browniana*, with an $X_1X_2X_3O$ condition; these may be forms in which a neo-Y has never developed (although the latter may have undergone an X-autosome fusion and later lost the neo-Y. The most complex situation occurs in the bisexual 'race' of *Heterocypris incongruens* (Bauer 1940); this 'species' also includes a parthenogenetic biotype consisting of females with twenty chromosomes. In the males six X's form a group at meiosis, and one of them is paired (probably by a chiasma) with a Y. Since there are only four pairs of autosomes, two kinds of sperms are formed, one with five chromosomes, the other with ten.

The small size of ostracod chromosomes makes a precise interpretation of these mechanisms difficult. Association of multiple X's into a pseudo-bivalent, pseudo-trivalent or pseudo-quinquevalent is obviously one factor involved. And it seems clear that in some genera X-autosome fusions have occurred, while in others there have been Y-autosome fusions. The tendency to formation of multiple sex-chromosome mechanisms seems to be an ancient and world-wide one in this group, since it occurs in two families and in both European and Australian species.

MULTIPLE SEX CHROMOSOMES IN STONE-FLIES

A most interesting series of sex-chromosome mechanisms exists in the Stone-flies (Plecoptera), but its evolutionary history can hardly be guessed at, on the basis of the available evidence. The behaviour of the sex chromosomes in this group at the first meiotic division seems to be highly peculiar, whether or not a 'multiple' mechanism exists. Certain species of Stone flies such as *Perla maxima*, *P. bipunctata* and *Acroneuria jezoensis* are simply XO in the males (Aubert and Matthey, 1943; Matthey 1946*b*,

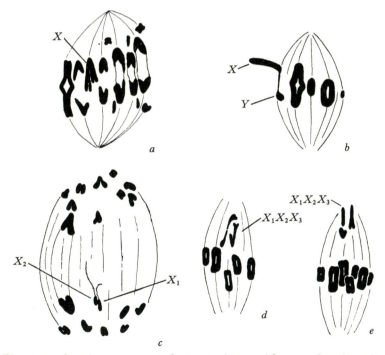

Fig. 17.20. Sex chromosomes at first metaphase and first anaphase in various species of stone flies (Plecoptera). (*a*) *Perla maxima* (XO); (*b*) *Perla immarginata* (XY); (*c*) *P. cephalotes* (X_1X_2O); (*d*) *Perlodes microcephala* ($X_1X_2X_3O$); (*e*) *P. jurassica* ($X_1X_2X_3O$). From Matthey and Aubert (1947) and Nakahara (1919), redrawn.

1947b; Matthey and Aubert 1947; Itoh 1933c); the X is a large metacentric element which is negatively heteropycnotic and lies in one-half of the first meiotic spindle. At this stage its centromere is nearest to the equator and its ends are directed towards one of the poles; at anaphase it moves in the direction of the centromere, i.e. 'backwards' through the equatorial region of the spindle.

Perla immarginata differs from the three species already mentioned in being XY in the male (Nakahara 1919). It is not quite clear whether the X and Y are united by chiasmata at the first meiotic division, but they form a bivalent or pseudo-bivalent and pass to opposite poles at first anaphase (presumably moving 'forwards'). Possibly the Y of this species is a neo-Y, but one cannot be certain.

Four other species of *Perla*, together with several species of *Isogenus* and *Isoperla*, are all X_1X_2O in the male (Matthey and Aubert, 1947; Junker 1923). In these forms the two X's seem to behave in the same general manner as those of the spiders, that is to say, they lie close together and usually parallel in one half of the spindle at the first metaphase. But their anaphase movement is 'backwards', just as in the XO species.

Finally, three species of *Perlodes* studied by Matthey (1946c) and Matthey and Aubert (1947) are $X_1X_2X_3O$ in the male. A peculiar association of the three sex chromosomes takes place in the prophase of the first meiotic division, but it is not clear whether they are held together by chiasmata. At first metaphase they lie in one half of the spindle, in a kind of 'zig-zag' configuration, possibly associated merely by their ends. At first anaphase they move 'backwards', as in the simpler mechanisms (Fig. 17.20).

As in the case of the spiders, it seems certain that the multiple X's of the Stone-flies (X_1X_2 or $X_1X_2X_3$) are not homologous. In most species of *Perlodes* all the three elements appear to differ in size, and two of them seem to be negatively heteropycnotic at first metaphase, while the third is not.

MULTIPLE SEX CHROMOSOMES IN NEMATODES

A whole series of multiple sex-chromosome mechanisms of the X_nO type are known in the nematodes (Goodrich 1914, 1916; Walton 1916, 1924). *Belascaris triquetra* and *Ganguleterakis spinosa* have X_1X_2O males, *Ascaris lumbricoides* is apparently $X_1X_2X_3X_4X_5O$, while *Contracaecum incurvum* has as many as eight different X elements ($X_1 \ldots X_8$) and a Y. Since Y's are only known in two species of nematodes (the other being *Belascaris mystax*) they are probably neo-Y's, due to recent X-autosome fusions, in both cases.

Obviously, the more complex sex-chromosome mechanisms in the

Nematoda have arisen from the simpler ones, and ultimately from the XO mechanisms which are of general occurrence in a number of families. But the exact way in which this has happened is unclear, although one must suspect that dissociation has played a major role, with gradual increase in the total amount of 'sex chromatin', either at the time of the dissociations, or by duplications occurring from time to time in phylogeny. Most of the multiple sex chromosomes known in nematodes occur in the family Ascaridae, in which various peculiar types of chromosome fragmentation take place in the somatic cells (see p. 54). Thus *Toxascaris canis* (Walton 1924) which shows twelve autosomal bivalents and $X_1 \ldots X_6$ at the first metaphase of the male (and presumably twelve autosomal bivalents and six X bivalents in the female) naturally has fertilization nuclei with $2n = 30$ and 36 in male and female embryos, respectively. But in the somatic cells each of these chromosomes fragments into 2, so that $2n\male = 60$ and $2n\female = 72$.

In every case where multiple sex chromosomes have been described in the nematodes, all the X's pass undivided as a group to one pole at first anaphase.

MISCELLANEOUS CASES

A few other 'X_nO' mechanisms may be briefly mentioned. One race of the coreid bug *Syromastes marginatus* is X_1X_2O in the male (Wilson 1909a, b) while another is $X_1X_2X_3O$ (Xavier 1945) and four Asiatic members of this family studied by Manna (1951b) are likewise X_1X_2O. In all these cases the multiple X's are very closely associated throughout the meiotic divisions in spermatogenesis.

Certain species of pyrrhocorid Cotton-stainers of the genus *Dysdercus* are also X_1X_2O in the males and $X_1X_1X_2X_2$ in the females. This is the situation in *D. mendesi* (Mendes 1947; Piza 1951) and in *D. koenigi* (Battaglia 1956). The cytology of the latter species was misinterpreted by Ray-Chaudhuri and Manna (1952), who mistook the X_1 and X_2 for an X and a Y. Some other species of *Dysdercus* are XO in the male (Piza 1947e).

The aphid *Euceraphis betulae* is $X_1X_2X_3X_4O$ in the male, the related species being XO (Shinji 1931); all the X's pass to the same pole at the first meiotic division and, as in other aphids (see p. 754) the second spermatocytes which lack sex chromosomes degenerate, so that only one kind of sperm is formed. One of the scale insects (*Matsucoccus gallicola*) even has an $X_1X_2X_3X_4X_5X_6O$ mechanism in the male (Hughes-Schrader 1948a). At first metaphase in the male the six X elements become arranged in a small ring, surrounded by the autosomal bivalents. They move as a unit to one pole at first anaphase and split in the second division. Since there are fourteen pairs of autosomes, the two kinds of sperm carry fourteen and twenty chromosomes, respectively.

GENERAL SIGNIFICANCE OF COMPLEX SEX-CHROMOSOME SYSTEMS

The multiple sex-chromosome mechanisms of the mantids, morabine grasshoppers, *Paratylotropidia*, *Karruacris browni*, fleas, *Potorous*, *Sorex araneus*, etc. may be compared with the complex structural heterozygosity found in *Oenothera*, *Rhoeo* and some other plant genera. In these plants, however, it is usual for the chromosomes concerned to form a ring of 4, 6, 8, 10, 12 . . . at meiosis; whereas in the animals mentioned above the sex chromosomes form a chain of three, and are incapable of forming a ring. In one species of animal, however, a sex-chromosome mechanism exists which does seem to take the form of a truly *Oenothera*-like ring at meiosis. This is the copepod *Diaptomus castor*, in which Matschek (1910) and Heberer (1932) demonstrated the existence of a ring of six chromosomes in oogenesis, in addition to fourteen bivalents. At the corresponding stage in the male there are seventeen bivalents and no ring. Since female hetero-gamety seems to be the rule in copepods (Beermann 1954; Rüsch 1960; Ar-Rushdi 1963), it is fairly clear that this mechanism has been derived from an XY one as a result of two successive translocations whereby two pairs of autosomes become involved in the sex-chromosome mechanism. Formally, this could be designated as an $X_1X_2X_3Y_1Y_2Y_3(\female)$: $X_1X_1X_2X_2X_3X_3(\male)$ system, the differential segments being presumably proximal or interstitial, and not terminal. There is some evidence that it only exists in one geographic race of the species, another race or group of populations not having a ring of chromosomes in oogenesis. The ring is probably held together by terminalized chiasmata, *Diaptomus* being a copepod with chiasmata in the female, unlike *Ectocyclops* (Beermann 1954).

An even more complex type of sex-chromosome mechanism has been described by Ogawa (1954) in the centipede *Otocryptops sexguttatus*, which has four X's and five Y's in the male ($2n\male = 15$, i.e. there are three pairs of autosomes). At the first meiotic division the nine sex chromosomes are associated in a chain configuration ($Y_1X_1Y_2X_2Y_3X_3Y_4X_4Y_5$) which orientates itself on the spindle so that the alternate members pass to opposite poles, the X's to one and the Y's to the other (Fig. 17.21). A significant difference from the *Oenothera* and *Rhoeo* rings and chains is that the chromosomes of the *O. sexguttatus* chain are of very different sizes and some of them are described as short rods (more probably they are J-shaped elements). Such a complex sex-chromosome mechanism could theoretically have arisen by four mutual translocations, starting with an XO system ('mantid mechanism') or from an XY mechanism by seven centric fusions, each involving the loss of a centromere ('grasshopper mechanism'); or perhaps by some combination of mutual translocations and centric fusions. A study of other centipedes and especially of other species of the genus *Otocryptops* makes it likely that centric fusions rather than mutual translocations have produced the sex-chromosome mechanism of *O. sexguttatus*. In the first place Ogawa (1961d) has described some individuals of this species which only have a chain of five sex chromosomes ($X_1Y_1X_2Y_2X_3$ or $Y_1X_1Y_2X_2Y_3$) in which the terminal members of the chain are acrocentric. And in the second place a fairly large number of species of centipedes are known to have XY males, whereas none are known with an XO mechanism (Ogawa 1953, 1961c; Putanna 1959). *O. curtus* ($2n\male = 32$) and *O. capillipedatus* ($2n\male = 22$) are both XY in the male sex (Ogawa 1961a, c). An unnamed species of the same genus with $2n\male = 23$ is XY_1Y_2; its sex chromosomes form a trivalent at meiosis (Ogawa 1955b). Finally, *O. rubiginosus* includes individuals which are X_1X_2Y ($2n\male = 27$), $X_1X_2Y_1Y_2$ ($2n\male = 26$) and either $X_1X_2X_3Y_1Y_2$ or $X_1X_2Y_1Y_2Y_3$ ($2n\male = 25$). The first show a trivalent, the second a chain of four sex chromosomes, the latter a chain of five (Ogawa 1961b). It is clear from the numbers of autosomes in the three types of males than an X-autosome fusion converted the X_1X_2Y condition into an $X_1X_2Y_1Y_2$ one, and that a further X-autosome or Y-autosome fusion (it is not certain which) led to the condition in which a chain of five sex chromosomes is found at meiosis. In this case only the terminal members of the chain are possibly acrocentrics, the three central ones being clearly equal-armed metacentrics, which is what we should expect if they had arisen by centric fusion. Analogy with *O. rubiginosus* and the fact that it has the lowest chromosome number known in the genus suggests that the chain of nine found in *O. sexguttatus* has been derived by centric fusions, although the sizes and shapes of the members of the chain may indicate that other structural changes have occurred in addition.

Since all these mechanisms of sex-chromosome complex heterozygosity in animals are confined to one sex, they differ from the *Oenothera* system in that no balanced lethals or 'Renner effects' are involved.

The even more complex sex-chromosome system of *Blaps polychresta* (Fig. 17.22) is, of course, quite different in principle from that of *Otocryptops* in that no chain is formed, and there is no mechanism of alternate segregation, although X's and Y's do travel to opposite poles.

Multiple sex chromosomes are an example of the tendency of living organisms to build up extremely elaborate systems which achieve a result that other species manage to arrive at in much simpler ways. Viewed as mechanisms of sex determination, the chain of nine chromosomes in

Otocryptops sexguttatus and the association of twelve X's and six Y's in *Blaps polychresta* simply do not 'make sense', when we know that in some other organisms the switch mechanism consists of only a very short section of one chromosome, perhaps a single genetic locus. Their main adaptive significance is, presumably, that they lead to fixation of a great deal of heterozygosity; although of course in one sex only. Another factor that may be of some importance is that a species with several X's $(X_1X_2\ldots)$ will have much genetic recombination within its 'sex-linked' genetic material, in the homogametic sex. Species such as the two cited above

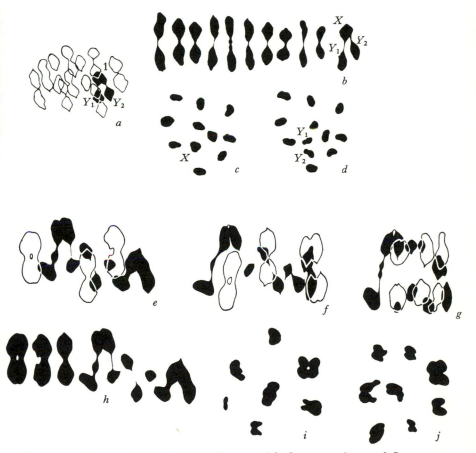

Fig. 17.21. Chromosomes of centipedes. *a* and *b*, first metaphases of *Otocryptops* sp. (XY_1Y_2); *c* and *d*, second metaphases of the same; *e, f, g, h*, first metaphases, in side view, of *Otocryptops sexspinosus*, showing the chain of nine sex chromosomes (in black in Figs. *e, f, g*); *i* and *j*, second metaphases with 7 and 8 chromosomes respectively. From Ogawa (1954, 1955).

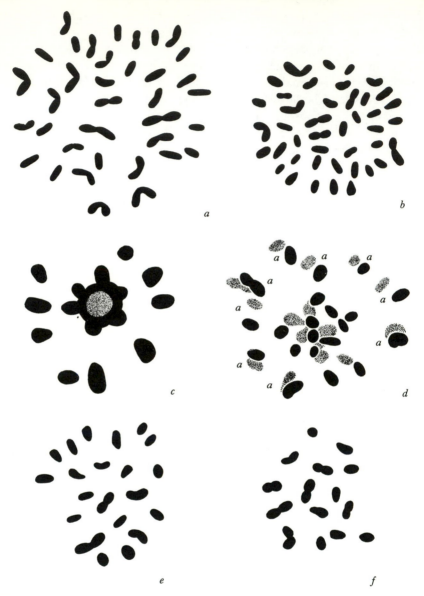

Fig. 17.22. Chromosomes of the beetle *Blaps polychresta*. *a*, spermatogonial metaphase (36 chromosomes); *b*, female somatic metaphase (42 chromosomes); *c*, first metaphase in the male (polar view) showing the mass of sex chromosomes surrounded by nine autosomal bivalents; *d*, very early first anaphase in polar view with one anaphase group in black, the other stippled (autosomes designated *a*); *e* and *f*, second metaphases, with 21 and 15 chromosomes respectively. From Guénin 1953.

probably achieve a considerable amount of adaptive polymorphism, and of heterotic interactions between X's and Y's in the heterogametic sex.

A few species of animals with complex sex-chromosome mechanisms are polymorphic for autosomal rearrangements or supernumerary chromosomes. The association may indicate a tendency for structural rearrangements to establish themselves relatively freely in the karyotype of the species in question. Thus the beetle *Chilocorus stigma*, with $X_1 X_2 Y$ males, is conspicuously polymorphic for autosomal fusions, the $X_1 X_2 Y$ fleas studied by Bayreuther (1954) are polymorphic for supernumerary chromosome regions and one of the $X_1 X_2 Y$ species of morabine grasshoppers is polymorphic for pericentric inversions and for a supernumerary chromosome. But the majority of species with complex sex-chromosome mechanisms show no evident chromosomal polymorphism, so that the association, where it occurs, may be fortuitous.

18

HAPLODIPLOIDY

> As bees are a peculiar and extraordinary kind of animal, so
> also their generation appears to be peculiar. That bees should
> generate without copulation is a thing which may be paralleled
> in other animals, but that what they generate should not be
> of the same kind is peculiar to them . . .
>
> ARISTOTLE: *De Generatione Animalium*

In a few groups of animals the males arise from unfertilized eggs, by a form
of parthenogenesis, while the females develop from fertilized eggs. Such a
genetic system may be referred to as *haplodiploidy, haploid parthenogenesis*
or *arrhenotoky* (male production), the latter term implying a contrast to
thelytoky, the type of parthenogenesis in which the progeny are all females
(see Ch. 19). Aristotle (see above) clearly grasped one essential difference
between arrhenotoky and thelytoky.

Males in haplodiploid species may be referred to as impaternate, since
they have no fathers. Females have two parents, but only three grand-
parents, five great-grandparents, etc. The formal genetics of haplo–diploid
species follows the same pattern as sex-limited inheritance with male
heterogamety; indeed, one way of looking at haplodiploidy is to consider
it as a genetic system in which there are only X-chromosomes with no Y and
no autosomes.

Haplodiploidy and thelytoky are in reality very different systems, even
though they sometimes coexist in the same group, and occasionally even in
the same species. The former is a mechanism which still involves genetic
recombination at each generation, while in the case of thelytoky recombination no longer occurs. In the terminology of Dobzhansky (1950*b*) a
haplodiploid population is still a 'Mendelian' one, albeit of a peculiar type
while a thelytokous population is non-Mendelian. Thelytoky is simply a
reproductive device which does not involve the development of any new

sex-determining mechanism, and in a species reproducing exclusively by thelytoky, males are absent altogether. Haplodiploidy, on the other hand, is a method of sex determination, as well as a reproductive system, the frequency of males in the population being detcrmined by the frequency with which unfertilized eggs are laid. It is thus characteristic of groups with haplodiploidy that the sex ratio may vary rather widely from species to species and from strain to strain within species, and also to some extent with various environmental factors, showing no particular tendency to conform to a fixed proportion of males in the population.

Thelytoky has arisen repeatedly in almost all the phyla of the animal kingdom, and can be induced experimentally in a number of ways. Arrhenotoky, on the other hand, is only known to have arisen about eight times in the whole evolutionary history of the Metazoa (probably six times in insects, once in arachnids and once in rotifers). This is not unexpected, since the origin of arrhenotoky – although the details are unknown – obviously involves the replacement of an original sex-determining mechanism by an entirely new one, an event which (as we have seen in the case of male and female heterogamety) can only have taken place under very special circumstances.

The four insect orders which show haplodiploidy (in some or all species) are the Hymenoptera, Homoptera, Thysanoptera and Coleoptera. In the Hymenoptera the whole order (apart from a few species which have become secondarily thelytokous) shows haplodiploidy, from the primitive sawflies to the most specialized parasitic and social wasps and bees. The mechanism is obviously of great antiquity in this case, going back to early Mesozoic or late Palaeozoic times, and any attempt to speculate as to its mode of origin would be fruitless, particularly since the evolutionary relationships of the Hymenoptera to other living and fossil groups of insects are not known with any degree of certainty.

In the Homoptera the majority of species (including all the members of the more primitive suborder Auchenorhyncha) have diploid males with a sex-chromosome mechanism of the usual kind (usually XO :XX). However, in the coccids of the tribe Iceryini and certain, at any rate, of the White-flies (family Aleurodidae) males are haploid and arise from unfertilized eggs. These must surely represent independent evolutionary origins of haplodiploidy. In addition, as we have already seen (Ch. 14) the Armoured scale insects (Diaspididae) have a unique type of genetic system in which males are haploid, but arise from fertilized eggs, the paternal chromosome set being eliminated early in development. Whether we should call the diaspidid system haplodiploidy is really a question of semantics. It is obviously akin in some respects to the situation in the cecidomyids of the subfamily Heteropezinae, in which the males have a haploid soma. Among the Coleoptera *Micromalthus debilis*, the sole member of the family Micromalthidae, of uncertain relationships, has male haploidy combined with a complicated life cycle involving paedogenetic reproduction. In addition some bark-beetles of the genus *Xyleborus* have quite independently acquired male haploidy. In the Thysanoptera (thrip insects) a number of species are known, either from genetic or cytologic evidence to show haplodiploidy and it is probable that this genetic system is found throughout the whole group, except for those species that reproduce by thelytoky (Shull 1917; Davidson and Bald 1931; Bournier 1956*a*, *b*).

In the mites and ticks (Arachnida, order Acarina) Schrader (1923*c*) first demonstrated male haploidy in a member of the Tetranychidae. Jary and Stapley (1936) and Putman (1939) showed that unmated females produced all-male progenies in *Histiostoma rostro-serratum* and *Phyllocoptes fockeui* respectively. Helle and Bolland (1967) showed that thirteen species of this family have haploid males. Later cytological and/or genetic studies have shown that haplodiploidy also exists in the Tarsonemidae (Pätau 1934, 1936; Cooper 1937, 1939), Harpyrhynchidae (Oliver and Nelson 1967; Moss, Oliver and Nelson 1968), Phytoseiidae (Hansell, Mollison and Putman 1964; Wysoki and Swirski 1968) and Dermanyssidae (Skaliy and Hayes 1949; Oliver, Camin and Jackson 1963; Oliver 1965*b*). In the Tyroglyphidae, Trombidiidae, Gamasidae (= Parasitidae), Ixodidae and Argasidae, however, the males are diploid and heterogametic (Sokolov 1934, 1945, 1954; Opperman 1935; Dutt 1954; Sharma and Joneja 1960; Oliver 1965*a*, 1966, 1967). In the Water-mites (Hydrachnellae) the males are diploid, but sex chromosomes have not been observed (Sokolow 1962*b*). As in the case of the Hymenoptera, a few species of mites such as *Chelyletus eruditus* have become thelytokous and males are unkown or extremely rare. Among the ticks, *Amblyomma rotundatum* is obligatorily thelytokous and *Haemaphysalis bispinosa* is represented in Australia by a triploid thelytokous biotype, although diploid and bisexual in some other areas (Oliver 1967).

The rotifers (phylum Trochelminthes) include a number of families among which the life-cycle varies considerably. In the group of the Bdelloidea males are unknown, and thelytoky is presumably the only method of reproduction (see p. 698). In the family Seisonidae (marine parasites of certain Crustacea) reproduction is apparently bisexual, males being present throughout the year and similar in size to the females. In the remaining families a complicated alternation of parthenogenetic and sexual generations occurs, the males (which are usually dwarf-sized and may be structurally degenerate) only being produced at certain seasons of the year or under particular environmental conditions. It is generally accepted that these males develop from unfertilized eggs which have undergone two maturation divisions and that they are haploid in constitution. However, the literature on the cytology of the Rotifera is in a rather chaotic state, there being so many points of disagreement between the various accounts that it is difficult to arrive at any firm conclusions.

When we say that the male of a species is haploid we mean that its germ-line nuclei contain half the number of chromosomes present in the corresponding diploid nuclei of the female. However, in insects most of the somatic tissue shows high degrees of endopolyploidy. The question, then, arises as to whether the final degree of endopolyploidy is the same in the male and female nuclei. Merriam and Ris (1954) attempted to answer this question by a study of nuclear size and DNA content of somatic nuclei in queen, workers and drone honey bees. They concluded that 'male tissues have about the same chromosome numbers as comparable female tissues' and that 'during growth, nuclei of the male undergo compensatory endomitosis so that in the adult the DNA is usually about the same as in the females. From an investigation of chromosome numbers alone in somatic tissues [which they do not appear to have carried out] it would be impossible to distinguish male from female'. However, it is doubtful whether their findings really justify this conclusion, except in the case of the intestinal cells (see Risler 1954). In the Malpighian tubules the majority of the nuclei in the queens appear to be 32-ploid, while most of those in the drones and workers seem to be 16-ploid. In the haplodiploid coccid *Steatococcus tuberculatus*, O'Brien (1956) found a complex situation. Hypodermal cells remain haploid in the male and diploid in the female (ratio 1:2); in the fat body most nuclei in both sexes are diploid (ratio 2:2); in the wax glands most male nuclei are tetraploid, while female nuclei are diploid (ratio 4:2); and in the Malpighian tubules most male nuclei are octoploid while female nuclei become 32-ploid (ratio 8:32). A generally similar situation appears to exist in the haplodiploid thysanopteran *Haplothrips* (Risler and Kempter 1962). Here also the hypodermis remains haploid in the male and diploid in the female, as do also the cells of the nervous system, tracheae and hind gut. Endopolyploidy occurs in the fat body, mid gut epithelium and Malpighian tubules, but in all of these the female nuclei have twice the chromosome number present in the male. However, the oenocyte nuclei are larger in the male than in the female and have presumably reached a higher degree of ploidy. We may thus conclude that in haplodiploid organisms each sex has its own characteristic developmental pattern of endopolyploidy, and that there is no overall regulatory process bringing the male somatic nuclei up to the same level of endopolyploidy as the female ones – indeed in certain tissues the level of endopolyploidy may even be *higher* in the male than in the female. The hypodermis and nerve cells usually retain the original haploid and diploid chromosome complements (Dreyfus and Breuer (1944) in the case of *Telenomus*, and Kerr (1948) in that of *Melipona*).

In view of these complex relationships it is perhaps fruitless to speculate as to whether mechanisms of dosage compensation are required in haplodiploid organisms. Thoday (in discussion on White 1964) suggested that in *Drosophila* 'the need for dosage compensation arises because there is one set of X genes and two sets of autosomal genes in the male, but two sets of each in the female'. But it is by no means certain that the need for dosage compensation does depend on this kind of genic balance; there is an alternative possibility, namely that it is the absolute dosage of 'sex-linked' loci (in haplodiploid organisms all loci can be regarded as sex-linked) in relation to the cytoplasm which is significant. Studies of the expression of mutant alleles in male individuals of Hymenoptera hypo- and hyperploid for the loci in question should be carried out. It may be that as far as many tissues are concerned, the male attains the same degree of ploidy as the female (which, as pointed out by Stern (1960a), would be equivalent to 'dosage compensation'). But this is obviously not so for cuticular structures derived from the hypodermis, which seems to always remain haploid in males.

As one might expect, the males in all haplodiploid groups produce haploid sperms by a type of spermatogenesis that does not involve any reduction in chromosome number. The males of Hymenoptera show several different types of spermatogenesis, but in none of these is there any synapsis or bivalent-formation. In what is probably the most widespread type of spermatogenesis in the order there is an abortive first meiotic division in which a non-nucleated cytoplasmic bud is eliminated at the end of a peculiar type of spindle, the spermatocyte nucleus never going into a true metaphase; a second meiotic division which is a simple mitosis then follows and gives rise to two equal spermatids that develop into sperms. This type of spermatogenesis has been reported in *Sirex* (Peacock and Greeson 1931) and *Habrobracon* (Torvik Greb 1935), following the early work of Mark and Copeland (1906) on *Vespa*. However, in *Polistes*, which is closely related to *Vespa*, Pardi (1947) failed to observe cytoplasmic buds, neither could they be found in the spermatogenesis of the sawflies *Pteronidea* (Sanderson 1933) and *Diprion* spp. (S. G. Smith 1941).

In four species of sphecoid wasps Whiting (1947) found no trace of the abortive first meiotic division and no cytoplasmic buds are given off prior to the equal mitotic division. No first meiotic division of any kind was observed in the bees *Osmia cornuta* and *Euplusia violacea* by Armbruster (1913) and Kerr (1948). Other bees do show the abortive first division, and in all bees (but not in other Hymenoptera) the second meiotic division

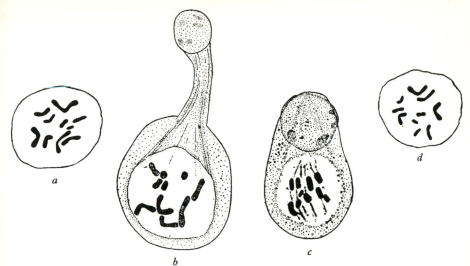

Fig. 18.1. Spermatogenesis of a haploid male *Habrobracon*. *a*, spermatogonial metaphase (10 chromosomes); *b* and *c*, stages showing the spermatocyte attached to a cytoplasmic 'bud' (possibly equivalent to a highly modified first division); *d*, metaphase of the single maturation division, which gives rise to two equal spermatids. From Torvik-Greb (1935).

involves an unequal cytoplasmic division, so that one functional spermatid and a small nucleated cell destined to degenerate are formed. Thus in bees each primary spermatocyte only gives rise to a single sperm (Sharma, Gupta and Kumbkarni 1961).

Various cytologists have at one time or another claimed that they had demonstrated sex chromosomes in Hymenoptera, but all these claims seem to have been refuted by later work or are inherently incredible. Thus Dozorcheva's (1936) description of a pair of sex chromosomes in the chalcid *Pteromalus puparum* has been discredited by later work (see Whiting 1940, 1945) and Dreyfus and Breuer's account of an *XY* pair in the females of the scelionid *Telenomus fariai*, in which they also reported an unequal division in the spermatogonia, is quite unconvincing. Manning (1949–50) described a hook-shaped body in the honey bee which he regarded as an *X*-chromosome, and his conclusions were supported by Kerr (1951*a*). The female was supposed to have fifteen pairs of autosomes and one *X* (i.e. thirty-one chromosomes altogether) males were supposed to have sixteen chromosomes, but to produce sperms with only fifteen. This cytological interpretation seems to have been decisively disproved. Females have 2n = 32, including two large hook-shaped chromosomes (Sanderson and Hall 1948); the body which fails to be included in the spermatic nuclei and which was interpreted as an X undoubtedly exists, but is Feulgen-negative and hence not a chromosome (Ris and Kerr 1952); and finally, sperm nuclei develop into male tissue in certain types of bee gynandromorphs (Rothenbuhler, Gowen and Park 1951).

In the oogenesis of the Hymenoptera synapsis and chiasma formation seem to take place quite normally in all except the thelytokous species (Doncaster 1906; Speicher 1936; S. G. Smith 1941). Thus, in all normally bisexual Hymenoptera, crossing-over occurs on the female side, but not in the males. The spermatogenesis of the diploid males of *Habrobracon* follows almost exactly the same course as that of the haploid ones, except that there are twenty chromosomes instead of ten at every stage. No synapsis takes place and the single maturation division is mitotic in type, It is thus clear that the type of spermatogenesis found in the Hymenoptera is genetically determined and is not merely an automatic consequence of haploidy.

The males of the haplodiploid mites, aleurodids and iceryine coccids all have meiosis replaced by a single mitotic division, each primary spermatocyte forming only two sperms. In the Thysanoptera, however, according to the account of Bournier (1956a, b) the situation is somewhat more complicated. The details are obviously very hard to make out on account of the small size of the nuclei, and it was not possible to actually count the chromosomes. The first maturation division, in spermatogenesis, gives two equal

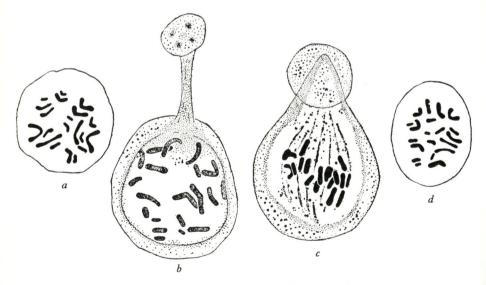

Fig. 18.2. Spermatogenesis of a biparental diploid male *Habrobracon*. The stages are the same as in the previous figure, and the only essential difference is that there are 20 chromosomes instead of 10. From Torvik-Greb (1935).

sized secondary spermatocytes. On the assumption that the male germ-line is haploid, this must be a mitotic division. The second maturation is obviously peculiar, two groups of chromosomes separate to opposite sides of the nucleus after the first telophase, and are visibly different in appearance. These groups pass into different-sized spermatids. In the large spermatids the nuclei are clear with diffuse chromosomes, while in the small cells the nuclei become pycnotic and degenerate. There seems to be one chromosome much larger than the others at the first maturation division. It is by no means certain that the number of chromosomes in the pycnotic mass after the second division is the same as in the nucleus of the functional spermatid. Obviously, further studies on the chromosome cycle of the Thysanoptera are needed before a rational interpretation can be arrived at. Some species have low chromosome numbers (e.g. $n = 10$ in *Taeniothrips simplex*), but others have quite high numbers, ranging up to $n = 50$ to 53 in *Aptinothrips rufua* (Bournier 1954; Pussard-Radulesco 1930).

A species of beetle (*Micromalthus debilis*), which belongs to the primitive suborder Archostemata (it is the sole species of the family Micromalthidae) has long been known to have haploid males (Scott 1936, 1938). It had earlier been shown by H. S. Barber (1913) that this species, which bores in rotten wood, has a very complicated life cycle, similar in many respects to that of the cecidomyids *Heteropeza* and *Miastor* (see p. 524). Some of the larvae reproduce paedogenetically, while others give rise to the sexual males and females. There are two main types of paedogenetic larvae, a viviparous type which produce only female offspring, and an oviparous type which give rise to males. The adult females are diploid ($2n = 20$), while the males are haploid ($n = 10$); no sex chromosomes have been observed at any stage.

The spermatogenesis of the males is anomalous. At a stage corresponding to first metaphase a 'unipolar' spindle is formed, to the diffuse end of which all the chromosomes are attached. This 'division' seems to be entirely abortive, since neither chromosomes nor cytoplasm divide. The second division is a normal mitosis, so that two sperms are formed from each primary spermatocyte.

Since *M. debilis* is the only species of its genus and family (which latter is a taxonomically isolated one) it would be useless to speculate as to the origin of its peculiar reproductive and chromosomal cycle. Sex determination in this species may possibly be environmentally controlled (Scott 1941) although the possibility of genetic factors being involved has not been excluded.

Two other species of beetles (*Xyleborus compactus* and *X. germanus*), which belong to the family Scolytidae, have recently been shown to reproduce by haplodiploidy (Entwistle, 1964; Takenouchi and Tagaki 1967). Females, which develop from fertilized eggs, have $2n = 16$; males, which arise from unfertilized eggs show $n = 8$ in the spermatogonia. An unusual feature is that there are apparently two maturation divisions in spermatogenesis, both mitotic in type.

From a genetic standpoint, haplodiploidy presents a number of problems which have been only partially resolved. Three major ones are: (1) the mechanism of sex determination, (2) the genetic structure of natural populations, (3) mechanisms of speciation. These questions have been discussed mainly in the case of the Hymenoptera; it does not necessarily follow that mechanisms of sex determination are similar in all haplodiploid groups, although one might expect that the genetic structure of natural populations would be of the same general type.

At first sight it would appear that neither the genic balance principle nor the 'dominant Y' mechanism could be operative in haplodiploid organisms where the females have the same chromosome set as the males but carry it in duplicate. The idea that sex determination in Hymenoptera might depend on some kind of balance between female-determining genes in the chromosomes and male-determining cytoplasmic factors seem very unlikely in view of the work on the degree of somatic endopolyploidy in the two sexes (see p. 684). The whole question was in a very uncertain state when the Whitings (1925, 1927) discovered that in the parasitic wasp *Habrobracon* (or *Bracon*) *juglandis* diploid males could be obtained by inbreeding (brother–sister mating). Apart from the case of *Mormoniella* (see p. 000) where diploid males were obtained in the offspring of triploid mothers, these are the only Hymenoptera in which diploid males have been obtained so far. The diploid males of *Habrobracon* were clearly biparental, i.e. derived from fertilized eggs and not by diploid parthenogenesis as might otherwise have been supposed) since they showed genetic characters inherited from both parent. This discovery naturally tended to discredit the idea that maleness in Hymenoptera was a direct consequence of haploidy. It was eventually found, as a result of genetic experiments with stocks carrying the gene *fused* (Whiting 1939, 1943*a*; Bostian 1939) that the female *Habrobracon* is heterozygous for a sex factor which is closely linked with *fused*. Thus femaleness seems to depend on heterozygosity for certain sex factors which may be designated X_1, X_2, X_3 . . . The normal haploid males are of several phenotypically different kinds, each carrying a different X factor, while the diploid males, which arise only on inbreeding, are homozygous for the X factor (e.g. X_1X_1, X_2X_2, X_3X_3 . . .). It is uncertain whether these X factors are actually a series of multiple alleles at a single locus. They may be very short differential regions or small duplications; at any rate they are not visible cytologically. Whiting (1943a) has demonstrated the existence of at least nine different X factors which lie on a linkage map which seems to have a genetic length of at least 350 cross-over units, and which must presumably have a very high chiasma frequency of about 7.0 (this point has not been checked cytologically, since Speicher (1936) only studied the later stages of oogenesis).

Apparently none of the combinations of X factors in *Habrobracon* lead to intersexuality. But some of the homozygous combinations have a rather low viability. The diploid males are almost, but not quite completely, sterile, but entirely male in appearance. In Japanese strains of *Habrobracon* the diploid males are reasonably fertile (Inaba 1939); when mated with normal diploid females they give rise to triploid daughters. Several types of genetic intersexuality are known in *Habrobracon*, but they do not seem to be particularly relevant to the normal mechanism of sex determination in the species; one mutation masculinizes female wasps, another feminizes males (Whiting 1943b). Even more interesting are the 'gynandroids' obtained by Whiting – individuals which are haploid throughout but which are derived from binucleate eggs so that the cells on the two sides of the body contain different X factors. Such individuals are basically male but have a streak of feminized tissue in the mid-line, where an ovipositor of some kind develops. Probably such individuals should be regarded as evidence for some kind of diffusible substances, produced under the influence of the X factors, which interact to switch development into a female pathway just as must happen normally in diploid cells heterozygous for sex factors.

Although the genetic mechanism controlling sex determination in *Habrobracon* is of considerable interest, it does not seem to be universal in the Hymenoptera. Thus in *Telenomus fariai* (family Scelionidae) Dreyfus and Breuer (1944) did not find any diploid males by cytological methods, although brother–sister mating seems to be the rule in this species, copulation taking place with the egg shells of the host, the bug *Triatoma* (the Scelionidae are egg-parasites). Also, in *Melittobia* spp. (Chalcidoidea) Schmieder and Whiting (1947) and Whiting (1947) found no diploid males after close inbreeding (see also Schmieder 1938). The possibility that they might be produced but be inviable was excluded, on account of the very high percentage egg hatch. The species of this genus reproduce mostly by matings between mothers and their sons or between sibs; it accordingly seems impossible that they could maintain a sex-determining mechanism of the *Habrobracon* type.

In another chalcidoid, *Mormoniella vitripennis*, inbreeding likewise does not lead to the production of diploid males. However, in this species some triploid females were obtained which, when unmated, gave rise to both haploid and diploid sons (Whiting 1960). The diploid males obtained in this species are hence uniparental and they are quite fertile, producing abundant diploid sperm. It mated to normal females they give rise to triploid daughters. Triploid females may, of course, be mated with either haploid or diploid males. In the former case the progeny are mostly inviable but a few daughters are produced. In the latter case a few sons were obtained from unfertilized eggs as well as many inviable eggs and some triploid and tetraploid daughters. Sex determination in *Mormoniella* is certainly not fully understood, but appears to depend solely on whether an egg is fertilized or not, regardless of ploidy. In a related chalcidoid wasp, *Pachycrepoideus dubius*, Whiting (1954) likewise failed to get diploid males after brother–sister mating. Populations of iceryine coccids are probably very inbred and in *I. purchasi* (see p. 506) self-fertilization usually occurs without any production of diploid males.

In the honey bee, Mackensen (1951, 1955) Rothenbuhler (1957) and Drescher and Rothenbuhler (1964) have produced evidence that sex is determined by a series of multiple complementary factors as in *Habrobracon*. Following inbreeding diploid males which are homozygous for X alleles are produced (Woyke 1963a, b, 1965, 1967); these diploid male larvae are eaten by the workers, but may be reared artificially. According to Kerr and Nielsen (1967), however, these individuals are partially feminized. The same workers inbred fifteen species of stingless bees (Meliponini) without obtaining any diploid males; they suggest that the X alleles of the evolutionarily advanced Apinae are recent in origin. Eleven different X alleles have been demonstrated in six inbred lines of honey bees. Rothenbuhler (1957) obtained some bees similar in principle to Whiting's gynandroids in *Habrobracon* i.e. mosaics of two different kinds of haploid tissue with feminized zones where areas carrying different sex factors meet.

If sex determination in many and perhaps most haplodiploid organisms depends neither on an X-autosome balance, a dominant Y, an X-cytoplasm balance or a *Habrobracon* type mechanism, what *does* it depend on? One plausible suggestion has been made by Da Cunha and Kerr (1957) and Kerr (1967). According to this hypothesis sex determination may depend on a series of genes m for maleness having non-cumulative effects and another series f for femaleness whose action is cumulative. Thus sex would be governed by the relationship $ff > mm = m > f$. In haploid individuals $m > f$ so that we get a male, while in diploids $ff > mm$ so that we get a female. In *Habrobracon* the X factors could be f genes which have lost their cumulative effects for female-determination in the homozygotes but which are still cumulative in the heterozygotes. The fact that triploid individuals of *Habrobracon* are females (Torvik-Greb 1935; Inaba 1939) seems to support this interpretation.

The population genetics of haplodiploid organisms has not yet been studied. It is clear that all gradations exist between outbreeding and inbreeding. Whatever the breeding system, selection against deleterious recessives should be extremely severe. Newly arisen mutations will be exposed to the full force of natural selection in the male sex from the very first. One might expect that this would lead to haplodiploid species having much less genetic polymorphism in their natural populations than organisms in which both sexes are diploid, in other words, the 'reservoir of hidden variability' should be small. This may, in fact, be so. However, sex-linked mutants which are not expressed in the male will constitute a special category. It may be that the apparently greater evolutionary plasticity of the female sex in the Hymenoptera (shown, *inter alia*, by the caste-polymorphism – environmentally controlled though it usually is* – of the female but not the male sex in social groups) is a consequence of this fact. Kerr (1951*b*) has reported two genes with female-limited effects in wild populations of bees belonging to the genera *Trigona* and *Melipona*. Polymorphism for chromosomal rearrangements such as inversions, which has been so important in furthering our understanding of genetic polymorphism in natural populations of Diptera and grasshoppers, does not seem to have been recorded in any haplodiploid groups.

Speciation mechanisms have also not been studied at all extensively in Hymenoptera and other haplodiploid groups. Chromosomal differences between related species seem to be similar in extent to those in groups where both sexes are diploid.

Various authors have speculated at one time or another on the possible occurrence of evolutionary polyploidy in groups with haplodiploidy. The evidence of chromosome numbers in coccids, mites and Hymenoptera seems rather unfavourable to such suggestions. Chromosome numbers seem to be very low and rather uniform in the two former groups. In the Hymenoptera we now have a reasonably large number of chromosome counts for the sawflies (Sanderson 1933; Comrie 1938; Peacock and Sanderson 1939; Smith 1941; Waterhouse and Sanderson 1958), ants (Hauschtek 1961, 1963*a*, *b*; Imai and Yosida 1964; Crozier 1968*a*, *b*, *c*, 1970), and bees (Kerr and Laidlaw 1956; Kerr and Araujo 1957; Kerr 1969). In the first group we have a range from $n = 6$ to $n = 21$, but there is really no indication of any polyploid series. *Diprion similis*, with $n = 14$ (S. G. Smith 1941) seemed at first to be a tetraploid by comparison with other species of the

---

* In almost all social Hymenoptera the castes are determined by the quantity and quality of the food they receive (trophogenic determination). But in the bee genus *Melipona*, Kerr (1950*a*, *b*) has presented evidence that the queen and worker castes are determined genetically, queens being heterozygous for several genes, each of which when homozygous, causes the individual to develop into a worker.

genus, having n = 7; but according to Smith (1960b) its karyotype has been derived by 'centric fission' (i.e. dissociation of metacentrics to give acrocentrics).

In ants the chromosome numbers (n) range from 4 (in a species of *Stenamma* studied by Hauschtek 1963a) to 26 and 27 (in various species of *Formica* and *Polyergus* studied by Imai) and 28 in *Myrmica sulcinodis*; the latter species may really be diplotetraploid rather than haplodiploid, since many cells in the testis showed 14 'chromosomes' (Hauschtek 1965). Apart from this, there is no evidence of polyploid series in ants and some large genera such as *Iridomyrmex* seem to show a graded series (e.g. n = 7, 8, 9, 14 in the latter genus). The primitive *Myrmecia ruginoda* has n = 15 (Crozier 1970). Chromosomes are generally metacentric in ants, but some forms have acrocentric chromosomes as well, so that the evolution of chromosome numbers has obviously been by fusion and dissociations, as in other groups. The wasp genus *Polistes* shows a graded series of chromosome numbers (n = 6, 9, 13, 14 and 21 in the species that have been investigated by Machida 1934 and Pardi 1947).

In bees chromosome numbers seem to be more uniform in the larger genera and even tribes; for example n = 9 in Meliponini, 12 in Bombini and 16 in Apini, Euglossini, Xylocopini and Megachilini. Several species of the tribe Trigonini from Africa and South America have n = 18 and are regarded by Kerr and Araujo (1957) as diplotetraploid by comparison with the closely related Meliponini with n = 9; this interpretation is particularly plausible in view of the fact that in two species of Trigonini the 18 chromosomes are synapsed in 9 pairs during the prophase of meiosis in the male, although desynapsis apparently occurs before metaphase. Kerr (1969) also regards *Melipona quinquefasciata*, with n = 18 as diplotetraploid by comparison with other species of the genus which have n = 9. There is even a possibility (but no certainty) that the Apini with n = 16 are diplotetraploid, since Thakar and Deodikar (1966) found *Apis florea* to have n = 8.

The idea that the Hymenoptera as a whole should be regarded as diplotetraploid rather than haplodiploid was put forward by Greenshields (1936) and revived by Maxwell (1958) but is not supported by any critical evidence; in fact, all the genetical work on Hymenoptera seems to negate it. S. G. Smith (1960b) rightly speaks of it as an 'obsolete theory'. We cannot know just how haplodiploidy arose in the Triassic Permian ancestors of the Hymenoptera, but the problem is not made any easier by the gratuitous assumption of a doubling of the chromosome number.

It is possible to consider the haplodiploid groups as possessing an $X_1 \ldots X_n O$ sex-chromosome mechanism, i.e. a genetic system with a multiple X but no Y and no autosomes. In fact, this way of expressing the situation may represent the mode of origin of haplodiploidy. It is a matter of general experience that haploid individuals (and in many cases even haploid patches of tissue) are inviable in *Drosophila* and grasshoppers (see p. 703). In Amphibia on the other hand, some haploid individuals of newts and frogs produced experimentally have been successfully reared through metamorphosis (Baltzer 1922; Fischberg 1948; Miyada 1960). They are, however, generally inviable and always weak and retarded in development. Some of the inviability of haploid tissues may be due to lethal alleles for which the diploid individuals are generally heterozygous. But in locust parthenogenesis the haploid tissues remain inviable even when derived from parents that must be completely homozygous (see p. 703), and the same is almost certainly true of *Drosophila*. We must therefore conclude that haploidy is lethal *per se* in most, but not all, groups of animals. That being so, it may well be that the haplodiploid mechanisms arose in evolution through loss of one chromosome at a time in the male, rather than by a sudden loss of a whole haploid set of autosomes. It may be significant that several of the haplodiploid groups have very low chromosome numbers. The iceryine coccids have n = 2 and the haplodiploid families of mites seem to invariably have n = 3 or 4. Thus in the ancestors of the modern mites and fluted scales, the final stage in the acquisition of haplodiploidy may have consisted merely in the loss from the male sex of one or two chromosome elements (which may well have been relatively small elements or perhaps large heterochromatic and genetically sub-inert). Brown (1964) makes the point that male haploidy is unlikely to have arisen in groups which depended to any extent on heterozygosity, and that in fact it is most likely to have arisen in species adapted to inbreeding.

The factors which control the sex ratio in species of haplodiploid animals are poorly understood. Some of them were reviewed by Flanders (1946). It is not clear whether all the eggs of an indivuidal female have an equal chance of being fertilized, or whether they may differ in respect of some cytoplasmic factors influencing the probability that they will be penetrated by a sperm. Mechanisms of the latter type seem very probable, on general grounds, but their existence would be extremely hard to prove.

19

THELYTOKY

'Dama, nascono in questo paese solamente galline senza gallo
alcuno ?'

BOCCACCIO: Decameron

The term thelytoky is used to designate those forms of reproduction in
which female individuals produce exclusively female progeny by a process
that does not involve fertilization. Botanists often use the term *apomixis* in
the same sense, but it seems preferable to employ this word with a more
restricted meaning, to designate one particular type of thelytoky (see p. 704).
The word 'fertilization' in the above definition must be understood in the
sense of genetic fertilization, i.e. fusion of nuclei coming from different
individuals. There are as a matter of fact a variety of types of thelytoky in
which the eggs only develop after penetration by a sperm; but in these cases
of *pseudogamy* or *gynogenesis* the sperm nucleus degenerates without fusing
with the egg nucleus so that it makes no genetic contribution (in the ordinary
sense) to the developing embryo. In some instances of pseudogamy it has
been demonstrated that inheritance is purely maternal and in others it may
be inferred to be so. The sperm in cases such as this may be derived from
the testis or ovotestis of a hermaphrodite or may even come from a male
individual of a different, but fairly closely related species (see pp. 724–41).

Thelytoky may be the sole mode of reproduction in a species or biotype
(*complete thelytoky*) or it may alternate with sexual reproduction in some
kind of more or less regular manner (*cyclical thelytoky* or *heterogony*),
dependent on the seasons or on environmental factors, as happens in many
parasitic worms, rotifers, cladocerans, aphids, gall wasps and a few cecidomyids.

From the genetic standpoint it is clear that species which reproduce by
cyclical thelytoky must exhibit genetic recombination, whereas in the case
of complete thelytoky there is no way in which mutations that have occurred
in two individuals, neither of which is ancestral to the other, can become

combined in a third. This seems to be a basic difference, regardless of whether sexual reproduction in any particular species with cyclical thelytoky is common or rare, regular or irregular in occurrence. Genetically speaking, the essential distinction is between complete thelytoky and all types of genetic systems involving fertilization. We shall deal with cyclical thelytoky in Chapter 20, since it really constitutes an entirely distinct type of genetic system.

The early literature on the various types of thelytoky was reviewed by Vandel (1931) and Suomalainen (1950) but several investigations of general importance on various aspects of thelytoky have been carried out more recently. Some of this more recent work has been reviewed by Peacock and Weidmann (1961), Suomalainen (1962), Narbel-Hofstetter (1964) and Uzzell (1970). Unfortunately there have been very few studies on the biometry of thelytokous populations and on the effects of selection on metrical characters. Thus a discussion of the genetic consequences of thelytoky still rests largely on inference from general theoretical considerations rather than on direct factual evidence (White 1970a).

Complete thelytoky is of significance in population genetics since it constitutes a highly distinct genetic system in which sexuality has been entirely abolished and recombination of genes in different individuals is no longer possible (although in some types of thelytoky, segregation of genes present in the same individual is still theoretically possible). The essential genetic feature would seem to be that it fixes particularly adaptive combinations of genes and perpetuates them *ad infinitum*, with no production of adaptively inferior individuals by segregation and recombination. Most thelytokous biotypes are probably very highly heterozygous. It is certainly not accidental that all the thelytokous Diptera whose polytene chromosomes have been studied have proved to be complex inversion heterozygotes (see p. 712). The importance of heterozygosity in parthenogenetic organisms has been stressed especially by White (1970a). Heterotic and epistatic interactions, whether within diploid karyotypes or in polyploid biotypes, are stabilized and 'frozen' by genetic systems in which recombination does not occur. Where segregation does not occur, the segregation load normally associated with heterosis is obviously zero. Heterozygosity of parthenogenetic biotypes may result from mutation and chromosomal rearrangement but some thelytokous forms are of hybrid origin and owe their heterozygosity to this cause.

Not *all* thelytokous mechanisms lead to great heterozygosity, however. Some of them must be expected to produce complete homozygosity. In these cases the evolutionary advantage of thelytoky presumably depends solely on its reproductive economy, or on the fact that it eliminates a male sex which is biologically unsatisfactory. But the cases of thelytoky which must be assumed to be associated with heterozygosity greatly outnumber those where homozygosity is to be expected, and some of the latter type are not fully authenticated by complete cytological evidence.

Since complete thelytoky has clearly arisen independently on many different occasions in the course of evolution, it is natural that the details should vary from case to case. However, the rigidity of all genetic systems in which there is no recombination of genes present in different individuals probably dooms them, in all instances, to a relatively brief evolutionary career. With the single exception of the bdelloid rotifers (Mayr 1963) no major group of metazoan animals (family, subfamily, tribe, etc. in the systematic hierarchy) consists entirely of thelotokous bioptypes, and essentially the same seems to be true of the plant kingdom. Among the Protozoa however, there are various groups, such as the spirotrichonymphid parasites of the termite genus *Stolotermes* (Day 1945; Cleveland and Day 1948) in which sexual reproduction and meiosis seem to be entirely absent. As far as the higher animals are concerned, it seems that within a very short time, on the geological time scale, thelytokous populations are likely to suffer extinction as a result of environmental changes to which they cannot respond by adaptive modifications, since there is little or no chance of their being genetically improved by natural selection. Only Omodeo (1953*a*) has claimed (on the basis of their geographic distribution) that certain thelytokous 'species' of oligochaetes probably date from mid-Tertiary times (see p. 733). Most thelytokous forms seem to be closely related to bisexual species which are yet extant: there seems to be no instance of a thelytokous species which is really isolated taxonomically, with no close relatives.

Some degree of genetic flexibility must still exist in thelytokous biotypes as a result of mutation. Darlington (1958) has assumed that thelytokous forms will retain the mutation rates they have inherited from their bisexual ancestors. But this is not necessarily so. In *Drosophila* genetic factors ('mutator genes') are known which raise the overall mutation rate (Ives 1950). Thus, once recombination has been abolished, there could still be selection for either higher or lower mutation rates. But there is no direct evidence on this point, fundamental though it is to an understanding of non-sexual genetic systems; we do not even know enough to predict whether we would expect selection for increased or decreased mutation rates.

The immediate advantages of thelytoky from the standpoint of reproductive economy are fairly obvious. By dispensing with the need for mating, it allows the whole of the adult life to be devoted solely to feeding and reproduction except, of course, in cases of pseudogamy. The prolificity of parthenogenetic organisms is thus expected to be higher than that of related bisexual forms; even if the mean number of offspring per female is the same, the potential increase in numbers per generation is double that of a bisexual population, since every individual is a female. However, the thelytokous arctic simuliid *Gymnopais* sp. lays only about twenty eggs compared with several hundreds in more southern bisexual species (Downes 1964).

More important, perhaps, are some of the advantages of thelytoky from the standpoint of *dispersal*. They may help to explain why some thelytokous biotypes have survived in spite of an inherent lack of adaptability. A single thelytokous female, transported by accident to a new locality or habitat can always give rise to a new colony if conditions are not too adverse. And, unlike the situation in bisexual species, no thelytokous population can be extinguished simply by being reduced to such a low density that there is little chance of the sexes meeting. This may be the reason why some thelytokous 'species' have wide distributions, although in small isolated colonies. Pseudogamous biotypes which depend on being fertilised by males of a closely related bisexual form do not, of course, enjoy any of these advantages; they are 'reproductive parasites' which can only exist or establish themselves in environments also occupied by the population which provides the males.

Clearly natural selection must continue to operate in thelytokous populations. But it will be a purely negative force, acting solely to eliminate unfit genotypes and will lack entirely the constructive, creative character it has in bisexual species where it constantly gives rise to new 'statistically improbable' combinations of alleles. In fact the evolution of the thelytokous populations must resemble the caricatures of the synthetic or genetic view of evolution that used to be held up to derision by some Lamarckian critics who had never grasped the essential nature of the action of selection on Mendelian populations. Natural selection may possibly tend to favour genotypes that are especially plastic, phenotypically, in the case of thelytokous populations (replacing genetic polymorphism by physiological adaptation). But there does not seem to be any definite evidence of this.

Treatment of thelytokous forms by standard taxonomic methods is frequently unsatisfactory. In some instances we may speak of a thelytokous 'species' in the sense of a *morphospecies*, the members of which all resemble one another very closely and differ from related bisexual forms. In fact such forms may show extreme morphological uniformity; far greater than in more variable bisexual species. The mantid *Brunneria borealis* (p. 714), the tettigoniid *Saga pedo* (p. 715), the grasshopper *Moraba virgo* (p. 717) and *Drosophila mangabeirai* (p. 713) all seem to be 'species' in this sense, and pose no nomenclatural problems. Each of these forms seems to represent a single origin of thelytoky and there has been no hybridization between bisexual and thelytokous biotypes.

It is quite different where thelytoky has arisen repeatedly in a group and where there is indirect evidence that occasional hybridization between males of bisexual forms and females of related thelytokous biotypes has occurred. In such instances we may have great complex assemblages of thelytokous forms, differing either in their gene sequences, chromosome numbers (degree of ploidy perhaps sometimes combined with aneuploidy) or precise cytogenetic mechanism of thelytoky. Such complicated situations simply do not lend themselves to the usual type of taxonomic treatments; if we recognize species as biological entities sharing a common gene pool, we must accept the consequences as far as these forms are concerned, namely that since they lack a gene pool, they really do not form species at all. Botanists have used the term *agamic complexes* to describe situations where a multitude of asexually-reproducing biotypes coexist; such agamic complexes seem to occur in animals, among earthworms, weevils, simuliid flies and in the brine shrimp, *Artemia*. Although some taxonomists have set up species and formal intraspecific categories in such groups, it may be that, as Stebbins (1950, p. 409) has put it, 'they are looking for entities which in the biological sense are not there' (although the protozoan parasites of the termites of the genus *Stolotermes*, referred to earlier, seem to form distinct species). Adopting a conservative viewpoint on the subject of parthenogenetic 'species', it would seem that rather less than a thousand such have been described in the animal kingdom, compared with a total of over a million bisexual species, i.e. that there are over a thousand bisexual genetic systems to every thelytokous one.

We shall discuss the evolutionary origin of thelytokous genetic systems later (p. 743). But it is desirable to point out here that agamic complexes do *not* necessarily demonstrate a capacity for divergent evolution within thelytokous systems, as some biologists (including the present author in earlier works) seem to have implied. More probably they are due to two causes, acting alone or in combination: (1) multiple polyphyletic origin of thelytoky from bisexual species 'preadapted' to this kind of evolutionary change of breeding system, (2) the building up of alloploid karyotypes by occasional mating of thelytokous females with males of related sexual biotypes (or with hermaphrodites functioning as males, in the case of the oligochaetes and planarians). In some instances autoploidy may also have played a part, but its role has probably been overestimated by some workers such as Christensen (1961) and Suomalainen (1962).

Mayr (1957, 1963) has discussed the general problem of morphological discontinuities between biotypes of asexual organisms. He concludes that 'the existing types are the survivors among a great number of produced forms, that the surviving types are clustered around a limited number of adaptive peaks, and that ecological factors have given the former continuum a taxonomic structure.' In the great majority of cases that have been investigated, morphologically distinct parthenogenetic biotypes are also chromosomally distinct either in ploidy or in respect of chromosomal rearrangements. They are hence *cytotypes* and we may solve the nomenclatural problem by simply referring to them as 'the triploid biotype', 'the 15-chromosome form', and so forth. A word of warning is in order here, however. From an originally diploid continuum, triploid biotypes may have evolved on several occasions, in different geographic areas. To speak of *the* triploid biotype in such cases would be wrong.

Experimental studies have shown that parthenogenetic reproduction can be induced in the eggs of most species by a wide variety of treatments such as pricking the egg with a needle, exposure to many kinds of chemical agents and so forth (Loeb 1913; Kawamura 1940; Tyler 1941; Beatty 1957). Discussion of artificial parthenogenesis is outside the scope of this book. But it has also been shown in a number of normally bisexual insect species that a small proportion of the eggs of virgin females hatch spontaneously (in *Drosophila* spp. by Stalker 1954; in phasmatids by Bergerard 1954, 1962; in Grouse locusts by Nabours and Forster 1929 and Nabours 1937; and in grasshoppers by a number of workers – see p. 702). This has been called 'occasional' or 'accidental' parthenogenesis (*tychoparthenogenesis* of some authors). In these cases the small percentage of unfertilized eggs that hatch can be increased by deliberate selection. Thus Carson (1962*b*, 1967*a*) was able to increase the frequency of tychoparthenogenesis in *Drosophila mercatorum* by selection from about 0.1% to approximately 6%. In this way he was able to obtain self-perpetuating thelytokous strains of what is normally a bisexual species. About 90% of the females obtained in this way appear to homozygous for all loci; they must arise through fusion of cleavage nuclei in pairs (or some equivalent endoreduplication of haploid nuclei). The remaining 10% are heterozygous for various loci, and must result from fusion of meiotic products (Carson, Wei and Niederkom 1969). Selection seems to have increased the frequency of two different mechanisms of parthenogenetic development. Tychoparthenogenesis is not, of course, confined to insects; it occurs for example in certain breeds of turkeys (Olsen and Marsden 1954; Poole and Olsen 1957; Poole 1959), where the offspring are apparently always male (restoration of diploidy, whatever the precise mechanism, presumably results in XX males and YY or OO embryos which would be inviable).

The term *facultative parthenogenesis* has been used to designate those cases where reproduction is normally bisexual but where a high percentage of the eggs will develop without fertilization. Thus in the phasmatid *Menexenus semiarmatus* 44% of unfertilized eggs hatch, and in another phasmatid, *Clitumnus extradentatus*, 66% do so (Bergerard 1954). To what extent unfertilized eggs do in fact occur and develop in such species under natural conditions is quite uncertain. Usually, facultative thelytoky gives rise to all-female progenies, and this is in fact what the word means. But in the ephemerid *Centroptilum luteolum*, Degrange (1956) has reported that batches of unfertilized eggs give rise to bisexual progenies ('facultative deuterotoky'). The chromosomal mechanism is not clear in this case, since Ephemeroptera are reported to have male heterogamety (see p. 607).

Artificial parthenogenesis, tychoparthenogenesis and facultative thelytoky all suggest that some capacity for parthenogenetic development is probably present in all eggs. It is tempting to suppose that biotypes which reproduce exclusively by thelytoky at the present time arose by a gradual process in which tychoparthenogenesis is an early stage, facultative thelytoky a somewhat later one; and that sexual reproduction has eventually become a rarity in such evolutionary lineages and finally disappeared altogether. Darlington (1958), on the other hand, has suggested that in some instances thelytoky may have arisen by a single mutation, and Hertwig (1920) claimed that she had actually observed such a mutation in the nematode *Rhabditis pellio*. As far as animals are concerned, most thelytokous mechanisms have probably arisen gradually; but in some instances the cytogenetic mechanism is such that a sudden origin seems more plausible (see p. 718). We shall deal later with the problem of the origin of polyploid parthenogenetic forms.

The cytological mechanism of 'accidental' thelytoky or tychoparthenogenesis has been studied in a number of species of grasshoppers, e.g. *Melanoplus differentialis* (King and Slifer 1934), *Locusta migratoria* (Bergerard and Seugé 1959), *Schistocerca gregaria* (Hamilton 1953, 1955), *Chorthippus longicornis* (Creighton and Robertson 1941) and *Romalea microptera* (Swann and Mickey 1947).

In all these Acrididae the proportion of unfertilized eggs which undergo development is very small. Such eggs seem to pass through two meiotic divisions and the embryos begin development with the haploid number of chromosomes. Diploid nuclei soon make their appearance, however, as a result of an endomitotic process (*sens. lat.* – i.e. two chromosomal replications between successive mitoses). Normal development depends on successful 'diploidization' and a great many embryos die or develop quite abnormally because of haploid/diploid mosaicism. In *Schistocerca gregaria* Hamilton was able to rear four successive generations of females parthenogenetically, but in *Locusta* Bergerard and Seugé found that partheno-produced females were always weak and usually unable to lay eggs.

The non-viability of haploid grasshopper embryos or ones with a considerable proportion of their cells and tissues still haploid does not seem to depend to any great extent on an 'uncovering' of recessive lethal genes, but is, rather, a consequence of the haploid condition as such. The proof of this is as follows: the viability of haploid tissues is no greater in the *second* and subsequent generations of parthenogenetic reproduction, even though lethal genes must all be eliminated in the first generation, because of the mode of diploidization (King and Slifer 1934). Bergerard and Seugé (1959) have shown that regions of partheno-produced embryos which are inhibited in development have a higher proportion of haploid cells than normally-developing regions.

In the Grouse locusts (Tetrigidae) accidental parthenogenesis has been studied in various species of *Apotettix*, *Paratettix* and *Tettigidae* (Nabours and Forster 1929; Robertson 1930, 1931). The individuals produced by parthenogenesis were almost invariably female, but a few males were obtained (in *A. eurycephalus* Nabours obtained thirteen XO males in about 5000 partheno-produced offspring, which presumably arose by non-disjunction and not by sex reversal, since females are XX).

Robertson believed that partheno-produced embryos in Grouse locusts arose from eggs in which the second meiotic division had been suppressed. He found cells in the developing embryos which contained the diploid number of chromosomes, but there was a marked tendency for the homologues to lie side by side in the metaphase plates and frequently diplochromosomes (i.e. four parallel chromatids more or less closely associated at the centromere) were seen. Robertson assumed that these parallel associations of homologues have persisted into the embryo from an abortive second meiotic division. If so, there would be a fundamental difference between the mechanism of accidental parthenogenesis in the grasshoppers (Acrididae) and Grouse locusts (Tetrigidae). It seems more probable that his interpretation was incorrect and that in the tetrigids, as in the true grasshoppers, partheno-produced embryos develop from haploidy to diploidy by a process of endomitosis and that the diplochromosomes are an indication of this.

In view of the apparently universal, even if rudimentary, capacity for parthenogenetic development, we may well ask why more species have not acquired thelytokous modes of reproduction, particularly in view of the reproductive economy involved. The barriers to this evolutionary development seem to be genetical rather than physiological. Many thelytokous biotypes must be so incapable of reacting to environmental changes by adaptive modifications (or at any rate they are so inefficient in making adaptive changes) that they can only survive if they are adapted to extremely narrow, constant and unchanging ecological niches such as dependence on a single food-plant or on a single type of prey. This may be one reason why thelytoky seems never to have been evolved in certain groups of characteristically mobile animals such as dragonflies, butterflies and most groups of vertebrates, whereas it has developed in some flightless bagworm moths (Psychidae) and in a few apterous mantids, phasmatids and grasshoppers; but another is almost certainly that in flightless insects the meeting of the sexes is a chancy business, requiring special ethological mechanisms which are liable to failure. Very mobile animals must almost necessarily be adapted to (or adaptable to) a wide variety of environments. Thelytoky is fairly widespread among non-mobile scale insects (Brown 1960b, 1965) but is unknown in the highly mobile Homoptera Auchenorrhyncha and has been reported in only a single species of the Heteroptera, the mirid *Campyloneura virgula* (Southwood and Leston 1959). It is found in a number of species of mites, but seems to be unknown in spiders, whose vagility is much greater.

Some thelytokous species are represented only by a few small widely-separated colonies, consisting of small numbers of individuals, surviving precariously in special habitats. The Mediterranean tettigoniid *Saga pedo* (Matthey 1941), the Australian eumastacid *Moraba virgo* (White, Cheney and Key 1963; White 1966b; White and Webb 1968) and the neotropical *Drosophila mangabeirai* (Carson 1962a) are certainly in this category, and the North American mantid *Brunneria borealis* probably is likewise (White 1948b). On the other hand, some of the thelytokous biotypes of the Brine shrimp *Artemia salina* and certain thelytokous oligochaetes form large, dense populations. Some arctic simuliid midges that are parthenogenetic form dense populations and have wide distributions (Downes 1962), being practically circumpolar species.

From a formal cytogenetic standpoint, we may draw a distinction between two main types of thelytoky: (1) *automixis* or *meiotic thelytoky*, in which meiosis still occurs in the developing oocyte, but is compensated for by a doubling of the chromosome number at some stage in the life cycle; and (2) *apomixis* or *ameiotic thelytoky*, in which meiosis has been entirely suppressed, the maturation division or divisions in the oocyte being mitotic in character (i.e. without any formation of bivalents). It would be logical to consider these two types of thelytoky separately, but since they may sometimes coexist in the same group, and since we shall need to discuss some highly interesting cases where it is not even known with certainty whether the genetic mechanism is automictic or apomictic, this logical arrangement is not possible. We shall accordingly consider some outstanding examples of both types of mechanism and then survey the known cases of thelytoky in animals, group by group.

AUTOMICTIC MECHANISMS

From a genetic standpoint there are a number of different modes of automixis, depending on the precise manner in which the chromosome number is restored; and some of these are largely equivalent to apomixis, so that the formal dichotomy between two systems is blurred by ones which are intermediate in their practical genetic consequences.

In the majority of automictic systems synapsis, chiasma formation and reduction proceed normally and four haploid nuclei are formed as a result of the two divisions, diploidy being then restored by fusion between two of these nuclei. However, in a number of instances, the chiasmata have not been seen with certainty and in the thelytokous forms of the enchytraeid genus *Fridericia* (as well as in sexual species of this genus) the bivalents are definitely achiasmatic according to Christensen (1961). Whether genetic crossing-over occurs in the oogenesis of *Fridericia* is, of course, unknown. In many organisms, including most species of *Drosophila*, the four post-meiotic nuclei in the egg lie in a row with an innermost one ('egg nucleus'), an outermost one and two in between. The genetic consequences will depend on *which* two of the four nuclei fuse. One possibility is fusion between the innermost nucleus and the one next to it, i.e. between 'second division sister nuclei'; this is sometimes referred to as fusion of the egg nucleus and the second polar body nucleus. Whether the first polar body nucleus undergoes the second meiotic division is genetically immaterial. This mode of restoration of the original chromosome number, following reduction, has been described in the facultatively parthenogenetic 'races' of the scale insects *Lecanium hesperidum* (Thomsen 1927, 1929) and *L. hemisphaericum* (Suomalainen 1940c)*. Other probable cases of the same mechanism are the diploid thelytokous biotype of the Brine shrimp *Artemia salina* at Sète (Artom 1931) and the thelytokous sawflies *Pristiphora pallipes* (Comrie 1938) and *Diprion polytomum* (Smith 1941). Inclusion of the two anaphase groups of the second meiotic division has also been reported in two thelytokous oribatid mites *Platynothrus peltifer* and *Thrypochthonius tectorum* (Taberly 1958b, 1960). In some parthenogenetic enchytraeid worms we have a mechanism which is genetically equivalent to fusion of second division sister nuclei: the second division is suppressed, but the chromatids fall

---

* Biotypes of these 'species' which reproduce by obligatory thelytoky are also known; they are apomictic, with a single mitotic maturation division in the egg.

asunder (Christensen 1960); this is what happens in the hexaploid *Cognettia glandulosa* (Fig. 19.1), in two species of *Mesenchytraeus* and in *Fridericia ratzeli*.

To understand the genetic results of this type of system we may consider the situation where the mother is heterozygous at a particular locus *Aa* (Fig. 19.2). The offspring will necessarily be homozygous (*AA* or *aa*) if the locus was proximal to the first chiasma, i.e. between the chiasma and the centromere. On the other hand, heterozygosity will be preserved if the chiasma occurs proximal to the locus in question. But even with proximal localization of chiasmata, such species will tend to get rid of heterozygosity much faster than it is generated by new mutations, i.e. they will tend towards complete homozygosity (unless chiasmata are *never* formed distally, in which case they will accumulate heterozygosity in the distal segments of their chromosomes, except when *pairs* of chiasmata can be formed proximally).

The genetic consequences are quite different if the fusion is regularly between the two central nuclei, i.e. between 'second division non-sisters'.

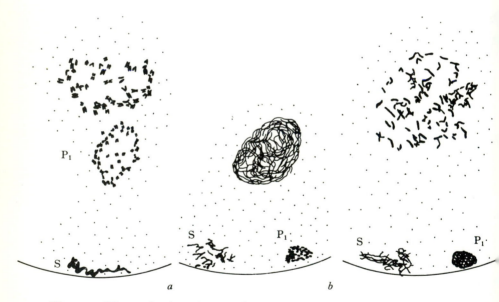

a   *b*

Fig. 19.1. The mechanism of restoration of the somatic number in the pseudogamic thelytokous enchytraeid worm *Cognettia glandulosa*. This species is a hexaploid which shows 54 chiasmatic bivalents at the first meiotic division. *a*, abortive second meiotic division with the chromatids beginning to fall asunder in the egg nucleus; *b*, interphase; *c*, metaphase of the first cleavage division with 108 chromosomes. P_1, the first polar body nucleus, S the genetically functionless sperm-nucleus which degenerates at the periphery of the egg. From Narbel-Hofstetter 1964, based on Christensen 1961.

I Fusion of Second Division Sister Nuclei

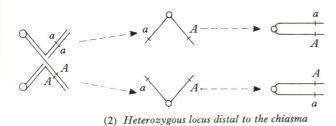

First metaphase   First anaphase   Second metaphase   Fusion nucleus

(1) *Heterozygous locus proximal to the chiasma*

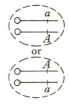

(2) *Heterozygous locus distal to the chiasma*

II Fusion of Second Division Non-Sister Nuclei

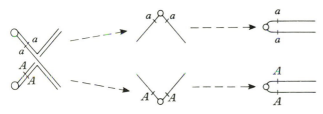

(1) *Heterozygous locus proximal to the chiasma*

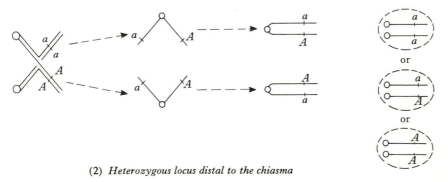

(2) *Heterozygous locus distal to the chiasma*

Fig. 19.2. Genetic consequences of meiotic thelytoky according to whether restoration of the diploid number is by fusion of second division sister nuclei or second division non-sister nuclei. Further explanation in the text.

In this case heterozygosity will be preserved for loci proximal to the first chiasma, but distal segments will become completely homozygous in course of time. As in the first case there will, theoretically, be a slow loss of heterozygosity, even for proximal loci, due to occasional formation of chiasmata very close to the centromere. But if the homozygotes formed in this way are inviable, heterozygosity for proximal regions may be preserved indefinitely, with only a minimal 'segregation load'. This is probably the situation in the dipteran *Lonchoptera dubia* (Stalker 1956b) which is a permanent heterozygote for a number of chromosomal inversions. In the diploid parthenogenetic race of the bagworm moth *Solenobia triquetrella* an XO or XY sex-chromosome constitution is maintained from generation to generation by fusion of the two central nuclei (see p. 721).

It will be appreciated that the genetic properties of automictic species will depend both on the mode of fusion of meiotic products (whether between second division sisters or second division non-sisters) and on the precise pattern of chiasma localization in the meiotic bivalents. No studies of the latter aspect appear to have been carried out on automictic species. It may be that these forms range from ones which are almost completely homozygous to ones that are as heterozygous as apomictic species. But the existence, in actuality, of these homozygous automictic biotypes must remain uncertain for the present, and it may be that they do not really occur, due to various types of strict chiasma localization.

Complete homozygosity for all loci will exist, however, in those cases of automictic thelytoky where restoration of the diploid number occurs in early ontogeny, by fusion of haploid cleavage nuclei in pairs or by endomitosis at the cleavage divisions. At least this will be so if the cleavage nuclei are derived from a single one of the four meiotic products; if they were derived from two or more of these nuclei heterozygosity or mosaicism could result.

Restoration of the somatic chromosome number by fusion of cleavage nuclei or endomitosis does not seem to be a common mechanism of thelytoky. It was formerly believed by Seiler and his school that this was the mechanism in both the diploid and the tetraploid thelytokous races of the psychid moth *Solenobia triquetrella* (Seiler 1949a; Seiler and Schäffer 1938, 1941). But later work (Seiler and Schäffer 1960; Seiler 1963) has shown that in both cases it is second-division non-sister nuclei which fuse to form the functional egg nucleus. Restoration of the chromosome number by fusion of cleavage nuclei in pairs or by endomitosis at the first cleavage division probably does occur in the thelytokous form of the White fly *Trialeurodes vaporariorum*, according to Thomsen (1927) and in the mite *Cheyletus eruditus* according to Peacock and Weidmann (1961). It definitely occurs in the Scale insect

Pulvinaria hydrangeae (Nur 1963*a*) and in an iceryine coccid, *Gueriniella serratulae* (Hughes-Schrader and Tremblay 1966).

Of all the thelytokous species of animals which have been investigated cytologically, these four are, then, the only ones (if indeed they have been correctly interpreted) in which we can be fairly certain that homozygosity is complete, and in which an interpretation of the adaptive advantage of thelytoky in terms of the fixation of heterosis is excluded.

In general, as we should expect, automictic biotypes are diploids, but some polyploids undoubtedly occur among earthworms (p. 732). There is no barrier to the development of evolutionary polyploidy in the thelytokous members of this group, because of the peculiar automictic mechanism which exists in most of them (Omodeo 1951*d*, *e*, 1952, 1955). In these forms doubling of the chromosome number occurs at a premeiotic oogonial division; synapsis then takes place in a nucleus which contains twice the somatic number of chromosomes. Multivalents are not formed, and in all probability pairing occurs only between genetically identical sister chromosomes. If this is so, a type of thelytoky which is formally automictic is genetically equivalent to apomixis, preserving heterozygosity unchanged from generation to generation. This kind of automictic mechanism is even compatible with odd-numbered polyploidy and we have the strange phenomenon of triploid and pentaploid biotypes which show only bivalents at meiosis. The same kind of mechanism occurs in some planarians (see p. 736), in the diploid thelytokous grasshopper *Moraba virgo* (see p. 718) and in some thelytokous salamanders and lizards.

APOMICTIC MECHANISMS

The cytology of apomictic mechanisms is, in general, very simple. There is no synapsis, no bivalents are formed and only a single maturation division occurs in the oocyte, which is an ordinary mitosis. Since no reduction of the chromosome number has occurred, no question of a compensatory doubling process arises. Any genetical heterozygosity which exists will be perpetuated from generation to generation and, in fact, as more and more mutations occur, heterozygosity must be expected to increase without limit. Assuming that they remain diploid in a cytological sense, apomictic forms should tend towards a state in which no allele is represented by an identical allele in the 'homologous' chromosome. To suggest that apomixis permits a species to exploit an unlimited variety of heterotic combinations is, perhaps, an exaggeration. But the adaptive combinations built up in such forms probably do have some physiological resemblances to the heterotic mechanisms of sexually reproducing species.

From a formal cytological standpoint, there are a few aberrant types of apomixis. In the parthenogenetic 'race' of the roach *Pycnoscelus surinamensis* (Matthey 1945*b*), which is diploid, the chromosomes apparently undergo synapsis but fall apart again at pachytene, no chiasmata being formed. The chromosomes are four-stranded following pachytene and undergo two mitotic divisions in rapid succession.

An even more peculiar apomictic system occurs in the turbellarian *Bothrioplana semperi* (Reisinger 1940). The egg capsules of this worm always contain two eggs which fuse to produce a single embryo. The nuclei of both these eggs simultaneously undergo two maturation divisions of a peculiar type and all the eight nuclei resulting from these divisions remain within the capsule and contribute to the future embryo, functioning as cleavage nuclei. A doubling of the chromosome number precedes each maturation division being followed by a compensatory reduction, but there is no true bivalent formation and no chiasmata occur, so that the system is genetically apomictic.

Apomictic systems do not in themselves, constitute a barrier to the development of evolutionary polyploidy. Whether apomictic biotypes become polyploid will consequently depend on whether polyploidy-producing mechanisms exist and on whether any polyploid genotypes which may arise can compete with the pre-existing diploid biotypes. Darlington (1958) has suggested that in a form like the triploid isopod *Trichoniscus coelebs* (Vandel 1928–40) the 'mutation' which established the triploid karyotype simultaneously gave rise to the apomictic oogenesis and the thelytokous mode of reproduction. Since all cytogenetic mechanisms are likely to be imperfect at their origin and only become improved as a result of a more or less long process of natural selection, it seems more probable that thelytoky arose and was perfected by selection before triploidy arose and the cytogenetic mechanism adapted to the increased chromosome number (rather than that the thelytokous biotype had to simultaneously improve the efficiency of both its apomixis and its triploidy). The question is, however, still an open one.

THELYTOKY IN DIPTERA

Particular interest attaches to studies of parthenogenesis in Diptera with polytene chromosomes in which inversions or other structural rearrangements can be identified. We give, in Table 19.1, a list of seventeen dipteran 'species' (in some instances complexes of biotypes) in which parthenogenesis appears to be the sole normal mode of reproduction; in six of these the polytene chromosomes do not appear to have been studied. A number of other thelytokous species of Diptera are known to exist, particularly in the family Chironomidae.

Several general features are immediately apparent. One is the highly

TABLE 19.1. *Thelytokous species of Diptera*

Family and Species	Type of Thelytoky	Ploidy	Inversion Polymorphism
Psychodidae			
Psychoda severini subsp. *parthenogenetica*[1]	?	?	?
Simuliidae			
Cnephia mutata[2]	?	3n	Over 700 biotypes heterozygous for different combinations of inversions.
Prosimulium ursinum[2]	?	3n	All individuals highly heterozygous cytologically.
P. macropyga[2]	?	3n	
Gymnopais sp.[2]	?	3n	
Chironomidae			
Limnophyes biverticillatus[3,4]	Ameiotic	2n	?
L. virgo[3,4]	Ameiotic	3n	One inversion
'Forms related to *L. virgo*'[3]	?	3n	?
Pseudosmittia arenaria[3,4]	Ameiotic	2n	Very complex structural heterozygosity
P. baueri	Ameiotic	2n	
P. sp.[4]	Ameiotic	3n	
Monotanytarsus boreoalpinus[5]	?	?	?
Lundstroemia parthenogenetica[5,7,8]	Ameiotic	3n	Two inversions and a deletion
Lonchopteridae			
Lonchoptera dubia[9] (four biotypes)	Automictic with fusion of meiotic products	2n	All four biotypes structurally heterozygous for inversions.
Chamaemyiidae			
Ochthiphila polystigma[10]	?	3n	?
Drosophilidae			
Drosophila mangabeirai[11,12,13]	Automictic with fusion of meiotic products	2n	Heterozygous for three inversions
Agromyzidae			
Phytomyza crassiseta[14]	?	2n and 3n	Several strains heterozygous for inversions

[1] Séguy 1950
[2] Basrur and Rothfels 1959
[3] Scholl 1956
[4] Scholl 1960
[5] Lindeberg 1958
[6] Freeman 1961
[7] Edward 1963
[8] Porter 1971
[9] Stalker 1956b
[10] Stalker 1956a
[11] Carson Wheeler and Heed 1957
[12] Carson 1962a
[13] Murdy and Carson 1959
[14] Block 1969

developed structural heterozygosity of these forms, all being heterozygous for one or more inversions and several being very complexly heterozygous.

The second general feature is that seven of the twelve 'species' that have been studied cytologically are triploids. In such cases it may be technically very difficult to determine just how the various cytological sequences are distributed over three homologues associated as a polytene 'chromosome'.

In the parthenogenetic chironomids belonging to the subfamily Ortho-cladiinae studied by Scholl (1956, 1960) there is apparently an abortive first meiotic division in the egg in which univalents attach to the spindle. A mitotic second division then occurs, both the resulting nuclei taking part in the formation of the embryo. Thus in this case no polar body nucleus is formed. An essentially identical mechanism seems to exist in the triploid Australian *Lundstroemia parthenogenetica* (subfamily Chironominae), which is heterozygous for two inversions and a deletion (Porter 1971).

In the North American *Lonchoptera dubia* (Stalker 1956b) it is certain that automixis occurs, with subsequent fusion of two of the four haploid products of meiosis. Since inversion heterozygosity is maintained from generation to generation, this may be always a union of the two central nuclei (derived from different second meiotic divisions), there being no chiasmata formed between the centromeres and the inverted regions. These conditions would give permanent inversion heterozygosity. Alternatively, if the pattern of fusion is variable, or if chiasmata are sometimes formed proximal to the inversions, the structurally homozygous individuals which would be produced may be inviable.

Lonchoptera dubia consists of four chromosomal races or biotypes which differ in their inversion heterozygosity. Within each race, however, there is absolute karyotypic uniformity. In many localities, two, three or even four races coexist. Thus they are almost certainly ecologically specialized to some extent, i.e. adapted to different ecological niches and not absolutely competitive. Stalker assumes that parthenogenesis developed gradually within an originally bisexual species, different gene sequences being trapped and fixed in the four biotypes which emerged from this process.

A much more complex 'super-species' is the North American simuliid *Cnephia mutata*, which has been investigated in Ontario by Basrur and Rothfels (1959). Two 'races' exist, a diploid bisexual one and a triploid thelytokous one; the two forms frequently coexist in the same stream, but there is good evidence that they do not interbreed. Except for some structural differences between the X and Y in male larvae, the diploid form is cytologically monomorphic. The triploid one, however, (which consists exclusively of females) shows extensive inversion-polymorphism and at one locality as many as seventy chromosome biotypes coexist (Rothfels *in litt.*), all differing in respect of inverted regions in two chromosomal elements. Basrur and Rothfels consider it likely that a reassortment of inversions by crossing-over and segregation occurs in the course of an automictic maturation process. They put forward two alternative interpretations of their data:

(1) Triploid *C. mutata* are of allopolyploid and highly 'polyphyletic' origin and perpetuated by ameiotic (apomictic) parthenogenesis, and (2) triploid *C. mutata* are autopolyploid and not necessarily 'polyphyletic' in origin; automixis was their mode of parthenogenetic reproduction, at least in the beginning, and loss of heterozygosis by crossing-over is balanced by heterosis of individuals carrying inversions. The former interpretation is not excluded, the latter is preferred . . .

Just how automixis could occur in a triploid, unless it involved premeiotic doubling of the chromosome number (in which case it would be equivalent to apomixis), is not clear. But until the maturation divisions in this most interesting species have been investigated, it seems futile to discuss the adaptive significance of chromosomal polymorphism in *Cnephia mutata*. Rothfels states (*in litt.*) that he had evidence of more than 700 different genotypes (representing permutations of a small number of inversions) in material ranging from Newfoundland to Alaska.

Three other boreal species of simuliids have been briefly reported by Basrur and Rothfels as thelytokous (all-female) triploids with complex inversion heterozygosity but no further details have been published. They also mention a species of *Prosimulium* which, paradoxically, includes triploid males as well as females; it is not clear whether this species is thelytokous (it is perhaps comparable to the triploid isopod *Trichoniscus coelebs* in which rare males are produced thelytokously).

Drosophila mangabeirai, the sole member of the genus known to reproduce by obligatory thelytoky, is a rare neotropical species; it is a diploid which is apparently always heterozygous for the same three inversions over a wide geographic area. Reproduction is definitely automictic in this case, with structural heterozygosity maintained by fusion on non-sister nuclei after the second meiotic division.

No cases of polyploidy higher than triploidy have been recorded in the Diptera. It is entirely possible that tetraploidy, pentaploidy, etc. would be relatively unworkable in organisms with polytene chromosomes in many of their somatic tissues.

In the parthenogenetic *Ochthiphila polystigma* triploidy has definitely been established but the polytene chromosomes have not been studied and nothing is known of the maturation divisions.

Since the heterozygous inversions of *Lonchoptera dubia* and *Drosophila mangabeirai* cover a substantial fraction of the total chromosome length we may inquire what happens to eggs in which chiasmata are formed between mutually inverted segments. Such chiasmata would give rise to dicentric and acentric chromatids. In bisexual species there is a precise mechanism whereby they pass into the polar nuclei, the future egg nucleus receiving only monocentric chromosomes; but this mechanism cannot very well operate in the cases of automictic parthenogenesis which depend on fusion of the two central nuclei following on meiosis, since it is precisely these central nuclei which will contain the dicentric and acentric strands derived from first division bridges. Egg- and embryo-mortality is relatively high in *L. dubia* and *D. mangabeirai* (at least 25% in the first and possibly as much as 40% in the second); it may be that some of this mortality is due to crossing-over between the mutually inverted segments. Another source of egg-mortality may be the occurrence of chiasmata between the centromere and the inverted region (which would lead to structurally homozygous egg nuclei in approximately 50% of cases when fusion of non-sister nuclei occurs after meiosis).

THELYTOKY IN THE ORTHOPTEROID ORDERS OF INSECTS

A few instances of thelytoky are known in most of the orthopteroid groups – one each in the roaches, mantids, crickets, grasshoppers and embiids, two in tettigoniids and an undetermined number in phasmatids (Chopard 1948).

The roach 'species' *Pycnoscelus surinamensis* includes diploid thelytokous biotypes with $2n = 34$ (Brazil, Australia, Thailand), 35 (Thailand), 37 (Indonesia), 39 (Brazil), 53 (U.S.A., Brazil), 54 (Panama, Jamaica) and 74 (Indonesia) according to Roth and Cohen (1968) and Roth (1970). Presumably, both polyploidy and aneuploidy have occurred. These thelytokous biotypes have apparently arisen, perhaps polyphyletically, from the bisexual species *P. indicus*, which itself shows two chromosome numbers ($2n\male\female = 35$, 36 and 37, 38). Matthey (1945b, 1948) showed that the eggs of the thelytokous females go through two maturation divisions, both mitotic in type, so that this is a case of apomixis. Females of *P. indicus* from Malaya are stated by Matthey to be facultatively parthenogenetic since virgins lay eggs which hatch successfully; but this is not true of the Hawaiian form (Roth and Willis 1956, 1961).

The only species of mantid known to be thelytokous (no males have ever been recorded) is the near-apterous *Brunneria borealis*, which is distributed from North Carolina to Texas, but is rare and local throughout its range. Several bisexual species of the genus are known from South America, but none occurs in Mexico or Central America. *B. borealis* is a diploid species (White 1948b), but the maturation divisions in the egg have not been studied. Somatic divisions show twenty-eight chromosomes, with no evident structural heterozygosity.

In the embiids (order Embioptera) one thelytokous species occurs in Sardinia, Corsica and some neighbouring Mediterranean islands (Stefani 1956); it also occurs in the Canary Islands and has been introduced into California (Ross 1944). Stefani has considered this as a 'race' of the bisexual species *Haploembia solieri*. However, since the eggs of the two forms are quite markedly dissimilar, and since the two differ in chromosome number, it seems far better to regard them as distinct species. The bisexual form has $2n\male = 19$, $2n\female = 20$, while the thelytokous one normally has $2n = 22$, although a few individuals seem to be triploid, with $3n = 33$. Stefani concludes, without any critical evidence, that the 22-chromosome individuals have four X's instead of the two that are present in a female of the bisexual form. There seems to be only one maturation division in the egg, so that thelytoky is probably apomictic in this case; not all the details are clear, however, from the published account. Stefani has put forward the interesting suggestion that thelytoky may have been adaptive in an insect whose males were severely parasitized by a gregarine that does not attack females. Thelytoky may be relatively common in the Embioptera, since a number of thelytokous African species have been reported by Ross (1961).

Four cases of thelytoky are known in the true Orthoptera, or Saltatoria – the cricket *Myrmecophila acervorum*, the tettigoniids *Saga pedo* and *Poecilimon intermedius* and the eumastacid grasshopper *Moraba virgo*. *Myrmecophila acervorum*, which inhabits ants' nests in Europe, is parthenogenetic in certain areas only, according to Schimmer (1909). Its cytology has not been investigated. The cytology of *Poecilimon intermedius*, which is parthenogenetic from Eastern Europe Russia across Siberia to the Altai and Tian-Shan mountains (Bei-Bienko 1954) is likewise unknown. This is a case of a thelytokous form whose range extends far beyond that of the rest of the genus (about eighty species of *Poecilimon* are known from the Mediterranean region, Balkans, Asia Minor and southern European Russia).

The large wingless tettigoniid, *Saga pedo*, which inhabits the countries north of the Mediterranean, from south Russia to central Spain, is a tetraploid thelytokous species (at least it is tetraploid in Switzerland – material from other areas has not been checked cytologically). The details of oogenesis have been studied by Matthey (1939a, 1941). The oogonial chromosome number is 68 (12 metacentrics and 56 acrocentrics); this is far higher than any other tettigoniid (Fig. 12.11) and approximately twice the chromosome numbers of three related bisexual species from Israel and Turkey, which have $2n\delta = 31$, 31, 33 (Matthey, 1946a, 1949b; E. Goldschmidt, 1946). No synapsis occurs in *S. pedo* and reproduction is apomictic, there being only a single maturation division in the egg. The cells of *S. pedo* are said to be, in general, no larger than those of its diploid, bisexual relatives.

The genus *Saga* (which includes about eleven species) is distributed over the Balkans, Syria, Palestine and Cyprus; like *Poecilimon intermedius*, *S. pedo* has hence extended its range far beyond that of the bisexual members of the genus. It seems, however, to be a relict species since the localities at which it occurs are very discontinuous and the individual populations of very small size. Presumably, after a period of successful dispersal, *S. pedo* has become extinct over the greater part of its range through inability to adapt to changing environments, surviving only in isolated colonies.

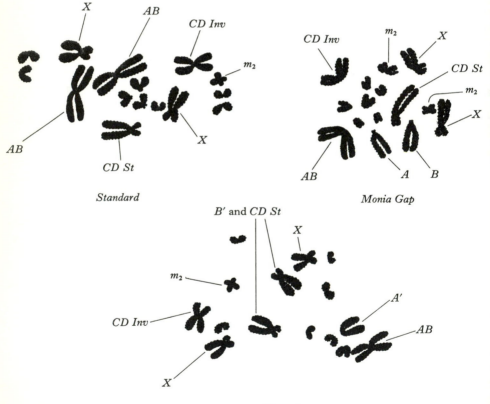

Fig. 19.3. Somatic metaphases from colcemid-treated ovarian follicle cells of *Moraba virgo*, showing the three karyotypes. The *B'* and *CD* (*Standard*) chromosomes are morphologically indistinguishable in the *Yatpool* karyotype.

The Australian eumastacid grasshopper *Moraba virgo* is the only one of about 250 species of the subfamily Morabinae to have become thelytokous (White, Cheney and Key 1963; White 1966*b*). It is a wingless species which occurs as isolated colonies living on various species of *Acacia* throughout a large area of western New South Wales and northern Victoria. Several karyotypes are known in the species (Fig. 19.3), but in each colony all the individuals are cytologically identical. There are always fifteen chromosomes in the somatic set (related bisexual species have $2n♀ = 18$). The 'standard' karyotype, which is widespread, includes a pair of large metacentric '*AB*' chromosomes, a smaller pair of '*CD*' elements which is always heterozygous for a pericentric inversion, a pair of *X*-chromosomes, and nine smaller chromosomes one of which is a small metacentric element ('m_1') that appears to have no homologue. At two localities additional rearrangements have been superimposed on this basic karyotype. In the Yatpool, Victoria, population all individuals are heterozygous for a translocation between one of the '*AB*' chromosomes and a small acrocentric element. And at Monia Gap, N.S.W., one of the *AB* metacentrics has undergone dissociation into two acrocentrics and there are only eight small chromosomes, which include m_2 and a slightly larger m_1 metacentric, which has clearly arisen by a centric fusion. Identification of the *CD* and *X*-chromosome pairs in *M. virgo* depends on comparison with karyotypes of

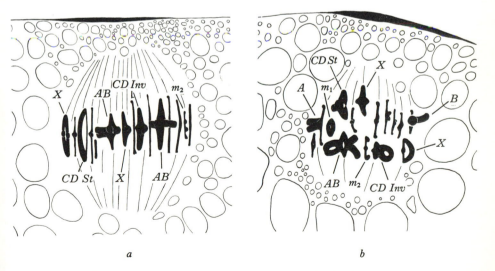

a   *b*

Fig. 19.4. First metaphases in side view, from egg of *Moraba virgo*. *a*, from an individual with the *Standard* karyotype; *b*, from one with the *Monia Gap* karyotype. From White, Cheney and Key (1963).

related bisexual species and is not absolutely certain (i.e. the pair here called 'CD' may in reality be the X).

The pericentric inversion in the CD pair and the m_2 chromosome, which are universal in the species, presumably date from a very early period in the phylogeny of *Moraba virgo*, and may have been handed down from 'bisexual times'. The translocation at Yatpool and the dissociation and fusion at Monia Gap are quite clearly more recent acquisitions; this conclusion is based on the premise that a parthenogenetic biotype without genetic segregation can never lose a chromosomal rearrangement in the course of its evolution – its degree of heterozygosity can be increased but not decreased (however it is barely possible, although highly unlikely, that a loss of heterozygosity could take place by 'back-mutation', i.e. through reinversion in the case of an inversion, refusion in the case of a dissociation, etc.) White and Webb (1968) have found, by tritiated thymidine autoradiography that *M. virgo* is heterozygous for late-replicating segments in the AB and CD pairs (i.e. one of the AB chromosomes and the short arm of the standard CD element are regularly late-labelling). At Monia Gap and Yatpool it is the late-labelling AB chromosome which has become involved in translocations, the normal-labelling one being unbroken.

The thelytoky of *M. virgo* is automictic, but of a type that is genetically equivalent to apomixis; there is a premeiotic doubling of the chromosome number in the oocyte (probably by some kind of endomitotic process), followed by a meiosis in which fifteen bivalents with chiasmata are formed. Apparently synapsis is restricted to sister chromosomes (which are exact molecular copies of one another), since all bivalents are structurally homozygous and no multivalents are ever formed (Fig. 19.4). Two polar bodies are formed, and the meiotic mechanism obviously ensures that structural and genic heterozygosity is perpetuated from generation to generation.

This mechanism of premeiotic doubling is such a radical departure from the normal that it is difficult to imagine its origin except as a single cytological mutation. The thelytoky of *M. virgo* is thus more likely to have arisen suddenly than through the stages of tychoparthenogenesis and facultative thelytoky. Possibly the bisexual population from which *M. virgo* arose was carrying the inversion in the CD, which may have imposed a heavy genetic load on the population. If so, the mechanism of thelytoky might have possessed an adaptive advantage.

Moraba virgo certainly arose from a bisexual ancestor having a karyotype like that of the Western Australian species 'P151' (White and Webb 1968). Two chromosomal rearrangements must have intervened: a fusion between the originally acrocentric X and one of the small autosomes (for which

M. virgo is homozygous) and the pericentric inversion in the CD chromosome, but the origin of the little m_2 metacentric is obscure. The X-autosome fusion in *virgo* implies that a neo-XY species or population must have existed as an intermediate stage between the $XO:XX$ species 'P151' and *virgo*. That neo-XY form is presumably now extinct. Since *virgo* is homozygous for the X-autosome fusion there does not seem any likelihood of its having arisen by hybridization, as many parthenogenetic forms have done; at least if any hybridization did occur it would have had to have been between two neo-XY taxa.

Whereas bisexual species of morabine grasshoppers generally show considerable polymorphism for colour pattern (green, grey and brown individuals), *M. virgo* is relatively constant in colour. Although restricted to a rather narrow range of food plants it has managed to adapt to life on a variety of species of *Acacia*. *Moraba virgo* is theoretically important since, apart from the Diptera with polytene chromosomes listed in Table 19.1, and some parthenogenetic lizards which have arisen by hybridization between species with different karyotypes (see p. 742), it is the only thelytokous species known to be a multiple heterozygote for chromosomal rearrangements. Structural heterozygosity may well exist in many other thelytokous species, but no special search for it has been made in most cases, and many thelytokous species have high chromosome numbers, small chromosomes or other conditions which would make pericentric inversions, unequal translocations and other theoretically observable rearrangements, difficult to detect in practice.

Thelytoky seems to be much more frequent in phasmatids (Stick insects) than in any of the other orthopteroid orders. The situation has been investigated in the species of the Mediterranean region (Bergerard 1958a, 1962), in some forms from India sent back to European laboratories (Cappe de Baillon and collaborators 1934–40), and in the New Zealand phasmatid fauna (Salmon 1955). However, in spite of much published information, the facts are not entirely clear, and the interpretation is still somewhat uncertain.

Most species of phasmatids are bisexual; but facultative thelytoky seems to be extremely widespread, i.e. most workers report a high percentage of hatching from unfertilized eggs (Hadlington and Shipp 1961; Bergerard 1962). In several instances we have a 'species' which has males in one geographic area but exists as an all-female thelytokous population elsewhere. Thus *Sipyloidea sipylus* is bisexual (but shows facultative parthenogenesis) in the Malay archipelago, its original home; in Madagascar, where it has been introduced, it is exclusively thelytokous. The mechanism of thelytoky is apparently apomictic, there being two maturation divisions in the egg (Pijnacker 1968). *Bacillus rossii* is parthenogenetic in the south of France;

the egg apparently undergoes two maturation divisions and eighteen bivalents are present at first metaphase (de Baehr 1970), so this is clearly a case of automixis. The embryos start development with a haploid chromosome complement, but doubling of the chromosome number by a process of endomitosis leads to a restoration of diploidy (Pijnacker 1969). Near Pisa, Italy, rare males occur, but these are fully fertile. Unmated females give all-female progenies, but ones that have been fertilized produce bisexual progenies with a normal sex ratio (Benazzi and Scali 1964). In north Africa 'B. rossii' is bisexual and north African males have been successfully mated with thelytokous females from southern France, producing bisexual progenies (Bergerard 1962, see also Montalenti and Fratini 1959).

In the facultatively thelytokous phasmatid *Clitumnus extradentatus* oogenesis is automictic, with bivalent formation and two meiotic divisions; restoration of the chromosome number takes place through 'diploidization' in the cleavage nuclei (Bergerard 1955, 1957, 1958a, 1962). In spite of the earlier claims of Cappe de Baillon and collaborators (Cappe de Baillon, Favrelle and de Vichet 1934a, b, 1935, 1937, 1938; Cappe de Baillon and de Vichet 1940), who thought that the high chromosome numbers of some *Carausius* strains indicated polyploidy, the thelytokous biotypes of phasmatids are probably all diploid. Cappe de Baillon *et al.* believed that the thelytokous *Carausius morosus* (somatic number 64) was a triploid, since some other species of the genus, such as the bisexual *C. juvenilis* have $2n = 42$. But Bergerard has shown that eggs of *C. morosus* give rise to intersexes and males if kept at high temperatures. Such males show thirty-two bivalents in their spermatogenesis (Pehani 1925; Bergerard 1962); the species is hence a diploid. A similar claim for triploidy was made by Cappe de Baillon and de Vichet (1940) in the case of the thelytokous *Leptynia hispanica* (somatic number 52–56) which may be contrasted with the bisexual *L. attenuata* ($2n = 36$). In this case no 'males' of *hispanica* have been studied, so that there is no direct evidence. But rather wide differences in chromosome numbers between closely related species seem to be rather characteristic of phasmatid genera (see, for example, the data of Hughes-Schrader (1947a) on *Isagoras*) so that one should not necessarily conclude that *L. hispanica* is a triploid simply on the basis of its chromosome number.

Pijnacker (1966) has shown that there are two mitotic maturation divisions in the egg of *Carausius morosus*. The thelytoky of this species is hence of the apomictic type.

The process of 'diploidization' observed by Bergerard in *Clitumnus* seems to involve 'blocked' metaphases with extra-wide chromosomes or diplochromosomes which resemble nuclei after colchicine-treatment. Clearly an extra replication cycle has occurred and a transformation from the haploid to the diploid condition will result. Individuals that have become diploid in this manner would, of course, be completely homozygous.

THELYTOKY IN LEPIDOPTERA

In the Lepidoptera, although tychoparthenogenesis has been recorded in a number of species, complete thelytoky seems to be known only in certain species of Bagworm moths (Psychidae), in which the females are flightless. In all the cases that have been investigated, the mechanism is an automictic one, although the details vary greatly. Since in Lepidoptera it is the female sex which is XY or XO, the cytogenetic mechanism of thelytoky obviously has to be one which retains this sex-determining heterozygosity and transmits it from generation to generation.

Solenobia triquetrella consists of three main 'races', a diploid bisexual one (DB), a diploid thelytokous one (DT) and a tetraploid thelytokous race (TT). The latter is widespread in Europe, both north and south of the Alps; in Switzerland it occurs particularly in areas which were covered by the Würm ice sheet (Seiler 1946). DB and DT occur north of the Alps, where they are broadly sympatric.

According to earlier accounts (Seiler and Schaeffer 1941) the restoration of the somatic chromosome number was supposed to be accomplished by a fusion of cleavage nuclei in pairs. It is now known that this is incorrect. The true mechanism in both the DT and TT races, involves the fusion of the two central nuclei resulting from meiosis (second division non-sisters) to form a '*Richtungs-Kopulations Kern*' (RKK) which becomes the functional nucleus that gives rise to the embryo (Seiler and Schäffer 1960; Seiler 1963). The inner nucleus ('true egg nucleus') also divides repeatedly, but only to give polyploid nuclei that degenerate in the yolk. The outermost polar-body nucleus degenerates without dividing. The 'RKK' is necessarily XO or XY, thus preserving the sex-chromosome heterozygosity and ensuring that the offspring are all females.

An RKK is also formed in the eggs of the bisexual race, which may accordingly be regarded as preadapted for thelytokous development. But if the egg is penetrated by a sperm the latter fuses with the innermost nucleus to form a true zygote nucleus which now undergoes normal cleavage, the RKK degenerating (Seiler 1959).

Another species of *Solenobia* has been studied by Seiler and especially by Narbel-Hofstetter (1950); it was earlier referred to as *S. pineti*, but is probably correctly named *S. lichenella*. A diploid bisexual form of this species occurs near Munich. No DT biotype is known in this case, but there is a TT biotype, with $4n = 118$–126, there being apparently some genuine aneuploidy. There are approximately sixty bivalents at the first meiotic division, but apparently some univalents may be present. The two halves of the first anaphase spindle collapse on one another and form the second-division spindle, a process referred to by Narbel-Hofstetter (1964) as *cataphase*.

In the central European form of another psychid moth, *Apterona helix* (= *Cochliotheca helix*), with 2n = 62, there are thirty-one bivalents at the first meiotic division, which has a normal anaphase (Narbel 1946). There is a slightly different type of cataphase (Fig. 19.5): second meiotic spindles (which are directly derived from the half-spindles of first anaphase), each with thirty-one chromosomes, come together side by side and fuse, so that a composite second metaphase plate contains the full diploid number of sixty-two chromosomes. Two telophase nuclei result from this second division and both of them function as cleavage nuclei, there being no polar body at all.

Clearly, the automictic system of *Apterona* is genetically equivalent to an apomixis. Heterozygosity will tend to increase in such a species. But the special peculiarity of this well-nigh unique system (but compare *Bothrioplana*, p. 738) is that each individual, since it arises from two nuclei which may be genetically different, must be expected to be a mosaic for all distal loci which undergo postreduction. In the Mediterranean region, *Apterona helix* is said to be bisexual. No tetraploid thelytokous race is known in this species.

Another European genus of psychids, *Luffia*, has been studied cytologically by Narbel-Hofstetter (1955, 1956, 1958, 1961, 1963a, b). *L. lapidella* is a bisexual diploid; however it seems to include strains which reproduce by pseudogamy and give all-female progeny. Such strains are necessarily strictly sympatric with the sexually reproducing biotype.

L. ferchaultella is a diploid thelytokous form. Actually, this name probably covers a multitude of biotypes which differ in external morphology (McDonogh 1943) and in the details of the maturation divisions in the egg (Narbel-Hofstetter 1961). 2n = 61, as in females of the related bisexual species *L. lapidella*, but there may be some variation of chromosome number. At first metaphase thirty-one bodies are present on the spindle. A first anaphase occurs, but the diploid number is restored, either by the formation of a restitution-spindle from the two half-spindles of the first division, or by fusion of the two completed second division spindles. In any case the first meiotic division is ineffective, the normal sequence of events being interrupted at some stage between first anaphase and second metaphase (probably at a characteristic stage in each strain). The second meiotic division results in the formation of a central diploid egg nucleus and a peripheral diploid polar body nucleus. This course of events is also followed in the pseudogamic

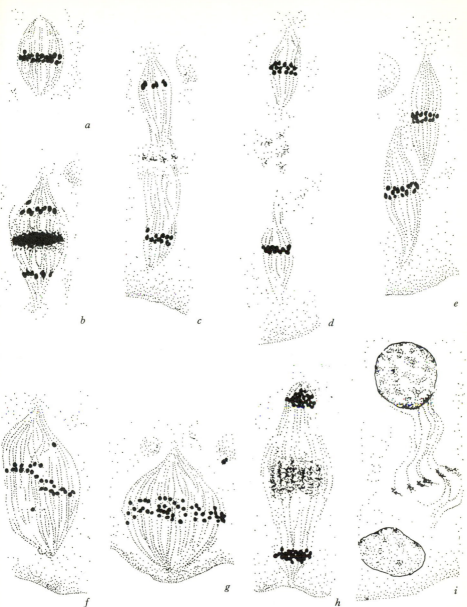

Fig. 19.5. Maturation divisions in the egg of *Apterona helix*. *a*, first metaphase; *b*, first anaphase, showing the eliminated bodies in the equatorial region of the spindle; *c*, interkinesis; *d*, second meiotic divisions both in metaphase; *e*, the two spindles approaching one another; *f*, fusion of the spindles; *g*, the diploid second metaphase; *h*, second anaphase; *i*, the two resulting nuclei. From Narbel-Hofstetter (1964).

race of *lapidella*, the only difference being the presence in the latter of a sperm nucleus which degenerates. Narbel-Hofstetter (1962) was successful in crossing males of *lapidella* with thelytokous females of *ferchaultella*; the offspring included parthenogenetic females, intersexes (presumably triploid) and some fertile males and females of the bisexual type.

THELYTOKY IN BEETLES

Considering the enormous number of species of Coleoptera, it is perhaps remarkable how few are thelytokous. Apart from *Micromalthus debilis*, which has a unique type of cyclical parthenogenesis (see p. 689), most of the known cases of thelytoky in beetles occur in the weevils (family Curculionidae). Isolated instances also occur in the families Ptinidae and Chrysomelidae.

The Spider beetle *Ptinus mobilis* is an interesting example of pseudogamy (Moore, Woodroffe and Sanderson 1956; Sanderson and Jacob 1957; Sanderson 1960). It is a triploid (3n = 27), which is presumably *XXX*. Apparently the oocyte nucleus undergoes an endomitotic doubling of chromosome number after the last oogonial division (as in *Moraba virgo*). A single maturation division then follows, there being twenty-seven bivalents ('pseudobivalents' according to Sanderson). However, the egg does not develop beyond the metaphase stage unless it is penetrated by a sperm of the related diploid bisexual species *P. clavipes* (2n = 18, *XY* in the male). No genetic fertilization occurs, however, and development is certainly thelytokous with maternal inheritance. The resemblance of this case to several that have been described in earthworms (see p. 732) is obviously very close.

The chrysomelid beetle *Adoxus obscurus* exists as a bisexual diploid in north America and an apomictic triploid in Europe (Suomalainen 1958b). There is no suggestion of pseudogamy in this case. A number of 'species' of another chrysomelid genus, *Calligrapha*, are apomictic tetraploids (Robertson 1964). A member of the Ciidae, *Cix fuscipes*, is a thelytokous diploid, but the maturation divisions in the egg were not studied. A few other instances of thelytoky are known in various families of beetles, but have not been investigated cytologically as yet.

A whole series of parthenogenetic races and species occur in the subfamilies Otiorrhynchinae, Brachyderinae, Cylindrorrhininae and Eremninae of the weevils (Curculionidae). They have been studied cytologically by Suomalainen (1940a, b, 1945, 1947, 1949, 1954a, 1955, 1958b, 1961, 1962, 1966a, 1969a), Seiler (1947), Takenouchi (1957a, b, 1959, 1961, 1963, 1964, 1965, 1966) and Mikulska (1949, 1951, 1953, 1960). We give a list of the

TABLE 19.2. *Polyploidy in parthenogenetic weevils* (Data from Suomalainen 1954a, 1955, 1969a and Takenouchi 1964)

Species	Canada	Finland	Poland	Germany	Austrian Alps	Switzer- land
*Otiorrhynchus niger**	–	–	–	–	3n & 4n	3n
*O. dubius**	–	4n	–	–	–	–
*O. scaber**	–	4n	3n & 4n	–	3n	3n & 4n
O. subcostatus	–	–	–	–	–	3n
O. singularis	–	3n	–	3n	–	3n
O. subdentatus	–	–	3n	–	4n	3n & 4n
*O. salicis**	–	–	3n	–	3n	3n†
O. sulcatus	–	3n	–	3n	–	3n
*O. rugifrons**	–	–	–	–	–	3n
*O. gemmatus**	–	–	–	–	3n	–
O. pauxillus	–	–	3n	–	–	3n
O. ovatus	3n	3n	–	–	–	–
*O. chrysocomus**	–	–	–	–	4n ?	3n
O. anthracinus	–	–	–	–	–	5n
O. ligustici	–	3n	–	–	–	–
O. proximus	–	–	3n	–	–	–
Peritelus hirticornis	–	–	–	–	–	3n & 4n
*Trachyphloeus bifoveolatus**	3n	3n	–	–	–	–
*Polydrusus mollis**	–	2n	2n	3n	–	3n
Eusomus ovulum	–	–	3n	–	–	–
Sciopithes obscurus	4n	–	–	–	–	–
Sciaphilus asperatus	3n	3n	–	3n	–	3n
Strophosomus melogrammus	3n	3n	–	3n	–	3n
Barynotus obscurus	–	4n	–	–	–	–
*B. squamosus**	3n	–	–	–	–	–
B. moerens	–	–	–	–	5n	3n
Tripiphorus carinatus	–	–	–	–	–	3n
T. curcullatus	–	–	–	–	–	4n
*Liophloeus tessulatus**	–	3n	–	–	–	–

Other parthenogenetic species of weevils (mainly Japanese):
 Catapionus gracilicornis (4n = 44 and 5n = 55) Takenouchi, 1957a, b
 Listroderes costirostris (3n = 28–41) Takenouchi 1957b; Sanderson 1956; Takenouchi, Yasue and Kawata 1970
 Pseudocneorhinus bifasciatus (4n = 44) Takenouchi 1957b
 Scepticus griseus (5n = 55) Takenouchi 1961
 Tropiphorus terricola (4n = 44) Takenouchi 1964
 Cyrtepistomus castaneus (3n = 30) Takenouchi 1970
 Blosyrus japonicus (6n = 61–66) Takenouchi 1970

* A diploid bisexual form is also known.   † Also triploid in Norway.

forty-four known thelytokous biotypes of this group in Table 19.2. It will be seen that only one (*Polydrosus mollis* in Poland and Finland) is a diploid; there are twenty-five triploid, twelve tetraploid, four pentaploid and one hexaploid strains. The basic chromosome number ($n = 11$) seems to be rather uniform throughout this group and is also found in the bisexual species, which have 2n = 22. However, the triploids *Liophloeus tessulatus*, *Eusomus ovulum* and the Australian form of *Listroderes costirostris* have 3n = 30 rather than 33 and *Strophosomus melanogrammus* has 3n = 34. The genetic system of these parthenogenetic weevils is always apomictic in type; only one maturation divison takes place in the egg. Since there is no synapsis, the full number of univalent chromosomes arrange themselves on the spindle of the maturation division. In many eggs this spindle is compound, i.e. there are several separate but adjoining metaphase plates, and in such cases each plate consists of either a haploid or a diploid set of univalents.

It is uncertain whether the polyploidy of the parthenogenetic weevils is autoploidy or alloploidy. Auto-triploidy could arise in an automictic species through fusion of three instead of two of the products of meiosis. But there is no real evidence that any of these thelytokous strains and species of weevils ever were automictic, and it is difficult to imagine autotriploidy arising in an apomictic species with a single mitotic maturation division in the egg, or even with two, as in *Pycnoscelus surinamensis*. However, it may be that the formation of several metaphase plates at the single maturation division has facilitated the origin of autopolyploid biotypes.

Allotriploidy could presumably have arisen as a result of copulation between parthenogenetic females and males of related bisexual biotypes. This would add an 'alien' haploid set to the pre-existing diploid karyotype. Possibly, some degree of heterosis would in most cases result from such an addition. If so, the large number of triploid species and strains of these weevils is satisfactorily explained. The few tetraploid forms may have arisen by a repetition of the process, or they may be autotetraploids. The pentaploid apomicts have most likely arisen as a result of mating between a tetraploid parthenogenetic female and a diploid male belonging to a different race or species. The main objection to regarding all these polyploid biotypes as autoploids is the difficulty in understanding how autoploidy could be so generally advantageous as to lead to the evolutionary success of the poly-ploids at the expense of the diploids (a difficulty which is considerable but perhaps not insuperable).

The striking biometrical differences found by Suomalainen (1961) between geographically remote populations of the same thelytokous 'species' may, as he has suggested, be due to evolution by mutation and selection. But they may also be due to polyphyletic origin of alloploid or partially alloploid taxa (for example, the populations at two localities may be *AAB* and *AAC* in composition, where *A*, *B* and *C* represent the haploid genomes of three original biotypes, of which *A* became thelytokous, *B* and *C* remaining bisexual).

The tendency to polyploidy in thelytokous weevils is not confined to

European species. Takenouchi (1961) has found a number of thelytokous species in Japan, including two pentaploids and one hexaploid. One of the pentaploids is regarded taxonomically as a geographic race of a widespread diploid bisexual species. The vast majority of species of weevils in all continents are, of course, strictly bisexual.

THELYTOKY IN SCALE INSECTS

The thelytokous systems of the weevils seem to be very uniform – apomixis, polyploidy and heterozygosity are their outstanding characteristics. By contrast, the mechanisms of thelytoky encountered in scale insects, although not known in detail, seem to be much more diverse. Brown (1965) has reported on twenty thelytokous species of the families Phoenicococcidae and Diaspididae (out of a total of 140 species studied by him). In addition there are a number of instances where a taxonomic 'species' includes both bisexual and thelytokous biotypes. Of all these forms at least nine seem to have apomictic mechanisms, while in four bivalents are seen in the egg at the first maturation division, so that some type of automictic system must be present. It is noteworthy that none of these apomictic coccids seem to be polyploids. Brown (1964) has argued in favour of the view that most coccid populations must be inbred and hence largely homozygous. But the very numerous apomictic forms that have evolved in this group would appear to be ones that are exploiting heterozygosity to the utmost. However, they are conservative in their chromosome numbers – it is the sexual members of the group rather than the thelytokous ones which exhibit unusually high or low chromosome numbers. Only one thelytokous coccid, *Leucaspis loewi*, shows a 'diploid' number of eleven, and is presumably heterozygous for a dissociation. Brown states that, in contrast to the situation in the armoured scale insects, most of the thelytokous members of the Lecaniidae (mealybugs) are automictic. In two species of *Lecanium* diploidy is restored by fusion of second division sister nuclei (Thomsen 1927, Suomalainen 1940c). In *Pulvinaria hydrangeae*, on the other hand, diploidization results from a fusion of the products of the first cleavage division (Nur 1963a).

THELYTOKY IN HYMENOPTERA

In the Hymenoptera, of course, thelytoky has arisen in the course of evolution from haplodiploid systems in which haploid eggs develop parthenogenetically into males. In other words, the potentiality to enter on the cleavage divisions without previous fertilization is already present in fully evolved form and all that is required is a change in the maturation divisions which will ensure that cleavage begins with the diploid rather than the haploid number of chromosomes.

Both automictic and apomictic thelytoky are known in the Hymenoptera. In the sawflies (Tenthredinidae) *Pristiphora pallipes* and the thelytokous race of *Diprion polytomum* are automictic (Comrie 1938; Smith 1941). On the other hand, in the sawfly *Strongylogaster macula* we have an apomictic mechanism (Peacock and Sanderson 1939).

The parasitic wasp *Nemeritis canescens* shows a modified form of automictic thelytoky (Speicher 1937). 2n = 22 and there are eleven bivalents at first metaphase. But both anaphase groups become included in a restitution nucleus and twenty-two chromosomes align themselves on the second metaphase spindle. At the second anaphase, however, eleven univalents pass to each pole (the two haploid sets, presumably). The chromosome number is restored to twenty-two by an endomitotic reduplication at the end of the second anaphase, so that the cleavage nuclei contain the diploid complement. This case is discussed in detail by Narbel-Hofstetter (1964), who compares it with that of *Bothrioplana* (p. 738).

THELYTOKY IN CRUSTACEA

The most famous instance of thelytoky in the Crustacea, and the one that has been most extensively studied, is that of the branchiopod 'Brine shrimp' *Artemia salina*. However, in spite of the large number of cytologists who have worked on this 'super-species' or 'agamic complex', some uncertainties remain. Owing to the fact that it is restricted to waters of high salinity, such as occur in certain lakes and salt pans, *Artemia* has a highly discontinuous distribution, being known from most of the Mediterranean countries, including those bordering on the Black Sea, North Africa and the western United States (Great Basin area, Gulf of California); it also occurs in Australia. *Artemia* eggs are probably carried vast distances in mud adhering to the feet of water-fowl (Barigozzi 1946), thus accounting for the wide dispersal of the 'species' and some of its biotypes.

The cytology of *Artemia* has been studied by Brauer (1894), Artom (1908, 1928, 1931), Gross (1932, 1935), Stella (1933), Barigozzi (1934, 1935, 1941, 1944, 1946, 1956), Haas and Goldschmidt (1946), E. Goldschmidt (1952) and Barigozzi and Tosi (1957, 1959), whose results are not entirely in agreement, and are critically discussed by Narbel-Hofstetter (1964).

The diploid bisexual 'race' of *Artemia*, occurs at many localities. Meiosis is normal in both sexes, with twenty-one bivalents, and eggs will not develop without fertilization. The female is apparently the heterogametic sex (Stefani 1963) and crossing-over between the X and Y takes place (Bowen 1965).

In the diploid thelytokous biotypes from a number of Mediterranean localities, twenty-one bivalents are formed in the egg and a first meiotic division is initiated in the same manner as in the bisexual form. Artom (1931) claimed that in material from Sète a first polar body is given off and that there is then a doubling of the chromosome number between the two maturation divisions; but his interpretation has been disputed by Barigozzi (1944) and Stefani (1960). Gross (1932) suggested that the diploid, tetraploid and 'octoploid' thelytokous biotypes at Margherita di Savoia were as described by Artom, except for the chromosome number in the polyploid forms. If so, the mechanism would be equivalent to a fusion of second division sister-nuclei.

In diploid thelytokous material from Cagliari, Sardinia, Stefani states that the first division spindle is parallel to the surface of the egg. It carries twenty-one bivalents. First anaphase is a long stage and eventually the spindle breaks in the middle and the two half-spindles come to lie side by side, the two anaphase groups forming the metaphase plate of a diploid second meiotic division whose spindle is perpendicular to the surface of the egg. One of the two diploid nuclei resulting from this second division is expelled as a polar body nucleus, while the other becomes the egg nucleus. This type of automixis would be liable to lead to genetic segregation of loci situated distal to the chiasma.

Most of the polyploid biotypes of *Artemia salina*, at any rate, are apomictic. There is a tetraploid form which shows eighty-four univalents at the single maturation division (Barigozzi 1944). A few triploid thelytokous individuals have also been found at one locality in Italy and odd-numbered polyploidy also occurs in some Palestinian strains, since E. Goldschmidt studied one with a somatic number of about 64 and another with about 108–109 chromosomes. According to the latter author, synapsis occurs in these polyploid strains; but the bivalents become resolved into their constituent chromosomes ('desynapsis') between diakinesis and metaphase. Thus the hyper-triploid form shows 40–45 bodies at diakinesis and the hyperpentaploid one 59–100. Barigozzi (1956) did not observe any synapsis in triploid individuals from Istria.

The most widespread biotypes of *Artemia* seem to be the diploid bisexual one and the diploid and tetraploid thelytokous strains, most of the others having been encountered only at single localities. However, many of the salt lakes from which *Artemia* has been studied seem to contain mixtures of two or more biotypes. The parthenogenetic ones normally produce all-female offspring, but some of them have occasionally been observed to give rise to a few males. It is obvious that the evolutionary history of this super-species has been extremely complex.

A classical case of polyploid parthenogenesis occurs in the isopod genus

Trichoniscus (Vandel 1928, 1934, 1940). There are a number of bisexual species of this genus in Europe, all with $2n = 16$. In central and southern France there exists a triploid thelytokous form ($3n = 24$) known as *coelebs*, which is regarded by Vandel as a 'race' of the diploid bisexual species *elizabethae*. The two forms are sympatric throughout a large area but do not seem to interbreed. In southern France *elizabethae* is found only in damp mountainous localities, while *coelebs* occurs in isolated colonies in the arid 'garrigues' of the Mediterranean region, an apparently precarious habitat for a crustacean whose life depends on the presence of water. Thus the parthenogenetic *coelebs* seem to have succeeded in establishing itself in regions where the bisexual members of the genus are unable to live. Parthenogenetic biotypes of *Trichoniscus* whose cytology is unknown also extend further north than any of the bisexual species and even reach the Baltic countries, Scandinavia and Iceland, so that they are possibly more resistant to cold as well as to drought.

The populations of *coelebs* are not entirely composed of females; there are a few males (1–2%) which are also triploid. The triploid females are apomictic; their chromosomes do not pair during oogenesis and there is only one maturation division which is mitotic in character. Thus the eggs which develop without fertilization possess the somatic number of 24 from the beginning. The establishment of the apomictic system must almost certainly have preceded the development of triploidy in this case; it is hardly likely to have been an automatic consequence of triploidy, as originally suggested by Vandel. Thus we must assume that a diploid biotype developed apomictic thelytoky and was later replaced by a triploid form. Perhaps this hypothetical diploid thelytokous *Trichoniscus* is extinct; but it may yet be found somewhere in Europe.

The triploid males which are occasionally met with constitute a problem. It is uncertain whether they have exactly the same karyotype as the triploid females. Their spermatogenesis is naturally anomalous; there is no chromosome pairing, but two maturation divisions occur, both of which are said to be mitotic in character. Thus the sperms of such males contain the somatic number of chromosomes (24). Triploid males will mate with triploid females, but their sperm is apparently not functional; they will usually not copulate with females of the diploid race.

Another case of 'geographic' thelytoky in isopods is that of *Nagara modesta* which includes a bisexual biotype on the island of Salajar and a thelytokous one on Christmas Island (Indian Ocean). The latter is probably apomictic, but whether it is polyploid is uncertain from the work of Hill (1948). Apomictic thelytoky occurs also in one 'race' of the ostracod *Heterocypris* (= *Cypris*) *incongruens*, according to Bauer (1940); this form is certainly a diploid.

THELYTOKY IN ARACHNIDA

In the mites (Acarina) numerous thelytokous forms have been recorded, but there are very few reports of thelytoky in the other groups of the Arachnida. However, the Brazilian scorpion *Tityus serrulatus* is almost certainly thelytokous (Matthiesen 1962). This species has 2n = 12 with two chromosomes smaller than the rest (Piza 1940) so that it is probably diploid.

Among the mites, thelytoky is well known in the predaceous species *Cheyletus eruditus*. Apparently the mechanism is an automictic one, there being two bivalents at the first maturation division; restoration of the diploid number is said to be by fusion of cleavage nuclei in pairs (Peacock and Weidmann 1961).

In two species of oribatid mites Taberly (1958b, 1960) has described a type of automictic thelytoky. Nine bivalents are clearly visible at the first metaphase. Restoration of the diploid number seems to be accomplished through inclusion of the two anaphase groups of the second meiotic division in a single nucleus (a process equivalent to fusion of second-division sister nuclei).

The tick *Haemaphysalis longicornis* (formerly *bispinosa*) reproduces parthenogenetically, but produces occasional functionless males; it has 33 chromosomes compared with $2n\mathcal{J} = 21$, $2n\female = 22$ in the bisexual form of *H. bispinosa* and *H. leporis palustris* and is probably a triploid (Bremner 1959; Oliver and Bremner 1968). Takenouchi, Shiitsu and Toshioka (1970) have shown that this is a case of apomictic thelytoky, with no synapsis in the oocyte nucleus.

THELYTOKY IN NEMATODES

Several root-knot nematodes of the family Heteroderidae reproduce by apomictic thelytoky combined with polyploidy (Triantaphyllou 1962, 1963, 1964). Some bisexual species of *Heterodera* and *Meloidogyne* probably have n = 9. *M. arenaria* includes apomictic biotypes with 36 chromosomes, which are probably tetraploid, and ones with 41–54 which are presumably hexaploid. *M. javanica*, with 43–48 chromosomes is likewise apomictic and probably pentaploid. The hexaploid strains of *M. arenaria* produce numbers of males which are probably functionless, as in the triploid *Trichoniscus coelebs* (see p. 730). *M. hapla* includes bisexual biotypes with n = 15, 16, and 17; some apomictic strains are probably hexaploid, with a somatic number of 45–46.

A number of forms of the genus *Rhabditis* studied by Belar (1923a, 1924) are thelytokous hermaphrodites which reproduce by pseudogamy. There is a single apomictic maturation division in the egg. Hertwig (1920) reported the origin by mutation of a pseudogamic apomictic strain of *R. pellio* which appeared suddenly in a sexually-reproducing stock.

THELYTOKY IN OLIGOCHAETES

The oligochaetes are, of course, fundamentally hermaphroditic animals and the thelytokous mechanisms that are fairly widespread in two of the main families (Lumbricidae and Enchytraeidae) have been derived from hermaphroditic genetic systems. To a considerable extent, the testes seem to have degenerated in thelytokous oligochaetes; but in a number of these species sperms are still formed and may even function in a pseudogamic manner, activating the eggs to development, even if they do not fertilize them in a genetic sense. However, the extent to which pseudogamy actually occurs in many of the thelytokous biotypes is rather uncertain; experimental studies seem indicated, rather than simple observation. The sexual species of earthworms are certainly cross-bred as a rule, but there seems to be no anatomical barrier to self-fertilization, which may sometimes occur.

In the only family of true earthworms that has been studied cytologically (the Lumbricidae) at least four genetic systems have been demonstrated (Muldal 1948, 1949, 1952; Omodeo 1951*a–f*, 1952, 1953*a, b*, 1955, 1962*a, b*). These are (1) diploid sexual species, (2) polyploid sexual species, (3) apomictic polyploids, (4) automictic polyploids. A single diploid thelytokous biotype (*Octolasium lacteum* from England and Switzerland) probably exists, but Omodeo has suggested that this form may actually be sub-triploid.

The most characteristic feature of the automictic thelytokous earthworms (which are far more numerous than the apomictic ones) is that the compensatory doubling of the chromosome number is a premeiotic one which occurs at the last oogonial division. This mitosis is entirely abortive, the abnormally short and condensed chromosomes all becoming included in a single nuclear membrane at the stage corresponding to telophase. Thus a triploid form with fifty-four chromosomes in its somatic cells and oogonia will have 108 elements in the leptotene oocyte nuclei. Since synapsis is restricted to pairs, the number of bivalents formed is the same as that of the somatic chromosomes (fifty-four in the example chosen). These bivalents show typical chiasmata and meiosis is entirely regular, two polar bodies being given off. Thus in this very unusual type of thelytoky even odd-numbered polyploids (3n, 5n . . .) show only bivalents in their oocytes, with no univalents or multivalents (Fig. 19.6). It is uncertain whether all bivalents are derived from the pairing of 'sister homologues' (i.e. chromosomes derived from the same parent chromosome at the premeiotic restitutional mitosis) or whether pairing of non-sister homologues occurs. The genetic consequences would obviously be somewhat different according to whether or not pairing of non-sister homologues takes place (Omodeo 1951c).

Dendrobaena rubida, including the 'forms' *typica, tenuis, subrubicunda, norvegicus* and *constricta* (whatever the precise nature of these 'forms') is a complex superspecies, including diploid ($2n = 34$) tetraploid, hexaploid and (probably) suboctoploid sexual biotypes and a triploid thelytokous one (Omodeo 1955). The geographic distribution of the various biotypes is complex; for example the hexaploid has been found in Italy and in Greenland and Iceland, but not in England (where only the diploid, triploid and tetraploid biotypes are known to occur). It is entirely possible that hexaploidy has been independently acquired in the arctic and temperate regions of Europe, since the diploid race is known from Turkey, Italy and Iceland. The triploid thelytokous biotype is only known from England (Muldal 1952); it has a 'premeiotic doubling' mechanism and shows fifty-one bivalents at meiosis. The sexual polyploids likewise show only bivalents at meiosis (and not multivalents), e.g. the hexaploids, with $6n = 102$ have fifty-one bivalents, like the triploid automictic race.

Dendrobaena octaedra is apparently always thelytokous. It differs from most thelytokous earthworms, however, in being apomictic, i.e. there is no premeiotic doubling of the chromosome number and no formation of bivalents. Most biotypes (from Iceland, Greenland and the Dolomites) seem to be hexaploid ($6n = 108$). But apparently aneuploid biotypes occur in Greenland and the Alps; some show 112–117 chromosomes, some 124, some 99. The exact status of these forms and their evolutionary relationships to one another is uncertain. *D. octaedra* is certainly an agamic complex, rather than a uniform morphospecies, since it includes forms with and without testes and accessory male organs or even with supernumerary non-functional testes (Omodeo 1953a). How far all this morphological variation is genotypic, how far phenotypic, and how it is distributed geographically, has not been determined. It may simply be a breakdown of developmental canalization in the absence of stabilizing selection (as suggested by Omodeo), but it seems more likely that it indicates the coexistence of numerous biotypes differing significantly from one another genetically, even if not in their visible cytology.

Some thelytokous earthworms possess testes, but the spermatogenesis is highly anomalous, the chromosome number being repeatedly doubled during the gonial divisions, so that giant cells with multipolar spindles are formed. In some instances no sperms are produced at all, in others a small number of abnormal and functionless sperms. The tetraploid race of

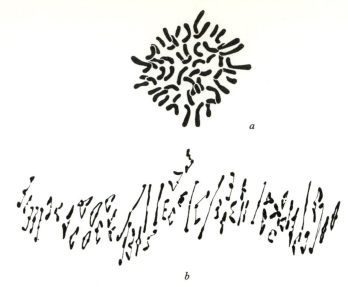

Fig. 19.6. Cytology of the triploid earthworm *Eiseniella tetraedra typica*. *a*, oogonial metaphase, showing 54 chromosomes; *b*, side view of first metaphase in the egg, showing 54 bivalents. From Omodeo (1952*a*) redrawn.

Allolobophora caliginosa trapezoides from Italy still seems to produce normal sperms and is most likely pseudogamous.

Eiseniella tetraedra is a particularly variable agamic complex which includes about eight forms differing in external morphology. Tetraploid biotypes have been found in England, Turkey and Calabria; in central Italy, on the other hand, triploid strains occur. All these various forms probably reproduce exclusively by thelytoky. Various degrees of reduction of the spermathecae are found and the male pores may be on the 13th segment (form *typica*) or the 15th (form *hercynia*).

There are a number of polyploid thelytokous species in the oligochaete family Enchytraeidae, but the usual cytogenetic mechanism is quite

TABLE 19.3. *Relation of polyploidy to mode of reproduction in Oligochaeta* (from Christensen 1961)

	Lumbricidae		Enchytraeidae		Total	
	Sexual	Parth	Sexual	Parth	Sexual	Parth
Diploid	34	1	46	1	80	2
Polyploid	8	16	27	14	35	30

different from that characteristic of the Lumbricidae. With the exception of the triploid *Lumbricillus lineatus* (3n = 39) only even-numbered polyploids have been reported (Christensen 1959, 1960, 1961). No premeiotic doubling of the chromosome number occurs in these species and bivalents are formed at the first meiotic division in the egg (Christensen 1961). Thelytoky in these species is hence automictic and restoration of the chromosome number takes place as a result of an ineffective second meiotic division, equivalent to fusion of second-division sister nuclei. In the diploid and tetraploid thelytokous species of *Mesenchytraeus* and *Cognettia glandulosa* the bivalents show clear chiasmata, but in a number of other thelytokous biotypes belonging to the genera *Fridericia* and *Achaeta* the bivalents resemble those of *Callimantis* (see p. 483); whether genetic crossing-over occurs is, of course, uncertain. Such achiasmatic bivalents also occur in some sexually-reproducing enchytraeids (see p. 477).

In *Lumbricillus lineatus* there is a peculiar apomictic mechanism. No pairing occurs and at the first anaphase the undivided univalents pass in approximately equal numbers to the two poles. The spindle then becomes V-shaped and the univalents at each pole split into their constituent chromatids, one set of daughter chromatids from each pole passing to the apex of the V-shaped spindle where they meet and form a triploid nucleus which becomes the functional egg nucleus (Christensen 1960).

THELYTOKY IN FLATWORMS

Some highly interesting instances of superspecies and agamic complexes based in part on pseudogamic reproductive mechanisms have been described in freshwater planarians by Benazzi, Benazzi-Lentati and Lepori in a long series of publications (Benazzi 1954*a*, *b*, 1955, 1957*a*, *b*, 1960, 1962; Benazzi and Benazzi-Lentati 1959*a*, *b*; Benazzi-Lentati 1957, 1960, 1961, 1966, 1970). These animals are all hermaphroditic, but self-fertilization apparently does not occur, since no eggs are obtained from unmated individuals. Apart from the cases of pseudogamy, thelytoky seems to be unknown in triclads, although it does occur in some rhabdocoels (Benazzi 1959).

The superspecies or species group *Dugesia gonocephala* consists of a number of forms from Europe, Africa and Asia. *D. gonocephala sensu stricto* is a widespread diploid species (n = 8) which reproduces sexually and has a normal meiosis in both spermatocytes and oocytes. Some individuals carry extra chromosomes, which are believed to be homologous to members of the regular set (i.e. they are not supernumerary chromosomes in the restricted sense in which we use that term). *D. etrusca* is likewise basically a diploid, sexual species, but in some strains there is considerable variation in chromosome number, due to extra chromosomes (2n up to 30 in some individuals). The extra elements may form bivalents at meiosis, but univalents and multivalents may also occur (some of the latter involved members of the regular chromosome complement, proving the homology of the 'extra' and normal elements). The extra chromosomes are inherited in a strictly maternal way (through the egg) since they are eliminated in the course of spermatogenesis. *D. ilvana* is apparently always diploid and reproduces sexually, while *D. sicula* and *D. cretica*, from Sicily and Crete respectively, although not fully investigated, seem to be hyper-diploid with some extra chromosomes.

The most complex form within this species group is *D. benazzii*, from Sardinia, Corsica and Capraia. Within this species three main biotypes have been found. The first is diploid, has normal spermatogenesis and oogenesis and reproduces sexually; in some populations polysomic individuals analogous to those of related species occur. The second biotype is somatically triploid ($3n = 24$), but the oocyte nuclei have undergone a doubling of the chromosome number and are hence hexaploid; at meiosis they show twenty-four bivalents. The primary spermatocytes are, however, diploid, since they have eliminated one haploid set; they form eight bivalents. This race reproduces pseudogamically with twenty-four maternal chromosomes, the sperm nucleus degenerating in the egg cytoplasm. Some individuals may have more than the usual number of chromosomes, but in all cases the sperms are haploid. The third biotype likewise reproduces by pseudogamy; it is tetraploid, i.e. the oocytes contain thirty-two chromosomes. No synapsis occurs in the egg and there is a single mitotic maturation division. One polar body is given off and later a cytoplasmic bud gets cut off, into which the sperm nucleus may migrate, having apparently accomplished its function of 'activating' the egg. The male germ-line is usually subtetraploid (i.e. there are fewer than 32 chromosomes) and in some cells it may be diploid ($2n = 16$), these numbers being due to chromosomal elimination in spermatogonia. However, a true meiosis does take place in spermatogenesis, and the sperms may contain any number of chromosomes between 8 and 16. It has been shown by Benazzi-Lentati and Bertini (1961) that in crosses between different biotypes of *Dugesia benazzii* the asynaptic character can be inherited through the male line, behaving hence as a dominant in inheritance. But a single individual may contain both synaptic and asynaptic oocytes.

Dugesia lugubris is likewise a complex of biotypes, differing in chromosome number and mode of reproduction. Biotype *A* is diploid ($2n = 8$) and reproduces sexually; *B* is triploid–hexaploid and pseudogamous like the analogous one in *D. benazzii*; its oocytes show twelve bivalents, the spermatocytes four bivalents, so that the eggs are triploid, the sperms being haploid. A variant of this biotype has a tetraploid soma and octoploid oocytes. Biotype *C* is a triploid which is asynaptic in the female line, the spermatocytes being diploid; biotype *D* is similar but tetraploid. Biotypes *E*, *F* and *G* seem to be all diploid, but their karyotypes are different and they seem to be reproductively isolated from one another and from the *A–D* group, so that they are best regarded as sibling species (Benazzi and Pulcinelli 1961). Benazzi (1960) and Benazzi-Lentati (1961, 1962) have carried out a number of crosses between the *A* and *C* biotypes of *D. lugubris* and have studied the inheritance of asynapsis and pseudogamic reproduction. Some of the hybrids laid both diploid and tetraploid eggs (Benazzi and Benazzi-Lentati 1959*a* and *b*). Later crosses in both *D. benazzii* and *D. lugubris* (Benazzi-Lentati 1970) have shown that the genetic determination of the various cytological and reproductive phenomena (polyploidy, asynapsis, ameiotic versus meiotic maturation, fertilization, pseudogamy) is complex and polygenic.

Polycelis nigra is also an assemblage of different biotypes. In Italian strains inheritance of colour is entirely maternal in all types of matings (Benazzi 1948*a*). Lepori (1949, 1950) has shown that this is because the sperm nucleus degenerates in the egg without contributing to the developing embryo. But penetration of the sperm into the egg is necessary to initiate development. One biotype is diploid with $2n = 16$. Another biotype, common in Italy, is triploid, with 24 chromosomes in the somatic cells, spermatogonia and primordial oogonia (Benazzi 1963). During oogenesis the chromosome number in the oogonia is doubled, so that at the first meiotic division there are 24 bivalents. After penetration of the egg by a sperm, two polar bodies are given off and the mature eggs contain 24 chromosomes. Two different types of spermatogenesis coexist in the same gonad; some spermatogonia undergo the same process of chromosome doubling as the oogonia and the spermatocytes derived from them show 24 bivalents and eventually give rise to tetraploid sperms with 24 chromosomes. Other spermatocytes, which have escaped the doubling process, have only 12 bivalents and give rise to diploid sperms. Either type of sperm can presumably 'activate' the eggs, but neither is functional in a genetic sense. No asynaptic biotypes of *Polycelis* seem to exist (Benazzi-Lentati 1970).

Scandinavian material of the *tenuis-nigra* complex of *Polycelis* has $2n = 12$ (Melander 1948, 1950*a*) but in some areas strains with $2n = 14$ occur, and there are heteroploid biotypes of *P. tenuis* in various parts of Europe, whose oocytes show about 17 to 22 bivalents.

Vegetative reproduction (by fission) seems to occur rather sporadically throughout this group of planarians. *Polycelis nigra* never undergoes fission, but in *P. felina* fission is common and in certain strains appears to be the only mode of reproduction. Fission also occurs in *Dugesia subtentaculata* and *D. cretica*. Many of these sterile, vegetatively reproducing planarians are apparently polyploids. Thus Dahm (1958) reports that *Phagocata vitta* includes a range of forms from triploids ($3n = 21$) to decaploids ($10n = 70$) and *Crenobia alpina* biotypes from tetraploids ($48 = 28$) up to 9-ploids ($9n = 63$).

The method of parthenogenetic development in the alloeocoel turbellarian *Bothrioplana semperi* seems to be entirely unique, and does not fall into any of the categories of thelytoky which we have discussed above. The egg capsules of this worm contain two nuclei, both of which contribute to the developing embryo (Reisinger 1940). Both these nuclei simultaneously undergo two maturation divisions of a peculiar type, and all the eight nuclei resulting from these divisions remain within the egg and contribute to the future embryo. The species seems to be a diploid, but an endomitotic doubling takes place just before the first maturation division, being followed by a reduction to the diploid state at the second division. If this case has been correctly interpreted, the meiotic divisions have been modified so that they function as embryonic cleavage divisions, all the products of meiosis becoming blastomeres (it will be recalled that in the psychid moth *Apterona helix* both the nuclei resulting from the second meiotic division function as cleavage nuclei).

THELYTOKY IN MOLLUSCA

The North American freshwater snail *Campeloma rufum* reproduces by apomictic thelytoky (Mattox 1937). It has the rather low chromosome number $2n = 12$, and is clearly a diploid. On the other hand, the apomictic form of *Paludestrina jenkinsi* from the British Isles is probably tetraploid by comparison with the thelytokous form from continental Europe (Rhein 1935; Peacock 1940; Sanderson 1940).

Parthenogenetic reproduction occurs in several Indian species of the freshwater snail genus *Melanoides* (Jacob 1954, 1957, 1958). In all of these the mechanism is an apomictic one, with two mitotic maturation divisions in the egg. *M. tuberculatus* has a diploid thelytokous 'race' ($2n = 32$) and a pentaploid or hexaploid one (90–94 chromosomes). *M. lineatus* and *M. scabra* are possibly both tetraploid forms (71–73 and 76–78 chromosomes). The polyploid biotype of *M. tuberculatus* and *M. lineatus* both produce rare males; the spermatogenesis of these individuals is 'normal' in the sense that two meiotic divisions occur. But the first metaphase shows a mixture of multivalents, bivalents and univalents and some chromosomal bridges occur at anaphase.

Jacob has interpreted the polyploid biotypes of these snails as auto-alloploids, but the evidence does not seem at all certain, although this interpretation is obviously not excluded. The rare males are presumably due to some type of non-disjunction. It is unfortunate that the large number and small size of the chromosomes seem to preclude a really detailed cytological investigation.

It is not clear why parthenogenesis seems to be absent from the hermaphroditic Pulmonata, whereas it is common in some other groups of hermaphroditic invertebrates such as the planarians and the oligochaetes. Possibly its absence is related to the barrier which seems to prevent the evolutionary establishment of polyploidy in pulmonate snails (see p. 458); polyploidy certainly occurs in some sexually reproducing planarians and it is relatively common in the non-thelytokous oligochaetes as well as almost universal in the thelytokous ones. Thelytoky may have been an evolutionary success in oligochaetes because it could be combined with polyploidy, and this may be impossible in the pulmonates, for 'histological' reasons of unknown nature (see p. 456).

THELYTOKY IN VERTEBRATES

About twenty-five species of vertebrates are now known to reproduce by constant thelytoky. These include three teleost fishes, two species of ambystomid salamanders and about twenty species of lizards, not all of which have been investigated cytogenetically (White 1970a). In the human species occasional meiotic thelytoky apparently leads to the production of dermoid ovarian teratomas which are incapable of an independent existence (Linder and Power 1970).

The little poeciliid freshwater fish species *Poecilia* (or *Mollienisia*) *formosa* from south Texas and northern Mexico consists entirely of females, which are apparently diploid (Hubbs and Hubbs 1932; Meyer 1938; Haskins, Haskins and Hewitt 1960; Kallman 1962). The details of oogenesis are not known so that we do not know whether automixis or apomixis occurs, but this is certainly a case of pseudogamy, since females have to be mated with a male belonging to one of several related bisexual species (*M. latipinna*, *M. sphenops*, *M. mexicana*) before they will reproduce. Penetration of the egg by a sperm is apparently necessary for development, but in the ordinary way no genetic fertilization takes place and inheritance is purely maternal. However occasionally true fertilization occurs and the offspring exhibit some paternal characters and a few rare males have been produced in this way (Haskins, Haskins and Hewitt 1960; Rasch, Darnell, Kallman and Abramoff 1965). These hybrid progeny are apparently triploid, having two maternal and one paternal chromosome set. A triploid population of *P. formosa* is known from one area of Mexico (Rasch, Prehn and Rasch 1970).

It is possible that *P. formosa* arose by hybridization between *P. latipinna* and *P. mexicana*, since it is heterozygous for two serum albumin alleles characteristic of those species (Abramoff, Darnell and Balsano 1968). It includes a number of different clones which show mutual graft-incompatibility; natural populations consist of mixtures of several such biotypes (Kallmann 1962). These histo-incompatibility clones may owe their origin to mutation or, conceivably, to occasional transfers of paternal chromosomes or segments of chromosomes to the genome of *P. formosa*. Kallmann (1964) considered that *P. formosa* was homozygous for its histo-incompatibility genes since grafts from normal females into hybrids between *formosa* and *sphenops* were not rejected. He interpreted this result as indicating that a haploid genome of *formosa* contains all the histo-incompatibility alleles that are present in the diploid. But the argument is invalid, as Rasch, Darnell, Kallman and Abramoff (1965) showed that these hybrids are triploid, i.e. they contain two genomes of *formosa*. As a form of hybrid origin in which mutations have presumably been able to establish themselves since the origin of its thelytokous reproductive system without becoming homozygous, *P. formosa* is presumably extensively heterozygous.

Two more poeciliid fishes that are genuinely thelytokous are triploid biotypes of *Poeciliopsis* living in the rivers of northwestern Mexico (R. J. Schultz 1967, 1969). One is believed to be a form of hybrid origin having two chromosome sets from *P. lucida* and one from *P. monacha* (*LLM*). Males of *P. lucida* provide sperms which activate the eggs of this form to development. The second triploid gynogenetic biotype seems to combine one chromosome set from *lucida* with two from *monacha*, i.e. it is *LMM* and is a 'reproductive parasite' on *monacha*.

In addition to these genuinely thelytokous forms of *Poeciliopsis* three diploid all-female biotypes are known. These are referred to by Schultz as *Cx*, *Cz* and *Fx*. They are probably all of hybrid origin. The first two depend for their reproduction on mating with males of *P. lucida*, while the third has a similar relationship to *P. occidentalis*, with which it coexists geographically. But in all these cases we are not dealing with gynogenesis, since the offspring show inheritance from both parents. The evidence indicates that the paternal chromosome set is not transmitted to the offspring, i.e. that these chromosomes are consistently rejected in oogenesis, only the maternal set being passed on (R. J. Schultz 1961, 1966). This most peculiar genetic mechanism, in which the male genome is merely 'borrowed' in each successive generation and never passed on, is referred to as *hybridogenesis* by Schultz (1969). It may be assumed that there is no crossing-over in oogenesis, but this is not certain. Although, strictly speaking, not a form of thelytoky, hybridogenesis involves all-female populations which must be highly heterozygous.

Gynogenetic thelytoky occurs also in certain North American salamanders of the *Ambystoma jeffersonianum* group (Uzzell 1963). There are two diploid bisexual species, *A. laterale* and *A. jeffersonianum* and two triploid thelytokous forms, *A. tremblayi* and *A. platineum* (Uzzell and Goldblatt 1967); apparently the former carries two chromosome sets of *laterale* and one of *jeffersonianum* (constitution *LLJ*), while the latter is *LLJ*. Both are hence of hybrid origin, as has been shown by a study of the serum proteins. The mechanism of thelytoky is automictic in these cases, since the oocyte nuclei show lampbrush elements which form bivalents and occasional quadrivalents (Macgregor and Uzzell 1964). A total of forty-two bivalents are normally present, which is equivalent to the hexaploid number of chromosomes ($2n = 28$ in the bisexual species). The oocytes undergo a premeiotic chromosomal replication, which is then followed by synapsis; whether pairing is normally restricted to sister chromosomes is not known. Apparently, sperm from a male of *A. jeffersonianum* or *A. laterale* is usually necessary to initiate development of the eggs, but the chromosomes of the sperm do not contribute to the individual. According to Macgregor and Uzell, the chiasma frequency of the pseudogamous forms is extraordinarily high, although if synapsis is almost invariably between sister chromosomes, these chiasmata would be genetically ineffective. In nature, *A. tremblayi* normally coexists with *A. laterale* and *A. platineum* with *A. jeffersonianum*, but at some localities the triploid thelytokous biotypes occur by themselves, so that gynogenesis can apparently be dispensed with.

It has been shown by numerous workers that certain species of the lizard genus *Cnemidophorus* in the southwestern U.S.A. and Mexico reproduce thelytokously (Minton 1958, Tinkle 1959; Maslin 1962, 1968; McCoy and Maslin 1962; Zweifel 1965; Pennock 1965). Lowe, Wright, Cole and Bezy (1970b) list nine parthenogenetic 'species' (some diploid, some triploid and one which includes both diploid and triploid biotypes).

The bisexual members of the genus *Cnemidophorus* mostly have $n = 23$, but the karyotypes of the various species differ significantly. The diploid thelytokous form *C. neomexicanus* is probably a stabilized hybrid between *C. tigris* and *C. inornatus*, since it seems to combine haploid sets derived from those species (Lowe and Wright 1966a, b). The triploid *C. perplexus* is probably closely related, having one *tigris* and two *inornatus* genomes.

The name *C. tesselatus* includes at least six largely allopatric partheno-genetic biotypes named *A* to *F*. Forms *C* to *F* are apparently stabilized diploid hybrids between *C. tigris* and *C. septemvittatus*; they differ in colour pattern, but each is very uniform in phenotype. Biotypes *A* and *B* are tri-ploids with one extra chromosome $(3n + 1)$; on the evidence of the karyotypes and of skin-grafting experiments, they are probably trihybrid, containing one genome from *tigris*, one from *septemvittatus* and one from *sexlineatus* (Wright and Lowe 1967*a*, *b*; Maslin 1967).

These forms seem to have arisen by repeated hybridizations between bisexual species, with loss of sexuality; the triploid biotypes probably as a result of occasional matings between bisexual parthenogenetic individuals and males of related bisexual species. The diploids are *XX* and the triploids *XXX*. The triploid *Cnemidophorus uniparens* has a premeiotic doubling of the chromosome number as in the parthenogenetic *Ambystomae* (Cuellar 1970). Lowe, Wright, Cole and Bezy (1970*a*) have reported occasional male hybrids between the bisexual *C. tigris* and the triploid parthenogenetic *C. sonorae*; these individuals were tetraploids and apparently had the sex-chromosome constitution *XXXY*. Tetraploid female hybrids probably also occur but would be difficult to distinguish phenotypically from normal *C. sonorae*.

Thelytokous reproduction occurs also in some lizards belonging to the super-species *Lacerta saxicola* in the Caucasus (Darevsky 1958, Darevsky and Kulikova 1961, 1964; Darevsky 1966; Darevsky and Danielyan 1968; Kupriyanova 1969). In this case the parthenogenetic biotypes (there are four of them) appear to be diploid $(2n = 38)$ and there is some evidence that they are automictic. Thelytokous females produce about 5–8% teratological offspring, some of which are males; it is not known whether these are due to polyploidy, aneuploidy, or mosaicism; they might be due to crossing-over producing homozygotes for lethal genes. The mechanism of parthenogenesis is consequently inefficient and possibly rather recent. Mating between bisexual and thelytokous biotypes of *L. saxicola* occurs in nature but pro-duces only sterile triploid $(3n = 57)$ female offspring with rudimentary gonads. When males of the bisexual form *L. saxicola valentini* were liberated in an area where the parthenogenetic forms *L. armeniaca* and *L. unisexualis* occurred naturally, both diploid (non hybrid) and triploid (hybrid) progeny were recovered, sometimes from the same oviduct.

The four parthenogenetic forms are all broadly sympatric in an area from which related bisexual lizard species are excluded; but narrow zones of overlap do occur on the edges. One of the four appears to be a direct derivative of the bisexual *L. defilippii*; the other three may be of hybrid origin.

At least five other species of lizards, belonging to several families (Agamidae, Chamaeleontidae and Geckonidae) are almost certainly thelytokous (White 1970a; Hall 1970).

EVOLUTIONARY ORIGIN OF THELYTOKY

That many types of automictic mechanisms which occur naturally could have arisen through natural selection for spontaneous development of eggs in which fusion of meiotic products has taken place is fairly obvious. There would, of course, have to be simultaneous selection increasing the frequency of such fusions and (in some instances at any rate) restricting fusion to second-division non-sister nuclei. This is clearly what has happened in *Lonchoptera dubia* and *Drosophila mangabeirai*; it is a sequence of events which has been repeated experimentally in *D. parthenogenetica* and *D. mercatorum*, species which normally reproduce bisexually (Stalker 1954; Carson 1962b).

The evolutionary origin of apomictic thelytoky is less clear. Have all apomictic species passed through a stage when they were automictic? No definite answer can be given, but it seems highly unlikely. Certainly, among the numerous species and strains of thelytokous weevils, no automictic ones are known. This suggests that in this group, at any rate, apomixis has repeatedly arisen from bisexuality, without passing through an intermediate automictic stage. It seems probable, on general principles, that thelytokous genetic mechanisms are too rigid and inflexible to be capable of the delicate evolutionary transformation from automixis to apomixis, which would necessarily involve a great many separate mutational steps if the new mechanism was to be efficient.

The various cases of 'occasional' parthenogenesis ('tychoparthenogenesis') which we have considered earlier, probably all involve automictic mechanisms. There is consequently some reason to think that automixis, where it has become established as the regular mode of reproduction, has done so by a gradual process of natural selection, starting with a bisexual population showing rare tychoparthenogenesis.

On the other hand, apomictic mechanisms (and automictic ones involving premeiotic doubling of the chromosome number) seem to represent such a radical departure from the pre-existing bisexual-meiotic situation that one can hardly imagine them arising through a gradual process extending over many generations. A particular lineage either is or is not apomictic – it can hardly be partly apomictic and partly bisexual. Thus apomixis, like alloploidy, seems to be an example of 'macromutation' in evolution. We know almost nothing about the liability of various genetic systems to undergo such mutations. The frequency of apomictic biotypes in certain genera of weevils may indicate a high tendency to undergo mutations abolishing meiosis in the egg. However, it may be that similar mutations are equally frequent in some other groups, but are unsuccessful in evolution for various reasons, and hence remain undetected. Even if successful apomictic mechanisms always originate by a macromutation, they may undergo modification and improvement later, by further mutation and selection, although in the clumsy inefficient way that must be characteristic of thelytokous systems in general. From a genetic standpoint there is no real difference between apomictic thelytoky and vegetative reproduction by budding or fission. Unfortunately, very little is known of the chromosome cytology of coelenterates, polyzoans, tunicates and other invertebrate groups in which vegetative reproduction is of common occurrence. More is known of the nuclear cytology of Protozoa; polyploidy seems to be widespread in the Spirotrichonymphidae, where reproduction is believed to be exclusively by vegetative means (Cleveland and Day 1958). There is, however, a basic difference between species which reproduce exclusively by vegetative means (a very small minority) and the much larger number in which vegetative and sexual reproduction coexist; the latter must be expected to have karyotypes which are compatible with a normal meiosis.

The idea that thelytoky is an 'escape from sterility' (Darlington 1939*a*, 1958, following Strasburger, 1905) need not be taken too literally. Even as far as plants are concerned (in which mechanisms of vegetative reproduction may permit the development of occasional sexual sterility), it has been rightly criticized by Gustafsson (1947) and Stebbins (1950). There are yet stronger reasons for rejecting it in the case of animals, in which 'sterility' simply cannot arise in the evolution of bisexual species. However, some types of thelytoky do undoubtedly permit the evolutionary development and exploitation of various genetic mechanisms (triploidy, pentaploidy etc.) which would lead to sterility if reproduction were sexual. And it also seems that many cases of thelytoky may have evolved in the first place as an 'escape' from various kinds of excessive or unbearable segregational genetic loads (or, to put it slightly differently, many instances of thelytoky have probably arisen by exploiting pre-existing heterotic mechanisms, genic or chromosomal).

If we take this view (that the main demogenetic significance of thelytoky in animals is that it fixes heterotic mechanisms, while getting rid of the genetic load represented by adaptively inferior homozygotes), we may well ask why it has not arisen more often, and in particular why it seems to be virtually or entirely absent in such groups as the Odonata, Heteroptera, spiders and many others. To say that thelytoky is rare because thelytokous species are evolutionary 'no-hopers' is not a sufficient answer; they have been successful in certain groups, many bisexual species are destined to fairly rapid extinction and the opportunism of the natural selection process cannot be expected to take account of future prospects. It would seem likely that the main obstacle to the development of thelytoky is the scarcity of sufficiently narrow, invariable ecological niches. The almost complete absence of thelytoky in groups of highly mobile animals, as well as from most groups of marine organisms, points strongly in this direction.

The main exception to the concept of thelytokous biotypes as highly heterozygous organisms which have shed their segregational load seems to be the few forms referred to on pp. 708–9 which must be completely homozygous in view of their mechanism of thelytoky (diploidization of cleavage nuclei). Here it would appear that the adaptive significance of thelytoky must have been in terms of reproductive advantage or because it enabled the species to get rid of a biologically unsatisfactory male sex (in most scale insects the adult males are very fragile, non-feeding and short-lived).

In some other cases (Diptera, *Moraba virgo*) there is a suggestion of a definite association of thelytoky and heterozygosity for chromosomal rearrangements. Lepidoptera, Crustacea, weevils, oligochaetes are not favourable material for the detection of most types of chromosomal rearrangements, so that thelytokous members of these and some other groups may well be structurally heterozygous in ways which escape notice. In a number of polyploid-thelytokous biotypes in such groups it has been suggested that aneuploidy exists (e.g. $4n - 2$, $5n - 4$, etc.). While some of these biotypes may, in fact, be aneuploid in the sense of lacking certain chromosomes or possessing others in excess, it is also possible that some supposed aneuploids are simply heterozygous for chromosomal fusions or dissociations.

One question which arises in connection with those thelytokous species which are cytologically heterozygous is as to whether the chromosomal rearrangements arose before or after the assumption of thelytokous reproduction. In all the thelytokous Diptera which are structurally heterozygous the rearrangements seem to be exclusively paracentric inversions, i.e. the type that is widespread in bisexual species. Pericentric inversions and translocations of various kinds which would diminish the fecundity of a bisexual species (but which, in general, should not do so in a thelytokous one) have not been recorded. Thus the evidence seems to point towards the conclusion that chromosomal rearrangements have not established themselves in these thelytokous strains of Diptera and that those which exist are an inheritance from 'bisexual times' (or, in the case of some triploid forms, may have been brought into the karyotype by the sperm of a related bisexual form).

However, the situation in *Moraba virgo* is clearly somewhat different, since the fusions, dissociations and translocations that have occurred at Monia Gap, Yatpool and Connor's Tank have clearly arisen in a form that was already thelytokous. The fact that far more rearrangements have not succeeded in establishing themselves in this species may be taken as evidence that the great majority of them are accompanied by deleterious position effects. Those that have been successful may have owed their establishment to chance; at any rate they seem to be confined to single colonies.

In Mendelian populations of sexually reproducing species the chromosomal rearrangements (inversions and translocations, including fusions and dissociations) that exist in a polymorphic state undoubtedly owe their survival to the fact that they act as cross-over suppressors, permitting the evolution of alternative coadapted complexes of genes. They cannot play this role in thelytokous species, where effective crossing-over has been suppressed throughout the entire chromosome complement. It is, thus, perhaps understandable that the evidence for the establishment of chromosomal rearrangements, *after* the assumption of thelytokous reproduction, is somewhat meager. We could only expect them to be successful if they had associated position effects which conferred a selective advantage on their carriers (such position effects are probably very rare or even non-existent). However, it is perhaps somewhat surprising that rather more structural heterozygosity has not been reported in thelytokous species; *Moraba virgo* is surely not the only such species in which we might expect to find rearrangements that had established themselves by chance. In particular, there seems to be almost no evidence for heterozygosity in respect of duplications and deficiencies in thelytokous species.

Since polyploidy seems to have been such a successful adjunct to thelytoky in weevils, earthworms and simuliids, we may well ask why such species as *Drosophila mangabeirai*, *Moraba virgo*, *Pycnoscelus surinamensis* and *Brunneria borealis* have remained diploid. Possible explanations include:

(1) absence of closely related bisexual biotypes to provide sperms, leading to alloploid combinations.

(2) a cytological limitation to the numbers of chromosomes that can be efficiently accommodated in the nuclei or on the spindles of certain cell-types.

(3) incompatibility of polyploidy with the mechanism of the maturation division or divisions in the egg.

(4) a simple absence of adaptive superiority of polyploid biotypes in these cases.

We may point out that there is one very obvious reason why alloploidy might not succeed in some instances of incipient thelytoky – namely that the genome brought into the egg by the foreign sperm will necessarily include 'genes for bisexuality', and these may prove to be at any rate partially dominant over the 'genes for thelytoky' in the original diploid karyotype.

We have suggested earlier that many thelytokous systems may have opportunistically exploited pre-existing heterotic mechanisms, and that their success has depended, in part at least, on their having shed the genetic load which heterosis entails in bisexual populations. Some authors have, however, gone further than this and suggested that as apomictic biotypes accumulate mutations, becoming more and more heterozygous in the process, they become more and more 'heterotic'. Whether one should use the term heterosis in this sense seems doubtful; but clearly, there will be selection, in all heterozygosity-preserving thelytokous systems, for newly mutated alleles which are adaptive in the heterozygous state.

In general, we have almost no information on the importance of mutation as a factor in the evolution of thelytokous forms. Suomalainen's evidence that geographically remote populations of parthenogenetic weevils taxonomically classified as the same 'species' (and having the same degree of ploidy) are widely different, biometrically, seems to point more to a polyphyletic origin of thelytoky or polyploidy (or both) than to mutational changes within a uniformly polyploid-parthenogenetic stock. Nevertheless, the author, in the second edition of this book, possibly went too far in suggesting that in a tetraploid apomict such as *Saga pedo* gene mutations would be almost completely recessive (i.e. ineffective) in the presence of three doses of the original allele. This may be so for the majority of mutations. But there will surely be many whose recessivity will not be complete, even in a tetraploid, i.e. they will not give rise to subtle biometric effects even if not to conspicuous phenotypic ones. One thing seems clear – populations of thelytokous insects seem to be almost completely monomorphic for colour, even where they belong to groups where colour polymorphism is usual.

It would seem, theoretically, as if mutation should lead to the existence of a large number of genetically different lineages within each thelytokous 'species'. In particular, one might expect that most types of automictic thelytoky, which provide the possibility of genetic segregation, should lead to a multiplicity of genetic strains. Although this seems undeniable from a theoretical standpoint, there is very little factual evidence for it. At least as far as animals are concerned, the agamic complexes that have been described seem due more to polyphyletic origin of either thelytoky or polyploidy, or both, than to subsequent evolutionary divergence by segregation or mutation. Only in the triploid simuliid *Cnephia mutata* does there seem to be clear evidence for segregation of different cytological biotypes in a thelytokous genetic system. Genetic divergence of thelytokous biotypes due to gene mutation must undoubtedly have occurred, but is naturally not evident cytologically.

Vandel (1931) and many subsequent authors have laid particular stress on what they call the geographic character of thelytoky. The distributional pattern shown in such genera as *Trichoniscus*, *Saga* and *Brunneria*, with the thelytokous forms extending their range far beyond that of the bisexual ones, is a rather characteristic one. Vandel assumed that in the case of *Trichoniscus* the triploid thelytokous form is more 'hardy' and therefore able to adapt to harsh environments. A more probable explanation would seem to be that these thelytokous forms have found it easier to extend their range just because every individual transported to a new locality is capable of founding a local colony. It is not entirely clear, however, how forms whose genetic system must be very inflexible manage to become adapted to new environments when they *do* get transported to them: the apparent ecological versatility in *space* seems to be at variance with their lack of evolutionary versatility in *time*. The solution of this apparent paradox is not obvious. Omodeo (1953a) has argued that the antiquity of some thelytokous biotypes is far greater than has generally been supposed, i.e. that the lack of adaptability of these forms to changing environmental conditions is to some extent a myth. His conclusion that the thelytokous systems of *Dendrobaena octaedra* and the triploid form of *Allolobophora caliginosa trapezoides* date from the Miocene is based on the facts of their geographic distribution. But *D. octaedra* seems to be a complex of hexaploid and aneuploid biotypes (see p. 733) rather than a uniform species, so it is quite possible that the existing biotypes have arisen polyphyletically at various times from some unknown and perhaps extinct diploid ancestor. Certainly the evidence for any particular time of origin of its genetic system is not compelling. In the case of *A. c. caliginosa*, since the diploid sexually-reproducing biotype still exists in England, at any rate, it seems rather unlikely that the triploid thelytokous biotype is of great antiquity and quite possible that it may be polyphyletic. It is quite likely that subterranean habitats have changed less during Tertiary time than above-ground ones, and hence that the genetic inflexibility of parthenogenetic genetic systems has not been the handicap that it would have been to the inhabitants of the world above ground. But there is no real evidence on this point and one has the feeling that the genetic evolution of earthworms has been extremely complex and that, in spite of the good beginning that has been made by Omodeo, we need vastly more basic cytological information from the whole geographical range of the species studied.

Thus the apparent ecological versatility of some thelytokous forms may simply reflect the fact that each widely successful thelytokous 'species' consists of a large number of genetically different biotypes, adapted to different ecological niches within the main area of occupation. Constantly penetrating, either through their own vagility or through accidental, passive dispersal, into new areas, some at least are likely to establish new colonies.

In the case of the parthenogenetic weevils, this apparently enabled them to colonize boreal environments in the wake of the retreating ice-sheets after the end of the glacial period. The same is probably true of the triploid thelytokous simuliids of the arctic (Downes 1962). Boreal and arctic environments probably provide relatively more really narrow, invariable ecological niches than the complex, shifting and changing habitats of warmer climates. There is also the fact that the short breeding season of arctic latitudes may render the finding of a mate a chancy proposition unless the population is extremely dense. The weevils and the simuliids are not the only insect groups to have parthenogenetic representatives in arctic environments; the mayfly *Baetis feminum* and the caddis-fly *Apatania zonella* are likewise thelytokous in the Canadian arctic (Downes 1964). We might perhaps expect desert and cave faunas to include a higher proportion of parthenogenetic forms than forest or meadow environments; but evidence on such questions seems to be lacking (Baccetti (1961) has recorded one instance of thelytoky in a cavernicolous orthopteran). We do not know what ecological factors keep *Drosophila mangabeirai* such a rare species in the forests of central America, but we may be reasonably certain that it occupies a niche which is narrowly restricted in some manner.

The adaptation of non-sexual genetic systems to the environments they occupy is one of the most important chapters of evolutionary theory. But its serious study has hardly begun. In spite of numerous investigations on cytological mechanisms of thelytoky, we are almost entirely without detailed knowledge of the biometry or ecology of *any* thelytokous form. Studies of the effects of artificial selection on thelytokous populations should be especially valuable. Wherever several cytologically different biotypes can be distinguished within an agamic complex it would be important to determine the exact extent to which they coexist in nature, or replace one another geographically or ecologically.

Although well-authenticated instances of pseudogamy exist in earthworms, beetles, vertebrates and several other groups, the phenomenon remains somewhat enigmatic and paradoxical. If the sperm in these cases does not contribute any chromosomes to the egg, what does it contribute? No answer can be given as yet to this question, and it is consequently quite uncertain whether the mechanism of activation of the egg in all cases is essentially the same. Theoretically, one would think that this kind of reproductive parasitism on a related bisexual biotype would be extremely restrictive to the spread and adaptation of the thelytokous biotype and that natural selection would be expected to do away with this dependence. But until the biochemical basis of the phenomenon is understood, speculation as to the extent to which it might be modified or abolished by natural selection is perhaps premature.

Several recent authors (e.g. R. J. Schultz 1969) have discussed the possibility that thelytokous genetic systems might in a few instances give rise to bisexual systems and that, in particular, some polyploid bisexual species (e.g. the South American anurans referred to on p. 438) may have evolved via a stage in which they were parthenogenetic. While this is conceivable in groups lacking morphologically differentiated sex chromosomes it is most unlikely to have occurred in any group with a well-developed sex-determining system depending on visibly different X- and Y-chromosomes.

20

CYCLICAL PARTHENOGENESIS

The basis of the alternation of generations as regards the idioplasm must therefore in all cases consist of a germ-plasm composed of ids of at least two different kinds, which ultimately take over the control of the organism to which they give rise.

AUGUST WEISMANN: *The germ plasm, a theory of heredity*, (trans. Parker and Rönnfeldt)

Animal species and groups with cyclical parthenogenesis (heterogony) seem to have combined the reproductive and dispersive advantages of thelytoky with the genetic advantages of sexuality. Since genetic recombination occurs in them from time to time, they are, from an evolutionary standpoint, more akin to sexual than to thelytokous forms. We must expect them to form definite biological species, rather than agamic complexes or sympatric clones. Their capacity to exploit the possibilities of heterosis is limited in the same way as that of ordinary sexual species. On the other hand, there is no reason to expect them to be evolutionary 'blind alleys', as thelytokous species undoubtedly are.

An alternation of generations does, however, pre-suppose two alternative ecological niches to which the sexual and thelytokous generations must be adapted; and these must, in general, coincide in space, although not in time. The species is dependent on two different food plants, two host species (in the case of a parasite) or an alternation of seasons.

Cyclical parthenogenesis may involve an alternation of thelytokous and bisexual generations, or of thelytokous and hermaphroditic ones. In the former case a cytogenetic tour-de-force is required, since a sex-determining mechanism has to be produced anew at each switch-over from thelytoky to bisexuality.

Cyclical parthenogenesis has been described in trematodes, rotifers, Cladocera, aphids, Gall wasps (Cynipidae), in a few Cecidomyidae and in the beetle *Micromalthus*. These cases exhibit a great diversity of cytogenetic detail and the ecological-evolutionary significance of the alternation of generations is certainly not the same in all of them. Thus in the trematodes the dispersive phase of the life cycle is in the parthenogenetic generations, while in the case of the larval parthenogenesis of certain Cecidomyidae it is the winged sexual generation which is alone capable of invading new habitats, the dispersive powers of the parthenogenetic (pedogenetic) larvae being very restricted.

In the digenetic trematodes (of which the sheep liver-fluke *Fasciola hepatica* is typical) there is an alternation between a hermaphroditic generation, parasitic in a vertebrate, and one or more generations of so-called miracidia, sporocysts and rediae, morphologically 'larval' forms which parasitize invertebrate hosts (mainly freshwater snails) and whose reproduction is apomictic. Eventually tailed larvae known as cercariae are produced, which reinfect the vertebrate host and develop into the 'adult' fluke. In this type of life cycle we find diploidy, an absence of any special sex chromosomes and no cytogenetic revolution or tour-de-force is needed to accomplish the change from apomixis back to sexuality, since both occur with the same karyotype.

The alternation of generations in the rotifers is between thelytokous and haplodiploid generations. It formally resembles the life cycle of the cynipid Gall wasps.

The life cycles of aphid species usually consist of a series of parthenogenetic summer generations, alternating with a single sexual generation which occurs during the colder part of the year. Such species are said to be *holocyclic*, since they have the 'whole' cycle; there are also some *anholocyclic* species which have lost the sexual part of the cycle, no doubt secondarily.

We may take *Tetraneura ulmi* as a typical representative of the holocyclic species. Its chromosome cycle has been described by Schwartz (1932). At the beginning of May (in Europe) the small female nymphs form galls on the leaves of the elm, inside which they become adult and produce by parthenogenesis about forty female offspring each. The wingless gall-making individuals are known as *fundatrices*; their offspring develop wings and make their way out of the galls. They then migrate to the roots of various species of grasses – whence their name of *emigrantes*. On this second host plant they give rise by parthenogenesis to several generations of *exules*. These latter finally give rise to *sexuparae*, winged forms that fly back to the elm, where they produce male and female *sexuales* (again by parthenogenesis). These sexuales pair, and from the fertilized eggs laid by the females emerge the gall-making fundatrices of the next year. All the fundatrices, emigrantes, exules and sexuparae are females, since the mechanism of thelytoky never seems to break down and produce a male by accident.

Schwartz showed that all the different types of thelytokous females of *Tetraneura ulmi* have 2n = 14, there being three long and four short pairs of chromosomes. The female sexuales also have 2n = 14, but the males have 2n = 13, so that there is an $XO:XX$ type of sex-determining mechanism, all the various types of females being XX and the single type of male being XO.

The details of the maturation processes in the parthenogenetically developing eggs of aphids have remained in doubt until quite recently, but seem to have been elucidated by Cognetti (1961a, b, c, 1962) and Pagliai (1961, 1962). Synapsis and bivalent formation seem to take place normally, at least in *Brevicoryne brassicae*, *Macrosiphum rosae*, *Myzodes persicae* and *Toxoptera aurantiae*. But the bivalents then resolve themselves into univalents 'without formation of a first division spindle. The single maturation division takes place with the diploid number of chromosomes and results in the casting out of a diploid polar body. Cognetti and his school call this process *endomeiosis*; they believe that genetic recombination by crossing-over occurs, and that this fact explains the responses to selection for winglessness in parthenogenetic lines of *M. persicae*.

The sexuparae differ from the other types of parthenogenetic females in that they produce two kinds of eggs which will develop into males and females, respectively. In *Tetraneura ulmi* each sexupara produces eggs of both kinds, but in some other species of aphids there are two kinds of sexuparae, male-producing and female-producing. Both kinds of eggs undergo a single maturation division, but whereas in the eggs which are destined to give rise to females all the chromosomes divide, in the eggs which will give rise to males the two X-chromosomes pair to form a bivalent which remains in the middle of the anaphase spindle after the other chromosomes have passed to the poles. One half of this XX bivalent then passes into the polar body nucleus, the other half remaining in the egg. Thus the XO condition in the male arises through the X-chromosome undergoing reduction at the single maturation division.

The spermatogenesis of the male aphids is of a highly characteristic and anomalous type. All the autosomes pair to form bivalents, so that at the metaphase of the first meiotic division there are six autosomal bivalents and the univalent X. The latter becomes stretched between the two poles at anaphase, appearing as a bipartite body lying in the long axis of the spindle.

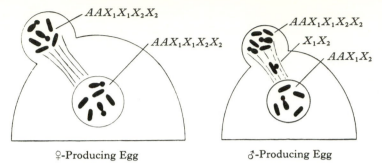

$\female$-Producing Egg $\male$-Producing Egg

Fig. 20.1. First meiotic division in male- and female-producing eggs of the aphid *Phylloxera caryaecaulis*. Based on the work of T. H. Morgan (1912).

Although it has the appearance of a body under considerable tension, it eventually passes into one of the daughter nuclei without dividing. Thus two kinds of secondary spermatocytes are formed, one possessing an X-chromosome, the other lacking one. The former receive far more cytoplasm than the latter, so that the minus-X spermatocytes are very much smaller than the plus-X ones. Only the latter undergo a second meiotic division, the smaller minus-X spermatocytes simply degenerating without forming sperms. Thus, although the males are XO, they are actually homogametic, producing only one kind of sperm. Ris (1942) claimed that the unequal division of the cytoplasm is actually caused by the X-chromosome, which is stretched in the axis of the spindle and prevents the cleavage furrow from cutting through the middle of the cell.

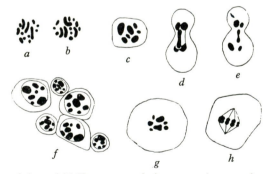

Fig. 20.2. Cytology of the aphid *Tetraneura ulmi*. *a*, somatic metaphase in male embryo (13 chromosomes); *b*, somatic metaphase in female embryo (14 chromosomes); *c*, first metaphase (polar view) in the male; *d* and *e*, first anaphases with the X-chromosome stretched on the spindle; *f*, a group of three large and three small interkinetic cells (the latter lack an X-chromosome and will degenerate without undergoing a second division; *g* and *h*, second meiotic divisions. From Schwartz (1932), redrawn.

The above account, based on *Tetraneura ulmi* and the species studied by Cognetti and Pagliai, is probably valid in essentials for other species of aphids except those which have become secondarily anholocyclic, having lost the sexual part of the cycle altogether. The meiosis of the males, in particular, always seem to be of the same peculiar type (Stevens 1905*b*, 1906; Morgan 1906, 1908, 1909*a*, *b*, 1912, 1915; Tannreuther 1907; de Baehr 1908, 1909, 1920; Honda 1921; Frolova 1924; Shinjii 1931; Suomalainen 1933; Lawson 1936). In some species such as *Phylloxera caryaecaulis* there is an X_1X_2O mechanism and in *Euceraphis betulae* the sex-chromosome mechanism is $X_1X_2X_3X_4O$ in the male, but the different X's all behave in spermatogenesis exactly like the single one in related species and all of them undergo reduction during the maturation of eggs destined to form males. Presumably aphids are precluded by their cyclical parthenogenesis and peculiar chromosome cycle from possessing Y-chromosomes; thus translocations between the X and an autosome, such as would give rise to neo-XY mechanisms in other groups, could hardly be expected to survive in aphids, whose X-chromosome must be presumed to be in a state of 'evolutionary isolation'.

Certain questions concerning the cytological cycles of the aphids remain unanswered. Thus, in *Tetraneura ulmi* it is not known why the eggs of a particular sexupara get rid of one X at meiosis (and hence produce males) while other eggs of the same individual retain both X's and give rise to females. In some other aphids, such as the species of *Phylloxera*, there are two kinds of sexuparae, each kind producing only male or only female offspring (Morgan). It is not known whether there is a cytological or genetic difference between the two kinds of sexuparae in these species; Morgan (1912) thought that a portion of one of the X's was eliminated during the maturation of eggs destined to give rise to male-producing sexuparae, but his observations have never been confirmed and Ris (1942) concluded that any elimination which does take place is analogous to the casting out of ribonucleoprotein in the eggs of moths and in the cleavage divisions of the mite *Pediculopsis* (see p. 57), and not the loss of a chromosome segment.

The details of the life-cycle vary greatly among the aphids. Some phylloxerans have only three generations in the course of the year, but most species of aphids have a much larger number of summer generations. In the adelgids the entire life-cycle extends over two years and may be extremely complicated. In some members of the subfamily Pemphiginae the sexual generation has been lost from the life-cycle, which has consequently become reduced to an endless succession of thelytokous generations.

The cyclical parthenogenesis of the Gall wasps (Hymenoptera, family

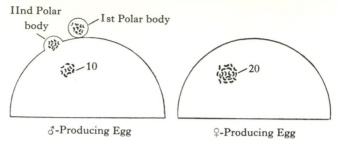

Fig. 20.3. Maturation process in the male- and female-producing eggs of the Gall-wasp *Neuroterus lenticularis*. In the first, two polar bodies are given off, in the second, none. Based on the work of Doncaster (1910, 1911).

Cynipidae) differs from that of the aphids in a number of respects. Typically, there are two generations a year, one thelytokous ('agamic'), the other bisexual. In the species of *Cynips*, which live on oaks, the thelytokous females develop within conspicuous galls during the summer months, emerge in winter and lay eggs in the buds which grow into simple incon-spicuous galls within which the individuals of the bisexual generation develop in the spring. In *Neuroterus lenticularis*, the fertilized eggs give rise to thelytokous ('agamic') females which, although penotypically alike, are of two genetic types, male-producers and female-producers (Doncaster 1910, 1911, 1916). In the oocytes of both types the chromosomes pair to form ten bivalents. In the eggs of male-producers (which are destined to give rise to haploid embryos) at least one and possibly two maturation divisions occur (Doncaster believed two, but Dodds (1939) thought that only one took place). The eggs of the female-producers seem to be unique in not giving off a polar body at all (Doncaster and Dodds are in agreement on this). Thus, if femaleness in *Neuroterus* depends on heterozygosity for a sex factor as in *Habrobracon* and the honey bee (see pp. 690–2) this hetero-zygosity could be transmitted by female-producers to their daughters.

The sexual generation of *Neuroterus* pair and produce diploid embryos which are all female and which will develop into male- and female-pro-ducers once more in the following spring. Doncaster showed that each sexual female always gives rise to either male-producers or female-pro-ducers but not both; so there are presumably two genetically different types of sexual female, outwardly similar.

In *N. lenticularis* the eggs of the sexual females must be fertilized or they will never develop, and all the eggs laid by sexual females give rise to thelytokous females. In *N. contortus*, however, Patterson (1928) bred 10 males and 231 females from eggs laid by sexual females which had mated.

No pairing took place between these rare males and the females of the agamic generation, so that the sexual instincts of the latter had apparently been lost. The distinction between male-producing and female-producing thelytokous females which is rigid in these species of *Neuroterus* breaks down in *Andricus operator*, where the same thelytokous female produces both sons and daughters. In other species an intermediate condition exists, each female producing progeny which are mainly of one sex, with a few exceptions. The whole system of unisexual progenies seems to depend on a special genetic mechanism which has developed in the Cynipidae and which reaches its final state in species like *Neuroterus lenticularis*, where progenies are strictly unisexual. As in the case of the aphids, a number of cynipids, belonging to several genera, have lost the sexual generation from the life cycle, males being unknown or very rare, and probably functionless.

Cyclic parthenogenesis has been known from the time of Weismann (1880) to occur in the Crustacea of the subclass Cladocera, but the cytogenetic mechanisms underlying it are still largely unknown. Thelytokous reproduction may go on for an indefinite number of generations, but under certain unfavourable environmental conditions males are produced; they are diploid and indistinguishable in karyotype from the females, sex chromosomes being unknown in this group (Mortimer 1936). The mechanism of thelytoky in Cladocera is apparently apomictic. Schrader (1926) concluded that there was no bivalent formation in the parthenogenetic eggs of *Daphnia*. However, Bacci, Cognetti and Vaccari (1961) have claimed that in the material of *D. pulex* which they examined there was some kind of synapsis in the parthenogenetic eggs, followed by a process of desynapsis before first metaphase. In certain biotypes of *Simocephalus exspinosus* and *Daphnia longispina* intersexes may be produced under some environmental conditions (Banta *et al.* 1939).

In certain nematodes there is a cyclical alternation of generations, but no parthenogenesis occurs, the successive generations being bisexual and hermaphroditic. We shall deal with these cases here, for the sake of convenience, since the behaviour of the sex chromosomes is somewhat similar to that of the X's in the aphids.

In *Angiostomum* (*Rhabdonema*) *nigrovenosum*, studied by Boveri (1911) and Schliep (1911), the male and female individuals are free-living, while the hermaphrodites are parasitic in the lung of a frog. Anatomically, these hermaphrodites are obviously modified females, their general structure being quite different from that of the males, although they produce sperms in an ovotestis. The females and hermaphrodites have a diploid set of twelve chromosomes. During oogenesis two meiotic divisions occur, so that the eggs are haploid and contain six chromosomes. The males are XO and have only eleven chromosomes. In the spermatogenesis of the free-living males two kinds of sperm are formed, with and without an X-chromosome, respectively. The latter kind seem, however, to be incapable of fertilizing the eggs, so that all eggs are inseminated by X-bearing sperms, and the hermaphrodites arising from these eggs are all XX, like the females. The spermatogenesis which goes on in the ovotestis follows a different course to that which occurs in the males. One X-chromosome becomes condensed faster than the other (it is not clear whether this is the paternal or the maternal X or whether it may be sometimes one and sometimes the other). No sex bivalent is formed during this meiosis but the two X's pass to opposite poles at first anaphase. All the sperms contain an X at first but in half of them it is extruded and lost at the time when the developing sperm separates from its residual cytoplasm. Thus the XX hermaphrodites produce one kind of egg but two kinds of sperm, and by the fertilization of these eggs, the XX and XO individuals of the sexual generation are produced once again.

The chromosome cycle of *Rhabdias fülleborni*, described by Dreyfus (1937), is very similar. The parasitic hermaphrodites which live in the lungs of toads have twelve chromosomes; these pair to form six bivalents in oogenesis, but in spermatogenesis the two X's do not undergo synapsis and both divide in the first meiotic division. Thus all second metaphases contain a haploid set of autosomes and two X's. At the second division one or both X's are lost from some cells, so that sperms with five and six chromosomes are formed. Thus the fertilized eggs which will develop into the males and females of the sexual generation have eleven and twelve chromosomes respectively.

21

CONCLUSIONS — GENETIC SYSTEMS AND EVOLUTIONARY PATTERNS

Nought may endure but Mutability
P. B. SHELLEY: *Mutability*
L'homme ne connaît le tout de rien
B. PASCAL: *Pensées*

We have now completed our survey of existing knowledge about the comparative cytogenetics of animals. Eventually the story of the chromosomal mechanism and its evolution will have to be entirely rewritten in molecular terms. But even the most spectacular successes of modern molecular biology, culminating in the elucidation of the coding mechanisms for amino acids that reside in DNA and RNA, have not led to any equivalent breakthrough in our understanding of chromosome morphology, behaviour and evolution. And a good deal of what has been learnt in recent years about the genetic systems of the Prokaryota seems rather irrelevant to the great problems of cytogenetics in the higher organisms. Meiotic crossing-over, synapsis, the mechanism and properties of chromosomal rearrangements, the transformation of 'ordinary' into 'giant' chromosomes in the amphibian oocyte and the polytene nuclei of the Diptera, the 'puffing' phenomenon, heteropycnosis and 'late replication' – all these are molecular phenomena, or have a molecular basis. But they represent a level or levels of organization above that represented by DNA biochemistry. A further level is represented by the cytogenetic structure of natural populations and a still higher one by the evolution of species.

The evidence we have examined in this book seems to establish three important principles. The first is that evolution is essentially a cytogenetic process, and that any attempt to interpret it solely in terms of taxonomy, ethology, ecology and distributional studies is bound to be a very weak and incomplete analysis. The second is that for a full understanding of genetic systems in evolution we need *both* population genetics *and* cytology, and that arguments based solely on mathematical genetic models or on the morphology and physiology of the cell are likely to present a partial and biased picture.

The third great generalization that emerges from this book is that the chromosomal mechanism itself varies in a great many ways from species to species and from group to group in the systematic hierarchy or, in other words, that it has undergone an evolution of its own, and that this somehow underlies the outwardly visible evolution of phenotypes. One way to interpret evolution (the palaeontologist's and the ecologist's way) is as a series of adaptive reactions on the part of living organisms to the ecological opportunities of the environment – a progressive filling of 'niches'. Such a viewpoint has much to commend it; but it clearly does not explain all the peculiar features of evolutionary patterns. To understand the processes involved in the differentiation of populations, species and higher categories, we need to know what is happening in the karyotype itself. This has been more generally appreciated by botanical writers (Manton 1950; Stebbins 1950; Darlington 1958, 1966; Grant 1964 and many others), in spite of the fact that the past history of higher plants is far less known than that of animals. The reason is, undoubtedly, that the evolution of higher plants (above the level of algae and fungi) is a story in which polyploidy has played a major role, so that botanical evolutionists have not been able to ignore cytogenetics as some of their zoological colleagues have done. Statements such as 'Speciation phenomena depend primarily on behavioural traits such as those that determine reproductive isolation, or a particular manner of niche exploitation and coexistence' (Brncic, Nair and Wheeler 1971) express only one aspect of a far more complex situation.

Evolution is, ultimately, a complex sequence of changes in the chromo-somal DNA. Some of the most eminent modern writers on evolutionary mechanisms have done no more than pay lip-service to this fact. We need to establish the chromosomes in the centre of our picture of evolution, as Darlington (1964) has rightly urged. Adaptation, isolation, competition, specialization, are all aspects of the evolutionary process. But there is another aspect which is concerned with such phenomena as the role of chromosomal rearrangements in generating systems of cytogenetic polymorphism (usually through heterozygote superiority) and genetic isolating mechanisms and hence speciation (as a result of heterozygote inferiority). There are also the evolutionary increases and decreases of chromosomal DNA, involving on the one hand duplication of genetic material and on the other hand its loss. And there is yet another aspect of evolutionary theory which is concerned with the unique cytogenetic systems of organisms – the various types of supernumerary chromosomes, the E-chromosomes of the Cecidomyidae, the haplodiploidy of many groups, the bizarre chromosome cycles of the coccids and some other groups, the strange modifications of meiosis in the dragonflies, the heterochromatinization and dosage compensation mechanisms of many organisms. There is also the whole subject of the evolution of sex and sex-determining mechanisms. These are phenomena which must have profoundly influenced the evolution of many groups. Yet some of the leading writers on the 'synthetic' or neo-Darwinian theory of evolution seem almost to leave them out of account, as if they were awkward facts which did not fit into their synthesis. Certainly, most of them could not have been predicted by any kind of mathematical genetics or model-building that is conceivable at present – a fact which should warn us that our vaunted 'synthesis' is no more than a preliminary framework, and that the final synthesis will be infinitely more complex; this is emphatically not a field in which progress will lead to simplification.

To a considerable extent, comparative cytogenetics deals with unique, unrepeatable events. The types of meiosis met with in Gall-midges, lice, Scale insects and dragonflies are all unique products of evolution. So, in all probability, are the chromosomal fragmentation phenomenon of the ascarid nematodes, and perhaps the 'fixed inactivation phenomenon' in the case of the mammalian X-chromosome. Haplodiploidy, polyteny, gene-amplification mechanisms associated with nucleoli, achiasmatic meiosis, holocentric chromosomes and most types of sex-chromosome mechanisms, on the other hand, have clearly been evolved independently several times. But most of these mechanisms have probably been 'unique events' in the particular groups and evolutionary lineages in which they have arisen. Cytogenetics therefore cannot endorse any view of evolution in which there is no place for unique, or at any rate very rare, events. Just what relation such cytogenetic inventions and innovations have had to the general course of evolution or to patterns of speciation in particular groups is not clear in most instances. But it seems extremely unwise to dismiss them as probably unimportant. Evolution certainly includes (in the words of Stebbins 1963) the projection 'into the geological time scale [of] the processes of mutation, genetic recombination, and natural selection as they are going on commonly and continually in all populations of organisms'. But cytogenetics forces us to the view that at all levels, rare, exceptional events and combinations of events play a role. Every chromosomal rearrangement, every evolutionary change in a karyotype, is a genetic revolution; and once it has occurred, a whole series of point mutations will be tried out in novel combinations in the new karyotype.

Within the past two decades a great variety of previously unsuspected cytogenetic mechanisms have been discovered. Some of these have been found in groups like the Armoured scale insects, which had not been previously investigated cytogenetically. But others such as the partial inactivation and heterochromatinization of one X-chromosome in female mammals, have been uncovered in groups that had been much studied by earlier workers. All this suggests that many surprising chromosomal mechanisms still remain to be discovered. The highly unusual chromosome cycles of the rodents *Ellobius lutescens* and *Microtus oregoni* and the achiasmatic male meiosis of certain thericleine grasshoppers show that such bizarre and unusual systems may turn up in groups whose chromosomal mechanisms were previously believed to be entirely orthodox. It may be objected by some that these peculiar cytogenetic mechanisms in single species are mere evolutionary curiosities, and that we should do better to concentrate our attention on more normal genetic and evolutionary mechanisms. But this is a field in which W. Bateson's dictum: 'Treasure your exceptions!' seems especially applicable. Our task is to wrest general principles from these unique phenomena.

In Chapter 13 we considered the evolution of the meiotic mechanism. In animals meiosis takes place during the formation of the gametes. And even within a single taxonomic group we find great variations in the morphology of the gametes and in the number that are produced by an individual. Thus within the grasshopper family Eumastacidae there are some subfamilies such as the Australian Morabinae in which the number of sperms is enormous (the female spermatheca being correspondingly large to receive them) while in other subfamilies such as the African Thericleinae the number of sperms is far smaller (perhaps by two orders of magnitude). Closely correlated with the difference in number of sperms is a difference in the size of the cellular elements involved. Morabine spermatocytes have little cytoplasm and the mean nuclear volume at pachytene in five species was 470 μm^3; thericleine spermatocytes have a large amount of cytoplasm and the mean nuclear volume of four species was 3390 μm^3. There is no corresponding disparity in the number and size of the eggs in these groups. The significance of the difference in sperm production is quite unknown, but it obviously involves the whole reproductive physiology of these insects and it is at least unlikely that it has had no effect on their genetic systems.

Certain relatively minor evolutionary changes in the meiotic mechanism, without any apparent immediate genetic consequences, may have in certain instances provided a future basis for rather far-reaching changes in the genetic mechanism. It is rather difficult to imagine just what the adaptive significance of the 'telescoping' of the diplotene and diakinesis stages of spermatogenesis in the mantids and thericleine grasshoppers (see Ch. 13) could be. But whatever the biochemical basis or adaptive significance of this 'cryptochiasmatic' meiosis, it seems to have furnished a basis from which achiasmatic mechanisms could be evolved. The pre-conditions for the development of achiasmatic meiosis in the Diptera, with their well-developed 'somatic pairing' phenomenon, were probably quite different. *A priori*, one would expect that if cryptochiasmatic meiosis has been an evolutionary success in one small group of African grasshoppers, it should have also evolved in some other groups of grasshoppers and grasshopper-like insects. But studies on dozens of subfamilies and hundreds of genera have failed to demonstrate it, or anything similar. Its real significance in the Thericleinae thus remains almost completely mysterious, since it almost certainly does not in itself lead to any increase or decrease in genetic recombination. In the mantids the extent to which the diplotene and diakinesis stages have been suppressed seems much more variable than in the thericleine grasshoppers, and at least one mantid genus (*Tithrone*) seems to show typical diplotene–diakinesis stages (Hughes-Schrader 1953*b*), perhaps as an evolutionary reversion.

Possibly, as suggested in Chapter 13, the extent to which the prophase stages of meiosis are extended in time (or telescoped and suppressed) depends on the extent of protein synthesis, preparatory to gamete-formation, which goes on at this time. The six months' long lampbrush diplotene stage in the yolky amphibian oocyte is probably merely an extreme case of mid-meiotic nuclear stasis during the growth phase of the cell. In spermatogenesis, groups with a long diffuse post-pachytene period may be ones which have much biosynthetic activity at this time; and ones with a very rapid and direct progression from pachytene to first metaphase (with no diplotene–diakinesis, or with these stages very abbreviated) may be ones which have already completed this biosynthesis at earlier stages. It has been shown by Henderson (1964), using tritiated uridine autoradiography, that in normal grasshopper spermatogenesis, the autosomes are active in RNA synthesis up to and including diakinesis, and again at the prophase of the second meiotic division, but not in the stages when the chromosomes are condensed; the univalent heteropycnotic X seems to be totally inactive as far as RNA synthesis is concerned at all stages of the male meiosis.

A large part of this book has been concerned with the role played by 'gross' chromosomal rearrangements (inversions, translocations of various kinds such as fusions and dissociations, duplications and deletions) in the genetics of natural populations and in speciation. A vast literature now exists on the significance of inversion polymorphism in natural populations of *Drosophilo* and some other organisms (see review of Sperlich 1967). Another line of investigation has been concerned with cytotaxonomy in the strict sense, i.e. with chromosomal differences between species and their relation to speciation. We have tried to synthesize and interrelate the evidence from these two kinds of study wherever possible. To some extent a parallel certainly exists; e.g. paracentric inversions commonly occur as polymorphisms in *Drosophila* populations and are also the commonest type of difference between the karyotypes of closely related species, while in the trimerotropine grasshoppers pericentric inversions play analogous roles both within and between species. But balanced polymorphism and speciation are basically antithetical phenomena since one results from the cohesive forces in natural populations, the other from divisive forces. As we have pointed out in Chapter 11, many types of cytotaxonomic differences that have arisen in the course of speciation, in particular fusions and dissociations which have led to changes in chromosome number, do not seem to exist as balanced polymorphisms in groups where they have undoubtedly established themselves repeatedly in phylogeny. Thus in the morabine grasshoppers a total of sixty fusions and dissociations are known to have reached fixation in the phylogeny of the group, but not a single population showing balanced polymorphism for either type of change is known (apart from the very narrow zones of overlap between races or species differing in chromosome number).

Where a parallelism exists in a group between intraspecific chromosomal rearrangements and interspecific ones we may legitimately conclude that balanced chromosomal polymorphism has given rise to cytotaxonomic differences between species through one chromosome sequence undergoing fixation in one incipient species while an alternative sequence undergoes fixation in a second incipient species (in certain cases one incipient species would retain both sequences, while the other becomes monomorphic for a single sequence). This general picture of a species with extensive chromosomal polymorphism giving rise to a number of species with less highly developed polymorphism (or with none at all) seems to be what has happened in certain species groups in *Drosophila*, e.g. the *repleta* group (Wasserman 1954–63). It was referred to in the second edition of this book as a 'collapse of polymorphism into monmorphism' and is undoubtedly a real evolutionary phenomenon. Presumably it occurs on an allopatric basis, with the descendant monomorphic (or less polymorphic) incipient species occupying different geographic areas, to which their chromosome sequences are adaptive.

But even in *Drosophila* this is fairly clearly not the whole story. It was estimated by Stone (1955) that fifty-eight chromosomal fusions have occurred in the phylogeny of the genus *Drosophila*. There is, however, no clear instance known of a *Drosophila* population polymorphic for this kind of change, unless we interpret the hybrid population of *D. a. americana* and *D. a. texana* in the south-central U.S.A. in this sense.

The conclusion we must draw from these facts (and a great many more instances of the same kind in beetles, mammals and in fact in almost every group of animals whose chromosomes have been studied) is that, in certain groups at any rate, fusions and dissociations which exist as cytotaxonomic differences between species have not been preceded by a condition of balanced polymorphism in an ancestral population. In other words, the acquisition of most fusions and dissociations (and some other types of chromosomal rearrangements in particular groups) is not a normal and usual part of 'phyletic' evolution, but is somehow directly (and perhaps causally) related to the process of speciation itself.

There is no reason to suppose, however, that this conclusion necessitates a belief in what has been called 'instantaneous speciation'. A good deal of confusion seems to exist in some published discussions of this question. Since each chromosomal rearrangement is, in origin, an instantaneous event and since some types of rearrangements undoubtedly reduce the fertility of the structural heterozygotes (but not the homozygotes) to a considerable extent, it is natural that adherents of theories of instantaneous speciation should have regarded chromosomal rearrangements as playing a special and primary role in the origin of new species. Mayr (1963) has criticized these views strongly, claiming that such evolutionists have relied on 'three separately valid observations: (1) that speciation is the origin of discontinuities, (2) that related species often differ in their chromosomal pattern, and (3) that certain chromosomal changes, such as reciprocal translocations, are partially isolating on account of considerable sterility among the heterozygotes' and naively combined them into a theory which not only runs 'into the same difficulties as all other attempts to establish new species through single individuals . . . but . . . is also contradicted by all the known facts of chromosomal variation.'

Surely, this is overstating the case. A belief that a chromosomal rearrangement which reduces the fecundity of the heterozygote by 5 or 10% (or even 25%) may play a special role in initiating phyletic divergence between an incipient species and the parental population does not imply an origin of the incipient species from a single individual, except in the sense that all genetic changes start in single individuals which transmit them to their progeny. And we have already seen that there are a good many 'known facts of chromosomal variation' which are favourable to the 'special role of chromosomal rearrangements' viewpoint, at least in certain groups of animals. We have discussed this question in Chapter 11, with special emphasis on the part played by chromosomal fusions and dissociations in the speciation of the morabine grasshoppers. It seems probable that similar mechanisms have operated in the case of some mammalian groups such as murid rodents and insectivores, whose vagility is of the same order of magnitude as that of wingless grasshoppers. But rearrangements causing an immediate and significant reduction in the fecundity of heterozygotes are much less likely to have played a role in the speciation of *Drosophila*, where individuals are far more mobile, i.e. liable to be carried considerable distances by air currents.

In *Drosophila* the survival and fixation in a natural population of a rearrangement which reduced the fertility of the heterozygote to a significant degree is highly unlikely. Nevertheless, even in *Drosophila* a few rearrangements which probably had this kind of effect (e.g. the pericentric inversion in chromosome 2 of the *montana* complex and some of the fusions and translocations in other species groups) certainly did undergo fixation in phylogeny. They most likely established themselves initially in geographically peripheral and rather strongly isolated populations and later came to play a special role as genetic isolating mechanisms. The paracentric inversion differences between closely related *Drosophila* species, however, are likely to have been more or less heterotic at the time of their initial establishment in the population, i.e. this type of cytotaxonomic difference must be derived from a former adaptive polymorphism. In general, however, heterotic polymorphisms will clearly be cohesive rather than divisive forces and will not play a primary role in speciation (although they may perhaps function in a secondary capacity if alternative polymorphic systems become established in different parts of the geographic range of a species). There is certainly no factual basis for Darlington's (1958) statement that *Drosophila robusta* is 'likely to split into species in the centre [of its distribution area] where it keeps its stock of inversion hybridity.' The speciation model according to which large polymorphic populations 'bud off' peripheral and more or less cytologically monomorphic isolates as a result of homoselection (Carson 1959) at the edge of the range is probably one that is frequently operative in *Drosophila* and many other groups. But the existence of numerous cases where species and incipient species share the same polymorphisms (e.g. in the *paulistorum* and *repleta* groups of *Drosophila*, in various groups of simuliids and in the grasshopper *Keyacris scurra* – see p. 257) proves that incipient species certainly do not always go through a period of monomorphism. They also establish the fact that species *can* split in the centre, even if the long-established balanced polymorphisms do not help them to do so. Just how far the 'stasipatric' speciation model of White, Blackith, Blackith and Cheney (1967) and White (1968a) or the theoretical model of Bazykin (1969) correspond to reality, either in wingless grasshoppers or in any other group, will only be decided by future work.

Although one may reasonably conclude that many fusions and dissociations have established themselves in populations of relatively sedentary species without the aid of heterosis, and that some rearrangements of this type may have played a special role in speciation, it is fairly clear that such rearrangements must not be of types which reduce the fecundity of the heterozygotes too drastically. This rules out most translocations, other than those involving interchanges of (1) minute terminal segments, (2) heterochromatic regions, or (3) whole arms of metacentrics (in the first and second cases aneuploidy may be compatible with viability and in the third aneuploidy will not occur if disjunction is of a regular zig-zag type). We have earlier spoken of rearrangements establishing themselves in partially isolated colonies of a species in spite of a reduction of the fecundity of the heterozygotes by 5 or 10%. A much greater reduction in fecundity (e.g. by 40 or 50%) will almost certainly doom a chromosomal rearrangement to rather rapid extinction, even in an isolated population of an extremely sedentary species. One piece of evidence supporting this view is the fact that very few tandem fusions seem to have established themselves in the evolution of groups in which centric fusions and dissociations have been repeatedly successful. If heterozygote inferiority were irrelevant to the situation there should have been almost as many tandem fusions as centric ones, at any rate where relatively short acrocentrics are involved.

A second line of evidence pointing in the same direction is the extreme rarity of mutual translocations (other than ones involving whole or virtually intact chromosome limbs) in such genera as *Drosophila* and *Chironomus*. Thus in the former genus Stone (1962) concluded that there are only three known cases of translocations being fixed in the evolution of the entire genus (and one of these instances, namely *Drosophila miranda*, is poorly documented and may not be genuine). This is in contrast to fifty-eight centric fusions, thirty-two pericentric inversions and some hundreds of paracentric ones. In fact the two undeniable cases of translocation (*sensu stricto*) in *Drosophila* – namely the transfer of the proximal region of the X to the fourth chromosome in *D. ananassae* (Kaufmann 1937; Kikkawa 1938) and the similar transfer in *D. tumiditarsus* (Hsiang 1949) both relate to segments that are at any rate mainly and perhaps entirely, heterochromatic.

We conclude, therefore, that critical evidence from well-studied groups does not favour John and Lewis' (1965) view that mutual translocations (other than fusions and dissociations) have played a significant and even a major role in animal speciation. This applies with especial force to translocations between acrocentric chromosomes. To suggest that 'interchanges in acrocentric chromosomes would be well suited' to serve as isolating mechanisms, because they will reduce the fecundity of heterozygotes so drastically, may be true in a formal sense. But to conclude that such rearrangements have been frequently successful in evolution is contrary to all the principles of mathematical population genetics and is contradicted by the evidence from *Drosophila* and other genera of Diptera whose polytene chromosomes have been carefully studied. The evidence that John and Lewis regard as supporting their viewpoint has been critically discussed in Chapter 15.

This question of the role of chromosomal rearrangements in animal speciation has been dealt with at some length because it is a critical one for the understanding of speciation processes and because a complete divergence exists between the extreme views of Mayr on the one hand and Darlington, Lewis and John on the other. Ten years ago the evidence on the cytogenetic processes involved in speciation rested very largely on *Drosophila*, a genus in which the overwhelming majority of the chromosomal differences between species are paracentric inversions that have rather clearly established themselves as a result of heterozygote superiority; the cytotaxonomic differences result from 'collapse into monomorphism' of former adaptive polymorphisms, Some cytotaxonomic differences in the trimerotropine grasshoppers must be interpreted in the same sense (see p. 248). But the evidence in favour of the view that many cytotaxonomic differences have arisen without passing through an adaptive polymorphism stage has been growing steadily. Even in *Drosophila* the fifty-eight fusions have most likely established themselves without benefit of heterosis.* But a balanced view of this whole question is badly needed. Clearly speciation *can* occur without the occurrence of chromosomal rearrangements. However, the very frequent existence in many groups (especially, but not exclusively, ones where the individuals are relatively sedentary) of cytotaxonomic differences between species that could not have arisen from balanced heterotic polymorphisms suggests that the role of particular kinds of chromosomal rearrangements in furnishing an impetus to the development of genetic isolating mechanisms between the incipient species may have been much greater in many groups than has been believed in the past.

The picture of the genetic and cytogenetic processes involved in speciation which emerges from recent work is hence a somewhat varied and complex one. The extent of vagility and the population structure of the group seems to play a most important role in determining what kinds of cytogenetic changes can establish themselves in populations and hence play a role

---

* Adaptive polymorphism for fusions or dissociations seems to depend entirely on distal localization of chiasmata and regularly alternate orientation of centromeres at meiosis. It is extremely unlikely that the latter condition could ever obtain with 100% regularity in an organism with achiasmatic meiosis. In accordance with this, there seems to be no record of an adpative polymorphism for a fusion or dissociation in any group of organisms where one sex is achiasmatic except for the peculiar case of the scorpions of such genera as *Tityus*, which is not properly understood (see p. 290).

in speciation. Sedentary, apterous animals with little or no capacity for passive disperal will exhibit one pattern; winged, migratory forms or ones liable to be carried considerable distances by wind or other agencies will be expected to show an entirely different one. But the number of individuals comprising a deme or local population will also be extremely important. The chromosomal differences between species of Equidae seem to be every bit as great as those between species of mice (see p. 448). Now 'horses' (i.e. zebras and wild asses of various species) are much more highly mobile than small rodents; but they exist, at any rate sometimes, as small local herds of a few individuals in which the establishment of a chromosomal fusion or dissociation by drift or as a result of homozygote superiority is quite conceivable, even if the fecundity of the heterozygote were somewhat reduced.

Obviously, however, even a highly significant coincidence in time between speciation and chromosomal rearrangements does not necessarily prove a causal relationship. Some types of rearrangements such as fusions or dissociations may, in certain groups, have facilitated, promoted or caused the kind of evolutionary divergence which constitutes speciation. But it is also possible that in some instances speciation (i.e. the formation of a small geographically and genetically isolated colony from a larger and more widespread species population) may simply have provided conditions favourable for the establishment and fixation of chromosomal rearrangements. Both kinds of relationships (rearrangements causing speciation and speciation favouring rearrangements) seem, *a priori*, plausible in groups with restricted vagility such as small mammals and wingless insects. Neither are they inevitably strict alternatives; during the period of speciation the population structure may favour rearrangements which later facilitate specific divergence and genetic isolation from the parent species.

Certain kinds of karyotypes seem to be generally characteristic of single species rather than of groups of species such as genera or families. This statement applies to neo-XY species in groups such as the Orthoptera which usually have an $XO:XX$ sex-chromosome system, and to species with multiple sex-chromosome mechanisms in almost all groups (although there are a few exceptions, such as the large group of mantid genera with X_1X_2Y males). Since it is hardly reasonable to suppose that $XO \rightarrow XY$ transformations in grasshoppers did not occur prior to the Pleistocene, we are almost forced to the conclusion that rearrangements of this type severely restrict the evolutionary potential of the species in which they occur – they may be short-term successes but are almost always long-term failures. 'Evolutionary potential' may be taken to include both capacity to survive and capacity to undergo speciation. It is perhaps especially the latter that seems to be lacking in forms that have acquired certain types of chromosomal rearrangements such as X-autosome fusions.

A more general problem in evolutionary theory is the overall long-term balance between increases and decreases in chromosome number. It is entirely possible that in a particular group, such as the genus *Drosophila*, a disequilibrium between the numbers of chromosomal fusions and dissociations may have existed in evolution. But when one considers the animal kingdom as a whole, or some large segment of it such as a phylum or class it is obvious that an approximate equality between the number of fusions and dissociations must have existed – at least this is so if we are right in thinking that losses and reduplications of whole chromosomes have not contributed, or have only made a very minor contribution to the process of evolutionary change in chromosome number.

The only alternative to the hypothesis that in all major groups of organisms fusions and dissociations have been about equally frequent would be to suppose that although fusions may have been more frequent (as the evidence from *Drosophila* and the Acrididae seems to suggest) they diminish the probability that species carrying them will leave descendent species. If species with numerous chromosomal fusions are often evolutionary blind-alleys, then the number of fusions occurring in evolution could be consider-ably in excess of the number of dissociations.

A fundamental aspect of chromosomal evolution is what we may call the *quantitative* one, namely the extent to which the basic haploid or diploid amount of DNA in the karyotype has changed throughout phylogeny, from species to species and group to group. So long as this could be explained away by the simple and convenient hypothesis of evolutionary polynemy (see p. 403) the significance of the evolution of DNA values appeared rather limited. But if we reject evolutionary polynemy, we are free to investi-gate more interesting alternatives. It is probable that the number of different polypeptides in a grasshopper or a mammal is somewhat greater than in a sponge or a coelenterate, and that the number of genetic loci is corres-pondingly increased. But the difference is hardly likely to be a tenfold one, as the difference in DNA values is (Table 1.3). And, on the other hand, it is highly unlikely that mammals (including man) have only a tenth or twentieth as many genetic loci (cistrons) as some salamanders such as *Necturus* and *Amphiuma*.

The extraordinarily high DNA values of these and other urodeles are a challenging problem for the cytogenetic evolutionist. Careful studies of the lampbrush chromosomes provide no support for the view that they might be due to polynemy. Nor is there any reason to suppose them to be caused by large blocks of inert or semi-inert heterochromatin. The total number of chromomeres in lampbrush chromosomes of *Triturus* species does not seem to be greatly, if at all, in excess of the number of bands in the polytene chromosomes of *Drosophila*.

The resolution of this paradox lies, we feel sure, in Callan's (1967) hypothesis of the organization of genetic units in chromosomes, or in whatever modification of this hypothesis may eventually be adopted. Once we have accepted the notion that each unit of encoded information consists of one 'master copy' followed by a series of 'slave repeats' which are active in RNA synthesis, then it follows that the evolution of DNA values is not (in the main, anyhow) an evolution of the number of different genetic units, but on the contrary an evolution of the average number of 'slave' units each 'master' unit possesses. But just why the master units of the salamander *Necturus* should have ten times as many slaves as those of the frog *Rana pipiens* is unclear. Once again, the whole question of cellular dimensions (Ch. 12) undoubtedly comes into the picture. Accurate measurements of cell sizes in comparable tissues of related organisms with very different DNA values seem to be badly needed. But we also need far more determinations of DNA values in a wide variety of species before their general significance can become apparent. There is a great lack of data for most of the invertebrate phyla, and many of the published determinations are in arbitrary units which cannot be converted to picograms. The extent of variation in DNA values is not really known for any group. We cannot, for example, say whether the rather wide variations found by Hughes-Schrader (1951, 1953a, 1957, 1958; see also Hughes-Schrader and Schrader 1956) in the DNA values of pentatomid, coccid and mantid species are characteristic of insect groups in general. In the acridid grasshoppers the work of John and Hewitt (1966b) suggested a rather narrower range of DNA values, but Kiknadze and Vysotskaya (1970) have found that some species of *Bryodema* and *Angaracris* have more than twice as much DNA per genome as *Eirenephilus longipennis*. Comparisons of DNA values in sibling species with and without considerable karyotypic differences should be especially significant. In the not-too-distant future, accurate determinations of the DNA content of individual chromosomes should become routine. Bailly (1967) has published one such study on the chromosomes of the salamander *Pleurodeles waltlii*.

Atkin *et al.* (1965) have published some data on the DNA values of nineteen species of reptiles, birds and mammals. The birds all had low values, the lizards and snakes intermediate ones and the mammals high values, there being only very minor differences between six mammalian species. Two species of Chelonia had values only slightly less than those of the mammals and considerably higher than those of the lizards and snakes;

somewhat surprisingly, the value for a species of alligator was close to that of the Chelonia.

Martin and Hayman (1965, 1967) have carried out a series of studies on karyotype evolution in Australian marsupials which involve accurate determinations of the amounts of DNA per diploid nucleus, as well as measurements of each chromosome arm. Taking *Macropus major* (n = 8) as a standard (DNA value – 100 units), *Protemnodon eugenii* (n = 8) has 98.8 units, *Macropus rufus* (n = 10) 108.2 units and *Protemnodon bicolor* (n = 5) 96.6 units (standard errors range from ±1.4 to ±2.0). It is tempting to conclude that the species with the highest chromosome number owes its high DNA value to 'gains' of material in the course of the two dissociations for which it presumably became homozygous at some stage of its phylogeny; and that the low DNA value of *P. bicolor* is due to loss of proximal regions of chromosomes in the course of three chromosomal fusions.

We have already indicated (pp. 403–5) the evidence against the 'polyneme' model of germ-line chromosome structure. Even if chromosomes in general are eventually shown to have a polyneme structure (which we believe is highly unlikely and probably impossible, on present evidence), two possibilities would still remain, which would be very different in their evolutionary consequences. On the first of these the degree of polynemy would be uniform, at least over major taxonomic groups and perhaps over all higher organisms. The second possibility is that which was favoured by the Schraders (see p. 403), namely that within individual groups such as families and genera exist species having different orders of polynemy in their germ-like chromosomes. If this were so, it would alter our views of evolution rather considerably, and an important task for evolutionists would be to identify the species or species-groups having different grades of polynemy within each genus or other taxonomic category being studied. Changes in degree of polynemy would be major evolutionary events and such changes would have to be studied in relation to speciation, evolutionary patterns and phylogeny. Although it seems that germ-line polynemy is (apart from temporary polyneme stages in a few parthenogenetic forms and in the coccid *Llaveiella*) a hypothesis that will not withstand close examination and that differential evolutionary polynemy, in particular, must be rejected, we feel that the evolutionary implications of such concepts need to be pointed out here, in case we should be shown eventually to be wrong in rejecting them.

It is clear that one can get a rough idea of the DNA value of a species from the volume of its chromosomes. Thus it is apparent that certain groups such as the grasshoppers and the urodeles will eventually be shown to have high DNA values in all their species. Conversely, there is almost certainly no species of bird in which the DNA value is really large. To this extent, DNA values are quite conservative in an evolutionary sense; probably more so than chromosome numbers, since we do have species in many groups that have undergone spectacular increases or decreases in chromosome number.

The reason for this conservatism in DNA values is not entirely clear, although there seems to be a closer correlation between DNA values and cell size than between amount of DNA and histological or anatomical complexity. Nevertheless, differences in DNA values of the order of 200 or 300% probably occur in many groups. Some rather surprising evolutionary changes in DNA values (up to a threefold difference) must have occurred in diploid tree frogs of the genus *Hyla* (see p. 438).

It is, of course, certain that, even if we adopt Callan's hypothesis, all differences in DNA values between related species are not due to changes in the number of slave sequences per genetic locus. Some of them are undoubtedly due to duplications and deletions of quite sizeable chromosome segments, probably in most instances heterochromatic ones.

The significance of Keyl's (1964, 1965*a, b*, 1966*a, b*) remarkable observations on single-locus duplication in *Chironomus thummi* populations (and in *C. thummi*, by comparison with the ancestral form, *C. piger*) is difficult to estimate until we know how far this phenomenon occurs in other species and other groups. Clearly, since the evolutionary separation of these two forms (sibling species or subspecies) there have been some hundreds, at least, of single locus duplications in *thummi* – but apparently none in *piger*. The result is a quite significantly greater DNA value in the nuclei of *thummi*. It is certainly possible that many of the differences between the DNA values of related species is due to the 'Keyl phenomenon', although evolutionary duplications and deletions of much longer regions (e.g. of heterochromatic segments) have undoubtedly contributed as well, in many instances.

A theoretically possible alternative to the above interpretation would be, of course, that many loci in *piger* have had their DNA content halved. But this seems unlikely since, from a comparison of the banding pattern of the polytene chromosomes of a number of *Chironomus* species, it was concluded that *piger* is phylogenetically older than *thummi* (Keyl 1957, 1962).

Duplication of DNA, on the scale that has occurred in *thummi*, cannot have been a general phenomenon in evolution without the existence of an antithetical process or processes of some kind; otherwise DNA values would have increased to fantastic levels from pre-Cambrian times to the present. If halving of the amount of DNA in a locus occurred from time to time we might expect to find that some loci of *thummi* would contain half as much DNA as the corresponding loci of *piger* – which seems not to be the case. If less than half a locus could be deleted we might expect to find some loci of *thummi* with three times, five times, six times . . . as much DNA as the homologous bands of *piger* – which is likewise not so. It may well be, of course, that these two sibling species represent a rather special case and that similar studies of other sibling pairs or species groups might reveal a different situation. One might find, for example, a pair of siblings where species *A* had some bands duplicated, quadruplicated, etc. by comparison with species *B*, and *B* had other bands duplicated by comparison with *A*.

Loss of chromosomal material by deletion undoubtedly occurs from time to time in evolution. It may be that periodic losses of substantial hetero-chromatic segments are a sufficient antithetical process to single-locus duplication – but this seems rather doubtful. Another possibility is, however, that species like *thummi*, which have undergone hundreds or thousands of locus-doubling events, have thereby acquired genetic systems which doom them as far as future progressive evolution is concerned. They may be blind alleys in the evolutionary process in the same way as high polyploids in most groups of plants are generally believed to be. If so, a rather novel view of the whole process of animal evolution would emerge; earlier ideas of 'racial senility' in some groups, currently viewed with disfavour, might have to be revived in a new form. It is fairly clear that there are a considerable number of loci in *thummi* which are polymorphic for DNA content – pre-sumably because of heterosis of the heterozygotes; the locus-duplications which have undergone fixation may have passed through a stage when they existed as heterotic polymorphisms.

It is not entirely clear whether the duplicated bands examined by Keyl in *Chironomus thummi* correspond in all respects to the complex loci of *Drosophila* genetics. At the morphological level a difference seems to exist; the complex loci of *Drosophila* are frequently resolvable into several bands in the salivary chromosomes, while the 'thick' bands of *C. thummi* are not. But this may not be a very fundamental distinction. Even in *Drosophila* the number of loci is sometimes greater than the number of visible bands and the single band $3C_7$ includes as many as ten genetically distinct pseudo-allelic loci (Green (1965)).

It is possible that within the next decade, biochemical techniques presently available for studying the DNA's and proteins of different species will enable us to obtain direct and objective measures of the extent of the genetic divergence between them. When combined with accurate deter-minations of the total amount of DNA in the karyotypes of numerous species, such biochemical data should not merely provide the basis for irrefutable phylogenies but should also throw entirely new light on numer-ous general problems of evolutionary mechanisms.

Information is already becoming available on the base ratios in the DNA's of different species. The ratios of adenine + guanine to cytosine + thymine $(A + G/C + T)$ and adenine + cytosine to guanine + thymine $(A + C/G + T)$ should equal unity; in fact most of the published values agree with this theoretical expectation (based on the Watson and Crick structure for DNA) within the limits of experimental error. However, studies on higher animals, plants and microorganisms have shown that there are systematic variations in the $A + T/C + G$ ratio, although different organs of the same species give the same ratio (Chargaff 1955). In most animals the above ratio is over 1; it ranges from 1.02 in a trypanosome up to 1.75 in the limpet *Patella*, 1.81 in a sea-urchin, *Psammechinus* and a reported 2.78 in the protozoan, *Tetrahymena* (Leslie 1961; Antonov and Belozerskii 1961; Antonov, Favorova and Belozerskii 1962). Most bacteria show values of less than unity, but some have values as high as 2.22 (Leslie 1961). Relatively closely related species of higher organisms may show fairly different ratios; for example, the mouse has been stated to have a value of 1.55, while the rat has one of 1.33. The significance of these differences is unknown; they have not been correlated with any differences in extent of heterochromatic segments, size of nucleolar structures, etc.

It has been shown by Britten and Kohne (1966, 1967, 1968) that the nuclear DNA's of most higher organisms contain a fraction which consists of up to a million copies of a short sequence of nucleotides. In the mouse this satellite DNA forms about 10–14% of the total quantity present; it has an $A + T/G + C$ ratio of about 1.94 as compared with 1.38 for the major fraction (Walker 1968). It had been shown earlier by Sueoka (1961) and Smith (1964) that crabs of the genus *Cancer* possess, in addition to a DNA of 'normal' composition, a minor component consisting almost entirely of adenine and thymine nucleotides. Some other Crustacea seem to lack this minor DNA component.

A full discussion of the significance of these minor DNA components would be outside the scope of this book. But the possibility that they are derived from chromosomal material recognized as heterochromatic by cytological techniques is an attractive one. The very general occurrence of 'ectopic' pairing between heterochromatic segments, the formation of the chromocentre in *Drosophila* polytene nuclei and various kinds of synaptic attractions between heterochromatic regions and chromosomes at meiosis all suggest that most types of heterochromatin contain extensive reduplication of nucleotide sequences.

Basically, there have always been two views of the heterochromatin–heteropycnosis problem. One of these regarded heterochromatin as a stuff, a substance or mixture of substances that differed permanently from euchromatin and manifested its character at various stages of the nuclear cycle by 'heteropycnosis', 'precocious condensation', 'differential staining', 'nucleic acid starvation' (Darlington and La Cour) or other special behaviour. This viewpoint finds expression in such unequivocal statements as 'the *Y* of *Drosophila melanogaster* is composed of heterochromatin'.

The second viewpoint emphasizes the process of heteropycnosis, which is regarded as a temporary condition or state which chromosomes or chromosome regions may assume, under appropriate circumstances. Heteropycnosis is seen as an essentially reversible process, which may be either positive or negative ('over-condensation' and 'under-condensation').

Clarification of this issue has come especially from studies of coccid cytology by Schrader, Hughes-Schrader and S. W. Brown (see p. 507) and of the *X*-chromosomes in female mammals by Ohno and other recent workers (see p. 594). Both these lines of investigation have shown that entire chromosomes may become positively heteropycnotic when genetically homologous chromosomes in the same nucleus remain non-heteropycnotic – and that this heteropycnosis involves genetic inactivation of many loci, at any rate. It has further been shown that pieces of mammalian autosomes translocated to *X*-chromosomes may become themselves inactivated ('spreading effect'). In the coccid case the chromosomes which become inactivated (in male embryos only) are an entire haploid genome which has been genetically active in the previous generation so that there can be no question of their being essentially inert.

The chemical basis of heteropycnosis has not yet been determined; but certain facts are known. Loewus, Brown and McLaren (1964), by studying the DNA's of male and female mealybug coccids, have shown that the DNA base ratios of the heteropycnotic and non-heteropycnotic chromosome sets are identical (as we might expect, if the integrity of the DNA is retained). A more probable difference might seem to be a replacement of the basic protein component by one of a different type, e.g. of a histone by a protamine, as is known to occur during salmon spermatogenesis (Stedman and Stedman 1947; Alfert 1956) – or a change in the type of histone present. However, Comings (1967) has shown that the histones of male and female mealy-bugs are virtually identical. One is thus forced to wonder whether the chemical basis of heteropycnosis does not depend in some way on the non-histone chromosomal proteins.

The term *heterochromatinization* (or heterochromatization) has been used by recent writers in several different senses, so that confusion is likely to result. When used in a developmental sense it is equivalent to the earlier term *heteropycnosis*. Thus we may say that grasshopper *X*-chromosomes undergo heteropycnosis during certain stages of mitosis and meiosis. The term *negative heteropycnosis* may be used to describe the 'under-condensed' conditions while *positive heteropycnosis* designates the 'over-condensed' state.

The second sense in which the term *heterochromatinization* has been used is to describe the changes which take place in a previously euchromatic segment when, as a result of a rearrangement, it is brought close to a hetero-chromatic region. This kind of heterochromatinization seems to be fre-quently irregular in manifestation and gives rise to variegation (the 'V-type position effects' of Lewis 1950). It involves cytological heteropycnosis and genetic activation.

A third kind of phenomenon which has been called *heterochromatinization* is the gradual evolutionary change of euchromatic segments into hetero-chromatic ones which has occurred, for example, in the Y-chromosomes of many neo-XY mechanisms. In this book we have restricted the term *heterochromatinization* to this kind of transformation which has come about as a result of a long term evolutionary process.

It seems probable that the developmental phenomenon (heteropycnosis), the position effect one and the evolutionary one (true heterochromatiniza-tion) all involve a switching-off of the genetic functions or at any rate a large part of them. But whereas in ordinary heteropycnosis the activity of the genes can be manifested in other tissues or at other stages of develop-ment, in evolutionary heterochromatinization the switching off is permanent and applies to all tissues.

There are a number of instances in which one or other of a number of chromosomes (but no more than one) becomes heteropycnotic at a particular stage. Particularly striking instances are the XX pair in the somatic chromo-somes of female mammals, and the 'variable D' *Comstockiella* systems of some scale insects (see p. 512). A formal explanation in these cases might be in terms of an extrachromosomal body or *episome* which fuses with one chromosome and determines its subsequent behaviour (being then un-available to fuse with other chromosomes); no direct evidence exists as to whether any such mechanism actually exists. In the case of the 'variable D' systems the 'choice' is between non-homologous chromosomes, any one of which may become a D-chromosome. In the mammalian X-chromo-some mechanism, if there is a single episome (or some equivalent mechan-ism) this would have to attach itself to the non-heteropycnotic X, rather than to the heteropycnotic one, since if there are more than two X's, all except one become heteropycnotic.

The condition of all the chromosomes in the sperm nucleus is undoubtedly equivalent to heteropycnosis. It has been known in a general way ever since the work of Muller and Settles (1927) that the genes are 'non-functional' in the spermatozoa of *Drosophila* and it is now known that RNA is not synthesized by the condensed chromosomes of the sperm head. Another example of a cell type that has a totally repressed genotype is the avian erythrocyte, whose nucleus does not synthesize RNA, DNA or protein (Allfrey and Mirsky 1951; Cameron and Prescott 1963).

A particularly interesting case of cells with many chromosomes heteropycnotic is provided by the cells of the germ-line in cecidomyids of both sexes, in which the S-chromosomes of both paternal and maternal origin are heavily condensed. One must presume that in these germ-line cells the only chromosomes which are present in the somatic tissues are mainly or totally repressed, genetically.

The genetic significance of negative heteropycnosis is not known. The undercondensed state of the X-chromosome in the spermatogonia of short-horned grasshoppers, crickets and some other groups may be an outward indication of a burst of biosynthetic activity which is the antithesis of the genetic repression associated with positive heteropycnosis.

The phenomenon of late replication of heterochromatic DNA seems to be a special manifestation of positive heteropycnosis. The X-chromosomes of short-horned grasshoppers are late-replicating in the male meiotic prophase, but not in the spermatogonia, when they are negatively heteropycnotic; neither do they ever seem to be late-replicating in female somatic cells, where they do not show either positive or negative heteropycnosis.

From an evolutionary standpoint we can imagine some kind of long-term circulation of genetic materials – increase in DNA through duplications, inactivation through heterochromatinization and loss of DNA as a result of deletions of heterochromatin. Such a picture seems to involve the idea that fully active genetic loci can be converted (by mutation) into partially or completely inactivated (i.e. heterochromatic) ones; but that the reverse process (conversion of inactivated loci into fully active ones) never occurs. Certainly there is no evidence for a process which would be the converse of evolutionary heterochromatinization. However, it is difficult to see how we could obtain evidence of such a process even if it did occur.

One important aspect of chromosomal evolution is, of course, the changes which have occurred in the total amount of genetic recombination (the entire range of the higher animals – from *Parascaris equorum* with n = 1 or 2 to *Lysandra atlantica* with about n = 220 – being through about two orders of magnitude). There have been two schools of thought on this subject of levels of recombination in evolution. One, which has its roots in Fisher's general theory of natural selection (see Fisher 1930) regards the amount of genetic recombination between individual polymorphic loci and the total amount of recombination in the entire karyotype, as being very directly under the control of natural selection. According to this view, selection constantly adjusts the amount of recombination between polymorphic loci, and the amount of recombination in a genetic system as a whole is adaptive to a particular kind of life cycle, population structure or environment (see Bodmer and Parsons 1962). There can be no doubt that by suitable genetic techniques of artificial selection it is possible to increase recombination in particular chromosome segments (Parsons 1958); there do not seem to have been any selection experiments in which the chiasma frequency of the whole nucleus was increased or decreased.

Fisher (1930) pointed out that, in the case of two polymorphisms Aa and Bb, if natural selection favours the AB and ab gametes and operates against Ab and aB gametes, then there will be selection for closer linkage between the two loci. The problem is complex, but a general model of a genetic system which leads to closer linkage by natural selection has been considered by Kimura (1956a). This raises the question why genotypes do not 'condense', i.e. why recombination remains at a relatively high level in most species. Turner (1967) attempted to answer this question mathematically and claimed to have shown that where three loci are considered, selection may favour looser rather than tighter linkage and that this effect becomes stronger as the number of loci increases. However, Lewontin (1971) has presented a rigorous mathematical demonstration that in a uniform environment genotypes *should* condense, or at least that restriction of recombination increases the mean fitness of the population. Clearly, recombination has been preserved in evolution because environments are *not* uniform, either in space or in time.

The opposite view to that of Fisher and his followers is vaguer and has had less articulate protagonists. Essentially, it stresses that the amount of recombination depends on many cellular processes and that it may be to some extent at any rate a fortuitous result of selection for chiasma frequencies and chiasma distributions which lead to meiotic efficiency.

A clear-cut decision between these alternatives is hardly possible at the present time and it seems at least possible that both are to some extent true. Interspecific comparisons are not very revealing; in other words if species A has 50% more recombination than species B, it is usually not possible to find an adaptive explanation in terms of life cycle or habitat. Intraspecific comparisons would be far more significant, but there have been very few studies of this type, e.g. of chiasma frequencies in central and peripheral populations of a species. Hewitt (1964) found that three species of British grasshoppers had significantly different chiasma frequencies (13.51, 15.39 and 18.04 per spermatocyte nucleus) but only *Chorthippus parallelus*, with an overall mean of 15.39, showed significant differences between populations (the range being from 14.69 to 16.32). It is well known that heterozygosity for inversions or the presence of supernumerary chromosomes or chromosome regions may increase the chiasma frequency of the other chromosomes in the nucleus. Nolte (1964) has presented some convincing evidence that in the Brown locust (*Locustana pardalina*) the process of 'gregarization' involves an increase in chiasma-frequency, so that individuals from dense swarms (and their descendants in laboratory stocks) have considerably higher chiasma frequencies than solitary ones. But both the mechanism and the adaptive significance (if any) of this phenomenon are somewhat obscure, and it is not really certain that the increase is due to natural selection. In a recent paper, Nolte (1968) has claimed that a pheromone increases chiasma frequency in locusts, but it seems doubtful if the evidence in this case is really reliable.

Several lines of evidence suggest that selectional control of chiasma frequencies may in some cases operate more in relation to the 'mechanical' than the 'genetic' functions of the chiasmata. One is that in different populations of the parthenogenetic grasshopper *Moraba virgo* which must have been long isolated from one another, the chiasma frequencies (in the egg, naturally) seem to be very constant and also similar to those of related bisexual species, i.e. they have remained highly canalized, in spite of the fact that the chiasmata can have no genetic function at all in this species (see p. 718).

The second is that in some species of grasshoppers (White, unpub.) and newts (Watson and Callan 1963) recombination has been reduced to a low level in one sex, by extreme chiasma-localization, without there having been any corresponding restriction of chiasma-distribution in the other sex; one would expect that if natural selection had favoured a low level of genetic recombination *per se*, it would have done so fairly equally in both sexes.

Hewitt's (1964, 1965) careful studies of the chiasma frequencies of four species of British grasshoppers do not seem to provide a basis for any general theory of the evolutionary determination of levels of genetic recombination. No geographic regularities were clearly shown and the species in mixed colonies did not show similar trends in chiasma frequency.

In theory it would seem highly desirable to know how levels of genetic recombination and structural heterozygosity are related to the 'breeding systems' (i.e. the coefficients of inbreeding) of natural populations (John 1965). But, unfortunately, almost nothing is known (objectively and accurately, that is) concerning the values of F (Sewall Wright 1922) in natural populations of animals; and more or less wild guesses concerning particular species or groups are more likely to lead to confusion than clarity.

If we have appeared somewhat blind to this aspect, it is not because we regard it as unimportant but simply because speculation in the virtual absence of scientifically valid data seems fruitless. For a good many species of animals where the population size is large and vagility considerable it is fairly clear that F will not differ significantly from zero. For some apterous sedentary organisms which form small local colonies it may possibly be large enough (in the range .01 to .1) to affect the population genetics of the species. But except for a very few forms where the life cycle is such as to enforce brother–sister or mother–son mating, or where a planktonic existence implies that $F = 0$, we simply do not have the basic information on which to base any theoretical discussions. Carson (1968) has discussed some of the genetic and evolutionary consequences of drastic fluctuations in the size of natural populations, from a theoretical standpoint. But locust species, which are outstanding examples of populations exhibiting spectacular 'flushes' and 'crashes', provide absolutely no evidence for his conclusions on species formation following population flushes, and may in fact be regarded as disproving his general thesis.

To a botanist it must appear rather strange that certain zoological taxonomists and evolutionists engage in extremely elaborate morphological and biometric studies designed to clear up systematic confusion in the groups they are dealing with without attempting even the most elementary chromosomal study. Thus Ehrlich (1961 – see also Ehrlich and Holm 1963) carried out a refined mathematical analysis of seventy-five characters in each of thirteen individuals of the butterfly genus *Euphydryas* in an attempt to assign apparently intermediate specimens to one or other of the two species *E. editha* and *E. chalcedona*. No attempt to determine the chromosome numbers was made; it is of course possible that the karyotypes of these species are indistinguishable – but the possibility that a difference exists would surely have been worth exploring, in connection with this investigation. And the despair of Ehrlich and Holm over taxonomic confusion in the case of the approximately 400 species of North American butterflies might be lessened if at any rate some of the taxonomically difficult complexes were subjected to cytologic analyses (an excellent beginning has been made by Maeki and Remington). But we must also warn the taxonomists that there do exist groups in which karyotypes are so uniform that their study is likely to be taxonomically unrewarding. Cytology is certainly a tool that deserves more attention by animal taxonomists than it has received in the past, and in certain cases it can furnish critical evidence on relationships that could not be obtained by any other method presently available; but it is not a magic key that will unlock all taxonomic problems, as some have supposed.

Mayr (1963 and elsewhere) has led the current revolt against the 'typological' view of the species – i.e. against the concept of an average or norm for each species from which variant populations and individuals deviate more or less. There can be no doubt that much of the typological thinking of the past went much too far and that in many respects 'population thinking' in evolutionary biology is a much more fruitful approach. However, we should beware of pushing the 'populationist' approach to extremes. Chromosome cytology emphasizes the fact that many species are polymorphic for inversions or other chromosomal rearrangements and that some species may be cytologically polymorphic in some areas and monomorphic in others. Nevertheless there must be many species of animals (probably the majority) which are not chromosomally polymorphic, so that they can be characterized by a karyotype consisting of n chromosomes of such-and-such lengths and containing a fixed amount of DNA. Thus the chromosome cytologist is bound to take a typological attitude, up to a point, when dealing with a number of species whose karyotypes differ but do not vary (or show little intraspecific variation).

The most extreme revolt against typological thinking is the attitude that species are largely artifacts of taxonomic procedure and that 'the generalization that organisms exist as distinct species is largely invalid' (Ehrlich and Holm 1963). In the light of the numerous examples cited in Chapters 11, 12, 15, 16 and 17 of this book it should be obvious that chromosome cytology provides innumerable examples of clear-cut differences between the karyotypes of species so similar morphologically that they are barely or not at all distinguishable by the methods of classical taxonomy, based on external characters. Ehrlich and Holm's (1962) peculiar attempts to belittle both natural selection and adaptation have been critically discussed by Stebbins (1963).

Cytotaxonomic studies on higher animals emphatically confirm the reality of biological species as distinct entities which generally do not intergrade. This is definitely so in the most detailed investigations of the *virilis*, *repleta* and *melanica* species groups of *Drosophila* by Stone (1955, 1962), Wasserman (1954, 1960, 1962a–d, 1963) and Stalker (1963, 1964a, b, c). The situation in the morabine grasshoppers seems to be essentially the same; there is not the least difficulty in assigning the great majority of individuals to a distinct species on the quite independent evidence of its karyotype *or* its male internal genitalia and other structures (e.g. egg guides of females). Only in a few instances in grasshoppers do we have large complexes of forms comparable to the *Drosophila paulistorum* case; further studies in these cases will no doubt settle the question as to what degree of evolutionary divergence the several 'chromosomal races' (as we provisionally call them) have reached. The situation in the Simuliidae and Chironomidae (Rothfels

1956; Dunbar 1958, 1959; Basrur 1959, 1962; Landau 1962; Keyl 1957*b*, 1961*b*, 1962; Wülker, Sublette and Martin 1968) while revealing many 'border-line cases' of chromosomal races, incipient species and recently evolved sibling species, certainly does not necessitate any abandonment or radical revision of the biological species concept. Neither, in our opinion, do the detailed studies of S. G. Smith (1962*b*, *c*, *d*) on the beetle genera *Chilocorus* and *Pissodes*. Taxonomic confusion does exist in the case of the African mice studied by Matthey (1963*b*, *c*, 1964*b*, 1965*b*, *c*) but here the number of individuals that have been subjected to critical cytological study is still very small. The exact status of the 'subspecies' of *Triturus cristatus* must await a detailed investigation of the lampbrush chromosomes of individuals from critical areas of contact or sympatry.

The 'synthetic' or neo-Darwinian theory of organic evolution represents a major stage in our understanding of the history of life on earth. But evolution is, as Patterson and Stone (1952) have said, 'the most comprehensive, fascinating and generalized theory of living organisms', and we must not expect that a few broad statements, however fundamental, represent any kind of final understanding. Rather than congratulating ourselves on the synthesis already achieved, we should be looking for new principles, new generalizations and, eventually, a new synthesis which will include all those unique cytogenetic evolutionary inventions which seem, for the time being, outside the scope of the currently accepted viewpoint. It seems safe to predict that any discussion of the broad mechanisms of evolution in, say, twenty-five years' time, will have to take far more account of the chromosomes themselves, as bodies composed of nucleic acids and proteins, and their relations to the rest of the cell at various stages of mitosis and meiosis. And, as indicated earlier, it will almost certainly lay more stress on the role of chromosomal rearrangements in initiating and promoting speciation. But, although we may expect the emphasis to change, we do not anticipate that any of the main conclusions of the current synthetic theory will be overthrown, as has been attempted, for example by Ehrlich and Holm in their superficial attacks on the biological species concept, natural selection and adaptation. In an earlier chapter (p. 450) we laid considerable stress on the concept of karyotypic orthoselection, as a major evolutionary principle. This, rather than any mystical 'orthogenetic' principle is surely the basis of most of the events of chromosomal evolution. The unique inventions of cytogenetic evolution only appear capricious, bizarre and unexpected because we cannot yet view them in relation to the whole of the molecular physiology of mitosis and meiosis in organisms in which they have occurred, The current explosive development of molecular biology has only just begun to have its impact in the evolutionary field.

As this develops, the results are bound to be somewhat unexpected. But we may be reasonably confident that they will amend, amplify and extend the major concepts of the first generation of 'synthetic evolutionists' rather than undermine or refute them altogether.

No reason exists for believing that the general principles of organic evolution which we can discern at the present time did not also operate in the past, at least ever since multicellular organisms arose. Thus, although the evolution of trilobites, ammonites and dinosaurs will never be interpretable *in detail* in cytogenetic terms, we can be reasonably certain that their chromosomal evolution followed the same general principles as we find in living forms. We can hence imagine, even if we cannot study, the grand pattern of the evolution of genetic systems as it has unfolded itself in the past, with its endless succession of mutations and chromosomal rearrangements, duplications and deletions, polymorphism and monomorphism, isolation and speciation, adaptive radiation, extinction and survival.

BIBLIOGRAPHY

ABDEL-HAMEED, F. and SHOFFNER, R. N. (1971). Intersexes and sex determination in chickens. *Science*, **172**: 963–4.

ABRAMOFF, P., DARNELL, R. M. and BALSANO, J. S. (1968). Electrophoretic demonstration of the hybrid origin of the gynogenetic teleost *Poecilia formosa*. *Amer. Nat.* **102**: 555–8.

ABRAMOWICZ, T. (1964). Occurrence of inversion in four natural populations of *Lasioptera rubi* Heeg. (Cecidomyiidae, Diptera). *Zool. Poloniae*, **14**: 269–85.

ACTON, A. B. (1955). Selective values of chromosome inversions in *Chironomus*. *Proc. Roy. Phys. Soc. Edinb.* **24**: 10–14.

ACTON, A. B. (1956). Crossing over within inverted regions in *Chironomus*. *Amer. Nat.* **90**: 63–5.

ACTON, A. B. (1957a). Chromosome inversions in natural populations of *Chironomus tentans*. *J. Genet.* **55**: 61–94.

ACTON, A. B. (1957b). Chromosome inversions in natural populations of *Chironomus dorsalis*. *J. Genet.* **55**: 261–75.

ACTON, A. B. (1957c). Sex chromosome inversions in *Chironomus*. *Amer. Nat.* **91**: 57–9.

ACTON, A. B. (1958). The gene contents of over-lapping inversions. *Amer. Nat.* **92**: 57–8.

ACTON, A. B. (1959). A study of the differences between widely separated populations of *Chironomus* (= *Tendipes*) *tentans* (Diptera). *Proc. Roy. Soc. Lond.* B, **151**: 277–96.

ACTON, A. B. (1961). An unsuccessful attempt to reduce recombination by selection. *Amer. Nat.* **95**: 119–20.

ACTON, A. B. (1962). Incipient taxonomic divergence in *Chironomus* (Diptera)? *Evolution*, **16**: 330–7.

ACTON, A. B. (1965). *Chironomus tentans* (Diptera): the giant chromosomes and taxonomic divergence. *Proc. XIIth Int. Congr. Entomol.* London, p. 245.

AEPPLI, E. (1951). Die Chromosomenverhältnisse bei *Dendrocoelum infernale* (Steinmann). Ein Beitrag zur Polyploidie im Tierreich. *Rev. Suisse Zool.* **58**: 511–18.

AEPPLI, E. (1952). Natürliche Polyploidie bei den Planarien *Dendrocoelum lacteum* (Müller) und *Dendrocoelum infernale* (Steinmann). *Z. indukt. Abst. Vererbl.* **84**: 182–212.

AGAR, W. E. (1911). The spermatogenesis of *Lepidosiren paradoxa*. *Quart. J. Micr. Sci.* **57**: 1–44.

AGAR, W. E. (1923). The male meiotic phase in two genera of marsupials (*Macropus* and *Petauroides*). *Quart. J. Micr. Sci.* **67**: 183–202.

AHMED, I. A. (1940). The structure and behaviour of the chromosomes of the sheep during mitosis and meiosis. *Proc. Roy. Soc. Edinb.* **60**: 260–70.

AIDA, T. (1921). On the inheritance of color in a freshwater fish, *Aplocheilus latipes* Temmick and Schlegel, with special reference to sex-linked inheritance. *Genetics*, **6**: 554–73.

ALBERTI, W. and POLITZER, G. (1923). Über den Einfluss der Röntgenstrahlen auf die Zellteilung. *Arch. mikr. Anat. Entwicklungsm.* **100**: 83–109.

ALBIZU DE SANTIAGO, C. (1968). Chromosome number in *Cephalocoema vittata* (Proscopiidae, Orthoptera). *J. Agric. Univ. Puerto Rico*, **52**: 351–5.

ALEXANDER, M. L. (1952). Gene variability in the *americana–texana–novamexicana* complex of the virilis group of *Drosophila*. *Univ. Texas Publ.* **5204**: 73–105.

ALFERT, M. (1956). Chemical differentiation of nuclear proteins during spermatogenesis in the salmon. *J. Biophys. Biochem. Cytol.* **2**: 109–14.

ALFERT, M. and GOLDSTEIN, N. (1955). Cytochemical properties of nucleoproteins in *Tetrahymena pyriformis*; a difference in protein composition between macro- and micronuclei. *J. Exp. Zool.* **130**: 403–21.

ALLFREY, V. and MIRSKY, A. E. (1952). Some aspects of the desoxyribonuclease activities of animal tissues. *J. Gen. Physiol.* **36**: 227–41.

ALLFREY, V., MIRSKY, A. E. and STERN, H. (1955). The chemistry of the cell nucleus. *Adv. Enzymol.* **16**: 411–500.

ALLISON, A. C. (1955). Aspects of polymorphism in man. *Cold Spr. Harb. Symp. Quant. Biol.* **20**: 239–55.

ALONSO, P. and PÉREZ-SILVA, J. (1965). Giant chromosomes in Protozoa. *Nature*, **205**: 313–14.

AMMERMANN, D. (1965). Cytologische und genetische Untersuchungen an dem Ciliaten *Stylonychia mytilus* Ehrenberg. *Arch. Protistenk.* **108**: 109–52.

ANDERSON, E. (1949). *Introgressive Hybridization*. New York: John Wiley.

ANDERSON, E. and HUBRICHT, L. (1938). Hybridization in *Tradescantia*. III. The evidence for introgressive hybridization. *Amer. J. Bot.* **25**: 396–402.

ANDERSON, W. W., DOBZHANSKY, TH. and KASTRITSIS, C. D. (1967). On the relationship of structural heterozygosity in the *X* and third chromosomes of *Drosophila pseudoobscura*. *Amer. Nat.* **101**: 89–92.

ANDREWARTHA, H. G. and BIRCH, L. C. (1954). *The Distribution and Abundance of Animals*. Chicago Univ. Press.

ANGELL, R., GIANNELLI, F. and POLANI, P. E. (1970). Three dicentric *Y*-chromosomes. *Ann. Hum. Genet.* **34**: 39–50.

ANKEL, W. E. (1930). Die atypische Spermatogenese von *Janthina* (Prosobranchia Stenoglossa). *Z. Zellforsch.* **11**: 242–326.

ANSLEY, H. R. (1957). A cytophotometric study of chromosome pairing. *Chromosoma*, **8**: 380–95.

ANTONOV, A. S. and BELOZERSKII, A. N. (1961). Comparative study of the nucleotide composition of the desoxyribonucleic acids of various vertebrate and invertebrate animals. *C.R. (Dokl.) Acad. Sci. U.R.S.S.* **138**: 1216–19.

ARAKAKI, D. T., VEOMETT, I., and SPARKES, R. S. (1970). Chromosome polymorphism in deer mouse siblings (*Peromyscus maniculatus*). *Experientia*, **26**: 425–6.

ARCOS-TERÁN, L. and BEERMANN, W. (1968). Changes of DNA replication behaviour associated with intragenic changes of the *white* region in *Drosophila melanogaster*. *Chromosoma*, **25**: 377–91.

ARMBRUSTER, L. (1913). Chromosomenverhältnisse bei der Spermatogenese solitärer Apiden (*Osmia cornuta* Latr.). Beiträge zur Geschlechtsbestimmungs-frage und zum Reduktionsproblem. *Arch. Zellforsch.* 11: 242–326.

ARRIGHI, F. E., MANDEL, M., BERGENDAHL, J. and HSU, T. C. (1970). Buoyant densities of DNA of mammals. *Biochem. Genet.* 4: 367–76.

ARRIGHI, F. E., SORENSON, M. W. and SHIRLEY, L. R. (1969). Chromosomes of the tree shrews (Tupaiidae). *Cytogenetics*, 8: 199–208.

AR-RUSHDI, A. H. (1963). The cytology of achiasmatic meiosis in the female *Tigriopus* (Copepoda). *Chromosoma* 13: 526–39.

ARTOM, C. (1908). La maturazione, le fecondazione e i primi stadi di sviluppo dell' uovo dell' *Artemia salina* Lin. di Cagliari. *Biologica*, I: 495–515.

ARTOM, C. (1928). Diploidismo e tetraploidismo dell' *Artemia salina. Verh. V Int. Kongr. Genet.* 1: 384–6.

ARTOM, C. (1931). L'origine e l'evoluzione della partenogenesi attraverso i differenti biotipi di une specie colletiva (*Artemia salina* L.) con speciale riferimento al biotipo diploide partenogenetico di Sète. *Mem. R. Accad. Ital. Fis. Mat. Nat.* 2: 1–57.

ASANA, J. J. (1934). Studies on the chromosomal of Indian Orthoptera. IV. The idiochromosomes of *Hierodula* sp. *Curr. Sci.* 2: 244–5.

ASANA, J. J. and KICHIJO, H. (1936). The chromosomes of six species of ant lions (Neuroptera). *J. Fac. Sci. Hokkaido Univ.* (VI) 5: 121–36.

ASANA, J. J. and MAHABALE, T. S. (1940). On the chromosomes of an agamid lizard *Calotes versicolor* Boulen. *Curr. Sci.* 9: 377–8.

ASANA, J. J. and MAHABALE, T. S. (1941). Spermatogonial chromosomes of two Indian lizards, *Hemidactylus flaviviridis* Rüppell and *Mabuya macularia* Blyth. *Curr. Sci.* 10: 494–5.

ASANA, J. J. and MAKINO, S. (1934). The idiochromosomes of an earwig, *Labidura riparia. J. Morph.* 56: 361–9.

ASANA, J. J. and MAKINO, S. (1937). The occurrence of V-shaped centrioles in the spermatocytes of some neuropteran insects. *Annot. Zool. Jap.* 16: 175–9.

ASANA, J. J., MAKINO, S. and NIIYAMA, H. (1940). A chromosomal survey of some Indian insects. III. Variations in the chromosome number of *Gryllotalpa africana*, due to the inclusion of supernumerary chromosomes. *J. Fac. Sci. Hokkaido Univ.* (VI) 7: 59–72.

ASHBURNER, M. (1967). Patterns of puffing activity in the salivary gland chromo-somes of *Drosophila*. I. Autosomal puffing patterns of a laboratory stock of *Drosophila melanogaster. Chromosoma*, 21: 298–428.

ASHBURNER, M. (1969a). Patterns of puffing activity in the salivary gland chromo-somes of *Drosophila*. II. The *X*-chromosome puffing patterns of *D. melanogas-ter* and *D. simulans. Chromosoma*, 27: 47–63.

ASHBURNER, M. (1969b). Patterns of puffing activity in the salivary gland chromo-somes of *Drosophila*. III. A comparison of the autosomal puffing patterns of the sibling species *D. melanogaster* and *D. simulans. Chromosoma*, 27: 64–85.

ASHBURNER, M. (1969c). Patterns of puffing activity in the salivary gland chromo-somes of *Drosophila*. IV. Variability of puffing patterns. *Chromosoma*, 27: 156–77.

ASHBURNER, M. (1970). Function and structure of polytene chromosomes during insect development. *Adv. Insect Physiol.* 7: 1–95.

ATCHLEY, W. R. and JACKSON, R. C. (1970). Cytological observations on spermato-genesis in four species of Mecoptera. *Canad. J. Genet. Cytol.* 12: 264–72.

ATKIN, N. B., MATTINSON, G., BEÇAK, W. and OHNO, S. (1965). The comparative DNA content of 19 species of placental mammals, reptiles and birds. *Chromosoma*, **17**: 1–10.

ATKIN, N. B. and OHNO, S. (1967). DNA values of four primitive Chordates. *Chromosoma*, **23**: 10–13.

ATKINS, L., BÖÖK, J. A., GUSTAVSON, K. H., HANSSON, O. and HJELM, M. (1963). A case of *XXXXY* sex chromosome anomaly with autoradiographic studies. *Cytogenetics*, **2**: 208–32.

ATKINS, L., TAFT, P. D. and DALAL, K. P. (1962). Asynchronous DNA synthesis of sex chromatin in human interphase nuclei. *J. Cell Biol.* **15**: 390–3.

AUBERT, J. and MATTHEY, R. (1943). Le problème des hétérochromosomes chez les Perles (Plécoptères). *Arch. J. Klaus Stift. Vererb.* **18**: 662–4.

AULA, P., GRIPENBERG, U., HJELT, L., KIVALO, E., LEISTI, J., PALO, J., VON SCHOULTZ, B. and SUOMALAINEN, E. (1967). Two cases with a ring chromosome in group E. *Acta Neurol. Scandinav.* **43**: suppl. 31, pp. 51–2.

AULA, P., NICHOLS, W. W. and LEVAN, A. (1968). Virus-induced chromosome changes. *Ann. N.Y. Acad. Sci.* **155**: 737–47.

AYRES, M., SAMPAIO, M. M., BARROS, R. M. S., DIAS, L. B. and CUNHA, O. R. (1969). A karyological study of turtles from the Brazilian Amazon region. *Cytogenetics*, **8**: 401–9.

BACCETTI, B. (1958). Notulae orthopterologicae. IX. Osservazioni cariologiche sulle *Dolichopoda* italiane. *Redia*, **43**: 315–27.

BACCETTI, B. (1961). Cariologia di popolazioni partenogenetiche e bisessuate di *Troglophilus cavicola* Koll. (Ins. Orthopt.) (Notulae Entomologicae XV). *Verh. XI Int. Kongr. Entomol. Wien*, **1**: 418–22.

BACCI, G., COGNETTI, G. and VACCARI, G. (1961). Endomeiosis and sex determination in *Daphnia pulex*. *Experientia*, **17**: 505–6.

BACHMANN, K. (1970). Specific nuclear DNA amounts in toads of the genus *Bufo*. *Chromosoma*, **29**: 365–74.

BACHMANN, K. (1972). Genome size in mammals. *Chromosoma*, **37**: 85–93.

BADR, F. M. and BADR, R. S. (1970). The somatic chromosomes of a wild population of rats: numerical polymorphism. *Chromosoma*, **30**: 465–75.

BAER, D. (1965). Asynchronous replication of DNA in a heterochromatic set of chromosomes in *Pseudococcus obscurus*. *Genetics*, **52**: 275–85.

BAILLY, S. (1967). Étude cytophotométrique de la teneur en acides nucléiques des chromosomes métaphasiques de l'amphibien urodèle *Pleurodeles waltlii* Michah. *Exp. Cell Res.* **48**: 549–56.

BAIMAI, V. (1969). Karyotype variation in *Drosophila birchii*. *Chromosoma*, **27**: 381–94.

BAJER, A. (1963). Observations on dicentrics in living cells. *Chromosoma*, **14**: 18–30.

BAJER, A. (1965). Sub-chromatid structure of chromosomes in the living state. *Chromosoma*, **17**: 291–302.

BAJER, A., HANSEN-MELANDER, E., MELANDER, Y. and MOLE-BAJER, J. (1961). Meiosis in *Cepaea nemoralis* studied by microcinematography. *Chromosoma*, **12**: 374–81.

BAJER, A. and JENSEN, C. (1969). Detectability of mitotic spindle microtubules with the light and electron microscope. *J. Microsc.* **8**: 343–54.

BAJER, A. and MOLE-BAJER, J. (1956). Cine-micrographic studies on mitosis in endosperm. II. Chromosome, cytoplasmic and brownian movements. *Chromosoma*, **7**: 558–607.

BAJER, A. and MOLE-BAJER, J. (1969). Formation of spindle fibres, kinetochore orientation and behaviour of the nuclear envelope during mitosis in endosperm. *Chromosoma*, **27**: 448–84.

BAKER, R. H. and KITZMILLER, J. B. (1963). Cytogenetic studies on *Anopheles punctipennis. Amer. Zoologist*, **3**: 535.

BAKER, R. H. and KITZMILLER, J. B. (1964*a*). Salivary gland chromosomes of *Anopheles punctipennis. J. Hered.* **55**: 9–17.

BAKER, R. H. and KITZMILLER, J. B. (1964*b*). The salivary gland chromosomes of *Anopheles aztecus. Rev. Inst. Salubr. Enferm. Trop. (Mexico)*, **24**: 43–54.

BAKER, R. J. (1967). Karyotypes of bats of the family Phyllostomidae and their taxonomic implications. *Southwestern Nat.* **12**: 407–28.

BAKER, R. J. and HSU, T. C. (1970). Further studies on the sex-chromosome systems of the American leaf-nosed bats (Chiroptera, Phyllostomatidae). *Cytogenetics*, **9**: 131–8.

BAKER, R. J. and MASCARELLO, J. T. (1969). Karyotypic analyses of the genus *Neotoma* (Cricetidae, Rodentia). *Cytogenetics*, **8**: 187–98.

BAKER, R. J. and PATTON, J. L. (1967). Karyotypes and karyotypic variation of North American vespertilionid bats. *J. Mammal.* **48**: 270–86.

BAKER, W. K. (1954). The anatomy of the heterochromatin. Data on the physical distance between breakage point and affected locus in a V-type position effect. *J. Hered.* **45**: 65–8.

BAKER, W. K. (1965). *Genetic Analysis*. Boston: Houghton Mifflin.

BAKER, W. K. (1968). Position-effect variegation. *Adv. in Genet.* **14**: 133–70.

BAKER, W. K. and REIN, A. (1962). The dichotomous action of *Y*-chromosomes on the expression of position-effect variegation. *Genetics*, **47**: 1399–407.

BALBIANI, E. G. (1881). Sur la structure du noyau des cellules cellulaires chez les larves de *Chironomus. Zool. Anz.* **4**: 637–41.

BALTZER, F. (1908). Über mehrpolige Mitosen bei Seeigeleirn. *Verh. Physik. Ges. Würzburg*, N.F. **39**: 291–330.

BALTZER, F. (1922). Über die Herstellung und Aufzucht eines haploiden *Triton taeniatus. Verh. Schweiz. Naturf. Ges. Bern.* **103**: 248–9.

BANERJEE, B. (1961). Chromosome morphology during the spermatogenesis of *Odontotermes redemanni* (Wasmann). *Caryologia*, **14**: 155–8.

BANERJEE, M. K. (1958). A study of the chromosomes during meiosis in twenty-eight species of Hemiptera (Heteroptera, Homoptera). *Proc. Zool. Soc. (Calcutta)*, **11**: 9–37.

BANTA, A. M., WOOD, T. R., BROWN, L. A. and INGLE, L. (1939). Studies on the physiology, genetics and evolution of some Cladocera. *Carneg. Inst. Wash. Publ.* **513**, 285 pp.

BANTOCK, C. (1961). Chromosome elimination in Cecidomyidae. *Nature*, **190**: 466–7.

BARBER, H. N. (1940). The suppression of meiosis and the origin of diplochromosomes. *Proc. Roy. Soc. Lond.* B, **128**: 170–85.

BARBER, H. N. (1941). Chromosome behavior in *Uvularia. J. Genet.* **42**: 223–57.

BARBER, H. S. (1913). The remarkable life-history of a new family (Micromalthidae) of beetles. *Proc. Biol. Soc. Wash.* **26**: 185–90.

BARIGOZZI, C. (1933*a*). Die Chromosomen-garnitur der Maulwurfsgrille und ihre systematische Bedeutung. *Z. Zellforsch.* **18**: 626–41.

BARIGOZZI, C. (1933*b*). L'unicità della specie 'Gryllotalpa gryllotalpa L.' e il suo ciclo biologico. *Boll. Lab. Zool. Portici*, **27**: 145–55.

BARIGOZZI, C. (1934). Diploidismo, tetraploidismo e octoploidismo nell' *Artemia salina* partenogenetica di Margherita di Savoia. *Boll. Soc. Ital. Biol. Sper.* **9**: 906–8.

BARIGOZZI, C. (1935). Il legame genetico fra i biotipi partenogenetici di *Artemia salina. Arch. Zool. (Ital.) Napoli,* **22**: 33–77.

BARIGOZZI, C. (1941). Cytogenetical analysis of two wild populations of *Artemia salina* in connexion with polyploidism. *Proc. VIIth Int. Genet. Congr. Edinb.* pp. 57–8.

BARIGOZZI, C. (1942). Sulla struttura dei cromosomi in nuclei iperploidi di *Gryllotalpa gryllotalpa* L. *Chromosoma,* **2**: 345–66.

BARIGOZZI, C. (1944). I fenomeni cromosomici delle cellule germinali in *Artemia salina* Leach. *Chromosoma,* **2**: 549–75.

BARIGOZZI, C. (1946). Über die geographische Verbreitung der Mutanten von *Artemia salina* Leach. *Arch. J. Klaus Stift. Vererbf.* **21**: 479–82.

BARIGOZZI, C. (1956). Différenciation des génotypes et distribution géographique d'*Artemia salina* Leach: données et problèmes. *Année Biol.* **33**: 241–51.

BARIGOZZI, C., DOLFINI, S., FRACCARO, M., RAIMONDI, G. R. and TIEPOLO, L. (1966). In vitro study of the DNA replication patterns of somatic chromosomes of *Drosophila melanogaster. Exp. Cell. Res.* **43**: 231–4.

BARIGOZZI, C. and PETRELLA, L. (1953). Number and behavior of chromosomes in *Aphiochaeta xanthina* Speiser. *Experientia,* **9**: 337.

BARIGOZZI, C. and SEMENZA, L. (1952). A preliminary note on the biology and chromosome cycle of *Aphiochaeta xanthina* Sp. *Amer. Nat.* **86**: 123–4.

BARIGOZZI, C. and TOSI, M. (1957). Nuovi dati sul numero cromosomico di *Artemia salina* anfigonica. *Convegno di Genetica* 1955. *Ricerca Scientifica,* **27**, suppl. 5 pp.

BARIGOZZI, C. and TOSI, M. (1959). New data on tetraploidy of amphigonic *A. salina* Leach and on triploids resulting from crosses between tetraploids and diploids. *Convegno di Genetica* 1957. *Ricerca Scientifica,* **29**, suppl. 6 pp.

BARKER, J. F. (1960). Variation of chiasma frequency in and between natural populations of Acrididae. *Heredity,* **14**: 211–14.

BARKER, J. F. (1966). Climatological distribution of a grasshopper supernumerary chromosome. *Evolution,* **20**: 665–7.

BARKER, K. R. and RIESS, R. W. (1966). An electron microscope study of spermateleosis in the hemipteran *Oncopeltus fasciatus. Cellule,* **66**: 39–54.

BARNES, H. F. (1935). On the gall midges injurious to the cultivation of willows. II. The 'shot hole' gall midges (*Rhabdophaga* spp.) *Ann. Appl. Biol.* **22**: 86–105.

BARNES, H. F. (1944). Investigations on the raspberry cane midge, 1943–1944. *J. Roy. Hort. Soc.* **69**: 370–5.

BARR, H. J. (1966). Problems in the developmental cytogenetics of nucleoli in *Xenopus. Nat. Cancer Inst. Monograph,* **23**: 411–24.

BARR, H. J. and ESPER, H. (1963). Nucleolar size in cells of *Xenopus laevis* in relation to nucleolar competition. *Exp. Cell Res.* **31**: 211–14.

BARR, H. J. and PLAUT, W. (1966). Comparative morphology of nucleolar DNA in *Drosophila. J. Cell Biol.* **31**: C17–22.

BARR, M. L. (1959). Sex chromatin. *Science,* **130**: 1302.

BARR, M. L. and BERTRAM, L. F. (1949). A morphological distinction between neurons of the male and female and the behavior of the nucleolar satellite during accelerated nucleoprotein synthesis. *Nature,* **163**: 676–7.

BARR, M. L. and CARR, D. H. (1960). Sex chromatin, sex chromosomes and sex anomalies. *Canad. Med. Assoc. J.* **83**: 979–86.

BARR, M. L. and CARR, D. H. (1962). Correlations between sex chromatin and sex chromosomes. *Acta Cytol.* **6**: 34–45.

BARSACCHI, G., BUSSOTTI, L., and MANCINO, G. (1970). The maps of the lampbrush chromosomes of *Triturus* (Amphibia, Urodela). IV. *Triturus vulgaris meridionalis*. *Chromosoma*, **31**: 255–79.

BARTON, D. E., DAVID, F. N. and MERRINGTON, M. (1965). The relative positions of the chromosomes in the human cell in mitosis. *Ann. Hum. Genet.* **29**: 139–46.

BASRUR, P. K. (1959). The salivary gland chromosomes of seven segregates of *Prosimulium* (Diptera: Simuliidae) with transformed centromere. *Canad. J. Zool.* **37**: 527–70.

BASRUR, P. K. (1962). The salivary gland chromosomes of seven species of *Prosimulium* (Diptera: Simuliidae) from Alaska and British Columbia. *Canad. J. Zool.* **40**: 1019–33.

BASRUR, V. R. (1957). Inversion polymorphism in the midge *Glyptotendipes barbipes* (Staeger). *Chromosoma*, **8**: 597–608.

BASRUR, V. R. and ROTHFELS, K. H. (1959). Triploidy in natural populations of the black fly *Cnephia mutata* (Malloch). *Canad. J. Zool.* **37**: 571–89.

BATTAGLIA, E. (1956). A new type of segregation of the sex chromosomes in *Dysdercus koenigii* Fabr. (Hemiptera–Pyrrhocoridae). *Caryologia*, **8**: 205–13.

BATTAGLIA, E. (1964). Cytogenetics of *B*-chromosomes. *Caryologia*, **17**: 245–99.

BATTAGLIA, E. and BOYES, J. W. (1955). Post reductional meiosis: its mechanism and causes. *Caryologia*, **8**: 87–134.

BAUER, H. (1931). Die Chromosomen von *Tipula paludosa* Meig. in Eibildung und Spermatogenese. *Z. Zellforsch.* **14**: 138–93.

BAUER, H. (1933). Die Wachsenden Oocytenkerne einiger Insekten in ihrem Verhalten zur Nuklealfärbung. *Z. Zellforsch.* **18**: 254–98.

BAUER, H. (1935). Der Aufbau der Chromosomen in den Speicheldrüsen von *Chironomus thummi* Kiefer (Untersuchungen an den Riesenchromosomen der Dipteren I). *Z. Zellforsch.* **23**: 280–313.

BAUER, H. (1936*a*). Beiträge zur vergleichenden Morphologie der Speicheldrüsenchromosomen (Untersuchungen an den Riesenchromosomen der Dipteren. II). *Zool. Jb. (Phys.)*, **56**: 239–56.

BAUER, H. (1936*b*). Structure and arrangement of salivary gland chromosomes in *Drosophila* species. *Proc. Nat. Acad. Sci. U.S.A.* **22**: 216–22.

BAUER, H. (1938). Die polyploide Natur der Riesenchromosomen. *Naturwiss.* **26**: 77–8.

BAUER, H. (1939). Chromosomenforschung (Karyologie und Cytogenetik). *Fortschr. Zool.* **4**: 584–97.

BAUER, H. (1940). Über die Chromosomen der bisexuellen und der parthenogenetischen Rasse des Ostracoden *Heterocypris incongruens* Ramd. *Chromosoma*, **1**: 620–37.

BAUER, H. (1945). Chromosomen und Systematik bei Chironomiden. *Arch. Hydrobiol.* **40**: 994–1008.

BAUER, H. (1946). Gekoppelte Vererbung bei *Phryne fenestralis* und die Beziehung zwischen Faktorenaustausch und Chiasmabildung. *Biol. Zentralbl.* **65**: 108–15.

BAUER, H. (1947). Karyologische Notizen. I. Über generative Polyploidie bei Dermapteren. *Z. Naturf.* **2b**: 63–6.

BAUER, H. (1952). Die Chromosomen im Soma der Metazoen. *Verh. deutsch. Zool. Ges.*, pp. 252–68.

BAUER, H. (1953). Die Chromosomenreduktion (Meiose). In: *Allgemeine Biologie* (ed.: M. Hartmann), 4 Aufl. Stuttgart: Gustav Fischer.

BAUER, H. (1967). Die kinetische Organisation der Lepidopterenchromosomen. *Chromosoma*, **22**: 101–25.

BAUER, H. and BEERMANN, W. (1952a). Die Polytänie der Riesenchromosomen. *Chromosoma*, 4: 630–48.

BAUER, H. and BEERMANN, W. (1952b). Chromosomale Soma-Keimbahn Differenzierung bei Chironomiden. *Naturwiss.* 39: 22–3.

BAUER, H. and BEERMANN, W. (1952c). Der Chromosomenzyklus der Orthocladiinen (Nematocera: Diptera). *Z. Naturf.* 7b: 557–63.

BAUER, H., DIETZ, R. and RÖBBELEN, C. (1961). Die Spermatocytenteilungen der Tipuliden. III. Das Bewegungsverhalten der Chromosomen in Translokationsheterozygoten von *Tipula oleracea*. *Chromosoma*, 12: 116–89.

BAUMGARTNER, W. J. (1929). Die Spermatogenese bei einer Grylle, *Nemobius fasciatus*. *Z. Zellforsch.* 9: 603–39.

BAYREUTHER, K. (1952). Extra-chromosomale Feulgenpositive Körper (Nukleinkörper) in der Oogenese der Tipuliden. *Naturwiss.* 39: 71.

BAYREUTHER, K. (1954). Die Chromosomen der Flöhe (Aphaniptera). *Naturwiss.* 41: 309.

BAYREUTHER, K. (1955). Holokinetische Chromosomen bei *Haematopinus suis* (Anoplura, Haematopinidae). *Chromosoma*, 7: 260–70.

BAYREUTHER, K. (1956). Die Oogenese der Tipuliden. *Chromosoma*, 7: 508–57.

BAYREUTHER, K. (1957). Extrachromosomales DNS-haltiges Material in der Oogenese der Flöhe. *Z. Naturf.* 12b.: 458–61.

BAYREUTHER, K. (1969). Die Cytogenetik zweier norddeutscher Populationen von *Nosopsyllus fasciatus* Bosc. (Aphaniptera). *Chromosoma*, 27: 20–46.

BAZYKIN, A. D. (1969). Hypothetical mechanism of speciation. *Evolution*, 23: 685–7.

BEATTY, R. A. (1957). *Parthenogenesis and Polyploidy in Mammalian Development*. Cambridge Univ. Press.

BEÇAK, M. L., BEÇAK, W. and RABELLO, M. N. (1966). Cytological evidence of constant tetraploidy in the bisexual South American frog, *Odontophrynus americanus*. *Chromosoma*, 19: 188–93.

BEÇAK, M. L., BEÇAK, W. and VIZOTTO, L. D. (1970). A diploid population of the polyploid amphibian *Odontophrynus americanus* and an artificial intraspecific triploid hybrid. *Experientia*, 26: 545–6.

BEÇAK, M. L., DENARO, L. and BEÇAK, W. (1970). Polyploidy and mechanisms of karyotypic diversification in Amphibia. *Cytogenetics*, 9: 225–38.

BEÇAK, W. (1967). Karyotypes, sex chromosomes and chromosomal evolution in snakes. *Venomous Animals and their Venoms*, 1: 53–95. New York: Academic Press.

BEÇAK, W. and BEÇAK, M. K. (1969). Cytotaxonomy and chromosomal evolution in Serpentes. *Cytogenetics*, 8: 247–62.

BEÇAK, W., BEÇAK, M. L., LAVALLE, D. and SCHREIBER, G. (1967). Further studies on polyploid amphibians (Ceratophrydidae). II. DNA content and nuclear volume. *Chromosoma*, 23: 14–23.

BEÇAK, W., BEÇAK, M. L. and NAZARETH, H. R. S. (1962). Karyotypic studies of two species of South American snakes (*Boa constrictor amarali* and *Bothrops jararaca*). *Cytogenetics*, 1: 305–313.

BEÇAK, W., BEÇAK, M. L. and NAZARETH, H. R. S. (1963). Karyotypic studies of South American snakes. *Proc. XIth Int. Genet. Congr.* 1: 136.

BEÇAK, W., BEÇAK, M. L., NAZARETH, H. R. S. and OHNO, S. (1964). Close karyological kinship between the reptilian suborder Serpentes and the class Aves. *Chromosoma*, 15: 606–17.

BEÇAK, W., BEÇAK, M. L. and OHNO, S. (1966). Intraindividual chromosomal polymorphism in green sunfish (*Lepomys cyanellus*) as evidence of somatic segregation. *Cytogenetics*, **5**: 313–20.

BEÇAK, W., BEÇAK, M. L., SCHREIBER, G., LAVALLE, D., and AMORIM, F. O. (1970). Interspecific variability of DNA content in Amphibia. *Experientia*, **26**: 204–6.

BECKER, H. J. (1959). Die Puffs der Speicheldrüsenchromosomen von *Drosophila melanogaster*. I. Beobachtungen zum Verhalten des Puffmusters im Normalstamm und bei zwei Mutanten, *Giant* und *Lethal-Giant-Larvae*. *Chromosoma*, **10**: 654–78.

BEDICHEK-PIPKIN, S. (1940*a*). Segregation and crossing-over in a 2, 3 translocation in *Drosophila melanogaster*. *Univ. Texas Publ.* **4032**: 73–125.

BEDICHEK-PIPKIN, S. (1940*b*). Multiple sex genes in the *X*-chromosome of *Drosophila melanogaster*. *Univ. Texas Publ.* **4032**: 126–56.

BEERMANN, S. (1959). Chromatin-Diminution bei Copepoden. *Chromosoma*, **10**: 504–14.

BEERMANN, W. (1952*a*). Chromomerenkonstanz und spezifische Modifikation der Chromosomenstruktur in der Entwicklung und Organdifferenzierung von *Chironomus tentans*. *Chromosoma*, **5**: 139–98.

BEERMANN, W. (1952*b*). Chromosomenstruktur und Zelldifferenzierung in der Speicheldrüse von *Trichocladius vitripennis*. *Z. Naturf.* **7b**: 237–42.

BEERMANN, W. (1954). Weibliche Heterogametie bei Copepoden. *Chromosoma*, **6**: 381–96.

BEERMANN, W. (1955). Geschlechtsbestimmung und Evolution der genetischen Y-Chromosomen bei *Chironomus*. *Biol. Zentralbl.* **74**: 525–44.

BEERMANN, W. (1956*a*). Nuclear differentiation and functional morphology of chromosomes. *Cold. Spr. Harb. Symp. Quant. Biol.* **21**: 217–32.

BEERMANN, W. (1956*b*). Inversionheterozygotie und Fertilität der Mannchen von *Chironomus*. *Chromosoma*, **8**: 1–11.

BEERMANN, W. (1961). Ein Balbiani-Ring als Locus einer Speicheldrüsenmutation. *Chromosoma*, **12**: 1–25.

BEERMANN, W. (1962). Riesenchromosomen. *Protoplasmatologia*, **6**: 1–161.

BEERMANN, W. (1967). Gene action at the level of the chromosome. In: *Heritage from Mendel* (ed.: R. A. Brink), pp. 179–201. Univ. Wisconsin Press.

BEERMANN, W. and BAHR, G. F. (1954). The submicroscopic structure of the Balbiani ring. *Exp. Cell. Res.* **6**: 195–201.

BEERMANN, W., HESS, O. and MEYER, G. F. (1963). Structure and function of the Y-heterochromatin in *Drosophila*. *Proc. 16th Int. Congr. Zool. Washington, D.C.* **4**: 283–8.

BEERMANN, W. and PEELING, C. (1965). H³-Thymidin-Markierung einzelner Chromatiden in Riesenchromosomen. *Chromosoma*, **16**: 1–21.

BEESON, G. E. (1960). Chromosome numbers of slugs. *Nature*, **186**: 257–8.

BEI-BIENKO, G. YA. (1954). Orthoptera, Tettigoniidae. Sub-family Phaneropterinae. *Fauna of the U.S.S.R.* **2**, part 2, pp. 301–4.

BĚLAŘ, K. (1923*a*). Über den Chromosomenzyklus von parthenogenetischen Erdnematoden. *Biol. Zentralbl.* **43**: 513–18.

BĚLAŘ, K. (1923*b*). Untersuchungen an *Actinophrys sol* Ehrenberg. I. Die Morphologie des Formwechsels. *Arch. Protistenk*, **46**: 1–96.

BĚLAŘ, K. (1924). Die Cytologie der Merospermie bei freilebenden *Rhabditis*-Arten. *Z. Zell. Gewebelehre*, **1**: 1–21.

BĚLAŘ, K. (1926). Der Formwechsel der Protistenkerne. *Ergebn. Zool.* **6**: 235–652.

794 BIBLIOGRAPHY

BĚLAŘ, K. (1929). Beiträge zur Kausalanalyse der Mitose. Untersuchungen an den Spermatocyten von *Chorthippus* (*Stenobothrus*) *lineatus* Panz. *Roux' Archiv Entwichlungsm.* **118**: 359–484.

BELLING, J. (1928). The ultimate chrommeres of *Lilium* and *Aloe* with regard to the number of genes. *Univ. Calif. Publ. Bot.* **16**: 153–70.

BELLING, J. (1931). Chiasmas in flowering plants. *Univ. Calif. Publ. Bot.* **16**: 311–38.

BELLING, J. (1933). Crossing-over and gene rearrangement in flowering plants. *Genetics*, **18**: 388–413.

BENAZZI, M. (1947). *Problemi biologici della sessualità.* Bologna: Cappelli.

BENAZZI, M. (1948a). Eredità materna nella planaria *Polycelis nigra. Mem. Soc. Tosc. Sci. Nat.* B, **55**: 6 pp.

BENAZZI, M. (1948b). Problemi genetici ed ormonici in rapporto alla ibridazione. *La Ricerca Scientifica,* **18**: 15 pp.

BENAZZI, M. (1954a). Un decennio di ricerche cariologiche sulle planarie. *La Ricerca Scientifica,* **24**: 1012–15.

BENAZZI, M. (1954b). La ginogenesi naturale (pseudogamia). *Problemi di Sviluppo,* pp. 38–64. Milano: Casa Editrice Ambrosiana.

BENAZZI, M. (1955). L'evoluzione della poliploidia nelle planarie appartenenti alla superspecie *Dugesia gonocephala. Rend. Accad. Naz. Lincei,* Ser. VIII, **18**: 527–33.

BENAZZI, M. (1957a). Introduzione alla analisi genetica dei biotipi cariologici di tricladi. *La Ricerca Scientifica,* **27**, suppl. pp. 9–24.

BENAZZI, M. (1957b). Cariologia di *Dugesia lugubris* (O. Schmidt) (Tricladida Paludicola). *Caryologia,* **10**: 276–303.

BENAZZI, M. (1959). I meccanismi riproduttivi nelle planarie; *Symp. Genet. Biol. Ital.* (*Atti Simp. Internaz. Biol. Sper.*) **9**: 16 pp.

BENAZZI, M. (1960). Evoluzione cromosomica e differenziamento razziale e specifico nei tricladi. *Accad. Naz. Lincei,* Quaderno **47**, *Evoluzione e Genetica,* pp. 273–297.

BENAZZI, M. (1962). I meccanismi riproduttivi nelle Planarie. *Symp. Genet. Biol. Ital.* **9**: 1–16.

BENAZZI, M. (1963). Il problema sistematico delle *Polycelis* del gruppo *nigratenuis* alla luce di ricerche citologiche e genetiche. *Monit. Zool. Ital.* **70–1**: 288–300.

BENAZZI, M. and BENAZZI-LENTATI, G. (1959a). Corredo cromosomico e meccanismi della ovogenesi e della fecondazione in ibridi fra biotipi diploidi e poliploidi di planarie. *La Ricerca Scientifica,* **29**, suppl. pp. 133–40.

BENAZZI, M. and BENAZZI-LENTATI, G. (1959b). Ricerche su ibridi tra i biotipi diploide e triplo-esaploide della planaria *Dugesia lugubris* (O. Schmidt). *Atti Soc. Tosc. Sci. Nat.* B, **65**: 45–57.

BENAZZI, M. and LEPORI, N. G. (1947). Le gonadi nei maschi dell'ibrido *Triton cristatus* ♀ × *Triton vulgaris* ♂. *Sci. Genet.* **3**: 1–18.

BENAZZI, M. and PULCINELLI, I. (1961). Analisi comparativa del cariogramma dei biotipi di *Dugesia lugubris. Atti Assoc. Genet. Ital.* **6**: 419–26.

BENAZZI, M. and SCALI, V. (1964). Modalità riproduttive della popolazione di *Bacillus rossius* (Rossi) dei dintorni di Pisa. *Rend. Accad. Naz. Lincei,* **36**: 311–14.

BENAZZI-LENTATI, G. (1957). Sul determinismo e sulla ereditarietà della aneuploidia in *Dugesia etrusca* Benazzi, planaria a riproduzione anfigonica. *Caryologia,* **10**: 352–87.

BENAZZI-LENTATI, G. (1960). Precoce separazione dei due cromatidi in cromosomi di *Dugesia lugubris* (O. Schmidt). *Caryologia*, **12**: 482–96.

BENAZZI-LENTATI, G. (1961). Considerazioni sul determinismo dei cicli cromosomici in ibridi di planarie. *Caryologia*, **14**: 271–7.

BENAZZI-LENTATI, G. (1962). Due modalità di sviluppo dello stesso tipo di uova in ibridi interraziali di planarie. *Acta Embryol. Morph. Exp.* **5**: 145–60.

BENAZZI-LENTATI, G. (1966). Amphimixis and pseudogamy in fresh-water triclads: experimental reconstruction of polyploid pseudogamic biotypes. *Chromosoma*, **20**: 1–14.

BENAZZI-LENTATI, G. (1970). Gametogenesis and egg fertilization in Planarians. *Int. Rev. Cytol.* **27**: 101–79.

BENAZZI-LENTATI, G. and BERTINI, V. (1961). Sul determinismo della 'asynapsi femminile' in ibridi di *Dugesia benazzii* (Tricladida, Paludicola). *Atti Soc. Tosc. Sci. Nat.* B, **68**: 83–112.

BENDER, M. A. and CHU, E. H. Y. (1963). The chromosomes of primates (Ch. 7). In: *Evolutionary and Genetic Biology of Primates* (ed.: J. Buettner-Janusch). New York: Academic Press.

BENDER, M. A. and METTLER, L. E. (1958). Chromosome studies of Primates. *Science*, **128**: 186–90.

BENIRSCHKE, K. (1964). Corrigendum. *Chromosoma*, **15**: 300.

BENIRSCHKE, K. (1967). Sterility and fertility of interspecific mammalian hybrids. In: *Comparative Aspects of Reproductive Failure* (ed.: K. Benirschke). New York: Springer-Verlag.

BENIRSCHKE, K. (1968). The chromosome complement and meiosis of the North American porcupine. *J. Hered.* **59**: 71–6.

BENIRSCHKE, K., Low, R. J., BROWNHILL, L. E., CADAY, L. B. and DE VENECIA-FERNANDEZ, J. (1964). Chromosome studies of a donkey-grevy zebra hybrid. *Chromosoma*, **15**: 1–13.

BENIRSCHKE, K., Low, R. J., SULLIVAN, M. M. and CARTER, R. M. (1964). Chromosome study of an alleged fertile mare mule. *J. Hered.* **55**: 31–8.

BENIRSCHKE, K., MALOUF, N., Low, R. J. and HECK, H. (1965). Chromosome complement: differences between *Equus caballus* and *Equus przewalskii* Poliakoff. *Science*, **148**: 387–3.

BENIRSCHKE, K., WURSTER, D. H., Low, R. J. and ATKIN, N. B. (1970). The chromosome complement of the aardvark, *Orycteropus afer*. *Chromosoma*, **31**: 68–78.

BENKERT, J. M. (1930). Chromosome number of the first form reproductive caste of *Reticulitermes flavipes* Kollar. *Proc. Pennsylvania Acad. Sci.* **4**: 97–9.

BENNETT, E. S. (1938). The origin and behavior of chiasmata. XIV. *Fritillaria chitralensis*. *Cytologia*, **8**: 443–51.

BENNETT, F. D. and BROWN, S. W. (1958). Life history and sex determination in the Diaspine scale, *Pseudaulacaspis pentagona* (Targ.) (Coccoidea). *Canad. Entomol.* **90**: 317–25.

BENNETT, J. H. (1958). The existence and stability of selectivity balanced polymorphism at a sex-linked locus. *Aust. J. Biol. Sci.* **11**: 598–602.

BENZER, S. (1955). Fine structure of a genetic region in bacteriophage. *Proc. Nat. Acad. Sci. U.S.A.* **41**: 334–54.

BERENDES, H. D. (1967). The hormone ecdysone as effector for specific changes in the pattern of gene activities of *Drosophila hydei*. *Chromosoma*, **22**: 274–93.

BERENDES, H. D. (1968). Factors involved in the expression of gene activity in polytene chromosomes. *Chromosoma*, **24**: 418–37.

BERENDES, H. D. (1970). Polytene chromosome structure at the submicroscopic level. I. A map of region X, 1-4E of *Drosophila melanogaster*. *Chromosoma*, **29**: 118–30.

BERENDES, H. D. and KEYL, H. G. (1967). Distribution of DNA in heterochromatin and euchromatin of polytene nuclei of *Drosophila hydei*. *Genetics*, **57**: 1–13.

BERENDES, H. D. and MEYER, G. F. (1968). A specific chromosome element, the telomere of *Drosophila* polytene chromosomes. *Chromosoma*, **25**: 184–97.

BERENDES, H. D., VAN BREUGHEL, F. M. A. and HOLT, TH. K. H. (1965). Experimental puffs in salivary gland chromosomes of *Drosophila hydei*. *Chromosoma*, **16**: 35–46.

BERGER, C. A. (1936). Observations on the relation between salivary-gland chromosomes and multiple chromosome complexes. *Proc. Nat. Acad. Sci. U.S.A.* **22**: 186–7.

BERGER, C. A. (1937). Additional evidence of repeated chromosome division without mitotic activity. *Amer. Nat.* **71**: 187–90.

BERGER, C. A. (1938). Multiplication and reduction of somatic chromosome groups as a regular developmental process in the mosquito, *Culex pipiens*. *Publ. Carnegie Instn.* **476**: 209–32.

BERGER, C. A. (1940). The uniformity of the gene complex in the nuclei of different tissues. *J. Hered.* **31**: 1–4.

BERGERARD, J. (1954). Parthénogenèse facultative de *Clitumnus extradentatus*. Br. (Phasmidae). *Bull. Soc. Zool. Fr.* **79**: 169–75.

BERGERARD, J. (1955). Mécanique cytologique de la parthénogénèse thelytoque facultative de *Clitumnus extradentatus* Br. (Phasmidae). *C.R. Acad. Sci. Paris*, **240**: 1143–5.

BERGERARD, J. (1957). Étude de la parthénogenèse facultative de *Clitumnus extradentatus*. *Bull. Biol. Fr. Belg.* **92**: 88–182.

BERGERARD, J. (1958a). La parthénogenèse des phasmides. A. Caractères généraux. *Bull. Biol. Fr. Belg.* **92**: 90–102.

BERGERARD, J. (1958b). La polyploidie somatique chez les animaux. *Année Biologique*, **34**: 119–32.

BERGERARD, J. (1962). Parthenogenesis in the Phasmidae. *Endeavour*, **21**: 137–43.

BERGERARD, J. and SEUGÉ, J. (1959). La parthénogenèse accidentelle chez *Locusta migratoria* L. *Bull. Biol. Fr. Belg.* **93**: 16–37.

BERLOWITZ, L. (1965a). Correlation of genetic activity, heterochromatinization and RNA metabolism. *Proc. Nat. Acad. Sci. U.S.A.* **53**: 68–73.

BERLOWITZ, L. (1965b). Analysis of histone *in situ* in developmentally inactivated chromatin. *Proc. Nat. Acad. Sci. U.S.A.* **54**: 476–80.

BERNSTEIN, N. and GOLDSCHMIDT, E. (1961). Chromosome breakage in structural heterozygotes. *Amer. Nat.* **95**: 53–6.

BERRY, R. O. (1939). Observations on chromosome elimination in the germ cells of *Sciara ocellaris*. *Proc. Nat. Acad. Sci. U.S.A.* **25**: 125–7.

BERRY, R. O. (1941). Chromosome bahavior in the germ cells and development of the gonad in *Sciara ocellaris*. *J. Morph.* **68**: 547–83.

BESSIS, M., BRETON-GORIUS, J. and THIÉRY, J. (1958). Centriole, corps de Golgi et aster des leucocytes. Etude au microscope électronique. *Rev. Hémat.* **13**: 363–86.

BEUTLER, E., YEH, M. and FAIRBANKS, V. F. (1962). The normal human female as a mosaic of X-chromosome activity: Studies using the gene for g-6-pd deficiency as a marker. *Proc. Nat. Acad. Sci. U.S.A.* **48**: 9–16.

BHATNAGAR, A. N. (1957). The spermatogonial chromosomes of the Indian Rat Snake *Ptyas mucosus* Linn. *Curr. Sci.* **26**: 154–5.

BHATNAGAR, A. N. (1958). Spermatogonial chromosomes of two lizards, *Hemidactylus brooki* Grey and *Varanus monitor* Linn. *Curr. Sci.* **27**: 504–5.

BHATNAGAR, A. N. (1959). Studies on the structure and behavior of chromosomes of *Oligodon arnensis* Shaw (Colubridae: Ophidia). *Cytologia*, **24**: 459–65.

BHATNAGAR, A. N. (1960a). Chromosomes of *Bungarus caeruleus* Schneider (Elapidae: Ophidia). *Cytologia*, **25**: 173–8.

BHATNAGAR, A. N. (1960b). Studies on the structure and behavior of chromosomes of two species of Colubrid snakes. *Caryologia*, **12**: 349–61.

BHATTACHARJEE, T. K. and MANNA, G. K. (1967). A cytological survey of Grylloidea. *Proc. 54th Ind. Sci. Congr.* **3**: 420.

BIANCHI, A. P. (1966). Note sulla cariologia di alcuni Efippigeridi (Insecta-Orthoptera). *Rend. Accad. Naz. Lincei*, Ser. VIII, **41**: 553–7.

BIANCHI, N. O., BEÇAK, W., DE BIANCHI, M. S. A., BEÇAK, M. L. and RABELLO, M. N. (1969). Chromosome replication in four species of snakes. *Chromosoma*, **26**: 188–200.

BIANCHI, N. O. and CONTRERAS, J. R. (1967). The chromosomes of the field mouse *Akodon azarae* (Cricetidae, Rodentia) with special reference to sex chromosome anomalies. *Cytogenetics*, **6**: 306–13.

BIANCHI, N. O. and MOLINA, J. O. (1967). DNA replication patterns in somatic chromosomes of *Leptodactylus ocellatus* (Amphibia, Anura). *Chromosoma*, **22**: 391–400.

BIANCHI, N. O., PAULETE-VANDRELL, J. and DE VIDAL RIOJA, L. A. (1969). Complement with 38 chromosomes in two South American populations of *Rattus rattus*. *Experientia*, **25**: 1111–2.

BICK, Y. A. E. and JACKSON, W. D. (1967a). A mammalian *XO* sex-chromosome system in the monotreme *Tachyglossus aculeatus* determined from leucocyte cultures and testicular preparations. *Amer. Nat.* **101**: 79–86.

BICK, Y. A. E. and JACKSON, W. D. (1967b). Karyotype of the monotremes *Ornithorhynchus anatinus* (Platypus) and *Tachyglossus aculeatus* (Echidna). *Nature*, **214**: 600–1.

BICK, Y. A. E. and JACKSON, W. D. (1967c). DNA content of monotremes. *Nature*, **215**: 192–3.

BIER, K. (1957). Endomitose und Polytänie in den Nährzellkernen von *Calliphora erythrocephala* Meigen. *Chromosoma*, **8**: 493–522.

BIER, K. (1958). Beziehung zwischen Wachstumsgeschwindigkeit, endometaphasischer Kontraktion und der Bildung von Riesenchromosomen in den Nährzellen von *Calliphora*. *Z. Naturf.* **13b**: 85–93.

BIER, K. (1959). Quantitative Untersuchungen über die Variabilität der Nährzellkernstruktur und ihre Beeinflussung durch die Temperatur. *Chromosoma*, **10**: 619–53.

BIER, K. (1960a). Über den Wechsel von polytäner und retikulärer Kernstruktur in den Nährzellen einiger Kalyptraten. *Verh. Dtsch. Zool. Gesell.* Bonn/Rhein: 298–304.

BIER, K. (1960b). Der karyotyp von *Calliphora erythrocephala* Meigen unter besonderer Berücksichtigung der Nährzellkern-chromosomen in gebündelten und gepaarten Zustand. *Chromosoma*, **11**: 335–64.

BIER, K., KUNZ, W. and RIBBERT, D. (1967). Struktur und Funktion der Oocytenchromosomen und Nukleolen sowie der Extra-DNS während der Oogenese panoistischer und meroistischer Insekten. *Chromosoma*, **23**: 214–54.

BIER, K., KUNZ, W. and RIBBERT, D. (1969). Insect oogenesis with and without Lampbrush chromosomes. In: *Chromosomes Today*, **2**, pp. 107–15. Edinburgh: Oliver and Boyd.

BIRNSTIEL, M. L., WALLACE, H., SIRLIN, J. L. and FISCHBERG, M. (1966). Investigations on the ribosomal cistrons associated with the nucleolar organizer of *Xenopus laevis. Nat. Cancer Inst. Monograph*, **23**: 431–48.

BLACKITH, R. E. and BLACKITH, R. M. (1966). The food of morabine grasshoppers. *Aust. J. Zool.* **14**: 877–94.

BLACKITH, R. E. and BLACKITH, R. M. (1969). Variation of shape and of discrete anatomical characters in the morabine grasshoppers. *Aust. J. Zool.* **17**: 697–718.

BLACKWOOD, M. (1956). The inheritance of *B*-chromosomes in *Zea mays. Heredity*, **10**: 353–66.

BLAIR, A. P. (1941). Variation, isolating mechanisms and hybridization in certain toads. *Genetics*, **26**: 398–417.

BLAIR, W. F. (1951). Interbreeding of natural populations of vertebrates. *Amer. Nat.* **85**: 9–30.

BLAYLOCK, B. G. (1965). Chromosomal aberrations in a natural population of *Chironomus tentans* exposed to chronic low-level radiation. *Evolution*, **19**: 421–9.

BLAYLOCK, B. G. (1966). Chromosomal polymorphism in irradiated natural populations of *Chironomus. Genetics*, **53**: 131–6.

BLAYLOCK, B. G. and KOEHLER, P. G. (1969). Terminal chromosome rearrangements in *Chironomus riparius. Amer. Nat.* **103**: 547–51.

BOCH, D. P. and BRACK, S. D. (1964). Evidence for the cytoplasmic synthesis of nuclear histone during spermiogenesis in the grasshopper *Chortophaga viridifasciata* (De Geer). *J. Cell Biol.* **22**: 327–40.

BLOCH, D. P. and GODMAN, G. C. (1955). A microphotometric study of the syntheses of desoxyribonucleic acid and nuclear histone. *J. Biophys. Biochem. Cytol.* **1**: 17–28.

BLOCH, D. P. and HEW, H. Y. C. (1960). Changes in nuclear histones during fertilization and early embryonic development in the pulmonate snail, *Helix aspersa. J. Biophys. Biochem. Cytol.* **8**: 69–81.

BLOCK, K. (1969). Chromosome variation in the Agromyzidae. II. *Phytomyza crassiseta* Zetterstedt – a parthenogenetic species. *Hereditas*, **62**: 357–81.

BODMER, W. F. and PARSONS, P. A. (1962). Linkage and recombination in evolution. *Adv. in Genet.* **11**: 1–100.

BOGART, J. P. (1967). Chromosomes of the South American amphibian family Ceratophridae, with a reconsideration of the taxonomic status of *Odontophrynus americanus. Canad. J. Genet. Cytol.* **9**: 531–42.

BOGART, J. P. (1968). Chromosome number difference in the Amphibian genus *Bufo*: the *Bufo regularis* species group. *Evolution*, **22**: 42–5.

BOLE-GOWDA, B. N. (1950). The chromosome study in the spermatogenesis of two lynx spiders (Oxyopidae). *Proc. Zool. Soc. Bengal* **3**: 95–107.

BOLE-GOWDA, B. N. (1952). Studies on the chromosomes and sex determining mechanism in four hunting spiders (Sparassidae). *Proc. Zool. Soc. Bengal*, **5**: 51–70.

BOLE-GOWDA, B. N. (1958). A study of the chromosomes during meiosis in twenty-two species of Indian spiders. *Proc. Zool. Soc. Bengal*, **11**: 69–108.

BONNET, P. (1934). Le gynandromorphisme chez les araignées. *Bull. Biol. Fr. Belg.* **68**: 167–87.

BONNEVIE, K. (1902). Über Chromatindiminution bei Nematoden. *Jena Z. Natur-wiss.* **36**: 275–88.

BÖÖK, J. A. (1940). Triploidy in *Triton taeniatus* Laur. *Hereditas*, **26**: 107–14.

BÖÖK, J. A. (1945). Cytological studies in *Triton*. *Hereditas*, **31**: 177–220.

BÖÖK, J. A., FRACCARO, M. and LINDSTEN, J. (1959). Cytogenetical observations in mongolism. *Acta paediat.* **48**: 453–68.

BÖÖK, J. A. and KJESSLER, B. (1964). Meiosis in the human male. *Cytogenetics*, **3**: 143–7.

BOOTHROYD, E. R. (1959). Chromosome studies on three Canadian populations of Altantic salmon, *Salmo salar* L. *Canad. J. Genet. Cytol.* **1**: 161–72.

BORGAONKAR, D. S. (1965). A list of chromosome numbers in the order Insectivora. *J. Hered.* **56**: 266, 291.

BARGAONKAR, D. S. (1966*a*). Addendum-chromosome numbers in the order Insectivora. *J. Hered.* **57**: 28.

BORGAONKAR, D. S. (1966*b*). A list of chromosome numbers in Primates. *J. Hered.* **57**: 60–4.

BORGAONKAR, D. S. (1967). Additions to the lists of chromosome numbers in the Insectivores and Primates. *J. Hered.* **58**: 211–13.

BORGAONKAR, D. S. and GOULD, E. (1968). Homozygous reciprocal translocation as a mode of speciation in *Microgale* Thomas 1883 (Tenrecidae – Insectivora). *Experientia*, **24**: 506–9.

BORING, A. M. (1913). The odd chromosome in *Cerastipsocus venosus*. *Biol. Bull. Woods Hole*, **24**: 125–32.

BORRAGÁN, P. J. and CALLAN, H. G. (1952). Chiasma formation in spermato-cytes and oocytes of the turbellarian *Dendrocoelum lacteum*. *J. Genet.* **50**: 449–54.

BOSEMARK, N. O. (1954). On accessory chromosomes in *Festuca pratensis*. I. Cyto-logical investigations. *Hereditas*, **40**: 346–76.

BOSEMARK, N. O. (1956). On accessory chromosomes in *Festuca pratensis*. IV. Cytology and inheritance of small and large accessory chromosomes. *Hereditas*, **42**: 235–60.

BOSEMARK, N. O. (1957). On accessory chromosomes in *Festuca pratensis*. V. Influence of accessory chromosomes on fertility and vegetative development. *Hereditas*, **43**: 211–35.

BOSTIAN, C. H. (1939). Multiple alleles and sex determination in *Habrobracon*. *Genetics*, **24**: 770–7.

BOURNIER, A. (1956*a*). Contribution à l'étude de la parthénogenèse des Thysanop-tères et de sa cytologie. *Arch. Zool. Exp. Gén.* **93**: 219–317.

BOURNIER, A. (1956*b*). Un nouveau cas de parthénogenèse arrhénotoque: *Liothrips oleae* Costa (Thysanoptera, Tubulifera). *Arch. Zool. Exp. Gen.* **93**, Notes et Revue: 135–141.

BOVERI, TH. (1887). Über Differenzierung der Zellkerne während der Fürchung des Eies von *Ascaris megalocephala*. *Anat. Anz.* **2**: 688–93.

BOVERI, TH. (1888). *Zellenstudien*, **2**. Jena: Fischer.

BOVERI, TH. (1892). Über die Entstehung des Gegensatzes zwischen den Gesch-lechtszellen und den somatischen Zellen bei *Ascaris meg*. *S.B. Ges. Morph. Physiol. Münch.* **8**: 114–25.

BOVERI, TH. (1909). Die Blastomerenkerne von *Ascaris megalocephala* und die Theorie der Chromosomenindividualität. *Arch. Zellforsch.* **3**: 181–268.

BOVERI, TH. (1910). Die Potenzen der *Ascaris*-Blastomeren bei abgeänderter Fürchung. *Festschr. f. R. Hertwig*, **3**: 131–214.

BOVERI, TH. (1911). Über das Verhalten der Geschlechtschromosomen bei Herma-
phroditismus. Beobachtungen an *Rhabditis nigrovenosa*. *Verh. phys.-med.
Ges. Würzb.* **41**: 83–97.

BOVEY, R. (1948). Un type nouveau d'hétérochromosomes chez un mammifère:
le trivalent sexuel de *Sorex araneus* L. *Arch. J. Klaus Stift. Vererbungsforsch,*
23: 507–10.

BOVEY, R. (1949). Les chromosomes des Chiroptères et des Insectivores. *Rev.
Suisse Zool.* **56**: 371–460.

BOWEN, R. H. (1922). Studies on insect spermatogenesis. IV. The phenomenon of
polymegaly in the sperm cells of the family Pentatomidae. *Proc. Amer. Acad.
Arts Sci.* **57**: 391–423.

BOWEN, S. T. (1963). The genetics of *Artemia salina*. II. *White*, a sex-linked mutation. *Biol. Bull. Woods Hole*, **124**: 17–23.

BOWEN, S. T. (1964). The genetics of *Artemia salina*. IV. Hybridization of wild
populations with mutant stocks. *Biol. Bull. Woods Hole*, **126**: 333–44.

BOWEN, S. T. (1965). Genetics of *Artemia salina* V. Crossing-over between the *X*
and *Y* chromosomes. *Genetics*, **52**: 695–710.

BOYES, J. W. (1952). A multiple sex-chromosome mechanism in a root maggot,
Hylemyia fugax. J. Hered. **43**: 195–9.

BOYES, J. W. (1953). Somatic chromosomes of higher Diptera. II. Differentiation
of Sarcophagid species. *Canad. J. Zool.* **31**: 561–576.

BOYES, J. W. (1954). Somatic chromosomes of higher Diptera. III. Interspecific
and intraspecific variation in the genus *Hylemya*. *Canad. J. Zool.* **32**: 39–63.

BOYES, J. W. (1958). Chromosomes in classification of Diptera. *Proc. Xth Int.
Congr. Entomol. Montreal*, **2**: 899–906.

BOYES, J. W. (1963). Somatic chromosomes of higher Diptera. VII. Sarcophagid
species in relation to their taxonomy. *Canad. J. Zool.* **41**: 1191–204.

BOYES, J. W. (1964). Somatic chromosomes of higher Diptera. VIII. Karyotypes
of species of Oestridae, Hypodermitidae, and Cuterebridae. *Canad. J. Zool.*
42: 599–604.

BOYES, J. W., COREY, M. J. and PATERSON, H. E. (1964). Somatic chromosomes of
higher Diptera. IX. Karyotypes of some Muscid species. *Canad. J. Zool.* **42**:
1025–36.

BOYES, J. W. and NAYLOR, A. F. (1962). Somatic chromosomes of higher Diptera.
VI. Allosome-autosome length of relations in *Musca domestica* L. *Canad. J.
Zool.* **40**: 777–84.

BOYES, J. W. and VAN BRINK, J. (1964). Chromosomes of Syrphidae. I. Variations
in karyotype. *Chromosoma* **15**: 579–590.

BOYES, J. W. and VAN BRINK, J. M. (1967). Chromosomes of Syrphidae. III. Karyotypes of some species in the tribes Milesiini and Myoleptini. *Chromosoma*, **22**:
417–55.

BOYES, J. W. and WILKES, A. (1953). Somatic chromosomes of higher Diptera.
I. Differentiation of Tachinid parasites. *Canad. J. Zool.* **31**: 125–65.

BRACHET, J. (1940). La localization de l'acide thymonucléique pendant l'oogenèse
et la maturation chez les amphibiens. *Arch. Biol.* **51**: 151–65.

BRANDHAM, P. E. (1969). Inversion heterozygosity and sub-chromatid exchange in
Agave stricta. Chromosoma, **26**: 270–86

BRAUER, A. (1893). Zur Kenntniss der Spermatogenese von *Ascaris. Arch. mikr.
Anat.* **42**: 153–213.

BRAUER, A. (1894). Zur Kenntniss der Reifung des parthenogenetisch sich entwickelnden Eies von *Artemia salina. Arch. mikr. Anat.* **43**: 162–222.

BREIDER, H. (1942). ZW-Männchen und WW-Weibchen bei *Platypoecilus maculatus*. *Biol. Zentralbl.* **62**: 187–95.

BRELAND, O. P. (1961). Studies on the chromosomes of mosquitoes. *Ann. Ent. Soc. Amer.* **54**: 360–75.

BREMNER, K. C. (1959). Observations on the biology of *Haemaphysalis bispinosa* Neumann (Acarina: Ixodidae) with particular reference to its mode of reproduction by parthenogenesis. *Aust. J. Zool.* **7**: 7–12.

BREUER, M. and PAVAN, C. (1955). Behaviour of polytene chromosomes of *Rhynchosciara* at different stages of larval development. *Chromosoma*, **7**: 371–86.

BREWEN, J. G. and BROCK, R. D. (1968). The exchange hypothesis and chromosome-type aberrations. *Mutation Res.* **6**: 245–55.

BREWEN, J. G. and PEACOCK, W. J. (1969a). Restricted rejoining of chromosomal subunits in aberration formation: a test for subunit dissimilarity. *Proc. Nat. Acad. Sci. U.S.A.* **62**: 389–94.

BREWEN, J. G. and PEACOCK, W. J. (1969b). The effect of tritiated thymidine on sister chromatid exchange in a ring chromosome. *Mutation Res.* **7**: 433–40.

BRIDGES, C. B. (1925a). Sex in relation to chromosomes and genes. *Amer. Nat.* **59**: 127–37.

BRIDGES, C. B. (1925b). Haploidy in *Drosophila melanogaster*. *Proc. Nat. Acad. Sci. U.S.A.* **11**: 706–10.

BRIDGES, C. B. (1930). Haploid *Drosophila* and the theory of genic balance. *Science*, **72**: 405–6.

BRIDGES, C. B. (1932). The genetics of sex in *Drosophila*. In *Sex and Internal Secretions*, Ch. 3, pp. 55–93. London: Baillère, Tindall and Cox.

BRIDGES, C. B. (1935a). The structure of salivary chromosomes and the relation of the banding to the genes. *Amer. Nat.* **69**: 59.

BRIDGES, C. B. (1935b). Salivary chromosome maps with a key to the banding pattern of the chromosomes of *Drosophila melanogaster*. *J. Hered.* **26**: 60–4.

BRIDGES, C. B. (1936). The Bar 'gene' a duplication. *Science*, **83**: 210–11.

BRIDGES, C. B. (1938). A revised map of the salivary gland *X*-chromosome of *Drosophila melanogaster*. *J. Hered.* **29**: 11–13.

BRIDGES, C. B. and P. N. (1939). A revised map of the right limb of the second chromosome of *Drosophila melanogaster*. *J. Hered.* **30**: 475–6.

BRIDGES, P. N. (1941a). A revised map of the left limb of the third chromosome of *Drosophila melangaster*. *J. Hered.* **32**: 64–5.

BRIDGES, P. N. (1941b). A revision of the salivary gland IIIR chromosome map of *Drosophila melanogaster*. *J. Hered.* **32**: 299–300.

BRIDGES, P. N. (1942). A new map of the salivary gland II L chromosome of *Drosophila melanogaster*. *J. Hered.* **33**: 403–8.

BRIEGER, F. G. and GRANER, E. A. (1943). On the cytology of *Tityus bahiensis* with special reference to meiotic prophase. *Genetics*, **28**: 269–74.

BRIEGER, F. G. and KERR, W. E. (1949). Sobre o comportamento mitotico e meiotico de cromosomas policentricos ou com ponto de inserção difuso. *Ann. Esc. Sup. Agric. L. de Queiroz*, **6**: 179–92.

BRINK, R. A. and COOPER, D. C. (1935). A proof that crossing-over involves an exchange of segments between homologous chromosomes. *Genetics*, **20**: 22–35.

BRINKLEY, B. R. and HUMPHREY, R. M. (1969). Evidence for subchromatid organization in marsupial chromosomes. I. Light and electron microscopy of X-ray-induced sidearm bridges. *J. Cell Biol.* **42**: 827–30.

BRITT, H. G. (1947). Chromosomes of digenetic Trematodes. *Amer. Nat.* **81**: 276–96.

BRITTEN, R. J. and KOHNE, D. E. (1968). Repeated sequences in DNA. *Science*, **161**: 529–40.

BRNCIC, J. D. (1953). Chromosomal variation in natural populations of *Drosophila guaramunu*. *Z. indukt. Abst. Vererbl.* **85**: 1–11.

BRNCIC, D. (1954). Heterosis and integration of the genotype in geographic populations of *Drosophila pseudoobscura*. *Genetics*, **39**: 77–88.

BRNCIC, D. (1955). Chromosomal variation in Chilean populations of *Drosophila immigrans*. *J. Hered.* **46**: 59–63.

BRNCIC, D. (1957). Chromosomal polymorphism in natural populations of *Drosophila pavani*. *Chromosoma*, **8**: 699–708.

BRNCIC, D. (1961a). Non-random association of inversions in *Drosophila pavani*. *Genetics*, **46**: 401–6.

BRNCIC, D. (1961b). Integration of the genotype in geographic populations of *Drosophila pavani*. *Evolution*, **15**: 92–7.

BRNCIC, D. (1970). Studies on the evolutionary biology of Chilean species of *Drosophila*. *Evol. Biol.* Suppl. Vol.: 401–36.

BRNCIC, D., NAIR, P. S. and WHEELER, M. R. (1971). Cytotaxonomic relationships within the *mesophragmatica* species group of *Drosophila*. *Studies in Genet.* VI. *Univ. Texas Publ.* **7103**: 1–16.

BRNCIC, D. and SOLAR, E. D. (1961). Life cycles and the expression of heterosis in inversion heterozygotes in *Drosophila funebris* and *Drosophila pavani*. *Amer. Nat.* **95**: 211–16.

BROOKS, J. L. (1950). Speciation in ancient lakes. *Quart. Rev. Biol.* **25**: 30–176.

BROWN, D. D. and DAWID, I. B. (1968). Specific gene amplification in oocytes. *Science*, **160**: 272–80.

BROWN, D. D. and GURDON, J. B. (1964). Absence of ribosomal RNA synthesis in the anucleolate mutant of *Xenopus laevis*. *Proc. Nat. Acad. Sci. U.S.A.* **51**: 139–47.

BROWN, D. D. and WEBER, C. S. (1968a). Gene linkage by RNA–DNA hybridization. I. Unique DNA sequences homologous to 4s RNA, 5s RNA and ribosomal RNA. *J. Mol. Biol.* **34**: 661–80.

BROWN, D. D. and WEBER, C. S. (1968b). Gene linkage by RNA–DNA hybridization. II. Arrangement of the redundant gene sequences for 28s and 18s ribosomal RNA. *J. Mol. Biol.* **34**: 681–97.

BROWN, S. W. (1957). Chromosome behavior in *Comstockiella sabalis* (Comstk.) (Coccoidea–Diaspididae). *Rec. Genet. Soc. Amer.* **26**: 362–3.

BROWN, S. W. (1958). The chromosomes of an *Orthezia* species (Coccoidea–Homoptera). *Cytologia*, **23**: 429–34.

BROWN, S. W. (1959). Lecanoid chromosome behavior in three more families of the Coccoidea (Homoptera). *Chromosoma*, **10**: 278–300.

BROWN, S. W. (1960a). Chromosome aberration in two aspidiotine species of the armored scale insects (Coccoidea–Diaspididae). *Nucleus*, **3**: 135–60.

BROWN, S. W. (1960b). Spontaneous chromosome fragmentation in the armored scale insects (Coccoidea–Diaspididae). *J. Morph.* **106**: 159–86.

BROWN, S. W. (1961). Fracture and fusion of coccid chromosomes. *Nature*, **191**: 1419–20.

BROWN, S. W. (1963). The Comstockiella system of chromosome behavior in the armored scale insects (Coccoidea: Diaspididae). *Chromosoma*, **14**: 360–406.

BROWN, S. W. (1964). Automatic frequency response in the evolution of male haploidy and other coccid chromosome systems. *Genetics*, **49**: 797–817.

BROWN, S. W. (1965). Chromosomal survey of the armored and palm scale insects (Coccoïdea: Diaspididae and Phoenicococcidae). *Hilgardia*, **36**: 189–294.

BROWN, S. W. (1966). Heterochromatin. *Science*, **151**: 417–25.

BROWN, S. W. (1967). Chromosome systems of the Eriococcidae (Coccoidea–Homoptera). I. A survey of several genera. *Chromosoma*, **22**: 126–50.

BROWN, S. W. and BENNETT, F. D. (1957). On sex determination in the diaspine scale insect *Pseudaulacaspis pentagona* (Targ.) (Coccoidea). *Genetics*, **46**: 510–23.

BROWN, S. W. and CLEVELAND, C. (1968). Meiosis in the male of *Puto albicans* (Coccoidea–Homoptera). *Chromosoma*, **24**: 210–32.

BROWN, S. W. and DE LOTTO, G. (1959). Cytology and sex ratios of an African species of armored scale insect (Coccoidea–Diaspididae). *Amer. Nat.* **93**: 369–79.

BROWN, S. W. and McKENZIE, H. L. (1962). Evolutionary patterns in the armored scale insects and their allies (Homoptera: Coccoidea: Diaspididae, Phoenicococcidae and Asterolecaniidae). *Hilgardia*, **33**: 141–70A.

BROWN, S. W. and NELSON-REES, W. A. (1961). Radiation analysis of a Lecanoid genetic system. *Genetics*, **46**, 983–1007.

BROWN, S. W. and NUR, U. (1964). Heterochromatic chromosomes in the coccids *Science*, **145**: 130–6.

BROWN, S. W., WALEN, K. H. and BROSSEAU, G. E. (1962). Somatic crossing-over and elimination of ring *X*-chromosomes of *Drosophila melanogaster*. *Genetics*, **47**: 1573–9.

BROWN, S. W. and ZOHARY, D. (1955). The relationship of chiasmata and crossing-over in *Lilium formosanum. Genetics*, **40**, 850–73.

BRUES, A. M. (1969). Genetic load and its varieties. *Science*, **164**: 1130–6.

BUDER, J. E. (1915). Die Spermatogenese von *Deilephila euphorbiae*. L. *Arch. Zellforsch.* **14**, 26–78.

BULLINI, L. and BIANCHI BULLINI, A. P. (1969). Nuovi dati sul corredo cromosomico degli Efippigeridi italiani (Orthoptera–Tettigonioidea). *Atti Accad. Naz. Lincei*, Ser. VIII. *Rendic. Sci. Fis. Mat. Nat.* **46**: 213, 216.

BULLOUGH, W. S. and LAURENCE, E. B. (1966). The diurnal cycle in epidermal mitotic division and its relation to chalone and adrenalin. *Exp. Cell Res.* **43**: 343–50.

BULTMANN, H. and CLEVER, V. (1969). Chromosomal control of foot pad development in *Sarcophaga bullata*. I. The puffing patterns. *Chromosoma*, **28**: 120–35.

BURCH, J. B. (1960a). Chromosome studies of aquatic pulmonate snails. *Nucleus*, **3**: 177–208.

BURCH, J. B. (1960b). Chromosomes of *Gyraulus circumstratius*, a freshwater snail. *Nature*, **189**: 497–8.

BURCH, J. B. (1961). The chromosomes of *Planorbarius corneus* (Linnaeus), with a discussion on the value of chromosome numbers in snail systematics. *Basteria*, **25**: 45–52.

BURCH, J. B. (1962). Cytotaxonomic studies of freshwater limpets (Gastropoda: Basommatophora). I. The European Lake Limpet *Acroloxus lacustri. Malacologia*, **1**: 55–72.

BURCH, J. B. (1963). Cytological studies of Planorbidae (Gastropoda: Basommatophora) II. The African subgenus *Bulinus* s.s. *Malacologia*, **1**: 387–400.

BURCH, J. B. (1964). Chromosomes of the succineid snail *Catinella rotundata. Occas. Pap. Mus. Zool. Mich.* no. 638: 8 pp.

BURCH, J. B. and HEARD, W. H. (1962). Chromosome numbers of two species of *Vallonia* (Mollusca, Stylommatophora, Orthurethra). *Acta Biol. Acad. Scient. Hungaricae*, **12**: 205–12.

BURISCH, E. (1963). Beiträge zur Genetik von *Megaselia scalaris* Loew (Phoridae). *Z. indukt Abst. Vererbl.* **94**: 322–30.

BUSH, G. L. (1962). The cytotaxonomy of the larvae of some Mexican fruit flies in the genus *Anastrepha* (Tephritidae, Diptera). *Psyche*, **69**: 87–101.

BUSH, G. L. (1966a). The taxonomy, cytology and evolution of the genus *Rhagoletis* in North America (Diptera, Tephritidae). *Bull. Mus. Comp. Zool. Harvard Univ.* **134**: 431–562.

BUSH, G. L. (1966b). Female heterogamety in the family Tephritidae (Acalyptratae, Diptera). *Amer. Nat.* **100**: 119–26.

BUSHNELL, R. J., BUSHNELL, E. P. and PARKER, M. V. (1939). A chromosome study of five members of the family Hylidae. *J. Tenn. Acad. Sci.* **14**: 209–15.

BUTOT, L. J. M. (1967). Phylogenetic position of Heterurethra (Gastropoda: Euthyneura) in the light of cytotaxonomy. *Genen en Phaenen*, **11**: 53–5.

BUXTON, P. A. (1940). *The Louse*. London: Arnold.

BUZZATI-TRAVERSO, A. (1950). Interspecific hybrids in the *obscura* group of *Drosophila*. *Program of the Columbus, Ohio Meeting, Soc. for the Study of Evolution.*

BUZZATI-TRAVERSO, A. and SCOSSIROLI, R. E. (1955). The '*obscura* group' of the genus *Drosophila*. *Adv. in Genet.* **7**: 47–92.

BYTINSKI-SALZ, H. (1934). Verwandtschaftsverhältnisse zwischen den Arten der Gattungen *Celerio* und *Pergesa* nach Untersuchungen über die Zytologie und Fertilität ihrer Bastarde. *Biol. Zentralbl.* **54**: 300–13.

BYTINSKI-SALZ, H. and GÜNTHER, A. (1930). Untersuchungen an Lepidopteren-hybriden. I. Morphologie und Cytologie einiger Bastarde der *Celerio hybr. galliphorbiae*-Gruppe. *Z. indukt. Abst. Vererbl.* **53**: 153–234.

CAIN, A. J. and SHEPPARD, P. M. (1954). The theory of adaptive polymorphism. *Amer. Nat.* **88**: 321–6.

CAIRNS, J. (1963a). The form and duplication of DNA. *Endeavour*, **22**: 141–5.

CAIRNS, J. (1963b). The chromosome of *Escherichia coli*. *Cold Spr. Harb. Symp. Quant. Biol.* **28**: 43–6.

CAIRNS, J. (1963c). The bacterial chromosome and its manner of replication as seen by autoradiography. *J. Mol. Biol.* **6**: 208–13.

CALLAN, H. G. (1941a). A trisomic grasshopper. *J. Hered.* **32**: 296–8.

CALLAN, H. G. (1941b). The sex-determining mechanism of the earwig, *Forficula auricularia*. *J. Genet.* **41**: 349–74.

CALLAN, H. G. (1942). Heterochromatin in *Triton*. *Proc. Roy. Soc. Lond.* B, **130**: 324–35.

CALLAN, H. G. (1949a). Chiasma interference in diploid, tetraploid and interchange spermatocytes of the earwig *Forficula auricularia*. *J. Genet.* **49**: 209–13.

CALLAN, H. G. (1949b). Chromosomes. *New Biol.* **7**: 70–88.

CALLAN, H. G. (1952). A general account of experimental work on amphibian oocyte nuclei. *Symp. Soc. Exp. Biol.* **6**: 243–55.

CALLAN, H. G. (1955). Recent work on the structure of cell nuclei. *Symp. on Fine Structure of Cells. IUBS Publ. Series B*, **21**: 89–109.

CALLAN, H. G. (1957). The lampbrush chromosomes of *Sepia officinalis* L., *Anilocra physodes* L. and *Scyllium catulus* and their structural relationship to the lampbrush chromosomes of Amphibia. *Pubbl. Staz. Zool. Napoli*, **29**: 329–46.

CALLAN, H. G. (1963). The nature of lampbrush chromosomes. In: *International Review of Cytology* (ed.: G. H. Bourne and J. F. Danielli), pp. 1–34. New York: Academic Press.

CALLAN, H. G. (1967). The organization of genetic units in chromosomes. *J. Cell Sci.* **2**: 1–7.

CALLAN, H. G. (1972). Replication of DNA in the chromosomes of eukaryotes. *Proc. Roy. Soc. Lond.* B, **181**: 19–41.

CALLAN, H. G. and JACOBS, P. A. (1957). The meiotic process in *Mantis religiosa* L. males. *J. Genet.* **55**: 200–16.

CALLAN, H. G. and LLOYD, L. (1956). Visual demonstration of allelic differences within cell nuclei. *Nature*, **178**: 355–7.

CALLAN, H. G. and LLOYD, L. (1960a). Lampbrush chromosomes. In: *New Approaches in Cell Biology* (ed: P.M.B. Walker). New York: Academic Press.

CALLAN, H. G. and LLOYD, L. (1960b). Lampbrush chromosomes of crested newts, *Triturus cristatus* (Laurenti). *Phil. Trans. Roy. Soc. Lond.* **243**: 135–219.

CALLAN, H. G. and MACGREGOR, H. C. (1958). Action of deoxyribonuclease on lampbrush chromosomes. *Nature*, **181**: 1479–80.

CALLAN, H. G. and MONTALENTI, G. (1947). Chiasma interference in mosquitoes. *J. Genet.* **48**: 119–34.

CALLAN, H. G. and SPURWAY, H. (1951). A study of meiosis in interracial hybrids of the newt, *Triturus cristatus*. *J. Genet.* **50**: 235–49.

CALLAN, H. G. and TAYLOR, J. H. (1968). A radioautographic study of the time course of male meiosis in the newt *Triturus vulgaris*. *J. Cell Sci.* **3**: 615–26.

CAMENZIND, R. (1962). Untersuchungen über bisexuelle Fortpflanzung einer paedogenetischen Gallmücke. *Rev. Suisse Zool.* **69**: 377–84.

CAMENZIND, R. (1966). Die Zytologie der bisexuellen und parthenogenetischen Fortpflanzung von *Heteropeza pygmaea* Winnertz, einer Gallmücke mit pädogenetischer Vermehrung. *Chromosoma*, **18**: 123–52.

CAMENZIND, R., and NICKLAS, R. B. (1968). The non-random segregation of *Gryllotalpa hexadactyla*. A micromanipulation analysis. *Chromosoma*, **24**: 324–35.

CAMERON, I. L. and PRESCOTT, D. M. (1963). RNA and protein metabolism in the maturation of the nucleated chicken erythrocyte. *Exp. Cell Res.* **30**: 609–12.

CAMPANA-ROUGET, Y. (1959). La teratologie des tiques I. *Ann. Parasit. Hum. Comp.* **34**: 209–60.

CANNON, H. G. (1922). A further account of the spermatogenesis of lice. *Quart. J. Micr. Sci.* **66**: 657–67.

CAPANNA, E. and CIVITELLI, M. V. (1964). Contributo alla conoscenza della cariologia dei Rinolofidi (Mammalia–Chiroptera). *Caryologia*, **17**: 361–71.

CAPANNA, E. and CIVITELLI, M. V. (1969). An endemic population of *Rattus rattus* (L.) with a 38 chromosome complement. *Mammal. Chrom. Newsl.* **10**: 220–2.

CAPANNA, E. and CIVITELLI, M. V. (1970). Chromosomal mechanisms in the evolution of Chiropteran karyotype: chromosomal tables of Chiroptera. *Caryologia*, **23**: 79–111.

CAPANNA, E., CIVITELLI, M. V. and NEZER, R. (1970). The karyotype of the Black Rat (*Rattus rattus* L.): Another population with a 38-chromosomes complement. *Experientia*, **26**: 422–5.

CAPPE DE BAILLON, P., FAVRELLE, M. and DE VICHET, G. (1934a). La parthénogenèse des Phasmes. *C.R. Acad. Sci. Paris*, **199**: 1069–70.

CAPPE DE BAILLON, P., FAVRELLE, M. and DE VICHET, G. (1934*b*). Parthénogenèse et variation chez les Phasmes. I. *Baculum artemis* Westw., *Carausius thieseni* n. sp. *Bull. Biol. Fr. Belg.* **68**: 109–66.

CAPPE DE BAILLON, P., FAVRELLE, M. and DE VICHET, G. (1935). Parthénogenèse et variation chez les Phasmes. II. *Carausius furcillatus* Pant., *Menexenus semiarmatus* Westw. *Bull. Biol. Fr. Belg.* **69**: 1–46.

CAPPE DE BAILLON, P., FAVRELLE M. and DE VICHET, G. (1937). Parthénogenèse et variation chez les Phasmes. III. *Bacillus rossii* Rossi, *Epibacillus lobipes* Luc., *Phobaeticus sinetyi* Br., *Parasosiba parva* Redt., *Carausius rotundato-lobatus* Br. *Bull. Biol. Fr. Belg.* **71**: 129–89.

CAPPE DE BAILLON, P., FAVRELLE, M. and DE VICHET, G. (1938). Parthénogenèse et variation chez les Phasmes. IV. Discussion des faits-conclusions. *Bull. Biol. Fr. Belg.* **72**: 1–47.

CAPPE DE BAILLON, P. and DE VICHET, G. (1935). Le mâle du *Clonopsis gallica* (Orthopt. Phasmidae). *Ann. Soc. Ent. Fr.* **104**: 259–72.

CAPPE DE BAILLON, P. and DE VICHET, G. (1940). La parthénogenèse des espèces du genre *Leptynia* Pant. (Orthopt. Phasmidae). *Bull. Biol. Fr. Belg.* **74**: 43–87.

CARLSON, E. A. (1959). Comparative genetics of complex loci. *Quart. Rev. Biol.* **34**: 33–67.

CARLSON, J. G. (1940). Immediate effects of 250 r of X-rays on the different stages of mitosis in neuroblasts of the grasshopper, *Chortophaga viridifasciata*. *J. Morph.* **66**: 11–23.

CARLSON, J. G. (1952). Microdissection studies of the dividing neuroblast of the grasshopper, *Chortophaga viridifasciata* (De Geer). *Chromosoma*, **5**: 199–220.

CARLSSON, G. (1966). Cytology of the black fly *Prosimulium frohnei* Sommerman. *Hereditas*, **55**: 73–8.

CARMODY, G., DIAZ COLLAZO, A., DOBZHANSKY, TH., EHRMAN, L., JAFFREY, I. S., KIMBALL, S., OBREBSKY, S., SILAGI, S., TIDWELL, T. and ULLRICH, R. (1962). Mating preferences and sexual isolation within and between the incipient species of *Drosophila paulistorum*. *Amer. Midl. Nat.* **68**: 67–82.

CARNOY, J. B. (1884). *La Biologie Cellulaire*. Fasc. 1. Lierre.

CAROTHERS, E. E. (1913). The Mendelian ratio in relation to certain orthopteran chromosomes. *J. Morph.* **24**: 487–511.

CAROTHERS, E. E. (1917). The segregation and recombination of homologous chromosomes as found in two genera of Acrididae (Orthoptera). *J. Morph.* **28**: 445–520.

CAROTHERS, E. E. (1921). Genetical behavior of heteromorphic homologous chromosomes of *Circotettix* (Orthoptera). *J. Morph.* **35**: 457–83.

CAROTHERS, E. E. (1926). The maturation divisions in relation to the segregation of homologous chromosomes. *Quart. Rev. Biol.* **1**: 419–35.

CAROTHERS, E. E. (1931). The maturation divisions and segregation of heteromorphic homologous chromosomes in Acrididae (Orthoptera). *Biol. Bull. Woods Hole*, **61**: 324–49.

CAROTHERS, E. E. (1941*a*). Interspecific grasshopper hybrids (*Trimerotropis citrina* × *T. maritima*); F₁, F₂ and backcrosses. *Proc. VIIth Int. Genet. Congr. Edinb.* p. 84.

CAROTHERS, E. E. (1941*b*). Interspecific hybridization in the Acrididae (*Trimerotropis citrina* × *T. maritima*). *Genetics*, **26**: 144.

CARR, D. H. (1965). Chromosome studies in spontaneous abortions. *Obstet. Gynec.* **26**: 308–26.

CARROLL, M. (1920). An extra dyad and an extra tetrad in the spermatogenesis of *Camnula pellucida* (Orthoptera): numerical variations in the chromosome complex within the individual. *J. Morph.* **34**: 375–455.

CARSON, H. L. (1944). An analysis of natural chromosome variability in *Sciara impatiens* Johannsen. *J. Morph.* **75**: 11–59.

CARSON, H. L. (1946). The selective elimination of inversion dicentric chromatids during meiosis in the eggs of *Sciara impatiens*. *Genetics,* **31**: 95–113.

CARSON, H. L. (1953). The effects of inversions on crossing-over in *Drosophila robusta*. *Genetics,* **38**: 168–86.

CARSON, H. L. (1955*a*). Variation in genetic recombination in natural populations. *J. Cell. Comp. Phys.* **45**, suppl. 2: 221–36.

CARSON, H. L. (1955*b*). The genetic characteristics of marginal populations of *Drosophila. Cold Spr. Harb. Symp. Quant. Biol.* **20**: 276–287.

CARSON, H. L. (1958*a*). The population genetics of *Drosophila robusta*. *Adv. in Genet.* **9**: 1–40.

CARSON, H. L. (1958*b*). Increase in fitness in experimental populations resulting from heterosis. *Proc. Nat. Acad. Sci. U.S.A.* **44**: 1136–41.

CARSON, H. L. (1959). Genetic conditions which promote or retard the formation of species. *Cold Spr. Harb. Symp. Quant. Biol.* **24**: 87–105.

CARSON, H. L. (1962*a*). Fixed heterozygosity in a parthenogenetic species of *Drosophila. Studies in Genet.* II, *Univ. Texas Publ.* **6205**: 55–62.

CARSON, H. L. (1962*b*). Selection for parthenogenesis in *Drosophila mercatorum*. *Rec. Genet. Soc. Amer.* **31**: 77.

CARSON, H. L. (1967*a*). Selection for parthenogenesis in *Drosophila mercatorum*. *Genetics,* **55**: 157–71.

CARSON, H. L. (1967*b*). Chromosomal polymorphism in altitudinal races of *Drosophila. Proc. Jap. Soc. Syst. Zool.* **3**: 10–16.

CARSON, H. L. (1968). The population flush and its genetic consequences. In: *Population Biology and Evolution* (ed.: R. C. Lewontin), pp. 123–37. Syracuse Univ. Press.

CARSON, H. L. (1969). Parallel polymorphisms in different species of Hawaiian *Drosophila. Amer. Nat.* **103**: 323–9.

CARSON, H. L. (1970). Chromosome tracers of the origin of species. *Science,* **168**: 1414–18.

CARSON, H. L. and BLIGHT, W. C. (1952). Sex chromosome polymorphism in a population of *Drosophila americana. Genetics,* **37**: 572.

CARSON, H. L., CLAYTON, F. and STALKER, H. D. (1967). Karyotypic stability and speciation in Hawaiian *Drosophila. Proc. Nat. Acad. Sci. U.S.A.* **57**: 1280–5.

CARSON, H. L., HARDY, D. E., SPIETH, H. T. and STONE, W. S. (1970). The evolutionary biology of the Hawaiian Drosophilidae. In: *Essays in Evolution and Genetics in Honor of Theodosius Dobzhansky* (ed.: M. K. Hecht and W. C. Steere), pp. 457–543. New York: Appleton-Century-Crofts.

CARSON, H. L. and HEED, W. B. (1964). Structural homozygosity in marginal populations of nearctic and neotropical species of *Drosophila* in Florida. *Proc. Nat. Acad. Sci. U.S.A.* **52**: 427–30.

CARSON, H. L. and STALKER, H. D. (1947). Gene arrangements in natural populations of *Drosophila robusta* Sturtevant. *Evolution,* **1**: 113–33.

CARSON, H. L. and STALKER, H. D. (1949). Seasonal variation in gene arrangement frequencies over a three-year period in *Drosophila robusta* Sturtevant. *Evolution,* **3**: 322–9.

CARSON, H. L. and STALKER, H. D. (1968). Polytene chromosome relationships in Hawaiian species of *Drosophila*. I, II and III. *Univ. Texas Publ.* **6818**: 335–80.

CARSON, H. L., WEI, I. Y. and NIEDERKORN, J. A. Jr. (1969). Isogenicity in parthenogenetic strains of *Drosophila mercatorum*. *Genetics*, **63**, 619–28.

CARSON, H. L., WHEELER, M. R. and HEED, W. B. (1957). A parthenogenetic strain of *Drosophila mangabeiri* Malogolowkin. *Univ. Texas Publ.* **5721**: 115–22.

CASPERSSON, T., FARBER, S., FOLEY, G. E., KUDYNOWSKI, J., MODEST, E. J., SIMONSSON, E., WAGH, V. and ZECH, L. (1968). Chemical differentiation along metaphase chromosomes. *Exp. Cell Res.* **49**: 219–22.

CASPERSSON, T. and SCHULTZ, J. (1938). Nucleic acid metabolism of the chromosomes in relation to gene reproduction. *Nature*, **142**: 294–5.

CASPERSSON, T. O. and SCHULTZ, J. (1939). Pentose nucleotides in the cytoplasm of growing tissues. *Nature*, **143**: 602–9.

CASPERSSON, T., ZECH, L., JOHANSSON, C. and MODEST, E. J. (1970). Identification of human chromosomes by DNA-binding fluorescent agents. *Chromosoma*, **30**: 215–27.

CASSAGNAU, P. (1966). Présence de chromosomes géants dans les glandes salivaires des *Neanura* (Collemboles suceurs). *C.R. Acad. Sci. Paris* D, **262**: 168–70.

CASSAGNAU, P. (1968a). Sur la structure des chromosomes salivaires de *Bilobella massoudi* Cassagnau (Collembola: Neanuridae). *Chromosoma*, **24**: 42–58.

CASSAGNAU, P. (1968b). L'évolution des pièces buccales et la polyténie chez les Collemboles. *C.R. Acad. Sci. Paris* D, **267**: 106–9.

CASSAGNAU, P. (1970). Inversions péricentriques et polygénotypisme dans une population de *Bilobella* (Collembole) du Mont Parnasse (Grèce). *C.R. Acad. Sci. Paris* D, **270**: 529–32.

CASTLEMAN, K. R. and WELCH, A. J. (1967). Match recognition in chromosome band structure. *Univ. Texas Labs. for Electronic and rel. Sci. Res. Tech. rep. no.* **31**: (65 pp.).

CASTRO-SIERRA, E. and WOLF, U. (1967). Replication patterns of the unpaired chromosome no. 9 of the rodent *Ellobius lutescens* Th. *Cytogenetics*, **6**: 268–75.

CASTRO-SIERRA, E. and WOLF, U. (1968). Studies on the male meiosis of *Ellobius lutescens* Th. *Cytogenetics*, **7**: 241–8.

CATCHESIDE, D. G. (1968). The control of genetic recombination in *Neurospora crassa*. *Replication and Recombination of Genetic Material*, pp. 216–26. Canberra: Aust. Acad. Sci.

CATTANACH, B. M. (1961a). A chemically-induced variegated type position effect in the mouse., *Z. indukt. Abst. Vererbl.* **92**: 165–82.

CATTANACH, B. M. (1961b). *XXY* mice. *Genet. Res.* **2**: 156–8.

CATTANACH, B. M. (1963). The inactive-*X* hypothesis and position effects in the mouse. *Genetics*, **48**: 884–5.

CATTANACH, B. M. (1964). Autosomal trisomy in the mouse. *Cytogenetics*, **3**: 159–66.

CAVALCANTI, A. G. L. (1948). Geographic variation of chromosome structure in *Drosophila prosaltans*. *Genetics*, **33**: 529–36.

CAVAZOS, L. F. (1951). Spermatogenesis of the horned lizard *Phrynosoma cornutum*. *Amer. Nat.* **85**: 373–9.

CAVE, M. D. (1968). Chromosome replication and synthesis of non-histone proteins in giant polytene chromosomes. *Chromosoma*, **25**: 392–401.

CELADA, F. and WELSHONS, W. J. (1963). An immunogenetic analysis of the male antigen in mice, utilizing animals with an exceptional chromosome constitution. *Genetics*, **48**: 139–51.

CHANDRA, H. S. (1962). Inverse meiosis in triploid females of the mealy bug, *Planococcus citri*. *Genetics*, **47**: 1441–54.

CHANDRA, H. S. (1963*a*). Cytogenetic studies following high dosage paternal irradiation in the mealy bug, *Planococcus citri*. I. Cytology of X_1 embryos. *Chromosoma*, **14**: 310–29.

CHANDRA, H. S. (1963*b*). Cytogenetic studies following high dosage paternal irradiation in the mealy bug, *Planococcus citri*. II. Cytology of X_1 females and the problem of lecanoid sex determination. *Genetics*, **14**: 330–46.

CHARGAFF, E. (1955). In: *The Nucleic Acids*, **I**. (ed.: E. Chargaff and J. N. Davidson), p. 307. New York: Academic Press.

CHARLTON, H. H. (1921). The spermatogenesis of *Lepisma domestica*. *J. Morph.* **35**: 381–423.

CHARNIAUX-COTTON, H. (1960). Sex determination. In: *The Physiology of Crustacea* (ed: T. H. Waterman.), Ch. 13. New York and London: Academic Press.

CHEISSIN, E. M. and OVCHINNIKOVA, L. P. (1964). A photometric study of DNA content in macronuclei and micronuclei of different species of *Paramecium*. *Acta Protozool.* **2**: 225–36.

CHEISSIN, E. M., OVCHINNIKOVA, L. P. and KUDRIAVTSEV, B. N. (1964). A photometric study of DNA content in macronuclei and micronuclei of different strains of *Paramecium caudatum*. *Acta Protozool.* **2**: 237–45.

CHEN, A. T. L. and FALEK A. (1969). Centromeres in human meiotic chromosomes. *Science*, **166**: 1008–10.

CHEN, T. R. and EBELING, A. W. (1966). Probable male heterogamety in the deep sea fish *Bathylagus wesethi* (Teleostei: Bathylagidae). *Chromosoma*, **18**: 88–96.

CHEN, T. R. and RUDDLE, F. H. (1970). A chromosome study of four species and a hybrid of the killifish genus *Fundulus* (Cyprinodontidae). *Chromosoma*, **29**: 255–67.

CHIARELLI, B. (1961). Chromosomes of the Orang-Utan. *Nature*, **192**: 285.

CHIARELLI, B. (1968). Chromosome polymorphism in the species of the genus *Cercopithecus*. *Cytologia*, **33**: 1–16.

CHICKERING, A. M. (1927). Spermatogenesis in the Belostomatidae. II. The chromosomes and cytoplasmic inclusions in the male germ cells of *Belostoma flumineum* Say, *Lethocerus americanus* Leidy and *Benacus griseus* Say. *J. Morph.* **44**: 541–607.

CHICKERING, A. M. (1932). Spermatogenesis in the Belostomatidae. III. The chromosomes in the male germ cells of a *Lethocerus* from New Orleans, Louisiana. *Pap. Mich. Acad. Sci.* **15**: 357–60.

CHICKERING, A. M. and BACORN, B. (1933). Spermatogenesis in the Belostomatidae. IV. Multiple chromosomes in *Lethocerus*. *Pap. Mich. Acad. Sci.* **17**: 529–33.

CHILDRESS, D. (1969). Polytene chromosomes and linkage group – chromosome correlations in the Australian Sheep Blowfly *Lucilia cuprina* (Diptera: Calliphoridae). *Chromosoma*, **26**: 208–14.

CHOPARD, L. (1948). La parthénogenèse chez les Orthopteroides. *Ann. Biol.* **52**: 15–22.

CHOWDAIAH, B. N., BAKER, R. H. and KITZMILLER, J. B. (1966). The salivary chromosomes of *Anopheles vestitipennis*. *Cytologia*, **31**: 144–52.

CHRISTENSEN, B. (1959). Asexual reproduction in the Enchytraeidae (Olig.). *Nature*, **184**: 1159–60.

CHRISTENSEN, B. (1960). A comparative cytological investigation of the reproductive cycle of an amphimictic diploid and a parthenogenetic triploid form of *Lumbricillus lineatus* (O.F.M.) (Oligochaeta, Enchytraeidae). *Chromosoma*, **11**: 365–79.

CHRISTENSEN, B. (1961). Studies on cyto-taxonomy and reproduction in the Enchytraeidae. With notes on parthenogenesis and polyploidy in the animal kingdom. *Hereditas*, **47**: 387–450.

CHRISTENSEN, B. (1966). Cytophotometric studies on the DNA content in diploid and polyploid Enchytraeidae (Oligochaeta). *Chromosoma*, **18**: 305–15.

CHU, E. H. Y. and BENDER, M. A. (1961). Chromosome cytology and evolution in primates. *Science*, **133**, 1399–1405.

CHU, E. H. Y. and GILES, N. H. (1957). A study of primate chromosome complements. *Amer. Nat.* **91**: 273–81.

CHU, E. H. Y. and SWOMLEY, B. A. (1961). Chromosomes of the lemurs. *Science*, **133**: 1925–6.

CHU, E. H. Y., THULINE, H. C. and NORBY, D. E. (1964). Triploid–diploid chimerism in a male tortoiseshell cat. *Cytogenetics*, **3**: 1–18.

CLARKE, E. (1959). Functional hermaphroditism and self-fertilization in a serranid fish. *Science*, **129**: 215–6.

CLAUS, G. (1956). La formule chromosomique du Gryllodea *Eneoptera surinamensis* De Geer et le comportement des chromosomes sexuels de cette espèce au cours de la spermatogenèse. *Ann. Sci. Nat. Zool.* Ser. II, **18**: 63–105.

CLAYTON, F. E. and WARD, C. L. (1954). Chromosomal studies of several species of Drosophilidae. *Univ. Texas Publ.* **5422**: 98–105.

CLAYTON, F. E. and WASSERMAN, M. (1957). Chromosomal studies of several species of Drosophilidae. *Univ. Texas Publ.* **5422**: 98–105.

CLEVELAND, L. R. (1949). The whole life cycle of chromosomes and their coiling systems. *Trans. Amer. Phil. Soc.* **39**: 1–100.

CLEVELAND, L. R. (1950a). Hormone-induced sexual cycles of flagellates. II. Gametogenesis, fertilization and one-division meiosis in *Oxymonas*. *J. Morph.* **86**: 185–214.

CLEVELAND, L. R. (1950b). Hormone-induced sexual cycles of flagellates. III. Gametogenesis, fertilization and one-division meiosis in *Saccinobaculus*. *J. Morph.* **86**: 215–28.

CLEVELAND, L. R. (1951). Hormone-induced sexual cycles of flagellates. VII. One-division meiosis and autogamy without cell division in *Urinympha*. *J. Morph.* **88**: 385–440.

CLEVELAND, L. R. (1957). Types and life cycles of centrioles of flagellates. *J. Protozool.* **4**: 230–41.

CLEVELAND, L. R. and DAY, M. (1958). Spirotrichonymphidae of *Stolotermes*. *Arch. Protistenk.* **103**: 1–53.

CLEVER, U. (1961). Genaktivitäten in den Riesenchromosomen von *Chironomus tentans* und ihre Beziehung zur Entwicklung. I. Genaktivierung durch Ecdyson. *Chromosoma*, **12**: 607–75.

CLEVER, U. (1966). Gene activity patterns and cellular differentiation. *Amer. Zool.* **6**: 33–41.

CLEVER, V. and KARLSON, P. (1960). Induktion von Puff-Veränderungen in den Speicheldrüsenchromosomen von *Chironomus tentans* durch Ecdyson. *Exp. Cell Res.* **20**: 623–6.

COCK, A. G. (1964). Dosage compensation and sex-chromatin in non-mammals. *Genet. Res.* **5**: 354–65.

COGGESHALL, R. E., YAKSTA, B. A. and SWARTZ, F. J. (1970). A cytophotometric analysis of the DNA in the nucleus of the giant cell, R-2, in *Aplysia*. *Chromosoma*, **32**: 205–12.

COGNETTI, G. (1961*a*). Citogenetica della partenogenesi negli Afidi. *Arch. Zool. Ital.* **46**: 89–122.

COGNETTI, G. (1961*b*). Endomeiosis in parthenogenetic lines of aphids. *Experientia,* **17**: 168–9.

COGNETTI, G. (1961*c*). Endomeiosi e selezione in ceppi partenogenetici di afidi. *Atti. Assoc. Genet. Ital.* **6**: 449–54.

COGNETTI, G. (1962). La partenogenesi negli afidi. *Boll. di Zool.* **29**: 129–47.

COGNETTI DE MARTIIS, L. (1922). Contributo alla conoscenza della spermatogenesi dei Rabdocelidi. *Arch. Zellforsch.* **16**: 249–84.

COHEN, M. M. and CLARK, H. F. (1967). The somatic chromosomes of five crocodilian species. *Cytogenetics,* **6**: 193–203.

COHEN, M. M. and GANS, C. (1970). The chromosomes of the order Crocodilia. *Cytogenetics,* **9**: 81–105.

COHEN, M. M., SHAW, M. W. and MACCLUER, J. W. (1966). Racial differences in the length of the human *Y*-chromosome. *Cytogenetics,* **5**: 34–52.

COLE, C. J. (1969). Sex chromosomes in teiid whiptail lizards (genus *Cnemidophorus*). *Amer. Mus. Novit.* no. **2395**: 1–14.

COLE, C. J. (1970). Karyotypes and evolution of the *spinosus* group of lizards in the genus *Sceloporus. Amer. Mus. Novit.* no. **2431**, 1–47.

COLE, C. J., LOWE, C. H. and WRIGHT, J. W. (1967). Sex chromosomes in lizards. *Science,* **155**: 1028–9.

COLEMAN, L. C. (1947). Chromosome abnormalities in an individual of *Chorthippus longicornis* (Acrididae). *Genetics,* **32**, 435–47.

COLEMAN, L. C. (1948). The cytology of some western species of *Trimerotropis* (Acrididae). *Genetics,* **33**: 519–28.

COLLIER, J. R. and McCANN COLLIER, M. (1962). The deoxyribonucleic acid content of the egg and sperm of *Ilyanassa obsoleta. Exp. Cell Res.* **27**: 553–9.

COLUZZI, M. (1964). Morphological divergences in the *Anopheles gambiae* complex. *Riv. Malariol.* **43**: 197–232.

COLUZZI, M. (1966). Osservazioni comparative sul cromosoma X nelle specie A e B del complesso *Anopheles gambiae. Rend. Accad. Naz. Lincei,* **40**: 671–8.

COLUZZI, M. (1968). Cromosomi politenici delle cellule nutrici ovariche nel complesso *gambiae* del genere *Anopheles. Parassitologia,* **10**: 179–84.

COLUZZI, M., CANCRINI, G. and DI DECO, M. (1970). The polytene chromosomes of *Anopheles superpictus* and relationships with *Anopheles stephensi. Parassitologia,* **12**: 101–12.

COLUZZI, M. and SABATINI, A. (1967). Cytogenetic observations on species A and B of the *Anopheles gambiae* complex. *Parassitologia,* **9**: 73–88.

COLUZZI, M. and SABATINI, A. (1968). Cytogenetic observations on species C of the *Anopheles gambiae* complex. *Parassitologia,* **10**: 155–6.

COLUZZI, M. and SABATINI, A. (1969). Cytogenetic observations on the salt water species, *Anopheles merus* and *Anopheles melas,* of the *gambiae* complex. *Parassitologia,* **11**: 177–87.

COMINGS, D. E. (1967). Histones of genetically active and inactive chromatin. *J. Cell Biol.* **35**: 699–708.

COMINGS, D. E. and MATTOCCIA, E. (1970). Studies of microchromosomes and a G–C rich DNA satellite in the quail. *Chromosoma,* **30**: 202–14.

COMINGS, D. E. and OKADA, T. A. (1970*a*). Whole mount electron microscopy of meiotic chromosomes and the synaptonemal complex. *Chromosoma,* **30**: 269–86.

COMINGS, D. E. and OKADA, T. A. (1970b). Mechanism of chromosome pairing during meiosis. *Nature*, **227**: 451–6.

COMINGS, D. E. and OKADA, T. A. (1970c). Association of chromatin fibers with the annuli of the nuclear membrane. *Exp. Cell Res.* **62**: 293–302.

COMINGS, D. E. and OKADA, T. A. (1970d). Association of nuclear membrane fragments with metaphase and anaphase chromosomes as observed by whole mount electron microscopy. *Exp. Cell Res.* **63**: 62–8.

COMRIE, L. C. (1938). Biological and cytological observations on tenthredinid parthenogenesis. *Nature*, **142**, 877–8.

COOKE, P. and GORDON, R. R. (1965). Cytological studies on a human ring chromosome. *Ann. Hum. Genet.* **29**: 147–50.

COOPER, K. W. (1937). Reproductive behavior and haploid parthenogenesis in the grass mite, *Pediculopsis graminum* (Reut.) (Acarina, Tarsonemidae). *Proc. Nat. Acad. Sci. U.S.A.* **23**: 41–4.

COOPER, K. W. (1939). The nuclear cytology of the grass mite *Pediculopsis graminum* (Reut.) with special reference to karyomerokinesis. *Chromosoma*, **1**: 51–103.

COOPER, K. W. (1941a). Bivalent structure in the fly *Melophagus ovinus* L. (Pupipara, Hippoboscidae). *Proc. Nat. Acad. Sci. U.S.A.* **27**: 109–14.

COOPER, K. W. (1941b). An investigation of the aberrant chromosome behavior in the male germ cells of flies parasitic on tropical bats and vultures (Diptera, section Pupipara). *Yearbook Amer. Phil. Soc. 1491*, pp. 122–7.

COOPER, K. W. (1944a). Invalidation of the cytological evidence for reciprocal chiasmata in the sex chromosome bivalent of male *Drosophila*. *Proc. Nat. Acad. Sci. U.S.A.* **30**: 50–4.

COOPER, K. W. (1944b). Analysis of meiotic pairing in *Olfersia* and consideration of the reciprocal chiasmata hypothesis of sex chromosome conjunction in male *Drosophila*. *Genetics*, **29**, 537–68.

COOPER, K. W. (1945). Normal segregation without chiasmata in female *Drosophila melanogaster*. *Genetics*, **30**: 472–84.

COOPER, K. W. (1946). The mechanism of non-random segregation of sex chromosomes in *Drosophila miranda*. *Genetics*, **31**: 181–94.

COOPER, K. W. (1949). The cytogenetics of meiosis in *Drosophila*. Mitotic and meiotic autosomal chiasmata without crossing-over in the male. *J. Morph.* **84**: 81–122.

COOPER, K. W. (1950). Normal spermatogenesis in *Drosophila*. In: *Biology of Drosophila* (ed.: M. Demerec), pp. 1–61. New York: J. Wiley.

COOPER, K. W. (1951). Compound sex chromosomes with anaphasic precocity in the male mecopteran *Boreus brumalis* Fitch. *J. Morph.* **89**: 37–58.

COOPER, K. W. (1959). Cytogenetic analysis of major heterochromatic elements (especially *Xh* and *Y*) in *Drosophila melanogaster*, and the theory of 'heterochromatin'. *Chromosoma*, **10**: 535–88.

CORDEIRO, A. R., TOWNSEND, J. I., PETERSEN, J. A. and JAEGER, E. C. (1959). Genetics of southern marginal populations of *Drosophila willistoni*. *Proc. Xth Int. Congr. Genet.* **2**: 58–9.

COREY, H. I. (1933). Chromosome studies in *Stauroderus* (an Orthopteran). *J. Morph.* **55**: 313–47.

COREY, H. I. (1938). Heteropycnotic elements of orthopteran chromosomes. *Arch. Biol.* **49**: 159–72.

COREY, H. I. (1939). Chromomere vesicles in orthopteran cells. *J. Morph.* **66**: 299–321.

CORIN-FREDERIC, J. (1969). The so-called aberrant sex-determining mechanisms in placental mammals: the special case of the sloth *Choelopus hoffmanni* Peters (Edentata, Xenarthra, family Bradypodidae). *Chromosoma*, **27**: 268–87.

CORNEO, G., GINELLI, E. and POLLI, E. (1958). Isolation of the complementary strands of a human satellite DNA. *J. Mol. Biol.* **33**: 331.

COSTELLO, D. P. (1970). Identical linear order of chromosomes in both gametes of the acoel turbellarian *Polychoerus carmelensis*: a preliminary note. *Proc. Nat. Acad. Sci. U.S.A.* **67**: 1951–8.

COUDRAY, Y., QUETIER, F. and GUILLE, E. (1970). New compilation of satellite DNA's. *Biochim. Biophys. Acta*, **217**, 259–67.

COURT BROWN, W. M., HARNDEN, D. G., JACOBS, P. A., MACLEAN, N. and MANTLE, D. J. (1964). Abnormalities of the sex-chromosome complement in man. *Med. Res. Council. Spec. Ser.* **305**.

COUSIN, G. (1934). Sur la fécondité naturelle et les caractères des hybrides issus du croisement de deux espèces de Gryllidés. (*Acheta campestris* L. et *bimaculata* De Geer). *C.R. Acad. Sci. Paris* **198**: 853–9.

COUSIN, G. (1941). Analyse biométrique d'une hybridation interspécifique chez les Gryllides. *Proc. VIIth Int. Genet. Congr. Edinb.* p. 90.

CRADDOCK, E. (1970). Chromosome number variation in a stick insect, *Didymuria violescens* (Leach). *Science*, **167**: 1380–2.

CRAIG, G. B. Jr., HICKEY, W. A. and VANDEHEY, R. C. (1960). An inherited male-producing factor in *Aedes aegypti*. *Science*, **132**: 1887–9.

CREIGHTON, M. and ROBERTSON, W. R. B. (1941). Genetic studies on *Chorthippus longicornis*. *J. Hered.* **32**: 339–41.

CRETSCHMAR, M. (1928). Das Verhalten der Chromosomen bei der Spermatogenese von *Orgyia thyellina* Btl. und *antiqua* L., sowie eines ihrer Bastarde. *Z. Zell-forsch* **7**: 290–399.

CRIPPA, M. (1964). The mouse karyotype in somatic cells cultured *in vitro*. *Chromosoma*, **15**: 301–11.

CROUSE, H. V. (1939). An evolutionary change in chromosome shape in *Sciara*. *Amer. Nat.* **73**: 476–80.

CROUSE, H. V. (1943). Translocations in *Sciara*; their bearing on chromosome behavior and sex determination. *Univ. Missouri Coll. Agr. Res. Bull.* **379**: 1–75.

CROUSE, H. V. (1947). Chromosome evolution in *Sciara*. *J. Hered.* **38**: 279–88.

CROUSE, H. V. (1954). X-ray breakages of lily chromosomes at first meiotic metaphase. *Science*, **119**: 485–7.

CROUSE, H. V. (1960a). The nature of the influence of X-translocation on sex of progeny in *Sciara coprophila*. *Chromosoma*, **11**: 146–66.

CROUSE, H. V. (1960b). The controlling element in sex chromosome behavior in *Sciara*. *Genetics*, **45**: 1429–43.

CROUSE, H. V. (1961a). Irradiation of condensed meiotic chromosomes in *Lilium longiflorum*. *Chromosoma*, **12**: 190–214.

CROUSE, H. V. (1961b). X-ray effects on sex of progeny in *Sciara coprophila*. *Biol. Bull. Woods Hole*, **120**: 8–10.

CROUSE, H. V. (1965). Experimental alterations in the chromosome constitution of *Sciara*. *Chromosoma*, **16**: 391–410.

CROUSE, H. V. (1968). The role of ecdysone in DNA-puff formation and DNA synthesis in the polytene chromosomes of *Sciara coprophila*. *Proc. Nat. Acad. Sci. U.S.A.* **61**: 971–8.

CROUSE, H. V. and KEYL, H.-G. (1968). Extra replications in the "DNA-puffs" of *Sciara coprophila*. *Chromosoma*, **25**: 357–64.

CROW, J. F. (1942). Cross fertility and isolating mechanisms in the *Drosophila mulleri* group. *Univ. Texas Publ.* **4228**: 53–67.

CROW, J. F. (1963). The concept of genetic load: a reply. *Amer. J. Hum. Genet.* **15**: 310–15.

CROW, J. F. and MORTON, N. E. (1960). The genetic load due to mother–child incompatibility. *Amer. Nat.* **94**: 413–19.

CROW, J. F., THOMAS, C. and SANDLER, L. (1962). Evidence that the segregation distortion phenomenon in *Drosophila* involves chromosome breakage. *Proc. Nat. Acad. Sci. U.S.A.* **48**: 1307–14.

CROZIER, R. H. (1968a). Interpopulation karyotype differences in Australian *Iridomyrmex* of the '*detectus*' group (Hymenoptera: Formicidae: Dolichoderinae) *J. Aust. Ent. Soc.* **7**: 25–7.

CROZIER, R. H. (1968b). Cytotaxonomic studies on some Australian dolichoderine ants (Hymenoptera: Formicidae). *Caryologia*, **21**: 241–59.

CROZIER, R. H. (1968c). The chromosomes of three Australian dacetine ant species (Hymenoptera: Formicidae). *Psyche*, **75**: 87–90.

CROZIER, R. H. (1970). Karyotypes of twenty-one ant species (Hymenoptera, Formicidae), with reviews of the known ant karyotypes. *Canad. J. Genet. Cytol.* **12**: 109–28.

CRUMPACKER, D. W. and KASTRITSIS, C. D. (1967). A new gene arrangement in the third chromosome of *Drosophila pseudoobscura. J. Hered.* **58**: 2–6.

CUELLAR, O. (1971). Reproduction and the mechanism of meiotic restitution in the parthenogenetic lizard *Cnemidophorus uniparens. J. Morph.* **133**: 139–66.

DA CUNHA, A. B. (1953). Chromosomal inversions with sex-limited effects. *Nature*, **172**: 815.

DA CUNHA, A. B. (1955). Chromosomal polymorphism in the Diptera. *Adv. in Genet.* **7**: 93–138.

DA CUNHA, A. B. (1960). Chromosomal variation and adaptation in insects. *Ann. Rev. Entomol.* **5**: 85–110.

DA CUNHA, A. B. (1972). Chromosome activities and differentiation. *Chromosomes Today*, **3**: 296–7.

DA CUNHA, A. B., BRNCIC, D. and SALZANO, F. M. (1953). A comparative study of chromosomal polymorphism in certain South American species of *Drosophila. Heredity*, **7**: 193–202.

DA CUNHA, A. B., BURLA, H. and DOBZHANSKY, TH. (1950). Adaptive chromosomal polymorphism in *Drosophila willistoni. Evolution*, **4**: 212–35.

DA CUNHA, A. B. and DOBZHANSKY, TH. 1954. A further study of chromosomal polymorphism in *Drosophila willistoni* in its relation to the environment. *Evolution*, **8**: 119–34.

DA CUNHA, A. B., DOBZHANSKY, TH., PAVLOVSKY, O. and SPASSKY, B. (1959). Genetics of natural populations. XXVIII. Supplementary data on the chromosomal polymorphism in *Drosophila willistoni* in its relation to the environment. *Evolution*, **13**: 389–404.

DA CUNHA, A. B. and KERR, W. E. (1957). A genetical theory to explain sex determination by arrhenotokous parthenogenesis. *Forma et Functio*, **1**: 33–6.

DA CUNHA, A. B., MORGANTE, J. S., PAVAN, C. and GARRIDO, M. C. (1968). Studies on the cytology and differentiation in *Sciaridae*. I. Chromosome changes induced by a gregarine in *Trichosia* sp. (Diptera, Sciaridae). *Caryologia*, **21**: 271–82.

DA CUNHA, A. B., PAVAN, C., MORGANTE, J. S. and GARRIDO, M. C. (1969). Studies on cytology and differentiation in Sciaridae. II. DNA redundancy in salivary gland cells of *Hybosciara fragilis* (Diptera). *Genetics*, **61**, suppl. 1: 335–49.

DAHM, A. G. (1958). *Taxonomy and ecology of five species groups in the family Planariidae*. Malmö: Nya Litografen.
DALLAI, R. and VEGNI TALLURI, M. (1969). A karyological study of three species of Scincidae (Reptilia). *Chromosoma*, **27**: 86–94.
D'ANCONA, U. (1945). Sexual differentiation of the gonad and the sexualization of the germ cells in teleosts. *Nature*, **156**: 603–4.
DANEHOLT, B. and EDSTRÖM, J. E. (1967). The content of deoxyribonucleic acid in individual polytene chromosomes of *Chironomus tentans*. *Cytogenetics*, **6**: 350–6.
DAREVSKY, I. S. (1958). Natural parthenogenesis in certain subspecies of *Lacerta saxicola* Eversmann. *Dokl. Akad. Nauk S.S.S.R.* **122**: 730–2.
DAREVSKY, I. S. (1966). Natural parthenogenesis in a polymorphic group of Caucasian rock lizards related to *Lacerta saxicola* Eversmann. *J. Ohio Herpetol. Soc.* **5**: 115–52.
DAREVSKY, I. S. and DANIELYAN, F. D. (1968). Diploid and triploid progeny arising from natural mating of parthenogenetic *Lacerta armeniaca* and *L. unisexualis* with bisexual *L. saxicola valentini*. *J. Herpetol.* **2**: 65–9.
DAREVSKY, I. S. and KULIKOWA, W. N. (1691). Natürliche Parthenogenese in der polymorphen Gruppe der kaukasischen Felseidechse (*Lacerta saxicola* Eversmann). *Zool. Jb.* (*Syst.*) **89**: 119–76.
DAREVSKY, I. S. and KULIKOWA, W. N. (1964). Natural triploidy in a polymorphic group of Caucasian rock lizards (*Lacerta saxicola* Eversmann) resulting from hybridization between bisexual and parthenogenetic forms of this species. *Dokl. Akad. Nauk S.S.S.R.* **158**: 202–5.
DARLINGTON, C. D. (1930). A cytological demonstration of genetic crossing-over. *Proc. Roy. Soc. Lond.* B, **107**: 50–9.
DARLINGTON, C. D. (1931). Meiosis. *Biol. Rev.* **6**: 221–64.
DARLINGTON, C. D. (1932a). The origin and behavior of chiasmata. V. *Chorthippus elegans*. *Biol. Bull. Woods Hole*, **63**: 357–67.
DARLINGTON, C. D. (1932b). *Recent Advances in Cytology*, 1st. ed. London: Churchill; Philadelphia: Blakiston.
DARLINGTON, C. D. (1934a). Anomalous chromosome pairing in the male *Drosophila pseudoobscura*. *Genetics*, **19**: 95–118.
DARLINGTON, C. D. (1934b). The origin and behavior of chiasmata. VII. *Zea mays* *Z. indukt. Abst. Vererbl.* **67**: 96–114.
DARLINGTON, C. D. (1935a). The internal mechanics of the chromosomes. I, II and III. *Proc. Roy. Soc. Lond.* B, **118**: 33–96.
DARLINGTON, C. D. (1935b). The time, place and action of crossing-over. *J. Genet.* **31**: 185–212.
DARLINGTON, C. D. (1936). Crossing-over and its mechanical relationships in *Chorthippus* and *Stauroderus*. *J. Genet.* **33**: 465–500.
DARLINGTON, C. D. (1937a). The biology of crossing-over. *Nature*, **140**: 759.
DARLINGTON, C. D. (1937b). *Recent Advances in Cytology*. 2nd ed. London: Churchill; Philadelphia: Blakiston.
DARLINGTON, C. D. (1939a). *Evolution of Genetic Systems*. 1st ed. Cambridge Univ. Press.
DARLINGTON, C. D. (1939b). The genetical and mechanical properties of the sex chromosomes. V. *Cimex* and the Heteroptera. *J. Genet.* **39**: 101–38.
DARLINGTON, C. D. (1939c). Misdivision and the genetics of the centromere. *J. Genet.* **37**: 341–64.
DARLINGTON, C. D. (1940a). The origin of isochromosomes. *J. Genet.* **39**: 351–61.

DARLINGTON, C. D. (1940*b*). The prime variables of meiosis. *Biol. Rev.* **15**: 307–22.

DARLINGTON, C. D. (1953). Polyploidy in animals. *Nature*, **171**: 191–4.

DARLINGTON, C. D. (1958). *Evolution of genetic systems.* Edinburgh and London: Oliver and Boyd.

DARLINGTON, C. D. (1964). Contending with evolution. *Sci. Progr.* **52**: 133–7.

DARLINGTON, C. D. (1966). The chromosomes as *we* see them. *Chromosomes Today*, **1** (ed.: C. D. Darlington and K. R. Lewis). pp. 1–6.

DARLINGTON, C. D. and DOBZHANSKY, TH. (1942). Temperature and 'sex ratio' in *Drosophila pseudoobscura. Proc. Nat. Acad. Sci. U.S.A.* **28**: 45–7.

DARLINGTON, C. D. and KEFALLINOU, M. (1957). Correlated chromosome aberrations at meiosis in *Gasteria. Chromosoma*, **8**: 364–70.

DARLINGTON, C. D. and LA COUR, L. (1940). Nucleic acid starvation of the chromosomes in *Trillium. J. Genet.* **40**: 185–213.

DARLINGTON, C. D. and LA COUR, L. F. (1942). *The Handling of Chromosomes.* London: Allen and Unwin.

DARLINGTON, C. D. and MATHER, K. (1949). *The Elements of Genetics.* London: Allen and Unwin.

DARLINGTON, C. D. and UPCOTT, M. B. (1941). The activity of inert chromosomes in *Zea mays. J. Genet.* **41**: 275–96.

DARLINGTON, C. D. and VOSA, C. G. (1963). Bias in the internal coiling direction of chromosomes. *Chromosoma*, **13**: 609–22.

DARLINGTON, C. D. and WYLIE, A. P. (1953). A dicentric cycle in *Narcissus. Heredity*, **6**, suppl.: 197–213.

DARTNALL, J. A. (1970). The chromosomes of some Tasmanian rodents. *Pap. Proc. Roy. Soc. Tasmania.* **104**: 79–80.

DAS, C. C. (1958). Studies on the structure and behavior of chromosomes of *Ranatra elongata* (Fabr.) (Nepidae: Hemiptera-Heteroptera). *Cellule*, **59**: 205–10.

DAS, C. C., KAUFMANN, B. P. and GAY, H. (1964). Histone-protein transition in *Drosophila melanogaster.* II. Changes during early embryonic development. *J. Cell Biol.* **23**: 423–30.

DASGUPTA, J. (1967). Meiosis in male tiger beetle *Cicindela catena* Fabr. (Cicindelidae: Coleoptera). *Sci. and Cult.* **33**: 491–3.

DASGUPTA, J. (1968). Analysis of male meiosis in *Cephalocoema canaliculata* Guerin (Orthoptera: Proscopiidae). *Cellule*, **67**: 129–36.

DASS, C. M. S. (1952). Meiosis in two members of the family Nepidae (Hemiptera–Heteroptera) *Caryologia*, **4**: 77–85.

DATTA, M. (1970). Reinvestigation of meiosis in the male goat, *Capra hircus* Linn., with special reference to chiasma formation in the sex and autosomal bivalents. *Cytologia*, **35**: 344–53.

DAVE, M. J. (1965). On unusual sex chromosomes found in two species of the Locustidae. *Cytologia*, **30**: 194–200.

DAVIDSON, G. (1962). The *Anopheles gambiae* complex. *Nature*, **196**: 907.

DAVIDSON, G. (1964). The five mating types in the *Anopheles gambiae* complex. *Riv. Malariol.* **43**: 167–83.

DAVIDSON, G., PATERSON, H. E., COLUZZI, M., MASON, G. F. and MICKS, D. W. (1967). The *Anopheles gambiae* complex. In: *Genetics of Insect Vectors of Disease* (ed. J. W. Wright and R. Pal), pp. 211–50. Amsterdam: Elsevier.

DAVIDSON, J. and BALD, J. G. (1931). Sex determination in *Frankliniella insularis* Franklin (Thysanoptera). *Aust. J. Exp. Biol. Med. Sci.* **8**: 139–42.

DAVIDSON, R. G., NITOWSKY, H. N. and CHILDS, B. (1963). Demonstration of two populations of cells in the human female heterozygous for glucose 6 phosphate dehydrogenase variants. *Proc. Nat. Acad. Sci. U.S.A.* **50**: 481–5.

DAVIDSON, W. M. and SMITH, D. R. (1954). A morphological sex difference in the polymorphonuclear neutrophil leukocytes. *Brit. Med. J.* **ii**: 43–4.

DAVIS, G. (1963). Autosomal behavior in claret-nondisjunctional *Drosophila females. Proc. XIth Internat. Congr. Genet.* **1**: 127–8.

DAVIS, H. S. (1908). Spermatogenesis in Acrididae and Locustidae. *Bull. Mus. Comp. Zool. Harvard Univ.* **53**: 59–158.

DAY, M. F. C. (1945). Concomitant speciation in *Euspironympha* (*gen. nov.*) and its host, *Stolotermes* (Isoptera). *Summaries of Theses, Harvard Univ.* pp. 39–42.

DEAVEN, L. L. and STUBBLEFIELD, E. (1969). Segregation of chromosomal DNA in chinese hamster fibroblasts *in vitro. Exp. Cell Res.* **55**: 123–35.

DE BAEHR, W. B. (1970). Über die Zahl der Richtungskörper in parthenogenetisch sich entwickelnden Eier von *Bacillus rossii. Zool. Jb.* (*Anat.*) **24**: 175–92.

DE BAEHR, W. B. (1908). Über die Bildung der Sexualzellen bei Aphididae. *Zool. Anz.* **33**: 507–17.

DE BAEHR, W. B. (1909). Die Oogenese bei einigen viviparen Aphididen und die Spermatogenese von *Aphis saliceti* mit besonderer Berücksichtigung der Chromosomenverhältnisse. *Arch. Zellforsch.* **3**: 269–333.

DE BAEHR, W. B. (1920). Recherches sur la maturation des oeufs parthénogénétiques dans l'*Aphis palmae. Cellule*, **30**: 315–53.

DEBAISIEUX, P. and GASTALDI, L. (1919). Les microsporidies parasites des larves de *Simulium. Cellule* **30**: 187–213.

DE CASTRO, N. M. (1963). Frequency variations of "drumsticks" of peripheral blood neutrophils in the rabbit in different alimentary conditions. *Acta Anat.* **52**: 341–368.

DE CASTRO, Y. G. P. (1946). Notas sôbre os cromossômios dos Proscopiidios. *Ann. Esc. Sup. Agric. L. de Queiroz*, **3**: 273–5.

DEDERER, P. H. (1907). Spermatogenesis in *Philosamia cynthia. Biol. Bull. Woods Hole*, **13**: 94–106.

DEGRANGE, C. (1956). La parthénogenèse facultative deutérotoque de *Centroptilum luteolum* (Müll.) (Ephemeroptère). *C.R. Acad. Sci. Paris*, **243**: 201–3.

DE LATTIN, G. (1951). Über die Bestimmung und Vererbung des Geschlechts einiger Oniscoideen (Crust., Isop.). I. Untersuchungen über die geschlechts-beeinflussende Wirkung von Farbfaktoren bei *Porcellio* und *Tracheoniscus. Z. indukt. Abst. Vererbl.* **84**: 1–37.

DE LATTIN, G. (1952). Über die Bestimmung und Vererbung des Geschlechts einiger Oniscoideen (Crust., Isop.) II. Zur Vererbung der Monogenie von *Cylisticus convexus* (DeG.). *Z. indukt. Abst. Vererbl.* **84**: 536–67.

DELBRÜCK, N. (1941). A theory of autocatalytic synthesis of polypeptides and its application to the problem of chromosome reproduction. *Cold Spr. Harb. Symp. Quant. Biol.* **11**: 33–7.

DE LESSE, H. (1952). Quelques formules chromosomiques chez les Lycaenidae (Lepidoptères Rhopalocères). *C.R. Acad. Sci. Paris*, **235**: 1692–4.

DE LESSE, H. (1953). Formules chromosomiques nouvelles chez les Lycaenidae (Lepid. Rhopal.). *C.R. Acad. Sci. Paris*, **237**: 1781–3.

DE LESSE, H. (1954). Formules chromosomiques nouvelles chez les Lycaenidae (Lepidoptères, Rhopalocères). *C.R. Acad. Sci. Paris*, **238**: 514–6.

DE LESSE, H. (1955). Une nouvelle formule chromosomique dans le groupe d'*Erebia tyndarus* Esp. (Lepidoptères, Satyrinae). *C.R. Acad. Sci. Paris,* **241**: 1505–7.

DE LESSE, H. (1956). Étude cytologique des *Lysandra* fixés par M. H. Beuret. *Mitt. Ent. Gesell. Basel* (N.F.), **6**: 77–80.

DE LESSE, H. (1957). Description de deux nouvelles espèces d'*Agrodiaetus* (Lep. Lycaenidae) separées à la suite de la decouverte de leurs formules chromosomiques. *Lambillionea,* **57**: 65–71.

DE LESSE, H. (1959a). Separation specifique d'un *Lysandra* d'Afrique du Nord a la suite de la decouverte de sa formule chromosomique (Lycaenidae). *Alexanor,* **1**: 61–4.

DE LESSE, H. (1959b). Caractères et repartition en France d'*Erebia aethiopellus* Hoffmsg. et *E. mnestra* Hb. *Alexanor,* **1**: 72–81.

DE LESSE, H. (1959c). Note sur deux espèces d'*Agrodiaetus* (Lep. Lycaenidae) recemment separées d'après leurs formules chromosomiques. *Lambillionea,* **59**: 5–10.

DE LESSE, H. (1960). Spéciation et variation chromosomique chez les Lépidoptères Rhopalocères. *Ann. Sci. Nat. Zool.* **12** Ser. 2: 1–223.

DE LESSE, H. (1964). Les nombres de chromosomes chez quelques *Erebia* femelles (Lep. Satyrinae). *Rev. Fr. Entomol.* **31**: 112–15.

DE LESSE, H. (1966a). Variation chromosomique chez *Agrodiaetus dolus* Hubner (Lep. Lycaenidae). *Ann. Soc. Entomol. Fr.* N.S., **2**: 209–14.

DE LESSE, H. (1966b). Formules chromosomiques de quelques Lépidoptères Rhopalocères d'Afrique Centrale. *Ann. Soc. Entomol. Fr.* N.S., **11**: 349–53.

DE LESSE, H. (1967). Les nombres de chromosomes chez les Lépidoptères Rhopalocères néotropicaux. *Ann. Soc. Entomol. Fr.* N.S., **3**: 67–136.

DE LESSE, H. (1969). Les nombres de chromosomes dans le groupe de *Lysandra coridon* (Lep. Lycaenidae). *Ann. Soc. Entomol. Fr.* N.S. **5**: 469–522.

DE LESSE, H. (1970). Les nombres de chromosomes dans le groupe de *Lysandra argester* et leur incidence sur sa taxonomie. *Bull. Soc. Entomol. Fr.* **75**: 64–8.

DE LESSE, H. and CONDAMIN, M. (1962). Formules chromosomiques de quelques Lépidoptères Rhopalocères du Sénégal. *Bull. I.F.A.N.* A, **24**: 464–73.

DE LESSE, H. and CONDAMIN, M. (1965). Formules chromosomiques de quelques Lépidoptères Rhopalocères du Sénégal et de Côte d'Ivoire. *Bull. I.F.A.N.* A, **27**: 1089–94.

DEMEREC, M. (1941). The nature of changes in the *white–Notch* region of the X-chromosome of *Drosophila melanogaster. Proc. VIIth Int. Genet. Congr.* pp. 99–103.

DEMEREC, M. (1942). The nature of the gene. In: *Cytology, Genetics and Evolution* (*Univ. Penna. Bicentenn. Conf.*).

DEMEREC, M. (1964). Homology and divergence in genetic material of *Salmonella typhimurium* and *Escherichia coli*. In: *Evolving Genes and Proteins. Symp. Inst. Microbiol. Rutgers State Univ.* pp. 505–510.

DEMEREC, M. (1965). Gene differentiation. *Nat. Cancer Inst. Monograph,* **18**: 15–20.

DEMEREC, M. and OHTA, N. (1964). Genetic analysis of *Salmonella typhimurium Escherichia coli* hybrids. *Proc. Nat. Acad. Sci. U.S.A.* **52**: 317–23.

DENNHÖFER, L. (1968). Die Speicheldrüsenchromosomen der Stechmücke *Culex pipiens.* I. Der normale Chromosomenbestand. *Chromosoma,* **25**: 365–76.

'DENVER REPORT' (1960). *Lancet,* **1**: 1063.

DEWEY, W. C. and HUMPHREY, R. M. (1962). Relative radiosensitivity of different phases in the life cycle of L-P 59 mouse fibroblasts and ascites tumor cells. *Radiat. Res.* **16**: 503–30.

DE WINIWARTER, H. (1927). Étude du cycle chromosomique chez diverses races de *Gryllotalpa gryll*. L. *Arch. Biol.* **37**: 515–72.

DE WINIWARTER, H. (1937). Les chromosomes du genre *Gryllotalpa*. *Cytologia*, Fujii Jubilee vol. pp. 987–94.

DIAZ, M. and PAVAN, C. (1965). Changes in chromosomes induced by micro-organism infection. *Proc. Nat. Acad. Sci. U.S.A.* **54**: 1321–8.

DIAZ, M., PAVAN, C. and BASILE, R. (1969). Effects of a virus and a microsporidian infections in chromosomes of various tissues of *Rhynchosciara angelae* (Nonato et Pavan 1951). *Rev. Brasil. Biol.* **29**: 191–206.

DIAZ, M. O. and SAEZ, F. A. (1968). DNA synthesis in the neo-X neo-Y sex-determination system of *Dichroplus bergi* (Orthoptera: Acrididae). *Chromosoma*, **24**: 10–16.

DICKSON, E., BOYD, J. B. and LAIRD, C. D. (1971). Sequence diversity of polytene chromosome DNA from *Drosophila hydei*. *J. Mol. Biol.* **61**: 615–27.

DIETZ, R. (1954). Multiple Geschlechtschromosomen bei dem Ostracoden *Notodromas monacha*. *Chromosoma*, **6**: 397–418.

DIETZ, R. (1955). Zahl und Verhalten der Chromosomen einiger Ostracoden. *Z. Naturf.* **10b**: 92–5.

DIETZ, R. (1956). Die Spermatocytenteilungen der Tipuliden. II Mitteilung. Graphische Analyse der Chromosomenbewegung während der Prometaphase I im Leben. *Chromosoma*, **8**: 183–211.

DIETZ, R. (1958). Multiple Geschlechtschromosomen bei den Cypriden Ostracoden, ihre Evolution und ihr Teilungsverhalten. *Chromosoma*, **9**: 359–440.

DIETZ, R. (1959). Centrosomenfreie Spindelpole in Tipuliden-Spermatocyten. *Z. Naturf.* **14b**: 749–52.

DIETZ, R. (1966). The dispensability of the centrioles in the spermatocyte divisions of *Pales ferruginea* (Nematocera). *Chromosomes Today*, **1** (ed.: C. D. Darlington and K. R. Lewis) pp. 161–6.

DIETZ, R. (1969). Bau und Funktion des Spindelapparats. *Naturwiss.* **56**: 237–48.

DIGBY, L. (1910). The somatic, premeiotic and meiotic nuclear divisions of *Galtonia candicans*. *Ann. Bot.* **24**: 727–58.

DOBZHANSKY, TH. (1934). Studies on hybrid sterility. I. Spermatogenesis in pure and hybrid *Drosophila pseudoobscura*. *Z. Zellforsch.* **21**: 169–223.

DOBZHANSKY, TH. (1935a). The Y-chromosomes of *Drosophila pseudoobscura*. *Genetics*, **20**: 366–76.

DOBZHANSKY, TH. (1935b). *Drosophila miranda*, a new species. *Genetics*, **20**: 377–91.

DOBZHANSKY, TH. (1936a). The persistence of the chromosome pattern in successive cell divisions in *Drosophila pseudoobscura*. *J. Exper. Zool.* **74**: 119–135.

DOBZHANSKY, TH. (1936b). Studies on hybrid sterility. II. Localization of sterility factors in *Drosophila pseudoobscura* hybrids. *Genetics*, **21**: 113–35.

DOBZHANSKY, TH. (1936c). L'effet de Position et la Théorie de l'Hérédité. *Actualités Sci. Industr.* no. **410**, 19 pp. Paris: Hermann.

DOBZHANSKY, TH. (1936d). Position effects on genes. *Biol. Rev.* **11**: 364–84.

DOBZHANSKY, TH. (1937a). Further data on the variation of the Y-chromosome in *Drosophila pseudoobscura*. *Genetics*, **22**: 340–6.

DOBZHANSKY, TH. (1937b). Further data on *Drosophila miranda* and its hybrids with *Drosophila pseudoobscura*. *J. Genet.* **34**: 135–51.

DOBZHANSKY, TH. (1937c). *Genetics and the Origin of Species*. 1st ed. New York: Columbia Univ. Press.

DOBZHANSKY, TH. (1939a). Microgeographic variation in *Drosophila pseudoobscura*. *Proc. Nat. Acad. Sci. U.S.A.* **25**: 311–14.

DOBZHANSKY, TH. (1939*b*). Mexican and Guatemalan populations of *Drosophila pseudoobscura*. *Genetics*, **24**: 391–412.

DOBZHANSKY, TH. (1940). Speciation as a stage in evolutionary divergence. *Amer. Nat.* **74**: 312–21.

DOBZHANSKY, TH. (1941*a*). *Genetics and the Origin of Species*. 2nd ed. New York: Columbia Univ. Press.

DOBZHANSKY, TH. (1941*b*). Discovery of a predicted gene arrangement in *Drosophila azteca*. *Proc. Nat. Acad. Sci. U.S.A.* **27**: 47–50.

DOBZHANSKY, TH. (1943). Genetics of natural populations. IX. Temporal changes in the composition of populations of *Drosophila pseudoobscura*. *Genetics*, **28**: 162–86.

DOBZHANSKY, TH. (1947). Genetic structure of natural populations. *Yearb. Carnegie Instn.* **46**: 155–65.

DOBZHANSKY, TH. (1948*a*). Chromosomal variation in populations of *Drosophila pseudoobscura* which inhabit northern Mexico. *Amer. Nat.* **82**: 97–106.

DOBZHANSKY, TH. (1948*b*). Genetic structure of natural populations. *Yearb. Carnegie Instn.* **47**: 193–203.

DOBZHANSKY, TH. (1950*a*). Genetics of natural populations. XIX. Origin of heterosis through natural selection in populations of *Drosophila pseudoobscura*. *Genetics*, **35**: 288–302.

DOBZHANSKY, TH. (1950*b*). Mendelian populations and their evolution. *Amer. Nat.* **84**: 401–18.

DOBZHANSKY, TH. (1951). *Genetics and the Origin of Species*. 3rd ed. New York: Columbia Univ. Press.

DOBZHANSKY, TH. (1956). Genetics of natural populations. XXV. Genetic changes in populations of *Drosophila pseudoobscura* and *Drosophila persimilis* in some localities in California. *Evolution*, **10**: 82–92.

DOBZHANSKY, TH. (1957*a*). Genetic loads in natural populations. *Science*, **126**: 191–4.

DOBZHANSKY, TH. (1957*b*). Genetics of natural populations. XXVI. Chromosomal variability in island and continental populations of *Drosophila willistoni* from Central America and the West Indies. *Evolution*, **11**: 280–93.

DOBZHANSKY, TH. (1957*c*). The *X*-chromosome in the larval salivary glands of hybrids *Drosophila insularis* × *Drosophila tropicalis*. *Chromosoma*, **8**: 691–8.

DOBZHANSKY, TH. (1958*a*). Evolution at work. *Science*, **127**: 1091–8.

DOBZHANSKY, TH. (1958*b*). Genetics of natural populations. XXVII. The genetic changes in populations of *Drosophila pseudoobscura* in the American southwest. *Evolution*, **12**: 385–401.

DOBZHANSKY, TH. (1963*a*). Species in *Drosophila*. *Proc. Linn. Soc. Lond.* **174**: 1–12.

DOBZHANSKY, TH. (1963*b*). Genetics of natural populations. XXXIII. A progress report on genetic changes in populations of *Drosophila pseudoobscura* and *Drosophila persimilis* in a locality in California. *Evolution*, **17**: 333–9.

DOBZHANSKY, TH. (1965). "Wild" and "domestic" species of *Drosophila*. *Proc. Symp. on the Genetics of Colonizing Species, Asilomar, Calif.*: 533–46.

DOBZHANSKY, TH. (1970). *Genetics of the Evolutionary Process*. New York: Columbia Univ. Press.

DOBZHANSKY, TH., ANDERSON, W. W. and PAVLOVSKY, O. (1966). Genetics of natural populations. XXXVIII. Continuity and change in populations of *Drosophila pseudoobscura* in the western United States. *Evolution*, **20**: 418–27.

DOBZHANSKY, TH. and BOCHE, R. D. (1933). Intersterile races of *Drosophila pseudoobscura* Frol. *Biol. Zentralbl.* **55**: 314–30.

DOBZHANSKY, TH., BURLA, H. and DA CUNHA, A. B. (1950). A comparative study of chromosomal polymorphism in sibling species of the *willistoni* group of *Drosophila*. *Amer. Nat.* **84**: 229–46.

DOBZHANSKY, TH. and DREYFUS, A. (1943). Chromosomal aberrations in Brazilian *Drosophila ananassae*. *Proc. Nat. Acad. Sci. U.S.A.* **29**: 301–5.

DOBZHANSKY, TH., EHRMAN, L., PAVLOVSKY, O. and SPASSKY, B. (1964). The superspecies *Drosophila paulistorum*. *Proc. Nat. Acad. Sci. U.S.A.* **51**: 3–9.

DOBZHANSKY, TH. and EPLING, C. (1944). Contributions to the genetics, taxonomy and ecology of *Drosophila pseudoobscura* and its relatives. *Publ. Carnegie Instn.* **554**, 183 pp.

DOBZHANSKY, TH. and EPLING, C. (1948). The suppression of crossing-over in inversion heterozygotes of *Drosophila pseudoobscura*. *Proc. Nat. Acad. Sci. U.S.A.* **34**: 137–41.

DOBZHANSKY, TH., HUNTER, A. S., PAVLOVSKY, O., SPASSKY, B. and WALLACE, B. (1963). Genetics of an isolated marginal population of *Drosophila pseudoobscura*. *Genetics*, **48**: 91–103.

DOBZHANSKY, TH., KRIMBAS, C. and KRIMBAS, M. G. (1960). Genetics of natural populations. XXX. Is the genetic load in *Drosophila pseudoobscura* mutational or balanced? *Genetics*, **45**: 741–53.

DOBZHANSKY, TH. and LEVENE, H. (1948). Genetics of natural populations. XVII. Proof of operation of natural selection in wild populations of *Drosophila pseudoobscura*. *Genetics*, **33**: 537–47.

DOBZHANSKY, TH. and PAVAN, C. (1943). Chromosome complements of some South American species of *Drosophila*. *Proc. Nat. Acad. Sci. U.S.A.* **29**: 268–375.

DOBZHANSKY, TH. and PAVLOVSKY, O. (1955). An extreme case of heterosis in a Central American population of *Drosophila tropicalis*. *Proc. Nat. Acad. Sci. U.S.A.* **41**: 289–95.

DOBZHANSKY, TH. and PAVLOVSKY, O. (1962). A comparative study of the chromosomes in the incipient species of the *Drosophila paulistorum* complex. *Chromosoma*, **13**: 196–218.

DOBZHANSKY, TH. and PAVLOVSKY, O. (1967). Experiments on the incipient species of the *Drosophila paulistorum* complex. *Genetics*, **55**: 141–56.

DOBZHANSKY, TH. and QUEAL, M. L. (1938). Genetics of natural populations. I. Chromosome variation in populations of *Drosophila pseudoobscura* inhabiting isolated mountain ranges. *Genetics*, **23**: 239–51.

DOBZHANSKY, TH. and SCHULTZ, J. (1931). Evidence for multiple sex factors in the *X*-chromosome of *Drosophila melanogaster*. *Proc. Nat. Acad. Sci. U.S.A.* **17**: 513–18.

DOBZHANSKY, TH. and SCHULTZ, J. (1934). The distribution of sex factors in the *X*-chromosome of *Drosophila melanogaster*. *J. Genet.* **28**: 349–86.

DOBZHANSKY, TH. and SOCOLOV, D. (1939). Structure and variation of the chromosomes in *Drosophila azteca*. *J. Hered.* **30**: 3–19.

COBZHANSKY, TH. and SPASSKY, B. (1941). Intersexes in *Drosophila pseudoobscura*. *Proc. Nat. Acad. Sci. U.S.A.* **27**: 556–62.

DOBZHANSKY, TH. and SPASSKY, B. (1959). *Drosophila paulistorum*, a cluster of species in *statu nascendi*. *Proc. Nat. Acad. Sci. U.S.A.* **45**: 419–28.

DOBZHANSKY, TH. and STURTEVANT, A. H. (1938). Inversions in the chromosomes of *Drosophila pseudoobscura*. *Genetics*, **23**: 28–64.

DOBZHANSKY, TH. and TAN, C. C. (1936a). A comparative study of the chromosome structure in two related species, *Drosophila pseudoobscura* and *D. miranda*. *Amer. Nat.* **70**: 47–8.

DOBZHANSKY, TH. and TAN, C. C. (1936*b*). Studies on hybrid sterility. III. A comparison of the gene arrangement in two species, *Drosophila pseudoobscura* and *Drosophila miranda*. *Z. indukt. Abst. Vererbl.* **72**: 99–114.

DODDS, K. S. (1939). Oogenesis in *Neuroterus baccarum* L. *Genetica*, **21**: 177–190.

DODSON, E. O. (1948). A morphological and biochemical study of lampbrush chromosomes of vertebrates. *Univ. Calif. Publ. Zool.* **53**: 281–314.

DONCASTER, L. (1906). On the maturation of the unfertilized egg and the fate of the polar bodies in the Tenthredinidae (Sawflies) *Quart. J. Micr. Sci.* **49**: 561–89.

DONCASTER, L. (1910). Gametogenesis of the gall fly, *Neuroterus lenticularis* (*Spathegaster baccarum*). I. *Proc. Roy. Soc. Lond.* B, **89**: 183–200.

DONCASTER, L. (1911). Gametogenesis and sex-determination of the gall fly, *Neuroterus lenticularis*. II. *Proc. Roy. Soc. Lond.* B, **82**: 88–113.

DONCASTER, L. (1916). Gametogenesis and sex-determination of the gall fly, *Neuroterus lenticularis* (*Spathegaster baccarum*). III. *Proc. Roy. Soc. Lond.* B, **89**: 183–200.

DONCASTER, L. and CANNON, H. G. (1919). On the spermatogenesis of the louse (*Pediculus humanus* and *P. capitis*) with observations on the maturation of the egg. *Quart. J. Micr. Sci.* **64**: 303–28.

DOTTRENS, E. (1959). Systématique des Corégones de l'Europe occidentale, basée sur une étude biométrique. *Rev. Suisse Zool.* **66**: 1–66.

DOWNES, J. A. (1962). What is an arctic insect? *Canad. Entomol.* **94**: 143–62.

DOWNES, J. A. (1964). Arctic insects and their environment. *Canad. Entomol.* **96**: 279–307.

DOZORCHEVA, R. L. (1936). The morphology of chromosomes in the ichneumon *Pteromalus puparum*. *C.R. Acad. Sci. U.R.S.S.* N.S. **3**: 339–42.

DRESCHER, W. and ROTHENBUHLER, W. C. (1964). Sex determination in the honey bee. *J. Hered.* **55**: 91–6.

DREYFUS, A. (1937). Contribução para o estudo do ciclo cromosomico e da determinação do sexo de *Rhabdias fülleborni*. *Bolm. Fac. Filos. Cïenc. S. Paulo*, Biol. Ger. no. **1**: 1–144.

DREYFUS, A. (1942). Estudos sôbre cromosomas de Gryllotalpidae brasileiros. I. Precessão, sincronismo e successão de cromosomãs sexuales. *Rev. Brasil. Biol.* **2**: 235–46.

DREYFUS, A. and BREUER, M. E. (1944). Chromosomes and sex determination in the parasitic hymenopteran *Telenomus fariai* (Lima). *Genetics*, **29**: 75–82.

DRISCOLL, C. J. and DARVEY, N. L. (1970). Chromosome pairing: effect of colchicine on an isochromosome. *Science*, **169**: 290–1.

DRONAMRAJU, K. R. (1960). Hypertrichosis of the pinna of the human ear, *Y*-linked pedigrees. *J. Genet.* **57**: 230–43.

DRONAMRAJU, K. R. (1964). *Y*-linkage in man. *Nature*, **201**: 424–5.

DRONAMRAJU, K. R. (1965). The function of the *Y*-chromosome in man, animals and plants. *Adv. in Genet.* **13**: 227–310.

DROUIN, J. A., SULLIVAN, C. R. and SMITH, S. G. (1963). Occurrence of *Pissodes terminalis* Hopp (Coleoptera: Curculionidae) in Canada: Life history, behaviour and cytogenetic identification. *Canad. Entomol.* **95**: 70–6.

DRUMMOND, F. H. (1933). The male meiotic phase in five species of marsupials. *Quart. J. Micr. Sci.* **76**: 1–11.

DRUMMOND, F. H. (1938). Meiosis in *Dasyurus viverrinus*. *Cytologia*, **8**: 343–52.

DUBININ, N. P. and NEMTSEVA, L. S. (1970). The phenomenon of "vesting" in ring chromosomes and its role in the mutation theory and understanding of the mechanism of crossing-over. *Proc. Nat. Acad. Sci. U.S.A.* **66**: 211–17.

DUBININ, N. P. and SIDOROV, B. N. (1935). The position effect of the *Hairy* gene. *Biol. Zh.* **4**: 555–68.

DUBININ, N. P., SOKOLOV, N. N. and TINIAKOV, G. G. (1936). Occurrence and distribution of chromosome aberrations in nature. *Nature,* **137**: 1035–6.

DUBININ, N. P., SOKOLOV, N. N. and TINIAKOV, G. G. (1937). Intraspecific chromosome variability. *Biol. Zh.* **6**: 1007–54.

DUBININ, N. P. and TINIAKOV, G. G. (1945). Seasonal cycles and the concentration of inversions in populations of *Drosophila funebris. Amer. Nat.* **79**: 570–2.

DUBININ, N. P. and TINIAKOV, G. G. (1946a). Natural selection and chromosomal variability in populations of *Drosophila funebris. J. Hered.* **37**: 39–44.

DUBININ, N. P. and TINIAKOV, G. G. (1946b). Structural chromosome variability in urban and rural populations of *Drosophila funebris. Amer. Nat.* **80**: 393–6.

DU BOIS, A. M. (1932a). A contribution to the embryology of *Sciara* (Diptera). *J. Morph.* **54**: 161–95.

DU BOIS, A. M. (1932b). Elimination of chromosomes during cleavage in the eggs of *Sciara* (Diptera). *Proc. Nat. Acad. Sci. U.S.A.* **18**: 352–6.

DU BOIS, A. M. (1933). Chromosome behavior during cleavage in the eggs of *Sciara coprophila* (Diptera) in relation to the problem of sex determination. *Z. Zellforsch.* **19**: 595–614.

DUNBAR, R. W. (1958). The salivary gland chromosomes of two sibling species of black flies included in *Eusimulium aureum* Fries. *Canad. J. Zool.* **36**: 23–44.

DUNBAR, R. W. (1959). The salivary gland chromosomes of seven forms of black flies included in *Eusimulium aureum* Fries. *Canad. J. Zool.* **37**: 495–525.

DUNBAR, R. W. (1965). Chromosome inversions as blocks to genetic exchange leading to sympatric speciation in black flies (Simuliidae, Diptera). *Proc. XIIth Int. Congr. Entomol. London:* 268–9.

DUNBAR, R. W. (1966). Four sibling species included in *Simulium damnosum* Theobald (Diptera: Simuliidae) from Uganda. *Nature,* **209**: 597–9.

DUNN, L. C. (1953). Variations in the segregation ratio as causes of variations of gene frequency. *Acta Genet. Statist. Med.* **4**: 139–47.

DU PRAW, E. J. (1970). *DNA and Chromosomes.* New York: Holt, Rinehart and Winston.

DURYEE, W. R. (1941). The chromosomes of the amphibian nucleus. *Univ. Pennsylvania Bicent. Conf. on Cytol. Genet. Evol.* pp. 129–141. Univ. Pennsylvania Press.

DURYEE, W. R. (1950). Chromosomal physiology in relation to nuclear structure. *Ann. N.Y. Acad. Sci.* **50**: 920–53.

DUTT, M. K. (1949). On the chromosomes of a cricket, *Liogryllus bimaculatus. Curr. Sci.* **18**: 411.

DUTT, M. K. (1954). Chromosome studies on *Rhipicephalus sanguineus* Latreille and *Hyalomma aegyptium* Newmann (Acarina: Ixodidae). *Curr. Sci.* **23**: 194–6.

DUTT, M. K. (1957). Cytology of three species of Coreid bugs with special reference to multiple sex chromosome mechanism. *Genetica,* **29**: 110–119.

ECKHARDT, R. A. and GALL, J. G. (1971). Satellite DNA association with heterochromatin in *Rhynchosciara. Chromosoma,* **32**: 407–27.

EDGAR, R. S. (1965). The bacteriophage chromosome. *Nat. Cancer Inst. Monograph,* **18**: 67–77.

EDSTRÖM, J. E. and BEERMANN, W. (1962). The base composition of nucleic acids in chromosomes, puffs, nucleoli and cytoplasm of *Chironomus* salivary gland cells. *J. Cell Biol.* **14**: 371–80.

EDWARD, D. H. D. (1963). The biology of a parthenogenetic species of *Lundstroemia* (Diptera: Chironomidae), with descriptions of the immature stages. *Proc. Roy. Entomol. Soc. Lond.* A, **38**: 165–70.

EDWARDS, J. H. (1970). The operation of selection. In: *Human Population Cytogenetics* (ed. R. A. Jacobs, W. H. Price and P. Law). pp. 241–262. Pfizer Medical Monographs no. 5, Edinburgh Univ. Press.

EDWARDS, J. H., HARNDEN, D. G., CAMERON, A. H., CROSSE, V. M. and WOLFF, O. H. (1960). A new trisomic syndrome. *Lancet,* **i**: 787–90.

EDWARDS, J. H., YUNCKEN, C., RUSHTON, D. I., RICHARDS, S. and MITTWOCH, U. (1967). Three cases of triploidy in man. *Cytogenetics,* **6**: 81–104.

EHRLICH, P. (1961). Has the biological species concept outlived its usefulness? *Syst. Zool.* **10**: 167–76.

EHRLICH, P. R. and HOLM, R. W. (1962). Patterns and populations. *Science,* **137**: 652–7.

EHRLICH, P. R. and HOLM, R. W. (1963). *The Process of Evolution.* New York: McGraw Hill.

EHRMAN, L. (1960). Genetics of hybrid sterility in *Drosophila paulistorum. Evolution,* **14**: 212–23.

EHRMAN, L. (1962). Hybrid sterility as an isolating mechanism in the genus *Drosophila. Quart. Rev. Biol.* **37**: 279–302.

EHRMAN, L. (1965). Direct observation of sexual isolation between allopatric and between sympatric strains of the different *Drosophila paulistorum* races. *Evolution,* **19**: 459–64.

EHRMAN, L. and WILLIAMSON, D. L. (1965). Transmission by injection of hybrid sterility to nonhybrid males in *Drosophila paulistorum*: preliminary report. *Proc. Nat. Acad. Sci. U.S.A.* **54**: 481–3.

EKBLOM, T. (1941). Chromosomenuntersuchungen bei *Salda littoralis* L., *Callocoris chenopodii* Fall., und *Mesovelia purcata* Muls. & Rey, sowie Studien über die Chromosomen bei verschiedenen Hemiptera–Heteroptera im Hinblick auf phylogenetische Betrachtungen. *Chromosoma,* **2**: 12–35.

ELSDALE, T. R., FISCHBERG, M. and SMITH, S. (1958). A mutation that reduces nucleolar number in *Xenopus laevis. Exp. Cell Res.* **14**: 642–3.

EMERSON, S. (1969). Linkage and recombination at the chromosome level. In *Genetic Organization,* **1** (ed.: E. W. Caspari and A. W. Ravin), pp. 267–360. New York and London: Academic Press.

EMERSON, S. and BEADLE, G. W. (1933). Crossing-over near the spindle fiber in attached-X chromosomes of *Drosophila melanogaster. Z. indukt. Abst. Vererbl.* **45**: 129–40.

EMMART, E. W. (1935). Studies on the chromosomes of *Anastrepha* (Diptera: Trypetidae). *Proc. Entomol. Soc. Wash.* **37**: 119–35.

ENTWISTLE, P. F. (1964). Inbreeding and arrhenotoky in the ambrosia beetle *Xyleborus compactus* (Eichh.) (Coleoptera: Scolytidae). *Proc. Roy. Entomol. Soc. Lond.* A, **39**: 83–8.

EPLING, C. (1944). The historical background. In: *Contributions to the Genetics, Taxonomy and Ecology of Drosophila pseudoobscura and its Relatives. Publ. Carneg. Inst.* **554**: 145–83.

EPLING, C. and LOWER, W. R. (1957). Changes in an inversion system during a hundred generations. *Evolutions,* **11**: 248–56.

EPLING, C., MITCHELL, D. F. and MATTONI, R. H. T. (1953). On the role of inversions in wild populations of Drosophila pseudoobscura. Evolution, **7**: 342–65.

EPLING, C., MITCHELL, D. F. and MATTONI, R. H. T. (1955). Frequencies of inversion combinations in the third chromosome of wild males of Drosophila pseudoobscura. Proc. Nat. Acad. Sci. U.S.A. **41**: 915–21.

EPLING, C., MITCHELL, D. F. and MATTONI, R. H. T. (1957). The relation of an inversion system to recombination in wild populations. Evolution, **11**: 225–47.

ERICKSON, J. (1965). Meiotic drive in Drosophila involving chromosome breakage. Genetics, **51**: 555–71.

ERICKSON, J. and ACTON, A. B. (1969). Spermatocyte granules in Drosophila melanogaster. Canad. J. Genet. Cytol. **11**: 153–68.

ESPER, H. and BARR, H. J. (1964). A study of the developmental cytology of a mutation affecting nucleoli in Xenopus embryos. Devel. Biol. **10**: 105–21.

EVANS, E. P., BRECKON, G. and FORD, C. E. (1964). An air-drying method for meiotic preparations from mammalian testes. Cytogenetics, **3**: 289–94.

EVANS, H. J. (1960). Supernumerary chromosomes in wild populations of the snail Helix pomatia L. Heredity, **15**: 129–38.

EVANS, H. J. (1961). Chromatid aberrations induced by gamma irradiation. I. The structure and frequency of chromatid interchanges in diploid and tetraploid cells of Vicia faba. Genetics, **46**: 257–75.

EVANS, H. J. (1962). Chromosome aberrations induced by ionising radiations. Int. Rev. Cytol. **13**: 221–321.

EVANS, H. J. (1966). Repair and recovery from chromosome damage after fractionated X-ray dosage. In: General Aspects of Radiosensitivity: Mechanisms of Repair, pp. 31–48. Vienna: Internat. Atomic Energy Auth.

EVANS, H. J., NEARY, G. J. and WILLIAMSON, F. S. (1959). The relative biological efficiency of single doses of fast neutrons and gamma rays on Vicia faba roots and the effects of oxygen. II. Chromosome damage: the production of micronuclei. Int. J. Radiat. Biol. **1**: 216–29.

EVANS, H. J. and SAVAGE, J. R. K. (1963). The relation between DNA synthesis and chromosome structure as resolved by X-ray damage. J. Cell Biol. **18**: 525–40.

EVANS, H. J., FORD, C. E., LYON, M. F. and GRAY, J. (1965). DNA replication and genetic expression in female mice with morphologically distinguishable X-chromosomes. Nature, **206**: 900–3.

EVANS, J. W. (1962). Evolution in the Homoptera: In: The Evolution of Living Organisms (ed.: G. W. Leeper). Melbourne. Univ. Press.

EVANS, W. L. (1954). Cytology of the grasshopper genus Circotettix. Amer. Nat. **88**: 21–32.

EXLINE, H. (1938). Gynandromorph spiders. J. Morph. **63**: 441–72.

FABERGÉ, A. C. (1942). Homologous chromosome pairing: the physical problem. J. Genet. **43**: 121–44.

FAHMY, O. G. (1949a). A new type of meiosis in Plastosciara pectiventris (Nematocera, Diptera) and its evolutionary significance. Proc. Egypt. Acad. Sci. **5**: 12–42.

FAHMY, O. G. (1949b). The mechanism of chromosome pairing during meiosis in male Apolipthisa subincana (Mycetophilidae, Diptera). J. Genet. **49**: 246–63.

FAHMY, O. G. (1952). The cytology and genetics of Drosophila subobscura. VI. Maturation, fertilization and cleavage in normal eggs and in the presence of the cross-over suppressor gene. J. Genet. **50**: 486–506.

FANKHAUSER, G. (1945). The effects of changes in chromosome number on amphibian development. *Quart. Rev. Biol.* **20**: 20–78.

FANKHAUSER, G. and HUMPHREY, R. R. (1950). Chromosome number and development of progeny of triploid axolotl females mated with diploid males. *J. Exp. Zool.* **115**: 207–50.

FASTEN, N. (1918). Spermatogenesis of the Pacific coast edible crab, *Cancer magister* Dana. *Biol. Bull. Woods Hole*, **34**: 277–306.

FAURÉ-FREMIET, E. (1957). Le macronucleus hétéromère de quelques Ciliés. *J. Protozool.* **4**: 7–17.

FEDERLEY, H. (1913). Das Verhalten der Chromosomen bei der Spermatogenese der Schmetterlingen *Pygaera anachoreta*, *curtula* und *pigra*, sowie einiger ihrer Bastarde. *Z. indukt. Abst. Vererbl.* **9**: 1–110.

FEDERLEY, H. (1914). Ein Beitrag zur Kenntnis der Spermatogenese bei Mischlingen zwischen Eltern verschiedener systematischer Verwandtschaft. *Öfvers. Finska Vetensk. Soc. Forh.* **56** (13): 1–28.

FEDERLEY, H. (1915). Chromosomenstudien an Mischlingen. I. Die Chromosomenkonjugation bei der Gametogenese von *Smerinthus populi* var. *astauti* × *populi*. *Öfvers. Finska Vetensk. Soc. Forh.* **57** (56): 1–36.

FEDERLEY, H. (1916). Chromosomenstudien an Mischlingen. III. Die Spermatogenese des Bastards *Chaerocampa porcellus* ♀ × *elpenor* ♂, *Öfvers. Finska Vetensk. Soc. Forh.* **58** (12): 1–17.

FEDERLEY, H. ((1928). Chromosomenverhältnisse bei Mischlingen. *Verh. V Int. Kongr. Genet. Berlin*, **1**: 271–93.

FEDERLEY, H. (1929). Über subletale und disharmonische Chromosomenkombinationen. *Hereditas*, **12**: 271–93.

FEDERLEY, H. (1931). Chromosomenanalyse der reziproken Bastarde zwischen *Pygaera pigra* und *P. curtula* sowie ihrer Rückkreutzungbastarde. *Z. Zellforsch.* **12**: 772–816.

FEDERLEY, H. (1933). Gibt es eine Geschlechtsumwandlung als Folge einer Spezieskreutzung? *Hereditas*, **18**: 91–100.

FEDERLEY, H. (1938). Chromosomenzahlen Finnländischer Lepidopteren. I. Rhopalocera. *Hereditas*, **24**: 221–69.

FEDERLEY, H. (1939). Geni e cromosomi. *Sci. Genet.* **1**: 186–205.

FEDERLEY, H. (1943). Zytologische Untersuchungen an Mischlingen der Gattung *Dicranura* (Lepidoptera). *Hereditas*, **29**: 205–54.

FEDERLEY, H. (1945). Die Konjugation der Chromosomen bei den Lepidopteren. *Soc. Sci. Fenn. Comment. Biol.* **9** (13): 1–12.

FEDYK, S. (1970). Chromosomes of *Microtus* (*Stenocranius*) *gregalis major* (Ognev, 1923) and phylogenetic connections between sub-arctic representatives of the genus *Microtus* Schrank, 1798. *Acta Theriologica, Bialowicza*, **15**: 143–152.

FERGUSON-SMITH, M. A. (1966). Sex chromatin, Klinefelter's syndrome and mental deficiency. In: *The Sex Chromatin* (ed.: K. L. Moore), pp. 277–315. London: Saunders.

FERREIRA, A. (1969). Chromosomal survey of some Australian Tettigoniids (Orthoptera–Tettigonioidea): Two species with neo-*XY* sex-determining mechanism. *Cytologia*, **34**: 511–22.

FERRIER, P. E., FERRIER, S. A. and BILL, A. H. (1968). A male pseudohermaphrodite with a dicentric *Y*-chromosome. Autoradiographic study. *Humangenetik*, **6**: 131–41.

FICQ, A. and PAVAN, C. (1961). Métabolisme des acides nucléiques et des protéines dans les chromosomes géants. *Path. Biol. Paris*, **9**: 756–7.

FIERS, W. and SINSHEIMER, R. L. (1962). The structure of the DNA of bacterio-phage ϕX174. III. Ultracentrifugal evidence for a ring structure. *J. Mol. Biol.* **5**: 424–34.

FISCHBERG, M. (1948). Experimentelle Auslösung von Heteroploidie durch Käl-tebehandlung der Eier von *Triton alpestris* aus verschiedenen Populationen. *Genetica*, **24**: 213–329.

FISCHBERG, M. and WALLACE, H. (1960). A mutation which reduces nucleolar number in *Xenopus laevis*. In: *The Cell Nucleus* (ed.: J. S. Mitchell), pp. 30–4. London: Butterworths.

FISHER, R. A. (1930). *The Genetical Theory of Natural Selection*. Oxford: Clarendon Press.

FLANDERS, S. E. (1946). Control of sex and sex-limited polymorphism in the Hymenoptera. *Quart. Rev. Biol.* **21**: 135–43.

FOGG, L. C. (1930). A study of chromatin diminution in *Ascaris* and *Ephestia*. *J. Morph.* **50**: 413–52.

FOGWILL, M. (1958). Differences in crossing-over and chromosome size in the sex cells of *Lilium* and *Fritillaria*. *Chromosoma*, **9**: 493–504.

FOOT, K. and STROBELL, E. C. (1914). The chromosomes of *Euschistus variolarius*, *Euschistus servus* and the hybrids of the F_1 and F_2 generations. *Arch. Zell-forsch.* **12**: 485–512.

FORD, C. E. (1970). The population cytogenetics of other mammalian species. In: *Human Population Cytogenetics* (*Pfizer Medical Monographs*, **5**), pp. 221–39. Edinburgh Univ. Press.

FORD, C. E. and HAMERTON, J. L. (1956). The chromosomes of man. *Nature*, **178**: 1020–3.

FORD, C. E. and HAMERTON, J. L. (1958). A system of chromosomal polymorphism in the common shrew (*Sorex araneus* L.). *Proc. XVth Int. Congr. Zool.* 177–9.

FORD, C. E., HAMERTON, J. L. and SHARMAN, G. B. (1957). Chromosome poly-morphism in the common shrew. *Nature*, **180**: 392–3.

FORD, C. E., JONES, K. W., MILLER, D. J., MITTWOCH, U., PENROSE, L. S., RIDLER, M. and SHAPIRO, A. (1959). The chromosomes in a patient showing both mongolism and the Klinefelter syndrome. *Lancet*, **i**: 709–10.

FORD, E. B. (1964). *Ecological Genetics*. London: Methuen (3rd edition 1971, London: Chapman and Hall).

FOX, D. P. (1966a). The effects of X-rays on the chromosomes of locust embryos. I. The early responses. *Chromosoma*, **19**: 300–16.

FOX, D. P. (1966b). The effects of X-rays on the chromosomes of locust embryos. II. Chromatid interchanges and the organization of the interphase nucleus. *Chromosoma*, **20**: 173–94.

FOX, D. P. (1967a). The effects of X-rays on the chromosomes of locust embryos. III. The chromatid aberration types. *Chromosoma*, **20**: 386–412.

FOX, D. P. (1967b). The effects of X-rays on the chromosomes of locust embryos. IV. Dose response and variation in sensitivity of the cell cycle for the induction of chromatid aberrations. *Chromosoma*, **20**: 413–41.

FOX, D. P. (1969a). The relationship between DNA value and chromosome volume in the Coleopteran genus *Dermestes*. *Chromosoma*, **27**: 130–44.

FOX, D. P. (1969b). DNA values in somatic tissues of *Dermestes* (Dermestidae: Coleoptera). I. Abdominal fat body and testis wall of the adult. *Chromosoma*, **28**: 445–56.

FOX, D. P. (1970a). A non-doubling DNA series in somatic tissues of the locusts *Schistocerca gregaria* (Forskål) and *Locusta migratoria* (Linn.) *Chromosoma*, **29**: 446–61.

Fox, D. P. (1970b). DNA values in somatic tissues of *Dermestes* (Dermestidae: Coleoptera). II. Malpighian tubules of the adult male. *Chromosoma*, **31**: 321–30.

Fox, D. P. (1972). DNA content of related species. *Chromosomes Today*, **3**: 32–7.

Fraccaro, M. (1962). Autosomal anomalies in man. In: *Chromosomes in Medicine* (ed.: J. L. Hamerton). *Little Club Clinic Dev. Med.* **5**: 184–206.

Fraccaro, M., Kaijser, K. and Lindsten, J. (1960). Chromosome abnormalities in father and mongol child. *Lancet*, **i**: 724–7.

Fraccaro, M. and Lindsten, J. (1964). The nature, origin and genetic implications of structural abnormalities of the sex chromosomes in man. In: *Cytogenetics of Cells in Culture* (ed.: R. J. C. Harris), pp. 97–110. New York and London: Academic Press.

Frederic, J. (1961). Contribution à l'étude du caryotype chez le poulet. *Arch. Biol.* **72**: 185–209.

Fredga, K. (1965a). New sex-determining mechanism in a mammal. *Nature*, **206**: 1176.

Fredga, K. (1965b). A new sex-determining mechanism in a mammal. Chromosomes of Indian Mongoose (*Herpestes auropunctatus*) *Hereditas*, **52**: 411–20.

Fredga, K. (1968a). Idiogram and trisomy of the water vole (*Arvicola terrestris* L.), a favourable animal for cytogenetic research. *Chromosoma*, **25**: 75–89.

Fredga, K. (1968b). Chromosomes of the masked shrew (*Sorex caecutiens* Laxm.). *Hereditas*, **60**: 269–71.

Fredga, K. (1970). Unusual sex chromosome inheritance in mammals. *Phil. Trans. Roy. Soc. Lond.* B, **259**: 15–36.

Freeman, P. (1961). The Chironomidae (Diptera) of Australia. *Austr. J. Zool.* **9**: 611–737.

Freire-Maia, N. (1952). Chromosomal variation in Brazilian domestic species of *Drosophila. Dros. Inf. Service*, **26**: 100.

Freire-Maia, N., Zanardini, I. F. and Freire-Maia, A. (1953). Chromosome variation in *Drosophila immigrans. Dusenia*, **4**: 303–11.

Friedländler, M. and Wahrman, J. (1966). Giant centrioles in neuropteran meiosis. *J. Cell Sci.* **1**: 129–44.

Friedländler, M. and Wahrman, J. (1970). The spindle as a basal body distributor. A study in the meiosis of the male silkworm moth, *Bombyx mori. J. Cell Sci.* **7**: 65–89.

Friesen, H. (1936). Spermatogoniales crossing over bei *Drosophila. Z. indukt. Abst. Vererbl.* **71**: 501–26.

Frizzi, G. (1947a). Cromosomi salivari in *Anopheles maculipennis. Sci. Genet.* **3**: 67–79.

Frizzi, G. (1947b). Determinazione del sesso nel genere *Anopheles. Sci. Genet.* **3**: 80–8.

Frizzi, G. (1947c). Salivary gland chromosomes of *Anopheles. Nature*, **160**: 226–7.

Frizzi, G. (1949). Genetica di popolazioni in *Anopheles maculipennis. Ric. Sci.* **19**: 544–52.

Frizzi, G. (1950). Studio sulla sterilità degli ibridi nel genere *Anopheles*. I. Sterilità nel incrocio fra *Anopheles mac. atroparvus* ed *Anopheles mac. typicus* e nel reincrocio dei cromosomi salivari. *Sci. Genet.* **3**: 260–70.

Frizzi, G. (1951). Dimorfismo cromosomico in *Anopheles maculipennis messeae. Sci. Genet.* **4**: 79–93.

FRIZZI, G. (1952). Nuovi contributi e prospettive di ricerca nel gruppo *Anopheles maculipennis* in base allo studio del dimorfismo cromosomico (ordinamento ad *X* invertito e tipico nel *messeae*). *Symp. Genet.* **3**: 231–65.

FRIZZI, G. (1953). Étude cytogénétique d'*Anopheles maculipennis* en Italie. Extension des réchèrches à d'autres espèces d'anopheles. *Bull. World Health Org.* **9**: 335–44.

FRIZZI, G. and CONTINI, C. (1962). Studio introduttivo citogenetico su alcune specie di Dixidae della Sardegna. *Boll. di Zool.* **29**: 621–33.

FRIZZI, G. and DE CARLI, L. (1954). Studio preliminare comparativo genetico e citogenetico fra alcune specie nordamericane di *Anopheles maculipennis* e l'*Anopheles maculipennis atroparvus* Italiano. *Symp. Genet. Pavia*, **2**: 184–206.

FROLOVA, S. L. (1924). Die Ei- und Samenreifung bei *Chermes strobilobius* und *Chermes pectinatae*. *Z. Zell.-u. Gewebel.* **1**: 29–56.

FROLOVA, S. (1929). Die Polyploidie einiger Gewebe bei Dipteren. *Z. Zellforsch.* **8**: 542–65.

FRÖST, S. (1959). The cytological behaviour and mode of transmission of accessory chromosomes in *Plantago serraria*. *Hereditas*, **45**: 191–210.

GABRUSEWYCZ-GARCIA, N. (1964). Cytological and autoradiographic studies in *Sciara coprophila* salivary gland chromosomes. *Chromosoma*, **15**: 312–44.

GABRUSEWYCZ-GARCIA, N. (1971). Studies in polytene chromosomes of Sciarids. I. The salivary chromosomes of *Sciara* (*Lycoriella*) *pauciseta* (II) Felt. *Chromosoma*, **33**: 421–35.

GABRUSEWYCZ-GARCIA, N. and KLEINFELD, R. G. (1966). A study of the nucleolar material in *Sciara coprophila*. *J. Cell. Biol.* **29**: 347–59.

GALGANO, M. (1933*a*). Studi intorno al comportamento della cromatina nella spermatogenesi di *Rana esculenta* L. Il numero e la forma dei cromosomi nel processo normale. *Arch. Ital. Anat. Embriol.* **31**: 1–86.

GALGANO, M. (1933*b*). Evoluzione dei spermatociti di I ordine e cromosomi pseudosessuali in alcune specie di anfibi. *Arch. Ital. Anat. Embriol.* **32**: 171–200.

GALL, J. G. (1952). The lampbrush chromosomes of *Triturus viridescens*. *Exp. Cell Res.* suppl. **2**: 95–102.

GALL, J. G. (1954). Lampbrush chromosomes from oocyte nuclei of the newt. *J. Morph.* **94**: 283–352.

GALL, J. G. (1955). Problems of structure and function in the Amphibian oocyte nucleus. *Symp. Soc. Exp. Biol.* **9**: 358–370.

GALL, J. G. (1956). On the submicroscopic structure of chromosomes. *Brookhaven Symp. Biol.* **8**: 17–32.

GALL, J. G. (1958). Chromosomal differentiation. McCollum-Pratt symposium on *The Chemical Basis of Development*, pp. 103–35. Baltimore: Johns Hopkins Press.

GALL, J. G. (1959). Macronuclear duplication in the ciliated protozoan *Euplotes*. *J. Biophys. Biochem. Cytol.* **5**: 295–308.

GALL, J. G. (1961). Centriole replication. A study of spermatogenesis in the snail *Viviparus*. *J. Biophys. Biochem. Cytol.* **10**: 163–93.

GALL, J. G. (1963*a*). Kinetics of deoxyribonuclease action on chromosomes. *Nature*, **198**: 36–8.

GALL, J. G. (1963*b*). Chromosomes and cytodifferentiation. In: *Cytodifferentiation and Macromolecular Synthesis* (ed.: M. Locke), pp. 119–43. New York: Academic Press.

GALL, J. G. (1963c). Chromosome fibers from an interphase nucleus. *Science*, **139**: 120–1.

GALL, J. G. (1966). Chromosome fibers studied by a spreading technique. *Chromosoma*, **20**: 221–33.

GALL, J. G. (1967). Octagonal nuclear pores. *J. Cell Biol.* **32**: 391–9.

GALL, J. G. (1968). Differential synthesis of the genes for ribosomal RNA during amphibian oogenesis. *Proc. Nat. Acad. Sci. U.S.A.* **60**: 553–60.

GALL, J. G. (1969). The genes for ribosomal RNA during oogenesis. *Genetics*, **61**: 121–32.

GALL, J. G. and CALLAN, H. G. (1962). H^3-uridine incorporation in lampbrush chromosomes. *Proc. Nat. Acad. Sci. U.S.A.* **48**: 562–70.

GALL, J. G., MACGREGOR, H. C. and KIDSTON, M. E. (1969). Gene amplification in the oocytes of Dytiscid water beetles. *Chromosoma*, **26**: 169–87.

GALLIEN, L. (1954). Démonstration de l'homogamétie du sexe mâle chez le Triton *Pleurodeles waltlii* Michah. par l'étude de la descendance d'animaux à sexe physiologique inversé après un traitement hormonal gynogène (benzoate d'oestradiol). *C.R. Acad. Sci. Paris*, **238**: 402–4.

GALLIEN, L. (1956). Consequences du changement du sexe pour la descendance d'un amphibien anoure *Xenopus laevis* Daudin. Femelles donnant une progéniture exclusivement mâle. *C.R. Acad. Sci. Paris*, D **241**: 998–1000.

GALLIEN, L., LABROUSSE, M. and LACROIX, J. C. (1966). Détection sur les chromosomes mitotiques et sur les chromosomes en ecouvillon (lampbrush) d'anomalies provoquées par l'irradiation de l'oeuf, chez l'amphibien urodèle *Pleurodeles waltlii* Michah. *C.R. Acad. Sci. Paris*, D **263**: 1984–7.

GASSNER, G. (1967). Synaptinemal complexes: recent findings. *J. Cell Biol.* **35**: 166A–167A.

GASSNER, G. (1969). Synaptinemal complexes in the achiasmatic spermatogenesis of *Bolbe nigra* Giglio-Tos (Mantoidea). *Chromosoma*, **26**: 22–34.

GATENBY, J. B. (1917). The degenerate (apyrene) sperm formation of moths as an index to the interrelations of the various bodies of the spermatozoon. *Quart. J. Micr. Sci.* **62**: 465–88.

GATES, R. R. (1953). Polyploidy and the sex chromosomes. *Acta Biotheoretica*, **11**: 27–44.

GAY, H. (1956). Nucleo-cytoplasmic relations in *Drosophila*. *Cold Spr. Harb. Symp. Quant. Biol.* **21**: 257–69.

GAY, H. (1959). In: *The Nature of the Materials of Heredity. Yearb. Carnegie Instn.* **58**: 440–9.

GAY, H., DAS, C. C., FORWARD, K. and KAUFMANN, B. P. (1970). DNA content of mitotically active condensed chromosomes of *Drosophila melanogaster*. *Chromosoma*, **32**: 213–23.

GEARD, C. R. and PEACOCK, W. J. (1969). Sister chromatid exchanges in *Vicia faba*. *Mutation Res.* **7**: 215–23.

GEITLER, L. (1937). Die Analyse des Kernbaus und der Kernteilung der Wasserläufer *Gerris lateralis* und *Gerris lacustris* (Hemiptera, Heteroptera) und die Somadifferenzierung. *Z. Zellforsch.* **26**: 641–72.

GEITLER, L. (1938). Über den Bau des Ruhekerns mit besonderer Berücksichtigung der Heteropteren und Dipteren. *Biol. Zentralbl.* **58**: 152–79.

GEITLER, L. (1939a). Die Entstehung der polyploiden Somakerne der Heteropteren durch Chromosomenteilung ohne Kernteilung. *Chromosoma*, **1**: 1–22.

GEITLER, L. (1939b). Das Heterochromatin der Geschlechtschromosomen bei Heteropteren. *Chromosoma*, **1**: 197–229.

GEITLER, L. (1940). Neue Untersuchungen über Bau und Wachstum des Zellkerns in Geweben. *Naturwiss.* **28**: 241–8.

GEITLER, L. (1941). Das Wachstum des Zellkerns in tierischen und pflanzlichen Geweben. *Ergebn. Biol.* **18**: 1–54.

GEITLER, L. (1953). *Endomitose und endomitotische Polyploidisierung. Protoplasmatologia Handbuch der Protoplasmaforschung*, **VIC**: 1–89. Berlin: Springer-Verlag.

GELEI, J. (1921). Weitere Studien über die Oogenese des *Dendrocoelum lacteum*. II. Die Längskonjugation der Chromosomen. *Arch. Zellforsch.* **16**: 88–169.

GELEI, J. (1922). Weitere studien über die Oogenese des *Dendrocoelum lacteum*. III. Die Konjugations-frage der Chromosomen in der Literatur und meine Befunde. *Arch. Zellforsch.* **16**: 299–370.

GEORGE, K. P. (1970). Cytochemical differentiation along human chromosomes. *Nature*, **226**: 80–1.

GERMAN, J. L. (1962). DNA Synthesis in human chromosomes. *Trans. N.Y. Acad. Sci.* **24**: 395–407.

GERMAN, J. (1964a). Identification and characterization of human chromosomes by DNA replication sequence. *Symp. Int. Soc. Cell Biol.* **3**: 191–207. New York: Academic Press.

GERMAN, J. (1964b). The pattern of DNA synthesis in the chromosomes of human blood cells. *J. Cell Biol.* **20**: 37–55.

GERSH, E. S. (1967). Genetic effects associated with band 3C1 of the salivary gland *X*-chromosome of *Drosophila melanogaster*. *Genetics*, **56**: 309–19.

GERSHENSON, S. (1928). A new sex ratio abnormality in *Drosophila obscura*. *Genetics*, **13**, 488–507.

GEYER-DUSZYŃSKA, I. (1959). Experimental research on chromosome elimination in Cecidomyidae (Diptera). *J. Exp. Zool.* **141**: 391–448.

GEYER-DUSZYŃSKA, I. (1961). Chromosome behavior in spermatogenesis of Cecidomyiidae (Diptera). *Chromosoma*, **11**: 499–513.

GEYER-DUSZYŃSKA, I. (1963). On the structure of the *XY* bivalent in *Mus musculus*. *Chromosoma*, **13**: 521–5.

GIANNELLI, F. (1963). The pattern of *X*-chromosome deoxyribonucleic acid synthesis in two women with abnormal sex chromosome complements. *Lancet*, **i**: 863–5.

GIARDINA, A. (1901). Origine dell'oocite e delle cellule nutrici nel *Dytiscus*. *Int. Monatschr. Anat. Physiol.* **18**: 418–84.

GIBSON, I. and HEWITT, G. M. (1970). Isolation of DNA from *B*-chromosomes in grasshoppers. *Nature*, **225**: 67–8.

GILBERT, C. W., MULDAL, S., LAJTHA, L. J. and ROWLEY, J. (1962). Time sequence of human chromosome duplication. *Nature*, **195**: 869–73.

GILCHRIST, B. M. and HALDANE, J. B. S. (1947). Sex linkage and sex determination in a mosquito, *Culex molestus*. *Hereditas*, **33**: 175–90.

GILES, N. H. JR. (1954). Radiation-induced chromosome aberrations in *Tradescantia*. In: *Radiation Biology*, **1**(2) (ed.: A. Hollaender), pp. 713–62. New York: McGraw-Hill.

GIMÉNEZ-MARTÍN, G., LÓPEZ-SÁEZ, J. F. and GONZÁLEZ-FERNÁNDEZ, A. (1963). Somatic chromosome structure (observations with the light microscope). *Cytologia*, **28**: 381–9.

GOIN, O. B., GOIN, C. J. and BACHMANN, K. (1967). The nuclear DNA of a caecilian. *Copeia* 1967, p. 233.

GOLDSCHMIDT, E. (1946). Polyploidy and parthenogenesis in the genus *Saga*. *Nature*, **158**: 587.

GOLDSCHMIDT, E. (1952). Fluctuation in chromosome number in *Artemia salina*. *J. Morph.* **91**: 111–34.

GOLDSCHMIDT, E. (1953). Multiple sex-chromosome mechanisms and polyploidy in animals. *J. Genet.* **51**: 434–40.

GOLDSCHMIDT, E. (1956a). Chromosomal polymorphism in a population of *Drosophila subobscura*, from Israel. *J. Genet.* **54**: 474–95.

GOLDSCHMIDT, E. (1956b). Structural polymorphism in the Israel race of *Drosophila subobscura. Bull. Res. Counc. Israel*, **5B**: 150.

GOLDSCHMIDT, E. (1958). Polymorphism and co-adaption in natural populations of *Drosophila subobscura. Proc. Xth Int. Congr. Entomol. Montreal*, **2**: 821–8.

GOLDSCHMIDT, R. (1931a). *Die sexuellen Zwischenstufen*. Berlin: J. Springer.

GOLDSCHMIDT, R. (1931b). Analysis of intersexuality in the gypsy moth. *Quart. Rev. Biol.* **6**: 125–42.

GOLDSCHMIDT, R. (1932). Untersuchungen zur Genetik der geographischen Variation. III. Abschliessendes über die Geschlechtsrassen von *Lymantria dispar* Roux. *Arch. Entwicklungsm. Organ.* **126**: 277–324.

GOLDSCHMIDT, R (1934). *Lymantria. Bibliogr. Genet.* **11**: 1–186.

GOLIKOWA, M. N. (1965). Der Aufbau des Kernapparates und die Verteilung der Nukleinsäuren und Proteine bei *Nyctotherus cordifornis* Stein. *Arch. Protistenk.* **108**: 191–216.

GOOCH, P. C. and FISCHER, C. L. (1969). High frequency of a specific chromosome abnormality in leukocytes of a normal female. *Cytogenetics*, **8**: 1–8.

GOODMAN, R. M., GOIDL, J. and RICHART, R. M. (1967). Larval development in *Sciara coprophila* without the formation of chromosomal puffs. *Proc. Nat. Acad. Sci. U.S.A.* **58**: 553–9.

GOODRICH, H. B. (1914). The maturation divisions in *Ascaris incurva. Biol. Bull. Woods Hole*, **27**: 147–50.

GOODRICH, H. B. (1916). The germ cells in *Ascaris incurva. J. Exp. Zool.* **21**: 61–100.

GORDON, M. (1937). Genetics of *Platypoecilus*. III. Inheritance of sex and crossing-over of the sex chromosomes in the platyfish. *Genetics*, **22**: 376–92.

GORDON, M. (1946). Interchanging genetic mechanisms for sex determination in fishes under domestication. *J. Hered.* **37**: 307–20.

GORDON, M. (1947). Genetics of *Platypoecilus maculatus*. IV. The sex-determining mechanism in two wild populations of the Mexican platyfish. *Genetics*, **32**: 8–17.

GORDON, M. (1951). Genetics of *Platypoecilus maculatus*. V. Heterogametic sex-determining mechanism in females of a domesticated stock originally from British Honduras. *Zoologica*, N.F. **37**: 91–100.

GORMAN, G. C. (1965). Interspecific karyotypic variation as a systematic character in the genus *Anolis* (Sauria: Iguanidae). *Nature*, **208**: 95–7.

GORMAN, G. C. (1968). The chromosomes of *Anolis chrysolepis scypheus* from a developing egg. *Herpetologica*, **24**: 263–4.

GORMAN, G. C. (1969). Chromosomes and evolution in the lizard genus *Anolis*. *Mammal. Chrom. Newsl.* **10**(1): 37.

GORMAN, G. C. and ATKINS, L. (1966). Chromosomal heteromorphism in some male lizards of the genus *Anolis. Amer. Nat.* **100**: 579–83.

GORMAN, G. C. and ATKINS, L. (1967). The relationships of the *Anolis* of the *roquet* species group (Sauria: Iguanidae). II. Comparative chromosome cytology. *Syst. Zool.* **16**: 137–43.

GORMAN, G. C. and ATKINS, L. (1968). Natural hybridization between two sibling species of *Anolis* lizards: chromosome cytology. *Science*, **159**: 1358–60.

GORMAN, G. C., ATKINS, L. and HOLZINGER, T. (1967). New karyotypic data on 15 genera of lizards in the family Iguanidae, with a discussion of taxonomic and cytological implications. *Cytogenetics*, **6**: 286–99.

GORMAN, G. C. and GRESS, F. (1970). Sex chromosomes of a Pygopodid lizard *Lialis burtonis*. *Experientia*, **26**: 206–7.

GORMAN, G. C., THOMAS, R. and ATKINS, L. (1968). Intra- and interspecific chromosome variation in the lizard *Anolis cristatellus* and its closest relatives. *Breviora, Mus. Comp. Zool. Harvard Univ.* **293**: 1–13.

GOTO, E. and YOSIDA, T. H. (1953). Sex chromosomes of *Rhaphidopalpa femoralis* Motsch. (Coleoptera, Chrysomelidae). *Kromosomo*, **17**–9: 674–6.

GÖTZ, W. (1965a). Beitrag zur Kenntnis der Inversionen, Duplikationen und Strukturtypen von *Drosophila subobscura* Coll. *Z. Vererbl.* **96**: 285–96.

GÖTZ, W. (1965b). Chromosomaler Polymorphismus in einem Muster von *Drosophila subobscura* aus Marokko, mit Darstellung der Heterozygotieverhältnisse aus Heterozygotiediagramm. *Z. Vererbl.* **97**: 40–5.

GÖTZ, W. (1967). Untersuchungen über den chromosomalen Strukturpolymorphismus in kleinasiatischen und persischen Populationen von *Drosophila subobscura* Coll. *Mol. Gen. Genet.* **100**: 1–38.

GOWEN, J. W. (1928). On the mechanism of chromosome behavior in male and female *Drosophila*. *Proc. Nat. Acad. Sci. U.S.A.* **14**: 475–7.

GOWEN, J. W. (1933). Meiosis as a genetic character in *Drosophila melanogaster*. *J. Exp. Zool.* **65**: 83–106.

GOWEN, J. W. and FUNG, S.-T. C. (1957). Determination of sex through genes in a major sex locus in *Drosophila melanogaster*. *Heredity*, **11**: 397–402.

GOWEN, J. W. and GAY, E. H. (1933). Eversporting as a function of the *Y*-chromosome in *Drosophila melanogaster*. *Proc. Nat. Acad. Sci. U.S.A.* **19**: 122–6.

GRAHAM, C. F. and MORGAN, R. (1966). Changes in the cell cycle during early amphibian development. *Devel. Biol.* **14**: 439–60.

GRANATA, L. (1910). Le cinesi spermatogenetiche di *Pamphagus marmoratus* (Burm). *Arch. Zellforsch.* **5**: 182–214.

GRANT, V. (1964). *The Architecture of the Germplasm*. New York: John Wiley.

GREEN, M. M. (1955). Pseudoallelism and the gene concept. *Amer. Nat.* **89**: 65–71.

GREEN, M. M. (1961). Phenogenetics of the *lozenge* loci in *Drosophila melanogaster*. II. Genetics of *lozenge-Krivshenko* (lz^K). *Genetics*, **46**: 1169–76.

GREEN, M. M. (1963). Pseudoalleles and recombination in *Drosophila*. In: *Methodology in Basic Genetics* (ed.: W. J. Burdette), pp. 279–90. San Francisco: Holden-Day.

GREEN, M. M. (1965). Genetic fine structure in *Drosophila*. *Proc. XIth Int. Genet. Congr. The Hague*, **2**: 37–49.

GREEN, M. M. and GREEN, K. C. (1949). Crossing-over between alleles at the *lozenge* locus in *Drosophila melanogaster*. *Proc. Nat. Acad. Sci. U.S.A.* **35**: 586–91.

GREEN, M. M. and GREEN, K. C. (1956). A cytogenetic analysis of the *lozenge* pseudoalleles in *Drosophila*. *Z. indukt. Abst. Vererbl.* **87**: 708–21.

GREENSHIELDS, F. (1936). Tetraploidy and Hymenoptera. *Nature*, **138**: 330.

GRELL, K. G. (1940). Der Kernphasenwechsel von *Stylocephalus* (*Stylorhynchus*) *longicollis* P. Stein. Ein Beitrag zur Frage der Chromosomenreduktion der Gregarinen. *Arch. Protistenk.* **94**: 161–200.

GRELL, K. G. (1949). Die Entwicklung der Makronucleusanlage im Exconjuganten von *Ephelota gemmipara* R. Herwig. *Biol. Zentrabl.* **68**: 289–312.

GRELL, K. G. (1950). Der Kerndualismus der Ciliaten und Suktorien. *Naturwiss.* **37**: 347–56.

GRELL, K. G. (1952). Der Stand unserer Kenntnisse über den Bau der Protisten-kerne. *Verh. deutsch. Zool. Ges. Freiburg.* pp. 212–51.

GRELL, K. G. (1953). Die Chromosomen von *Aulacantha scolymantha* Haeckel. *Arch. Protistenk.* **99**: 1–54.

GRELL, K. G. (1956). Protozoa and algae. *Ann. Rev. Microbiol.* **10**: 307–28.

GRELL, K. G. (1958). Untersuchungen über die Fortpflanzung und Sexualität der Foraminiferen. II. *Rubratella intermedia. Arch. Protistenk.* **102**: 291–308.

GRELL, K. G. (1962). Morphologie und Fortpflanzung der Protozoen (Einschliesslich Entwicklungsphysiologie und Genetik). *Fortschr. Zool.* **14**: 1–85.

GRELL, R. F. (1962*a*). The chromosome. *J. Tenn. Acad. Sci.* **37**: 43–53.

GRELL, R. F. (1962*b*). A new hypothesis on the nature and sequence of meiotic events in the female of *Drosophila melanogaster. Proc. Nat. Acad. Sci. U.S.A.* **48**: 165–72.

GRELL, R. F. (1964*a*). Chromosome size at distributive pairing in *Drosophila melanogaster* females. *Genetics,* **50**: 151–66.

GRELL, R. F. (1964*b*). Distributive pairing: the size-dependent mechanism for regular segregation of the fourth chromosomes in *Drosophila melanogaster. Proc. Nat. Acad. Sci. U.S.A.* **52**: 226–32.

GRELL, R. F. (1965). Chromosome pairing, crossing-over and segregation in *Drosophila melangaster. Int. Sympos. Genes and Chromosomes – Structure and Function. Nat. Cancer Inst. Monograph,* **18**: 215–42.

GRELL, R. F. (1967). Pairing at the chromosomal level. *J. Cell Physiol.* suppl. 1 to **70**: 119–46.

GRELL, R. F. (1969). Meiotic and somatic pairing. Chapter 6, pp. 361–492. In: *Genetic Organization,* **1** (ed.: E. W. Caspari and A. W. Ravin). New York: Academic Press.

GRELL, S. M. (1946*a*). Cytological studies in *Culex*. I. Somatic reduction divisions. *Genetics,* **31**: 60–76.

GRELL, S. M. (1946*b*). Cytological studies in *Culex*. II. Diploid and meiotic divisions. *Genetics,* **31**: 77–94.

GRIFFEN, A. B. and BUNKER, M. C. (1964). Three cases of trisomy in the mouse. *Proc. Nat. Acad. Sci. U.S.A.* **52**: 1194–8.

GRIFFEN, A. B. and BUNKER, M. C. (1967). Four further cases of autosomal primary trisomy in the mouse. *Proc. Nat. Acad. Sci. U.S.A.* **58**: 1446–52.

GRIFFITHS, R. B. (1941). Triploidy (and haploidy) in the newt, *Triturus viridescens,* induced by refrigeration of fertilized eggs. *Genetics,* **26**: 69–88.

GRINBERG, M. A., SULLIVAN, M. M. and BENIRSCHKE, K. (1966). Investigation with tritiated thymidine of the relationship between the sex chromosomes, sex chromatin and the drumsticks in the cells of the female nine-banded armadillo, *Dasypus novemcinctus. Cytogenetics,* **5**: 64–74.

GRINCHUK, T. M. (1967). A study of the polymorphism of polytene chromosomes of the Adirondack black fly *Prosimulium hirtipes* (Diptera, Simuliidae) indigenous to the Leningrad region. *Genetica,* no. **1**: 165–72.

GRIPENBERG, U. (1967). The cytological behavior of a human ring chromosome. *Chromosoma,* **20**: 284–9.

GROPP, A., TETTENBORN, U. and VON LEHMANN, E. (1970). Chromosomenvaria-tionen vom Robertson'schen Typus bei der Tabakmaus *M. poschiavinus* und ihren Hybriden mit der Laboratoriumsmaus. *Cytogenetics,* **9**: 9–23.

GROPP, A., TETTENBORN, U. and LÉONARD, A. (1970). Identification of acrocentric chromosomes involved in the formation of fusion metacentrics in mice. Proposal for nomenclature of *M. poschiavinus* metacentrics. *Experientia,* **26**: 1018–9.

GROSS, F. (1932). Untersuchungen über die Polyploidie und die Variabilität bei *Artemia salina*. *Naturwiss.* **20**: 962–7.

GROSS, F. (1935). Die Reifungs- und Furchungsteilungen von *Artemia salina* in Zusammenhang mit dem Problem des Kernteilungs-mechanismus. *Z. Zellforsch.* **23**: 522–65.

GROSSBACH, U. (1968). Cell differentiation in the salivary glands of *Camptochironomus tentans* and *C. pallidivittatus. Ann. Zool. Fenn.* **5**: 37–40.

GROSSBACH, U. (1969). Chromosomen-Aktivität und biochemische Zelldifferenzierung in den Speicheldrüsen von *Camptochironomus. Chromosoma,* **28**: 136–87.

GRUMBACH, M. M., MARKS, P. A. and MORISHIMA, A. (1962). Erythrocyte glucose 6-phosphate dehydrogenase activity and *X*-chromosome polysomy. *Lancet,* **i**: 1330–2.

GRUMBACH, M. M. and MORISHIMA, A. (1962). Sex chromatin and the sex chromosomes; on the origin of sex chromatin from a single *X*-chromosome. *Acta Cytol.* **6**: 46–60.

GRUMBACH, M. M., MORISHIMA, A. and TAYLOR, J. H. (1963). Human sex chromosome abnormalities in relation to DNA replication and heterochromatinization. *Proc. Nat. Acad. Sci. U.S.A.* **49**: 581–9.

GRÜNEBERG, H. (1937). The position effect proved by a spontaneous reinversion of the *X*-chromosome in *Drosophila melanogaster. J. Genet.* **34**: 169–89.

GRÜNEBERG, H. (1969). Threshold phenomena versus cell heredity in the manifestation of sex-linked genes in mammals. *J. Embryol. Exp. Morph.* **22**: 145–79.

GUÉNIN, H. A. (1948). La formule chromosomique de *Blaps mortisaga* L. *Experientia,* **4**: 221.

GUÉNIN, H. A. (1949). L'evolution de la formule chromosomique dans le genre *Blaps* (Coleopt. Tenebr.). *Rev. Suisse Zool.* **56**: 336.

GUÉNIN, H. A. (1952). Hétérochromosomes de Cicindèles. *Rev. Suisse Zool.* **59**: 277–82.

GUÉNIN, H. A. (1953). Les chromosomes sexuels multiples de *Blaps polychresta* Forsk. (Col. Tenebr.). *Rev. Suisse Zool.* **60**: 462–6.

GUÉNIN, H. A. (1956). Le complexe hétérochromosomique de *Caenoblaps nitida* Schüst (Col. Ténébr. Blaptinae). *Rev. Suisse Zool.* **63**: 298–303.

GUÉNIN, H. A. (1957). Contribution à la connaissance cytologique des scorpions: les chromosomes de *Pandinus imperator* Koch. *Rev. Suisse Zool.* **64**: 349–53.

GUÉNIN, H. A. (1961). Contribution à la connaissance cytologique des scorpions: les chromosomes de *Buthus occitanus* Amor. *Vie et Milieu,* **12**: 89–96.

GUNSON, M. M., SHARMAN, G. B. and THOMSON, J. A. (1968). The affinities of *Burramys* (Marsupialia: Phalangeroidea) as revealed by a study of its chromosomes. *Aust. J. Sci.* **31**: 40–1.

GUPTA, M. L. (1964). Chromosome number and sex-chromosome mechanism in fifteen species of the Indian praying mantids. *Curr. Sci.* **33**: 369–70.

GUPTA, M. L. (1966a). Chromosome number and sex-chromosome mechanism in some more species of the Indian mantids. *Experientia,* **22**: 457.

GUPTA, M. L. (1966b). A preliminary account of the meiotic mechanism in nineteen species of the Indian mantids. *Res. Bull. Panjab Univ.* **17**: 421–2.

GUPTA, Y. (1964). Chromosomal studies in some Indian Lepidoptera. *Chromosoma,* **15**: 540–61.

GUSTAFSSON, Å. (1947). Apomixis in higher plants. Part III. Biotype and species formation. *Lunds Univ. Årsskr.* **43** (12): 183–370.

GUSTAVSSON, I., FRACCARO, M., TIEPOLO, L. and LINDSTEN, J. (1968). Presumptive X-autosome translocation in a cow: preferential inactivation of the normal X-chromosome. *Nature*, **218**: 183–4.

GUSTAVSSON, I. and SUNDT, C. O. (1966). Chromosome complex of the family Canidae. *Hereditas*, **54**: 249–54.

GUSTAVSSON, I. and SUNDT, C. O. (1968). Karyotypes in five species of deer (*Alces alces* L., *Capreolus capreolus* L., *Cervus elaphus* L., *Cervus nippon nippon Temm.* and *Dama dama* L.). *Hereditas*, **60**: 233–48.

GUSTAVSSON, I. and SUNDT, C. O. (1969). Three polymorphic chromosome systems of centric fusion type in a population of Manchurian Sika Deer (*Cervus nippon hortulorum* Swinhoe). *Chromosoma*, **28**: 245–54.

GUTHERZ, S. (1907). Kenntniss der Heterochromosomen. *Arch. Mikr. Anat. Entwicklungsm.* **69**: 491–514.

GUYÉNOT, E. and DANON, M. (1953). Chromosomes et ovocytes des Batraciens. *Rev. Suisse Zool.* **60**: 1–129.

GUYÉNOT, E. and NAVILLE, A. (1929). Les chromosomes et la réduction chromatique chez *Drosophila melanogaster* (cinèses somatiques, spermatogénèse, ovogénèse). *Cellule*, **39**: 25–82.

HAAS, G. and GOLDSCHMIDT, E. (1946). A decaploid strain of *Artemia salina*. *Nature*, **158**: 329.

HACKMAN, W. (1948). Chromosomenstudien an Araneen mit besonderer Berücksichtigung der Geschlechtschromosomen. *Acta Zool. Fenn.* **54**: 1–101.

HADLINGTON, P. and SHIPP, E. (1961). Diapause and parthenogenesis in the eggs of three species of Phasmatodea. *Proc. Linn. Soc. N.S.W.* **86**: 268–79.

HADORN, E., RUCH, F. and STAUB, M. (1964). Zum DNS Gehalt in Speicheldrüsenkernen mit "übergrössen Riesenchromosomen" von *Drosophila melanogaster. Experientia*, **20**: 566.

HALDANE, J. B. S. (1922). Sex ratio and unisexual sterility in hybrid animals. *J. Genet.* **12**: 101–9.

HALDANE, J. B. S. (1931). The cytological basis of genetical interference. *Cytologia*, **3**: 54–65.

HALDANE, J. B. S. (1936). A search for incomplete sex-linkage in man. *Ann. Eug.* **7**: 28–57.

HALDANE, J. B. S. (1957). The conditions for coadaptation in polymorphism for inversions. *J. Genet.* **55**: 218–25.

HALDANE, J. B. S. (1958). The cost of natural selection. *J. Genet.* **55**: 511–24.

HALDANE, J. B. S. (1962). Conditions for stable polymorphism at an autosomal locus. *Nature*, **193**: 1108.

HALFER, C., TIEPOLO, L., BARIGOZZI, C. and FRACCARO, M. (1969). Timing of DNA replication of translocated Y-chromosome sections in somatic cells of *Drosophila melanogaster. Chromosoma*, **27**: 395–408.

HALKKA, L. and HALKKA, O. (1968). RNA and protein in nucleolar structures of dragonfly oocytes. *Science*, **162**: 803–5.

HALKKA, O. (1956). Studies on mitotic and meiotic cell division in certain Hemiptera under normal and experimental conditions. *Ann. Acad. Sci. Fenn.* AIV, **32**, 1–80.

HALKKA, O. (1959). Chromosome studies on the Hemiptera Homoptera Auchenorrhyncha. *Ann. Acad. Sci. Fenn.* AIV, **43**: 1–71.

HALKKA, O. (1964). Recombination in six homopterous families. *Evolution*, **18**: 81–8.

HALKKA, O., HALKKA, L. and NYHOLM, M. (1969). Lamellate bodies in the nuclei of dragonfly (Odonata) oocytes. *Z. Zellforsch.* **94**: 534–41.

HALKKA, O. and SKARÉN, U. (1964). Evolution chromosomique chez genre *Sorex*: nouvelle information. *Experientia*, **20**: 314–15.

HALKKA, O., SKARÉN, V. and HALKKA, L. (1970). The karyotypes of *Sorex isodon* Turov and *S. minutissimus* Zimm. *Ann. Acad. Sci. Fenn.* AIV, **161**: 1–5.

HALL, W. P. (1970). Three probable cases of parthenogenesis in lizards (Agamidae, Chamaeleontidae, Geckonidae). *Experientia*, **26**: 1271–3.

HAMERTON, J. L. (1968). Significance of sex chromosome derived heterochromatin in mammals. *Nature*, **219**: 910–11.

HAMERTON, J. L., GIANNELLI, F., COLLINS, F., HALLETT, J., FRYER, A., McGUIRE, HULLIGER, L., TAYLOR, A. and LANG, E. M. (1961). Somatic chromosomes of the gorilla. *Nature*, **192**: 225–8.

HAMERTON, J. L., GIANELLI, F., COLLINS, F., HALLETT, J., FRYER, A., McGUIRE, V. M. and SHORT, R. V. (1969). Non-random X-inactivation in the female mule. *Nature*, **222**: 1277–8.

HAMILTON, A. G. (1953). Thelytokous parthenogenesis of four generations in the Desert Locust *Schistocerca gregaria* Forsk. (Acrididae). *Nature*, **172**: 1153–4.

HAMILTON, A. G. (1955). Parthenogenesis in the Desert Locust (*Schistocerca gregaria* Forsk) and its possible effect on the maintenance of the species. *Proc. Roy. Entomol. Soc. Lond.* A, **30**: 103–14.

HANSELL, R. I. C., MOLLISON, M. M. and PUTMAN, W. L. (1964). A cytological demonstration of arrhenotoky in three mites of the family Phytoseiidae. *Chromosoma*, **15**: 562–7.

HARD, W. L. (1939). The spermatogenesis of the lycosid spider *Schizocosa crassipes* (Walckenaer). *J. Morph.* **65**: 121–50.

HARDY, D. E. (1965). Evolution and genetics of Hawaiian *Drosophilidae*. *Proc. XIIth Int. Congr. Entomol. London*, p. 119.

HARMAN, M. T. (1920). Chromosome studies in Tettigidae. II. Chromosomes of *Paratettix* BB and CC and their hybrid BC. *Biol. Bull. Woods Hole*, **38**: 213–31.

HARNDEN, D. G. and JACOBS, P. A. (1961). Cytogenetics of abnormal sexual development in man. *Brit. Med. Bull.* **17**: 206–12.

HARNDEN, D. G., MILLER, O. J. and PENROSE, L. S. (1960). The Klinefelter-mongolism type of double aneuploidy. *Ann. Hum. Genet.* **24**: 165–9.

HARTL, D. L. (1968). Evidence for dysfunctional sperm production in Segregation Distorter (SD) males of *Drosophila melanogaster*. *Genetics*, **60**: 187.

HARTL, D., HIRAIZUMI, Y. and CROW, J. F. (1967). Evidence for sperm dysfunction as the mechanism of Segregation Distortion in *Drosophila melanogaster*. *Proc. Nat. Acad. Sci. U.S.A.* **58**: 2240–5.

HARRIS, P. (1962). Some structural and functional aspects of the mitotic apparatus in sea-urchin embryos. *J. Cell Biol.* **14**: 475–87.

HARTMANN, M. (1909). Polyenergide Kerne. Studien über multiple Kernteilungen und generative Chromidien bei Protozoen. *Biol. Zentralbl.* **29**: 481–7, 491–506.

HARTMANN, M. (1953). *Allgemeine Biologie*. Stuttgart: Gustav Fischer Verlag.

HARTMANN-GOLDSTEIN, I. J. (1967). On the relation between heterochromatization and variegation in *Drosophila*, with special reference to temperature-sensitive periods. *Genet. Res.* **10**: 143–60.

HASKINS, C. P., HASKINS, E. F. and HEWITT, R. E. (1960). Pseudogamy as an evolutionary factor in the Poeciliid fish *Mollienisia formosa*. *Evolution*, **14**: 473–83.

HAUSCHKA, T. S. and HOLDRIDGE, B. A. (1962). A cytogenetic approach to the Y linked histoincompatibility antigen in mice. *Ann. N. Y. Acad. Sci.* **1**: 12–22.

HAUSCHTECK, E. (1959). Über die Zytologie der Parthenogenese und der Geschlechtsbestimmung bei der Gallmücke *Oligarces paradoxus* Mein. *Experientia,* **15**: 260–4.

HAUSCHTECK, E. (1961). Die Chromosomen von fünf Ameisenarten. *Rev. Suisse Zool.* **68**: 218–23.

HAUSCHTECK, E. (1962). Die Zytologie der Paedogenese und der Geschlechtsbestimmung einer heterogonen Gallmücke. *Chromosoma,* **13**: 163–82.

HAUSCHTECK, E. (1963*a*). Die Chromosomen einiger in der Schweitz vorkommender Ameisenarten. *Vjschr. naturf. Ges. Zürich,* **107**: 213–20.

HAUSCHTECK, E. (1963*b*). Chromosomes of Swiss ants. *Proc. XIth Int. Genet. Congr.* **1**: 140.

HAUSCHTECK, E. (1965). Halbe haploide chromosomenzahl im Hoden von *Myrmica sulcinodis* Nyl. (Formicidae). *Experientia,* **21**: 323–5.

HAYASHI, M., HAYASHI, M. N. and SPIEGELMAN, S. (1964). DNA circularity and the mechanism of strand selection in the generation of genetic messages. *Proc. Nat. Acad. Sci. U.S.A.* **51**: 351–9.

HAYATA, I., SHIMBA, H., KOBAYASHI, T. and MAKINO, S. (1970). Preliminary accounts on the chromosomal polymorphism in the field mouse *Apodemus giliacus,* a new form from Hokkaido. *Proc. Japan. Acad.* **46**: 567–71.

HAYMAN, D. L. and MARTIN, P. G. (1965*a*). An autoradiographic study of DNA synthesis in the sex chromosomes of two Marsupials with an XX/XY_1Y_2 sex-chromosome mechanism. *Cytogenetics,* **4**: 209–18.

HAYMAN, D. L. and MARTIN, P. G. (1965*b*). Sex-chromosome mosaicism in the Marsupial genera *Isoodon* and *Perameles. Genetics,* **52**: 1201–6.

HAYMAN, D. L. and MARTIN, P. G. (1965*c*). Supernumerary chromosomes in the marsupial *Schoinobates volans* (Kerr). *Aust. J. Biol. Sci.* **18**: 1081–2.

HAYMAN, D. L. and MARTIN, P. G. (1969). Cytogenetics of marsupials. In: *Comparative Mammalian Cytogenetics* (ed.: K. Benirschke), pp. 191–217. New York: Springer-Verlag.

HAYMAN, D. L., MARTIN, P. G. and WALLER, P. F. (1969). Parallel mosaicism of supernumerary chromosomes and sex chromosomes in *Echymipera kalabu* (Marsupialia). *Chromosoma,* **27**: 371–80.

HEBERER, B. (1932). Die Spermatogenese der Copepoden. II. *Z. Zool.* **142**: 191–253.

HEBERER, G. (1937). *X*-chromosomen und Spermiengrösse. Untersuchungen an einheimischen und tropischen Orthopteren. *Z. indukt. Abst. Vererbl.* **73**: 479–82.

HEDDLE, J. A. (1965). Randomness in the formation of radiation-induced chromosome aberrations. *Genetics,* **52**: 1329–34.

HEDDLE, J. A. and BODYCOTE, D. J. (1968). The strandedness of chromosomes (Abstract) *J. Cell Biol.* **39**: 60*a*.

HEDDLE, J. A. and TROSKO, J. E. (1966). Is the transition from chromosome to chromatid aberrations the result of the formation of single-stranded DNA? *Exp. Cell Res.* **42**: 171–7.

HEDDLE, J. A., WOLFF, S., WHISSEL, D. and CLEAVER, J. (1967). Distribution of chromatids at mitosis. *Science,* **158**: 929–31.

HEGNER, R. W. (1914). Studies on germ cells. I and II. *J. Morph.* **25**: 375–509.

HEINONEN, L. and HALKKA, O. (1967). Early stages of oogenesis and metabolic DNA in the oocytes of the house cricket, *Acheta domesticus. Ann. Med. Exp. Biol. Fenn.* **45**: 101–9.

HEITZ, E. (1928). Das Heterochromatin der Moose. I. *Jb. wiss. Bot.* **69**: 762–818.

HEITZ, E. (1929). Heterochromatin, Chromocentren, Chromomeren. *Ber. Deutsch. Bot. Ges.* **47**: 274–84.

HEITZ, E. (1931). Die Ursache der gesetzmässigen Zahl, Lage, Form and Grösse pflanzlicher Nukleolen. *Planta* **12**: 775–844.

HEITZ, E. (1933). Die somatische Heteropyknose bei *Drosophila melanogaster* und ihre genetische Bedeutung. *Z. Zellforsch.* **20**: 237–87.

HEITZ, E. (1934). Über α und β Heterochromatin sowie Konstanz und Bau der Chromomeren bei *Drosophila. Biol. Zentralbl.* **54**: 588–609.

HEITZ, E. and BAUER, H. (1933). Beweise für die Chromosomennatur der Kernschleifen in den Knäuelkernen von *Bibio hortulanus* L. *Z. Zellforsch.* **17**: 67–82.

HELENIUS, O. (1952). The mode of bivalent orientation in the Hemiptera. *Hereditas*, **38**: 420–34.

HELLE, W. and BOLLAND, H. R. (1967). Karyotypes and sex determination in spider mites (Tetranychidae). *Genetica*, **38**: 43–53.

HELWIG, E. R. (1929). Chromosomal variations correlated with geographical distribution in *Circotettix verruculatus* (Orthoptera). *J. Morph.* **47**: 1–36.

HELWIG, E. R. (1933). The effect of X-rays upon the chromosomes of *Circotettix verruculatus* (Orthoptera). *J. Morph.* **55**: 265–312.

HELWIG, E. R. (1938). The frequency of reciprocal translocations in irradiated germ cells of *Circotettix verruculatus* (Orthoptera). *Arch. Biol.* **49**: 143–58.

HELWIG, E. R. (1941). Multiple chromosomes in *Philocleon anomalus* (Orthoptera: Acrididae). *J. Morph.* **69**: 317–27.

HELWIG, E. R. (1942). Unusual integrations of the chromatin in *Machaerocera* and other genera of the Acrididae (Orthoptera). *J. Morph.* **71**: 1–33.

HELWIG, E. R. (1955). Spermatogenesis in hybrids between *Circotettix verruculatus* and *Trimerotropis suffusa* (Orthoptera: Oedipodidae). *Univ. Colorado Studies Ser. Biol.* **10**: 49–64.

HELWIG, E. R. (1958). Cytology and taxonomy. *Bios*, **29**: 59–71.

HENDERSON, S. A. (1961). The chromosomes of the British Tetrigidae (Orthoptera). *Chromosoma*, **12**: 553–72.

HENDERSON, S. A. (1964). RNA synthesis during male meiosis and spermiogenesis. *Chromosoma*, **15**: 345–66.

HENDERSON, S. A. (1965). Chromosome behavior in diploid and tetraploid cells of *Sphodromantis gastrica* and its bearing on chromosome evolution in the mantids. *Chromosoma*, **16**: 192–221.

HENDERSON, S. A. (1966). Time of chiasma formation in relation to the time of deoxyribonucleic acid synthesis. *Nature*, **211**: 1043–7.

HENDERSON, S. A. (1967a). The salivary gland chromosomes of *Dasyneura crataegi* (Diptera: Cecidomyiidae). *Chromosoma*, **23**: 38–58.

HENDERSON, S. A. (1967b). A second example of normal coincident endopolyploidy and polyteny in salivary gland nuclei. *Caryologia*, **20**: 181–6.

HENDERSON, S. A. (1969). Chromosome pairing, chiasmata and crossing-over. Chapter 14, pp. 326–57. In: *Handbook of Molecular Cytology* (ed.: A. Lima-de-Faria). Amsterdam and London: North Holland.

HENDERSON, S. A. (1970). Sex chromosomal polymorphism in the earwig *Forficula. Chromosoma*, **31**: 139–64.

HENDERSON, S. A. and KOCH, C. A. (1970). Co-orientation stability by physical tension: a demonstration with experimentally interlocked bivalents. *Chromosoma*, **29**: 207–16.

HENDERSON, S. A., NICKLAS, R. B. and KOCH, C. A. (1970). Temperature-induced orientation instability during meiosis: an experimental analysis. *J. Cell Sci.* **6**: 323–50.

HENDERSON, S. A. and PARSONS, T. (1963). The chromosomes of eleven species of Tipulid. *Caryologia*, **16**: 337–46.

HENNIG, W., HENNIG, I. and STEIN, H. (1970). Repeated sequences in the DNA of *Drosophila* and their localization in giant chromosomes. *Chromosoma*, **32**: 31–63.

HENNIG, W. and WALKER, P. M. B. (1970). Variations in the DNA from two rodent families (Cricetidae and Muridae). *Nature*, **225**: 915–19.

HERREROS, B. and GIANNELLI, F. (1967). Spatial distribution of old and new chromatid subunits and frequency of chromatid exchanges in induced human lymphocyte endoreduplication. *Nature*, **216**: 286–7.

HERSHEY, A. D., BURGI, E. and INGRAHAM, L. (1963). Cohesion of DNA molecules isolated from phage Lambda. *Proc. Nat. Acad. Sci. U.S.A.* **49**: 748–55.

HERTWIG, O. (1890). Vergleich der Ei- und Samenbildung bei Nematoden. Eine Grundlage für cellulare Streitfragen. *Arch. mikr. Anat.* **36**: 1–138.

HERTWIG, P. (1920). Abweichende Form der Parthenogenese bei einer Mutation von *Rhabditis pellio*. Eine experimentell Cytologische Untersuchung. *Arch. mikr. Anat.* **94**: 303–34.

HESS, O. (1965). Struktur-differenzierung im *Y*-Chromosom von *Drosophila hydei* und ihre Beziehungen zu Gen-Aktivitäten. III. Sequenz und Lokalisation der Schleifenbildungsorte. *Chromosoma*, **16**: 222–48.

HESS, O. and MEYER, G. F. (1963*a*). Chromosomal differentiations of the lampbrush type formed by the *Y*-chromosome in *Drosophila hydei* and *Drosophila neohydei*. *J. Cell Biol.* **16**: 527–39.

HESS, O. and MEYER, G. F. (1963*b*). Artspezifische funktionelle differenzierungen des *Y*-Heterochromatins bei *Drosophila*-Arten der *D. hydei*-Subgruppe. *Port. Acta Biol.* **7**: 29–46.

HEWITT, G. M. (1964). Population cytology of British grasshoppers. I. Chiasma variation in *Chorthippus brunneus*, *Chorthippus parallelus* and *Omocestus viridulus*. *Chromosoma*, **15**: 212–30.

HEWITT, G. M. (1965). Population cytology of British grasshoppers. II. Annual variation in chiasma frequency. *Chromosoma*, **16**: 579–690.

HEWITT, G. M. and JOHN, B. (1965). The influence of numerical and structural chromosome mutations on chiasma conditions. *Heredity*, **20**: 123–35.

HEWITT, G. M. and JOHN, B. (1967). The *B*-chromosome system of *Myrmeleotettix maculatus* (Thunb.). III. The statistics. *Chromosoma*, **21**: 140–62.

HEWITT, G. M. and JOHN, B. (1968). Parallel polymorphism for supernumerary segments in *Chorthippus parallelus* (Zetterstedt). I. British populations. *Chromosoma*, **25**: 319–42.

HEWITT, G. M. and JOHN, B. (1970*a*). The *B*-chromosome system of *Myrmeleotettix maculatus* (Thunb.). IV. The dynamics. *Evolution*, **24**: 169–80.

HEWITT, G. M. and JOHN, B. (1970*b*). Parallel polymorphism for supernumerary segments in *Chorthippus parallelus* (Zetterstedt). IV. Ashurst revisited. *Chromosoma*, **31**: 198–206.

HICKEY, W. A. and CRAIG, G. B. JR. (1966*a*). Distortion of sex ratio in populations of *Aedes aegypti*. *Canad. J. Genet. Cytol.* **8**: 260–78.

HICKEY, W. A. and CRAIG, G. B. Jr. (1966*b*). Genetic distortion of sex ratio in a mosquito, *Aedes aegypti*. *Genetics*, **53**: 1177–96.

HILDRETH, P. E. (1965). Doublesex, a recessive gene that transforms both males and females of *Drosophila* into intersexes. *Genetics*, **51**: 659–78.

HILL, R. (1948). Parthenogenese bei *Nagara modesta* Dollf. (Isopoda). *Chromosoma*, **3**: 232–56.

HINDLE, E. (1919). Sex inheritance in *Pediculus humanus* var. *corporis*. *J. Genet.* **8**: 267–77.

HINDLE, E. and PONTECORVO, G. (1942). Mitotic divisions following meiosis in *Pediculus corporis* males. *Nature*, **149**: 668.

HINEGARDNER, R. (1968). Evolution of cellular DNA content in teleost fishes. *Amer. Nat.* **102**: 517–23.

HINTON, T. (1942). A comparative study of certain heterochromatic regions in the mitotic and salivary gland chromosomes of *Drosophila melanogaster*. *Genetics*, **27**: 119–27.

HINTON, T. (1945). A study of chromosome ends in salivary gland nuclei of *Drosophila*. *Biol. Bull. Woods Hole*, **88**: 144–65.

HINTON, T. and ATWOOD, K. C. (1941). Terminal adhesions of salivary gland chromosomes in *Drosophila*. *Proc. Nat. Acad. Sci. U.S.A.* **27**: 491–6.

HINTON, T. and SPARROW, A. H. (1941). The non-random occurrence of terminal adhesion in salivary chromosomes of *Drosophila*. *Genetics*, **26**: 155.

HIRAI, H. (1951). The chromosomes of two species of the Oligotomidae (Embioptera, Insecta). *Kromosomo*, **8**: 321–3.

HIRAI, H. (1955a). Chromosome studies in the Neuroptera. II. The chromosomes of 13 species of Japanese Hemerobioidea. *Kromosomo*, **29**: 999–1004.

HIRAI, H. (1955b). Cyto-taxonomical studies of the Japanese Neuroptera. II. The Hemerobioidea. *Misc. Rep. Yamashina's Inst. Ornith. Zool.* **7**: 293–6.

HIRAI, H. (1955c). Chromosome studies in the Neuroptera. III. Chromosomes of seven species of the Myrmeleonoidea. *Zool. Mag. (Tokyo)*, **64**: 370–4.

HIRAI, H. (1956). Compound sex-chromosomes with distance painting in the male neuropteran *Plethosmylus decoratus*. *Anot. Zool. Japan*, **29**: 155–60.

HIRAIZUMI, Y., SANDLER, L. and CROW, J. F. (1960). Meiotic drive in natural populations of *Drosophila melanogaster*. III. Populational implications of the *segregation-distorter* locus. *Evolution*, **14**: 433–44.

HIRAIZUMI, Y. and WATANABE, S. S. (1969). Aging affect on the phenomenon of segregation distortion in *Drosophila melanogaster*. *Genetics*, **63**: 121–31.

HOCHMAN, B. (1971). Analysis of chromosome 4 in *Drosophila melanogaster*. II. Ethyl methanesulfonate induced lethals. *Genetics*, **67**: 235–52.

HOLLIDAY, R. (1962). Mutation and replication in *Ustilago maydis*. *Genet. Res.* **3**: 472–86.

HOLLIDAY, R. (1964). A mechanism for gene conversion in fungi. *Genet. Res.* **5**: 282–304.

HOLLIDAY, R. (1968). Genetic recombination in fungi. In: *Replication and Recombination of Genetic Material* (ed.: W. J. Peacock and R. D. Brock), pp. 157–74. Canberra: Aust. Acad. Science.

HONDA, H. (1921). Spermatogenesis of Aphids: the fate of the smaller secondary spermatocytes. *Biol. Bull. Woods Hole*, **40**: 349–69.

HOTTA, Y., ITO, M. and STERN, H. (1966). Synthesis of DNA during meiosis. *Proc. Nat. Acad. Sci. U.S.A.* **56**: 1184–91.

HOWARD, A. and PELC, S. R. (1951). Nuclear incorporation of P32 as demonstrated by autoradiographs. *Exp. Cell Res.* **2**: 178–87.

HOWARD, E. F. and PLAUT, W. (1968). Chromosomal DNA synthesis in *Drosophila melanogaster*. *J. Cell Biol.* **39**: 415–29.

HOWELL, W. M. and DUCKETT, C. R. (1971). Somatic chromosomes of the lamprey, *Ichthyomyzon gagei* (Agnatha: Petromyzonidae). *Experientia*, **27**: 222–3.

HSIANG, W. (1949). The distribution of heterochromatin in *Drosophila tumiditarsus*. *Cytologia* **15**: 149–52.

HSU, T. C. (1952). Chromosomal variation and evolution in the *virilis* group of *Drosophila*. *Univ. Texas Publ.* **5204**: 35–72.

HSU, T. C. (1963). Longitudinal differentiation of chromosomes and the possibility of interstitial telomeres. *Exp. Cell Res.* suppl. **9**: 73–85.

HSU, T. C. (1964). Mammalian chromosomes *in vitro*. XVIII. DNA replication sequence in the Chinese hamster. *J. Cell Biol.* **23**: 53–62.

HSU, T. C. (1969). Robertsonian fusion between homologous chromosomes in a natural population of the Least Cotton Rat, *Sigmodon minimus* (Rodentia, Cricetidae). *Experientia*, **25**: 205–6.

HSU, T. C. and ARRIGHI, F. E. (1968). Chromosomes of *Peromyscus* (Rodentia, Cricetidae). I. Evolutionary trends in twenty species. *Cytogenetics*, **7**: 417–46.

HSU, T. C., BAKER, R. J. and UTAKOJI, T. (1968). The multiple sex-chromosome system of American leaf-nosed bats (Chiroptera, Phyllostomidae). *Cytogenetics*, **7**: 27–38.

HSU, T. C., DEWEY, W. C. and HUMPHREY, R. M. (1962). Radiosensitivity of the cells of Chinese hamster *in vitro* in relation to the cell cycle. *Exp. Cell Res.* **27**: 441–52.

HSU, T. C. and LIU, T. T. (1948). Microgeographic analysis of chromosomal variation in a Chinese species of *Chironomus* (Diptera). *Evolution*, **2**: 49–57.

HSU, T. C. and REARDEN, H. H. (1965). Further karyological studies on Felidae. *Chromosoma*, **16**: 365–71.

HSU, T. C., REARDEN, H. H. and LUQUETTE, G. F. (1963). Karyological studies of nine species of Felidae. *Amer. Nat.* **97**: 225–34.

HUANG, C. C. (1968). Karyologic and autoradiographic studies of the chromosomes of *Rattus* (*Mastomys*) *natalensis*. *Cytogenetics*, **7**: 97–107.

HUANG, C. C. and CLARK, H. F. (1967). Chromosome changes in cell lines of the box turtle (*Terrapene carolina*) grown at two different temperatures. *Canad. J. Genet. Cytol.* **9**: 449–61.

HUANG, C. C., CLARK, H. F. and GANS, C. (1967). Karyological studies on fifteen forms of Amphisbaenians (Amphisbaenia – Reptilia). *Chromosoma*, **22**: 1–15.

HUBBS, C. L. and HUBBS, L. C. (1932). Apparent parthenogenesis in nature, in a form of fish of hybrid origin. *Science*, **76**: 628–30.

HUBBS, C. L., HUBBS, L. C. and JOHNSON, R. E. (1943). Hybridization in nature between species of catastomid fishes. *Contr. Lab. Verteb. Biol. Univ. Mich.* **22**: 1–76.

HUBBS, C. L. and MILLER, R. R. (1943). Mass hybridization between two genera of Cyprinid fishes in the Mohave desert, California. *Pap. Mich. Acad. Sci.* **28**: 343–78.

HUBERMAN, J. A. and ATTARDI, J. G. (1966). Isolation of metaphase chromosomes from HeLa cells. *J. Cell Biol.* **31**: 95–105.

HUBERMAN, J. A. and RIGGS, A. D. (1966). Autoradiography of chromosomal DNA fibers from Chinese hamster cells. *Proc. Nat. Acad. Sci. U.S.A.* **55**: 599–606.

HUBERMAN, J. A. and RIGGS, A. D. (1968). On the mechanism of DNA replication in mammalian chromosomes. *J. Mol. Biol.* **32**: 327–41.

HUETTNER, A. F. (1930). The spermatogenesis of *Drosophila melanogaster*. *Z. Zellforsch.* **11**: 615–37.

HUETTNER, A. F. (1934). Octoploidy and diploidy in *Miastor americana*. *Anat. Rec.* **58**, suppl.: 8.

HUGHES-SCHRADER, S. (1924). Reproduction in *Acroschismus wheeleri* Pierce. *J. Morph.* **39**: 157–205.

HUGHES-SCHRADER, S. (1925). Cytology of hermaphroditism in *Icerya purchasi* (Coccidae). *Z. Zellforsch.* **2**: 264–92.

HUGHES-SCHRADER, S. (1927). Origin and differentiation of the male and female germ cells in the hermaphrodite of *Iceryi purchasi* (Coccidae). *Z. Zellforsch.* **6**: 509–40.

HUGHES-SCHRADER, S. (1930*a*). Contribution to the life history of the iceryine coccids, with special reference to parthenogenesis and haploidy. *Ann. Entomol. Soc. Amer.* **23**: 359–80.

HUGHES-SCHRADER, S. (1930*b*). The cytology of several species of iceryine coccids, with special reference to parthenogenesis and haploidy. *J. Morph.* **50**: 475–95.

HUGHES-SCHRADER, S. (1931). A study of the chromosome cycle and the meiotic division figure in *Llaveia bouvari* – a primitive coccid. *Z. Zellforsch.* **13**: 742–69.

HUGHES-SCHRADER, S. (1935). The chromosome cycle of *Phenacoccus* (Coccidae). *Biol. Bull. Woods Hole,* **69**: 462–8.

HUGHES-SCHRADER, S. (1940). The meiotic chromosomes of the male *Llaveiella taenechina* Morrison and the question of the tertiary split. *Biol. Bull. Woods Hole,* **78**: 312–37.

HUGHES-SCHRADER, S. (1942). The chromosomes of *Nautococcus schraderae* Vays. and the meiotic division figures of male Llaveine coccids. *J. Morph.* **70**: 261–99.

HUGHES-SCHRADER, S. (1943*a*). Meiosis without chiasmata in diploid and tetraploid spermatocytes of the mantid *Callimantis antillarum* Saussure. *J. Morph.* **73**: 111–41.

HUGHES-SCHRADER, S. (1943*b*). Polarization, kinetochore movements and bivalent structure in the meiosis of male mantids. *Biol. Bull. Woods Hole,* **85**: 265–300.

HUGHES-SCHRADER, S. (1944). A primitive coccid chromosome cycle in *Puto* sp. *Biol. Bull. Woods Hole,* **87**: 167–76.

HUGHES-SCHRADER, S. (1947*a*). The 'pre-metaphase stretch' and kinetochore orientation in phasmids. *Chromosoma,* **3**: 1–21.

HUGHES-SCHRADER, S. (1947*b*). Reversion of *XO* to *XY* sex-chromosome mechanism in a phasmid. *Chromosoma,* **3**: 52–65.

HUGHES-SCHRADER, S. (1948*a*). Cytology of coccids (Coccoidea–Homoptera). *Adv. in Genet.* **2**: 127–203.

HUGHES-SCHRADER, S. (1948*b*). Expulsion of the sex chromosome from the spindle in spermatocytes of a mantid. *Chromosoma,* **3**: 257–70.

HUGHES-SCHRADER, S. (1950). The chromosomes of mantids (Orthoptera: Manteidae) in relation to taxonomy. *Chromosoma,* **4**: 1–55.

HUGHES-SCHRADER, S. (1951). The desoxyribonucleic acid content of the nucleus as a cytotaxonomic character in mantids (Orthoptera: Mantoidea). *Biol. Bull. Woods Hole,* **100**: 178–87.

HUGHES-SCHRADER, S. (1953*a*). The nuclear content of desoxyribonucleic acid and interspecific relationships in the mantid genus *Liturgousa* (Orthoptera: Mantoidea). *Chromosoma,* **5**: 544–54.

HUGHES-SCHRADER, S. (1953*b*). Supplementary notes on the cytotaxonomy of mantids (Orthopteroidea: Mantoidea). *Chromosoma,* **6**: 79–90.

HUGHES-SCHRADER, S. (1955). The chromosomes of the giant scale *Aspidoproctus maximus* Louns. (Coccoidea–Margarodidae) with special reference to asynapsis and sperm formation. *Chromosoma*, **7**: 420–38.

HUGHES-SCHRADER, S. (1957). Differential polyteny and polyploidy in Diaspine Coccids (Homoptera: Coccoidea). *Chromosoma*, **8**: 709–18.

HUGHES-SCHRADER, S. (1958). The DNA content of the nucleus as a tool in the cytotaxonomic study of insects. *Proc. Xth Int. Congr. Entomol. Montreal*, **2**: 935–44.

HUGHES-SCHRADER, S. (1959). On the cytotaxonomy of Phasmids (Phasmatodea). *Chromosoma*, **10**: 268–77.

HUGHES-SCHRADER, S. (1963). Hermaphroditism in an African coccid, with notes on other margarodids (Coccoidea–Homoptera). *J. Morph.* **113**: 173–84.

HUGHES-SCHRADER, S. (1969). Distance segregation and compound sex chromosomes in mantispids (Neuroptera: Mantispidae). *Chromosoma*, **27**: 109–29.

HUGHES-SCHRADER, S. and MONAHAN, D. F. (1966). Hermaphroditism in *Icerya zeteki* Cockerell, and the mechanism of gonial reduction in iceryine coccids (Coccoidea: Margarodidae Morrison). *Chromosoma*, **20**: 15–31.

HUGHES-SCHRADER, S. and RIS, H. (1941). The diffuse spindle attachment of coccids as verified by the mitotic behavior of induced fragments. *J. Exp. Zool.* **87**: 429–56.

HUGHES-SCHRADER, S. and SCHRADER, F. (1956). Polyteny as a factor in the chromosomal evolution of the Pentatomini (Hemiptera). *Chromosoma*, **8**: 135–51.

HUGHES-SCHRADER, S. and SCHRADER, F. (1961). The kinetochore of the Hemiptera. *Chromosoma*, **12**: 327–50.

HUGHES-SCHRADER, S. and TREMBLAY, E. (1966). *Gueriniella* and the cytotaxonomy of iceryine coccids (Coccoidea: Margarodidae). *Chromosoma*, **19**: 1–13.

HUMPHREY, R. M. and BRINKLEY, B. R. (1969). Ultrastructural studies of radiation-induced chromosome damage. *J. Cell Biol.* **42**: 745–53.

HUMPHREY, R. R. (1942). Sex reversal and the genetics of sex determination in the axolotl (*Amblystoma mexicanum*). *Anat. Rec.* **84**, suppl.: 465.

HUMPHREY, R. R. (1945). Sex determination in ambystomid salamanders: a study of the progeny of females experimentally converted into males. *Amer. J. Anat.* **76**: 33–66.

HUNGERFORD, D. A., CHANDRA, H. A., SNYDER, R. L. and UIMER, F. A. Jr. (1966). Chromosomes of three Elephants, two Asian (*Elephas maximus*) and one African (*Loxodonta africana*). *Cytogenetics*, **5**: 243–6.

HUNGERFORD, D. A., SHARAT CHANDRA, H. and SNYDER, R. L. (1967). Somatic chromosomes of a Black Rhinoceros (*Diceros bicornis* Gray 1821). *Amer. Nat.* **101**: 357–8.

HUSKINS, C. L. (1941). The coiling of chromonemata. *Cold Spr. Harb. Symp. Quant. Biol.* **9**: 13–17.

HUSKINS, C. L. and SMITH, S. G. (1935). Meiotic chromosome structure in *Trillium erectum* L. *Ann. Bot.* **49**: 119–50.

HUSKINS, C. L. and WILSON, G. B. (1938). Probable causes of the changes in direction of the major spiral in *Trillium erectum* L. *Ann. Bot.* N.S. **2**: 281–91.

HUSTED, L. and BURCH, P. R. (1946). The chromosomes of Polygyrid snails. *Amer. Nat.* **80**: 410–29.

HUSTED, L. and RUEBUSH, T. K. (1940). A comparative study of *Mesostoma ehrenbergii ehrenbergii* and *Mesostoma ehrenbergii wardii*. *J. Morph.* **67**: 387–410.

IKEDA, K. and MAKINO, S. (1936). Studies on the sex and chromosomes of the oriental human blood fluke, *Schistosomum japonicum* Katsurada. *J. Fac. Sci. Hokkaido Imp. Univ.* **5**: 57–71.

IMAI, H. and YOSIDA, T. H. (1964). Chromosome observations in Japanese ants. *Ann. Rep. Nat. Inst. Genet.* **15**: 64–6.

INABA, A. (1953). Cytological studies in molluscs. I. Chromosomes in Basommato-phoric Pulmonata. *J. Sci. Hiroshima Univ.* B, **14**: 221–8.

INABA, A. (1959*a*). Cytological studies in molluscs. II. A chromosome survey in the Stylommatophoric Pulmonata. *J. Sci. Hiroshima Univ.* B, **18**: 71–94.

INABA, A. (1959*b*). On the chromosomes of a nudibranch gastropod, *Dendrodoris* (*Dendrodoris*) *nigra* (Stimpson). *J. Sci. Hiroshima Univ.* B, **18**: 95–8.

INABA, F. (1939). Diploid males and triploid females of the parasitic wasp *Habrobracon pectinophorae* Watanabe. *Cytologia*, **9**: 517–23.

INOUYÉ, S. (1964). Organization and function of the mitotic spindle. In: *Primitive Motile Systems in Cell Biology* (ed.: R. D. Allen and N. Kamiya), pp. 549–94. New York: Academic Press.

IRIKI, S. (1932). Studies on amphibian chromosomes VII. On the chromosomes of *Megalobatrachus japonicus*. *Sci. Rep. Tokyo Bunrika Daig.* **B1**: 61–126.

IRWIN, J. O. (1965). A theory of the association (overlap) of chromosomes in karyotypes illustrated by Dr. Patricia Jacobs' data. *Ann. Hum. Genet.* **28**: 361–7.

ITOH, H. (1933*a*). The chromosomes of *Chauliodes japonicus* McLachlan. *Zool. Mag.* (*Tokyo*), **45**.

ITOH, H. (1933*b*). The chromosomes of *Protohermes grandis* Thunberg. *Zool. Mag.* (*Tokyo*), **45**.

ITOH, H. (1933*c*). The sex chromosomes of a stone fly, *Acroneuria jezoensis* Okamoto (Plecoptera). *Cytologia*, **4**: 427–43.

ITOH, H. (1934). Chromosomal variation in the spermatogenesis of a grasshopper, *Locusta danica* L. *Jap. J. Genet.* **10**: 115–34.

ITOH, M., SASAKI, M., and MAKINO, S. (1970). The chromosomes of some Japanese snakes with special regard to sexual dimorphism. *Jap. J. Genet.* **45**: 121–8.

IVANOV, V. G. (1967). Chromosome set of *Ellobius talpinus* Pall. *Tsitologia*, **9**: 879–83.

IVANOV, V. G. and FEDOROVA, V. G. (1970). Sex heteromorphism of chromosomes in *Lacerta strigata* Eichwald. *Tsitologia*, **12**: 1582–5.

IVES, P. T. (1947). Second chromosome inversions in wild populations of *Drosophila melanogaster*. *Evolution*, **1**: 42–7.

IVES, P. T. (1950). The importance of mutation rate genes in evolution. *Evolution*, **4**: 236–52.

JACKSON, L. G. and ELLEM, K. A. O. (1968). The karyotype of the Australian longnosed bandicoot (*Perameles nasuta*). *Cytogenetics*, **7**: 183–9.

JACKSON, W. F. and CHEUNG, D. S. M. (1967). Distortional meiotic segregation of a supernumerary chromosome producing differential frequencies in the sexes in the short-horned grasshopper *Phaulacridium vittatum*. *Chromosoma*, **23**: 24–37.

JACOB, F. and WOLLMAN, E. (1957). Analyse des groupes de liaison génétique de differentes souches donatrices d'*Escherichia coli* K12. *C.R. Acad. Sci. Paris*, **245**: 1840–3.

JACOB, J. (1954). Allopolyploidy in the Melaniid snails. *Nature*, **174**: 1061.

JACOB, J. (1957). Cytological studies of Melaniidae (Mollusca) with special reference to parthenogenesis and polyploidy. I. Oögenesis of the parthenogenetic species of '*Melanoides*' (Prosobranchia-Gastropoda). *Trans. Roy. Soc. Edinb.* **63**: 341–52.

JACOB, J. and SIRLIN, J. L. (1963). Electron microscope studies on salivary gland cells. I. The nucleus of *Bradysia mycorum* Frey (Sciaridae), with special reference to the nucleolus. *J. Cell Biol.* **17**: 153–65.

JACOBS, P. A., BAIKIE, A. G., COURT BROWN, W. M., MACGREGOR, T. N., MACLEAN, N. and HARNDEN, D. G. (1959). Evidence for the existence of the human "super-female". *Lancet*, **ii**: 423–5.

JACOBS, P. A., BRUNTON, M., MELVILLE, M. M., BRITTAIN, R. P. and McCLEMONT, W. F. (1965). Aggressive behavior, mental subnormality and the *XYY* male. *Nature*, **208**: 1351–2.

JACOBS, P. A. and ROSS, A. (1966). Structural abnormalities of the *Y*-chromosome in man. *Nature*, **210**: 352–4.

JANDE, S. S. (1959a). Chromosome number and sex mechanism in twenty-seven species of Indian Heteroptera. *Res. Bull. (N.S.) Panjab Univ.* **10**: 215–17.

JANDE, S. S. (1959b). Chromosome number and sex mechanism in nineteen species of Indian Heteroptera. *Res. Bull. (N.S.) Panjab Univ.* **10**: 415–17.

JANSSENS, F. A. (1900). Rapprochements entre les cinèses polliniques et les cinèses sexuelles dans le testicule des Tritons. *Anat. Anz.* **17**: 520–4.

JANSSENS, F. A. (1909). Spermatogénèse dans les batraciens. V. La théorie de la chiasmatypie, nouvelle interprétation des cinèses de maturation. *Cellule*, **25**: 387–411.

JANSSENS, F. A. (1924). La chiasmatypie dans les insectes. *Cellule*, **34**: 135–359.

JARY, S. G. and STAPLEY, J. H. (1936). Investigations on the insect and allied pests of cultivated mushrooms. VI. Observations on the Tyroglyphid mite *Histiostoma rostro-serratum* Megnin. *J. S-E. Agric. Coll. Wye*, no. **38**: 67–74.

JAWORSKA, H. and LIMA-DE-FARIA, A. (1969). Multiple synaptinemal complexes at the region of gene amplification in *Acheta. Chromosoma*, **28**: 309–27.

JOHN, B. (1957). *XY* segregation in the Crane Fly *Tipula maxima. Heredity*, **11**: 209–15.

JOHN, B. (1965). Review of E. Mayr: *Animal Species and Evolution. Heredity*, **19**: 744–5.

JOHN, B. and HEWITT, G. M. (1963). A spontaneous interchange in *Chorthippus brunneus* with extensive chiasma formation in an interstitial segment. *Chromosoma*, **14**: 638–50.

JOHN, B. and HEWITT, G. M. (1965a). The *B*-chromosome system of *Myrmeleotettix maculatus* (Thunb.) I. The mechanics. *Chromosoma*, **16**: 548–78.

JOHN, B. and HEWITT, G. M. (1965b). The *B*-chromosome system of *Myrmeleotettix maculatus* (Thunb.). II. The statics. *Chromosoma*, **17**: 121–38.

JOHN, B. and HEWITT, G. M. (1966a). A polymorphism for heterochromatic supernumerary segments in *Chorthippus parallelus. Chromosoma*, **18**: 254–71.

JOHN, B. and HEWITT, G. M. (1966b). Karyotype stability and DNA variability in *Chromosoma*, **20**: 155–72.

JOHN, B. and HEWITT, G. M. (1968). Patterns and pathways of chromosome evolution within the Orthoptera. *Chromosoma*, **25**: 40–74.

JOHN, B. and HEWITT, G. M. (1969). Parallel polymorphism for supernumerary segments in *Chorthippus parallelus* (Zetterstedt) III. The Ashurst population. *Chromosoma*, **28**: 73–84.

JOHN, B. and HEWITT, G. M. (1970). Inter-population sex chromosome polymorphism in the grasshopper *Podisma pedestris*. I. Fundamental facts. *Chromosoma*, **31**: 291–308.

JOHN, B. and LEWIS, K. R. (1957). Studies on *Periplaneta americana*. I. Experimental analysis of male meiosis. *Heredity*, **11**: 1–9.

JOHN, B. and LEWIS, K. R. (1958). Studies on *Periplaneta americana*. III. Selection for heterozygosity. *Heredity*, **12**: 185–97.

JOHN, B. and LEWIS, K. R. (1959). Selection for interchange heterozygosity in an inbred culture of *Blaberus discoidalis* (Serville). *Genetics*, **44**: 251–67.

JOHN, B. and LEWIS, K. R. (1960a). Chromosome structure in *Periplaneta americana*. *Heredity*, **15**: 47–54.

JOHN, B. and LEWIS, K. R. (1960b). Nucleolar controlled segregation of the sex bivalent in beetles. *Heredity*, **15**: 431–9.

JOHN, B. and LEWIS, K. R. (1965a). *The Meiotic System. Protoplasmatologia*, **VIf1**. Vienna and New York: Springer-Verlag.

JOHN, B. and LEWIS, K. R. (1965b). Genetic speciation in the grasshopper *Eyprepocnemis plorans. Chromosoma*, **16**: 308–44.

JOHN, B. and LEWIS, K. R. (1968). *The Chromosome Complement. Protoplasmatologia*, **VIa**. Vienna: Springer-Verlag.

JOHN, B. and LEWIS, K. R. (1969). *The Chromosome Cycle. Protoplasmatologia*, **VIb**. Vienna and New York: Springer-Verlag.

JOHN, B., LEWIS, K. R. and HENDERSON, S. A. (1960). Chromosome abnormalities in a wild population of *Chorthippus brunneus. Chromosoma*, **11**: 1–20.

JOHN, B. and QURAISHI, H. B. (1964). Studies on *Periplaneta americana*. IV. Pakistani populations. *Heredity*, **19**: 147–56.

JOHN, B. and SHAW, D. D. (1967). Karyotype variation in dermestid beetles. *Chromosoma*, **20**: 371–85.

JOHNSON, G. (1957). Étude de la spermatogénèse chez l'Oniscoïde hermaphrodite, *Rhyscotoides legrandi. C.R. Acad. Sci. Paris*, **244**: 1087–9.

JONES, A. W. (1944). *Macrostomum hustedi* n.sp.; a morphological and cytological study of a rhabdocoel turbellarian. *J. Morph.* **75**: 347–59.

JONES, A. W. (1945). Studies in cestode cytology. *J. Parasitol.* **31**: 213–235.

JONES, K. W. (1970). Chromosomal and nuclear location of mouse satellite DNA in individual cells. *Nature*, **225**: 912–15.

JUCCI, C. (1948). Physiogenetics in silkworms *Proc. VIIIth Int. Genet. Congr. Stockholm*, pp. 286–97.

JUDD, B. H. (1955). Direct proof of a variegated-type position effect at the *white* locus in *Drosophila melanogaster. Genetics*, **40**: 739–44.

JUDD, B. H. (1962). An analysis of mutations confined to a small region of the X-chromosome of *Drosophila melanogaster. Science*, **138**: 990–1.

JUDD, B. H. (1964). The structure of intralocus duplication and deficiency chromosomes produced by recombination in *Drosophila melanogaster*, with evidence for polarized pairing. *Genetics*, **49**: 253–65.

JUNKER, H. (1923). Cytologische Untersuchungen an den Geschlechtsorganen der halbzwittrigen Steinfliege *Perla marginata. Arch. Zellforsch.* **17**: 185–259.

KAHLE, W. (1908). Die Pädogenesis der Cecidomyiden. *Zoologica*, **21**: 1–80.

KAHN, J. (1962). The nucleolar organizer in the mitotic chromosome complement of *Xenopus laevis. Quart. J. Micr. Sci.* **103**: 407–9.

KAHN, J. (1964). Cytotaxonomy of ticks. *Quart. J. Micr. Sci.* **105**: 123–7.

KALLMAN, K. D. (1962). Population genetics of the gynogenetic teleost, *Mollienisia formosa* (Girard). *Evolution*, **16**: 497–504.

KALLMAN, K. D. (1964). Homozygosity in a gynogenetic fish. *Genetics*, **50**: 260–1.

KALLMAN, K. D. (1965). Sex determination in the teleost *Xiphophorus maculatus. Genetics*, **52**: 450–1.

KARNKOWSKA, Z. (1932). Les chromosomes dans la spermatogénése de la puce. *C.R. Soc. Biol. Paris.* **110**: 670–1.

KASTRITSIS, C. D. (1966a). A comparative chromosome study in the incipient species of the *Drosophila paulistorum* complex. *Chromosoma*, **19**: 208–22.

KASTRITSIS, C. D. (1966b). Cytological studies on some species of the *tripunctata* group of *Drosophila. Studies in Genetics III, Univ. Texas Publ.* **6615**: 413–74.

KASTRITSIS, C. D. (1967). A comparative study of the chromosomal polymorphs in the incipient species of the *Drosophila paulistorum* complex. *Chromosoma*, **23**: 180–202.

KASTRITSIS, C. D. and CRUMPACKER, D. W. (1967). Gene arrangements in the third chromosome of *Drosophila pseudoobscura*. *J. Hered.* **58**: 113–29.

KASTRITSIS, C. D. and DOBZHANSKY, TH. (1967). *Drosophila pavlovskiana*, a race or a species? *Amer. Midl. Nat.* **78**: 244–7.

KASTURI BAI, A. R. and SUGANDHI, M. R. (1968). Chromosome study of *Haltica caerulea* Oliver. *Caryologia*, **21**: 161–6.

KATAYAMA, H. (1939a). On the chromosomes of *Hybris subjacens* Walk. (Neuroptera: Ascalaphidae). *Jap. J. Genet.* **15**: 75–7.

KATAYAMA, H. (1939b). The sex chromosomes of a may-fly, *Ameletus costalis* Mats. (Ephemerida). *Jap. J. Genet.* **15**: 139–44.

KAUFMAN, T. C., SHEN, M. W. and JUDD, B. H. (1969). The complementation map of mutations in a small region of the X-chromosome of *Drosophila melanogaster*. *Genetics*, **61**: 30–1.

KAUFMANN, B. P. (1933). Interchange between X- and Y-chromosomes in attached-X females of *Drosophila melanogaster*. *Proc. Nat. Acad. Sci. U.S.A.* **19**: 830–8.

KAUFMANN, B. P. (1934). Somatic mitoses of *Drosophila melanogaster*. *J. Morph.* **56**: 125–55.

KAUFMANN, B. P. (1936). A terminal inversion in *Drosophila ananassae*. *Proc. Nat. Acad. Sci. U.S.A.* **22**: 591–4.

KAUFMANN, B. P. (1937). Morphology of the chromosomes of *Drosophila ananassae*. *Cytologia* Fujii Jubilee vol., pp. 1043–5.

KAUFMANN, B. P. (1940). The nature of hybrid sterility – abnormal development in eggs of hybrids between *Drosophila miranda* and *Drosophila pseudoobscura*. *J. Morph.* **66**: 197–213.

KAUFMANN, B. P. (1954). Chromosome aberrations induced in animal cells by ionizing radiations. In: *Radiation Biology*, **1** (2): (ed.: A. Hollaender), pp. 627–711. New York: McGraw-Hill.

KAUFMANN, B. P. and GAY, H. (1969). The capacity of the fourth chromosome of *Drosophila melanogaster* to establish end-to-end contacts with the other chromosomes in salivary-gland cells. *Chromosoma*, **26**: 395–409.

KAUFMANN, B. P. and IDDLES, M. K. (1963). Ectopic pairing in salivary gland chromosomes of *Drosophila melanogaster*. I. Distributional patterns in relation to puffing. *Port. Acta Biol.* A, **7**: 225–248.

KAUFMANN, B. P., McDONALD, M. R., GAY, H., WILSON, K., WYMAN, R. and OKUDA, N. (1948). Organization of the chromosome. *Yearb. Carnegie Instn.* **47**: 144–55.

KAWAGUCHI, E. (1928). Zytologische Untersuchungen am Seidenspinner und seinen Verwandten. I. Gametogenese von *Bombyx mori* L. und *B. mandarina* M. und ihre Bastarde. *Z. Zellforsch.* **7**: 519–52.

KAWAMURA, K. (1960). Studies on cytokinesis in neuroblasts of the grasshopper, *Chortophaga viridifasciata* (De Geer). *Exp. Cell Res.* **21**: 1–18.

KAWAMURA, T. (1940). Artificial parthenogenesis in the frog. III. The development of the gonads in triploid frogs and tadpoles. *J. Sci. Hiroshima Univ.* B, **8**: 117–64.

KAWAMURA, T. (1951a). Reproductive ability of triploid newts, with remarks on their offspring. *J. Sci. Hiroshima Univ.* B, **12**: 1–10.

KAWAMURA, T. (1951b). The offspring of triploid males of the frog, *Rana nigromaculata*. *J. Sci. Hiroshima Univ.* B, **12**: 11–20.

KAWAMURA, T. (1952). Studies on hybridization in amphibians. IV. Hybrids between *Hynobius nebulosus* Schlegel and *Hynobius naevius* (Schlegel). *J. Sci. Hiroshima Univ.* B, **13**: 139–49.

KAWAMURA, T. (1953). Studies on hybridization in amphibians. V. Physiological isolation among four *Hynobius* species. *J. Sci. Hiroshima Univ.* B, **14**: 73–116.

KAWAMURA, T. and YOKOTA, R. (1959). The offspring of sex-reversed females of *Rana japonica* Guenther. *J. Sci. Hiroshima Univ.* B, **18**: 31–38.

KAYANO, H. (1956a). Cytogenetic studies in *Lilium callosum*. I. Three types of supernumerary chromosomes. *Mem. Fac. Sci. Kyushu Univ.* E, **2**: 47–52.

KAYANO, H. (1956b). Cytogenetic studies in *Lilium callosum*. II. Preferential segregation of a supernumerary chromosome. *Mem. Fac. Sci. Kyushu Univ.* E, **2**: 53–60.

KAYANO, H. (1957). Cytogenetic studies in *Lilium callosum*. III. Preferential segregation of a supernumerary chromosome in EMC's. *Proc. Jap. Acad.* **33**: 553–8.

KAYANO, H. (1960a). Chiasma studies in structural hybrids. III. Reductional and equational separation in *Disporum sessile*. *Cytologia*, **25**: 461–7.

KAYANO, H. (1960b). Chiasma studies in structural hybrids. IV. Crossing-over in *Disporum sessile*. *Cytologia*, **25**: 468–75.

KAYANO, H. (1962). Cytogenetic studies in *Lilium callosum*. V. Supernumerary *B*-chromosomes in wild populations. *Evolution*, **16**: 246–53.

KEENAN, R. D. (1932). The chromosomes of *Sphenodon punctatum*. *J. Anat.* **67**: 1–17.

KENNEDY, J. A. (1969). The karyotypes of some Australian rodents (Rodentia: Muridae). *Aust. J. Zool.* **17**: 465–71.

KERKIS, J. J. (1933). Development of gonads in hybrids between *Drosophila melanogaster* and *Drosophila simulans*. *J. Exp. Zool.* **66**: 477–509.

KERNAGHAN, R. P. and EHRMAN, L. (1970). An electron microscopic study of the etiology of hybrid sterility in *Drosophila paulistorum*. I. Mycoplasma-like inclusions in the testes of sterile males. *Chromosoma*, **29**: 291–304.

KERR, W. E. (1948). Estudos sôbre o gênero *Melipona*. *An. Esc. Sup. Agric. L. de Queiroz*, **5**: 181–276.

KERR, W. E. (1950a). Evolution of the mechanism of caste determination in the genus *Melipona*. *Evolution*, **4**: 7–13.

KERR, W. E. (1950b). Genetic determination of castes in the genus *Melipona*. *Genetics*, **35**: 143–52.

KERR, W. E. (1951a). Sex chromosomes in honey bee. *Evolution*, **5**: 80–1.

KERR, W. E. (1951b). Bases para o estudo da genética de populações dos Hymenoptera em geral e dos Apinae sociais em particular. *An. Esc. Sup. Agric. L. de Queiroz*, **8**: 219–354.

KERR, W. E. (1967). Genetic structure of the populations of Hymenoptera. *Ciencia e Cultura*, **19**: 39–44.

KERR, W. E. (1969). Some aspects of the evolution of social bees. *Evol. Biol.* **3**: 119–75.

KERR, W. E. and ARAUJO, V. DE P. (1957). Contribuição ao estudo citológico dos Apoidea. *Garcia de Orta, Revista da Junta das Missões Geográficas e de Investigações do Ultramar*, **5**: 431–3.

KERR, W. E. and LAIDLAW, H. H. JR. (1956). General genetics of bees. *Adv. in Genet.* **8**: 109–53.

KERR, W. E. and NIELSEN, R. A. (1967). Sex determination in bees (Apinae). *J. Apic. Res.* **6**: 3–9.

KEUNEKE, W. (1924). Über die Spermatogenese einiger Dipteren. *Z. Zellforsch.-u Gewebel.* **1**: 357–412.

KEY, K. H. L. (1968). The concept of stasipatric speciation. *Syst. Zool.* **17**: 14–22.

KEYL, H.-G. (1956). Beobachtungen über die ♂-Meiose der Muschel *Sphaerium corneum. Chromosoma,* **8**: 12–17.

KEYL, H.-G. (1957*a*). Zur Karyologie der Hydrachnellen (Acarina). *Chromosoma,* **8**: 719–29.

KEYL, H.-G. (1957*b*). Untersuchungen am Karyotypus von *Chironomus thummi.* I. Karte der Speicheldrüsen-Chromosomen von *Chironomus thummi thummi* und die cytologische Differenzierung der Subspezies *Ch. th. thummi* und *C. th. piger. Chromosoma,* **8**: 739–56.

KEYL, H.-G. (1960*a*). Die cytologische Diagnostik der Chironomiden. II. Diagnosen der Geschwisterarten. *Chironomus acidophilus* n. sp. und *Ch. uliginosus* n. sp. *Arch. Hydrobiol.* **57**: 187–95.

KEYL, H.-G. (1960*b*). Chromosomenumbau und Evolution in der Gattung *Chironomus. Verh. deutsch. Zool. Ges.,* pp. 280–3.

KEYL, H.-G. (1960*c*). Erhöhung der chromosomalen Replikationsrate durch Microsporidieninfektion in Speicheldrüsenzellen von *Chironomus. Naturwiss.* **47**: 212–13.

KEYL, H.-G. (1961*a*). Chromosomenevolution bei *Chironomus.* I. Strukturabwandlungen an Speicheldrüsen-Chromosomen. *Chromosoma,* **12**: 26–47.

KEYL, H.-G. (1961*b*). Die cytologische Diagnostik der Chironomiden III. *Arch. Hydrobiol.* **58**: 1–6.

KEYL, H.-G. (1962). Chromosomenevolution bei *Chironomus.* II. Chromosomenumbauten und phylogenetische Beziehungen der Arten. *Chromosoma,* **13**: 496–514.

KEYL, H.-G. (1963*a*). Crossing-over bei Bastarden von *Chironomus thummi piger* × *C. thummi thummi. Chromosoma,* **13**: 588–99.

KEYL, H.-G. (1963*b*). DNS-Konstanz im Heterochromatin von *Glyptotendipes. Exp. Cell Res.* **30**: 245–7.

KEYL, H.-G. (1964). Verdopplung des DNS-Gehalts kleiner Chromosomenabschnitte als Faktor der Evolution. *Naturwiss.* **51**: 46–7.

KEYL, H.-G. (1965*a*). Duplikationen von Unterheiten der chromosomalen DNS während der Evolution von *Chironomus thummi. Chromosoma,* **17**: 139–80.

KEYL, H.-G. (1965*b*). A demonstrable local and geometric increase in the chromosomal DNA of *Chironomus. Experientia,* **21**: 191–3.

KEYL, H.-G. (1966*a*). Increase of DNA in chromosomes. *Chromosomes Today,* **1** (ed.: C. D. Darlington, and K. R. Lewis), pp. 99–101. Edinburgh and London: Oliver and Boyd.

KEYL, H.-G. (1966*b*). Lokale DNS-Replikationen in Riesenchromosomen. In: *Probleme der biologischen Reduplikation, pp. 55–69. Berlin, Heidelberg and New York*: Springer-Verlag.

KEYL, H.-G. and HÄGELE, K. (1966). Heterochromatin-proliferation an den Speicheldrüsenchromosomen von *Chironomus melanotus. Chromosoma,* **19**: 223–30.

KEYL, H.-G. and PELLING, C. (1963). Differentielle DNS-Replikation in den Speicheldrüsen-chromosomen von *Chironomus thummi. Chromosoma,* **14**: 347–59.

KEZER, J. (1962). The chromosome number of the European cave Salamander *Proteus anguinus. Bioloski Vestnik,* **10**: 45–8.

KEZER, J. (1970). Observations on salamander spermatocyte chromosomes during the first meiotic division. *Dros. Inf. Serv.* **45**: 194–200.

Kiauta, B. (1967). Considerations on the evolution of the chromosome complement in Odonata. *Genetica*, **38**: 430–46.

Kiauta, B. (1968). Variation in size of the dragonfly *m*-chromosome, with considerations on its significance for the chorogeography and taxonomy of the order Odonata, and notes on the validity of the rule of Reinig. *Genetica*, **39**: 64–74.

Kiauta, B. (1969*a*). Sex chromosomes and sex-determining mechanisms in Odonata, with a review of the cytological conditions in the family Gomphidae, and references to the karyotypic evolution in the order. *Genetica* **40**: 127–57.

Kiauta, B. (1969*a*). Autosomal fragmentations and fusions in Odonata and their evolutionary implications. *Genetica*, **40**: 158–80.

Kiblisky, P. (1969*a*). Patterns of 7 species of leaf-nosed bats of Venezuela (Chiroptera: Phyllostomidae). *Experientia*, **25**: 1203–4.

Kiblisky, P. (1969*b*). The chromosomes of the hispid cotton rat, *Sigmodon hispidus*, from two localities in Venezuela. *J. Mammal.* **50**: 810–11.

Kichijo, H. (1934). The chromosomes of some neuropterous insects of the family Chrysopidae. *J. Fac. Sci. Hokkaido Univ.* VI, **3**: 55–65.

Kichijo, H. (1941*a*, *b*). Cited from Makino, 1951 (originals not seen).

Kichijo, H. (1942). Chromosomes of *Oligotoma japonica* (Embioptera). *Jap. J. Genet.* **18**: 196–7.

Kihlman, B. A. and Hartley, B. (1967). 'Sub-chromatid' exchanges and the 'folded fibre' model of chromosome structure. *Hereditas*, **57**: 289–94.

Kikkawa, H. (1938). Studies on the genetics and cytology of *Drosophila ananassae*. *Genetica*, **20**: 458–516.

Kiknadze, I. I. and Vysotskaya, L. V. (1970). Measurement of DNA mass per nucleus in the grasshopper species with different numbers of chromosomes. *Tsitologia*, **12**: 1100–7.

Kimura, M. (1956*a*). A model of a genetic system which leads to closer linkage by natural selection. *Evolution*, **10**: 278–87.

Kimura, M. (1956*b*). Rules for testing stability of a selective polymorphism. *Proc. Nat. Acad. Sci. U.S.A.* **42**: 336–40.

Kimura, M. (1960). Genetic load of a population and its significance in evolution. *Jap. J. Genet.* **35**: 7–33.

Kimura, M. (1962). A suggestion on the experimental approach to the origin of supernumerary chromosomes. *Amer. Nat.* **96**: 319–20.

Kimura, M. and Kayano, H. (1961). The maintenance of supernumerary chromosomes in wild populations of *Lilium callosum* by preferential segregation. *Genetics*, **46**: 713–19.

King, J. C. (1947). A comparative analysis of the chromosomes of the *guarani* group of *Drosophila*. *Evolution*, **1**: 48–62.

King, R. L. (1923). Heteromorphic homologous chromosomes in three species of *Pseudotrimerotropis* (Orthoptera: Acrididae). *J. Morph.* **38**: 19–63.

King, R. L. (1931). Chromosomes of three species of Mantidae. *J. Morph.* **52**: 523–33.

King, R. L. (1950). Neo-*Y* chromosome in *Hypochlora alba* and *Mermiria intertexta* (Orthoptera: Acrididae). *J. Morph.* **87**: 227–57.

King, R. L. and Beams, H. W. (1934). Somatic synapsis in *Chironomus* with special reference to the individuality of the chromosomes. *J. Morph.* **56**: 577–86.

King, R. L. and Beams, H. W. (1938). The multiple chromosomes of *Paratylotropidia brunneri* Scudder (Orthoptera: Acrididae). *J. Morph.* **63**: 289–300.

King, R. L. and Slifer, E. H. (1934). Insect development. VIII. Maturation and early development of unfertilized grasshopper eggs. *J. Morph.* **56**: 603–20.

KIT, S. (1961). Equilibrium sedimentation in density gradients of DNA preparations from animal tissues. *J. Mol. Biol.* **3**: 711–16.

KITCHIN, R. M. (1970). A radiation analysis of a Comstockiella chromosome system: destruction of heterochromatic chromosomes during spermatogenesis in *Parlatoria oleae* (Coccoidea: Diaspididae). *Chromosoma*, **31**: 165–97.

KITZMILLER, J. B. (1953). Mosquito genetics and cytogenetics. *Rev. Brasil. Malariol. Doen. Trop.* **5**: 285–359.

KITZMILLER, J. B. (1966). The salivary gland chromosomes of *Anopheles algeriensis*. *Riv. Malariol.* **45**: 51–59.

KITZMILLER, J. B. and BAKER, R. H. (1963). The salivary chromosomes of *Anopheles freeborni*. *Mosquito News* **23**: 254–61.

KITZMILLER, J. B. and BAKER, R. H. (1965). The salivary chromosomes of *Anopheles earlei*. *Canad. J. Genet. Cytol.* **7**: 275–83.

KITZMILLER, J. B., BAKER, R. H. and CHOWDAIAH, B. N. (1966). The salivary gland chromosomes of *Anopheles neomaculipalpus*. *Caryologia*, **19**: 1–12.

KITZMILLER, J. B. and FRENCH, W. L. (1961). Chromosomes of *Anopheles quadrimaculatus*. *Amer. Zool.* **1**: 366.

KITZMILLER, J. B., FRIZZI, G. and BAKER, R. H. (1967). Evolution and speciation within the *maculipennis* complex of the genus *Anopheles*. In: *Genetics of Insect Vectors of Disease* (ed.: J. W. Wright and R. Pal), pp. 151–210. Amsterdam: Elsevier.

KLASSEN, W., FRENCH, W. L., LAVEN, H. and KITZMILLER, J. B. (1965). The salivary chromosomes of *Anopheles quadrimaculatus* Say. *Mosquito News*, **25**: 328–34.

KLINGER, H. P. (1963). The somatic chromosomes of some Primates (*Tupaia glis, Nycticebus coucang, Tarsius bancanus, Cercocebus aterrimus, Symphalangus syndactylus*). *Cytogenetics*, **2**: 140–51.

KLINGER, H. P., LINDSTEN, J., FRACCARO, M., BARRAI, I. and DOLINAR, Z. J. (1965). DNA content and area of sex chromatin in subjects with structural and numerical aberrations of the *X*-chromosome. *Cytogenetics*, **4**: 96–116.

KLINGSTEDT, H. (1931). Digametie beim Weibchen der Trichoptere *Limnophilus decipiens* Kol. *Acta Zool. Fenn.* **10**: 1–69.

KLINGSTEDT, H. (1934). Chromosomenstudien an Neuropteren. I. Ein Fall von heteromorphen Chromosomenpaaren als Beispiel vom Mendeln der Chromosomen. *Mem. Soc. Fauna Flora Fenn.* **10**: 3–11.

KLINGSTEDT, H. (1937*a*). On some tetraploid spermatocytes in *Chrysochraon dispar* (Orth.) *Comm. Biol. Helsingf.* **12**: 194–209.

KLINGSTEDT, H. (1937*b*). Chromosome behavior and phylogeny in the Neuroptera. *Nature*, **139**: 468–9.

KLINGSTEDT, H. (1939). Taxonomic and cytological studies on grasshopper hybrids. I. Morphology and spermatogenesis of *Chorthippus bicolor* Charp. and *Ch. biguttulus* L. *J. Genet.* **37**: 389–420.

KLUSS, B. C. (1962). Electron microscopy of the macronucleus of *Euplotes eurystomus*. *J. Cell Biol.* **13**: 462–5.

KNÜLLE, W. (1954). Gynandromorphie bei *Erigone vagans spinosa* Camb. (Micryphantidae: Araneae). *Zool. Anz.* **152**: 219–27.

KOBEL, H. R. (1962). Heterochromosomen bei *Vipera berus* L. (Viperidae, Serpentes). *Experientia*, **18**: 173–4.

KOBEL, H. R. and VAN BREUGEL, F. M. A. (1968). Observations on *ltl* (lethal tumorous larvae) of *Drosophila melanogaster*. *Genetica*, **38**: 305–27.

KOLLER, P. C. (1935). The internal mechanics of the chromosomes. IV. *Proc. Roy. Soc. Lond.* B, **118**: 371–97.

KOLLER, P. C. (1936a). The genetical and mechanical properties of the sex chromosomes. II. Marsupials. *J. Genet.* **32**: 451–72.

KOLLER, P. C. (1936b). The origin and behavior of chiasmata, XI. *Dasyurus* and *Sarcophilus*. *Cytologia*, **7**: 82–103.

KOLLER, P. C. (1937). The genetical and mechanical properties of the sex chromosomes. III. Man. *Proc. Roy. Soc. Edinb.* **57**: 194–214.

KOLLER, P. C. (1938). The genetical and mechanical properties of the sex chromosomes. IV. The golden hamster. *J. Genet.* **36**: 177–95.

KOLLER, P. C. (1941a). The genetical and mechanical properties of the sex chromosomes. VII. *Apodemus sylvaticus* and *A. hebridensis*. *J. Genet.* **41**: 375–90.

KOLLER, P. C. (1941b). The genetical and mechanical properties of the sex chromosomes. VIII. The cat (*Felis domestica*). *Proc. Roy. Soc. Edinb.* B, **61**: 78–94.

KOLLER, P. C. and DARLINGTON, C. D. (1934). The genetical and mechanical properties of the sex chromosomes. I. *Rattus norvegicus* male. *J. Genet.* **29**: 159–73.

KOLTZOV, N. K. (1934). The structure of the chromosomes in the salivary glands of *Drosophila*. *Science*, **80**: 312–13.

KOMAI, T. and ISHIHARA, T. (1956). On the origin of the male tortoiseshell cat. *J. Hered.* **47**: 287–91.

KORSCHELT, E. (1884). Über die eigentümliche Bildung in den Zellkernen der Speicheldrüse von *Chironomus plumosa*. *Zool. Anz.* **7**: 189–94.

KOSSWIG, C. (1939). Die Geschlechtsbestimmungsanalyse bei Zahnkarpfen. *Rev. Fac. Sci. Univ. Istanbul*, **4**: 239–70.

KOZLOVSKY, A. N. (1970). Chromosome polymorphism in eastern-Siberian populations of the common shrew, *Sorex araneus* L. *Tsitologia*, **12**: 1459–64.

KRACZKIEWICZ, Z. (1935a). Études cytologiques sur l'oogénèse et la diminution de la chromatine dans les larves paedogénétiques de *Miastor metraloas* Meinert (Diptera). *Folia Morph. Warsz.* **6**: 1–37.

KRACZKIEWICZ, Z. (1935b). Nouvelles recherches sur l'oogénèse et la diminution dans les larves paedogénétiques de *Miastor metroloas*. *C.R. Soc. Biol. Paris*, **119**: 1201–2.

KRACZKIEWICZ, Z. (1936). De la difference entre les premiers stades de maturation des oocytes parthénogénétiques et des oocytes "sexués" de *Miastor metraloas* Meinert (Diptera). *C.R. Soc. Biol. Paris*, **123**: 879–82.

KRACZKIEWICZ, Z. (1937). Recherches cytologiques sur le cycle évolutif de *Miastor metraloas*. *Cellule*, **46**: 57–74.

KRACZKIEWICZ, Z. (1938). La spermatogénèse chez *Miastor metraloas*. *C.R. Soc. Biol. Paris*, **127**: 1143–6.

KRACZKIEWICZ, (1950). Recherches cytologiques sur les chromosomes de *Lasioptera rubi* Heeg. (Cecidomyidae). *Zool. Polon.* **5**: 73–117.

KRACZKIEWICZ, Z. (1966). Premiers stades de l'oogénèse de *Rhabdophaga saliciperda* (Cecidomyiidae, Diptera). *Chromosoma*, **18**: 208–29.

KRAEPELIN, K. (1905). Die geographische Verbreitung der Scorpione. *Zool. Jb. Abt. Syst.* **22**: 321–64.

KRIMBAS, K. B. (1960). Ta didyma eidi Gryllotalpa en Helladi. Kylfarologiki kai morfologiki ereyna, 27 pp. Athens: privately printed.

KRIVSHENKO, J. D. (1963). The chromosomal polymorphism of *Drosophila busckii* in natural populations. *Genetics*, **48**: 1239–58.

KROEGER, H. (1960). The induction of new puffing patterns by transplantation of salivary gland neuclei into egg cytoplasm of *Drosophila*. *Chromosoma*, **11**: 129–45.

KROEGER, H. (1963a). Experiments on the extranuclear control of gene activity in Dipteran polytene chromosomes. *J. Cell Comp. Phys.* **62**: suppl. 1: 45–59.

KROEGER, H. (1963b). Chemical nature of the system controlling gene activities in insect cells. *Nature*, **200**: 1234–5.

KROEGER, H. (1964). Zellphysiologische Mechanismen bei der Regulation von Genaktivitäten in den Riesenchromosomen von *Chironomus thummi*. *Chromosoma*, **15**: 36–70.

KUNZ, W. (1967). Lampenbürstenchromosomen und multiple Nukleolen bei Orthopteren. *Chromosoma*, **21**: 446–62.

KUNZ, W. (1969). Die Entstehung multipler Oocytennukleolen aus akzessorischen DNS-Körpern bei *Gryllus domesticus*. *Chromosoma*, **26**: 41–75.

KUNZ, W., TREPTE, H. H. and BIER, K. (1970). On the function of the germ line chromosomes in *Wachtliella persicariae* (Cecidomyiidae). *Chromosoma*, **30**: 180–92.

KUNZE-MÜHL, E. and SPERLICH, D. (1955). Inversionen und chromosomale Strukturtypen bei *Drosophila subobscura* Coll. *Z. indukt. Abst. Vererbl.* **87**: 65–84.

KUPKA, E. (1948). Chromosomale Verschiedenheiten bei schweizerischen Coregonen (Felchen). *Rev. Suisse Zool.* **55**: 285–93.

KUPKA, E. (1950). Die Mitosen- und Chromosomenverhältnisse bei der grossen Schwebrenke, *Coregonus wartmanni* (Bloch), des Attersees. *Öst. Zool. Z.* **2**: 605–23.

KUPRIYANOVA, L. A. (1969). Karyological analysis of lizards of the subgenus *Archaeolacerta*. *Tsitologia* **11**: 803–814.

KURNICK, N. B. and HERSKOWITZ, I. W. (1962). The estimation of polyteny in *Drosophila* salivary gland nuclei based on determination of desoxyribonucleic acid content. *J. Cell. Comp. Phys.* **39**: 281–99.

KUSCHAKEWITSCH, S. (1913). Studien über den Dimorphismus der männlichen Geschlechtselemente bei den Prosobranchia. I. *Arch. Zellforsch.* **10**: 237–323.

KUSHNIR, T. (1948). Chromosomal evolution in the European mole cricket. *Nature*, **161**: 531–2.

KUSHNIR, T. (1952). Heterochromatic polysomy in *Gryllotalpa gryllotalpa* L. *J. Genet.* **50**: 361–83.

KUWADA, Y. and NAKAMURA, T. (1934). Behavior of chromonemata in mitosis. II. Artificial unravelling of coiled chromonemata. *Cytologia*, **5**: 244–7.

LA COUR, L. F. (1953). The *Luzula* system analyzed by X-rays. *Heredity*, **6**, suppl. vol.: 77–81.

LA COUR, L. F. and PELC, S. R. (1958). Effect of colchicine on the utilization of labelled thymidine during chromosomal reproduction. *Nature*, **182**: 506–8.

LA COUR, L. F. and PELC, S. R. (1959). Effect of colchicine on the utilization of labelled thymidine during chromosomal reproduction. *Nature*, **183**: 1455–6.

LA COUR, L. F. and RUTISHAUSER, A. (1954). X-ray breakage experiments with endosperm. I. Sub-chromatid breakage. *Chromosoma*, **6**: 696–709.

LACROIX, J.-C. (1967). Obtention de femelles trisomiques fertiles chez l'amphibian urodèle *Pleurodelus waltlii* Michah. *C.R. Acad. Sci. Paris* D, **264**: 85–8.

LACROIX, J.-C. (1968a). Étude descriptive des chromosomes en écouvillon dans le genre *Pleurodeles* (Amphibien, urodèle). *Ann. Embryol. Morphog.* **1**: 179–202.

LACROIX, J.-C. (1968b). Variations expérimentales ou spontanées de la morphologie et de l'organization des chromosomes en écouvillon dans le genre *Pleurodeles* (Amphibien, Urodèle). *Ann. Embryol. Morphog.* **1**: 205–48.

LACROIX, J.-C. (1970). Mise en évidence sur les chromosomes en écouvillon de *Pleurodeles poireti* Gervais, Amphibien urodèle, d'une structure liée au sexe, identifiant le bivalent sexuel et marquant le chromosome *W. C. R. Acad. Sci. Paris*, **171**: 102–4.

LANCEFIELD, D. E. (1929). A genetic study of crosses of two races or physiological species of *Drosophila obscura. Z. indukt. Abst. Vererbl.* **52**: 287–317.

LANDAU, R. (1962). Four forms of *Simulium tuberosum* (Lundstr.) in Southern Ontario: a salivary gland chromosome study. *Canad. J. Zool.* **40**: 921–39.

LANG, K. (1953). *Apseudes hermaphroditicus* n. sp. A hermaphroditic tanaid from the Antarctic. *Ark. Zool.* [2] **4** (18): 341–50.

LANTZ, L. A. (1947). Hybrids between *Triturus cristatus* Laur. and *Triturus marmoratus* Latr. *Proc. Zool. Soc. Lond.* **117**: 247–58.

LANTZ, L. A. and CALLAN, H. G. (1954). Phenotypes and spermatogenesis of interspecific hybrids between *Triturus cristatus* and *Triturus marmoratus*. *J. Genet.* **52**: 165–85.

LARK, K. G. (1967). Non random segregation of sister chromatids in *Vicia faba* and *Triticum boeoticum. Proc. Nat. Acad. Sci. U.S.A.* **58**: 352–9.

LARK, K. G. (1969). Sister chromatid segregation during mitosis in polyploid wheat. *Genetics*, **62**: 289–305.

LARK, K. G., CONSIGLI, R. A. and MINOCHA, H. C. (1966). Segregation of sister chromatids in mammalian cells. *Science*, **154**: 1202–5.

LAWSON, C. A. (1936). A chromosome study of the aphid *Macrosiphum solanifolii. Biol. Bull. Woods Hole*, **70**: 288–307.

LAY, D. M. and NADLER, C. F. (1969). Hybridization in the rodent genus *Meriones*. I. Breeding and cytological analyses of *Meriones shawi* (♀) × *Meriones lybicus* (♂) hybrids. *Cytogenetics*, **8**: 35–50.

LEA, D. E. (1946). *Actions of Radiations on Living Cells*. Cambridge Univ. Press.

LEBEDEFF, G. A. (1938). Intersexuality in *Drosophila virilis* and its bearing on sex determination. *Proc. Nat. Acad. Sci. U.S.A.* **24**: 165–72.

LE CALVEZ, J. (1947). Morphologie et comportement des chromosomes dans la spermatogénèse de quelques Mycetophilides. *Chromosoma*, **3**: 137–65.

LE CALVEZ, J. (1949). Données caryologiques sur l'Embioptère *Monotylota ramburi* Enderl. *C.R. Acad. Sci. Paris*, **229**: 245–6.

LE CALVEZ, J. (1950). Recherches sur les Foraminifères. II. Place de la méiose et sexualité. *Arch. Zool. Exp. Gen.* **87**: 211–44.

LE CALVEZ, J. and CERTAIN, P. (1950). Données caryologiques sur quelques Pulmonés basommatophores. *C.R. Acad. Sci. Paris*, **231**: 794–5.

LÉCHER, P. (1967). Cytogénétique de l'hybridation expérimentale et naturelle chez l'isopode *Jaera* (*albifrons*) *syei* Bocquet. *Arch. Zool. Exp. Gén.* **108**: 633–98.

LEE, M. R. and ZIMMERMAN, E. G. (1969). Robertsonian polymorphism in the cotton rat, *Sigmodon fulviventer. J. Mammal.* **50**: 333–9.

LEJEUNE, J., DUTRILLAUX, B., LAFOURCADE, J., BERGER. R., ABONYI, D. and RETHORÉ, M. O. (1968). Endoreduplication du bras long du chromosome 2 chez une femme et sa fille. *C.R. Acad. Sci. Paris*, **266**: 24–6.

LEJEUNE, J., GAUTIER, M. and TURPIN, R. (1959). Étude des chromosomes somatiques de neuf enfants mongoliens. *C.R. Acad. Sci. Paris*, **248**: 1721–2.

LENNOX, B. (1961). Indirect assessment of number of *X*-chromosomes in man, using nuclear sexing and colour vision. *Brit. Med. Bull.* **17**: 196–9.

LEPORI, N. G. (1949). Ricerche sulla ovogenesi e sulla fecondazione nella planaria *Polycelis nigra* Ehrenberg con particolare riguardo all'ufficio del nucleo spermatico. *Caryologia*, **1**: 280–95.

LEPORI, N. G. (1950). Il ciclo cromosomico, con poliploidia, endomitosi e ginogenesi, in popolazioni italiane di *Polycelis nigra* Ehrenberg. *Caryologia*, **2**: 301–24.

LESLIE, I. (1961). Distribution of purines and pyrimidines in nucleic acids. In: *Biochemists' Handbook* (ed.: Cyril Long), pp. 198–201. London: Spon.

LESTON, D. (1957). Cytotaxonomy of Miridae and Nabidae (Hemiptera). *Chromosoma*, **8**: 609–16.

LESTON, D. (1958). Chromosome number and the systematics of Pentatomorpha (Hemiptera). *Proc. Xth Int. Entomol. Congr. Montreal*, **3**: 911–18.

LEVAN, A. (1940). Meiosis of *Allium porrum*, a tetraploid species with chiasma localization. *Hereditas*, **26**: 454–62.

LEVAN, A., FREDGA, K. and SANDBERG, A. A. (1965). Nomenclature for centromeric position of chromosomes. *Hereditas*, **52**: 201–20.

LEVAN, A. and HAUSCHKA, T. S. (1953). Endomitotic reduplication mechanisms in ascites tumors of the mouse. *J. Nat. Cancer Inst.* **14**: 1–43.

LEVAN, A., HSU, T. C. and STICH, H. F. (1962). The idiogram of the mouse. *Hereditas*, **48**: 677–87.

LEVENE, H. (1953). Genetic equilibrium when more than one ecological niche is available. *Amer. Nat.* **87**: 331–3.

LEVITAN, M. (1951). Selective differences between males and females in *Drosophila robusta*. *Amer. Nat.* **85**: 385–8.

LEVITAN, M. (1952). A ring chromosome in wild-caught *Drosophila*. *Rec. Genet. Soc. Amer.* **21**: 44.

LEVITAN, M. (1954). Additional evidence of position effects in natural populations. *Genetics*, **39**: 979.

LEVITAN, M. (1955). Studies of linkage in populations. I. Association of second chromosome linkages in *Drosophila robusta*. *Evolution*, **9**: 27–41.

LEVITAN, M. (1957). Natural selection for linked gene arrangements. *Anat. Rec.* **127**, 430.

LEVITAN, M. (1958). Non-random associations of inversions. *Cold Spr. Harb. Symp. Quant. Biol.* **23**: 251–68.

LEVITAN, M. (1961). Proof of an adaptive linkage association. *Science*, **134**: 1617–9.

LEVITAN, M. (1963). A maternal factor which breaks paternal chromosomes. *Nature*, **200**: 437–8.

LEVITAN, M., CARSON, H. L. and STALKER, H. D. (1954). Triads of overlapping inversions in *Drosophila robusta*. *Amer. Nat.* **88**: 113–14.

LEVITAN, M. and MONTAGU, A. (1971). *Textbook of Human Genetics*. Oxford: Clarendon Press.

LEVITAN, M. and SALZANO, F. N. (1959). Studies of linkage in populations. III. An association of linked inversions in *Drosophila guarumunu*. *Heredity*, **13**: 243–8.

LEVITAN, M. and SCHILLER, R. (1963). Further evidence that the chromosome breakage factor in *Drosophila robusta* involves a maternal effect. *Genetics*, **48**: 1231–8.

LEWIS, E. B. (1945). The relation of repeats to position effects in *Drosophila melanogaster*. *Genetics*, **30**: 137–66.

LEWIS, E. B. (1950). The phenomenon of position effect. *Adv. Genet.* **3**: 73–115.

LEWIS, E. B. (1951). Pseudoallelism and gene evolution. *Cold Spr. Harb. Symp. Quart. Biol.* **16**: 159–74.

LEWIS, E. B. (1955). Some aspects of position pseudoallelism. *Amer. Nat.* **89**: 73–89.

LEWIS, E. B. (1963). Genes and developmental pathways. *Amer. Zool.* **3**: 33–56.

LEWIS, E. B. (1964). Genetic control and regulation of developmental pathways. In: *Role of Chromosomes in Development*. New York: Academic Press.

LEWIS, E. B. (1967). Genes and gene complexes. In: *Heritage from Mendel* (ed.: R. A. Brink), pp. 17–47. Univ. Wisconsin Press.

LEWIS, E. B. and GENCARELLA, W. (1952). Claret and non-disjunction in *Drosophila melanogaster*. *Rec. Genet. Soc. Amer.* **21**: 44–5.

LEWIS, F. J. W., HYMAN, J. M., MACTAGGART, M. and POULDING, R. H. (1963). Trisomy of autosome 16. *Nature*, **199**: 404.

LEWIS, H. (1951). The origin of supernumerary chromosomes in natural populations of *Clarkia elegans*. *Evolution*, **5**: 142–57.

LEWIS, K. R. and JOHN, B. (1957a). Studies on *Periplaneta americana*. II. Interchange heterozygosity in isolated populations. *Heredity*, **11**: 11–22.

LEWIS, K. R. and JOHN, B. (1957b). Bivalent structure in *Periplaneta americana*. *Nature*, **179**: 973.

LEWIS, K. R. and JOHN, B. (1957c). The organization and evolution of the sex multiple in *Blaps mucronata*. *Chromosoma*, **9**: 69–80.

LEWIS, K. R. and JOHN, B. (1959). Breakdown and restoration of chromosome stability following inbreeding in a locust. *Chromosoma*, **10**: 589–618.

LEWIS, K. R. and JOHN, B. (1963). *Chromosome Marker*. London: Churchill.

LEWIS, K. R. and JOHN, B. (1966). The meiotic consequences of spontaneous chromosome breakage. *Chromosoma*, **18**: 287–304.

LEWONTIN, R. C. (1965). Selection in and of populations. In: *Ideas in Modern Biology* (ed.: J. A. Moore). Garden City, New York: Natural History Press.

LEWONTIN, R. C. (1971). The effect of genetic linkage on the mean fitness of a population. *Proc. Nat. Acad. Sci. U.S.A.* **68**: 984–6.

LEWONTIN, R. C. and DUNN, L. C. (1960). The evolutionary dynamics of a polymorphism in the house mouse. *Genetics*, **45**: 705–22.

LEWONTIN, R. C. and KOJIMA, K. (1960). The evolutionary dynamics of complex polymorphisms. *Evolution*, **14**: 458–72.

LEWONTIN, R. C. and WHITE, M. J. D. (1960). Interaction between inversion polymorphisms of two chromosome pairs in the grasshopper *Moraba scurra*. *Evolution*, **14**: 116–29.

LEZZI, M. (1965). Die Wirkung von DNAse auf isolierte Polytänchromosomen. *Exp. Cell Res.* **39**: 289–92.

LI, C. C. (1955). The stability of an equilibrium and the average fitness of a population. *Amer. Nat.* **89**: 281–95.

LIGHT, S. F. (1942). The determination of the castes of social insects. *Quart. Rev. Biol.* **17**: 312–26.

LIMA-DE-FARIA, A. (1948). *B*-chromosomes of rye at pachytene. *Port. Acta. Biol.* A, **2**: 167–74.

LIMA-DE-FARIA, A. (1949a). The structure of the centromere of the chromosomes of rye. *Hereditas*, **35**: 77–85.

LIMA-DE-FARIA, A. (1949b). Genetics, origin and evolution of kinetochores. *Hereditas*, **35**: 422–44.

LIMA-DE-FARIA, A. (1952). Chromomere analysis of the chromosome complement of rye. *Chromosoma*, **5**: 1–68.

LIMA-DE-FARIA, A. (1956). The role of the kinetochore in chromosome organization. *Hereditas*, **42**: 85–160.

LIMA-DE-FARIA, A. (1959a). Incorporation of tritiated thymidine into meiotic chromosomes. *Science*, **130**: 503–4.

LIMA-DE-FARIA, A. (1959b). Differential uptake of tritiated thymidine into hetero- and euchromatin in *Melanoplus* and *Secale*. *J. Biophys. Biochem. Cytol.* **6**: 457–66.

LIMA-DE-FARIA, (1962). Metabolic DNA in *Tipula oleracea*. *Chromosoma*, **13**: 47–59.

LIMA-DE-FARIA, A., BIRNSTIEL, M. and JAWORSKA, H. (1969). Amplification of ribosomal cistrons in the heterochromatin of *Acheta*. *Genetics*, **61** suppl.: 145–59.

LIMA-DE-FARIA, A. and JAWORSKA, H. (1968). Late DNA Synthesis in heterochromatin. *Nature*, **217**: 138–42.

LIMA-DE-FARIA, A. and MOSES, M. J. (1966). Ultrastructure and cytochemistry of metabolic DNA in *Tipula*. *J. Cell Biol.* **30**: 177–92.

LIMA-DE-FARIA, A., NILSSON, B., CAVE, D., PUGA, A. and JAWORSKA, H. (1968). Tritium labelling and cytochemistry of extra DNA in *Acheta*. *Chromosoma*, **25**: 1–20.

LIMA-DE-FARIA, A., REITALU, J. and BERGMAN, S. (1961). The pattern of DNA synthesis in the chromosomes of man. *Hereditas*, **47**: 695–704.

LIN, T. P. (1954). The chromosomal cycle in *Parascaris equorum* (*Ascaris megalocephala*): oogenesis and diminution. *Chromosoma*, **6**: 175–98.

LINDEBERG, B. (1958). A parthenogenetic race of *Monotanytarsus boreoalpinus* Th. (Dipt., Chironomidae) from Finland. *Ann. Ent. Fenn.* **24**: 35–8.

LINDER, D. and POWER, J. (1970). Further evidence for postmeiotic origin of teratomas in the human female. *Ann. Hum. Genet.* **34**: 21–30.

LOEB, J. (1913). *Artificial parthenogenesis and Fertilization*. Chicago Univ. Press.

LOEWUS, M. W., BROWN, S. W. and McLAREN, A. D. (1964). Base ratios in DNA in male and female *Pseudococcus citri*. *Nature*, **203**: 104.

LONDON CONFERENCE (1963). The London conference on the normal human karyotype. *Cytogenetics*, **2**: 264–8.

LONGLEY, A. E. (1927). Supernumerary chromosomes in *Zea mays*. *J. Agric. Res.* **35**: 769–84.

LONGLEY, A. E. (1956). The origin of diminutive *B*-type chromosomes in maize. *Amer. J. Bot.* **43**: 18–22.

LOOMIS, E. C. and STONE, B. F. (1970). Gynandromorphs of *Boophilus* ticks (Acarina: Ixodidae). *J. Aust. Entomol. Soc.* **9**: 68–70.

LORKOVIĆ, Z. (1941). Die Chromosomenzahlen in der Spermatogenese der Tagfalter. *Chromosoma*, **2**: 155–91.

LORKOVIĆ, Z. (1949). Chromosomen-Vervielfachung bei Schmetterlingen und ein neuer Fall fünffacher Zahl. *Rev. Suisse Zool.* **56**: 243–9.

LORKOVIĆ, Z. (1950). Neue ostasiatische Arten und Rassen der Gattung *Leptidea* nebst Nomenklaturberichtigungen. *Glasnik Hrv. Priridosl. Drustvo, Zagreb* 2B, 2–3: 57–76.

LORKOVIĆ, Z. (1968). Karyologischer Beitrag zur Frage der Fortpflanzungsverhältnisse südeuropäischer Taxone von *Pieris napi* (L.) (Lep., Pieridae). *Biol. Glasnik*, **21**: 95–136.

LOW, R. J. and BENIRSCHKE, K. (1969). The replicating pattern of the chromosomes of *Pan troglodytes*. *Proc. 2nd Int. Congr. Primatol., Atlanta, Ga.* **2**: 92–102.

LOWE, C. H. and WRIGHT, J. W. (1966a). Evolution of parthenogenetic species of *Cnemidophorus* (Whiptail lizards) in western North America. *J. Arizona Acad. Sci.* **4**: 81–7.

LOWE, C. H. and WRIGHT, J. W. (1966b). Chromosomes and karyotypes of Cnemidophorine teiid lizards. *Mammal. Chrom. Newsl.* no. **22**: 199–200.

LOWE, C. H., WRIGHT, J. W., COLE, G. J. and BEZY, R. L. (1970a). Natural hybridization between the Teiid lizards *Cnemidophorus sonorae* (parthenogenetic) and *Cnemidophorus tigris* (bisexual). *Syst. Zool.* **19**: 114–27.

Lowe, C. H., Wright, J. W., Cole, C. J. and Bezy, R. L. (1970*b*). Chromosomes and evolution of the species groups of *Cnemidophorus* (Reptilia: Teiidae). *Syst. Zool.* **19**: 128–41.

Lubs, H. A. and Ruddle, F. H. (1971). Chromosome polymorphism in American negro and white populations. *Nature*, **233**: 134–6.

Ludwig, W. (1950). Zur Theorie der Konkurrenz. Die Annidation (Einnischung) als fünfter Evolutionsfaktor. *Neue Ergeb. Probleme Zool., Klatt Festschr.* pp. 516–37.

Lueken, W. (1966). Autosomale weibchendeterminierende Realisatoren bei *Armadillidium nasutum* (Isopoda Terrestria). *Z. Vererbl.* **97**: 345–52.

Luria, S. E. (1959). The reproduction of viruses: a comparative survey. In: *The Viruses*, **1**, (ed.: F. M. Burnet and W. M. Stanley), pp. 549–68. New York: Academic Press.

Lwoff, A. (1943). *L'Évolution Physiologique: Étude des Pertes de Fonctions chez les Microorganismes*. Paris: Hermann.

Lyon, M. F. (1961). Gene action in the *X*-chromosome of the mouse (*Mus musculus* L). *Nature*, **190**: 372–3.

Lyon, M. F. (1962). Sex chromatin and gene action in the mammalian *X*-chromosome. *Amer. J. Hum. Genet.* **14**: 135–48.

Lyon, M. F. (1963). Attempts to test the inactive-*X* theory of dosage compensation in mammals. *Genet. Res.* **4**: 93–103.

Lyon, M. F. (1966*a*). Lack of evidence that inactivation of the mouse *X*-chromosome is incomplete. *Genet. Res.* **8**: 197–204.

Lyon, M. F. (1966*b*). *X*-chromosome inactivation in mammals. In: *Advances in Teratology* (ed.: D. H. M. Woollam), pp. 93–103. London: Logos Press.

Lyon, M. F. (1968). Chromosomal and subchromosomal inactivation. *Ann. Rev. Genet.* **2**: 31–52.

McCarthy, B. J. and Hoyer, B. H. (1964). Identity of DNA and diversity of messenger RNA molecules in normal mouse tissues. *Proc. Nat. Acad. Sci. U.S.A.* **52**: 915–22.

McCarthy, M. D. (1945). Chromosome studies on eight species of *Sciara* (Diptera) with special reference to chromosome changes of evolutionary significance. *Amer. Nat.* **79**: 104–21.

McClintock, B. (1932). A correlation of ring-shaped chromosomes with variegation in *Zea mays. Proc. Nat. Acad. Sci. U.S.A.* **18**: 677–81.

McClintock, B. (1933). The association of non-homologous parts of chromosomes in the mid-prophase of meiosis in *Zea mays. Z. Zellforsch.* **19**: 191–237.

McClintock, B. (1934). The relation of a particular chromosomal element to the development of the nucleoli in *Zea mays. Z. Zellforsch.* **21**: 294–328.

McClintock, B. (1938). The production of homozygous deficient tissues with mutant characteristics by means of the aberrant mitotic behavior of ring-shaped chromosomes. *Genetics*, **23**: 315–76.

McClintock, B. (1939). The behavior in successive nuclear divisions of a chromosome broken at meiosis. *Proc. Nat. Acad. Sci. U.S.A.* **25**: 405–16.

McClintock, B. (1940). The relation of a particular chromosomal element to the development of the nucleoli in *Zea mays. Z. Zellforsch.* **21**: 294–335.

McClintock, B. (1941*a*). The stability of broken ends of chromosomes in *Zea mays. Genetics*, **26**: 234–82.

McClintock, B. (1941*b*). Spontaneous alterations in chromosome size and form in *Zea mays. Cold Spr. Harb. Symp. Quant. Biol.* **9**: 72–81.

McCLINTOCK, B. (1942). The fusion of broken ends of chromosomes following nuclear fusion. *Proc. Nat. Acad. Sci. U.S.A.* **28**: 458–63.

McCLINTOCK, B. (1951). Chromosome organization and genic expression. *Cold Spr. Harb. Symp. Quant. Biol.* **16**: 13–47.

McCLUNG, C. E. (1902). The spermatocyte divisions of the Locustidae. *Kansas Univ. Sci. Bull.* **1**: 185–238.

McCLUNG, C. E. (1914). A comparative study of the chromosomes in orthopteran spermatogenesis. *J. Morph.* **25**: 651–749.

McCLUNG, C. E. (1917). The multiple chromosomes of *Hesperotettix* and *Mermiria*. *J. Morph.* **29**: 519–605.

McCLUNG, C. E. (1928). Differential chromosomes of *Mecostethus gracilis*. *Z. Zellforsch.* **7**: 756–78.

McCLUNG, C. E. (1932). Multiple chromosomes in the Orthoptera. *Actas Congr. Int. Biol. Montevideo*, pp. 1831–48.

McCLUNG, C. E. (1941). The tetramite of orthopteran spermatocytes. *J. Morph.* **69**: 575–84.

McCLUNG, C. E. and ASANA, J. J. (1933). The chromosomes of *Schizodactylus monstrosus*. *J. Morph.* **55**: 185–91.

McCLURE, H. M., BELDEN, K. H., PIEPER, W. A. and MACKSON, C. B. (1969). Autosomal trisomy in a chimpanzee: resemblance to Down's syndrome. *Science*, **165**: 1010–12.

McCOY, C. J. Jr. and MASLIN, T. P. (1962). A review of teiid lizard *Cnemidophorus cozumelus* and the recognition of a new race, *Cnemidophorus cozumelus rodecki*. *Copeia*, 1962, pp. 620–7.

McDONOGH, R. S. (1943). The variation in adult structure of *Luffia ferchaultella* (Stephens) (Lepidoptera, Psychidae). *Trans. Roy. Entomol. Soc. Lond.* **93**: 149–72.

MACGREGOR, H. C. (1968). Nucleolar DNA in oocytes of *Xenopus laevis*. *J. Cell Sci.* **3**: 437–44.

MACGREGOR, H. C. and CALLAN, H. G. (1962). The actions of enzymes on lampbrush chromosomes. *Quart. J. Micr. Sci.* **103**: 173–203.

MACGREGOR, H. C. and KEZER, J. (1970). Gene amplification in oocytes with eight germinal vesicles from the tailed frog *Ascaphus truei* Stejneger. *Chromosoma*, **29**: 189–206.

MACGREGOR, H. C. and UZZELL, T. M. Jr. (1964). Gynogenesis in salamanders related to *Ambystoma jeffersonianum*. *Science*, **143**: 1043–5.

MACHATTIE, L. A. and THOMAS, C. A. Jr. (1964). DNA from bacteriophage lambda: molecular length and conformation. *Science*, **144**: 1142–4.

MACHIDA, J. (1934). The spermatogenesis of the three species of *Polistes* (Hymenoptera). *Proc. Imp. Acad. Tokyo*, **10**: 515–18.

McINTOSH, A. J. and SHARMAN, G. B. (1953). The chromosomes of some species of marsupials. *J. Morph.* **93**: 509–31.

McINTOSH, J. R., HEPLER, P. K. and VAN WIE, D. G. (1969). Model for mitosis. *Nature*, **224**: 659–63.

MACKENSEN, O. (1951). Viability and sex determination in the honey bee (*Apis mellifica* L.). *Genetics*, **36**: 500–9.

MACKENSEN, O. (1955). Further studies on a lethal series in the honey bee. *J. Hered.* **46**: 72–4.

MacKNIGHT, R. H. (1939). The sex-determining mechanism in *Drosophila miranda*. *Genetics*, **24**: 180–201.

MacKNIGHT, R. H. and COOPER, K. W. (1944). The synapsis of the sex chromosomes of *Drosophila miranda* in relation to their directed segregation. *Proc. Nat. Acad. Sci. U.S.A.* **30**: 384–7.

McKusick, V. A. (1964). *On the X-chromosome of Man.* Washington, D. C.: Amer. Inst. Biol. Sci.

McKusick, V. A. (1970). Human genetics. *Ann. Rev. Genet.* **4**: 1–46.

McLaren, A. and Walker, P. M. B. (1969). Correspondence of certain fractions of mouse DNA. *Nature,* **221**: 771–2.

Maclean, N. (1962). The drumsticks of polymorphonuclear leucocytes in sex chromosome abnormalities. *Lancet,* **i**: 1154–8.

Maden, B. E. H. (1968). Ribosome formation in animal cells. *Nature,* **219**: 685–9.

Maeda, T. (1930). On the configurations of gemini in the pollen mother-cells of *Vicia faba* L. *Mem. Coll. Sci. Kyoto Imp. Univ.* (B), **5**: 89–123.

Maeki, K. (1953). Cytological studies of some Japanese butterflies (Lepidoptera–Rhopalocera). *Ann. Stud. Kwansei Gakuin Univ.* **1**: 67–70.

Maeki, K. (1958a). A comparative study of chromosomes in 15 species of the Japanese Pieridae (Lepidoptera–Rhopalocera). *Jap. J. Genet.* **33**: 349–55.

Maeki, K. (1958b). On the cytotaxonomical relationship in *Leptidea* (Lepidoptera–Rhopalocera). *Jap. J. Genet.* **33**: 283–5.

Maeki, K. (1961). A study of chromosomes in thirty-five species of the Japanese Nymphalidae (Lepidoptera–Rhopalocera). *Jap. J. Genet.* **36**: 137–46.

Maeki, K. and Remington, C. L. (1960a). Studies of the chromosomes of North American Rhopalocera. I. Papilionidae. *J. Lepidopt. Soc.* **13**: 193–203.

Maeki, K. and Remington, C. L. (1960b). Studies of the chromosomes of North American Rhopalocera. II. Hesperiidae, Megathymidae and Pieridae. *J. Lepidopt. Soc.* **14**: 37–57.

Maeki, K. and Remington, C. L. (1961a). Studies of the chromosomes of North American Rhopalocera. III. Lycaenidae, Danaidae, Satyrinae, Morphinae. *J. Lepidopt. Soc.* **14**: 127–47.

Maeki, K. and Remington, C. L. (1961b). Studies of the chromosomes of North American Rhopalocera. IV. Nymphalinae, Charaxidinae, Libytheinae, *J. Lepidopt. Soc.* **14**: 179–201.

Magenis, R. E., Hecht, F. and Lovrien, E. W. (1970). Heritable fragile site on chromosome 16: probable localization of haptoglobin locus in man. *Science,* **170**: 85–7.

Maguire, M. P. (1967). Evidence for homologous pairing of chromosomes prior to meiotic prophase in maize. *Chromosoma,* **21**: 221-31.

Maguire, M. P. (1968). Nomarski interference contrast resolution of sub-chromatid structure. *Proc. Nat. Acad. Sci. U.S.A.* **60**: 533–6.

Maillet, P. L. and Folliott, R. (1965). Sur les ultrastructures chromosomiques de la méiose chez *Philaenus spumarium* L. mâle (Homoptère "Cercopidae"). *C.R. Acad. Sci. Paris,* **260**: 3486–9.

Mainx, F. (1951a). Die Bedeutung von Chromosomenaberrationen in natürlichen Populationen. *Port. Acta Biol.* ser. A, Goldschmidt vol., pp. 563–92.

Mainx, F. (1951b). Die Verbreitung von Chromosomendislokationen in natürlichen Populationen von *Liriomyza urophorina* Mik. *Chromosoma,* **4**: 521–34.

Mainx, F. (1958). Population genetics of *Drosophila* species on the Liparic Islands. *Proc. Xth Int. Genet. Congr.* **2**: 175.

Mainx, F. (1959). Die Geschlechtsverhältnisse der Phoride *Megaselia scalaris* und das Problem einen alternativen Geschlechtsbestimmung. *Z. Vererbl.* **90**: 251–6.

Mainx, F. (1962). Ein neuer Modus der genotypischen Geschlechsbestimmung. *Biol. Zentralbl.* **81**: 335–40.

Mainx, F. (1964a). Mosaikbildungen bei *Megaselia scalaris* durch doppelte Befruchtung. *Z. Vererbl.* **95**: 222–5.

MAINX, F. (1964b). The genetics of *Megaselia scalaris* Loew (Phoridae): a new type of sex determination in Diptera. *Amer. Nat.* **98**: 415–30.

MAINX, F., FIALA, Y. and VON KOGERER, E. (1956). Die geographische Verbreitung der chromosomalen Strukturtypen von *Liriomyza urophorina* Mik. *Chromosoma*, **8**: 18–29.

MAINX, F., KOSKE, TH. and SMITAL, E. (1953). Untersuchungen über die chromosomale Struktur europäischer Vertreter der *Drosophila obscura*-Gruppe. *Z. indukt. Abst. Vererbl.* **85**: 354–72.

MAINX, F., KUNZE, E. and KOSKE, T. (1953). Cytologische Untersuchungen an Lunzer Chironomiden. *Öst. Zool. Z.* **4**: 33–44.

MAKINO, S. (1931). The chromosomes of *Diestrammena japonica* Karny (an orthopteran). *Zool. Mag.* **43**: 635–46.

MAKINO, S. (1932a). An unequal pair of idiochromosomes in the tree cricket *Oecanthus longicauda* Mats. *J. Fac. Sci. Hokkaido Univ.* VI, **2**: 1–36.

MAKINO, S. (1932b). Notes on the chromosomes of *Rana temporaria* L. and *Bufo sachalinensis* Nikolski. *Proc. Imp. Acad. Tokyo.* **8**: 23–5.

MAKINO, S. (1934). The chromosomes in *Hynobius leechii* and *H. nebulosus*. *Trans. Sapporo Nat. Hist. Soc.* **13**: 351.

MAKINO, S. (1935). The chromosomes of *Cryptobranchus allegheniensis*. *J. Morph.* **58**: 573–83.

MAKINO, S. (1939a). The chromosomes of three species of *Hynobius* belonging to the *naevius* group. *Zool. Mag., Tokyo*, **51**: 729–33.

MAKINO, S. (1939b). On the tetraploid spermatocytes produced by irradiation in *Podisma mikado* (Acrididae). *Jap. J. Genet.* **15**: 80–2.

MAKINO, S. (1941). Studies on the murine chromosomes. I. Cytological investigations of mice included in the genus *Mus. J. Fac. Sci. Hokkaido Imp. Univ.* VI, **7**: 305–80.

MAKINO, S. (1943). The chromosome complexes in goat (*Capra hircus*) and sheep (*Ovis aries*) and their relationship. *Cytologia*, **13**: 39–54.

MAKINO, S. (1944). Karyotypes of domestic cattle, zebu and domestic water buffalo. *Cytologia*, **13**: 247–64.

MAKINO, S. (1949). A chromosomal survey in some asiatic species of the Muridae, with special regard to the relationship of the chromosomes upon taxonomy. *Cytologia*, **15**: (153–60).

MAKINO, S. (1951). *An Atlas of the Chromosome Numbers in Animals*. Ames: Iowa State College Press.

MAKINO, S. (1955). Notes on the cytological feature of male sterility in the mule. *Experientia*, **11**: 224.

MAKINO, S. (1956). *A Review of the Chromosome Numbers in Animals* (revised ed.). Tokyo: Hokuryukan Publ. Co.

MAKINO, S. and ASANA, J. J. (1948). A sexual difference in the chromosomes of two species of Agamid lizards. *Chromosoma*, **3**: 208–19.

MAKINO, S. and MOMMA, E. (1949). An idiogram study of the chromosomes in some species of reptiles. *Cytologia*, **15**: 96–108.

MAKINO, S., NIIYAMA, H. and ASANA, J. J. (1938). On the supernumerary chromosomes in the mole cricket, *Gryllotalpa africana* de Beauvois, from India. *Jap. J. Genet.* **14**: 272–7.

MAKINO, S., SOFUNI, T. and SASAKI, M. S. (1963). A revised study of the chromosomes in the horse, the ass and the mule. *Proc. Jap. Acad.* **39**: 176–81.

MAKINO, S., UDAGAWA, T. and YAMASHINA, Y. (1956). Karyotype studies in birds. A comparative study of the chromosomes in the Columbidae. *Caryologia*, **8**: 275–93.

MALOUF, N. and SCHNEIDER, T. G. (1965). Karyotype of *Felis aurata*. *Mammal. Chrom. Newsl.* no. **15**: 107.

MANCINO, G. (1963). La minuta struttura di un tratto del lampbrush chromosome XII di *Triturus cristatus cristatus*. *Rend. Accad. Naz. Lincei*, **34**: 65–7.

MANCINO, G. (1965). Le mappe dei cromosomi lampbrush di *Triturus alpestris apuanus* e *T. helveticus helveticus* (Anfibi Urodeli) – riassunto. *Boll. di Zool.* **32**: 539–40.

MANCINO, G. (1966). Le mappe dei cromosomi lampbrush di *Triturus vulgaris meridionalis* (Anfibi Urodeli). *Atti Soc. Tosc. Sci. Nat.* **73**: 3–4.

MANCINO, G. and BARSACCHI, G. (1965). Le mappe dei cromosomi "lampbrush" di *Triturus* (Anfibi Urodeli). I. *Triturus alpestris apuanus*. *Caryologia*, **18**: 637–65.

MANCINO, G. and BARSACCHI, G. (1966*a*). Le mappe dei cromosomi "lampbrush" di *Triturus* (Anfibi Urodeli). II. *Triturus helveticus helveticus*. *Riv. Biol.* **19**: 311–51.

MANCINO, G. and BARSACCHI, G. (1966*b*). Cariologia di *Salamandrina perspicillata* (Anfibi Urodeli). *Boll. di Zool.* **33**: 251–67.

MANCINO, G. and BARSACCHI, G. (1969). The maps of the lampbrush chromosomes of *Triturus* (Amphibia Urodela). III. *Triturus italicus*. *Ann. Embryol. Morph.* **2**: 355–77.

MANCINO, G., BARSACCHI, G. and NARDI, I. (1969). The lampbrush chromosomes of *Salamandra salamandra* (L.) (Amphibia, Urodela). *Chromosoma*, **26**: 365–87.

MANCINO, G., NARDI, I. and BARSACCHI, G. (1970). Spontaneous aberrations in lampbrush chromosome XI from a specimen of *Triturus vulgaris meridionalis* (Amphibia, Urodela). *Cytogenetics*, **9**: 260–71.

MANCINO, G. and SCALI, V. (1961). Anomalie spermatogenetiche in un esemplare di *Triturus helveticus*. *Arch. Zool. Ital.* **46**: 149–66.

MANGE, E. J. (1961). Meiotic drive in natural populations of *Drosophila melanogaster*. VI. A preliminary report on the presence of segregation-distortion in a Baja California population. *Amer. Nat.* **95**: 87–96.

MANNA, G. K. (1950). Multiple sex-chromosome mechanism in a reduviid bug *Conorhinus rubrofasciatus* (De Geer). *Proc. Zool. Soc. Bengal*, **3**: 155–61.

MANNA, G. K. (1951*a*). Multiple sex-chromosome mechanism in a reduviid bug, *Conorhinus rubrofasciatus* (De Geer) *Proc. Zool. Soc. Bengal*, **3**: 155–61.

MANNA, G. K. (1951*b*). A study of the chromosomes during meiosis in forty-three species of Indian Heteroptera. *Proc. Zool. Soc. Bengal*, **4**: 1–116.

MANNA, G. K. (1954). A study of chromosomes during meiosis in fifteen species of Indian grasshoppers. *Proc. Zool. Soc. Bengal*, **7**: 39–58.

MANNA, G. K. (1957). Sex mechanism in *Dysdercus*. *Curr. Sci.* **26**: 187.

MANNA, G. K. (1958). Cytology and interrelationships between various groups of Heteroptera. *Proc. Xth Int. Congr. Entomol. Montreal*, **2**: 919–34.

MANNA, G. K. (1962). A further evaluation of the cytology and interrelationships between various groups of Heteroptera. *Nucleus*, **5**: 7–28.

MANNA, G. K. (1967). Cytological analyses of the sex chromosome from testis cells of grasshopper – a review. *Nucleus*, **10**: 140–58.

MANNA, G. K. (1969). Some aspects of chromosome cytology. *Proc. 56th Indian Sci. Congr.* **2**: 1–30.

MANNA, G. K. and BHATTACHARJEE, T. K. (1964). Studies of gryllid chromosomes. II. Chromosomal polymorphisms in *Pteronemobius taprobanensis* (Walk.); and chromosome morphology of *Loxoblemmus* sp. *Cytologia*, **29**: 196–206.

MANNA, G. K. and CHATTERJEE, K. (1963). Polymorphic sex chromosomes in *Euprepocnemis* sp. I. The meiosis in the *XO* type male and in the neo-*X* and neo-*Y* type male. *Nucleus*, **6**: 121–34.

MANNA, G. K. and MAZUMDER, S. C., (1967a). Evolution of caryotype in an interesting species of grasshopper, *Tristria pulvinata* Uvarov. *Cytologia*, **32**: 236–47.

MANNA, G. K. and MAZUMDER, S. C. (1967b). Qualitative aspects of X-ray-induced sex-chromosome aberrations in grasshoppers. *Nucleus*, **10**: 128–39.

MANNA, G. K. and SMITH, S. G. (1959). Chromosomal polymorphism and inter-relationships among bark weevils of the genus *Pissodes* Germar. *Nucleus*, **2**: 179–208.

MANNING, F. J. (1949–50). Sex determination in the honey bee. *Microscope*, **7**: 175–80, 209–11, 237–41 and 259; **8**: 63–6.

MANTON, I. (1945). Chromosome length at early meiotic prophases in *Osmunda*. *Ann. Bot.* **9**: 155–78.

MANTON, I. (1950). *Problems of Cytology and Evolution in the Pteridophyta*. Cambridge Univ. Press.

MANTON, I. and SLEDGE, W. A. (1954). Observations on the cytology and taxonomy of the pteridophyte flora of Ceylon. *Phil. Trans. Roy. Soc. Lond.* B, **238**: 127–85.

MANTON, I. and SMILES, J. (1943). Observations on the spiral structure of somatic chromosomes in *Osmunda* with the aid of ultraviolet light. *Ann. Bot.* **7**: 195–212.

MARGOT, A. (1946). Démonstration de l'absence d'hétérochromosomes morphologiquement differenciés chez deux espèces de sauriens, *Anguis fragilis* L. et *Lacerta vivipara* Jacquin. *Rev. Suisse Zool.* **53**: 555–96.

MARK, E. L. and COPELAND, M. (1906). Some stages in the spermatogenesis of the honey bee. *Proc. Amer. Acad. Arts Sci.* **42**: 101–12.

MARKS, G. E. (1957). Telocentric chromosomes. *Amer. Nat.* **91**: 223–32.

MARTIN, J. (1962). Interrelation of inversions in the midge *Chironomus intertinctus* (Diptera: Nematocera). I. A sex-linked inversion. *Aust. J. Biol. Sci.* **15**: 666–73.

MARTIN, J. (1963). The cytology and larval morphology of the Victorian representatives of the subgenus *Kiefferulus* of the genus *Chironomus* (Diptera: Nematocera). *Aust. J. Zool.* **11**: 301–22.

MARTIN, J. (1964). Morphological differences between *Chironomus intertinctus* Skuse and *C. paratinctus* sp. nov. with descriptions and a key to the subgenus *Kiefferulus* (Diptera: Nematocera). *Aust. J. Zool.* **12**: 279–87.

MARTIN, J. (1965). Interrelation of inversion systems in the midge *Chironomus intertinctus*. II. A non-random association of linked inversions. *Genetics*, **52**: 371–83.

MARTIN, J. (1966). Female heterogamety in *Polypedilum nubifer* (Diptera: Nematocera). *Amer. Nat.* **100**: 157–9.

MARTIN, J. (1967). Meiosis in inversion heterozygotes in Chironomidae. *Canad. J. Genet. Cytol.* **9**: 255–68.

MARTIN, J. H. D. (1969). Chromosomes of some native Muridae in Queensland. *Queensl. J. Agric. Animal Sci.* **26**: 125–41.

MARTIN, P. G. and HAYMAN, D. L. (1965). A quantitative method for comparing the karyotypes of related species. *Evolution*, **19**: 157–61.

MARTIN, P. G. and HAYMAN, D. L. (1966). A complex sex chromosome system in the hare-wallaby *Lagorchestes conspicillatus* Gould. *Chromosoma*, **19**: 159–75.

MARTIN, P. G. and HAYMAN, D. L. (1967). Quantitative comparisons between the karyotypes of Australian marsupials from three different superfamilies. *Chromosoma*, **20**: 290–310.

MASLIN, T. P. (1962). All-female species of the lizard genus *Cnemidophorus*, Teiidae. *Science*, **135**: 212–13.

MASLIN, T. P. (1967). Skin grafting in the bisexual teiid lizard *Cnemidophorus sexlineatus* and in the unisexual *C. tesselatus*. *J. Exp. Zool.* **166**: 137–50.

MASLIN, T. P. (1968). Taxonomic problems in parthenogenetic vertebrates. *Syst. Zool.* **12**: 219–31.

MATHAI, C. K., OHNO, S. and BEUTLER, E. (1966). Sex-linkage of the glucose 6-phosphate dehydrogenase gene in the family Equidae. *Nature*, **210**: 115–16.

MATHER, K. (1933a). Interlocking as a demonstration of genetical crossing-over during chiasma formation. *Amer. Nat.* **67**: 476–9.

MATHER, K. (1933b). The relation between chiasmata and crossing-over in diploid and triploid *Drosophila melanogaster*. *J. Genet.* **27**: 243–59.

MATHER, K. (1935a). Meiosis in *Lilium*. *Cytologia*, **6**: 354–80.

MATHER, K. (1935b). Reductional and equational separation of the chromosomes in bivalents and multivalents. *J. Genet.* **30**: 53–78.

MATHER, K. (1936). The determination of position in crossing-over. I. *Drosophila melanogaster*. *J. Genet.* **33**: 207–35.

MATHER, K. (1938). Crossing-over. *Biol. Rev.* **13**: 252–92.

MATHER, K. (1953). The genetical structure of populations. *Symp. Soc. Exp. Biol.* **7**: 66–95.

MATHER, K. and LAMM, R. (1935). The negative correlation of chiasma frequencies. *Hereditas*, **20**: 65–70.

MATHER, K. and STONE, L. H. A. (1933). The effect of X-radiation upon somatic chromosomes. *J. Genet.* **28**: 1–24.

MATHER, W. B. (1961). Chromosomal polymorphism in *Drosophila rubida* Mather. *Genetics*, **46**: 799–810.

MATHER, W. B. (1963). Patterns of chromosomal polymorphism in *Drosophila rubida*. *Amer. Nat.* **97**: 59–64.

MATSCHEK, H. (1910). Über Eireife und Eiablage bei Copepoden. *Arch. Zellforsch.* **5**: 37–111.

MATTHEY, R. (1931). Chromosomes de reptiles sauriens, ophidiens et cheloniens. L'évolution de la formule chromosomiale chez les sauriens. *Rev. Suisse Zool.* **38**: 117–86.

MATTHEY, R. (1932a). Les chromosomes et la systématique zoologique. *Rev. Suisse Zool.* **39**: 229–37.

MATTHEY, R. (1932b). Les chromosomes de l'amphisbénien acrodonte *Trogonophis wiegmanni* Kaupp. *Arch. Zool. Exp. Gen.* **74**: 193–204.

MATTHEY, R. (1932c). Chromosomes de sauriens: la cinèse hétérotypique et la réduction chromatique chez *Gerrhonotus scincicauda* (= *G. caeruleus* Wiegm.). *C.R. Soc. Biol. Paris*, **110**: 158–9.

MATTHEY, R. (1933). Nouvelle contribution à l'étude des chromosomes chez les sauriens. *Rev. Suisse Zool.* **40**: 281–318.

MATTHEY, R. (1934a). La formale chromosomiale de *Lacerta vivipara* Jacquin. *C.R. Soc. Biol. Paris*, **117**: 315–16.

MATTHEY, (1934b). La formule chromosomiale du Kangourou *Macropus parryi* Benn. *C.R. Soc. Biol. Paris*, **117**: 407–8.

MATTHEY, R. (1936). La formule chromosomiale et les hétérochromosomes chez les *Apodemus* européens. *Z. Zellforsch. Mikr. Anat.* **25**: 501–15.

MATTHEY, R. (1939a). La formule chromosomiale de la sauterelle parthénogénétique *Saga serrata* Fabr. et de l'*Ephippigera vitium*. *C.R. Soc. Biol. Paris*, **132**: 369–70.

MATTHEY, R. (1939b). La loi de Robertson et la formule chromosomiale chez deux Lacertiens: *Lacerta ocellata* Daud., *Psammodromus hispanicus* Fitz. *Cytologia*, **10**: 32–9.

MATTHEY, R. (1941). Étude biologique et cytologique de *Saga pedo* Pallas (Orthoptères: Tettigoniidae). *Rev. Suisse Zool.* **48**: 91–102.

MATTHEY, R. (1945a). L'évolution de la formule chromosomiale chez les vertébrés. *Experientia*, **1**: 50–6 and 78–86.

MATTHEY, R. (1945b). Cytologie de la parthénogénèse chez *Pycnoscelus surinamensis* L. *Rev. Suisse Zool.* **52**: 1–109.

MATTHEY, R. (1946a). Démonstration du caractère géographique de la parthénogénèse de *Saga pedo* Pallas et de sa polyploidie, par comparaison avec les espèces bisexuées *S. ephippigera* Fisch. et *S. gracilipes* Uvar. *Experientia*, **2**: 260–1.

MATTHEY, R. (1946b). Troisième contribution à l'étude des chromosomes chez les Perles. *Bull. Soc. Vaud. Sci. Nat.* **63**: 297–8.

MATTHEY, R. (1946c). Communications préliminaires sur les chromosomes des Plécoptères. IV, Formules chromosomiques des Perlodidae et évolution génerale des hétérochromosomes. *Experientia*, **8**: 497–8.

MATTHEY, R. (1947a). Encore les hétérochromosomes des *Apodemus*. *Arch. J. Klaus Stift. Vererbf.* **22**: 85–92.

MATTHEY, R. (1947b). Les chromosomes sexuels des Plécoptères. V. *Perla abdominalis* Burm., *P. baetica* Rambur, *P. bipunctata* Pict. *Rev. Suisse Zool.* **54**: 259.

MATTHEY, R. (1947c). Quelques formules chromosomiales. *Sci. Genet.* **3**: 23–32.

MATTHEY, R. (1948). La formule chromosomiale de *Pycnoscelus surinamensis* L. Race bisexuée et race parthénogénétique. Existence probable d'une parthénogénèse diploide facultative. *Arch. J. Klaus Stift. Vererbf.* **23**: 217–20.

MATTHEY, R. (1949a). La formule chromosomique et la méiose chez *Apteromantis bolivari* Werner (Mantidae, Amelinae). *Arch. J. Klaus Stift. Vererbf.* **24**: 114–23.

MATTHEY, R. (1949b). Les chromosomes de *Saga cappadocica* Werner (Orthoptera, Tettigoniidae). *Arch. J. Klaus Stift. Vererbf.* **25**: 44–6.

MATTHEY, R. (1949c). Les chromosomes sexuels géants de *Microtus agrestis* L. et le mécanisme d'association des hétérochromosomes. *Rev. Suisse Zool.* **56**: 337.

MATTHEY, R. (1949d). *Les Chromosomes des Vertébrés*. Lausanne: Rouge.

MATTHEY, R. (1950a). Les chromosomes sexuels géants de *Microtus agrestis* L. *Cellule*, **53**: 163–84.

MATTHEY, R. (1950b). La formule chromosomique et le type de digamétie chez *Bittacus italicus* Müll. (Mecoptera). *Arch. J. Klaus Stift. Vererbf.* **25**: 607–11.

MATTHEY, R. (1951). Systématique et critères cytologiques. *Bull. Soc. Vaud. Sci. Nat.* **65**: 111–20.

MATTHEY, R. (1952a). Chromosomes de Muridae (Microtinae et Cricetinae). *Chromosoma*, **5**: 113–38.

MATTHEY, R. (1962b). Chromosomes sexuels multiples chez un rongeur (*Gerbillus pyramidum* Geoffroy). *Arch. J. Klaus Stift. Vererbf.* **27**: 163–6.

MATTHEY, R. (1953a). Les chromosomes des Muridae. *Rev. Suisse Zool.* **60**: 225–83.

MATTHEY, R. (1953b). A propos de la polyploidie animale: réponse à un article de C. D. Darlington. *Rev. Suisse Zool.* **60**: 466–71.

MATTHEY, R. (1953c). La formule chromosomique et le problème de la détermination sexuelle chez *Ellobius lutescens* Thomas (Rodentia–Muridae–Microtinae). *Arch. J. Klaus Stift. Vererbf.* **28**, 65–73.

MATTHEY, R. (1954a). Nouvelles recherches sur les chromosomes des Muridae. *Caryologia*, **6**: 1–44.

MATTHEY, R. (1954b). Un cas nouveau de chromosomes sexuels multiples dans le genre *Gerbillus* (Rodentia–Muridae–Gerbillinae). *Experientia*, **10**: 464–5.

MATTHEY, R. (1955). Nouveaux documents sur les chromosomes des Muridae. Problèmes de cytologie comparée et de taxonomie chez les Microtinae. *Rev. Suisse Zool.* **62**: 163–206.

MATTHEY, R. (1956). Cytologie comparée des Muridae. L'origine des Ellobii. *Experientia*, **12**: 337.

MATTHEY, R. (1957). Cytologie comparée et taxonomie des Chamaeleontidae (Reptilia–Lacertilia). *Rev. Suisse Zool.* **64**: 709–32.

MATTHEY, R. (1958a). Un nouveau type de détermination du sexe chez les mammifères *Ellobius lutescens* Th. et *Microtus* (*Chilotus*) *oregoni* Bachm. (Muridés–Microtinés). *Experientia*, **14**: 240–1.

MATTHEY, R. (1958b). Les chromosomes des mammifères euthériens: liste critique et essai sur l'évolution chromosomique. *Arch. J. Klaus Stift. Vererbf.* **33**: 253–297.

MATTHEY, R. (1959). Formules chromosomiques de 'Muridae' et 'Spalacidae'. La question du polymorphisme chromosomique chez les mammifères. *Rev. Suisse Zool.* **66**: 175–209.

MATTHEY, R. (1961). La formule chromosomique et la position systématique de *Chamaeleo gallus* Günther (Lacertilia). *Zool. Anz.* **166**: 153–9.

MATTHEY, R. (1962). Études sur les chromosomes d'*Ellobius lutescens* Th. (Mammalia–Muridae–Microtinae). *Cytogenetics*, **1**: 180–95.

MATTHEY, R. (1963a). Polymorphisme chromosomique intraspécifique et intraindividual chez *Acomys minous* Bate (Mammalia–Rodentia–Muridae). *Chromosoma*, **14**: 468–97.

MATTHEY, R. (1963b). Polymorphisme intraspécifique chez un mammifère *Leggada minutoides* Smith (Rodentia–Muridae). *Rev. Suisse Zool.* **70**: 173–90.

MATTHEY, R. (1963c). Cytologie comparée et polymorphisme chromosomique chez des *Mus* africains appartenant aux groupes *bufo-triton* et *minutoides* (Mammalia–Rodentia). *Cytogenetics*, **2**: 290–322.

MATTHEY, R. (1964a). Études sur les chromosomes d'*Ellobius lutescens* (Mammalia–Muridae–Microtinae). II. Informations complémentaires sur les divisions méiotiques. *Rev. Suisse Zool.* **71**: 401–10.

MATTHEY, R. (1964b). Evolution chromosomique et spéciation chez les *Mus* du sous-genre *Leggada* Gray 1837. *Experientia*, **20**: 657–65.

MATTHEY, R. (1964c). La formule chromosomique et la position systématique de *Pitymys tatricus* Kratochvil (Rodentia–Microtinae). *Z. Säugetierk.* **29**: 235–42.

MATTHEY, R. (1964d). La signification des mutations chromosomiques dans les processus de spéciation. Etude cytogénétique du sous-genre *Leggada* Gray (Mammalia–Muridae). *Arch. Biol.* **75**: 169–206.

MATTHEY, R. (1965a). Le problème de la détermination du sexe chez *Acomys selousi* de Winton. Cytogénétique du genre *Acomys* (Rodentia, Muridae). *R. Suisse Zool.* **72**: 119–44.

MATTHEY, R. (1965b). Un type nouveau de chromosomes sexuels multiples chez une souris africaine du groupe *Mus* (*Leggada*) *minutoides* (Mammalia–Rodentia). Mâle: X_1X_2Y. Femelle X_1X_2/X_1X_2. *Chromosoma*, **16**: 351–64.

MATTHEY, R. (1965c). Cytogénétique des *Mus* africains du sous-genre *Leggada*. Etude de 35 exemplaires provenant de l'Afrique du Sud, de la Côte d'Ivoire et du Gabon. *Caryologia*, **18**: 161–79.

MATTHEY, R. (1965d). Etudes de cytogénétique sur des Murinae africains appartenant aux genres *Arvicanthis, Praomys, Acomys* et *Mastomys* (Rodentia). *Mammalia*, **29**: 228–249.

MATTHEY, R. (1966a). Cytogénétique et taxonomie des rats appartenant au sous-genre *Mastomys* Thomas (Rodentia–Muridae). *Mammalia*, **30**: 105–19.

MATTHEY, R. (1966b). Présence dans une population congolaise de *Mus* (*Leggada*) *triton* Th. de femelles hétérozygotes pour une déletion caractérisée par la suppression du bras court de l'un des chromosomes × metacentriques. *Z. Vererbl.* **97**: 361–9.

MATTHEY, R. (1966c). Nouvelles contributions à la cytogénétique des *Mus* africains du sous-genre *Leggada*. *Experientia*, **22**: 400–1.

MATTHEY, R. (1967). Cytogénétique des *Leggada*: (1) La formule chromosomique de *Mus* (*Leggada*) *bufo* Th., (2) Nouvelles données sur la déletion portant sur le bras court d'un *X* chez *Mus* (*Leggada*) *triton* Th. *Experientia*, **22**: 133.

MATTHEY, R. (1968a). Cytogénétique et taxonomie du genre *Acomys*: *A. percivali* Dollman et *A. wilsoni* Thomas, espèces d'Abyssinie. *Mammalia*, **32**: 621–7.

MATTHEY, R. (1968b). Les chromosomes de *Leggadina hermannsburgensis hermannsburgensis* Waite et le problème des Muridae australiens. *Experientia*, **24**: 1160–1.

MATTHEY, R. (1970). Nouvelles données sur la cytogénétique et la spéciation des *Leggada* (Mammalia–Rodentia–Muridae). *Experientia*, **26**: 102–3.

MATTHEY, R. and AUBERT, J. (1947). Les chromosomes des Plécoptères. *Bull. Biol. Fr. Belg.* **81**: 202–46.

MATTHEY, R. and MEYLAN, A. (1961). Le polymorphisme chromosomique de *Sorex araneus* L. (Mamm. Insectivora). Etude de deux portées de 5 et 9 petits. *Rev. Suisse Zool.* **68**: 223–7.

MATTHEY, R. and VAN BRINK, J. (1956a). La question des hétérochromosomes chez les Sauropsidés. I. Reptiles. *Experientia*, **12**: 53–5.

MATTHEY, R. and VAN BRINK, J. (1956b). Note préliminaire sur la cytologie chromosomique comparée des Caméléons. *R. Suisse Zool.* **63**: 241–6.

MATTHEY, R. and VAN BRINK, J. M. (1957). Sex chromosomes in Amniota. *Evolution*, **12**: 163–5.

MATTHEY, R. and VAN BRINK, J. (1960). Nouvelle contribution à la cytologie comparée des Chamaeleontidae (Reptilia–Larcertilia). *Bull. Soc. Vaud. Sc. Nat.* **67**: 333–48.

MATTHIESEN, F. A. (1962). Parthenogenesis in scorpions. *Evolution*, **16**: 255–6.

MATTOX, N. T. (1937). Oogenesis of *Campeloma rufum*, a parthenogenetic snail. *Z. Zellforsch.* **27**: 455–64.

MATUSZEWSKI, B. (1960). Chromosome behavior during oogenesis in the gall midge *Mikiola fagi* Hart. (Cecidomyidae, Diptera). *Bull. Acad. Polon. Sci.* Cl. II, **8**: 101–3.

MATUSZEWSKI, B. (1962). Oogenesis in *Mikiola fagi* Hart. (Cecidomyiidae: Diptera). *Chromosoma*, **12**: 741–811.

MATUSZEWSKI, B. (1964). Polyploidy and polyteny induced by a Hymenopteran parasite in *Dasyneura urticae* (Diptera, Cecidomyiidae). *Chromosoma*, **15**: 31–5.

MATUSZEWSKI, B. (1965). Transition from polyteny to polyploidy in salivary glands of Cecidomyiidae. *Chromosoma*, **16**: 22–34.

MAXWELL, D. E. (1958). Sawfly cytology with emphasis on the Diprionidae (Hymenoptera: Symphyta). *Proc. Xth Int. Congr. Entomol.* 1956, **2**: 961–78.

MAYNARD SMITH, J. (1966). Sympatric speciation. *Amer. Nat.* **100**: 637–50.

MAYR, E. (1940). Speciation phenomena in birds. *Amer. Nat.* **74**: 249–78.

MAYR, E. (1942). *Systematics and the Origin of Species*. New York: Columbia Univ. Press.

MAYR, E. (1947). Ecological factors in speciation. *Evolution*, **1**: 263–88.

MAYR, E. (1954). Change of genetic environment and evolution. In: *Evolution as a Process* (ed.: J. Huxley, A. C. Hardy and E. B. Ford), pp. 157–80. London: Allen and Unwin.

MAYR, E. (1957). Species concepts and definitions. In: *The Species Problem* (ed.: E. Mayr). *Amer. Assoc. Adv. Sci. Pub.* no. 50, pp. 371–88.

MAYR, E. (1963). *Animal Species and Evolution*. Harvard: Belknap Press.

MAZIA, D. (1961). Mitosis and the physiology of cell division. In: *The Cell*, **3** (ed.: J. Brachet and A. E. Mirsky), pp. 77–421. New York: Academic Press.

MAZIA, D. and DAN, K. (1952). The isolation and biochemical characterization of the mitotic apparatus of dividing cells. *Proc. Nat. Acad. Sci. U.S.A.* **38**: 826–38.

MEAD, G. W. (1960). Hermaphroditism in archibenthic and pelagic fishes of the order Iniomi. *Deep Sea Res.* **6**: 234–5.

MECHELKE, F. (1961). Das Wandern des Aktivitätsmaximums im BR-locus von *Acricotopus lucidus* als Modell für die Wirkungsweise eines Komplexen Locus. *Naturwiss.* **48**: 29.

MELANDER, Y. (1948). Cytological studies on Scandinavian flatworms belonging to Tricladida Paludicola. *Proc. VIIIth Int. Genet. Congr. Stockholm*, pp. 625–6.

MELANDER, Y. (1950a). Studies on the chromosomes of *Ulophysema öresundense*. *Hereditas*, **36**: 233–55.

MELANDER, Y. (1950b). Accessory chromosomes in animals, especially in *Polycelis tenuis*. *Hereditas*, **36**: 261–96.

MELANDER, Y. (1959). The mitotic chromosomes of some cavicorn mammals (*Bos taurus* L., *Bison bonasus* L. and *Ovis aries* L.). *Hereditas*, **45**: 649–64.

MELANDER, Y. and KNUDSEN, O. (1953). The spermiogenesis of the bull from a karyological point of view. *Hereditas*, **39**: 505–17.

MELLAND, A. M. (1942). Types of development of polytene chromosomes. *Proc. Roy. Soc. Edinb.* B **61**: 316–27.

MENDES, L. O. T. (1947). Sôbre a meiose de *Dysdercus mendesi* Bloete (1937), Hemiptera–Pyrrhocoridae. *Bragantia*, **7**: 243–56.

MENDES, L. O. T. (1949). Observacões citologicas em *Dysdercus*. Cadeias de cromossômios em tecido somático de *Dysdercus mendesi* Bloete (Hemiptera– Pyrrhocoridae). *Bragantia*, **9**: 53–7.

MENON, P. S. (1955). On the multiple sex-chromosome mechanism in a lygaeid. *Oxycarenus hyalinipennis* (Costa). *Experientia*, **11**: 483–6.

MENZEL, M. Y. and PRICE, J. M. (1966). Fine structure of synapsed chromosomes in F_1 *Lycopersicon esculentum–Solanum lycopersicoides* and its parents. *Amer. J. Bot.* **53**: 1079–86.

MERRIAM, R. W. and RIS, H. (1954). Size and DNA content of nuclei in various tissues of male, female and worker honeybees. *Chromosoma*, **6**: 522–38.

MESA, A. (1956). Los cromosomas de algunos acridoideos uruguayos. *Agros, Revta. no.* **141** *de la Asociación de Estudiantes de Agronomía, Montevideo*, pp. 32–45.

MESA, A. (1960). Cariología de una nueva especie uruguaya del género *Scotussa* (Orthoptera, Catantopidae). *Rev. Soc. Uruguayana Entomol.* **4**: 87–94.

MESA, A. (1961). Morfologia falica y cariología de *Neuquenina fictor* (Rehn) (Orthoptera–Acridoidea). *Comm. Zool. Mus. Hist. Nat. Montevideo*, **5**: 1–11.

MESA, A. (1962a). Los cromosomas de *Eurotettix lilloanus* Lieb. (Orthoptera: Catantopidae). *Acta Zool. Lilloana*, **18**: 99–104.

MESA, A. (1962b). Cariología de *Dichroplus bergi* Stal (Orthoptera–Acrididae). *Rev. Agric. Piracicaba*, **37**: 41–9.

MESA, A. (1963a). Acerca de la cariología de Ommexechidae. *Rev. Soc. Uruguayana Entomol.* **5**: 37–43.

MESA, A. (1963b). Mecanísmo cromosomico de determinación sexual poco frecuente en *Scotussa daguerrei* Lieb. (Orthoptera–Acrididae). *Rev. Soc. Entomol. Argentina,* **26**: 119–24.

MESA, A. (1964a). Los cromosomas de *Pachyossa* sp. (Orthoptera, Ommexechidae). *Rev. Soc. Uruguayana Entomol.* **6**: 49–54.

MESA, A. (1964b). Cariología de *Scotussa delicatula* Lieb. *An. II Congr. Latino-Amer. Zool. S. Paulo,* **1**: 3–7.

MESA, A. (1965). Caryology of four Chilean species of Gryllacridoids of the genus *Heteromallus* (Orthoptera: Gryllacridoidea: Rhaphidophoridae). *Occas. Pap. Mus. Zool. Mich.* no. 640 (13 pp.).

MESA, A. (1970a). The chromosomes of two species of the genus *Australotettix* Richards (Gryllacridoidea: Macropathinae). *J. Aust. Entomol. Soc.* **9**: 7–10.

MESA, A. (1970b). Cytogenetic and evolutionary studies on Macropathinae. Unpublished Ph.D. thesis: Univ. Melbourne.

MESA, A. and BRAN, E. J. (1964). Acerca de los cromosomas de *Eneoptera surinamensis. An. II. Congr. Latino-Amer. Zool. S. Paulo,* **1**: 9–16.

MESA, A. and DE MESA, R. S. (1967). Complex sex-determining mechanisms in three species of South American grasshoppers (Orthoptera, Acridoidea). *Chromosoma,* **21**: 163–80.

MESA, A., FERREIRA, A. and DE MESA, R. S. (1968). The karyotype of some Australian species of Macropathinae (Gryllacridoidea–Rhaphidophoridae). *Chromosoma,* **24**: 456–66.

MESA, A. FERREIRA, A. and DE MESA, R. S. (1969). The chromosomes of three species of Gryllacridids (Gryllacridoidea, Rhaphidophoridae, Macropathinae). *Caryologia,* **22**: 149–59.

MESELSON, M. and STAHL, F. W. (1958). The replication of DNA in *Escherichia coli. Proc. Nat. Acad. Sci. U.S.A.* **44**: 671–82.

METZ, C. W. (1914). Chromosome studies on the Diptera. I. A preliminary survey of five different types of chromosome groups in the genus *Drosophila. J. Exp. Zool.* **17**: 45–59.

METZ, C. W. (1916a). Chromosome studies on the Diptera. II. The paired association of chromosomes in the Diptera and its significance. *J. Exp. Zool.* **21**: 213–79.

METZ, C. W. (1916b). Chromosome studies on the Diptera. III. Additional types of chromosome groups in the Drosophilidae. *Amer. Nat.* **50**: 587–99.

METZ, C. W. (1922). Chromosome studies on the Diptera. IV. Incomplete synapsis in *Dasyllis grossa. Biol. Bull. Woods Hole,* **43**: 253–66.

METZ, C. W. (1926). Chromosome studies on *Sciara* (Diptera). I. Difference between the chromosomes of the two sexes. *Amer. Nat.* **60**: 42–56.

METZ, C. W. (1933). Monocentric mitosis with segregation of chromosomes in *Sciara* and its bearing on the mechanism of mitosis. *Biol. Bull. Woods Hole,* **54**: 333–47.

METZ, C. W. (1934). Evidence indicating that in *Sciara* the sperm regularly transmits two sister sex chromosomes. *Proc. Nat. Acad. U.S.A.* **20**: 31–6.

METZ, C. W. (1935). Structure of the salivary gland chromosomes in *Sciara. J. Hered.* **26**: 177–88.

METZ, C. W. (1937). Small deficiencies and the problem of genetic units in the giant chromosome. *Genetics,* **22**: 543–56.

METZ, C. W. (1938a). Chromosome behavior, inheritance and sex determination in *Sciara. Amer. Nat.* **72**: 485–520.

METZ, C. W. (1938*b*). Observations on evolutionary changes in the chromosomes of *Sciara* (Diptera). *Publ. Carnegie Instn.* **501**: 275–94.

METZ, C. W. (1939). Observations on the nature of minute chromosomal differences in species hybrids and in ordinary wild strains of *Sciara*. *Genetics*, **24**: 105.

METZ, C. W. (1941). Species hybrids, evolutionary chromosome changes and the mechanism of chromosome rearrangement in *Sciara*. *Proc. VIIth Int. Genet. Congr. Edinb.* pp. 215–18.

METZ, C. W. (1947). Duplications of chromosome parts as a factor in evolution. *Amer. Nat.* **81**: 81–103.

METZ, C. W. (1959). Chromosome behavior and cell lineage in triploid and mosaic salivary glands of species hybrids in *Sciara*. *Chromosoma*, **10**: 515–34.

METZ, C. W. and ARMSTRONG, L. S. (1961). Observations on deficient chromosome groups in developing *Sciara* larvae. *Growth*, **25**: 89–106.

METZ, C. W. and LAWRENCE, E. G. (1938). Preliminary observations on *Sciara* hybrids. *J. Hered.* **29**: 179–86.

METZ, C. W., MOSES, M. and HOPPE, E. (1926). Chromosome behavior and genetic behavior in *Sciara*. I. Chromosome behavior in the spermatocyte divisions. *Z. indukt. Abst. Vererbl.* **42**: 237–70.

METZ, C. W. and NONIDEZ, J. F. (1921). Spermatogenesis in the fly, *Asilus sericeus* Say. *J. Exp. Zool.* **32**: 165–85.

METZ, C. W. and NONIDEZ, J. F. (1923). Spermatogenesis in *Asilus notatus* Wied. (Diptera). *Arch. Zellforsch.* **17**: 438–49.

METZ, C. W. and NONIDEZ, J. F. (1924). The behavior of the nucleus and chromosomes during spermatogenesis in the robber fly, *Lasiopogon bivittatus*. *Biol. Bull. Woods Hole*, **46**: 153–64.

METZ, C. W. and SCHMUCK, M. L. (1931*a*). Differences between the chromosome groups of soma and germ-line in *Sciara*. *Proc. Nat. Acad. Sci. U.S.A.* **15**: 862–6.

METZ, C. W. and SCHMUCK, M. L. (1931*b*). Studies on sex determination and the sex chromosomes in *Sciara*. *Genetics*, **16**: 225–53.

MEVES, F. (1902). Über oligopyrene und apyrene Spermien und über ihre Entstehung, nach Beobachtungen an *Paludina* und *Pygaera*. *Arch. Mikr. Anat.* **61**: 1–84.

MEYER, G. F. (1963). Die Funktionsstrukturen des *Y*-Chromosoms in den Spermatocytenkernen von *Drosophila hydei*, *D. neohydei*, *D. repleta* und einigen anderen *Drosophila*-Arten. *Chromosoma*, **14**: 207–55.

MEYER, G. F. (1964). A possible correlation between the submicroscopic structure of meiotic chromosomes and crossing-over. *Proc. IIIrd Europ. Conf. Electr. Microsc.* pp. 461–2.

MEYER, G. F. and HESS, O. (1965). Struktur-Differenzierung im *Y*-Chromosome von *Drosophila hydei* und ihre Beziehungen zu Gen-Aktivitäten. II. Effekt der RNS-Synthese-Hemmung durch Actinomycin. *Chromosoma*, **16**: 249–70.

MEYER, H. (1938). Investigations concerning the reproductive behavior of *Mollienisia* 'formosa'. *J. Genet.* **36**: 327–66.

MEYLAN, A. (1960). Contribution a l'étude du polymorphisme chromosomique chez *Sorex araneus* L. (Mamm. Insectivora). (Note préliminaire). *Rev. Suisse Zool.* **67**: 258–61.

MEYLAN, A. (1964). Le polymorphisme chromosomique de *Sorex araneus* L. (Mamm.–Insectivora). *Rev. Suisse Zool.* **71**: 903–83.

MEYLAN, A. (1965*a*). Répartition géographique des races chromosomiques de *Sorex araneus* L. en Europe (Mamm.–Insectivora). *Rev. Suisse Zool.* **72**: 636–46.

MEYLAN, A. (1965b). La formule chromosomique de *Sorex minutus* L. (Mammalia–Insectivora). *Experientia*, **21**: 268–9.

MEYLAN, A. (1967). Formules chromosomiques et polymorphisme robertsonien chez *Blarina brevicauda* (Say) (Mammalia: Insectivora). *Canad. J. Zool.* **45**: 1119–27.

MICHIE, D. (1953). Affinity: a new genetic phenomenon in the house mouse. Evidence from distant crosses. *Nature*, **171**: 26–7.

MIKULSKA, I. (1949). Cytological studies upon genus *Otiorrhynchus* (Curculionidae, Coleoptera) in Poland. *Experientia*, **5**: 473–5.

MIKULSKA, I. (1951). The chromosome number in *Otiorrhynchus salicis* Ström (Curculionidae, Coleoptera) in Poland. *Bull. Acad. Polon. Sci.* Cl. II, 1950, 269–76.

MIKULSKA, I. (1953). The chromosomes of the parthenogenetic and thelytokian Weevil *Eusomus ovulum* Germ. (Curculionidae Coleoptera). *Bull. Acad. Polon. Sci.* Cl. II, 1951, 293–307.

MIKULSKA, I. (1960). New data to the cytology of the parthenogenetic and thelytokian weevil *Eusomus ovulum* Germ. (Curculionidae, Coleoptera) from Poland. *Cytologia*, **25**: 322–33.

MILLER, D. D. (1939). Structure and variation of chromosomes in *Drosophila algonquin*. *Genetics*, **24**: 699–708.

MILLER, D. D. (1941). Interspecific hybrids involving *Drosophila athabasca*. *Genetics*, **26**: 161.

MILLER, D. D. and ROY, R. (1964). Further data on Y-chromosome types in *Drosophila athabasca*. *Canad. J. Genet. Cytol.* **6**: 334–48.

MILLER, O. L. (1965). Fine structure of lampbrush chromosomes. *Int. Symp. Genes and Chromosomes – Structure and Function*: National Cancer Inst. Monograph no. 18, pp. 79–99.

MILOVIDOV, P. (1949). *Physik und Chemie des Zellkernes. Protoplasma Monographien*, **20**, part 2. Berlin: Borntraeger.

MINOUCHI, O. (1934). Über die überzähligen Chromosomen der Orthopteren. I. Über die überzähligen Chromosomen in der Spermatogenese von *Trixalis nasuta*. *Z. Zellforsch.* **20**: 707–29.

MINOUCHI, O. (1935). Über die Chromosomen des Bohnenkäfers *Zabrotes subfasciatus* Boh. (Bruchidae). *Z. Zellforsch.* **23**: 361–74.

MINTON, S. A. Jr. (1958). Observations on amphibians and reptiles of the Big Bend region of Texas. *Southwestern Nat.* **3**: 28–54.

MIRSKY, A. E. and RIS, H. (1947a). Isolated chromosomes. *J. Gen. Physiol.* **31**: 1–6.

MIRSKY, A. E. and RIS, H. (1947b). The chemical composition of isolated chromosomes. *J. Gen. Physiol.* **31**: 7–18.

MIRSKY, A. E. and RIS, H. (1949). Variable and constant components of chromosomes. *Nature*, **163**: 666–7.

MIRSKY, A. E. and RIS, H. (1951a). The desoxyribonucleic acid content of animal cells and its evolutionary significance. *J. Gen. Physiol.* **34**: 451–62.

MIRSKY, A. E. and RIS, H. (1951b). The composition and structure of isolated chromosomes. *J. Gen. Physiol.* **34**: 472–92.

MISRA, A. B. (1937a). The chromosomes of an earwig, *Forficula scudderi* Borm. *Jap. J. Genet.* **13**: 171–4.

MISRA, A. B. (1937b). A note on the chromosomes of *Acrydium japonicum*. *Jap. J. Genet.* **13**: 175–6.

MITTAL, O. P. (1960). Chromosome number and sex mechanism in twenty species of Indian spiders. *Res. Bull. (N.S.) Panjab Univ.* **11**: 245–7.

MITTWOCH, U. (1967). *Sex Chromosomes*. New York and London: Academic Press.

MIYADA, S. (1960). Studies on haploid frogs. *J. Sci. Hiroshima Univ.* B, **19**: 1–56.

MOENS, P. B. (1966). Segregation of tritium-labeled DNA at meiosis in *Chorthippus*. *Chromosoma*, **19**: 277–85.

MOENS, P. B. (1968a). The structure and function of the synaptinemal complex in *Lilium longiflorum* sporocytes. *Chromosoma*, **23**: 418–51.

MOENS, P. B. (1968b). Fine structure aspects of chromosome pairing at meiotic prophase. *Genetics*, **60**: 205–6.

MOENS, P. B. (1969). Multiple core complexes in grasshopper spermatocytes and spermatids. *J. Cell Biol.* **40**: 542–51.

MOFFETT, A. A. (1936). The origin and behavior of chiasmata. XIII. Diploid and tetraploid *Culex pipiens*. *Cytologia*, **7**: 184–97.

MOHN, N. and SPIESS, E. B. (1963). Cold resistance of karyotypes in *Drosophila persimilis* from Timberline of California. *Evolution*, **17**: 548–63.

MOHR, O. L. (1923). A genetic and cytological analysis of a section deficiency involving four units of the *X*-chromosome in *Drosophila melanogaster*. *Z. indukt. Abst. Vererbl.* **32**: 108–232.

MOHR, O. L. and EKER, R. (1934). The grasshopper *Tachycines asynamorus*, a new laboratory animal for cytological purposes. *Cytologia*, **5**: 384–90.

MOMMA, E. (1941). The chromosomes of two species of *Euscyrtus* (Gryllodea, Podoscyrtinae). *Jap. J. Genet.* **17**: 307–9.

MOMMA, E. (1943). A karyogram study on eighteen species of Japanese Acrididae (Orthoptera). *J. Fac. Sci. Hokkaido Univ.* **9**: 59–69.

MONESI, V. (1962). Autoradiographic study of DNA synthesis and the cell cycle in spermatogonia and spermatocytes of mouse testis, using tritiated thymidine. *J. Cell Biol.* **14**: 1–18.

MONESI, V., CRIPPA, M. and ZITO-BIGNAMI, R. (1967). The stage of chromosome duplication in the cell cycle as revealed by X-ray breakage and ^{3}H-thymidine labeling. *Chromosoma*, **21**: 369–86.

MONTALENTI, G. (1946). Sui cromosomi e il numero dei chiasmi nella spermatogenesi di *Psychoda* sp. (Dipt., Psychodidae). *Rend. Accad. Naz. Lincei Ser.* VIII, **1**: 120–2.

MONTALENTI, G. (1947). L'asetto cromosomico e l'interferenza dei chiasmi nella spermatogenesi di *Simulium equinum* L. (Diptera, Simuliidae). *Rend. Accad. Naz. Lincei* Ser. VIII, **2**: 471–474.

MONTALENTI, G. (1960). Alcune considerazioni sull' evoluzione della determinazione del sesso. *Accad. Naz. Lincei – Quaderno* **47** *– Evoluzione e Genetica*, pp. 153–81.

MONTALENTI, G. (1961). La variabilità del rapporto sessi e il suo valore selettivo. *Atti Assoc. Genet. Ital.* **6**: 405–8.

MONTALENTI, G. and FRATINI, L. (1959). Observations on the spermatogenesis of *Bacillus rossius* (Phasmoidea). *Proc. XVth Int. Congr. Zool. Lond.* pp. 749–50.

MONTALENTI, G. and VITAGLIANO-TADINI, G. (1960). Osservazioni sull'idoneità (fitness) di alcuni incroci di *Asellus*. *Atti Assoc. Genet. Ital.* **5**: 207–16.

MONTALENTI, G. and VITAGLIANO-TADINI, G. (1963). Variability of sex ratio and sex-determining mechanism in *Asellus*. *Proc. XIth Int. Genet. Congr., The Hague*, **1**: 186–7.

MOORE, B. P., WOODROFFE, G. E. and SANDERSON, A. R. (1956). Polymorphism and parthenogenesis in a ptinid beetle. *Nature*, **177**: 847–8.

MOORHEAD, P. S. (1954). Chromosome variation in giant forms of *Drosophila montana*. *Univ. Texas Publ.* **5422**: 106–29.

MOORHEAD, P. S. and DEFENDI, V. (1963). Asynchrony of DNA synthesis in chromosomes of human diploid cells. *J. Cell Biol.* **16**: 202–9.

MORESCALCHI, A. (1962). Osservazioni sul corredo cromosomico di *Rana esculenta* L. (nota preliminare). *Boll. di Zool.* **29**: 601–9.

MORESCALCHI, A. (1963*a*). Il corredo cromosomico e il problema dei cromosomi sessuali in *Bufo viridis* Laur. *Rend. Accad. Sci. Fis. Mat. Soc. Naz. Sci. Lett. Arti, Napoli,* **30**: 236–240.

MORESCHALCHI, A. (1963*b*). Conferma della presenza di eterocromosomi in *Xenopus laevis* Daudin. *Rend. Accad. Sci. Fis. Mat. Soc. Naz. Sci. Lett. Arti, Napoli,* **30**: 310–14.

MORESCALCHI, A. (1964). Il corredo cromosomico di *Discòglossus pictus* Otth: cromosomi sessuali, spiralizzazione cromosomica e zone eterocromatiche. *Caryologia,* **17**: 327–45.

MORESCALCHI, A. (1967*a*). Note citotassonomiche su *Ascaphus truei* Stejn. *Atti Soc. Peloritana, Sci. Fis. Mat. Nat.* **13**: 23–30.

MORESCALCHI, A. (1967*b*). Le relazioni tra il cariotipo di Anuri Displasioceli. I. Il corredo cromosomico di alcuni Ranidae. *Caryologia,* **20**: 65–85.

MORESCALCHI, A. (1968*a*). Initial cytotaxonomic data on certain families of amphibious Anura (Displasiocoela after Noble). *Experientia,* **24**: 280–3.

MORESCALCHI, A. (1968*b*). Hypothesis on the phylogeny of the Salientia, based on karyological data. *Experientia,* **24**: 964.

MORESCALCHI, A. (1968*c*). The karyotypes of two specimens of *Leiopelma hochstetteri* Fitz. (Amphibia Salientia). *Caryologia,* **21**: 37–46.

MORESCALCHI, A. (1970). Karyology and vertebrate phylogeny. *Boll. di Zool.* **37**: 1–28.

MORGAN, L. V. (1933). A closed *X*-chromosome in *Drosophila melanogaster*. *Genetics,* **18**: 250–83.

MORGAN, T. H. (1906). The male and female eggs of the phylloxerans of the hickories. *Biol. Bull. Woods Hole,* **10**: 201–6.

MORGAN, T. H. (1908). The production of two kinds of spermatozoa. *Proc. Soc. Exp. Biol. N. Y.* **5**: 56–7.

MORGAN, T. H. (1909*a*). Sex determination and parthenogenesis in phylloxerans and aphids. *Science,* **29**: 234–7.

MORGAN, T. H. (1909*b*). A biological and cytological study of sex determination in phylloxerans and aphids. *J. Exp. Zool.* **7**: 239–352.

MORGAN, T. H. (1912). The elimination of the sex chromosomes from the male-producing eggs of phylloxerans. *J. Exp. Zool.* **12**: 379–98.

MORGAN, T. H. (1915). The predetermination of sex in phylloxerans and aphids. *J. Exp. Zool.* **19**: 285–321.

MORGAN, T. H., BRIDGES, C. B. and STURTEVANT, A. H. (1925). The genetics of *Drosophila*. *Bibliogr. Genet.* **2**: 1–262.

MORGAN, T. H., REDFIELD, H. and MORGAN, L. V. (1943). The constitution of the germinal materials in relation to heredity. *Carnegie Inst. Wash. Yearb.* **42**: 171–4.

MORGAN, W. P. (1928). A comparative study of the spermatogenesis of five species of earwigs. *J. Morph.* **46**: 241–73.

MORISHIMA, A., GRUMBACH, M. M. and TAYLOR, J. H. (1962). Asynchronous duplication of human chromosomes and the origin of sex chromatin. *Proc. Nat. Acad. Sci. U.S.A.* **48**: 756–63.

MORITA, J. (1932). Existence of small and large atypical spermatozoa in *Viviparus*. *Folia Anat. Jap.* **10**: 35–51.

MORRIS, J. McL. (1953). The syndrome of testicular feminization in male pseudohermaphrodites (82 cases). *Amer. J. Obstet. Gynecol.* **65**: 1192–211.

MORTIMER, C. (1936). Experimentelle und cytologische Untersuchungen über den Generationswechsel der Cladoceren. *Zool. Jb. (Allg. Zool.)* **56**: 323–88.

MORTON, J. E. (1964). *Molluscs* (3rd ed.). London: Hutchinson.

MORTON, N. E., CROW, J. F. and MULLER, H. J. (1956). An estimate of the mutational damage in man from data on consanguineous marriages. *Proc. Nat. Acad. Sci. U.S.A.* **42**: 855–63.

MOSES, M. J. (1956). Chromosomal structures in crayfish spermatocytes. *J. Biophys. Biochem. Cytol.* **2**: 215–17.

MOSES, M. J. (1958). The relation between the axial complex of meiotic prophase chromosomes and chromosome pairing in a salamander (*Plethodon cinereus*). *J. Biophys. Biochem. Cytol.* **4**: 633–8.

MOSES, M. J. (1968). The synaptinemal complex. *Ann. Rev. Genet.* **2**: 363–412.

MOSES, M. J. (1969). Structure and function of the synaptonemal complex. *Genetics*, **61** suppl.: 41–51.

MOSES, M. J. and COLEMAN, J. R. (1964). Structural patterns and the functional organization of chromosomes. In: *The Role of Chromosomes in Development* (ed.: M. Locke), pp. 11–50. New York: Academic Press.

MOSES, M. J. and YERGANIAN, G. (1952). Desoxypentose nucleic acid (DNA) content and cytotaxonomy of several Cricetinae (hamsters). *Genetics*, **37**: 607–8.

MOSS, J. P. (1966). The adaptive significance of B-chromosomes in rye. In: *Chromosomes Today*, **I** (ed.: C. D. Darlington and K. R. Lewis), pp. 15–23. London and Edinburgh: Oliver and Boyd.

MOSS, W. W., OLIVER, J. H. JR. and NELSON, B. C. (1968). Karyotypes and developmental stages of *Harpyrhynchus novoplumaris* sp. n. (Acari: Chelytodea: Harpyrhychidae), a parasite of North American birds. *J. Parasitol.* **54**: 377–92.

MOURAD, A. M. and MALLAH, G. S. (1960). Chromosomal polymorphism in Egyptian populations of *Drosophila melanogaster*. *Evolution*, **14**: 166–70.

MUKHERJEE, A. S. and BEERMANN, W. (1965). Synthesis of ribonucleic acid by the *X*-chromosomes of *Drosophila melanogaster* and the problem of dosage compensation. *Nature*, **207**: 785–6.

MUKHERJEE, B. B. (1965). Cytological aspect of the *X*-chromosome differentiation in normal and abnormal cells. *Canad. J. Genet. Cytol.* **7**: 189–201.

MUKHERJEE, B. B. and SINHA, A. K. (1964). Single-active-*X*-hypothesis: cytological evidence for random inactivation of *X*-chromosomes in a female mule complement. *Proc. Nat. Acad. Sci. U.S.A.* **51**: 252–8.

MULDAL, S. (1948). No title. *John Innes Hort. Inst. 39th Ann. Rep.* pp. 21–2.

MULDAL, S. (1949). Cytotaxonomy of British earthworms. *Proc. Linn. Soc. Lond.* **161**: 116–18.

MULDAL, S. (1952). The chromosomes of the Earthworms. I. The evolution of polyploidy. *Heredity*, **6**: 55–76.

MULDER, M. P., VAN DUIJN, P. and GLOOR, H. J. (1968). The replicative organization of DNA in polytene chromosomes of *D. hydei*. *Genetica*, **39**: 385–428.

MULLER, H. J. (1925). Why polyploidy is rarer in animals than in plants. *Amer. Nat.* **59**: 346–53.

MULLER, H. J. (1932*a*). Some genetic aspects of sex. *Amer. Nat.* **66**: 118–38.

MULLER, H. J. (1932*b*). Further studies on the nature and causes of gene mutations. *Proc. VIth Int. Genet. Congr. Ithaca*, **1**: 213–55.

MULLER, H. J. (1938). The re-making of chromosomes. *Collecting Net, Woods Hole*, **13**: 181–95 and 198. (Reprinted in *Studies in Genetics, the Selected papers of H. J. Muller*. Indiana Univ. Press, 1962).

MULLER, H. J. (1940*a*). Bearings of the *Drosophila* work on systematics. In: *The New Systematics* (ed.: J. Huxley), pp. 185–268. Oxford: Clarendon Press.

MULLER, H. J. (1940b). Analysis of the process of structural change in chromosomes of *Drosophila. J. Genet.* **40**: 1–66.

MULLER, H. J. (1950a). Our load of mutations. *Amer. J. Hum. Genet.* **2**: 111–76.

MULLER, H. J. (1950b). Evidence for the precision of genetic adaptation. *Harvey Lectures.* Ser. XLIII, pp. 165–229.

MULLER, H. J. (1956). On the relation between chromosome changes and gene mutations. *Brookhaven Symp. in Biol.* **8**: 126–47.

MULLER, H. J. (1959). The mutation theory re-examined. *Proc. Xth Int. Genet. Congr.* **1**: 306–17.

MULLER, H. J. and HERSKOWITZ, I. H. (1954). Concerning the healing of chromosome ends produced by breakage in *Drosophila melanogaster. Amer. Nat.* **88**: 177–208.

MULLER, H. J. and PONTECORVO, G. (1940a). Recombinants between *Drosophila* species the F_1 hybrids of which are sterile. *Nature,* **146**: 199.

MULLER, H. J. and PONTECORVO, G. (1940b). The artificial mixing of incompatible germplasms in *Drosophila. Science,* **92**: 418 (see also *Science,* **92**: 476).

MULLER, H. J. and PONTECORVO, G. (1941). Recessive genes causing interspecific sterility and other disharmonies between *Drosophila melanogaster* and *Drosophila simulans. Genetics,* **27**, 157.

MULLER, H. J. and PROKOFYEVA, A. A. (1935). The individual gene in relation to the chromomere and the chromosome. *Proc. Nat. Acad. Sci. U.S.A.* **21**: 16–26.

MULLER, H. J., PROKOFIEVA, A. A. and KOSSIKOV, K. V. (1936). Unequal crossing-over in the Bar mutant as a result of duplication of a minute chromosome section. *C. R. Acad. Sci. U.R.S.S.* **1**: 87–8.

MULLER, H. J. and SETTLES, F. (1927). The non-functioning of the genes in spermatozoa. *Z. indukt. Abst. Verebl.* **43**: 285–312.

MÜNTZING, A. (1945). Cytological studies of extra chromosome fragments in rye. II. Transmission and multiplication of standard fragments and isofragments. *Hereditas,* **31**: 457–77.

MÜNTZING, A. (1950). Accessory chromosomes in rye populations from Turkey and Afghanistan. *Hereditas,* **36**: 507–9.

MÜNTZING, A. (1957). Frequency of accessory chromosomes in rye strains from Iran and Korea. *Hereditas,* **43**: 682–5.

MÜNTZING, A. (1958). A new category of chromosomes. *Proc. Xth Int. Genet. Congr.* **1**: 453–67.

MÜNTZING, A. (1963). Effects of accessory chromosomes in diploid and tetraploid rye. *Hereditas,* **49**: 371–426.

MÜNTZING, A. (1966). Accessory chromosomes. *Bull. Bot. Soc. Bengal,* **20**: 1–15.

MURDY, W. H. and CARSON, H. L. (1959). Parthenogenesis in *Drosophila mangabeirai* Malog. *Amer. Nat.* **93**: 355–63.

NABOURS, R. K. (1937). Methoden und Ergebnisse bei der Züchtung von Tetriginae. *Abderhalden's Handb. Biol. Arbeitsmeth.* **9**: (3): 1309–65.

NABOURS, R. K. and FORSTER, M. E. (1929). Parthenogenesis and the inheritance of colour patterns in the Grouse Locust *Paratettix texanus* Hancock. *Biol. Bull. Woods Hole,* **54**: 129–55.

NADLER, C. F. and BLOCK, M. H. (1962). The chromosomes of some North American chipmunks (Sciuridae) belonging to the genera *Tamias* and *Eutamias. Chromosoma,* **13**: 1–15.

NAISSE, J. (1969). Rôle des neurohormones dans la différenciation sexuelle de *Lampyris noctiluca. J. Insect Physiol.* **15**: 877–92.

NAKAHARA, W. (1919). A study on the chromosomes in the spermatogenesis of the stone-fly *Perla immarginata* Say, with special reference to the question of synapsis. *J. Morph.* **32**: 509–22.

NAKAMURA, K. (1927). Preliminary notes on Reptilian chromosomes. I. The chromosomes of some snakes. *Proc. Imp. Acad. Tokyo,* **3**: 296–7.

NAKAMURA, K. (1928). On the chromosomes of a snake (*Natrix tigrina*). *Mem. Coll. Sci. Kyoto Imp. Univ.* B, **4**: 1–8.

NAKAMURA, K. (1932). Studies on reptilian chromosomes. III. Chromosomes of some geckos. *Cytologia.* **3**: 156–68.

NAKAMURA, K. (1935). Studies on reptilian chromosomes. VI. Chromosomes of some snakes. *Mem. Coll. Sci. Kyoto Imp. Univ.* B, **10**: 361–402.

NANKIVELL, R. N. (1967). A terminal association of two pericentric inversions in first metaphase cells of the Australian grasshopper *Austroicetes interioris* (Acrididae). *Chromosoma,* **22**: 42–68.

NARASIMHA RAO, M. V. and AMMERMANN, D. (1970). Polytene chromosomes and nucleic acid metabolism during macronuclear development in *Euplotes*. *Chromosoma,* **29**: 246–54.

NARBEL, M. (1946). La cytologie de la parthénogénèse chez *Apterona helix* Sieb. (Lepidoptera: Psychides). *Rev. Suisse Zool.* **53**: 625–81.

NARBEL-HOFSTETTER, M. (1950). La cytologie de la parthénogénèse chez *Solenobia sp.* (*lichenella* L.?) (Lépidoptères, Psychides). *Chromosoma,* **4**: 56–90.

NARBEL-HOFSTETTER, M. (1955). La pseudogamie chez *Luffia lapidella* Goeze (Lépid. Psychide). *Rev. Suisse Zool.* **62**: 224–9.

NARBEL-HOFSTETTER, M. (1956). La cytologie des *Luffia* (Lépid. Psych.): Le croisement de l'espèce parthénogénétique avec l'espèce bisexuée. *Rev. Suisse Zool.* **63**: 203–8.

NARBEL-HOFSTETTER, M. (1958). Le sort des polocytes dans les oeufs de *Luffia* (Lépidoptère Psychide). *Arch J. Klaus Stift. Vererbf.* **33**: 57–62.

NARBEL-HOFSTETTER, M. (1961). Cytologie comparée de l'espèce parthénogénétique *Luffia ferchaultella* Steph. et de l'espèce bisexuée *L. lapidella* Goetze (Lepidoptera, Psychidae). *Chromosoma,* **12**: 505–52.

NARBEL-HOFSTETTER, M. (1962). Le croisement des espèces parthénogénétique et bisexuée chez *Luffia* (Lépidoptère Psychide). Les élevages et leur résultats. *Rev. Suisse Zool.* **69**: 465–79.

NARBEL-HOFSTETTER, M. (1963a). Cytologie de la pseudogamie chez *Luffia lapidella* Goeze (Lépid. Psych.). *Chromosoma,* **13**: 623–45.

NARBEL-HOFSTETTER, M. (1963b). La répartition géographique des trois formes cytologiques de *Luffia* (Lépidoptère, Psychide). *Mitt. Schweiz. Ent. Ges.* **36**: 275–88.

NARBEL-HOFSTETTER, M. (1964). *Les Altérations de la Méiose chez les Animaux Parthénogénétiques.* Protoplasmatologia **VI/F/2**, 190 pp. Vienna: Springer-Verlag.

NAVASHIN, M. (1932). The dislocation hypothesis of evolution of chromosome numbers. *Z. indukt Abst. Vererbl.* **63**: 224–31.

NAVILLE, A. (1932). Les bases cytologiques de la théorie du "crossing-over": Étude sur la spermatogénèse et l'ovogénèse des Calliphorinae. *Z. Zellforsch.* **16**: 440–70.

NAVILLE, A. and DE BEAUMONT, J. (1933). Recherches sur les chromosomes des Nevroptères. *Arch. Anat. Mikr.* **29**: 199–234.

NAVILLE, A. and DE BEAUMONT, J. (1934). Les chromosomes des Panorpes. *Bull. Biol. Fr. Belg.* **68**: 98–107.

NAVILLE, A. and DE BEAUMONT, J. (1936). Recherches sur les chromosomes des Nevroptères. Deuxieme partie. *Arch. Anat. Micr.* **32**: 271–302.

NEARY, G. J. and EVANS, H. J. (1958). Chromatid breakage by irradiation and the oxygen effect. *Nature,* **182**: 890–1.

NEBEL, B. R. (1932). Chromosome studies in the Tradescantiae. II. The direction of coiling of the chromonema in *Tradescantia reflexa* Raf., *T. virginiana* L., *Zebrina pendula* Schnitzl and *Rhoeo discolor* Hance. *Z. Zellforsch.* **16**: 285–304.

NEBEL, B. R. (1936). Chromosome structure. X. An X-ray experiment. *Genetics,* **21**: 605–14.

NEBEL, B. R. and RUTTLE, M. L. (1936). Chromosome structure. IX. *Tradescantia reflexa* and *Trillium erectum. Amer. J. Bot.* **23**: 652–63.

NELSON-REES, W. A., KNIAZEFF, A. J., BAKER, R. J. and PATTON, J. L. (1968). Intraspecific chromosome variation in the bat, *Macrotus waterhousii* Gray. *J. Mammal.* **49**: 706–12.

NES, N. (1963). The chromosomes of *Chinchilla lanigera. Acta Vet. Skand.* **4**: 128–35.

NEUFFER, M. G., JONES, L. and ZUBER, M. S. (1968). The mutants of maize. *Spec. Publ. Crop. Sci. Soc. Amer., Madison, Wis.* 74 pp.

NEWBY, W. W. (1942). A study of intersexes produced by a dominant mutation in *Drosophila virilis,* Blanco stock. *Univ. Texas Publ.* **4228**: 113–45.

NEWCOMER, E. H. (1957). The mitotic chromosomes of the domestic fowl. *J. Hered.* **48**: 227–34.

NEWMAN, L. J. (1967). Meiotic chromosomal aberrations in wild populations of *Podophyllum peltatum. Chromosoma,* **22**: 258–73.

NICKLAS, R. B. (1959). An experimental and descriptive study of chromosome elimination in *Miastor* spec. (Cecidomyidae; Diptera). *Chromosoma,* **10**: 301–36.

NICKLAS, R. B. (1960). The chromosome cycle of a primitive cecidomyid – *Mycophila speyeri. Chromosoma,* **11**: 402–18.

NICKLAS, R. B. (1961a). Recurrent pole-to-pole movements of the sex chromosome during prometaphase I in *Melanoplus differentialis* spermatocytes. *Chromosoma,* **12**: 97–115.

NICKLAS, R. B. (1961b). The relationship between DNA content and alternative meiotic patterns in certain Discocephalinids (Pentatomidae; Heteroptera). *J. Biophys. Biochem. Cytol.* **9**: 486–90.

NICKLAS, R. B. (1963). A quantitative study of chromosomal elasticity and its influence on chromosome movement. *Chromosoma,* **14**: 276–95.

NICKLAS, R. B. (1965). Chromosome velocity during mitosis as a function of chromosome size and position. *J. Cell Biol.* **25**: Mitosis suppl.: 119–35.

NICKLAS, R. B. (1967). Chromosome micromanipulation. II. Induced reorientation and the experimental control of segregation in meiosis. *Chromosoma,* **21**: 17–50.

NICKLAS, R. B. (1970). Mitosis. In: *Advances in Cell Biology,* **2** (ed: D. M. Prescott, L. Goldstein and E. H. McConkey). New York: Appleton-Century-Crofts.

NICKLAS, R. B. and JAQUA, R. A. (1965). X-chromosome DNA replication developmental shift from synchrony to asynchrony. *Science,* **147**: 1041–3.

NICKLAS, R. B. and KOCH, C. A. (1969). Chromosome micromanipulation. III. Spindle fiber tension and the reorientation of mal-oriented chromosomes. *J. Cell Biol.* **43**: 40–50.

NICKLAS, R. B. and STAEHLY, C. A. (1967). Chromosome micromanipulation. I. The mechanics of chromosome attachment to the spindle. *Chromosoma,* **21**: 1–16.

NICOLETTI, B. (1968). Il controllo genetico della meiosi. *Atti Assoc. Genet. Ital.* **13**: 1–71.

NICOLETTI, B., TRIPPA, G. and DE MARCO, A. (1967). Reduced fertility in SD males and its bearing on segregation distortion in *Drosophila melanogaster*. *Atti Accad. Naz. Lincei*, **43**: 383–92.

NIGON, V. and BOVET, P. (1955). La teneur des gamètes en acide désoxyribonucléique chez *Parascaris equorum* Goeze. *C.R. Soc. Biol. Paris*, **149**: 129–30.

NIIYAMA, H. (1937). The problem of male heterogamety in the Decapod Crustacea, with special reference to the sex chromosomes in *Plagusia dentipes* de Haan and *Eriocheir japonicus* de Haan. *J. Fac. Sci. Hokkaido Imp. Univ., Zool.* **5**: 283–95.

NIIYAMA, H. (1938). The *XY*-chromosomes of the shore crab *Hemigrapsus sanguineus* de Haan. *Jap. J. Genet.* **14**: 33–8.

NIIYAMA, H. (1941). The *XO* type of sex chromosome found in *Ovalipes punctatus* de Haan (Crustacea: Decapoda). *Jap. J. Genet.* **17**: 41–5.

NIIYAMA, H. (1951). The chromosomes of a hermit crab, *Eupagurus ochotensis*, showing the greatest number so far found in animals. *Kromosomo*, **8**: 315–17.

NIIYAMA, H. (1959). A comparative study of the chromosomes in Decapods, Isopods, and Amphipods, with some remarks on cytotaxonomy and sex determination in the crustacea. *Mem. Fac. Fish. Hokkaido Univ.* **7**: 1–60.

NIIYAMA, H. (1962). On the unprecedentedly large number of chromosomes of the crayfish *Astacus trowbridgei* Stimpson. *Annot. Zool. Japan*, **35**: 229–33.

NIIYAMA, H. (1966). The chromosomes of two species of edible crabs (Brachyura, Decapoda, Crustacea). *Bull. Fac. Fish. Hokkaido Univ.* **16**: 201–5.

NIKOLEI, E. (1961). Vergleichende Untersuchungen zur Fortpflanzung heterogoner Gallmücken unter experimentellen Bedingungen. *Z. Morph. Ökol. Tiere*, **50**: 281–329.

NISHIKAWA, S. (1962). A comparative study of the chromosomes in marine Gastropods, with some remarks on cytotaxonomy and phylogeny. *J. Shimonoseki Coll. Fish.* **11**: 539–76.

NIYAMASENA, S. G. (1940). Chromosomen und Geschlecht bei *Bilharzia mansoni*. *Z. Parasitenk.* **11**: 690–701.

NODA, S. (1968). Achiasmate bivalent formation by parallel pairing in PMC's of *Fritillaria amabilis*. *Bot. Mag. Tokyo*, **81**: 344–5.

NOGUSA, S. (1955a). Chromosome studies in Pisces. IV. The chromosomes of *Mogrunda obscura* (Gobiidae), with evidence of male heterogamety. *Cytologia*, **20**: 11–18.

NOGUSA, S. (1955b). Chromosome studies in Pisces V. Variation of the chromosome number in *Acheilognathus rhombea* due to multiple-chromosome formation. *Annot. Zool. Japan*, **28**, 249–55.

NOGUSA, S. (1957). The chromosomes of the Japanese lancelet, *Branchiostoma belcheri* (Gray), with special reference to the sex chromosomes. *Annot. Zool. Japan*, **30**: 42–6.

NOGUSA, S. (1960). A comparative study of the chromosomes in fishes with particular considerations on taxonomy and evolution. *Mem. Hyogo Univ. Agric.* **3**: 1–62.

NOLTE, D. J. (1939). A comparative study of seven species of Transvaal Acrididae, with special reference to the chromosome complex. *J. Entomol. Soc. S. Afr.* **2**: 196–260.

NOLTE, D. J. (1964). The nuclear phenotype of locusts. *Chromosoma*, **15**: 367–88.

NOLTE, D. J. (1968). The chiasma-inducing pheromones of locusts. *Chromosoma*, **23**: 346–58.

NONIDEZ, J. F. (1915). Estudios sobre las celulas sexuales. I. Los cromosomas goniales y las mitosis de maduración en *Blaps lusitanica* y *B. waltli*. *Mem. R. Soc. Esp. Hist. Nat.* **10**: 149–90.

NONIDEZ, J. F. (1920). The meiotic phenomena in the spermatogenesis of *Blaps*, with special reference to the *X* complex. *J. Morph.* **34**: 69–117.

NORDENSKIÖLD, H. (1951). Cyto-taxonomical studies in the genus *Luzula*. I. Somatic chromosomes and chromosome numbers. *Hereditas*, **37**, 325–55.

NORDENSKIÖLD, H. (1961). Tetrad analysis and the course of meiosis in three hybrids of *Luzula campestris*. *Hereditas*, **47**: 203–38.

NORDENSKIÖLD, H. (1962). Studies of meiosis in *Luzula purpurea*. *Hereditas*, **48**: 503–19.

NOUJDIN, N. I. (1944). The regularities of the influence of heterochromatin on mosaicism. *Zh. Obshch. Biol.* **5**: 357–88.

NOVITSKI, E. (1946). Chromosome variation in *Drosophila athabasca*. *Genetics*, **31**: 508–24.

NOVITSKI, E. (1947). Genetic analysis of an anomalous sex ratio condition in *Drosophila affinis*. *Genetics*, **32**: 526–34.

NOVITSKI, E. (1961). Chromosome breakage in inversion heterozygotes. *Amer. Nat.* **95**: 250–1.

NOVITSKI, E. (1964). An alternative to the distributive pairing hypothesis in *Drosophila*. *Genetics*, **50**: 1449–51.

NOVITSKI, E. and DEMPSTER, E. R. (1958). An analysis of data from laboratory populations of *Drosophila melanogaster*. *Genetics*, **43**: 470–9.

NOVITSKI, E., PEACOCK, W. J. and ENGEL, J. (1965). Cytological basis of sex ratio in *Drosophila pseudoobscura*. *Science*, **148**: 516–17.

NUÑEZ, O. (1962). Cytology of Collembola. *Nature*, **194**: 946–7.

NUR, U. (1961). Meiotic behavior of an unequal bivalent in the grasshopper *Calliptamus palaestinensis* Bdhr. *Chromosoma*, **12**: 272–9.

NUR, U. (1962*a*). A supernumerary chromosome with an accumulation mechanism in the lecanoid genetic system. *Chromosoma*, **13**: 249–71.

NUR, U. (1962*b*). Sperm, sperm bundles and fertilization in a mealy bug *Pseudococcus obscurus* Essig (Homoptera: Coccoidea). *J. Morph.* **111**: 173–99.

NUR, U. (1962*c*). Population studies of supernumerary chromosomes in a mealy bug. *Genetics*, **47**: 1679–90.

NUR, U. (1963*a*). Meiotic parthenogenesis and heterochromatinization in a soft scale, *Pulvinaria hydrangeae* (Coccoidea: Homoptera). *Chromosoma*, **14**: 123–39.

NUR, U. (1963*b*). A mitotically unstable supernumerary chromosome with an accumulation mechanism in a grasshopper. *Chromosoma*, **14**: 407–22.

NUR, U. (1965). A modified Comstockiella chromosome system in the olive scale insect *Parlatoria oleae* (Coccoidea: Diaspididae). *Chromosoma*, **17**: 104–20.

NUR, U. (1966*a*). Harmful supernumerary chromosomes in a mealy bug population. *Genetics*, **54**: 1225–38.

NUR, U. (1966*b*). The effect of supernumerary chromosomes on the development of mealy bugs. *Genetics*, **54**: 1239–49.

NUR, U. (1966*c*). Nonreplication of heterochromatic chromosomes in a mealy bug, *Planococus citri* (Coccoidea: Homoptera). *Chromosoma*, **19**: 439–48.

NUR, U. (1967). Chromosome systems in the Eriococcidae (Coccoidea: Homoptera). II. *Gossyparia spuria* and *Eriococcus araucariae*. *Chromosoma*, **22**: 151–63.

NUR, U. (1968*a*). Endomitosis in the mealy bug, *Planococcus citri* (Homoptera: Coccoidea). *Chromosoma*, **24**: 202–9.

NUR, U. (1968*b*). Synapsis and crossing-over within a paracentric inversion in the grasshopper *Camnula pellucida*. *Chromosoma*, **25**: 198–214.

NUR, U. (1969*a*). Mitotic instability leading to an accumulation of B-chromosomes in grasshoppers. *Chromosoma*, **27**: 1–19.

NUR, U. (1969b). Harmful *B*-chromosomes in a mealy bug: additional evidence. *Chromosoma*, **28**: 280–97.

NUR, U. (1970). Translocations between eu- and heterochromatic chromosomes and spermatocytes lacking a heterochromatic set in male mealy bugs. *Chromosoma*, **29**: 42–61.

NUR, U. and CHANDRA, H. S. (1963). Interspecific hybridization and gynogenesis in mealy bugs. *Amer. Nat.* **97**: 197–202.

NUR, U. and NEVO, E. (1969). Supernumerary chromosomes in the cricket frog *Acris crepitans. Caryologia*, **22**: 97–102.

O'BRIEN, R. (1956). Deoxyribose nucleic acid in a haplo-diploid species (*Steatococcus tuberculatus Morrison*). *Chromosoma*, **8**: 229–59.

OFFERMANN, C. A. (1936). Branched chromosomes as symmetrical duplications. *J. Genet.* **32**: 103–16.

OGAWA, K. (1950). Chromosome studies in the Myriapoda. I. The chromosomes of *Thereuonema hilgendorfi* Verhoeff (Chilopoda) with special regard to the post-reductional separation of the sex chromosomes. *Jap. J. Genet.* **25**: 106–10.

OGAWA, K. (1952). Chromosome studies in the Myriapoda. VI. A study on the sex-chromosomes in allied species of Chilopods. *Annot. Zool. Japan*, **25**: 434–40.

OGAWA, K. (1953). Chromosome studies in the Myriapoda. V. A chromosomal survey in some Chilopods with a cyto-taxonomic consideration. *Jap. J. Genet.* **28**: 12–18.

OGAWA, K. (1954). Chromosome studies in the Myriapoda. VII. A chain-association of the multiple sex chromosomes found in *Otocryptops sexspinosus* (Say). *Cytologia*, **19**: 265–72.

OGAWA, K. (1955a). Chromosome studies in the Myriapoda. VIII. Behavior of X-ray induced chromosome fragments of *Thereuonema hilgendorfi. Zool. Mag. Tokyo*, **64**: 167–70.

OGAWA, K. (1955b). Chromosome studies in the Myriapoda. X. An *X-2 Y* type of sex-determining mechanism found in *Otocryptops* sp. *Annot. Zool. Japan*, **28**: 244–8.

OGAWA, K. (1955c). Chromosome studies in the Myriapoda. IX. The diffuse kinetochores verified by X-ray-induced chromosome fragments in *Esastigmatobius longitarsis* Verhoeff. *Zool. Mag. Tokyo*, **64**: 291–3.

OGAWA, K. (1961a). Chromosome studies in the Myriapoda. XII. On the sex chromosomes of *Otocryptops capillipedatus* Takakuwa. *Jap. J. Zool.* **13**: 63–8.

OGAWA, K. (1961b). Chromosome studies in the Myriapoda. XIII. Three types of the sex chromosomes found in *Otocryptops rubiginosus* (L. Koch). *Jap. J. Genet.* **36**: 122–8.

OGAWA, K. (1961c). Chromosome studies in the Myriapoda. XIV. The chromosomes of four species of Chilopods. *Zool. Mag. Tokyo*, **70**: 171–5.

OGAWA, K. (1961d). Chromosome studies in the Myriapoda. XV. On individually different three karyotypes found in *Otocryptops sexspinosus* (Say) (Chilopoda) (Preliminary report). *Zool. Mag. Tokyo*, **70**: 176–9.

OGAWA, K. (1961e). Chromosome studies in the Myriapoda. XVI. The chromosomes of five species of Chilopods. *Zool. Mag. Tokyo*, **70**: 203–6.

OGUMA, K. (1921). The idiochromosomes of the mantis. *J. Coll. Agric. Univ. Hokkaido*, **10**: 1–27.

OGUMA, K. (1934). Studies on the sauropsid chromosomes. II. The cytological evidence proving female heterogamety in the lizard (*Lacerta vivipara*). *Arch. Biol. Paris*, **45**: 27–46.

OGUMA, K. (1937a). Studies on sauropsid chromosomes. III. The chromosomes of the soft-shelled turtle, *Amyda japonica* (Temminck and Schleg.), as additional proof of female heterogamey in the Reptilia. *J. Genet.* **34**: 247–64.

OGUMA, K. (1937b). Absence of the Y-chromosome in the vole, *Microtus montebelli* Edw. with supplementary remarks on the sex-chromosomes of *Evotomys* and *Apodemus*. *Cytologia*, Fujii Jubilee vol., pp. 796–808.

OGUMA, K. (1946). Karyotype and phylogeny of the mantids. *Kromosomo*, **1**: 1–5.

OGUMA, K. and ASANA, J. J. (1932). Additional data to our knowledge of the dragon-fly chromosome with a note on occurrence of XY-chromosome in the ant lion (Neuroptera). *J. Fac. Sci. Hokkaido Univ.* Ser. VI, **1**: 133–42.

OHMACHI, F. (1935). A comparative study of chromosome complements in the Gryllodea in relation to taxonomy. *Bull. Mie Coll. Agric. For.* no. 5, pp. 1–48.

OHMACHI, F. and UESHIMA, N. (1955). On the chromosome complements of three nearly related species of *Loxoblemmus arietulus* Saussure (Preliminary note). *Bull. Fac. Agric. Mie Univ.* **10**: 21–31.

OHMACHI, F. and UESHIMA, N. (1957a). A study on local variation of chromosome complements in *Scapsipedus aspersus* (Orthoptera:Gryllodea). *Bull. Fac. Agric. Mie Univ.* **14**: 43–9.

OHMACHI, F. and UESHIMA, N. (1957b). Variation of chromosome complements in *Euscyrtus hemelytrus* de Haan (Orthoptera:Gryllodea). *Bull. Fac. Agric. Mie Univ.* **15**: 1–13.

OHNO, S. (1963). Dynamics of the condensed female X-chromosome. *Lancet*, **i**: 273–4.

OHNO, S. (1965). Direct handling of germ cells. In: *Human Chromosome Methodology* (ed.: J. J. Yunis). New York: Academic Press.

OHNO, S. (1967). Sex chromosomes and sex-linked genes. *Monographs on Endocrinology*, **1**. Berlin, Heidelberg and New York: Springer-Verlag.

OHNO, S., BEÇAK, W. and BEÇAK, M. L. (1964). X-autosome ratio and the behavior pattern of individual X-chromosomes in placental mammals. *Chromosoma*, **15**: 14–30.

OHNO, S. and CATTANACH, B. M. (1962). Cytological studies of an X-autosome translocation in *Mus musculus*. *Cytogenetics*, **1**: 129–40.

OHNO, S., CHRISTIAN, L. C. and STENIUS, C. (1962). Nucleolus-organizing micro-chromosomes of *Gallus domesticus*. *Exp. Cell Res.* **27**: 612–14.

OHNO, S., JAINCHILL, J. and STENIUS, C. (1963). The creeping vole (*Microtus oregoni*) as a gonosomic mosaic. The OY/XY constitution of the male. *Cytogenetics*, **2**: 232–9.

OHNO, S., KAPLAN, W. D. and KINOSITA, R. (1957). Heterochromatic regions and nucleolus organizers in chromsomes of the mouse, *Mus musculus*. *Exp. Cell Res.* **13**: 358–64.

OHNO, S., KAPLAN, W. D. and KINOSITA, R. (1959a). On the end-to-end association of the X- and Y-chromosomes of *Mus musculus*. *Exp. Cell Res.* **18**: 282–90.

OHNO, S., KAPLAN, W. D. and KINOSITA, R. (1959b). Formation of the sex chromatin by a single X-chromosome in liver cells of *Rattus norvegicus*. *Exp. Cell Res.* **18**: 415.

OHNO, S., KITTRELL, W. A., CHRISTIAN, L. C., STENIUS, C. and WITT, G. A. (1963). An adult triploid chicken (*Gallus domesticus*) with a left ovotestis. *Cytogenetics*, **2**: 42–9.

OHNO, S. and MAKINO, S. (1961). The single-X nature of sex chromatin in man. *Lancet*, **i**: 78–9.

OHNO, S., MAKINO, S., KAPLAN, W. D. and KINOSITA, R. (1961). Female germ cells of man. *Exp. Cell Res.*, **24**: 106–10.

OHNO, S., MURAMOTO, J., CHRISTIAN, L. and ATKIN, N. B. (1967). Diploid-tetraploid relationship among old-world members of the fish family Cyprinidae. *Chromosoma*, **23**: 1–9.

OHNO, S., POOLE, J. and GUSTAVSSON, I. (1965). Sex-linkage of erythrocyte glucose 6-phosphate dehydrogenase in two species of wild hares. *Science*, **150**: 1737–8.

OHNO, S., STENIUS, C. and CHRISTIAN, L. (1966). The *XO* as the normal female of the creeping vole (*Microtus oregoni*). In: *Chromosomes Today*, **1** (ed.: C. D. Darlington and K. R. Lewis), pp. 182–7. Edinburgh and London: Oliver and Boyd.

OHNO, S., STENIUS, C. H., CHRISTIAN, L. C., BEÇAK, W. and BEÇAK, M. L. (1964). Chromosomal uniformity in the avian class Carinatae. *Chromosoma*, **15**: 280–8.

OHNO, S., STENIUS, C., FAISST, E. and ZENZES, M. T. (1965). Post-zygotic chromosomal rearrangements in rainbow trout (*Salmo irideus* Gibbons). *Cytogenetics*, **4**: 117–29.

OHNO, S., WEILER, C., POOLE, J., CHRISTIAN, L. and STENIUS, C. (1966). Autosomal polymorphism due to pericentric inversions in the deer mouse (*Peromyscus maniculatus*) and some evidence of somatic segregation. *Chromosoma*, **18**: 177–87.

OHNO, S., WOLF, U. and ATKIN, N. B. (1968). Evolution from fish to mammals by gene duplication. *Hereditas*, **59**: 169–87.

OHNUKI, Y. (1968). Structure of chromosomes. I. Morphological studies of the spiral structure of human somatic chromosomes. *Chromosoma*, **25**: 402–28.

OKSALA, T. (1943). Zytologische Studien an Odonaten. I. Chromosomenverhältnisse bei der Gattung *Aeschna* mit besonderer Berücksichtigung der postreduktionellen Teilung der Bivalente. *Ann. Acad. Sci. Fenn.* **A4**: 1–65.

OKSALA, T. (1944). Zytologische Studien an Odonaten. II. Die Entstehung der meiotische Präkozität. *Ann. Acad. Sci. Fenn.* **A4**: 1–37.

OKSALA, T. (1945). Zytologische Studien an Odonaten. III. Die Ovogenese. *Ann. Acad. Sci. Fenn.* **A4**: 1–32.

OKSALA, T. (1948). The concept and mechanism of chromosome reduction. *Hereditas*, **34**: 104–12.

OKSALA, T. (1952). Chiasma formation and chiasma interference in the Odonata. *Hereditas*, **38**: 449–80.

OLIVER, C. P. (1940). A reversion to wild type associated with crossing over in *Drosophila melanogaster*. *Proc. Nat. Acad. Sci. U.S.A.* **26**: 452–4.

OLIVER, J. H. Jr. (1965*a*). Cytogenetics of ticks (Acari: Ixodoidea). II. Multiple sex chromosomes. *Chromosoma*, **17**: 323–7.

OLIVER, J. H. Jr. (1965*b*). Karyotypes and sex determination in some dermanyssid mites (Acarina: Mesostigmata). *Ann. Entomol. Soc. Amer.* **58**: 567–73.

OLIVER, J. H. (1967). Cytogenetics of acarines. In: *Genetics of Insect Vectors of Disease* (ed.: J. Wright and R. Pal). Amsterdam: Elsevier.

OLIVER, J. H. Jr. (1966). Cytogenetics of ticks (Acari: Ixodoidea). I. Karyotypes of the two *Ornithodoros* species (Argasidae) restricted to Australia. *Ann. Entomol. Soc. Amer.* **59**: 144–7.

OLIVER, J. H. and BREMNER, K. C. (1968). Cytogenetics of ticks III. Chromosomes and sex determination in some Australian hard ticks (Ixodidae). *Ann. Entomol. Soc. Amer.* **61**: 837–44.

OLIVER, J. H. Jr., CAMIN, J. H. and JACKSON, R. C. (1963). Sex determination in the snake mite *Ophionyssus natricis* (Gervais) (Acarina: Dermanyssidae). *Acarologia*, **5**: 180–4.

OLIVER, J. H. Jr. and NELSON, B. C. (1967). Mite chromosomes: an exceptionally small number. *Nature*, **214**: 809.

OLIVIERI, G. and BREWEN, J. G. (1966). Evidence for non-random rejoining of chromatid breaks and its relation to the origin of sister chromatid exchanges. *Mutation Res.* **3**: 230–6.

OLSEN, M. W. and MARSDEN, S. J. (1954). Natural parthenogenesis in turkey eggs. *Science*, **120**: 545–6.

OMODEO, P. (1951a). Corredo cromosomico e spermatogenesi aberrante in *Allolobophora caliginosa trapezoides*. *Boll. di Zool.* **18**: 27–39.

OMODEO, P. (1951b). Problemi zoogeografici ed ecologici relativi a lombrichi peregrini con particolare riguardo al ripo di riproduzione ed alla struttura cariologica. *Boll. di Zool.* **18**: 117–22.

OMODEO, P. (1951c). Problemi genetici connessi con la poliploidia di alcuni lombrichi. *Boll. di Zool.* **18**: 123–9.

OMODEO, P. (1951d). Raddoppiamento del corredo cromosomico nelle cellule germinali femminili di alcuni lombrichi. *Atti Accad. Fisiocritici Siena* Ser. XII, **19**: fasc. 3, 4 pp.

OMODEO, P. (1951e). Il fenomeno della restituzione premeiotica in lombrichi partenogenetici. *Boll. Soc. Ital. Biol. Sper.* **27**: 1292–3.

OMODEO, P. (1951f). Gametogenesi e sistematica interspecifica come problemi conessi con la poliploidia nei Lumbricidae *Mem. Soc. Tosc. Sci. Nat.* B, **58** 12 pp.

OMODEO, P. (1952). Cariologia dei Lumbricidae. *Caryologia*, **4**: 173–275.

OMEDEO, P. (1953a). Specie e razze poliploidi nei lombrichi. *Convegno di Genetica, Ricerca Scientifica* **23**, suppl.: 43–9.

OMODEO, P. (1953b). Considerazioni sulla poliploidia nei lombrichi. *Boll. Soc. Ital. Biol. Sper.* **29**: 1437–9.

OMODEO, P. (1955). Cariologia dei Lumbricidae. II. Contributo. *Caryologia*, **8**: 135–78.

OMODEO, P. (1962a). Oligochètes des alpes. *Mem. Mus. Civ. Stor. Nat. Verona*, **10**: 71–96.

OMODEO, P. (1962b). Oligochètes de l'Afghanistan. II. *Atti Accad. Fisiocritici Siena*. Ser. XIII, **11**, 17 pp.

OPPERMANN, E. (1935). Die Entstehung der Riesenspermien von *Argas columbarum* Shaw (*reflexus* F.). *Z. mikr.-anat. Forsch.* **37**: 538–60.

ORIAN, A. J. E. and CALLAN, H. G. (1957). Chromosome numbers of gammarids. *J. Mar. Biol. Assoc. U.K.* **36**: 129–42.

ORLOV, V. N. and ALENIN, V. P. (1968). Karyotypes of some species of shrews of the genus *Sorex* (Insectivora, Soricidae). *Zool. Zh.* **47**: 1071–4.

ORTIZ, E. (1951). Los cariotipos de *Gryllotalpa gryllotalpa* (L.) de la península ibérica. *Bol. Soc. Esp. Hist. Nat.* **49**: 153–8.

ORTIZ, E. (1958). El valor taxonómico de las llamadas razas cromosómicas de *Gryllotalpa gryllotalpa* (L.). *Publ. Inst. Biol. Aglicades*, **27**: 181–94.

ORTIZ, E. (1969). Chromosomes and meiosis in Dermaptera. *Chromosomes Today*, **2** (ed.: C. D. Darlington and K. R. Lewis), pp. 33–43. Edinburgh and London: Oliver and Boyd.

ÖSTERGREN, G. (1945). Parasitic nature of extra chromosome fragments. *Bot. Notiser*, pp. 157–63.

ÖSTERGREN, G. (1947). Heterochromatic B-chromosomes in *Anthoxanthum*. *Hereditas*, **33**: 261–96.

Östergren, G. (1950). Considerations on some elementary features of mitosis. *Hereditas*, **36**: 1–18.

Östergren, G. and Wakonig, T. (1954). True or apparent sub-chromatid breakage and the induction of labile states in cytological chromosome loci. *Bot. Notiser*, pp. 357–75.

Ott, J. (1968). Nachweis natürlicher reproduktiver Isolation zwischen *Sorex gemellus* sp. n. und *Sorex araneus* Linnaeus 1758 in der Schweiz. *Rev. Suisse Zool.* **75**: 53–75.

Ottonen, P. O. (1966). Salivary gland chromosomes of six species in the IIIS-1 group of *Prosimulium* Roub. (Diptera: Simuliidae). *Canad. J. Zool.* **44**: 677–701.

Oura, G. (1936). A new method of unravelling the chromonema spirals. *Z. wiss. Mikr.* **53**: 36–7.

Ovchinnikova, L. P., Selivanova, G. V. and Cheissin, E. M. (1965). Photometric study of the DNA content in the nuclei of *Spirostomum ambiguum* (Ciliata, Heterotricha). *Acta Protozool.* **3**: 69–78.

Owen, A. R. G. (1949). A possible interpretation of apparent interference across the centromere found by Callan and Montalenti in *Culex pipiens. Heredity*, **3**: 357–67.

Owen, A. R. G. (1953). A genetical system admitting of two distinct stable equilibria under natural selection. *Heredity*, **7**: 97–102.

Pagliai, A. (1961). L'endomeiosi in *Toxoptera aurantiae* (Boyer de Foscolombe) (Homoptera, Aphididae). *Rend. Accad. Naz. Lincei* Ser. VIII, **31**: 455–7.

Pagliai, A. (1962). La maturazione dell'uovo partenogenetico e dell'uovo anfigonico in *Brevicoryne brassicae. Caryologia*, **15**: 537–44.

Pai, S. (1949). Sexuelle Stufen und Chromosomengrösse der Keimzellen von *Chirocephalus nankinensis. Exp. Cell Res.* suppl. **1**: 143–5.

Painter, R. H. (1930). Observations on the biology of the hessian fly. *J. Econ. Entomol.* **23**: 326–8.

Painter, T. S. (1914). Spermatogenesis in spiders. *Zool. Jb.* **38**: 509–76.

Painter, T. S. (1922). Studies in mammalian spermatogenesis. I. The spermatogenesis of the Opossum (*Didelphys virginiana*). *J. Exp. Zool.* **35**: 13–38.

Painter, T. S. (1925). A comparative study of the chromosomes of mammals. *Amer. Nat.* **59**: 385–409.

Painter, T. S. (1933). A new method for the study of chromosome rearrangements and the plotting of chromosome maps. *Science*, **78**: 585–6.

Painter, T. S. (1934a). A new method for the study of chromosome aberrations and the plotting of chromosome maps. *Genetics*, **19**: 175–88.

Painter, T. S. (1934b). The morphology of the *X*-chromosome in salivary glands of *Drosophila melanogaster* and a new type of chromosome map for this element. *Genetics*, **19**: 448–69.

Painter, T. S. (1934c). Salivary chromosomes and the attack on the gene. *J. Hered.* **25**: 455–76.

Painter, T. S. (1935). The morphology of the third chromosome in the salivary gland of *Drosophila melanogaster* and a new cytological map of this element. *Genetics*, **20**: 301–26.

Painter, T. S. (1939). The structure of salivary gland chromosomes. *Amer. Nat.* **75**: 315–30.

Painter, T. S. and Biesele, J. J. (1966a). The fine structure of the hypopharyngeal gland cell of the honey bee during development and secretion. *Proc. Nat. Acad. Sci. U.S.A.* **55**: 1414–19.

Painter, T. S. and Biesele, J. J. (1966b). Endomitosis and polyribosome formation. *Proc. Nat. Acad. Sci. U.S.A.* **56**: 1920–5.

PAINTER, T. S. and COLE, L. J. (1943). The genetic sex of pigeon–ring dove hybrids as determined by their sex chromosomes. *J. Morph.* **72**: 411–40.

PAINTER, T. S. and TAYLOR, A. N. (1942). Nucleic acid storage in the toad's egg. *Proc. Nat. Acad. Sci U.S.A.* **28**: 311–17.

PANELIUS, S. (1968). Germ line and oogenesis during paedogenetic reproduction in *Heteropeza pygmaea* Winnertz (Diptera: Cecidomyiidae). *Chromosoma,* **23**: 333–45.

PANELIUS, S. (1971). Male germ-line, spermatogenesis and karyotypes of *Heteropeza pygmaea* Winnertz (Diptera: Cecidomyiidae). *Chromosoma,* **32**: 295–331.

PANITZ, R. (1964). Hormonkontrollierte Genaktivitäten in den Riesenchromosomen von *Acricotopus lucidus. Biol. Zentralbl.* **83**: 197–230.

PANSHIN, I. B. (1935). New evidence for the position effect hypothesis. *C.R. Acad. Sci. U.R.S.S.* N.S. **4**: 85–8.

PARDI, L. (1947). Richerche sui Polistini. VIII. La spermatogenesi di *Polistes gallicus* (L.) e di *Polistes* (*Leptopolistes*) *omissus* (Weyrauch). *Sci. Genet.* **3**: 14–22.

PARDON, J. F. and RICHARDS, B. M. (1972). The molecular structure of nucleohistone. *Chromosomes Today,* **3**: 38–46. London and Edinburgh: Longmans.

PARDUE, M. L. and GALL, J. G. (1970). Chromosomal localization of mouse satellite DNA. *Science,* **168**: 1356–8.

PARDUE, M. L., GERBI, S. A., ECKHARDT, R. A. and GALL, J. G. (1970). Cytological localization of DNA complementary to ribosomal RNA in polytene chromosomes of Diptera. *Chromosoma,* **29**: 268–90.

PARISER, K. (1927). Die Cytologie und Morphologie der Triploiden Intersexe des rückgekreuzten Bastards von *Saturnia pavonia* L. und *S. pyri* Schiff. *Z. Zellforsch.* **5**: 415–47.

PARKES, A. S. (1963). The sex ratio in human populations. In: *Man and his Future* (Ciba Foundation Volume, ed.: G. Wolstenholme). London: Churchill.

PARSHAD, R. (1956). Cytological studies in *Heteroptera.* I. The behavior of chromosomes during the male meiosis of *Ranatra sordidula* with general considerations on the evolution of compound sex chromosomes in Nepidae. *Caryologia,* **8**: 349–60.

PARSHAD, R. (1957). Cytological studies in Heteroptera. VI. Chromosome complement and meiosis in twenty-six species of the Pentatomoidea, Lygaeoidea and Coreoidea with a consideration of the cytological bearing on the status of these super-families. *Res. Bull. Panjab Univ. Zool.* No. 133, pp. 521–59.

PARSHAD, R. (1958). Structure of the heteropteran kinetochore. *Cytologia,* **23**: 25–32.

PARSONS, P. A. (1958). Selection for increased recombination. *Amer. Nat.* **92**: 255–6.

PARSONS, P. A. (1959). Possible affinity between linkage groups V and XIII of the house mouse. *Genetica,* **29**: 304–11.

PASTERNAK, J. (1964). Chromosome polymorphism in the blackfly *Simulium vittatum* (Zett.). *Canad. J. Zool.* **42**: 135–58.

PÄTAU, K. (1934). Ein neuer Fall haploider Parthenogenese. *Naturwiss.* **22**: 648–9.

PÄTAU, K. (1935). Chromosomenmorphologie bei *Drosophila melanogaster* und *Drosophila simulans* und ihre genetische Bedeutung. *Naturwiss.* **23**: 537–43.

PÄTAU, K. (1936). Cytologische Untersuchungen an der haploid-parthenogenetischen Milbe *Pediculoides ventricosus* Newp. Berl. *Zool. Jb.* (*Zool.*) **56**: 277–322.

PÄTAU, K. (1941). Cytologische Nachweis einer positiven Interferenz über das Centromer. *Chromosoma,* **2**: 36–63.

PÄTAU, K. (1948). *X*-segregation and heterochromasy in the spider *Aranea reaumuri*. *Heredity*, **2**: 77–100.

PÄTAU, K., SMITH, D. W., THERMAN, E., INHORN, S. L. and WAGNER, H.P. (1960). Multiple congenital anomaly caused by an extra autosome. *Lancet*, **i**: 790–3.

PÄTAU, K., THERMAN, E., SMITH, D. W. and DE MARS, R. I. (1961). Trisomy for chromosome no. 18 in man, *Chromosoma*, **12**: 280–5.

PATHAK, S. (1970). The karyotype of *Mus playthrix* Bennet (1832), a favourable mammal for cytogenetic investigation. *Mammal. Chrom. Newsl.* **11**: 105.

PATTERSON, J. T. (1928). Sex in the Cynipidae; and male-producing and female-producing lines. *Biol. Bull. Woods Hole*, **54**: 201–11.

PATTERSON, J. T. (1942). *Drosophila* and speciation. *Science*, **95**: 153–9.

PATTERSON, J. T. and CROW, J. F. (1940). Hybridization in the *mulleri* group of *Drosophila*. *Univ. Texas Publ.* **4032**: 251–6.

PATTERSON, J. T. and GRIFFEN, A. B. (1944). A genetic mechanism underlying species isolation. *Univ. Texas Publ.* **4445**: 212–23.

PATTERSON, J. T. and STONE, W. S. (1952). *Evolution in the Genus Drosophila*. New York: Macmillan.

PATTERSON, J. T., STONE, W. S. and BEDICHEK, S. (1937). Further studies on *X*-chromosome balance on *Drosophila*. *Genetics*, **22**: 407–26.

PATTERSON, J. T., STONE, W. S. and GRIFFEN, A. B. (1940). Evolution of the *virilis* group in *Drosophila*. *Univ. Texas Publ.* **4032**: 218–50.

PATTERSON, J. T., STONE, W. S. and GRIFFEN, A. B. (1942). Genetic and cytological analysis of the *virilis* species group. *Univ. Texas Publ.* **4228**: 162–200.

PATTON, J. L. (1969*a*). Chromosome evolution in the pocket mouse, *Perognathus goldmani* Osgood. *Evolution*, **23**: 645–62.

PATTON, J. L. (1969*b*). Karytypic variation in the pocket mouse, *Perognathus penicillatus* Woodhouse (Rodentia–Heteromyidae). *Caryologia*, **22**: 351–8.

PATTON, J. L. and SOULE, O. H. (1967). Natural hybridization in pocket mice, genus *Perognathus*. *Mammal. Chrom. Newsl.* **8** (4): 263–4.

PAULETE-VANDRELL, J. (1970). Reduction of chromosome number in Brazilian *Rattus rattus rattus*. *Mammal. Chrom. Newsl.* **11**: 99–101.

PAVAN, C. (1946). Chromosomal variation in *Drosophila nebulosa*. *Proc. Nat. Acad. Sci. U.S.A.* **32**: 137–45.

PAVAN, C. (1965). Synthesis (of session on chromosome structure). *Proc. XIth Int. Genet. Congr. The Hague*, **2**: 335–42.

PAVAN, C. (1967). Chromosomal changes induced by infective agents. *Triangle*, **8**: 42–8.

PAVAN, C. and BASILE, R. (1966). Chromosome changes induced by infections in tissues of *Rhynchosciara angelae*. *Science*, **151**: 1556–8.

PAVAN, C. and DA CUNHA, A. B. (1968). Chromosome activities in normal and in infected cells of Sciaridae. *The Nucleus: Seminar on Chromosomes*, pp. 183–96.

PAVAN, C. and DA CUNHA, A. B. (1969). Chromosomal activities in *Rhynchosciara* and other Sciaridae. *Ann. Rev. Genet.* **3**: 425–50.

PAVAN, C. and PERONDINI, A. L. P. (1966). Changes in chromosomes induced by infection in cells of *Sciara*. *Genetics*, **54**: 353.

PAVAN, C. and PERONDINI, A. L. P. (1967). Heterozygous puffs and bands in *Sciara ocellaris* Comstock (1882). *Exp. Cell Res.* **48**: 202–5.

PAVAN, C., PERONDINI, A. L. P. and PICARD, T. (1969). Changes in chromosomes and in development of cells of *Sciara ocellaris* induced by microsporidian infections. *Chromosoma*, **28**: 328–45.

PAYNE, F. (1909). Some new types of chromosome distribution and their relation to sex. *Biol. Bull. Woods Hole.* **18**: 174–9.

PAYNE, F. (1910). The chromosomes of *Acholla multispinosa. Biol. Bull. Woods Hole,* **18**: 174–9.

PAYNE, F. (1912*a*). A further study of the chromosomes of the Reduviidae. II. The nucleolus in the young oocytes and the origin of the ova in *Gelastocoris. J. Morph.* **23**: 331–47.

PAYNE, F. (1912*b*). The chromosomes of *Gryllotalpa borealis* Burm. *Arch. Zellforsch.* **9**: 141–8.

PAYNE, F. (1916). A study of the germ cells of *Gryllotalpa borealis* and *Gryllotalpa vulgaris. J. Morph.* **28**: 287–327.

PCHAKADZE, G. (1930). Karyologische Untersuchungen an Trichopteren. *Arkh. Anat. Gist. Embr.* **9**: 227–31.

PEACOCK, A. D. (1940). A possible case of polyploidy in a prosobranch mollusc. *Nature,* **146**: 368.

PEACOCK, A. D. and GRESSON, R. A. R. (1931). Male haploidy and female diploidy in *Sirex cyaneus.* F. (Hymen.) *Proc. Roy. Soc. Edinb.* **51**: 97–103.

PEACOCK, A. D. and SANDERSON, A. R. (1939). The cytology and the thelytokous parthenogenetic sawfly *Thrinax macula. Trans. Roy. Soc. Edinb.* **59**: 647–60.

PEACOCK, A. D. and WEIDMANN, U. (1961). Recent work on the cytology of animal parthenogenesis. *Przeglad Zoologiczny,* **5**: 5–27.

PEACOCK, W. J. (1961). Sub-chromatid structure and chromosome duplication in *Vicia faba. Nature,* **191**, 832–3.

PEACOCK, W. J. (1963). Chromosome duplication and structure as determined by autoradiography. *Proc. Nat. Acad. Sci. U.S.A.* **49**: 793–801.

PEACOCK, W. J. (1965*a*). Chromosome replication. *Int. Symp. Genes and Chromosomes – Structure and Function*: *Nat. Cancer Inst. Monograph,* **18**: 101–2.

PEACOCK, W. J. (1965*b*). Non-random segregation of chromosomes in *Drosophila* males. *Genetics,* **51**: 573–83.

PEACOCK, W. J. (1970). Replication, recombination and chiasmata in *Goniaea australasiae* (Orthoptera: Acrididae). *Genetics,* **65**: 593–617.

PEACOCK, W. J. and ERICKSON, J. (1964). An indicator of polarity in the spermatocyte? *Dros. Inf. Ser.* **39**: 107–8.

PEACOCK, W. J. and ERICKSON, J. (1965). Segregation distortion and regularly nonfunctional products of spermatogenesis in *Drosophila melanogaster. Genetics,* **51**: 313–28.

PEARSON, N. E. (1929). The structure and chromosomes of three gynandromorphic katydids (*Amblycorypha*). *J. Morph.* **47**: 531–47.

PEARSON, P. L. and BOBROW, M. (1970). Definite evidence for the short arm of the *Y*-chromosome associating with the *X*-chromosome during meiosis in the human male. *Nature,* **226**: 959–61.

PEARSON, P. L., BOBROW, M. and VOSA, C. G. (1970). Technique for identifying *Y*-chromosomes in human interphase nuclei. *Nature,* **226**: 78–80.

PEGINGTON, C. J. and REES, H. (1967). Chromosome size in Salmon and Trout. *Chromosoma,* **21**: 475–7.

PEHANI, H. (1925). Die Geschlechtszellen der Phasmiden. Zugleich ein Beitrag zur Fortpflanzungsbiologie der Phasmiden. *Z. wiss. Zool.* **125**: 167–238.

PELLING, C. (1959). Chromosomal synthesis of ribonucleic acid as shown by incorporation of uridine labelled with tritium. *Nature,* **184**: 655–6.

PELLING, C. (1964). Ribonukleinsäure-Synthese der Riesenchromosomen. Autoradiographische Untersuchungen an *Chironomus tentans. Chromosoma,* **15**: 71–122.

PELLING, C. (1966). A replicative and synthetic chromosomal unit – the modern concept of the chromomere. *Proc. Roy. Soc. Lond.* B, **164**: 279–89.

PELLING, C. and BEERMANN, W. (1966). Diversity and variation of the nucleolar organizing regions in chironomids. *Nat. Cancer Inst. Monograph*, **23**: 393–408.

PENNOCK, L. A. (1965). Triploidy in parthenogenetic species of the Teiid Lizard genus *Cnemidophorus*. *Science*, **149**: 539–40.

PENNOCK, L. A., TINKLE, D. W. and SHAW, M. W. (1969). Minute Y-chromosome in the lizard genus *Uta* (family Iguanidae). *Cytogenetics*, **8**: 9–19.

PENROSE, L. S., ELLIS, J. R. and DELHANTY, J. D. A. (1960). Chromosomal translocations in mongolism and in normal relatives. *Lancet*, **ii**: 409–10.

PENROSE, L. S. and STERN, C. (1958). Reconsideration of the Lambert pedigree (Ichthyosis hystrix gravior). *Ann. Hum. Genet.* **22**: 258–83.

PENTZOS-DAPONTE, A. and SPERLICH, D. (1966). Vitalitätsprüfungen an intra- und interlokalen Heterozygoten von *Drosophila subobscura*. *Z. Vererbl.* **97**: 46–54.

PERJE, M. (1948). Studies on the spermatogenesis of *Musca domestica*. *Hereditas*, **34**: 207–32.

PERKOWSKA, E., MACGREGOR, H. C. and BIRNSTIEL, M. L. (1968). Gene amplification in the oocyte nucleus of mutant and wild type *Xenopus laevis*. *Nature*, **217**: 649–50.

PERONDINI, A. L. P. and PERONDINI, D. R. (1966). Sex chromatin in somatic cells of the *Philander opossum quica* (Temminck, 1827), Marsupialia). *Cytogenetics*, **5**: 28–33.

PERREAULT, W. J., KAUFMANN, B. P. and GAY, H. (1968). Similarity in base composition of heterochromatic and euchromatic DNA in *Drosophila melanogaster*. *Genetics*, **60**: 289–301.

PERROT, J. L. (1934). La spermatogénèse et l'ovogénèse du Mallophage *Goniodes stylifer*. *Quart. J. Micr. Sci.* **76**: 353–77.

PESSON, P. (1941). Description du mâle de *Pulvinaria mesembryanthemi* Vallot et observations biologiques sur cette espèce. (Hemipt. Coccidae). *Ann. Soc. Ent. Fr.* **110**: 71–7.

PETRUKHINA, T. E. (1966). Chromosomal polymorphism in the population of flies of the species *Eusimulium latipes latipes* (Diptera, Simuliidae). *Genetica*, **12**: 78–84.

PETZELT, CH. (1970). RNS- und Proteinsynthese im Ablauf der Spermatocytenteilungen von *Pales ferruginea* (Nematocera). *Chromosoma*, **29**: 237–45.

PHILIP, U. (1942). An analysis of chromosomal polymorphism in two species of *Chironomus*. *J. Genet.* **44**: 129–42.

PICCINI, E. and STELLA, M. ã1970). Some avian karyograms. *Caryologia*, **23**: 189–202.

PIJNACKER, L. P. (1966). The maturation divisions of the parthenogenetic stick insect *Carausius morosus* Br. (Orthoptera, Phasmidae). *Chromosoma*, **19**: 99–112.

PIJNACKER, L. P. (1968). Oogenesis in the parthenogenetic stick insect *Sipyloidea sipylus* Westwood (Orthoptera, Phasmidae). *Genetica*, **38**: 504–15.

PIJNACKER, L. P. (1969). Automictic parthenogenesis in the stick insect *Bacillus rossius* Rossi (Cheleutoptera, Phasmidae). *Genetica*, **40**: 393–9.

PIZA, S. DE T. (1939). Comportamento dos cromossômios na primeira divisão do espermatocito do *Tityus bahiensis*. *Sci. Genet.* **1**: 255–61.

PIZA, S. DE T. (1940). Poliploidia natural em *Tityus bahiensis* (Scorpiones) associada a aberraçoes cromossômicas espontaneas. *Rev. Biol. Hyg. S. Paulo*, **10**: 143–55.

PIZA, S. DE T. (1943a). Meiosis in the male of the Brazilian scorpion, *Tityus bahiensis*. *Rev. Agric. S. Paulo*, **18**: 249–76.

PIZA, S. DE T. (1943b). Cromossômios em Proscopiidae (Orthoptera–Acridoidea). *Rev. Agric. S. Paulo*, **18**: 347–50.

PIZA, S. DE T. (1943c). A propósito de meiose do *Tityus bahiensis*. *Rev. Agric. S. Paulo*, **18**: 351–69.

PIZA, S. DE T. (1944). A case of spontaneous end-to-end permanent union of two non-homologous chromosomes in the Brazilian scorpion *Tityus bahiensis*, accompanied by irregularities in pairing. *Rev. Agric. S. Paulo*, **19**: 133–47.

PIZA, S. DE T. (1945). Comportamento do heterocromossômio em alguns ortópteros do Brazil. *An. Esc. Sup. Agric. L. de Queiroz*, **2**: 174–207.

PIZA, S. DE T. (1946a). Uma nova modalidade de sexo-determinacão no grilo sul-americano *Eneoptera surinamensis*. *An. Esc. Agric. L. de Queiroz*, **3**: 339–46.

PIZA, S. DE T. (1946b). Soldadura par uma das extremidades de dois cromossômios homólogos do *Tityus*. *An. Esc. Agric. L. de Queiroz*, **3**: 339–46.

PIZA, S. DE T. (1947a). Interessante comportamento dos cromossômios na espermatogenese do escorpião *Isometrus maculatus* de Geer. *An. Esc. Sup. Agric. L. de Queiroz*, **4**: 177–82.

PIZA, S. DE T. (1947b). Uma raca cromossômica natural de *Tityus bahiensis* (Scorpiones–Buthidae). *An. Esc. Sup. Agric. L. de Queiroz*, **4**: 183–92.

PIZA. S. DE T. (1947c). Notas sôbre cromossômios de alguns escorpiões brasileiros. *An. Esc. Sup. Agric. L. de Queiroz*, **3**: 169–76.

PIZA, S. DE T. (1947d). Breve noticia sobre a espermatogenese de *Lutosa brasiliensis* Brunner (Tettigonoidea–Stenopelmatidae). *An. Esc. Sup. Agric. L. de Queiroz*, **4**: 202–8.

PIZA, S. DE T. (1947e). Cromossômios de *Dysdercus* (Hemiptera–Pyrrhocoridae). *An. Esc. Sup. Agric. L. de Queiroz*, **4**: 209–16.

PIZA, S. DE T. (1948a). Primeiras observaçoes sôbre os cromossômios de *Tityus trivittatus* Krpln. (Scorpiones–Buthidae). *Rev. Agric. S. Paulo*, **24**: 177–80.

PIZA, S. DE T. (1948b). Uma nova raça cromossômica natural de *Tityus bahiensis* (Scorpiones–Buthidae). *Rev. Agric. S. Paulo*, **24**: 181–6.

PIZA, S. DE T. (1948c). Variações cromossômicas do *Tityus bahiensis* de São Joaquim. *Rev. Agric. S. Paulo*, **24**: 187–94.

PIZA, S. DE T. (1949). 'Ouro Prêto', nova e interessante raça cromossômica de *Tityus bahiensis* (Scorpiones–Buthidae). *Ciência e Cultura*, **2**: 57–9.

PIZA, S. DE T. (1950a). Variações cromossômicas do *Tityus bahiensis* de Ribeirão Preto. *Ciência e Cultura*, **2**: 57–9.

PIZA, S. DE T. (1950b). Observações cromossômicas em escorpiões brasileiros. *Ciência e Cultura*, **2**: 202–6.

PIZA, S. DE T. (1951). Interpretacão do typo sexual de *Dysdercus mendesi* Bloete (Hemiptera–Pyrrhocoridae) *Genet. Iberica*, **3**: 107–12.

PIZA, S. DE T. (1952). Primeiras observaçoes sôbre os cromossômios do *Tityus metuendus* Pocock (Scorpiones, Buthidae). *Sci. Genet.* **4**: 162–7.

PIZA, S. DE T. (1957a). The chromosomes of *Rhopalurus* (Scorpiones–Buthidae). *Canad. Entomol.* **89**: 565–8.

PIZA, S. DE T. (1957b). Comportamento dos cromossômios na espermatogenese de *Microtomus conspicillaris* (Drury). *Rev. Agric.* **32**: 53–64.

PIZA, S. DE T. (1958). Normally dicentric insect chromosomes. *Proc. Xth. Int. Congr. Entomol.* **2**: 945–51.

PLAUT, W. (1968). On DNA replication in polytene chromosomes. In: *Replication and Recombination of Genetic Material*, pp. 87–92. Canberra: Aust. Acad. of Science.

PLAUT, W. and NASH, D. (1964). Localized DNA synthesis in polytene chromosomes and its implications. In: *The Role of Chromosomes in Development* (ed.: M. Locke), pp. 113–35. New York: Academic Press.

POLANI, P. E., BRIGGS, J. H., FORD, C. E., CLARKE, C. M. and BERG, J. M. (1960). A mongol girl with 46 chromosomes. *Lancet*, **i**: 721–4.

POLICANSKY, D. and ELLISON, J. (1970). 'Sex ratio' in *Drosophila pseudoobscura*: spermiogenic failure. *Science*, **169**: 886–7.

POLLISTER, A. W. (1939). Centrioles and chromosomes in the atypical spermatogenesis of *Vivipara*. *Proc. Nat. Acad. Sci. U.S.A.* **25**: 189–95.

POLLISTER, A. W. and POLLISTER, P. F. (1943). The relation between centriole and centromere in atypical spermatogenesis of Viviparid snails. *Ann. N.Y. Acad. Sci.* **45**: 1–48.

PONSE, K. (1942). Sur la digamétie du crapaud hermaphrodite. *Rev. Suisse Zool.* **49**: 185–9.

POOLE, H. K. (1959). The mitotic chromosomes of parthenogenetic and normal turkeys. *J. Hered.* **50**: 151–4.

POOLE, H. K. and OLSEN, M. W. (1957). The sex of parthenogenetic turkey embryos. *J. Hered.* **48**: 217–18.

PORTER, D. L. (1971). Oogenesis and chromosomal heterozygosity in the thelytokous midge, *Lundstoemia parthenogenetica* (Diptera, Chironomidae). *Chromosoma*, **32**: 333–42.

POTTER, I. C. and ROTHWELL, B. (1970). The mitotic chromosomes of the Lamprey, *Petromyzon marinus* L. *Experientia*, **26**: 429–30.

POULSON, D. F. and METZ, C. W. (1938). Studies on the structure of nucleolus forming regions and related structures in the giant salivary gland chromosomes of Diptera. *J. Morph.* **63**: 363–95.

POWERS, P. B. A. (1942). Metrical studies on spermatogonial chromosomes of Acrididae (Orthoptera). *J. Morph.* **71**: 523–76.

PRABHOO, N. R. (1961). A note on the giant chromosomes in the salivary glands of *Womersleya* sp. (Collembola: Insecta). *Bull. Entomol. Madras*, **2**: 21–2.

PRESCOTT, D. M. (1962). Nucleic acid and protein metabolism in the macronuclei of two ciliated Protozoa. *J. Histochem. Cytochem.* **10**: 145–53.

PRESCOTT, D. M. (1966). The synthesis of total macronuclear protein, histone and DNA during the cell cycle in *Euplotes eurystomus*. *J. Cell Biol.* **31**: 1–9.

PRESCOTT, D. and BENDER, M. A. (1963). Autoradiographic study of chromatid distribution of labelled DNA in two types of mammalian cells *in vitro*. *Exp. Cell Res.* **29**: 430–42.

PRESCOTT, D. M. and KIMBALL, R. F. (1961). Relation between RNA, DNA and protein synthesis in the replicating nucleus of *Euplotes*. *Proc. Nat. Acad. Sci. U.S.A.* **47**: 686–93.

PREVOSTI, A. (1964). Chromosomal polymorphism in *Drosophila subobscura* populations from Barcelona (Spain). *Genet. Res.* **5**: 27–38.

PROKOFIEVA, A. A. (1934). On the chromosome morphology of certain Pisces. *Cytologia*, **5**: 498–506.

PROKOFIEVA, A. A. (1935*a*). Morphologische Struktur der Chromosomen von *Drosophila melanogaster*. *Z. Zellforsch.* **22**: 225–62.

PROKOFIEVA, A. A. (1935*b*). The structure of the chromocenter. *Cytologia*, **6**: 438–43.

PROKOFIEVA, A. A. (1935*c*). The structure of the chromocenter. *Dokl. Akad. Nauk. U.S.S.R.* N.S. **2**: 498–9.

PROKOFIEVA-BELGOVSKAYA, A. A. (1937*a*). The structure of the *Y*-chromosome in the salivary glands of *Drosophila melanogaster*. *Genetics*, **22**: 94–103.

PROKOFIEVA-BELGOVSKAYA, A. A. (1937b). Inert regions in the distal ends of chromosomes of *Drosophila melanogaster*. *Bull. Acad. Sci. U.R.S.S.* Ser. Biol., **3**: 719–24.

PROKOFIEVA-BELGOVSKAYA, A. A. (1938). The inert region in the subterminal part of the *X*-chromosome of *Drosophila melanogaster*. *Bull Acad. Sci. U.R.S.S.* Ser. Biol. pp. 97–103.

PROKOFIEVA-BELGOVSKAYA, A. A. (1941). Cytological properties of inert regions and their bearing on the mechanisms of mosaicism and chromosome rearrangement. *Dros. Inf. Serv.* **15**: 35–6.

PROKOFIEVA-BELGOVSKAYA, A. A. (1947). Heterochromatization as a change of chromosome cycle. *J. Genet.* **48**: 80–98.

PUSSARD-RADULESCO, E. (1930). Recherches biologiques et cytologiques sur quelques Thysanoptères. *Ann. Epiphyties*, **16**: 103–88.

PUTMAN, W. L. (1939). The plum nursery mite (*Phyllocoptes fockeui* Nal. and Trt.) *Ann. Rep. Ent. Soc. Ontario*, **70**: 33–9.

PUTTANNA, C. R. (1959). Cytological studies of Indian chilopods. I. The chromosomes of some species of Scolopendridae (Myriapoda: Chilopoda). *Chromosoma*, **10**: 179–83.

RABELLO, M. N. (1970). Chromosomal studies in Brazilian Anurans. *Caryologia*, **23**: 45–59.

RAI, K. S. (1963). A comparative study of mosquito karyotypes. *Ann. Entomol. Soc. Amer.* **56**: 160–9.

RAI, K. S. and CRAIG, G. B. (1961). A study of the karyotypes of some mosquitoes. *Genetics*, **46**: 891.

RAICU, P., BRATOSIN, S. and HAMAR, M. (1968). Study on the karyotype of *Spalax leucodon* Nordm. and *S. micróphthalmus* Guld. *Caryologia*, **21**: 127–35.

RAIKOV, I. B. (1969). The macronucleus of Ciliates. In: *Research in Protozoology*, **3** (ed.: T. T. Chen), pp. 1–128. Oxford: Pergamon Press.

RAMBOUSEK, F. (1912). Cytologische Verhältnisse der Speicheldrüsen der *Chironomus*-Larve. *Sitzungsb. Königl. Böhm. Ges. Wiss., Math-Naturw. Kl.* (cited from Beermann, 1962).

RANDOLPH, L. F. (1941). Genetic characteristics of the *B*-chromosomes in maize. *Genetics*, **26**: 608–31.

RAO, M. V. N. and PRESCOTT, D. M. (1967). Micronuclear RNA synthesis in *Paramoecium caudatum*. *J. Cell Biol.* **33**: 281–5.

RAO, S. R. V. (1957). Notes on the chromosome structure of *Eurybrachis*. *Genet. Iberica*, **9**: 283–92.

RAO, S. R. V. (1958). The kinetochore problem in Hemiptera. *Curr. Sci.* **27**: 303–4.

RAO, T. R. (1934). Chromosomal aberrations occurring in unirradiated grasshoppers. *Half-yearly J. Mysore Univ.* **7**: 1–12.

RAO, T. R. (1937). A comparative study of the chromosomes of eight genera of Indian Pyrgomorphinae (Acrididae). *J. Morph.* **61**: 223–55.

RASCH, E., DARNELL, R. M., KALLMAN, K. D. and ABRAMOFF, P. (1965). Cytophotometric evidence for triploidy in hybrids of the gynogenetic fish, *Poecilia formosa. J. Exp. Zool.* **160**: 155–70.

RASCH, E. M., PREHN, L. M. and RASCH, R. W. (1970). Cytogenetic studies of *Poecilia* (Pisces). II. Triploidy and DNA levels in naturally occurring populations associated with the gynogenetic teleost *Poecilia formosa* (Girard). *Chromosoma*, **31**: 18–40.

RAUSCH, R. L. and RAUSCH, V. R. (1965). Cytogenetic evidence for the specific distribution of an Alaskan marmot, *Marmota broweri* Hall and Gilmore (Mammalia: Sciuridae). *Chromosoma*, **16**: 618–23.

RAYBURN, M. F. (1917). Chromosomes of *Nomotettix*. *Kansas Univ. Sci. Bull.* **10**: 267–70.

RAY-CHAUDHURI, S. P. and BOSE, I. (1948). Meiosis in diploid and tetraploid spermatocytes of *Atractomorpha* sp. (Acrididae) *Proc. Zool. Soc. Bengal.* **1**: 1–12.

RAY-CHAUDHURI, S. P. and GUHA, A. (1952). *X* and neo-*Y* mechanism of sex determination in the grasshopper *Thisiocetrus pulcher*. *Nature*, **169**: 78–9.

RAY-CHAUDHURI, S. P. and MANNA, G. K. (1951). Chromosome evolution in wild populations of Acrididae. I. On the occurrence of a supernumerary chromosome in *Aiolopus* sp. *Proc. Ind. Acad. Sci.* **34**: 55–61.

RAY-CHAUDHURI, S. P. and MANNA, G. K. (1952). A new type of segregation of the sex chromosomes in the meiotic divisions of the cotton stainer, *Dysdercus koenigii* (Fabr.) *J. Genet.* **51**: 191–7.

RAY-CHAUDHURI, S. P. and MANNA, G. K. (1955). Evidence of a multiple sex-chromosome mechanism in a pyrrhocorid bug, *Physopelta schlanbuschi* Fabr. *Proc. Zool. Soc. Calcutta*, **8**: 65–77.

RAY-CHAUDHURI, S. P., SINGH, L. and SHARMA, T. (1972). The evolution of sex chromosomes and the formation of W chromatin in snakes. *Chromosomes Today*, **3**: 298. London and Edinburgh: Longmans.

RAYLE, R. E. and GREEN, M. M. (1968). A contribution to the genetic fine structure of the region adjacent to *white* in *Drosophila melanogaster*. *Genetica*, **39**: 497–507.

REED, T. E., SIMPSON, N. E. and CHOWN, B. (1963). The Lyon hypothesis. *Lancet*, **ii**: 467–8.

REES, H. (1964). The question of polyploidy in the Salmonidae. *Chromosoma*, **15**: 275–9.

REES, H. and EVANS, G. M. (1966). A correlation between the localization of chiasmata and the replication pattern of chromosomal DNA. *Exp. Cell Res.* **44**: 161–4.

REES, H. and JAMIESON, A. (1954). A supernumerary chromosome in *Locusta*. *Nature*, **173**: 43.

REGNART, H. C. (1933). Additions to our knowledge of chromosome numbers in the Lepidoptera. *Proc. Univ. Durham Phil. Soc.* **9**: 79–83.

REHN, J. A. G. (1948). The acridoid family Eumastacidae (Orthoptera). A review of our knowledge of its components, features and systematics, with a suggested new classification of its major groups. *Proc. Acad. Nat. Sci. Philad.* **100**: 77–139.

REIG, O. A. and BIANCHI, N. O. (1969). The occurrence of an intermediate didelphid karyotype in the short-tailed opossum (genus *Monodelphis*) *Experientia* **25**: 1210–11.

REIG, O. A. and KIBLISKY, P. (1969). Chromosome multiformity in the genus *Ctenomys* (Rodentia, Octodontidae). *Chromosma*, **28**: 211–44.

REISINGER, E. (1940). Die cytologische Grundlage der parthenogenetischen Dioogonie. *Chromosoma*, **1**: 531–53.

REISMAN, L. E., KASAHARA, S., CUNG, C. Y., DARNELL, A. and HALL, B. (1966). Anti-mongolism. Studies in an infant with a partial monosomy of the 21 chromosome. *Lancet*, **ii**: 394–5.

REITALU, J. (1970). Observations on the behavioral pattern of the sex chromosome complex during spermatogenesis in man. *Hereditas*, **64**: 283–90.

REITBERGER, A. (1939). Die Cytologie des pädogenetischen Entwicklungszyklus der Gallmücke *Oligarces paradoxus* Mein. *Chromosoma*, **1**: 391–473.

RENDEL, J. M. and KELLERMAN, G. M. (1955). Deoxyribonucleic acid content of marsupial cells. *Nature*, **176**: 829.

REVELL, S. H. (1947). Controlled *X*-segregation at meiosis in *Tegenaria*, *Heredity*, **1**: 337–47.

REVELL, S. H. (1955). A new hypothesis for 'chromatid' changes. *Proc. Radiobiol. Symp. Liège* 1954 (ed.: Z. M. Bacq and P. Alexander). London: Butterworth.

REVELL, S. H. (1959). The accurate estimation of chromatid breakage and its relevance to a new interpretation of chromatid aberrations induced by ionizing radiations. *Proc. Roy. Soc. Lond.* B, **150**: 563–89.

REVELL, S. H. (1963). Chromatid aberrations – the generalized theory. In: *Radiation-induced Chromosome Aberrations*. (*ed.*: S. Wolfe). pp. 41–72. Columbia Univ. Press.

REVELL, S. H. (1966). Evidence for a dose-squared term in the dose-response curve for real chromatid discontinuities induced by X-rays and some theoretical consequences thereof. *Mutation Res.* **3**: 34–53.

REVERBERI, G. and PITOTTI, M. (1942). Il ciclo biologico e la determinazione fenotipica del sesso di *Ione thoracica* Montagu, Bopiride parassita di *Calianassa laticauda* Otto. *Pubbl. Staz. Zool. Napoli*, **19**: 111–84.

REYNOLDS, J. P. (1938). Sex determination in a 'bisexual' strain of *Sciara coprophila*. *Genetics*, **23**: 203–20.

RHEIN, A. (1935). Diploide parthenogenesis bei *Hydrobia jenkinsi* Smith (Prosobranchia). *Naturwiss.* **23**: 100.

RHOADES, M. M. (1938). On the origin of a secondary trisome through the doubling of a half-chromosome fragment. *Genetics*, **23**: 163–4.

RHOADES, M. M. (1940). Studies of a telocentric chromosome in maize with reference to the stability of its centromere. *Genetics*, **25**: 483–520.

RHOADES, M. M. (1950). Meiosis in maize. *J. Hered.* **41**: 58–67.

RHOADES, M. M. (1961). Meiosis. In: *The Cell*, **3** (ed.: J. Brachet and A. E. Mirsky). New York: Academic Press.

RHOADES, M. M. and KERR, W. E. (1949). A note on centromere organization. *Proc. Nat. Acad. Sci. U.S.A.* **35**: 129–32.

RHOADES, M. M. and VILKOMERSON, H. (1942). On the anaphase movement of chromosomes. *Proc. Nat. Acad. Sci. U.S.A.* **28**: 433–6.

RIBBANDS, C. R. (1941). Meiosis in Diptera. I. Prophase associations of non-homologous chromosomes and their relation to mutual attraction between centromeres, centrosomes and chromosome ends. *J. Genet.* **41**: 411–42.

RIBBERT, D. (1967). Die Polytänchromosomen der Borstenbildungszellen von *Calliphora erythrocephala*. Unter besonderer Berücksichtigung der geschlechtsgebundenen Strukturheterozygotie und des Puffmusters während der Metamorphose. *Chromosoma*, **21**: 296–344.

RIBBERT, D. and BIER, K. (1969). Multiple nucleoli and enhanced nucleolar activity in the nurse cells of the insect ovary. *Chromosoma*, **27**: 178–97.

RICHTER, H. and HÜNDLER, M. (1955). Die geographische Verbreitung der chromomalen Strukturtypen von *Drosophila subobscura* Coll. *Z. indukt. Abst. Vererbl.* **87**: 85–92.

RILEY, R. (1958). Chromosome pairing and haploids in wheat. *Proc. 10th Int. Genet. Congr.* **2**: 234–5.

RILEY, R. (1960). The diploidization of polyploid wheat. *Heredity*, **15**: 407–29.

RILEY, R. (1966). The genetic regulation of meiotic behaviour in wheat and its relatives. *Proc. 2nd Int. Wheat Genet. Symp.* **2**: 395–408.

RILEY, R. and LAW, C. N. (1965). Genetic variation in chromosome pairing. *Adv. in Genet.* **13**: 57–114.

RINGERTZ, N. R., ERICSSON, J. L. E. and NILSSON, O. (1967). Macronuclear chromatin structure in *Euplotes*. *Exp. Cell Res.* **48**: 97–117.

RINGERTZ, N. R. and HOSKINS, G. C. (1965). Cytochemistry of macronuclear reorganization. *Exp. Cell Res.* **38**: 160–79.

RIS, H. (1942). A cytological and experimental analysis of the meiotic behavior of the univalent X-chromosome in the bearberry aphid *Tamalia* (=*Phyllaphis*) *coweni* (Ckll.) *J. Exp. Zool.* **90**: 267–326.

RIS, H. (1945). The structure of meiotic chromosomes in the grasshopper and its bearing on the nature of 'chromosomes' and 'lampbrush' chromosomes. *Biol. Bull. Woods Hole*, **89**: 242–57.

RIS, H. (1956). A study of chromosomes with the electron microscope. *J. Biophys. Biochem. Cytol.* **2**, suppl.: 385–92.

RIS, H. (1957). Chromosome structure. *Symp. Chem. Basis of Hered.*, pp. 23–62. Baltimore: Johns Hopkins Press.

RIS, H. (1966). Fine structure of chromosomes. *Proc. Roy. Soc. Lond.* B, **164**: 246–57.

RIS, H. (1967). Ultrastructure of the animal chromosome. In: *Regulation of Nucleic Acid and Protein Biosynthesis*, pp. 11–21. Amsterdam: Elsevier.

RIS, H. and KERR, W. E. (1952). Sex determination in the honey-bee. *Evolution*, **6**: 444–5.

RIS, H. and KLEINFELD, R. (1952). Cytochemical studies on the chromatin elimination in *Solenobia* (Lepidoptera). *Chromosoma*, **5**: 363–71.

RIS, H. and MIRSKY, A. E. (1949). Quantitative cytochemical determination of desoxyribonucleic acid with the Feulgen nucleal reaction. *J. Gen. Physiol.* **33**: 125–46.

RISHIKESH, N. (1959*a*). Chromosome behavior during spermatogenesis of *Anopheles stephensi sensu stricto*. *Cytologia*, **24**: 447–58.

RISHIKESH, N. (1959*b*). Morphology and development of the salivary glands and their chromosomes in the larvae of *Anopheles stephensi sensu stricto*. *Bull. World Health Org.* **20**: 47–61.

RISLER, H. (1954). Die Somatische Polyploidie in der Entwicklung der Honigbiene (*Apis mellifica* L.) und die Wiederherstellung der Diploidie bei den Drohnen. *Z. Zellforsch.* **41**: 1–78.

RISLER, H. and KEMPTER, E. (1962). Die Haploidie der Männchen und die Endopolyploidie in einigen Geweben von *Haplothrips* (Thysanoptera). *Chromosoma*, **12**: 351–61.

RITOSSA, A. (1962). A new puffing pattern induced by temperature shock and DNP in *Drosophila*. *Experientia*, **18**: 571.

RITOSSA, F. M., ATWOOD, K. C., LINDSLEY, D. L. and SPIEGELMAN, S. (1966). On the chromosomal distribution of DNA complementary to ribosomal and soluble RNA. *Nat. Cancer Inst. Monograph*, **23**: 449–72.

RITOSSA, F. M. and SPIEGELMAN, S. (1965). Localization of DNA complementary to ribosomal RNA in the nucleolus organizer region of *Drosophila melanogaster*. *Proc. Nat. Acad. Sci. U.S.A.* **53**: 737–45.

ROBERTS, B. (1968). DNA granule synthesis in the giant foot pad nuclei of *Sarcophaga bullata*. *J. Cell Biol.* **39**: 112A.

Roberts, P. A., Kimball, R. F. and Pavan, C. (1967). Response of *Rhynchosciara* chromosome to a microsporidian infection: increased polyteny and generalized puffing. *Exp. Cell. Res.* **47**: 408–22.

Robertson, J. G. (1964). Effect of supernumerary chromosomes on sex ratio in *Calligrapha philadelphica* L. (Coleoptera: Chrysomelidae). *Nature,* **204**: 605.

Robertson, W. R. B. (1916). Chromosome studies. I. Taxonomic relationships shown in the chromosomes of Tettigidae and Acrididae. V-shaped chromosomes and their significance in Acrididae, Locustidae and Gryllidae: chromosomes and variation. *J. Morph.* **27**: 179–331.

Robertson, W. R. B. (1917). A deficient supernumerary accessory chromosome in a male of *Tettigidea parvipennis. Kansas Univ. Sci. Bull.* **10**: 275–83.

Robertson, W. R. B. (1930). Chromosome studies, V. Diploidy and persistant chromosome relations in partheno-produced Tettigidae (*Apotettix eurycephalus* and *Paratettix texanus*). *J. Morph.* **50**: 209–57.

Robertson, W. R. B. (1931). On the origin of partheno-produced males in Tettigidae (*Apotettix* and *Paratettix.*) *Genetics,* **16**: 353–6.

Robinson, R. (1958). Genetic independence of four mutants in the Syrian hamster. *Nature,* **183**: 125–6.

Rocchi, A. (1967). Sulla presenza di cromosomi soprannumerari in una popolazione di *Asellus coxalis. Caryologia,* **20**: 107–13.

Rodman, T. C. (1967). The larval characteristics and salivary gland chromosomes of a tumorigenic strain of *Drosophila melanogaster. J. Morph.* **115**: 419–46.

Rohde, R. A. (1965). Congenital chromosomal syndromes. A model for parthogenesis. *Calif. Med.* **103**: 249–53.

Roman, H. (1947*a*). Mitotic non-disjunction in the case of interchanges involving the *B*-type chromosome in maize. *Genetics,* **32**: 391–409.

Roman, H. (1947*b*). Directed fertilization in maize. *Proc. Nat. Acad. Sci. U.S.A.* **34**: 46–52.

Roman, H. and Ullstrup, A. J. (1951). The use of A-B translocations to locate genes in maize. *Agron. J.* **43**: 450–4.

Ross, E. S. (1944). A revision of the Embioptera or web-spinners of the new world. *Proc. U.S. Nat. Mus.* **94**: 401–504.

Ross, E. S. (1961). Parthenogenetic African Embioptera. *Wasmann. J. Biol.* **18**: 297–304.

Roth, L. E. and Daniels, E. W. (1963). Electron microscope studies of mitosis in amebae. I. The giant ameba *Pelomyxa carolinensis. J. Cell Biol.* **12**: 57–78.

Roth, L. M. (1970). Evolution and taxonomic significance of reproduction in Blattaria. *Ann. Rev. Entomol.* **15**: 75–96.

Roth, L. M. and Cohen, S. H. (1968). Chromosomes of the *Pycnoscelus indicus* and *P. surinamensis* complex. *Psyche,* **75**: 54–76.

Roth, L. M. and Willis, E. R. (1956). Parthenogenesis in cockroaches. *Ann. Entomol. Soc. Amer.* **49**: 195–204.

Roth, L. M. and Willis, E. R. (1961). A study of bisexual and parthenogenetic strains of *Pycnocelus surinamensis* (Blattaria: Epilamprinae). *Ann. Entomol. Soc. Amer.* **54**: 12–25.

Roth, T. F. (1966). Changes in the synaptenemal complex during meiotic prophase in mosquito oocytes. *Protoplasma,* **61**: 346–86.

Rothenbuhler, W. C. (1957). Diploid male tissue as new evidence on sex determination in honey bees. *J. Hered.* **48**: 160–8.

ROTHENBUHLER, W. C., GOWEN, J. W. and PARK, O. W. (1951). Androgenesis in gynandromorphic honeybees (*Apis mellifera* L.). *Genetics*, **36**: 573.

ROTHFELS, K. H. (1950). Chromosome complement, polyploidy and supernumeraries in *Neopodismopsis abdominalis* (Acrididae). *J. Morph.* **87**: 287–316.

ROTHFELS, K. H. (1956). Black flies: siblings, sex and species grouping. *J. Hered.* **47**: 113–22.

ROTHFELS, K., ASPDEN, M. and MOLLISON, M. (1963). The *W*-chromosome of the Budgerigar *Melopsittacus undulatus. Chromosoma*, **14**: 459–67.

ROTHFELS, K. H. and BASRUR, P. K. (1960). The interrelations of *Prosimulium hirtipes* (Fries) and allied European and North American species (Diptera: Simuliidae). A salivary gland chromosome study. *Verh. XI. Int. Kongr. Entomol. Wien.* **1**: 17–25.

ROTHFELS, K. H. and DUNBAR, R. W. (1953). The salivary gland chromosomes of the black fly *Simulium vittatum* Zett. *Canad. J. Zool.* **31**: 226–41.

ROTHFELS, K. H. and FAIRLIE, T. W. (1957). The non-random distribution of inversion breaks in the midge *Tendipes decorus. Canad. J. Zool.* **35**: 221–63.

ROTHFELS, K. and FREEMAN, M. (1966). The salivary gland chromosomes of three North American species of *Twinnia* (Diptera: Simuliidae). *Canad. J. Zool.* **44**: 937–45.

RUBENSTEIN, I., THOMAS, C. A. Jr. and HERSHEY, A. D. (1961). The molecular weights of T2 bacteriophage DNA and its first and second breakage products. *Proc. Nat. Acad. Sci. U.S.A.* **47**: 1113–22.

RÜCKERT, J. (1892). Zur Entwicklungsgeschichte des Ovarialeies bei Selachiern. *Anat. Anz.* **7**: 107–158.

RUDKIN, G. T. (1964). The proteins of polytene chromosmes. *The Nucleohistones* (ed.: Bonner and Ts'o), pp. 184–92. San Francisco: Holden Day.

RUDKIN, G. T. (1965*a*). The structure and function of heterochromatin. *Proc. XIth Int. Genet. Congr. The Hague*, **2**: 359–74.

RUDKIN, G. T. (1965*b*). Nonreplicating DNA in giant chromosomes. *Genetics*, **52**: 470.

RUDKIN, G. T. and CORLETTE, S. L. (1957). Disproportionate synthesis of DNA in a polytene chromosome region. *Proc. Nat. Acad. Sci. U.S.A.* **43**: 964–8.

RUEBUSH, T. K. (1937). The genus *Dalyellia* in America. *Zool. Anz.* **119**: 237–56.

RUEBUSH, T. K. (1938). A comparative study of Turbellarian chromosomes. *Zool. Anz.* **122**: 321–9.

RUMPLER, Y. (1970). Étude cytogénétique du *Lemur catta. Cytogenetics*, **9**: 239–44.

RUMPLER, Y. and ALBIGNAC, R. (1969). Existence d'une variabilité chromosomique intraspécifique chez certains Lémuriens. *C.R. Soc. Biol. Paris*, **163**: 1989–92.

RÜSCH, M. E. (1960). Untersuchungen über Geschlechtsbestimmungsmechanismen bei Copepoden. *Chromosoma*, **11**: 419–32.

RUSSELL, L. B. (1961). Genetics of mammalian sex chromosomes. *Science*, **133**: 1795–803.

RUSSELL, L. B. (1963). Mammalian *X*-chromosome action: inactivation limited in spread and in region of origin. *Science*, **140**: 976–8.

RUSSELL, L. B. (1964*a*). Genetic and functional mosaicism in the mouse. In: *Role of chromosomes in Development* (ed.: M. Locke), pp. 153–81. New York: Academic Press.

RUSSELL, L. B. (1964*b*). Another look at the single-active-*X* hypothesis. *Trans. N.Y. Acad. Sci.* Ser. II. **26**: 726–36.

RUSSELL, L. B. and BANGHAM, J. W. (1959). Variegated-type position effects in the mouse. *Genetics*, **44**: 532.

RUSSELL, L. B. and CHU, E. H. Y. (1961). An *XXY* male in the mouse. *Proc. Nat. Acad. Sci. U.S.A.* **47**: 571–5.

RUTISHAUSER, A. (1960). Telocentric fragment chromosomes in *Trillium grandiflorum. Heredity,* **15**: 241–6.

SACHS, L. (1952). Polyploid evolution and mammalian chromosomes. *Heredity,* **6**: 357–64.

SAEZ, F. A. (1956). Caso extraordinario de un ortóptero acridido con ocho cromosomas diploides y mecanismo sexual *XY. Biologica, Santiago–Chile,* **22**: 27–30.

SAEZ, F. A. (1957). An extreme karyotype in an orthopteran insect. *Amer. Nat.* **91**: 259–64.

SAEZ, F. A. (1963). Gradient of the heterochromatinization in the evolution of the sexual system 'neo-*X*–neo-*Y*'. *Port. Acta Biol.* A, **7**: 111–38.

SAEZ, F. A. and BRUM, N. (1959). Citogenética de anfibios anuros de América del Sur. *Anales Fac. Med. Montevideo,* **44**: 414–23.

SAEZ, F. A. and BRUM, N. (1960). Chromosomes of South American amphibians. *Nature,* **175**: 945.

SAEZ, F. A. and BRUM-ZORILLA, N. (1966). Karyotype variation in some species of the genus *Odontophrynus* (Amphibia–Anura). *Caryologia,* **19**: 55–63.

SAEZ, F. A. and DIAZ, A. (1958). Sistema sexual neo-*X* and -*Y* en *Xyleus laevipes* (Orthoptera, Romaleinae). *Arch. Soc. Biol. Montevideo* **23**: 13–27.

SAEZ, F. A. and DIAZ, A. (1960). Neo-*X* neo-*Y* system of sex determination in *Xyleus laevipes* (Orthoptera: Romaleinae).

SAEZ, F. A., ROJAS, P. and DE ROBERTIS, E. (1935). Le problème des chromosomes sexuels des amphibiens. *C.R. Soc. Biol. Paris,* **120**: 368.

SAEZ, F. A., ROJAS, P., and DE ROBERTIS, E. (1936). Untersuchungen über die Geschlechtszellen der Amphibien (Anuren). Der meiotische Prozess bei *Bufo arenarum. Z. Zellforsch.* **24**: 725–77.

SAEZ, F. A. and ZORILLA, B. DE N. (1963). Cytogenetics of South American amphibians. *Proc. Xth Int. Genet. Congr.* **1**: 141.

SALMON, J. T. (1955). Parthenogenesis in New Zealand stick insects. *Trans. Roy. Soc. New Zealand,* **82**: 1189–92.

SALOMON, R., KAYE, A. M. and HERZBERG, M. (1969). Mouse nuclear satellite DNA 5-methylcytosine content, pyrimidine isoplith distribution and electron microscopic appearance. *J. Mol. Biol.* **43**: 581–92.

SALTET, P. (1959). La formule chromosomique de *Dolichopoda linderi* Duf. (Orth. Rhaphidophoridae). *C.R. Acad. Sci. Paris,* **248**: 851–3.

SALTET, P. (1960). La formule chromosomique de *Dolichopoda palpata* et *D. bolivari* (Orth. Rhaphidophoridae). *C.R. Acad. Sci. Paris,* **250**: 2612–14.

SALTET, P. (1967). Les Dolichopodes de corse (Orth. Rhapidophoridae). I. Étude cytologique préliminaire. *Bull. Soc. Hist. Nat. Toulouse,* **103**: 265–8.

SANDERSON, A. R. (1933). The cytology of parthenogenesis in the Tenthredinidae. *Genetica,* **14**: 321–451.

SANDERSON, A. R. (1940). Maturation in the parthenogenetic snail *Potamopyrgus jenkinski* Smith, and in the snail *Peringia ulvae* (Pennant). *Proc. Zool. Soc. Lond.* A, **110**: 11–15.

SANDERSON, A. R. (1956). Maturation in the parthenogenetic weevil *Listroderes costirostris* Schonh. (*obliquus* Gull.) *Proc. XIVth Int. Congr. Zool. Copenhagen,* p. 185.

SANDERSON, A. R. (1960). The cytology of a diploid bisexual spider beetle, *Ptinus clavipes* Panzer and its triploid gynogenetic form *mobilis* Moore. *Proc. Roy. Soc. Edinb.* **67**: 333–50.

SANDERSON, A. R. and HALL, D. W. (1948). The cytology of the honey bee, *Apis mellifica* L. *Nature*, **162**: 34–5.

SANDERSON, A. R. and JACOB, J. (1957). Artificial activation of the egg in a gyno-genetic spider beetle. *Nature*, **179**: 1300.

SANDERSON, K. E. and DEMEREC, M. (1965). The linkage map of *Salmonella typhimurium*. *Genetics*, **51**: 897–913.

SANDLER, L. and HIRAIZUMI, Y. (1959). Meiotic drive in natural populations of *Drosophila melanogaster*. II. Genetic variation at the Segregation-Distorter locus. *Proc. Nat. Acad. Sci. U.S.A.* **45**: 1412–22.

SANDLER, L. and HIRAIZUMI, Y. (1960a). Meiotic drive in natural populations of *Drosophila melanogaster*. IV. Instability at the Segregation-Distorter locus. *Genetics*, **45**: 1269–87.

SANDLER, L. and HIRAIZUMI, Y. (1960b). Meiotic drive in natural populations of *Drosophila melanogaster*. V. On the nature of the SD region. *Genetics*, **45**: 1671–89.

SANDLER, L., HIRAIZUMI, Y. and SANDLER, I. (1958). Meiotic drive in natural populations of *Drosophila melanogaster*. I. The cytogenetic basis of segregation-distortion. *Genetics*, **43**: 233–50.

SANDLER, L., LINDSLEY, D. D., NICOLETTI, B. and TRIPPA, G. (1968). Mutants affecting meiosis in natural populations of *Drosophila melanogaster*. *Genetics*, **60**: 525–58.

SANDLER, L. and NOVITSKI, E. (1957). Meiotic drive as an evolutionary force. *Amer. Nat.* **91**: 105–10.

SANNOMIYA, M. (1962). Intra-individual variation in number of *A*- and *B*-chromosomes in *Patanga japonica*. *Chrom. Inf. Serv. Tokyo*, **3**: 30–32.

SANNOMIYA, M. and KAYANO, H. (1968). Local variation and year-to-year change in frequencies of *A*- and *B*-chromosomes in natural populations of some grasshopper species. *Proc. XIIth Int. Genet. Congr.* **2**: 116–17.

SARÀ, M. (1950). Sui cromosomi di *Telmatoscopus albipunctatus*, con alcuni dati su quelli di *Telmatoscopus ustulatus* (Diptera, Psychodidae). *Caryologia*, **3**: 204–10.

SASAKI, M. (1965). Meiosis in a male with Down's syndrome. *Chromosoma*, **16**: 637–51.

SASAKI, M. and ITOH, M. (1967). Preliminary notes on the karyotype of two species of turtles, *Clemys japonica* and *Geoclemys reevesii*. *Chrom. Inf. Serv. Tokyo*, **8**: 21–2.

SASAKI, M. and MAKINO, S. (1965). The meiotic chromosomes of man. *Chromosoma*, **16**: 637–51.

SASAKI, M. and MAKINO, S. (1967). Sex-chromosome abnormalities in man: a review. *Gunma Symp. on Endocrin.*, **4**: 3–22.

SATO, I. (1932). Chromosome behavior in the spermatogenesis of urodele amphibia *Diemyctylus* (*Triturus*) *pyrrhogaster* (Boie). *J. Sci. Hiroshima Univ.* B, **2**: 33–47.

SATO, I. (1936). On the chromosomes in spermatogenesis of the scorpion *Buthus martensii* Karsch. *Zool. Mag. Tokyo*, **48**: 956–7.

SAX, K. (1938). Chromosome aberrations induced by X-rays. *Genetics*, **23**: 494–516.

SAX, K. (1941). Types and frequencies of chromosomal aberrations induced by X-rays. *Cold Spr. Harb. Symp. Quant. Biol.* **9**: 93–101

SAX, K. and MATHER, K. (1939). An X-ray analysis of progressive chromosome splitting. *J. Genet.* **37**: 483–90.

SCHELLENBERG, A. (1913). Das accessorische Chromosom in den Samenzellen der Locustide *Diestrammena marmorata* de Hahn. *Arch. Zellforsch.* **11**: 489–514.

SCHIMMER, F. (1909). Beitrag zu einer Monographie der Gryllodengattung *Myrmecophila* Latr. *Z. wiss. Zool.* **93**: 409–534.

SCHLIEP, W. (1911). Das Verhalten des Chromatins bei *Angiostomum* (*Rhabdonema*) *nigrovenosum*. *Arch. Zellforsch.* **7**: 87–138.

SCHMID, W. (1962). DNA replication patterns of the heterochromosomes in *Gallus domesticus*. *Cytogenetics*, **1**: 344–52.

SCHMID, W. (1963). DNA replication patterns of human chromosomes. *Cytogenetics*, **2**: 175–93.

SCHMID, W., SMITH, D. W. and THEILER, K. (1965). Chromatinmuster in verschiedenen Zelltypen und Lokalisation von Heterochromatin auf Metaphase-Chromosomen bei *Microtus agrestis, Mesocricetus auratus, Cavia cobaya* und beim Menschen. *Arch. J. Klaus Stift. Vererbf.* **40**: 35–49.

SCHMIEDER, R. G. (1938). The sex ratio in *Melittobia chalybii* Ashmead, gametogenesis and cleavage in females and in haploid males (Hymenoptera: Chalcidoidea). *Biol. Bull. Woods Hole*, **70**: 256–66.

SCHMIEDER, R. G. and WHITING, P. W. (1947). Reproductive economy in the Chalcidoid wasp *Melittobia*. *Genetics*, **32**: 29–37.

SCHMUCK, M. L. (1934). The male chromosome group of *Sciara pauciseta*. *Biol. Bull. Woods Hole*, **66**: 224–7.

SCHMUCK, M. L. and METZ, C. W. (1932). Maturation divisions and fertilization in the eggs of *Sciara coprophila*. *Proc. Nat. Acad. Sci. U.S.A.* **18**: 349–52.

SCHOLL, H. (1955). Ein Beitrag zur Kenntnis der Spermatogenese der Mallophagen. *Chromosoma*, **7**: 271–4.

SCHOLL, H. (1956). Die Chromosomen parthenogenetischer Mücken. *Naturwiss.* **43**: 91–2.

SCHOLL, H. (1960). Die Oogenese einiger parthenogenetischer Orthocladiinen (Diptera). *Chromosoma*, **11**: 380–401.

SCHRADER, F. (1921). The chromosomes of *Pseudococcus nipae*. *Biol. Bull. Woods Hole* **40**: 259–70.

SCHRADER, F. (1923a). The origin of the mycetocytes in *Pseudococcus*. *Biol. Bull. Woods Hole*, **45**: 279–302.

SCHRADER, F. (1923b). A study of the chromosomes in three species of *Pseudococcus*. *Arch. Zellforsch.* **17**: 45–62.

SCHRADER, F. (1923c). Haploidie bei einer Spinnmilbe. *Arch. mikr. Anat.* **79**: 610–22.

SCHRADER, F. (1926). The cytology of pseudo-sexual eggs in a species of *Daphnia*. *Z. indukt. Abst. Vererbl.* **40**: 1–27.

SCHRADER, F. (1929). Experimental and cytological investigations of the life cycle of *Gossyparia spuria* and their bearing on the problem of haploidy in males. *Z. wiss. Zool.* **134**: 149–79.

SCHRADER, F. (1931). The chromosome cycle in *Protortonia primitiva* (Coccidae) and a consideration of the meiotic division apparatus in the male. *Z. wiss. Zool.* **138**: 386–408.

SCHRADER, F. (1932). Recent hypotheses on the structure of spindles in the light of certain observations in Hemiptera. *Z. wiss. Zool.* **142**: 520–39.

SCHRADER, F. (1940a). The formation of tetrads and the meiotic mitoses in the male of *Rhytidolomia senilis* Say (Hemiptera, Heteroptera). *J. Morph.* **67**: 123–42.

SCHRADER, F. (1940b). Touch-and-go pairing in chromosomes. *Proc. Nat. Acad. Sci. U.S.A.* **26**: 634–6.

SCHRADER, F. (1941a). The spermatogenesis of the earwig *Anisolabis maritima* Bon. with reference to the mechanism of chromosomal movement. *J. Morph.* **68**: 123–41.

SCHRADER, F. (1941*b*). Heteropycnosis and non-homologous association of chromosomes in *Edessa irrorata* (Hemiptera, Heteroptera). *J. Morph.* **69**: 587–604.

SCHRADER, F. (1944). *Mitosis: the Movements of the Chromosomes in Cell Division.* New York: Columbia Univ. Press.

SCHRADER, F. (1945*a*). Regular occurrence of heteroploidy in a group of Pentatomidae (Hemiptera). *Biol. Bull. Woods Hole*, **88**: 63–70.

SCHRADER, F. (1945*b*). The cytology of regular heteroploidy in the genus *Loxa* (Pentatomidae – Hemiptera). *J. Morph.* **76**: 157–77.

SCHRADER, F. (1946*a*). The elimination of chromosomes in the meiotic chromosomes of *Brachystethus rubromaculatus* Dallas. *Biol. Bull. Woods Hole*, **90**: 19–31.

SCHRADER, F. (1946*b*). Autosomal elimination and preferential segregation in the harlequin lobe of certain Discocephalini (Hemiptera). *Biol. Bull. Woods Hole*, **90**: 265–90.

SCHRADER, F. (1947). The role of the kinetochore in the chromosomal evolution of the Heteroptera and Homoptera. *Evolution.* **1**: 134–42.

SCHRADER, F. (1953). *Mitosis: the Movement of Chromosomes in Cell Division.* (2nd ed.) New York: Columbia Univ. Press.

SCHRADER, F. (1960*a*). Cytological and evolutionary implications of aberrant chromosome behavior in the harlequin lobe of some Pentatomidae (Heteroptera). *Chromosoma*, **11**: 103–28.

SCHRADER, F. (1960*b*). Evolutionary aspects of aberrant meiosis in some Pentatominae (Heteroptera). *Evolution*, **14**: 498–508.

SCHRADER, F. and HUGHES-SCHRADER, S. (1926). Haploidy in *Icerya purchasi*. *Z. wiss. Zool.* **128**: 182–200.

SCHRADER, F. and HUGHES-SCHRADER, S. (1931). Haploidy in Metazoa. *Quart. Rev. Biol.* **6**: 411–38.

SCHRADER, F. and HUGHES-SCHRADER, S. (1956). Polyploidy and fragmentation in the chromosomal evolution of various species of *Thyanta* (Hemiptera). *Chromosoma*, **7**: 469–96.

SCHRADER, F. and HUGHES-SCHRADER, S. (1958). Chromatid autonomy in *Banasa* (Hemiptera: Pentatomidae). *Chromosoma*, **9**: 193–215.

SCHRADER, F. and LEUCHTENBERGER, C. (1950). A cytochemical analysis of the functional interrelationships of various cell structures in *Arvelius albopunctatus* (De Geer). *Exp. Cell Res.* **1**: 421–52.

SCHREIBER, G. and MEMORIA, J. M. P. (1957). Algunos aspectos do problema do polimorfismo cromossômico e ecologia em anofelinos brasileiros. *Rev. Bras. Malar.* **9**: 101–3.

SCHREIBER, G. and PELLEGRINO, J. (1950). Eteropicnosi di autosomi come possibile mecanismo di speciazione. *Sci. Genet.* **3**: 215–26.

SCHREINER, A. and SCHREINER, K. E. (1905). Über die Entwicklung der männlichen Geschlechtszellen von *Myxine glutinosa* (L.). *Arch. Biol.* **21**: 183–357.

SCHULTZ, J. (1936). Variegation in *Drosophila* and the inert chromosome regions. *Proc. Nat. Acad. Sci. U.S.A.* **22**: 27–33.

SCHULTZ, J. and CATCHESIDE, D. G. (1937). The nature of closed *X*-chromosomes in *Drosophila melanogaster*. *J. Genet.* **35**: 315–20.

SCHULTZ, J. and DOBZHANSKY, TH. (1933). Triploid hybrids between *Drosophila melanogaster* and *Drosophila simulans*. *J. Exp. Biol.* **65**: 73–82.

SCHULTZ, J. and REDFIELD, H. (1951). Interchromosomal effects on crossing-over in *Drosophila*. *Cold Spr. Harb. Symp. Quant. Biol.* **16**: 175–97.

SCHULTZ, R. J. (1961). Reproductive mechanism of unisexual and bisexual strains of the viviparous fish *Poeciliopsis*. *Evolution*, **15**: 302–25.

SCHULTZ, R. J. (1966). Hybridization experiments with an all-female fish of the genus *Peociliopsis*. *Biol. Bull. Woods Hole*, **130**: 415–29.

SCHULTZ, R. J. (1967). Gynogenesis and triploidy in the viviparous fish *Poeciliopsis*. *Science*, **157**: 1564–7.

SCHULTZ, R. J. (1969). Hybridization, unisexuality and polyploidy in the teleost *Poeciliopsis* (Poeciliidae) and other vertebrates. *Amer. Nat.* **103**: 605–19.

SCHWARTZ, D. (1953). The behavior of an X-ray-induced ring chromosome in maize. *Amer. Nat.* **87**: 19–28.

SCHWARTZ, D. (1955). Studies on crossing over in maize and *Drosophila*. *J. Cell. Comp. Physiol.* suppl. 2, **45**: 171–88.

SCHWARTZ, H. (1932). Der Chromosomenzyclus von *Tetraneura ulmi* De Geer. *Z. Zellforsch.* **15**: 645–86.

SCHWARZACHER, H. G. and SCHNEDL, W. (1966). Position of labelled chromatids in diplochromosomes of endo-reduplicated cells after uptake of tritiated thymidine. *Nature*, **209**: 107–8.

SCOTT, A. C. (1936). Haploidy and aberrant spermatogenesis in a Coleopteran, *Micromalthus debilis* Le Conte. *J. Morph.* **59**: 485–515.

SCOTT, A. C. (1938). Paedogenesis in the Coleoptera. *Z. Morph. Ökol. Tiere*, **33**: 633–653.

SCOTT, A. C. (1941). Reversal of sex production in *Micromalthus*. *Biol. Bull. Woods Hole*, **81**: 420–31.

SCOTT, C. D. and GREGORY, P. W. (1965). An *XXY* trisomic in an intersex of *Bos taurus*. *Genetics*, **52**: 473–4.

SEARLE, A. G. (1962). Is sex-linked *Tabby* really recessive in the mouse? *Heredity*, **17**: 297.

SEARS, E. R. (1952*a*). The behavior of isochromosomes and telocentrics in wheat. *Chromosoma*, **4**: 551–62.

SEARS, E. R. (1952*b*). Misdivision of univalents in common wheat. *Chromosoma*, **4**: 535–50.

SEARS, E. R. (1966). Chromosome mapping with the aid of telocentrics. *Proc. IInd. Int. Wheat Genet. Symp.*; *Hereditas suppl. vol.* **2**: 370–81.

SEARS, E. R. and CAMARA, A. (1952). A transmissible dicentric chromosome. *Genetics*, **37**: 125–35.

SÉGUY, E. (1950). *La Biologie des Diptères*. Paris: Paul Lechevalier.

SEILER, J. (1914). Das Verhalten der Geschlechtschromosomen bei Lepidopteren. *Arch. Zellforsch.* **13**: 159–269.

SEILER, J. (1923). Geschlechtschromosomenuntersuchungen an Psychiden. IV. Die Parthenogenese der Psychiden. *Z. indukt. Abst. Vererbl.* **31**: 1–99.

SEILER, J. (1946). Die Verbreitungsgebiete der verschiedenen Rassen von *Solenobia triquetrella* (Psychidae) in der Schweiz. *Rev. Suisse Zool.* **53**: 529–33.

SEILER, J. (1947). Die Zytologie eines parthenogenetischen Rüsselkäfers *Otiorrhynchus sulcatus* F. *Chromosoma*, **3**: 88–109.

SEILER, J. (1949*a*). L'intersexualité chez les Lépidoptères. *XIIIᵉ Congr. Int. Zool. Paris*, pp. 155–61.

SEILER, J. (1949*b*). Das Intersexualitätsphänomen. *Experientia*, **5**: 425–38.

SEILER, J. (1958). Die Entwicklung des Genitalapparates bei triploiden Intersexen von *Solenobia triquetrella* F.R. (Lepid. Psychidae). Deutung des Intersexualitätsphänomens. *Roux' Arch. Entwickm.* **150**: 199–372.

SEILER, J. (1959). Untersuchungen über die Entstehung der Parthenogenese bei *Solenobia triquetrella* F.R. (Lepidoptera, Psychidae). I. Mitteilung. Die Zytologie der bisexuellen *S. triquetrella*, ihr Verhalten und ihr Sexualverhältnis. *Chromosoma*, **10**: 73–114.

SEILER, J. (1963). Untersuchungen über die Enstehung der Parthenogenese bei *Solenobia triquetrella* F. R. (Lepidoptera, Psychidae). *Z. Vererbl.* **94**: 29–66.

SEILER, J. (1965). Sexuality as developmental process. *Proc. XIth. Int. Genet. Congr.* **2**: 199–207.

SEILER, J. and SCHÄFFER, K. (1938). Der Chromosomenzyklus einer diploid-parthenogenetischen. *Solenobia triquetrella. Arch. exper. Zellforsch.* **22**: 215–16.

SEILER, J. and SCHÄFFER, K. (1941). Der Chromosomenzyklus einer diploid-parthenogenetischen *Solenobia triquetrella. Rev. Suisse Zool.* **48**: 537–40.

SEILER, J. and SCHÄFFER, K. (1960). Untersuchungen über die Entstehung der Parthenogenese bei *Solenobia triquetrella* F.R. (Lepidoptera, Psychidae). II. Analyse der diploidparthenogenetischen *S. triquetrella*, Verhalten, Aufzuchtresultate und Zytologie. *Chromosoma*, **11**: 29–102.

SENNA, A. (1911). La spermatogenesi di *Gryllotalpa vulgaris* Latr. *Monit. Zool. Ital.* **22**: 65–77.

SESHACHAR, B. R. (1937). The chromosomes of *Ichthyophis glutinosus* (Linn.). *Cytologia*, **8**: 327–30.

SESHACHAR, B. R. (1939). The spermatogenesis of *Uraeotyphlus narayani. Cellule*, **48**: 61–76.

SESHACHAR, B. R. (1944). The chromosomes of *Genophis carnosus* Bedd. *J. Mysore Univ.* B, **5**: 1–3.

SESHACHAR, B. R. and BAGGA, S. (1962). Chromosome number and sex-determining mechanism in the dragonfly *Hemianax ephippiger* (Burmeister). *Cytologia*, **27**: 443–9.

SETO, T., KEZER, J. and POMERAT, C. M. (1969). A cinematographic study of meiosis in salamander spermatocytes *in vitro. Z. Zellforsch.* **94**: 407–24.

SETO, T., POMERAT, C. M. and KEZER, J. (1964). The chromosomes of *Necturus maculosus* as revealed in cultures of leukocytes. *Amer. Nat.* **98**: 71–8.

SHARMA, G. P. (1950). Spermatogenesis in the spider, *Plexippus paykulli. Res. Bull. East. Panjab Univ* **5**: 67–80.

SHARMA, G. P. (1956). Sex mechanism in *Dysdercus* (Pyrrhocoridae-Heteroptera). *Curr. Sci.* **25**: 360.

SHARMA, G. P., GUPTA, B. L. and KUMBKARNI, C. G. (1961). Cytology of spermatogenesis in the honey-bee, *Apis indica. J. Roy. Micr. Sci.* **79**: 337–51.

SHARMA, G. P., GUPTA, M. L. and RANDHAWA, G. S. (1967). Polysomy in *Chrotogonus trachypterus* (Blanchard) (Orthoptera: Acridoidea: Pyrgomorphidae) from three more populations and its possible role in animal speciation. *Res. Bull. (N.S.) Panjab Univ.* **18**: 157–63.

SHARMA, G. P., JANDE, S. S., GREWAL, M. S. and CHOPRA, R. N. (1958). Cytological studies on the Indian spiders. II. *Res. Bull. Panjab Univ.* **156**: 255–69.

SHARMA, G. P., JANDE, S. S. and TANDON, K. K. (1959). Cytological studies on the Indian spiders. IV. *Res. Bull. (N.S.) Panjab Univ.* **10**: 73–80.

SHARMA, G. P. and JONEJA, M. G. (1960). Centromere in the sex chromosome of the males of *Hyalomma aegyptium* and *Rhipicephalus sanguineus* (Acarina: Ixodidae). *Curr. Sci.* **29**: 437–8.

SHARMA, G. P. and PARSHAD, R. (1955). The morphology of chromosomes in *Laccotrephes maculatus* Fabr. (Hemiptera–Heteroptera). *Res. Bull. Panjab Univ.* **72**: 67–72.

SHARMA, G. P., PARSHAD, R. and BEDI, T. S. (1962). Breakdown of the meiotic stability in *Chrotogonus trachypterus* (Blanchard) (Orthoptera: Acridoidea: Pyrgomorphidae). *Res. Bull. (N.S.) Panjab Univ.* **13**: 281–308.

SHARMA, G. P., PARSHAD, R. and HANDA, R. (1962). Meiosis in two species of *Palamnaeus* (Scorpiones–Scorpionidae). *Res. Bull. (N.S.) Panjab Univ.* **13**: 85–9

SHARMA, G. P., PARSHAD. R. and JONEJA, M. G. (1959). Chromosome mechanism in the males of three species of scorpions (Scorpiones–Buthidae). *Res. Bull. (N.S.) Panjab Univ.* **10**: 197–207.

SHARMA, G. P., PARSHAD, R. and SEHGAL, P. (1956). Meiosis without chiasmata in *Periplaneta americana. Nature*, **178**: 1004–5.

SHARMA, G. P., PARSHAD, R. and SEHGAL, P. (1959). Cytological analysis of the male meiosis in *Periplaneta americana. J. Genet.* **56**: 1–7.

SHARMA, G. P. and SINGH, S. (1957). Cytological studies on the Indian spiders. *Res. Bull. Panjab Univ.* **120**: 389–93.

SHARMA, G. P., SUD, G. C. and PARSHAD, R. (1957). Chromosome complement and the sex mechanism in *Dysdercus cingulatus* (Pyrrhocoridae–Heteroptera). *Caryologia*, **9**: 278–85.

SHARMA, G. P., TANDON, K. K. and GREWAL, M. S. (1960). Cytological studies on the Indian spiders. V. *Res. Bull. (N.S.) Panjab Univ.* **11**: 201–6.

SHARMA, T. (1961). A study on the chromosomes of two Lycosid spiders. *Proc. Zool. Soc. Calcutta*, **14**: 33–8.

SHARMA, T. (1963). A study on the chromosomes of three gryllids. *Proc. Zool. Soc. Calcutta*. **16**: 45–9.

SHARMAN, G. B. (1952). The cytology of Tasmanian short horned grasshoppers (Orthoptera: Acridoidea). *Pap. Roy. Soc. Tasm.* **86**: 107–22.

SHARMAN, G. B. (1956). Chromosomes of the common shrew. *Nature*, **177**: 941–2.

SHARMAN, G. B. (1961). The mitotic chromosomes of marsupials and their bearing on taxonomy and phylogeny. *Aust. J. Zool.* **9**: 38–60.

SHARMAN, G. B. (1971). Late DNA replication in the paternally derived X-chromosome of female Kangaroos. *Nature*, **230**: 231–2.

SHARMAN, G. B. (1970). Reproductive physiology of marsupials. *Science*, **167**: 1221–8

SHARMAN, G. B. and BARBER, H. N. (1952). Multiple sex chromosomes in the marsupial *Potorous. Heredity*, **6**: 345–55.

SHARMAN, G. B., McINTOSH, A. J. and BARBER, H. N. (1950). Multiple sex chromosomes in the marsupials. *Nature*, **166**: 996.

SHAW, D. D. (1970). The supernumerary segment system of *Stethophyma*. I. Structural basis. *Chromosoma*, **30**: 326–43.

SHAW, M. W. and KROOTH, R. S. (1964). The chromosomes of the Tasmanian Rat-kangaroo (*Potorous tridactylis apicalis*). *Cytogenetics*, **3**: 19–33.

SHELLHAMMER, H. S. (1969). Supernumerary chromosomes of the harvest mouse, *Reithrodontomys megalotis. Chromosoma*, **27**: 102–8.

SHEPPARD, P. M. (1953). Polymorphism, linkage and the blood groups. *Amer. Nat.* **87**: 283–94.

SHEPPARD, P. M. (1955). Genetic variability and polymorphism: synthesis. *Cold Spr. Harb. Symp. Quant. Biol.* **20**: 271–5.

SHIMBA, H. and ITOH, M. (1969). On the chromosomes of the shrew *Sorex unguiculatus. J. Fac. Sci. Hokkaido Univ.* VI, **17**: 263–5.

SHINJI, O. (1931). The evolutional significance of the chromosomes of the *Aphididae. J. Morph.* **51**: 373–433.

SHORT, R.B.(1957). Chromosomes and sex in *Schistosomatium douthitti* (Trematoda: Schistosomatidae). *J. Hered.* **48**: 2–6.

SHIWAGO, P. J. (1924). The chromosome complex in the somatic cells of male and female of the domestic chicken. *Science*, **60**: 45–6.

SHULL, A. F. (1917). Sex determination in *Anthothrips verbasci. Genetics*, **2**: 480–8.

SIMON, R. C. and DOLLAR, A. M. (1963). Cytological aspects of speciation in two North American teleosts, *Salmo gairdneri* and *Salmo clarki lewisi. Canad. J. Genet. Cytol.* **5**: 43–9.

SIMPSON, G. G. (1944). *Tempo and Mode in Evolution.* New York: Columbia Univ. Press.

SIMPSON, G. G. (1960). The evolution of life. In: *Evolution After Darwin,* **1** (ed.: Sol Tax). Univ. Chicago Press.

SIMPSON, G. G. (1970). Uniformitarianism. An inquiry into principle, theory and method in geohistory and biohistory. *Essays in Evolution and Genetics in Honor of Theodosius Dobhaznsky. Evolutionary Biology,* suppl. vol., pp. 43–96. New York: Appleton-Century-Crofts.

SINGH, L., SHARMA, T. and RAY-CHAUDHURI, S. P. (1970). Multiple sex chromosomes in the common Indian Krait, *Bungarus caeruleus* Schneider. *Chromosoma,* **31**: 386–91.

SINGH, R. P. and McMILLAN, D. B. (1966). Karyotypes of three subspecies of *Peromyscus. J. Mammal.* **47**: 261–6.

SINSHEIMER, R. L. (1959). A simple stranded deoxyribonucleic acid from bacteriophage ø X 174. *J. Mol. Biol.* **1**: 43–53.

SKALIY, P. and HAYES, W. J. (1949). The biology of *Liponyssus bacoti* (Hirst 1913) Acarina: Liponyssidae). *Amer. J. Trop. Med.* **29**: 759–72.

SKARÉN, U. and HALKKA, O. (1966). The karyotype of *Sorex caecutiens* Laxmann. *Hereditas,* **54**: 376–8.

SKINNER, D. M., BEATTIE, W. G., KERR, M. S. and GRAHAM, D. E. (1970). Satellite DNA's in Crustacea: two different components with the same density in neutral CsCl gradients. *Nature,* **227**: 837–9.

SLACK, H. D. (1938*a*). Chromosome numbers in *Cimex. Nature,* **142**: 358.

SLACK, H. D. (1938*b*). The association of non-homologous chromosomes in Corixidae (Hemiptera, Heteroptera). *Proc. Roy. Soc. Edinb.* **58**: 192–212.

SLACK, H. D. (1939*a*). The chromosomes of *Cimex. Nature,* **143**, 78.

SLACK, H. D. (1939*b*). Structural hybridity in *Cimex* L. *Chromosoma,* **1**: 104–18.

SLIZYNSKI, B. M. (1944). A revised map of salivary gland chromosome IV. *J. Hered.* **35**: 322–4.

SLIZYNSKI, B. M. (1945). 'Ectopic' pairing and the distribution of heterochromatin in the *X*-chromosome of salivary gland nuclei of *Drosophila melanogaster. Proc. Roy. Soc. Edinb.* **62**: 114–19.

SLIZYNSKI, B. M. (1950). *Chironomus* versus *Drosophila. J. Genet.* **50**: 77–8.

SLIZYNSKI, B. M. (1964*a*). Chiasmata in spermatocytes of *Drosophila melanogaster. Genet. Res.* **5**: 80–4.

SLIZYNSKI, B. M. (1964*b*). Cytology of the *XXY* mouse. *Genet. Res.* **5**: 328–9.

SMITH, S. G. (1941). A new form of spruce sawfly identified by means of its cytology and parthenogenesis. *Sci. Agric.* **21**: 244–305.

SMITH, S. G. (1944). The reproduction of the nucleus. *Sci. Agric.* **24**: 491–501.

SMITH, S. G. (1945). Heteropycnosis as a means of diagnosing sex. *J. Hered.* **36**: 194–6.

SMITH, S. G. (1952). The cytology of some tenebrionid beetles (Coleoptera). *J. Morph.* **91**: 325–64.

SMITH, S. G. (1953*a*). A pseudo-multiple sex-chromosome mechanism in an Indian gryllid. *Chromosoma,* **5**: 555–73.

SMITH, S. G. (1953*b*). Chromosome numbers of Coleoptera. *Heredity,* **7**: 31–48.

SMITH, S. G. (1956*a*). Extreme chromosomal polymorphism in a Cocinellid beetle. *Experientia,* **12**: 52.

SMITH, S. G. (1956b). Animal cytology and cytotaxonomy. *Proc. Genet. Soc. Canada*, **3**: 57–64.

SMITH, S. G. (1956c). The status of supernumerary chromosomes in *Diabrotica* after a lapse of fifty years. *J. Hered.* **47**: 157–64.

SMITH, S. G. (1957a). Comparative cytology of Chilocorini (Coleoptera). *Proc. Genet. Soc. Canada*, **2**: 42.

SMITH, S. G. (1957b). Chromosomal evolution in *Chilocorus stigma*: an exception to "Robertson's law". *Rec. Genet. Soc. Amer.* **26**: 396.

SMITH, S. G. (1958). Animal cytology and cytotaxonomy. *Proc. Genet. Soc. Canada*, **3**: 57–64.

SMITH, S. G. (1959). The cytogenetic basis of speciation in Coleoptera. *Proc. Xth Int. Genet. Congr.* **1**: 444–50.

SMITH, S. G. (1960a). Chromosome numbers of Coleoptera II. *Canad. J. Genet. Cytol.* **2**: 66–88.

SMITH, S. G. (1960b). Cytogenetics of insects. *Ann. Rev. Entomol.* **5**: 69–84.

SMITH, S. G. (1962a). Unique incompatibility system in a hybrid species. *Science*, **168**: 36–7.

SMITH, S. G. (1962b). Tempero-spatial sequentiality of chromosomal polymorphism in *Chilocorus stigma* Say (Coleoptera: Coccinellidae). *Nature*, **193**, 1210–11.

SMITH, S. G. (1962c). Cytogenetic pathways in beetle speciation. *Canad. Entomol.* **94**: 941–55.

SMITH, S. G. (1962d). Chromosomal polymorphism and inter-relationships among bark weevils of the genus *Pissodes* Germar: an amendment. *Nucleus*, **5**: 65–6.

SMITH, S. G. (1963). Cytotaxonomy: the phylogenetic implications of incompatibility systems. *Canad. J. Genet. Cytol.* **5**: 108.

SMITH, S. G. (1965a). Cytological species-separation in Asiatic *Exochomus* (Coleoptera: Coccinellidae). *Canad. J. Genet. Cytol.* **7**: 363–173.

SMITH, S. G. (1965b). Heterochromatin, colchicine and karyotype. *Chromosoma*, **16**: 162–5.

SMITH, S. G. (1966). Natural hybridization in the Coccinellid genus *Chilocorus*. *Chromosoma*, **18**: 380–406.

SMITH, S. G. and EDGAR, R. S. (1954). The sex determining mechanism in some North American Cicindelidae (Coleoptera). *Rev. Suisse Zool.* **61**: 657–67.

SMITH, S. G. and TAKENOUCHI, Y. (1962). A unique incompatibility system in a hybrid species. *Science*, **138**: 36–7.

SMITH-STOCKING, H. (1936). Genetic studies on selective segregation of chromosomes in *Sciara coprophila* Lintner. *Genetics*, **21**: 421–43.

SMITH-WHITE, S. G., PEACOCK, W. J., TURNER, B. and DEN DULK, G. M. (1693). A ring chromosome in man. *Nature*, **197**: 102–3.

SNOW, M. H. L. and CALLAN, H. G. (1969). Evidence for a polarized movement of the lateral loops of newt lampbrush chromosomes during oogenesis. *J. Cell Sci.* **5**: 1–25.

SOKOLOW, I. (1913). Über die Spermatogenese der Skorpione. *Arch. Zellforsch.* **9**: 399–432.

SOKOLOV, I. I. (1934). Untersuchungen über die Spermatogenese bei den Arachniden. V. Über die Spermatogenese der Parasitidae (= Gamasidae, Acari). *Z. Zellforsch.* **21**: 1–42.

SOKOLOV, I. I. (1945). On karyological study of some species of Acari and the problem of sex determination in the group. *Izvest. Akad. Nauk U.S.S.R. Otdel. Biol. Nauk* no. 6, pp. 654–63.

SOKOLOV, I. I. (1954). The chromosome complexes of Acari and their significance for systematics and phylogeny. *Trudy Leningradskovo Obschchestva Estestvoispit.* **72**: 124–58.

SOKOLOV, I. I. (1960). Studies on nuclear structures in spiders (Araneina) I. Karyological peculiarities in spermatogenesis. *Vopr. Cytol. i Protistol.* pp. 160–86.

SOKOLOV, I. I. (1962*a*). Studies on nuclear structures in spiders (Araneina). II. The sex chromosomes. *Tsitologia*, **4**: 617–625.

SOKOLOW, I. (1962*b*). Einige karyologische Beobachtungen an *Limnochares aquatica* L. (Hydrachnellae, Acari). *Zool. Anz.* **168**: 302–9.

SOLARI, J. (1965). Structure of the chromatin in the sea urchin sperm. *Proc. Nat. Acad. Sci. U.S.A.* **53**: 503–11.

SOLDATOVIĆ, B., ŽIVKOVIĆ, S., SAVIĆ, J. and MILOSEVIĆ, M. (1967). Vergleichende Analyse der Morphologie und der Anzahl der Chromosomen zwischen werschiedenen Populationen von *Spalax leucodon* Nordmann, 1840. *Z. Säugetier.* **32**: 238–46.

SONNEBORN, T. M. (1940). The relation of macronuclear regeneration in *Paramecium aurelia* to macronuclear structure, amitosis and genetic determination. *Anat. Rec.* **78**, suppl.: 53–4.

SONNEBORN, T. M. (1946). Inert nuclei: inactivity of micronuclear genes in variety 4 of *Paramecium aurelia*. *Genetics*, **31**: 231.

SONNEBORN, T. M. (1947). Recent advances in the genetics of *Paramecium* and *Euplotes*. *Adv. in Genet.* **1**: 263–358.

SONNEBORN, T. M. (1954). Gene-controlled, aberrant nuclear behaviour in *Paramecium aurelia*. *Microbial Genet. Bull.* **11**: 24–5.

SORSA, M. (1969). Ultrastructure of puffs in the proximal part of chromosome 3R in *Drosophila melanogaster*. *Ann. Acad. Sci. Fenn.* AIV, **150**: 1–21.

SOTELO, J. R. and WETTSTEIN, R. (1966). Association of nucleolus and sex chromosomes in Gryllidae spermatocytes. *Nat. Cancer Inst. Monograph*, **23**: 77–89.

SOUTHERN, D. I. (1967). Pseudo-multiple formation as a consequence of prolonged non-homologous chromosome association in *Metrioptera brachyptera*. *Chromosoma*, **21**: 272–84.

SOUTHERN, D. I. (1969). Stable telocentric chromosomes produced by centric misdivision in *Myrmeleotettix maculatus* (Thunb). *Chromosoma*, **26**: 140–7.

SOUTHERN, E. M. (1970). Base sequence and evolution of guinea-pig α-satellite DNA. *Nature*, **227**: 794–8.

SOUTHWOOD, T. R. E. and LESTON, D. (1959). *Land and Water Bugs of the British Isles.* London: Warne.

SPARKES, R. S. and ARAKAKI, D. T. (1966). Intrasubspecific and intersubspecific chromosomal polymorphism in *Peromyscus maniculatus* (deer mouse). *Cytogenetics*, **5**: 411–18.

SPEICHER, B. R. (1936). Oogenesis, fertilization and early cleavage in *Habrobracon*. *J. Morph.* **61**: 453–71.

SPEICHER, B. R. (1937). Oogenesis in a thelytokous wasp, *Nemeritis canescens*. *J. Morph.* **61**: 453–71.

SPENCER, W. P. (1940*a*). Subspecies, hybrids and speciation in *Drosophila hydei* and *Drosophila virilis*. *Amer. Nat.* **74**: 157–79.

SPENCER, W. P. (1940*b*). Levels of divergence in *Drosophila* speciation. *Amer. Nat.* **74**: 299–311.

SPERLICH, D. (1961). Untersuchungen über den chromosomalen Polymorphismus einer Population von *Drosophila subobscura* auf den Liparischen Inseln. *Z. Vererbl.* **92**: 74–84.

SPERLICH, D. (1964). Chromosomale Strukturanalyse und Fertilitatsprüfung an einer Marginalpopulation von *Drosophila subobscura*. *Z. Vererbl.* **95**: 73–81.

SPERLICH, D. (1967). Populationsgenetik. *Fortschr. Zool.* **18**: 223–78.

SPERLICH, D. and FEUERBACH, H.-M. (1966). Ist der chromosomale Strukturpolymorphismus von *Drosophila subobscura* stabil oder flexibel? *Z. Vererbl.* **98**: 16–24.

SPIESS, E. (1950). Experimental populations of *Drosophila persimilis* from an altitudinal transect of the Sierra Nevada. *Evolution*, **4**: 14–33.

SPIESS, E. B. (1957). Relation between frequencies and adaptive values of chromosomal arrangements in *Drosophila persimilis*. *Evolution*, **11**: 84–93.

SPIESS, E. B. (1961). Chromosomal fitness changes in experimental populations of *Drosophila persimilis* from Timberline in the Sierra Nevada. *Evolution*, **15**: 340, 351.

SPURWAY, H. (1953). Genetics of specific and subspecific differences in European newts. *Symp. Soc. Exp. Biol.* **7**: 200–37.

SPURWAY, H. and CALLAN, H. G. (1950). Hybrids between some members of the Rassenkreis *Triturus cristatus*. *Experientia*, **6**: 95–6.

SPURWAY, H. and HALDANE, J. B. S. (1954). Genetics and cytology of *Drosophila subobscura*. IX. An autosomal recessive mutant transforming homogametic zygotes into intersexes. *J. Genet.* **52**: 208–25.

SRIKANTAPPA, L. and ASWATHANARAYANA, N. W. (1970). Analysis of male meiosis in *Nala lividipes* (Dufour) (Labiduridae–Dermaptera). *Cytologia*, **35**: 354–358.

SRIKANTAPPA, L. and RAJASEKARASETTY, M. R. (1969–70). Contributions to the cytology of Dermaptera. *J. Mysore Univ.* (N.S.) B, **23**: 27–34.

SRIVASTAVA, M. L. D. (1957). Compound sex-chromosome mechanism and regularly occurring meiotic aberrations in the spermatogenesis of *Macropygium reticulare* (Pentatomidae–Hemiptera). *Cellule*, **58**: 259–72.

STADLER, L. J. (1931). The experimental modification of heredity in crop plants. I. Induced chromosomal irregularities. *Sci. Agric.* **11**: 557–72.

STAHL, F. W. (1967). Circular genetic maps. *J. Cell Comp. Physiol.* **70**, suppl: 1–12.

STAIGER, H. (1954). Der Chromosomendimorphismus beim Prosobranchier *Purpura lapillus* in Beziehung zur Ökologie der Art. *Chromosoma*, **6**: 419–478.

STAIGER, H. (1955). Reziproke Translokationen in natürlichen Populationen von *Purpura lapillus* (Prosobranchia). *Chromosoma*, **7**: 181–97.

STAIGER, H. and BOCQUET, C. (1956). Les chromosomes de la super-espèce *Jaera marina* (F.) et de quelques autres Janiridae (Isopodes asellotes). *Bull. Biol. Fr. Belg.* **90**: 1–32.

STALKER, H. D. (1940). Chromosome homologies in two subspecies of *Drosophila virilis*. *Proc. Nat. Acad. Sci. U.S.A.* **26**: 575–8.

STALKER, H. D. (1954). Parthenogenesis in *Drosophila*. *Genetics*, **39**: 4–34.

STALKER, H. D. (1956a). A case of polyploidy in Diptera. *Proc. Nat. Acad. Sci. U.S.A.* **42**: 194–8.

STALKER, H. D. (1956b). On the evolution of parthenogenesis in *Lonchoptera* (Diptera). *Evolution*, **10**: 345–59.

STALKER, H. D. (1960a). Relationship of meiotic drive and inversion association in *Drosophila paramelanica*. *Rec. Genet. Soc. Amer.* **29**: 94–5.

STALKER, H. D. (1960b). Chromosomal polymorphism in *Drosophila paramelanica* Patterson. *Genetics*, **45**: 95–114.

STALKER, H. D. (1961). The genetic systems modifying meiotic drive in *Drosophila paramelanica*. *Genetics*, **46**: 177–202.

STALKER, H. D. (1963). Cytotaxonomy in the *Drosophila melanica* species group. *Proc. Xth Int. Genet. Congr.* **1**: 139–40.

STALKER, H. D. (1964*a*). The salivary gland chromosomes of *Drosophila nigromelanica*. *Genetics*, **49**: 883–93.

STALKER, H. D. (1964*b*). Chromosomal polymorphism in *Drosophila euronotus*. *Genetics*, **49**: 669–87.

STALKER, H. D. (1964*c*). The evolutionary relationships of six species of *Drosophila* as determined by photographic mapping techniques. *Genetics*, **50**: 289–90.

STALKER, H. D. (1965). The salivary chromosomes of *Drosophila micromelanica* and *Drosophila melanica*. *Genetics*, **51**: 327–507.

STALKER, H. D. (1966). The phylogenetic relationships of the species in the *Drosophila melanica* group. *Genetics*, **53**: 327–42.

STALKER, H. D. and CARSON, H. L. (1948). An altitudinal transect of *Drosophila robusta* Sturtevant. *Evolution*, **2**: 295–305.

STEBBINS, G. L. (1945). Evidence for abnormally slow rates of evolution with particular reference to the higher plants and the genus *Drosophila*. *Lloydia*, **8**: 84–102.

STEBBINS, G. L. (1950). *Variation and Evolution in Plants*. Columbia Univ. Press.

STEBBINS, G. L. (1963). Perspectives – I. *Amer. Scientist*, **51**: 262–70.

STEBBINS, G. L. (1966). *Processes of Organic Evolution*. Englewood Cliffs, N.J.: Prentice-Hall.

STEDMAN, E. and STEDMAN, E. (1947). The chemical nature and functions of the components of cell nuclei. *Cold Spr. Harb. Symp. Quant. Biol.* **12**: 224–36.

STEFANI, R. (1956). Il problema della partenogenesi in '*Haploembia solieri*' Ramb. (Embioptera–Oligotomidae). *Atti Accad. Naz. Lincei*, Ser. VIII, **5**. Sez. III, pp. 127–201.

STEFANI, R. (1959). Secondo contributo alla conoscenza della cariologia negli insetti Embiotteri: il corredo cromosomico nelle specie dell'Europa meridionale. *Rend. Acad. Lincei*, Ser. VIII. *Sci. fis. mat. nat.* **26**: 396–99.

STEFANI, R. (1960). L'*Artemia salina* partenogenetica a Cagliari. *Riv. di Biol.* **53**: 463–91.

STEFANI, R. (1963). La digametia femminile in *Artemia salina* Leach e la costituzione del corredo cromosomico nei biotipi diploide anfigonico e diploide partenogenetico. *Caryologia*, **16**: 625–36.

STEINBERG, A. G. (1936). The effects of autosomal inversions on crossing over in the *X*-chromosome of *Drosophila melanogaster*. *Genetics*, **21**: 615–24.

STEINBERG, A. G. (1937). Relations between chromosome size and effects of inversions on crossing-over in *Drosophila melanogaster*. *Proc. Nat. Acad. Sci. U.S.A.* **23**: 54–6.

STEINBERG, A. G. and FRASER, F. C. (1944). Studies on the effects of *X*-chromosome inversions on crossing-over in the third chromosome of *Drosophila melanogaster*. *Genetics*, **29**: 83–103.

STEINITZ-SEARS, L. M. (1963). Cytogenetic studies bearing on the nature of the centromere. *Genetics Today. Proc. XIth Int. Gent. Congr. The Hague*, **1**: 123.

STELLA, E. (1933). Phenotypical characteristics and geographical distribution of several biotypes of *Artemia salina* L. *Z. indukt. Abst. Vererbl.* **65**: 412–46.

STEOPOE, I. (1925). La spermatogénèse chez *Nepa cinerea*. *C.R. Soc. Biol. Paris*, **92**: 1436–8.

STEOPOE, I. (1927). La spermatogénèse chez *Ranatra linearis*. *C.R. Soc. Biol. Paris*, **96**: 1030–1.

STEOPOE, I. (1931). La spermatogénèse chez *Nepa cinerea*. *Ann. Sci. Univ. Jassy*, **16**: 611–54.

STEOPOE, I. (1939). Nouvelles recherches sur la spermatogénèse chez *Gryllotalpa vulgaris* de Roumanie. *Arch. Zool. Exp. Gen.* **80**: 445–64.

STERN, C. (1929). Über die additive Wirkung multipler Allele. *Biol. Zentralbl.* **49**: 261–90.

STERN, C. (1931). Zytologisch-genetische Untersuchungen als Beweise für die Morgansche Theorie des Faktoren-austauschs. *Biol. Zentralbl.* **51**: 547–87.

STERN, C. (1957). The problem of complete Y-linkage in man. *Amer. J. Hum. Genet.* **9**: 147–66.

STERN, C. (1960*a*). Dosage compensation – development of a concept and new facts. *Canad. J. Genet. Cytol.* **2**: 105–18.

STERN, C. (1960*b*). *Principles of Human Genetics.* San Francisco: Freeman.

STERN, C. (1965). Synthesis. *Proc. XIth Int. Genet. Congr. The Hague,* **2**: 221–6.

STERN, C., CENTERWALL, W. R. and SARKAR, S. S. (1964). New data on the problem of Y-linkage of hairy pinnae. *Amer. J. Hum. Genet.* **16**: 455–71.

STEVENS, N. M. (1905*a*). Studies in spermatogenesis with special reference to the 'accessory chromosome'. *Publ. Carnegie Instn.* **36**, pt. I: 3–32.

STEVENS, N. M. (1905*b*). A study of the germ cells of *Aphis rosae* and *Aphis oenotherae. J. Exp. Zool.* **2**: 313–33.

STEVENS, N. M. (1906). Studies on the germ cells of Aphids. *Publ. Carnegie Instn.* **51**: 1–28.

STEVENS, N. M. (1908*a*). The chromosomes in *Diabrotica vittata, Diabrotica soror,* and *Diabrotica* 12-*punctata. J. Exp. Zool.* **5**: 453–70.

STEVENS, N. M. (1908*b*). A study of the germ cells of certain Diptera, with reference to the .heterochromosomes and the phenomena of synapsis. *J. Exp. Zool.* **5**: 359–74.

STEVENS, N. M. (1909). Further studies on the chromosomes of the Coleoptera. *J. Exp. Zool.* **6**: 101–13.

STEVENS, N. M. (1912). Supernumerary chromosomes and synapsis in *Ceuthophilus* sp. *Biol. Bull. Woods Hole,* **22**: 231–8.

STEVENS, W. L. (1936). The analysis of interference. *J. Genet.* **32**: 51–64.

STEVENSON, A. C. and BOBROW, M. (1967). Determinants of sex proportions in man, with consideration of the evidence concerning a contribution from X-linked mutations to intra-uterine death. *J. Med. Genet.* **4**: 190–221.

STICH, H. F. (1962). Variations of the deoxyribonucleic acid (DNA) content in embryonal cells of *Cyclops strenuus. Exp. Cell Res.* **26**: 136–43.

STOHLER, R. (1928). Cytologische Untersuchungen an den Keimdrüsen der mitteleuropäischen Kröten (*Bufo viridis* Laur., *B. calamita* Laur. und *B. vulgaris* Laur.). *Z. Zellforsch.* **7**: 400–75.

STONE, W. S. (1942). The l x^B factor and sex determination. *Univ. Texas Publ.* **4228**: 146–52.

STONE, W. S. (1949). The survival of chromosomal variation in evolution. *Univ. Texas Publ.* **4920**: 18–21.

STONE, W. S. (1955). Genetic and chromosomal variability in *Drosophila. Cold Spr. Harb. Symp. Quant. Biol.* **20**: 256–69.

STONE, W. S. (1962). The dominance of natural selection and the reality of superspecies (species groups) in the evolution of *Drosophila. Univ. Texas Publ.* **6205**: 507–538.

STONE, W. S., GUEST, W. C. and WILSON, F. D. (1960). The evolutionary implications of the cytological polymorphisms and phylogeny of the virilis group of *Drosophila. Proc. Nat. Acad. Sci. U.S.A.* **46**: 350–61.

STONE, W. S. and PATTERSON, J. T. (1947). The species relationships in the *virilis* group. *Univ. Texas Publ.* **4720**: 157–60.

STRASBURGER, E. (1905). Die Apogamie der Eualchemillen und allgemeine Gesichtspunkte die sich aus ihr ergeben. *Jb. wiss. Bot.* **41**: 88–164.

STREISINGER, G. and BRUCE, V. (1960). Linkage of genetic markers in phages T_2 and T_4. *Genetics*, **45**: 1289–96.

STREISINGER, G., EDGAR, R. S. and DENHARDT, G. H. (1964). Chromosome structure in phage T_4. I. Circularity of the linkage map. *Proc. Nat. Acad. Sci. U.S.A.* **51**: 775–9.

STRICKBERGER, M. W. and WILLS, C. J. (1966). Monthly frequency changes of *Drosophila pseudoobscura* third chromosome gene arrangements in a California locality. *Evolution*, **20**: 592–602.

STUMM-ZOLLINGER, E. (1953). Vergleichende Untersuchungen über die Inversion-häufigkeit bei *Drosophila subobscura* in Populationen der Schweitz und Süd-westeuropas. *Z. indukt. Abst. Vererbl.* **85**: 382–407.

STUMM-ZOLLINGER, E. and GOLDSCHMIDT, E. (1959). Geographical differentiation of inversion systems in *Drosophila pseudoobscura*. *Evolution*, **13**: 89–98.

STURTEVANT, A. H. (1920). Intersexes in *Drosophila simulans*. *Science*, **51**: 325–7.

STURTEVANT, A. H. (1921). Genetic studies on *Drosophila simulans*. I. Introduction: hybrids with *Drosophila melanogaster*. *Genetics*, **5**: 488–500.

STURTEVANT, A. H. (1925). The effects of unequal crossing-over at the *Bar* locus in *Drosophila*. *Genetics*, **10**: 117–47.

STURTEVANT, A. H. (1929). The *claret* mutant type of *Drosophila simulans*: a study of chromosome elimination and of cell lineage. *Z. wiss. Zool.* **135**: 323–55.

STURTEVANT, A. H. (1945). A gene in *Drosophila melanogaster* that transforms females into males. *Genetics*, **30**: 297–9.

STURTEVANT, A. H. (1946). Intersexes dependent on a maternal effect in hybrids between *Drosophila repleta* and *D. neorepleta*. *Proc. Nat. Acad. Sci. U.S.A.* **32**: 84–7.

STURTEVANT, A. H. and BEADLE, G. W. (1936). The relations of inversions in the *X*-chromosome of *Drosophila melanogaster* to crossing-over and disjunction. *Genetics*, **21**: 554–604.

STURTEVANT, A. H. and DOBZHANSKY, TH. (1936a). Geographical distribution and cytology of 'sex-ratio' in *Drosophila pseudoobscura* and related species. *Genetics*, **21**: 473–90.

STURTEVANT, A. H. and DOBZHANSKY, TH. (1936b). Inversions in the third chromosome of wild races of *Drosophila pseudoobscura* and their use in the study of the species. *Proc. Nat. Acad. Sci. U.S.A.* **22**: 447–52.

SUGIYAMA, M. (1933). Behavior of the sex chromosomes in the spermatogenesis of Japanese earwig, *Anisolabis marginalis*. *J. Fac. Sci. Tokyo Univ.* IV, **3**: 177–82.

SUOMALAINEN, E. (1933). Der Chromosomenzyklus bei *Macrosiphum pisi* Kalt. *Z. Zellforsch.* **19**: 583–94.

SUOMALAINEN, E. (1940a). Polyploidy in parthenogenetic Curculionidae. *Hereditas*, **26**: 51–64.

SUOMALAINEN, E. (1940b). Beiträge zur Zytologie der parthenogenetischen Insekten. *Ann. Acad. Sci. Fenn.* A, **54**: 1–143.

SUOMALAINEN, E. (1940c). Beiträge zur Zytologie der parthenogenetischen Insekten. II. *Lecanium hemisphaericum*. *Ann. Acad. Sci. Fenn.* A, **57**: 1–30.

SUOMALAINEN, E. (1945). Zu den Chromosomenverhältnissen und dem Artibil-dungsproblem bei parthenogenetischen Tieren. *Sitzungsb. Finn. Akad. Wiss.* pp. 181–201.

SUOMALAINEN, E. (1946). Die Chromosomenverhältnisse in der Spermatogenese einiger Blattarien. *Ann. Acad. Sci. Fenn.* **4** (10): 1–60.

SUOMALAINEN, E. (1947). Parthenogenese und Polyploidie bei Rüsselkäfern (Curculionidae). *Hereditas*, **33**: 425–56.

SUOMALAINEN, E. (1949). Parthenogenesis and polyploidy in the weevils, Curculionidae. *Ann. Ent. Fenn.* **14**, suppl.: 479–82.

SUOMALAINEN, E. (1950). Parthenogenesis in animals. *Adv. in Genet.* **3**: 193–253.

SUOMALAINEN, E. (1953). The kinetochore and the bivalent structure in the Lepidoptera. *Hereditas,* **39**: 88–96.

SUOMALAINEN, E. (1954a). Zur Zytologie der parthenogenetischen Curculioniden der Schweiz. *Chromosoma,* **6**: 627–55.

SUOMALAINEN, E. (1954b). Das Zentromer und die Bivalentenstruktur bei den Schmetterlingen. *Atti IX Congr. Int. Genet., Caryologia* vol. suppl.: pp. 780–1.

SUOMALAINEN, E. (1955). A further instance of geographical parthenogenesis and polyploidy in the weevils, Curculionidae. *Arch. Soc. Vanamo,* **9**: suppl. 350–4.

SUOMALAINEN, E. (1958a). On polyploidy in animals. *Proc. Finn. Acad. Sci. Letters,* **1**: 105–19.

SUOMALAINEN, E. (1958b). Polyploidy in parthenogenetic beetles. *Proc. Xth Int. Congr. Genet. Montreal,* **2**: 283.

SUOMALAINEN, E. (1961). On morphological differences and evolution of different polyploid parthenogenetic weevil populations. *Hereditas,* **47**: 309–41.

SUOMALAINEN, E. (1962). Significance of parthenogenesis in the evolution of insects. *Ann. Rev. Entomol.* **7**: 349–66.

SUOMALAINEN, E. (1963). On the chromosomes of the Geometrid moths *Cidaria. Proc. XIth Int. Genet. Congr.* **1**: 137–8.

SUOMALAINEN, E. (1965). On the chromosomes of the Geometrid moth genus *Cidaria. Chromosoma,* **16**: 166–84.

SUOMALAINEN, E., (1966a). The first known case of polyploidy in a parthenogenetic curculionid native of America. *Hereditas,* **57**: 213–16.

SUOMALAINEN, E. (1966b). Achiasmatische Oogenese bei Trichopteren. *Chromosoma,* **18**: 201–7.

SUOMALAINEN, E. (1969a). Evolution in parthenogenetic Curculionidae. *Evol. Biol.* **3**: 261–96.

SUOMALAINEN, E. (1969b). On the sex chromosome trivalent in some Lepidoptera females. *Chromosoma,* **28**: 298–308.

SUOMALAINEN, E. and HALKKA, O. (1963). The mode of meiosis in the Psyllina. *Chromosoma,* **14**: 498–510.

SUOMALAINEN, H. O. T. (1952a). Occurrence and phylogenetic significance of a true sex chromosome bivalent in a lace wing, *Boriomyia nervosa* F. (Neuroptera, Hemerobiidae) *Arch. Soc. Vanamo,* **7**: 45–9.

SUOMALAINEN, H. O. T. (1952b). Localization of chiasmata in the light of observations on the spermatogenesis of certain Neuroptera. *Ann. Zool. Soc. Vanamo,* **15**: 1–104.

SUTHERLAND, G. R. (1970). Chromosome abnormalities in newborn babies. *Aust. J. Ment. Retard.* **1**: 77–81.

SUTTON, E. (1940). Terminal deficiencies in the *X*-chromosome of *Drosophila melanogaster. Genetics,* **25**: 628–35.

SUTTON, E. (1943). Bar eye in *Drosophila melanogaster*: a cytological analysis of some mutations and reverse mutations. *Genetics,* **28**: 97–107.

SUZUKI, S. (1950). Spiders with extremely low or high chromosome numbers. *Jap. J. Genet.* **25**: 221–2.

SUZUKI, S. (1951). Cytological studies in spiders. I. A comparative study of the chromosomes in the family Argiopidae. *J. Sci. Hiroshima Univ.* B, **12**: 67–98.

SUZUKI, S. (1952). Cytological studies in spiders. II. Chromosomal investigation in the twenty-two species of spiders belonging to the four families Clubionidae, Sparassidae, Thomisidae and Oxyopidae, which constitute Clubionoidea, with special reference to sex chromosomes. *J. Sci. Hiroshima Univ.* B, **13**: 1–52.

SUZUKI, S. (1954). Cytological studies in spiders. III. Studies on chromosomes of fifty-seven species of spiders belonging to seventeen families, with general considerations on chromosomal evolution. *J. Sci. Hiroshima Univ.* B. **15**: 23–136.

SUZUKI, S. and OKADA, A. (1950). A study on the chromosomes of a spider, *Heteropoda venatoria*, with special reference to X_1, X_2 and X_3 chromosomes. *J. Sci. Hiroshima Univ.* B, **11**: 29–44.

SVÄRDSON, G. (1945). Chromosome studies on Salmonidae. *Rep. Swed. State Inst. Freshw. Fish. Res. Drottningholm*, **23**: 1–151.

SVÄRDSON, G. (1957). The Coregonid problem. VI. The Palaearctic species and their intergrades. *Rep. Swed. State Inst. Freshw. Fish Res. Drottningholm*, **38**: 357–84.

SWANN, M. F. and MICKEY, G. H. (1947). Parthenogenetic grasshoppers and their bearing upon polyploidy and sex determination. *Proc. Louisiana Acad. Sci.* **10**: 78–92.

SWANSON, C. P. (1947). X-ray and ultra-violet studies on pollen tube chromosomes. II. The quadripartite structure of the prophase chromosomes of *Tradescantia*. *Proc. Nat. Acad. Sci. U.S.A.* **33**: 229–32.

SWANSON, C. P. (1957). *Cytology and Cytogenetics*. Englewood Cliffs, N.J.: Prentice-Hall.

SWANSON, C. P., MERZ, T. and YOUNG, W. J. (1967). *Cytogenetics*. Englewood Cliffs, N.J.: Prentice-Hall.

SWIFT, H. (1950). The constancy of deoxyribosenucleic acid in plant nuclei. *Proc. Nat. Acad. Sci. U.S.A.* **36**: 643–54.

SWIFT, H. (1962). Nucleic acids and cell morphology in Dipteran salivary glands. In: *The Molecular Control of Cellular Activity* (ed.: J. M. Allen), pp. 73–125. New York: McGraw-Hill.

TABERLY, G. (1958a). Les nombres chromosomiques chez quelques espèces d'Oribates (Acariens). *C.R. Acad. Sci. Paris*, **246**: 3284–5.

TABERLY, G. (1958b). La cytologie de la parthénogenèse chez *Platynothrus peltifer* (Koch) (Acarien, Oribate) *C.R. Acad. Sci. Paris*, **247**: 1655–7.

TABERLY, G. (1960). La régulation chromosomique chez *Thrypochthonius tectorum* (Berlese), espèce parthénogénétique d'Oribate (Acarien): un nouvel exemple de mixocinèse. *C.R. Acad. Sci. Paris*, **250**: 4200–1.

TAKAGI, N. and FUJIMAKI, Y. (1966). Chromosomes of *Sorex shinto saevus* Thomas and *Sorex unguiculatus* Dobson. *Jap. J. Genet.* **41**: 109–13.

TAKENOUCHI, Y. (1957a). On a parthenogenetic weevil, *Catapionus gracilicornis* Roelofs. *Zool. Mag.* **66**: 198–205.

TAKENOUCHI, Y. (1957b). Polyploidy in some parthenogenetic weevils (a preliminary note). *Annot. Zool. Jap.* **30**: 38–41.

TAKENOUCHI, Y. (1959). Some oecological observations on three species of curculionid weevils, with special reference to parthenogenetic reproduction. *J. Hokkaido Gakugei Univ.* **10**: 297–339.

TAKENOUCHI, Y. (1961). The cytology of bisexual and parthenogenetic races of *Scepticus griseus* Roelofs (Curculionidae: Coleoptera). *Canad. J. Genet. Cytol.* **3**: 237–41.

TAKENOUCHI, Y. (1963). A further investigation on the chromosomes in twenty-three species of weevils (Curculionidae, Coleoptera). *J. Hokkaido Gakugei Univ.* **13**: 160–74.

TAKENOUCHI, Y. (1964). A preliminary note on the chromosomes of four partheno-genetic weevils (Brachyrhininae) in Canada. *Jap. J. Genet.* **39**: 74–9.

TAKENOUCHI, Y. (1965). Chromosome survey in thirty-four species of bisexual and parthenogenetic weevils of Canada. *Canad. J. Genet. Cytol.* **7**: 663–87.

TAKENOUCHI, Y. (1966). Tetraploid and pentaploid races of the Japanese partheno-genetic weevil, *Catapionus gracilicornis* Roelofs (Curculionidae, Coleoptera). *Annot. Zool. Jap.* **39**: 47–54.

TAKENOUCHI, Y. (1968). A modification of the Xyy_p sex-determining mechanism in a lady-bird beetle, *Epilachna pustulosa* Kono (Coccinellidae: Coleoptera). *Chrom. Inf. Serv.* **9**: 25–7.

TAKENOUCHI, Y. (1969a). The Xyy_p sex-determining mechanism found in *Scepticus insularis* Roelofs (Curculionidae: Coleoptera). *Jap. J. Genet.* **44**: 189–90.

TAKENOUCHI, Y. (1969b). A chromosomal study in a weevil, *Deporaus pacatoides* Voss (Curculionidae: Coleoptera): an additional instance of the Xyy_p sex-determining mechanism. *Japan. J. Genet.* **44**: 379–80.

TAKENOUCHI, Y. (1969c). On the compound sex chromosomes in *Diocalandra* sp. (Curculionidae: Coleoptera). *Chrom. Inf. Serv.* **10**: 4–6.

TAKENOUCHI, Y. (1970). Further studies of the chromosomes of Japanese weevils (Coleoptera: Curculionidae). *Canad. J. Genet. Cytol.* **12**: 273–7.

TAKENOUCHI, Y. and MURAMOTO, N. (1969). Chromosome numbers of Heteroptera. *J. Hokkaido Univ. Educ.* IIB, **20**: 1–15.

TAKENOUCHI, Y., SHIITSU, T. and TOSHIOKA, S. (1970). A chromosome study of two parthenogenetic Ticks, *Haemaphysalis bispinosa* Neumann and *H. longi-cornis* Neumann (Acarina: Ixodidae). *J. Hokkaido Univ. Educ.* IIB, **20**: 45–50.

TAKENOUCHI, Y. and TAGAKI, K. (1967). A chromosome study of two parthenogene-tic Scolytid beetles. *Annot. Zool. Jap.* **40**: 105–10.

TAKENOUCHI, Y., YASUE, Y. and KAWATA, K. (1970). A chromosome survey on the parthenogenetic weevil, *Listroderes costirostris* Schönherr (Curculionidae: Coleoptera) in Okayama prefecture. *Zool. Mag.* **79**: 71–9.

TAN, C. C. (1935). Salivary gland chromosomes in the two races of *Drosophila pseudoobscura*. *Genetics*, **20**: 392–402.

TAN, C. C. (1937). The cytological maps of the autosomes in *Drosophila pseudo-obscura*. *Z. Zellforsch.* **26**: 439–61.

TANAKA, Y. (1953). Genetics of the silkworm, *Bombyx mori*. *Adv. Genet.* **5**: 240–324.

TANNREUTHER, G. W. (1907). History of the germ cells and early embryology of certain aphids. *Zool. Jb.* (Anat.), **24**: 609–42.

TAYLOR, A. L. and THOMAN, M. S. (1964). The genetic map of *Escherichia coli* K12. *Genetics*, **50**: 659–77.

TAYLOR, E. W. (1965). Brownian and saltatory movements of cytoplasmic granules and the movement of anaphase chromosomes. *Proc. IVth Int. Congr. Rheology*, part 4, pp. 175–191. New York: Interscience.

TAYLOR, J. H. (1957). The time and mode of duplication of chromosomes. *Amer. Nat.* **91**: 209–21.

TAYLOR, J. H. (1958a). Sister chromatid exchanges in tritium-labelled chromosomes. *Genetics*, **43**: 515–29.

TAYLOR, J. H. (1958b). The organization and duplication of genetic material. *Proc. Xth Int. Genet. Congr. Montreal*, **1**: 63–78.

TAYLOR, J. H. (1960a). Nucleic acid synthesis in relation to the cell division cycle. *Ann. New York Acad. Sci.* **90**: 409–21.

TAYLOR, J. H. (1960b). Asynchronous duplication of chromosomes in cultured cells of Chinese hamster. *J. Biophys. Biochem. Cytol.* **7**: 455–63.

TAYLOR, J. H. (1963*a*). DNA synthesis in relation to chromosome reproduction and the reunion of breaks. *J. Cell Comp. Physiol.* suppl. 1 to vol. **62**: 73–86.

TAYLOR, J. H. (1963*b*). Control mechanisms for chromosome reproduction in the cell cycle. *Int. Soc. Cell. Biol.* **2**: 161–77.

TAYLOR, J. H. (1965). Distribution of tritium-labelled DNA among chromosomes during meiosis. I. Spermatogenesis in the grasshopper. *J. Cell Biol.* **25**: 57–67.

TAYLOR, J. H. (1966). The duplication of chromosomes. In: *Probleme der biologischen Reduplikation*, pp. 9–26. Berlin: Springer-Verlag.

TAYLOR, J. H. (1968). Rates of chain growth and units of replication in DNA of mammalian chromosomes. *J. Mol. Biol.* **31**: 579–94.

TAYLOR, J. H. and McMASTER, (1954). Autoradiographic and microphotometric studies of desoxyribose nucleic acid during microgametogenesis in *Lilium longiflorum*. *Chromosoma*, **6**: 489–521.

TAYLOR, J. H., WOODS, P. S. and HUGHES, W. L. (1957). The organization and duplication of chromosomes as revealed by autoradiographic studies using tritium-labelled thymidine. *Proc. Nat. Acad. Sci. U.S.A.* **43**: 122–8.

TAYLOR, K. M., HUNGERFORD, D. A., SNYDER, R. L. and ULMER, F. A. Jr. (1968). Uniformity of karyotypes in the Camelidae. *Cytogenetics*, **7**: 8–15.

TAZIMA, Y. (1954). Mechanisms of the sex determination in the silkworm *Bombyx mori*. Proc. IXth Int. Genet. Congr. pp. 958–60. *Caryologia*, **6**, suppl. vol.

TAZIMA, Y. (1964). *The Genetics of the Silkworm*. London: Logos Press and Academic Press.

TEISSIER, G. (1954). Conditions d'équilibre d'un couple d'alleles et superiorité des hétérozygotes. *C.R. Acad. Sci. Paris*, **238**: 621–3.

THAKAR, C. V. and DEODIKAR, G. B. (1966). Chromosome number in *Apis florea* Fab. *Curr. Sci.* **35**: 186.

THERMAN, E., PATAU, K., SMITH, D. W. and DEMARS, R. I. (1961). The *D* trisomy syndrome and *XO* gonadal dysgenesis in two sisters. *Genetics*, **13**: 193–204.

THODAY, J. M. (1951). The effect of ionizing radiations on the broad bean root. Part IX. Chromosome breakage and the lethality of ionizing radiations to the root meristem. *Brit. J. Radiol.* N.S. **24**: 572–6, 622–8.

THODAY, J. M. (1958). Effects of disruptive selection: the experimental production of a polymorphic population. *Nature*, **181**: 1124–5.

THODAY, J. M. and BOAM, T. B. (1959). Effects of disruptive selection II. Polymorphism and divergence without isolation. *Heredity*, **13**: 205–18.

THODAY, J. M. and GIBSON, J. B. (1962). Isolation by disruptive selection. *Nature*, **193**: 1164–6.

THOMPSON, C. (1911). The spermatogenesis of an orthopteran *Ceuthophilus latebricola* Scudder, with special reference to the accessory chromosome. *Ann. Rep. Mich. Acad. Sci.* **13**: 97–104.

THOMPSON, M. W. (1965). Genetic implications of heteropyknosis of the *X*-chromosome. *Canad. J. Genet. Cytol.* **7**: 202–13.

THOMSEN, M. (1927). Studien über die Parthenogenese bei einigen Cocciden und Aleurodiden. *Z. Zellforsch.* **5**: 1–116.

THOMSEN, M. (1929). Sex determination in *Lecanium*. *Trans. IVth Int. Congr. Entomol. Ithaca*, pp. 18–24.

THOMSON, J. A. (1969). The interpretation of puff patterns in polytene chromosomes. *Curr. Mod. Biol.* **2**: 333–8.

THOMSON, J. A. and GUNSON, M. M. (1970). Developmental changes in the major inclusion bodies of polytene nuclei from larval tissues of the blowfly, *Calliphora stygia*. *Chromosoma*, **30**: 193–201.

THORPE, W. H. (1930). Biological races in insects and allied groups. *Biol. Rev.* **5**: 177–212.

THORPE, W. H. (1940). Ecology and the future of systematics. In: *The New Systematics* (ed.: J. Huxley). Oxford: Clarendon Press.

THORPE, W. H. (1945). The evolutionary significance of habitat selection. *J. Anim. Ecol.* **14**: 67–70.

THROCKMORTON, L. H. (1962). The problem of phylogeny in the genus *Drosophila*. *Univ. Texas Publ.* **6205**: 207–343.

THROCKMORTON, L. H. (1965). Similarity versus relationship in *Drosophila*. *Syst. Zool.* **14**: 221–36.

TINKLE, D. W. (1959). Observations on the lizards *Cnemidophorus tigris, Cnemidophorus tesselatus* and *Crotaphytus wislezeni*. *Southwestern Nat.* **4**: 195–200.

TJIO, J. H. and LEVAN, A. (1954). Chromosome analysis of three hyperdiploid ascites tumors of the mouse. *Lunds Univ. Årsskr.* N.F. **2**: 50, no. 15.

TJIO, J. H. and LEVAN, A. (1956). The chromosome number of man. *Hereditas*, **42**: 1–6.

TOBARI, Y. N. and KOJIMA, K.-I. (1967). Selective modes associated with inversion karyotypes in *Drosophila ananassae*. I. Frequency-dependent selection. *Genetics*, **57**: 179–88.

TODD, N. B. (1970). Karyotypic fissioning and canid phylogeny. *J. Theor. Biol.* **26**: 445–80.

TOKUNAGA, CH. and HONJI, Y. (1956). Crossing-over in the autosome of the male of *Aphiochaeta xanthina* Speiser. *Kobe Coll. Studies*, **2**: no. 3: 1–9.

TOKUYASU, K. T., PEACOCK, W. J. and HARDY, R. W. (1972). Dynamics of spermiogenesis in *Drosophila melanogaster*. I. Individualization process. *Z. Zellforsch.* **124**: 479–506.

TOLIVER, A. and SIMON, E. H. (1967). DNA synthesis in 5-Bromouracil tolerant HeLa cells. *Exp. Cell Res.* **45**: 603–17.

TOMLIN, S. G. and CALLAN, H. G. (1951). Preliminary account of an electron microscope study of chromosomes from newt oocytes. *Quart. J. Micr. Sci.* **92**: 221–4.

TORVIK-GREB, M. (1935). The chromosomes of *Habrobracon*. *Biol. Bull. Woods Hole*, **68**: 25–34.

TOSI, M. (1959). The chromosome sets of *Gryllotalpa* L. and their geographical distribution. *Caryologia*, **12**: 189–98.

TOWNSEND, J. I. (1952). Genetics of marginal populations of *Drosophila willistoni*. *Evolution*, **6**: 428–42.

TOYOFUKU, Y. (1957). Chromosomal polymorphism found in natural populations of *Drosophila*. *Jap. J. Genet.* **32**: 229–33.

TOYOFUKU, Y. (1958). Further notes on the chromosomal polymorphism in natural populations of *Drosophila* in Hokkaido. VII. *Annot. Zool. Jap.* **31**: 103–8.

TRIANTAPHYLLOU, A. C. (1962). Oogenesis in the root-knot nematode *Meloidogyne javanica*. *Nematologica*, **7**: 105–13.

TRIANTAPHYLLOU, A. C. (1963). Polyploidy and parthenogenesis in the root-knot Nematode *Meloidogyne arenaria*. *J. Morph.* **113**: 489–99.

TRIANTAPHYLLOU, A. C. (1964). Chromosomal forms and reproductive patterns in the root-knot nematode *Meloidogyne hapla*. *Genetics*, **50**: 291–2.

TROEDSSON, P. H. (1944). The behavior of the compound sex chromosomes in the females of certain Hemiptera Heteroptera. *J. Morph.* **75**: 103–47.

TROSKO, J. E. and BREWEN, J. G. (1966). Cytological observations on the strandedness of mammalian metaphase chromosomes. *Cytologia*, **31**: 208–12.

TROSKO, J. E. and WOLFF, S. (1965). Strandedness of *Vicia faba* chromosomes as revealed by enzyme digestion studies. *J. Cell Biol.* **26**: 125–35.

TRUJILLO, J. M., STENIUS, CH., CHRISTIAN, L. C. and OHNO, S. (1962). Chromosomes of the horse, the donkey and the mule. *Chromosoma*, **13**: 243–48.

TSCHERMAK-WOESS, E. (1963). *Strukturtypen der Ruhekerne von Pflanzen und Tieren. Protoplasmatologia*, **1**. Vienna: Springer-Verlag.

TURNER, J. P. (1930). Division and conjugation in *Euplotes patella* Ehrbg., with special reference to the nuclear phenomena. *Univ. Calif. Publ. Zool.* **33**: 193–258.

TURNER, J. R. G. (1967). Why does the genotype not congeal? *Evolution*, **21**: 645–56.

TURPIN, R. and LEJEUNE, J. (1965). *Les Chromosomes Humains (Caryotype normal et variations pathologiques)*. Paris: Gauthier-Villars.

TWITTY, V. C. (1964). Fertility of *Taricha* species hybrids and viability of their offspring. *Proc. Nat. Acad. Sci. U.S.A.* **51**: 156–61.

TYLER, A. (1941). Artificial parthenogenesis. *Biol. Rev.* **16**: 291–335.

UCHIDA, J. A., RAY, M., McRAE, K. N. and BESANT, D. F. (1968). Familial occurrence of trisomy 22. *Amer. J. Hum. Genet.* **20**: 107–18.

UDAGAWA, T. (1952). Karyogram studies in birds. I. Chromosomes of five Passeres. *Cytologia*, **17**: 311–16.

UDAGAWA, T. (1953). Karyogram studies in birds. II. The chromosomes of three species belonging to the Columbidae, Ardeidae and Alcidae. *Annot. Zool. Jap.* **26**: 28–31.

UDAGAWA, T. (1955). Karyogram studies in birds. V. The chromosomes of five Passerine birds. *Annot. Zool. Jap.* **28**: 19–25.

UDAGAWA, T. (1957). Karyogram studies in birds. IX. The chromosomes of five species of Turdidae. *J. Fac. Sci. Hokkaido Univ.* VI, **13**: 338–43.

UESHIMA, N. (1957). Chromosomal polymorphism in a species of field cricket, *Anaxipha pallidula* (Orthoptera: Gryllodea). *J. Nagoya Jogakuin Coll.* **4**: 78–83.

UESHIMA, N. (1963). Chromosome behavior of the *Cimex pilosellus* complex (Cimicidae: Hemiptera). *Chromosoma*, **14**: 511–21.

UESHIMA, N. (1967). Supernumerary chromosomes in the human bed bug, *Cimex lectularius* Linn. (Cimicidae: Hemiptera). *Chromosoma*, **20**: 311–31.

UHL, C. H. (1965). A model of sister-strand crossing-over. *Nature*, **206**: 1003–5.

ULLERICH, F. H. (1961). Achiasmatische Spermatogenese bei der Skorpionsfliege *Panorpa* (Mecoptera). *Chromosoma*, **12**: 215–32.

ULLERICH, F. H. (1963). Geschlechtschromosomen und Geschlechtsbestimmung bei einigen Calliphorinen (Calliphoridae, Diptera). *Chromosoma*, **14**: 45–110.

ULLERICH, F. H. (1966). Karyotyp und DNS-Gehalt von *Bufo bufo, B. viridis, B. bufo* × *B. viridis* und *B. calamita* (Amphibia, Anura). *Chromosoma*, **18**: 316–42.

ULLERICH, F. H. (1967). Weitere Untersuchungen über Chromosomenverhältnisse und DNS-Gehalt bei Anuren (Amphibia). *Chromosoma*, **21**: 345–68.

ULLERICH, F. H. (1970). DNS-Gehalt und Chromosomenstruktur bei Amphibien. *Chromosoma*, **30**: 1–37.

ULLERICH, F. H., BAUER, H. and DIETZ, R. (1964). Geschlechtsbestimmung bei Tipuliden (Nematocera, Diptera). *Chromosoma*, **15**: 591–605.

ULRICH, H. (1962). Generationswechsel und Geslechtsbestimmung einer Gallmücke mit viviparen Larven. *Verh. Deutsch. Zool. Ges. (Wien) 1962*, pp. 139–59.

UNNÉRUS, V., FELLMAN, J. and DE LA CHAPELLE, A. (1967). The length of the human Y-chromosome. *Cytogenetics*, **6**: 213–27.

UPCOTT, M. (1937). The external mechanics of the chromosomes VI. The behaviour of the centromere at meiosis. *Proc. Roy. Soc. Lond.* B, **124**: 336–61.

URBANI, E. (1969). Cytochemical and ultrastructural studies of oogenesis in the Dytiscidae. *Monit. Zool. Ital.* **3**: 55–87.

URBANI, E. and RUSSO-CAIA, S. (1964). Osservazioni citochimiche e autoradiografiche sul metabolismo degli acidi nucleici nella oogenesi di *Dytiscus marginalis* L. *Rend. Ist Sci. Univ. Camerino,* **5**: 19–50.

UTAKOJI, T. (1967). The karyotype of *Microtus montebelli*. *Mammal. Chrom. Newsl.* **8**: 283.

UYENO, T. and MILLER, R. R. (1971). Multiple sex chromosomes in a Mexican cyprinodont fish. *Nature,* **231**: 452–3.

UZZELL, T. M. (1963). Natural triploids in salamanders related to *Ambystoma jeffersonianum*. *Science,* **139**: 113–14.

UZZELL, T. M. (1970). Meiotic mechanisms of naturally occurring unisexual vertebrates. *Amer. Nat.* **104**: 433–45.

UZZELL, T. M. and GOLDBLATT, S. M. (1967). Serum proteins of Salamanders of the *Ambystoma jeffersonianum* complex, and the origin of the triploid species of this group. *Evolution,* **21**: 345–54.

VAARAMA, A. (1953). Chromosome fragmentation and accessory chromosomes in *Orthotrichum tenellum. Hereditas,* **39**: 305–16.

VALKANOV, A. (1938). Cytologische Untersuchungen über die Rhabdocoelen. *Jb. Univ. Sofia (Phys.-Math. Fak.)* **34**: 321–402.

VALLÉE, L. (1959). Recherches sur *Triturus blasii* de l'Isle, hybride naturel de *Triturus cristatus* Laur. × *Triturus marmoratus* Latr. *Mem. Soc. Zool. Fr.* **31**: 1–96 (of reprint).

VAN BRINK, J. M. (1959). L'expression morphologique de la digametie chez les sauropsidés et les monotrèmes. *Chromosoma,* **10**: 1–72.

VANDEL, A. (1928). La parthénogenèse géographique: contribution a l'étude biologique et cytologique de la parthénogenèse naturelle. I. *Bull. Biol. Fr. Belg.* **62**: 164–281.

VANDEL, A. (1931). *La Parthénogenèse*. Paris: Doin Frères.

VANDEL, A. (1934). La parthénogenèse géographique. II. Les mâles triploides d'origine parthénogénétique de *Trichoniscus (Spiloniscus) elizabethae* Herold. *Bull. Biol. Fr. Belg.* **68**: 419–63.

VANDEL, A. (1940). La parthénogenèse géographique. IV. Polyploidie et distribution géographique. *Bull. Biol. Fr. Belg.* **74**: 94–100.

VANDEL, A. (1941). Recherches sur la génétique et la sexualité des isopodes terrestres. VI. Les phenomènes de monogénie chez les Onïscoïdes. *Bull. Biol. Fr. Belg.* **75**: 316–63.

VANDENBERG, S. G., MCKUSICK, V. A. and MCKUSICK, A. B. (1962). Twin data in support of the Lyon hypothesis. *Nature,* **194**: 505–6.

VENDRELY, R. and VENDRELY, C. (1948). La teneur du noyau cellulaire en acide désoxyribonucléique à travers les organes, les individus et les espèces animales. *Experientia,* **4**: 434–6.

VENDRELY, C. and VENDRELY, R. (1949). Sur la teneur individuelle en acide désoxyribonucléique des gametes d'oursins *Arbacia* et *Paracentrotus*. *C.R. Soc. Biol. Paris,* **143**: 1386–7.

VENKATANARASIMHAIAH, C. B. and RAJASEKARASETTY, M. R. (1964). Contributions to the cytology of Indian scorpions. Chromosomal behavior in the male meiosis of *Palamnaeus gravimanus*. *Caryologia,* **17**: 195–201.

VETTURINI, M. (1968). Cromosomi e mutanti di mais. *Maydica,* **13**(8): 1–63.

VIALLI, M. (1967). La quantità di ADN per nucleo negli eritrociti di *Latimeria*. *Rendic. Ist. Lombardo Cl. Sci. Mat. Nat.* **91**: 680–6.

VINCENT, W. S. (1965). The Nucleolus. In: *Genetics Today*, **2** (ed.: S. J. Geerts), pp. 343–58. Oxford: Pergamon Press.

VINCENT, W. S., HALVORSON, H. O., CHEN, H. R. and SHIN, D. (1968). Ribosomal RNA cistrons in single and multinucleolate oöcytes. *Biol. Bull. Woods Hole*, **135**: 441.

VIRKKI, N. (1954). Akzessorische Chromosomen bei zwei Käfern, *Epicometis hirta* Poda und *Oryctes nasicornis* L. (Scarabeidae). *Ann. Acad. Sci. Fenn.* AIV, **26**: 1–19.

VIRKKI, N. (1961). Non-conjugation and late conjugation of the sex chromosomes in the beetles of the genus *Alagoasa* (Chrysomelidae: Alticinae). *Ann. Acad. Sci. Fenn.* AIV, **54**: 1–22.

VIRKKI, N. (1962a). On the cytology of some neotropical elaterids (Coleoptera), with special reference to the neo-XY of the Pyrophorini. *Ann. Acad. Sci. Fenn.* AIV, **61**: 1–21.

VIRKKI, N. (1962b). Chromosomes of certain Meloid beetles from El Salvador. *Ann. Acad. Sci. Fenn.* AIV, **57**: 1–11.

VIRKKI, N. (1963). High chromosome number and giant postreductional sex chromosomes in the beetle *Walterianella venusta* Schaufuss (Chrysomelidae, Alticinae). *J. Agr. Univ. Puerto Rico*, **47**: 154–63.

VIRKKI, N. (1964). On the cytology of some neotropical Chrysomelids (Coleoptera). *Ann. Acad. Sci. Fenn.* AIV, **75**: 1–24.

VIRKKI, N. (1965). Chromosomes of certain Anthribid and Brentid beetles from El Salvador. *Ann. Acad. Sci. Fenn.* AIV, **83**: 1–11.

VIRKKI, N. (1967a). Orientation and segregation of asynaptic multiple sex chromosomes in the male *Omophoita clerica* Erickson (Coleoptera: Alticidae). *Hereditas*, **57**: 275–88.

VIRKKI, N. (1967b). Chromosome relationships in some North American Scarabaeoid beetles, with special reference to *Pleocoma* and *Trox. Canad. J. Genet. Cytol.* **9**: 107–25.

VIRKKI, N. (1967c). Rapid allocyclic changes in the centric and an arm segment of the X-chromosome of *Omphoita superba* Weise (Coleoptera, Alticidae). *Hereditas*, **58**: 262–4.

VIRKKI, N. (1968a). Regular segregation of seven asynaptic sex chromosomes in the male of *Asphaera daniela* Bechyné (Coleoptera, Alticidae). *Caryologia*, **21**: 47–51.

VIRKKI, N. (1968b). A chiasmate sex quadrivalent in the male of an Alticid beetle, *Cyrsylus volkameriae* (F.) *Canad. J. Genet. Cytol.* **10**: 898–907.

VIRKKI, N. (1970). Sex chromosomes and karyotypes of the Alticidae (Coleoptera). *Hereditas*, **64**: 267–82.

VIRKKI, N. and PURCELL, C. M. (1965). Four pairs of chromosomes: the lowest number in Coleoptera. *J. Hered.* **56**: 71–4.

VITAGLIANO, G. (1947). La spermatogenesi e la distribuzione dei chiasmi in *Asellus aquaticus. Pubbl. Staz. Zool. Napoli*, **21**: 164–82.

VITAGLIANO-TADINI, G. (1958). Il probabile significato biologico della monogenia. *Rend. Accad. Naz. Lincei*, ser. VIII, **24**: 562–6.

VITAGLIANO-TADINI, G. (1963). La variabilità del rapporto sessi in *Asellus aquaticus* e la sua determinazione oligogenica *Rend. Accad. Naz. Lincei*, ser. VIII, **34**: 573–82.

VON PFALER-COLANDER, E. (1941). Vergleichend-karyologische Untersuchungen an Lygaeiden. *Acta Zool. Fenn.* **30**: 1–119.

VORHIES, C. T. (1908). The development of the nuclei in the spinning gland cells of *Platyphylax designatus* Walker (Trichopteran). *Biol. Bull. Woods Hole*, **15**: 54–64.

VORONTSOV, N. N. and RAJABLI, S. I. (1967). Chromosome complements and cyto-genetic differentiation of two forms within the superspecies *Ellobius talpinus*. *Tsitologia*, **9**: 848–52.

VOSA, C. G. (1968). A method to reveal sub-chromatids in somatic chromosomes. *Caryologia*, **21**: 381–3.

VOSA, C. G. (1970). Heterochromatin recognition with fluorochromes. *Chromosoma*, **30**: 366–72.

WAGONER, D. E. (1965). The linkage group-karyotype relationship in the house fly (*Musca domestica* L.) *Genetics*, **52**: 482–3.

WAHRMAN, J. (1954*a*). Evolutionary changes in the chromosome complement of the Amelinae (Orthoptera: Mantodea). *Experientia*, **10**: 176–7.

WAHRMAN, J. (1954*b*). Cytological polymorphism and chromosomal evolution in mantids. *Proc. IXth Int. Genet. Congr. Bellagio*, pp. 683–4.

WAHRMAN, J. (1966). A carabid beetle with only eight chromosomes. *Heredity*, **21**: 154–9.

WAHRMAN, J., GOITEIN, R. and NEVO, E. (1969*a*). Mole rat *Spalax*, evolutionary significance of chromosome variation. *Science*, **164**: 82–4.

WAHRMAN, J., GOITEIN, R. and NEVO, E. (1969*b*). Geographic variation of chromo-some forms in *Spalax*, a subterranean mammal of restricted mobility. In: *Comparative Mammalian Cytogenetics* (ed.: K. Benirschke). New York: Springer-Verlag.

WAHRMAN, J. and O'BRIEN, R. (1956). Nuclear content of DNA in chromosomal polymorphism in the genus *Ameles* (Orthoptera: Mantoidea). *J. Morph.* **99**: 259–70.

WAHRMAN, J. and RITTE, U. (1963). Crossing-over in the sex bivalent of male mammals. *Proc. XIth Int. Genet. Congr.* **1**: 125.

WAHRMAN, J. and ZAHAVI, A. (1955). Cytological contributions to the phylogeny and classification of the rodent genus *Gerbillus*. *Nature*, **175**: 600.

WAHRMAN, J. and ZAHAVI, A. (1958). Cytogenetic analysis of mammalian sibling species by means of hybridization. *Proc. Xth Int. Genet. Congr.* **2**: 304–5.

WALD, H. (1936). Cytologic studies on the abnormal development of the eggs of the claret mutant type of *Drosophila simulans*. *Genetics*, **21**: 264–81.

WALEN, K. H. (1963). The pattern of DNA synthesis in the chromosomes of the marsupial *Potorous tridactylis*. *Proc. XIth Int. Genet. Congr. The Hague*, **1**: 106.

WALEN, K. H. (1964). Somatic crossing over in relationship to heterochromatin in *Drosophila melanogaster*. *Genetics*, **49**: 905–23.

WALEN, K. H. (1965). Spatial relationships in the replication of chromosomal DNA. *Genetics*, **51**: 915–21.

WALKER, T. G. (1962). Cytology and evolution in the fern genus *Pteris* L. *Evolution*, **16**: 27–43.

WALLACE, B. (1948). Studies on "sex ratio" in *D. pseudoobscura*. I. Selection and "sex ratio". *Evolution*, **2**: 189–217.

WALLACE, B. (1953). On coadaptation in *Drosophila*. *Amer. Nat.* **87**: 343–58.

WALLACE, B. (1958). The comparison of observed and calculated zygotic distribu-tions. *Evolution*, **12**: 113–14.

WALLACE, B. (1970). *Genetic Load: Its Biological and Conceptual Aspects*. Englewood Cliffs, N.J.: Prentice-Hall.

WALLACE, C. and FAIRALL, N. (1968). Chromosome analysis in the Kruger national park; a rare translocation chromosome in the Kudu. *S. Afr. J. Med. Sci.* **33**: 113–18.

WALLACE, H. and BIRNSTIEL, M. L. (1966). Ribosomal cistrons and the nucleolar organizer. *Biochim. Biophys. Acta*, **114**: 296–310.

WALLACE, J. (1960). The development of anucleolate embryos of *Xenopus laevis*. *J. Embryol. Exp. Morph.* **8**: 405–13.

WALLACE, M. E. (1953). Affinity: a new genetic phenomenon in the house mouse. Evidence from within laboratory stocks. *Nature*, **171**: 27–8.

WALLACE, M. E. (1958). Experimental evidence for a new genetic phenomenon. *Phil. Trans. Roy. Soc. Lond.* B, **241**: 211–54.

WALTERS, M. S. (1970). Evidence on the time of chromosome pairing from the preleptotene spiral stage in *Lilium longiflorum* "Croft". *Chromosoma*, **29**: 375–418.

WALTON, A. C. (1916). *Ascaris canis* (Werner) and *Ascaris felis* (Goeze): a taxonomic and cytological comparison. *Biol. Bull. Woods Hole*, **31**: 364–71.

WALTON, A. C. (1924). Studies on Nematode gametogenesis. *Z. Zell.-u. Gewebel.* **1**: 167–239.

WARD, C. L. (1949). Karyotype variation in *Drosophila*. *Univ. Texas Publ.* **4920**: 70–9.

WARD, C. L. (1952). Chromosome variation in *Drosophila melanica*. *Univ. Texas Publ.* **5204**: 137–57.

WARMKE, H. E. (1946). Sex determination and sex balance in *Melandrium*. *Amer. J. Bot.* **33**: 648–59.

WARTERS, M. (1944). Chromosomal aberrations in wild populations of *Drosophila*. *Univ. Texas Publ.* **4445**: 129–74.

WASSERMAN, A. O. (1957). Factors affecting interbreeding in sympatric species of spadefoots (genus *Scaphiopus*). *Evolution*, **11**: 320–38.

WASSERMAN, A. O. (1970). Polyploidy in the common tree toad *Hyla versicolor* LeConte. *Science*, **167**: 385–6.

WASSERMAN, M. (1954). Cytological studies of the *repleta* group. *Univ. Texas Publ.* **5422**: 130–52.

WASSERMAN, M. (1960). Cytological and phylogenetic relationships in the *repleta* group of the genus *Drosophila*. *Proc. Nat. Acad. Sci. U.S.A.* **46**: 842–59.

WASSERMAN, M. (1962a). Cytological studies of the *repleta* group of the genus *Drosophila*. III. The *mercatorum* subgroup. Univ. Texas Publ. **6205**: *Studies in Genet.* **2**: 63–71.

WASSERMAN, M. (1962b). Cytological studies of the *repleta* group of the genus *Drosophila*. IV. The *hydei* subgroup. *Univ. Texas Publ.* **6205**, *Studies in Genet.* **2**: 73–83.

WASSERMAN, M. (1962c). Cytological studies of the *repleta* group of the genus *Drosophila*. V. The *mulleri* subgroup. *Univ. Texas Publ.* **6205**, *Studies in Genet.* **2**: 85–117.

WASSERMAN, M. (1962d). Cytological studies of the *repleta* group of the genus *Drosophila*. VI. The *fasciola* subgroup. *Univ. Texas Publ.* **6205**, *Studies in Genet.* **2**: 119–34.

WASSERMAN, M. (1963). Cytology and phylogeny of *Drosophila*. *Amer. Nat.* **97**: 333–52.

WASSERMAN, M. and WILSON, F. D. (1957). Further studies on the *repleta* group. *Univ. Texas Publ.* **5721**: 132–56.

WATERHOUSE, F. L. and SANDERSON, A. R. (1958). Geographical color polymorphism and chromosome constitution in sympatric species of saw flies. *Nature*, **182**: 477.

WATSON, I. D. and CALLAN, H. G. (1963). The form of bivalent chromosomes in newt oocytes at first metaphase of meiosis. *Quart. J. Micr. Sci.* **104**: 281–95.

WATSON, J. D. and CRICK, F. H. C. (1953). The structure of DNA. *Cold Spr. Harb. Symp. Quant. Biol.* **18**: 123–31.

WEBB, G. C. (1972). Unpublished Ph.D. thesis, University of Melbourne.

WEBB, G. C. and WHITE, M. J. D. (1970). A new interpretation of the sex-determining mechanism of the European earwig, *Forficula auricularia. Experientia*, **26**: 1387.

WEBER, F. (1966). Beitrag zur Karyotypanalyse der Laufkäfergattung *Carabus* L. (Coleoptera). *Chromosoma*, **18**: 467–76.

WEILER, C. and OHNO, S. (1962). Cytological confirmation of female heterogamety in the African Water Frog (*Xenopus laevis*). *Cytogenetics*, **1**: 217–23.

WEISMANN, A. (1880). Beiträge zur Naturgeschichte der Daphnoiden, VI, VII. *Z. wiss. Zool.* **33**: 55–270.

WEISS, I. (1958). Vergleichend Kernvolumetrische Untersuchungen an *Cricetus cricetus, Cricetulus barabensis griseus* und dem angeblich tetraploiden *Mesocricetus auratus* (Gold-hamster). *Z. mikr. anat. Forsch.* **64**: 44–58.

WELSHONS, W. J. (1958). The analysis of a pseudoallelic recessive lethal system at the *Notch* locus of *Drosophila melanogaster. Cold Spr. Harb. Symp. Quant. Biol.* **23**: 171–6.

WELSHONS, W. J. (1965). Analysis of a gene in *Drosophila. Science*, **150**: 1122–9.

WELSHONS, W. J. and VON HALLE, E. S. (1962). Pseudoallelism at the *Notch* locus in *Drosophila. Genetics*, **47**: 743–59.

WENDROWSKY, V. (1928). Über die Chromosomenkomplexe der Hirudineen. *Z. Zellforsch.* **8**: 153–75.

WENRICH, D. H. (1917). Synapsis and chromosome organization in *Chorthippus* (*Stenobothrus*) *curtipennis* and *Trimerotropis suffusa* (Orthoptera) *J. Morph.* **29**: 479–516.

WERNER, O. S. (1927). The chromosomes of the Indian runner duck. *Biol. Bull. Woods Hole*, **52**: 330–72.

WERNER, Y. L. (1956). Chromosome numbers of some male Geckos (Reptilia: Gekkonoidea). *Bull. Res. Counc. Israel*, **5B**: 319.

WESTERGAARD, M. (1940). Studies on cytology and sex determination in polyploid forms of *Melandrium album. Dansk Bot. Ark.* **10**, no. 5: 1–131.

WESTERGAARD, M. (1953). Über den Mechanismus der Geschlechtsbestimmung bei *Melandrium album. Naturwiss.* **40**: 253–60.

WESTERGAARD, M. (1958). The mechanism of sex determination in dioecious flowering plants. *Adv. in Genet.* **9**: 217–81.

WESTERGAARD, M. (1964). Studies on the mechanism of crossing-over. I. Theoretical considerations. *C.R. Trav. Lab. Carlsberg*, **34**: 359–405.

WESTERMAN, M. (1968). The effect of X-irradiation on male meiosis in *Schistocerca gregaria* (Forskål). II. The induction of chromosome mutations. *Chromosoma*, **24**: 17–36.

WESTERMAN, M. (1969). Parallel polymorphism for supernumerary segments in *Chorthippus parallelus* (Zetterstedt). II. French populations. *Chromosoma*, **26**: 7–21.

WETTSTEIN, R. and SOTELO, J. R. (1965). Fine structure of meiotic chromosomes. The elementary components of metaphase chromosomes of *Gryllus argentinus. J. Ultrastr. Res.* **13**: 367–81.

WETTSTEIN, R. and SOTELO, J. R. (1967). Electron microscope serial reconstruction of the spermatocyte I nuclei at pachytene. *J. Microsc.* **6**: 557–76.

WHARTON, L. T. (1942). Analysis of the *repleta* group of *Drosophila. Univ. Texas Publ.* **4228**: 23–52.

WHARTON, L. T. (1943). Analysis of the metaphase and salivary chromosome morphology within the genus *Drosophila. Univ. Texas Publ.* **4313**: 282–319.

WHARTON, L. T. (1944). Interspecific hybridization in the repleta group. *Univ. Texas Publ.* **4445**: 175–93.

WHITE, M. J. D. (1933). Tetraploid spermatocytes in a locust, *Schistocerca gregaria. Cytologia*, **5**: 135–9.

WHITE, M. J. D. (1935*a*). Eine neue form von Tetraploidie nach Röntgenbestrahlung. *Naturwiss.* **23**: 390–1.

WHITE, M. J. D. (1935*b*). The effects of X-rays on mitosis in the spermatogonial divisions of *Locusta migratoria* L. *Proc. Roy. Soc. Lond.* B, **119**,: 61–84.

WHITE, M. J. D. (1938). A new and anomalous type of meiosis in a mantid, *Callimantis antillarum* Sanssure. *Proc. Roy. Soc. Lond.* B, **125**: 516–23.

WHITE, M. J. D. (1940*a*). The heteropycnosis of sex chromosomes and its interpretation in terms of spiral structure. *J. Genet.* **40**: 67–82.

WHITE, M. J. D. (1940*b*). The origin and evolution of multiple sex-chromosome mechanisms. *J. Genet.* **40**: 303–36.

WHITE, M. J. D. (1940*c*). A translocation in a wild population of grasshoppers. *J. Hered.* **31**: 137–40.

WHITE, M. J. D. (1941*a*). The evolution of the sex chromosomes. I. The XO and X_1X_2Y mechanisms in praying mantids. *J. Genet.* **42**: 143–72.

WHITE, M. J. D. (1941*b*). The evolution of the sex chromosomes. II. The X-chromosome in the Tettigonidae and Acrididae and the principle of 'evolutionary isolation' of the X. *J. Genet.* **42**: 173–90.

WHITE, M. J. D. (1945). *Animal Cytology and Evolution.* (1st ed.) Cambridge Univ. Press.

WHITE, M. J. D. (1946*a*). The cytology of the Cecidomyidae (Diptera). I. Polyploidy and polyteny in salivary gland cells of *Lestodiplosis* spp. *J. Morph.* **78**: 201–19.

WHITE, M. J. D. (1946*b*). The cytology of the Cecidomyidae (Diptera). II. The chromosome cycle and anomalous spermatogenesis of *Miastor. J. Morph.* **79**: 323–70.

WHITE, M. J. D. (1946*c*). The evidence against polyploidy in sexually reproducing animals. *Amer. Nat.* **80**: 610–19.

WHITE, M. J. D. (1946*d*). The spermatogenesis of hybrids between *Triturus cristatus* and *T. marmoratus* (Urodela). *J. Exp. Zool.* **102**: 179–207.

WHITE, M. J. D. (1947). The cytology of the Cecidomyidae (Diptera). III. The spermatogenesis of *Taxomyia taxi. J. Morph.* **80**: 1–24.

WHITE, M. J. D. (1948*a*). The cytology of the Cecidomyidae (Diptera). IV. The salivary gland chromosomes of several species. *J. Morph.* **82**: 53–80.

WHITE, M. J. D. (1948*b*). The chromosomes of the parthenogenetic mantid *Brunneria borealis. Evolution*, **2**: 90–3.

WHITE, M. J. D. (1949). A cytological survey of wild populations of *Trimerotropis* and *Circotettix* (Orthoptera, Acrididae). I. The chromosomes of twelve species. *Genetics*, **34**: 537–63.

WHITE, M. J. D. (1950*a*). Cytological studies on gall midges. *Univ. Texas Publ.* **5007**: 1–80.

WHITE, M. J. D. (1950*b*). Cytological polymorphism in natural populations of grasshoppers. *Yearb. Amer. Phil. Soc. 1949*, pp. 183–5.

White, M. J. D. (1951a). A cytological survey of wild populations of *Trimerotropis* and *Circotettix* (Orthoptera, Acrididae). II. Racial differentiation in *T. sparsa*. *Genetics*, **36**: 31–53.

White, M. J. D. (1951b). Cytogenetics of orthopteroid insects. *Adv. in Genet.* **4**: 267–330.

White, M. J. D. (1951c). Structural heterozygosity in natural populations of the grasshopper *Trimerotropis sparsa*. *Evolution*, **5**: 376–94.

White, M. J. D. (1951d). Cytological polymorphism and racial differentiation in grasshopper populations. *Yearb. Amer. Phil. Soc. 1950*, pp. 158–60.

White, M. J. D. (1951e). Evolution of cytogenetic mechanisms in animals. A chapter in: *Genetics in the Twentieth Century*. New York: Macmillan.

White, M. J. D. (1951f). Structural heterozygosity in natural populations of the grasshopper *Trimerotropis sparsa*. *Evolution*, **5**: 376–94.

White, M. J. D. (1953). Multiple sex-chromosome mechanisms in the grasshopper genus *Paratylotropidia*. *Amer. Nat.* **87**: 237–44.

White, M. J. D. (1954). An extreme form of chiasma localization in a species of *Bryodema* (Orthoptera, Acrididae). *Evolution*, **8**: 350–8.

White, M. J. D. (1954). *Animal Cytology and Evolution*. (2nd ed.) Cambridge Univ. Press.

White, M. J. D. (1956). Adaptive chromosomal polymorphism in an Australian grasshopper. *Evolution*, **10**: 298–313.

White, M. J. D. (1957a). Cytogenetics of the grasshopper *Moraba scurra*. I. Meiosis of interracial and interpopulation hybrids. *Aust. J. Zool.* **5**: 285–304.

White, M. J. D. (1957b). Cytogenetics of the grasshopper *Moraba scurra*. II. Heterotic systems and their interaction. *Aust. J. Zool.* **5**: 305–37.

White, M. J. D. (1957c). Cytogenetics of the grasshopper *Moraba scurra*. IV. Heterozygosity for 'elastic constrictions'. *Aust. J. Zool.* **5**: 348–54.

White, M. J. D. (1957d). Some general problems of chromosomal evolution and speciation in animals. *Surv. Biol. Progr.* **3**: 109–47.

White, M. J. D. (1957e). An interpretation of the unique sex-chromosome mechanism of the rodent *Ellobius lutescens* Thomas. *Proc. Zool. Soc. Calcutta, Mookerjee Memor. Vol.* pp. 113–14.

White, M. J. D. (1958). Restrictions on recombination in grasshopper populations and species. *Cold Spr. Harb. Symp. Quant. Biol.* **23**: 307–17.

White, M. J. D. (1959a). Speciation in animals. *Aust. J. Sci.* **22**: 32–9.

White, M. J. D. (1959b). Telomeres and terminal chiasmata – a reinterpretation. *Univ. Texas Publ.* **5914**: 107–11.

White, M. J. D. (1960). Are there no mammal species with *XO* males – and if not, why not? *Amer. Nat.* **94**: 301–4.

White, M. J. D. (1961a). The role of chromosomal translocations in urodele evolution in the light of work on grasshoppers. *Amer. Nat.* **95**: 315–21.

White, M. J. D. (1961b). Cytogenetics of the grasshopper *Moraba scurra*. VI. A spontaneous pericentric inversion. *Aust. J. Zool.* **9**: 784–90.

White, M. J. D. (1962). A unique type of sex-chromosome mechanism in an Australian mantid. *Evolution*, **16**: 75–85.

White, M. J. D. (1963). Cytogenetics of the grasshopper *Moraba scurra*. VIII. A complex spontaneous translocation. *Chromosoma*, **14**: 140–5.

White, M. J. D. (1964). Cytogenetic mechanisms in insect reproduction. *Insect Reproduction, symp. no. 2, Roy. Entomol. Soc. Lond.* pp. 1–12.

White, M. J. D. (1965a). Chiasmatic and achiasmatic meiosis in African Eumastacid grasshoppers. *Chromosoma*, **16**: 271–307.

WHITE, M. J. D. (1965b). Sex chromosomes and meiotic mechanisms in some African and Australian mantids. *Chromosoma*, **16**: 521–47.

WHITE, M. J. D. (1966a). A case of spontaneous chromosome breakage at a specific locus occurring at meiosis. *Aust. J. Zool.* **14**: 1027–34.

WHITE, M. J. D. (1966b). Further studies on the cytology and distribution of the Australian parthenogenetic grasshopper, *Moraba virgo*. *Rev. Suisse Zool.* **73**: 383–98.

WHITE, M. J. D. (1967). Karyotypes of some members of the grasshopper families Lentulidae and Charilaidae. *Cytologia*, **32**: 184–9.

WHITE, M. J. D. (1968a). Models of speciation. *Science*, **159**: 1065–70.

WHITE, M. J. D. (1968b). Karyotypes and nuclear size in the spermatogenesis of grasshoppers belonging to the subfamilies Gomphomastacinae, Chininae and Biroellinae (Orthoptera, Eumastacidae). *Caryologia*, **21**: 167–79.

WHITE, M. J. D. (1969). Chromosomal rearrangements and speciation. *Ann. Rev. Genet.* **3**: 75–98.

WHITE, M. J. D. (1970a). Heterozygosity and genetic polymorphism in parthenogenetic animals. *Evol. Biol.: Essays in Evolution and Genetics in honor of Theodosius Dobzhansky*, pp. 237–62.

WHITE, M. J. D. (1970b). Asymmetry of heteropycnosis in tetraploid cells of a grasshopper. *Chromosoma*, **29**: 51–61.

WHITE, M. J. D. (1970c). Karyotypes and meiotic mechanisms of some eumastacid grasshoppers from East Africa, Madagascar, India and South America. *Chromosoma*, **30**: 62–97.

WHITE, M. J. D. (1970d). Cytogenetics of speciation. *J. Aust. Entomol. Soc.* **9**: 1–6.

WHITE, M. J. D. (1971a). The chromosomes of *Hemimerus bouvieri* Chopard (Dermaptera). *Chromosoma*, **34**: 183–9.

WHITE, M. J. D. (1971b). Unpublished data.

WHITE, M. J. D. (1971c). The value of cytology in taxonomic research on Orthoptera. *Proc. Int. Study Conf. on Current and Future Problems of Acridology.*

WHITE, M. J. D. and ANDREW, L. E. (1960). Cytogenetics of the grasshopper *Moraba scurra*. V. Biometric effects of chromosomal inversions. *Evolution*, **14**: 284–92.

WHITE, M. J. D. and ANDREW, L. E. (1962). Effects of chromosomal inversions on size and relative viability in the grasshopper *Moraba scurra*. In: *The Evolution of Living Organisms*, pp. 94–101. Melbourne Univ. Press.

WHITE, M. J. D., BLACKITH, R. E., BLACKITH, R. M., and CHENEY, J. (1967). Cytogenetics of the *viatica* group of Morabine grasshoppers. I. The 'coastal' species. *Aust. J. Zool.* **15**: 263–302.

WHITE, M. J. D., CARSON, H. L. and CHENEY, J. (1964). Chromosomal races in the Australian grasshopper *Moraba viatica* in a zone of geographic overlap. *Evolution*, **18**: 417–29.

WHITE, M. J. D. and CHENEY, J. (1966). Cytogenetics of the *cultrata* group of Morabine grasshoppers. I. A group of species with XY and X_1X_2Y sex-chromosome mechanisms. *Aust. J. Zool.* **14**: 821–34.

WHITE, M. J. D. and CHENEY, J. (1972). Cytogenetics of a group of morabine grasshoppers with XY and X_1X_2Y males. *Chromosomes Today*, **3**: 177–96.

WHITE, M. J. D., CHENEY, J. and KEY, K. H. L. (1963). A parthenogenetic species of grasshopper with complex structural heterozygosity (Orthoptera: Acridoidea). *Aust. J. Zool.* **11**: 1–19.

WHITE, M. J. D. and CHINNICK, L. J. (1957). Cytogenetics of the grasshopper *Moraba scurra*. III. Distribution of the 15- and 17-chromosome races. *Aust. J. Zool.* **5**: 338–47.

WHITE, M. J. D. and KEY, K. H. L. (1957). A cytotaxonomic study of the *pusilla* group of species in the genus *Austroicetes* Uv. *Aust. J. Zool.* **5**: 56–87.

WHITE, M. J. D., KEY, K. H. L., ANDRÉ, M. and CHENEY, J. (1969). Cytogenetics of the *viatica* group of Morabine grasshoppers. II. Kangaroo Island populations. *Aust. J. Zool.* **17**: 313–28.

WHITE, M. J. D., LEWONTIN, R. C. and ANDREW, L. E. (1963). Cytogenetics of the grasshopper *Moraba scurra*. VII. Geographic variation of adaptive properties of inversions. *Evolution,* **17**: 147–62.

WHITE, M. J. D., MESA, A. and MESA, R. (1967). Neo-XY sex-chromosome mechanisms in two species of Tettigonioidea. *Cytologia,* **32**: 190–9.

WHITE, M. J. D. and MORLEY, F. H. W. (1955). Effects of pericentric rearrangements on recombination in grasshopper chromosomes. *Genetics,* **40**: 604–19.

WHITE, M. J. D. and NICKERSON, N. H. (1951). Structural heterozygosity in a very rare species of grasshopper. *Amer. Nat.* **85**: 239–46.

WHITE, M. J. D., and WEBB, G. C. (1968). Origin and evolution of parthenogenetic reproduction in the grasshopper *Moraba virgo* (Eumastacidae: Morabinae). *Aust. J. Zool.* **16**: 647–71.

WHITEHOUSE, H. L. K. (1963*a*). A theory of crossing-over and gene conversion involving hybrid DNA. *Proc. XIth Int. Genet. Congr. The Hague,* **2**: 87–8.

WHITEHOUSE, H. L. K. (1963*b*). A theory of crossing-over by means of hybrid deoxyribonucleic acid. *Nature,* **199**: 1034–40.

WHITEHOUSE, H. L. K. (1965). Crossing-over. *Sci. Progr.* **53**: 285–96.

WHITEHOUSE, H. L. K. (1966). An operator model of crossing-over. *Nature,* **211**: 708–13.

WHITEHOUSE, H. L. K. (1970). The mechanism of genetic recombination. *Biol. Rev.* **45**: 265–315.

WHITEHOUSE, H. L. K. and HASTINGS, P. J. (1965). The analysis of genetic recombination on the polaron hybrid DNA model. *Genet. Res.* **6**: 27–92.

WHITING, A. R. (1927). Genetic evidence of diploid males in *Habrobracon*. *Biol. Bull. Woods Hole,* **53**: 438–49.

WHITING, P. W. (1939). Multiple alleles in sex determination of *Habrobracon*. *J. Morph.* **66**: 323–55.

WHITING, P. W. (1940). Sex linkage in *Pteromalus*. *Amer. Nat.* **74**: 377–9.

WHITING, P. W. (1943*a*). Multiple alleles in complementary sex determination of *Habrobracon. Genetics,* **28**: 365–82.

WHITING, P. W. (1943*b*). Intersexual females and intersexuality in *Habrobracon. Biol. Bull. Woods Hole,* **85**: 238–43.

WHITING, P. W. (1945). The evolution of male haploidy. *Quart. Rev. Biol.* **20**: 231–60.

WHITING, P. W. (1947). Some experiments with *Melittobia* and other wasps. *J. Hered.* **38**: 11–20.

WHITING, P. W. (1954). Comparable mutant eye colors in *Mormoniella* and *Pachycrepoideus* (Hymenoptera: Pteromalidae). *Evolution,* **8**: 135–47.

WHITING, P. W. (1960). Polyploidy in *Mormoniella. Genetics,* **45**: 949–70.

WHITING, P. W. and WHITING, A. R. (1925). Diploid males from fertilized eggs in Hymenoptera. *Science,* **62**: 437.

WHITTEN, J. M. (1965). Differential deoxyribonucleic acid replication in the giant foot-pad cells of *Sarcophaga bullata. Nature,* **208**: 1019–21.

WHITTEN, J. M. (1969). Coordinated development in the foot pad of the fly *Sarcophaga bullata* during metamorphosis: changing puffing patterns of the giant cell chromosomes. *Chromosoma,* **26**: 215–44.

WHITTEN, M. J. (1965). Chromosome numbers in some Australian leafhoppers (Homoptera Auchenorrhyncha). *Proc. Linn. Soc. N.S.W.* **90**: 78–85.

WHITTEN, M. J. and TAYLOR, W. C. (1969). Chromosomal polymorphism in an Australian leafhopper (Homoptera: Cicadellidae). *Chromosoma*, **26**: 1–6.

WHITTINGHILL, M. (1937). Induced crossing-over in *Drosophila* males and its probable nature. *Genetics*, **22**: 114–29.

WHITTINGHILL, M. (1947). Spermatogonial crossing-over between the third chromosomes in the presence of the curly inversions of *Drosophila melanogaster*. *Genetics*, **32**: 608–14.

WHITTINGHILL, M. (1955). Cross-over variability and induced crossing-over. *J. Cell. Comp. Physiol.* **45**, suppl. 2: 189–220.

WICKBOM, T. (1943). Cytological studies in the family Cyprinodontidae. *Hereditas*, **29**: 1–24.

WICKBOM, T. (1945). Cytological studies on Dipnoi, Urodela, Anura and *Emys*. *Hereditas*, **31**: 241–6.

WICKBOM, T. (1949). A new list of chromosome numbers in Anura. *Hereditas*, **35**: 242–5.

WICKBOM, T. (1950*a*). The chromosomes of *Pipa pipa*. *Hereditas*, **36**: 363–6.

WICKBOM, T. (1950*b*). The chromosomes of *Ascaphus truei* and the evolution of the anuran karyotypes. *Hereditas*, **36**: 406–18.

WILLIAMS, R. J. (1956). *Biochemical Individuality*. New York: John Wiley.

WILLIAMSON, D. L. and EHRMAN, L. (1967). Induction of hybrid sterility in non-hybrid males of *Drosophila paulistorum*. *Genetics*, **55**: 131–40.

WILSON, E. B. (1905*a*). Studies on chromosomes. I. The behavior of the idiochromosomes in Hemiptera. *J. Exp. Zool.* **2**: 371–405.

WILSON, E. B. (1905*b*). Studies on chromosomes II. The paired microchromosomes, idiochromosomes and heterotropic chromosomes in Hemiptera. *J. Exp. Zool.* **2**: 507–45.

WILSON, E. B. (1906). Studies on chromosomes. III. The sexual difference of the chromosome groups in Hemiptera, with some considerations on the determination and inheritance of sex. *J. Exp. Zool.* **3**: 1–40.

WILSON, E. B. (1907). Notes on the chromosome group of *Metapodius* and *Banasa*. *Biol. Bull. Woods Hole*, **12**: 303–13.

WILSON, E. B. (1909*a*). Studies on chromosomes. V. The chromosomes of *Metapodius*. A contribution to the hypothesis of the genetic continuity of chromosomes. *J. Exp. Zool.* **6**: 147–205.

WILSON, E. B. (1909*b*). The female chromosome groups in *Syromastes* and *Pyrrhocoris*. *Biol. Bull. Woods Hole*, **16**: 199–204.

WILSON, E. B. (1909*c*). Studies on chromosomes. IV. The accessory chromosome in *Syromastes* and *Pyrrhocoris*, with a comparative review of the types of sexual difference of the chromosome groups. *J. Exp. Zool.* **6**: 69–99.

WILSON, E. B. (1910). Studies on chromosomes. VI. A new type of chromosome combination in *Metapodius*. *J. Exp. Zool.* **9**: 53–78.

WILSON, E. B. (1911). Studies on chromosomes. VII. A review of the chromosomes of *Nezara*; with some more general considerations. *J. Morph.* **22**: 71–110.

WILSON, E. B. (1912). Studies on chromosomes. VIII. Observations on the maturation phenomenon in certain Hemiptera and other forms, with considerations on synapsis and reduction. *J. Exp. Zool.* **13**: 345–431.

WILSON, E. B. (1913). A chromatoid body simulating an accessory chromosome in *Pentatoma*. *Biol. Bull. Woods Hole*, **24**: 392–411.

WILSON, E. B. (1925). *The Cell in Development and Heredity*. New York: Macmillan.

WILSON, E. B. (1931). The distribution of sperm-forming material in scorpions. *J. Morph.* **52**: 429–83.

WILSON, E. B. (1932). Polyploidy and metaphase patterns. *J. Morph.* **53**: 443–71.

WILSON, G. B., SPARROW, A. H. and POND, V. (1959). Sub-chromatid rearrangements in *Trillium erectum*. I. Origin and nature of configurations induced by ionizing radiation. *Amer. J. Bot.* **46**: 309–16.

WIMBER, D. E. and PRENSKY, W. (1963). Autoradiography with meiotic chromosomes of the male newt (*Triturus viridescens*) using H³-thymidine. *Genetics*, **48**: 1731–8.

WIMBER, D. E. and STEFFENSEN, D. M. (1970). Localization of 5s RNA genes on *Drosophila* chromosomes by RNA–DNA hybridization. *Science*, **170**: 639–41.

WINGE, Ö. (1923). Crossing-over between the *X*- and *Y*-chromosomes in *Lebistes*. *J. Genet.* **13**: 201–27.

WINGE, Ö. (1932). The nature of sex chromosomes. *Proc. VIth Int. Genet. Congr. Ithaca*, **1**: 343–55.

WINGE, Ö. (1934). The experimental alteration of sex chromosomes into autosomes and vice versa, as illustrated by *Lebistes*. *C.R. Trav. Lab. Carlsberg*, **21**: 1–49.

WINGE, Ö. and DITLEVSEN, E. (1947). Color inheritance and sex determination in *Lebistes*. *Heredity*, **1**: 65–83.

WITSCHI, E. (1923). Über die genetische Konstitution der Froschzwitter. *Biol. Zbl.* **43**: 83–96.

WODSEDALEK, J. E. (1916). Causes of sterility in the mule. *Biol. Bull. Woods Hole*, **30**: 1–56.

WOLF, B. E. (1960). Zur Karyologie der Eireifung und Fürchung bei *Chloeon dipterum* L. (Bengtsson) (Ephemerida, Baetididae). *Biol. Zentralbl.* **79**: 153–98.

WOLF, E. (1941). Die Chromosomen in der Spermatogenese einiger Nematoceren. *Chromosoma*, **2**: 192–246.

WOLF, E. (1946). Chromosomenuntersuchungen au Insekten. *Z. Naturf.* **1**: 108–9.

WOLF, E. (1950). Die Chromosomen in der Spermatogenese der Dipteren *Phryne* und *Mycetobia. Chromosoma*, **4**: 148–204.

WOLFE, S. L. and JOHN, B. (1965). The organization and ultrastructure of meiotic chromosomes in *Oncopeltus fasciatus. Chromosoma*, **17**: 85–103.

WOLFF, S. (1961). Radiation genetics. In: *Mechanisms in Radiobiology* (ed.: M. Errera and A. Forssberg). New York: Academic Press.

WOLFF, S. (1969). Strandedness of chromosomes. *Int. Rev. Cytol.* **25**: 279–96.

WOLFF, S. and HEDDLE, J. A. (1968). Some chromosome studies with tritiated thymidine. In: *Replication and Recombination of Genetic Material* (ed.: W. J. Peacock and R. D. Brock), pp. 105–13. Canberra: Aust. Acad. of Sci.

WOLFF, S. and LUIPPOLD, H. E. (1964). Chromosome splitting as revealed by combined X-ray and labelling experiments. *Exp. Cell Res.* **34**: 548–56.

WOLSTENHOLME, D. R. (1965). The distribution of DNA and RNA in salivary gland chromosomes of *Chironomus tentans* as revealed by fluorescence microscopy. *Chromosoma*, **17**: 219–29.

WOLSTENHOLME, D. R. (1966*a*). Electron microscope identification of the interphase chromosomes of *Amoeba proteus* and *Amoeba discoides* using autoradiography: with some notes on helices and other nuclear components. *Chromosoma*, **19**: 449–68.

WOLSTENHOLME, D. R. (1966*b*). Direct evidence for the presence of DNA in interbands of *Drosophila* salivary gland chromosomes. *Genetics*, **53**: 357–60.

WONG, S. K. and THORNTON, I. W. B. (1966). Chromosome numbers of some psocid genera (Psocoptera). *Nature*, **211**: 214–15.

WOODARD, J., GELBER, B. and SWIFT, H. (1961). Nucleoprotein changes during the mitotic cycle in *Paramecium aurelia*. *Exp. Cell Res.* **23**: 258–64.

WOODARD, J., GOROVSKY, M. and SWIFT, H. (1966). DNA content of a chromosome of *Trillium erectum*: effect of cold treatment. *Science*, **151**: 215–16.

WOODARD, J. and SWIFT, H. (1964). The DNA content of cold treated chromosomes. *Exp. Cell Res.* **34**: 131–7.

WOODS, P. S. and SCHAIRER, M. U. (1959). Distribution of newly synthesized deoxyribonucleic acid in dividing chromosomes. *Nature*, **183**: 303–5.

WOYKE, J. (1963*a*). Drone larvae from fertilized eggs of the honeybee. *J. Apic. Res.* **2**: 19–24.

WOYKE, J. (1963*b*). What happens to diploid drone larvae in a honeybee colony. *J. Apic. Res.* **2**: 73–5.

WOYKE, J. (1965). Genetic proof of the origin of drones from fertilized eggs of the honeybee. *J. Apic. Res.* **4**: 7–11.

WOYKE, J. (1967). Diploid drone substance – cannibalism substance. *Proc. XXIth Int. Apic. Congr. Maryland*, pp. 57–8.

WRIGHT, J. W. and LOWE, C. H. (1967*a*). Evolution of the alloploid parthenospecies *Cnemidophorus tesselatus* (Say). *Mamm. Chrom. Newsl.* **8**: 95–6.

WRIGHT, J. W. and LOWE, C. H. (1967*b*). Hybridization in nature between partheno-genetic and bisexual species of whiptail lizards (genus *Cnemidophorus*). *Amer. Mus. Novit.* **2286**: 1–36.

WRIGHT, S. (1922). Coefficients of inbreeding and relationship. *Amer. Nat.* **56**: 330–8.

WRIGHT, S. (1941). On the probability of fixation of reciprocal translocations. *Amer. Nat.* **75**: 512–22.

WÜLKER, W., SUBLETTE, J. E. and MARTIN, J. (1968). Zur Cytotaxionomie nordamerikanischer *Chironomus*-Arten. *Ann. Zool. Fenn.* **5**: 155–8.

WURSTER, D. H. (1969). Cytogenetic and phylogenetic studies in Carnivora. In: *Comparative Mammalian Cytogenetics* (ed.: K. Benirschke), pp. 310–29. New York: Springer-Verlag.

WURSTER, D. H. and BENIRSCHKE, K. (1968*a*). The chromosomes of the Great Indian Rhinoceros (*Rhinoceros unicornis* L.). *Experientia*, **24**: 511.

WURSTER, D. H. and BENIRSCHKE, K. (1968*b*). Chromosome studies in the super-family Bovoidea. *Chromosoma*, **25**: 152–71.

WURSTER, D. H. and BENIRSCHKE, K. (1970). Indian Muntjak, *Muntiacus muntjak*: a deer with a low diploid chromosome number. *Science*, **168**: 1364–6.

WURSTER, D. H., BENIRSCHKE, K. and NOELKE, N. (1968). Unusually large sex chromosomes in the Sitatunga (*Tragelaphus spekei*) and the Blackbuck (*Antilope cervicapra*). *Chromosoma*, **23**: 317–23.

WYATT, I. J. (1961). Pupal paedogenesis in the Cecidomyiidae (Diptera). I. *Proc. Roy. Entomol. Soc. Lond.* **36**: 133–43.

WYSOKI, M. and SWIRSKI, E. (1968). Karyotypes and sex determination of ten species of phytoseiid mites (Acarina: Mesostigmata). *Genetica*, **39**: 220–8.

XAVIER, A. DA C. M. (1945). Cariologia comparada de alguns Hemipteros Heteropteros (Pentatomideos e Coreideos). *Mem. Est. Mus. Zool. Univ. Coimbra*, no. 163, pp. 1–105.

YAMAMOTO, H. (1970). Heat shock induced puffing changes in Balbiani rings. *Chromosoma*, **32**: 171–90.

YAMAMOTO, T. (1963). Induction of reversal in sex differentiation of *YY* zygotes in the medaka *Oryzias latipes*. *Genetics*, **48**: 293–306.

YAMAMOTO, T. (1964*a*). The problem of the viability of *YY* zygotes in the medaka, *Oryzias latipes*. *Genetics*, **50**: 45–58.

YAMAMOTO, T. (1964*b*). Linkage map of sex chromosomes in the medaka, *Oryzias latipes*. *Genetics*, **50**: 59–64.

YAMASHINA, Y. (1942). A revised study of the chromosomes of the muscovy duck, the domestic duck and their hybrids. *Cytologia*, **12**: 163–9.

YAMASHINA, Y. (1943). Studies on sterility in hybrid birds. IV. Cytological researches in hybrids in the family Phasianidae. *J. Fac. Sci. Hokkaido Univ.* VI, **8**: 307–86.

YAMASHINA, Y. (1944). Karyotype studies in birds. I. Comparative morphology of chromosomes in seventeen races of domestic fowl. *Cytologia*, **13**: 271–96.

YAMASHINA, Y. (1950). The chromosomes of some birds belonging to the Gressores, Pygopodes and Alectorides. *Kromosomo*, **7**: 283–88.

YAMASHINA, Y. (1951). Studies on the chromosomes in twenty-five species of birds. *Papers f. Coordinating Comm. Research Genetics*, **2**: 27–38.

YASMINEH, W. G. and YUNIS, J. J. (1970). Localization of mouse satellite DNA in constitutive heterochromatin. *Exp. Cell Res.* **59**: 69–75.

YEAGER, C. H., PAINTER, T. S. and YERKES, R. M. (1940). The chromosomes of the chimpanzee. *Science*, **91**: 74–5.

YONENAGA, Y., FROTA-PESSOA, O. and LEWIS, K. R. (1969). Karyotypes of seven species of Brazilian bats. *Caryologia*, **22**: 63–80.

YONG, H. S. (1968). Karyotype of four Malayan rats (Muridae, genus *Rattus* Fischer). *Cytologia*, **33**: 174–80.

YONG, H. S. (1969). Karyotypes of Malayan rats (Rodentia–Muridae, genus *Rattus* Fischer). *Chromosoma*, **27**: 245–67.

YONG, H. S. (1972). Population cytogenetics of the Malayan House Rat. *Rattus rattus diardii. Chromosomes Today*, **3**: 223–7. London and Edinburgh:

YOSIDA, T. H. (1953). Multiple sex chromosome mechanism in *Rhaphidopalpa femoralis* (Coleoptera, Chrysomelidae). *Ann. Rep. Nat. Inst. Genet. Japan*, **3**: 35.

YOSIDA, T. H. (1957). Sex chromosomes of the tree frog, *Hyla arborea japonica*. *J. Fac. Sci. Hokkaido Univ.* Zool. **13**: 352–58.

YOSIDA, T. H. and AMANO, K. (1965). Autosomal polymorphism in laboratory bred and wild Norway rats, *Rattus norvegicus*, found in Misima. *Chromosoma*, **16**: 658–67.

YOSIDA, T. H., MORIGUCHI, Y., KANG, Y. S. and SHIMAKURA, K. (1967). Population survey of no. 1 chromosome polymorphism of Black Rats (*Rattus rattus*) collected in Japan and Korea. *Ann. Rep. Nat. Inst. Genet. Misima*, **17**: 61–3.

YOSIDA, T. H., NAKAMURA, A. and FUKAYA, T. (1965). Autosomal polymorphism in *Rattus rattus* (L.) collected in Kusudomari and Misima. *Chromosoma*, **16**: 70–8.

YOSIDA, T. H. and TSUCHIYA, K. (1969). Scientific expedition for the study of rodents to south east Asia and Oceania. III. Chromosomal polymorphism and new karyotypes of Black Rat, *Rattus rattus*, collected in South east Asia and Oceania. *Ann. Rep. Nat. Inst. Genet. Misima*, **19**: 11–12.

YOUNG, W. J., MERZ, T., FERGUSON-SMITH, M. A. and JOHNSTON, A. W. (1960). Chromosome number of the chimpanzee, *Pan troglodytes. Science*, **131**: 1672–3.

YUNIS, J. J. and YASMINEH, W. G. (1970). Satellite DNA in constitutive heterochromatin of the guinea pig. *Science*, **168**: 263–5.

ZAHAVI, A. and WAHRMAN, J. (1957). The cytotaxonomy, ecology and evolution of the Gerbils and Jirds of Israel (Rodentia: Gerbillinae). *Mammalia*, **21**: 341–80.

ZASLAVSKI, V. A. (1963). Hybrid sterility as a factor limiting propagation of allopatric species. *Dokl. Akad. Nauk S.S.S.R.* **149**: 470–1.

ZELENY, C. (1921). The direction and frequency of mutation in the Bar eye series of multiple allelomorphs in *Drosophila. J. Exp. Zool.* **34**: 203–33.

ZIMMERING, S., BARNABO, J. M., FEMINO, J. and FOWLER, G. L. (1970). Progeny: sperm ratios and Segregation-distorter in *Drosophila melanogaster. Genetica*, **41**: 61–4.

ZIMMERING, S., SANDLER, L. and NICOLETTI, B. (1970). Mechanisms of meiotic drive. *Ann. Rev. Genet.* **4**: 409–36.

ZIMMERMAN, E. C. (1938). Cryptorhynchinae of Rapa. *Bernice P. Bishop Mus. Bull.* **151**: 1–75.

ZIMMERMAN, E. G. (1970). Karyology, systematics and chromosomal evolution in the rodent genus, *Sigmodon. Publ. Mus. Michigan State Univ.* **4**, no. 9: 385–454.

ZIMMERMAN, E. G. and LEE, M. R. (1968). Variation in chromosomes of the cotton rat, *Sigmodon hispidus. Chromosoma*, **24**: 243–50.

ZIMMERMANN, A. M. (1960). Physico-chemical analysis of the isolated mitotic apparatus. *Exp. Cell Res.* **20**: 529–47.

ZIRKLE, R. E. (1957). Partial-cell irradiation. *Adv. in Biol. and Med. Phys.* **5**: 104–46.

ZWEIFEL, R. G. (1965). Variation in and distribution of the unisexual lizard, *Cnemidophorus tesselatus. Amer. Mus. Novit.* **2235**: 1–49.

INDEX